ADVANCES IN ENERGY, ENVIRONMENT AND MATERIALS SCIENCE

PROCEEDINGS OF THE INTERNATIONAL CONFERENCE ON ENERGY, ENVIRONMENT AND MATERIALS SCIENCE (EEMS 2015), GUANGZHOU, P.R. CHINA, 25–26 AUGUST 2015

Advances in Energy, Environment and Materials Science

Editors

Yeping Wang
School of Mechanical Engineering. Tongji University, Shanghai, China

Jianhua Zhao
College of Materials Science and Engineering, Chongqing University, Chongqing, China

CRC Press
Taylor & Francis Group
Boca Raton London New York

CRC Press is an imprint of the
Taylor & Francis Group, an **informa** business

A BALKEMA BOOK

Published by:
CRC Press/Balkema
P.O. Box 447, 2300 AK Leiden, The Netherlands
e-mail: Pub.NL@taylorandfrancis.com
www.crcpress.com – www.taylorandfrancis.com

First issued in paperback 2020

© 2016 by Taylor & Francis Group, LLC
CRC Press/Balkema is an imprint of the Taylor & Francis Group, an informa business

No claim to original U.S. Government works

Typeset by V Publishing Solutions Pvt Ltd., Chennai, India

ISBN 13: 978-0-367-73737-5 (pbk)
ISBN 13: 978-1-138-02931-6 (hbk)

Visit the Taylor & Francis Web site at
http://www.taylorandfrancis.com

and the CRC Press Web site at
http://www.crcpress.com

Table of contents

Preface   xv

EEMS2015 organizing committee   xvii

Energy science and technology

A reactive power control method of reducing wind farm power loss   3
G.F. Zhang, J.J. Sun, Y.Y. Ge, S.H. Li, F. Sun & C.J. Xie

Research on the construction of information service platforms for electricity
market large data   9
J.M. Wang, M. Chao, G. Lin, C.C. Gao & D.N. Liu

Application of fieldbus control system on AC/DC microgrid   15
Y.H. Zhang & J. Chang

The regulate and control method research of wind acceptance regarding power
peak adjustment   19
T. Zhang, J.J. Sun, C. Wang, Q. Zhang & G. Wang

A research on wind and thermal power joint ancillary service market behaviors   25
A. Li, L. Wang & X.G. Zhao

Research on peak-regulation pricing and compensation mechanism of wind
power in China   29
D. Zhao, L. Wang & X.G. Zhao

A method to extend fast charging lead-acid battery cycle life for electric vehicles   33
Y.Z. Zhang, H.J. Wang, L. Wang & X.G. Zhao

Study of solar energy technology applied in oil fields   37
W. Qiu, H. Huang, X. Xu & R. Wang

Research on energy storage optimal configuration of renewable energy generation under
different control objectives   41
B. Liu, Y. Meng, S.Y. Zhou & T. Lin

A new electric power emergency command system for electric grid   45
X.M. Jin, X.D. Chen, T. Li, S. Jiang, Y.H. Cui, W.T. Sun & J. Zhang

Experimental study on the performance of 15 kW OTEC system   49
F.Y. Chen, L. Zhang, W.M. Liu, L. Liu & J.P. Peng

Construction solutions on data center of PV power systems of Qinghai Province based
on big data technologies   53
Y. Yun

Dynamic simulation on a micro-grid system consisting of photovoltaic power, gas combined
heat-and-power and battery   57
H. Kuronuma, N. Kobayashi & H.Y. Huang

Numerical simulation analysis on the tower foundation deformation of the high voltage
transmission line caused by iron ore mining and filling   61
B.S. Xu, C. Chen, Q. Yan, S. Dong & S.W. Liu

Application on the some nuclear power engineering of the hydro-fracturing technique
X.B. Wang & M. Zhang
67

Improvements on frequency capture range and stability of multi-phase output charge
pump Phase-Locked Loop
J. Xiao, Y. Chen & Y.Z. Qiu
73

Research on deployment scheme of the power synchronization network management system
Z.C. Xing, L. Teng, Q. Gao, M.X. Wang & Y. Wang
77

Study on the key identification technologies of 360° full-range intelligent inspection
in electric power communication equipment room
S.Z. Yang & H.L. Zeng
85

Environmental analysis and monitoring

Correlations between enzymes and nutrients in soils from the *Rosa sterilis* S.D. Shi planting
bases located in karst areas of Guizhou Plateau, China
J.L. Li, H. Yang, X.D. Shi, M.Y. Fan, L.Y. Li, J.W. Hu & C.C. Li
91

Progress on the combined effects of PPCP residues in drinking water
H.-h. Gao, L.-t. Qin, L.-l. Huang, Y.-p. Liang, L.-Y. Mo & H.-h. Zeng
97

The progress of constructed wetland soil organic carbon mineralization
X.F. Li, Y.L. Ding, S.Y. Bai, L. Sun & J.J. You
101

The economic analysis of the household Combined Heat and Power system
X. Zhou, W.Q. Xu, Y. Shi, J. Wang & X. Wu
105

Isolation and identification of biphenyl-biodegrading strain
W.J. Liang, M.F. Niu, H. Liu & Y. Yan
109

Study on the eliminating outburst technology and effect with increasing permeability
and forced gas drainage by drilling hole cross the seam in outburst seam
B.Q. Yuan & Y.G. Zhang
113

Hydrocarbon expulsion threshold from lacustrine mudstone and shale: a case study
from Dongying Depression, China
Z.H. Chen, M. Zha & C.X. Jiang
119

Research progress on purification of oil spill coast based on slow release fertilizers
J. Meng & X.L. Zheng
125

Improved design and experimental research on sampling pipes for inductive environmental
gas monitoring
Y.L. Jiang & G. Li
131

Investigation and evaluation of water quality in the Qingshitan reservoir in Guilin city
L.Y. Liang, Y.P. Liang & H.H. Zeng
135

Investigation and evaluation of water quality of Qingshitan reservoir surrounding
agricultural wetland
Z.J. Wu, Y.P. Liang & H.H. Zeng
141

Effect of nitrogen fertilizer on the adsorption of cadimum in typical Chinese soils
X.H. Li & R.C. Guo
145

Effects of nano-devices on growth and major elements absorption
of hydroponic lettuce
Y.J. Wang, R.Y. Chen, H.C. Liu, S.W. Song, W. Su & G.W. Sun
151

Induced sysmatic resistance of potato to common scab by MeJA
G.B. Deng, Q. Deng, G.S. Zhang, E.J. Chen & X.L. Chen
155

Research on soil moisture dynamic under negative pressure irrigation
X.D. Shi, Q.Y. Zhou, Y. Wang, C.Y. Cao, G.Y. Jin & Y.F. Liu
161

Analysis and research on the mixed problem of oil field sewage 165
Z. Y. Tao, R.S. Zhao, G.M. Li & S. Zhao

Effect of different potassium levels on the growth and photosynthesis of sweet sorghum
seedlings under salt stress 169
H. Fan, J. Zhang, X.K. Liu, K. Dong, Y.P. Xu & L. Li

The trend and relationship analysis of precipitation and relative humidity under the climate
change in Haihe River Basin from 1961 to 2010 175
M.J. Yang, Z. Y. Yang, S. Y. Zhang, Y.D. Yu, Z.L. Yu, D.M. Yan & S.N. Li

Chaotic characterization of multiphase mixing effects by computational homology
and the 0–1 test for chaos 181
J.W. Huang, J.X. Xu, S.B. Wang & J.H. Hu

Application effects of reclaimed water reuse technology on pharmaceutical wastewater 187
W.J. Liang & Y. Yan

Pesticide usage and environmental protection 191
G.E. Khachatryan & N.I. Mkrtchyan

Review of the mechanisms of low pressure non-Darcy phenomena of shale gas 195
Y.X. Cui, Z.M. Hu & Q. Li

Numerical study of biofouling effects on the flow dynamic characteristics of blade sections 199
T. Zhang, S.-s. Min & F. Peng

The practical application of AHP to select a zone of population evacuation—take
Xinzhou District of Wuhan as an example 203
J. Qin

Research on technical reforms of nitrogen and phosphorus removal process
in T Sewage Company 209
H.X. Shen

The development status of medical waste category, management and disposal in China 213
Z.B. Bao, H.J. Teng, D.C. Jin, F. Yang, X.Y. Liu & Y. Li

Accessing LUCC and ecosystem service value in Weigan River Basin 219
X.N. Qiao, Y.J. Yang & H.B. Zhang

Comparing different kinds of materials for adsorption of pollutants in wastewater 223
P. Wang & J.R. Chen

The classification of medical waste and its disposal technology 227
Z.B. Bao, H.J. Teng, D.C. Jin, J.X. Peng, N. Wu & L. Yang

Experimental analysis on the mechanical characteristic of single-fractured rock masses under
the freeze–thaw condition 233
Y.N. Lu, X.P. Li & X.H. Wu

Effect of irrigation frequency on crop water productivity and economic benefit
of oasis jerusalem artichoke (*Helianthus Tuberosus. L*) 237
Z. Y. Bao, H.J. Zhang & S.X. Chai

Numerical study of Zhuanghe coastal water using a two-dimensional finite volume method 241
Y. Y. Xu, M.L. Zhang & Y. Qiao

Spatial analysis of drought in Haihe River Basin from 1961 to 2010 based on PDECI and SPI 247
M.J. Yang, Z. Y. Yang, Y.D. Yu, G.Q. Dong & Z.L. Yu

The ground surface displacement of shallow buried circular cavity in a soft layered
half-space impacted by SH wave 253
Y.B. Zhao, H. Qi, X.H. Ding & D.D. Zhao

Discussion on the design of steel skeleton of great span greenhouse in cold region 259
L. Zhai & D.-h. Liu

Research on spatiotemporal changes of water level corresponding to different flood
frequency to ENSO in the Pearl River Delta 265
H.Y. Qiao, Q. Jia & Y. Xu

Effect of large flood in 2012 on river planform change in the Inner Mongolia
reaches of the Yellow River 271
S. Yu, K. Wang & L. Shi

Test of hysteretic freezing characteristics of clay during freezing and thawing 275
D.Y. Li & B. Liu

Color/Turbidity detection of water supply pipeline cleaning and its application 283
X.L. Xu

Study on the influence of ecological protection measures on water conservation capacity
in the Sanjiangyuan Region 291
J.Q. Zhai, Y. Zhao, H.H. Li, Q.M. Wang & K.N. Chen

Study on the ground subsidence caused by pipe jacking 299
X.K. Bao, C.H. Huang & H.H. Liu

Experimental study of cyclic loading history influence on small strain shear modulus
of saturated clays in Wenzhou 303
X.B. Li, C. Gu, X.Q. Hu & G.Q. Fu

Application on geophysical monitoring technology of geological structure detection
in Jinfeng coal mine 309
Z.H. Lu, X.P. Lai, Y.Z. Zhang, P.F. Shan & X.J. Yue

Architectural environment and equipment engineering

Stress monitoring in the ARMA model of Changchun subway
construction applications 317
X.-Q. Zhang, J.-K. Zhang, M.-S. Wang & G.-D. Yang

Simulating analysis and research on transformer winding deformation fault 321
Y. Chen, Q. Peng, J. Tang & Y.H. Yin

Study on the equipment maintenance support efficiency evaluation based on the gray
correlation analysis method 327
X. Zhao, M. Guo & Y.J. Ruan

Effect of polyurea reinforced masonry walls for blast loads 331
J.G. Wang, H.Q. Ren, X.Y. Wu & C.L. Cai

Development and optimization of an improved vacuum assisted conventional extraction
process of epigoitrin from *Radix Isatdis* using orthogonal test design 339
Y.Q. Wang, Z.F. Wu, J.P. Lan, X. Wang & M. Yang

Calculation of key parameters on laser penetrating projectile steel 345
G.F. Song, L.C. Li, S.G. Wang & F.J. Xia

The research on temperature control of mass concrete 349
J. Deng, Y.B. Li & S.L. Wu

The heat-insulating property evaluation and application of high permeability-high
strength concrete materials 353
Y.-z. Tan, Y.-x. Liu & P.-y. Wang

Research on the injection characteristics of biogas engine 357
Z.H. Fang, K. Wang, Y. Sun & G.J. Guo

Study on relationship of land use & landscape pattern and water quality
in Dahuofang reservoir watershed 361
R.C. Guo & X.H. Li

Theoretical and experimental study of volumetric heat transfer coefficient
for Direct-Contact Heat Exchanger 367
J.W. Huang, S.B. Wang & J.H. Hu

Research on dynamic analysis of pod vibration based on ANSYS/LS-DYNA 375
J.Y. Jiang & S.M. Li

The Influence of different testing feed velocity on the dynamic modulus of elasticity of lumber 379
P.F. Zhang, W. Zhang, Q.W. Zhang, H.L. He & Z. Jin

The mechanism of residents' participation of architectural heritage conservation 385
X.Y. Ma

Analysis of displacements and temperature increments of the half space subjected
to a flat cylindrical heat source 391
J.C.-C. Lu, W.-C. Lin & F.-T. Lin

Research on high-rise residential district environment landscape problems and countermeasures 397
Y. Sun & G.L. Gao

Regulation mode analysis and selection of power plant boiler fan 401
Y.F. Li, J. Zhao, S.G. Hu & H.L. Jia

Optimization of the matching performances between pump and valve in the boom
converging loop of a PFC excavator 405
W.H. Jia, D.S. Zhu, C.B. Yin & H. Liu

The role of micro/nano-structure in the complex wettability of butterfly wing 411
G. Sun & Y. Fang

Reinforcement simulation for ramp bridge 417
Y. Li & B. Pan

Single-variable model for a dynamic soil structure 421
X. Chen, Y.X. Yan & X.J. Hou

Study of influencing factors on the strain localization formation of saturated clay under
the plane strain condition 427
X.Y. Gu & T.Q. Zhou

The experiment and study of concrete inorganic protective preparation in cold regions 433
S.Z. Wang

Research on the effect of the durability of arch foot concrete of the Xixihe bridge
as the reinforcement corrosion 437
K. Li

Factors affecting the rapid excavation speed of rock roadways based on the blasting method 441
G.-h. Li, J. Li, X. Wu & J.-p. Wu

Effect of high strength reinforced steel and axial compression ratio on the hysteretic behavior
of rectangular bridge piers 445
X. Rong, D.D. Xu & P. Liu

The application on fuzzy neural control in boiler liquid level control system 451
H.X. Cheng, G.Q. Zhang, J. Li, L. Cheng & L.L. Kong

Simulation on hydraulic fracturing propagation using extended finite element method 455
D.X. Wang, B. Zhou & S.F. Xue

Experimental study of the pre-split crack effect on the stress wave propagation caused
by explosive 459
J.J. Shi, X. Wei, H.M. An & H.L. Meng

Research on ultimate strength of confined concrete with circular stirrups 465
M.Y. Lu & D.S. Gu

Ground motion of half space with ellipse inclusion and interfacial crack for SH waves 469
X.H. Ding, H. Qi, Y.B. Zhao & D.D. Zhao

The performance study of notch-stud connections of Timber-Concrete Composite beam 475
G.J. He, H.Z. Xiao, L.P. Chen & L. Li

Vibration test and finite element analysis of a timber construction with curbwall 481
F.F. Qian, W.D. Lu, W.Q. Liu, X.W. Cheng, D.P. Lv & X.X. Liu

Studies on internal forces of shield lining segments with different design models 487
K.W. Ding & Y.G. Wang

Constitutive relationship of super high-strength concrete filled steel box 493
X.M. Chen, J. Duan & Y.G. Li

Offshore wind turbine aerodynamic damping analysis and apply in semi-integrated
analysis method 497
B. Wang, W.H. Wang, X. Li & Y. Li

Interaction gesture recognition for highway 3D space alignment design 505
L.D. Long, X.S. Fu, H.L. Zhu & Y.Q. Wang

Research on the united tension formulas of short bridge suspenders for two typical kinds
of boundary conditions 513
J.B. Liao, H.L. Liu & G.W. Tang

Application of Probabilistic Fracture Mechanics in fatigue life evaluation
of crane beam 519
G. Xue & Y.T. Meng

Failure analysis of large, low head water pump units 525
B.Y. Qiu, J.Y. Cao & Y.Z. Yang

Study for seismic behavior of dovetail mortise-tenon joints with gap damaged 531
K. Feng, K. Kang, X.J. Meng, J. Hou & T.Y. Li

Finite element analysis of the flange sealing of pipes under the condition of vibration 537
Z.R. Yang, D.Y. Zhang, L. Sun & G.J. Wang

A method for parameter optimization of locking dowel base on the orthogonal experiment 543
S.-t. Chen, W.-x. Zhang, X.-l. Du & Y.-h. Zhang

Research on crack propagation process of concrete using photoelastic coatings 549
H.B. Gao, D.S. Song & S.J. Ouyang

Environmental materials

Synthesis and adsorption of OMMT/AHL grafted PAA superabsorbent composite 555
Y.F. Xu, H.X. Zhao, G.P. Chen, E.X. Lian & Y.L. Ma

Distribution characteristics of saturated hydrocarbon and UCM, organic matter sources
and oil-gas indicative significance of core P7327 in the Chukchi Sea 561
Q.Y. Zhao, R.H. Chen, H.S. Zhang & B. Lu

Absorption kinetics of CO_2 in MEA promoted K_2CO_3 aqueous solutions 569
W.H. Si, C.L. Mi & D. Fu

Kinetics in Mn (II) ion catalytic ozonation of Papermaking Tobacco Slice wastewater 573
S.Y. Chen, Y.M. Li & L.R. Lei

Formula optimization of dust coagulation agent for application in open-pit iron
ore mine blasting 581
E.W. Ma, Q.S. Wu, X.Y. Wu, Q.C. Lang & Z.J. Cui

Properties of hemihydrate phosphogypsum and application to solidified material 587
C.Q. Liu, Q.L. Zhao, K.J. Zhou & J.Q. Li

Study on chemical oxidation of acetone gas 593
W.X. Zhao, H. Kang & A.L. Ren

The research on the non-isothermal crystallization process of POM fiber 597
H. Tan, Y. Wang, X.X. Hu, J.Q. Huang & C. Li

Preparation of composite flocculant and its application in urban recycled
water treatment 601
C.G. Xue, F.T. Wang & C.Y. Tong

Effects of cadmium on calcium and manganese uptake, and the activity of tonoplast
proton pumps in pakchoi roots 607
G.W. Sun, Z.J. Zhu & X.Z. Fang

ABA's antioxidant protection role in salt-stressed cells of bloom-forming cyanobacteria
Microcystis aeruginosa 613
E.J. Chen, H.F. Xue, Y.J. He, X.Q. Zhang, X.L. Chen & G.B. Deng

Synthesis of highly monodisperse Polystyrene microspheres and assembly
of the polystyrene colloidal crystals 619
C. Yang, L.M. Tang, Q.S. Li, L.T. Feng, A.L. Bai, H. Song & Y.M. Yu

Thermogravimetric analysis of Polyoxymethylene fiber used for cement
concrete modification 625
J.H. Huang, Y. Wang, X.X. Hu & C. Li

Synthesis of uniform zirconium oxide colloidal particles by hydrothermal method
and the influence factor analysis 629
A.L. Bai, H. Song, Q.S. Li, Y.Q. Wang, C. Yang & Y.M. Yu

An electro-thermal model for Li-ion battery under different temperatures 635
P. Chen, F. Sun, H. He & W. Huo

Capacity degradation of commercial LiFePO$_4$ cell at elevated temperature 641
W.P. Cao, J. Li & Z.B. Wu

Oxygen Reduction Reaction on sulfur doped graphene by density functional study 647
J.P. Sun & X.D. Liang

Study of thermodynamics on the formation of tetrahydrofuran hydrate
by React Calorimeter (RC1e) 653
X.D. Dai, X.M. Hu, K. Liu, H.Y. Li, Y. Liang & W. Xing

In-situ synthesis of LSCF-GDC composite cathode materials for Solid Oxide Fuel Cells 657
J. Li, Z. Wu, N. Zhang, D. Ni & K. Sun

Ni-Co/AlMgO$_x$ catalyzed biodiesel production from Waste Cooking Oil in supercritical CO$_2$ 665
M.Y. Xi, W.Y. Zhang, M.N. Cui, X.B. Chu & C.Y. Xi

Special wettability of locust wing and preparation of biomimetic polymer film
by soft lithography 669
G. Sun & Y. Fang

Novel graphene-based composites for adsorption of organic pollutants in wastewater treatment 673
M.Y. Zhou & J.R. Chen

Corrosion inhibition of nitrite in existing reinforced concrete structures 677
J. Liu, J.L. Yan, Y.H. Dai & Y.S. Li

Binding mechanism of chloride ions in mortar 681
J. Liu, J.L. Yan, Y.H. Dai & Y.S. Li

Computer applications

Multi-channel PM$_{2.5}$ sampler based on intelligent embedded software control system 687
D.Z. Hua, W. Huang, Z.H. Liu, X. Cao, H.J. Ye & W.P. Liu

Structure optimization design based on the numerical simulation of subway station deep
foundation in open cut   691
L.L. Tan, M. Li & M.L. Wu

Numerical calculation on Stress Intensity Factor in rock using Extended Finite Element Method   695
B. Sun, B. Zhou & S.F. Xue

Yinger Learning Dynamic Fuzzy Neural Network algorithm for the three stage inverted pendulum   701
P. Zhang, G.D. Gao, X. Zhang & W. Chen

Simulation of ET_0 based on *PSO* and *LS-SVM* methods   707
B. Ju, H. Liu & D. Hu

Evapotranspiration (ET) prediction based on least square support vector machine   715
J.P. Liu, W. Wang & J.J. Zhou

A method of high maneuvering target dynamic tracking   721
J.-p. Yang, W.-t. Liang & J. Wang

A novel semantic data model for big data analysis   727
Y. Li, K. Li, X.-T. Zhang & S.-B. Wang

Research on the power allocation algorithm in MIMO-OFDM systems   731
X.C. Lin & Y.X. Zu

Analysis and research of android security system   735
L. Zheng & Y.J. Liu

Feature fused multi-scale segmentation method for remote sensing imagery   741
T.Q. Chen, J.H. Liu, Y.H. Wang, F. Zhu, J. Chen & M. Deng

Performance study on PWM rectifier of electric vehicle charger   745
L.X. Qu

Analysis and research on iOS security system   751
L. Zheng & D.D. Li

Design and implementation of the anti-cancer diet system based on .net   757
Q. Li & Z. Xiang

Non-Intrusive Load Identification based on Genetic Algorithm and Support of Vector Machine   761
Y.L. Ma, B. Yin, X.P. Ji & Y.P. Cong

A fast algorithm combined with SSIM and adjacent coding information for HEVC
intra prediction   765
T. Yan, X.X. Xiao, X.C. Zhang, H. Liu & X.S. Zhang

Persuasive tech in keeping chronic patients' willingness in health self-management   769
Y.Y. Guo, M.G. Yang, J. Hu & L.K. Tao

Modeling and simulation of the air fuel ratio controlling the biogas engine based
on the fuzzy PID algorithm   773
Z.H. Fang, Y. Sun, K. Wang, G.J. Guo & J.L. Wang

Evaluation of software realization algorithms of industrial building operation life   777
Z.T. Vladimirovna & S.P. Nikolaevich

Anomaly detection based on the dynamic feature of network traffic   781
Y.X. Zhang, S.Y. Jin, Y.Z. Wang & Y.X. Wang

Application of computer simulation in the elastic-plastic seismic response
of bottom frame structure   791
H.Y. Deng & B.T. Sun

The application of simulation method in one dimension pipeline flow based
on elastic pipeline   797
X.D. Xue, P.F. Cao, J. Shao, L.J. Duan, J.N. Zhang & H. Lan

On designing MAC—ROUTE cross layer to support hybrid antennas
in wireless ad hoc networks   803
Y.J. Zhang & Y. Li

Detecting anomalies in outgoing long-wave radiation data by a window average
martingale method   809
J. Zeng & L.P. Chen

Human detection system based on Android   815
Q. Tian, S. Lin, Y. Wei & W.W. Fei

Research on attack model and security mechanism of RFID system   819
T. Wang, S.S. Wu, Q. Li, H.F. Luo & R.Q. Liu

Wear feature extraction for diamond abrasives based on image processing techniques   823
Y.F. Lin, F. Wu & C.F. Fang

Detection the deviation from process target using VP loss chart   829
C.-M. Yang & S.-F. Yang

Optical rotation calculations on chiral compounds with HF and DFT methods   833
L.R. Nie, J. Yu, X.X. Dang, W. Zhang, S. Yao & H. Song

Decision-making method of strategy intelligence based on the lifecycle model
for technology equipages   839
N. Ma, L. Zhang, J. Zhao, Y. Xue & H.N. Guo

Logical model of outdoor lighting soft interface TALQ protocol   845
J. Liang & D.H. Jia

The matlab simulation of the frequency hopping transmission station   849
N. Wang & Z.G. Song

Author index   853

Preface

This collection of papers from the International Conference on Energy, Environment and Materials Science (EEMS2015, Guangzhou, China, 25–26 August 2015), aims to present novel and fundamental advances in the areas of energy, and environmental and materials sciences. The 161 accepted papers (from 400 submissions) are divided into five chapters:

– Energy Science and Technology
– Environmental Analysis and Monitoring
– Architectural Environment and Equipment Engineering
– Environmental Materials
– Computer Applications

In preparing EEMS2015, we would like to thank the contributors, reviewers, committee members and CRC Press/Balkema (Taylor & Francis Group) for their support of EEMS 2015.

The Organizing Committee of EEMS2015

EEMS2015 organizing committee

CONFERENCE CHAIRS

Prof. Bachir Achour, *University Mohamed Khider of Biskra, Algeria*
Prof. Li Xie, *Tongji University, China*

TECHNICAL COMMITTEE CO-CHAIRS

Professor L. Hench, *Imperial College London, UK*
Sir C.J. Humphreys, *University of Cambridge, UK*
Professor L. Jiang, *Chinese Academy of Sciences, Beijing, China*

TECHNICAL PROGRAM COMMITTEE

Prof. Yulin Zhong, *South China University of Technology, China*
Prof. Jun Zhang, *Xiangtan University, China*
Prof. Min Xie, *Tongji University, China*
Prof. Peijiang Zhou, *Wuhan University, China*
Prof. Tao Zhu, *China University of Mining and Technology, China*
Prof. Xiaojun Zhou, *Southwest Jiaotong University, China*
Prof. Xiangjun Yu, *Kunming University, P.R. China*
Prof. Tiehong Wu, *Inner Mongolia University, China*
Prof. Haiyun Wang, *China Three Gorges University, China*
Prof. Yonggang Yang, *Beijing Institute of Graphic Communication, China*
Prof. Donglai Xie, *South China University of Technology, China*
Prof. Guangqun Huang, *China Agricultural University, China*
Prof. Johnson Carl, *Portland Community College, US*
Assistant Prof. Miguel Ângelo Dias Azenha, *University of Minho, Portugal*
Dr. Ching Yern Chee, *Faculty of Engineering University of Malaya, Malaysia*
Prof. Chang Li, *University of Science and Technology Liaoning, China*
Prof.Guochang Zhao, *Shenyang Aerospace University, China*
Prof. Hong Liang, *Yunnan University, China*
Prof. Weisheng Shi, *Kunming University of Science and Technology, China*
Prof. Mulan Zhu, *Xiamen University of Technology, China*
Prof. Xinxian Zhai, *Henan Polytechnic University, China*
Prof. Manxia Zhang, *Dalian Maritime University, China*
Prof. Taihua Yang, *Shanghai University of Electric Power, China*
Prof. Hai Fan, *Shandong Normal University, China*
Prof. Shaochun Chen, *Fuzhou University, China*
Prof. Youyuan Wang, *Nanchang Hangkong University, China*
Prof. Ruitao Peng, *Xiangtan University, China*
Prof. Junli Han, *Lnner Mongolia University of Science and Technology, Mongolia*
Prof. Zhonghong Chen, *China University of Petroleum, China*
Prof. Liya Xu, *Liming Vocational University, China*

Energy science and technology

A reactive power control method of reducing wind farm power loss

Guanfeng Zhang
Electric Power Research Institute of Liaoning Electric Power Company, Shenyang, China

Junjie Sun
School of Electric Engineering, Wuhan University, Wuhan, China

Yangyang Ge, Shenghui Li, Feng Sun & Cijian Xie
Electric Power Research Institute of Liaoning Electric Power Company, Shenyang, China

ABSTRACT: For wind farms consisting of double-fed wind turbines with centralized compensation device SVC, a method was proposed based on a minimum of reactive power compensation, utilization of installed reactive power compensation device SVC and double-fed unit's reactive ability to achieve reactive power optimization. According to actual reactive power generation capability of Double-fed wind turbines and reactive power compensation device SVC, wind Farm automatic voltage control system distributes reactive power with a network Loss minimum of the objective function, this allocation method can play DFIG reactive power regulation capability and can reduce the loss of wind farms. The simulation results show the proposed scheme can effectively reduce the wind farm power losses and improve voltage stability margin.

1 INTRODUCTION

With wind power capacity growing, stabilizing influence of reactive power on the wind farm has become increasingly obvious security, the accident of wind farm voltage collapse was caused due to windfarm reactive configured properly, and caused huge economic losses (Chi, 2007). Currently, most wind turbines use four-quadrant power electronic converters to connect to the grid, achieve decoupling control of active and reactive power by the converters, its own reactive power regulation capability can be applied to wind farm reactive power regulation (Wang, 2011; Jia, 2010). Moreover, wind farms equipped with reactive power compensation equipment, can achieve reactive power control, through cooperative control of DFIG and reactive power compensation device (Zhu, 209).

For wind farm reactive power control strategies, numerous studies have been done at home and abroad. Wind turbine reactive power output sequence was determined to achieve reactive power optimization through reactive power sensitivity different locations outlet buses of wind turbines (Li, 2005), but this method does not consider DFIG reactive power regulation. Literature (Chen, 2009) analyses the reactive power ultimate limit of DFIG, and achieve reactive power optimization control to the losses of the DFIG based on VSCF electrical equivalent circuit, this method only considered reactive loss of wind turbines, not the loss of

centralized compensation device, cannot achieve the overall optimization of reactive power.

This article provides a reactive power control method of reducing the power consumption of the wind farm, makes full use of fans nonfunctional force and considers SVC own loss, achieving a coordinated control of wind farm between SVC reactive power compensation equipment and wind turbines, and minimizes the wind farm power loss.

2 DOUBLE-FED WIND TURBINE POWER OPERATING CHARACTERISTICS

In order to obtain wind turbine active and reactive power coupling relationship, mathematical model for wind turbine was analyzed first. DFIG stator side using rotation practices of the generator, rotor side using the motor rotation practices, the equivalent circuit shown in Figure 1.

The following equation was calculated according to the equivalent circuit:

$$\begin{cases} \dot{E}_m = \dot{I}_m \cdot jX_m \\ \dot{U}_s = \dot{E}_m - (R_s + jX_{s\sigma})\dot{I}_s \\ \dfrac{\dot{U}_r}{s} = \dot{E}_m + \left(\dfrac{R_r}{s} + jX_{r\sigma}\right)\dot{I}_r \\ \dot{I}_r = \dot{I}_s + \dot{I}_m \end{cases} \tag{1}$$

where $\dot{E}_m, \dot{U}_s, \dot{U}_r$ was gap magnetic field induced electromotive force, stator and rotor voltage of

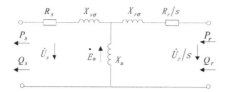

Figure 1. DFIG equivalent circuit.

DFIG respectively, $\dot{I}_m, \dot{I}_s, \dot{I}_r$ was excitation current, the stator and rotor current value respectively, $X_m, X_{s\sigma}, X_{r\sigma}$ was magnetizing inductance and the stator and rotor leakage reactance respectively, X_s, X_r was stator and rotor side reactance value after conversion respectively, s was the slip of DFIG, P_s, P_r, Q_s, Q_r was stator and rotor of active and reactive power respectively; By the formula $X_s = X_{s\sigma} + X_m, X_r = X_{r\sigma} + X_m$, rotor side variable have been converted to the stator.

The stator voltage and current RMS is expressed in the form

$$\begin{cases} \dot{U}_s = U_s + j0 \\ \dot{I}_s = I_{sP} + jI_{sQ} \end{cases} \quad (2)$$

where U_s was stator voltage RMS, I_{sP}, I_{sQ} was the active and reactive component of the stator current.

The rotor side current is based on the above formula:

$$\dot{I}_r = \frac{P_s X_s - Q_s R_s}{3 X_m U_s} - j\frac{3U_s^2 + P_s R_s + Q_s R_s}{3 X_m U_s} \quad (3)$$

Without taking into account reactive power consumes of the system, active and reactive power injected the system are:

$$\begin{cases} P_x = P_s + P_r \\ Q_x = Q_S \end{cases} \quad (4)$$

DFIG rotor maximum current value was set as $I_{r\max}$, which is 150% of the rated current value generally. Stator reactive power can be equivalent to the output of a single wind turbine reactive power. So, a single DFIG reactive power output limit formula is: (Shen, 2003)

$$\begin{cases} -\sqrt{\dfrac{9I_{r\max}^2 U_s^2 X_m^2}{X_s^2} - P_s^2} - \dfrac{3U_s^2}{X_s^2} \leq Q_{reg} \\ \leq \sqrt{\dfrac{9I_{r\max}^2 U_s^2 X_m^2}{X_s^2} - P_s^2} - \dfrac{3U_s^2}{X_s^2} \\ P_W \dfrac{\sqrt{1-\lambda_L^2}}{\lambda_L} \leq Q_{reg} \leq P_W \dfrac{\sqrt{1-\lambda_H^2}}{\lambda_H} \end{cases} \quad (5)$$

where λ was the power factor of wind turbine, power factor range of wind turbines under normal working conditions was $\lambda_L \leq \lambda \leq \lambda_H$, Q_{reg} was each wind turbine maximum output capacity in real time. As can be seen from the formula (5), when wind turbine power factor reaches the maximum, it will lose the ability to regulate reactive power; Under certain wind speed (active issue of the stator certain), the stator reactive and absorptive capacity is asymmetric.

3 REACTIVE POWER COMPENSATION DEVICE SVC OPERATING CHARACTERISTIC ANALYSIS

Typical SVC can be divided into Thyristor Controlled Reactor (TCR), Thyristor Switched Reactor (TSR) and Thyristor Switched Capacitor (TSC). TCR single-phase equivalent circuit shown in Figure 2, Shunt reactor dynamic control reactive from minimum to maximum within the range by

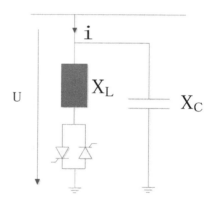

Figure 2. SVC (TCR) single-phase equivalent circuit.

Table 1. The loss results of reactive compensation devices.

No.	Type	Capacity (Mvar)	Loss value (MW)	Losses as a percentage of capacity
1	TCR	−15	0.336	2.24%
2	SVG	±4 Mvar	0.155	3.10%
3	SVG	±5	0.17	3.40%
4	SVG	±5 Mvar	0.18	3.60%
5	SVG	±7 Mvar	0.209	2.98%
6	SVG	±20 Mvar	0.35	3.50%
7	SVG	±5 Mvar	0.299	5.98%
8	SVG	±5 Mvar	0.185	3.70%
9	SVG	±3 Mvar	0.097	3.20%

controlling bidirectional thyristor, SVC is equivalent to a variable shunt reactors, because SVC device also includes transformers, filters, cooling systems, will have a greater power loss during the operation, Table 1 is the loss results of reactive power compensation device SVC and SVG (static synchronous compensator).

4 WIND FARM REACTIVE POWER REGULATION AND OPTIMIZATION

Wind Farm automatic voltage control system features is to ensure wind farms outlets reactive demand for the grid, then sets and distributes wind farm reactive power demand. The entire wind farm reactive power control principle is shown in Figure 3.

This paper implements reactive power control of wind farm level via a layered approach, divided into reactive setting layer and reactive power distribution layer. First, compare the actual voltage and reference voltage given by Power Dispatching Center at control point in the reactive tuning layer, and get the change of reactive power demand for wind farms, then sends reactive demand to reactive distribution layer to determine each wind Turbine and SVC reactive power setpoint.

4.1 *Windfarm reactive power demand value calculation*

The amount of reactive power compensation is calculated via wind farms and outlets voltage and reactive power relations:

$$\Delta U = \Delta Q / S_{sc} \qquad (6)$$

$$S_{sc} = \left(U_{now} - U_{last}\right) \Big/ \left(\sum Q_{now} \Big/ U_{now} - \sum Q_{last} \Big/ U_{last}\right) \qquad (7)$$

In Equation (6), (7), ΔQ was reactive power variation, S_{sc} was Short circuit capacity of the system bus side, $\sum Q_{now}$, $\sum Q_{last}$, was last Total Reactive and current total reactive respectively, U_{last}, U_{now} was last bus voltage, current bus voltage.

4.2 *Wind farm level of reactive power optimization*

The entire wind farm reactive power was distributed based on loss minimum, reactive power optimization of the entire wind farm setting program flowchart is shown in Figure 4.

1. When wind turbines generally reactive output capacity Q_{Wref} More than total reactive power demand-ΔQ, at this time SVC reactive power compensation is not required, calculating Q_{ref} though the reactive setting layer, then Distribute Q_{ref} according to the individual wind turbine reactive power capability.

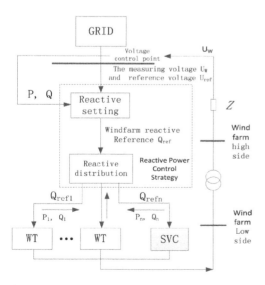

Figure 3. Control diagram of DFIG based wind farm.

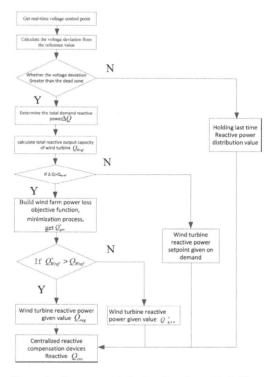

Figure 4. Strategy optimization flowchart of windfarm reactive power control.

5

2. When $Q_{Wref} < \Delta Q < (Q_{Wref} + Q_{SVCmax})$, Wind turbines and SVC take part in reactive power regulator at the same time, in this paper, in order to achieve the coordination and control between the wind farm reactive power compensation equipment and wind turbines, and wind farm power loss is minimized, the objective function between the wind farm loss and wind turbine reactive is built. Under certain wind speed, wind turbine active power has been set, total power consumption:

$$\Delta P_{\Sigma}(S,V) = P_0(V) + P_S(S,V) + P_{LL}(S,V) + P_{svc}$$
$$= \left(\frac{V}{V_N}\right)^2 \cdot P_{N0} + \left(\frac{S}{S_N}\right)^2 \cdot \left(\frac{V_N}{V}\right) \cdot P_{NS}$$
$$+ \left(\frac{S}{S_N}\right)^2 \cdot \left(\frac{V_N}{V}\right) \cdot P_{NL} + P_{svc}$$

(8)

where ΔP_{Σ} was wind farms total power loss, P_0 was transformer load loss, P_S was transformer short-circuit loss, P_{LL} was Line Loss, V was Transformer operating voltage value, V_N was Transformer rated voltage, S was transformer operation apparent power value, S_N was transformer rated apparent power value, P_{N0} was transformer rated load loss, P_{NS} was transformer rated short-circuit loss, P_{NL} was Line Rated power loss, P_{svc} was power loss of reactive power compensation device.

The total power loss ΔP_{Σ} is set as the objective function, since equipment power consumption of all the series in a set of utility line is:

$$P = I^2 R = \left(S_N / \sqrt{3} U_N\right)^2$$
$$R = \left(P_{gen} \times n / \cos\phi / k / \sqrt{3} U_N\right)^2 R$$

(9)

where $\cos\phi = \lambda$, n is the total number of units of wind turbines, k is the percentage of the rated voltage of the actual voltage.

Power loss of reactive power compensation device P_{svc} is 2%–3% of rated capacity, which is set to a known quantity in accordance with the actual situation of the wind farm referring to Table 1.

So equation (8)–(9) shows that, wind farms total power loss ΔP_{Σ} only has relation with reactive power of each wind turbine-Q_W, the equation (6) can be transformed to the objective function of wind turbines Reactive Dispatching:

$$T = \Delta P_{\Sigma}(S,V) = \Delta P_{\Sigma}(Q_{gen})$$
$$= P_0(V) + P_S(S,V) + P_{LL}(S,V) + P_{svc}$$

(10)

Constraints are equation (5) and $\Delta Q = Q^*_{Wref} + Q_{SVC} - \Delta Q_{\Sigma}$, where ΔQ_{Σ} was the total wind farm reactive power loss, it can use wind turbine capacity reactive representation.

The objective function T was made minimum treatment, wind generator reactive power value Q^*_{gen} is obtained when the existence of minimum of the objective function, overall output capacity of wind turbines was $Q^*_{Wref} = \Sigma^n_{k=1} Q^*_{gen}$, then compare overall actual wind turbine reactive power output capability Q_{Wref} with Q^*_{Wref}, if $Q_{Wref} \leq Q^*_{Wref}$, wind turbine reactive given was Q_{reg}, Centralized Reactive Power Compensation Device reactive given was $Q_{svc} = \Delta Q - Q_{Wref} - Q_T - Q_{LL}$; if $Q_{Wref} > Q^*_{Wref}$, wind turbine reactive given was Q^*_{gen}, Centralized Reactive Power Compensation Device reactive given was $Q^*_{svc} = \Delta Q - Q^*_{Wref} - Q_T - Q_{LL}$, where Q_T and Q_{LL} was Transformers and reactive power loss, respectively.

3. If $|Q_{SVCmax} + Q_{Wref}| \leq |\Delta Q|$, wind Farm automatic voltage regulator control system will give alarm prompt transformer taps.

5 OPERATOR CASES

In this paper, a wind farm in Liaoning was simulated, test and verify the effectiveness of the wind farm reactive control strategy. Wind farm system wiring shown in Figure 5.

The wind farm installed 33 sets of 1.5 MW double-fed induction generator, fan outlet voltage was 690 V, each fan collector through four 35 kV line access in 220 kV substation, and the –10 MVA SVC was installed at the 35 KV bus.

In this paper, to doubly-fed wind farm, take net loss minimum as target to optimize reactive power. Under different wind speeds, wind farm operation as shown in Figure 6–8.

Through simulation calculation shows, when the SVC reactive power compensate alone, there is a network problem of excessive consumption, the

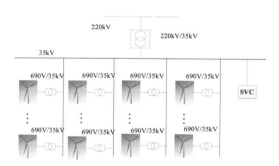

Figure 5. The wiring diagram of sample system.

6

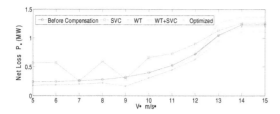

Figure 6. Doubly fed wind farm losses changes under different wind speeds.

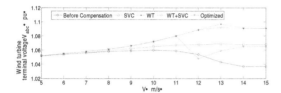

Figure 7. Wind turbine terminal voltage changes under different wind speeds.

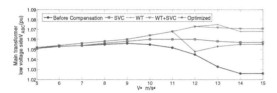

Figure 8. Wind farm main transformer low-side voltage changes under different wind speeds.

largest up 1.369 MW; the wind turbines compensate alone, terminal voltage is too high, especially when the fan is running at high speed, can reach 1.097 pu; After optimization, reducing the wind farm network losses and reduce the 1.369 MW to 1.197 MW, improved voltage stability margin,

the maximum terminal voltage decreases from 1.097 pu to 1.08 pu.

6 CONCLUSION

This paper analyzes the wind farm reactive power control principles and reactive power compensation device SVC operating characteristics, proposes a wind farm reactive power control optimization algorithm, which take the wind farm internal power loss as the objective, optimizes the wind farm internal reactive power control strategy on the basis of ensuring the safe and stable operation of wind farms, effectively reduce the wind farm internal reactive loss and improve voltage stability margin.

REFERENCES

Chen Ning, et al. Strategy for reactive power control of wind farm for improving voltage stability in wind power integrated region [J]. Proceedings of the CSEE, 2009, 29(10): 102–106.

Chi Yong-ning, et al. Study on impact of wind power integration on power system [J]. Power System Technology, 2007, 31(3): 77–81.

Jia Jun-chuan, et al. Novel reactive power optimization control strategy for doubly fed induction wind power generation system [J]. Proceedings of the CSEE, 2010, 30(30): 87–92.

Li Jing, et al. Research on subsection and layer control strategy of doubly-fed variable speed wind turbine [J]. Power System Technology, 2005, 29(9): 15–21.

Shen Hong, et al. Variable-speed wind power generation using doubly fed wound rotor induction machine a comparison with alternative schemes [J]. Power System Technology, 2003, 27(11): 60–63.

Wang Cheng-fu, et al. Coordinate var-control during fault of wind power system [J]. Electric Power Automation Equipment, 2011, 31(9): 14–21.

Zhu Xue-ling, et al. Research on the compensation of reactive power for wind farms [J]. Power System Protection and Control, 2009, 37(16): 68–72.

Research on the construction of information service platforms for electricity market large data

Junmei Wang, Ma Chao & Guo Lin
State Grid Chongqing Electric Power Corporation, Chongqing, China

Chuncheng Gao
Nari Group Corporation/State Grid Electric Power Research Institute, Nanjing, Jiangsu Province, China

Dunnan Liu
North China Electric Power University, Beijing, China

ABSTRACT: Electric power large data is an inevitable course for the power industrial technology innovation in energy revolution. Moreover, it is the value form leap of the next generation intelligent power system in the large data era. There are many different kinds of business information of electricity market design, which contains rich information. In order to excavate and apply the biggest value of these information effectively, this paper was based on the information composition of electricity market large data. Besides, construction of information service platforms for electricity market large data was researched on the aspects of application service and consulting service.

1 INTRODUCTION

It is not only a technical progress, but also a great change that involves the development idea, management system and technical route of the whole power system in the large data era. Moreover, it is the value form leap of the next generation intelligent power system in the large data era. The two core lines of power electric power large data are reshaping the core value of electric power and transforming the development mode of electric power (Yan, 2013). The electricity market business involves various kinds of information, such as power transaction data and electrical coal market data. The amount of data is very huge, while the distribution of information is scattered. It is a large amount of work for a single market member to collect and store these data. It needs to put in special and huge manpower and time to obtain effective information related to its business (Jiang, 2007).

Thus, it is responsible for power market organizers to offer just, fair and open information service for intra-regional generation enterprises and companies with large electricity consumption. Also, they have an obligation to build an intra-regional unified information service platform for large data power market, helping market members to find problems, optimize operation and increase benefit (Shi, 2000; Ye, 2009). This paper would design the modes of application service and consulting service to offer application service and consulting service for market members. Consequently, we can help them to get the picture of the market operation situation, grasp market trends, reduce operational costs, participate in trade reasonably and increase the whole benefit.

2 THE INFORMATION COMPOSITION OF ELECTRICITY MARKET LARGE DATA

Electricity market large data contains multi-source heterogeneous information source sets. As the publisher of large data, transaction center offers information service to four objects (power grid, plant, large customers and the public). The data cross association between the above five objects is shown in Figure 1.

There are two main kinds of data in electricity market large data: power grid data sources and public information source for the government and society.

Power grid data sources include the information mastered by transaction center, power grid, plant and large customers. As the information publisher, transaction center masters comprehensive information. And all the data related to trade can be obtained by it. Most of the information mastered by power grid, plant and large customers can

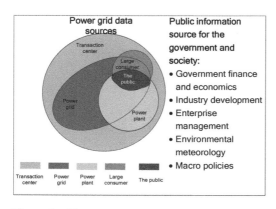

Figure 1. The construction of the information of electricity market large data.

be offered by transaction center. However, some information is private, which is mastered by them respectively. While the information mastered by the public is open.

Outer cataloging data sources are government social public information sources, including government finance and economics, industry development, enterprise management, environmental meteorology, macro policies and so on. This kind of information is obtained from the outside of power grid.

2.1 Government finance and economics

Government finance and economics data are reflected by three indexes, which can be expressed by data from direct observation.

1. Regions GDP: It is the total values of goods and services produced in a country or a region during a given period (a quarter or a year)
2. Consumer Price Index (CPI): It is a macroeconomic indicators which reflects the changes of the price level of goods and services purchased by households.
3. Producer Price Index (PPI): It is an index to measure the average changes of ex-factory prices of manufacturers.

2.2 Industry development

Industry production total value growth rate: According to the industrial category, industry is divided into the first, second and third industry, to calculate the total ratio of industry production total value with the previous year.

Industrial structure: It is the composition, connection and proportion relations of the industries.

2.3 Enterprise management

Profitability: Profitability indicators mainly include the operating profit ratio, the profit margin of costs, earnings cash guarantee multiple, rate of return on total assets, return on net assets and return on capital. In practice. Listed companies often use earnings per share, dividend per share, price-earning ratio and net assets per share to evaluate the profitability.

Debt paying ability: It refers to the ability to repay the long-term debt and short-term debt by enterprises with their assets. Cash payment ability and the ability to repay its debt, it is the key to the healthy survival and development.

Development ability: It is used to analyze development ability mainly considering the following eight indicators: operating income growth rate, hedging and proliferating ratios, rate of capital accumulation, total assets growth rate, business profit growth rate, technology investment ratio, the average growth rate of business income in three years and the average growth rate of assets in three years.

2.4 Environmental meteorology

Temperature: The change of temperature has certain effect to the electricity use of users.

Precipitation: Hydro power capacity depends on the amount of precipitation.

Natural disasters: The data of natural disasters (earthquake, frozen) can provide early warning for safe operation of the power system.

2.5 Macro policies

Macro policy information service is mainly provided to the user in the form of announcement or web link. It mainly includes the national policies related to energy, low carbon, energy saving and environmental protection policy.

Based on the above various information sources, we can construct information service platform. The platform provides two kinds of services, the mode of application services and consulting services.

3 APPLICATION SERVICE MODE OF INFORMATION SERVICE PLATFORMS FOR ELECTRICITY MARKET LARGE DATA

Three kinds of information service mode are provided.

1. User's free retrieval application service mode
2. Platform information publishing service mode

3. Flexible and interactive self-defined package and intelligent pushing service mode.

3.1 *Platform information publishing service mode*

The information service platform for electricity power large data mainly includes five modules, respectively being introduced as follows:

1. Power market reform

   It mainly sets three topics, which mainly introduces the review of the history of power market reform, the relevant laws and regulations and the situation of current market transactions. At the same time, it can also introduce the situation of foreign power market construction, mainly including California electricity market, the PJM electricity market, the Nordic electricity market, etc.

2. Information publishing

   It mainly sets three topics, including the news broadcasts related with power market transaction, the situation of current electricity trading (trading price and trading electric quantity, etc.), power market annual (quarterly) report, etc. This part is the basic module of the electric power large data information service, which mainly provides the fundamental information of market.

3. Information push

   It mainly sets two topics, including information service subscription and mobile client intelligent information push. The information service subscription is divided into two categories: subscription by keywords and subscription by information type. Mobile client intelligent information push can take the form of Wechat subscription number to push power market information service to users. Push content mainly includes: the power market transaction data of this day, the industry policy, the international electricity market memorabilia and etc.

4. Member service

   Information service platforms for electricity market large data provide paying members customized service, including consulting platform and customized package. By consulting platform, we can achieve real-time online contact with related experts and scholars, the user can also customize the content of package service according to their own preferences.

5. Links to sites

   This part mainly provides convenience for the users to search other relevant information. And links to sites mainly includes grid companies, generation enterprises, emission right transaction platforms such as carbon emission, overseas power market information service platform and etc.

3.2 *Flexible and interactive self-defined package and intelligent pushing service mode*

1. Flexible and interactive self-defined package mode

   Users can only passively accept the type and content of certain packages. Facing to all types of power market users, it develops basic module respectively for the user to choose from. For each type of user, it provides the content customization of fixed module number respectively. And the content of additional customized information belongs to the category of value-added services.

   Government: the trade/industry development analysis prediction module, saving power and electric quantity index statistics module, industry/region energy efficiency analysis module, energy efficiency per unit GDP analysis module, industry load characteristic analysis module and etc.

   Power grid: electric quantity balance analysis module, electricity demand analysis module, purchasing electricity auxiliary decision analysis module and etc.

   Generation: fairness analysis module, power market bidding strategy analysis module and etc.

   Big users: electricity consumption behavior analysis module, power supply security analysis module, the economic benefits of direct electricity purchase analysis module and etc.

   Based on the above modules, users can choose the information type that they want to get freely according to the actual needs. According to user's choice, large data information service center provides unique package information. The main characteristics of flexible and interactive self-defined package mode are:

   1. Set optionally. According to actual needs, the user can choose time limit of package (a month, half a year, a year), type of package information (basic information, transaction data, analysis report, etc.), package delivery time, etc. Information types, information amount is completely decided by the users.
   2. Change optionally. The user can change package content according to the actual demand situation. And the following month would act.
   3. Replace optionally. The user can convert package within different information content which is equivalent to each other according to the actual demand situation.

2. Intelligent pushing service mode

   According to the user's habit of customizing and retrieving, it makes frequency analysis and correlation analysis to recorded data. On the basis of the analysis results, it pushes the concerned information to users. Learning methods are:

1. Learning directly. It provides operable friendly interface, and allows users to actively put forward suggestion of modification and maintenance of user mode.
2. Feedback learning. They can learn by the all previous feedback suggestion from users' retrieving results.
3. The historical study: Through the analysis of users' historical searching records, after a period of accumulation, we can discover the potential law of user requirements.
4. Observational learning. For client search tools and systems, by making full use of the advantage of combining client tools and environment, we can observe from more aspects and get the characteristics information that is relevant to the user.
5. Reasonable speculation. According to the type of user registration information, we can roughly infer the information that users care about. For example, power grid company customers focus more on power trading information. Power plant customers more concern about the price bidding information. Users concern about peak and valley price and other information which is closely linked with electricity consumption, etc.

4 RESEARCH ON THE MODE OF CONSULTING SERVICE FOR ELECTRICITY MARKET LARGE DATA

4.1 *Intelligence consulting service mode*

First of all, according to the content of power market large data, it build market analysis knowledge

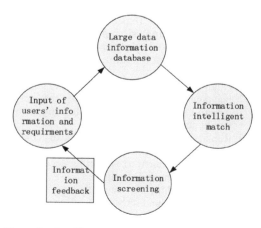

Figure 2. Intelligent consulting service mode.

base of systematization. Or in view of the problems or theme, it builds knowledge base. And then it realizes the fast matching between users' consultation content and knowledge base, to provide users with consulting results.

Intelligence consulting service mode is mainly for the behavior of ordinary users browsing the basic information. This kind of information has large amount but low value, and can be obtained widely. Therefore, all users can search the content of the information they need by the form of intelligent search.

4.2 *Half intelligence consulting service mode*

First of all, according to the content that users need to consult, it retrieves a correlation content from large data platform artificially. And then the professional staffs integrate the content, feedback to the user in the form of the consulting report.

Half intelligence consulting service mode is mainly aimed at the demand for specific information from member users, such as power market development analysis report, power market users' behavior analysis, etc. This kind of consulting reports has high value. They can guide users to make scientific and reasonable management strategies. Therefore, only paying members have privilege to require to provide relevant consulting report. Moreover, different types of membership grade is for different consulting report.

4.3 *Artificial training service mode*

Electricity market large data information service platform builds network training function module, to offer network classroom service for users' consulting.

Artificial training service mode belongs to the highest level of power market large data consulting services, which provides core member of the information service to consulting and training. Users can not only master power market development trends, but also can accept the 'one to one' precise service. Large data information operators can according to customers' actual condition, tailor for customers' behavior strategy to participate in power market transaction, and put forward risk aversion strategies.

5 CONCLUSIONS

This paper built information service module through analyzing the information category composition of electricity market large data and the attention to all kinds of information from each market main body. Moreover, based on the above original information and processed information service module, we built three kinds of power market large data information application service modes and three kinds of consulting service modes. They involved the demand of each main body participating in the power market. Also, they can change along with the demand of the market main body. We can make the most application of the value of power market large data, according to electricity market large data information service platforms achieved by the model in this paper.

REFERENCES

Jiang Jianjian, Kang Chongqing, Xia Qing: Key Information and Rationality Criteria of Electricity Trade. Automation of Electric Power Systems. 2007, 05:34–39.

Shi Yongli: Research of electric power market operation modes. Modern Electric Power. 2000, 04:94–100.

Yan Long-chuan, Li Ya-xi, Li Bin-chen, Zhao Zi-yan: Opportunity and Challenge of Large data for the Power Industry. Electric Power Information Technology. 2013, 04:1–4.

Ye Qing, Tuo Guangzhong, Zhang Yong: Transaction Information Publication System of Power Market and Its Operation Mode. North China Electric Power. 2009, 12:1–4.

Advances in Energy, Environment and Materials Science – Wang & Zhao (Eds)
© 2016 Taylor & Francis Group, London, ISBN 978-1-138-02931-6

Application of fieldbus control system on AC/DC microgrid

Yunhui Zhang & Jiang Chang

School of Mechanical and Electrical Engineering, Shenzhen Polytechnic, Shenzhen, China

ABSTRACT: Microgrid is becoming a hot spot for the development of world energy system, but also opens up a new direction for the use of renewable energy. When the microgrid control system makes the decision, it depends on acquisition and transmission of the information more than the traditional grid. At the same time, the response characteristics of microgrid equipment have higher requirements on the reliability and real-time communication. In the paper, the advanced FCS technology is used in AC/DC hybrid microgrid, and it makes full use of the advantage of the fieldbus control technology in microgrid communication and data acquisition. It can achieve high precision, high reliability, rapid data acquisition and monitoring for the microgrid, and then improve the operation stability of microgrid.

1 INTRODUCTION

New energy application market is starting in China, concern about the microgrid has been rising. In the 12th five-year plan on renewable energy development, it has been proposed that 30 new energy microgrid demonstration projects will be built in 2015. Although total size of the projects is not big, they are significant. Future development potential of microgrid is huge.

As a new research area, microgrid has some characteristics that are different from the traditional power grid in many aspects (Wang et al. 2005, Wang & Li. 2010, Puttgen et al. 2003). For example, inverter is taken as grid connection interface for the distributed power supply in microgrid, while synchronous generator is often used in traditional power grid, so different operation control and management mode need to be adopted. Because of the existence of distributed power, bidirectional flow of energy becomes possible, so bidirectional flow of information must be realized for corresponding scheduling and protection. These characteristics result in microgrid control system, compared with the traditional power grid, relies more on acquisition and transmission of information when making decisions, while response characteristic of the equipments in microgrid demand for more real-time and reliability of the communication.

Therefore, the establishment of a perfect microgrid communication system is the basic requirement for the microgrid operation control and management. It is advantageous to obtain real-time, accurate and reliable information, can realize the system's centralized monitoring and comprehensive dispatch, and the system optimization control, make the system run more safe, economical, and reliable.

In AC/DC hybrid microgrid, master-slave control mode is the most common, and the data signals from all field equipments are transmitted to control room. However, it will inevitably lead to disadvantages such as low reliability, poor real-time. Since all field nodes are connected with the master control machine, the choice of it determines the configuration of the whole field equipment, the openness is poor. With the development of control, computer, network, module integration technology, Fieldbus Control System (FCS) is appeared in the late 1980's. As the basis of digital communication network, it establishes the communication between the production process and the control equipment, and the connection between the field, the control equipment and the higher control management levels. It is not only a grass-roots network, but also an open, new distributed control system. On the one hand, the solution based on open, standardized, overcomes the defects caused by the closed system; on the other hand, it becomes a new distributed structure, the control function completely into the site.

Open, dispersive, and digital communication are the most obvious advantages of fieldbus system. Because of its advantages, the advanced FCS technique. is used in AC/DC hybrid microgrid in this paper, make full use of the advantages of FCS Technology in microgrid control communication and data collection, to enhance stability and economic benefits for the microgrid operation, and more convenient for maintenance, operation, monitoring.

2 FIELDBUS CONTROL SYSTEM (FCS) AND ITS FEATURES

IEC gives the fieldbus definition (Yang. 2010, Li. 2013): Field bus is a technology which applied to the production site, and bidirectional, serial and multi node digital communication is realized between field equipments, and control devices. Fieldbus broke the structure of the traditional control system, and formed a new distributed control system, that is fieldbus control system. It is a new generation control system based on the base type pneumatic instrument control system, the electric unit combined with analog instrument control system, centralized digital control system, distributed control system.

FCS changes single scattered measure and control devices into network nodes, taking the fieldbus control network as the transmission link, each network node can be connected to the network system and control system which can communicate with each other and complete the control task.

The characteristics of FCS are as follows:

1. Bus connection type:
   Each field device is a network node on the bus, and the bus type network connection is used, the connection is simple and can be expanded;
2. Openness of the system:
   Interconnection and information exchange can be realized between different devices. Equipment with similar performance from different manufacturers can replace each other;
3. Digital and intelligent:
   Fieldbus instruments have digital communication capability, and digital communication and network connection are used between the equipments to replace analog signal transmission in the traditional measurement and control system. Digital communication is advantageous to improve the information quantity of communication transmission. Digital calculation can improve control accuracy;
4. Certainty and timeliness of communication:
   FCS provides the communication mechanism for the transmission of data, and provides the time management function to meet the real-time requirements of the control system;
5. Environmental adaptability:
   As the bottom level of the factory network, field bus system works in the front of the production site with bad working environment, and it has environment adaptability for different working conditions.

3 APPLICATION OF FCS IN HYBRID MICROGRID

3.1 Functional requirements for AC/DC hybrid microgrid system

The research object in this paper is hybrid microgrid system in Shenzhen Polytechnic, AC bus for three-phase 380 V, DC bus using DC600V. Distributed power (DG) on the AC bus is composed of 10 kW monocrystalline silicon, 10 kW polycrystalline silicon, 3 kW amorphous silicon, 3 kW wind turbine and so on. The DG of the DC bus is made of 4 kW silicon and 4 kW polycrystalline silicon. The energy storage system is composed of 50 kWh lithium iron phosphate battery and 50 kW 10 s super capacitor, and connects the bus and energy storage system to the grid through PCS 50 kW cabinet. There are user loads and adjustable simulation loads on AC bus and DC bus. The system uses the modular design, DC microgrid and AC microgrid, AC/DC hybrid microgrid can be implemented separately. Through flexible combination of modules, the microgrid system can be realized for various functions and occasions, and can meet the needs of many applications.

The main functions of the system are as follows:

1. The information can be real-time interactive among the modules of the system;
2. Real-time parameters and running status information, alarming information of the PV, wind power, AC/DC load and hybrid energy storage subsystem can be acquired in real time. Monitoring comprehensive information of microgrid system, including microgrid system frequency, public connection point voltage, distribution exchange power, statistics and analysis in many aspects, to realize microgrid full control;
3. Predict the DG power generation and load demand, develop an operation plan. According to collected current, voltage, power and other information, real-time adjustment of the operation plan, control the starting and stopping of DG, load and energy storage device, guarantee stability of microgrid voltage and frequency, and provide related protection function for the system;
4. Implement the DG regulation of the microgrid, the charge & discharge control of the hybrid energy storage and load control, and the transient power balance and the low frequency load shedding of the microgrid and the realization of the transient safe operation of the microgrid;
5. Through real-time comprehensive monitoring of the microgrid, achieve the optimal control of DG, energy storage device and load energy according to the power and load characteristics when the microgrid is connected to the grid, from the network operation and state transitions, so to achieve the safe and stable operation of microgrid, it needs to improve the energy utilization of microgrid;
6. From the point of view of the security and economic operation of the microgrid, the microgrid can be coordinating optimal dispatched, and

the microgrid can accept the adjustment control command of the higher power distribution network.

3.2 *System design of microgrid control network*

Because of the distributed characteristic of microgrid, massive amounts of control data and the flexible and changeable control mode, it is difficult to realize flexible and effective dispatch by centralized control method with unified judgment and scheduling by the dispatching center. So the control power is dispersed to the microgrid element, the distributed coordination control method which is based on the microgrid dispatching by the components to change the running state will effectively solve these problems. Thus, FCS technology is the core technology of microgrid control, but also the key to improve energy efficiency, reduce costs and improve the reliability.

In order to meet the functional requirements of the system, the internal information of the microgrid is divided into three layers: Field intelligent node control level, MGCC microgrid center control level and remote monitoring level. Communication structure diagram is as shown in Figure 1.

The fieldbus structure based on Modbus TCP is adopted in the field intelligent node control level. Field intelligent node provides Ethernet interface.

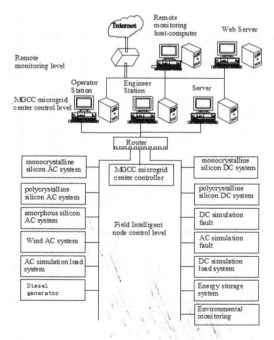

Figure 1. System communication structure diagram.

Modbus TCP industrial ethernet protocol, which has been widely used in industrial field, is used to connect the field equipment network with the central control level of MGCC microgrid, and realize data exchange. MGCC center control level including engineer station, operator station, etc. Modbus TCP protocol is adopted to connect the layer between monitoring computer, and the layer and upper management information network are connected, data exchange is realized, and the monitoring and scheduling is easier. The remote monitoring level includes the monitoring PC and Web server, etc., so as to realize the scheduling between the microgrid and the centralized management of information.

1. Field intelligent node control level

The microgrid includes many subsystems, scattered in the site, and the whole microgrid control system is divided into distributed power, energy storage unit, load simulation, AC/DC load system. Therefore, multiple field intelligent nodes are respectively used to collect real-time operation parameters, such as PV array output current and voltage, inverter operation state parameters, inverter output grid electricity and electrical energy quality parameters, environmental parameters. In the whole system, intelligent nodes also act as a local controller role.

The control nodes, the node and the MGCC center control level, are connected by the redundancy ethernet switch, and the field level control network is isolated from the upper level network.

On the intelligent node control level, the real-time parameters are processed by the arithmetic computation, and form corresponding local transient control strategies, also transfers real-time data through industrial Ethernet to the MGCC microgrid center control level. The overall control strategy of the microgrid is formed, and the feedback information and control commands of the MGCC microgrid center control level are transmitted to the distributed controller to ensure the stable operation of the microgrid. Therefore, the monitoring node should not only have a strong data acquisition and processing power, communication ability, and the stability and real-time requirements are very high.

Each field intelligent nodes use M44MAD series PLC as the core of Modbus TCP distributed controller. The controller has ethernet communication module, supports the TCP/IP protocol, and has IP address itself. M44MAD series PLC can be remote debugged and redeveloped through Modbus TCP industrial ethernet. The PLC can communicate with PC through many kinds of fieldbus, including CAN and Modbus/TCP.

2. MGCC microgrid center control level

MGCC microgrid center control level has one or more computer, printer, etc., including the operator station, engineer station, data management station (server) and other equipment. As human-machine interface for the whole microgrid, it mainly completes the monitoring, control and operation of the controlled equipment, responsible for the data collection, verification, and record of the alarm information, and the historical data recording and other functions. This monitoring platform can directly control every single controller of the microgrid and collect the working status of each device through the ethernet. Through the network, MGCC energy management software for microgrid can complete data acquisition, running control, power dispatch, and other functions for the subsystem, and all the controller can realize programming, debugging, download.

3. Remote monitoring level

Remote monitoring level has one or more computer, printer, etc., including host-computer and Web server, achieve centralized management of the scheduling and information between the microgrid. For remote monitoring level, host-computer monitoring interface is prepared by Kingview, including data acquisition, data display, data storage, data query and alarm prompting modules. Data acquisition module has been listening to the TCP/IP network interface monitoring center; receive the real time data from MGCC transmission through Ethernet. According to received packet identification, data display module display data from each node. Data storage module use Server SQL database to save data. The data query module has the function querying the historical data according to time and equipment number, and can draw the data result into curve shape, and visually identify the changing trend of the data. Alarm module has hierarchical alarm function, when the data exceeds the threshold, pop up device alarm dialog box.

Web server using the standard Internet browser can read all kinds of information equipment, Modify the configuration of the device and view the history of the fault records, but also the diagnostic function of the system equipment, and it can realize the upgrade of automation system or intelligent device in the Ethernet environment, greatly simplifying the renewal and rebuilding of the control system.

3.3 *Basic communication process*

The basic flow of data communication among different layers is as follows:

Step 1: Collect the operating parameters and equipment operating parameters such as voltage, current, temperature, air pressure, wind speed, wind direction, etc. The above parameters are collected by field intelligent node control level (PLC), then packaged into a Modbus TCP packet which is sent to the engineer station;

Step 2: After the engineer station received TCP Modbus data frame, it sends confirmation message to the intelligent node which sent communication packets, to fulfill a complete Modbus communications;

Step 3: Engineer station will transmit Modbus TCP data frame to the host on the remote monitoring level.

4 SUMMARY

It is inseparable from the data acquisition system to form a correct and real-time control strategy for microgrid system. Microgrid with traditional data acquisition is through the ways such as analog, then its anti-interference ability is weak, the transmission process has the signal attenuation, the data precision is not high. Therefore, the data acquisition characteristics of the microgrid and topological structure of industry ethernet network, which achieves conformance data communication from the field to the control layer until the management level, are used in this paper. Based on FCS, realize microgrid real time numerical data acquisition and monitoring, so as to realize high precision, high reliability, fast data acquisition for microgrid. The system can realize the comprehensive monitoring of information, and is easy to maintain, and has the on-line, real-time function.

REFERENCES

Li Zhengjun. 2013. Fieldbus and industrial Ethernet and its application technology. Beijing: Machinery Industry Press.

Puttgen, H.B., P.R. Macgrego, F.C. Lambert. 2003. Distributed generation: semantic hype or the dawn of a newera. IEEE Power Energy Magazine 1:22–29.

Wang Cheng-shan, Li Peng. 2010. Development and Challenge of Distributed Generation, Microgrid and Smart Distribution System. Automation of Electric Power Systems 2(34):10–14.

Wang Jian, Li Xing-yuan, Qiu Xiao-yan. 2005. Power System Research on Distributed Generation Penetration. Automation of Electric Power Systems 9(24):90–97.

Yang Xian-hui. 2009. Network control system-Field Bus Technology (Second Edition). Beijing: Tsinghua University Press.

Advances in Energy, Environment and Materials Science – Wang & Zhao (Eds)
© 2016 Taylor & Francis Group, London, ISBN 978-1-138-02931-6

The regulate and control method research of wind acceptance regarding power peak adjustment

Tao Zhang
Electrical Power Research Institute of Liaoning Electric Power Co. Ltd., Shenyang, China

Junjie Sun
School of Electric Engineering, Wuhan University, Wuhan, China

Chao Wang, Qiang Zhang & Gang Wang
Electrical Power Research Institute of Liaoning Electric Power Co. Ltd., Shenyang, China

ABSTRACT: Traditional calculation method of the ability to wind acceptance, without considering the scheduling operation mode of actual power grid, is lack of common practicability. Under the premise of considering peak regulation constraints, sensitivity analysis on the important boundary conditions which affect the ability to wind acceptance, relying on the D5000 technical support system, reading operation data of the generator mode in power grid. The theory and method are proposed, by which wind acceptance index of provincial grid can be calculated, and then optimization scheduling model which coordinate power grid scheduling plan with accept ability of wind power is constructed, thus realizing the coordinated operation of the conventional power and wind power is realized, reducing the impact of wind power to the grid, and making the calculated results more accurately reflect the actual acceptance of wind power. The run data proves the calculation method of the ability to wind acceptance is effective.

1 INTRODUCTION

In the state supports the renewable energy industry policy regulation and encouraged, wind power in China got rapid development (Feng, 2011; Xie, 2011; Zhang, 2011). China has reached 60 of 830 MW wind power grid capacity, capacity of 1, 00.4 billion kWh in 2012. New wind power installed capacity of up to 14 000 MW, wind power has been over nuclear power become the third largest after coal and hydropower main power supply (Xie, 2012; Hou, 2012; Geng, 2011; Liu, 2011; Zhang, 2012; Huang, 2010). With the rapid growth of wind power installed capacity, large-scale wind power grid given problem more and more prominent, all regions have appeared different degree of wind power brownouts phenomenon, research on the acceptance ability of power grid wind power in the industry. The traditional calculation method of wind power acceptance ability, mainly according to the factors that may influence the system load, combined with the wind load characteristics to define (Zhang, 2012; Wang, 2011; Wang, 2010; Yao, 2010; Sun, 2011; Li, 2011). The calculation method is very adaptive to the theoretical analysis, but in determining the actual power grid wind power acceptance, without considering the actual

operation mode of the power grid, so lack of common utility. At present, the method of calculating the acceptance ability of wind power has been studied: The literature (Li, 2010) was established to accommodate wind power capacity for peak objective function model, in the calculation of the capacity of the wind power, the uncertainty of the intermittent energy is considered, but the influence of the power side is not considered; The literature (Han, 2010) proposed a method for calculating wind power acceptance capacity constraints to the peak load capacity of power network, quantitatively researching on the impact of wind power on power peaking, but not related to the actual power grid dispatching operation mode; The literature (Zheng, 2010) studied the optimal dispatching method of wind power to improve the acceptance of wind power, what basis on considering grid security and stability constraints, but this method is only applicable to optimization of wind power operation strategy, and does not apply to other units in the arrangement of grid short boot mode; The literature (Zha, 2012) built the simulation platform which based on the timing analysis of Delphi, under the condition of the transmission power of the tie line, the load curve is adjusted, the improvement of wind power acceptance ability in the new load time series curve, however, due to not considering the scheduling scheme in the unit on the boot mode, the calculation accuracy is affected. Therefore, in order to meet the requirements of the rapid development of wind power, in strengthening the power grid construction, the method of calculating the wind power acceptance ability of the actual operation mode of the power network and the power grid is needed, not only can meet the peak power constraint, but also can maximize reflect the wind power acceptance ability of the actual power grid.

On the basis of the above research, this paper presents a method of calculating the acceptance of wind power based on the operation mode of scheduling. This method is based on the D5000 intelligent power grid dispatching technology support system, and the operation data of the short time power mode of the power grid is read, considering power balance equations and contact line adjustment coefficient, the peaking capability of the unit comprehensive constrained, on an actual power system boot way to properly adjust, achieving the coordinated operation of conventional power and wind power, reduce the impact of wind power on power grid operation, so that the results can more accurately reflect the actual acceptance of wind power grid.

2 PRINCIPLE ANALYSIS OF WIND POWER ACCEPTANCE

2.1 *Wind power acceptance space under the constraint of peaking*

In the analysis of power system active power balance, on the one hand to satisfy the peak load capacity of power demand, on the other hand also consider valley load time grid unit can reach the minimum output. In consideration of the valley period of load peak power constraint based, the difference between the value of load and the system in all the units minimum output electricity can be supplied by wind power, the difference of power is the wind power of the receiving space. Under the constraint of power peaking wind power maximum acceptance space and load forecasting value, tie line, minimum unit output and reserve capacity, calculation formula is:

$$P_{max} = P_{pre} + P_{line} - P_{Gmin} - P_R \qquad (1)$$

In the formula: represents the peaking capacity under the constraint of maximum wind power acceptance space; represents the maximum load; represents contact line plan represents minimum output of conventional units; represents spinning reserve capacity.

By formula (1) can be seen, when the ability of the conventional unit clipping small and power load demand is also very hour, the smaller the force and the load demand is also very small, the smaller the wind power acceptance space, when the peak wind receiving space constraint is negative, it should limit the output of wind power to meet peak constraint.

This article mainly concerns about the peak load restraint multi variable correlation, put forward the formation principle of wind power acceptance space. Highlight the peak load restraint the lowest margin, the selected month peak to valley difference, as a typical, launches the analysis, program and the time scale in typical day in 96 time node as a benchmark. Taking the provincial power grid for a typical day operation of the power grid as an example, as shown in Figure 1 the principle of peak under the constraint of wind power accommodation space.

Figure 1 is the provincial power grid a typical day, the maximum adjustable output, full bore power supply load and the province of the minimum adjustable output curve. In Figure 1, A curve is the maximum adjustable output curve, the C curve is the province's minimum adjustable output curve, and the B curve is the full diameter of the load curve.

Where: the maximum adjustable output curve by the plan of the maximum direct water, fire, nuclear power units, the tie line is received by the electric power (both are converted to the power supply side) and the maximum power station is added to the power and the spinning reserve; Provincial minimum adjustable output curve by the plan of the minimum of direct regulating water, fire and nuclear power unit output, tie line by electric power were converted to the supply side) and the smallest local power plant Internet power can be obtained by adding; full bore power load curve by the full aperture load statistics obtained.

Figure 1. Schematic diagram of wind power acceptance under peak regulating constraint.

From the angle of the operation of the power grid, wind power with the change of the wind speed is changeable. However, such stochastic variations can be grid accepted depending on current grid peak shaving means wind Internet space size, including with conventional power supply, load minimum output power difference. The accurate index value of the wind power is the target of this article. In Figure 1, B curve and C curve of the space for the wind power. The maximum/minimum output of the unit and the load peak and valley changes are reflected by the formula (1). What can be seen from Figure 1, the time period of the load is the smallest of the power grid acceptance of the wind power, which the main time of wind power.

2.2 *Beyond the peak wind power dispatching constraint*

Actual power grid wind power to accept ability essentially reflects to abandon the wind amount of wind power. When the wind power forecasting power exceeds the maximum wind power, the measures to limit the wind power output are required. Two typical wind power wind conditions are listed below.

1. Off-peak load of wind power
   When the thermal power units because of the heating or the stability of the system operation constraints lead to valley period of load output large, in order to ensure the system active power balance must be removal unit minimum adjustable output curve (including wind turbine output) was higher than that of low load curve to ensure the safe and stable operation of the system. In Figure 2, the shadow part is the output of the power grid during the load and the low period.
2. Peak load of the wind power
   When the wind power output is very large, there is a possibility that the minimum adjustable output curve (including wind turbine output)

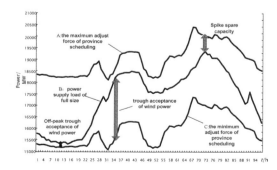

Figure 2. Curves of abandon wind power and others during low load period.

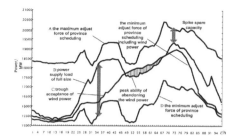

Figure 3. Curves of abandon wind power and others during peak load period.

is higher than that of the load peak curve. As shown in Figure 3 if the unit minimum adjustable output higher than this moment full bore load, in order to ensure the system active power balance must be removal unit minimum adjustable output curve (including wind turbine output) is higher than that of the peak load curve of parts to ensure the safe and stable operation of the system. In Figure 3, the shadow part is the output of the power grid of the load peak period.

3 CALCULATION METHOD OF WIND POWER ACCEPTANCE CAPACITY BASED ON THE OPERATION MODE OF DISPATCHING

3.1 *Calculation process design*

Based on the operation modes of the acceptance of wind power capacity calculation method is considering the unit output, contact line adjustment factor, system spinning production capacity, local hydro thermal power electric power to the Internet. According to the operation mode of the power network, the reasonable peak mode is determined by the peak load forecasting. Then in the peak load time of electricity under the premise of network loss and the Internet power plant, the network loss rate based on industry standard selection. Determining the load low moment smallest boot mode that on the basis of participate in all kinds of power peaking capacity analysis and meet the Electricity Regulatory Bureau approved the minimum start-up mode, in order to obtain the minimum power supply network, combined with the low load, and peak earning power. Based on the study of wind farm cluster effect, conclude the power grid safety electric capacity, so as to provide technical support for the reasonable arrangement of wind power planning and scheduling strategy.

The calculation process of the capacity of wind power based on the operation mode of scheduling is shown in Figure 4.

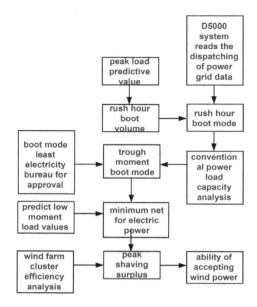

Figure 4. Design drawing of calculation process.

3.2 Calculation method

1. Calculation principle
   1. According to the maximum load of the month, the minimum load, the minimum power of the power plant, the power of the power plant and the power to start the power, the power of the wind power is calculated on a monthly basis;
   2. According to the power balance of the power grid, during the low load time peak pressure, prone to abandon wind phenomenon, so adopt the principle of load low peak shaving, calculate the wind power acceptance;
   3. In the peak load period, according to the actual power grid dispatch load statistics, the power to participate in the power balance;
   4. In the off-peak load, Directing thermal power, nuclear power, hydropower, wind power, ting line transmission power and local Internet power in peak shaving;
   5. According to the peak time of power grid peaking capacity calculate the acceptance capacity of wind power in the month.
2. Calculation steps
   1. The calculation of peak output of the unit
      The power to turn on the Internet is the sum of the power, nuclear power and hydropower calculate by formula (2).

$$p'_{\text{Gpower max}} = p'_{\text{Gf max}}(1-\beta_f) + p_{\text{Gnmax}}(1-\beta_n) + p_{\text{Gwmax}}(1-\beta_w)$$

(2)

In the formula: represents direct power grid power combined (not considering the rotation of the spare capacity and the limited capacity of the unit); represents direct power to adjust the total power (not considering the spinning reserve and the unit output is limited); represents coal-fired power plant auxiliary power rate; represents straight nuclear power boot combined; represents the rate of nuclear power plant auxiliary power; represents direct the waterworks boot combined; represents the rate of power plant auxiliary power.

According to the sum of the contact line output and the power of the Internet power is power of the network, conclude formula (3).

$$p_{\text{power max}} = (p_{\text{line max}} + p_{\text{Gpower max}})(1-\alpha)$$ (3)

In the formula: represents peak grid for power; represents tie line output; represents the loss rate of Net.

According to power balance, Full bore loads in rush hour is the sum of Network for power and local electric power grid power, conclude formula (4).

$$p_{\text{L max}} = p_{\text{power max}} + p_{\text{G max}}$$ (4)

In the formula: represents peak load; represents the local fire power grid peak power.

Considering the spinning reserve and the limited capacity of the unit, use formula (5) to calculate the capacity of the direct adjustment.

$$p_{\text{Gf max}} = p'_{\text{Gf max}} + p_{\text{G}} + p_{\text{Glim}}$$ (5)

In the formula: represents spinning reserve; represents unit output capacity.

Formula (2)–formula (5) can be adjusted directly to the power of the starting capacity, as shown in formula (6).

$$p_{\text{Gf max}} = \frac{p_{\text{L max}} - p_{\text{Gpower max}} - (p_{\text{line max}} + p_{\text{Gn max}}(1-\beta_n) + p_{\text{Gw max}}(1-\beta_w))(1-\alpha)}{(1-\alpha)(1-\beta_h)} + p_o + p_{\text{Glim}}$$

(6)

2. Calculation of the low output power of the unit
   According to the minimum starting mode of the thermal power unit, determine the power generating capacity on the peak time and starting capacity of thermal power units in the trough.

   According to formula (7) calculate other boot capacity thermal power unit on the peak time.

$$p_{Gf\,max} = p_{G\,max} - p_{Gh\,max} \qquad (7)$$

According to other boot capacity thermal power unit and peak rate, use formula (8) to calculate other starting capacity of thermal power units in the trough $p_{Gf\,min}$.

$$p_{Gfe\,min} = p_{Gfe\,max}(1 - \delta) \qquad (8)$$

In the formula, represents the other peak power unit rate.

Therefore, total capacity of thermal power generating units at low point $p_{Gh\,min}$ is:

$$\begin{aligned} p_{Gf\,min} &= p_{Gh\,min} + p_{Gf\,else\,min} \\ &= p_{Gh\,min} + p_{Gf\,else\,max}(1 - \delta) \qquad (9) \end{aligned}$$

3. Calculation of peak load capacity of power network

Direct power on the Internet is the sum of thermal power, nuclear power and hydro power of the Internet in the trough, use formula (10) to calculate.

$$\begin{aligned} p_{Gpower\,min} = {}& p_{Gf\,min}(1 - \beta_f) \\ &+ p_{Gn\,min}(1 - \beta_n) + p_{Gw\,min}(1 - \beta_w) \end{aligned}$$
$$(10)$$

In the formula: represents direct power grid power combined; represents direct power to adjust the total power; represents straight nuclear power boot combined; represents direct the waterworks boot combined.

According to Network for power is the sum of tie line output and direct power on the Internet, conclude formula (8).

$$p_{power\,min} = (p_{line\,min} + p_{Gpower\,min})(1 - \alpha) \quad (11)$$

In the formula: represents trough for electric power network; represents tie line output.

Using formula (12) to calculate peak load capacity when the load is low.

$$\begin{aligned} p_{peak} = {}& p_{L\,min} - p_{power\,min} - p_{else\,min} \\ = {}& p_{L\,min} - p_{else\,min} - (p_{line\,min} \\ &+ p_{Gf\,min}(1 - \beta_f) + p_{Gn\,min}(1 - \beta_n) \\ &+ p_{Gw\,min}(1 - \beta_w))(1 - \alpha) \end{aligned}$$
$$(12)$$

4. Calculation of wind power acceptance

According to the wind power installed to provide network for power equal to the load capacity, conclude formula (13).

$$p_{Gwind}(1 - \beta_{wind})(1 - \alpha)\varepsilon = p_{peak} \qquad (13)$$

4 ANALYSIS OF WIND POWER ACCEPTANCE CAPABILITY OF LIAONING POWER GRID

4.1 Liaoning power grid load status

Liaoning power grid is the main power grid, which has less power than the coal, and no gas, fuel and other power generating units fast peak load capacity of power network is relatively weak. At the same time, Liaoning power grid thermal power units due to poor coal quality and equipment defects will always exist, put great pressure on grid trough peak shaving. In addition, grid heating unit of Liaoning is more, in the winter heating period, heating units need to heat, the output can be adjusted range limited, therefore, Winter is the most difficult period of peak of Liaoning power grid, and the reverse peak winter wind power output has a greater difficulty of power peaking.

4.2 Computational boundary conditions

In the case of the system power supply structure, the whole network of wind power accommodation affected by load size, water seasons and dry seasons, the heating period and non heating period and other multiple factors, the ability of the wind power system has obvious seasonal characteristics. Therefore, in this paper, we consider the factors such as the starting mode and load characteristics of different months, 2013~2015 in Liaoning power grid is the monthly peak load balance, According peak earnings to calculate the ability of wind power consumptive each month. In the process of load balance, considering the full aperture load, local hydrothermal Internet power, and tie line by electric capacity, system spinning reserve capacity and additional units to participate in peaking factors.

4.3 Analysis of wind power acceptance capability of Liaoning power grid

Based on the 2.2 section, the method of calculating the wind power acceptance capability based on the operation mode of the scheduling is proposed, the principle of scheduling the daily scheduling of unit output, on the basis of meeting the minimum boot mode, combined with the annual load forecasting and peak and valley difference of each level, concluded Liaoning wind power accommodation. The results are shown in Table 1.

1. From the installed wind power seasonal average consumption perspective, 2013~2015 annual average wind power installed capacity for consumption were 2 810 MW, 2 750 MW and 2 290 MW.
2. From the balance results can be seen, 2013~2015 Liaoning power grid wind power installed in winter the lack of absorptive

capacity, summer can consume large amount of wind power, which installed wind power consumptive capacity is almost 0 on January and February, to force acceptance of wind power, it is necessary to break the minimum starting unit; Power grid acceptance of wind power reached the maximum on September and October.

5 CONCLUSION

To accommodate wind power dispatching accurate guidance and planning under the constraints of power peaking, this article mainly considers the global change of operation parameters, given power grid peak shaving methods of wind power capacity calculation principle of the accepted, and the establishment of the corresponding wind power scheduling index calculation model and method of acceptance. The solving process of peaking in the parameter of conventional thermal power units, output and standby, the same month the typical daily load peak and valley value and wind power admissible capacity is a key parameter modeling and computing, an accurate analysis of the peak load and wind power to accept the numerical relationship between amount of dispatch. Through the analysis of a provincial power grid actual examples, using this model to analyze and calculate that wind power acceptance ability index value of the peak load capacity of power network, example, and the results verify the index for the rationality of the evaluation of the power grid peak shaving means, and guide the intermittent wind power access mode of power grid dispatching safe and stable operation.

REFERENCES

Feng Li-min, FAN Guo-ying, Zheng Tai-yi, et al. Design of wind power dispatch automation system in Jilin power grid [J]. Automation of Electric Power Systems, 2011, 35(11): 39–42.

Geng Jing, Yan Zheng, Jiang Chuan-wen, et al. The studies of peak regulation transaction considering environment costs [J]. Power System Protection and Control, 2011, 39(20): 111–114.

Han Zi-fen, Chen Qi-juan. Wind power dispatch model based on constraints [J]. Automation of Electric Power Systems, 2010, 34(2): 89–92.

Hou Ting-ting, Lou Su-hua, Zhang Zi-hua, et al. Capacity optimization of corollary thermal sources transmitted with large-scale clustering wind power [J]. Transactions of China Electrotechnical Society, 2012, 27(10): 255–260.

Huang Xue-liang, Liu Zhi-ren, Zhu Rui-jin, et al. Impact of power system integrated with large capacity of variable speed constant frequency wind turbines [J]. Transactions of China Electrotechnical Society, 2010, 25(4): 142–150.

Li Xiao-ming, Rong Shi-yang, Li Xiao-long, et al. Analysis on maximum wind power penetration into power grid [J]. Hebei Electric Power, 2011, 30(1): 31–35.

Li Zhi, Han Xue-shan, Yang Ming, et al. Power system dispatch considering wind power grid integration [J]. Automation of Electric Power Systems, 2010, 34(19): 15–18.

Liu Yang-yang, Jiang Chuan-wen, Li Lei, et al. Peak regulation right trading model considering DSM [J]. Power System Protection and Control, 2011, 39(9): 38–43.

Lu Shun, Quan Cheng-hao, Liu Jian, et al. The analysis of wind dissolve technical support system [C] // The third paper of power quality and flexible transmission.

Sun Rong-fu, Zhang Tao, Liang Ji. Evaluation and application of wind power integration capacity in power grid [J]. Automation of Electric Power Systems, 2011, 35(4): 70–74.

Wang Rui, Gu Wei, Sun Rong, et al. Analysis on wind power penetration limit based on probabilistically optimal power flow [J]. Power System Technology, 2011, 35(12): 214–217.

Wang Zhi-ming, Su An-long, Lu Shun. Analysis on capacity of wind power integrated into Liaoning power grid based on power balance [J]. Automation of Electric Power Systems, 2010, 34(3): 86–88.

Xie Guo-hui, Li Qiong-hui, Gao Chang-zheng. The research of wind dissolve ability based on Balmorel model [J]. Energy Technology and Economics, 2011, 23(5): 29–33.

Xie Jun, Zhang Xiao-hua, Wu Fu-xia, et al. Peaking cost allocation using cooperative game theory and engineering concept [J]. Power System Protection and Control, 2012, 40(11): 16–22.

Yao Jin-xiong, Zhang Shi-qiang. Analysis on capacity of wind power integration into grid based on peak load regulation [J]. Power System and Clean Energy, 2010, 27(7): 25–28.

Zha Hao, Shi Wen-hui. The research of probability optimization coordinating wind dissolve ability [J]. Power System Protection and Control, 2012, 40(22): 14–17.

Zhang Kun, Wu Jian-dong, Mao Cheng-xiong, et al. Optimal control of energy storage system for wind power generation based on fuzzy algorithm [J]. Transactions of China Electrotechnical Society, 2012, 27(10): 236–240.

Zhang Li-zi, Fan Peng-fei, Ma Xiu-fan, et al. Timing planning method for investment of clustering wind farms considering risk of peak load regulation [J]. Proceedings of the CSEE, 2012, 32(7): 14–18.

Zhang Ming-li, Li Qing-chun, Zhang Nan. The evaluation method research of wind dissolve ability based on multi-objective economics peak shaving model [J]. Northeast Electric Power Technology, 2011, 22(9): 23–25.

Zheng Tai-yi, Feng Li-min, Wang Shao-ran, et al. An optimized wind power dispatching method considering security constraints in the power grid [J]. Automation of Electric Power Systems, 2010, 34(15): 71–75.

Advances in Energy, Environment and Materials Science – Wang & Zhao (Eds)
© 2016 Taylor & Francis Group, London, ISBN 978-1-138-02931-6

A research on wind and thermal power joint ancillary service market behaviors

Ang Li
School of Economics and Management, North China Electric Power University, Huilongguan Changping District, Beijing, China

Ling Wang
State Grid Liaoning Electric Power Co. Ltd., Benxi Power Supply Company, Benxi, China

Xingang Zhao
School of Economics and Management, North China Electric Power University, Huilongguan Changping District, Beijing, China

ABSTRACT: Wind and thermal power joint ancillary service is an effective and efficient way to guarantee the safe and stable operation of electricity systems, to reduce service cost and improve the integration of wind power. In order to make a further research on wind and thermal power joint ancillary service market behaviors, this paper, based on the evolutionary game theory, and in the condition of bounded rationality and incomplete information, simulates the transaction behaviors to reach equilibrium of wind and thermal power vendors in the ancillary service market and the process of their gaming. Thereby, the market behaviors of both wind and thermal power vendors in ancillary service market in different conditions are analyzed. The findings are expected to provide a basis for the macro-control of wind and thermal power joint ancillary service market.

1 INTRODUCTION

1.1 *Background*

China's current ancillary service work by starting up/shutting down thermal power or hydropower units, controlling the power and load of traditional generation to ensure the safety of grid operating and stability of electricity supply. With the rapid development of wind power in China, the problem of wind power integration has received widespread attention from the society and power dispatch departments, but serious problems of wind curtailment still exist in some areas especially in valley periods. The increasing installed capacity of wind power has made it more difficult to solve the problem of wind power integration. Simultaneously, the startup/shutdown cost of wind power units is much lower than traditional thermal power units. Renewable energy connection to the gird makes it more difficult for traditional thermal power units to provide ancillary services. In some places, problems began to arise that traditional ancillary services only cannot ensure the safety and stability of electricity supply. Therefore, wind and thermal power joint ancillary service is an efficient and effective way to resolve and adjust overcapacity to provide a good market and institutional environment.

1.2 *Literature review*

Because of the features: volatility, uncertainty and not being able to be peak-regulated (Tu 2009), wind power connected to the grid mainly affects the peak-regulation and the standbys of ancillary services, with little effect on other ancillary services (Hannele 2008). With the increase of wind power installed capacity, the effect caused by the instability of wind power load is more and more significant. It is more difficult for thermal power units to provide corresponding ancillary services. This will surely affect the fairness of generating vendors (Li 2013). Zhang (2010) through the simulation of chronological load curve and wind power output sequence curves in planning years analyzed the influence of large-scale wind power system on peak-regulation from a new perspective. Bai (2009) brought out a framework of promoting efficient wind power integration ancillary services. He (2013) proposed several sharing and compensation systems of ancillary service costs caused by wind power accessed. Zhang (2005) compared different ancillary service models and conducted extensive research on pricing and market system problems.

In addition, current researches on renewable energy storage and prediction technologies have made it possible for wind power vendors to join the ancillary service market. To be close to the actual operating situation of the system in the future, Wang (2011) introduced the interval value estimate of maximum wind power units' capacity under a certain confidence probability. Liu (2009) conducted a weighted average combined forecasting model of wind power farm output. Liu (2013) introduced the integrated learning method and conducted a dynamic adjustment of weight distribution integrated learning method for wind power prediction.

This paper, based on the evolutionary game theory, simulates the behaviors of wind and thermal power vendors in the joint ancillary service market, studies on the strategies wind and thermal power vendors choose to reach the equilibrium under the assumption of bounded rationality and incomplete information and on how to achieve the optimum benefit.

2 ASSUMPTIONS AND PARAMETERS DEFINITIONS

2.1 Model assumptions

This model has 4 assumptions:

Assumption 1: Both thermal and wind power vendors are bounded rationality.
Assumption 2: Assumptions are restricted by incomplete information.
Assumption 3: Wind power vendors are able to provide ancillary services themselves.
Assumption 4: In the wind and thermal power joint ancillary service market, the capacity can be completely integrated.

2.2 Parameters definitions

Parameters used in modeling and their economic definitions are given as follows:

Q_1-Conventional generating capacity of thermal power vendors
Q_2-Conventional generating capacity of wind power vendors
P-Electricity price
C-Generating cost of thermal power vendors and cost of providing ancillary services for themselves
C'-Unit cost of thermal power units to provide ancillary services
C''-Costs of wind power units to provide ancillary services
α-Capacity of thermal power units corresponding to ancillary services needed by wind power vendors.
β-Capacity of wind power units corresponding to ancillary services needed by wind power vendors.
a-Compensation of thermal power vendors for providing unit ancillary service.
Of which: $C' \geq C''$, $\beta \geq \alpha$.

3 BASIC MODEL

Replicator dynamics actually describe the dynamic differential equations that the frequency of a particular strategy used in a population. According to the principle of evolution, if the adaptive value and payment of a strategy is higher than the average, this strategy will develop in the population. That is, the survival of the fittest is reflected that the growth rate $\frac{1}{X_k}\frac{dx_k}{dt}$ of this strategy is greater than zero, which can be calculated as follows:

$$\frac{1}{X_k}\frac{dx_k}{dt} = [u(k,s) - u(s,s)], k = 1, 2 \ldots K \qquad (1)$$

In which, x represents the ratio using strategy k in one population; $u(k,s)$ represents the adaptive value using strategy k; $u(s,s)$ represents average adaptive value; k stands for a different strategy.

Table 1 is the payoff matrix of thermal and wind power vendors in the ancillary services market strategy gaming.

In this 2×2 asymmetric game payoff matrix, π and Π represent profits of thermal power vendors and wind power vendors. The payoffs of wind and thermal power vendors are:

$$\{\pi_1, \Pi_1\} = \{Q_1 P + \alpha a - C - \alpha C', Q_2 P - \alpha a\}$$

$$\{\pi_2, \Pi_2\} = \{Q_1 P - C, (Q_2 - \beta)P - \beta C''\}$$

$$\{\pi_3, \Pi_3\} = \{(Q_1 + \alpha)P - C, 0\}$$

$$\{\pi_4, \Pi_4\} = \{(Q_1 + \alpha)P - C, (Q_2 - \beta)P - \beta C''\}$$

If the proportion of ancillary services thermal power vendors choose to sell is x, the proportion they choose not to sell will be $1-x$. Similarly, when the proportion of ancillary services wind power vendors buy from thermal power vendors is y, the proportion they choose not to buy is $1-y$.

Then, the fitness thermal power vendors choose to sell their ancillary services is:

Table 1. Payoff matrix.

	Wind power vendors	
	Buying ancillary services	Not buying ancillary services
Thermal power vendors		
Selling ancillary services	π_1, Π_1	π_2, Π_2
Not selling ancillary services	π_3, Π_3	π_4, Π_4

$$\mu_T(S,J) = y(Q_1P + \alpha a - C - \alpha C')$$
$$+ (1-y)(Q_1P - C) \tag{2}$$

The fitness thermal power vendors choose not to sell their ancillary services is:

$$\mu_T(NS,J) = y(Q_1P + \alpha P - C)$$
$$+ (1-y)(Q_1P + \alpha P - C) \tag{3}$$

The average fitness of thermal power vendors is:

$$\bar{\mu}_T = x \cdot \mu_T(S,J) + (1-x) \cdot \mu_T(NS,J) \tag{4}$$

Therefore, the dynamic replication thermal power vendors choose to sell their ancillary services is:

$$\frac{dx}{dt} = x[u_T(S,J) - \bar{u}_s] = x(1-x)[y(\alpha a - \alpha C') - \alpha P] \tag{5}$$

Similarly, the dynamic replication wind power vendors choose to buy ancillary services is:

$$\frac{dy}{dt} = y[\mu_W(B,J) - \bar{u}_b]$$
$$= y(1-y)[x(Q_2P - \alpha a) + \beta C'' + \beta P - Q_2P] \tag{6}$$

Differential equations (5) and (6) describe the group dynamic of evolution system, and represent the adjusting speed of thermal and wind power vendors towards different trading strategies. If and only if the replicated dynamic equation is 0, the game will come to a relatively stable equalization. Making $dx/dt = 0$ and $dy/dt = 0$, four equilibrium results are obtained: $O(0,0)$, $A(1,0)$, $B(1,0)$, $C(1,1)$. If and only if $P + C' < a < \frac{\beta}{\alpha}(P + C'')$, D is the fifth equilibrium. The equation is:

$$D\left(\frac{Q_2P - \beta P - \beta C''}{Q_2P - \alpha a}, \frac{P}{a - C'}\right)$$

The Jacobian of the replication dynamic process is:

$$J = \begin{vmatrix} (1-2x)[y(\alpha a - \alpha c') - \alpha P] & x(1-x)(\alpha a - \alpha c') \\ y(1-y)(Q_2P - \alpha a) & (1-2y)[x(Q_2P - \alpha a) \\ & + \beta C'' + \beta P - Q_2P] \end{vmatrix}$$

In Table 2, for different equilibrium, if the sign of Jacobian is positive but the symbol of the trace is negative, the corresponding equilibrium will be stable. If the sign of Jacobian is positive and the symbol of the trace is positive too, the corresponding equilibrium will be unstable. Besides, if the sign of Jacobian is negative, the equilibrium will be a saddle point.

Table 2. Local stability of evolutionary game.

Sign of Jacobian	Sign of the trace	Stability of corresponding equilibrium
+	−	Stable equilibrium
+	+	Unstable equilibrium
−		Saddle point

1. If the parameters do not meet $P + C' < a < \frac{\beta}{\alpha}(P + C'')$, $O(0,0)$ is the only stable equilibrium, A and B are saddle points, C is unstable equilibrium. In this condition, the vendors' market strategy will be stable that thermal power vendors do not sell and wind power vendors do not buy ancillary services.
2. If $P + C' < a < \frac{\beta}{\alpha}(P + C'')$, the corresponding Jacobian of equilibrium D is:

$$J = \begin{vmatrix} 0 & \dfrac{(Q_2P - \beta P - \beta C'')}{(Q_2P - \alpha a)^2}(\beta P + \beta C'' - \alpha a)(\alpha a - \alpha C') \\ \dfrac{P(a - C' - P)}{(a - C')^2}(Q_2P - \alpha a) & 0 \end{vmatrix}$$

In this case, equilibrium $D\left(\frac{Q_2P - \beta P - \beta C''}{Q_2P - \alpha a}, \frac{P}{a - C'}\right)$ is a saddle point; $O(0,0)$ and $C(1,1)$ are stable equilibriums; A and B are unstable equilibriums. The strategy of thermal and wind power vendors premising on bounded rationality will be stable in the circumstances that thermal power vendors sell while wind power vendors buy ancillary services and thermal power vendors do not sell while wind power vendors do not buy according to their long-time learning and experience. The dynamic evolution game process of thermal and wind power vendors is shown in Figure 1.

In the dynamic evolution game process of thermal and wind power vendors, the area of quadrilateral OADB stands for the probability that thermal power vendors do not sell and wind power vendors do not buy ancillary services. The larger the area OADB is, the greater possibility that thermal and wind power vendors will choose this strategy. The area of quadrilateral OADB is shown as follows:

$$S_{OADB} = \frac{1}{2}\left(\frac{Q_2P - \beta P - \beta C''}{Q_2P - \alpha a} + \frac{P}{a - C'}\right)$$

With other parameters constant, the greater the capacity of thermal power vendors providing ancillary services, (i.e. the smaller α or C' and S_{OADB} is), the greater preference thermal

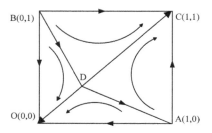

Figure 1. The dynamic evolution game process of thermal and wind power vendors.

and wind power vendors will have to cooperative market strategies. Analogously, the smaller the capacity of wind power vendors providing ancillary services (i.e. the greater β or C'' and S_{OADB} is), the more willing thermal and wind power vendors will be to choose cooperative market strategies. According to $\frac{\partial S_{OADB}}{\partial Q_2}$, S_{OADB} is decreasing with the increase of Q_2. That is, under the condition of $P + C' < a < \frac{\beta}{\alpha}(P + C'')$, the possibility of wind and thermal power vendors' cooperation is decreasing with the increase of the wind power generating capacity. According to $\frac{\partial S_{OADB}}{\partial a}$, in the case of $P + C' < a < \frac{\beta}{\alpha}(P + C'')$ and $a \geq C' + \frac{Q_2 P \sqrt{P} - \alpha C' \sqrt{P}}{\alpha \sqrt{P} + \sqrt{\alpha(Q_2 P - \beta P - \beta C'')}}$, if other parameters are constant, S_{OADB} is increasing with the increasing of parameter α. That is, the increase of compensation will lead to an increasing possibility of thermal and wind power vendors' cooperation. According to $\frac{\partial S_{OADB}}{\partial P}$, when $P > \frac{\alpha a}{Q_2} + \frac{\sqrt{(Q_2 - \beta)(a - C')\alpha a - (a - C')\beta C'' Q_2}}{Q_2}$, if other parameters are constant, S_{OADB} is decreasing with the increasing of parameter P. That is, with the rising of electricity prices, the possibility of thermal and wind power vendors' cooperation in ancillary service market is decreased.

4 RESULTS

When the parameters meet the condition of $P + C' > a$, which means the compensation thermal power vendors received from providing ancillary services for wind power vendors is lower than the opportunity and operating cost of units providing ancillary services, the thermal power vendors will not volunteer to provide ancillary services for wind power vendors. When the parameters meet $a > \frac{\beta}{\alpha}(P + C'')$ that is when wind power vendors buy ancillary services from thermal power vendors, if the compensation paid to thermal power vendors is higher than the cost wind power vendors providing ancillary services themselves, the wind power vendors will not buy ancillary services from thermal power vendors voluntarily. Only in the condition

that $P + C' < a < \frac{\beta}{\alpha}(P + C'')$, will it be possible that transactions between wind and thermal power vendors happen in the joint ancillary service market. In that case, their business possibility is related to the generating capacity of wind power, the compensation level of ancillary services, current prices of electricity and the capability of wind and thermal power vendors have to provide ancillary services.

ACKNOWLEDGMENT

This article is supported by "Science and technology projects funding by State Grid Liaoning electric power co., LTD. Benxi Power Supply Company" (Grant NoFZJS1400824, FZJS1400825), "Science and technology project funding by State Grid Liaoning electric power co., LTD" (Grant No.5222 JJ14001F) and "Science and technology project funding by State Grid Corporation".

REFERENCES

Bai Jianhua, Xin Songxu et al. 2009. Analysis on the Planning and Operation Problems of Large-scale Wind Power Development in China. *Electric Power Technologic Economics*, 2: 004.

He Yang, Hu Junfeng, Yan Zhitao, et al. 2013. Compensation Mechanism for Ancillary Service Cost of Grid-Integration of Large-Scale Wind Farms. *Power System Technology*, 37(12): 3552–3557.

Hannele H., Michael M., Brendan K. et al. 2008. Using standard deviation as a measure of increased operational reserve requirement for wind power. *Wind Engineering*, 32(4): 355–378.

Li Yu. 2013. A Discussion on the Transaction Systems of Wind and Thermal Power Units' Startup/shutdown Peak-regulated Ancillary Services Based on a Large-scale Wind Power Grid. *Inner Mongolia Science Technology & Economics*, (16): 75–77.

Liu Chun, Fan Gaofeng, Wang Weisheng et al. 2009. A combination forecasting model for wind farm output power. *Power System Technology*, 33(13): 74–79.

Liu Kewen, Pu Tianjiao, Zhou Haiming, et al. 2013. A Short Term Wind Power Forecasting Model Based on Combination Algorithms. *Proceedings of the CSEE*, (34).

Qian W., Lizi Z., Guohui X. 2011. Probabilistic Calculation and Confidence Interval Estimation of Several Wind Farm Penetration Limit. *Acta Energiae Solaris Sinica*, 4: 020.

Tu Qiang. Existing applications of wind power forecasting technology and relevant suggestions. 2009. *Power System and Clean Energy*, 25(10): 4–9.

Zhang N., Zhou T., Duan C., et al. 2010. Impact of Large-Scale Wind Farm Connecting With Power Grid on Peak Load Regulation Demand. *Power System Technology*.

Zhang Baohui, Wang Yongli, Tan Lunnong et al. 2005. A new transaction mechanism for ancillary service market based on bilateral reliability bidding of both power supply and demand. *Power System Technology*, 29(23): 50–55.

Advances in Energy, Environment and Materials Science – Wang & Zhao (Eds)
© 2016 Taylor & Francis Group, London, ISBN 978-1-138-02931-6

Research on peak-regulation pricing and compensation mechanism of wind power in China

Di Zhao
School of Economics and Management, North China Electric Power University,
Huilongguan Changping District, Beijing, China

Ling Wang
State Grid Liaoning Electric Power Co. Ltd., Benxi Power Supply Company, Benxi, China

Xingang Zhao
School of Economics and Management, North China Electric Power University,
Huilongguan Changping District, Beijing, China

ABSTRACT: Wind power is an important renewable energy. However, wind curtailment occurs occasionally due to the large volatility of wind. Peak-regulation ancillary service is the most important way to adjust load balance, promote grid-connected of wind power generation, and reduce wind curtailment. This paper analyzes the recent status quo of wind power and ancillary services in China. So the study of peak-regulation is crucial. And also proposes the model which can price and compensate the peak-regulation ancillary services of wind power generation in China, in accordance with the principle of "compensation for costs and reasonable revenues". Then calculate the price and compensation of peak-regulation in the four different types of wind resource areas in China.

1   INTRODUCTION

1.1   *Background*

Recent years, wind energy as a kind of clean, renewable energy, gradually became an important strategic choice for countries to develop clean energy. With the increasing attention Chinese government pays to renewable energy development, as the fastest-growing and most promising way of renewable energy generation, the degree of concern about wind power development also increases [Wang Ruogu, 2011]. Wind power generation has the advantages of energy-saving and lower cost. However, due to the active output of wind power is volatile, intermittent and random, in order to maintain the balance of power in real time and to ensure the safe and stable operation of power system, hydropower, thermal power and other conventional power need to provide for integration of wind power with a lot of auxiliary services, such as peak-regulation, frequency-modulation, voltage-regulation and standby, and others. This increases the burden of ancillary services of thermal power and hydropower generation and results in the increasing of energy consume and generation cost for conventional units involve in power system ancillary services. The phenomenon of

wind curtailment occurs occasionally. Therefore, the study of peak-regulation pricing and compensation mechanism of wind power generation is crucial for promoting integration of wind power, as well as ensuring safe and stable operation of the power system.

1.2   *Literature review*

Actually, there were many scholars who had analyzed related issues about peak-regulation pricing and compensation mechanism. Shen Shuangjing (2003) studied the pricing and correlative economic compensation for peak adjustment of ancillary services in power market. The models, with the goal of the max sum of power energy income and peak capacity income, are presented to optimize combination of units. He mainly introduced the cost of deep peak-regulation, but it did not analyze the cost of start and stop peak-regulation. Hu Jianjun, Hu Feixiong (2009) introduced the idea of peak-regulation capacity liability regime and proposed a new method based on equivalent available load rate indicators to compensate for the peak-regulation capacity. Wang Ruogu et al. (2011) proposed an estimate method of hydropower peak-regulation compensation. This method analyzed the water loss due to the hydropower output below the low-bound of the fee peaking internal, and then the water loss can be converted to profit loss, which can serve as the hydropower peaking compensation. Xie Jun, Li Zhenkun et al. (2013) according to the peaking value of generators which is quantified by using the proposed economic dispatch model, based on Shapley value and peaking mileage concepts, two peaking cost compensation methods are proposed. Wang Mei (2013) adopts K-means cluster analysis to classify the units based on the unit capacity and the highest peaking rate indicators, and then puts forward a new auxiliary service compensation model. Cui Qiang et al. (2015) proposed a stagger peak-valley Time-of-Use (TOU) price mechanism, in which heavy energy-consuming enterprises are impelled to participate in the peak load regulation under the Demand Side Management (DSM). Lv Quan et al. (2013) introduced the operation mechanism, technical characteristics and available scopes of peak regulation by cogeneration units. Moreover, the application and economic efficiency of this peak regulation mode are forecasted. Zhu Zhiling et al. (2011) summarizes the key factors which impact on wind power integration, including system regulation capacity, transmission capacity, technical performance of wind power, wind power dispatch levels, etc. Then some measures are proposed to improve wind power integration in China. Poul Alberg Østergaard (2006) analyses the possibilities for integrating even more wind power using new power balancing strategies that exploit the possibilities given by the existence of CHP plants, as well as the potential impact of heat pumps used for district heating and installed for integration purposes.

This paper analyzes the recent status quo of wind power and ancillary services in China. And also proposes the model which can price and compensate the peak-regulation ancillary services of wind power generation in China, in accordance with the principle of "compensation for costs and reasonable revenues".

2 MODEL OF PRICING AND COMPENSATING ON PEAK-REGULATION

In the electricity market, all main participants in the power system is to achieve the objective of maximizing benefits, in the premise of ensuring the safe and stable operation of the power system. For example, when the load is low, generator units need to reduce output in order to achieve load balance. This would reduce the generating efficiency of generator units, if the power plants' economy is affected due to providing the peak-regulation ancillary services. Therefore, to ensure the enthusiasm

of providing peak-regulation ancillary services, it is crucial to provide some compensation for the wind power plants which provide peak-regulation ancillary services, to ensure that their cost is covered or even get a reasonable profit.

Currently, peak-regulation ancillary services can be divided to paid peak-regulation and free peak-regulation, which is called basic peak-regulation. Unlike basic peak-regulation, paid peak-regulation can be divided into deep peak-regulation and start and stop peak-regulation. The way of peak-regulation, which active output of the generator set accounts for less than 50% of the rated capacity is called deep peak-regulation. The way of peak-regulation, which needs to achieve stopping the generator units according to the peak-regulation needs of power grid and starting the generator units in 24 hours, is called start and stop peak-regulation. So, the compensation of peak-regulation can also be divided into deep peak-regulation compensation and start and stop peak-regulation compensation.

The installed capacity of wind turbines is generally about 2 MW, so it is generally to use the wind farm as a unit to provide peak regulation ancillary services for the power grid. The cost of running peak regulation capacity may be replaced by its equivalent opportunity cost. The generation capacity of wind farm reduces since it provides peak regulation ancillary services. Then reducing amount of generation capacity can decrease the revenue of electric energy, the decreased revenue is just the opportunity cost, which the wind farm provide running peak regulation capacity. The paid peak regulation can be made up of the deep peak regulation and start and stop peak regulation.

2.1 Deep peak-regulation

A certain wind farm i losses a part of capacity due to providing peak regulation ancillary services in a certain time Δt. Thus, we can use the revenue difference before and after peak regulation as the opportunity cost loss of deep peak regulation. In another words, it is the compensation of deep peak regulation.

1. When it provides maximum output $P_{i,\max}$, its revenue is $G_{Tp_{i,\max}}$

$$G_{Tp_{i,\max}} = P_{i,\max}\Delta t\left(\rho_E - C_i\right)\left(1 - P_{i,FOR}\right) \quad (1)$$

    ρ_E—Electricity price in Δt
    C_i—Cost of generation at the output level
    $P_{i,FOR}$—Short-term forced outage rate of the wind farm.

2. When it provides output P_i, its revenue is G_{Tp_i}

$$G_{Tp_i} = P_i\Delta t\left(\rho_E - C_i\right)\left(1 - P_{i,FOR}\right) \quad (2)$$

3. So, there is a revenue difference ΔG_i before and after the peak regulation in a certain wind farm i in a certain time Δt.

$$\Delta G_i = G_{Tp_{i,\max}} - G_{Tp_i}$$
$$= \left(P_{i,\max} - P_i\right)\Delta t\left(\rho_E - C_i\right)\left(1 - P_{i,FOR}\right) \quad (3)$$

2.2 Start and stop peak-regulation

A certain wind farm i losses all the capacity due to providing start and stop peak regulation ancillary services in a certain time Δt. Thus, the loss of electricity during this period multiplied by the corresponding price is the opportunity cost of start and stop peak regulation. The calculating method is equal to the revenue before the deep peak regulation in the maximum output case. It can be shown as follows:

$$G_{Tp_{i,\max}} = P_{i,\max}\Delta t\left(\rho_E - C_i\right)\left(1 - P_{i,FOR}\right) \quad (4)$$

ρ_E—Electricity price in Δt
C_i—Cost of generation at the output level
$P_{i,FOR}$—Short-term forced outage rate of the wind farm.

3 EMPIRICAL ANALYSIS

By July 2009, the NDRC (National Development and Reform Commission) announce "A notice on improving the wind power feed-in tariff policy". According to the announcement, NDRC established the principle of regional benchmarking feed-in tariff. As seen from the Table 1, the whole nation was divided into four types of wind resource areas. The wind power in-grid prices in the four areas were formulated based on the situation of wind energy resources and wind power projects construction.

Therefore, this article will be based on the four types of wind resource areas, calculates their different peak regulation price and compensation of wind power.

3.1 Peak-regulation price of wind power

First, we can calculate the peak-regulation price of wind power in Category I. We can see that in Category I, $\rho_E = 83.57$ \$/MWh. Assuming profit rate is 8%. So, we can conclude that $\rho_E - C_i = 6.6856$ \$/MWh. In China, the Short-term forced outage rate of the wind farm is usually equal to 0.05. So, the peak-regulation price of wind power can be calculated:

$$P_{tf} = 6.6856 \times (1 - 0.05) = 6.3513 \text{ \$/MWh}$$

Table 1. Benchmark feed-in tariffs for China's onshore wind power.

Resource zone	Benchmark feed-in tariff	Area coverage
Category I	83.57 \$/MWh	Inner Mongolia autonomous except: Tongliao, Chifeng, xing'anmeng, Hulunbeier; Xinjiang uygur autonomous: Urumqi, Yale, Changing hui autonomous prefecture, Karamay, Shihezi
Category II	88.49 \$/MWh	Hebei province: Zhangjiakou, Chengde; Inner Mongolia autonomous: Tongliao, Chifeng, xing'anmeng, Hulunbeier; Gansu province: Zhangye, Jiayuguan, Jiuquan
Category III	95.04 \$/MWh	Jilin province: Baicheng, Songyuan; Heilongjiang province: Jixi, Shuangyashan, Qitaihe, Suihua, Yichun, Daxinganling region, Gansu province except: Zhangye, Jiayuguan, Jiuquan, Xinjiang autonomous region except: Urumqi, Yale, Changing, Karamay, Shihezi, Ningxia Hui autonomous region
Category IV	99.96 \$/MWh	Other areas of China not mentioned above

Sources: Yuanxin Liu et al. 2015 [11]

Table 2. Peak-regulation price of wind power in China.

	Category I	Category II	Category III	Category IV
Peak-regulation price	6.3513 \$/MWh	6.7252 \$/MWh	7.223 \$/MWh	7.597 \$/MWh

We can also calculate the peak-regulation price of wind power in the other three wind resource areas using the same method. The result is shown in Table 2.

3.2 Peak-regulation compensation of wind power

Based on the last part, we can conclude that a certain wind farm i can get compensation ΔG_i if it is providing deep peak-regulation.

$$\Delta G_i = 6.3513\left(P_{i,\max} - P_i\right)\Delta t \quad (5)$$

Table 3. Compensation of peak-regulation in China.

	Compensation of peak-regulation	
	Deep peak-regulation	Start and stop peak-regulation
Category I	$6.3513\,(P_{i,\max} - P_i)\,\Delta t$	$6.3513\,P_{i,\max}\,\Delta t$
Category II	$6.7252\,(P_{i,\max} - P_i)\,\Delta t$	$6.7252\,P_{i,\max}\,\Delta t$
Category III	$7.233\,(P_{i,\max} - P_i)\,\Delta t$	$7.233\,P_{i,\max}\,\Delta t$
Category IV	$7.597\,(P_{i,\max} - P_i)\,\Delta t$	$7.597\,P_{i,\max}\,\Delta t$

And a certain wind farm i can get compensation ΔG_i if it providing strat and stop peak-regulation.

$$\Delta G_i = 6.3513 P_{i,\max}\Delta t \tag{6}$$

So according to the four types of wind resource areas in China, we can calculate the compensation of deep, start and stop peak regulation in the different wind resource areas in China. The result is shown in Table 3.

4 CONCLUSION

Ancillary service is the most important way to ensure the balance of electricity supply and demand, protect the safe and stable operation of power system and improve quality of power. With the expanding scale of the grid-connected of wind power, peak-regulation pricing and compensation mechanism as an important way to adjust load balance, promote grid-connected of wind power generation and reduce wind curtailment, whether it can be fully provided is a decisive factor to the capacity that system can accept wind power. This paper analyzes the recent status quo of wind power and ancillary services in China. And also proposes the model which can price and compensate the peak-regulation ancillary services of wind power generation in China, in accordance with the principle of "compensation for costs and reasonable revenues". Then calculate the price and compensation of peak-regulation in the four different types of wind resource areas in China. The prices of peak-regulation are 6.3513 $/MWh, 6.7252 $/MWh, 7.223 $/MWh, 7.597 $/MWh. China's wind power is still under development, in order to achieve safe and economic operation of wind power, it is necessary to carry out the reform of the electricity market, and improve peak-regulation pricing and compensation mechanism

of wind power generation. However, measures to implement still need further study.

ACKNOWLEDGMENT

This article is supported by "Science and technology project funding by State Grid Liaoning electric power co., LTD. Benxi Power Supply Company" (Grant NoFZJS1400824, FZJS1400825), "Science and technology project funding by State Grid Liaoning electric power co., LTD" (Grant No. 5222 JJ14001F) and "Science and technology project funding by State Grid Corporation".

REFERENCES

Cui Qaing, Wang Xiuli, Wang Weizhou. 2015. Stagger Peak Electricity Price for Heavy Energy-Consuming Enterprises Considering Improvement of Wind Power Accommodation. *Power System Technology*, 39(4): 946–951.

Hu Jianjun, Hu Feixiong. 2009. A novel compensation mechanism of remunerative peak load regulation under energy-saving dispatch framework. *Automation of Electric Power Systems*, 33(10): 16–18.

Jin Dan, Ding Kun, He Shien. 2011. Discussion on active power regulation mechanism of wind power generation in Denmark and its enlightment to China. *Electric Power Environmental Protection*, 27(4): 50–53.

Lv Quan, Chen Tianyou. 2013. Review and Perspective if Integrating Wind Power into CHP Power System for Peak Regulation. *Electric Power*, 46(11): 129–136.

Poul Alberg Østergaard. 2006. Ancillary services and the integration of substantial quantities of wind power. *Applied Energy*, (83): 451–463.

Shen Shuangjing. 2003. Pricing and correlative economic compensation for peak adjustment of ancillary services in power market. Shanghai Jiao Tong University: Shanghai Jiao Tong University. 1–72.

Wang Ruogu, Wang Jianxue et al. 2011. A cost analysis and practical compensation method for hydropower units peaking service. *Automation of Electric Power Systems*, 35(23): 41–46.

Wang Mei. 2013. Research on Peak-regulation Ancillary Service Compensation and Trade Mechanism for Promoting Wind Power Utilization. North China Electric Power University: North China Electric Power University. 1–70.

Xie Jun, Li Zhenkun. 2013. Peaking Value Quantification and Cost Compensation for Generators. *Transaction of China Electrotechnical Society*, 28(1): 271–276.

Yuanxin Liu et al. 2015. The industrial performance of wind power industry in China. *Renewable and Sustainable Energy Reviews* 43: 644–655.

Zhu Lingzhi, Chen Ning, Han Hualing. 2011. Key Problems and Solutions of Wind Power Accommodation. *Automation of Electric Power Systems*, 35(22): 29–34.

Advances in Energy, Environment and Materials Science – Wang & Zhao (Eds)
© *2016 Taylor & Francis Group, London, ISBN 978-1-138-02931-6*

A method to extend fast charging lead-acid battery cycle life for electric vehicles

Yuzhuo Zhang
School of Economics and Management, North China Electric Power University, Beijing, China

Haojun Wang
Yancheng Power Supply Bureau, Jiangsu Yancheng, China

Ling Wang
State Grid Liaoning Electric Power Co. Ltd., Benxi Power Supply Company, Benxi, China

Xingang Zhao
School of Economics and Management, North China Electric Power University, Huilongguan Changping District, Beijing, China

ABSTRACT: Study of making electric vehicles becomes the focus of scholars as the development of new energy. The study of lead-acid batteries is an important factor, which affects the development of electric vehicles. In order to increase the lead-acid battery charging efficiency, the lead-acid battery charging and discharging mechanism is based on a fast charge mode. The experiment proved that this method could effectively improve the charging efficiency, shortened the charging time and increased battery cycle life. This paper provides a new method for greatly improving future electric vehicle fast charging life.

1 INTRODUCTION

Due to the shortage of fuel consumption resulting in energy, environment increasingly destroyed. Environmentally friendly and energy-efficient electric vehicles will be better solutions to solve these problems [Genchang Wu, 2009]. As the core of electric vehicles, how to extend battery cycle life becomes a problem to overcome for scholars. Now the main types of battery, which are acid batteries, nickel cadmium batteries, nickel metal hydride batteries, and lithium-ion batteries, have advantages and disadvantages. Lead-acid batteries with its low cost, large capacity, no memory effect and mature technology advantages become the first choice of electric vehicle batteries [Abudura A, 2001] [Jinghong Zhong, 2006] [Yonghua Song] [Chaowei Duan, 2013]. Meanwhile, the main reason for affecting the life of lead-acid battery is power supply, thus the use of reasonable charging method has important practical significance [Jingjin Chen, 2004].

Juan Hao (2010) mentioned two traditional charging methods in the constant current charging and constant voltage charging. Although the control circuit methods are simple and easy to implement, but the charge cycle is too long, technical tend to a single charge, and the battery would cause unnecessary harm due to which battery life is affected, thus there are inevitable limitations. Based on the principle stage constant current battery charge, the battery depth of discharge cannot be distinguished and it is difficult to control, so it affects the battery life. According to the law proposed by Maas, fast charging method is of complexity structure and high cost. Several techniques above are developing relatively mature, but the lead-acid battery itself nonlinear dispersion and complexity of the various methods are not well satisfied the rapid development of modern electric vehicle charging requirements. It is not ideal. Therefore, on the basis of conventional charging methods, this paper further proposes new charging model. The results of using the charging circuit PSCAD simulation show that the new method can greatly improve fast charging rate and shorten the charging time.

2 LEAD-ACID BATTERY WORKS

Lead-acid battery is a secondary power supply. The positive plate active material is lead dioxide (Pb), through the water in the sulfur acid molecules, unstable material lead hydroxide (PbOH) is produced by the combination of water and a small amount of lead dioxide. Because lead ions are on the anode, hydroxide ions is present in the solution, so the positive lack of electrons. A chemical reaction occurs, when lead-acid battery charges, the negative plate of lead and sulfuric acid electrolyte. It can get lead ions. Due to the transfer of lead ions in the electrolyte, the negative plate excess of two electrons can be seen. When opening, after a series of chemical reactions, the excess electrons of the negative electrode plate and the positive electrode plate e deletion with a potential difference between the two plates. According to the theory above, the reaction obtained positive and negative electrodes of lead-acid batteries and the total reaction is as follows:

Negative reactions:

$$Pb + HSO_4^- \rightleftharpoons PbSO_4 + H^+ + 2e$$

Cathode reaction:

$$PbO_2 + 3H^+ + HSO_4^- + 2e \rightleftharpoons PbSO_4 + 2H_2O$$

Battery overall reaction:

$$Pb + PbO_2 + 2H^+ + HSO_4^- \rightleftharpoons 2PbSO_4 + 2H_2O$$

Lead-acid battery charging process, in a nutshell, is to produce electrical energy into chemical energy and stored. When charging, at the negative electrode, the formation of metallic lead is due to the reduction of lead sulfate of metallic lead, and its speed is far greater than the speed of lead sulfate to form. At the positive electrode, speed of lead sulfate oxidized to lead dioxide is also accelerated and turns into lead dioxide. In order to avoid loss of water, when to maintain, ionized water needs to be added on a regular basis.

Internal battery reaction is as follows:

Positive:

$$PbSO_4 - 2e + 2H_2O = PbO_2 + H_2SO_4 + 2H^+$$
$$H_2O - 2e = 2H^+ + \frac{1}{2}O_2$$

Negative:

$$PbSO_4 + 2e + 2H^+ = Pb + H_2SO_4$$
$$2H^+ + 2e = H_2$$

The charging process can be divided into three stages, namely efficient phase, mixed phase, and gas phase precipitation. The main role of the high-stage is to convert Pb to $PbSO_4$ and PbO_2. The charge acceptance rate, which is the ratio of the electrical energy converted to electrochemical for reserve and charging from the electromechanical, is about 100%. In the mixing stage, charge

acceptance rate gradually decreases as a result of water electrolysis and the main reactions occur at the same time. We can know when the battery voltage and the acid concentration is no longer growing, the battery is considered is full. Gas precipitation stage, the battery is already full, then will self-discharge and water electrolysis reaction.

In summary, there are many factors that can influence the battery fast charge, such as the different levels of activity of the active substance plate, the electrolyte concentration and temperature, thus the charge will make a big difference. Different put some state, using and saving time will affect the battery charging. Since the charge current curve of the nonlinear characteristic, the charging time with the charging process decreases exponentially.

3 THE BASIC PRINCIPLE OF BATTERY CHARGE AND DISCHARGE

Rapid charging technology is based on three basic laws of the battery charge, namely, three laws by Maas made. These three laws are as follows.

The first Law: A battery can be in any given current discharge, and then its charge acceptance rate and discharge capacity of C is inversely proportional to the square root, namely:

$$a = \frac{K}{\sqrt{C}} \tag{1}$$

Formula k is a proportional constant.
And because:

$$a = \frac{I_o}{C} \tag{2}$$

Therefore,

$$I_o = K\sqrt{C} \tag{3}$$

It can be seen that for the same discharge current, power and charge acceptance is proportional. To release more power, the stronger charge is acceptance.

The second law: In any depth of discharge, battery charge acceptance ratio and the logarithm of discharge current for both linear relationship. Available:

$$a = \frac{K}{\sqrt{C}} \log(KI_d) \tag{4}$$

The formula shows that the discharge rate and depth of discharge affect battery charge acceptance rate. According to the second law, Charge acceptance rate will decrease as the battery long time of discharge in small current.

The third Law: The battery charging acceptance current is the sum of each discharge rate under current. That is:

$$I_t = I_1 + I_2 + I_3 + \dots \tag{5}$$

$$a_t = \frac{I_t}{C_t} \tag{6}$$

where I_t is the total for the current; C_t is total power emitted; a_t is total charge acceptance rate.

Release by the battery discharge can make all the battery and the charging acceptance current increases. Thus, to discharge before and after the battery in charging can increase the charge acceptance rate.

Charging technology has a big breakthrough for ever because of Mass three laws as the theory basis of quick charge technology. In the process of quick charge battery, a short stop charging and join the discharge pulse in it.

4 NEW FAST CHARGE MODE

4.1 Lead-acid battery model

Based on the understanding of current various kinds of charging model and on the basis of summing up predecessors' experience, we propose the circuit model shown in Figure 1 and Figure 2. The charging circuit comprises two main parts, the main one is the charging circuit was shown in the Figure 1, and the charging control circuit was shown in the Figure 2.

Charge controls the main route back three components:

1. Lead-acid battery equivalent model circuit.

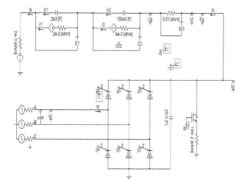

Figure 1. Charge and discharge of the main circuit.

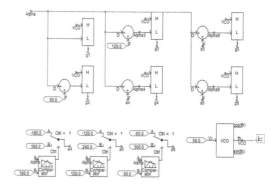

Figure 2. Pulse control loop.

2. Figure 3 is a three-phase bridge rectifier circuit: It is composed of six sent 60 degrees each pulse and it makes six thyristor conduction of the circuit, so as to control and regulate output voltage value.
3. IGBT control circuit is shown in Figure 4: This circuit controls the stop and reverses the lead-acid battery charging and discharging process.

4.2 Select new rapid charge mode data

1. The choice of the negative pulse amplitude has a very important significance.
   Negative pulse is mainly used to remove the polarization effect. If the amplitude value choice is very small, it does not remove the polarization effect. If the amplitude of choice is very big, it will shorten the life of the battery plate. Therefore, we will set multiple sets of parameters in the process of testing and the amplitude can vary widely.
2. Select the pulse width
   1. Before the stop time: Whenever the battery after a period of time to stop, polarization effect weakened gradually, ohm polarization immediately disappear, concentration polarization will gradually diminish over time. The longer the intermittent time, depolarization effect is more obvious. But because this is a quick charge, the interval time should not be too long, so we should choose the right down time.
   2. Negative pulse of time: Because the dynamic characteristic of the rechargeable battery, according to the different types of batteries, they need the negative pulse time is not the same. Specific analysis requires specific circumstances. Generally, select between 100 ms to 600 ms. If the time is too short, the depolarization will not achieve the effect, but if the time is too long, it will increase the amount of discharge, the battery charge is negative.

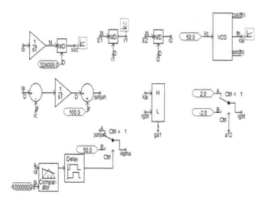

Figure 3. Controlled three-phase bridge rectifier circuit.

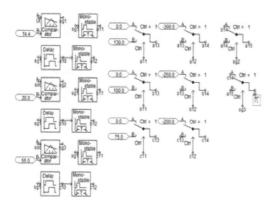

Figure 4. IGBT control circuit.

3. The stop time: this time is not appropriate for too long, the longer the terminal voltage will increase more time.

Depolarization pulse charging method has been developed into several kinds of modes. Such as constant current, constant periodic pulse fast charging method, constant current, set gas rate pulse charging, constant current, constant voltage pulse fast charging.

3. Determine the size of the charge current

The first stage takes 1–2 times the rated capacity of the battery as a large current during charging current value. Various stages of the selection of current required to charge according to each stage in the electricity into the battery SOC decision.

5 CONCLUSIONS

This paper focuses on the technology fast charging electric vehicles. The charging process parameters, such as the charge level among the variation width of the positive pulse amplitude, the amplitude of the positive pulse, the negative pulse duration, the amplitude of the negative pulse, etc. We explore the theoretical and experimentally study these parameters and propose a new fast—charge mode. This paper presents the design studies for its fast charge mode efficient, rapid, non-destructive, and so on. It can be as the quick charger to provide a reliable reference and theoretical basis for the research on electric vehicles.

ACKNOWLEDGMENT

This article is supported by "Science and technology project funding by State Grid Liaoning electric power co., LTD. Benxi Power Supply Company" (Grant No. FZJS1400823 and FZJS1400824), and "Science and technology project funding by State Grid Liaoning electric power Co, Ltd.," (Grant No. 5222JJ14001F).

REFERENCES

Abudura A., Kenichi S. 2001. Present status and future prospects of electric vehicles. Advanced Technology of Electrical Engineering and Energy 20(1):49–53.

Chaowei Duan, Lei Zhang et al. 2013. Design of the pulse-type fast speed charging system of lead-acid battery for electric car. Process Automation Instrumentation 34(7):75–77.

Genchang Wu. 2009. Study on selection of the electric vehicle batteries and the charging methods. *China Science and Technology Information* 23:144–155.

Jinghong Zhong, Chengning Zhang et al. 2006. The research about quick pulse charge system of lead-acid battery for electric vehicle. Power Supply Technology 30(6):504–506.

Jingjin Chen, Ningmei Yu. 2004. Study on the multi-stage constant current charging method for VRLA battery system. Power Supply Technology 28(1):32–33.

Juan Hao, Qiang Li, Jianhua Yue. 2010. Discussion on charge pattern for electric vehicle charging station. *Inner Mongolia Electric Power* 28(6):7–9.

Yonghua Song, Yuexi Yang et al. 2011. Present Status and Development Trend of Batteries for Electric Vehicles. Power System Technology 35(4):1–7.

Advances in Energy, Environment and Materials Science – Wang & Zhao (Eds)
© 2016 Taylor & Francis Group, London, ISBN 978-1-138-02931-6

Study of solar energy technology applied in oil fields

W. Qiu, H. Huang, X. Xu & R. Wang
Petroleum Exploration and Production Research Institute, SINOPEC, Beijing, China

ABSTRACT: Solar energy is a clear renewable energy. It is applied in oilfields exploration, production, storage, and transportation to reduce cost and carbon emissions. Solar energy used for generation and low-temperature heat supply is widely used in auxiliary facility in oil fields. Besides, solar energy plays an important part in large projects in exploration, storage, and transportation. Through collectors for vapor generation is used in heavy oil development instead of vapor generated by fuel. Solar energy as single heater source or auxiliary heater system has been widely used in petroleum transportation. Combined solar energy with other energy to solve the influence of radiation change will help to increase the application of solar energy in petroleum engineering.

1 INTRODUCTION

Solar energy is an important clean renewable energy. It has been one of good choices to solve lack of resource problem, environmental pollution, and climate change. Wide use of solar energy can reduce the pollution effectively and the reliance of traditional resources (Yan, Y.F. et al. 2012, Li, K. & He, F.N. 2009). Recently, the application of solar energy is mostly focused on power generation and heating (Arif, H. & Zeyad, A. 2011, Shi, J.L. 2013). According to the prediction of global resource structure from Europe United Research Centre, solar electrical energy generation technology will develop faster after 2030, and replace traditional resource step by step (Hu, Z.P. 2008).

Low voltage electrical technology has been applied in petroleum exploration and development as logistic support and auxiliary measures. In 2001, a solar electrical energy station was put into operation in the offshore platform in Shengli oilfields to supply power for navigation lights, fog horn, and living facilities (Lang, X.C. & Wang, M.X. 2001). Amoco Company used solar battery for casing anti-corrosion by catholic protection (Ralph, S.Jr. & Horkondee, J. 1979). Solar energy is also fully used in Jidong oilfield offshore artificial island of petroleum exploration to supply light and heat (Cao, X.C. et al. 2009). Solar energy is also widely used in heavy oil thermal recovery and oil storage and transportation.

2 GETTING STARTED SOLAR ENERGY APPLICATION ON HEAVY OIL THERMAL RECOVERY

Underground viscosity of heavy oil is over 50 mPa.s or relative density is higher than 0.92 at room temperature. The worldwide proved reserves of heavy oil is more than 2100×10^8 m^3 and the production is over 900×10^4 m^3/d. Thermal enhanced oil recovery such as steam injection is a common way for heavy oil recovery. Steam injection concludes cyclic steam stimulation and steam flooding. The steam stimulation technology is the most popular method in developing the heavy oil and also the main method for thermal production of heavy oil in our country. Inject amount of steam into the well first, then shut-in the well for a period of time. When the vapor expands to the oil column, open the well for production. Steam flooding heats oil formation by continuous steam injection of high dryness to reduce oil viscosity and the injected steam turns to hot fluid driving oil to productive wells where oil is to be developed.

Both these two methods need amounts of steam. According to some research results, the best injection rate of these two methods is usually higher than 100 t/d, even to 400 t/d for some special oilfields (Wu, X.D. et al. 2007, Ling, J.J. et al. 1996). The vapor temperature is at 100–250°C (Yang, B. et al. 2012, Mozaffari, S. et al. 2013). Steam is usually generated by steam generators, which will cost a lot and release greenhouse gases such as CO_2, causing pollution to the environment (Zhang, X.K. et al. 2007). The solar steam generator can figure out these problems.

Figure 1 shows the structure of common trough solar vapor generator. Water absorbs the solar energy from vacuum tube collector array mirror field and generates steam at the outlet. After separation of steam and water, water will flow to the recycle pump again and steam will be used in the petroleum production. The scale of the mirror field is depended on the steam temperature and rate that is required. For example, a vacuum tube

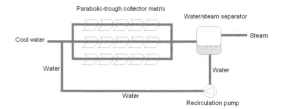

Figure 1. Schematic of direct steam generation using solar power.

collector array mirror field has 7 columns with total collector length 980 m (including pre-heated length 294 m, evaporation length 490 m and super-heat section length 196 m). The steam generated by the field can reach 290°C and the superheat section 411°C (Fraidenraich, N. et al. 2013).

The solar steam generator applied in the industry includes DC type, intermediate filler type and recirculation type (Du, J.L. et al. 2013). Among these, DC type has simplest structure with lowest initial cost, but the parameters are hard to control. The recirculation type costs less with easy control of parameters (Liao, W.C. & Ke X.F. 2008).

It is analyzed that steam generated trough solar vapor generator can meet the demand of heavy oil thermal recovery. However, the oilfield needs to fit the following requirements. The sunlight is sufficient and the climate is suitable with stronger solar radiation and lower wind speed. The dust content is low. And an enough plain space is also needed for solar collectors.

Glass Point Solar Company fixed the first solar recovery device in California in 2011. At the same year, Bright Source Energy Company and Shevron Corporation built a 29 MW solar steam generator at Coalinga oilfield in California (Palmer, D. & O'Donnell, J. 2014). Oman Amal West oilfields used solar thermal recovery technology to get stream 50 t per day. After one year operation, it was shown that using solar thermal recovery in the desert in Mid-east is feasible. The system operated time was 98.6% and it can operate in the desert. Heavy oil resource in China is distributed in Xinjiang, Liaohe, Shengli and so on, which has good conditions for using solar energy for thermal recovery.

3 SOLAR ENERGY IN OIL STORAGE AND TRANSPORTATION

Produced oil needs to be stored nearby and then transported to plants by long pipelines, oil truck, or ship. Pipelines are the main way of crude oil transportation with large volume and low cost.

However, due to high viscosity of heavy oil it is hard to transport at room temperature. As a result, it needs to take measures such as heating, adding light oil or drag reducer and so on to reduce viscosity and friction loss, and save the power consumption (Guo, J.L. & Han, Q.S. 2011).

Heavy oil transported by heating pipeline is an important way, including heat treated and pre-heated transportation (Sun, W.M. 2003.). The former way is to heat oil before transportation and then cool it down to transportation temperature at same rate in steady state or cool it at low rate in the pipelines. How to control heating and cooling temperature is the key point of this technology. The latter way pre-heats the pipe by electric heating belts, thermal station, or hot fluid and takes measure for insulation to maintain crude oil flowing at high temperature. Steam-hot water heating and electric heating are two main methods for heating crude oil.

Traditional heating methods use fuel and electricity causing high cost. Since 1990s, solar energy heating technology has been developed fast and put into practice. Use of solar energy to heat crude oil for transportation includes directly and indirectly heating (Jia, Z.W. & Song C.M. 2009). Directly heating is to heat the crude oil in the solar collectors with high effectiveness. However, the collector is hard to wash due to high viscosity of crude oil. And temperature difference is too high to control, causing crude oil burning in serious conditions. The heat from solar energy of the indirectly heating method is transferred by media. Then the media enters the heat exchanger and heats the crude oil. It improves the safety and stability of the system. The common solar energy heating process is shown in Figure 2 (Jia, Z.W. & Yang Q.M. 2009).

At normal radiation, the recycle pump starts, and hot water from the tank flows to the heat exchanger. Crude oil heated in the exchanger flows to the pipeline. Before entering to the pipeline net, a heating furnace turns on automatically according to crude oil temperature. If the temperature is lower than the required value, crude oil will enter into the furnace to be heated.

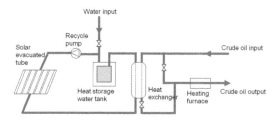

Figure 2. Process flow diagram of crude oil heating using solar power.

The temperature of crude oil needs to be heated to 50°C~55°C from 25°C~30°C before entering the next test station. It is required that the temperature of collector outlet has to reach 70°C~80°C. So, solar collectors must have high thermal efficiency, stable operation, and high resistance to pressure and freeze resistance in winter. There are three kinds of solar collectors including flat plate solar collector, all-glass solar vacuum tubular collector, and heat-pipe type vacuum collector. The flat plate solar has a higher thermal efficiency with smaller temperature difference between water inlet temperature and ambient temperature. As the difference gets larger, the efficiency gets lower fast. Vacuum collectors are influenced by temperature and have a high thermal efficiency even when the temperature is higher than 90°C (Yang, X.F. et al. 1997). Flat plate collectors have strong pressure resistance, stable operation, and bad freezing resistance, which are just the contrary to the all-glass solar vacuum tubular collectors. Heat-pipe type vacuum collectors are characterized with strong pressure and freezing resistance and stable operation (Zhu, M. et al. 2006). Each kind of collector has its own characters. And the choice of collectors should depend on the detailed conditions in the field.

At present, solar energy to heat oil has been widely used in oil fields, especially, for the areas with long-time sunshine and large solar radiation. The average daylight hour for Jiangsu oilfields every year is about 2200 h and the solar radiation is 6000 MJ/m². Liaohe oilfields are 2750 h sunshine and 7000 MJ/m² solar radiation. Both of these two fields have good conditions for using solar energy (Hu, J.Y. & Hua, X.P. 2009 & Wang, X.S. et al. 2004). In 2006, Wanglongzhuang oilfield in Jiangsu applied 30 sets heat-pipe type vacuum collectors with the total area 60 m² to assist the electricity heat system for heating. Good results were shown. Between July and September with enough solar radiation, electricity heat system may be stopped and the other time the solar system and electricity system can be used in turns. It saved 4×10^4 kWh after using the system in one year. Libao oilfield built 262 m² solar collectors in 2008 for heating oil in a 25 m³ storage tank with an electricity heater inside (Wu, M.J. & Wang, C.L. 2010). It changed the traditional heating mode and reduced a lot of cost.

Liaohe oilfields built a heating system with solar collection 392 m². It occupied 792 m². And the heat exchanging area reached 55 m² due to promotion of exchanging tube structure. When it has good sunshine condition, solar energy can be used separately to heat the circulated fluid and the rest of the energy will be stored in the heat accumulator. When the sunshine is not good enough, the water jacket heater can be used. Since using the system, it saves 30% gas consumption every day (He, Z.N. 2014).

4 DISCUSSION

With the development of technology and high requirement of environment protection, solar energy has applied in petroleum industry more and more. However, there are two challenges. One is that solar radiation periodic changes by time and season. And the second challenge is that when it's cloudy or rainy or foggy and so on, the solar radiation will be affected.

Periodical or temporary changes of solar radiation have influence on steam production. However, it was studied that the ultimate recovery is not sensitive for the changes of steam rejection rate. These changes only affects seasonal recovery (Heel A.V. et al 2010). It is also recommended that solar system combined with coal units for steam generation (Chen H.P. 2013). If the solar radiation is weak, coal units are used only for heavy oil recovery.

For crude oil storage and transportation, the widely used methods are combination of solar energy and electricity for heating and combination solar energy and heat pump for heating (Peng Z. 2008, Wang X.S. et al. 2010). At good conditions for solar radiation, electricity heaters or heat pumps don't start up. And when it gets worse, electricity heaters or heat pumps will supply energy to make up the weakness of slow start-up and un-stability of the solar energy system. So, solar energy fluctuation has little influence for the production if proper measures are taken.

5 CONCLUSION

1. Solar energy is clean and regenerative, which can save energy and protect environment. Solar energy for power generation and low-temperature heating has been widely used in the oilfield production and life, such as office warming, lights and low-voltage power supply for equipments, and so on.
2. Steam injection is one of those important methods for heavy oil production. Trough solar vapor generator can replace some of coal and fuel for steam generation. Application of the technology has just been started.
3. Solar energy as a sole heat resource or auxiliary heat resource is used for crude oil storage and transportation. There are many types of solar collectors and mature pipeline design methods. This technology has been popularized and applied in many oilfields.
4. Solar radiation changes by time and season and reduces a lot at bad weather, which effects heating performance. As a result, solar energy

should be applied in areas with enough sunlight and strong annual average solar radiation and combined with other heating methods to keep the heat supply stable.

REFERENCES

Arif, H. & Zeyad, A. 2011. A key review on present status and future directions of solar energy studies and applications in Saudi Arabia[J]. *Renewable and Sustainable Energy Reviews*, 15: 5021–5050.

Cao, X.C. et al. 2009. Application on solor energy in Jidong oilfield exploration and development[J]. *Energy Conservation Technology*, 27(3): 281–283.

Chen, H.P. 2013. Optimal design of a solar aided coal-fired thermal power generation system[J]. Thermal Power Generation, 42(8): 13–16, 33.

Du, J.L. et al. 2013. Experimental study of trough type solar thermal technology used in heavy oil production[J]. *Solar Energy*, 1: 27–32.

Fraidenraich, N. et al. 2013. Analytical modeling of direct steam generation solar power plants[J]. *Solar Energy*, 98: 511–522.

Guo, J.L. & Han, Q.S. 2011. Research and application of energy-saving technology for viscous crude pipeline transportation[J]. *Sino-global Energy*, 16(4): 101–103.

He, Z.N. 2014. Application of solar heating system for raw petroleum during its piping transport[J]. *Energy Procedia*, 48: 1173–1180.

Heel, A.V. et al. 2010. The impact of daily and seasonal cycles in solar-generated steam on oil recovery[C]. SPE 129225.

Hu, J.Y. & Hua, X.P. 2009. Application of solar energy auxiliary electric heating technology in crude oil storage and transportation[J]. *Petroleum Engineering Construction*, 12: 24–25.

Hu, Z.P. 2008. Energy structure variation and alternative energy development prospect at home and abroad[J]. *China Electical Equipment Industry*, (10): 29–38.

Jia, Z.W. & Song C.M. 2009. Discussion on solar energy-saving system used for crude oil transmissing[J]. *Petroleum Engineering Construction*, 35(4): 57–59.

Jia, Z.W. & Yang Q.M. 2009. Discussion on energy conservation of crude oil conveying system using solar power[J]. *Oil-Gasfield Surface Engineering*, 28(3): 5–7.

Lang, X.C & Wang, M.X. 2001. Application of solar energy in offshore production platform[J]. *Ocean Technology*, 20(4): 73–75.

Li, K. & He, F.N. 2009. Analysis on mainland China's solar energy distribution and potential to utilize solar energy as an alternative energy source[J]. Progress in Geography, 29(9): 1049–1054.

Liao, W.C. & Ke X.F. 2008. Application of solar energy in medium & high temperature steam generation[J]. *Energy Conservation Technology*, 26(4): 328–331.

Ling, J.J. et al. 1996. The effect of steam injection rate on the performance of steam drive[J]. *Petroleum Exploration and Development*, 23(1): 66–68.

Mozaffari, S. et al. 2013. Numerical modeling of steam injection in heavy oil reservoirs[J]. *Fuel*, 112: 185–192.

Palmer, D. & O'Donnell, J. 2014. Construction, Operations and Performance of the First Enclosed Trough Solar Steam Generation Pilot for EOR Applications[C]. *SPE-169745-MS*.

Peng Z. 2008. Crude oil storage and transportation system using solar-electrical heating[J]. Oil-Gasfield Surface Engineering, 27(6): 58–59.

Ralph, S.Jr. & Horkondee, J. 1979. Application of solar energy to producing operations of oil and gas field [J]. *Journal of Petroleum Technology*, 2: 151–154.

Shi, J.L. 2013. Review of renewable energy development in China[J]. *Automation Panorama*, 6: 28–32.

Sun, W.M. 2003. Review of heavy oil pipeline technology[J]. *Oil-Gasfield Surface Engineering*, 22(5): 23–24.

Wang, X.S. et al. 2004. Applied research of solar energy for heating crude oil of transportation[J]. *Oil & Gas Storage and Transportation*, 23(7): 41–45.

Wang X.S. et al. 2010. New crude oil transportation system using solar energy and heat pump heating[J]. Oil-Gasfield Surface Engineering, 29(3): 55–56.

Wu, M.J. & Wang, C.L. 2010. Research and application of energy conservation technique of surface technology in Libao Oilfield[J]. *Complex Hydrocarbon Reservoirs*, 3(4): 82–84.

Wu, X.D. et al. 2007. Steam stimulated wells gas injection orthogonal optimization design[J]. *Petroleum Drilling Techniques*, 35(3): 1–4.

Yan, Y.F. et al. 2012. Application and utilization technology of solar energy[J]. Acta Energiae Solaris Sinica, 33(S): 37–56.

Yang, B. et al. 2012. Research on optimized multiple thermal fluids stimulation of offshore heavy oil reservoirs[J]. *Petroleum Geology and Engineering*, 26(1): 54–56.

Yang, X.F. et al. 1997. Performance computation of a few kinds of solar collectors and comparison among them[J]. *Journal of Chongqing Jianzhu University*, 19(5): 98–100.

Zhang, X.K. et al. 2007. Study of modification for thermal recovery steam generator for oil fields[J]. *Industrial Boiler*, 27(3): 8–10.

Zhu, M. et al. 2006. Collectors of solar heating system for crude oil transportation[J]. *Energy Technology*, 27(1): 10–12.

Advances in Energy, Environment and Materials Science – Wang & Zhao (Eds)
© 2016 Taylor & Francis Group, London, ISBN 978-1-138-02931-6

Research on energy storage optimal configuration of renewable energy generation under different control objectives

B. Liu & Y. Meng
Shanghai Electric Power Design Institute Co. Ltd., Shanghai, China

S.Y. Zhou & T. Lin
School of Electrical Engineering Wuhan University, Wuhan, China

ABSTRACT: The output power of renewable energy power plants (wind farms and solar power plants) fluctuates. In order to smooth power fluctuation or tracking power generation schedule, energy storage devices are often configured in power system. In this paper, charging and discharging characteristics of battery is analyzed as representative of energy storage devices and an energy storage system model is established. For different operation modes, an optimal configuration model is studied under two different control objectives, which mean tracking power generation schedule and smoothing the output fluctuations. The optimization objective is the minimization of energy storage devices investment costs, and the decision variables are the rated power and rated capacity of energy storage devices. For a practical case, Matlab is used to calculate the optimal solution.

1 INTRODUCTION

In recent years, fossil energy shortages and environmental pollution have become increasingly serious. To relieve the stress of the energy and environmental problems, countries all over the world are developing clean renewable energy. Among all sorts of renewable energy, wind power and solar photovoltaic power develop most rapidly. At present, China leads the world in capacity of all installed wind turbines.

Although wind power generation and solar photovoltaic power generation help to save energy and protect the environment, the output power will change with weather conditions such as the wind speed, and the output cannot maintain stability like that of traditional power generation to operate precisely according to generation schedule. In addition, wind power farms and photovoltaic power plants in large scale accessing to power grid may affect the stability and security of the system, and bring inconvenience to users.

In order to smooth the power fluctuations of wind farms and photovoltaic plants, energy storage devices can be installed on the outlet side of wind farms and photovoltaic plants. Energy storage devices can limit the renewable energy power fluctuations to some degree, and reduce the impact on power system and users. However, the installation of energy storage devices will increase the costs, so the research on the configuration of energy storage devices in new energy power generation system is significant.

In (Hu, 2012), the annual energy balance method is proposed for the independent wind power generation system, using batteries to balance the power difference between the power generating system and the power consumption for capacity configuration. This method is also applied to wind power and solar power hybrid generation systems with simple algorithm, and it applies well to engineering. In (Wang, 2008), peak integration method is proposed, taking the maximum absolute value of the integration of batteries charging and discharging on the entire time axis as the capacity of batteries. In (Liu, 2008), maximum negative integration method is proposed, taking the maximum value of integration in the interval of batteries discharging as the capacity of the battery.

However, the methods mentioned above fail to take an account of the internal characteristics of the battery, such as the state of charge. The capacity of batteries determined by methods above may not supply smooth power in wind power and photovoltaic power generation system, due to restriction of the operating conditions of batteries.

In this paper, a model is established with the internal characteristics of batteries taken into account for certain amount of wind power and photovoltaic power generation. There are two control objectives: (1) tracking generation schedule; (2) smoothing output power fluctuations to the predetermined range. The decision variables are the rated power and rated capacity of energy storage devices, and the optimization objective is the minimization of energy storage devices investment. Sequential enumeration method is used to get the optimal solution.

2 WIND POWER AND PHOTOVOLTAIC POWER GENERATION SYSTEM

Wind power and photovoltaic power generation system usually consist of wind turbines, solar photovoltaic cells, energy storage devices, inverters and the load. When wind power and photovoltaic power supply the load sufficiently, the energy storage devices charges to store the excess energy, otherwise the energy storage devices discharges to ensure smooth and continuous power supply.

2.1 Model of wind power generation

The output power of wind power generation is influenced by a number of factors, whereas it can be attributed to wind speed only with secondary factors ignored, as shown below (Bowden, 1983).

$$P_W = \begin{cases} 0, v \leq v_i, v \geq v_o \\ \dfrac{v^3 - v_i^3}{v_R^3 - v_i^3} P_R, v_i \leq v \leq v_o \\ P_R, v_R \leq v \leq v_o \end{cases} \quad (1)$$

In equation 1, P_W is the output power, v_i is the cut-in wind speed, v_o is the cut-out wind speed, v_r is the rated wind speed, P_R is the rated output power.

2.2 Model of solar photovoltaic generation

In practical engineering, the output power of photovoltaic cells can be described in simplified model, namely, that the output power of the photovoltaic cells is only related with solar radiation and ambient temperature values. The equation is shown as below (Niu, 2010).

$$P_{PV} = P_{std} G_{AC} \frac{1 + k(T_c - T_r)}{G_{std}} \quad (2)$$

where P_{PV} is the output power of photovoltaic cells, G_{AC} is the light intensity; P_{std} is the maximum test power under standard test conditions; G_{std} is the light intensity under standard test conditions; k is the power temperature coefficient; T_c is the working temperature of photovoltaic cell; T_r is the reference temperature.

2.3 Model of energy storage devices

The remaining energy of batteries at moment t is related with that at moment $t-1$, charging and discharging during interval [$t-1$, t] and its own energy attenuation. The charging and discharging

process of it can be described as follows (Ding, 2012; Ding, 2011; Ding, 2011):

When charging, there is:

$$P_S(t) \geq 0 \quad (3)$$

$$E(t) = (1 - \sigma)E(t-1) + \eta P_S(t)\Delta t \quad (4)$$

When discharging, there is:

$$P_S(t) \leq 0 \quad (5)$$

$$E(t) = (1 - \sigma)E(t) + P_S(t)\Delta t / \eta \quad (6)$$

where $E(t)$ is the remaining energy of batteries at moment t; $P_S(t)$ is the charging/discharging power of batteries during interval [$t-1$, t]; σ is its self-discharging rate; η is the efficiency of its charging and discharging process; Δt is the time window for calculation. To simplify the problem, ideal case is taken into consideration, i.e. self-discharging rate σ is taken as 0, the charging and discharging efficiency η is taken as 100%. Therefore, the charging and discharging process can be unified expressed as:

$$E(t) = E(t-1) + P_S(t)\Delta t. \quad (7)$$

3 MODEL OF OPTIMAL CONFIGURATION OF ENERGY STORAGE DEVICES

In order to record the remaining energy during the process of charging and discharging, and calculate the charging and discharging power value of power stations at different time, annual data of wind farms and photovoltaic power plants is needed as known condition. In this paper, energy storage devices are assumed to be installed on the outlet side of wind farms and photovoltaic power plants to meet the specific requirements of control objectives. And there are two control objectives: a. tracking generation schedule; b. smoothing power fluctuation to predetermined range.

3.1 The decision variables and objective function

Economic and rational allocation of power and capacity of batteries are very necessary. The decision variables are the rated power and rated capacity of energy storage devices (Ma, 2015; Xie, 2012), and the optimization objective is the minimization of energy storage devices investment, i.e.

$$\min f = aP_R + bE_R \quad (8)$$

where a and b respectively represent the unit price of battery power and battery capacity, and the corresponding units are Yuan (¥)/(kW) and Yuan (¥)/(MWh); P_R is the rated power of batteries, and

E_R is the rated capacity of batteries, f is the investment cost of energy storage devices. In the objective function contains, both the rated power and the rated capacity are taken into consideration.

3.2 Constraints

Constraints include the self constraints of the energy storage devices and constraints under different control objectives. Self constraints of the energy storage devices constraints:

1. limit of the charging and discharging power,

$$-P_R \leq P_S(t) \leq P_R \qquad (9)$$

2. limits of the State of Charge (SOC) of energy storage devices. State of Charge (SOC) represents the ratio of the remaining capacity of the energy storage devices and its rated capacity, and SOC can be expressed in percentage, ranging from (Bowden, 1981). Expressions of SOC can be written as:

$$SOC = (E_{ini} + \sum P_S(t)\Delta t)/E_R \qquad (10)$$

where E_{ini} is the initial capacity of energy storage devices. The SOC limit is:

$$SOC_{min} \leq SOC \leq SOC_{max} \qquad (11)$$

For different control objectives, there are different constraints. Tracking generation schedule is required to meet the constraints of power balance, so the sum of wind power, photovoltaic power, and the power of the energy storage devices should keep equal with the power generation schedule, namely:

$$P_W(t) + P_{PV}(t) - P_{GS}(t) = P_S(t) \qquad (12)$$

where $P_{GS}(t)$ is the scheduled power; and smoothing power fluctuations to the predetermined range has to satisfy the constraints below:

$$|(P_W(t) + P_{PV}(t) - P_S(t)) - (P_W(t-1) + P_{PV}(t-1) - P_S(t-1))| \leq \delta \qquad (13)$$

where δ is the range of power fluctuation during period Δt.

3.3 Optimization method

In this paper, sequential enumeration method is adopted to solve the model. For example, under control objective a, first the power differences between wind power, photovoltaic power, and scheduled power generation need to be calculated.

The rated power of batteries equals to the maximum absolute value of the power difference. Then charging and discharging power of batteries during each time window needs to sum up to get the maximum positive value and negative absolute value. And then rated capacity can be calculated according to limits of SOC. Finally, the minimization of investment cost can be calculated. The method is analogous under objective b. When the power difference is within the limit, power of batteries equals 0, otherwise it equals the maximum of the absolute value of the power difference plus or minus δ.

4 CASE ANALYSIS

In this paper, data of the case is from annual data of wind farms and solar photovoltaic plants in a certain area. The rated wind power is 90 MW, and the rated photovoltaic power is 50 MW. After computing, the annual wind power generation curve are shown in Figure 1.

The generation schedules of wind power generation, photovoltaic power generation and wind power and photovoltaic power hybrid generation are not the same. Due to space limitations, only wind power and its generation schedule are shown below as examples in Figure 1 and Figure 2. The limit range of power fluctuations is shown in Table 1.

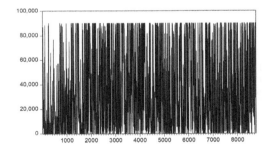

Figure 1. Annual wind power curve.

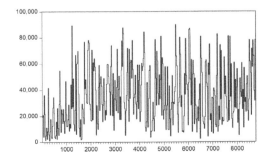

Figure 2. Annual wind power GS curve.

Table 1. Limit range of power fluctuations.

Operation mode	ΔP in 1 hour
Wind power generation	Rated power/3
Photovoltaic power generation	Rated power
Wind power and photovoltaic power hybrid generation	Rated power/3

Table 2. Parameters of lithium iron phosphate battery.

Power unit cost a	1500 Yuan/kW
Capacity unit cost b	3500 Yuan/(MWh)
SOC range	[0.2,1]
σ	0
η	100%
Δt	1 h
E_{ini}	$0.5\,E_R$

Table 3. Results of optimal configuration.

Operation mode	Control objective a			Control objective b		
	Rated power/kW	Rated capacity/MWh	Investment/ 10^7 Yuan	Rated power/kW	Rated capacity/MWh	Investment/ 10^7 Yuan
Wind + battery	8.2773×10^4	1.4076×10^3	12.9086	5.5292×10^4	1.0971×10^3	8.6778
PV + battery	2.3769×10^4	1.1295×10^2	3.6049	1.5521×10^4	2.2435×10^2	2.4067
Wind + PV + battery	8.8453×10^4	1.6011×10^3	13.8283	3.4792×10^4	1.0461×10^3	5.5849

In this paper, lithium iron phosphate battery is taken as an example, and its parameters are shown in Table 2.

According to the acquired data of wind power and photovoltaic power, as well as generation schedule and the predetermined power fluctuation range, the rated power, rated capacity and investment are calculated under different operation modes for different control objectives. By use of Matlab, results of optimal configuration are shown in Table 3.

According to the results, conclusion can be drawn that: under control objective a, photovoltaic power generation costs least, while wind power and photovoltaic power hybrid generation costs most; under control objective b, photovoltaic power generation costs least, while wind power generation costs most.

5 CONCLUSION

In this paper, method of energy storage devices configuration is studied for wind power and photovoltaic power generation system. Energy storage devices model is established with the internal characteristics of battery taken into account. The optimization objective is the minimization of energy storage devices investment costs, and the decision variables are the rated power and rated capacity of energy storage devices. For different control objectives, Matlab is used to get the optimal solution under different operating modes for the practical case.

REFERENCES

Bowden, G.J. & Barker, P.R. 1983. Weibull distribution function and wind power statistics. *Wind Engineering*, 7(2): 85–98.

Ding, Min & Zhang, Yingyuan 2011. Economic operation optimization for microgrids including na/S battery storage. proceedings of the CSEE, 31(4): 7–14 (in Chinese).

Ding, Ming & Lin, Gende 2012. A Control strategy for hybrid energy storage systems. *Proceedings of the CSEE*, 32(7): 1–6 (in Chinese).

Ding, Ming & Xu, Ningzhou 2011. Modeling of BESS for smoothing renewable energy output fluctuation. *Automation of Electric Power Systems*, 35(2): 66–72 (in Chinese).

Hu, Guozhen & Duan, Shanxu 2012. Sizing and cost analysis of photovoltaic generation system based on vanadium redox battery. *Transactions of China Electrotechnical Society*, 35(5): 260–267 (in Chinese).

Liu, Zhihuang & Yang, Yimin 2008. Model of optimal design for wind/photovoltaic hybrid power system. *Computer Engineering and Design*, 29(18): 4836–4838 (in Chinese).

Ma, Suliang & Ma, Huimeng 2015. Capacity configuration of the hybrid energy storage system based on bloch spherical quantum genetic algorithm. *Proceedings of the CSEE*, 35(3): 592–596 (in Chinese).

Niu, Ming & Huang, Wei 2010. Research on economic operation of grid-connected microgrid. *Power System Technology*, 34(11): 38–42 (in Chinese).

Wang, X.Y. 2008. Determination of battery storage capacity in energy buffer for wind farm. *IEEE Trans on Energy Conversion*, 123(3): 868–878.

Xie, Shixiao 2012. A chance constrained programming based optimal configuration method of hybrid energy storage system. *Power System Technology*, 36(5): 79–83 (in Chinese).

A new electric power emergency command system for electric grid

Xiaoming Jin, Xiaodong Chen & Tie Li
Liaoning Electric Power Co. Ltd., State Grid Corporation of China, Shenyang, Liaoning, China

Shan Jiang & Yuhang Cui
Tellhow Software Co. Ltd., Nanchang, Jiangxi, China

Wentao Sun & Jian Zhang
Liaoning Electric Power Co. Ltd., State Grid Corporation of China, Shenyang, Liaoning, China

ABSTRACT: The modern power system is growing rapidly in the context of constant social development and economic growth, as a result, power grid tends to run in more complex modes. To resolve the problems, such as lack of effective prejudgment for power accidents, and unablility to effectively and quickly define the power equipment state, a new electric power emergency command system is proposed. The system has been designed based on "One Center, Two Host Lines, Three Links and Multiple Platforms". It cores the unified command for system security, realizes the closed-loop control for emergency disposal of risks and failures, ensures a well analyzed put-in-place control, and achieves the linkage of overall system. Depending on this system, the operating staff in charge of grid control can be of great capacity in control of volume growth and accidents handling, besides on-time disposal capacity against grid accidents.

1 INTRODUCTION

Electric Power Emergency Command System constitutes a new need with higher demands for the power management, which is mainly applicable for aid supports to commanders when a vital accident occurs. The said aid supports can be divided into three parts: first, collect and display various information associated with the system; second, make data summary and statistics; third, provide an assistant decision-making via an intelligent method based on those above. Until now, most of power emergency commanding systems have been at Phase I or Phase II, that is, data collection and demonstration phase, which can only help decision-makers to produce some statements and statistical reports. In details, decision-making still relies on commander's past experiences.

At present, the grid system applies single servers in provinces and local areas which work for accidents monitoring and multiple-departments, so there exist a quantity of problems related to data sharing. Upon the above analysis, the shortcomings are as follows: one lies at wide application and uncompleted functions. The reason is that the system used in a variety of departments can only summarize common characteristics of dispatching departments and take less consideration on their actual status. The other lies at poor extensibility and difficult post-maintenance. First, the system serves

a variety of departments, so it features for weak data access and poor data security, which may bring great difficulty for second functional use. Second, such weak data access and high data sharing rate may result in re-development difficulties, even risks of high development cost and maintenance cost.

In electric power system, the factors that may affect grid's safe and stable operation cover many aspects, such as grid's running mode, flow status, equipment status, etc. In addition, how a personnel can obtain these data timely and accurately is vital important. The secure and stable operation of Liaoning Power Grid lies at a mass of scattered system platforms and unsatisfied data interactivity. To change this status, a new method herein, according to power grid emergency standards, has been proposed to realize system's centralized management, data integration and smooth communications, besides on-time monitoring of sudden accidents, prediction and modeling, failure forecasting, and accident handling. The final is for integrated, intellectual, and optimized control.

2 TECHNICAL REQUIREMENTS

2.1 Technical problems that need to be solved

The problem needed to be solved lies in how to transform and upgrade the traditional mode to a modern one, namely, from the extensive mode to an intelligentized mode (Liu, 2010). Depending on this system, the operating staff in charge of grid control can be of great capacity in control of volume growth and accidents handling, besides on-time disposal capacity against grid accidents.

Modern power system is growing rapidly in the context of constant social development and economic growth, as a result, power grid tends to run in more complex modes (Gantz, 2011). System's stable operation is subject to a variety of factors, such as current distribution, equipment state, weather situation, etc. The result is that the system can only play a supporting role in conclusion of power flow and decision-making of power supports and has defects of longer response time, strong subjective and poor systematicness. Besides those, it is lack of the effective prejudgment for power accidents, and unable to effectively and quickly define the power equipment state, accident influence scope and flow control mode especially in the case of cascading faults.

2.2 *The situation in Liaoning electric grid*

The new proposed method has been designed in Liaoning electric grid in the northeast part of China, with more than ten sets of different intelligent grid distribution systems, such as real time data monitoring, equipment management, online system analysis, failure data collection, etc (Zhao, 2013; Demchenko, 2012). Hence, the data that has been collected and managed can be synthetically computed, matched, and analyzed for the purpose of intelligent and all-round comprehensive analysis made available for grid accidents. Meanwhile, the real time failure data system can help the personnel in charge of control and regulation to obtain and understand completed and accurate key data within the shortest time and accidents trend, besides in favor of aid decision making as well. Such an upgrading from the traditional extensive mode to a modern intelligent system can highly help the personnel in charge of grid control and regulation to boost their capacities in grid control and accidents disposal, furthermore, help power system boost its capacity in real time emergency processing against grid accidents.

3 SYSTEM STRUCTURE AND FUNCTIONS

One importance against blackouts becomes vital to improve grid service quality other than daily operation, particularly, in special periods such as earthquake, thunderstorms, lingering high temperatures in summers, and yearly seasonal typhoons. The present system that specializes in grid management and control is unable to service as an information platform in case of meeting blackouts. Hence a new system has been proposed that features for technical countermeasures against grid (emergency) operation and platform building via a variety of analysis of recently major blackouts at home and abroad, sudden-onset disaster and its characteristics, as well as the theoretical foundation and the existing problems.

3.1 *System structure*

The system has been designed based on "One Center, Two Host Lines, Three Links and Multiple Platforms". Concretely to say, it cores the unified command for system security, realizes the closed-loop control for emergency disposal of risks and failures, ensures a well analyzed put-in-place control, and achieves the linkage of overall system. The system targets at leveling up the operation, focuses at the running and equipment situation, seizes two hosts including risk control and default disposal, implements the risk management, keeps the linkage efficiency under control. The purpose is to establish a uniform platform for the whole system. The work-flow of the system is shown in Figure 1.

3.2 *System functions*

1. Mass data acquisition in system provides a search function of leap-type data query (Wigan, 2013). A centralized control mode system is established without switching different systems for data query and information retrieval, which depends on multi-format, mass data acquisition, and message-switching technique.
2. Comprehensive analysis can be made on problems related to grid security. This system is considered as a comprehensive information-based forecast system against risks and disasters, which depends on Java's SOA (service-oriented platform building

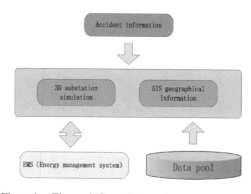

Figure 1. The work-flow of the system.

technology) to perform grid's security analysis. Grid security is an organic integrity, which can be divided into information security, operation security and infrastructure security, whereas only more research on operation security and infrastructure security in the past. Figure 2 shows 3D simulation of transformer substation picture in Liaoning electric grid.

3. A leap-type and accessible data exchange platform can be established to realize integration. XML-based data exchange platform is built across different regional networks, departments, platforms, and databases. This platform automatically extracts system-required data across heterogeneous platforms and databases as the standard XML-format data exchange protocol. And data is stored in power dispatching III area's center public resource database based on XML-standard unloading format and compression transmission technology to conduct data flow, data exchange between different systems to realize integration.

4. Instant information can be transmitted safely, quickly, and stably in combination with mobile information technology. System's automatically distributing work, sending message to look for someone and sending message to phones can be realized by comprehensively using database JOB technology, GSM/MODEM, SOCKET SMS technology.

Emergency Command Project is a web-based approach besides with Java client-side and IOS client-side. Concrete thinking shown in Figure 3.

Figure 2. 3D simulation of transformer substation picture in Liaoning electric grid.

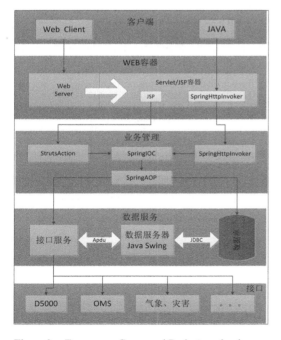

Figure 3. Emergency Command Project mechanism.

4 SYSTEM SOFTWARE

4.1 *Introduction to spring*

Spring, an open-source framework is created to decrease complexity of enterprises' application development. One of framework's main advantages lies in its layered architecture which can give you prior option of components, meanwhile provide integrated framework for J2EE application development. Spring Framework's functions can be applied to any J2EE servers, mostly of which also can be used to non-managed environment. Spring Framework's core lies in supporting reusable business and data access objects which not tied to specific J2EE service. Such objects can be reusable among different J2EE environments (Web or EJB), standalone applications and test environments. Here is mainly to introduce Spring's technologies AOP and IOC.

Spring AOP: Spring AOP (Aspect-Oriented Programming) module directly integrates aspect-oriented programming function into Spring Framework upon configuration management characteristic. As a result, it is easy to make any object in Spring Framework support AOP. And transaction management service is provided for objects in Spring's applications by Spring AOP module. Using Spring AOP can integrate declarative transaction management into applications without depending on EJB (Enterprise Java Bean).

4.2 *Introduction to web container*

WEB container: helps the application components herein (such as JSP, SERVLET) achieve interactive

Table 1. Advantages of the new method.

	Analysis and calculation of grid's security and stability	Grid mode adjustment, output of power plant and load adjustment, etc
Traditional operation mode	15 minutes, off-line computation in technical support rooms	10 minutes, decision-making upon operating experience after operators' querying procedures
New method	Auto starting on-line computation to obtain a consequence and auto prompting grid's weakness	Auto generation assistant decision-making

at environment variable interface without any interference. All these can be realized by WEB server such as TOMCAT, WEBLOGIC, WEB-SPHERE, etc. Interfaces provided herein strictly abide by the WEB APPLICATION standards in J2EE specification. WEB server in accordance with the above standard is considered as the WEB container in J2EE.

4.3 Introduction to spring HTTP invoke

Spring HTTP invoker is a remote invocation model in Spring Framework, performing HTTP-based remote invocation (it can penetrate a firewall), and use JAVA's serialization for data transfer. It is a bit similar to but different from WebService that client-sides can easily invoke objects on remote servers just as invoking local objects.

4.4 Introduction to struts

Struts architecture respectively maps Model, View and Controller to components in Web to implement MVC design pattern's concept. Model is composed by Action represented system state and business logic, and definition labels provided by struts and JSP realize View, besides, Controller is in charge of control flow, completed by Action Servlet and Action Mapping.

Beneficial effects that the system brings to Power Grids are listed on tables below for contrast of dispatching emergency command system between before and after using this method when accidents occur.

5 CONCLUSIONS

In modern electric power system, the factors that may affect grid's safe and stable operation cover many aspects. For example, the grid's running mode, flow status, equipment status, etc. In particular, how a personnel can obtain these data timely and accurately is vital important. The secure and stable operation of Liaoning Power Grid lies at a mass of scattered system platforms and unsatisfied data interactivity. To solve these problems, a new power emergency command system is proposed-according to power grid emergency standards-to realize system's centralized management, data integration and smooth communications. Furthermore, on-time monitoring of sudden accidents, prediction and modeling, failure forecasting and accident handling. The final function is for integrated, intellectual, and optimized control.

REFERENCES

Demchenko Y., Zhao Zhiming, Grosso P., et al. Addressing big data challenges for scientific data infrastructure [C]//2012 IEEE 4th International Conference on Cloud Computing Technology and Science. Taipei, Taiwan: IEEE, 2012: 614–617.

Gantz J., Reinsel D. Extracting value from chaos [J]. Proceedings of IDC iView, 2011: 1–12.

Liu Zhenya. Smart grid technology [M]. Beijing: Chinese Electric Power Press, 2010 (in Chinese).

Wigan M.R., Clarke R. Big data's big unintended consequences [J]. IEEE, 2013, 46(6): 46–53.

Zhao Guodong, Yi Huanhuan, Mi Wanjun, et al. Historic opportunity in the era of big data [M]. Beijing: Tsinghua University Press, 2013 (in Chinese).

Advances in Energy, Environment and Materials Science – Wang & Zhao (Eds)
© 2016 Taylor & Francis Group, London, ISBN 978-1-138-02931-6

Experimental study on the performance of 15 kW OTEC system

Fengyun Chen
Harbin Engineering University, Harbin, China
First Institute of Oceanography State Oceanic Administration, Qingdao, China

Liang Zhang
Harbin Engineering University, Harbin, China

Weimin Liu, Lei Liu & Jingping Peng
First Institute of Oceanography State Oceanic Administration, Qingdao, China

ABSTRACT: The cycle efficiency calculation method for the Ocean Thermal Energy Conversion (OTEC) is obtained through the theoretical analysis and model establishment of the equipment in the OTEC system. A 15 kW OTEC plant is built using Rankine cycle and ammonia as working medium to research the performance of the OTEC plant under different operating conditions. Then the relationship between cycle efficiency, turbine efficiency and seawater temperature difference are obtained through the analysis of the experimental data, and the heat transfer performance of the evaporator and condenser is obtained. The results show that within the scope of the test conditions, the system achieves 15 kW power rating system, and the maximum efficiency of turbine about 73% when the temperature difference is 19.7°C.

1 INTRODUCTION

An Ocean Thermal Energy Conversion (OTEC) plant is basically a heat engine that utilizes the temperature difference between the warm surface seawater and deep cold seawater to drive a turbine to produce electricity (Lavi, 1980). In Rankine cycle the low boiling point working fluid is vaporized by the warm surface seawater in the evaporator. The vaporized fluid expands to do work in a turbine and generate electricity. Then the low pressure exhaust is condensed in the condenser by cold deep seawater. The condensed working fluid is pumped back to the evaporator and the cycle is repeated. Figure 1 shows a schematic diagram of a closed cycle OTEC plant (Uehara, 1990).

Ocean thermal conversion energy is abundant in south China sea. The theory reserves of ocean thermal conversion energy close to China's offshore is about $14.4 \times 10^{21} \sim 15.9 \times 10^{21}$ J and the total installed capacity which could be utilized is $17.47 \times 10^8 \sim 18.22 \times 10^8$ kW, and 90% of the reserves distribute in the south China sea (Wang, 2009). In the face of environmental pollution and gradually reduce of fossil fuel, the utilization of the OTEC can not only promote the development of the marine economy and coastal defense construction, but also solve the power shortage of coastal and island area. The construction of the OTEC plant could supply power and fresh water for the oil production in offshore engineering, and achieve the purpose of energy conservation and emission reduction.

The development of OTEC will contribute to alleviating the pressure of the energy, improving our country's industrial structure, and also environmental protection. As a result, the ocean thermal energy as a kind of clean and renewable energy, the development and utilization of the ocean thermal energy has important practical significance to the sustainable development of national economy and the improvement of people's living standard.

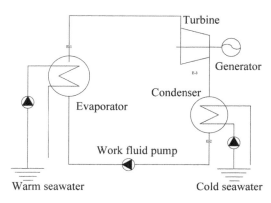

Figure 1. Schematic diagram of a Rankine cycle OTEC plant.

2 SYSTEM DESCRIPTION

2.1 Principle of Rankine's cycle

In this paper the OTEC system uses Rankine cycle and ammonia as working medium in this system. The system is composed of evaporator, turbine, generator, condenser, pump etc. The liquid ammonia evaporates in the evaporator where it gains heat from the warm surface seawater. Ammonia vapor enters into the turbine and expands thereby delivering mechanical work. The working fluid at the exit of the turbine is condensed in the condenser by cold deep seawater. The working fluid is saturated liquid at the exit of the condenser, it is then pumped into the evaporator, in this way the cycle is continued. The experimental system is showed in Figure 2.

The T-S diagram of Rankine cycle is showed in Figure 3. In the T-S diagram 1–2 means the progress of conversion of heat into power in turbine, 2–3 means the exothermic process from the working fluid at the exit of the turbine to cold deep seawater. 3–4 is the process of working medium through the pump, 4–1 is the evaporation of working medium in the evaporator.

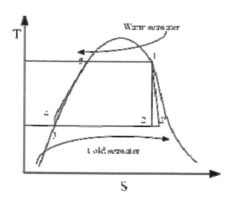

Figure 2. Diagram of the experimental system.

Figure 3. The T-S diagram of Rankine cycle.

2.2 Thermodynamic analysis

Thermodynamic analysis in each component is conducted in this study. For the cycle performance simulation, the following assumption are made: pressure drop in pipes and heat losses to the environment in the condenser, evaporator, generator, turbine and pump are neglected; The system remains uniform flow condition.

Evaporator: under this process, the liquid ammonia is heated at constant pressure. Equation of heat balance is:

$$Q_E = m\ (h_1 - h_4) \tag{1}$$

Condenser: The working fluid at the exit of the turbine is condensed by cold deep seawater. Equation of heat balance of condenser is:

$$Q_C = m\ (h_2 - h_3) \tag{2}$$

Turbine: the turbine converts thermal energy of ammonia vapor into mechanical work. The work produced by turbine is:

$$W_t = m\ (h_1 - h_2) \tag{3}$$

Pump: The process is assumed as isentropic process. The work consumed by the pump is:

$$W_P = m\ (h_4 - h_3) \tag{4}$$

Thermal efficiency of the cycle is defined on the basis of the first law of thermodynamic as the ratio of net power output to the heat transferred from the warm surface seawater to the working fluid in the evaporator.

$$\eta_t = \frac{(h_1 - h_2) - (h_5 - h_4)}{h_1 - h_4} \tag{5}$$

where Q_E = heat exchange amount through evaporator; m = system mass flow rate; and h = specific enthalpy; Q_C = heat exchange amount through condenser; W_t = work produced by turbine; W_p = work consumed by pump; η = thermal efficiency.

3 SELECTION AND CALCULATION OF SYSTEM COMPONENTS

3.1 Selection of heat exchanger

Due to the temperature difference between warm surface seawater and cold deep seawater is small, so the area of heat exchanger needs to be larger. Heat exchanger is an important equipment in thermal conversion process, so choosing a heat exchanger which has high performance can improve system efficiency, and greatly reduce the cost of investment. Commonly used form of heat exchangers include shell-and-tube heat exchanger, plate heat exchanger, plate-fin heat exchanger etc. Whereby the shell-and-tube heat exchanger is widely used and the technology has been perfected. However coefficient of shell-and-tube heat exchanger is small, result in increasing of the temperature difference, the volume of the heat exchanger, and then additional cost. Design of the ocean thermal energy conversion system is based on two factors: high heat transfer efficiency and low cost (compact size).

At present the structure of plate heat exchanger includes dismountable-mountable type and brazed plate heat exchangers, the compression capability of the dismountable-mountable plate is strong, however it is expensive. Currently heat exchangers adopt brazed plate heat exchanger in case of operating pressure greater than 1.0 MPa. Considering the. But considering the cost of brazed plate heat exchanger is large, in the developed OTEC plant, dismountable-mountable plate heat exchangers are chose. In order to increase the compression capability, use silicone rubber gasket and coated pressure tight sealant. In pressure testing experiment could be 2.0 MPa. Considering the corrosive performance of seawater, SOS316L is chose as plate heat exchanger material, material of E as gasket.

3.2 Calculation model of heat exchanger

The heat transfer coefficient is calculated and compared with the experimental data.

Calculation equation of heat transfer coefficient of the evaporator is:

$$Q = c_p m \Delta t = K A \Delta t_m \quad (6)$$

Thus the heat transfer coefficient is:

$$K = c_p m \Delta t / (A \Delta t_m) \quad (7)$$

In the evaporator, the fluids flow in counter flow, the temperature difference is called the log mean temperature difference.

$$\Delta t_m = \frac{(t_{win} - t_{cout}) - (t_{wout} - t_{cin})}{\ln \frac{(t_{win} - t_{cout})}{(t_{wout} - t_{cin})}} \quad (8)$$

where c_p = specific heat at constant pressurer; m = system mass flow rate; Δt = temperature drop of warm surface water through the evaporator; Δt_m = log mean temperature difference of the working fluid in evaporator; K = heat transfer coefficient; and A = heat transfer area.

Calculation equation of heat transfer coefficient of the condenser is:

$$Q = c_p m \Delta t = K A \Delta t_m \quad (9)$$

Thus the heat transfer coefficient of the condenser is:

$$K = c_p m \Delta t / (A \Delta t_m) \quad (10)$$

In the condenser, the working fluid and the cold deep seawater flow in counterflow, the temperature difference is called the log mean temperature difference.

$$\Delta t_m = \frac{(t_{win} - t_{cout}) - (t_{wout} - t_{cin})}{\ln \frac{(t_{win} - t_{cout})}{(t_{wout} - t_{cin})}} \quad (11)$$

where c_p = specific heat at constant pressurer; m = system mass flow rate; Δt = temperature rise of cold deep seawater through the condenser; Δt_m = log mean temperature difference of the working fluid in condenser; K = heat transfer coefficient; and A = heat transfer area of the condenser.

4 RESULTS AND DISCUSSION

In the present paper, based on the experimental data, the effect of the major thermodynamic parameter on the system performance for Rankine cycle is analysed. In experiment process get the data by changing the temperature difference between warm and cold seawater. Figure 4 shows the variation of turbine power output versus the temperature difference between warm and cold seawater. It is apparent from Figure 4 that turbine power output

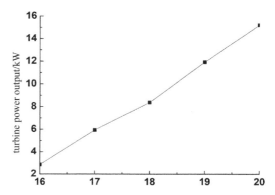

Figure 4. Variation of turbine power output versus the temperature difference between warm and cold seawater.

increase with the increase in temperature difference between warm and cold seawater. The plant achieve the rated power when the temperature difference between warm and cold seawater is 19.7°C.

Figure 5 shows the variation of turbine efficiency versus the temperature difference between warm and cold seawater. It is apparent from Figure 5 that turbine efficiency increase with the increase in temperature difference between warm and cold seawater. Turbine efficiency reaching the maximum, about 73%, when turbine operate under the rated conditions.

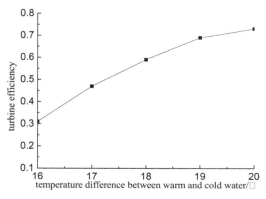

Figure 5. Variation of turbine efficiency versus the temperature difference between warm and cold seawater.

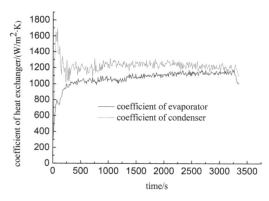

Figure 6. Variation of coefficient of heat exchangers versus over time.

Coefficient of the heat exchanger under different working conditions is obtained and heat transfer performance is tested by experiment. Figure 6 show the variation of coefficient of the heat exchanger versus over time under the conditions: the temperature of cold water is 20°C at inlet, temperature deference is 2.71°C, mass flow-rate of cold water is 129 m³/h, the temperature of warm water is 40°C at inlet, temperature deference is 2.03°C, mass flow-rate of warm water is 125.3 m³/h, mass flow-rate of ammonia is 1.27 m³/h. Coefficient of heat exchanger can be up to about 1700 W/(m²·h). When the heat exchanger is normally running, the coefficient of the heat exchangers are between 1200 W/(m²·h) and 1300 W/(m²·h).

5 CONCLUSION

Through the theoretical analysis and experiment of the 15 kW OTEC plant, the following results are obtained:

1. Turbine efficiency reaching the maximum, about 73%, when the temperature difference between warm and cold seawater is 19.7°C.
2. Coefficient of heat exchanger can be up to about 1700 W/(m²·h). When the heat exchanger is normally running coefficient is between 1200 and 1300 W/(m²·h).

REFERENCES

Huang Liang Min. The Overview of China's Sustainable Development [M]. Beijing: Science Press, Ocean Development and Management, 2007.

Lavi A. Ocean thermal energy conversion: a general introduction. Energy 1980; 5: 469–480.

Liu Qi. Study on China's New Energy Development [J]. Power System and Clean Energy, 2010, 1, 26(1): 1–2.

Uehara H, Ikegami Y. Optimization of a closed-cycle OTEC system. Trans ASME Journal of Solar Energy Engineering 1990; 112: 112–256.

Wang Chuan Kun. Analysis Methods and Reserves Evaluation of Ocean Energy Resources [M]. Beijing: China Ocean Press, 2009.

Wang Xun. Economic and environmental benefits of ocean thermal energy conversion [J]. Marine Sciences, 2008, 32(11): 84–87.

Advances in Energy, Environment and Materials Science – Wang & Zhao (Eds)
© 2016 Taylor & Francis Group, London, ISBN 978-1-138-02931-6

Construction solutions on data center of PV power systems of Qinghai Province based on big data technologies

Yu Yun
Nanjing College of Information Technology, Nanjing, China

ABSTRACT: With the rapid development of construction of PV plants in Qinghai province, the need for constructing a data center for monitoring and analyzing generation data and environment data of PV plants has become more and more emergent for the sake of the safety and efficiency of PV power generation and dispatching. In this paper, a total solution on constructing a data center of PV Power System in Qinghai province based on Big Data Technologies was put forward by presenting the logical architecture, network topology, physical architecture, and determination of storage configuration. With the support of Big Data technology, the data center can be built as a safe, scalable, efficient, trusted platform for the data and business management of the PV plants. And it will be also profitable for the power grid companies who are responsible for the dispatching and planning of the PV systems within their operation area.

1 INTRODUCTION

By the year 2014, the total installed capacity of PV power systems in Qinghai province of China have reached 4,120 MW. With the rapid development of construction of PV plants in Qinghai province, the need for constructing a data center for monitoring and analyzing generation data and environment data of PV plants has become more and more emergent for the sake of the safety and efficiency of PV power generation and dispatching.

For each PV power system, the measured values, e.g. voltage, current, output power, and so on, we may care can reach up to 20,000 data points. In sum, the designed data capacity for the data center should be set to 2,500,000 data points. To store and handle such large scale of data, the storage capacity for the data center is 100 TB for the first stage. Besides the observed data which can be organized in form of structured data, the non-structured data of design document, monitoring VCR data and so on is also very vital for the efficiency analysis of PV power system for the data center. The data center will be designed providing services for the related system like dispatching system, power trading system, economic analysis system and so on. So, the data with data center should be easy to be accessed for those outside systems.

To meet the needs above, the 'Big Data' technologies (Viktor M. et al. 2013) represented by Hadoop/Spark technology (Srinath P. & Thilina G. 2013, Holden K. et al. 2015) are naturally addressed for providing a safe, efficient, robust solution for the constructing data center of PV systems.

According to the heavy job of large scale of observed data collection within PV plants passed on to the site of data center far away, the communication solutions should also be addressed.

2 SYSTEM ARCHITECTURE

2.1 Logical architecture

The data flow of the data center can be depicted as the following. The data collection devices installed on-site collecting observed data, which was passed to data center through communication channels. After data analysis, the results will be pushed to display frontend such as desktop PC, smart phone, and display screen. According to the data flow, the data center of PV systems in Qinghai province can be logically divided into 4 layers: infrastructure layer, platform layer, application layer, and display layer, which is depicted as Figure 1.

1. Infrastructure Layer provides basic run-time environment for the business logics of the whole system. Apache Hadoop software library is introduced as the base framework for reliable, scalable, distributed computing. The distributed file system—*HDFS*, the batch processing framework—*Map/Reduce* and the coordination service for distributed applications—*Zookeeper* (Flavio J. & Benjamin R. 2013) form the core layer of Hadoop. Above the core layer, *Hbase* (Nick D. & Amandeep K. 2012) designed for structured storage and Hive—a data warehouse infrastructure that provides data summarization

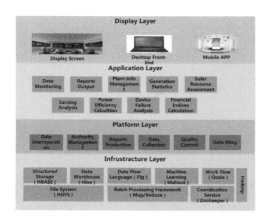

Figure 1. Logical architecture.

and ad hoc querying are based on distributed file system, providing supports for data interoperation mid-ware and data collection mid-ware in the platform layer. The data flow language—*Pig* is an execution framework for parallel computation built upon Map/Reduce and Zookeeper. *Mahout* (Sean O. et al. 2010)—a scalable machine learning and data mining library and *Oozie* (Hortonworks. 2015)—a workflow scheduler system to manage Hadoop jobs provide programming API for the modules of data mining and data quality control in the platform layer.

2. Platform Layer is a collection of business midware providing common use functions for business application modules. The platform layer is running upon the Hadoop services framework. The design objectives are to implement good abstractions for application functions, reduce the dependency between application interfaces and infrastructure layer, and ensure the scalability and reusability. The modules in platform layer includes: 1) Authority Management: a mid-ware which provides a series of management features, such as authority service, log management, and user management. The module provides support for the corresponding UI modules in application layer. 2) Data mining: a mid-ware enables features of data statistics and data mining. One part of the module is designed to be run as backgrounder in the state of batch processing to support off-line tasks. Another part of the module can be run within the application containers, accepting data analysis requests from application layer and returning analysis results. 3) Data interoperation: a mid-ware designed to support on-line data query features. It accepts real-time query requests from application layer then submit to database and return the query results. 4) Report production: a mid-ware for report

production by template definition and template replacement. The report system is designed for cross-platform and compatible with excel and xml file formats. 5) Data collection: performing interactions with communication servers to implement persistence operation of on-line monitoring data and statistics data. 6) Data quality control: a module implementing the function of data quality control by filtering and annotation to data based on the domain knowledge and statistic algorithms. It is responsible for the integrity and validity of data.

3. Application Layer contains all of the features shown in UI. Based on BS structure, the application layer is deployed in application container and run in Web browser. The main features include: on-line data monitoring, devices management, reports output, Earnings calculation of generation and generation efficiency analysis.

4. Display Layer renders the application data through display screen, desktop PC and mobile app.

2.2 *Network topology*

Figure 2 depicted the network topology of data center of PV power systems in Qinghai province, which is deployed in 3 security zones: PV Plants, internal zone of master platform, and external zone of master platform.

1. An on-site communication terminal is deployed in PV plant obtaining operation data of the PV system. There are 3 ways for the communication terminal to obtain the operation data from PV system according to different types of data: the device running data are fetched from the existing

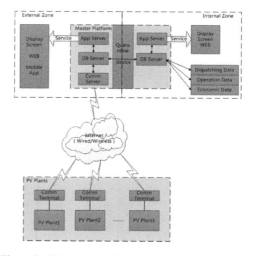

Figure 2. Network topology.

production monitoring system on-site and sent through COM port using the 101 power communication protocol; the climate observation data are directly sent from weather station through Modbus-RTU protocol; the generated energy data are directly sent through DLT645 protocol to the communication terminal. Then, the collected data in the on-site communication terminal are sent through VPN network to the remote master station of data center.

2. The master platform of the data center is deployed in both the external zone and the internal zone. There are quarantine devices for single direction data transport from internal zone to external zone. The devices only support the legal data transport through the TNS protocol, which is used by Oracle. The devices can provide the filtering, analysis and detection services to SQL query, blocking the malice access to platform, guaranteeing the safety of data content and platform. The communication server cluster, the distributed server cluster, Oracle server, and application server cluster are deployed in the external zone. The distributed server cluster supports the large data storage and application server cluster supports the internet information access. Oracle server and application server are also deployed in internal zone. Oracle server provides the storage service for dispatching data, sales data, and economic data. The application server provides the information service for intranet user.

2.3 *Physical architecture*

The physical architecture of data center is depicted in Figure 3. The whole data center platform is divided into external system and internal system by quarantine devices. The external system is the core part of the whole data center. The communication server cluster, Hadoop cluster, Oracle server, and application server cluster are deployed in it communicating though 10 Gbytes network. The external system is designed to support data processing, core business logic running, and UI service. To obtain the ability of collecting, persisting, and analyzing the TB to PB level data in the near future; Hardtop technologies are introduced in the system. Through Hadoop framework, coordination of distributed applications, distributed file system, distributed database, data warehouse, and data mining service are implemented. To implement real-time duplex communication between remote master platform and on-site communication terminals, communication server cluster are deployed supporting the input of more than 2,500,000 data values. The detail devices of external system are enumerated below:

1. Communication server cluster: PC servers, supporting active standby, providing the data collection service and NTP time service for the system. The communication server cluster should support 150 data channels and 2,500,000 data values.
2. Hadoop cluster: adopting Hadoop framework, to provide distributed business database, data warehouse, and application services.
3. Oracle server: supporting the data transport between external and internal system through quarantine devices using the oracle TNS protocol. It is deployed as the mirror of internal database.
4. Application server cluster: providing frontend services, supporting the interactions through Web browser and mobile terminals.
5. 10 Gbytes network system: providing data communication channels of high efficiency.
6. NAT switcher: providing Network Address Transformation service to support the information distribution and data collection.
7. GPS chronometer: receiving standard time from GPS system and providing time service to the system.

The internal system is designed for internal data (such as energy data) collection, internal web information distribution service which is a supplement for the external system providing some mirror services for the latter. It can also provide data service for other internal business systems. The detail devices configuration is listed below:

1. Oracle server: internal data collection and data analysis.
2. Application server: supporting internal web application service and UI service.
3. Workstations: internal information terminals.
4. Gbytes Network: providing the internal data communications.

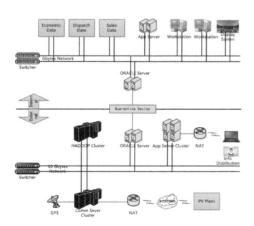

Figure 3.   Physical architecture.

3 DETERMINE OF STORAGE CONFIGURATION

To determine the storage configuration of the data center, the data size of collection and analysis jobs should be calculated. The data types been collected and analyzed includes: PV plants production monitoring data, dispatching data, energy production data from sales data, statistics data, testing data, and economic analysis data.

1. PV plants production monitoring data. By 2014, the total installed capacity of Qinghai province is 4120 MW. According to the statistics data, there are 400–600 data values for 1 MW installed capacity. Through comprehensive consideration, in the first phrase, the data center is designed to contain 2,500,000 data values. For each 15 minutes, these data values will be refreshed once. From engineering practice we know that each data values take up room for 128 bytes, that is to say, the growing data size for a year is about 11.21 GB. So the total data size is 56 GB for 5 years.
2. Dispatching data. The dispatching data of PV plants includes running data of inverters, climate data, and output power data with the total of 100 data values for each plant and 15,000 data values for 150 plants. Set the data refreshing frequency to once for 15 minutes. Then the growing data size is 44.85 GB for a year and the total data size is 224.25 GB for 5 years.
3. Energy production from sales data. Relatively smaller size from sales data can be estimated as 3 GB for a year and 15 GB for the total 5 years.
4. Statistics Data. They come from the statistical operation oriented to multi-topics analysis on the detail business data. The number of planned entity tables of statistics data is estimated to 200. Each entity tables contains: time field (8 bytes), unit (16 bytes), 5 business dimensions (5 attributes for each dimensions, 8 bytes for each attribute), 10 data items (10 bytes for each items). So the total data size of each record in each entity table is 164 bytes. The total data size for each entity table is 23.69 GB for 5 years. And the summed data size for all of the entity tables is 4.6 TB.
5. Testing data: 1 GB for each plant and 150 GB for total.
6. Economic analysis data: 200 GB extra data for reserved size.

Table 1 enumerated the whole calculations of storage configuration. The total data size of data center for 5 years is 62 TB. Available storage spaces = (original data size * number of replications +

Table 1. Calculation of storage configuration.

Item	Unit	Amount
Compression ratio	%	33%
Original size	TB	60
Number of replications		3
Reserved size		1
Extension rate	%	25%
Disk redundancy	%	30%

reserved size) * compression ratio * (1+extension rate) * (1+disk redundancy) = 128.7 TB.

4 CONCLUSION

With the rapid development of construction of PV plants in Qinghai province, the need for constructing a data center for monitoring and analyzing generation data and environment data of PV plants has become more and more emergent for the sake of the safety and efficiency of PV power generation and dispatching.

In this paper, a total solution on constructing a data center of PV Power System in Qinghai province based on Big Data Technologies was put forward by presenting the logical architecture, network topology, physical architecture, and determination of storage configuration. With the support of Big Data technology, the data center can be built as a safe, scalable, efficient, trusted platform for the data and business management of the PV plants. And it will be also profitable for the power grid companies who are responsible for the dispatching and planning of the PV systems within their operation area.

REFERENCES

Flavio J. & Benjamin R. 2013. *ZooKeeper: Distributed Process Coordination*. Sebastopol: O'Reilly.
Holden K. et al. 2015. *Learning. Spark*. Sebastopol: O'Reilly.
Hortonworks. *Apache Oozie*. 2015. oozie.apache.org.
Nick D. & Amandeep K. 2012. *Hbase in Action*. Shelter Island: Manning Publication.
Sean O. et al. *Mahout in Action*. 2010. Shelter Island: Manning Publication.
Srinath P. & Thilina G. 2013. *Hadoop MapReduce Cookbook*. Birmingham: Packt Publishing Limited.
Viktor M. & Kenneth C. 2013. *Big Data, A Revolution That Will Transform How We Live, Work, and Think*. Boston: Houghton Mifflin Harcourt.

Advances in Energy, Environment and Materials Science – Wang & Zhao (Eds)
© *2016 Taylor & Francis Group, London, ISBN 978-1-138-02931-6*

Dynamic simulation on a micro-grid system consisting of photovoltaic power, gas combined heat-and-power and battery

Hideaki Kuronuma & Noriyuki Kobayashi
Department of Chemical Engineering, Nagoya University, Nagoya, Japan

Hongyu Huang
Guangzhou Institute of Energy Conversion, Chinese Academy of Sciences, Guangzhou, China

ABSTRACT: In this study, dynamic analysis of a micro smart grid network system with solar and heat management in a smart grid network system has been investigated. The micro grid network system has been evaluated with considering demanded power changing at different time and weather and the benefits has been estimated.

1 INTRODUCTION

Renewable energy and conservation has become one of the hottest research topics during these days. Researches are looking for ways to increase energy efficiency and ensure a sustainable energy supply that will benefit to realize the energy sustainable society in the future (Meysam, 2012). And now, there is a new way to leverage technology that is generating interest and action—the smart grid. The smart grid is a type of electrical grid, which attempts to predict and intelligently respond to the behavior and actions of all electric power users connected to it—suppliers, consumers and those that do both—in order to efficiently deliver reliable, economic, and sustainable electricity services. The smart grid will help: improve the reliability, capacity and capability of the electricity network; ensure a safe, cost-effective, and environmentally sustainable energy supply; enable faster response and resolution to outages; create a resilient, open and dynamic information network (Colson, 2013; Huang, 2012; Kevin, 2012).

In recent years, a lot of researches relating to the smart grid system with distribution management systems have been carried out (Meysam, 2012; Colson, 2013; Huang, 2012; Kevin, 2012; Feng, 2013; Yun, 2014). Meysam et al (Meysam, 2012) illustrated the smart grid and micro-grid development in the recent decades, and showed that the challenges for the successful realization of smart grid includes the integration of renewable energy resources, real time demand response, and management of intermittent energy sources. Colson et al (Colson, 2013) proposed a comprehensive real-time micro-grid power management and control with distributed agents, and the system and formulations presented demonstrate the viability and capability of decentralized agent based control for micro-grids. Kevin et al (Kevin, 2012) analyzed the cost effective of smart grid network combing power and communication network. Yun et al (Yun, 2014) introduced the development of smart distribution management system for real-time predictive operation in distribution systems, the system consists of device level for the real time data acquisition and the server level for the data related to the voltage, current, faults, power quality, and load profiles of the network.

Each respective element technology in smart grid system (such as solar photovoltaics, gas engine, and various forms of distributed generation, thermal storage, secondary battery, and many more) are being developed. Furthermore, the authors are interested in optimizing the heat and electricity in the smart grid network, it is necessary to investigate the solar and heat management in a micro grid.

In this study, in order to contribute to a green innovation and optimize the heat and electricity in the smart grid network, we tried to develop a simulation model of a micro smart grid system that can predict the needed function of respective element technology according to demand of end user at different weather conditions, it is necessary to investigate the feasibility of this micro smart grid system. We proposed a combined complex smart grid (as shown in Fig. 1) consisting of lots of micro grids including gas engine (heat), solar power plant, thermal storage, end users (such as LED, secondary battery, factory, etc.). The number of installed and the size of each facility of the smart grid and place of installation has been optimized by modeling simulation, and the benefits due to energy balance calculation has been estimated.

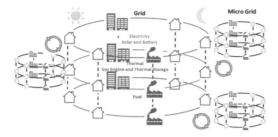

Figure 1. Optimization of heat and electricity supply in a smart grid network.

2 MODELING

At the first stage, model of a micro smart grid system with real time change and management of solar and heat was carried out to establish the evaluation and calculation method of fuel and electricity amount and the benefit effects.

2.1 Model and approach

Model of a micro smart grid system is shown in Figure 2, which is consisted of photovoltaics from isolation, a gas engine fueled by methane, and a secondary battery for electricity storage. The image of a micro smart grid model is shown in Figure 2. In the simulation, the following conditions are assumed.

- If solar power is not enough, replenish from the gas engine which has a maximum output of 35 kW grid-power;
- The excess of solar power is stored and used at night;
- If the total power supply is not enough at high consumption, more gas engine is set in parallel;
- The difference of solar radiation in sunny, cloudy day is considered.

Solar power = solar radiation × area of solar
panels × number × DC-AC conversion
efficiency × module conversion efficiency
× (1 − temperature correction factor)
× (1 − other losses)/3600   (1)

The approach of the micro smart grid system is illustrated as follows. A non-steady state mode is used by Visual Modeler because insolation for photovoltaics is a non-steady parameter. Units in the micro grid system are added by C program language in the analysis system as show in Figure 2.

All energy forms including electricity from photovoltaics and gas engine, methane fuel, thermal heat from gas engine are all transformed as hydrogen energy as shown in Figure 3.

Table 1. Specifications of equipments.

Solar panel	Product	Sanyo Electric HIT-B200J01
	Module conversion efficiency	17%
	Size	1,319m × 0,894m (14kg)
	Numbers	16 pieces (3.2kW total equivalent)
	Temperature correction factor (loss)	12% (June~August)
	Other losses	5%
Power conditioner	Product	Sanyo Electric SSI-TL27A2
	DC-AC conversion efficiency	94.5%
	Power consumption	5% of the solar power consumption
	Standby power at night	3W
Secondary battery	Product	Energy Farm Light EF-2
	Capacity	2500Wh
	Charge efficiency	90%
	Discharge efficiency	90%

Table 2. Specifications of gas engine.

Fuel type	Methane
Power output	35kW
Power generation efficiency	34%
Exhaust heat recycle efficiency	51%
Overall thermal efficiency	85%

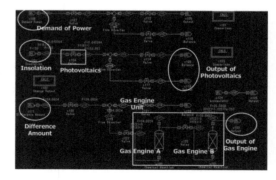

Figure 2. Model of a micro smart grid system.

The calculation flow of the micro smart grid system is shown in Figure 4. The insolation is non-steady and changes with time at every 5 seconds, as the electricity from photovoltaics also changes with time corresponding. If the electricity supply is not enough, the gas engine works and began to supply electricity. If the electricity is in excess, it is stored as electricity or thermal storage.

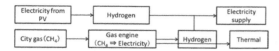

Figure 3. Every energy forms transformed as hydrogen energy.

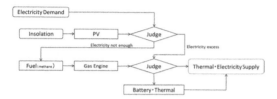

Figure 4. Calculation flow of the micro grid system.

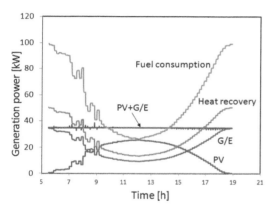

Figure 5. Simulation results (sunny day).

2.2 Result and discussion

Based on the above conditions, the thermal or battery recycle effect of the grid power consumption has been calculated. The results of sunny day and cloudy day are shown in Figures 5 and 6.

The horizontal axis of the graph is real time of a day; the vertical axis of the graph is electricity power. The black line is the demanded fuel in the micro smart grid system; the red line is electricity from gas engine; the blue line is the recycled thermal energy; the green line is electricity from solar power; the purple line is the sum electricity amount of solar power and gas engine; the brown line the demanded power. When there is no solar power output from the middle of the night until the morning, replenish from secondary battery or the gas engine. If solar power exceeds the power consumption during the day, the excess is stored with secondary battery and used at night.

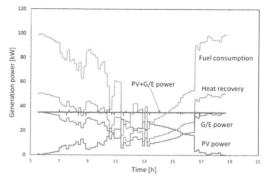

Figure 6. Simulation results (cloudy day).

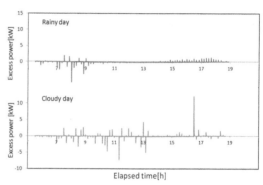

Figure 7. Excess amount of electricity with time.

As shown in Figures 5 and 6, the demanded fuel decreases when the solar power increases because of the higher insolation. It is possible to reduce the fuel consumption with the introduction of solar power and secondary battery. In the case of cloudy, the fuel consumption is changed dramatically because of the non-steady isolation at cloudy day.

From the above results, we estimated the amount of recycled thermal energy. The amount of recycled thermal energy is defined as Equation (2):

$$\text{Recycled thermal energy} = \text{Demanded fuel} \times \text{Heat recycle efficiency} \qquad (2)$$

In the sunny day, the amount of recycled thermal energy is about 1.36 GJ during the time period of 5:30 to 19:00, which is averaged about 28 kW. On the other hand, the amount of recycled thermal energy is about 1.56 GJ, corresponding as 32 kW in the cloudy day.

In order to obtain feasibility of the micro smart grid system, it is necessary to know the lacked amount of electricity at these weather conditions

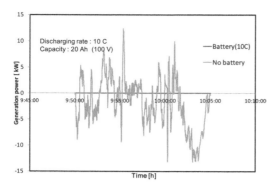

Figure 8. Excess amount of electricity with time (added battery).

which is show in Figure 7. The positive value means the lacked amount of electricity between the demanded electricity and supplied electricity. The negative value means the excess amount of electricity between the supplied electricity and demanded electricity. In case of sunny day, the electricity balance between supplies and demand can be kept, whereas, the lack between supplies and demand at cloudy increases. Figure 8 is the result for added battery. It was possible to supply stability with the electric power by using this battery (discharging rate 10 C, capacity 20 Ah).

3 CONCLUSIONS

In this study, in order to contribute to a green innovation and optimize the heat and electricity in the smart grid network, we tried to develop a simulation model of a micro smart grid system that can predict the needed function of respective element technology according to demand of end user at different weather conditions. A combined complex smart grid was proposed, at the first stage, model of a micro smart grid system with real time change was carried out to establish the evaluation and calculation method of complex smart grid and estimate the benefits effect.

It was able to model the case of a micro smart grid system introducing the equipments of solar power generation and gas engine power generation with thermal storage and secondary battery for electricity storage. The simulation was carried out to estimate the energy consumption and recycled thermal energy by the balance calculation. The benefit effects of sunny day, cloudy day are 1.36 GJ and 1.56 GJ, equal as 28 kW and 32 kW respectively.

In case of sunny day, the electricity balance between supplies and demand can be kept. Whereas, the lack between supplies and demand at cloudy increases. The lacked amount of electricity between supplies and demand is maximum at 12 kW at cloudy, in order to solve this shortfall, a maximum battery at 20 Ah is needed.

REFERENCES

Colson C.M., Nehrir M.H. Comprehensive real-time microgrid power management and control with distributed agents. IEEE Transitions on Smart Grid, March, 2013.
Feng G., Luis H., Robert M. Comprehensive real-time simulation of the smart grid. IEEE Transactions on Industry Applications, March/April 2013.
Huang H.Y., Kobayashi N.Y., Yukita K.Z., Cheng Y., Guo F.H., He Z.H., Yuan H.R. Predictive Simulation of Smart Grid Network System. Soft Computing and Intelligent Systems (SCIS) and 13th International Symposium on Advanced Intelligent System (ISIS), 20–24 Nov, 2012.
Kevin M., Juan A.O., Chris D. Combining power communication network simulation for cost-effective smart grid analysis.
Meysam S., Chin K.G., Chee W.T. A review of recent development in smart grid and micro-grid laboratories. IEEE International Power Engineering and Optimization Conference, 6–7 June 2012.
Yun S.Y., Choo C.M., Kwon S.C., Song K. Development of smart distribution management system for predictive operation of power distribution systems. 21st International Conference on Electricity Distribution, 6–9 June 2014.

Advances in Energy, Environment and Materials Science – Wang & Zhao (Eds)
© 2016 Taylor & Francis Group, London, ISBN 978-1-138-02931-6

Numerical simulation analysis on the tower foundation deformation of the high voltage transmission line caused by iron ore mining and filling

Bangshu Xu, Cheng Chen & Qin Yan
School of Civil Engineering, Shandong University, Jinan, China

Shan Dong
Limited by Share Ltd., Metallurgical Design Institute of Shandong Province, Jinan, China

Shiwu Liu
Fujian Yongfu Engineering Consultants Ltd., Fujian, China

ABSTRACT: ±800 kv high voltage transmission line project from Shanghai temple to Shandong province needs to cross the Shenghong iron ore mining area. Shenghong iron mine has not yet been constructed. It has yet to determine the distribution of underground iron roughly depending on the survey report. In order to analyze the tower foundation deformation of the high voltage transmission line caused by iron goaf, the paper adopts Flac3D software to simulate the process of iron ore mining and filling. The main conclusions are: (1) Excavation of the sloping iron mine causes the surface final vertical displacement a "parabolic" distribution and the horizontal displacement curve a "horizontal s" distribution. (2) During the excavation, the maximum subsidence offsets to the goaf slowly. In the influence range of the excavation, the trend of the parabolic in the incremental interval is almost consistent with the trend of the iron. The shallower depth of the iron, the smaller depth of the surface it could cause. The horizontal displacement shows symmetrical distribution on both sides of the iron. The horizontal displacement curve changes from "straight line" distribution to "parabolic" distribution slowly, and finally shows in "horizontal s" distribution. (3) The surface vertical displacement and the horizontal displacement at the location of the tower foundation are very small, the maximum different vertical displacement and different horizontal displacement satisfy the requirement of the standard.

1 INTRODUCTION

So far, study on the surface buildings caused by iron ore mining and filling was quite mature. Zhang et al. studied the home buildings caused by mining and filling by utilizing numerical simulation method (Zhang et al. 2005). Wang et al. analyzed surface subsidence caused by ore mining through three-dimensional finite element method of porous media solid-liquid coupling (Wang et al. 2010). Wang et al. had a research on the impact of the key isolation layer under the filling shaft to the surface subsidence by using numerical simulation method (Wang et al. 2010). But there was little research on the surface subsidence caused by sloping ore. It is hard to express clearly about the surface deformation rule through the mechanics-physics model as the complexity of the ore geological conditions. However, the simulation method can make the flaw up. On the basis of the seniors' research, the surface and surface buildings deformation rule during the sloping ore mining and filling are studied

with Flac3D, with the Shenghong iron ore as the background.

2 ENGINEERING BACKGROUND

The project of ±800 kv high voltage transmission line is between the Shanghai temple and the Shandong province. The line begins at the junction of the Liangshan and Dongping, and the terminal is located at the south of the Yuning, by way of Dongping, Wenshang, Ningyang. The line has to cross the Yangdian iron and Shenghong iron goaf as the path corridor restriction in the churchyard of Weishang. The mining area is alluvial plain geomorphology; ground elevation is about 50 m; the ore shows in monocline or stratiform distribution. The trend of the ore is nearly perpendicular to the line path. The overall occurrence is 256° ∠45~55°. The occurrence is relatively stable and buried under 44.3 m. Ore output level is 4.30 ~ −610.70 m and the design mining elevation is −40 m ~ −400 m.

The ore above –40 m is reserved as protecting pillar. The design of the underground ore adopts the shaft development scheme and shallow-hole shrinkage stoping method. The middle design height is 60 m and the iron ore is divided into 7 middle levels includes –40 m, –100 m, –160 m, –220 m, –280 m, –340 m, and –400 m. On the basis of characteristics of mineral geology, mining method, mining scheme and so on, adopting the method of numerical simulation, analyzing the deformation and movement rules of the surface, solving the deformation characteristic value, making a judge whether the ore area could be used as the architectural place and the degree of harm to the tower foundation.

3 NUMERICAL SIMULATION

3.1 Modeling

The quasi-three dimensional geologic model is established according to geological section map of the geological prospecting line in the geologic report. When establishing the model, the iron mine is simplified that nonuniform thickness ore is equivalent to equal thick iron ore, so equivalent thickness of iron ore I-3 is 2.5 m; equivalent thickness of iron ore I-5 is 4.0 m; equivalent thickness of iron ore I-6

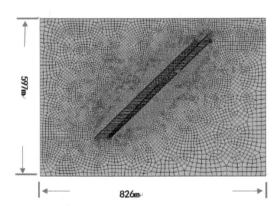

Figure 1. Numerical mesh model.

is 3.5 m. In order to consider the boundary effect of the model, left and right boundary of the model are taken 200 m that the influence range of ore, and the surface is taken as upper boundary, lower boundary is 597 m, and width is 826 m. The mesh model is established through Abaqus software, and then imported to the Flac3D software. The model is shown in Figure 1.

3.2 Physical and mechanical parameters

According to the geological prospecting report, 70 m under the surface is quaternary loose rock, and the other is basement. Backfill parameters refer to Zhang et al., and iron mine parameters refer to the geological prospecting report (Zhang et al. 2009). The physical and mechanical parameters are shown in Table 1.

3.3 Process of mining and filling

The first phase section height of the ore is 60 m and is divided into seven middles that include –40 m, –100 m, –160 m, –220 m, –280 m, –340 m, and –400 m. Among them, –100 m middle is the level of return air and the other is the mining level. Generally, iron ore is mined from –400 m level to –340 m, –280 m, –220 m, –160 m, –100 m level upwardly. Accordingly, the ore is divided into six mining levels including diagram 1, diagram 2, diagram 3, diagram 4, diagram 5, diagram 6 and are shown in Figure 2.

3.4 Result and analysis

3.4.1 Analysis of surface displacement
In the Figure 3, iron mine is mined into six levels, and every level acts as a construction step. Analyzing the result of the six construction steps, it could be seen from the Figure 3 (a) to Figure 3 (b) that the settlement first occurs in roof position as the mining and filling of the ore, then gradually spreads to the surface. Surface vertical displacement increases gradually during the constructions, and maximum surface subsidence offsets to the goaf slowly. After the mining, vertical displacement reaches to the maximum value. Due to the

Table 1. Physical and mechanical parameters.

Stratum	Density (Kg/m³)	Elastic modulus (GPa)	Poisson ratio	Cohesion (MPa)	Internal friction angle (°)	Strength tension (MPa)
Quaternary	1960	0.05	0.32	0.05	28	0.02
Basement	2650	1.6	0.22	1.25	35	1.28
Ore	3460	12	0.26	14.9	37.9	12.2
Backfill	2000	0.6	0.3	0.75	30	1.20

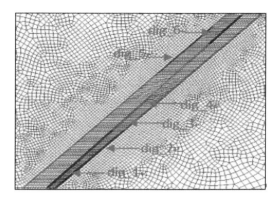

Figure 2.   Mining and filling sequences.

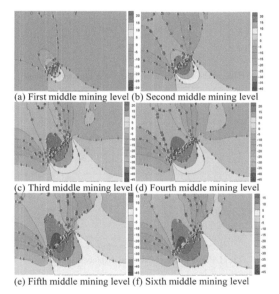

(a) First middle mining level (b) Second middle mining level

(c) Third middle mining level (d) Fourth middle mining level

(e) Fifth middle mining level (f) Sixth middle mining level

Figure 3.   Vertical displacement isoline map of different middle mining levels (unit mm).

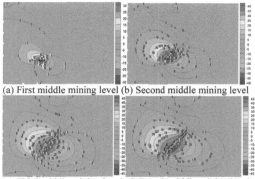

(a) First middle mining level (b) Second middle mining level

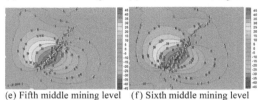

(c) Third middle mining level (d) Fourth middle mining level

(e) Fifth middle mining level   (f) Sixth middle mining level

Figure 4.   Horizontal displacement isoline map of different middle mining levels (unit mm).

action of overlying rock gravity, the goaf deforms, moves and breaks gradually during the mining. The goaf area increases, and settlement spreads to the surface. At the same time, a moving belt forms in a certain range of the surface. The buildings in the moving belt may be influenced by the mining.

It could be seen in the Figure 4, the horizontal displacement of soil caused by different mining levels shown in symmetrical distribution. Above the ore, horizontal displacement turns to right. Under the mine, horizontal displacement turns to left, because rock is disturbed and rock stress releases after the mining. However, the strength of the filling cannot reach the rock strength,

under the action of active earth pressure, rock moves to the goaf, causing the rock above the ore moved to the bottom right corner and the mine under the mine moved to the top left corner. The horizontal displacement spreads to the surface with the goaf increasing. As the dip angle of the mine, the horizontal displacement above the ore spreads faster than the horizontal displacement under the ore. So it could be seen in the Figure 4 (f), after the construction, the surface horizontal displacement basically shows the trend that moves to the right.

It could be seen in Figure 5 (a) that the vertical displacement shows in "Parabolic" distribution, parabola apex gradually moves downward and shifts to the goaf with the area of the goaf increasing, and the trend of the settlement curve is almost consistent with the trend of ore in the parabolic incremental interval. The maximum displacement occurs at the surface where X equals to 250 m, while the surface projection interval corresponds to the range of the mine distributed between X equals to 240 m to X equals to 320 m. So, in the process of the sloping ore mining and filling, the deepest depth of the ore corresponds to the range of the surface that has the largest settlement. It could be seen in the Figure 5 (b), the surface horizontal displacement changes from the "Straight line" distribution to "Parabolic" distribution. After analysis, in the beginning of the mining, ore hanging wall declines, footwall raises, making footwall soil moves negatively, and horizontal displacement

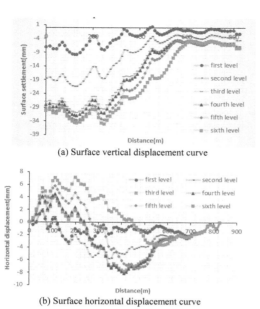

(a) Surface vertical displacement curve

(b) Surface horizontal displacement curve

Figure 5. Surface displacement curve changes during the construction steps.

is both negative. It could be seen in the Figure 4, positive horizontal displacement caused by the hanging wall spreads to surface faster than foot-wall, making the surface of the projection interval corresponding to the hanging wall moving positively and "Parabolic" distribution converts to "Horizontal S" distribution. So, the trend and angle of the sloping mine have a great impact on the surface displacement.

3.4.2 Tower foundation displacement analysis

The tower foundation is located at the surface where X equals to 370 m. It could be figured out in the Figure 6 that (a) the vertical displacement increases with the area of the goaf increasing. Diagram 1 and diagram 2 construction steps have great impact on the displacement of tower foundation. The slope of the curve changes fastest, because the quantities of the ore in the diagram 1 and diagram 2 are the most, and the other four construction steps are less than them. Accordingly, it has little disturbance to the surrounding rock, and surface settlement correspondingly reduces. The final settlement reaches 25.8 mm. The trend of horizontal displacement at the location of the tower foundation is equal to the surface. In the previous five constructions steps, the location of the tower foundation moves negatively of x-axis and the displacement increases firstly and then decreases. In the construction of step 6, the horizontal displacement at the location of the tower foundation moves to the positive direction

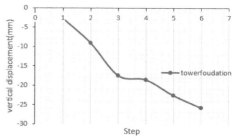

(a) Vertical displacement curve of tower foundation changes during the construction steps

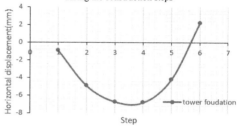

(b) Horizontal displacement curve of tower foundation changes during the construction steps

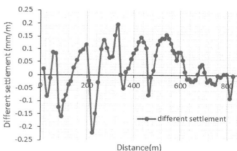

(c) Different value of vertical displacement curve

(d) Different value of horizontal displacement curve

Figure 6. Displacement curve of tower foundation.

of x-axis. The final horizontal displacement value reaches 25.8 mm positively of x-axis. According to the provision of "coal mining regulation under buildings, railways and water-bodies" published by Coal Industry Press, the allowed deformation values of building are got (Coal Industry Press, 2000):

Slope: $i = \pm 3$ mm/m
Curvature: $k = \pm 0.2 \times 10^{-3}$
Horizontal displacement: $\varepsilon = \pm 2$ mm/m

It could be seen in the Figure 6 (c) and Figure 6 (d) the maximum different value of vertical displacement reaches 0.25 mm/m, which is less than the specified value 3 mm/m. The maximum different value of horizontal displacement reaches 0.15 mm/m, which is less than the code value 2 mm/m. In conclusion, iron ore mining has little impact on the tower foundation deformation of the high voltage transmission line.

4 CONCLUSION

The paper describes the studies of the surface deformation rules caused by iron ore mining and filling. The main conclusions are:

1. The surface vertical displacement caused by sloping ore mining and filling shows in "Parabolic" distribution. During the mining, parabola vertex displacement increases gradually and offsets to the goaf with the goaf area increasing. The trend of the parabola is almost the same with the trend of the ore.
2. The surface horizontal displacement increases firstly and then decreases with the goaf area increasing, which shows in "Straight line" distribution and then to "Parabolic" distribution, finally changing into a "Horizontal S" distribution. Different angles and angle of the ore lead to different horizontal displacement of the ore.
3. The maximum vertical displacement of the tower foundation reaches 25.8 mm, and the final horizontal displacement reaches 2.2 mm.

The maximum different horizontal displacement reaches 0.15 mm/m, and the maximum different vertical displacement reaches 0.25 mm/m. According to coal mining regulation under buildings, railways, and water-bodies, tower foundation deformation is in the range of specified requirement. In conclusion, ore mining and filling has little impact on the deformation of tower foundation of the high voltage transmission line.

ACKNOWLEDGMENTS

Financial supports for this paper provided by National Science Foundation of China (No. 50909056) and Provincial Science Foundation of Shandong Province (No. ZR2014EEM014, ZR2014EEM029), and Science and Technology Project Plan in 2015, Ministry of Housing and Urban-Rural Development of China (No. 2015-K5-004) are gratefully acknowledged.

REFERENCES

Coal Mining Regulation under Buildings, Railways and Water-bodies, Coal Industry Press, 2000.06.
Wang P., Xu M.G., Li H.Q. Numerical simulation analysis on surface caused by mining in filling mining method and caving method [J]. Industrial Safety and Environmental Protection. 2010(10):50–52.
Wang X.J., Fang S.Y., Liu J.X. Numerical simulation analysis on surface subsidence controlled by key isolation layer in filling shift [J]. Metal Mine. 2010(10):13–16.
Zhang M.H., Gao Q., Zhai S.X. Design optimization and numerical simulation analysis on mining and filling of Jinchuan second ore [J]. Metal Mine. 2009(11):28–31.
Zhang T., Xu M.G., Ouyang Z.H. Numerical simulation analysis on a black iron ore mining and filling [J]. Gold. 2005(11):24–27.

Advances in Energy, Environment and Materials Science – Wang & Zhao (Eds)
© 2016 Taylor & Francis Group, London, ISBN 978-1-138-02931-6

Application on the some nuclear power engineering of the hydro-fracturing technique

Xianbin Wang & Ming Zhang
National Nuclear Power Planning Design and Research Institute, Beijing, China

ABSTRACT: Initial earth stress condition is the important parameter used in the nuclear engineering design of the outlet tunnel, and the hydro-fracturing technical is the ideal method on measuring the earth stress. Based on the practical engineering, this paper introduces the principles and methods of the hydro-fracturing method, the results show that: the maximum horizontal principle stress is 3.92–6.07 MPa, and the lateral principle stress is 2.75–4.92 MPa. The direction of maximum horizontal principle stress is NE33°, and the structural stress is main. Earth stress has some benefits on the stability of the tunnel, which cannot produce the rock burst phenomenon, and it can propose the important information for the design of tunnel.

1 INTRODUCTION

Now in China, water supply and drainage tunnel is the most applied scheme in the nuclear power plants. The arrangement of the grotto principal axis is influenced by the condition of original crustal stress. And for the spatial arrangement of tunnel, the forecast of adjoining rock stability, and design of lining, the results of crustal stress test is very significant. According to the in-situ stress test method issued by International Society for Rock Mechanics, there are many in-situ stress test methods, including the drill hole diameter changing measurement, strain measurement, stress recover measurement, and hydro-fracturing technique. Comparing with the other three measuring methods, hydro-fracturing technique has many advantages, like simplicity of operator, short test period, and so on. So the limitation of the point stress state and the nonuniformity of the geological conditions could be avoided. And the error caused by the choice of rock elastic parameter can be also avoided (Li, 2006). So in the rock mass stress measurement, hydro-fracturing technique cannot be matched by other kinds of the measurements.

2 THE FUNDAMENTAL PRINCIPLE HYDRAULIC FRACTURING TECHNIQUE

Hydraulic fracturing technique is a kind of in-situ stress test methods, which the crustal stress is calculated according to the pressure of the fracture conditional curve. And the fundamental principle is based on these three hypotheses as follows: (1) the surrounding rock is isotropous and elastic; (2) the

fluid in the rock is under the Darcy law; (3) one of the principal divection is the vertical direction which is according to the gaging hole in the vertical direction (Peng, 2006 & Liu, 1999).

3 PROJECT EXAMPLE

3.1 *Project profile*

There is one nuclear power plant planning and constructing two 1000 MW pressurized water reactor nuclear power unit. The water intake system is gravity type water supply tunnel. Each unit has one water supply tunnel, which the diameter and length of the tunnel is 4.8 m and 1200 m, the space between the two tunnels is 36 m, and the elevation of the tunnel bottom is −15 m. The rock which the tunnels went across is ignimbrite, and hydraulicfracturing technique is applied to measure the crustal stress parameter.

3.2 *Measuring method of the hydraulic fracturing technique*

3.2.1 *Measuring equipment*

There are three parts for the measuring equipment: (1) Packer system of the drilling and pressure-bearing section, which is constituted by two parkers. And between these two parkers, there is a space for pressure-bearing section; (2) compression system which is included in heavy pressure fluid pump with mass flow and push-pull valve; (3) record system which is included in functions recorder, pressure transducer, pressure gage, and so on.

In the Figure 1, the length of the low-duty packer is 3.4 m with 1.2 m rubber sleeve. And the length of the hydraumatic section is 1.0 m.

Figure 1. The low-duty packer.

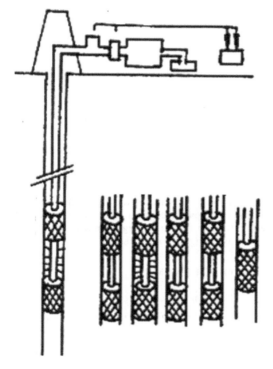

Figure 2. The geo-stress measurement routine of hydraulic fracturing.

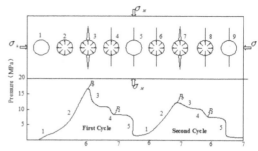

Figure 3. The conditional curve of hydraulic fracturing.

3.2.2 *Measurement procedure*

The single tube compression system is applied in this test. The push-pull value on the orifices is applied to control the flow direction of the liquid, and load to the packer. The measurement procedure is showed in the Figure 2, and the fracturing conditional curve is depicted in Figure 3.

Before the hydraulic fracturing test, the permeable rate and the gradient of the drill hole must be checked. The leakproofness of the drill pipe is also checked and the measurement procedure is as follows:

1. Down and seal the packer: Two packers are arranged in the selected section, and let the packers expand for the space of the compression (the pressure of the down and seal the packer in this test is 4 MPa);

2. Water injection and inflating: After driving the change-over valve by drill pipe, inflating to the fracturing section by water injection, so the pressure on the wall of hole increases gradually.
3. Fracturing palisades: Under enough pressure, the cracks occur in the direction of the least resistance on the wall of hole and expand in the direction which is perpendicular to minimum principal stress direction. Accordingly, the pressure decreases fast due to the fracture of the rock, when the pumping pressure gets the critical bursting pressure.
4. Turn off the pump: after the pressure pump is turned off, the pumping pressure decreases fast. When the pressure decreases to the pressure of critical closure state, it is called closing pressure P_s;
5. Pressure relief: open the pressure value and release the pressure for letting the cracks closed.
6. Re-open: repeat the steps from 2) to 5) until getting the reasonable parameters.
7. Blocking: after the test, the liquid in the packer is expelled and the packer is shrunk to the original state.
8. Recording the direction of cracks: Applying the directional impression device, the length and the direction are recorded.

3.2.3 *Parameters of the pressure*

Pressure parameters P_b, P_s, P_r, and P_0 are the basis for calculating the crustal stress of the hydraulic fracturing, which is confirmed by the fracturing conditional. Averagely, the peak value of the first circulation pressurization curve is selected as the bursting pressure P_b. And the inflection point of the rising section of circulation pressurization curve is selected as reopening pressure P_r. Due to the rising section of circulation pressurization curve is steep. So the point where is the departure from the straight as the reopening pressure, which is depicted in the Figure 4. For the instantaneous shutoff pressure, it is selected from the inflection point of the declining part of fracturing cyclic curve. To the cases whose inflection point is not clear, the tangent method and double tangent method could be applied, which is depicted in the Figure 5. The pore water pressure P_0 of the rock mass could be got from the pore water pressure gauge. When the testing depth is very little, the hydrostatic pressure could be regarded as pore water pressure. And in this test, the hydrostatic pressures of different sections are regarded as the pore water pressure, which the formula is $P_0 = \gamma_{water} H$: where the γ_{water} is the volume-weight

of water, and the H is the depth of the rock (Chen. 2004; An, 2004).

The formula for calculating the principal stress of the vertical direction is $\sigma_v = \gamma H$, where the γ is the unit weight which is 27, and H is the height.

3.2.4 Analysis of the test results

The depth of the testing drill hole is 84 m, and the diameter is 75 mm. In the range from 29.7 m to 79.3 m, 7 crustal stress trial curves were got, which the testing results are showed in the Table 1. And the representative records of the geostress survey are depicted in the Figure 6.

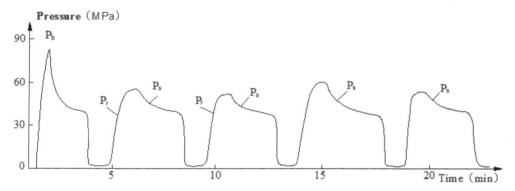

Figure 4.　Confirm P_s and P_r according to cyclic curve of pressure.

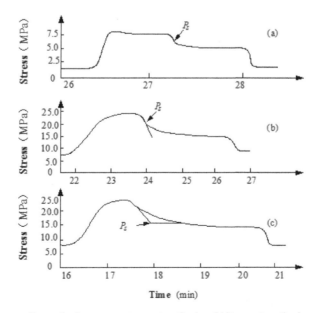

Figure 5.　Confirm P_s according to bathmometry, tangent method and bitangent method.

Table 1. The outcome of drilling hydraulic fracturing method.

Depth/ m	Bursting pressure P_b/MPa	Reopening pressure P_r/MPa	Closing pressure P_s/MPa	Pore pressure P_0/MPa	Waterhead pressure P_H/MPa	Maximum horizontal principal stress σ_H/MPa	Minimum horizontal principal stress σ_h/MPa	Geostatic stress σ_v/MPa	Side-pressure coefficient $\lambda = \sigma_H/\sigma_v$	Direction of the maximal horizontal principal stress
29.7	6.28	4.69	2.78	0.30	0.30	3.95	3.08	0.80	4.92	
38.7	5.1	4.03	2.52	0.39	0.39	3.92	2.91	1.04	3.75	
43.3	6.27	4.53	2.86	0.43	0.43	4.48	3.29	1.17	3.83	
52.3	4.82	4.13	2.81	0.52	0.52	4.82	3.33	1.41	3.42	32°
61.3	8.1	5.36	3.21	0.61	0.61	4.88	3.82	1.66	2.95	
74.8	7.88	6.3	3.7	0.75	0.75	5.55	4.45	2.02	2.75	33°
79.3	6.84	4.92	3.4	0.79	0.79	6.07	4.19	2.14	2.84	

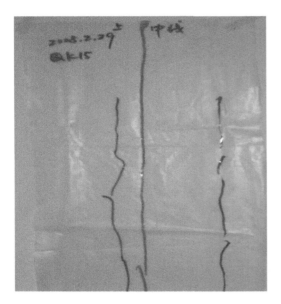

Figure 6. The broken crack record of 19.3 m.

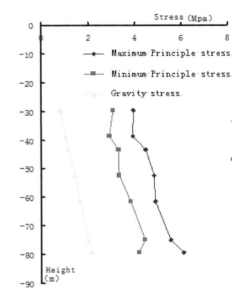

Figure 7. The curve stress changing with depth.

In this test, all the pressure-time history curves are according to the theoretical curves. And the records of the cracks are clear. So the data of this test are reliable. In the testing range of the depth, the value of the maximum horizontal principal stress is in the range of 3.92~6.07 MPa, and the direction of the maximum horizontal principal stress is about NE33° that is influenced by regional tectonics. The coefficient of horizontal pressure λ is in the range of 2.75~4.92.

The maximum horizontal principal stress, the minimum horizontal principal stress, and the gravity stress all have the tendency which is increasing with the development of the depth, which is depicted in the Figure 7.

According to the uniaxial compressive strength of the rock ($Rc = 80~90$ MPa), the σ_{max} is 6.07 MPa, Rc/σ_{max} is in the range of 12.92~14.54, which are all lager than 7.

4 CONCLUSION

1. The direction of the maximum horizontal principal stress is about NE33° which is influenced by regional tectonics;
2. The maximum horizontal principal stress, the minimum horizontal principal stress, and the gravity stress all have the tendency which is increasing with the development of the depth;

3. The lithology in the area is rigid rock, which the Rc/σ_{max} are all larger than 7. The crustal stress is helpful to the stability of the tunnel; and
4. The results tested by hydraulic fracturing reflect the condition of the crustal stress better, which could be regarded as the guidance of the similar project.

REFERENCES

An Qimei & Ding Lifeng. 2004. Research of crust stress measurement and its application with hydraulic fracturing in zhouning Hydro-power, Rock and Soil Mechanics: 1672–1676.

Chen Kuiying. 2004. Application of Geostatic stress measurements by hydraulic splitting to chilling tunnel. *Technology of Highway and Transport*: 71–74.

Li Jinsuo & Peng Hua. 2006. Application of hydraulic fracturing in-situ stress measurements in tunneling along the dali-Lijiang Railway, Yunnan, *Geological Bulletin of China*: 644–648.

Luo Chaowen & Liu Yunfang. 1999. Three Dimensional Geostress Test in Underground Plant Area of Shuibuya Project, *Journal of Yangtze River Scientific Research Institute:* 45–47.

Peng Hua & Cui Wei. 2006. Hydrofracturing In-Situ Stress Measurements of the Water Diversion Area in the First Stage of the South-North Water Diversion Project (Western Line), *Journal Of Geomechanics*: 182–190.

Advances in Energy, Environment and Materials Science – Wang & Zhao (Eds)
© 2016 Taylor & Francis Group, London, ISBN 978-1-138-02931-6

Improvements on frequency capture range and stability of multi-phase output charge pump Phase-Locked Loop

Jian Xiao, Yue Chen & Yanzhang Qiu

School of Electronic and Control Engineering, Chang'an University, Xi'an, P.R. China

ABSTRACT: A Multi-phase Output Charge Pump PLL (Phase-Locked Loop) with wide frequency capture range used for high speed series pixel data recovery is designed and implemented in this paper. The input clock can be automatically detected and divided into three segments to achieve the high frequency capture range. The PLL is used to sample the high-speed video signal and realized up to UXGA format digital visual signal transform, SMIC 0.18 μm CMOS process is used for verification, testing results show that the RMS jitter of the PLL output clock is 24ps and peak-to-peak jitter is less than 200ps at the condition of input 1.65Gbps pixel data signals.

1 INTRODUCTION

With the development of flat panel display technology in recent years, the speed and stability of tranceiver for serial uncompressed data transmission are required to be increased. The image quality and visual resolution can be significantly improved by serial uncompressed data tranceiver technology. The EMI interference during transmission and high-speed digital video signal transmission for long distances can been implemented through the use of advanced coding algorithms [3,4] (DDWG 1999 & HDMI Licensing LLC 2004). Even the single channel data transfer rates can achieve 1.65GHz even for the DVI or HDMI Single Link Data transmission requirements it is of importance for the design of long-distance and high-speed digital video signal receiver [6,9] (Lee et al.1998 & Zukui 2002).

Although the circuits of uncompressed high-speed serial data transceiver based on over-sampling technology are simple in structure, the complexity of designing PLL used in data recovery is much increased. In order to achieve high-speed serial data recovery, a multi-phase output charge pump PLL was designed in this paper by using 2.5 times divider method and differential VCO of six delay units, realizing over-sampling clock of 12 equal phases offset [1,8] (Xiaoping et al. 2007 & Cheng et al. 2004). The V-I converter with two-stage amplifier and a source follower is designed to improve the current matching and the response speed [5,10] (Rogers et al. 1998 & Moon et al. 2001). The input clock signal was detected using frequency detection circuit, automatically shifting among the three VCO and parameters of PLL loop. Therefore, the VCO operates at the corresponding clock frequency range, not only resulting in an improvement

on frequency capture range, but also satisfying the PLL loop parameter requirements. Compared with traditional methods, this design scheme can greatly reduce design complexity of the PLL.

2 SYSTEM FRAME AND PRINCIPLE

The basic structure of conventional charge pump PLL consists of four parts: phase detector (PFD), Charge Pump (CP), loop filter (LPF) and a Voltage Controlled Oscillator (VCO). In order to achieve a wide frequency capture range, an input frequency detector and logic control circuit are so designed in the system that the frequency of the input clock signal can be divided into three portions. The loop filter and VCO circuit are distributed into three groups, processing the corresponding frequency signal, respectively [7] (Liao et al. 2013). The whole block diagram of PLL was shown in Figure 1[2] (Christian et al. 2014). The 25 MHz~165 MHz reference clock signal was used as the input of PLL (Fref), the output is the phase locked clock signal with 12 equal phase gap (ck_ph [0:11]). The external reference clock signal is fed into the phase detector with the pre-charged structure to achieve

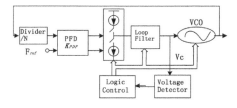

Figure 1. Block diagram of PLL.

high-speed no dead zone phase and frequency detecting. The charge pump circuit uses differential input structure with unity gain amplifier, and the charge pump current is controlled by the logic control circuit. The 6-stage differential delay unit is used in VCO to output 12 clock signals with equal phase gap (ck_ph [0:11]), where the ck_ph0 was divided by 2.5 and fed back to the input of PFD, and the center frequencies of the three groups correspond to the center of three frequency bands, respectively. The control voltage Vc outputted from the filter was detected using detection circuit. The charge pump current was switched and the corresponding frequency bands of VCO chosen by using the logic control circuit. Thus the PLL can operate at a clock frequency of the corresponding segment in order to achieve 50 ~ 412 MHz frequency capture range. The second order closed loop transfer function of the charge pump PLL can be written as:

$$H(s) = \frac{\frac{I_P}{2\pi}\left(R + \frac{1}{sC}\right)K_0}{s + \frac{I_P}{2\pi}\left(R + \frac{1}{sC}\right)K_0/N} \quad (1)$$

Natural frequency and damping factor are expressed as:

$$\omega_n = \sqrt{\frac{I_P K_o}{2\pi NC}} \quad (2)$$

$$\zeta = \frac{\omega_n RC}{2} = \frac{R}{2}\sqrt{\frac{I_P K_o C}{2\pi N}} \quad (3)$$

3 SCHEMATIC IMPLEMENT OF PLL

The PFD is an important part of the phase-locked loop, that compare frequency and phase of the system input signal with the internal feedback signal. Its linearity, resolution, bandwidth and sensitivity of PFD directly affect the performances of the system, as shown in Figure 2. The phase and frequency detector (pre-PFD) with pre-charged structure was used, in which the RSD Flip-Flop of con-PFD was substituted by the node of the pre-charge circuit, resulting in many advantages such as high speed, simple in structure, short delay path and small parasitic capacitance. In order to eliminate dead zone of the charge pump circuit, the pulse width outputted from locked PFD is approximately set to 500ps, thus the up and down signal could maintain enough effective level to turn on the charge pump switches. The pulse signal not only must keep appropriate width, because too large pulse width may cause the up and down current mismatching resulting in jitter generation, but also enough width was required to fully switch up and down.

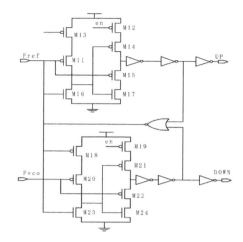

Figure 2.   PFD schematic.

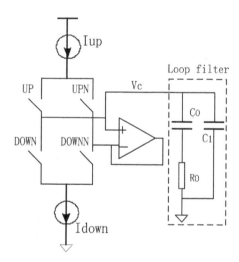

Figure 3.   Charge pump schematic.

A buffer module composed of inverter and transmission gate was designed to acquire the up, down and their complementary signals. The equal width and strict synchronization of the output pulse signal (UPN, UP, DOWNN, DOWN) were achieved by adjusting the size of MOS transistors and the ratio of NMOS to PMOS transistors.

The charge pump and loop filter circuit were shown in Figure 3. Current steering charge pump is designed for improving switching time. The current is mirrored into charge path and discharge path. IUP and IDOWN are cas-code structure and its bias voltage supplied by a low voltage cascode structure. When the pull-up signal is activated, the source current flows into the loop filter, so that the output voltage of the loop filter rises, which forces a higher oscillation frequency. When the pull down signal is activated, the current is discharged, and force the oscillator frequency decrease. Ideally, the pull-up and pull-down currents should be exactly equal. This can be accomplished by introducing feedback into the biasing scheme with a single op-amp [9] (Zukui 2002). The op-amp senses the output voltage and compares it with the voltage at the drains of the mirror transistors. The design forces the currents through both the NMOS and PMOS transistors to be almost exactly equal, regardless of the output voltage. Transient simulations are used to extract the charge pump current. Further analysis shows that the clock feed through of the differential pair generate more contribution on spur at the high frequency conditions. We can adjust the W/L ratio of the input NMOS and PMOS transistor with consideration of process parameters to make the spur charges canceled each other. The charge pump current directly affects the PLL loop parameters. In order to ensure suitable PLL loop parameters, the charge pump current Iup and Idown are adjusted through the logic control circuit to meet the requirements.

The dotted box in Figure 3 is the loop filter, a small capacitor is paralleled with the RC filter to form a second older filter circuit, it can effectively suppress the output voltage variations. A resistor R0 in series with C0 that can introduce a zero is used to cancel a pole for stability in PLL loop. C1 is used to eliminate the ripple of control voltage caused by the RC filter.

The VCO circuit includes a V-I converter and a current controlled ring oscillator, the total gain of VCO is the product of the two stages. Current controlled ring oscillator implemented with 6 stages delay units as shown in Figure 4. VCO output frequency is:

$$\frac{\sqrt{uC_{ox}(W/L)I_D}}{C_{load}} \tag{4}$$

The gain of VCO is:

$$K_{VCO} = \frac{uC_{ox}}{2}\left(\frac{W}{L}\right)\frac{1}{C_{load}} \tag{5}$$

I_D is the bias current of the delay unit, C_{load} is the all output terminal capacitance of 6 stages delay units. The output of the oscillator is sent to next stage circuit with a buffer between them. In order to match with each delay unit, the other 5 stages delay units are connected to the same buffer. We can change the resonance frequency by changing the value of parasitical capacitance include the nod capacitance of MOSFET and the load capacitance etc. The V-I converter transforms the Vc voltage of loop filter output into the bias voltage to generate the bias current for the delay unit Id, thus the control

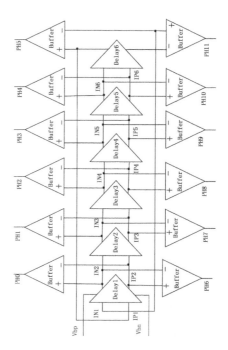

Figure 4. Frame of voltage oscillator.

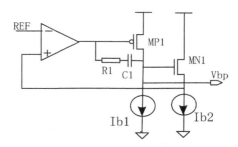

Figure 5. V-I converter schematic.

voltage can control the oscillation frequency of the VCO. The designed V-I circuit is shown in Figure 5, the bias voltage Vbn is provided directly by the loop filter, the VCO bias voltage Vbp is generated from two stage amplifier, the first stage amplifier use differential structure with PMOS load and NMOS input transistor, the transistor that generate tail current is biased on Vbn as same as Ib1 and Ib2. The same size NMOS transistors is used for current match, R1 and C1 are used to implement the miller compensation of two stage amplifier in order to ensure the loop stability. The output of the amplifier is feedback to the in-phase terminal of amplifier through the source following amplifier composed of MN1 and Ib1, thus improved the stability and response speed of the output bias.

The frequency divider used a 2.5 times division. In order to achieve 3X oversampling, the traditional method is two times the frequency of the reference clock firstly, and then five divider to get the 2.5 times division, the shortcomings of this approach are obvious, caused more system noise, circuit complexity, power consumption and large chip area consumption. An improved 2.5 times divider for the chare pump, PLL, was presented in this paper. It is simple circuit structure, very low power consumption and small chip area as well as the low costs.

4 SIMULATION AND TEST

The system was simulated by using Hspice-RF provided by Synopsys. The VCO jitter and phase noise simulation is deeply analyzed as the VCO noise is the main source of noise, the main program of simulation is shown as follows:

.HBOSC TONE = 0.412 g nharms = 30 probe-
    node = va1p, va1n, 2.5 +fspts = 40, 900 meg,
    110 meg .phasenoise v(va1p,va1n) dec 100 10
    1E11 method = 2 carrierindex = 1
.probe phasenoise phnoise
.probe phasenoise phnoise jitter
.meas phasenoise rj rmsjitter phnoise from 10 to
    10 meg units = sec

The result shows that the phase noise is 95dBc/Hz at 100 kHz frequency offset on the condition of 412 MHz center frequency of the VCO, the jitter is less than 20ps for 1μs. The designed PLL is used in the visual interface receiver chip design and taped out.

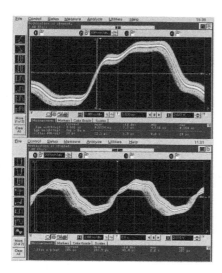

Figure 6.   Eye diagram of system clock signal.

Figure 6 is the eye diagrams of PLL output clock signal as external system clock signal is165 MHz. It can be seen from the testing results, the clock signal RMS jitter is 24ps, peak-to-peak jitter is 194ps.

5 CONCLUSIONS

The charge pump PLL with 12-phase clock output used for high-speed serial uncompressed data recovery was implemented. The frequency divider method of 2.5 times is used, 6 stages differential delay cell is adopted for the VCO, obtaining 12 equal phase offset clock. The frequency detector automatically detects and segment frequency of the input clock signal to achieve a large frequency capture range. Hspice RF tools is used to simulate and estimate the jitter and phase noise performances, simulation results show that the VCO phase noise is 95dBc/Hz as the frequency offset is 100 kHz at 412 MHz. SMIC 0.18μm CMOS mixed signal process was used for tape out and verification, testing results can satisfy the design requirements.

REFERENCES

Bai Xiaoping, Zhang Hongwei. Study of Digital Video Interface (DVI) Architecture Design in Computer System [C]. ICCCAS, 2007:1200–1203.

Christian H, Christian H, Ulrich H. Stability Analysis of a Charge Pump Phase-Locked Loop Using Autonomous Difference Equations [J]. IEEE Transactions on Circuits and Systems, 2014, 61(9):2569–25.

Digital Visual Interface (DVI) 1.0 Specification [S]. Digital Display Working Group, 1999.

High-Definition Multimedia Interface (HDMI) Specification Version R1.1 [S]. HDMI Licensing LLC, 2004.

John Rogers, Calvin Plett, Foster Dai. Integrated Circuit Design for High-Speed Frequency Synthesis [M]. Artech House Publishers, 2006.

Lee K, Shin Y, Kim S, et al., 1.04GBd Low EMI Digital Video Interface System Using Small Swing Serial Link Technique [J]. IEEE Journal of Solid-State Circuits, 1998, 33(5): 816–823.

Te-Wen Liao; Jun-Ren Su. Chung-Chih Hung. Ring-VCO based low noise and low spur frequency synthesizer [C]. Circuits and Systems (ISCAS), 2013 IEEE International Symposium, 2013:1861–1864.

W.C. Cheng, M. Pedram. Chromatic Encoding: a Low Power Encoding Technique for Digital Visual Interface [J]. IEEE Transactions on Consumer Electronics, 2004, 50(1): 320–328.

Ye Zukui. Design and Implementation of Digital Visual Interface Reciever [D]. Tainan:Tainan University of Technology, 2002.

Yongsam Moon, Deog-Kyoon Jeong, and Gijung Ahn. A 0.6–2.5-GBaud CMOS Tracked 3 Oversampling Transceiver with Dead-Zone Phase Detection for Robust Clock/Data Recovery [J]. IEEE Journal of Solid-State Circuits, 2001, 36(12): 1974–1983.

Advances in Energy, Environment and Materials Science – Wang & Zhao (Eds)
© 2016 Taylor & Francis Group, London, ISBN 978-1-138-02931-6

Research on deployment scheme of the power synchronization network management system

Zechao Xing, Ling Teng, Qiang Gao, Miaoxin Wang & Yang Wang
Information and Communication Department, China Electric Power Research Institute, Beijing, China

ABSTRACT: In order to solve the problem that the power synchronous network management is short of comprehensive unified management, the passage discusses the key technologies of the power synchronization network management based on the characteristics of telecommunication. Considering the standpoint of electric power industries, we came up with four schemes, which was third-party network central deployment, original factory network central deployment, third-party network dispersed deployment and original factory network dispersed deployment, and compared with the four schemes from the aspects of initial investment, operation costs and technical superiority, it turned out that the third-party network central deployment was the most suitable scheme for national network management. At the end, we gave specific use of the scheme, taking North China as an example.

1 INTRODUCTION

Power frequency synchronization network plays an important role in monitoring control and operation management, fault analysis for power system (Wang, 2009). Frequency synchronization network can provide accurate unified reference synchronization frequency to the grid service that is a powerful safeguard for the normal operation of the power grid (Wang, 2009; Chen, 2012). In order to meet the needs of electric power enterprise vigorous development, high standards, high precision synchronization network is constantly developing.

With the continuous expansion of the scale of network synchronization, the equipment type and quantity are increased that lead to the complexity of the whole network is increasing. The different developers developed a variety of network management system by using different technology and most of them use the respective management protocol (Liu, 2004; Wang, 2006). This situation will inevitably cause that network protocols are not compatible, management information can't be exchanged and the network is lack of comprehensive management of the entire network and other issues.

Therefore, in order to solve the problem that the network protocol in the synchronization network management is not uniform and the synchronization network management is lack of comprehensive management methods, this passage analyses the key technology in the management of power network synchronization based on the present situation of electric power synchronization network, and does an-depth research in the deployment scheme, finally puts forward the best deployment plan to solve the problem.

2 THE PRESENT SITUATION AND ANALYSIS OF POWER SYNCHRONIZATION NETWORK

The current power frequency synchronization network only deploys the original manufacturers and uses localized operation management mode which is lack of a unified monitoring method. According to statistics, the number of whole network equipment management (including local maintenance terminal) is 38 sets: the northeast division has 1 set, the east china division has 1 set, province companies have 36 sets. Only 16 provinces deploy the original factory management system, synchronization equipment, and the remaining 11 provinces have not achieved the management company.

The clocks in frequency synchronization network are 405 units, and 176 of them have the network management which account 43.46% of the total number, more than half of the clocks did not achieve the management. All of the management of frequency synchronization network that have been deployed, is the original manufacturer of network management. The original manufacturer network management manufacturers have 4 represents: Huawei, Xintai, Datang Telecom, Titan; the mainstream manufacturers have 3 represents: Huawei, Datang Telecom, Titan. The number of equipment they managed accounted for 99.4% in the total number of pipe equipment.

From the above results, all of manufacturers can't manage the equipments by a unified, effective way, and they can't manage other equipments of other manufacturers. They are lack of unified maintenance management means for synchronous devices in the whole network. Therefore, there is an urgent need to study a reasonable power synchronization network management scheme, unified deployment, implementation of synchronization the equipment of different manufacturers to maintain a unified network management system or management terminal.

3 RESEARCH ON POWER SYNCHRONIZATION NETWORK MANAGEMENT SYSTEM TECHNOLOGY

3.1 *Analysis of power synchronization network management level*

Power synchronization network management system, can be divided into three levels according the network or the equipment different, as shown in Figure 1.

In the Synchronous management in power network, equipment monitoring is also known as network element management which is founda-

tion of professional network management system and integrated network management. The professional network management, that is on a single network management, power synchronization network management is one of them. The integrated network management, network management oriented integrated communication network, is also known as for unified network management. In the power system, the State Grid remote terminal maintenance management system (Terminal Management System, TMS) is an integrated network management system.

3.2 *Research on the management mode of power synchronous network management system*

The management mode of the network management system mainly includes the centralized management mode, management mode of the distributed and hierarchical and distributed management mode.

3.2.1 *Centralized management mode*
For smaller networks, due to the limited amount of information in the management contents and transmission, network geographical distribution is relatively concentrated, administrative requirements for centralized management, management personnel are less, usually by centralized management, whereby a network management center implementation of the management of the whole network, as shown in Figure 2.

3.2.2 *Distributed management mode*
Distributed management is the network management that according to some classification method divide the network into a plurality of "regional management", and each of the management areas set up a management center which is responsible for the management of the regional network, the function of centralized management mode of

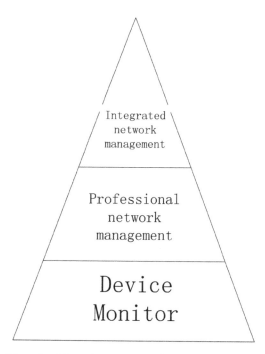

Figure 1. Network management level.

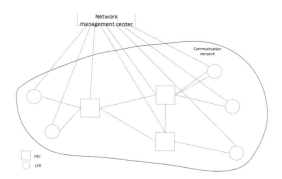

Figure 2. Centralized network communication management.

distribution to the various regional management center, as shown in Figure 3.

3.2.3 Hierarchical distributed management mode

The distribution pattern of the classification actually is the synthesis of the above two models. A larger scale network is managed by hierarchical and distributed mode that divide network management system into several levels, level depends on the size of the network management, and each upper level management of a number of a junior, superior in centralized, lower on the function of distribution, in order to achieve the optimal allocation of network management, as shown in Figure 4.

3.3 Research on interface protocol of power synchronous network management system

We need to build a unified Network management system for multi vendor synchronization equipment. And the equipment manufacturers or maintenance terminal and professional network management protocol is to study and identify technology. The interface protocol of network management system mainly in the Q3 protocol, Common Object Request Broker Architecture (CORBA), Simple Network Management Protocol (SNMP).

The Q3 protocol is the Telecommunications Management Network (TMN) interface specification. And th rich description can support

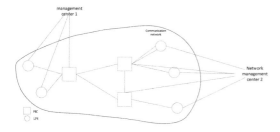

Figure 3. Distributed network communication management.

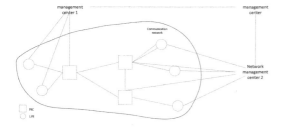

Figure 4. Hierarchical distributed network management mode.

the alarm management, configuration management and performance management (Liu, 2003; Li, 1999; Liao, 2002). But there are obvious deficiencies, because it to communicate with the CLNS protocol that requires running platform installed third-party software provides the protocol stack, and interoperability between different product requirements are relatively high, which increased the project difficulty; Q3 interface modeling is based on the fine-grained modeling, so the modeling and development are more complex, design, development and test cost is high; Q3 interface operation expenses of the larger, on the operation platform and network system hardware requirements higher.

CORBA protocols: CORBA can like Q3 interface as reliable support alarm, configuration and performance management, in the aspect of modeling belongs to the coarse-grained modeling, and development, testing compare to Q3 is simple, on a variety of underlying communication protocol completely shielded (Lu, 2006; Fang, 2002). The CORBA interface also has some disadvantages, such as the need to buy third party CORBA software platform, the cost is high.

SNMP protocol: SNMP protocol as the TCP/IP protocol stack in the network management protocol, is widely used in computer network management. The advantages of SNMP protocol is that the protocol is simple and reliable, and does not need the third party software support, development and maintenance is simple, and the agreement is continuous development and function in the continuous strengthening (Yan, 2008; Li, 2008; Zheng, 2012).

In the power synchronization network deployment scheme, due to the development and maintenance of SNMP protocol is simple, and it do not need to third party software support, SNMP is usually used as the synchronous equipment north to interface protocol; due to the CORBA protocols on a variety of underlying communication protocols are completely shielded, CORBA protocols is usually used as original equipment manufacturers of network management of the north to the interface protocol.

4 RESEARCH ON DEPLOYMENT SCHEME OF NETWORK MANAGEMENT SYSTEM

4.1 Network management system deployment plan

The deployment of network management system directly affect the system of management efficiency and investment cost, while it is closely related to access and network evolution and TMS. Therefore, the construction of synchronous network management system to carry out research on the pre deployment mode is an important part of. Through the study of the synchronization network management system technology, combined with the characteristics of electric power communication itself proposed synchronous network management to deploy four modes: third party network deployment, the original manufacturer network centralized deployment and the third party network management decentralized deployment, the original manufacturer network distributed deployment.

Mode one: third party network management centralized deployment

This mode that the equipment of different manufacturers is managed by third party network synchronization management, achieve the unifying network monitoring cross equipment manufacturers, and the third party network in centralized deployment mode, is adopted to the centralized model, including two levels of network management system, which backbone network third party network management system the company responsible for the synchronous backbone network, network management, provincial network management system responsible for network management Moto synchronization network.

Mode two: the original manufacturers centralized deployment of network management

This mode is that the original equipment management monitor and configure the operation of the network, and the equipment management in the province's centralized deployment. They are responsible for all synchronization network equipment of province. And, in province of synchronization network equipment is not a unified brand, there may be multiple sets of equipment management.

Mode third: three party network management decentralized deployment

This mode is that the third party network synchronization management integrated manage the network equipment of different manufacturers, to achieve cross equipment manufacturers of the unified network monitoring and the third party network management respectively in backbone, provinces, cities and counties three deployment. They are responsible for management backbone, provinces, cities and counties synchronization network equipment.

Mode four: original vendor network decentralized deployment

This model is that the original equipment management monitor and configure the the network, and the equipment management respectively deploy in backbone, provinces, cities and counties. They are responsible for all the synchronization network devices management backbone, provinces, cities and counties.

4.2 Research on deployment scheme of network management system

Following these four kinds of deployment for deployment on the network are studied from three aspects: initial investment, operation and maintenance cost and technical advantages of.

1. Initial investment

In order to get a more comprehensive comparison of four deployment models about the initial investment, we study from the development cost network rectification costs and cost of construction three aspects, as shown in Table 1.

In the Table 1 "+" is basic units for investment funds. The initial investment of four kinds of models are compared. Due to model four (original manufacturer network scattered deployment) some units have been with decentralized deployment of EMS, these units do not need to build network management, so the construction cost is low, that lead to the model four have obvious advantages in upfront investment. And mode III (third party network scattered deployment) due to the need for each province new third party network, so the construction of more expenses, investment in the early needs the most. And modes one and two due to the adoption of the centralized deployment mode, construction costs is lower, so the upfront investment also has certain advantages compared to the model three.

2. Operation and maintenance costs and difficulty

Mode one (third party network centralized deployment mode): using this mode only need sign with a third party manufacturers maintenance contracts, and operation and maintenance personnel only need to master a third party network management application functions that can complete the daily network operation and maintenance; third party network have access to TMS, for centralized deployment of third party network only maintain a third party network northbound interface. The operation and maintenance costs of the method is relatively low.

Table 1. Comparison of the initial investment of four deployment modes.

Evaluating indicator	Mode one	Mode two	Mode three	Mode four
Cost of development	+	++	+	++
Network rectification	+	+	+	+
Construction cost	+++	+++	++++++	+
Comprehensive cost	+++++	++++++	++++++++	++++

Mode two (original manufacturer network centralized deployment mode): before device without brand focus, even in the provincial centralized deployment mode of the original manufacturer network management there are two or more original manufacturer of network management, coordination of multiple vendors, maintenance costs are relatively high, and the need for multiple manufacturers of network equipment operation, is not conducive to the unified management; original manufacturer network access after TMS and need to maintain multiple original network north to the interface and the interface protocol may have several, requiring a higher ability of operation and maintenance personnel. The operation and maintenance costs and the difficulty of the way is higher.

Model three (third party network decentralized deployment): the use of the required signing with different levels of multiple third-party manufacturers maintenance contracts, operation and maintenance personnel also need to acquire multiple sets of third party network management application function before they can complete the daily network operation and maintenance. Third party network access to TMS, because the decentralized deployment is required to maintain multiple sets of third party network north interface. The operation and maintenance costs and the difficulty of the way is high.

Mode four (original manufacturer network dispersed deployment): distributed deployment mode of the original manufacturer network management will exist in different levels, each province at least has 10 sets of the original manufacturer network management, operation and maintenance is required to coordinate multiple vendors, maintenance costs are relatively high, and the need to operate in multi vendor network management equipment, is not conducive to the unified management. After the dispersion of the deployment of the original manufacturer network have access to TMS and need to maintain multiple original network north to the interface and the interface protocol may have several, requiring a higher ability of operation and maintenance personnel. The operation and maintenance costs and the difficulty of the way is higher.

In the aspects of maintenance costs and difficulty, the mode one (third party network centralized deployment mode) is relatively low.

3. Technological advantage

In the aspect of network control ability, due to the centralized management of centralized deployment mode which is dispersed equipment, have very strong centralized management and control capabilities, and they can able to statistical analysis synchronization network MTIE and TDEV and FREQ and performance data, and scattered deployment can only on the part of the equipment performance according to the analysis, so in the aspect of the ability to manage and control network, model one and two are stronger than modes three and four.

In the aspect of the ability of technology standardization, network management from the original manufacturers generally do not have the mature interface to the north and if they plan to access to the TMS, network management need upgrade. And the north interface protocol of the original manufacturer network management exits many types, technical standards ability is weak. Although the third party management on the equipment of access has normalized, but for the pattern of scattered construction is bound to occur, many manufacturers multi protocol, technology standardization is low. Mode one (third party network centralized deployment) just need to develop one set of third-party network management interface protocol program, and through a protocol to transmission of several kinds of data, that lead to there are fewer protocol and protocol conversion less workload. In addition, for devices within a network changes do not impact on the TMS access, as long as the devices are connected to the third party network management, network management interface without need to change can be submitted to change the device information. So in the technical standardization ability, mode one is stronger than mode two, three and four.

In terms of technology, mode one (third party network centralized deployment) has obvious advantage in the aspect of network control ability and standard technical capabilities.

In summary, the advantages and disadvantages of the four deployment patterns are shown in Table 2.

Therefore, in the process of synchronous network management evolution and development, it is recommended for synchronous network management system centralized deployment mode, at the same time, considering the construction of third-party integrated network management that is mode one (third party network centralized deployment mode) which have centralized management and control ability, reduce the cost of operation and maintenance of late, and easy access to the TMS system.

4.3 Implementation of network management deployment

Taking North China synchronous network management as an example, the implementation of

Table 2. The comparison of network management deployment.

Evaluating indicator	Mode one	Mode two	Mode three	Mode four
Initial investment	Big	Medium	Big	Small
Operation and maintenance costs	Small	Medium	Medium	High
Technological superiority	Strong	Weak	Medium	Weak
Comprehensive choice	Superior	Good	Good	Medium

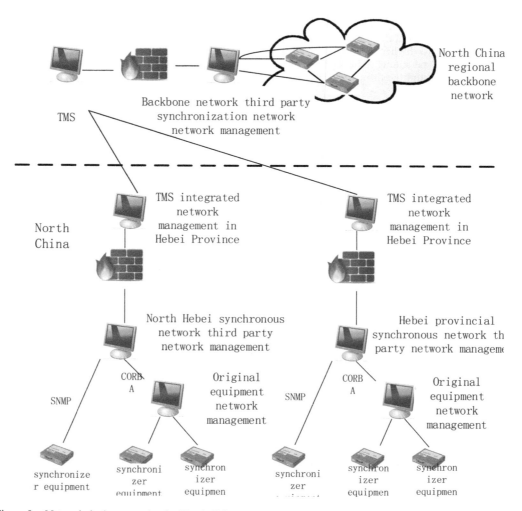

Figure 5. Network deployment plan for North China.

network management deployment is given, as shown in Figure 5.

The scheme is based on third party network centralized management deployment scheme which divided into two-level architecture: backbone network and provincial network. In North China area backbone synchronization network and provincial synchronization network, we use third-party synchronization network management and equipment. The management system of Jibei Province manage the network at the provincial level and municipal synchronization device and the original manufacturer device synchronization network management. The provincial third-party synchronization network management system obtain synchronization information by two ways: through SNMP acquisition device synchronization information, through the CORBA protocol to collect equipment management information, and all the synchronization information transfer through the electric power dispatch data network. Provincial third-party synchronization network synchronization information through network data transfer to provincial TMS integrated network management, integrated management of provincial TMS the information synchronization via an integrated data network passed to the national grid TMS integrated network management. For the security of the network, the network firewall is needed built between the scheduling data network and the comprehensive data network.

Synchronous information in the backbone network is also transferred by dispatching data network to the backbone of the third-party synchronization network, and then the information is transferred through the integrated data network to the national grid TMS integrated network management.

5 CONCLUSION

The construction of the network management system plays a pivotal role in the development of the synchronous network. This passage analyzed from the point of supporting power synchronization network efficient operation, based on the key technology of power synchronization complete network management system. We got the conclusion: the third party network deployment is the most suitable for the optimal power synchronization network deployment program. And based on the example of North China, we described practical application of the third party network management. The construction of power synchronization network management will continue to experience the process that identifying problems, solving problems and drawing lessons and gaining experience, understanding of the law, and gradually improving. As a power synchronization network application system, there will be a more broad application prospects.

ACKNOWLEDGEMENTS

Fund Project: the China State Grid Corp technology project the key technology research and application of the power synchronization key technology based on the time frequency fusion.

REFERENCES

Chen Bao Ren. Research on the construction scheme of the power network integrated [J]. Power Survey and design, 2012, (2):57–62.

Fang Ning, Shen Zhuo Wei. CORBA based network management [J]. computer engineering and application, 2002, 09:142–146.

Li Bo. Research and implementation of key technologies of network management system based on SNMP protocol [D]. Beijing Jiaotong University, 2008.

Li Tian Jian, Zeng Wen Fang. Analysis and Prospect of application of [J]. Network management protocol of the computer system, 1999, 05:17–20.

Liao Jian Xin, Zhang, Wang Jing, Meng Hui Yang. Mobile intelligent network management system analysis and design [J]. Modern telecommunication technology, 2002, 05:22–24+41.

Liu Li Na. The development of power system communication integrated network management system [J]. Power system communications, 2004,25(7):5–7.

Liu Tong Na. The power communication network management protocol [D]. North China Electric Power University (Hebei), 2003.

Lu Yan, Peng to offer. Research and application of CORBA in the network management [J]. Modern electronic technology, 2006, 10:47–49.

Wang Yu Dong, Yu Tian Qing. The research on communication power system time synchronization network [J]. The communication of electric power system, 2009, 30(7): 64–67.

Wang Wu Chao. Power integrated network management system for communication network construction of [J]. Power system communication, 2006, 27(5): 16–20.

Yan Yi. SNMP protocol in the campus network management application [D]. Chongqing University, 2008.

Zheng Xiao Ping. Research and implementation of network management platform based on SNMP protocol [D]. Shanghai Jiao Tong University, 2012.

Advances in Energy, Environment and Materials Science – Wang & Zhao (Eds)
© 2016 Taylor & Francis Group, London, ISBN 978-1-138-02931-6

Study on the key identification technologies of 360° full-range intelligent inspection in electric power communication equipment room

Shuzhen Yang
College of Engineering, Shanghai Second Polytechnic University, Shanghai, China

Huilan Zeng
Shanghai Sidun Information Technology Co. Ltd., Shanghai, China

ABSTRACT: The intelligent inspection of electric power communication equipment room is one of the means to insure devices' reliability, to enhance the technology capability of the intelligent inspection of electric power communication equipment room. The applications of intelligent inspection identification are emerging. Due to the increased demand of visual angle and scope requirements and the accuracy of Intelligent inspection, this paper presents the key identified technologies of intelligent inspection, which suits for electric power communication equipment room, including image acquisition applications based on RFID, automatic inspection route design based on tour demand, automatic inspection precise positioning based on image identification, full angle signal receiving application and application of intelligent patrol management system.

1   INTRODUCTION

With the rapid development of enterprise information, the number of information equipment increase very fast and the information network coverage becomes bigger and bigger. The number of equipment ervender of electric power enterprises is numerous; servers, computer terminals and software upgrade frequently, management and maintenance workload increases, the technical level and depth increases, information management and maintenance work become more and more difficult. Intelligent monitoring system for industry communication room in the market trend and development situation is of concern.

In order to detect equipment problems and troubleshooting problems and to ensure the reliability of the equipment, the communications room equipment needs to be visited regularly. The inspection in electric power equipment room is an important part of the maintenance work of Information systems. With the rapid development of information room, inspection work becomes more complex, if you continue to use traditional manual means of inspection for information systems, something will be missed in the inspection and it will become more and more difficult with costs rising. There is an urgent need to use advanced modern technology to promote the efficiency of inspection work to further improve the level of management and supervision, thus, the management of power enterprise information equipment

room becomes more scientific, efficient, refined, and standardized (Liu et al. 1989, Ma et al. 2014). In this paper, we carry out the study on the key identification technology of a full-range intelligent inspection in telecom equipment room based on its characteristics and difficulties.

2   ANALYSIS ON THE FULL-RANGE INTELLIGENT INSPECTION DEMAND OF POWER AND COMMUNICATION EQUIPMENT ROOM

The full-range intelligent inspection demands of communication equipment room are reflected in the following aspects (Shi 2009, Zhao 2012):

Whole-process: the inspection is controllable and under control in the whole process;

Scientification: the scientification of inspection of data analysis;

Standardization: inspection items are classified into three levels based on the inspection points corresponding to the ID card, relevant equipment under inspection points as well as items to be inspected of each equipment; management standardization of inspection;

Full-range: full range instead of local intelligent inspection & identification in communication equipment room;

High-performance: ensure that inspections are fully-implemented; operations are simple, user-friendly, easy-to-use, convenient, and quick;

inspections are of centralized management and flexible allocation.

3 GETTING STARTED KEY TECHNOLOGIES OF FULL-RANGE INTELLIGENT INSPECTION OF POWER COMMUNICATION EQUIPMENT ROOM

The development of equipment inspection technologies has experienced the Artificial Age, the Information Flow Age, and the Intelligence Age, the application of equipment and technologies by using of on-site data acquisitor etc. Though remote video identification has emerged in Intelligence Age (Tai 2008). As shown in the Figure 1 below key technologies and main contents are described as follows:

3.1 *Image acquisition application based on Radio Frequency Identification (RFID)*

Using RFID and Image Intelligent Identification, video imaging technology can be combined with equipment ID and information for the rapid positioning of equipment conditions.

RFID, Radio Frequency Identification, commonly known as electronic tag, is a kind of non-contact automatic identification technology, which identifies target objects and acquires relevant data through radio-frequency signals without artificial intervention and can be used in a variety of harsh environments. RFID can identify objects moving at high speed and multiple tags simultaneously, and is easy to operate. Radio Frequency Identification can identify the stationary or moving objects automatically and exchange data by using radio frequency

signal and its spatial coupling and transmission properties (Lang 2006, Xiong 2008, Hu et al. 2014). It is well known that the previous image acquisitions are transmitted by cables & wires; it is difficult for the realization of subtleness & artisticness, inconvenient for installation & maintenance, and of weak resistance to electrical interference, troublesome investigation of failures and so on (Xue et al. 2014). The basic working principle of RFID technology: after tags have entered the magnetic field, the receiving reader sends out radio frequency signals, which send out the product data stored in chips (passive or passive tags) by energies obtained by induced currents; reader, after reading and decoding, transfers data to the Central Information System (CIS) for data processing.

The RFID-based Image Acquisition Application makes use of the Radio Frequency technology to transmit images collected by image sensor to RF transmitting terminals with Wireless Radio Frequency technology, and then the RF receiving terminals receive electromagnetic waves containing images and send images to PC for display through USB chips. The application is featured by low cost, low power consumption, stable performance, etc., and has achieved the target of transmitting images acquired on-site to PC through RF chips in order to monitor all situations in real time.

3.2 *Automatic inspection route design based on equipment rooms' demand of inspection*

Combined with technical applications of RFID and Image Identification & Target Positioning etc., the inspection route shall be designed in conjunction with the inspection demands of equipment room and shall cover all key routes; with timing management, the free inspection & monitoring shall be controlled in the range of 360 degrees of freedom.

Among them, objects to be inspected and contents needed mainly include:

Server. Includes CPU utilization ratio, memory usage, disk (or tablespace) usage (%), and server logs etc.
Network equipment. Mainly includes,
Firewall: including power LEDs, TRUST port LEDs, UNIRUST port LEDs, event warnings, HA LEDs, and Ethernet LEDs.
Switchboard: includes power LEDs, fan LEDs, optical fiber module LEDs, RJ-45 port LEDs, and panel LEDs.
Devices in equipment room. Mainly includes,
UPS: including online time, battery life, input voltage, output voltage, battery status, indoor temperature, system logs, LEDs, mainframe power wiring, UPS battery leaking, and battery pack DC output power 48V–56V.

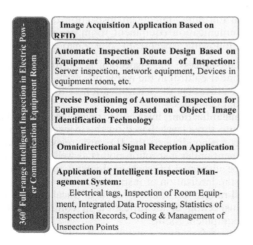

Figure 1. The key technologies and solutions.

Air-conditioner: includes starting, refrigeration, operation, temperature, humidity, etc.

Power supply: check that all power supply plugs are free of abnormal phenomenon of burning, heating etc.; all power supply circuits, including power lines, UPS mainframe & UPS battery lines and UPS power supply lines, switchboard internal power supply, air conditioner power lines are free of the abnormal phenomenon of burning, heating etc.; LEDs of voltage, current and power supply as well as other panel LEDs are normal, etc.

Sanitary status, windows & doors: clean and locked.

Fire extinguishers: check for leakage, etc.

3.3 *Precise positioning of automatic inspection for equipment room based on Object Image Identification technology*

The intelligent inspection identification of equipment room is closely linked with the precise positioning of automatic inspection of equipment room based on the Object Image Identification technology. The precise positioning of target plays an important role in object identification, image understanding & analysis. Target identification has been successfully applied in many areas, such as judgment of mineral reserves and oil rock stratum structure, etc,. In general, there are two ways for the identification & positioning of targets:

Positioning of target by information distributed in space or time, that is, model identification; the model identification system based on statistical methods consists of four parts: data acquisition, pre-processing, feature extraction & selection, and classification & decision.

Target positioning by means of matching. Image Matching Algorithm can be divided into two categories: the first is the Statistical Matching Algorithm, mainly includes Mean Absolute Difference Measure, Mean Squared Deviation Measure as well as the matching algorithm based on Gray Correlation Measure; the second is the matching algorithm based on features, that is, features such as boundaries, textures, entropies, energies etc. are extracted from images; those features can be used for rapid matching.

3.4 *Omnidirectional signal reception*

By using the 360° rotating device, omnidirectional infrared RC wireless signal repeater dn other devices, the signal reception angle is omnidirectional 360°; every corner of equipment room is covered so as to achieve the control of high sensitivity and high accuracy. The signal reception angle is adjustable and shall satisfy the different requirements of direction and omnidirection.

3.5 *Application of intelligent inspection management system*

Electrical tags: composed of coupled components and chips; each tag has a unique electronic code fitted on the equipment to identify target objects; relevant information is stored in tags and can be updated.

Inspection of room equipment for the reading and writing of information on tags.

a. Identification of inspection point: read the ID number of RFID card installed in specified location and match them with the ID number stored in the inspection plan, prompting inspection items under respective inspection point and contents required to be inspected so as to prevent omission and mis-inspection. b. Read information of RFID card and browse inspection items in order.

Integrated data processing.

a. preparation of inspection plan; b. maintenance of inspection data; inspectors can supplement the processing scheme and conditions for failures found in inspection; c. Summary of inspections;

Statistics of inspection records. collect information required based on information of time, inspection point, defect etc,. and provide chart display.

The first is data of the day, which summarizes and indicates details of the inspection results of the day, provides staged display based on ID points, equipment and items as well as fuzzy inquiry and group inquiry in line with multiple conditions. The second is man-hour statistics, which provides the time consumed by designated inspection in selected equipment room and provides chart display. The third is defect statistics which provides information specifying whether defects occurred in designated equipment room within selected time and conditions of defects occurred as well as detailed data tables, and provides chart display.

Coding & management of inspection points. The system records the inspection contents involved in inspections, for instance, the inspection for network equipment room includes UPS mainframe, UPS batteries, air conditioners; the inspection of server includes small equipment, PC servers etc.

Management of inspection point: records and manages the information of inspection points, including equipment room in which inspection points are located, contents of inspections etc. Inspection points shall be corresponding to the RFID card ID arranged on-site one by one; users can prepare the inspection plan based on information of inspection points and guiding inspectors' work.

4 SUMMARY

Combined with features and difficulties of the communication room inspections, the communication rooms' full range of intelligent inspection technology to identify the application requirements, and through to inspect the recognition technology research, is put forward to meet the communications room all 360 degree intelligent inspection recognition key technology and solution. The program is scientific, standardized, efficient, practical, and the whole range of the whole process and other characteristics.

ACKNOWLEDGMENT

This work was supported by Shanghai Committee of Science and Technology, China (Grant No. 13521103605) and Leading Academic Discipline Project of Shanghai Second Polytechnic University (No. XXKYS1403, No. XXKPY1311).

REFERENCES

Hu, Z.W. Wang, Y. Li, X. & Liu, D.M. 2014. Research and Utilization on RFID Label in Substation Inspection. *5th International Conference, CloudComp 2014, Guilin, China, October 19–21, 2014, Revised Selected Papers:* 232–239.

Lang, W.M. 2006. *RFID technology principle and application.* Beijing: China Machine Press.

Liu, Y.L.& Zhang, B.H. 2006. Application of the electric power telecommunication station monitor and control system in Yichang Electric Power Bureau. *Central China Electric Power* 19(1): 28–30.

Ma, Y.X. Liu, Q.D. & Bao, H. 2014. Design and implementation of substation intelligent patrol background management system. *Power System Protection and Control* 10: 125–129.

Shi, G. 2009. *Application Research on substation intelligent inspection system in Chengde Power Supply Company. Baoding: North China Electric Power University.*

Tai, B. 2008. Application of intelligent inspection system on inspection of substation equipment, *Guangdong Power Transmission Technology* 1: 21–23.

Xiong, C.R. 2008. Research and design of intelligent data collection terminal based on the RFID. *Electrical Automation* 30(5): 54–56, 65.

Xue, J. Zheng, W.B. & Zhang, Z.H. 2014. Development of Intelligent Substation Inspection System Based on Image Monitoring. *Electric Power Information and Communication Technology* 7: 21–23.

Environmental analysis and monitoring

Advances in Energy, Environment and Materials Science – Wang & Zhao (Eds)
© *2016 Taylor & Francis Group, London, ISBN 978-1-138-02931-6*

Correlations between enzymes and nutrients in soils from the *Rosa sterilis* S.D. Shi planting bases located in karst areas of Guizhou Plateau, China

Jieling Li, Hao Yang, Xuedan Shi, Mingyi Fan & Lingyun Li
Institute of South China Karst, Guizhou Normal University, Guiyang, Guizhou, China

Jiwei Hu & Chaochan Li
Guizhou Provincial Key Laboratory of Information System of Mountainous Areas and Protection of Ecological Environment, Guizhou Normal University, Guiyang, Guizhou, China

ABSTRACT: This paper reports the contents of soil Organic Matter (OM), Total Nitrogen (TN), Total Phosphorus (TP), Total potassium (TK), Hydrolyzable Nitrogen (HN), Available Phosphorus (AP), available potassium (AK), pH level, and the activities of urease, catalase and invertase from the two *Rosa sterilis* S.D. Shi planting bases located in karst areas of Guizhou Plateau. The relationships between enzyme activities and nutrient contents for soils were also investigated. The results show that catalase and OM have significant positive correlation at 0.05 level, while invertase and TN have a significant positive correlation at 0.01 level. Therefore, the activities of soil enzymes could be used as a potential indicator of soil fertility for these planting bases. The results demonstrate that the planting of *Rosa sterilis* is beneficial to improving soil organic matter content and curbing the degradation of the surface soil quality in karst areas.

1 INTRODUCTION

Soil is the loose stuff on the surface of the lithosphere, which can provide necessary nutrients for plants, the base material of land plants, important place for the exchange of material and energy in the terrestrial ecosystems. Tabatabai et al. indicated that soil enzymes are the core of soil ecosystems (Tabatabai et al., 2002). Soil enzyme is metabolic power of soil organisms, playing a biological catalytic action for materials circulation and energy flow. They participate in all biological and chemical progress, while enzymes have an important effect on essential elements circulation and transformation of plant growing, such as C, N, P in soil (Bruce A.C., 2005). The planting can influence the content of soil enzymes directly or indirectly, and living roots of plants have an effect on soil enzymes (Cao et al., 2003). In other earlier investigations on the essence of soil fertility, enzyme activities are used as auxiliary indices of soil fertility (Cao et al., 2003; Zhang et al., 2013; De La Paz Jimenez M. et al., 2002). Therefore, soil enzyme activities are potential indicators to maintain soil fertility, which can affect the strength of soil nutrients transformation (Liu et al., 2011). They are the important parameters for evaluating the soil fertility.

The Guizhou karst region is located in one of the three world famous karst development zones in the center of the east Asia. The nutrients lose quickly under the particular external conditions, such as high mountains and steep slopes, warm climate and abundant rainfall, because carbonate rock has the features of easy penetration, poor soil forming ability (Liu & Wolfgang D, 2007). The conservation of soil has always been an important issue in karst areas. *Rosa sterilis* S.D. Shi, as an endemic species in Guizhou, is widely studied for the aim of food and medicinal development. It is suitable for stony soil in karst areas because of its well-developed root systems, strong ability of water and soil conservation. In recent years, many investigations have been concentrated on morphological characteristics, cuttage and tissue culture propagation, analysis of aroma components and pharmacological properties, and resistance to powdery mildew, however the study on soil fertility about *Rosa sterilis* orchard has rarely been reported (Zheng et al., 2013).

Wudang District (26°45′N, 106°58′E) and Pingba County (26°28′N, 106°17′E) were selected as the representative planting bases in the Guizhou karst region (Fig. 1). Firstly, they are located in the middle area of Guizhou Province with the subtropical plateau monsoon climate and the yellow zonal soil. Secondly, the planting age of Wudang is 3–4 years, while that of Pingba is 7–8 years. Finally, they are key areas of rocky desertification affected counties listed by the national ministry of science and technology, and a large scale cultivation of *Rosa sterilis* has been developed.

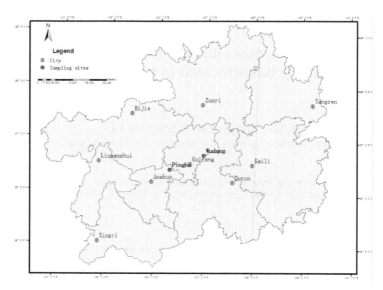

Figure 1. The location of two *Rosa sterilis planting bases* in Guizhou province.

The present study was aimed to analyze the activities of three enzymes and the contents of Organic Matter (OM), Total Nitrogen (TN), Total Phosphorus (TP), total potassium (TK), Hydrolyzable Nitrogen (HN), Available Phosphorus (AP), available potassium (AK), pH level in soils from Wudang and Pingba, and to discuss the possibility of enzyme activities as a indicator of soil fertility. The results of this investigation could be used for soil quality management to reduce nutrient loss in karst areas.

2 MATERIALS AND METHODS

2.1 *Sampling collection and analysis*

Soil samples were collected during July, 2014, using a compass to determine the slope and aspect, and a hand-global position system to determine the sampling location. According to the "S" shaped sampling method, 5–10 sample points were selected in each study area. Sampling depth was about 0–40 cm with the top-down sampling, and each soil sample was mixed, respectively. After air-drying and eliminating impurities, soils were ground with agate mortar to 0.149 mm (100 mesh) and screened for analysis. Each sample was determined with three replicates.

2.2 *Determination of nutrient contents and soil enzyme activities*

The OM of soil was determined by potassium dichromate-external heating method, TN was

determined by semi-micro Kjeldahl method, and TK was determined by sodium hydroxide melt-Flame photometry method. TP was determined by ClO_4-H_2SO_4-Mo-Sb colorimetry, while HN, AP and AK were determined by alkali diffusion method, 0.5 mol/L sodium bicarbonate extract-Mo-Sb colorimetry, 0.1 mol/L ammonium acetate extract-flame photometric method, respectively. pH was determined by a glass electrode using a 5:1 water-to-soil ratio (wt/wt) (Institute of soil science of Chinese academy of sciences, 1978). Catalase was determined by $KMnO_4$ method, urease was determined by phenol-sodium hypochlorite colorimetric method, and invertase was determined by 3, 5–2 salicylic acid colorimetry method (Guan, 1986). Activities of the three enzymes were expressed by $0.2 \ mol \cdot L^{-1} \ KMnO_4 \ ml \cdot (g \cdot 20 \ min)^{-1}$, (NH_3)-N $mg \cdot g^{-1} \cdot d^{-1}$, glucose $mg \cdot g^{-1} \cdot d^{-1}$ (the difference between test and blank value), respectively.

2.3 *Statistical analysis*

In this investigation, sample location map was made using ArcGIS 10.0, and data analysis was conducted with SPSS 19.0.

3 RESULTS AND DISCUSSION

3.1 *Status of soil nutrients*

The results for the analysis of soil nutrients are listed in Table 1. They were analyzed according

Table 1. Descriptive analysis for soil fertility of study areas.

Study area	Items	Minimal value	Maximal value	Mean value	CV
Wudang	TN	1.24	3.00	2.09	40.60%
	OM	1.67	3.00	2.31	26.59%
	TP	0.55	1.03	0.79	25.42%
	TK	1.05	1.41	1.21	14.33%
	HN	2.15	3.00	2.55	16.34%
	AP	0.65	1.04	0.76	21.50%
	AK	0.47	0.71	0.59	17.60%
	pH	2.00	3.00	2.60	21.07%
	Urease	0.17	1.25	0.49	89.89%
	Catalase	0.81	1.02	0.90	9.41%
	Invertase	13.36	34.80	22.86	37.53%
Pingba	TN	1.23	3.00	2.46	27.70%
	OM	1.92	3.00	2.46	16.88%
	TP	1.35	2.25	1.87	16.21%
	TK	1.12	1.59	1.38	14.12%
	HN	2.97	3.00	3.00	0.38%
	AP	2.06	3.00	2.85	12.31%
	AK	0.77	1.57	1.00	26.55%
	pH	2.00	3.00	2.71	17.98%
	Urease	0.53	0.87	0.74	16.89%
	Catalase	0.70	0.98	0.81	13.38%
	Invertase	14.23	32.91	24.36	27.46%

*The unit of TN, OM, OP, TK is $g \cdot kg^{-1}$, the unit of HN, AP, AK is $mg \cdot kg^{-1}$, the unit of catalase is $0.2\ mol \cdot L^{-1}\ KMnO_4\ ml \cdot (g \cdot 20\ min)^{-1}$, the unit of invertase is glucose $mg \cdot g^{-1} \cdot d^{-1}$, the unit of urease is $NH_3\text{-}N\ mg \cdot g^{-1} \cdot d^{-1}$.

to the Second National Soil Survey Classification Standard of Soil Nutrients. The contents of TN in the Wudang and Pingba planting bases are both with the top grade with the mean values of 2.09 $g \cdot kg^{-1}$ and 2.46 $g \cdot kg^{-1}$, respectively. The samples from Wudang with the top grade and the second grade account for 40% of the total, respectively. And the other samples belong to the third grade. The coefficient of variation (CV) for TN in Wudang is 40.60%. There are 87.5% of the samples from Pingba with the top grade. The CV for TN in soils from Pingba is 27.70%. The contents of TP for all soils from Wudang are with the third grade with the mean value of 0.79 $g \cdot kg$-1. The soils from Wudang with the second grade account for 40% of the total, while soils with the top, third and fourth grade account for 20% of the total. All soils in Pingba belong to the top grade. Thus, there is an obvious difference between soils from both areas for the content of TP and the grade in Pingba is significantly higher than that in Wudang. The values of CV for TP in soils from Wudang and Pingba are 16.21% and 25.42%, respectively. For OM, TK, HN and AK, the mean values for both planting bases belong to the sixth grade. The mean values of TN,

Table 2. Discrepancy of soil nutrients in karst areas.

	TN	OM	TP	TK	HN	AP	AK	pH	Urease	Catalase	Invertase
Wudang	2.09 (0.27)	2.31 (0.37915)	0.79 (0.09)	1.21 (0.07782)	2.55 (0.1866)	0.76 (0.07308)	0.59 (0.04675)	2.60 (0.24495)	0.49 (0.19602)	0.90 (0.03808)	22.86 (3.83656)
Pingba	2.46 (0.15684)	2.46 (0.25803)	1.87 (0.11448)	1.38 (0.0739)	3.00 (0.00429)	2.85 (0.13275)	1.00 (0.0999)	2.71 (0.18443)	0.74 (0.04703)	0.81 (0.04073)	24.36 (2.52819)
Guiyang (yellow soil)[a]	0.126	4.64	0.073	0.73	26	0.0315	/	5.1	/	/	/

*The unit of TN, OM, OP, TK is $g \cdot kg^{-1}$, the unit of HN, AP, AK is $mg \cdot kg^{-1}$. The value in () is the standard deviation. Wudang contain 5 soils, and Pingba contain 8 soils.
a: The Guiyang experimental dates in the table were collected from literature [11].

TP, TK, AP in Wudang and Pingba are higher than those in Guiyang (yellow soil) (Soil survey Office of Guizhou Province, 1994), and the mean values of OM and HN in both study areas are lower than those in Guiyang (yellow soil), as shown in Table 2.

Except for catalase, the indices (Table 2) for soils from Pingba are higher than those for soils from Wudang, especially for TP, AP and AK. According to independent sample T test, the contents of TP and AP are of extremely significant difference ($P < 0.01$) and the contents of AK and HN are of significant difference ($P < 0.05$). The differences of both areas may be caused by the planting age. The soil surface vegetation will be mineralized over the time. Coupled with the long term management (such as fertilization), more nutrients are accumulated. All of the above factors can change the physical and chemical characteristics of soil, water content and soil temperature. The status of nutrients in Pingba is better than that in Wudang. Compared with the Second National Soil Survey Classification Standard, the nutrients status for both areas are significantly poor. Thus, the conservation of soil nutrients is urgent in these karst areas.

As shown in Table 3, TN and HN have a significant correlation ($P < 0.05$), and the significant correlation ($P < 0.05$) was found between OM and pH. AP and HN have a significant correlation ($P < 0.05$), while there are significant correlations ($P < 0.01$) between TP and HN and between AP and AK. AP has a significant correlation with AK ($P < 0.01$). It is clear that these indices have direct or indirect correlation between each other. Scientific fertilization and strengthened human management are needed for development of the planting bases in karst areas.

3.2 Soil enzyme activities Relationship between soil nutrient contents and enzyme activities

The soil enzymes participate in the transformation processes of the soil nutrients elements directly.

To a certain degree, they can reflect the dynamic changes of the soil nutrients (Liang et al., 2013). The average activities of urease and invertase in soils from Pingba are higher than those in soils from Wudang, while the average activity of catalase in soils from Wudang is higher than that in soils from Pingba (Table 2). The values of CV for urease and invertase for Bingba are lower than those for Wudang (Table 1). The large coefficients of variation indicate a great degree of dispersion. Thus, there is a significant positive correlation between the stability of soil enzyme activities and the planting age.

3.3 Relationship between soil nutrient contents and enzyme activities

The catalysis of single enzyme is specific, and different enzymes have a fixed effect on one of the soil matrix. Soil enzymes reflect the transformation processes of organic compounds (Guan et al., 1984). A range of previous investigations indicated that there is a close relationship between the activities of soil enzyme and the contents of nutrients, because soil enzymes participate in the transformation of nutrients (Paz Jimenez M.D., 2002; Liu, 2011). As shown in Table 3, pH and catalase both have significant positive correlation with OM at 0.05 level. Guan et al. found that catalase can destroy the hydrogen peroxide generated by biochemical reactions in soil and reduce the damage to organism (Guan et al., 1984). As a kind of reductase in soil, the activity of catalase characterizes the transformation rate of organic matter (Lv et al., 2009). Invertase and TN have a significant positive correlation at 0.01 level, while TN and HN have a significant positive correlation at 0.05 level. Invertase can split sucrose into glucose and fructose. It is clear that invertase can increase the soluble nutrients in the soil. There is

Table 3. The coefficients of correlation between soil enzyme activities and fertility factors.

	TN	OM	TP	TK	HN	AP	AK	pH	Urease	Catalase	Invertase
TN	1	0.341	0.36	0.154	0.660*	0.326	0.33	0.251	0.049	0.005	0.764**
OM		1	0.271	0.346	0.435	0.143	0.353	0.658*	−0.083	0.595*	0.381
TP			1	0.342	0.746**	0.923**	0.781**	0.245	0.369	−0.248	0.353
TK				1	0.123	0.438	0.392	0.291	0.402	−0.033	0.052
HN					1	0.682*	0.561	0.477	0.118	−0.162	0.575
AP						1	0.740**	0.239	0.454	−0.428	0.264
AK							1	0.325	0.289	0.096	0.475
pH								1	−0.135	0.449	0.532
Urease									1	−0.306	−0.021
Catalase										1	0.337
Invertase											1

**Indicated 0.05 significant levels, **indicated 0.01 significant levels.

a close relationship between the activity of invertase and transformation of OM (Lv et al., 2009). Urease, as a kind of hydrolase, can catalyze the urea into ammonia. And the activity of urease can reflect the capacity of N supply in soil (Chen et al., 1985). OM is the source of C and N in soil nutrients (Zhou, 1987). The results show that urease has no significant correlation with the soil nutrients, probably because the content of soil organic matter is too low.

The results obtained in this investigation demonstrate that the activities of soil enzyme and nutrients have a close relationship. The OM not only affect catalase but also invertase. Soil enzymes play an important role in soil biochemical processes and formation of soil fertility, especially for the decomposition of OM and the activation process of N and P.

4 CONCLUSION

The soil nutrients and enzyme activities grow with the age of *Rosa sterilis* planting bases in karst areas, except for catalase. The plant of *Rosa sterilis* is beneficial to the restoration of vegetation and the rise in soil fertility. Soil enzymes are more sensitive to the change of soil environment compared with the soil nutrients, which are easily determined (Zhou, 1987). And soil enzymes can not only reflect the bioactivity of soil, but also indicate the transformation rate of soil nutrients (Guan, 1986). The main factors affecting the activities of catalase and invertase are OM and TN. Urease has no significant correlation with the soil nutrients. Previous investigations showed that OM is the source of C and N for soil nutrients. Urease is related to the transformation of C and N (Chen et al., 1985). Consequently, the content of OM is the key indicator impacting the soil enzyme activities of the planting bases in karst areas. And in this investigation, the activities of these three kinds of soil enzymes are not enough to evaluate the soil fertility, which should be improved in the future.

ACKNOWLEDGEMENT

This research was funded by Program for Changjiang Scholars and Innovative Research Team in University (No. RT1227), Agricultural Science and Technology Project of Guizhou Province (No. Qiankehe NY [2015] 3022-1) and Guizhou Normal University (Graduate Student Innovation Fund, 2014 (25)).

REFERENCES

Bruce, A.C. 2005. Enzyme activities as a component of soil biodiversity: A review. *Pedobiologia* 49(6): 637–644.

Cao, H. Sun, H. Yang, H. Sun, B. & Zhao, Q.G. 2003. A review soil enzyme activity and its indicate for soil quality. *Chinese Journal of Applied Environmental Biology* 9(1): 105–109.

Chen, E.F. Zhou, L.K. Qiu, F.Q. Yan, C.S. & Gao, Z.Q. 1985. Study on the essence of soil fertility. *Acta Pedologica Sinica* 22(2): 113–119.

De La Paz Jimenez, M. De La Horra, A. Pruzzo, L.R. Palma, M. 2002. Soil quality: A new index based on microbiological and biochemical parameters. *Biology and Fertility of Soils* 35(4): 302–306.

Guan, S.Y. 1986. *Soil enzyme and its research methods.* Beijing: Agricultural Press.

Guan, S.Y. Shen, G.Q. Meng, S.P. Yao, Z.H. & Min, J.K. 1984. Enzyme activity in main soil in china. *Acta Pedologica Sinica* 2(4): 368–381.

Institute of soil science of Chinese academy of sciences. 1978. *Soil physical and chemical analysis*: 132–136. Shanghai: Shanghai Scientific and Technical Publishers.

Liang, Y. Yang, H. Cao, J.H. Bo, Q.Z. Li, L. Fang, P.J. & Wang, K.R. 2013. Change of soil nutrient and enzyme activities under different land use. *Journal of Guangxi Normal University: Natural Science Edition* 31(1): 125–129.

Liu, S.J. Xia, X. Chen, G.M. Mao, D. Che, S.G. & Li, Y.X. 2011. Study Progress on Functions and Affecting Factors of Soil Enzymes. *Chinese Agricultural Science Bulletin* 27(21): 1–7.

Liu, Z.H. & Wolfgang D. 2007. *Karst dynamics and environment.* Beijing: Geology Publishing House.

Lv, C.H. Zheng, F.L. & An, S.S. 2009. The characteristics of soil enzyme activities and nutrients during vegetation succession. *Agricultural Research in the Arid Areas* 27(2): 227–232.

Soil survey Office of Guizhou Province. 1994. *The soil of Guizhou Province.* Guiyang: Guizhou science and Technology Press.

Tabatabai, M.A. & Dick, W.A. 2002. Enzymes in soil: Research and developments in measuring activities. In: Burns, R.G. & Dick, R.P. (Eds.), *Enzymes in the Environment*: Chapter 21: 567–596. New York, USA: Marcel Dekker.

Zeng, X.J. Liu, D.Q. & Zhu, S.M. 2005. Effects of different concentration of atrazine on soil catalase activity under three soil fertilities. *Hunan Agricultural Sciences* (6): 33–35.

Zhang, X.C. Zheng, F.L. An, J. & Yang, W.G. 2013. Relationship between soil enzyme activities and soil nutrient of a sloping farmland in the black soil region. *Journal of Arid Land Resources and Environment* 27(11): 106–110.

Zheng, Y. Yao, X.P. Jian, G. Guo, Z.D. & He. L.Y. 2013. The Current Situation and Outlook of Research and Development of *Rosa sterilis. Guizhou Forestry Science and Technology* 41(2): 62–64.

Zhou, L.K. 1987. *Soil enzymology*: 116–267. Beijing: Science Press.

Advances in Energy, Environment and Materials Science – Wang & Zhao (Eds)
© 2016 Taylor & Francis Group, London, ISBN 978-1-138-02931-6

Progress on the combined effects of PPCP residues in drinking water

Huan-huan Gao, Li-tang Qin, Liang-liang Huang, Yan-peng Liang,
Ling-Yun Mo & Hong-hu Zeng
Mining and Metallurgy and Environmental Science Experiment Center, College of Environmental Science and
Engineering, Guilin University of Technology, Guilin, Guangxi, China

ABSTRACT: Pharmaceuticals and Personal Care Products (PPCPs) that are widely used in the process of human production and life are discharged into surface water and groundwater. Thus, this is considered to be a new type of environmental pollutants and has drawn increased attention of people. Related research has mainly focused on the characterization of PPCP pollution to the environment and its impact. However, it is only confined to the detection and investigation of their types and levels. Only a few studies are available on the removal of PPCP from the water environment, as well as the study on the joint toxicity of mixture pollutants. The present study focuses on the reviews of the influence of PPCPs in drinking water on human health, and explores the impact of the joint toxicity of these pollutants on the ecological environment by using the sensitive biological indicator. Moreover, in order to protect the safety of drinking water, we put forward the detection and control methods of drugs in drinking water in accordance with the "new environmental law".

1 INTRODUCTION

In recent years, with the high-speed development and production of national economy, and living consumption, various kinds of pollutants are generated and discharged into the environment. Low-dose mixture pollutants in the environment are ubiquitous. Pharmaceuticals and Personal Care Products (PPCPs), which include beast medicine, agricultural medicine, human medicine, cosmetics, and other chemical substances, are closely related to human health (Daughton et al. 1999). As the current sewage treatment process is not designed for PPCPs, some PPCP substances in sewage treatment plants have not been removed effectively, so that PPCPs and their metabolites lead to numerous barriers such as biological degradation into natural water bodies, and even some groundwater supplies, polluting the water. They are also adsorbed on the activated sludge, and through fertilizers and other agricultural production activities eventually, they enter the environment. It has grown up to be a new type of pollutants in the water environment (Waiser et al. & Escher et al. 2011, Kaplan et al. 2013 & Jia et al. 2009). PPCPs present in the surface water and groundwater affect the safety of drinking water directly (Dai et al. 2015), and, therefore, their impact on human health is of particular concern.

Zhou et al. (2008) investigated the occurrence and source distributions of PPCPs in the Beijing North Canal. It was found that more than 60% of PPCP load in this water area is from new untreated emissions due to the presence of unlawful discharge of untreated sewage with not enough sewage treatment plant capacity. In the outskirts of Beijing, the sewage collection rate is only 50 percent and a large number of sewage enters directly into the aquatic environment, posing a threat to the surface water. Sun et al. (2014) studied 50 kinds of pharmaceuticals and personal care products in the wastewater treatment plant located in Xiamen, China, and found that 38 substances were detected in the wastewater treatment plant influent, among which the concentration of acetaminophen was highest, reaching up to 2963.5 ng/l. Thus, we should gradually increase the capacity of sewage treatment plants, and minimize or prevent the discharge of untreated sewage and wastewater directly into the environment, which is one of the effective methods for controlling PPCPs in the water environment.

China is the world's largest producer and exporter of chemical raw materials. Seventy percent of total drug production is accounted by antibiotic drug. In Western countries, antibiotic production accounts only for 30% of the total drug production (Ge et al. 2011). And in the past, these chemical raw materials industries in the production process consumed high energy consumption and generated heavy pollution at the expense of the environment in China, but now they have become the world's low-cost APIs "foundry". Some chemical and pharmaceutical companies in the wastewater discharge process have no discharge standards and other issues, which pollute numerous water bodies in China with pharmaceutical residues. Therefore, the overall levels of antibiotics and detection frequencies of China's surface water environment were higher than those of other countries. PPCPs are different from traditional Persistent Organic Pollutants (POPs) and do not have "toxic" "persistent" "accumulation" and "mobility" characteristics. Most of the PPCPs have strong polarity, are soluble in water and have weak volatility, which means that they will be released through the aqueous phase transfer and the food chain diffusion in the environment. Despite the short half-life and low concentration of PPCPs, human activities generate and discharge PPCPs continuously into the environment and present a "persistent" state. The substances are called as "virtual persistent chemical substances" by the scientists. Due to the environmental pollutants usually existing in the form of mixture, it becomes difficult to remove these pollutants by using the conventional treatment process. Some PPCP substances remain in the groundwater or drinking water and are probably harmful to human health. The study on the combined effects of PPCP mixtures on the environment is relatively scarce. Therefore, the study on the combined effects of PPCP mixture pollutants on the ecological environment has an important theoretical and practical significance.

PPCPs present in water substances is currently detected by using physical and chemical methods, such as high-performance liquid chromatography and tandem mass spectrometry, but rarely using biological detection methods. Liu et al. (2006) used freshwater luminescent bacteria, such as *Vibrio qinghaiensis* sp. Q67, as the indicator organism, and established the microplate luminescence testing method for determining the toxicity of environmental pollutants. This method can be used for the determination of the joint toxicity of pollutants in the water environment, and mixture joint toxicity models can be constructed. Nonetheless, most research is focused on pesticides, part of the heavy metal compounds and ionic liquid on the joint toxicity of luminescent bacteria, but studies of PPCP substances on the joint toxicity of luminescent bacteria are rare. Concentration Addition (CA) and Independent Action (IA) are two basic models for toxicity prediction, and the combined effects of mixture contain addition, antagonism, and synergy. However, CA and IA cannot predict the interaction between mixture substances. Evaluating synergy between chemical compounds that are related to the environment is a major challenge (Vasquez et al. 2014). Therefore, we should strengthen joint toxicity studies of these mixtures and develop a more comprehensive model.

2 THE PRESENT SITUATION AND THE COMBINED EFFECTS OF PPCPS IN DRINKING WATER

The main sources of drinking water are from surface water sources and groundwater, and groundwater is typically used as surface water supply. It is buried in the underground rock voids and can flow, so its management and monitoring are not easy. Sewage from living water, industries, hospitals, pharmaceutical companies, and aquaculture farms contains various kinds of pollutants. The sewage is transported to the wastewater treatment plant by means of the municipal pipe network process. Some sewage containing medicinal compounds will flow into the groundwater and thus pollute the water. This is one of the reasons why pharmaceutical residues are detected in drinking water.

2.1 *PPCPs in the drinking water survey*

Researchers have detected a variety of water bodies. Antibiotics have been detected in most rivers in China. Pearl River in Guangzhou section is seriously affected with higher concentrations of sulfa drugs and bisphenol A. Let us take bisphenol A for example: the highest level reached 782 ng/l in the Huangpu River, Suzhou River, and the Pearl River (Zhao et al. 2009). And bisphenol A concentrations up to 432 ng/l were detected in the Huangpu River upper untreated water used for drinking purpose, which is very likely due to untreated water contamination (Wang et al. 2012). It was pointed out that the drinking water in Nanjing in December 2014 had tetracycline, oxytetracycline, chlortetracycline and other six kinds of antibiotics, but the latest 106 national indicators of drinking water quality standards do not include antibiotics monitoring indicators.

Li et al. (2014) surveyed five waterworks and a groundwater sample in Beijing, China in 2013 to test 12 kinds of PPCPs in drinking water, and found that except tetracycline, trimethoprim, and 17β-estradiol, the other nine kinds of drugs had different levels of detection. Wang (2012) researched the drinking water sources in Anhui section and detected antibiotics, sulfonamide, and tetracycline at altered levels of detection in running water, groundwater, and surface water.

2.2 *Joint toxicity of PPCP mixture substances*

Contaminants in the environment are not of single type, but are in the form of mixture, and of the original type of PPCP pollutants as well. These mixtures can produce a joint toxicity action on aquatic organisms, soil, and human health itself. Zhu et al. (2014) studied the concentration levels of four antibiotics in the Qinghe River in Beijing, and aquatic organisms and pollution characteristics. They found that the coefficient of the harmful effect of ofloxacin, ciprofloxacin, and norfloxacin on algae and aquatic plants was larger than 1, and implied that these three kinds of antibiotics increased the environmental risks of algae and aquatic plants in Qinghe water. Yu et al. (2014) studied antibiotics wastewater characteristics and acute toxicity, and found that wastewater toxic can be attributed to the combined effects of toxic compounds. They assessed the toxicity of wastewater on *Scenedesmus obliquus* and *fees Vibrio* in a Chinese pharmaceutical factory, and found that the sewage treatment plant influent samples are toxic to these two indicator organisms. Although the entire wastewater treatment process has a high removal efficiency, antibiotic residues in water are still at high concentrations. Given the combined effects of mixture pollutants, the biggest challenge is the experimental design and analysis results of pharmaceutical joint actions in the ecosystem. In addition, the action of pharmaceutical transformation products such as biodegradation, photodegradation products, and metabolites on humans and animals has scarcely been addressed by toxicity studies.

Johnson et al. (2015) evaluated ciprofloxacin, sulfamethoxazol, trimethoprim, and erythromycin levels in some rivers in Europe, and found that with the widespread use of antibiotics, antibiotic concentrations in rivers continued to increase ranging from 0.1 to 1 μg/l in the maximum exposure area. Besides, toxicity of the mixture may enhance this risk as currently there is a lack of direct toxicity measurement of antibiotics in the water test method. Liu et al. (2008) used chlorophenol mixtures to study the joint toxicity on freshwater luminescent bacteria Q67, and found the acute toxicity of these substances on photobacterium, which was related to the number of chlorine atoms and chlorophenol toxicity of binary and ternary mixtures that were obviously higher than that of a single compound. An et al. (2014) studied three kinds of quorum-sensing inhibitors with typical sulfa antibiotic joint toxicity on *Vibrio fischeri*. They found that the combined effects of furan ketone compounds with sulfa antibiotic mixture were synergistic and additive under the binary toxicity ratio. They also found that joint actions of pyrrole ketone and pyrrole compounds with sulfa antibiotics were additive and antagonism, respectively. These studies indicate that there may be some kind of interaction between mixture pollutants that makes the joint toxicity to increase or decrease. Therefore, the exploration of the joint toxicity of PPCPs in drinking water is necessary.

3 DISCUSSION AND RESEARCH PROSPECTS

Pharmaceutical residues have attracted widespread national attention through their trace detection in drinking water in many countries. Some scholars have previously undertaken research on how to effectively remove PPCPs. T. Alvarino et al. (2015) studied the removal of PPCPs from the supernatant using the sewage nitrification anammox process by increasing the hydraulic retention time and PPCP molecules into the sludge particles. And they showed that by increasing the rate of nitrosation, ibuprofen, bisphenol A and triclosan can be removed, and the removal of erythromycin is closely related to the anaerobic ammonia oxidation reaction rate.

The new "environmental law" for the first time has included "protect public health" into general and clear "conservation priority" principle. Water is the basis of life and directly related to human health. In support of government policies, strengthening the water monitoring is significant in order to more comprehensively control the drinking water quality. The state should gradually improve the drinking water quality standards in the type of monitoring indicators. Investigation of water pollutants in the environment primarily involves the survey of PPCPs. Overall, levels of these substances and detection frequencies are high and should still be of important concern (Wang et al. 2014). At present, we should optimize and upgrade PPCP detection and analysis technology, and optimize the use of these technologies to research contaminants in drinking water. In addition, research on the joint toxicity of PPCP compounds in drinking water and establishment of a reliable toxicity prediction model is quite urgent.

ACKNOWLEDGMENT

This work was supported by the National Natural Science Foundation of China (21407032), the Scientific and Technological Research Projects of Guangxi Universities (ZD2014059), and the Guangxi Higher Education Innovation Team and Excellent High-level Academic Projects. The authors gratefully acknowledge their financial support.

REFERENCES

Alvarino, T., S. Suarez, E. Katsou, et al. Removal of PPCPs from the sludge supernatant in a one stage nitration/anammox process [J]. Science Direct, 2015, 68: 701–709.

An, Q.Q., Yao, Z.F., Gu, Y.F., et al. Preliminary joint toxicity and its mechanism of antibiotics and sulfa quorum sensing inhibitors on luminescent bacteria [J]. Environmental Chemistry. 2014, 33(12): 2068–2075.

Dai, G.H., Wang, B., Huang, J., et al. Occurrence and source apportionment of pharmaceuticals and personal care products in the Beiyun River of Beijing, China [J]. Chemosphere. 2015, 119: 1033–1039.

Daughton, C.G., Ternes, T.A. Pharmaceuticals and personal care products in the environment: Agents of subtle change? Environmental Health Perspectives, 1999, 107: 907–938.

Escher, B.I., Baumgartner, R., Koller, M., et al. Environmental Toxicology and risk assessment of pharmaceuticals from hospital wastewater. Water Research, 2011, 45(1): 75–92.

Ge, L., Chen, J.W., Zhang, S.Y., et al. Photodegradation of fluoroquinolone antibiotic gatifloxacin in aqueous solutions [J]. Chin Sci Bull, 2010, 55: 1495–1500.

Jia, A., Hu, J.Y., Sun, J.X., et al. Pharmaceuticals and personal care products in the environment [J]. Advance in chemistry. 2009, 3: 389–399.

Johnson Andrew, C., Virginie Keller, Egon Dumont, et al. Assessing the concentrations and risks of toxicity of the antibiotics ciprofloxacin, sulfamethoxazole, trimethoprim and erythromycin in European rivers [J]. Science of the Total Environment. 2015, 511, 747–755.

Kaplan, S. Review: Pharmacological Pollution in Water. Critical Reviews in Environmental Science and Technology, 2013, 43(10): 1074–1116.

Li, X.F., Yuan, S.Y., Jiang, X.M., et al. Detect 6 samples of drinking water and 12 species of pharmaceuticals and personal care products by Liquid chromatography-tandem mass spectrometry [J]. Environmental Chemistry. 2014, 33(9): 1573–1580.

Liu, B.Q., Ge, H.L., Liu, S.S. Determination of environmental pollutants on *Qinghai Vibrio* microplate luminescence intensity inhibit chemiluminescence [J]. Eco toxicology. 2006, 1(2): 186–191.

Liu, Z.T., Li, Z.X., Li, Z., et al. Chlorophenols joint toxicity on freshwater Q67 of luminescent bacteria [J]. Environmental Science, 2008, 21(2): 115–119.

Sun, Q., Lv, M., Hu, A.Y., et al. Seasonal variation and removal of pharmaceuticals and personal care products in a wastewater treatment plant in Xiamen, China [J]. Journal of Hazardous Materials. 2014, 277: 69–75.

Vasquez, M.I., A. Lambrianides, M. Schneider, et al. Environmental side effects of pharmaceutical cocktails: What we know and what we should know [J]. Journal of Hazardous Materials, 2014, 279: 169–189.

Waiser, M.J., Humphries, D., Tumber, V., et al. Effluent-dominated streams. Part 2: Presence and possible effects of pharmaceuticals and personal care products in Wascana Creek, Saskachewan, Canada. Environmental Toxicology and Chemistry, 2011, 30(2): 508–519.

Wang, D., Sui, Q., Zhao, W.T., et al. The environmental research drugs and personal care products in China Surface Water [J]. Chinese Science Bulletin. 2014, 59(9): 743–751.

Wang, H.X., Zhou, Ying, Wang, X., et al. Shanghai Water Environment screening and evaluation of major phenolic pollutants [J]. Fudan University (Medical Sciences). 2012, 39(03): 231–237.

Wang, Y. Distribution of contamination in drinking water sources and 9 antibiotics section in Anhui sewage. Hefei: a master's degree thesis of Anhui Agricultural University. 2012.

Yu, X., Zuo, J.E., Tang, X.Y., et al. A combined evaluation of the characteristics and acute toxicity of antibiotic wastewater [J]. Ecotoxicology and Environmental Safety. 2014, 106: 40–45.

Yu, X., Zuo, J.E., Tang, X.Y., et al. Toxicity evaluation of pharmaceutical wastewater using the alga Scenedesmus obliquus and the bacterium Vibrio fischeri [J]. J. Hazard. Mater. 2014, 266: 68–74.

Zhao, J.L., Ying, G.G., Wang, L., et al. Determination of phenolic endocrine disrupting chemicals and acidic pharmaceuticals in surface water of the Pearl Rivers in South China by gas chromatography-negative chemical ionization-mass spectrometry [J]. Sci Total Environ, 2009, 407(2): 962–974.

Zhou, X.F., Zhang, Y.L., Dai, C.M. Research advance in drinking water treatment to remove pharmaceuticals and personal care products [J]. Environment and Health, 2008, 25(11): 1024–1027.

Zhu, L., Zhang, Y., Qu, X.D., et al. Beijing Qinghe water bodies and aquatic organisms in vivo antibiotic pollution characteristics [J]. Environmental Science. 2014, 27(2): 139–146.

Advances in Energy, Environment and Materials Science – Wang & Zhao (Eds)
© 2016 Taylor & Francis Group, London, ISBN 978-1-138-02931-6

The progress of constructed wetland soil organic carbon mineralization

Xuefen Li, Yanli Ding, Shaoyuan Bai, Lin Sun & Junjie You
Guangxi Scientific Experiment Center of Mining, Metallurgy and Environment, Guilin University of Technology, Guilin, China

ABSTRACT: Water, temperature, soil depth, and redox potential have certain effects on soil organic carbon mineralization in wetland. Besides, wetland and wetland vegetation of different types also have an impact on soil organic carbon mineralization. Wetland clog is not only associated with hydraulic load and pollution load, but also connected with the organic carbon mineralization rate.

1 INTRODUCTION

Constructed wetlands can also be used to deal with a variety of wastewater, including urban wastewater, initial rainwater, agricultural runoff, industrial wastewater, paper wastewater, mine drainage, landfill seepage, and wastewater treatment plant effluent treatment. They usually consist of a gravel matrix, in which water-loving plants are grown and through which wastewater flows (M.F. Shamim, et al. 2013). Over time, the gravel matrix becomes clogged as a result of particulate accumulation and biofilm growth. Constructed wetland is one of the wastewater treatment technologies according to the synergistic effect of the natural wetland ecosystem involving the physical, chemical, and biochemical reaction, generally by the constructed wetland substrates and growth in the water plant, which constitutes a unique soil (matrix)–plant–microbial ecosystem. The artificial wetland matrix is a major storage site of wetland ecosystem, and soil organic carbon mineralization mainly refers to the soil natural organic matter and exogenous organic matter, such as plant litter, residual root, and organic materials. Under the action of microorganisms, releasing carbon dioxide to the atmosphere (CO_2) in the decomposition process (Huang Y. et al. 2002) is a key step and an important part of the soil ecosystem carbon cycle. In terrestrial ecosystems, wetlands are an important part of the carbon cycle in biochemical processes. Soil organic carbon mineralization is an important part of the internal and external material circulation related to the wetland system. It has an important significance for the regional carbon distribution and carbon balance (Chimner R.A., et al. 2003).

Accurate identification of the "mineralizable" soil C pool is essential, as it is an important component in modeling soil C dynamics to changing environmental factors (Alessandro Saviozzi. et al. 2014). Soil microbial decomposer of organic matter is the main microbial stability of organic carbon in the soil. Some studies have shown that microorganisms synthesize the formation of organic carbon in promoting the role (Kogel Knabner I. 2002; Lutzow M., et al. 2006; Six J., et al. 2006; Wolters V. 2000). The rate at which the soil organic matter is degraded to CO_2 depends mainly on the nature of its interactions with the soil matrix [3]. Among soil properties, texture was found to influence the rate of the first phase of the C mineralization process. The fine-texture of urban soils showed the lowest rate constants of C mineralization and is consequently, the lowest C losses. Lastly, results also indicate that the extent of the easily mineralizable organic C may affect the mineralization rate of the slow C pool in urban soils (Alessandro Saviozzi. et al. 2014). Research conducted in other countries has focused on the use of the model of changing trend of soil organic carbon for prediction (Ise T, Dunn A.L., et al. 2008; Weindorf D.C., et al. 2010).

2 EFFECT OF VARIOUS CONDITIONS ON THE MINERALIZATION OF ORGANIC CARBON

Many researchers have studied the effects of various conditions on the mineralization of organic carbon. Yang Jisong (2008), who used a closed culture method, studied the wetland soil organic carbon mineralization dynamics of *Deyeuxia angustifolia*, and discussed the effects of temperature and moisture conditions on organic carbon mineralization. The results showed that the organic carbon mineralization in wetland soil organic carbon during the early stage of culture (0~2 d) mineralization rate is higher, the mineralization rate decreased, the total amount of mineralization increased by 60%~210% (75% WHC) and 30%~200% (flooding) at the temperature rise of 10 degrees Celsius. The first-order kinetic equation can well describe the mineralization of soil organic carbon dynamics of wetland, and shows a parabolic correlation between organic carbon mineralization and soil depth obviously. Therefore, the soil depth and temperature have a significant effect on the soil organic carbon mineralization of wetland, and the effect of soil moisture on the mineralization of organic carbon was not significant. Yang Gairen et al. (2009) studied the effects of water content on the redox potential and carbon mineralization of wetland sediments, and found that the relationship between mineralization rates and redox potentials (Eh) was well fitted with second-order parabolic equations ($P < 0.05$). Mineralization rates and accumulative amount of organic C showed a positive correlation with Eh up to 300 mV. However, a significant negative correlation was observed when Eh increased above 300 mV. Shao Xuexin. et al. (2011) studied the Hangzhou Bay Wetland Natural tidal flat and wetland reclamation soil organic carbon content and its distribution pattern, to reveal the effects of vegetation succession, the invasion of alien species and reclamation on soil organic carbon distribution. Zhu Sixi. et al. (2014) attempted to measure the contents of Substrate Organic Carbon (SOC) in four different seasons in a full-scale constructed wetland to investigate the effects of plant diversity on the seasonal dynamics of SOC. Guiping Fu (2013) studied the constructed wetland system operated under a high hydraulic loading condition over a period of 1 year until the wetland medium was completely clogged. He found that among all the components of organic matter, labile organic matter and fulvic acids were the leading factors causing wetland clogging, with the former playing the most prominent role in the process.

3 INFLUENCE OF HYDRAULIC LOADING AND ORGANIC CARBON MINERALIZATION ON THE CLOGGING OF WETLAND

3.1 Influence of hydraulic loading

Matrix clogging is one of the common problems of wetland wastewater treatment system, which will shorten the life of wetlands and removal efficiency

to reduce pollutants. The essence of the blockage effective porosity reduction process mainly involves the matrix layer with unfiltered accumulation material and permeability coefficient decrease (Zhu, 2009). It also includes 4 aspects: inorganic plugging, plugging, microbial organic and hydraulic loading (Ni, 2012). As CO_2 is trapped and nitrification is inhibited, O_2 consumption can solely be attributed to C mineralization (Marco, 2007). According to the study by Xu Qiaoling et al. (2014), hydraulic load, organic load and suspended solids load are the three factors of constructed wetland clogging that influenced the degree of different bindings. The organic load is the main factor affecting the substrate clogging. Under the same total organic matter content and total oxygen demand situation, with low hydraulic load, a high concentration of organic matter of the wetland system is more prone to clogging than with high hydraulic load, which results in low organic matter concentrations of wetland system. Anna Pedescoll (2011) studied the effect of two types of primary treatment (Hydrolitic Up-Flow Sludge blanket (HUS)) reactor and conventional settling) and the flow regime (batch and continuous) on clogging development in subsurface flow constructed wetlands, and determined clogging indicators (e.g. accumulated solids, hydraulic conductivity and drainable porosity) in an experimental plant with three treatment lines. Song Zhixin (2014) discussed the relationship between the flow distribution and hydraulic loading, the matrix material, layering different size matrix fill charge, and plant root distribution.

Table 1 (Wang, 2014) and Table 2 (Xiong, 2011) present the study of the blockage of the constructed wetland under different hydraulic loading conditions. The results also show that the hydraulic load has a great effect on substrate clogging in constructed wetlands. The low hydraulic load matrix is not easy to be blocked, and the content of accumulation in the matrix decreases with the increase in the substrate depth.

3.2 Organic carbon mineralization

The degree of constructed wetland is proportional to the pollution load of wetland. However, in

Table 1. Predictions on blockage with different hydraulic loading conditions.

Hydraulic loading Cm/d	Content of organic matter‖g/kg	Content of accumulated organic matter‖g/kg	The capacity of organic matter‖g/kg	The accumulated rate of organic matter‖g/(kg·m)
50	4.361	3.509	2.315	0.5013
100	5.5753	4.721	1.103	0.6744
150	6.676	5.824	0	0.832
200	16.73	15.88	0	2.268

Table 2. Accumulated matter content in vertical-flow constructed wetlands.

Hydraulic loading (m/d)	Accumulated matter content Mg/L^{-1}		
	Organic	Inorganic	Total
0.3	997	10911	11908
0.6	10045	55355	65400
0.9	9840	69803	79643
1.2	1202	18398	19600

recent years, Knowles in his survey found that the French 53 artificial wetlands pollutant load was 17 g/m²d and the hydraulic loading was 0.06 m/d, with no jam phenomenon. He also found that the pollutant load was low (3 g/m²d) and the hydraulic load was high (0.3 m/d) in 21 individual workers wetland in Germany, resulting in the phenomenon of serious congestion. With no obvious blocking of French wetland, another study has found that the organic carbon mineralization rate reached 60% in the bed body filler, and the system showed a good performance of the hydraulic load (Paul, 2010). Therefore, blocking associated with the constructed wetland soil organic carbon mineralization rate, ensuring sufficient organic carbon mineralization, is an effective way to solve the problem of clogging.

4 CONCLUSION AND PROSPECT

Many scholars at home and abroad have studied on different ecological systems of soil microbial biomass carbon, nitrogen content, the transformation and its influencing factors. However, there are only a few reports on the parameters calculated according to the special model, which allowed us to detect in the constructed wetland soils a pool of easily biodegradable organic compounds that was exhausted in a very short period of time, and in the wetland system, the organic carbon mineralization of microorganisms has a great impact on the substrate clogging.

ACKNOWLEDGMENTS

This work was funded by the National Natural Science Foundation of China (No. 51408147, 51168012, 41404116) and the Guangxi Natural Science Foundation (No. 2014GXNSFBA118234), and supported by the project of high-level innovation team and outstanding scholar in Guangxi colleges and universities.

REFERENCES

Alessandro Saviozzi, Giacomo Vanni, Roberto Cardelli. 2014. Carbon mineralization kinetics in soils under urban environment. Applied Soil Ecology, 73 (2014) 64–69.

Alessandro Saviozzi, Giacomo Vanni, Roberto Cardelli. 2014. Carbon mineralization kinetics in soils under urban environment. Applied Soil Ecology, 73 64–69.

Anna Pedescoll, Angélica Corzo, et al. 2011. The effect of primary treatment and flow regime on clogging development in horizontal subsurface flow constructed wetlands: An experimental evaluation. water research 4 5:3579–3589.

Chimner R.A., Cooper D.J., 2003. Carbon dynamics of pristine and hydro logically modified fens in the southern Rocky Mountains. Canadian Journal of Botany, 81: 477–491.

Guiping FuJiahong Zhang, et al., 2013. Medium clogging and the dynamics of organic matter accumulation in constructed wetlands. Ecological Engineering 60:393–398.

Ise T., Dunn A.L., Wofsy S.C., et al. 2008. High sensitivity of peat decomposition to climate change through water-table feedback. Nature Geo science, 1(11):763–766.

Kogel Knabner I., 2002. The macromolecular organic composition of plant and microbial residues as inputs to soil organic matter [J]. Soil Biology & Biochemistry, 34:139–162.

Lutzow M., Kogel-Knabnerl, Eksehrnitt K, et al. 2006. Stabilization of organic matter in temperate soils: mechanisms and their relevance under different soil conditions-are view [J]. European Journal of soil Science, 57(4):426–445.

Marco Grigatti Manuel Dios Pe'rez, et al. 2007. A standardized method for the Determination of the intrinsic carbon and nitrogen mineralization capacity of natural organic matter sources. Soil Biology & Biochemistry, 39 (2007) 1493–1503.

Ni Zheng, Wang Cui-hong, Xie Ke-jun, guo Qi, Yu Qing-yi, Song Nan. 2012 Prevention and renew method for system blockage in artificial wetland. Hunan Agricultural Sciences, (17): 73–76.

Paul Knowles, Gabriela Dotro, Jaime Nivala, et al. 2010. Clogging in subsurface-flow treatment wetlands: Occurrence and contributing factors [J], Ecological Engineering, 37(2), 99–112.

Shamim, M.F., M. Bencsi, R.H., Morris, M.I., Newton. 2013. MRI measurements of dynamic clogging in porous systems using sterilised sludge. Microporous and Mesoporous Materials 178:48–52.

Shao Xuexin, Yang Wenying, Wu Ming, Jiang Keyi. 2011. Soil organic carbon content and its distribution pattern in Hangzhou Bay coastal wetlands. Chinese Journal of Applied Ecology, Mar 22(3):658–664.

Six J., Frey S.D., Thiet R.K., et al. 2006. Bacterial and fungal contributions to carbon Sequestration imago-ecosystems. Soil Science Society America Journal, 70:555–569.

Song Zhixin, ding Yanli, Xie Qinglin, Bai Shao yuan, You Shaohong. 2014. Research Progress in Relationship between Flow Field Distribution and Clogging in Subsurface Flow Constructed Wetlands. wetland science, 12(5):677–682.

Wang Xiaomao, Xu Qiaoling, Cui Lihua, Li Guowan. 2014. Effects of Hydraulic Loading of Vertical Flow Constructed Wetland Clogging. Agricultural Science & Technology, 15(11): 2030–2034.

Weindorf D.C., Zhu Y. 2010. Spatial variability of soil properties at Capulin Volcano, New Mexico, USA: Implications for sampling strategy. Pedosphere, 20(2): 185–197.

Wolters V. 2000. Invertebrate control of soil organic matter stability. Biology and Fertility of Soil, 31:1–19.

Xiong Zuofang, Zhou Yunxin, Xian Ping Chen Weiqiang Wu Qi. 2011. Hydraulic load on soil clogging on vertical flow constructed wetland. Technology of Water Treatment. 37(7):34–36.

Xu Qiaoling, Cui Lihua, Zhang Ling. 2014. Load working together on soil clogging on vertical flow constructed wetland. Environmental Science & Technology, 37(6 N):1–5.

Yang Gai-ren, Tong Cheng-li, Xiao He-ai, Wu Jin-shui. 2009. Effects of Water Content on Redox Potential and Carbon Mineralization of Wetland Sediments. Environmental Science. 30(8):2382–2386.

Yang Jisong, Liu Jingshuang, Sun Lina. 2008. Effects of temperature and soil moisture on wetland soil organic carbon mineralization. Chinese Journal of Ecology, 27(1): 38–42.

Zhu Sixi, Chang Jie, Ge Ying. et al. 2014. Effects of plant diversity on substrate organic carbon dynamics in a vertical subsurface flow constructed wetland. Guangdong Agricultural Science. 1:46–51.

Zhu Jie, Chen Hong-bin. 2009. Discussion on constructed wetlands clogging. China Water & Waste water, 25(6):24–28.

Advances in Energy, Environment and Materials Science – Wang & Zhao (Eds)
© *2016 Taylor & Francis Group, London, ISBN 978-1-138-02931-6*

The economic analysis of the household Combined Heat and Power system

Xiang Zhou, Weiqing Xu, Yan Shi, Jia Wang & Xian Wu
School of Automation Science and Electrical Engineering, Beihang University, Beijing, China

ABSTRACT: In this paper, a method for the economic analysis of natural gas Combined Heat and Power (CHP) system was proposed. The CHP system could not only provide living hot water, but also supply electricity for families. To evaluate the economy of the CHP heating device, a model for the economic analysis of CHP heating devices was also proposed. Based on this model, the heating device of the CHP system was compared with electric water heaters, gas water heaters, and heat pump water heaters. It was found that the new heating device was more economic than the other three types of heaters. This paper can be used as a reference for the optimization of the device.

1 INTRODUCTION

With the rapid development of economy, people's demand for hot water has become increasingly larger (Tian, 2012). In the past, there were different types of heater products. Nowadays, the development of household water heater has made a great progress, from the initial electric water heater, gas water heater to solar heater and air source heat pump water heater (Lu, 2014). With the more and more serious energy crisis, "low carbon" has gradually become the first choice when people choose home appliances (He, 2012).

Among the above four types of heater products, electric water heaters and gas water heaters are cheaper than solar or heat pump water heaters (Feng, 2012). But solar and heat pump water heaters are superior to electric or gas water heaters in energy saving (Kuang, 2003). Compared with other conventional systems, the CHP system has obvious advantages. For example, the system can reduce energy loss and realize cascade utilization of energy (Ye, 2008). Furthermore, the CHP system can strengthen environmental protection and improve air quality (Wang, 2006).

To improve the economy of heating device, a small CHP system was proposed in this paper, which could provide not only hot water, but also electricity simultaneously. Therefore, the small natural gas CHP system will become the development trend in the future. To illustrate the advantages of the device, its economy was studied, and this paper can be used as a reference for the optimization of the device.

2 THE WORKING PRINCIPLE OF THE CHP SYSTEM

Natural gas CHP system is an advanced technology for energy saving. It uses the combustion of natural gas to produce high-grade electric energy, and low-grade thermal energy, which is discharged from the power equipment, is fully utilized for heating. This system realizes the energy cascade utilization (Xu, 2003). A schematic diagram of the system is shown in Figure 1.

As shown in Figure 1, natural gas acts as the heat source. It expands in the engine and then generates electricity in the alternator, providing power for household appliances. At the same time, by recycling waste heat from the engine's cooling system, the cooling system and the exhaust gas, the low grade heat can be finally utilized for the production of hot water.

In the conventional system, natural gas is always used to burn directly for heating the water. But the CHP system uses the waste thermal energy for heating the water, and it realizes energy cascade utilization.

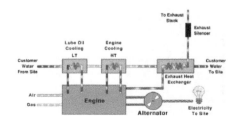

Figure 1. Schematic of the cogeneration system.

This system can not only meet the domestic hot water needs, but also provide electricity. The overall rate of energy utilization is improved.

3 METHOD OF CALCULATION

In this paper, the CHP device is mainly oriented to families, providing household hot water and a part of the power.

Supposing the demand of hot water for every family is 150 L per day, the initial temperature of water is 15°C, and the temperature of hot water is 60°C, the total quantity of heat for heating the water can be calculated as follows:

$$Q = Cm\Delta T = 4.2 \times 150 \times 45 = 28350 \; kJ \qquad (1)$$

Table 1 presents the efficiency of different water heaters, which can be used to calculate the consumption of electricity or natural gas for producing 150 L water (Xia, 2014).

The CHP device can produce not only hot water, but also electricity. The electricity produced can be sold to the state grid or used by the household. Due to the unknown thermal efficiency of the cogeneration device, the annual running cost of the CHP device can be obtained according to different thermal efficiencies, and compared with

the electric water heater, gas water heater and heat pump water heater. Cost comparison for different thermal efficiencies is shown in Figure 2.

Besides, the formula for calculating the annual running cost of the cogeneration device is given as follows:

$$A = B - C \qquad (2)$$

where A is the annual running cost; B is the price of natural gas; C is the price of electricity.

As shown in Figure 2, when the thermal recovery efficiency of the CHP device reaches 20%, the annual operating cost can be equal to that of the electric water heater. When the thermal recovery efficiency reaches 40%, the annual running cost can be equal to that of the gas water heater. When the thermal recovery efficiency reaches 80%, the annual running cost can be equal to that of the heat pump water heater. Thus, when the thermal efficiency of the CHP device reaches 50%, the annual running cost is reduced to a very low level. According to the experience, the thermal recovery efficiency of the CHP system can be more than 50% (Wang, 2002; Shao, 2014; Li, 2014), so the running cost of the device can meet the requirements of the market.

As the acquisition cost of the water heaters is different, the total cost of the different water heaters can be calculated, and compared with the cost of

Table 1. Comparison of several water heaters.

Category	Calorific value	Efficiency/(%)	Consumption of electricity (natural gas)	Running cost RMB per day
Electric water heater	3600 (kJ/kwh)	95 (heat)	8.29 kwh	4.145
Gas water heater	38500 (kJ/m³)	80 (heat)	0.92 m³	2.1
Heat pump water heater	3600 (kJ/kwh)	400 (heat)	1.96 kwh	0.98
CHP device	38500 (kJ/m³)	30 (electricity) +50 (heat)	1.47 m³	0.99

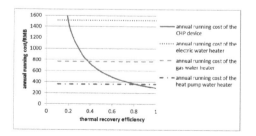

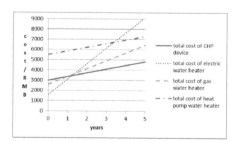

Figure 2. Comparison of the annual running cost of different water heaters.

Figure 3. Comparison of the total cost of different water heaters.

Table 2. Comprehensive performance of various water heaters.

Category	Acquisition cost/yuan	Volume	Energy saving	Span/year	Economy
Electric water heater	1600	Large	Low	10	Bad
Gas water heater	2600	Small	Medium	15	Medium
Heat pump water heater	5500	Large	High	15	Good
CHP device	3000	Medium	High	15	Good

the CHP device, where the heat recovery efficiency of the CHP device is about 50%.

The total cost can be calculated as follows:

$$Q = D + A \times N \qquad (3)$$

where Q is the total cost; D is the acquisition cost; A is the annual running cost; and N is the number of years. The comparison is shown in Figure 3.

From Figure 3, it can be seen that the total cost of the CHP device will drop below that of all kinds of water heaters after a working period of 2.5 years. This can be attributed to its low-cost operation, which has the best economic benefit. Among the four devices, heat pump water heater has the most expansive acquisition cost and the mildest increasing trend, while on the contrary, electric water heater has the cheapest acquisition cost and the highest running cost. Compared with the above two water heaters, the CHP device and gas water heater have an obvious advantage in the total cost. Taking the very low running cost of the cogeneration device into consideration, it is fit for the household, and has a good market prospect.

4 ANALYSIS OF COMPREHENSIVE PERFORMANCE

Each water heater has its own advantages and disadvantages, and their comprehensive performance needs to be considered when a water heater is chosen. The comprehensive performance of various water heaters was compared with the CHP device, and the results are summarized in Table 2 (Cai, 2006; Liu, 2015; Li, 2015; Yin, 2014; Xiao, 2003; Deng, 2014).

From Table 2, the acquisition cost of the CHP device is found to be 45% lower than that of the heat pump water heater. And the acquisition cost of the CHP device is similar to that of the gas water heater. Compared with the electric water heater, the CHP device has a great advantage in energy saving. Besides, it has some other advantages such

as moderate volume, convenient installation, long service span, and good economic benefit. So, the CHP device has a very good market value.

5 CONCLUSION

In this paper, the CHP device was compared with electric water heater, gas water heater, heat pump water heater, respectively. Their running cost and the total cost were analyzed, and their comprehensive performances were compared. In conclusion, the acquisition cost of the CHP device is lower than that of the heat pump water heater. Compared with the heat pump water heater, there are few differences in energy saving. The CHP device has more advantages in energy saving than the electric water heater and the gas water heater. In the long run, the cost of the CHP device is lower than that of the electric water heater and the gas water heater. Therefore, the comprehensive performance of the CHP device is more economical.

REFERENCES

Cai Ying, Qin Chaokui. Technological Progress of Gas Water Heater [J]. Gas and Heat. 2006(11).
Caiyou Wang, Weijie Cao. Cogeneration device of natural gas engine [J]. Shanghai Energy Conservation. 2002.
Daoming Xia. Research on multi energy water heater [J]. Solar Energy. 2014.
Feng Baile, He Haixia. Analysis on the energy conservation of household air-source heat pump water heater [J]. Shanxi Architecture. 2012(24).
Huaguo Shao, Changyuan Xiong. Discussion on the Development of Combined Heat and Power Turbines in China [J]. Hydropower and New Energy. 2014(09).
Li Pei, Wang jian. Caculation of energy comprehensive utilization efficiency of distributed natural gas energy system. Heating Ventilation and Air Conditioning. 2014.
Li Yang. Analysis of solar water heaters and other kinds of water heater for advantages and disadvantages [J]. China Anti-Counterfeiting Report. 2015(02).

Lu qing. How to choose household water heater [J]. China Public Science. 2014(01).

Naiqiang Xu, Lichang Lei. Gas turbine cogeneration—an effective form of energy utilization [J]. Movable Power Station and Vehicle. 2003.

Nanfang Xiao. Several advantages of gas water heater. Modern appliances. 2003.

Shengnan Yin. Analysis of the characteristics for air source heat pump water heater [J]. Journal of Green Science and Technology. 2014(09).

Tian Yajie, Liu Rong. Market Survey and Analysis of Gas Water Heaters in Beijing [J]. Gas & Heat.2012(05).

Wang Zhenming. New development of CHP and distributed energy in China [J]. Journal of Shenyang Institute of Engineering (Natural Science). 2006(01).

Xiaojin Liu, Chao Xu. Economic analysis of the auxiliary heating solar heat pump hot water system [J]. Energy Research and Utilization. 2015(01).

Xiuyin Deng. The overview about selection of water heater [J]. Education Teaching Forum. 2014(22).

Ye Weizhong. Energy Saving and Development of Distributed Energy [J]. Journal of Shenyang Institute of Engineering (Natural Science). 2008(02).

Yincheng He. The Study on the International Competitiveness of China's Household Electrical Appliance Industry in the Perspective of Low-carbon Economy. Anhui University. 2012(10).

Yuhui Kuang. Solar Heat Pump Water Heater [J]. Solar Energy. 2003(04).

Advances in Energy, Environment and Materials Science – Wang & Zhao (Eds)
© *2016 Taylor & Francis Group, London, ISBN 978-1-138-02931-6*

Isolation and identification of biphenyl-biodegrading strain

Wenjuan Liang, Mingfen Niu & Huan Liu
School of Municipal and Environmental Engineering, Shenyang Jianzhu University, Shenyang, Liaoning, China

Yong Yan
School of Civil Engineering, Shenyang Jianzhu University, Shenyang, Liaoning, China

ABSTRACT: Using biphenyl as selective medium, one biphenyl-degrading strain which could use biphenyl as sole carbon source was isolated from the sewage sludge of biological treatment process in sewage treatment plant, named G-2. Through observing the morphological characteristics of the strain and analyzing 16S rRNA gene sequence, the phylogenetic tree was constructed. The results showed that strain G-2 could tolerate biphenyl concentration of 2 g/L, and had high sequence homology with Paracoccus. Preliminarily, G-2 was classified into *Paracoccus*. This paper provides theoretical basis for biodegradation of biphenyl, which makes for the enrichment and perfection of the library on biphenyl degrading bacteria.

1 INTRODUCTION

Biphenyls are the natural components of coal tar, crude oil, and natural gas. Because of the incomplete combustion of coal and petroleum, biphenyls can be emitted into the environment through the form of exhaust gas, causing long-term pollution. Biphenyls are instable, and can be gradually chloridized to Polychlorinated Biphenyls (PCBs) in the presence of certain appropriate catalysts. Polychlorinated Biphenyls (PCBs) are one of 12 persistent organic pollutants, and a kind of serious carcinogenic and teratogenic agents. Microbial degradation is the most promising method for biphenyls pollution abatement. Previous studies of biphenyl degrading strains included *Achromobacter* (Abramowicz 1990), *Sphingomonas*, *Comamonas* (Potrawfke 1998), *Burkhoderia* (Gibson 1973), *Beijerincksa* (Bumpus & Tien 1985) and white rot fungus (Zeddel 2001) etc., and these strains had a certain degree of degradation to biphenyls and biphenyl derivatives. But the degradation rates were generally low, and the tolerance concentrations of biphenyls were not very high, either. For example, strain *Achrombacter* sp. BP3 was isolated from a long-term oil contaminated soil (Dong 2008), whose highest concentration of benzene degradation was only 50 mg/L; strain *Bacillus megaterium* LB07 was isolated from the mountain soil (Yang & Fan 2009), whose highest concentration of benzene degradation was only 150 mg/L. Therefore, selection of biphenyl-degrading bacteria with strong tolerance capability and degradation rates has significances to ecological restoration of biphenyl contaminated sites.

In this study, one strain was isolated from the activated sludge of biological treatment process in sewage treatment plant, which could gain rapid growth using biphenyls as sole carbon source. Through the analysis of 16S rRNA gene sequence, the strain was identified by molecular means.

2 MATERIAL AND METHOD

2.1 *Species source and culture medium*

The strain was from activated sludge of biochemical treatment in wastewater treatment plant.

Basic culture medium: KH_2PO_4 2.93 g, K_2HPO_4 5.87 g, $MgSO_4$ 0.3 g, $FeSO_4$ 0.01 g, NaCl 2.0 g, $(NH_4)_2SO_4$ 5.0 g, distilled water 1000 mL, pH 7.0.

Selective culture medium: KH_2PO_4 2.93 g, K_2HPO_4 5.87 g, $MgSO_4$ 0.3 g, $FeSO_4$ 0.01 g, NaCl 2.0 g, $(NH_4)_2SO_4$ 5.0 g, biphenyl 0.5 g, distilled water 10 ml, the concentration of biphenyl for 0.5 g/L, pH 7.0.

LB medium: peptone 10 g, yeast extract 5 g, NaCl 10 g, distilled water 10 ml, pH 7.0. Liquid culture medium was added with agar (the concentration of 20 g/L) to prepare LB solid medium. LB culture medium is used for the amplification and cultivation of the purified strain.

2.2 *Isolation and morphology observation of degrading bacteria*

10 mL activated sludge suspension was inoculated into 40 ml selective medium, and the temperature in incubator was set to 30°C. Every 5 days 20% of inoculation amount was transferred to fresh

medium for continuous cultivation. This experiment transferred to 4 phases.

1 ml enrichment bacteria solution was added into 9 ml sterilized normal saline in test tube, making different concentrations of bacteria solution by 10 fold dilution method. 0.1 ml bacteria solution of three dilution degrees (10^{-5}, 10^{-6}, 10^{-7}) was coated on LB medium plates, which were placed in an incubator at 37°C for 3 ~ 5 days.

Single colony which grows well was selected in different dilution plates, and continuous streaked in medium plate until strains were purified. The purified strain is inoculated on the LB inclined medium.

The isolated strain was Gram stained and observed by the microscope.

2.3 16S rRNA gene sequence analysis

2.3.1 DNA extraction of degrading bacteria
DNA of degrading bacteria was extracted according to Biospin Bacteria Genomic DNA extraction kit (Baidu Wenku). 20 mg/mL RNAase was included into the extraction process to avoid the interference of RNA (Meng & Peng 2014).

2.3.2 PCR amplification
In the PCR reaction system (see Table 1), negative control used 1 μL 16S-free-H_2O instead of template DNA. The forward and reverse primer were E-coli27F(5′-AGAGTTTGATCCTGGCTCAG-3′) and E-coli492R (5′-TACCTTGTTACGACT-3′) respectively.

PCR amplification procedures in this study were as follows: pre-denature at 95°C for 2 min, then 95°C 1 min, 55°C 1 min, 72°C 2 min, for 30 cycles, finally extended at 72°C for 2 min.

PCR products were detected by agarose gel electrophoresis. Electrophoresis conditions: voltage of 120 V, the sample solution for 6 × Loading Buffer, sample proportion: DNA solution 10 μL + 6 × loading buffer 2 μL, and sample volume of each lane for 20 μL, Marker was λ Hind III, electrophoresis time for 50 min.

2.3.3 16S rRNA gene sequence analysis and phylogenetic tree construction
Sequencing: the target DNA was sent to Sangon Biotech Ltd. for 16S rRNA gene sequencing.

Phylogenetic tree analysis: homologous alignment of obtained DNA sequences were detected in nucleotide database of National Biotechnology Information Center (NCBI) GenBank, and phylogenetic tree was constructed using the software MEGA v.4.

3 RESULTS AND ANALYSIS

3.1 Morphological features

After enrichment and domestication, one strain named G-2 could grow quickly on biphenyl as sole carbon source in aerobic conditions.

The colony morphology of strain G-2 on the LB solid medium plate was shown in Figure 1-a. Through Gram staining, G-2 was gram-positive bacterium, and the microscopic observation was shown in Figure 1-b.

The colony morphology and microscopy observation results of strain G-2 were summarized in Table 2.

3.2 16S rRNA gene sequence analysis

3.2.1 Total DNA extraction
DNA agarose gel electrophoresis of strain G-2 was shown in Figure 2.

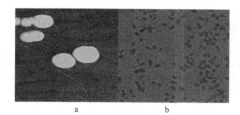

a   b

Figure 1. Bacterial colony morphology (a) and Gram stain microscopy photo (b) of strain G-2.

Table 1. PCR reaction system.

Reaction system	Contents
Template DNA	1 μL
dNTP mix (10 mM)	0.5 μl
Forward primer (10 μM)	0.5 μl
Reverse primer (10 μM)	0.5 μl
10*Taq reaction Buffer	2.5 μl
DNA polymerase (5 μ/μl)	0.2 μl
16S-free H_2O	44.8 μl
Total system	50.0 μl

Table 2. Morphological features of strain G-2.

Strain	G-2
Gram stain	G^+
Bacterial shape	Rod
Colony size	Rod
Colony shape	Round
Colony edge	Smooth
Colony color	White
Colony surface	Smooth, transparent, plump
Colony side	Highly convex

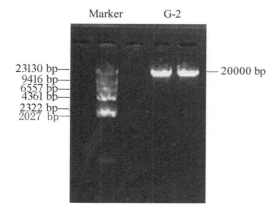

Figure 2. DNA electrophoretogram of the degrading bacteria G-2.

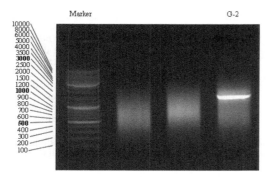

Figure 3. Electrophoretogram of PCR products.

In Figure 2, there appeared only one band of about 20 kb (two sample points). The band was clear without any unpurposed protein, RNA or single nucleotide bands, indicating that the total DNA was extracted successfully and could be the template for PCR amplification.

3.2.2 PCR amplification of 16S rRNA gene
The purified DNA fragments were recovered and used as template for 16S rRNA gene PCR (Qi 2007). The electrophoretogram of the amplified product was shown in Figure 3.

Figure 3 showed electrophoretogram of PCR products. The 16S rRNA gene sequence was amplified, whose length was about 1.5 kb.

3.2.3 Gene sequencing
Target DNA bands were cut from electrophoresis results of PCR products, and required DNA fragments were recovered by uniq-10 column DNA gel recovery kit for sequencing. The sequence size of G-2 16S rDNA was 1418 bp.

3.2.4 16S rRNA gene sequence analysis and phylogenetic tree construction
The gene sequence of about 1.5 kb was uploaded to the NCBI GenBank database, using the blast tool to find similar sequences. The sequence of degrading bacteria was compared with the known similar gene sequences by means of homologous alignment (see Table 3).

The 16S rDNA sequences of strains which were highly homologous with G-2 were downloaded from GenBank, multiple sequence comparison was done by Clustal 1.8, and the phylogenetic tree was constructed using MEGA v.4 software. Phylogenetic tree is shown in Figure 4.

As shown in Figure 4, strain G-2 has highest homology with *Paracoccus denitrificans* strain 381, *Paracoccus versutus* strain ATCC 25364 and *Paracoccus pantotrophus* strain ATCC 35512, thus strain G-2 can be classified into *Paracoccus*.

It was reported that the degradation of PCBs depended on the presence of the biphenyl, and biphenyl was often used as co-metabolic substrates. Some biphenyl degrading bacteria can also degrade PCBs. This study investigates the taxonomic status of strain G-2, and provides theoretical basis for the study of biphenyl degradation.

The current research related to *Paracoccus* focused more on ammonia oxidation and aerobic

Table 3. Comparison of 16S rDNA sequences between G-2 and related bacteria.

Strain	Homology
Paracoccus pantotrophus strain ATCC 35512	98%
Paracoccus denitrificans strain 381	98%
Paracoccus versutus strain ATCC 25364	97%
Paracoccus halophilus strain HN-182	96%
Paracoccus solventivorans strain ATCC 700252	96%

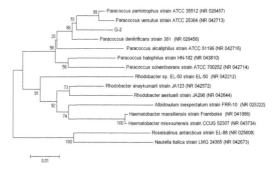

Figure 4. Phylogenetic tree of G-2 based on 16S rDNA sequence.

denitrification than on degradation of PAHs. For example, Zhang et al reported a strain of *Paracoccus* sp. could oxide a series of PAHs, such as anthracene, phenanthrene, fluoranthene, fluorene, flexion, and pyrene (Zhang & Kallimanis 2004). Mao Jian reported *Paracoccus* mediation in PAHs contaminated soil, and found that *Paracoccus* were dominant bacteria in pollution restoration (Mao & Luo 2009). This study found that *Paracoccus* G-2 could utilize biphenyl as sole carbon source, which provided a basis for the bioremediation of biphenyl and PCBs contaminated sites.

4 CONCLUSION AND DISCUSSION

Through enrichment and domestication, one strain was isolated from the activated sludge of the sewage treatment plant, which could use biphenyl as sole carbon source for rapid growth. Through observing the morphological features of the strain and analyzing 16S rRNA gene sequence. Preliminarily, G-2 was classified into *Paracoccus*.

This preliminary results showed that strain G-2 could tolerate biphenyl concentration of 2 g/L. The degradation characteristics of the strain need further study. This study can provide a theoretical basis for biphenyl biodegradation, and it is beneficial for the enrichment and perfection of the library on biphenyl degradation bacteria.

REFERENCES

Abramowicz, A. 1990. Aerobic and anaerobic biodegradation of PCBs: A review. Crit. Rev. Biotechnol 10(3): 241–248.

Baidu Wenku. Biospin bacterial genomic dna extraction kit instructions [EB/OL]. http://wenku.baidu.com/view/e1ea217a31b765ce050814fa.html.

Bumpus, J.A. & Tien, M. 1985. Oxidation of persistent environmental pollutants by a white rot fungus. Science 228(4706): 1434–1436.

Dong, X.J. 2008. Isolation and characterization of biphenyldegrading bacterium Achromobacter sp. BP3 and cloning of catabolic genes. Nanjing: Master Dissertat Agricultural University.

Gibson, D. 1973. Initial reaction in the oxidation of ethylbenzene by Pseudomonas putida. Biochemistry 12(8): 1520–1528.

Mao, J. & Luo M. 2009. Biodegradation of PAHs by Paracoccus aminovorans HPD-2 in contaminated soil. Soils 41(3): 448–453.

Meng, H. & Peng, Z. 2014. Compared different isolated RNA method from blood. Journal of Tarim University 26(1): 46–49.

Potrawfke, T. 1998. Mineralization of low-chlorinated biphenyls by Burkholderia sp. strain LB400 and by a two membered consortium upon directed interspecies transfer of chlorocatechol pathway genes. Applied Microbiology and Biotechnology 50(4): 440–446.

Qi, Y. 2007. Study on isolation and biodegrading characteristics of chlorobenzoate-degrading bacteria. Tianjin: Tianjin University.

Yang, J. & Fan Y.L. 2009. Screening of a biphenyl-degrading strain and determination of the degradation conditions. Chinese Journal of Microbiology 29(3): 37–42.

Zeddel, A. 2001. PCB congener selective biodegradation by the white rot fungus pleurotus ostreatus in contaminated soil. Chemosphere 43(2): 207–215. http://wenku.baidu.com/view/e1ea217a31b765ce050814fa.html.

Zhang, H.M. & Kallimanis, A. 2004. Isolation and characterization of novel bacteria degrading polycyclic aromatic hydrocarbons from polluted Greek soils. Applied Microbiologyand Biotechnology 65(1): 124–131.

Advances in Energy, Environment and Materials Science – Wang & Zhao (Eds)
© *2016 Taylor & Francis Group, London, ISBN 978-1-138-02931-6*

Study on the eliminating outburst technology and effect with increasing permeability and forced gas drainage by drilling hole cross the seam in outburst seam

Benqing Yuan & Yinggui Zhang
China Coal Technology Engineering Group Chongqing Research Institute, Chongqing, China
National Key Laboratory of Gas Disaster Detecting, Preventing and Emergency Controlling, Chongqing, China

ABSTRACT: To the question of the difficult gas extraction and the long elimination of the outburst low permeability coal seam, taking the specific gas geology conditions of 6-2 coal seam in a mine of huainan mining area for example, the theoretical analysis and engineering practice of the pressure grouting in the deep and shallow hole and the loose blasting in the medium-length hole are used. The grouting technology in the deep and shallow hole blocking off the fissures of the roadway and the deep and shallow drill sites, and it solves the question of the extraction from the roadway surrounding rock crack with high-pressure grouting of the drilling hole. The loose blasting in the medium-length hole makes the coal fracture sufficiently develop, and the fracture network resulting from blasting makes the degree of pressure relief of the coal in the test area increase significantly, and the coal seam permeability is effectively increased. The engineering practice results show that the borehole gas extraction concentration could be increased by 7.4~10.5 times by grouting in the deep and shallow hole, and the average extraction scalar quantity has increased 9.2 times after the loose blasting in the medium-length hole, and the time of drainage up to the standard is 85 d, it shortened four months than traditional method. The technology guarantees the normal alternate mining.

1 OUTLINE

As the depth of mining increases, seam gas pressure and gas content increase, risk of coal and gas outburst also increases correspondingly, coal seam permeability decreases. (Zhou, 1999; Yuan, 2004) Practice shows that, protection layer mining can greatly reduce or eliminate outburst danger as the most effective regional measure for preventing coal and gas. But because of the geological condition of coal mine, when tunneling in outburst coal without protection layer mining, outburst prevention measures for strip mining gas region of gas pre-drained seam gateway by constructing crossing boreholes with rock roadway in roof and floor are used (Yu, 2011). At the same time, the effect of outburst prevention measures for strip mining gas region of gas seam gateway is effected by coal seam permeability, borehole layout and other factors, and poor pre-drained effect and long drainage time will make alternate mining tight. In order to improve the effect of pre-drained, hydraulic fracturing, hydraulic flushing and other pressure releasing and permeability improvement are applied for reducing intensity and times of outburst effectively in some mines at home and abroad (Lyu, 2013; Liu, 2008; Li, 2010; Tong, 2011), but there are some limitations

in these measures, for example, hydraulic flushing has less effect for crossing borehole downward on permeability improvements.

Based on this, it is of extreme importance to study the technology of enhanced gas drainage by permeability improvement of outburst coal under the concrete gas geological conditions of mines. The test coal seam is 6-2 coal seam in a mine of huainan mining area which using the comprehensive control measures of grouting in the deep and shallow hole and loose blasting to improve concentration and flow rate of gas drainage. The technology can improve drainage rate of low permeability coal seam, reduce drainage time, guarantee safety quick tunneling in outburst coal and normal alternate mining.

2 GENERAL SITUATION OF TEST AREA

The maximum actual measured raw gas pressure of 6-2 coal seam in a mine of huainan mining area is 3.2 MPa, gas content is 7.02 m^3/t which has strong outburst hazard. Permeability of 6-2 coal seam is very poor and the average permeability coefficient is 0.00068 m^2/MPa·d, concentration of gas pre-drained seam gateway by constructing crossing boreholes is so low that drainage concentration of more than 60% of the drillings is less than 5%. Pre-drained effect is so poor and drainage time is so long that drainage standard time is more than 120 days. The average thickness of 6-2 coal seam at machinery roadway of test face is 3 m and the dip is 2~8°.

Horizontal distance between floor drainage roadway and machinery roadway of test face is 10 m and the vertical distance between floor drainage roadway and 6-2 coal seam is 10~15.6 m.

Lithology of 6-2 coal seam floor range 0 to 6.8 m is composed of sandy mudstone, thin sand mud interbed and mudstone, and beyond 6.8 m lithology is thin sand mud interbed which thickness is 19.5 m where floor drainage roadway is layout in with fractured surrounding rock and development crack. Each 60 m there is a drilling field arranged in floor drainage roadway to construct crossing boreholes.

3 DRAINAGE TECHNOLOGY AND EFFECT OF GROUTING IN THE DEEP AND SHALLOW HOLE

3.1 Theory analysis of grouting sealing under pressure

Grouting under pressure is that fractures in rocks are filled by slurry to exclude water and gas in it under pressure. Practice shows that: grouting under pressure can block the fractures around roadway and drillings. Slurry can not only make the fractures splitting extend and fill the pore to increase diffusion range with grouting pressure, but also go deep into rock microfissure and generate cohesion to consolidate with solid particles through distributing as branch after slurry concreting under permeation pressure gradient. Leaking channel is blocked effectively and gas drainage effect is improved. (Wang, 2014; Zong, 2013) Grouting under pressure impels deformation of broken rocks transforms from brittle into ductility, lateral deformation and axial deformation tends to coordination, stability of broken rocks strengthens and it has strong plasticity which can form bearing structure to adapt deformation easily.

3.2 Technology of grouting in the shallow hole

Rocks around the roadway are damaged seriously and fractures in the rocks develop gradually to transfixion, at last, relief ring is formed when floor drainage roadway tunneling in thin sand mud interbed. Blocking relief ring imprecisely will make effect of borehole short-circuit and draining poor. But occluding layer is formed in shallow surrounding rock of roadway and drilling field by grouting in the shallow hole to prevent the problem of the slurry leaking from cracks in the roadway when grouting in the deep hole and sealing the hole.

Layout parameters of grouting in the shallow hole: distance of holes bottom is 4 m and depth of hole is 2 m. Grouting pipe of 20 mm diameter is used, as shown in Figure 1. Water-cement ratio of slurry is 1:1 and grouting pressure is more than 2 MPa.

3.3 Technology of grouting in the deep hole

Double as deep grouting holes, crossing pre-drained boreholes are grouted at intervals one hole to block fractures around deep rocks further. After grouting, the hole is drilled again to be used for draining. Profile of grouting in the deep hole is shown as Figure 1.

Technology of grouting in the deep hole: after constructed ahead 4 m by drill bit of 153 mm diameter and core barrel of 108 mm diameter is put in, the borehole is grouted to strengthen core barrel. After slurry concreting, the hole is drilled again by drill bit of 94 mm diameter to the place in front of 6-2 coal seam 2 m and PVC pipe of one inch diameter is put in. Screw the flange which welded grouting pipe of 0.5 inch diameter and output-pipe of one inch diameter to core barrel and attach the output-pipe to PVC pipe, close the gate and grout with high pressure. Water-cement ratio of slurry is more than 0.7:1 and grouting pressure is more than 4 MPa. After all the deep holes are grouted, drainage borehole can be constructed.

3.4 Effect analysis

According to field data, gas scalar increment of a single hole is not obvious, but gas drainage concentration has changed greatly after grouting in the deep and shallow hole, as shown in Figure 2. Gas drainage concentration 17 days ago in the figure is the data before grouting.

Gas drainage concentrations in different gas drillings have been improved in different degree after grouting, as shown in Figure 2. The average concentration of gas drainage of No.1 drainage hole is

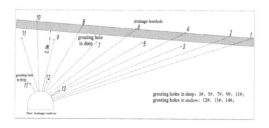

Figure 1. Profile of grouting in the deep and shallow hole.

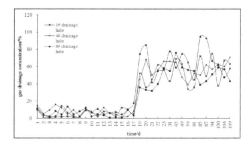

Figure 2. Changes of gas drainage concentration on one hole before and after grouting in the deep and shallow hole.

increased from previous 5.5% to 49.5% after grouting, which is 9 times before grouting. The average concentration of gas drainage of No.6 drainage hole is increased from previous 6.3% to 66.2% after grouting which is 10.5 times before grouting; The average concentration of gas drainage of No. 9 drainage hole is increased from previous 7.8% to 57.4% after grouting which is 7.4 times before grouting.

4 ANTIREFLECTION TECHNOLOGY AND EFFECT OF MEDIUM AND DEEP HOLE LOOSE BLASTING

4.1 Relief antireflection principle of medium and deep hole loose blasting

Objective of medium and deep hole loose blasting is to make redistribution of coal stress, increase coal fractures and change coal structure, which can realize relief and antireflection, improve permeability coefficent in coal seam and gas drainage rate, decrease risks of coal seam outburst[4-7]. Under explosive stress wave, detonation gas and gas pressure, relief antireflection technology with medium and deep hole loose blasting make stress concentration and energy concentration exist in coal, new fractures form and original fractures extend and develop, interconnected fracture network form within a certain range centered at blasting holes and transfixion cracks form between blasting holes and drainage boreholes. Under blasting shock wave and stress wave, stress in coal decreases, massive adsorption gas desorb and pours into new fractures, coal permeability around blasting holes increases which create advantages for draining free gas.

4.2 Design of medium and deep hole loose blasting

Gas drainage boreholes of 113 mm diameter are constructed in floor drainage roadway drilling field

under the concrete gas geological conditions of test face. Drilling control area is outward 15 m of contour line in two sides of the coal roadway to be tunneled and final coal point spacing is 5 m × 5 m.

Blasting holes are constructed after drainage boreholes. Setting parameter of blasting holes: every 5 m there are two or three blasting holes arranged and borehole end is outward 15 m of contour line in two sides of the coal roadway to be tunneled. Blasting holes of 94 mm diameter in two sides of the roadway are constructed to internal 2 m of coal roof, the one in the middle of roadway are constructed to internal 1 m of roof. Arrangement of blasting holes is shown as Figure 3. Distance between borehole end location of blasting holes and drainage boreholes is more than 2.5 m.

4.3 Charging and sealing technology of medium and deep hole loose blasting

In order to prevent hole collapsing, PVC pipe of 1.5 inch diameter whose upper connected with an empty pipe of 75 mm diameter and 0.5 m length is put in the hole to verify depth before charging. Then hole depth is recorded and charging length is determined.

Gas drainage water gel explosive grain is applied to the medium and deep hole loose blasting, which is 75 mm diameter, 1 m length and 5 Kg/m weight. Powdered charge of each blasting hole is not more than six. Profile of blasting hole arrangement is shown in Figure 4.

There is a hole of 6 mm diameter on front end of each grain because of the heavy grains. Based on the inclination of blasting holes, 6~9 antiskid wires are choked in the hole wall forming barb to fix grains. Charge with handhold pipe to the hole every time 2 grains, and the nozzle and output-pipe are put in blasting hole together at last.

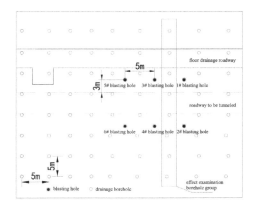

Figure 3. Drillings layout plan of medium and deep hole loose blasting.

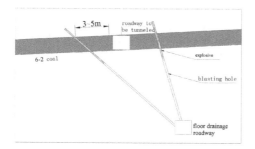

Figure 4. Drillings layout profile of medium and deep hole loose blasting.

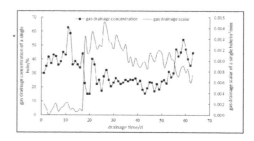

Figure 5. Changes of drilling gas concentration and scalar before and after blasting.

Sealing begins as soon as grain is installed. Sealing uses cement grout and sealing length is more than 12 m. Sealing step: ① Sealing uses polyurethane in the range of 2 m to orifice after grouting pipes of 4 m are put in the hole. ② The PVC output-pipe is put to the explosive. ③ Water-cement ratio of slurry is 0.7:1 when grout with pneumatic grouting pump. Stop grouting and tangle up output-pipe if slurry returns from output-pipe. ④ Blast after time of slurry solidification is more than 24 h.

4.4 Effect analysis

Count and analyze the change law of drilling gas concentration and scalar in affected area before and after blasting, as shown in Figure 5. Gas concentration and scalar 17 days ago in the figure is the data before grouting. Arrangement of investigation drillings is shown in Figure 3.

The Figure 5 shows:

1. Medium and deep hole loose blasting plays a great role in antireflection for coal seam. The average gas scalar in drilling groups is 0.0012 m³/min before blasting, and first increases sharply and then decreases slowly after blasting. The average gas scalar is 0.011 m³/min which increases in 9.2 folds.
2. The average gas concentration in drilling groups before blasting is 41.2%, and first decreases and then increases after blasting which is less than the data before blasting a little. The range of gas concentration is 15% to 54% and the average is 27.4%.
3. In a word, medium and deep hole loose blasting with water gel explosive can make fractures in the coal develop full and improve the relief degree of coal in test field greatly which make antireflection effect remarkable.
4. It will spend 85 d that drainage rate of one evaluation unit with grouting in the deep and shallow hole and medium and deep hole loose

blasting exceeding 35% which shortened four months than traditional method by drainage statistics of test area.

5 THE MAIN CONCLUSION

1. Grouting in the deep and shallow hole and medium and deep hole loose blasting can improve effect of gas drainage greatly by theory verification and field test.
2. The problem which is the slurry leaking from cracks in the roadway when grouting sealing is solved and gas drainage concentration has been increased 7.4–10.5 fold because occluding layer is formed in shallow surrounding rock of roadway and drilling field by grouting in the shallow hole, fractures of rocks in deep are blocked by grouting in the deep hole further.
3. After medium and deep hole loose blasting with water gel explosive is used, fractures in the coal develop full and improve the relief degree of coal in test field greatly which make Antireflection effect remarkable. In comparison to no blasting, the average gas drainage scalar is increased by 9.2 times.
4. Crossing pre-drained borehole in outburst coal with grouting in the deep and shallow hole and medium and deep hole loose blasting is safe and reliable, simple process, technically feasible, in which the grouting in the deep and shallow hole is a key measure to improve extraction concentration, and the increased coal seam permeability with loose blasting in the medium-length hole is the basis measure that improves the drainage quantity and shortens the drainage time. It is worth to popularize in outburst control of high outburst seam.

REFERENCES

Zhou Shining, Lin Baiquan. The Theory of Gas Flow and Storage in Coal Seams [M]. Beijing: China Coal Industry Publishing House, 1999.

Yuan Liang. Theory and Technology of Gas Drainage and Capture in Soft Multiple Coal Seams of Low Permeability [M]. Beijing: China Coal Industry Publishing House, 2004.

State Administration of Work Safety. Coal and Gas Outburst Provisions [M]. Beijing: China Coal Industry Publishing House, 2009, 4.

Yu Yeting, Yu Tao, Deng Zhong, ect. Research on Outburst Elimination Effect of Enhanced Gas Drainage by Permeability Improvement of Outburst Coal Seam [J]. Journal of Anhui Institute of Architecture & Industry, 2011, 19(6): 47–50.

Lyu Pengfei, Dou Xiaoxing, Zhu Tonggong, ect. Application on Permeability Improved Technology with Deep Borehole Energy Accumulation Blasting in Coal Seam [J]. Coal Science and Technology, 2013, 41(12): 35–38.

Liu Zegong, Cai Feng, Xiao Yingqi. Numerical Simulation and Analysis of Effect of Stress Release and Permeability Improvement in Coal Seams by Deep-hole Presplitting Explosion [J]. Journal of Anhui University of Science and Technology (Natural Science), 2008, 28(12): 16–20.

Li Jiangxin, Lin Baiquan, Li Guoqi, ect. Theory and Practice Of Pressure Releasing and Permeability Improvement with Long Hole Loose Blasting [J]. Safety in Coal Mine, 2010: 52–54.

Tong Bi, Yu Tao, He Shanlong. Cut Shot, Pressure Releasing, Permeability Improvement Reinforced Gas Drainage Technology with Varied Diameter Borehole in Seam Section of Borehole Through Strata [J]. Coal Engineering, 2011, 60–62, 66.

Wang Zhaofeng, Wu Wei. Analysis on Major Borehole Sealing Methods of Mine Gas Drainage Boreholes [J]. Coal Science and Technology, 2014, 42(6): 31–34, 103.

Zong Yijiang, Han Lijun, Han Guilei. Mechanical Characteristics of Confined Grouting Reinforcement for Cracked Rock Mass [J]. Journal of Mining & Safety Engineering, 2013, 30(4), 483–488.

Advances in Energy, Environment and Materials Science – Wang & Zhao (Eds)
© *2016 Taylor & Francis Group, London, ISBN 978-1-138-02931-6*

Hydrocarbon expulsion threshold from lacustrine mudstone and shale: a case study from Dongying Depression, China

Zhonghong Chen, Ming Zha & Chunxue Jiang
School of Geoscience, China University of Petroleum, Qingdao, Shandong, China

ABSTRACT: The new parameters to reflect hydrocarbon expulsion potential, which defined as Hydrocarbon Expulsion Amount (HEA) and Hydrocarbon Expulsion Saturation (HEA) were used to determine the hydrocarbon expulsion threshold of mudstone and shale. the calculation of the two parameters are based on the Rock Eval results including soluble hydrocarbon (S_1), pyrolysed hydrocarbon (S_2) and Total Organic Carbon content (TOC). HEA is defined as Sp/TOC (Sp is the amount of hydrocarbon expulsion, and TOC is the total organic carbon content) and hydrocarbon expulsion, and HES is calculated from the equation $\Phi \cdot HES \cdot \rho_o/\rho_s = Sp/TOC$ (Where: Φ represents porosity of mudstone or shale, %; ρ_o and ρ_s are density of crude oil and density of mudstone or shale (g/mL), respectively). The results of hydrocarbon expulsion threshold study of Well N38 in Dongying depression show that the critical amount for the initial hydrocarbon expulsion is 300 mg/g; the critical saturation for the initial hydrocarbon expulsion is 8.56% (with error considered, it is between 2.5% and 9.5%). Type III organic matter cannot reach the hydrocarbon expulsion threshold, thus is basically invalid organic matter to generate and expel hydrocarbon. The relations between HES, Tmax and VRo are: $HES = 0.0495TOC^{1.7913}$, $HES < 0.0757T_{max} - 32.404$, and $HES < 4.4505VRo-2.3607$.

1 INTRODUCTION

An important link in and effective means for the research of hydrocarbon expulsion from mudstone and shale is to determine the critical state of the process, only when meeting the critical state, can hydrocarbon be discharged from mudstone and shale.

Researchers have put forward different concepts to characterize the critical state of hydrocarbon expulsion, for instance, critical saturation of hydrocarbon expulsion (Dickney, 1975), critical saturation quantity of hydrocarbon expulsion (McAuliffe, 1979), hydrocarbon expulsion threshold (Pang et al, 1997), migration threshold (Wang et al.,1995). But due to different research perspectives, methods and study conditions, they haven't reached a full consensus in how to determine the critical state of hydrocarbon expulsion. The critical saturation of hydrocarbon expulsion commonly accepted is 1%–20%, 10% on average. Due to capillary force, oil and gas unable to be discharged from source rock smoothly, form multiphase fluid with pore water. According to the seepage theory of multiphase fluids in porous media, the premise of oil and gas discharge from source rock is that their phase saturation exceeds their minimum phase seepage threshold. The minimum saturation of oil and gas in source rock to move is called critical saturation of oil and gas. The oil and gas generated by source rock occupied the pore center firstly, when the quantity of generated hydrocarbon is enough to connect into a conduit, more hydrocarbons generated later can overcome the capillary resistance and flow outside along the conduit. In this sense, the quantity of residual hydrocarbon just before the oil and gas migration is called the critical saturation. McAullife (1978) proposed that oil and gas generated by source rock were adsorbed by kerogen network firstly, they must meet the adsorption before overcome capillary resistance and source rock adsorption resistance to flow outside. When the oil and gas quantity generated is enough to meet the adsorption need of kerogen, more oil and gas generated later can overcome capillary resistance and move outside along the kerogen network. In this case, the increasing hydrocarbon concentration in the source rock and the capillary resistance in kerogen network make up the forces behind oil and gas discharge.

Combining the previous viewpoints with the recent study, the authors think oil and gas must meet the need of adsorption by hydrocarbon expulsion channel, and form a continuous network of hydrocarbon expulsion before discharge. This is to say oil and gas must satisfy the critical saturation of hydrocarbon expulsion amount. The saturation reaching this amount in the pore network is the critical saturation. Obviously the critical amount and critical saturation depend on the nature of source rock and geological conditions, and are different under different environments.

Hydrocarbon's expulsion from mudstone and shale includes two consecutive processes: the release of hydrocarbon from kerogen and expulsion out of source rock. Thus enough hydrocarbon in the source rock is necessary to meet the adsorption of the kerogen network and the rock before they can form considerable expulsion. Therefore, hydrocarbon expulsion threshold actually is a potential critical state, only when the mudstone and shale reached the hydrocarbon expulsion threshold, can effective hydrocarbon discharge take place, otherwise, the mudstone and shale can be regarded as invalid source rock.

Dongying Depression is a typical negative rift tectonic unit located at the north-eastern part of the Bohai Bay basin, eastern China. Large set of dark mudstone and shale are deposited in the Member 3 (Es_3) of the Paleogene Shahejie Formation was deposited in this depression (Chen & Zha, 2006). The N-38 Well is located in the Niuzhuang sag, the south part of Dongying Depression. A total of 275 samples were analyzed via Rock-Eval pyrolysis. Mean random vitrinite reflectance (VR_O) was measured on 138 selected mudstone and shale samples.

2 SELECTION OF PARAMETER FOR HYDROCARBON EXPULSION THRESHOLD

2.1 Previous parameters

Because hydrocarbon generation threshold is mainly controlled by thermal maturity of organic matter, hydrocarbon generation threshold is comparatively easy to determine, and there are also a number of parameters to characterize it. As hydrocarbon expulsion is influenced by many factors, and the amount of expelled hydrocarbon and hydrocarbon expulsion threshold is difficult to determine and figure out. From Rock Eval pyrolysis analysis, several parameters including S_1, S_2 and Tmax can be obtained. S1 represents dissolved hydrocarbon quantity in the rock, S_2 is the pyrolysis hydrocarbon quantity measured at high temperature, and ($S_1 + S_2$) indicates residual hydrocarbon generation potential at present. Tmax is peak temperature corresponding to S_2. TOC is the residual total organic carbon content, In previous studies, the parameters S_1/TOC and ($S_1 + S_2$)/TOC, called hydrocarbon generation potential index, were often used to determine the hydrocarbon expulsion threshold (Zhou and Pang, 2002; Pang et al., 2003). The moment this index reduced was

proposed to represent the time hydrocarbon began to discharge from mudstone, and the corresponding depth is the hydrocarbon expulsion threshold depth.

Residual hydrocarbon generation potential $S_1 + S_2$ is the difference between original hydrocarbon generation potential of source rocks and the expelled hydrocarbon quantity. The $S_1 + S_2$ is affected by both of the original hydrocarbon generation potential and the expelled hydrocarbon quantity. Thus, the $S_1 + S_2$ is not directly related to the hydrocarbon expulsion, and the value of $S_1 + S_2$ is not in linear relationship with expelled hydrocarbon quantity. The less residual hydrocarbon amount doesn't necessarily mean more expelled hydrocarbon quantity either. In addition, when using the ($S_1 + S_2$)/TOC to reflect the expelled quantity of hydrocarbon, the mudstone or shale samples that for pyrolysis experiment should be with similar kerogen type and similar evolution state, and the more samples examined, the better the actual geological conditions can be reflected (Chen et al., 2004).

2.2 The new parameters calculated method

Using geochemical data obtained from pyrolytic experiment (a relationship between the test data of limited samples and corresponding logging response could be established in areas with few samples, then the relationship can be used to obtain TOC from logging data), the hydrocarbon generation and hydrocarbon expulsion quantity (expressed as So and Sp, respectively, mg/g.rock) and other various parameters can be calculated (14).

Then the new parameter "Hydrocarbon Expulsion Amount" (HEA) can be defined as Sp/TOC, the hydrocarbon expulsion potential index can reflect the capacity of hydrocarbon expulsion from mudstone and shale. Hydrocarbon expulsion threshold can be determined directly from the distribution of Sp/TOC profile with burial depth.

Through comparison of the different geochemical parameters in the profile of mudstone and shale, it is found that Sp/TOC curves show a convex type curve shape, lowest value point at 2950 m and highest value point at 3070 m. The parameter S_1/TOC (hydrocarbon index, decreases with the hydrocarbon discharge) shows small change overall, but display obvious increase and high values in the lower of Es_3; the ($S_1 + S_2$)/TOC shows an overall increasing trend from the shallow to the deep (Fig. 1). It is inferred that the significant increase of S_1/TOC and ($S_1 + S_2$)/TOC in the lower Es_3 has two reasons: on the one hand, the layer has higher TOC (an average of 4.0%); on the other hand, the large set of mudstone (more than 50 m thick

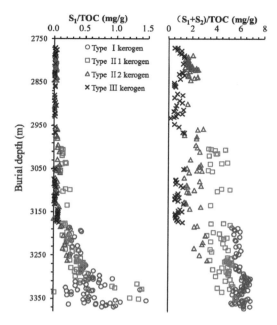

Figure 1. The distribution of the S_1/TOC and $(S_1 + S_2)$/TOC profiles in the Paleogene of N-38 Well, Dongying depression.

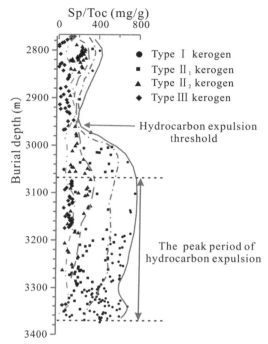

Figure 2. Using the Sp/TOC profile to determine the hydrocarbon expulsion threshold: an example from the Paleogene of N-38 Well, Dongying depression.

in single layer) blocked the smooth expulsion of hydrocarbon in large quantity. The comparative analysis shows that the proposed parameter Sp/TOC can be used to characterize hydrocarbon expulsion threshold (Fig.2).

2.3 Determination of hydrocarbon expulsion threshold

Figure 2 shows that in the overall Sp/TOC profile the lowest value of Sp/TOC is 200 mg/g at 2950 m, and the Sp/TOC increases gradually until the peak occurred at 3070 m where the highest HEA was 800 mg/g.

Since before the hydrocarbon expulsion the necessary adsorption by kerogen network and the source rocks is must be satisfied, as the mudstones just begin to enter the hydrocarbon expulsion threshold when the Sp/TOC reached 300 mg/g, and form considerable hydrocarbon discharge in a certain scale. Thus, the 2950 m was regarded as the depth of hydrocarbon expulsion threshold, and the 3070 m was regarded as the depth corresponding to the peak of the hydrocarbon expulsion. Analysis of thermal evolution of this interval shows that the Ro of these mudstones at 2950 m reached 0.5% and that at 3070 m is about 0.65%, which is consistent with the results of analysis on the hydrocarbon expulsion threshold.

The hydrocarbon saturation when the hydrocarbon is discharged can be calculated by the Sp/TOC value as following equation.

$$\Phi \cdot HES \cdot \rho_o / \rho_s = Sp/TOC \qquad (1)$$

where: Φ represents porosity of mudstone or shale, %; ρ_o and ρ_s are density of crude oil and density of mudstone or shale (g/mL), respectively.

The density of mudstone or shale can be directly obtained from density logging (Den). After the analysis on crude oil in the tested wellblock and conversion of them to the underground state of high temperature and high pressure, the oil density is estimated at 0.8 g/mL in the studied area; through well logging response and the related geological analysis, the initial porosity (Φ_0) of mudstone or shale that close to the surface was supposed to be 62%, then the porosity at depth H is calculated using the following equation:

$$\Phi = 0.62 e^{-0.000624H} \qquad (2)$$

The Figure 3 shows the distribution of the HES profiles in the Paleogene of N-38 Well, Dongying depression. The results show oil saturation of the studied mudstone and shale at 2950 m and

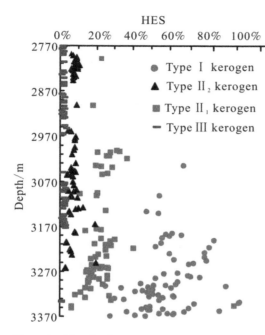

Figure 3. The distribution of the HES profiles in the Paleogene of N-38 Well, Dongying depression.

3070 m the values of HES are 8.56% and 64.42%, respectively.

3 MAIN CONTROLLING FACTORS FOR HYDROCARBON EXPULSION THRESHOLD

The hydrocarbon expulsion process is a very complex geological process because it was affected and controlled by a number of geological factors, such as type and amount of organic matter, thermal evolution degree, lithology structure and thickness of mudstone or shale, the distribution of pore fluid pressure inside the mudstone and shale, pore structure and development of micro-cracks. Among them, the organic matter type, the material base for source rock is the most important factor. Source rocks with different types of organic matter differ a lot in hydrocarbon expulsion process.

The Figure 2 shows that HEA curves of various types of organic matter are basically consistent. At 2950 m, type I and type II organic matter both entered hydrocarbon expulsion threshold, while type III of organic matter showed very limited capacity of hydrocarbon generation and expulsion and does not enter hydrocarbon expulsion threshold at this depth, thus is basically invalid organic matter to generate and expel hydrocarbon. At peak

of hydrocarbon generation, different types of organic matter show disparities in the capacity of hydrocarbon expulsion. At 3070 m, values of HEA for type I, type II 1, type II 2 and type III kerogen is 800 mg/g, 600 mg/g, 380 mg/g and 160 mg/g respectively, and the corresponding values of HES

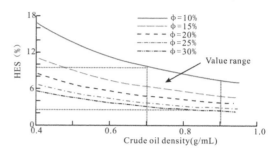

Figure 4. Analysis on the effect of porosity of mudstone and crude oil density on the HES in the estimation.

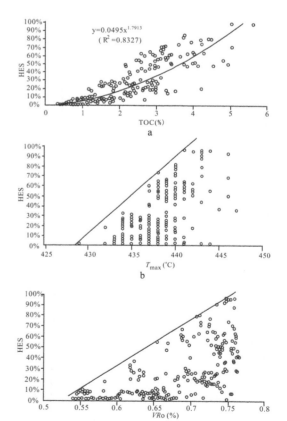

Figure 5. The effects of TOC, T_{max} and VRo to the HES.

are 64.42%, 29.86%, 13.1% and 1.59%, respectively (Fig. 3).

The calculation results are inevitably influenced by the selected parameters. In the actual calculation, Sp/TOC, TOC and ρs can be directly obtained, but Φ and ρo are needed to be calculated indirectly, thus they have a certain error range. Figure 3 shows when the hydrocarbon expulsion threshold is 300 mg/g, the HES value is influenced by Φ and ρo. It is estimated with the error taken into consideration that the critical oil saturation when the hydrocarbon is initially to be expelled from the mudstone and shale is between 2.5% and 9.5% (Fig. 4).

The Figure 5 shows the relations between the well relation between the HES and TOC.

$$HES = 0.0495TOC^{1.7913} \ (R^2 = 0.83) \tag{3}$$

The analysis results suggested that the correlations between HES and is not good, but there is still certain regularity: the values of HES increase with the increasing of T_{max} and on the whole, and distribute inside the fan area (Fig. 5b and 5c), there are following relations:

$$HES < 0.0757T_{max} - 32.404 \tag{4}$$

$$HES < 4.4505VRo - 2.3607 \tag{5}$$

ACKNOWLEDGEMENTS

This study was co-supported by the National Natural Science Foundation of China (Grants No. 41272140), the Fundamental Research Fund for the Central Universities, and the China Post-doctoral Science Foundation (2015M571227). We thank the Shengli Oil Company of Sinopec for their collaboration in this study.

REFERENCES

Chen, Z.H., Zha, M. 2006. Over-pressured fluid compartment and hydrocarbon migration and accumulation in Dongying Depression. Acta Sedimentologica Sinica, 24, 607–615.

Chen, Z.H., Zha, M., Jin, Q. et al., 2004. An investigation on generation and expulsion of hydrocarbon from source rocks of the Shahejie formation in the Well Niu-38, Dongying depression. Scientia Geologica Sinica, 39, 356–366.

Pang, X.O. 1997. Basic concept of hydrocarbon expulsion threshold and its research significance and application. Geoscience, 11, 510–521.

Pang, X.Q., Zuo, S.J., Jin, Z.J. 2003. Inversion modeling of the hydrocarbon amount destroyed by tectonic event in the Jiyang Depression. Petroleum Exploration and Development, 30 (3): 56–60.

Wang, X.Z., Zhou, D.X., Wang, X.J. 1995. A distribution of residual dead oil content in source rocks and its migration threshold. Petroleum Exploration and Development, 22 (2): 17–23.

Zhou, J., Pang, X.Q. 2002. A method for calculating the quantity of hydrocarbon generation and expulsion. Petroleum Exploration and Development, 29 (1): 24–27.

Advances in Energy, Environment and Materials Science – Wang & Zhao (Eds)
© 2016 Taylor & Francis Group, London, ISBN 978-1-138-02931-6

Research progress on purification of oil spill coast based on slow release fertilizers

Juan Meng
Ocean University of China, Qingdao, Shandong, China
Qingdao University of Technology, Qingdao, Shandong, China

Xilai Zheng
Ocean University of China, Qingdao, Shandong, China

ABSTRACT: This review states slow release fertilizers' effective promotion in petroleum bioremediation on oil spill coasts, and its advantages of high efficiency, environmental friendliness and low cost. Important factors influencing nutrient release mechanism in beach environment are discussed, including structure, temperature, moisture, microorganism, pH value and salinity. It is highlighted that three preconditions are necessary to purify oil spill coast using slow release fertilizers: abundant and available petroleum-degrading microorganisms, appropriate nutrient background concentration, and adequate dissolved oxygen supply. Although slow release fertilizers have made significant progress in purifying oil spill coast, challenges still exist, including fertilizers' great behavioral differences when applied to different beach environments, the data inconsistency between laboratory and field, etc. In future research, we shall take full account of and simulate complicated and diverse factors of the open beach environment, and design fertilizer formulas for different site environments.

1 INTRODUCTION

With increasingly wide utilization of petroleum, oil spill accidents happen frequently around the world (Bei & Dong 2010). Oil spill pollution has become a worldwide environmental problem. When the oil spill is brought to the beach by wind and wave, its pollution and damage as well as the difficulty to clean it are much larger than on the sea (Shi et al. 2013). The beach polluted by oil spill can be disposed by physical method, chemical method and bioremediation. Bioremediation is praised highly because of its advantages of being environmentally friendly and low cost, etc. Currently, bioremediation has become an important approach for purifying beaches polluted by oil spill. Its basic measures mainly include: inoculating microorganisms, adding nutritive salts (mainly including N, P), changing environmental factors (such as content of dissolved oxygen, pH, etc.) and adding surfactant, etc (Xia et al. 2007). As proved by existing researches, in most beaches which are seriously polluted by oil spill, lacking of nutritive salts (in particularly N, P) is an important factor which limits biodegradation (Prichard & Costa 1991). However, in actual beach environment, the nutrients of common water-soluble nutritive salts added are prone to being washed away by the tide

and wave (Wrenn et al. 1997, Bragg 1994). To solve this problem, and maintain the concentration of nutritive salts within ideal range, slow release fertilizer gradually enters into people's vision. Moreover, it is increasingly applied to purifying beaches polluted by oil spill (Bragg 1994). This paper summarizes the performance of slow release fertilizers, the factors influencing slow release of nutrients in beach environment as well as the progress of the researches on purifying beaches polluted by oil spill by using slow release fertilizers. Also, it discusses the feasibility and deficiency of applying slow release fertilizers to beaches polluted by oil spill. In addition, this paper focuses on pointing out some problems existing in actual application, and proposes some suggestions for future research direction.

2 SLOW RELEASE FERTILIZER

Generally, the available nutrients of slow release fertilizer can release slowly by changing its chemical components or coating semi-watertight or watertight matters on the surface (Wu & Chen 2003). Researching and developing new environmental-friendly fertilizers and slow release fertilizers is listed as the priority theme in China's National

Outline for Medium and Long Term S&T Development (2006–2020). The approaches for preparing slow release fertilizers mainly include biological chemistry approach, physical approach and chemical approach. The fertilizers produced by means of physical coating method, i.e. Coated Fertilizer (CF) has become one of the main research orientations of slow release fertilizers by far (Wu & Chen 2003). The researches on coated fertilizers both home and abroad are the most extensive and in-depth. The summary in this paper also focuses on relevant researches on coated fertilizers.

3 SLOW RELEASE PERFORMANCE

Compared with common urea, slow release fertilizers effectively provide nutrients in a continuous manner. Therefore, this can provide theoretical support for its application to beaches polluted by oil. Zhang et al. (2005) coated urea by organic and inorganic film coating materials to prepare slow release fertilizer. The research results show that the curve of the accumulated leaching amount of its NH_4^+—N is close to S shape, which indicates that nutrient release experiences the process of slow→ fast→ slow. The leaching speed of nutrients finally slows down, which indicates that this coated slow release fertilizer can supply nutrients stably and sustainably. The larger the organic constituent content in the formula is, the better the overall slow release effect of this coated urea ammonium nitrogen will be. Cao et al. (2007) had made similar results. For urea coated with one to three layers of glycerol ester of rosin films, the larger the coating rate is, the better the fertilizer's slow release performance will be. In fact, the film forming advantage of most macromolecular organic compounds lies in uniformity and compactness. Therefore, the slow release effect of organic films is generally superior to that of inorganic films. The research results from Gu et al. (2013) also prove this by regional contrast tests on slow release fertilizers coated by different coating technologies. During the critical growth period of the new root of rice seedling, sulfur coated urea basically dissolves, leaving only the empty shells; 2/3 of the slow release particles coated with double films of sulfur and macromolecular polymer dissolves; only 1/3 of the slow release particles coated with resin dissolves. This indicates that the particles coated with resin have the best slow release effect and longer release period.

Adding materials with adsorbability may effectively reinforce the slow release performance of slow release fertilizers. For the slow release fertilizer prepared by the coating material of bamboo charcoal, the leaching amount of K^+ is less than conventional urea. Moreover, it is significantly different from urea (Lu et al. 2008). The role of auxiliary is also of great importance. If the urea is separately coated by paraffin as the coating material, it basically has no influence on the slow release performance of urea. However, if paraffin is used as sealing agent to seal the coated urea (the dosage of paraffin is approximate 1% of the mass of the coated urea), the slow release performance of urea will be increased significantly. All the sealed coated fertilizers are leached for 6 times, and the accumulated dissolution rate is less than 6% (Cao et al. 2007). This indicates that paraffin is not an effective coating material. However, if it is added into the formula as an auxiliary, it can significantly increase the slow release performance of slow release fertilizers.

In conclusion, different types of slow release fertilizers are greatly different in terms of slow release performance. Even the same kind of slow release fertilizer may have different slow release performances. This is related to the factors such as selecting design parameters for fertilizer formula, whether auxiliary is added as well as the environment where the fertilizer performance is evaluated. To further explore the nutrient release mechanism, in-depth exploration should be carried out on the effect from the influence factors on slow release of nutrients.

4 THE MAIN FACTORS INFLUENCING THE NUTRIENT RELEASE OF SLOW RELEASE FERTILIZERS IN BEACH ENVIRONMENT

The main factors influencing the nutrient release of slow release fertilizers include the structure of the slow release fertilizer, temperature, moisture, microorganism, pH value and salinity, etc.

4.1 Structure

For coated fertilizers, the structural characteristics of the coating, such as the material characteristics of the coating, thickness, coating area, porosity and the sinuosity of pores, etc. will all affect the permeability of the coating. The smaller the film permeability is, the longer the time for controlled releasing the nutrients of the fertilizer will be. In addition, the type of fertilizer core nutrient also affects the controlled release effect of coated fertilizer (Wu & Chen 2003). When a slow release fertilizer is applied to seawater system with inferior buffer capacity, the effect of nitrate nitrogen is superior to ammoniacal nitrogen. This may be because that acid metabolin is generated amid the process that microorganisms biodegrade by using slow release fertilizer with ammoniacal nitrogen fertilizer core. Such metabolin may inhibit bio-degradation of crude oil (Wrenn et al. 1994). The result is opposite in case of salt marsh soil environment. Under such condition, the slow release effect of ammoniacal nitrogen fertilizer core is superior to nitrate nitrogen. To achieve the same bioremediation effect, the amount of ammoniacal nitrogen required is only as much as 20% of nitrate nitrogen. This may be because that nitrate nitrogen is not prone to being washed away from the environment due to the strong adsorbability between it and organic matters (Jackson & Pardue 1999).

4.2 Temperature

Temperature is closely related to the release rate of the nutrients in coated controlled release fertilizers (Cheng et al. 2013). The results from researches on applying slow release fertilizer to remediating oil spill site show that ambient temperature has a huge influence on the selection for fertilizer type. Meanwhile, when using common water-soluble urea and sulfur coated urea slow release fertilizer to conduct bioremediation for crude oil pollution on low-energy beach, if the ambient temperature is higher than 15°C, the slow release fertilizer can provide nutrients continuously and effectively, and its effect is superior to common water-soluble urea. However, when the ambient temperature is low, the release rate for the nutrients of the slow release fertilizer is very slow; it cannot effectively provide nutrients. Under such condition, the adding effect of water-soluble nutritive salts is superior to that of slow release fertilizer. This is because that low-temperature environment affects the permeability of sulfur coating, thus slowing down the release rate for the nutrients of the slow release fertilizer (Lee & Trembley 1993).

4.3 Moisture

Under certain temperature, with the increase of soil moisture content, the nutrient release rate will accelerate (Wu & Chen 2003, Zou 2007). However, there are inconsistent conclusions from some researches. According to the experiments on the accumulated nitrogen release rate of macromolecular slow release fertilizer under different moisture contents, under 25°C and 35°C, after cultivating for 34d, the accumulated nitrogen release rates as disposed in different soil moisture contents present: 60% > 40% > 80% > 100%. During the experiment interval, with the increase of moisture content, the accumulated nitrogen release rate rises before falls (Cheng et al. 2013). This indicates that when the soil moisture content is excessively high, the nitrogen release rate of macromolecular slow release fertilizer will slow down. This may be

because that soil urease and phosphatase are less active under excessively high soil moisture content, thus limiting the release of macromolecular slow release fertilizer and phosphorus.

4.4 Soil microorganisms

The slow release fertilizer and microorganisms in soil mutually affect each other. Increase of soil microorganisms will accelerate the dissolution rate of the nutrients (Jin et al. 2007). This may be because the microorganisms or various enzymes in soil participate in the degradation process of slow release fertilizer. However, this inference still needs to be further proved by microbial experiment. On the other hand, slow release fertilizer will provide nutrient substances simultaneously to accelerate the reproduction and growth of microorganisms (Hu et al. 2011). The increase of various bacteria is conducive to degrading the crude oil in beach environment polluted by oil spill (Song et al. 2004).

4.5 Influences from pH value and salinity

There are few researches on the influence from pH value and salinity on the release performance of slow release fertilizer, and the conclusions are not always consistent. Some researches show that when the pH value of the soil is in the range of 5~8, the soil acidity and alkalinity change will hardly affect the release rate of the nutrients in SCU (sulfur coated urea) (Wu & Chen 2003). However, other researches show that pH value has a huge influence on the nitrogen release performance of SCU and formaldehyde urea. Seawater medium which is near neutral is conducive to releasing nitrogen in SCU, alkalescent environment is more conducive to releasing of formaldehyde urea, in the salinity range (24.00, 32.00), salinity has no significant influence on the nitrogen release of the two fertilizers (Wu et al. 2010).

5 RESEARCHES ON PURIFYING BEACHES POLLUTED BY OIL SPILL BY USING SLOW RELEASE FERTILIZERS

5.1 Successful cases

After Exxon Vadez oil spill, people managed to eliminate the oil spill pollution within short time by using bioremediation technology, which initiated purifying oil spill pollution by using bioremediation (Bragg 1994). Later, relevant researches were carried out in laboratories and sites flourishingly. Many researches show that adding nutritive salts plays a remarkable promoting role in disposing oil spill pollution (Oh 2001, Fernandez et al. 2006). Currently, relevant researches also increase in China. Cui et al. (2013) simulated beach environment polluted by oil spill, and conducted contrast tests of multiple disposals. As indicated by the results, compared with disposal by natural weathering, after a 28d disposal period, the degradation rate of petroleum hydrocarbon increases by 24.70% and 26.40% if disposed by adding compound bacteria solution and soluble nutritive salts; the degradation rate of petroleum hydrocarbon increases by 56.88% and 59.61% if disposed by adding compound bacteria solution and slow release fertilizer. This indicates that slow release fertilizer can coordinate compound bacteria solution to strengthen biodegradation effect, and the slow release fertilizer's promoting effect for biodegradation is superior to soluble nutritive salts.

5.2 *Application premise*

In fact, certain premise conditions are required in order to purify beaches polluted by oil spill by using slow release fertilizer. Firstly, it needs to confirm whether there are suitable, abundant and available microorganisms at the site of oil spill pollution (Ronald & Don 2005). In marine environment, there are over 200 kinds of microorganisms which can degrade oil. They belong to 70 categories, including 40 categories of bacteria (Song et al. 2004). Existing researches show that the marine environment samples all contain microorganisms which can degrade oil spill population. Furthermore, the quantity of microorganisms near oil spill area is obviously larger than pollution-free sea area. This is because that the occurrence of oil spill provides abundant carbon sources, which promotes the reproduction and growth of microorganisms (Ronald & Don 2005). In case that the degrading microorganisms are insufficient at the site, adding exogenous microorganisms may be considered. Exogenous microorganisms have been proved to be successful in some researches (Fernandez et al. 2006, Cui et al. 2013). However, other researches show that exogenous microorganisms are not always effective; sometimes, it may cause serious hidden danger to environmental safety (Neralla et al. 1995).

Secondly, before determining to adopt bioremediation by adding slow release nutritive salts, it needs to investigate the background concentration of the nutrients at the oil spill site. When conducting bioremediation by using slow release fertilizer, if the nitrogen contained in the sludge exceeds 100 mol/L, then it's unnecessary to add external assistant nutrients (Oudot et al. 1998). This is because that high concentration nitrogen can only promote circulation of N and the activity of nitrobacteria, and has little effect for promoting petroleum degradation (Dunena et al. 1998).

Whether slow release fertilizer can be used by microorganisms to purify the beach polluted by oil spill is limited by dissolved oxygen in the environment on site. Oxygen is of great importance to biodegradation of oil spill. The effect of biodegradation under anaerobic condition is much slower than that under aerobiotic condition. According to the onsite bioremediation experiment on the mangrove forest polluted by oil spill at Australian Glastone Port, supplying oxygen by air compressor, coordinated by applying nutrient substances, significantly stimulates the growth of indigenous microorganisms, and improves the using effect of nutritive salts. Compared with an oil pollution site which is not disposed by bioremediation, during oxygen supply period, the quantity of alkane degrading bacteria in the mangrove forest increases by 1000 times, and the quantity of aromatic hydrocarbon degrading bacteria increases by 100 times (Ramsay et al. 2000).

5.3 *Problems*

To realize the goal of continuously providing nutrients in a more satisfactory manner, currently, people engage in preparing oleophilic slow release fertilizer. Because it can adsorb on the oil layer and retain for some time at intertidal zone, and it is not prone to being washed away by sea water, oleophilic organic fertilizer Inipol EAP-22 has presented satisfactory application effects at some arenaceous coasts (Prichard & Costa 1991). However, as proved by other researches, it can hardly play a promoting role when applied to a low-energy arenaceous coast in Canada (Lee & Levy 1989). When disposing the residual oil spill at intertidal zone, Inipol EAP-22 is more effective than common water-soluble urea. However, when the water migration on supralittoral zone is limited, it has no advantage in terms of promoting oil biodegradation (Sveum et al. 1994). This indicates that even the same slow release fertilizer may have different effects when it is applied to different beach environments.

According to the bioremediation experiment on coarse sand—gravel sandbeach polluted by oil spill, during the 120d remediation period, although the nitrogen and phosphorus contents of the slow release fertilizer added is significantly larger than that in the control system (only adding crude oil), its biodegradation constant ks is slightly less than the control system. This indicates that although the slow release fertilizer system can provide high level nutrients in a long time, it has a little effect on degradation. This may be because indigenous microorganisms cannot directly make use of the nutritive elements therein, and the nutrients released in slow release fertilizer may even form high ammonia-nitrogen environment which may restrain the growth of microorganisms (Liang et al. 2012).

In the pilot scale test on biodegradation of heavy crude oil in Tianjin intertidal zone at circum-Bohai-Sea, the degradation rate of the system crude oil added with microbial agents and nutritional agents is 72.2%. This indicates that jointly using the two can significantly promote degradation of crude oil. In the test on Dalian "7·16" oil spill accident, the degradation rate by jointly using the two for 42 days is only 39.88%. This data is not ideal compared with the data in pilot scale test. The main reason is because the remediating time started from September 3, when the air temperature was low in Dalian sea area, which had a huge influence on the bioremediation effect (Wu et al. 2011). The activity of microorganisms will reduce in low-temperature environment. Meanwhile, the release rate of the nutrients in the slow release fertilizer is too slow to effectively provide nutrients (Lee & Trembley 1993).

As indicated by the aforesaid researches, applying slow release fertilizer to purifying beaches polluted by oil spill has achieved some achievements. However, when applied to different beach environments, different fertilizers and even the same fertilize may have greatly different behaviors. Therefore, suitable slow release fertilizers should be designed and selected for different environments. Sometimes, a slow release fertilizer may have satisfactory effect in laboratory or pilot scale test, but it has common effect on site. This is mainly because that the beach is an open environment, where the site disposal effect may reduce due to the influence from temperature, salinity, wind and wave as well as tide, etc.

6 CONCLUSION AND PROSPECT

Slow release fertilizer can provide nutrients continuously in a long time. Therefore, it can be well applied to purifying beaches polluted by oil spill, and can effectively promote microorganisms to biodegrade petroleum. Featured by being low cost and environmentally friendly, it has broad application prospect. Currently, laboratory and pilot scale tests on purifying oil pollution by using slow release fertilizers have been gradually carried out both home and abroad. Satisfactory effects have been achieved. However, the nutrient release of slow release fertilizer is affected by multiple factors, which results in greatly different slow release behaviors. Therefore, relevant experiments should be carried out in an in-depth manner. In addition, there are few researches on the sites of beaches polluted by oil spill. The site environment of beaches polluted by oil spill is complicated and changeable, its temperature, salinity, pH value, wind and wave, tide and ocean current, etc. are all important factors which affect the slow release effect of slow

release fertilizer. Therefore, the subsequent main research direction should be designing the formulas of slow release fertilizers for different site environments, and surveying its purifying effect as well as influence factors.

REFERENCES

Bei, S.J. & Dong, Y. 2010. Revelation of oil spill in Mexico Bay. *China Maritime Safety* 19(6): 6–8.
Bragg, J.R. 1994. Effectiveness of biodegradation for the Exxon valdez oil spill. *Nature* 368: 413–418.
Cao, Z.H., Wang, Y.M., Guo, H.X., Tang, H. 2007. Preparation of rosin-glyceride-coated fertilizer and study on its slow release property. *Science & Technology in Chemical Industry* 15(3): 1–4.
Cheng, D.D., Zhao, G.Z., Liu, Y.Q., Hao, S.Q. 2013. Influences of soil temperature and moisture on nutrients release of polymeric slow release fertilizer and soil enzyme activity. *Journal of Soil and Water Conservation* 27(6): 216–225.
Cui, Z.S., Li, Q., Gao, W., Yang, B.J., Han, B., Zhang, K.Y., Zhou, W.J., Yang, G.P., Zheng, L. 2013. Applicability of composite bacterial culture in bioremediation of simulated oil-polluted beach. *Chinese Journal of Applied and Environmental Biology* 19(2): 324–329.
Dunena, K., Jenriings, E., Hettenbachs. 1998. Nitrogen cycling and nitic oxide emissions in oil impacted prairie soils. *Bioremediation Journal* 11(3): 195–208.
Fernandez, A.P., Vila, J., Garrido, F.J.M., Grifoll, M., Lema, J.M. 2006. Trials of bioremediation on a beach affected by the heavy oil spill of the prestige. *Journal of Hazardous Materials* 137(3): 1532–1531.
Gu, A.J., Tang, M.L., Lu, H., Zhou, S.L., Xia, W., Wang, H. 2013. Different coating slow release fertilizers' application effect in rice. *Shanghai Agricultural Science and Technology* (4): 119–141.
Hu, K., Wang, L.B., Du, H.L. 2011. Efficiency of mixed the microbial agent and slow-release fertilizer on reclaimed soil's microbial ecology. *Journal of Soil and Water Conservation* 25(5): 86–93.
Jackson, W.A. & Pardue, J.H. 1999. Potential for enhancement of biodegradation of crude oil in Louisiana salt marshes using nutrient amendments. *Water, Air, and Soil Pollution* 109(2): 343–355.
Jin, X.K., Gao, J.F., Liu, Y.Q. 2007. Slow-release property of a macromolecular slow-release fertilizer. *Chemical Research* 18(1): 61–63.
Lee, K. & Levy, E.M. 1989. Enhancement of the natural biodegradation of condensate and crude oil on beaches of Atlantic Canada. In *Proceedings of 1989 oil spill conference*: 479–486. Washington DC: American Petroleum Institute.
Lee, K. & Trembley, G.H. 1993. Bioremediation: application of slow-release fertilizers on low energy substrates. In *Proceedings of the 1993 Oil Spill Conference*: 449–454. Washington DC: American Petroleum Institute.
Liang, S.K., Zhou, P., Li, G.R., Chen, Y., Guo, L.G., Yang, S.M., Wu, L. 2012. Field-scale bioremediation of coarse sand-gravel beach contaminated by oil spill. *Journal of Tianjin University* 45(4): 343–348.

Lu, Y.Q., Zhang, X.H., Wang, D.Q., Liang, M.N., Ji, R.L., Tong, X.W., Jin, Y., Zhu, Y.N. 2008. Leaching characteristics of bamboo-charcoal coated urea and common nitrogen fertilizers. *Journal of Guilin University of Technology* 28(3): 363–369.

Neralla, S., Write, A., Weaver, R.W. 1995. Microbial inoculants and fertilization for bioremediation of oil in wetlands. In Hinchee, R.E.(ed.), *Bioaugmentation for Site Remediation*: 33–38. Columbus, OH: Battelle Press.

Oh, Y.S. 2001. Effects of nutrients on crude oil biodegradation in the upper intertidal zone. *Marine Pollution Bulletin* 42(12): 1367–1372.

Oudot, J., Merlin, F.X., Pinvidic, P. 1998. Weathering rates of oil components in a bioremediation experiment in estuarine sediments. *Marine Environmental Research* 45(2): 113–125.

Prichard, P.H. & Costa, C.F. 1991. EPA's Alaska oil spill bioremediation project. *Environmental Science and Technology* 25(3): 372–379.

Ramsay M.A., Swannell R.P.J., Shipton. W.A., Duke N.C., Hill R.T. 2000. Effect of bioremediation on the microbial community in oiled mangrove sediments. *Marine Pollution Bulletin* 41: 413–419.

Ronald, L.C. & Don, L.C. 2005. *Bioremediation: principles and applications*. UK: Cambridge University Press.

Shi, G.B., Li, X.L., Wang, Y.F. Liu, F.J., Liu, G.Q. 2013. Technology research on oil spill removal of the typical shoreline. *China Water Transport* 13(5): 89–94.

Song, Z.W., Xia, W.X., Cao, J. 2004. Biodegradation and bioremediation of petroleum contaminants in seawater. *Chinese Journal of Ecology* 23(3): 99–102.

Sveum, P., Faksness, L.G., Ramstad, S. 1994. Bioremediation of oil-contaminated shorelines: The role of carbon in fertilizers. In Hinchee, R.E. (ed.), *Hydrocarbon Bioremediation*: 163–174. Boca Raton, FL: CRC Press Inc.

Wrenn, B.A., Haines, J.R., Venosa, A.D., Kadkhodayan, M., Suidan, M.T. 1994. Effects of nitrogen source on crude oil biodegradation. *Journal of Industrial Microbiology* 13(5): 279–286.

Wrenn, B.A., Suidan, M.T., Strohmeier, K.L., Eberhart, B.L., Wilson, G.J., Venosa, A.D. 1997. Nutrient transport during bioremediation of contaminated beaches: evaluation with lithium as a conservative tracer. *Water Research* 31(3): 515–524.

Wu, L., Liang, S.K., Wang, X.L., Song, D.D., Zhang, G.C., Pang, B., Jin, H.J., Chen, Y. 2010. Study on nutrient release characteristics of two slow-release fertilizers in seawater and their abilities to enhance biodegradation of petroleum hydrocarbons. *Environmental chemistry* 29(3): 455–461.

Wu, L., Li, G.R., Chen, Y., Qu, L., Liang, S.K., Li, F.X., Wang, S., Li, Q. 2011. Research on bioremediation technology and its application of "7·16" oil spill accident scene in Dalian. In *2011 China Environmental Science Society academic annual meeting*: 2852–2866. Beijing: China Environmental Science Society.

Wu, Z.J. & Chen, L.J. 2003. *Slow controlled release fertilizers: principle and application*. Beijing: Science Press.

Xia, W.X., Li, J.C., Song, Z.W., Zheng, X.L., Lin, H.T., Lin, G.Q. 2007. Application of bioremediation agents in the cleaning of oil polluted beach. *Chinese Journal of Environmental Engineering* 1(8): 9–14.

Zhang, Y.L., Zou, H.T., Yu, N. 2005. Study on a preparation of slow-release organic and inorganic material coated urea. *Chinese Journal of Soil Science* 36(2): 55–58.

Zou, H.T. 2007. *Study on the manufacture of environmental friendly coated and slow-release fertilizer and its nutrient control and release mechanism*: 35–44. Shenyang: Shenyang agricultural university.

Advances in Energy, Environment and Materials Science – Wang & Zhao (Eds)
© *2016 Taylor & Francis Group, London, ISBN 978-1-138-02931-6*

Improved design and experimental research on sampling pipes for inductive environmental gas monitoring

Yalong Jiang & Gai Li
College of Civil and Environmental Engineering, Anhui Xinhua University, Hefei, Anhui, P.R. China

ABSTRACT: In order to improve the detection sensitivity to hazardous gases, inductive environmental gas monitoring sampling pipes are improved and relevant parameters are calculated. This paper makes use of CO and CO_2 from the combustion of cotton ropes to simulate hazardous gases in the environment and compares the detection effects before and after improvement. The experiment shows that after improvement of the sampling pipe, the response time of the detection system to hazardous gases (CO_2 and CO) is greatly shortened, from 400 s and 439 s to 55 s and 73 s successively. Moreover, the detection value of gas concentration has been improved, the maximum value of CO_2 reaches 400 ppm and that of CO 90 ppm.

1 INTRODUCTION

In recent years, the hazardous gases mentioned above have begun to seriously endanger human life as their emissions have increased along with social development, production and daily life. Thus, it is increasingly important to monitor their concentrations (Chai, 2012). For example, SF_6 has been extensively applied to high and ultra-high voltage breakers in the power industry due to its superior performance. However, such factors as differences in manufacturing and installation quality and the aging of equipment result in leakage which causes danger to the normal operation of equipment and the safety of workers (Guo et al., 2011). Furthermore, pollution resulting from interior decoration has become one of the most serious social issues. Due to the danger to the human body, it is of great importance for health protection to monitor the concentration of hazardous gases (methanal, benzene and ammonia et al.). According to statistics, over 2.8 million people (more than half of whom are adolescents and children) die from interior decoration pollution annually world-wide (Wen, 2008). Additionally, gas detection has been widely used (Fang et al., 2014; Yang and Wang, 2010; Li et al., 2008; Zhao, 2013; Hou et al., 2015; Liang et al., 2013) in other circumstances such as preservation of cultural relics (Li et al., 2014) and coal mining (Wei, 2010).

Conventional detection of hazardous environmental gases operates passively wherein the detection system detects gases and give off an alarm. Gases spread slowly at first and arrive at the detection system after the passage of some time, so it is hard to accomplish early detection. Nevertheless, an inductive detection system can extract and measure air actively, which greatly improves the detection sensitivity. The design of the sampling pipe is critical, since failures in improving detection sensitivity and early detection result from unreasonable design.

2 THEORETICAL ANALYSIS

It is quite sophisticated to calculate the parameters of the sampling pipe strictly with two dimensional fluid theory. In view of the engineering applications, many actual flow problems are deemed to be one-dimensional and a one-dimensional solution is adopted in this paper.

The sampling pipe is a PVC tube 10 m in length and 25 mm in diameter. 4 small holes are on the side of the pipe; their diameter is 5 mm and distance between them is 20 cm.

Without loss of generality, it is supposed that the diameter of the main sampling pipe is D. The small holes are round and burr-free, the diameter is d and the external pressure is barometric pressure P. In order to analyze any hole, it is supposed that downstream pressure is P_i and upstream pressure is P_i^*, then, the flow rate in the main sampling pipe is U_i as shown in Figure 1.

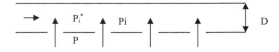

Figure 1. Sampling pipe parameter calculation.

It can be obtained from the momentum equation that:

$$P_i - P_i^* = \rho\left(U_i^2 - U_{i+1}^2\right) \qquad (1)$$

It can be obtained from the energy equation that:

$$P_i - P_i^* = \frac{2}{\rho}\left(U_i^2 - U_{i+1}^2\right) \qquad (2)$$

After the correction factor k is introduced (the value of k can be experimentally determined), the equation is established as below:

$$P_i^* - P_i = k_i\rho\left(U_{i+1}^2 - U_i^2\right) \qquad (3)$$

In case that the fluid is regarded as incompressible, the continuity equation is obtained as follows:

$$D^2 U_i = D^2 U_{i+1} + d_i^2 U_i^* \qquad (4)$$

where, U_i^* = the gas flow rate in the hole.
It can be obtained that:

$$U_i^* = C_0\sqrt{\frac{2}{\rho}\left[\left(P_i^* + P_i\right)/2 - P_a\right]} \qquad (5)$$

where C_0 = the flow coefficient and is generally 0.62, $(P_i^* + P_i)/2$ = the average pressure in the ith hole.
Given that $a = \frac{kC_0 d_i^2}{D^2}$, the following 3 equations can be obtained from formulas (3)~(5):

$$P_i^* = P_i + aU_i\sqrt{8\rho(P_i - P_a) + a^2} \qquad (6)$$

$$U_i^* = C_0\sqrt{\frac{2}{\rho}\left\{P_i - P_a + aU_i\left[8\rho(P_i - P_a)\right]^{\frac{1}{2}}\right\}} \qquad (7)$$

$$U_{i+1} = U_i - \frac{d_i^2}{D^2}U_i^* \qquad (8)$$

The upstream pressure difference of the ith hole and downstream pressure difference of the $i+1$th hole is shown below:

$$P_i^* = P_{i+1} + \lambda_{i+1}\frac{\rho l U_{i+1}^2}{2D} \qquad (9)$$

wherein, the friction coefficient is $\lambda = 64/\mathrm{Re}$ when $\mathrm{Re} \leq 2100$, or $\lambda = 0.3164/\mathrm{Re}^{0.25}$ when $\mathrm{Re} > 2100$.
The air flow rates in the main pipe and at the entrance of the holes can be obtained from the calculation formula based on the geometric parameters of the sampling pipe.

Table 1. Gas flow rate before and after improvement.

	U			
i	U_i^* (m/s)	U_i (m/s)	$U_i^{*\prime\prime}$ (m/s)	U_i' (m/s)
1	29.79	3.81	28.6	0.36
2	29.42	2.63	28.4	0.27
3	29.21	1.47	28.2	0.15
4	29.15	0.31	28.1	0.06

It is supposed that the air density is $\rho = 1.2$ kg/m^3, air viscosity coefficient is $\mu = 1.8 \times 10^{-5}$ Pa·s and correction factor is $k = 1/2$. The value of a and Re_1 are calculated first, then, $U_1^* = 29.79$ m/s, $U_1 = 3.81$ m/s and $P_1^* = 98629.3$ Pa based on formulas (7)~(9).
Other parameters can be calculated in the same way. The values of U_2^*, U_3^* and U_4^* are 29.4 m/s, 29.2 m/s and 29.1 m/s, respectively; those of U_2, U_3 and U_4 are 2.63 m/s, 1.47 m/s and 0.31 m/s, respectively. Before improvement, the values of $U_1^{*\prime\prime}$, $U_2^{*\prime\prime}$, $U_3^{*\prime\prime}$ and $U_4^{*\prime\prime}$ are 28.6 m/s, 28.4 m/s, 28.2 m/s and 28.1 m/s, respectively; and those of U_1', U_2', U_3' and U_4' are 0.36 m/s, 0.27 m/s, 0.15 m/s and 0.06 m/s, respectively. Parameters before and after improvement are compared in Table 1.
It can be seen that the air flow rates in the pipe and holes and the flow have been improved by using the improved by-pass sampling pipe.

3 EXPERIMENT AND ANALYSIS

The structural diagram of the detection system is shown in Figure 2. The high-power motor (RG125-19/12 N) is connected with the sampling pipe through the pedestal, the air sample to be detected passes through the sampling pipe and flows into the air cavity, the other end of the air cavity is connected with the gas detector which can be driven by the built-in low-power pump to extract air from the detection area. The working voltage of the sampling unit's motor is 12 V and the power is 5 W.
In order to test the performance of the improved system in detecting gas concentration, CO and CO_2 generated form the combustion of cotton ropes are used for verification. 60 pieces of 30 cm long cotton ropes are hung on a holder, lit and immediately extinguished to maintain continuous smoking. Detection results before improvement are shown in Figure 3.
The experiment time was 1200s, in which the CO can barely flow through the pipe into the gas detector due to lack of power. Although a little CO_2 flows into the detection system, the maximum

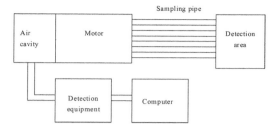

Figure 2. Structural diagram of induction environment gas monitoring system.

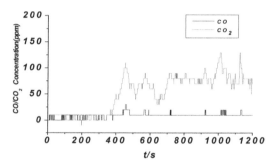

Figure 3. Experimental results before improvement.

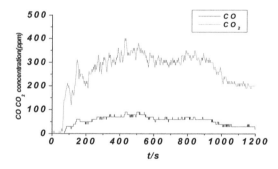

Figure 4. Experimental results after improvement.

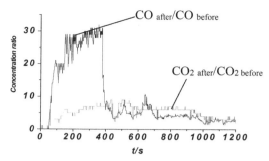

Figure 5. Concentration ratio of CO_2 and CO before and after improvement.

The system continuously analyzes the sampled concentration of CO and CO_2, but issues an alarm signal only in the case that the gas concentration detected or the growth rate of the gas concentration exceeds the given threshold value. The experiment results show that the gas concentrated detected after improvement is obviously higher than that before improvement, in other words, the measured value truthfully reflects the actual gas concentration in the detection area.

The concentration ratio of CO_2 and CO before and after improvement is shown in Figure 5. We can see that the concentration of CO has at least been tripled on average and the maximum value is up to 32 times higher. That of CO_2 has at least been quadrupled and the maximum value is up to 9 times higher. Moreover, the response time to CO has been shortened 6.01 times and the response time to CO_2 has been shortened 7.27 times.

4 CONCLUSIONS

Along with demands of social development, monitoring of hazardous environmental gas concentration takes on an increasing significance. This paper improves the conventional design of the sampling pipe in inductive environmental gas monitoring so as to improve the sensitivity of the detection system and compares the system response performance to gases before and after improvement in the experiments. The experimental results show that the maximum concentration value of CO_2 rises from 130 ppm to 400 ppm and that of CO from a low value to 90 ppm. Particularly, the concentration of CO is raised at least 3 times with a maximum of up to 32 times; while that of CO_2 is raised at least 4 times with a maximum of up to 9 times. Additionally, the response time to CO is shortened 6.01 times from 439 s to 73 s and the response time to CO_2 is shortened 7.27 times from 400 s to 55 s. Therefore, the optimized design of the sampling

value of gas concentration detected is only 130 ppm. The primary problem is the delay time in system response of as long as 400 s.

Under identical external conditions, the experiment above was repeated with the improved sampling pipe, and the experimental results in Figure 4 show that the concentration of CO and CO_2 have been improved. The maximum value of CO_2 reached up to 400 ppm and that of CO 90 ppm. Additionally, the response time of the detection system to CO_2 has been shortened from 400 s to 55 s, and to CO from 439 s to 73 s.

pipe can facilitate monitoring the concentration of hazardous environmental gases more efficiently.

ACKNOWLEDGMENT

This study was sponsored by the Key Research Program of the Department of Education of Anhui Province for Natural Science (KJ2014A101).

REFERENCES

Chai, S.C. 2012. Research on multi-gas sensor measurement method in closed environment. *Instrument Technique and Sensor* 4:76–78.

Fang, J.J., He, L.L., Xu, L.J., et al. 2014. Overview of source and detection assessment of foul gas in naval vessel environment. *Journal of Navy Medicine* 35(2):164–167.

Guo, L.M., Zhao, H.M., Lu, Y.P., et al. 2011. Design of environment online intelligent monitoring system on leakage of SF_6. *Instrument Technique and Sensor* 8:76–78.

Hou, N.N., Jin, Z., Sun, B., et al. 2015. New strategy for rapid detection of the stimulants of persistent organic pollutants using gas sensor based on 3-D porous single-crystalline ZnO nanosheets. *IEEE Sensors Journal* 15(7):3668–3674.

Li, H., Wang, W.L., Zhao, F.Y., et al. 2014. Passive sampling—ion chromatography for corrosive gases monitoring at Emperor Qin'S terracotta warriors and horses site. *Sciences of Conservation and Archaeology* 26(4):54–61.

Li, J., Zuo, Y., Shi, C.O., et al. 2008. Simultaneous determination of acidic gas in micro environment by ion chromatograph. *Environmental Chemistry* 27(5):658–661.

Liang, J.J., Yang, J.H., Tang, Z.L., et al. 2013. Study on implementation methods for butanone gas detection under circumstance of high humidity. *Transducer and Microsystem Technologies* 32(3):65–67,71.

Wei, H. 2010. An non-dispersive infrared (NDIR) gas sensor applicative for mine safety. *Journal of Minjiang University* 31(2):97–99.

Wen, W.W. 2008. Development of an indoor toxic gas monitoring system. XiAn: *Xi'An University of Science & Technology*.

Yang, L., Wang, L. 2010. Study on detection of toxic gases in traffic environment. *Computer Engineering and Applications* 46(8):202–204.

Zhao, Y., Piao, R.G., Wang H. 2013. Multi-channel gas detection system design based on Lab VIEW. *Instrument Technique and Sensor* 4:37–40.

Advances in Energy, Environment and Materials Science – Wang & Zhao (Eds)
© 2016 Taylor & Francis Group, London, ISBN 978-1-138-02931-6

Investigation and evaluation of water quality in the Qingshitan reservoir in Guilin city

Liying Liang, Yanpeng Liang & Honghu Zeng
College of Environmental Science and Engineering, Guilin University of Technology, Guilin, China
The Guangxi Talent Highland for Hazardous Waste Disposal Industrialization, Guilin, China

ABSTRACT: In order to investigate water quality variations at different spaces in it, the modern experiment instruments were used to analyze 7 water quality indicators measured at 31 different sites. The research assessed water pollution state and factors, and proposed some protective measures and schemes to improve the water quality. Based on the water quality indicators monitoring results of 31 sampling sites, in which shows that pH in Qingshitan reservoir is weakly alkaline. The value of pH ranges from 7.01 to 8.00 in west lake, which it ranges from 7.60 to 8.00 in east lake. The scope of reservoir Dissolved Oxygen (DO) ranges from 8.07 to 9.10 mg/L. The Total Phosphorus (TP) content of the reservoir ranges from 0.048 to 0.070 mg/L. Reservoir of total organic carbon content is high, ranging from 6.22 to 12.83 mg/L. Total nitrogen content ranges from 0.189 to 0.368 mg/L. According to the environmental quality standard of surface water GB3838-2002, total nitrogen surpasses standard value, which indicates that the reservoir water quality belongs to class III. Based on the lake eutrophication evaluation and classification standard, the water quality is intermediate level of eutrophication. Therefore, reinforcing the prevention and control of industrial and domestic pollution of upstream region are expected to improve the water quality of the Qingshitan reservoir.

1 INTRODUCTION

The Qingshitan reservoir is located in Guilin, Guangxi Zhuang autonomous region, Lingchuan county green lion tam town, 30 km distance to Guilin, 18 km distance to Lingchuan county town. The main functions of reservoir are irrigation, water supply, flood control, power generation, and tourism, etc. The control catchment area of reservoir reach is 474 km²; and the total capacity reach is 600 million m³; usable capacity is 405 million m³. It is the fourth in the Guangxi District reservoirs; and it is also the biggest reservoir in north Guangxi, which design of irrigation area reach 29240 ha and total control area is about 984 km² (Chen et al. 2013).

With the rapid development of industry, agriculture, and urbanization, the aqueous ecosystem in China has been severely decayed, particularly due to the water quality problem (Sun et al. 2011). National surface water quality report in August 2013 showed that in 749 sections of 406 rivers across the country, I–III water quality section shared 69%, VI–V shared 23%, and below V shared 8% (CMENC 2013). At present, the eutrophication of lakes and reservoirs has been widely regarded as a kind of environmental pollution problems. Eutrophication of water bodies is a dynamic development process, the initial performance to a large number of growth of algae and aquatic plants, the enhancement of

the photosynthesis, increase the dissolved oxygen content in water, zooplankton production increase; And to the late show is the algae blooms, the water transparency and dissolved oxygen levels drop, eventually led to the deterioration of water quality, a large number of other aquatic organisms die. Many lakes and reservoirs in China are faced with different degrees of eutrophication threats, such as the Dianchi lake, Taihu, Chaohu and already in a state of eutrophication (Jin 2001). Survey 135 representative reservoirs evaluation in China, found mesotrophic and eutrophic nutritional reservoir capacity of the number and proportion in large, medium, and small reservoirs are in dominant position, Most obviously, the lake library has different degree of eutrophication phenomenon, and above all in mesotrophic and eutrophic level. The reservoir scale eutrophication degree also from 11.2 in 1993 growth to 37%, in 2004, up nearly 26% (Meng et al. 2007).

It is necessary to investigate and evaluate water quality in the Qingshitan reservoir. In recent years, owing to the development of tourism and water tank keep fishing, and the surrounding farmland increase in the amount of pesticide used, more and more waste was produced, water ecological function was seriously damaged, which resulted in more contaminated water in the Qingshitan reservoir. And the reservoir as Guilin in many parts of the water supply, water quality testing the water quality index evaluation quality is especially important. The purpose of this paper is to reflect that the Qingshitan reservoir full system total nitrogen, total phosphorus, total organic carbon, such as the status of the water quality index, the water source of Guilin in order to accurately evaluate the water environment and ecological security, to make scientific decision for prevention and control of pollution to provide reliable technical support.

2 WATER SAMPLES OF ACQUISITION AND PROCESSING

2.1 The distribution of the sample point selection

In the study, sample sites were designed by using cross section method and reservoirs sampling technical guidance. The samples were collected in December 28, 2014, using GPS to fix sampling points. The data of The Qingshitan reservoir water quality were collected from 31 monitoring sites. The west lake of sampling points from 1 to 23 and the east lake of sampling points from 24 to 31 (see Fig. 1). The samples were taken back by sealing brown glass bottles and determined in laboratory used national standard method.

2.2 The main experimental instruments and equipment

TOC Analyzer (Analytik Jena, Multi N/C® 3100 a new generation of total organic carbon analyzer),

Figure 1. Distribution of the sampling points.

Inductive Coupled Plasma Optical Emission Spectrometer (PerKinElmer, Optima 7000DV, ICP-OES) Portable Dissolved Oxygen Meter, Portable pH Meter, Thermometer.

2.3 The determination of water quality index

Monitoring indicators included water temperature, Dissolved Oxygen (DO), pH, Total Nitrogen (TN), Total Phosphorus (TP), Total Organic Carbon (TOC), Total Carbon (TC). Among them, the temperature, DO, pH, were in situ tested by thermometer, portable dissolved oxygen meter and Portable pH meter. The other indicators were tested in laboratory.

2.4 The methods

The results were produced using precision instrument measured data directly. In addition to field measurement indicators, the rest of the laboratory measurement indexes through the use of precision measuring instrument. Adding 10 ml of water samples into 10 ml of centrifugal pipe, and then manual sampling, the values were measured by using TOC analyzer and ICP analyzer.

2.5 The standards

Standards for Drinking Water quality, Environmental quality standard of surface water GB3838-2002.

3 THE RESULT AND DISCUSSION

3.1 The change of water temperature

In the two lakes, the temperature range is not large. In the 23 sampling points of the west lake, the upper average temperature is 20.64°C; the middle is 20.01°C; the bottom is 20.03°C. 8 sampling points of the east lake, the upper average temperature is 20.29°C; the middle is 20.16°C, the lower is 20.46°C. Temperature changes in the scope of meet the upper > the middle > the lower overall.

3.2 The change of pH value

There are many effects of water quality indicators, and the pH is one of them. It plays an important role in evaluating index of eutrophication (Sui et al. 2011). In 23 points of the west lake, the value of pH were tested, and the mean value reach 7.30. The pH value varied between 7.01 and 8.00. And eight testing sites in the east lake, the mean value of pH reach 7.84. The pH value varied between 7.60 and 8.00, which is slightly higher than the average of the west lake. The water quality is slightly alkaline, but still in conformity with the Environmental quality standard of surface water GB3838-2002. This scope shall not be less than 6.5 and no more than 8.5. Alkaline water is more suitable to the growth of blue-green algae. It is influential to sharpen the reservoir eutrophication process. The variation of value of pH shows apparently between different sampling points.

3.3 The change of dissolved oxygen

Dissolved Oxygen (DO) is an important index of water can maintain ecological balance, and also is the primary productivity of lake and the comprehensive reflection of hydrodynamic conditions. Research the DO and seasonal change of layer to recognize the eutrophication process of lake has important significance() In 23 points of the west lake, the mean value of upper water dissolved oxygen is 9.10 mg/L, the middle is 8.62 mg/L, the lower at 8.58 mg/L. Three layers of dissolved oxygen value change are obvious. East lake 8 sampling points, the upper water dissolved oxygen mean value is 8.30 mg/L, the middle is 8.44 mg/L, the lower is 8.07 mg/L. The east lake three layers water dissolved oxygen average lower than that of the west lake. Two different sampling point value of dissolved oxygen in the lake change is more obvious, and dissolved oxygen value range meet the upper > the middle > the lower. According to the water environment quality standard, the reservoir water quality belongs to class I based on content of dissolved oxygen.

3.4 The changes of total phosphorus

The content of phosphorus has an important influence in the process of eutrophication. With the increase of total phosphorus concentration, the water quality of eutrophication in reservoir will be enhanced (Wang et al. 2006). The Qingshitan reservoir variation in content of Total Phosphorus (TP) in the west lake is from 0.048 to 0.069 mg/L, with a mean of 0.054 mg/L. The content of Total Phosphorus (TP) in the east lake change ranges from 0.049 to 0.070 mg/L, with a mean of 0.053 mg/L. The average content of total phosphorus in the east lake is slightly higher than the west lake. As show in the Figure 2, the changes

of total phosphorus are not obvious between each sample points, and changes not regular between each layer. According to the lake eutrophication evaluation and classification standard (Table 1), the water quality of the Qingshitan reservoir is the intermediate level of eutrophication. According to the water environment quality standard, reservoir water quality belongs to III class.

3.5 The change of total organic carbon

Total organic carbon is made from carbon to represent how much organic material in water, including Dissolved Organic Carbon (DOC) and Particulate Organic Carbon (POC). It reflects the process of organic pollutants in water pollution (Teng et al. 2014). The average of total organic carbon of the upper water is 6.24 mg/L in the reservoir of the west lake; the middle is 7.84 mg/L; lower level of 12.11 mg/L. The average of total organic carbon is 12.83 mg/L in the east lake upper class water, the middle is 10.60 mg/L, the lower is 11.04 mg/L. Observing the data, the total organic carbon in the west lake is lower than that of the east lake. Observing the data, the total organic carbon in the west lake is lower than that of the east lake. The People's Republic of China issued by the ministry of construction on September 28, 1999, the water quality standard of drinking water purification (CJ 94-1999), TOC inspection standards shall not exceed 4 mg/L. But it did not make provisions to drinking water sources.

3.6 The change of the total carbon

Total carbon content changes greatly between each point. Upper class water in the west lake is from 3.07 to 17.87 mg/L; the middle change from 6.54 to 27.07 mg/L; the lower change from 7.19 to 22.44 mg/L. The total carbon content of upper class water at east lake changes from 5.65 to 45.55 mg/L; the middle changes from 5.91 to 21.92 mg/L; the lower changes from 8.90 to 27.59 mg/L.

3.7 The change of total nitrogen

Nitrogen and its various nitrides is one of the main characterization indexes, which is the main

Table 1. Lakes eutrophication evaluation and classification standard.

Total phosphorus (TP/mg/L)	0.001	0.004	0.01	0.025	0.05	0.1	0.2	0.6	0.9	1.3
Score	10	20	30	40	50	60	70	80	90	100
Nutrition level	Dystrophic		Mesotrophic		Eutrophic					

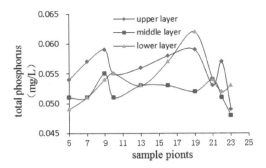

Figure 2. Parts of the sample points of west lake total phosphorus content changes.

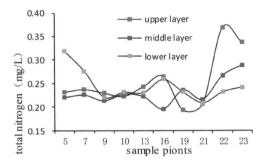

Figure 3. Parts of the sample points of west lake total nitrogen content changes.

research content of eutrophication (Wang et al. 2001). Nitrogen is an important nutrient element of phytoplankton growth, also is the main factor causing eutrophication (Andriedx F. et al. 1997). West lake water of total nitrogen content in the reservoirs ranges from 0.189 to 0.368 mg/L, with a mean of 0.242 mg/L. Total nitrogen content in the east lake water change ranges from 0.198 to 0.394 mg/L, with a mean of 0.237 mg/L. The average of total nitrogen content in West Lake is slightly higher than that of the east lake. From the Figure 3, it shows that total nitrogen content changes evidently between each site. According to the surface water environment quality standard GB3838-2002, only consider the index of total nitrogen, the Qingshitan reservoir water quality belongs to class I.

4 CONCLUSIONS

Through the investigation, the Qingshitan reservoir water quality condition overall is good. Water temperature and pH changed little in space. Water temperature and pH value meet the water environmental quality standards. The value of dissolved oxygen meets the water quality standard which is class I. Total phosphorus content on the high side, in the level, belongs to the class III water. Total nitrogen content is up to the water quality standard, which belongs to class II. On the whole, the Qingshitan reservoir water quality pertains to class III. Strengthening the environmental protection is a duty that can't wait any more, and making a significant contribution to the people of reservoir area. For the Qingshitan reservoir water quality, and put forward the following suggestions:

1. Accelerate the construction of reservoir upstream of sewage treatment facilities, and reduce the sewage directly discharged into the reservoir.
2. To increase the construction of agricultural ecological sections.
3. Strengthen the implementation of reduce the production and the detergent with phosphorus.
4. Strictly control the amount of water tank fish and perfect all kinds of measures.

ACKNOWLEDGMENTS

The authors thank the Guangxi Key Laboratory of Environmental Pollution Control Theory and Technology for the research assistance and the financial supports from the National Natural Science Foundation of China (No. 51268008, 21207024), the Key Project of Chinese Ministry of Education (No. 210170, JiaoJiSi [2010]114), the Program for Excellent Talents in the Guangxi Higher Education Institutions (No. GuiiaoRen [2010]65), The Guangxi Talent Highland for Hazardous Waste Disposal Industrialization.

REFERENCES

Andriedx F. et al (1997). A two-year survey of phosphorus speciation in the sediments of the Bay of Seine. Continental Shelf Research. 17(10): 1229–1245.
Chinese National Environmental Monitoring Center (CNEMC) (2013). National surface water quality monthly (August) in, Test Report. (in chinese)
Hongjie Gao et al. 2015. Chemometrics data of water quality and environmental heterogeneity analysis in Pu River, China. Environ Earth Sci (2015) 73:5119–5129.
Hongming Meng et al. 2007. The main reservoir eutrophication status evaluation. Journal of henan normal university (natural science edition), 2007, 35(2): 133–137.

Jing Wang et al. 2006. The total phosphorus distribution, migration and control of in the Miyun reservoir. Water conservancy of Jilin. 280(8):01–04.

Lei Chen et al. 2013. Investigation and evaluation Guilin Qingshitan reservoir water quality present situation. Agricultural and science in Guangdong 40(5):160–164.

Lianghuang Sui et al. 2011. A research of aquaculture water PH, dissolved oxygen, and between chlorophyll Journal of environmental engineering 5(6):1201–1208.

Lijia Wang et al. 201. The pollution and eutrophication problems of Dahuofang reservoir water quality. Liaoning urban and rural environmental science and technology 21(6):45–47.

Mingde Teng et al. 2014. Total organic carbon spatial distribution characteristics of Wan feng lake. Environmental Monitoring of China 30 (2):35–39.

Sun H., et al (2011) Water quality assessment of Yellow River based on multivariate statistical analysis. J Agro-Environ Sci 30(6):1193–1199.

Xiangcan Jin. 2001. Control and management technology of lake eutrophication. Beijing.

Xiaoyi Cheng et al. 2013. Modeling of seasonal vertical variation of dissolved oxygen and its impacts on water environment in Shahe Reservoir within Tianmuhu Reservoir. Lake Sciences 2013, 25 (6):818–826.

Advances in Energy, Environment and Materials Science – Wang & Zhao (Eds)
© *2016 Taylor & Francis Group, London, ISBN 978-1-138-02931-6*

Investigation and evaluation of water quality of Qingshitan reservoir surrounding agricultural wetland

Zhongjian Wu, Yanpeng Liang & Honghu Zeng
College of Environmental Science and Engineering, Guilin University of Technology, Guilin, China
The Guangxi Talent Highland for Hazardous Waste Disposal Industrialization, Guilin, China

ABSTRACT: Six water quality parameters which including pH, dissolved oxygen, total phosphorus, total nitrogen, total organic carbon, total carbon of Qingshitan reservoir surrounding agricultural wetland are tested. The results show that pH of Qingshitan reservoir surrounding agricultural wetland water present weak alkalinity, and the variation range of the total phosphorus content in streams, ditches, ponds respectively are 0.049~0.120 mg/L, 0.062~0.114 mg/L, 0.055~0.139 mg/L; the variation range of total nitrogen content in streams, ditches, ponds are 0.353~1.880 mg/L, 1.070~1.640 mg/L, 0.633~2.64 mg/L; the variation range of dissolved oxygen content in streams, ditches, ponds are 8.490~9.210 mg/L, 8.690~8.840 mg/L, 8.330~10.450 mg/L; the variation range of total organic carbon content in streams, ditches, ponds are 8.180~17.040 mg/L, 13.980~27.490 mg/L, 8.050~24.490 mg/L; the variation range of total carbon content in streams, ditches, ponds are 14.170~46.060 mg/L, 26.940~34.870 mg/L, 17.540~54.93 mg/L. According to Environmental Quality Standard for Surface Water (GB3838-2002), the water quality status of the streams and ditches reached the third water quality standard as a whole, but the total nitrogen exceeded this standard, and total nitrogen exceeding standard rate is 33%. The water quality of the ponds reached the fourth water quality standard. It is found that the loss of chemical fertilizers and pesticides in agricultural production and discharge of household garbage and waste water and feces of livestock are the main factors that influenced the water quality in pollution sources investigation. Based on the assessment, some suggestion on the agricultural wetland water protection and management are put forward.

1 INTRODUCTION

Guilin is a national historical and cultural city and the world famous tourism city. The Lijiang River is the soul of the Guilin scenery. Qingshitan reservoir is located in Qingshitan town, Lingchuan County, Guilin City, Guangxi Province. It is a comprehensive utilization of large-scale reservoir, which is mainly irrigation, both Lijiang tourism hydrating, power generation, aquaculture, flood control, etc (Chen et al. 2013). The water quality of reservoir reached Environmental Quality Standard for Surface Water Class II in the mid-90s in last century. After the mid-90s, due to large-scale cage fish culture and tourism resort development in the reservoir, especially on the edge of the West Lake, living garbage and sewage of the Gongping town were discharged arbitrarily, which made the surrounding environment of the reservoir polluted and leaded to a progressive decline in water quality of the reservoir (Chen 2013). Some studies show that the main indicators of water quality monitoring of Qingshitan present a rising tendency year by year, and water quality of reservoir is gradual deterioration (Yin 2012).

In recent years, with the development of economic construction, urban and rural construction and dense population to the city, Guilin is facing great pressure in urban water supply, tourism environment and ecological environment in (Wen et al. 2006). Therefore, the water qualities of Qingshitan reservoir directly affect people's health in Guilin, economic development and social stability. Thus, there is an important practical significance in evaluating the water quality of Qingshitan reservoir.

Water quality of Qingshitan reservoir surrounding agricultural wetland water is closely related to Qingshitan reservoir. In order to know the causes of the declining water quality of Qingshitan reservoir, we collected 21 water samples from Qingshitan reservoir surrounding agricultural wetland water, and detected the 6 water quality parameters, such as pH, dissolved oxygen, total phosphorus, total nitrogen, total organic carbon, and total carbon. To evaluate the water quality of Qingshitan reservoir surrounding agricultural wetland water with reference to Environmental Quality Standard for Surface Water (GB3838-2002), and provide a scientific basis for the water resources utilization and water pollution control.

2 RESEARCH METHODS

2.1 Sample collection

In order to fully understand the water pollution sources of Qingshitan reservoir, the water quality of the reservoir surrounding agricultural wetlands are investigated. Due to the complicated terrain of the study area, and ponds, rivers and ditches distribution is not uniform, the sampling mainly along the highway around the reservoir, according to the village cloth. Select the ditch, spring water, village ponds, and river which empties into the reservoir. Sampling time is November 22, 2014. The sampling points used GPS, a total of 13 sampling sites, and 21 water samples are collected, and the sampling volume of 3.5 L, and we collected water samples with brown bottles. The sampling distribution is shown in Figure 1 (sample point 1~13) and the corresponding place names of each sample point is shown in Table 1.

2.2 Determination of water quality parameters

Water quality determination includes field measurement parameters: pH and temperature;

Figure 1. Sampling point distributions.

Laboratory measurement parameters: Dissolved Oxygen (DO), Total Phosphorus (TP), Total Nitrogen (TN), Total Organic Carbon (TOC), Total Carbon (TC).

2.3 Main experimental apparatus

TOC Analyzer (Analytik Jena, Multi N/C ® 3100 a new generation of total organic carbon analyzer); Inductive Coupled Plasma Optical Emission Spectrometer (PerKinElmer, Optima 7000 DV, ICP-OES); Portable Dissolved OxygenMeter; Portable pH Meter, Thermometer.

2.4 Evaluation method

The water quality of water samples is evaluated according to Environmental Quality Standard for Surface Water (GB3838-2002) Class III. In this study, TOC evaluation of surface water in place of COD. Total organic carbon is the total carbon of organic carbon in water, and it is a comprehensive index of total content of organic matter in water, including benzene, pyridine and other aromatic hydrocarbons. Therefore, TOC can completely reflect the pollution level of organic matter in water (Wang et al. 2001). The related research showed that there is a linear correlation between TOC and COD in surface water, a correlation coefficient of 4 surface water was 0.40~0.93 (Sun et al. 2013). The correlation coefficient was 0.90, which was used for TOC evaluation in this research.

3 THE DETECTION RESULTS

3.1 Water quality detection result of streams and springs

A total of 6 water samples of three rivers and three springs are detected, and pH, DO, TP are not exceeded. TN of Xinjiang and Shangtianxin two streams did not conform to Environmental Quality Standard for Surface Water (GB3838-2002) Class III, the standard-exceeding rates of them are 33%. The variation range of TOC is 8.18~17.04 mg/L, the average of TOC is 11.93 mg/L, and the variation range of TC is 14.17~46.06 mg/L, the average of TC is 26.20 mg/L. The results of the detection of water samples are shown in Table 2.

Table 1. The corresponding place names of the sampling points.

Sampling site No.	1	2	3	4	5	6	7	8~10	11	12	13
Place names	Xin Jiang	Mei Zi	Nong Tan	Tian Xin	Da Qiao	Si Ji	Wu Mei	Gong Ping town	Huang village	Ping long zi	Song shu jiang

Table 2. The water quality detection results of the ditches.

Place names	pH	DO (mg/L)	TP (mg/L)	TOC (mg/L)	TN (mg/L)	TC (mg/L)
Zhou village	9.00	8.84	0.062	14.49	1.240	36.87
Xinjiang	8.80	8.69	0.114	13.98	1.070	26.94

3.2 Water quality detection result of ditches

The water quality of the two ditches in the Zhou Village and Xinjiang is detected, according to the national standard of PH in the Water quality standard for Surface Water (GB3838-2002) is 6~9. Zhou village and Xinxiang ditches' pH significantly higher than the national standard limit, is weak alkaline. Other water quality parameters detection is shown in Table 3. It shows that the water quality status of ditches reached the third water quality standard as a whole, but the total nitrogen exceeded this standard.

3.3 Water quality detection result of ponds

According to Water quality standard for Surface Water (GB3838-2002) Class III, pH national standard limit is 6~9, DO national standard is greater than 5 mg/L. A total of 12 water samples of ponds are detected, and the variation range of pH is 7.30~9.70, the average of pH is 8.25; the variation range of DO is 8.13~10.45 mg/L, the average of DO is 8.70 mg/L; the variation range of TP is 0.055~0.139 mg/L, the average of TP is 0.090 mg/L; the variation range of TOC is 8.05~24.49 mg/L, the average of TOC is 14.48 mg/L; the variation range of TN is 0.286~2.640 mg/L, the average of TN is 1.144 mg/L; the variation range of TC is 17.54~54.93 mg/L, the average of TC is 32.87 mg/L. There are 5 points of the pH that are not consistent with Water quality standard for Surface Water (GB3838-2002) Class III, the standard-exceeding rates of them are 42%, the water quality is weakly alkaline; DO is in conformity with Water quality standard for Surface Water (GB3838-2002) Class III. There are 8 points of the TN that are not consistent with Water quality standard for Surface Water (GB3838-2002) Class III, the standard-exceeding rates of them are 67%. There are 3 points of the TOC that are not consistent with Water quality standard for Surface Water (GB3838-2002) Class III, the standard-exceeding rates of them are 25%. The results of the detection of water samples are shown in Table 4. In terms of pH, DO, TP, TN, TOC, TC, we evaluate water quality in accordance with Environmental Quality Standard for Surface Water (GB3838-2002), it turn

Table 3. The water quality detection results.

Detection items	Detection result ($\bar{x} \pm s$)	Over standard Number (n (%))	National standard
pH	8.40 ± 0.41	0 (0)	6~9
DO	8.74 ± 0.40	0 (0)	≧5
TP	0.075 ± 0.030	0 (0)	≦0.2
TN	0.807 ± 0.700	2 (33)	≦1.0
TOC	11.93 ± 3.29	0 (0)	≦18
TC	26.20 ± 12.36	–	–

Note: except for pH dimensionless, over standard number of units are EA, the other units are mg/L.

Table 4. The water quality detection results.

Detection items	Detection result ($\bar{x} \pm s$)	Over standard Number (n (%))	National standard
pH	8.25 ± 0.68	5 (42)	6~9
DO	8.70 ± 0.60	0 (0)	≧5
TP	0.090 ± 0.030	0 (0)	≦0.2
TN	1.144 ± 0.600	8 (67)	≦1.0
TOC	14.48 ± 4.68	3 (25)	≦18
TC	32.87 ± 9.94	–	–

Note: except for pH dimensionless, over standard number of units are EA, the other units are mg/L.

out that ponds water quality belong to the fourth water quality standard.

4 DISCUSSIONS

Contrast in agricultural wetland in three different water pollution, pollution degree from large to small is ponds, ditches, streams, and springs. The possible cause of this phenomenon is that pond is relatively stable water, water exchange slower, and there are a large number of livestock farms; poultry feed and manure flow into the pond, and make pond water quality worse. The stream is flowing continuously and changing frequently that the water pollution is not too serious. Ditches are easy to be affected by human activities on both sides of the road; the pesticide bags and garbage fall into the ditch and make the deterioration of water quality. Ditches pollution situation is between ponds quality and streams quality.

Qingshitan reservoir surrounding agricultural wetland water' pH, total phosphorus, total nitrogen, total organic carbon and total carbon are significantly higher. The water quality of drinking water is affected by many factors, such as the local soil, seasonal climate, human industrial activities and domestic sewage; livestock can penetrate into the water through a variety of ways and cause water pollution (Ruan et al. 2010). The field investigation found that the living sewage and garbage were discharged directly by residents of Qingshitan. It will lead to new threats for water quality of reservoir. There are also some livestock and poultry farms in the shore of Qingshitan reservoir. Thus, it can produce a great deal of sewage, and without any treatment measures, it is an important factor affecting the water quality of the reservoir. The reason for the high total phosphorus concentration is that a large number of available phosphorus and compound fertilizer are applied in agricultural production. The characteristic of this kind of fertilizer is that it has a fast effect but the loss is quick, which is easily flowed to water environment and caused the pollution of the streams and ponds (Gao et al. 2006). Due to the lack of awareness of environmental protection, the extensive use of chemical fertilizers, pesticides, and indiscriminate discharge of the sewage and garbage, resulting in a variety of pollutants flow into the water, the water' pH, total phosphorus, total nitrogen, total organic carbon are significantly higher.

5 CONCLUSIONS

Qingshitan reservoir surrounding agricultural wetland can flow directly or indirectly into the Qingshitan reservoir, thus affecting the quality of the Qingshitan reservoir water quality. After considerable investigation may come to a conclusion that the loss of chemical fertilizers and pesticides in agricultural production and discharge of household garbage and waste water and feces of livestock are the main factors that influenced the water quality in pollution sources investigation. The water quality of Qingshitan reservoir surrounding agricultural wetland is maintained between the third and the fourth water quality standard, and the water quality situation may even worsen in summer. Thus, the future evolution tendency is not optimistic, and water pollution is a big potential threat. According to the current development trend of water quality, the reservoir water supply security will be affected. According to the Qingshitan reservoir surrounding agricultural wetland pollution situation, put forward the following suggestions:

1. To strengthen the comprehensive improvement of the surrounding environment of the Qingshitan, strictly control the emissions of pollutants and carry out thorough investigation of the hazards of water environmental violations.
2. To strengthen the environmental awareness of the Qingshitan nearby villagers. The majority of people are the most effective environmental supervisors, and the hazards should be introduced to villagers that the reservoir pollution may have a huge negative impact for the villagers' life. Standard legal system and the act of

endangering the reservoir must be investigated for legal responsibility according to the law.
3. To take the corresponding environmental policies according to different sources of pollution. Pesticide and chemical fertilizer in the farmland are important ways to pollute the water quality of the reservoir, and the government can limit the amount of fertilizer to reduce the excess supply of nitrogen and phosphorus.
4. In view of the livestock industry in water pollution control, we should take the livestock manure harmless treatment and resource utilization (Liu et al. 2010).

ACKNOWLEDGMENTS

The authors thank the Guangxi Key Laboratory of Environmental Pollution Control Theory and Technology for the research assistance and the financial supports from the National Natural Science Foundation of China (No. 51268008, 21207024), the Key Project of Chinese Ministry of Education (No. 210170, JiaoJiSi [2010]114), the Program for Excellent Talents in the Guangxi Higher Education Institutions (No. Gui iaoRen [2010]65), The Guangxi Talent Highland for Hazardous Waste Disposal Industrialization.

REFERENCES

Bin Ruan, Ru feng, et al. 2010. Current situation and Countermeasure of livestock manure pollution in China. Guangdong Agricultural Sciences 37(6):213–216.
Hui Liu, Lingyun Wang, et al. 2010. Current situation and Countermeasure of livestock manure pollution in China. Guangdong Agricultural Sciences 37(6):213–216.
Lei Chen, Jianping Qian, et al. 2013. Investigation and evaluation of water quality of Qingshitan reservoir in Guilin city. Guangdong Agricultural Sciences 40(5):160–164.
Liyang Sun, Zhipeng Yao, et al. 2013. Study on the relationship between TOC and COD in surface water. China Environmental Monitoring 29(2):125–130.
Mingzhong Chen. 2013. The technology and managemengt measures of water source protection and ecological restoration in Qingshitan reservoir. Technology and Enterprise (10):140–140.
Yanhong Wen, Anyou He. 2006. The upper reaches of the Lijiang River Qingshitan reservoir eutrophocation and its comprehensive treatment. Guangxi Fishery Science and Technology (3):20–25.
Yang Gao, Jinzhong Zhou, et al. 2006. Study on the environmental pollution and protection of Sand River Reservoir. Chinese Agricultural Science Bulletin 28(8):503–506.
Yuanyuan Yin. 2012. Water quality change trend and evaluation of the Qinghsitan reservoir. Instrumentation and Analysis (2):38–40.
Zhiguo Wang, Guogang Li. 2001. Study on the correlation between TOC and COD in water. China Environmental Monitoring 15(1):1–3.

Effect of nitrogen fertilizer on the adsorption of cadimum in typical Chinese soils

Xuhui Li & Ruichao Guo

Key Laboratory of Environment Change and Water-Land Pollution Control (University of Henan Province), College of Environment and Planning, Henan University, Kaifeng, China

ABSTRACT: Adsorption of Cd was of great importance for its environmental behaviors. Nitrogen fertilizers were often used to intensify the efficiency of phytoremediation. Aiming for exploring the influences of nitrogen fertilizers on Cd adsorption, batch experiments were carried out in typical Chinese soils. Cd was strongly adsorbed by typical Chinese soils. Phaeozem had the highest capacity of Cd adsorption, followed with moisture soil and krasnozem. Soil OC contents and soil pH value were key factors affecting Cd adsorption in soils tested. The application of nitrogen fertilizer, especially for NH_4^+-N, lowered down the adsorption of Cd, which meant that the bioavailability of Cd was enhanced. Our study suggested that NH_4^+-N was suitable to be used as nitrogen fertilizer in phytoremediation.

1 INTRODUCTION

Rapid urbanization is increasing environmental contamination in these areas with intensive industrial and agricultural activities result in increased use of pesticides, fungicides, and chemical fertilizers (Thawornchaisit & Polprasert, 2009) and increased use of wastewater irrigation and sewage sludge (Muchuweti *et al.* 2006). These activities result in increased heavy metal contamination of soils. Cadmium (Cd) is a heavy metal of great concern in agricultural ecosystems because of its high toxicity to animals and human health. Cd toxicity especially affects human health rather than plants and animals because of its longevity and the accumulation in organs through consumption of Cd-contaminated foods (Cheng & Gobas, 2007). Chronic exposure to human beings of cadmium through ingestion and/or inhalation results in severe damage to kidneys and lungs and can lead to other pathological symptoms such as itai-itai disease (Phillipp, 1980). It is one of the toxic metals under scrutiny by the US Environmental Protection Agency (USEPA). Cd enters agricultural soils through addition of sewage sludge, composts, chemical fertilizers, and wastewater irrigation (Muchuweti *et al.*, 2006; Zhou *et al.*, 2006). In China, Cd contamination has been considered as a serious problem for food safety (Bao *et al.*, 2013). According to the China's Soil Pollution State Bulletin (the Ministry of Environmental Protection of PRC & the Ministry of Land and Resources of PRC, 2014), the rate of Cd exceeding the standard was more than 7.0%.

The adsorption process of pollutants in soil affects its transportation (Wilde *et al.*, 2008) and bioavailability and hence is of great importance to environmental regulation and pollution control (Virag & Kiss, 2009). Freundlich models and/or Langmuir models were often used to fit adsorption isotherms of Cd. The main factors affecting Cd adsorption/desorption are soil organic matter content, pH values, amounts of clay minerals, Fe and Mn oxides, and calcium carbonate (Tipping *et al.*, 2003).

Fertilizers are widely used to improve the performance of phytoremediation, which is considered as an effective technique for the remediation Cd contamination soils. However, the interaction between fertilizer and Cd has not been studied systematically. Hence, the present study is aiming for exploring the influences of nitrogen fertilizers on Cd adsorption in typical Chinese soils.

2 MATERIALS AND METHODS

2.1 Soil characterization

Phaeozem, moisture soil and krasnozem, three typical soils distributed in northeastern, central, and southwestern China, were used in the present study. Phaeozem was collected from the Northeast Institute of Geography and Agroecology, Chinese Academy of Sciences, the moisture soil was collected from Henan University and krasnozem was collected from Yunan Academy of Forestry. Soil samples were collected from the surface layer (0–20 cm depth), air-dried, and sieved through a 2-mm, nylon-fiber sieve to remove stones, plant roots, and other large particles.

The organic C contents were 1.91%, 0.89%, and 0.33% for phaeozem, moisture soil, and krasnozem, respectively. The responding pH values were 6.96, 8.78, and 5.10, respectively. The organic C content was determined using $K_2Cr_2O_7$ method. The soil pH value was measured with a glass electrode in a 1:2.5 soil/water suspension using 1 M $CaCl_2$ solution.

2.2 Adsorption experiments

Adsorption experiments were carried out under the guidance of the standard batch equilibration method (OECD 2000). The previous study showed that 24 hrs was long enough to reach equilibrium for adsorption of Cd, so the adsorption experimental period was 24 hrs. The experiments were carried out at the original pH value of each soil at 25 ± 0.5 °C. 0.25 g of soil samples were weighed into 40 mL plastic centrifuge tubes, and 20 mL 0.01 M $CaCl_2$ solution containing different concentrations of Cd (0.30, 1.00, 1.50, 2.00, 2.50, 3.00 and 3.50 mg/L) were introduced. A blank (without soil) was also maintained to assure the quality control of the experiments. The samples were centrifuged at 2,500 rpm for 30 min after having been horizontally shaken in dark by an orbital shaker at 140 rpm for 24 hrs. The suspensions were filtered through a 0.25 μm filter membrane and used for Cd concentration determination.

The experiments of Cd adsorption affected by nitrogen fertilizers were carried out in the Cd solution containing different amount of NH_4Cl or $NaNO_3$. The concentrations of nitrogen were set as 250 and 500 N mg/L.

All of the experiments were carried out in triplicates. And the concentration of Cd was determined by ICP-AES (Thermo Fisher iCAP 6000 Series).

2.3 Data processing and statistical analysis

All data were analyzed by Excel 2007 (Microsoft, USA) and Matlab 7.0 (Mathworks, USA). Adsorption amount Cd in adsorption experiments was calculated as equations (1):

$$C_s = \frac{V}{M_s(C_i - C_e)} \tag{1}$$

where C_s (mg/kg) represents the adsorbed amount of a pollutant on soil per unit, V (L) is the volume of solution added, M_s (g) is the weight of soil samples, C_i (mg/L) is the initial concentration, C_e (mg/L) is the equilibrium concentration in the liquid phase.

The linear model and the Langmuir model are theoretical models based on distribution theory and monolayer adsorption theory, respectively. The Freundlich model is an empirical expression that encompasses the heterogeneity of the surface and exponential distribution of the sites and their energies. The adsorptions Cd were in turns described by the linear model, the Freundlich model and the Langmuir model:

$$C_s = K_d C_e \tag{2}$$

$$C_s = K_f C_e^{\frac{1}{n}} \tag{3}$$

$$C_s = \frac{C_{max}(K_L C_e)}{1 + K_L C_e} \tag{4}$$

where K_d (kg/L) is the distribution coefficient, K_f is the Freundlich constant, 1/n represents energy distribution of adsorption sites, C_{max} (mg/kg) is the maximum adsorption capacity of an adsorbent and K_L (L/mg) is the Langmuir constant related to the affinity of soil. K_{oc} (L/kg) was the K_d value corrected for organic carbon content (OC%) by calculating:

$$K_{OC} = \frac{K_d}{OC\%} \tag{5}$$

3 RESULTS AND DISCUSSION

3.1 Cd adsorption in typical Chinese soils

The parameters of the linear model, the Freundlich model and the Langmuir model for the adsorption of Cd were listed in Table 1.

Cd was strongly adsorbed by the soils studied and linear, Freundlich and Langmuir models can describe the adsorption isotherms accurately. Compared with the linear and Langmuir model, the Freundlich

Table 1. Parameters of linear, Freundlich and Langmuir models for Cd adsorption.

Soil type	Linear		Friendrich			Langmuir		
	K_d	R^2	K_f	1/n	R^2	C_{max}	K_L	R^2
S1	182.61	0.96	189.78	0.76	0.99	476	0.62	0.99
S2	64.15	0.96	88.35	0.60	0.99	198	0.84	0.97
S3	14.46	0.98	14.15	1.06	0.96	14373	0.001	0.96

*S1, S2 and S3 represented for phaeozem, moisture soil and krasnozem, respectively.

model had a better simulation in low concentrations for Cd adsorption, suggested by higher related coefficients. The adsorption isotherms of Cd in different soils were presented in Figure 1.

As shown in Table 1 and Figure 1, the capacity on Cd adsorption varied with soil types. Phaeozem showed the highest Cd adsorption capacity, followed with moisture soil and krasnozem. This might be caused by the different soil OC contents. The Soil Organic Matter (SOM) was the most important component for many biological, physicochemical and chemical processes. The soil OC content was about fifth time higher than moisture soil, and triple of krasnozem. However, the order of K_{OC} values was quite different. The K_{OC} values (calculated from Equation 5) were 9885.16, 19440.95, and 1339.31 L/kg for phaeozem, moisture soil and krasnozem, respectively. Obviously, the K_{OC} values reduced with the decrement of soil pH values, indicating pH value was another key factor influenced the Cd adsorption in soils. As pH

value might change the characteristic and surface charge of soil organic matter, thus affecting the adsorption of Cd (Li *et al.*, 2011).

The shape of the adsorption isotherm is an important characteristic, because it provides information about the adsorption mechanisms. 1/n values were less than 1.0 for Cd adsorption in phaeozem and moisture soil, which suggested that the adsorption isotherms were the L shape. The percentage of adsorption decreased with an increase in the initial concentration. The 1/n value for Cd adsorption in krasnozem was higher than 1.0, indicating the adsorption of Cd in krasnozem was difficult. In the Frendlich model, the nonlinearity factor 1/n is related to the energy distribution of an adsorption site and has been related to different adsorption sites and characters of SOM. The 1/n values of adsorption of Cd on moisture soil were lower than those on phaeozem and krasnozem, indicating that there was a more heterogenous adsorption site energy distribution for moisture soil. This might be caused by both the different maturation degree of SOM and soil pH values.

3.2 Cd adsorption as affected by NH_4^+-N

The addition of NH_4^+-N reduced the adsorption of Cd in soils. However, the influence varied with soil types. The parameters of the linear model, the Freundlich model, and the Langmuir model for the adsorption of Cd are listed in Table 2.

The influences of NH_4^+-N on Cd adsorption in typical Chinese soils were illustrated in Figure 2. The addition of NH_4^+-N decreased the adsorption of Cd in phaeozem. The K_d values reduced more than 50% under treatment with NH_4^+-N concentrations of 250 mg/L. However, the K_d valued reduced a little when the NH_4^+-N concentration increased to 500 mg/L form 250 mg/L.

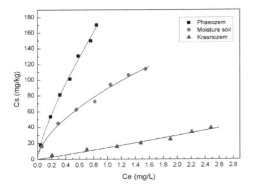

Figure 1. Cd adsorption isotherms in typical Chinese soils.

Table 2. Parameters of linear, Freundlich and Langmuir models for Cd adsorption under different NH_4^+-N concentrations.

Soil type	Conc.	Linear		Friendrich			Langmuir		
		K_d	R^2	K_f	1/n	R^2	C_{max}	K_L	R^2
S1	CK	182.61	0.96	189.78	0.76	0.99	476	0.62	0.99
	250	89.53	0.98	103.85	0.78	0.98	423	0.32	0.97
	500	83.23	0.98	101.88	0.74	0.99	369	0.39	0.99
S2	CK	64.15	0.96	88.35	0.60	0.99	198	0.84	0.97
	250	34.08	0.97	51.83	0.65	0.99	171	0.45	0.99
	500	19.22	0.79	40.98	0.46	0.94	79.2	1.20	0.97
S3	CK	14.46	0.98	14.15	1.06	0.96	14373	0.001	0.96
	250	13.31	0.98	16.37	0.84	0.98	158	0.11	0.99
	500	6.26	0.87	12.01	0.56	0.95	34	0.58	0.98

*S1, S2 and S3 represented for phaeozem, moisture soil and krasnozem, respectively.

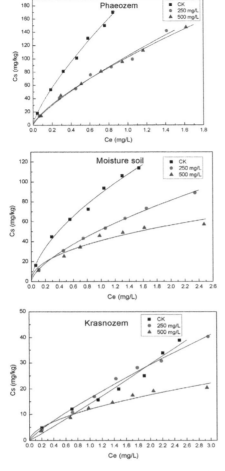

Figure 2. Effect of NH_4^+-N on adsorption of Cd in typical Chinese soils.

For moisture soil, the influence of NH_4^+-N on Cd adsorption was greater. Comparing with CK, the K_d values reduced 47% and 72% in treatments with NH_4^+-N concentrations of 250 and 500 mg/L, respectively. The influence of NH_4^+-N on Cd adsorption in krasnozem was reflected in the variation in 1/n value. With slight influence on K_f values, the 1/n values reduced from 1.06 to 0.56 with NH_4^+-N concentrations increased from 0 to 500 mg/L.

3.3 Influence of NO_3^--N on Cd adsorption

Although NO_3^--N decreased the adsorption of Cd, the influences were much slighter than those of NH_4^+-N. The parameters of the linear model, the Freundlich model and the Langmuir model for the adsorption of Cd are listed in Table 2.

The adsorption isotherms of Cd in different soils were given in Figure 3. As suggested by Table 3 and Figure 3, for Cd adsorption in krasnozem, the addition of NO_3^--N increased K_f value but decreased the 1/n value. This might be explained by the different pH values of soils. The pH value of krasnozem was lower than 7.0, indicating the surface charge of krasnozem was positive. Because Cd was introduced as cation, the addition of NO_3^--N neutralized part of the surface positive charge, thus enhancing the adsorption of Cd, especially in treatments with low Cd concentration.

The application of both NH_4^+-N and NO_3^--N inhibited the adsorption of Cd, thus increasing the bioavailability of Cd. Furthermore, the influence of NH_4^+-N was more significant, suggesting that NH_4^+-N was a suitable nitrogen fertilizer using in

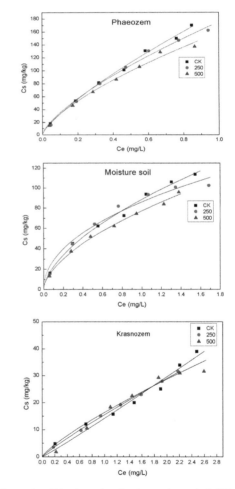

Figure 3. Cd adsorption isotherms in typical Chinese soils under different NO_3^--N concentrations.

Table 3. *Parameters of linear, Freundlich and Langmuir models for Cd adsorption under different NO_3^--N concentrations.

Soil type	Conc.	Linear		Friendrich			Langmuir		
		K_d	R^2	K_f	$1/n$	R^2	C_{max}	K_L	R^2
S1	CK	182.61	0.96	189.78	0.76	0.99	476	0.62	0.99
	250	163.69	0.96	175.01	0.70	0.99	356	0.90	0.99
	500	151.35	0.96	158.46	0.68	0.98	296	1.03	0.99
S2	CK	64.15	0.96	88.35	0.60	0.99	198	0.84	0.97
	250	52.41	0.84	85.83	0.48	0.96	144	1.62	0.98
	500	57.15	0.96	77.01	0.59	0.99	153	1.05	0.98
S3	CK	14.46	0.98	14.15	1.06	0.96	14373	0.001	0.96
	250	13.86	0.99	15.47	0.90	0.99	224	0.07	0.99
	500	12.65	0.84	16.32	0.81	0.95	102	0.19	0.96

*S1, S2 and S3 represented for phaeozem, moisture soil and krasnozem, respectively.

phytoremediation of Cd contamination. This might be caused by that both Cd^{2+} and NH_4^+-N were positive charge, and the competition for similar adsorption sites reduced the adsorption of Cd.

4 CONCLUSION

Cd was strongly adsorbed by the soils studied and linear, Freundlich and Langmuir models can describe the adsorption isotherms accurately.

SOC contents and soil pH values were key factors affected the adsorption of Cd in soils. Phaeozem had the highest capacity of Cd adsorption, followed with moisture soil and krasnozem.

Both NH_4^+-N and NO_3^--N increased the bioavailability of Cd. As the increasement of Cd bioavailability of NH_4^+-N was greater, it is recommended to be used as nitrogen fertilizer in phytoremediation.

ACKNOWLEDGMENT

This work was financially supported by the National Natural Science Foundation of China (grant Nos. 41430637, 41201520 and 41201494) and Jilin Provinical Science & Technology Department (grant No. 20130522178 JH).

REFERENCES

Bao, Y.Y., Wan, Y., Zhou, Q.X., Bao, Y.J., Li, W.M., Liu, Y.X., 2013. Competitive adsorption and desorption of oxytetracycline and cadmium with different input loadings on cinnamon soil. Journal of Soils and Sediments, 13: 364–374.

Cheng, W.W.L. and Gobas, F., 2007. Assessment of human health risks of consumption of cadmium contaminated cultured oysters. Human and Ecological Risk Assessment 13, 370–382.

Li, X., Zhou, Q., Wei, S., Ren, W., Sun, X., 2011. Adsorption and desorption of carbendazim and cadmium in typical soils in northeastern China as affected by temperature. Geoderma 160: 347–354.

Ministry of Environmental Protection of PRC & Ministry of Land and Resources of PRC, 2014. China's Soil Pollution State Bulletin. Beijing. (in Chinese).

Muchuweti, A., Birkett, J.W., Chinyanga, E., Zvauya, R., Scrimshaw, M.D., Lester, J.N., 2006. Heavy metal content of vegetables irrigated with mixtures of wastewater and sewage sludge in Zimbabwe: Implications for human health. Agriculture Ecosystems & Environment 112, 41–48.

OECD, Ed. 2000. OECD guidelines for the testing of chemicals. Adsorption/desorption using a batch equilibrium method OECD Test Guideline. Paris, OECD Publications.

Phillipp, R., 1980. Cadminum in the environment and cancer registration. Journal of Epidemiology and Community Health 34, 151–151.

Thawornchaisit, U. & Polprasert C., 2009. Evaluation of phosphate fertilizers for the stabilization of cadmium in highly contaminated soils. Journal of Hazardous Materials 165, 1109–1113.

Tipping, E., Rieuwerts, J., Pan, G., Ashmore, M.R., Lofts, S., Hill, M.T.R., Farago, M.E., Thornton, I., 2003. The solid-solution partitioning of heavy metals (Cu, Zn, Cd, Pb) in upland soils of England and Wales. Environmental Pollution 125, 213–225.

Virag, D. and Kiss, A., 2009. Comparative study of accessibility of distinctive pesticides. Journal of Environmental Science and Health Part B-Pesticides Food Contaminants and Agricultural Wastes 44, 69–75.

Wilde, T.d., Mertens, J., Spanoghe. P., Ryckeboer, J., Jaeken, P., Springael, D., 2008. Sorption kinetics and its effects on retention and leaching. Chemosphere 72, 509–516.

Zhou, Q.X., Zhang, Q.R., Sun, T.H., 2006. Technical innovation of land treatment systems for municipal wastewater in northeast China. Pedosphere 16, 297–303.

Advances in Energy, Environment and Materials Science – Wang & Zhao (Eds)
© *2016 Taylor & Francis Group, London, ISBN 978-1-138-02931-6*

Effects of nano-devices on growth and major elements absorption of hydroponic lettuce

Y.J. Wang, R.Y. Chen, H.C. Liu, S.W. Song, W. Su & G.W. Sun
College of Horticulture, South China Agricultural University, Guangzhou, China

ABSTRACT: A hydroponic experimental study was performed to determine the effects of nano-devices on the growth and major elements absorption of hydroponic lettuce with and without nano-devices. The results show that the fresh and dry weights of shoots and roots were significantly enhanced by nano-devices and the contents of soluble sugar, soluble protein, and Vc were increased in treatment with nano-devices. Nano-devices could improve the shoot and root absorption of N, P, and K. Moreover, the variation trend of N, P, K contents in nutrient solution was in accordance with the absorption amounts of lettuce plants as evidenced by the consumptions of major elements of the treatment group were higher than those of the control group in nutrient solution. Taken together, the results suggest that nano-devices could increase the absorption of N, P, K elements in lettuce, and further enhance the yield of lettuce.

1 INTRODUCTION

Since the 90s of 20th century, nano technology developed rapidly and made great achievements in the areas of ceramics, catalysis, biology, medicine, and so on. It is likely to bring the third industrial revolution to human in the 21st century and is considered to be the "key" of solving many scientific flaws (Behzadi et al. 2012, Chen et al. 2012). Nano-device (including nano-network, nano-ceramic chip, nano-film) belongs to nano-structure material, and is prepared by calcining carriers or ceramic coating with nano-material. Nano-devices have the advantages of insolubility in water, no toxicity, easy use, low cost, reuse, and security and reliability (Li et al. 2014). Study of Liu et al. (2008) showed that water treated with nano-devices had smaller molecular group and higher activity. Nano-devices could produce many special physical, chemical, and biological effects, which could change the role behavior of water and other substances related to organisms, such as the effect of increasing water solubility, improving cell biological permeability of water (Liu et al. 2007). There were positive effects in vegetables production (Sun et al. 2010, Wang et al. 2013, Li et al. 2014) and crop cultivation (Lu et al. 2002, Zhou et al. 2010) by using water treated with nano-devices for soaking seed or watering plant, which promotes obviously the growth and metabolism of organisms. Our previous experiment showed that nano-devices had obviously improved the yield and quality of hydroponic lettuce and significantly increased the vigor of root. The best amount of

usage in Guangzhou in autumn and winter is 2.80 cm²/L (Li et al. 2013). Root is the most active organ of nutrient absorption, which relate closely to crop yield (Li et al. 1992). Absorption of major elements such as nitrogen, phosphorus and potassium is the foundation of crop growth and development. However, the effects of nano-devices on growth and major elements absorption of hydroponic lettuce have been rarely reported. Therefore, this experiment aimed to study the effects of nano-devices on the growth and major elements absorption of hydroponic lettuce, so as to provide a theoretical guidance for the application of nanotechnology in vegetable production.

2 MATERIALS AND METHODS

2.1 Materials

The main constituents of nano-device used in the experiment are TiO_2 and ZnO coated on plastic film, and the area is 14 cm × 12 cm, shown in Figure 1.

Figure 1. The front and back image of nano-devices.

2.2 Experimental design

There were 2 treatments with and without nano-devices in the experiment: CK: 0 cm²/L, T: 2.8 cm²/L, the unit of cm²/L represents the area of nano-devices in per liter nutrient solution. Each treatment was repeated in triplicate and 18 plants of lettuce (cv. Italian) were in each repeated with random arrangement in bubble chamber (92 cm × 53 cm × 8 cm) with 30 L 1/2 Hoagland nutrient solution formula (pH ≈ 6.2). The experiment was carried out in the greenhouse of the College of Horticulture of South China Agricultural University during October 23rd to December 19th in 2014. The seedlings cultivated in perlite with 3 leaves were transplanted (November 15th). Meanwhile, nano-devices with required amount were soaked in the nutrient solution, and the nutrient solution in the bubble chamber was not replaced until harvest. Before nano-devices put into the bubble chamber, it took 20 minutes to be irradiated in the sun. After lettuce transplanted, the nano-devices also need to be taken out of nutrient solution in the bubble chamber and exposed under sunlight for 20 minutes every 10 days. The samples of lettuce were tested after harvest, and samples of nutrient solution were tested at the 10th, 15th, 20th, 25th, and 30th days after transplanting, respectively.

2.3 Measurement and methods

The fresh weight of shoot was weighted with a balance; the shoot were de-enzymed at 105°C for 15 minutes in oven, dried at 80°C until the weight was constant, and then the dry weight was measured; Vc content was measured with molybdic blue method (Li 2000); Soluble protein contents and soluble sugar contents were measured with Li's (2000) methods; N, P, K contents in lettuce were measured with Bao's (2000) methods; the major elements in nutrient solution were measured by the College of Natural Resources and Environment of SCAU.

2.4 Statistical analysis of data

Statistical analysis was performed with SPSS17.0, and single factor analysis was performed with Duncan methods. The figures were made by Excel 2007.

3 RESULTS

3.1 Effects of nano-devices on yield of hydroponic lettuce

The fresh and dry weights of shoots and roots of treatment with nano-devices were significantly

Table 1. Effects of nano-devices on yield of hydroponic lettuce.

Treatments	CK	T
Shoot fresh weight (g/plant)	108.04 ± 17.70	157.80 ± 13.32*
Shoot dry weight (g/plant)	5.71 ± 0.96	8.47 ± 0.76*
Root fresh weight (g/plant)	9.90 ± 3.02	15.68 ± 4.65*
Root dry weight (g/plant)	0.48 ± 0.12	0.78 ± 0.17*

Note: The * in the same range indicates significant difference ($P < 0.05$). Following tables are as the same.

Table 2. Effects of nano-devices on quality of hydroponic lettuce.

Treatments	CK	T
Contents of soluble sugar (mg/g)	7.54 ± 0.50	11.14 ± 0.19*
Contents of soluble protein (mg/g)	4.99 ± 0.25	7.15 ± 0.62*
Contents of Vc (mg/100 g)	6.90 ± 0.67	7.95 ± 1.06

higher than those of the control (Table 1). Compared with CK, the shoot fresh and dry weights of T treatment were increased by 46.06% and 48.34%, respectively, while the root fresh and dry weights of T treatment were increased by 58.38% and 62.50%, respectively. In summary, nano-devices could increase the yield of hydroponic lettuce.

3.2 Effects of nano-devices on quality of hydroponic lettuce

The contents of soluble sugar, soluble protein, and Vc were significantly increased in treatment with nano-devices (Table 2). Compared with CK, soluble sugar contents and soluble protein contents were increased by 47.75% and 43.29%, respectively, while the contents of Vc had no significant difference.

3.3 Effects of nano-devices on major elements absorption of hydroponic lettuce

Compared with CK, contents of N, P, and K of lettuce shoot were increased significantly by 15.7%, 34.1%, 50.8%, respectively and those of root were increased significantly by 12.8%, 33.5%, 53.2%, respectively (Table 3). Therefore, nano-devices could improve the major elements absorption in shoot and root of hydroponic lettuce.

Table 3. Effects of nano-devices on major elements absorption of hydroponic lettuce.

Treatments	CK	T
Contents of N in shoot (mg/g)	30.90 ± 0.01	$35.76 \pm 0.05*$
Contents of P in shoot (mg/g)	4.22 ± 0.01	$5.66 \pm 0.04*$
Contents of K in shoot (mg/g)	42.87 ± 0.06	$64.63 \pm 0.07*$
Contents of N in root (mg/g)	31.14 ± 0.04	$35.72 \pm 0.03*$
Contents of P in root (mg/g)	7.55 ± 0.02	$10.08 \pm 0.04*$
Contents of K in root (mg/g)	27.69 ± 0.12	$42.41 \pm 0.17*$

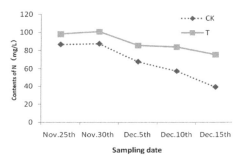

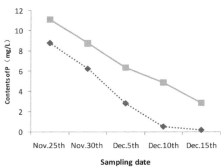

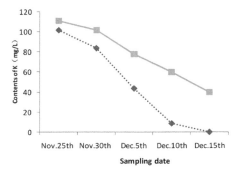

Figure 2. Effects of nano-devices on major elements in nutrient solution.

3.4 Effects of nano-devices on major elements in nutrient solution

It demonstrated that major elements in nutrient solution were gradually reducing with date (Fig. 2). The consumptions of major elements contents in nutrient solution of the treatment group were higher than those of the control group at each sampling stage, especially at the middle and later stage (After December 5th). Compared with CK, the consumptions of N, P, and K in the treatment group were increased by 31.96%, 89.37%, and 85.53%, respectively, at 25th day after transplanting (December 10th).

4 DISCUSSION AND CONCLUSION

Nitrogen, phosphorus, and potassium are the essential elements of plants, in which absorption directly affects the growth of plants. Under normal circumstances, dissolving force of minerals in water is constant under certain temperature, but there was evidence that water activated with nano-devices had higher solubility (Cao et al. 2006), which suggested that nano-devices changed the arrangement of water molecules and energy state. Therefore, with the rapid absorption of water, large amounts of N, P, and K were carried into the plant body to synthesize chloroplasts and mitochondria, so as to transform rapidly into bio-energy starch granule, thus increasing the potential of plant roots absorption of nutrients and moisture (Qian et al. 2010), and promoting the growth and development of plants. In this experiment, it was suggested that nano-devices had effects on the enhancement of both fresh and dry weights of shoot and root, and improved the quality to a certain extent. Zheng et al. (2005), working with spinach, have verified that TiO_2 nano-particles could increase the activity of several enzymes and promote the adsorption of nitrate, which accelerated the transformation of inorganic nitrogen into organic nitrogen. The results show that the major elements consumption of treatment group was significantly higher than that of control group, which suggested more elements in treatment nutrient solution which were absorbed by plants. The absorption of N, P, and K in both shoot and root could be significantly promoted by nano-devices, which were corroborated by the changes of major elements in the nutrient solution. In conclusion, nano-devices could significantly improve the yield and quality of hydroponic lettuce; promote the absorption of major elements in shoot and root.

ACKNOWLEDGMENT

This work was supported by the Project of China Agriculture Research System (Project No. CARS-25-C-04).

REFERENCES

Bao, S.D. 2000. Soil chemical analysis. Beijing: *China Agriculture Press.*

Behzadi, S. & Imani, M. & Yousefi, M. & Galinetto, P. & Simchi, A. & Amiri, H. & Stroeve, P. & Mahmoudi, M. 2012. Pyrolytic carbon coating for cytocompatibility of titanium oxide nanoparticles: a promising candidate for medical applications. *Nanotechnology,* 23(4), 45102.

Cao, Y.J. & Liu, An.X. & Liao, Z.W. & Li, Y.H. 2006. Preliminary study on effects of nano-materials on phosphor nutrition of maize. *Ecology and Environment,* 15(5), 1072–1074.

Chen, Y.X. & Tao, Y.B. & Liu, Y. 2012. Survey on the application of nano materials and technology in water treatment. *Journal of Green Science and Technology,* (11), 144–146.

Li, G.L. & Chen, R.Y. & Liu, H.C. & Song, S.W. & Sun, G.W. 2013. Effects of nano-devices on growth and physiological characteristics of hydroponic lettuce. *Journal of Shenyang Agricultural University,* 44(5), 656–659.

Li, G.L. & Chen, R.Y. & Liu, H.C. & Song, S.W. & Sun, G.W. 2014. Effects of nano-devices on growth, quality and activities of enzymes in nitrogen metabolism of hydroponic lettuce. *Key Engineering Materials,* 609–610, 1453.

Li, G.L. & Chen, R.Y. & Liu, H.C. & Song, S.W. & Sun, G.W. 2014. Effect of nano-devices on growth and quality of lettuce. *China Vegetables,* (03), 26–29.

Li, H.S. 2000. Principles and techniques of plant hysiological biochemical experiment. Beijing: *Higher Education Press.*

Li, J. 2000. Study on molybdic blue method of L-VC test by spectrometry. *Food Science,* 21(8).

Li, Y.S. & Feng, L.P. & Guo, M.L. & Han, X.X. 1992. Studies on the growth characteristics of root system and its relation with cultural practices and yield in cotton (*G. hirsutum* L.) II The effects of cultural practices on the growth of root system and its relation with above ground parts and yield of cotton. *Acta Gossypii Sinica,* (02), 59–66.

Liu, An.X. & Lu, Q.M. & Cao, Y.J. & Liao, Z.W. & Xu, Q.H. 2007. Effects of composite nanomaterials on rice growth. *Plant Nutrition and Fertilizer Science,* (02), 344–347.

Liu, An.X. & Liao, Z.W. 2008. Effects of nano materials on water clusters. *Journal of Anhui Agricultural Science,* (36), 15780–15781.

Lu, C.M. & Zhang, C.Y. & Wen, J.Q. & Wu, G.R. & Tao, M.X. 2002. Effect and mechanism of nano materials on the germination and growth of soybean. *Soybean Science,* (03), 168–171.

Qian, Y.F. & Shao, C.H. & Qiu, C.F. & Chen, X.M. & Li, S.L. & Zuo, W.D. & Peng, C.R. 2010. Primarily study of the effects of nanometer carbon fertilizer synergist on the late rice. *Acta Agriculturae Boreali-Sinica,* (S2), 249–253.

Sun, G.W. & Chen, H.X. & Chen, R.Y. & Liu, H.C. & Song, S.W. & Liao, Z.W. 2010. Effects of different treatments on growth and quality of Chinese Cabbage. *Nanoscience & Nanotechnology,* 7(5), 21–24.

Wang, Y.J. & Zheng, Y.J. & Sun, G.W. 2013. Effect of nano-devices on growth and quality of pea sprouts. *Vegetables,* (11), 7–9.

Zheng, L. & Hong, F.S. & Lu, S.P. & Liu, C. 2005. Effect of nano-TiO$_2$ on strength of naturally aged seeds and growth of spinach. *Biol Trace Elem Res.* 104: 83–91.

Zhou, S.B. & He, L.J. & He, L.H. 2010. Effects of treated water of nano-devices on waxy corn growth and physiological changes. *Journal of Maize Sciences,* (01), 87–89.

Advances in Energy, Environment and Materials Science – Wang & Zhao (Eds)
© 2016 Taylor & Francis Group, London, ISBN 978-1-138-02931-6

Induced sysmatic resistance of potato to common scab by MeJA

Guobin Deng
Yunnan Academy of Biodiversity, Southwest Forestry University, Kunming, Yunnan Province, China

Qing Deng & Gensong Zhang
Yunnan Chechuan Biotechnology Co. Ltd., Kunming, Yunnan Province, China

Erjuan Chen & Xiaolan Chen
School of Life Sciences, Yunnan University, Kunming, Yunnan Province, China

ABSTRACT: Effect of methyl jasmonate (MeJA) on the resistance of potato to common scab was studied in this paper. To elucidate the physiological mechanism of resistance induced by MeJA, the contents of superoxide anion free radical and MDA, and different pigments (chlorophyll a, b and carotenoid) were measured. Also, the activities of pathogen-resistance related Phenylalanine Ammonia Lyase (PAL) and anti-oxygen enzymes Polyphenol Oxidase (PPO), Peroxidase (POD) and Catalase (CAT) were analyzed. The results showed that MeJA can effectively inhibit the growth rate of seedlings, increase the contents of chlorophyll a and b, decrease the MDA level, and promote the activities of enzyme PAL and anti-oxygen enzymes POX, POD and CAT. Thus, MeJA can induce the system resistance of potato to common scab and could be used as anti-pathogenic agent for potato common scab control.

1 INTRODUCTION

1.1 *Common scab of potato caused by Streptomyces spp*

Common scab is an important disease of potato and taproot vegetables caused by actinobacteria belong to genus *Streptomyces*, including *S. scabies* and other closely related species (Goyer et al.,1998; Loria et al., 2006; Boucheck-Mechiche et al., 2000). They caused raised, superficial or pitted necrotic lesions on tuber surface which affect marketability of tubers and result in economic losses (Hill and Lazarovits, 2000). Common scab-inducing *Streptomyces* spp. are known for being phenotypically and genetically diverse (Healy et al., 1999), which make it hard to employ conventional methods to control its burst and harmful effluence on the yield.

1.2 *Jasmonates and their function in plant disease resistance*

Jasmonates are a family of compounds derived from Jasmonic Acid (JA) with varying biological activities, and MeJA is the volatile methyl ester of JA (Acosta and Farmer, 2008). Plant responds to microorganisms through the JA-mediated defense pathway or the salicylate-mediated defense pathway (Thaler et al., 2012). Necrotrophic pathogens are those that kill plant cells as part of the infection process and include most fungi and oomycetes as well as some bacteria. Plant defense against pathogens by JA signaling pathway is known as long-term ISR (Induced Systemic Resistance) response, including the production of antimicrobial metabolites (phytoalexins), antimicrobial peptides or proteins, hydrolytic enzymes that directly target the pathogen (chitinases and gulcanases), and enzymes that produce toxic reactive oxygen species (Balbi and Devoto, 2008; Bari and Jones, 2009; Ballare, 2011).

1.3 *Control of common scab of potato caused by streptomyces spp*

Management of common scab of potato is problematic with no single reliable control treatment available. Chemical and physical methods were adopted to control common scab of potato, such as foliar application of certain substituted phenoxy, benzoic, and picolinic acids to growing potato plants (McIntosh et al., 1988), adjustment of soil pH and crop rotation (Waterer, 2002) soil amendments (Mishra and Srivastava, 2004), excess irrigation during tuber formation and agrochemicals on scab (Loria et al., 1997) when applied shortly after crop emergence and that disease suppression was associated with diminished sensitivity of tubers from treated plants to thaxtomin A, a toxin produced by pathogenic *Streptomyces spp.* (Tegg et al., 2008; 2012; Thompson et al., 2013).

For the diversity and complex constitutions of scab-inducing pathogens and its genetic variation, chemical and traditional agricultural managements hardly control the development of disease. Biological methods, which can induce the systemic defense mechanism in host plants, may be a new strategy for its control.

Here we report that MeJA can efficiently induce the defense of potato seedlings to common scab during germ microtubes production in greenhouses. The possible mechanism of induced sysmatic resistance by MeJA was also analyzed.

2 MATERIALS AND METHODS

2.1 Planting material

Potato seedlings and seed tubers of the variety Shepody was used in this study, which is sensitive to scab-inducing *Streptomyces* spp.

2.2 MeJA treatment

After 15 days of the seedlings were transplanted into the medium in the greenhouse, seedlings were divided into three groups. Group I is the control, without any treatment. Group II is the MeJA treatment, in which the seedlings will be sprayed by MeJA at a final concentration of 25 ppm. Group III is the MeJA treatment group at a final concentration of 50 ppm. MeJA treatment was conducted on the day15 and day 22, a growth stage before tube induction at underground part of seedlings, equal volume of water was used for control group. Each group has three repeats. Each repeat has a area of 3 m*1 m.

2.3 Measurements of physiological parameters

2 weeks after MeJA treatment, the growth of seedling was compared by measuring the height of seedlings and the average length of nodes. And the physiological parameters were measured using the second leaves of seedlings in each group. The content of chlorophyll and carotenoid were extracted and measured by absorption measurements at

proper wavelength. Phenylalanine Ammonia Lyase (PAL) activity was determined by following the method of Dickerson et al. (1984). Polyphenol Oxidase (PPO), Peroxidase (POD) and Catalase (CAT) activities were determined using the method of Chance and Maehly (1955) and Wu (2007). The method for measurement of superoxide radicals O2- and MDA is described by Wu (2007).

2.4 Measurements of disease parameters

After 45 days of planting, the seed tubes of each group were harvested and disease parameters like scab index and scab incidence were measured. The tubers scored individually for scab occurrence according to the scale of Singhai (2011).

2.5 Statistical analysis

All experiments were performed in 3 replicates. Data presented in this study are presented in means-Standard Deviation (SD).

3 RESULTS

3.1 Effect of MeJA treatment of the scab incidence and index of seed tubes

As shown by Table 1, Scab incidence and scab index of seed tubes dramatically decreased after MeJA folia application during the seedling growth early stage (2 weeks after planting). 25 ppm MeJA treatment resulted in a drop of scab index from 44.6% of control to 7.0%, and of scab incidence from 69.2% to 18.3%. To our surprise, 50 ppm MeJA treatment only caused an 18.7% and 29.1% decline of scab index and incidence, respectively. These results indicate that exogenous MeJA can effectively induce the sysmatic resistance of potato to common scab disease.

3.2 The effects of MJ treatments on the growth of potato seedlings

To determine whether the folia application of MeJA have effect on the growth of young seedlings, the

Table 1. Effect of MeJA treatment of the scab index of tubes.

Treatment	Total number of tubes	Number of infected tubes	Scab incidence (%)	Scab index (%)
Control group	156	108	69.2%	44.6%
25 ppm MeJA	104	19	18.3%**	7.0%**
50 ppm MeJA	161	67	40.1%*	25.9%*

*means significant difference by two-tail test; **means highly significant difference by two-tail test.

Average height of aboveground part of seedlings and length of inter-nodes were measured.

Dwarf response was obvious after the MeJA treatment, as seen from Figure 1, sharp decrease of average height of seedlings was observed from 44 cm in control to 18 cm of group I seedlings after 25 ppm MeJA treatment. The decrease of average height of seedlings resulted in the decline of internodes length (11 cm vs. 5 cm). Dwarf seedlings and shortened internodes help to elevate the anti-lodging and other stress-resistance capability, which may be part of the mechanism MeJA induced.

3.3 Effects of MeJA treatment on the pigments content of potato seedlings leaves

Chlorophyll a and b, as well as caroteniod, of Seedlings were measured 2 weeks after treatment. MeJA induced the synthesis of chlorophyll a, b and carotenes.

Figure 2 shows the contents of three types of pigments. Chlorophyll a climbs from 1000 to 1780 ng/g, while chlorophyll b from 340 to 700 ng/g.

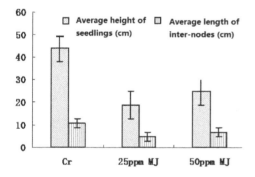

Figure 1. Effects of MeJA treatment on the growth of potato seedlings.

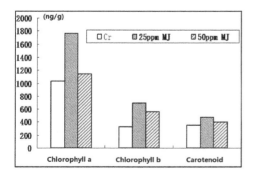

Figure 2. Effects of MeJA treatment on the pigments content of potato seedlings leaves.

The caroteniod contend increased gently compared to chloroyphyll. These results are consistant with the dark-green outlook of seedling with MeJA treatment.

3.4 Effect of MeJA on the activity of superoxide radical and antioxidant enzymes

Phenylalanine Ammonia Lyase (PAL) is the first enzyme of the general phenylpropanoid metabolism and controls a key branch point in the biosynthetic pathways of flavanoid phytoalexins which are mostly antimicrobial compounds (Bowles, 1990). Reactive Oxigen Species (ROS) burst is a typical characteristic of cells under stress. And MDA is the oxidant product of membrane lipid under stress conditions. Plants evolve elaborate mechanism to maintain the balance of production of scavange of ROS, Polyphenol Oxidase (PPO), Peroxidase (POD) and Catalase (CAT) are three key enzymes involving in ROS scavenge. To unveil the underline mechanism of MeJA induced common scab resistance of potato, we detected the content of superoxide anion free radical and MDA, and the activity of PAL, PPO and CAT.

In group II (25 ppm MeJA), although the superoxide anion free radicals level doesn't change much in the seedlings of group II, MDA contents decreased from a average value of 0.69 to 0.55 mg/g_{FW}. And all the enzymes measured have increased activity, they are 131%, 38.8%, 140% and 269%, for PAL, PPO, POD and CAT respectively, higher than those of control.

In the case of group III (50 ppm MeJA), the superoxide anion free radicals level and MDA decreased from a average value of 167.7 to 46.6 mg/L and from 0.69 to 0.41 mg/g_{FW}. Most notably, POD activity gets doubled and reaches a value as high as 613.3 U/g_{FW}.

4 DISCUSSION

Induced Systemic Resistance (ISR) is a state of enhanced defensive capacity triggered by specific contact stimuli whereby the plant's active defenses are potentiated against subseqauent pathogen challenge. The resistance responses are usually systemic but localized forms also exist and effective against a broad range of pathogens (Sticher et al., 1997). MeJA has been reported function in ISR (Induction of systemic resistance) in many plants, such as tomato, tobacco and peppers (Ballare, 2011). Our results showed that it can also function in potato and enhance the resistance of seedlings to common scab caused by pathogen Streptomyces spp, with folia application before the tube formation stage.

Table 2. Effects of MeJA on the activity of superoxide radical and antioxidant enzymes of potato seedling leaves.

Treatment	Control	25 ppmMeJA	50 ppmMeJA
O2⁻ content (mg/L)	167.7 ± 10.3	167.4 ± 12.4	46.6 ± 6.5
MDA (mg/g_{FW})	0.69 ± 0.07	0.55 ± 0.06	0.41 ± 0.06
PAL (U/g_{FW})	67.0 ± 4.8	$154.8 \pm 6.2^{**}$	74.9 ± 5.4
PPO (U/g_{FW})	134.5 ± 15.3	$186.7 \pm 23.4^{**}$	162.5 ± 10.5
POD (U/g·min)	125.2 ± 21.1	$300.0 \pm 47.2^{**}$	$613.3 \pm 43.2^{**}$
CAT (U/g·min)	65.2 ± 6.0	$240.5 \pm 55.0^{**}$	$103.7 \pm 39.9^{*}$

*means significant difference by two-tail test; **means highly significant difference by two-tail test.

By inhibition of inter-nodes elongation and increase the synthesis of chlorophyll a and b (Fig. 1 and Table 1), MeJA promotes the stress resistance capability of potato seedlings and decrease the incidence of common scab of seed tubes underground. The signals from the aboveground leaves and stems to underground roots and tubes remains unknown and worthy of further research. But exogenous MeJA may also function via changing the hormone composition in the seedlings such as auxin and Gibberellins (GA) (Yang et al., 2012), since they are main regulators involving in elongation of stem and the growth rate of aboveground part of seedlings.

MeJA may induce the expression of defense genes throughout the whole plant associated with the production of chitinase, POX, PPO and PAL and expression of stress-related proteins (Timmusk and Wagner, 1999). PAL activities was higher in the treated group seedlings, as shown by Table 2, which was required for biosynthesis of phenolica in response to pathogen infection that serve as precursors for biosynthesis of lignin. Increase in PPO, POD and CAT activities are ofter associated with modulation of active oxygen species, which may play a direct or indirect role in reducing pathogen viability and spread. Also, MeJA may also regulate plant cell wall reinforcement process and the formation of mechanical barrier such as lignin and cell wall cross linking proteins.

Chemicals were used extensively for plant disease management in the past and resulted in multiple ill effects like polluting the environment and disturbing the ecological balances. ISR exploiting by the natural defense machinery of plants is an alternative, non-conventional and eco-friendly approach for plant protection. And as efficient ISR inducer, MeJA is a prospectively anti-pathogenic agent, which is synthesized by plant itself and has versatile function in development and stress resistance.

But it is important to realize that the concentration used for folia management and the timing of spray application should be very deliberation and essential for the result. High level of exogenous MeJA may lead to undesirable effects, as shown by our study.

ACKNOWLEDGEMENTS

This research was supported by NSFC (NO. 30960036 and NO. 31260283) and YNUY (NO. 201421).

REFERENCES

Acosta I.F., and Farmer E.E. 2008. Jasmonates. In The Arabidopsis Book (*The American Society of Plant Biologists*), pp. 1–13.

Balbi, V., and Devoto, A. 2008. Jasmonate signalling network in *Arabidopsis thaliana:* crucial regulatory nodes and new physiological scenarios. *New Phytol.* 177, 301–318.

Ballare, C.L. 2011. Jasmonate-induced defenses: A tale of intelligence, collaborators and rascals. *Trend Plant Sci.* 16:249–257.

Bari, R., and Jones, J.D. 2009. Role of plant hormones in plant defence responses. *Plant Mol. Biol.* 69,: 473–488.

Boucheck-Mechiche, K., Gardan, L., Normand, P. 2000. DNA relatedness among strains of *Streptomyces* pathogenic to potato in France: description of three new species, *S. europaeiscabiei sp* nov, and *S. stelliscabiei sp.* nov associated with common scab, and *S. reticuliscabiei sp* nov associated with netted scab. *Int. J. Syst. Evol. Microbiol.* 50:91–99.

Chance, B., Maehly, A.C. 1955. Assay of catalases and peroxidases. *Methods in Enzymology* 2:764–755.

Dickerson, D.P., Pascholati, S.F., Hagerman, A.E., et al. 1984. Phenylalanine ammonia-lyase and hydroxycinnamate. *Physiological and Molecular Plant Pathology* 25:11–123.

Goyer, C., Vachon, J., Beaulieu, C. 1998. Pathogenicity of Streptomyces scabies mutants altered in thaxtomin a production. *Phytopathology* 88:442–445.

Healy FG, Bukhalid RA, Loria R. 1999. Characterization of an insertion sequence element associated with genetically diverse plant pathogenic *Streptomyces* spp. *J Bacteriol* 181(5):1562–1568.

Loria, R., Kers, J., Joshi, M. 2006. Evolution of plant pathogenicity in Streptomyces. *Annu. Rev. Phytopathol,* 44:467–487.

McIntosh, A.H. Bateman, G.L. and Chamberlain, K. 1988. Substituted benzoic and picolinic acids as foliar sprays against potato common scab. *Annals of Applied Biology* 112(3):397–401.

Mishra, K.K., and Srivastava, J.S. 2004. Soil amendments to control common scab of potato. *Potato Research* 47:101–109.

Singhai, P.K., Sarma, B.K., and Srivastava J.S. 2011. Phenolic acid content in potato peel determines natural infection of common scab caused by *Streptomyces* spp. *Biological Control.* 57:150–157.

Sticher, L., Mauch-Mani, B., Metraux, J.P. 1997. Systemic acquired resistance. *Annual Review of Phytopathology* 35:235–270.

Tegg, R.S., Corkrey, R. and Wilson, C.R. 2012. Relationship Between the Application of Foliar Chemicals to Reduce Common Scab Disease of Potato and Correlation with Thaxtomin A Toxicity. *Plant Disease* 96(1):97–103.

Tegg, R.S., Gill, W.M. Thompson, H.K., et al. 2008. Auxin induced resistance to common scab disease of potato linked to inhibitionof thaxtomin A toxicity. *Plant Disease* 92(9):1321–1328.

Thaler, J.S., Humphrey, P.T., and Whiteman, N.K. 2012. Evolution of jasmonate and salicylate signal crosstalk. *Trends Plant Sci* 17:260–270.

Thompson, H.K., Tegg, R.S., Davies, N.W., et al. 2013. Determination of optimal timing of 2,4-dichlorophenoxy-acetic acid foliar applications for common scab control in potato. *Annals of Applied Biology* 163(2): 242–256.

Waterer, D. 2002. Impact of high soil pH on potato yields and grade losses to common scab. *Canadian Journal of Plant Science* 82(3):583–586.

Wu, Z.X., Gan, N.Q., Huang, Q., Song, L.R. 2007. Response of *Microcystis* to copper stress. *Environmental Pollution* 147:324–330.

Yang, D.L. et al. 2012. Plant hormone jasmonate prioritizes defense over growth by interfering with gibberellin signaling cascade. *Proc. Natl. Acad. Sci* 109: E1192–E1200.

Advances in Energy, Environment and Materials Science – Wang & Zhao (Eds)
© *2016 Taylor & Francis Group, London, ISBN 978-1-138-02931-6*

Research on soil moisture dynamic under negative pressure irrigation

Xiande Shi, Qingyun Zhou, Yan Wang, Chengyue Cao, Guoyu Jin & Yunfei Liu
College of Water Conservancy Engineering, Tianjin Agricultural University, Tianjin, China

ABSTRACT: Negative pressure irrigation is a novel water-saving irrigation. The irrigation process does not require the external pressure equipment for water extraction. The soil suction was used to absorb water automatically from the source of water, which elevation is lower than the irrigation device. A new irrigation material named porous ceramic (ceramic, USA) combined with advanced negative pressure irrigation could keep the soil moisture always within the most suitable state for crop growth. Negative pressure irrigation could achieve the water-saving and the energy-saving target. In this paper, experiments were conducted to research soil moisture movement under negative pressure irrigation with porous ceramic douche. The results were showed as following. Negative pressure irrigation system using the porous ceramic douche could perform well when the inner negative pressure reached −15 cm. With the infiltration time, the soil moisture surround the porous ceramic douche became more and more big, and the soil moisture content increased gradually. Eventually, it reached the maximum soil moisture content value and fluctuated around the maximum soil moisture content value. The soil moisture was automatically controlled under negative pressure irrigation, which could always maintain soil moisture content in a certain range.

1 INTRODUCTION

Today's world is short of water resources and available water resources are gradually reduced. The existing irrigation methods, such as drip irrigation sprinkler irrigation, etc., saved water, however, consumed a large amount of energy. Those irrigation methods were also unable to improve the regulation of soil moisture state. Negative pressure irrigation technology can meet the requirements of crop's water demand and save water and energy without any power conditions. The difference of pressure between the soil metrics potential and irrigation pressure head was used to drive water into the soil under negative pressure irrigation. The whole negative pressure irrigation system must meet the airtight condition. Therefore, the application of negative pressure irrigation was affected by material properties of water supply device. The porous ceramic imported from the United States are porous material with water permeable and airtight characteristics. Combined the negative pressure irrigation with the porous ceramic douche, the automatic regulation of the soil moisture could be obtained. So, negative pressure irrigation was water saving and energy saving irrigation.

Negative pressure irrigation is based on theory of soil hydrodynamic. As long as the negative pressure water head does not exceed a certain range, the negative pressure irrigation can perform well. Even if the water elevation is lower than the water intake elevation, the negative pressure irrigation is also completely feasible (Livingston, 1908, 1918). Soil texture, water supply head, irrigation douche (Kato et al., 1982), and water quality are the main factors that affect the soil moisture movement under the negative pressure irrigation (Jiang 2006; Liu 2002).

In 2001, automatic water-controlling irrigation system on vegetable cultivation was established under negative pressure irrigation (Liu 2002). The results showed that the negative pressure irrigation system could perform according to the soil moisture. It irrigated when the soil was short of water, so the system had the function of self-regulation. Just because the irrigation system still needs water-carrying and pressurized equipment which would lead to the cost increased, negative pressure irrigation had not been promoted. In 2004, a new type of negative pressure irrigation system was proposed by Lei et al. (Jiang, 2006). Soil moisture dynamics were researched under different pressure (H = 0.5 m (positive pressure), 0 m (no pressure), −1 m, 2 m, −3 m, −4 m) under negative pressure irrigation using clay pipe. The experimental analysis showed that the system could automatically absorb water relying on the soil suction from water source whose elevation is lower than the soil elevation in a certain range. Negative pressure irrigation system has no water pressure equipment with energy saving and cost low. Therefore, it will be a wide application prospect.

Jiang et al. (2005) studied how different emitter materials (pottery and fiber) and different kinds of soil texture (clay loam and sandy loam) affect flow at the same water head (−0.5 m) under negative pressure irrigation system. The results showed that fiber emitter had better connectivity and higher output flow than clay douche.

The new irrigation douche materials, porous ceramic (ceramic, USA), combined with advanced negative pressure regulation could make the soil moisture always in the most suitable condition for crop's growth. The irrigation system could achieve the goal of saving water and energy. The porous ceramic cylinder contains a lot of capillary, but it still has tubular characteristics with the function of water permeable, airtight, and water delivery. An airtight system with the negative pressure irrigation was established. Theoretically, automatic adjustment irrigation could be realized. In this paper, the porous ceramic cylinder whose diameter is 1 cm (ceramic, USA) was used in negative pressure irrigation. Under the negative pressure head of 15 cm, the soil wetting front distance, the soil moisture content, and the cumulative infiltration were observed and recorded. The effects of porous ceramic cylinder on soil moisture dynamic were researched under negative pressure irrigation.

2 MATERIALS AND METHODS

2.1 Experimental site

The experiment was conducted at Agricultural Water-saving Irrigation Experimental Center by Tianjin Agricultural University. The experimental center is located in Xiqing district (N 39°08′, E116°57′, altitude 5.49 m), Tianjin, China. The site is in a typical warm temperate continental monsoon climate zone with a mean annual temperature of 11.6°C. The mean annual precipitation is 586 mm. The groundwater table is 2.6–3.7 m below the ground surface.

2.2 Materials

The experimental setup consists of organic glass soil box, water supply device, piezometer tube, and water tank as showed in Figure 1. The organic glass box is 20 cm high. The length and width are both 30 cm. On one side of the central part, a hole at the central part was bored 10 cm from the bottom of the soil box. The diameter of the hole is 1.2 cm for connecting with the porous ceramic cylinder. Markov bottle was used for infiltration water supply and controlling the negative pressure head. The experimental soil was collected from Yangcenzi village, Tianjin, China.

The experimental soil bulk density is 1.68 g/cm³. The saturated moisture content is 22.15% (mass percent). The soil physical and chemical properties are shown in Table 1. The soil was crushed and layered each 5 cm in the soil box. The water tank and markov bottle were connected by the rubber tube. The height between the liquid level and the centre axis of the porous ceramic cylinder is 15 cm. It means the negative pressure value is −15 cm. The piezometer tube was directly connected with the porous ceramic cylinder to ensure the whole system was airtight.

The porous ceramic cylinder was used as douche of negative pressure irrigation. The air entry value of porous ceramic cylinder in experiment is 0.1 MPa. The saturated hydraulic conductivity is 3.11×10^{-5} cm/s. The maximum pore size of the experimental ceramic cylinder is 6.0×10^{-6} m.

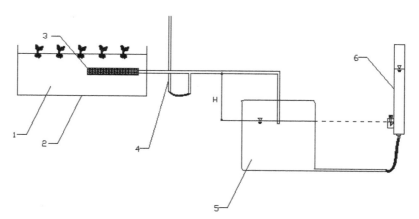

Figure 1.   The experimental setup of negative pressure irrigation with porous ceramic cylinder douche.
(1. Soil for experiment 2. Soil box for experiment 3. The Clay pipe 4. Piezometer tube 5. Water tank 6. The markov bottle).

Table 1. The soil physical and chemical properties.

Soil depth (cm)	Soil salinity (g/kg)	pH	Organic matter (g/kg)	N (mg/kg)	P (mg/kg)	K (mg/kg)
0–20	4.4	8.4	4.6	70	18.3	336.7
20–40	3.7	8.6	16.7	140	14.6	255.4

2.3 Design of experiment and measurements

Three replicates were respectively carried out at −15 cm pressure under negative pressure irrigation. The diameter of the experimental porous ceramic cylinder is 1 cm and buried at 10 cm soil depth on soil box. The soil volumetric water content was measured at 10 cm soil depth. The measured items include piezometer tube, amount water of infiltration, the horizontal and vertical wetting front distance and soil electrical conductivity at the horizontal distance of 0 cm and 5 cm. In the first two days, those items were measured every two hours. Then the soil water content was measured at 1 day interval until the end of the experiment.

Soil volumetric water content, soil temperature, and soil electrical conductivity was measured by three parameters monitoring system (WET-2, Delta, UK). The gravimetric sampling technique was used to calibrate the WET-2 display unit at the end of the experiment. The horizontal and vertical wetting front distances were measured by ruler.

3 RESULTS

3.1 Variation of soil wetting front

Figure 2 described the change of the horizontal soil wetting front surround the clay pipe when the distance between the porous ceramic cylinder and the water level of the tank is 15 cm. From the diagram, it can be seen that the soil wetting front didn't form at the beginning of the experiment. With the passage of infiltration time, the horizontal soil wetting front distance was gradually increased linearly, and the linear correlation coefficient is 0.9547 closed to 1. That is consistent with the results of Liu (2001) and Liang (2011). After 10 days infiltration, the wetting shape was a circle and the radius, i.e. the horizontal soil wetting front distance is about 10 cm. The average increasing distance of horizontal soil wetting front is about 1 cm at the soil surface.

The variation of the vertical soil wetting front distance surround the porous ceramic cylinder was showed in Figure 3. The change law of the vertical soil wetting front is similar to the horizontal soil wetting front, and the linear correlation coefficient is 0.958. On the 10th day, the vertically soil wetting front distance reached to 10 cm. Indicating that the

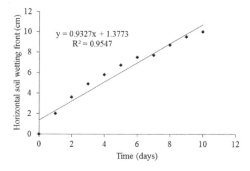

Figure 2. Variation of horizontal soil wetting front distance under negative pressure irrigation with porous ceramic cylinder douche.

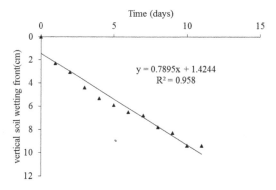

Figure 3. Variation of vertical soil wetting front distance under negative pressure irrigation with porous ceramic cylinder douche.

soil wetting front surround porous ceramic cylinder became larger under negative pressure irrigation with the porous ceramic cylinder douche when negative pressure was 15 cm. The average increasing distance of vertical soil wetting front is about 1 cm similar to that of horizontal soil wetting front distance.

3.2 Distribution of soil moisture content

Figure 4 showed the variation of soil volumetric moisture content surrounding the porous ceramic cylinder. The soil moisture content nearby the porous ceramic cylinder and the place whose

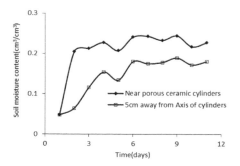

Figure 4. Variation of soil moisture content under negative pressure irrigation with porous ceramic cylinder douche.

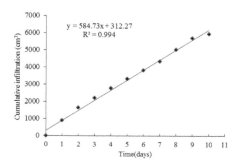

Figure 5. Variation of cumulative infiltration under negative pressure irrigation with porous ceramic cylinder douche.

horizontal distance is 5 cm apart from the porous ceramic cylinder were measured. Soil volumetric moisture content nearby the porous ceramic cylinder changed faster than that of 5 cm away. After a period of infiltration, the two soil moisture contents wander up and down slightly at the maximum. The value of soil water content nearby the porous ceramic cylinder was bigger than that of 5 cm away. The value of soil moisture content nearby the porous ceramic cylinder and 5 cm away about were 0.25 cm^3/cm^3 and 0.18 cm^3/cm^3, respectively. The soil moisture content besides the porous ceramic cylinder increased with infiltration time under negative pressure irrigation. It changed slightly in a certain range after a certain period indicating that the negative pressure irrigation model had self-regulatory function, which is different from other forms of irrigation. The closer the soil near the porous ceramic cylinder, the faster the moisture content changed, the earlier the constant state reached.

3.3 Variation of cumulative infiltration

In the experiment, the water level changed in Markov bottle was observed. The length of Markov bottle is 6 cm, and width is 5 cm. According to observed falling water level, the amount of water absorbed by soil in soil box was calculated. As it was shown in Figure 5, the amount of irrigation water increased linearly with infiltration. It can be concluded that the amount of water absorbed by soil had a good correlation with infiltration time and correlation coefficient is 0.994 very close to 1. The results proved that the speed of water seepage from the porous ceramic cylinder to the soil was constant under negative pressure irrigation. The results were consistent with Liu (2001).

4 CONCLUSION

In this paper, the law of soil moisture movement and soil wetting front distance under negative pressure irrigation with porous ceramic cylinder douche based on the experiment were analyzed. The results showed that negative pressure irrigation system could be achieved when the negative pressure is 15 cm. The soil moisture content increased with infiltration and reached a maximum value. The soil moisture content wandered up and down slightly at the maximum value and maintained at a certain range. So, the soil moisture state could be adjusted under negative pressure irrigation with porous ceramic cylinder douche.

ACKNOWLEDGMENTS

Thanks for funding support of College students' innovative entrepreneurial training project.

REFERENCES

Jiang Peifu, Lei Tingwu, F. Bralts Vincent, et al. The effect of soil texture and emitter material on the water flow and soil water transport of negative pressure irrigation [J]. Journal of Agricultural Engineering, 2006, 22(4):19–22.

Jiang Peifu. Study on the principle and experiment of negative pressure irrigation technology [D]. Beijing: China Agricultural University, 2006.

Kato Z., Tejima, S. Theory and fundamental studies on subsurface method by use of negative pressure [J]. Transaction of JSIDER, 1982, 101:46–54.

Liu Mingchi. Establishment and application of automatic pressure irrigation water vegetable cultivation system [D]. Beijing: Chinese Academy of Agricultural Sciences, 2001.

Liang Jintao. Study on the characteristics of soil moisture movement in negative pressure irrigation [D]. Shanxi: Taiyuan University of Technology, 2011.

Livingston, B.E. A method for controlling plant moisture [J]. Plant World, 1908, 11:39–40.

Livingston, B.E. Porous clay cones for the auto-irrigation of potted plants [J]. Plant world, 1918, 21:202–208.

Advances in Energy, Environment and Materials Science – Wang & Zhao (Eds)
© *2016 Taylor & Francis Group, London, ISBN 978-1-138-02931-6*

Analysis and research on the mixed problem of oil field sewage

Zhongye Tao, Rusong Zhao, Guimei Li & Sha Zhao
Beijing Institute of Petrochemical Technology, Beijing, China

ABSTRACT: With the continuous development of oil production in the domestic oil field, a large amount of sewage is produced. Due to the water saving resources, pollution prevention, environmental protection and other factors, Sewage reinjection treatment has become increasingly important. In the process of oilfield reinjection often occur in mixed injection after mixing, poor water quality, not standard phenomenon. For this purpose, this paper has done some research, putting forward some suggestions for the practical industrial process.

1 INTRODUCTION

With the development of oil field in China, the large area of water flooding, and oil recovery, are expanded, producing large amount of sewage every year. Just the eastern part of an oil field on the daily sewage treatment capacity reached 83.8×10^4 m^3 (Zhong, 2013). Behind the national "ecological environment" "water ten" and the enterprise itself requirements, there will be more and more important in the treatment of sewage reinjection. In the process of oilfield sewage reinjection, because the location of the oil wells and water wells or other reasons often occur in different blocks of the mixed injection problem of sewage treatment. In this paper, some related research about sewage mixed problem have been carried out, and a number of engineering proposals are put forward.

2 EXPERIMENTAL SECTION

2.1 *Experimental apparatus and reagents*

pH500 pH detector (First Clean Corporation), ZZW-II/P the multi parameter test instrument of Water quality (Voight Testing Technology Co., Ltd, Zhengzhou, China), 501S Super constant temperature sink (Shanghai Yuejin Medical Instrument Factory, Shanghai, China), DelsaTM Nano C Particle Aanlyzer (Beckman Coulter.Inc), Water treatment A, B agent (Laboratory synthesis).

2.2 *Experimental method*

Experimental water samples were collected from a sewage treatment station in an oilfield in the eastern part of china. Investigating and analyzing of mixed phenomenon and characteristics of water samples were carried out in different sewage stations. Using different combination order, the advantages and disadvantages of the "mixed after treatment" or "treatment after mixed" were studied. Meanwhile, the stability of mixed water samples was investigated, and the effect of mixing time was studied, too.

2.3 *Water quality index and analysis method*

The final result of the sewage treatment needs to meet the requirements of the (water quality standard practice for analysis of oilfield injecting waters in clastic reservoirs) (SY/T 5329-2012) and the oil field Management Bureau. The S ion, Fe ion and DO (dissolved oxygen) were detected by SY/T 5329-2012, 5.8, 5.10 and 5.7 respectively, which attached the attention of Sewage treatment station. And the particle size and Zeta potential of water samples was studied to explore the effect of mixing time.

3 RESULTS AND DISCUSSION

3.1 *Mixed phenomenon of different water samples*

Collecting water samples from 1#, 2#, 3# Sewage station each 500 ml at temperature 21°C. Fe^{2+}, total Fe, S^{2-} and pH of water samples were detected. According to the actual water injection mode, taking 1# and 2#, 1# and 3# water samples, such as the proportion of mixed to 500 ml. Fe^{2+}, total Fe, S^{2-}, pH and other characteristics of the mixed water samples were detected after the fully mixed. Data indexes of different water samples and mixed water samples, See Table 1.

It Can be seen from Table 1: The experimental collection of water samples was weakly acidic. 1#, Fe ion content is low, but the content of S^{2-} is

Table 1. Data indexes of different water samples and mixed water samples.

Water samples	Fe^{2+} (mg/l)	Total Fe (mg/l)	S^{2-} (mg/l)	pH
1#	1.42	1.88	0.83	6.60
2#	3.56	5.08	0.45	6.27
3#	11.6	15.8	0.12	6.43
1#/2#	2.34	3.14	0.14	6.41
1#/3#	4.91	7.06	0	6.59

Table 2. Water quality data of "mixed after treatment".

Water samples	Fe^{2+} (mg/l)	Total Fe (mg/l)	S^{2-} (mg/l)	pH
1*#	0.10	0.45	0	6.45
2*#	0.10	1.38	0	6.36
1*#/2*#	0.10	1.68	0	6.37

*Treated water samples.

Table 3. Water quality data of "mixed after treatment".

Water samples	Fe^{2+} (mg/l)	Total Fe (mg/l)	S^{2-} (mg/l)	pH
1#	1.42	1.88	0.83	6.60
2#	3.56	5.08	0.45	6.27
1#/2#	2.34	3.14	0.14	6.41
1*#/2*#	0.10	0.20	0	6.77

*Treated water samples.

the highest. The situation of 3# water samples is contrary to it. The 2# water samples are between the two.

After the fully mixed, S^{2-} ions in 1# and 2# mixed water samples were involved in the reaction, and its content is lower than the original. The change of S^{2-} ion content in 1# and 3# mixed water sample was more obvious, the content of which is directly reduced to zero. S^{2-} ion is more active than Fe ion, and is a priority to participate in the reaction, needing to pay attention to it.

3.2 Water treatment sequence combination

In the face of oil field sewage mixing problem, the combination of water treatment sequence should be studied. That is to deal with two kinds of means "mixed after treatment" or "treatment after mixed". Two ways of dealing with each have their advantages and disadvantages. The way of "mixed after treatment" is the actual operation of the process, there is no project to transform the problem, but there is often a mixed problem of water quality which is not up to the standard. However, the way of "treatment after mixed" avoids the need of dealing with the problem two times effectively. Low cost. But there may be a need for engineering reform. Collecting water samples from 1# and 2# Sewage station each 500 ml at temperature 21°C, equal proportion mixed to 500 ml after treatment. Record the change of water quality, data see Table 2.

Collecting water samples from 1# and 2# Sewage station each 500 ml at temperature 21°C, then equal proportion mixed to 500 ml. Treating the mixed water samples, and Fe^{2+}, total Fe, S^{2-} content and pH value of the mixed water samples were recorded. Data see Table 3.

Seeing from Table 2, Table 3, the two combination methods are ideal for S^{2-} processing. The content of total Fe was not decreased but increased after the way of "mixed after treatment", which shows that the iron removal is the key problem in the process of the late period. In the process of mixing with water, there was the water self purification reaction, which can reduce the post processing

load and reduce the cost of expenditure. At the same time, the removal rate of iron, pH value is also higher than the previous one, after which the water treatment process is more useful.

3.3 Mixed water sample stability

In the transport process of the mixed oil field Sewage. Due to transport pipe, DO, SRB, IB, and other factors, the stability of mixed water samples will be changed. The 1# and 2# mixed water samples were treated in the treatment at 21°C, and then the processed sample is deposited in a number of sample bottles. Besides, monitoring the change of water sample, which was in the sample bottle, the change of indicators is shown in Figure 1.

From the above it can be seen that S^{2-} ion content of mixed water samples has been maintained at 0 mg/l. The total Fe content in the process of experiencing a period of decline and then showing a state of fluctuation. Fe^{2+} content in the 1–2 hours also appeared to fluctuate. After 2 hours, the stability of Fe^{2+} and S^{2-} was good. The whole of these reflect the complexity of the water quality of oil field Sewage and the problem of the poor stability of the process. Therefore, in the process of water treatment it needs to cooperate with the use of deoxidizing agent, bactericide, scale inhibitor, and corrosion inhibitor.

3.4 Effect of different mixing time

In the process of mixing 1# and 2# water samples, there are both the "water self purification" effect and the external impetus. In a constant temperature

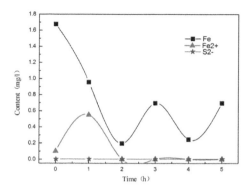

Figure 1.   The stability of mixed water samples.

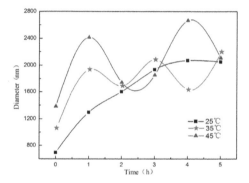

Figure 2.   Time variation of particle size of mixed water samples with different temperatures.

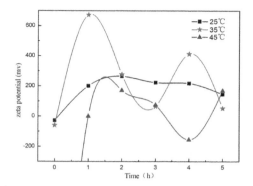

Figure 3.   Time variation of Zeta potential of mixed water samples with different temperatures.

water bath at 25°C, 35°C, and 45°C, the particle size and Zeta potential of the mixed water samples were measured, respectively. As shown below.

From Figure 2, seeing that the particle size of the mixed water samples increased with the increase of time, and the higher the temperature was, the higher the Oscillating wave was. This is because of the molecular heat movement. The change of particle size of the mixed water samples of the three temperatures in the 2–3 hour period showed the Consistency, which shows that the water treatment is less affected by temperature in this time.

At the same time, it can be found from Figure 3 that the Zeta potential value of mixed water samples was greatly influenced by temperature. The Zeta potential values of the mixed water samples of the three temperature segments in the 2–3 hour period were also demonstrated to show the Consistency, and the absolute value of Zeta potential is small. Zeta potential value is usually used as an index to characterize the stability of the dispersion system, describing the degree of attraction or repulsion between particles. In solution, the absolute value of Zeta potential is lower, the greater the attraction between particles, the more prone to aggregation. At 2–3 hours time, the absolute value of Zeta potential of the mixed water sample is relatively low, that is, the relatively unstable point. This point can be selected in the Sewage treatment.

4   CONCLUSIONS

1#, 2#, 3# Sewage treatment station water samples are weak acid. Different Sewage treatment station water samples will have the "self-cleaning" effects when mixed. In mixed water samples, the S^{2-} ion is more active than Fe, and the priority is to participate in the reaction.

Two kinds of water treatment combination means both have their advantages and disadvantages. But, the way of "treatment after mixed" uses the self reaction in the process of mixing water samples, reducing the post processing load. And compared with the actual operation of the "mixed after treatment" approach to the more effective and more beneficial to the back of the water treatment process.

Mixed water samples have the problem of stability, and in the process of transportation, water quality stability is often poor. Therefore, in the process of water treatment it needs to cooperate with the use of deoxidizing agent, bactericide and scale inhibitor etc.

2–3 hours after mixing with different treatment stations, the particle size and Zeta potential of the mixed water samples were less affected by temperature. And the consistency was good. At 2–3 hours time, the absolute value of Zeta potential of the mixed water sample is relatively low, that is, the relatively unstable point. Therefore, the mixed Sewage treatment is suitable for 2–3 hours after mixing of water samples.

REFERENCES

Bottero J.Y., Bruant M., Cases J.M. 1988, Adsorption of Nonionic Polyacrylamide on Montmorillonite: Relation between Adsorption, Potential Turbidity, Enthalpy of adsorption Data and 13C-NMR in Aqueous Solution. Colloid & Interface Sci., 124(2): 515–527.

Hao Y. & Haber S. 1998, Electrophoretic motion of a charged spherical particle normal to a planar dielectric wall. International Journal of Multiphase Low, 24: 793–824.

Logan B.E. & Kilps J.R. 1995, Fractal dimensions of aggregates formed in different fluid mechanical environments. Water Research. 29(2):443–453.

Robert J.H. 1981, Zeta potential in colloid science principles and applications. New York: San Francisco.

Rubing zhong. 2013, New progress of shengli oilfield reinjection sewage treatment technology. Chemical Engineering Of Oil & Gas, 20(1):86–90.

Russell A.S., Scales P.J., Underwood S.M. 1995, An electrophoretic investigation of the relaxation term in electrokinetic theory. Langmuir, (11):1112–1115.

Sorbie K.S. & Phil D. 1991, Polymer-Improved Oil Recovery. Florida, Blackie: 246–340.

Zaltoun A. 1992, Stabilization of Montmorillonite Clay in Porous Media by High-Molecular-Weight Polymers. SPE Prodn. Engng. 7(2):160–166.

Advances in Energy, Environment and Materials Science – Wang & Zhao (Eds)
© 2016 Taylor & Francis Group, London, ISBN 978-1-138-02931-6

Effect of different potassium levels on the growth and photosynthesis of sweet sorghum seedlings under salt stress

Hai Fan, Jie Zhang, Xiaokun Liu, Ke Dong, Yaping Xu & Li Li
Key Laboratory of Plant Stress Research, Shandong Normal University, Jinan, Shandong, P.R. China

ABSTRACT: In order to investigate the effect of potassium on the growth of sweet sorghum [*Sorghum bicolor* (L.) Moench] seedlings under salt stress and the optimum potassium level under salinity. Seedlings of sweet sorghum were supplied with 0.1 mmol L^{-1}, 3 mmol L^{-1}, 6 mmol L^{-1} and 9 mmol L^{-1} potassium fertilizers under 0% or 0.6% NaCl conditions. Then the growth parameters of the seedlings including fresh and dry weight, photosynthetic gas exchange, leaf chlorophyll *a* fluorescence as well as the content of MDA and relative electric conductance were determined. Results showed that plant growth and photosynthesis decreased under salt stress. Furthermore, with the increase of potassium level, plant fresh and dry weight, net Photosynthetic rate (Pn), Transpiration rate (Tr), stomatal conductance (Gs), the maximum photochemical efficiency of PSII (PHIPo), quantum yield under light (φPSII), the driving force on light absorption basis (D.F), the active reaction centers per cross section (RS/CSm), trapped energy per cross section (TRo/CSm) and PQ size of PS II (Sm) all increased and reached the peak at about 6–9 mmol L^{-1} potassium level. On the other hand, the close extent of PS II Reaction center (Vj) and the content of MDA as well as relative electric conductance all decreased. It is found that potassium at 6–9 mmol L^{-1} was beneficial to the growth of sweet sorghum, under both control and saline conditions. Though sweet sorghum grew much better under 9 mmol L^{-1} potassium level under salinity, it is not significant compared with 6 mmol L^{-1} potassium level.

1 INTRODUCTION

Soil salinity is a world-wide ecological problem especially in arid and semi-arid areas, which affects the growth and yield of glycophytes. Generally salt stress comprises two effects on plants: osmotic stress and ionic stress, which act in collusion to impair plant growth, disturb ion homeostasis, reduce photosynthesis, and affect many key physiological processes (Zhu 2001, Munns 2002). Only halophytes and some salt-tolerant glycophytes can live in saline soils.

Sweet sorghum is a variety of sorghum which is reported to be both drought tolerant and salt tolerant, It contains abundant sugar in its stalk which can be converted to fuel alcohol with an estimated production of 3 tones per ha land. Sweet sorghum is a very promising energy plant in alleviating energy crisis (Reddy et al. 2005). It is now widely cultivated in North China, both as energy crop and fodder, yet its nutrition requirement under salinity has seldom been studied, except for nitrogen. We found that the plant demand for nitrogen is reduced under salinity for the sake of photoinhibition (Fan et al. 2013).

Potassium is an essential mineral element for all plants. It participates in osmotic adjustment and

the movement of guard cells. Being cofactor of more than 50 enzymes, it is essential for the enzyme activity, especially the enzymes for photosynthesis, respiration, cell elongation, and protein synthesis. It is especially important under saline conditions, since NaCl stress typically reduces plant potassium uptake by competitive Na/K absorption. The plant potassium retaining capacity is regarded as one of the most important index of salt tolerance (Anschütz et al. 2014). In this paper, we planted sweet sorghum in plastic pots under saline or non-saline conditions and supplied them with different potassium levels in order to study the effect of different potassium levels on the growth, photosynthesis, chlorophyll fluorescence as well as other indices of sweet sorghum and to find out the optimum nitrogen level for the photosynthesis and growth of sweet sorghum, and to lay the basis for sweet sorghum cultivation in saline soils.

2 MATERIALS AND METHODS

2.1 Plant culture

Seeds of sweet sorghum (*Sorghum bicolor* (L.) Moench, cv. Jitianza 2) were provided by Shandong Academy of Agriculture Science. After soaking water for 12 h, the seeds were sown in plastic pots filled with clean river sand and watered with tap water until 2-leaf stage. After that the plants were cultivated in green house at 35°C/25°C (day/night), 60%RH and 1000 µmol photon m^{-2} s^{-1}, and irrigated daily with Hoagland nutrient solution of different potassium levels (0.1 mmol L^{-1}, 3 mmol L^{-1}, 6 mmol L^{-1} and 9 mmol L^{-1} potassium). After 15 more days, the plants of each concentration were treated with or without 0.6% NaCl respectively, each with 3 replications. 20 days later, the growth parameters, photosynthetic gas exchange, chlorophyll fluorescence as well as MDA and membrane permeability were determined.

2.2 Photosynthetic gas exchange measurement

The net Photosynthetic rate (Pn), Transpiration rate (Tr), stomatal conductance (Gs) and intercellular CO_2 concentration (Ci) were recorded with a Ciras-2 Portable Photosynthetic system (PP systems, Haverhill, USA). The PAR was 1000 µmol m^{-2} s^{-1}.

2.3 PSII Quantum yield (ΦPSII) and photochmical quenching (qP) measurement

ΦPSII and qP were measure with a FMS-2 modulated chlorophyll fluorometer (Hansatech, Kings Lynn, UK). Samples of intact leaves were light adapted for 30 min before the measurements, then the steady state fluorescence under sun light (Fs) was taken, afterwards the leaf was illuminated with saturated pulse of 15000 µmol m^{-2} s^{-1} for 0.7 s to get Fm′, followed by 5 s of far-red light under darkness to measure Fo′, ΦPSII was calculated as (Fm′-Fs)/ Fm′ and qP as (Fm′-Fs)/(Fm′-Fo′).

2.4 Chlorophyll a fluorescence transient measurement

Chlorophyll a fluorescence transient was measured with a handy PEA analyzer (Hansatech, Kings Lynn, UK). Samples of intact leaves were inserted in leaf clips and dark-adapted for 30 min before the measurements and then illuminated with light (3000 µmol m^{-2} s^{-1}, 650 nm peak wavelength) for 1 s provided by an array of three light emitting diodes. The data was then calculated using a JIP test software Biolyzer 3.0 (developed by Ronald Rodriguez in the Laboratory of Bioenergetics, University of Geneva) to obtain the following parameters: PHI (Po), RC/CSm, TRo/CSm, Sm, Vj and D.F. Where PHIPo, the maximum quantum yield of primary photochemistry; Tro/CSm, trapped energy per cross section; RC/CSm, density of Reaction Centers (RC) per cross section; Sm, Area/($F_M - F_o$), normalized complementary area above the OJIP

curve; Vj, relative variable fluorescence at J-step; D.F, Driving force at the light absorption basis (Strasser et al. 2004).

2.5 Malondialdehyde (MDA) and membrane permeability measurement

MDA content was measured by the spectrometric method of Lee and Csallany (Lee & Csallany 1987). Membrane permeability was expressed as relative electric conductance.

2.6 Seedling fresh weight and dry weight measurement

The seedlings were removed from the plastic pots and washed with fresh water, and then the surface water was adsorbed with tissue paper. The shoot and root was then separated and the fresh weight was measured immediately. After heated at 60°C for 2d, the dry weight was measured.

2.7 Data analysis

The data were analyzed with SPSS 20 by one-way Anova with Bonferroni's correction.

3 RESULTS

3.1 The growth of sweet sorghum seedlings

The effect of salt and potassium level on the growth status was shown in Figure 1. It is obvious that 0.6% NaCl seriously affected the growth of sweet sorghum, no matter how much potassium

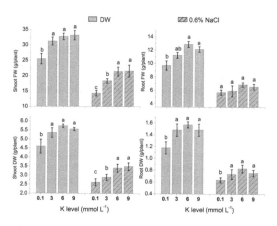

Figure 1. Effect of different potassium levels on the growth of sweet sorghum under salt stress. Different letters in the figure indicates significant difference at P < 0.05 significance level between treatments. The following are the same.

was supplied. With the increase of potassium concentration, the fresh weight and dry weight all increased significantly, especially in the shoot. At least 3 mmol L^{-1} potassium was adequate for its growth under control and 6 mmol L^{-1} at under salinity. Higher than 6 mmol L^{-1} potassium did not exert significant difference under either saline or non-saline conditions. Moreover, 0.1 mmol L^{-1} was not adequate for sweet sorghum growth, yet the growth reduction was only 22% and 20% in shoot FW and DW compared to that grown under 6 mmol L^{-1} potassium level. This phenomenon implies that the plant has a mechanism to utilized effectively when the environmental potassium nitrogen is low.

3.2 The photosynthetic gas exchange of sweet sorghum

0.6% NaCl treatment significantly reduced Pn, Gs and Tr while increased Ci (Fig. 2). On the other hand, supplying sweet sorghum with 6–9 mmol L^{-1} potassium significantly increased Pn, Gs, and Tr, meanwhile lowered Ci, under both saline and non-saline conditions (p < 0.05). Increase of Pn and Tr was closely related to the increase of Gs, this was based on the fact that potassium was involved in the osmotic potential regulation and thus increasing potassium supply promoted stomatal opening (Gs). Limitation of potassium may increase stomtal resistance and then restrict photosynthesis. Potassium also regulated the enzymes of carbon assimilation such as Rubisco. Thus adequate potassium supply increased photosynthesis and transpiration.

3.3 ΦPSII and qP

Quantum yield of PSII (ΦPSII) and photochemical quenching (qP) represent the ability of light conversion of PSII. Under low potassium level, salt stress seriously reduced ΦPSII and qP (Fig. 3). However, with the increase of potassium level, the differences of ΦPSII and qP between CK and 0.6% were not significant any more, implying supplying plants with potassium may alleviate salt injury under salinity.

3.4 Chlorophyll a fluorescence transient

The photochemical activity of the PSII can be measured by JIP-test (Fig. 4). Under CK and 0.6% NaCl treatments, PHIPo, RC/CSm, Tro/CSm and D.F all increased with potassium level. On the contrary, Vj decreased with potassium concentration. Implying sufficient potassium can increase the photochemical efficiency by increasing the number of active reactions centers, increasing the trapped energy of the leave section and promote the quantum-driven electron movement through the electron transport chain and meanwhile increasing the size of PQ sink and the extent of reaction centers' openness.

3.5 MDA and membrane permeability

MDA is a product of membrane lipid peroxidation resulted from Reactive Oxygen

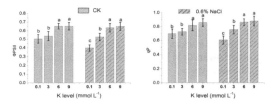

Figure 3. Changes of φPSII and qP under different potassium and NaCl levels.

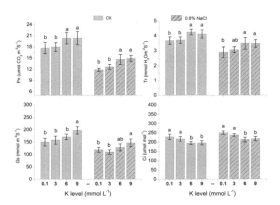

Figure 2. The changes of Pn, Tr, GS and Ci under different potassium and NaCl levels.

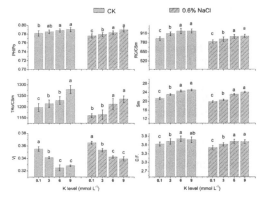

Figure 4. Changes of PHIPo, RC/CSm, Tro/CSm, Sm Vj and D.F under different potassium and NaCl levels.

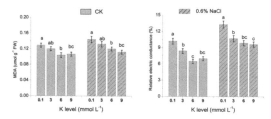

Figure 5. Changes of MDA and membrane permeability under different potassium and NaCl levels.

Species (ROS) attack. Under salt stress, more ROS are generated than can be scavenged by antioxidant enzymes and substances, which leads to the generation of MDA (Fig. 5). MDA not only is an indicator of membrane lipid peroxidation, but also reacts with deoxyadenosine and deoxyguanosine in DNA, forming DNA adducts (Marnett 1999). Membrane lipid peroxidation increases membrane permeability and causes leakage of nutritive substances from the cell (Shabala 2000), which leads to disturbance of ion homeostasis and metabolism (Munns & Tester 2008). Studies have shown that insufficient potassium reduces the efficiency of photosynthesis, and the plant can no longer use all of the sunlight striking its leaves for energy. The excess solar energy begin to accelerate the formation of ROS (Wang et al. 2015). It can be seen from Figure 5 that 0.6% NaCl increased relative electric conductance and increased potassium supply lowered MDA production and membrane permeability.

4 CONCLUSIONS

Potassium is absorbed by plants in larger amounts only secondary to nitrogen. One of the major roles of potassium is in regulating water use efficiency in the plant. If the plant is lack of potassium, the stomates will only open partially and be slower in closing. Potassium is also vital in carbohydrate metabolism. If a plant is deficient in potassium, photo-synthesis will decline while the plant respiration rate increases. This causes the plant's carbohydrate supply to decrease. Furthermore, potassium participates in the transportation of carbohydrates and activates enzymes to metabolize carbohydrates for the manufacture of amino acids and proteins. Potassium promotes cell division and growth by helping to move starches and sugars between plant parts. Adequate potassium supply adds stalk and stem stiffness, increases drought, cold and disease resistance, and gives plumpness to grain and seed (Mengel & Kirkby 2001).

Soil salinization is one of the abiotic factors limiting the distribution and growth of most glycophytes (Zhu 2001, Munns 2002). Salt stress lowers soil water potential, causes ionic stress and oxidative stress to the plant. Apart from these, salt stress generally reduces K uptake by the plant through competitive Na/K absorption. Potassium is one of the most important essential element to plants and saline soil is generally lack of it. Retaining of potassium in the plant under salinity is regarded as one of the most important property of salt-tolerant plants. It is not difficult to imagine that increased potassium supply under salinity may alleviate plant salt injury. As was shown in Figure 1, 3, 6 and 9 mmol L^{-1} potassium significantly increased the fresh weight and dry weight of sweet sorghum. However, the growth of sweet sorghum was not always positively correlated with potassium level, the effect of 9 mmol L^{-1} had almost the same effect as 6 mmol L^{-1}.

The role of potassium in photosynthesis relies on three aspects: promotion of stomatal aperture, light energy conversion and CO_2 assimilation. Thus potassium shortage influences all the steps and hence lower photosynthetic rate and bigger chances of photodamage (Wang et al. 2015). Figure 2 shows the increase of Pn is positively correlated with Gs. Figure 3 shows that the quantum yield and photochemical conversion efficiency of PSII is significantly promoted by higher potassium concentration under both saline and non-saline conditions, while Figure 4 shows that potassium and increase the number of active reaction centers, increase the light trapping ability and increase the size of PQ pool as well as the openness of reaction centers. As a result, sufficient potassium implies more solar energy can be converted to chemical energy and stored in the photosynthates through CO_2 assimilation, while insufficient potassium supply induced higher MDA and membrane permeability.

ACKNOWLEDGEMENTS

This work was financially supported by the National Key Technology Research and Development Program of the Ministry of Science and Technology of China (2009BADA7B05).

REFERENCES

Anschütz U., Becker D. & Shabala S. 2014. Going beyond nutrition: regulation of potassium homoeostasis as a common denominator of plant adaptive responses to environment. J. plant physiol. 171(9): 670–687.

Fan H., Meng G., Cheng R., *et al.* 2013. Effect of different nitrogen levels on the photosynthesis and growth of sweet sorghum seedlings under salt stress. Adv Mater Res. 726–731: 4352–4357.

Lee H.S. & Csallany S. 1987. Measurement of Free and Bound Malondialdehyde in Vitamin E-Deficient and -Supplemented Rat Liver Tissues. Lipids 22: 104–107.

Mengel K. & Kirkby E.A. 2001. Principles of Plant Nutrition. Dordrecht: Kluwer Academic Publishers.

Munns R. 2002. Comparative physiology of salt and water stress. Plant Cell Environ 25: 239–250.

Munns R. & Tester M. 2008. Mechanisms of salinity tolerance. Annu. Rev Plant Biol. 59: 651–681.

Reddy B.V.S., Ramesh S., Sanjana R.P., *et al.* 2005. Sweet sorghum—A potential alternative raw material for bio-ethanol and bio-energy. International Sorghum and Millets Newsletter 46: 79–86.

Strasser R.J., Tsimilli-Michael M. & Srivastava A. 2004. Chlorophyll *a* Fluorescence. Advances in Photosynthesis and Respiration. In GC Papageorgiou and Govindjee (ed). Chlorophyll a Fluorescence: A Signature of Photosynthesis: 321–362. Netherlands: Springer.

Shabala S. 2000. Ionic and osmotic components of salt stress specifically modulate net ion fluxes from bean leaf mesophyll. Plant Cell Environ. 23(8): 825–837.

Wang X.G., Zao X.H., Jiang C.J., *et al.* 2015. Effects of potassium deficiency on photosynthesis and photoprotection mechanisms in soybean. J Integr Agr. 14: 856–863.

Zhu J.K. 2001. Plant salt tolerance. Trends Plant Sci. 6: 66–71.

Advances in Energy, Environment and Materials Science – Wang & Zhao (Eds)
© 2016 Taylor & Francis Group, London, ISBN 978-1-138-02931-6

The trend and relationship analysis of precipitation and relative humidity under the climate change in Haihe River Basin from 1961 to 2010

M.J. Yang, Z.Y. Yang, S.Y. Zhang, Y.D. Yu, Z.L. Yu, D.M. Yan & S.N. Li
State Key Laboratory of Simulation and Regulation of the River Basin Water Cycle,
China Institute of Water Resources and Hydropower Research (IWHR), Beijing, P.R. China
Water Resources Department, China Institute of Water Resources and Hydropower Research, Beijing, P.R. China

ABSTRACT: Under the background of climate change, the tendency, mutability and periodicity of the annual precipitation of 47 meteorological stations in Haihe River Basin (HRB) and the surrounding area and their annual relative humidity were analyzed, by various statistical methods, including moving average method, Mann-Kendall test, Yamamoto diagnose, Maximum Entropy method, and Wavelet analysis. Also, a research about the relationship between precipitation and relative humidity was made with the support of SPSS. The results show that: (1) From 1961 to 2010, the decreasing trends of the annual precipitation and annual relative humidity in HRB are both significant, with a trend of −1.8426 mm/a and −0.0607%/a separately. (2) The mutation years of annual precipitation concluded from M-K test and Yamamoto diagnose are similar, both around 1976. However, the mutation year of the annual relative humidity are different with different methods. The M-K method shows the mutation occurred in 2002, while the Yamamoto diagnose indicates it happened in 1976, same as the annual precipitation. (3) The oscillation period of the two elements are both around 26 years by Wavelet analysis. But the results of Maximum Entropy method contain more than one period; the most significant of which is 9 years. (4) The regression analysis shows the change of relative humidity has a 2 10-days hysteresis after the precipitation.

1 INTRODUCTION

Under the background of climate change, agricultural development in China is facing a magnificent challenge. Climate change aggravates the agricultural disaster by affecting the occurrence of the extreme weather, and the society and economy will suffer stronger blows from these disastrous events than the normal evolution of regional climate (Wang, et al., 2009, Min, et al., 2008, Yang, 2008). Two of the most significant factors that affect the agriculture in water level are precipitation and relative humidity which determines the best place for the growth of different crops and the appropriate time to sow the seed. Haihe River Basin is one of the most crucial winter wheat producing region in China, with the climate varying with seasons and the evolution of water resources is of high complication. In recent decades, many hydrology researchers in China have launched lots of broad and detailed researches about the mechanism of how the climate and water resources affect the agriculture in HRB, and achieved a series of progresses (Yan, et al., 2013, Yang, et al., 2013, Li, et al., 2012). Studies also proved that the precipitation in Haihe River Basin has decreased in the

recent 50 years (Du, et al., 2014, Chen, et al., 2011), as well as the relative humidity.

This article, based on the daily precipitation data and daily relative humidity data for 50 years and the support of GIS, Matlab and SPSS, tests the trend, mutability and periodicity of the annual precipitation and annual relative humidity with the methods of moving average, Mann-Kendall (Wang, et al., 2010), Yamamoto (Zhao, et al., 2006), Maximum Entropy (Dagbegnon, et al., 2015) and Wavelet Analysis (Shujiang, et al., 2007) and gains the temporal relationship between precipitation and relative humidity. It will provide scientific support to the improvement of agricultural cropping pattern and, on the other hand, this article also includes a tentative research in the field of water cycle (Zhang, et al., 2008) that could act as a reference for agriculture development in HRB.

2 OVERVIEW OF STUDY AREA

Haihe River Basin (HRB), between 112°E~120°E and 35°N~43°N, located in the west of Bohai sea, crosses eight provinces including the entire Beijing and Tianjin, the most part of Hebei province, small part of Shanxi province, Henan province, Liaoning province, Shandong province and the Inner Mongolia Autonomous Region. The areas covers a total area of 317,800 km², in which mountain area occupies 189,000 km², 59.5 percent of the total area, and the rest 129,000 km² is the flat area, account for 40.5 percent of the total area. The terrain consists of three types, including plateau, mountain and plain, with elevation rising from southeast to northwest. The western part of HRB is Shanxi plateau and Taihangshan Mountains; the Inner Mongolia Plateau and Yanshan Mountains located in the northern part; the eastern and southeastern part of HRB is the North China Plain.

The HRB is mainly composed of three rivers: Haihe River, Luanhe River and Tuhaimajiahe River, the major climate of HRB is the warm and semi-humid continental monsoon climate, making Haihe River Basin one of the main granary of China and the main character is that the winter is cold and dry while the summer is hot and humid. The annual mean precipitation from 1961 to 2010 is 521 mm, concentrating in the rain season (summer), which makes summer flood and inundation frequent. The annual mean relative humidity from 1961 to 2010 is 59.6%.

3 DATA AND METHODS

3.1 *Data sources*

The daily precipitation data from 1961 to 2010 in 47 meteorological stations in and around HRB is obtained from *China Meteorological Data Sharing Service System* and the basic station information are listed in the Table 1 with Annual Mean Precipitation (AMP) and Annual Mean Relative Humidity (AMRH).

3.2 *Data processing*

In this article, two main representative parameters (precipitation and relative humidity) were chosen as the researching objects to reflect the climate change. The main methods in processing data are as follows: (1) moving average method, using the moving average of time series to identify the variation trend. (2) Mann-Kendall test, a non-parameter method to test the mutation of climatic factors. (3) Yamamoto diagnose, using the ratio of signal to noise to ascertain the mutation of climatic factors. (4) Wavelet analysis, a kind of enhanced Fourier analysis method that could get the characteristics of signal in a partial scale. (5) Maximum Entropy method, a kind of self-regression model that the peak of spectrum density curve shows the periodicity. With the support of GIS, three main aspects including the tendency, mutability

Table 1. The selected 20 stations in HRB and the AMP (mm) and AMRH (%).

Code	Station	Long/°	Lat/°	AMP	AMRH
53898	Anyang	114.400	36.050	564.076	65.914
54518	Bazhou	116.383	39.117	506.728	60.635
54602	Baoding	115.517	38.850	519.758	61.298
54511	Beijing	116.467	39.800	549.606	56.882
54423	Chengde	117.950	40.983	520.822	55.431
53487	Datong	113.333	40.100	371.904	52.005
54714	Dezhou	116.317	37.433	518.439	63.584
54736	Dongying	118.667	37.433	578.784	64.863
54208	Duolun	116.467	42.183	376.032	60.910
54308	Fengning	116.633	41.217	458.316	53.733
53564	Hequ	111.150	39.383	462.074	54.527
54906	Heze	115.433	35.250	476.141	73.735
53463	Huhehaote	111.683	40.817	401.492	53.500
53391	Huade	114.000	41.900	314.228	40.922
54405	Huailai	115.500	40.400	378.350	50.682
54624	Huangye	117.350	38.367	588.688	62.644
54725	Huimin	117.533	37.483	561.144	65.644
54823	Jinan	116.983	36.600	702.894	57.611
54326	Jianpingxian	119.700	41.383	461.522	52.219
54539	Leting	118.883	39.433	603.262	66.118
54705	Nangong	115.383	37.367	476.334	64.171
54449	Qinhuangdao	119.517	39.850	637.688	61.905
54436	Qinglong	118.950	40.400	694.734	59.346
54606	Raoyang	115.733	38.233	518.172	63.905
54808	Shenxian	115.667	36.233	539.922	68.930
53698	Shijiazhuang	114.417	38.033	526.232	60.778
54454	Suizhong	120.350	40.350	619.368	62.552
53772	Taiyuan	112.550	37.783	441.930	58.961
54534	Tangshan	118.150	39.667	608.976	61.115
54527	Tianjin	117.067	39.083	536.526	61.711
54623	Tianjintanggu	117.717	39.050	578.644	63.987
54311	Weichang	117.750	41.933	435.102	55.684
53593	Weixian	114.567	39.833	400.898	55.888
54213	Wengniuteqi	119.017	42.933	349.092	47.630
53480	Wulanchabu	113.067	41.033	358.930	52.091
53588	Wutaishan	113.517	38.950	747.306	65.496
53663	Wuzhai	111.817	38.917	421.052	57.812
53986	Xinxiang	113.883	35.317	574.022	67.277
53798	Xingtai	114.500	37.067	516.286	61.858
53975	Yangcheng	112.400	35.483	597.804	60.701
53478	Youyu	112.450	40.000	414.862	59.081
53787	Yushe	112.983	37.067	544.278	57.860

and periodicity of both precipitation and relative humidity revealing the characteristics of climate change were analyzed. In addition, a tentative research in the field of water cycle was made using regression analysis method based on the SPSS that could provide a scientific support to the improvement of agricultural cropping pattern.

4 RESULTS AND DISCUSSION

4.1 *Tendency*

47 meteorological stations in and around the HRB were chosen and processed using GIS to get the Annual Precipitation (AMP) and Annual Relative Humidity (AMRH) from 1961 to 2010 statistically (Fig. 1). Figure 1 (a) shows the variation trend of annual precipitation; Figure 1 (b) illustrates the annual relative humidity. From the 5 years moving average of annual departure, it shows that both AMP and AMRH has a decreasing trend during the last 50 years. The precipitation drops sharply in the late 1970s and 1990s while the relative humidity declines consistently since the early 1990s, with the declining gradients of each element are -1.8426 mm/a and -0.0607%/a respectively and indicates that the reduction of precipitation in HRB will lead to a severe agricultural irrigation water shortage, while the drying air would exacerbate the severe condition for the crops' growth.

4.2 *Mutation*

4.2.1 *M-K test*

According to the M-K statistical curves (Fig. 2), the x-axis value of the intersection point of UF and UB illustrates the year of mutation. From Figure 2 (a), the crossover point around 1976 demonstrates that the mutation of AMP happens in 1976, that means the requirement and supplement of water in the agricultural irrigation pattern changed in the year 1976 in HRB. However, the situation reflected in Figure 2 (b) is entirely different that, in 2002, relative humidity mutation occurs, before that there are no remarkable change of the air moisture.

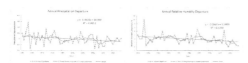

(a). Variation trend of AMP (b). Variation trend of AMRH

Figure 1. The 5 years moving average of annual departure and the linear trend of both AMP and AMRH.

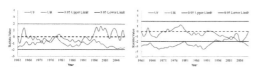

(a). Mutation period of AMP (b). Mutation period of AMRH

Figure 2. The mutation period of both AMP and AMRH using M-K method.

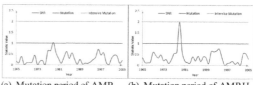

(a). Mutation period of AMP (b). Mutation period of AMRH

Figure 3. The mutation period of both AMP and AMRH using Yamamoto method.

4.2.2 *Yamamoto diagnose*

As a comparison and compensation, in Figure 3, the results of Yamamoto diagnose have distinctions with the conclusion of the M-K method above. In Figure 3 (a), the mutation period of AMP occurred in 1976, corresponding with M-K test, further verifying the precipitation in HRB experienced a sudden and profound change in 1976. But, compared with the M-K method, as it is showed in Figure 3 (b), the statistic value of relative humidity climbing over 1 in 1975 and peaked to 2 in 1976 indicates that the moisture quantity in air has a similar mutation condition with precipitation.

4.3 *Periodicity*

4.3.1 *Wavelet analysis*

According to the Figure 4, the periods of AMP and AMRH are gained by Wavelet analysis. Form the distribution of the intensity of wavelet coefficient in various time scale, it is illustrated clearly that during the last 50 years, both AMP (Fig. 4 (a)) and AMRH (Fig. 4 (b)) has the most significant periodicity of 26 years, with the signal intensity reaching the peak. In conclusion, from 1961 to 2010, the AMP has a 26 years oscillating period while the AMRH has a 26 years oscillating period in HRB.

4.3.2 *Maximum entropy method*

In Figure 5, the periodicity of AMP and AMRH in HRB from 1961 to 2010 was extracted using Maximum Entropy method. From Figure 5 (a, b), the spectrum density curves of AMP and AMRH has more than one peak, among which the most significant period of AMP and AMRH both are 9 years. Moreover, the second distinctive periods of AMP and AMRH are 5 years and 14 years respectively. Therefore, from 1961 to 2010, the AMP has an oscillating period of 9 years and 5 years while the AMRH has an oscillating period of 9 years and 14 years in HRB.

4.4 *Regression analysis*

Water supplement influences crop growth in two aspects: (1) soil moisture affects the ability of root absorbing water, (2) air moisture affects

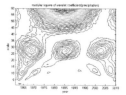
(a). Periodicity of AMP

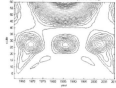
(b). Periodicity of AMRH

Figure 4.   The periodicity of both AMP and AMRH using wavelet analysis.

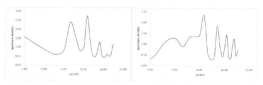

(a). Periodicity of AMP   (b). Periodicity of AMRH

Figure 5.   The periodicity of both AMP and AMRH using maximum entropy method.

the photosynthsis, respiration and evapotranspiration of crops. Therefore, as a reference, it will help improving the productivity if the correlation between the precipitation and relative humidity was figured out to make it clear of the local water cycle. Figure 6 shows the variation trend of interannual mean 10-day precipitation and interannual mean 10-day relative humidity in HRB. As portrayed in the Figure 6, it is clear that P and RH has a significant consistancy, with a synchronous increase and decrease trend, meanwhile, the variation of RH has a silght hysteresis after the precipitation. To further analyze the degree of hysteresis and find the accurate lagging time, the measured RH was lagged for 0~4 10-days to get the Pearson correlation coefficient between P and RH in each lagging time and then was portrayed in Figure 7. As illustrated in Figure 7, when the lagging time reaches to 2 10-days, the value of Pearson correlation coefficient comes to the peak, 0.907, where the consistancy is the highly conspicuous.

To ascertain the relationship between those two factors, precipitation was considered as an independent variable, with other meteorologic parameter closely referring to relative humidity (temperature (T/°C), wind speed (W/m*s^{-1}) and radiation (R/h)) as independent variable as well, and the relative humidity was taken as dependent variable. On SPSS, the multiple linear regression equations before and after 2 10-days hysteresis were established. The regression equations before and after hysteresis are as follows:

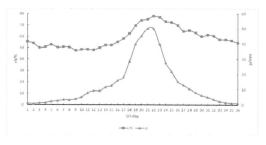

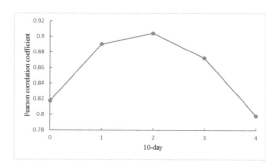

Figure 6.   Correlation analysis of P and RH in 10-day scale.

Figure 7.   Pearson correlation coefficient after different hysteresis.

Before Hysteresis:

$$RH = 113.362 + 0.047P + 0.677T - 9.126W - 0.538R \quad (1)$$

$$R^2 = 0.984, \ p = 0.000$$

After Hysteresis:

$$RH = 108.755 + 0.11P + 0.649T - 7.43W - 0.542R \quad (2)$$

$$R^2 = 0.989, \ p = 0.000$$

In Table 2, there are the significance value of each regression parameters showing that: the before hysteresis, P = 0.193, it is not significant for not reaching the significant standard ($\alpha = 0.05$); after hysteresis, P = 0.001, it reflects a significance by attaining the significant standard ($\alpha = 0.05$). Figure 8 expresses the fitting results between the simulated value from the regression model and the measured value, concluding that model considering hysteresis after the rainoff events behaves better than that considering hysteresis before the event, especially at the peak. It further identified the exist of 2 10-days hysteresis that may be because of the

Table 2. The significance value of each parameter.

	Constant	P	T	W	R
Sig. Before Hysteresis	0.000	0.193	0.000	0.000	0.000
Sig. After Hysteresis	0.000	0.001	0.000	0.000	0.000

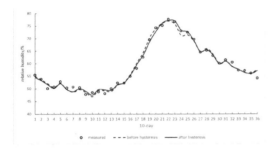

Figure 8. Simulation effect before and after 2 10-day hysteresis.

water storage ability of underlying surface that makes the variation of air moisture unsensitive to the rain.

5 CONCLUSIONS

Based on the researches of AMP and AMRH by statistic methods, it can be concluded that:

1. Both AMP and AMRH has decreasing trend during the last 50 years with a gradient of −1.8426 mm/a and −0.0607%/a respectively.
2. The mutation periods of precipitation showed by the M-K curve and Yamamoto curve are both in 1976, however, the results have some distinctions when it comes to the relative humidity with the mutation periods in 2002 and 1976 separately.
3. According to the results of Wavelet Analysis and Maximum Entropy method, there are 9 years and 26 years oscillating periodicity of AMP and AMRH occurs in HRB from 1961 to 2010.
4. After the regression analysis, as showed in the formulas, the influence of precipitation in relative humidity is not very remarkable, but, the conclusion is definite that the variation of RH has a 2 10-days hysteresis after P which means the underlying surface has the ability to alleviate the variation intensity of air moisture right after the rainfall event.

ACKNOWLEDGEMENTS

This study is financially supported by the Project of the Ministry Of Water Resources Public Welfare Industry Special Scientific Research (201401001).

REFERENCES

Chen M. & Wu Y.X. 2011. Distribution of rainstorm centers and processes under different weather systems in Haihe River Basin. *Journal of Hohai University (natural science edition)* (03): 237–241.
Dagbegnon C.S.D. & Vijay P.S. 2015. Retrieving vegetation growth patterns from soil moisture, precipitation and temperature using maximum entropy. *Ecological Modelling* 309–310(10–24): 10–21
Du H., Xia J., Zeng S. et al. 2014. Variations and statistical probability characteristic analysis of extreme precipitation events under climate change in Haihe River Basin, China. *Hydrological Processes* 28(3): 913–925.
Li L.X., Yan D.H., Qin T.L. et al. 2012. Drought variation in Haihe river basin from 1961 to 2010. *Journal of Arid Land Resources and Environment* 26(11): 61–67.
Min S. & Qian Y.F. 2008. Regionality and persistence of extreme precipitation events in China. *Advances in Water Science* (06): 763–771.
Shujiang K. & Henry L. 2007. Wavelet analysis of hydrological and water quality signals in an agricultural watershed. *Journal of Hydrology* 338(1–2): 1–14.
Wang Z.F. & Qian Y.F. 2009. Frequency and intensity of extreme precipitation events in China. *Advances in Water Science* (01): 1–9.
Wang Z.L., Chen X., Hao Z.C. et al. 2010. Long-team Trend and Jump Change for Major Climate Processes over the Area of Huaibei. *Journal of Irrigation and Drainage* 29(05): 52–56.
Yan D.H., Yuan Z., Yang Z.Y. et al. 2013. Spatial and temporal changes in drought since 1961 in Haihe River Basin. *Advances in Water Science* 24(01): 34–41.
Yang J.H., Jiang Z.H., Wang P.X. 2008. Temporal and spatial distribution characteristics of extreme precipitation events in china. *Climate and environmental research* (01): 75–83.
Yang Z.Y., Yuan Z., Yan D.H. et al. 2013. Study of spatial and temporal distribution and multiple characteristics of drought and flood in Huang-Huai-Hai River Basin. *Advances in Water Science* 24(05): 618–625.
Zhang Y., Zhang Z.H., Yao F.Q. et al. 2008. Analysis on time lag effect of soil temperature compared to meteorological factors. *Research of agricultural modernization* 29(04): 468–470.
Zhao F.F. & Xu Z.X. 2006. Long-term trend and jump change for major climate processes over the upper yellow river basin. *Acta Meteorologica Sinica* 64(02): 52–56.

Advances in Energy, Environment and Materials Science – Wang & Zhao (Eds)
© 2016 Taylor & Francis Group, London, ISBN 978-1-138-02931-6

Chaotic characterization of multiphase mixing effects by computational homology and the 0–1 test for chaos

Junwei Huang
Faculty of Mechanical and Electrical Engineering, Yunnan Agricultural University, Yunnan, P.R. China

Jianxin Xu, Shibo Wang & Jianhang Hu
State Key Laboratory of Complex Nonferrous Metal Resources Clean Utilization,
Kunming University of Science and Technology, Yunnan, P.R. China

ABSTRACT: Mixing technology is widely used in engineering. This paper introduces 0–1 test for chaos and Poincare sections, for the characterization of multiphase mixing effects based on computational homology and image analysis. Experimental results show that the chaotic behavior in multiphase mixing process can be decided by calculating the median (K) of parameter $K\text{-}c$ approaching asymptotically to zero or one, and the value of parameter K may give a good quantification of multiphase mixing effects. Finally, it is illustrated to verify the theoretical analysis according to Poincare sections and the largest Lyapunov exponents. Particularly, by Poincare sections of the first Betti numbers, the abnormal mixing effect could be distinguished graphically. It provides a good theoretical and methodological support for optimization design of mixing equipments.

1 INTRODUCTION

Mixing is a major problem in engineering. Usually good mixing is desirable and is ensured by stirring or somehow generating shear or turbulence, but of course followed by diffusion at the molecular level. If mixing precedes a relatively fast chemical reaction, the extent of mixing will inevitably affect the course of the reaction (Ivlevaa et al. 2011). Recently, statistical and deterministic chaos analysis techniques have been used to analyze the nonlinear behavior in multiphase mixing system (Paglianti et al. 2000; Briens & Ellis 2000; Robert et al. 2004; Shoji 2004; Liu et al. 2006). In this paper, based on the reaction of $CH4 + ZnO$ with molten salt system (Ao et al. 2008), the quantification of gas-liquid-solid three phase mixing effects is investigated by hydrodynamic modeling. A new technique for quantifying the efficiency of multiphase flow mixing has been proposed in (Xu et al. 2011), the zeroth and first betti numbers time series could give a good quantification of mixing homogeneity and inhomogeneity respectively. However, it is pointed out an interesting fact that the zeroth Betti number series in all the experiments and the first Betti number series in some experiments approximately follow the classic statistical bell curve (Normal Distribution). In order to verify the accuracy of our optimization experimental model and obtain the dynamic characteristics of multiphase

flow mixing effects, some deeper statistical and deterministic chaos analysis of multiphase mixing dynamics should be studied.

The main aim of the present study is to analyze the multiphase mixing effects from view point of chaotic dynamic theory. Histograms as a method of statistical analysis are usually introduced to study the multiphase flow (Chiti et al. 2011; Haam & Brodkey 2000), however, Betti numbers histograms of multiphase mixing is analyzed for the first time. For the deterministic chaos analysis, many testing methods have been proposed to decide whether a dynamic system is regular or chaotic, such as calculating the maximum Lyapunov exponent, observing the poincare section, calculating the power spectral and so on (Solomon et al. 2003; Johnssona et al. 2000; Shi et al. 2008). Recently, a new test named "the 0–1 test for chaos" has been developed to detect the presence of low-dimensional chaos in time series (Gottwald & Melbourne 2004, 2005, 2008, 2009; Falconer & Gottwald 2007). The input is the time series of a relevant variable and output is zero or one. This method has the advantage to be easy to implement and does not need the underlying equations, the reconstruction of the phase space, or the dimension of the actual system. The test has been applied successfully on theoretical time series (Sun et al. 2010; Ascani et al. 2008), both with and without noise, from various dynamical systems (van der

Pol, Kortweg de Vries, Lorenz, the logistic map), as well as on experimental data (Falconer et al. 2007). This paper is organized as follows. In section 2, we briefly describe the multiphase flow mixing model and the 0–1 test for chaos. In section 3, The 0–1 test is applied to characterize multiphase mixing effects and compare with the methods of computing the maximum Lyapunov exponent and poincare section. In section 4, we present our conclusions.

2 MODELING AND METHODOLOGY

In order to investigate the complex behavior often observed in a molten salt three phase flowing bed reactor, an experimental multiphase mixing model of gas agitated reactors stirred by top lance gas injection is built. As is pointed out in (Xu et al. 2011), the chemical reactor is a 5-L transparent cylindrical glass vessel with diameter 17.2 cm and height 26.8 cm. It is used to simulate a molten salt three phase flowing-bed reactor. The lance diameter is 9.6 mm. The range of top submerged lance lengths is set at 50–80 mm. Water is used to simulate the liquid phase (molten salt) in chemical reactors, nitrogen is used to simulate the gas phase (CH4), black polystyrene particles with diameter 0.45 mm are used to simulate the solid phase catalyst (ZnO). Nitrogen is forced downwards as a jet from a pipe (lance) that could be moved vertically. Nitrogen, solid particles and water come together in the chemical reactor to form a particular flow pattern in such a way. Figure 1 shows the experimental set-up of multiphae mixing system. A camera can be used to gain the patterns at the speed of 30 frames per second and take 10,000 images in each experiment, in order to eliminate any bias due to reflection, subtracting from each image an

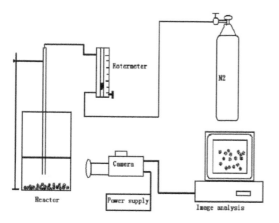

Figure 1. Experimental set-up of multiphase flow mixing system.

initial image, then binarization of these patterns are carried out in Matlab software environment. Since black polystyrene particles are quite different from the transparent nitrogen and water in color, the nitrogen bubbles in these patterns will disappear after binarization. In the end, Betti numbers of the binary images are calculated by CHomP free software.

We have developed a new Betti numbers method based on computable homology and image analysis (Xu et al. 2011; Kaczynski et al. 2004), for quantifying the mixing effects. The zeroth Betti numbers are used to estimate the numbers of pieces in the patterns, leading to a useful parameter to characterize the mixture homogeneity. The first Betti numbers are introduced to characterize the nonhomogeneity of the mixture. The mixing efficiency can be characterized by the Betti numbers for binary images of the patterns. In the end, we found an interesting fact that the zeroth Betti number series in all the experiments and the first Betti number series in some experiments approximately follow the classic statistical bell curve (Normal Distribution). On the basis, the deterministic chaos analysis of multiphase mixing dynamics are further discussed by the 0–1 test for chaos method.

3 THE 0–1 TEST ALGORITHM FOR CHAOS

Given an observation $\phi(j)$ for $j = 1, \dots N$, $\phi(j)$ represents an one-dimensional observable data set. For $c \in (0, \pi)$, we compute the translation variables. $p_c(n) = \sum_{j=1}^{n} \phi(j)\cos jc$, $q_c(n) = \sum_{j=1}^{n} \phi(j)\sin jc$ $n = 1, 2, 3, \dots, N$. To determine the growth of p_c and q_c it is convenient to look at the mean square displacement, defined as $M_c(n) = \lim_{N \to \infty} \frac{1}{N} \sum_{j=1}^{N} [(p_c(j+n) - p_c(j))^2 + (q_c(j+n) - q_c(j))^2]$. Note that this definition requires $n \ll N$. In practice we find that $n_{cut} = N/10$ yields good results. The test for chaos is based on the growth rate of $M_c(n)$ as a function of n. The next step is to estimate the asymptotic growth rate K-c (Ascani et al. 2008). We now present an alternative method for determining K-c from the mean square displacement. Form the vectors $\xi = (1, 2, 3, \dots, n_{cut})$ and $\Delta = (M_c(1), M_c(2), M_c(3), \dots, M_c(n_{cut}))$. Given vectors x, y of length q, we define covariance and variance in the usual way: $\text{cov}(x, y) = \frac{1}{q} \sum_{j=1}^{q} (x(j) - \bar{x})(y(j) - \bar{y})$, where $\bar{x} = \frac{1}{q} \sum_{j=1}^{q} x(j)$

$\text{var}(x) = \text{cov}(x, x)$ Now define the correlation coefficient $K - c = corr(\xi, \Delta) = \frac{\text{cov}(\xi, \Delta)}{\sqrt{\text{var}(\xi)\text{var}(\Delta)}} \in [-1, 1]$, K-c = 0 for regular dynamics and K-c = 1 for chaotic dynamics.

4 RESULTS AND DISCUSSION

4.1 Deterministic chaos analysis of multiphase flow mixing effects

By the 0–1 test for chaos, the chaotic behavior in multiphase mixing process can be decided by calculating the parameter K-c approaching asymptotically to zero or one, K-$c = 1$ stands for chaotic dynamic, and K-$c = 0$ stands for nonchaotic dynamic. As pointed out in (Gottwald & Melbourne 2009), the median is robust against outliers associated with resonances. K-c should be taken as the median rather than the mean. In Figure 2, (a) the median (K) of K-c of the zeroth Betti numbers time series at $l = 8$ cm; $Q = 1000$ L/h is 0.9937, it is very close to 1, for chaotic behavior; (b) the median (K) of K-c of the first Betti numbers time series at $l = 8$ cm; $Q = 1000$ L/h is 0.9971, for chaotic behavior; (c) the median (K) of K-c of the zeroth Betti numbers time series at $l = 5$ cm; $Q = 1500$ L/h is 0.9727, also for chaotic behavior; (d) the median (K) of K-c of the first Betti numbers time series at $l = 5$ cm; $Q = 1500$ L/h is 0.9952; (e) the median (K) of K-c of the zeroth Betti numbers time series at $l = 6$ cm; $Q = 1000$ L/h is 0.9769; (f) the median (K) of K-c of the first Betti numbers time series at $l = 6$ cm; $Q = 1000$ L/h is 0.9942. All the medians (K) of K-c are close to one, however, the median value of parameter K-c may give a

good quantification of multiphase mixing effects. In these experiments, the median value (K) of K-c at the case of $l = 8$ cm; $Q = 1000$ L/h is largest with 0.9937 and 0.9971, which has been confirmed as the best mixing condition. We have tried to compute the median (K) of K-c for every case more times, but there is very small fluctuation.

In Figure 3, we find that all the median (K) of K-c of the first Betti numbers are very close to one, for the chaotic behavior, although these cases do not follow the normal distribution. It is noticed that another interesting feature of the 0–1 test for chaos is that the inspection of the dynamics of the (p,q)-trajectories provides a simple visual test of whether the underlying dynamics is chaotic or nonchaotic (Cafagna & Grassi 2008). If the behavior of (p, q)-trajectories is Brownian, the underlying dynamics is chaotic; if the behavior is bounded, the underlying dynamics is non-chaotic. Clearly seen in Figure 4, the behaviors of all the (p, q)-trajectories are Brownian, which indicates the chaotic dynamics.

4.2 Poincare section and Lyapunov exponents

An even more indepth analysis can be obtained with the use of a Poincare section, which is introduced by Poincare for examining the motion of

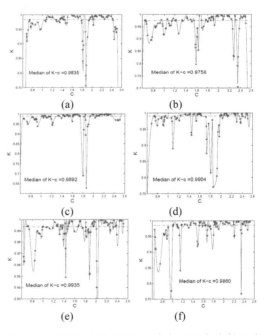

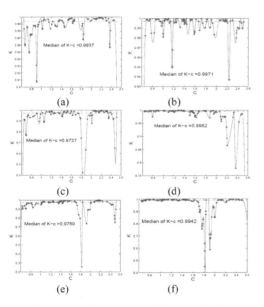

Figure 2. Plots of K versus c of the zeroth (left) and first (right) Betti numbers at $l = 8$ cm; $Q = 1000$ L/h (a,b), $l = 5$ cm; $Q = 1500$ L/h (c,d) and $l = 6$ cm; $Q = 1000$ L/h (e,f). Red dashed lines stand for the median K.

Figure 3. Plots of K versus c of the zeroth (left) and first (right) Betti numbers at $l = 5$ cm; $Q = 1900$ L/h (a,b), $l = 5.5$ cm; $Q = 2000$ L/h (c,d) and $l = 6.5$ cm; $Q = 2000$ L/h (e,f). Red dashed lines stand for the median K.

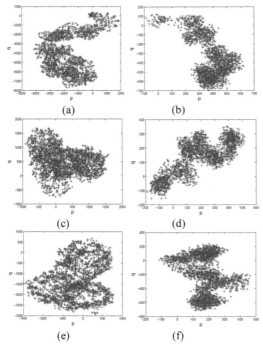

Figure 4. The dynamics of the translation components (p,q) of the zeroth (left) and first (right) Betti numbers time series. Figure (a,b) for $l = 8$ cm; $Q = 1000$ L/h; Figure (c,d) for $l = 5$ cm; $Q = 1500$ L/h; Figure (e,f) for $l = 6$ cm; $Q = 1000$ L/h.

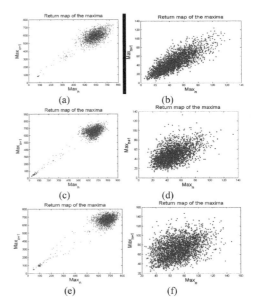

Figure 5. Poincare sections plots of the zeroth (left) and first (right) Betti numbers. Figure (a,b) for $l = 8$ cm; $Q = 1000$ L/h; Figure (c,d) for $l = 5$ cm; $Q = 1500$ L/h; Figure (e,f) for $l = 6$ cm; $Q = 1000$ L/h.

dynamical systems. We can use this method to investigate the dynamical behavior of multiphase mixing system. Considering the Poincare section of the steady-state image (Shi et al. 2008).

1. For a periodic case, the Poincare section is only one fixed point or a few discrete points in plane.
2. For a quasi-periodic case, the Poincare section is a closed curve in plane.
3. For a chaotic case, the Poincare section is some dense points with the similar fractal structure.

Figure 5 and Figure 6 show that the Poincare sections are some dense points with the similar fractal structure, however, there are some different structures, especially, the Poincare sections of the first Betti numbers time series for the better mixing effects are quite different from ones for the abnormal mixing effects, such as $l = 5$ cm; $Q = 1900$ L/h, $l = 5.5$ cm; $Q = 2000$ L/h, and $l = 6.5$ cm; $Q = 2000$ L/h. By the Poincare sections of the first Betti numbers time series, graphically at least, the abnormal mixing effect could be distinguished.

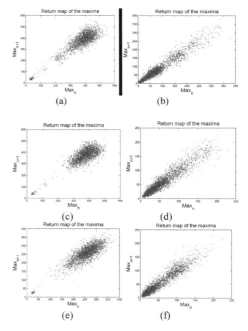

Figure 6. Poincare sections plots of the zeroth (left) and first (right) Betti numbers. Figure (a,b) for $l = 5$ cm; $Q = 1900$ L/h; Figure (c,d) for $l = 5.5$ cm; $Q = 2000$ L/h; Figure (e,f) for $l = 6.5$ cm; $Q = 2000$ L/h.

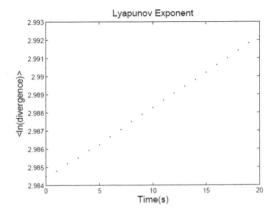

Figure 7. Lyapunov exponent profile for the chaotic zeroth Betti numbers time series ($l = 8$ cm; $Q = 1000$ L/h).

The usual test of whether a deterministic dynamical system is chaotic or nonchaotic involves the calculation of the maximal Lyapunov exponents (Solomon et al. 2003). As is known to us, a positive maximal Lyapunov exponent indicates chaos. In this study, the maximum Lyapunov exponent is calculated using Rosenstein method (Rosenstein et al. 1993). In Figure 7, a practical method (Rosenstein et al. 1993) is used to calculate the maximum Lyapunov exponent, the maximum Lyapunov exponent is easily and accurately calculated using a least-squares fit to the "average" line defined by plot of ln (divergence) versus time. This method is easy to implement and fast because it uses a simple measure of exponential divergence. The dashed line has a slope equal to the theoretical value of maximum Lyapunov exponent, the clear presence of a positive slope affords one the qualitative confirmation of a positive exponent. Finally, all the results indicate a positive exponent, which means chaotic.

5 CONCLUSIONS

In this study, characterization of multiphase flow mixing effects is investigated by the 0–1 test for chaos. Chaotic dynamic behaviors of multiphase mixing are then characterized by 0–1 test. The median (K) of parameter K-c calculated by correlation method is approaching asymptotically to one, which indicates a chaotic state, and the median values (K) of parameter K-c may give a good description of the multiphase mixing effects. Finally, Poincare sections, as well as the maximum Lyapunov exponents succeed in detecting the presence of chaos in the zeroth and the first betti numbers time series for quantifying the efficiency of multiphase

mixing. Particularly, using Poincare sections and fitting Normal distribution of the first Betti numbers time series, graphically at least, the abnormal mixing effects could be distinguished. The present analysis of multiphase mixing efficiency, emphasizes that the statistical and chaotic characteristics of the Betti numbers time series based on algebraic topology are important for a deeper understanding of multiphase flow mixing dynamics.

ACKNOWLEDGMENTS

This work is supported by the Special Preliminary Research of 973 Plan, China (2014CB460605), the National Natural Science Foundation of China (Project Nos. 51406071), and the Natural Science Foundation of Yunnan Province (2013FB020).

REFERENCES

Ao Xian quan, Wang Hua, Wei Yong gang. Novel method for metallic zinc and synthesis gas production in alkali molten carbonates [J]. Energy Conversion and Management, 2008, 49: 2063–2068.

Ascani F. et al. Detection of low-dimensional chaos in quasi-periodic time series: the 0–1 test [J]. Tech. Rep., Santa Fe Institute Complex Systems Summer School, 2008.

Briens Lauren A., Ellis N. Time-series analysis approach for the identification of flooding/loading transition in gas-liquid stirred tank reactors [J]. Chem. Eng. Sci., 2000, 55(23): 5793–5802.

Cafagna D., Grassi G. Fractional-order Chua's circuit: time-domain analysis, bifurcation, chaotic behavior and test for chaos [J]. International Journal of Bifurcation and Chaos in Applied Sciences and Engineering, 2008, 18: 615–639.

Chiti F., Bakalis S., Bujalski W., Barigou M., Eaglesham A., Nienow Alvin W. Using positron emission particle tracking (PEPT) to study the turbulent flow in a baffled vessel agitated by a Rushton turbine: Improving data treatment and validation [J]. Chem. Eng. Res. Des. 2011, 89(10):1947–1960.

CHomP. http://chomp.rutgers:edu/software

Falconer I., Gottwald G.A., Melbourne I., Wormnes K. Application of the 0–1 test for chaos to experimental data [J]. SIAM Journal on Applied Dynamical Systems, 2007, 6: 395–402.

Gottwald G.A., Melbourne I. Testing for chaos in deterministic systems with noise [J]. Physica D, 2005, 212: 100–110.

Gottwald G.A., Melbourne I. A new test for chaos in deterministic systems [J]. Proceedings of The Royal Society of London, 2004, 460: 603–611.

Gottwald G.A., Melbourne I. Comment on reliability of the 0–1 test for chaos [J]. Physical Review E, 2008, 77: 028201.

Gottwald G.A., Melbourne I. On the Implementation of the 0–1 Test for Chaos [J]. SIAM J. Applied Dyn. Syst, 2009, 8: 129–145.

Haam S.J., Brodkey R.S. Laser Doppler anemometry measurements in an index of refraction matched column in the presence of dispersed beads, Part I [J]. Int. J. Multiphas. Flow, 2000, 26: 1419–1438.

Ivlevaa T.P., Merzhanova A.G., Rumanova E.N., Vaganovaa N.I., Campbellb A.N., Hayhurstb A.N. When Do Chemical Reactions Promote Mixing? [J]. Chem. Eng. J. 2011, 168: 1–14.

Johnssona F., Zijerveldb R.C., Schoutenb J.C., Van den Bleekb C.M., Lecknera B. Characterization of fluidization regimes by time-series analysis of pressure fluctuations [J]. Int. J. Multiphas. Flow, 2000, 26: 663–715.

Kaczynski T., Mischaikow K., Mrozek M. Computational Homology [J]. Applied Mathematical Sciences, Springer-Verlag, New York, 2004, 157.

Liu Ming-Yan, Qiang Ai-Hong, Sun Bing-Feng. Chaotic characteristics in an evaporator with a vapor–liquid–solid boiling flow [J]. Chem. Eng. Process, 2006, 45(1): 73–78.

Paglianti A., Pintus S., Giona M. Time-series analysis approach for the identification of flooding/loading transition in gas–liquid stirred tank reactors [J]. Chem. Eng. Sci, 2000, 55: 5793–5802.

Robert Z., Wilhemus J.M., Kruijf T.D., Van Der Hagen H.J.J. Investigating the Nonlinear Dynamics of Natural-Circulation, Boiling Two-Phase Flows [J]. Nucl. Technol, 2004, 146: 244–256.

Rosenstein M.T., Collins J.J., De Luca C.J. A practical method for calculating largest Lyapunov exponents from small data sets [J]. Physica D, 1993, 65: 117–134.

Shi Pei-Ming, Liu Bin, Hou Dong-Xiao. Chaotic motion of some relative rotation nonlinear dynamic system [J]. Acta Physica Sinica, 2008, 57(3): 1321–1328.

Shoji M. Studies of boiling chaos: a review [J]. Int. J. Heat Mass Transfer. 2004, 47(6–7): 1105–1128.

Sun Ke-Hui, Liu Xuan and Zhu Cong-Xu. The 0–1 test algorithm for chaos and its applications [J]. Chin. Phys. B, 2010, 19: 110510.

Thomas H. Solomon, Brian R. Wallace, Nathan S. Miller, Courtney J.L. Spohn. Lagrangian chaos and multiphase processes in vortex flows [J]. Commun. Nonlinear Sci. Numer. Simul, 2003, 8: 239–252.

Xu Jianxin, Wang Hua, Fang Hui. Multiphase mixing quantification by computational homology and imaging analysis [J]. Appl. Math. Model., 2011, 35(5): 2160–2171.

Application effects of reclaimed water reuse technology on pharmaceutical wastewater

Wenjuan Liang
School of Municipal and Environmental Engineering, Shenyang Jianzhu University, Shenyang, Liaoning, China

Yong Yan
School of Civil Engineering, Shenyang Jianzhu University, Shenyang, Liaoning, China

ABSTRACT: Using coagulation, ultrafiltration and reverse osmosis treatments as the core technologies of the water reuse treatment, the treatment effects of small test and pilot test on the effluent of secondary biological treatment in one pharmaceutical factory were analyzed. The results of small experiment showed that, the optimal dosage was 15–20 ppm, transmembrane pressure in ultrafiltration was stable, and electrical conductivity of reverse osmosis stabilized at 600 µS/cm. The results of pilot test showed that, the turbidity of the clarifier effluent was lower than 10 NTU, and the removal rate of COD was 15.7%–48.8%. Ultrafiltration system operated steadily, whose transmembrane pressure stabilized at 0.4–0.6 kg/cm². The turbidity of water produced is stable, which was controlled around 0.17–0.28 NTU. The removal for COD was not very effective. The reverse osmosis operated steadily, and the desalination effect was gradually optimized. The application of the coagulation, ultrafiltration and reverse osmosis on pharmaceutical wastewater reuse project was effective.

1 INTRODUCTION

Pharmaceutical wastewater has low biodegradability and poor water quality (Dong 2008), which is one kind of industrial wastewater which is the most difficult to be treated (Gibson 1973). A large number of high concentration of organic polluted wastewater can be produced in such processes as fermentation, filtration, ion exchange, concentration, esterification and purification, etc (Yang & Shang 2007). Duel-membrane technology is one of the hot spots in the international development and engineering application (You & Deng 2014). Ultrafiltration can remove most turbidity and organic compounds in the wastewater (Fan 2012), which can reduce the pollution of the reverse osmosis membrane (Song 2012). Reverse osmosis membrane has good desalination effects (Li & Gao 2014). The high-quality effluent can be directly reused in the pharmaceutical industry (Dong 2013), meanwhile, the concentrated water backflows to the conventional process (Yan 2012), achieving zero emissions of wastewater (Chen 2013) and cleaner production (Zeng 2008).

One pharmaceutical factory in Shenyang is a large pharmaceutical company, main producing VC products, whose daily processing capacity was 30,000 tons of wastewater. The wastewater treatment process: pretreatment—hydrolysis, acidification—BAF—UNITANK, and the effluent can meet the discharge standard. In order to save water resources and reuse the effluent as industrial circulating cooling water after being deeply treated (Liu 2015), taking the effluent of secondary biological treatment in one pharmaceutical factory as the object of study, taking coagulation, ultrafiltration and reverse osmosis treatments as the core technologies of the water reuse small experiment and pilot test, the application effects of treatment process were analyzed.

2 TECHNOLOGICAL PROCESS, MATERIALS AND EQUIPMENTS

2.1 Technological process

The reclaimed water reuse engineering of secondary biological treatment effluent in one pharmaceutical factory was studied. Technological process: raw water tank—dosing—mechanical stirring clarification pool—buffering tank—ultrafiltration—water storage tank—reverse osmosis.

The product water was collected regularly for monitoring.

2.2 Experiment materials

Coagulation equipment is mechanical stirring clarification pool. Coagulant is Polyaluminium Chloride (PAC), and coagulant aid is Polyacrylamide (PAM).

2.3 Equipments

1. 1 set of water feed pump, delivering pre-process water, overcoming the pressure drop and the pipeline resistance from multi-media filter.
2. 2 multi-media filters with one preparing, Ø2200 × H1800 mm.
3. 1 set of backwash pump, flow 90 m³/h, water head 28 m.
4. 1 set of security filter, processing capacity 33 m³/h, within 12 spray type PP filter, filter material is SS304.
5. 15 sets of ultrafiltration membrane modules, Seville SV1060B type, capacity 29 m³/h.
6. 1 set of UF backwash pump, flow 60 m³/h, water head 22 m.
7. 1 set of UF water producing tank, PE material, volume 20 m³.
8. 1 set of reverse osmosis water feed pump, pump flow 28 m³/h, water head 29 m, power 4 kW.
9. 1 set of reverse osmosis security filter, shell material SS304, filtration precision 5 μm.
10. 1 set of reverse osmosis high pressure pump, flow 28 m³/h, water head 176 m, power 22 kW.
11. Reverse osmosis membrane modules, American Hyde LFC3-LD type, designed water yield capacity 17 m³/h, system recovery rate 70%.

3 RESULTS AND ANALYSIS

3.1 Small test

3.1.1 Coagulation test

To determine the optimal dosage of coagulants, beaker experiment was carried out. When the dosage was 15–20 ppm (Fig. 1), the effluent COD could obtain a removal rate of 32%, and the increase of dosage did not improve the clarified water. Therefore, the optimal dosage of coagulants was 15–20 ppm.

3.1.2 Ultrafiltration test

In the case of adding no flocculants, the change of Transmembrane Pressure (TMP) with time is shown in Figure 2. TMP was not stable. After adding flocculants (Fig. 3), TMP was stable in 1.00–1.10 kg/cm².

3.1.3 Reverse osmosis test

The raw water was heavy salt water with salt content of above 5000 ppm, and osmotic pressure about 5 kg/cm². Concentrated water's discharge flow was 6 L/min, the circulating flow rate was 80 L/min, and pure water yield increased from 8 L/min at the start of operation to 4.8 L/min after 25 min. The electrical conductivity increased initially and drop subsequently, and finally stabilized at 600 μS/cm (Fig. 4).

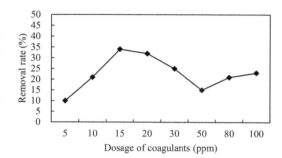

Figure 1. The effluent COD removal rate changes with the dosage of coagulants.

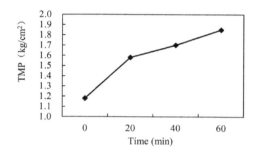

Figure 2. Transmembrane pressure changes without flocculants.

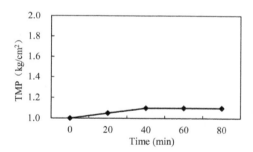

Figure 3. Transmembrane pressure changes when adding flocculants.

3.2 Pilot test

3.2.1 Operation analysis of mechanical stirring clarification pool

The effluent turbidity and COD of mechanical stirring clarification pool are respectively shown in Figure 5 and Figure 6.

The turbidity removal efficiency of raw water in mechanical stirring clarification pool was excellent. Even if the turbidity of raw water was as high as 70 NTU, the effluent turbidity can also stabilize below 10 NTU. The average removal rate was 83.1%

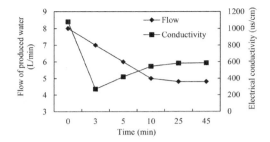

Figure 4. Reverse osmosis water flow and conductivity change change curve with time.

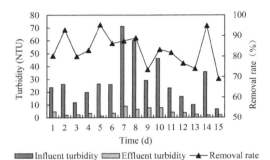

Figure 5. Influent and effluent turbidity and removal rate.

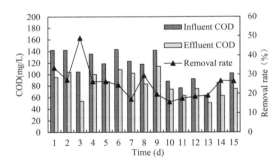

Figure 6. Influent and effluent COD and removal rate.

through coagulation. The peak values of removal rate had a good correlation with influent turbidity. The removal rate of COD was between 15.7%–48.8%, indicating that the content of insoluble organic matters in the raw water was large, and coagulation could effectively remove them. Coagulation provided powerful guarantees for ultrafiltration and reverse osmosis's stable operation.

3.2.2 Operation analysis of ultrafiltration

The water yield of ultrafiltration membrane was stably at 2 m³/h, and the concentrated water yield

was 12 L/min. Transmembrane pressure fluctuated slightly with the running time, and was stable around 0.4–0.6 kg/cm². The overall stability of the ultrafiltration system was good.

Figure 7 and Figure 8 respectively describe the effluent turbidity and COD in ultrafiltration treatment. The water turbidity of the ultrafiltration system was stable around 0.17–0.28 NTU; the removal rate was 85.79%–97.45%, and the average was 90.85%. The removal efficiency of turbidity was very good, and the removal rate was relatively stable, which could provide stable conditions for subsequent reverse osmosis system. Ultrafiltration system was not efficient for COD removal, and the overall removal rate was not higher than 60%, which was mainly caused by a large number of small molecular soluble COD in the influent.

3.2.3 Operation analysis of reverse osmosis

The water yield was 11 L/min; the concentrated water discharge was 6 L/min; the circulating water was 80 L/min; the recovery rate was 66%; and the working pressure was 11.7 kg/cm². There was no such phenomena as membrane fouling or concentration polarization.

Figure 9 is a continuous monitoring data of electrical conductivity of the reverse osmosis effluent.

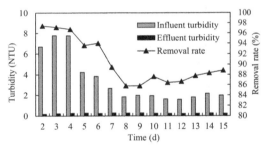

Figure 7. Influent and effluent turbidity and removal rate of ultrafiltration.

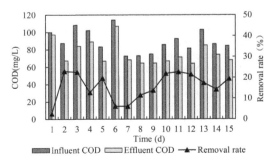

Figure 8. Influent and effluent COD and removal rate of ultrafiltration.

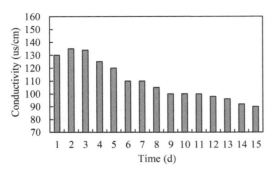

Figure 9. Effluent conductivity change of reverse osmosis.

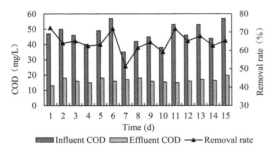

Influent COD   Effluent COD   Removal rate

Figure 10. Influent and effluent COD concentration and removal rate of reverse osmosis.

Electrical conductivity decreased gradually, desalination effect was gradually optimized.

Influent and effluent COD concentrations in reverse osmosis are shown in Figure 10.

The COD average removal rate of reverse osmosis was 64.6%, and the peak value was 72.34%. The effluent COD was 13–20 mg/L with an average of 16.4 mg/L. After reverse osmosis treatment, partial water could be used as the circulating cooling water.

4 CONCLUSION AND DISCUSSION

The ultrafiltration and reverse osmosis treatment effects of reclaimed water reuse small and pilot tests on the effluent of secondary biological treatment in one pharmaceutical factory were analyzed. The results of small experiment showed that, the optimal dosage was 15–20 ppm, transmembrane pressure in ultrafiltration was stable, and electrical conductivity of reverse osmosis stabilized at 600 μS/cm. The results of pilot test showed that,

the turbidity of the clarifier effluent was lower than 10 NTU, and the removal rate of COD was 15.7%–48.8%. Ultrafiltration system operated steadily, whose transmembrane pressure stabilized at 0.4–0.6 kg/cm². The turbidity of water produced was controlled around 0.17–0.28 NTU. The removal for COD was not very effective. The reverse osmosis operated steadily, and the desalination effect was gradually optimized. The application of the coagulation, ultrafiltration and reverse osmosis on pharmaceutical wastewater reuse project was effective.

REFERENCES

Chen H.B. 2013. Membrane bioreactor technology for treatment and reuse of mixed domestic wastewater from residential areas. Journal of Tongji University (Natural Science) 41(2): 247–252+288.

Dong K. 2013. Design of slaughterhouse wastewater treatment and water reuse. Changchun: Changchun University 24–69.

Dong X.J. 2008. Isolation and characterization of biphenylde-grading bacterium Achromobacter sp. BP3 and cloning of catabolic genes. Nanjing: Master Dissertat Agricultural University.

Fan J.H. 2012. Effect of quality-based pretreatment on enhanced treatment of pharmaceutical wastewater. China Water & Wastewater 28(23): 34–37.

Gibson, D. 1973. Initial reaction in the oxidation of ethylbenzene by Pseudomonas putida. Biochemistry 12(8): 1520–1528.

Li Y.F. & Gao Y. 2014. Review on advanced oxidation processes for the degradation of disinfection by-product formation potential. Technology of Water Treatment 40(5): 1–4+9.

Liu S.N. 2015. Treatment of electroplating wastewater by coupling vacuum membrane distillation and membrane gas absorption. Technology of Water Treatment 41(3): 68–70+75.

Song X. 2012. The present situation and research progress in the advanced treatment of pharmaceutical wastewater. Guangzhou Chemical Industry 40(12): 29–31.

Yan J.L. 2012. Application study on advanced treatment of textile wastewater by integrated ozonation-biological aerated filter process. Chengdu: South China University of Technology 21–39.

Yang Q. & Shang H.T. 2007. Pilot studies of UF-RO for municipal wastewater treatment. Membrane Science and Technology 27(3): 71–74.

You W.J. & Deng X.T. 2014. Recycled water technological design of a cement plant in Shanxi Province. Industrial Water Treatment 34(12): 87–90.

Zeng H.C. 2008. Treatment of textile wastewater using ultratfiltration and reverse osmosis dualmembrane system. Chinese Journal of Environmental Engineering 2(8): 1021–1025.

Advances in Energy, Environment and Materials Science – Wang & Zhao (Eds)
© 2016 Taylor & Francis Group, London, ISBN 978-1-138-02931-6

Pesticide usage and environmental protection

G.E. Khachatryan & N.I. Mkrtchyan
A.I Alikhanyan Science National Laboratory, Yerevan Physics Institute, Yerevan, Armenia

ABSTRACT: The consequences of interaction of the bacterium-pesticide by the way of example of the growth of some soil aerobic bacteria on different media containing preparation actara (insecticide) or only its active component thiamethoxam are presented. It's shown that the presence of actara in the nutrient meat-peptone agar leads to restriction or complete interruption of the growth of some soil strains. The presence of actara in the solid minimal salt medium containing actara, as the only source of carbon and nitrogen allows observing slow growth, which is more restricted on the medium with pure thiamethoxam. During the growth on actara-containing minimal medium the halo around of the colonies/plaques of some of strains were revealed, that appears after few consequent inoculations on the minimal media, and disappears after few intermediate inoculations on meat-peptone agar media. Some other phenomena are discussed also.

1 INTRODUCTION

One of the most important issues of modern ecology—is a constant problem of environmental pollution by various xenobiotics and their effects on living organisms and humans. This issue is the focus of researchers worldwide. Meanwhile, in our view, a separate and very important issue could be the consequences of a large-scale (and inevitable!) use of pesticides in agroindustry on the microbial community of the soil. The microbial community interpreted like certain system called on for and able to provide balance the possible imbalances generated in the soil after getting into the toxic compounds. And it's, like, a self-evident process. However, be aware that sudden changes typical of urban civilization, as a rule, have a negative impact on the nature. It seems to us that the things in fact are much more complicated, that the role of microorganisms is much more significant, and that this problem is not adequately discussed in the scientific literature (Kirk et al. 2004).

The authors believe that the reason for the difficulty of how to assess the processes, and long-term effects of pesticide exposure, since comprehensiveness of these processes. Several of environmentalists-agrobiologists have long warned of the possible danger, noting, in particular, the changes in the composition of microbial communities that are inherent de-facto to certain types of soil, as a result of the use of pesticides (Kiryushin 1996, Byzov et al. 1989).

If we recall that the decomposition of chemical compounds entering to the soil continues, sometimes for years, at least for us, is becoming clear that this fact is demanded more attention. As the decisive argument is not superfluous to recall the unfortunate history of DDT, and the fact that about 80 years (estimates vary greatly even for the natural environment) (NPIC, Wang 2010) is required for his semi-decomposition in nature. Given that the quantity and quality of pesticides (leaving aside the problem of accidents, release into the environment of xenobiotics of the other nature) is constantly growing and changing; the fact that the initial chemicals can be subjected to the lot of transformations getting more dangerous under the natural conditions; that natural conditions can vary dramatically even in small areas, etc. Humanity can be interested in one particular question: how much the claimed properties of each pesticide, even passed testing in certain circumstances, correspond to the actual state of affairs.

Not going deep into the intricacies of estimates given in different databases and official documents that easily can be found in Net, which show that the problem of simultaneous influence of multiple factors is a vital issue, is possible to conclude, that there is no way to determine exactly how much and where the substance used will be appeared, and what kind of physical, chemical, and biological factors promoting the process of decomposition (humidity, light exposure, possibilities of microbial community or a vegetative cover on destruction of toxic compound) and with what efficiency will act.

The huge amount of pesticides is produced over the world, and we randomly have chosen, as an example, a single one, with high effectiveness by its action and with minimal toxicity, according to tests, insecticide—a commercial preparation

"Actara", whose active component is the neonicotinoid thiamethoxam. The only criterion for the selection was the fact of sufficient solubility in water, as for the fast visualization of the effects of influence we wanted to work using critical concentrations of the pesticide. In addition, the high solubility allows for performing of comparative experiments on the row of concentrations.

The purpose of this report is to present to the attention of scientific community some results obtained by the group of radiation biophysics of A.I. Alikhanyan National Science Laboratory (Yerevan Physics Institute), while studying the interaction of bacterium-pesticide by the way of example of interaction of number of aerobic bacteria with insecticide actara and separately with its active ingredient—thiamethoxam.

2 USED MATERIALS AND METHODOLOGY

About of 150 identified and unidentified strains of aerobic microorganisms of different genus, including pseudomonades, bacillus, yeasts has been tested during the work. Most of them—is the soil microorganisms isolated by us from different soils of Armenia. Passage of inoculums carried out by glass spatula, by streaking or by the use of bacteriological brush. All the used chemicals were of high purity. Technical agar of Japan production thoroughly washed with large amounts of distilled water. Commercial preparation "Actara" was of Syngenta (Basel, Switzerland) production. Thiamethoxam was purified in conditions of our laboratory from the mentioned preparation "Actara", and for comparison, the thiamethoxam, purchased from the company Changzhou Elly Industrial Import @ Export Co., Ltd. (Changzhou, China) were used. For the simplicity onwards the locally purified thiamethoxam will be marked as T_L, and purchased from Chaina—T_C.

In the present work the concentration of the used pesticide close to the limit of its solubility (about 200 mg/%), or 50 mg/% based on pure thiamethoxam.

Passages were performed on a Meat-Peptone Agar (MPA), on a Minimal salt Medium (MM), comprising: agar—2%, $K_2HPO_4 \times 3H_2O$—0,15%, KH_2PO_4—0,05%, $MgSO_4 \times 7H_2O$—0,02%, and on the media, in which the actara (MM + A) or thiamethoxam (MM + T) as the only source of carbon and nitrogen were added.

During the experiments on the nutrient media the pure MPA was used as the control, in the case of experiments on minimal media—MM.

The results of growth on the MPA as in the case of control, as in the case of actara or thiamethoxam adding were recorded through two days after the passage, in the case of minimal media—through a week or ten days.

All experiments were repeated at least three times in several replications to achieve a statistical confidence.

The results presented on the figures were obtained by the passage of inoculums on the surface of the agar dishes using the bacteriological brush, using the same set of cultures that have been selected from the total number of tested strains for the reasons of visualization.

3 RESULTS AND DISCUSSION

Earlier we discussed the possibility of biodegradation of the insecticide actara by some aerobic bacteria isolated from the different soils, particularly from the soil of Nubarashen burial ground of toxic compound, located near the capital of Armenia—Yerevan city, as well as some of the consequences of the presence of insecticide in the growth medium on testing bacteria (Mkrtchyan 2011).

Briefly touch on the reasons that prompted us to take this step. Due to the certain circumstances the soil on territory of the burial ground was contaminated by large amounts of the number of toxic compounds, mainly by DDT. We were interested, how this fact is affected to the local microbial community. And, if the soil microorganisms in that place are survived, how they are tolerant to the presence of pesticides being in routine use. Therefore, the following logics of the research was formed: to estimate the resistance to pesticides (in this case—actara) microbial cultures that were collected in our laboratory, from various sources (including—from museums of other institutions, from the soils, sludge, etc.); to compare the obtained data with the results obtained when tested microorganisms isolated from the soil of Nubarashen's burial ground site.

The first and most important question, which begs from the outset of the study—a question whether any microorganisms are able to grow on agar with actara, as the only source of carbon and nitrogen? The answer was unambiguous—are capable, and during the multiple inoculations and the long time. Should understand what is the source of nutrition—thiamethoxam or additional components of the preparation? It was found that additional components play a critical role in the growth process, since the intensity of colony formation in the presence of actara was much higher than that of the pure thiamethoxam (Fig. 1B).

Comparison of Figure 1B and Figure 1C testifies, that the process of purification of thiamthoxam from the actara was enough successful, as almost half

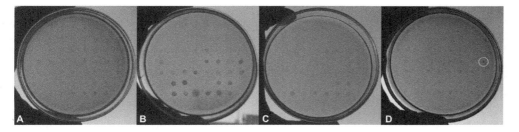

Figure 1. Growth of the tested cultures of bacteria on the minimal media: A—MM, B—MM + A, C—MM + T_L, D—MM + T_C.

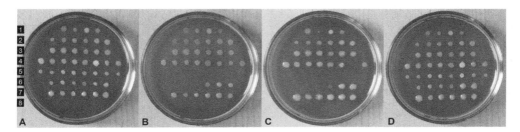

Figure 2. Growth of the tested cultures of microorganizms on nutrient environments: A—MPA, B—MPA with the actara, C—MPA with the thiamethoxam, purified from actara, D—MPA with the thiamethoxam of firm production. The arrangement of the strains is the same, as on Figure 1.

of the cultures inoculated on the MM + T_L medium does not grow. On the MM + T_C the growth of cultures is a little bit weaker, but the general number of full-grown strains is practically identical.

Unfortunately, the photos (especially black and white) do not reflect all completeness of the image since the traces of the rods of bacteriological brush on the agar surface are similar to the views of the plaques (Fig. 1A). For the obviousness we have made, one might say, transparent photos: agar plate against a daylight. However, even in this case the picture remained uncertain, therefore we have marked the unique colony on Figure 1D, which has shown so convex picture, so that it dismisses the possibility to doubt.

From the very beginning of the researches the number of interesting phenomena was revealed, the main of which was more tolerance of cultures isolated from the Nubarashen's burial ground to the presence of actara in the MM (MM + A), than cultures stored in our collection of microorganisms. The seven of them grew very well, and more 11, showing a moderate growth on the medium MM + A (Fig. 1B), did not grew up on the medium of the MPA + A (Fig. 2B). The replacement of actara to thiamethoxam of local purification added two more strains, whose growth was deteriorated (Fig. 2C). It was great surprise for us when the use of thiamethoxam of firm production led to a view,

similar to the control (Fig. 2A and 2D), as all of them grew well in its presence.

On Figure 1 the results of growth of the investigated strains on media MM, MM + T_L, and MM + T_C are shown. Apparently, from Figures we can observe the number of the amazing phenomena: some of the cultures, which are not grown on MPA + A medium, enough well grow on MC + A medium and on the contrary (compare the lines 4, 5, 6 on Fig. 2B and 1B).

It is necessary to notice, that growth of cultures on thiamethoxam, slowed more down, than on actara. The presented photos are made on 8th day after passage. But even after 14 days the strains on the control dishes were seeing only in track quantities (Fig. 1C and 1D).

Another very interesting phenomenon was founded out: so-called "halo"—obviously visible ring-nimbus around of colonies/plaques of some of investigated strains, which appears during the growth on MM + A (Fig. 3). The halo is not revealed right after passage, but after 2–3 consecutive passages on MM + A. Return steps with intermediate passages on MPA led to reduction of visible effect by the minimum media containing actara. After three consecutive passages on MPA the phenomenon disappeared practically completely, and again was shown after consecutive passages on the medium with actara.

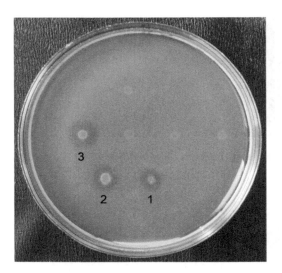

Figure 3. Formation of halo on MC + A medium. Counter-clockwise from below: 1—after the first passage; 2—after two passages; 3—after three; other plaques—the strains that not forming halo.

4 CONCLUSION

In the work the consistent pattern of growth of various strains of the soil microorganisms on the nutrient and minimal media containing insecticide actara or its active component—thiamethoxam were discussed. It is shown, that from almost one and a half hundreds tested cultures of the aerobic microorganisms, only seven, isolated from the soil of the Nubarashen's burial ground, had a capablity of good growth on the solid media containing actara, as the only source of carbon and nitrogen. Thus, around the colonies of some of them the accurately distinguishable halo was formed.

Specified seven strains were capable to grow on actara-containing medium at multiple passages during the long time. It is revealed eleven more cultures, not capable to grow on the nutrient media in the presence of actara, but which grew well on the minimal medium containing insecticide. Any of these cultures did not maintain repeated passage on MM + T_L media.

Despite of variety of the ideas, regarding the nature of halo, which are based on preliminary studies of this phenomenon and on the base of which further researches will be carried out, at the given stage we would like to avoid unsubstantiated reasoning.

Authors express their gratitude to Dr. V. Ghavalyan for purification of thiamethoxam from actara, and to Dr. V. Harutyunyan for continuous support of the works on given theme.

REFERENCES

Byzov, B.A. et al. 1989. In Microorganisms and soil protection, D.G. Zvyagintsev (eds), *Microbiological aspects of soil contamination by pesticides:* 86–129. Moscow, Moscow university press (in rus).

Kirk J., et al. 2004. Methods of studying soil microbial diversity. *Journal of Microbiological Methods* 58: 169–188.

Kiryushin V.I. 1996. *Ecological bases of agriculture.* Moscow: Kolos.

Mkrtchyan N.I. et al. 2011. The growth of some soil aerobic bacteria in the presence of insecticides Actara and Confidor *Biol. J. of Armenia*, 3: 6–11 (in rus).

National Pesticide Information Center—http://npic.orst.edu/factsheets/ddtgen.pdf.

Rodríguez R.A. and Gary A. Toranzos 2003. Stability of bacterial populations in tropical soil upon exposure to Lindane, *Int. Microbiol* 6: 253.

Wang G. et al. 2010. Co-metabolism of DDT by the newly isolated bacterium, *pseudoxanthomonas* sp. Wax. *Brazilian Journal of Microbiology* 41: 431–438.

Advances in Energy, Environment and Materials Science – Wang & Zhao (Eds)
© *2016 Taylor & Francis Group, London, ISBN 978-1-138-02931-6*

Review of the mechanisms of low pressure non-Darcy phenomena of shale gas

Yaxing Cui, Zhiming Hu & Qi Li
Research Institute of Porous Flow and Fluid Mechanics, Langfang, Hebei, China

ABSTRACT: The shale gas flow is generally divided into continuous flow, slip flow, transition flow and free molecular flow. Among those, slip flow, transition flow and free molecular flow are non-Darcy at low pore pressure. Different from sandstone reservoirs, shale reservoirs have complex pore structures. Shale gas flow at low pressure or in micropores/nanopores is significant to the exploitation. Make clear the mechanisms of low pressure non-Darcy flow and review existing relevant researches.

1 INTRODUCTION

1.1 *Gas flow in the field*

In the process of exploitation of shale gas, it is widely accepted that shale gas flow can be divided into three processes: seepage, diffusion, and desorption. When pore pressure is low enough, slippage effect will occur because of the particularity of gas flow.

The mechanism of shale gas flow is more complex than gas flow in sandstone in consequence of low permeability, low porosity, and complex structure of shale reservoirs.

1.2 *Gas flow in the laboratory*

Laboratory measurements can obtain the curve that describes apparent permeability that varies with pressure or the Knudsen number (Fig. 1, Fig. 2).

The Knudsen number is dimensionless. It is defined as:

$$K_n = \lambda/d \tag{1}$$

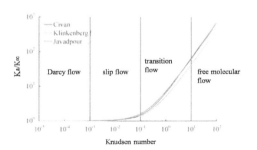

Figure 1.   Apparent permeability coefficient and Knudsen number change curve.

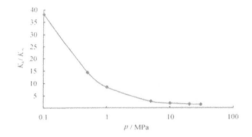

Figure 2.   Apparent permeability coefficient and pressure change curve.

where λ is defined as:

$$\lambda = \frac{\kappa T}{\sqrt{2}\pi\delta^2 p} \tag{2}$$

in which d = the diameter of pores, λ = the average mean-free-path of the gas, T = temperature, κ = the Boltzmann constant, δ = the collision diameter of the gas molecule, p = pressure.

The curve shows the nonlinear relationship between Knudsen number and apparent permeability coefficient. It is clear that the nonlinear part appears when slip flow plays a major role. This nonlinear part is the consequence of slippage effect. When the Knudsen number is larger or pore pressure is lower, the nonlinear phenomenon is more obvious, which is caused by free molecular flow occurring, namely, diffusion. It doesn't reflect the influence of desorption in the curve. The reason is that desorption has a close relationship with shale gas reserves rather than flow processes.

During the whole process, gas flow can be divided into several parts. The mechanism of each part is different. Non-Darcy flow at low pore pressure mainly consists of slip flow, transition flow and free molecular flow. Many researches have been conducted to explain the mechanisms of non-Darcy flow parts. It plays a great role to make the mechanisms of non-Darcy flow at low pressure clear to the exploitation of shale gas.

2 DIFFUSION MECHANISM

2.1 Diffusion categories

Shale gas reservoirs are mainly composed of non-organic matter, organic matter, natural fracture, and hydraulic fracture. In different porous media, diffusive gas flow follows different mechanisms. According to the Knudsen number (Thimons, 1973), diffusion flow can be divided into three kinds—Fick diffusion, Knudsen diffusion, and surface diffusion.

When K_n is higher than 10, the diffusion is considered as Knudsen diffusion. It mainly occurs in the nanometer pores of the shale matrix. When K_n is lower than 0.1, the diffusion is considered as Fick diffusion which occurs in micropores and microfractures. When K_n is between the two numbers, diffusion is considered as transition diffusion which consists of Fick diffusion and Knudsen diffusion. Surface diffusion usually occurs in solid kerogen. Adsorption gas moves along the surface of pore canal wall.

2.2 Fick diffusion

Fick diffusion is driven by concentration gradient. Shale gas diffuses from the matrix to the fracture system. Fick diffusion can be divided into quasi-steady-state diffusion and unsteady-state diffusion (Kuuskraa, 1985). Quasi-steady-diffusion is described by Fick's first law. Concentration is only related to the distance and has nothing to do with time. Diffusion coefficient based on Fick's first law can be expressed as:

$$D = -\frac{J}{dC/dx} \tag{3}$$

where J = diffusion flux, D = diffusion coefficient, dC/dx = concentration gradient.

Actually gas concentration will vary with time in the process of gas reservoir development. Thus, taking unsteady-state diffusion, namely Fick's second law to describe diffusion process is more accurate. The diffusion coefficient is usually regarded as a constant for the convenience of solving unsteady diffusion equation. The unsteady diffusion is expressed as:

$$\frac{\partial C}{\partial t} = \frac{DAZ_{SC}RT_{SC}}{p_{SC}} \frac{\partial^2 C}{\partial x^2} \tag{4}$$

where A = area, Z_{SC} = gas compressibility factor in standard conditions, R = universal gas constant, T_{SC} = standard temperature, p_{SC} = standard pressure, C = molar concentration, t = time.

By the view of microstructure, the diffusion coefficient of Fick diffusion can be expressed as

$$D = \frac{1}{3}\sqrt{\frac{8RT}{\pi M}} \frac{\kappa T}{\sqrt{2}\pi\sigma^2 p} \tag{5}$$

where M = molar mass, σ = effective molecular diameter.

There are a lot of influence factors of the diffusion coefficient, such as gas component properties, petrophysical properties, pore fluid properties, temperature, and pressure. In order to make the laboratory measured diffusion coefficient more close to the actual value of rocks in the field, many studies have been conducted to modify diffusion coefficient. Eyring et al. proposed temperature correction according to statistical thermodynamics. Zhang et al. (Zhang, 1996) proposed the relationship between diffusion coefficient and the media pore radius. Li et al. (Li, 2001) proposed the transformation relation between diffusion coefficient of saturated samples and dry samples and temperature correction of saturated sample diffusion coefficient.

2.3 Knudsen diffusion

When the pore diameter is far less than the mean free path of gas molecules, Knudsen diffusion occurs. Gas molecules move through the channel in the form of a single molecule free movement. Knudsen diffusion coefficient is defined as (Ziarani, 2012):

$$D_K = \frac{2r}{3}\left(\frac{8ZRT}{\pi M}\right)^{1/2} \tag{6}$$

where D_K = Knudsen diffusion coefficient, r = pore radius, Z = gas compressibility factor.

2.4 Transition diffusion

When the Knudsen number is between 0.1 and 10, the two kinds of diffusion occurs at the same time. Generally, the harmonic average of the two diffusion coefficient is taken as transition diffusion coefficient, which can be expressed as:

$$D_T = \frac{DD_k}{D + D_k} \tag{7}$$

where D_T = transition diffusion coefficient, D = Fick diffusion coefficient, D_K = Knudsen diffusion coefficient.

2.5 *Conclusions*

In shale reservoir, diffusion properties and the value of diffusion coefficient are closely related to the pore scale. In initial exploitation, diffusion in pores with different diameters is mainly Fick diffusion. The diffusion coefficient values are almost the same. In later period, diffusion mechanism and diffusion coefficient values are influenced by the pore scale as gas pressure and concentration decrease during the exploitation.

3 SLIP FLOW MECHANISM

3.1 *Slippage effect*

Gas flow can be divided into several parts based on Knudsen number (Frauk, 2010). When K_n is lower than 0.01, it is continuous flow. When K_n is between 0.01 and 0.1, it is slip flow. When K_n is larger than 10, it's free molecular flow. When K_n is between 0.1 and 10, slip flow and free molecular flow occur simultaneously, it is called transition flow.

Slip flow is a special phenomenon of gas flow in low permeability reservoirs. Many studies have proved the existence of slippage effect in the shale gas flow. Wu et al. (Wu, 2005) and Zhu et al. (Zhu, 2007) all obtained the curve that describes the relationship between flux and difference of squares gradient of pressure by experiment. The curve isn't coincident with Darcy law, which indicates slippage effect occurs during gas flow. Because of slippage effect, the tangential speed of gas molecules relative to shell wall surface in porous media is not equal to zero when gas flow through a low permeability rock sample.

Klinkenberg proposed an expression for slippage effect when single-phase gas flows in porous media. It is expressed as:

$$K_g = K_\infty \left(1 + \frac{b}{p}\right) \tag{8}$$

where b is the gas slippage factor, it is expressed as:

$$b = \frac{4c}{r}\lambda p \tag{9}$$

where K_g = gas permeability, namely, apparent permeability, K_∞ = absolute permeability, p = average pressure of import and export pressure, c = scale factor.

Florence put forward a theoretical model and obtained the generalized Darcy expression (Florence, 2007). It is expressed as:

$$V = -\frac{K_\infty}{\mu}\left(1 + \frac{b}{p}\right)\nabla p \tag{10}$$

where V = gas flow velocity, μ = gas viscosity.

As previously mentioned, Zhu et al. (Zhu, 2007) proposed that the theoretical foundation of the Klinkenberg equation only applies to the situation that K_n is between 0.01 and 0.1. The control equation of flow is still the classic N-S equation. When K_n is larger than 0.1, the continuum hypothesis and the Klinkenberg equation are no longer valid (Fig. 3).

3.2 *Slippage effect boundaries*

The boundary whether to consider the effects of slippage effect is still no final conclusion. Many scholars have put forward their own views.

For single-phase gas flow, Zhu et al. (Zhu, 2007) suggested the boundary is that reservoir permeability is lower than 0.1×10^{-3} μm^2 and the pore pressure is lower than 1.5 MPa through low permeability core of Sulige field experiments. Gao et al. (Gao 2011) proposed that slippage effect appears when the pore pressure of shale gas reservoirs reaches a relatively low level (e.g. lower than 10 MPa). Zhang et al. (Zhu, 2006) proposed that there are a lot of influence factors of slippage effect. The boundary is close to these factors, such as pressure, temperature, pore structures of porous media, the kind of gas and so on.

The situation will be different from single-phase gas slip when gas flow is under the condition of rocks with irreducible water. The variation trend of the degree of slippage effect changing with factors hasn't reached a consistent conclusion. According to the Klinkenberg theory, slippage effect under the condition of irreducible water existing will be more obvious when temperature rises, overburden pressure increases and irreducible water saturation

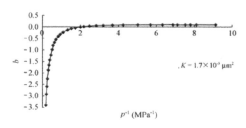

Figure 3. The slippage factor of the low permeability core and average pore pressure relation curve.

increases. Jones et al. (Jones, 1980) and Kewen et al. (Kewen, 2001) also reached a conclusion that the gas slippage factor would increase as temperature rises or gas effective permeability decreases. While Rose et al. (Rose, 1949) and Rushing et al. (Rushing, 2003) obtained opposite results that the gas slippage factor decreases with irreducible water saturation, temperature and overburden pressure increasing through experiments.

3.3 *Conclusions*

Slippage effect is a common phenomenon in shale gas flow. It is generally ignored when studying gas flow in sandstone, while it plays an important role in shale gas flow. The Klinkenberg equation and the gas slippage factor are usually used to describe slippage effect. As for the boundaries that considering slippage effect in gas flow, results of single-phase gas can be applied to production. When it is gas-water phase, the results are strongly influenced, which still needs to explore.

4 CONCLUSIONS

As pressure or the Knudsen number varies, shale gas flow will go through several stages. When pressure is high, the flow is non-Darcy flow with high velocity. When pressure is relatively low, the flow is also non-Darcy flow that includes slip flow and diffusion.

Diffusion usually divided into three kinds according to the Knudsen number. In micro fractures and micro pores, it is Fick diffusion. In initial exploitation, Fick diffusion occupies the main position. Shale gas diffuses from micro fractures or pores, and seepages when in fractures or connected pores. In later period of exploitation, Knudsen diffusion and transition diffusion are gradually obvious.

The slippage effect has been discovered for a long time. The Klinkenberg equation is a widely used way to describe this phenomenon. How to judge when the slippage effect appears and when should the slippage effect be taken into account

hasn't reached consensus yet. While gas-water phase is common in the field, further studies need to be conducted.

For shale reservoirs' complex structure, non-Darcy flow at low pressure is a considerable part in shale gas flow, so making clear the mechanisms of diffusion and slip flow is significant to the development of shale gas and utilization of residual gas.

REFERENCES

Faruk Civan. 2010. Effective correlation of apparent gas permeability in tight porous media. Transport in Porous Media 82(2):375–384.

Florence, F.A. et al. 2007. Improved permeability prediction relations for low permeability sands. Rocky Mountain Oil & Gas Technology Symposium.

Gao Shusheng et al. 2011. Impact of slippage effect on shale gas well productivity. Natural Gas Industry 31(4):55–58.

Jones, F.O. & Owens, W.W. 1980. A laboratory study of low-permeability gas sands. JPT 1631-1640.

Kewen, L. & Horne, R.N. 2001. Gas slippage in two-phase flow and the effect of temperature. SPE 68778.

Kuuskraa, V.A. & Sedwick, K. 1985. Technically recoverable Devonian shale gas in Ohio, West Virginia, and Kentucky. SPE-MS14503.

Li Haiyan et al. 2001. Methods on the study of gas diffusion coefficient. Petroleum Exploration and Development 28(2):33–35.

Rose, W.D. 1949. Permeability and gas-slippage phenomena. API Drilling and Production Practice 209–217.

Rushing, J.A. et al. 2003. Measurement of the two-phase gas slippage phenomenon and its effect on gas relative permeability in tight gas sands. SPE 84297.

Thimons, E.D. & Kissell, F.N. 1973. Diffusion of methane through coal. Fuel 52(4):274–280.

Wu Ying et al. 2005. New calculation method of Kelinberg constant and non-Darcy coefficient for low permeable gas reservoirs. Natural Gas Industry 25(5):78–80.

Zhang Jun & Guo Ping. 2006. Research on gas slippage effect of low permeability and tight gas reservoir. Fault-Block Oil & Gas Field 13(3):54–56.

Ziarani, S.A. & Aguilera, R. 2012. Knudsen's permeability correction for tight porous media. Transp. Porous. Med 91(1):239–260.

Zhu Guangya et al. 2007. A research of impacts of gas flow slippage effect in low permeability gas reservoir. Natural Gas Industry 27(5):44–47.

Advances in Energy, Environment and Materials Science – Wang & Zhao (Eds)
© *2016 Taylor & Francis Group, London, ISBN 978-1-138-02931-6*

Numerical study of biofouling effects on the flow dynamic characteristics of blade sections

Tao Zhang, Shao-song Min & Fei Peng
Department of Naval Engineering, University of Naval Engineering, Wuhan, China

ABSTRACT: Generally blade surfaces are of polished metal and have no antifouling provision, which makes them susceptible to biofouling, yet its effects on the hydrodynamic performance of blade sections are rarely studied. The present work aims at quantifying the effects and providing a deep insight into the physical mechanism by means of Computational Fluid Dynamics (CFD). The simulation employs the SST k-ω turbulence model and is carried out on NACA 4424 airfoils. Being selected as the study subject from fouling community, barnacles are directly modeled at geometry level instead of the well-known wall function methods. The results show that flow separation occurs earlier and the separation region becomes larger around airfoils with increase of the fouling height. Fouling on airfoils leads to a significant reduction of lift-drag ratio (C_L/C_D) and hence would result in a remarkable decrease of propulsive efficiency of propellers. Once the height of fouling exceeds a certain limit, however, the increase of the height will have little effect on C_L/C_D.

1 INTRODUCTION

The settlement and subsequent growth of flora and fauna on surfaces exposed in aquatic environments is termed biofouling (Schultz 2004). Biofouling begins to occur immediately after a ship is immersed in water and will continue to accumulate throughout a ship at sea until a cleaning and repainting process is performed. It is well established that fouling on ships increases the surface roughness of the hull, which in turn, caused increased frictional resistance and fuel consumption and decreased top speed and range (Schultz 2011). Moreover, in the case of war vessels and other types of ships in which extreme speed is essential, the occurrence of fouling may result in the loss of advantages for which great sacrifice has been made (Woods Hole Oceanographic Institute, 1952).

Generally propeller blade surfaces are of polished metal and have no antifouling provision, which makes them susceptible to fouling. The 25th International Towing Tank Conference (ITTC) pointed out that in its report by far for propeller roughness the biggest cause is fouling and a small roughness increase of the propeller can causes large increases in the required power (ITTC, 2008).

In the papers above, although fouling effects on performance of blade sections have been seen research efforts, the flow details have not been yet obtained due to limitation of experimental methods. On aspect of numerical simulation study, there are some problems with application of turbulence models and near-wall treatment methods. The aim of the present study is to settle those problems and investigate the impact of fouling on the hydrodynamic performance of blade sections by utilizing CFD method.

This paper is organized as follows: the numerical approach will be described in Section 2, followed by results and discussion in Section 3, and conclusions will be drawn at the end.

2 NUMERICAL APPROACH

2.1 *Fouling modeling*

Barnacles are dominant organism of fouling community on propellers and the actual distribution on a ship propeller. In the experiment of Orme et al. (2001), barnacles were modeled by setting small conical shapes on the surface of airfoils, which is shown in Figure 1. As the 2-D shape of cones is triangle and a 2-D airfoil is examined here, it is simplified by making triangle-shape spikes modeling the barnacles as shown in Figure 2.

In the experiment of Orme et al. (2001), two variables are used to describe the fouling cases: height and density. Fouling height is defined as the size of barnacles on the airfoil surface, and fouling density is defined as the number of specimens per square meter. To investigate the effect of fouling height alone, the density is set to be constant. Furthermore, the control case of a smooth airfoil is set up for comparison purposes.

Figure 1.   Experimental model from Orme et al. (2001).

Figure 2.   Present research model of NACA 4424 airfoil.

Table 1.   Computational fouling cases.

Case	Height (mm)	Total spikes on each side of the foil
NoFoul	0.0	0
MinHe	0.7	7
MedHe	3.2	7
MaxHe	5.7	7

With these thoughts in mind, the computational fouling cases can be set up in Table 1. Different fouling cases of MinHe, MedHe and MaxHe are studied at a fixed fouling density, and Nofoul is the presentation of the control case.

2.2   Numerical modeling

2.2.1   Mathematical formulation

The simulation is conducted by using a commercial Computational Fluid Dynamics package FLUENT. The Reynolds-averaged continuity and momentum equations for incompressible flows in tensor notation and Cartesian coordinates are summarized below:

$$\frac{\partial \overline{u}_i}{\partial x_i} = 0 \tag{1}$$

$$\frac{\partial (\rho \overline{u}_i)}{\partial t} + \frac{\partial (\rho \overline{u}_i \overline{u}_j)}{\partial x_j} = -\frac{\partial \overline{p}}{\partial x_j} + \frac{\partial}{\partial x_j}\left(\mu \frac{\partial \overline{u}_i}{\partial x_j} - \rho \overline{u'_i u'_j} \right) \tag{2}$$

where ρ is the fluid density, $\overline{u}_j$ is averaged components of the velocity vector, $\rho \overline{u'_i u'_j}$ is the Reynolds stresses, $\overline{p}$ is mean pressure and μ is the dynamic viscosity.

The Reynolds number (Re) based on the coming velocity and chord length for the flow is about 4.6×10^5 as the fluid density ρ and dynamic viscosity μ equal to 1.225 kg/m^3 and 1.7894×10^{-5} kg/(m·s) respectively. Therefore, the flow can be considered as low Re turbulent and thus low Re turbulence model is required. The Shear-Stress Transport (SST) k-w turbulence model is developed to effectively blend the robust and accurate formulation of the k-w model in the near-wall region with free-stream independence of the k-ε model in the far field. Moreover, the near-wall flow is resolved directly with the integration method instead of the wall function based on empirical formulas.

$$\frac{\partial (\rho k)}{\partial t} + \frac{\partial}{\partial x_i}(\rho k u_i) = \frac{\partial}{\partial x_j}\left(\Gamma_k \frac{\partial k}{\partial x_j} \right) + G_k - Y_k + S_k \tag{3}$$

$$\frac{\partial (\rho w)}{\partial t} + \frac{\partial (\rho w u_i)}{\partial x_i} = \frac{\partial}{\partial x_j}\left(\Gamma_w \frac{\partial w}{\partial x_j} \right) + G_w - Y_w + S_w D_w \tag{4}$$

where k is the turbulent kinetic energy, w is the specific dissipation rate. In these equations, G_k represents the generation of turbulence kinetic energy due to mean velocity gradients. G_w represents the generation of w. Γ_k and Γ_w represent the effective diffusivity of k and w, respectively. Y_k and Y_w represent the dissipation of k and w due to turbulence. S_k and S_w are user-defined source terms.

2.2.2   Grid and meshing

Considering the geometry irregularity of the airfoil surface under fouling conditions, hybrid grid

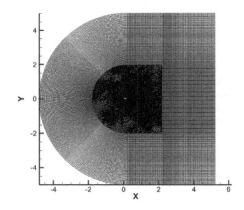

Figure 3.   The whole computational zone and mesh.

method is adopted: unstructured grid is used in the inner C-region near the airfoil and structured grid is used in the outer C-region far away from the airfoil. The circle radius of inner and outer C-region is $10c$ and $25c$, which equal to the length of the side of inner and outer squares beyond the trailing edge respectively. The term c refers to the chord length and equals to 0.2 m. Near-wall grid is arranged clustering to the surface of the airfoil to capture the boundary layer flow field accurately, especially the viscous sublayer. The whole computational zone and local mesh distribution for NACA 4424 airfoils are displayed in Figure 3 and Figure 4.

3   RESULTS AND DISCUSSIONS

3.1   *Lift and drag*

The variation of the differences of C_L/C_D with AOA is shown in Figure 5 for different fouling

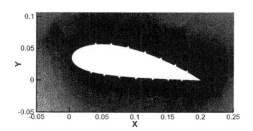

Figure 4.   Local amplified grids for the NACA 4424.

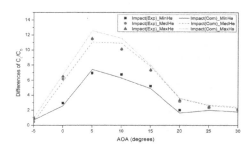

Figure 5.   Results of C_L/C_D for NACA 4424 airfoil under each fouling case at different AOA.

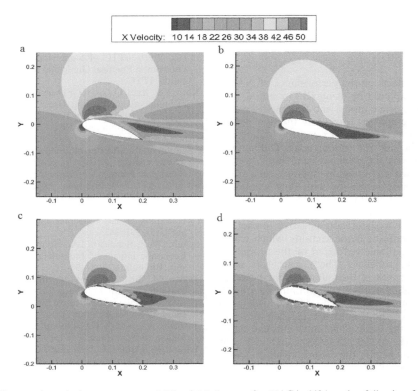

Figure 6.   Streamwise velocity contours at AOA of 15 degrees for NACA 4424 under following fouling cases: (a) Nofoul; (b) MinHe; (c) MedHe; (d) MaxHe.

cases, and compared with the corresponding computation results. It can be seen that the results of C_L/C_D from CFD are in fair agreement with that obtained by the experiment. The impact of fouling rises almost linearly with increase of AOA before the stall angle, however, when AOA exceeds the stall angle, the impact declines gradually, especially when the AOA increases greater than 20 degrees, the impact becomes very slight. The C_L/C_D results of MaxHe case are nearly the same as MedHe case although the fouling height of the former is almost twice that of the latter. Therefore, it can be inferred that once the height exceeds a certain limit, the increase of the height has little effect on C_L/C_D. Nevertheless, the critical value of fouling height needs further research to determine.

3.2 *Velocity contours*

The streamwise velocity (x-velocity) contours for various fouling cases are shown in Figure 6. As seen clearly, flow separation occurs earlier and the separation region becomes larger around the airfoils with increase of fouling height. For the case of higher fouling height, high velocity region of upper surface of the airfoil is smaller, thus the negative pressure of the upper surface becomes smaller resulting in lower lift.

4 CONCLUSION

The fouling effects on the performance of NACA 4424 foil is explored by using CFD method. The SST k-w turbulence model and fine mesh around the airfoil is employed to capture the flow details near the airfoil, especially around the fouling elements and beyond the trailing edge. The method proposed avoids the weakness and limitation of the well-known wall function methods, such as the non-applicability in dealing with flow separation and dependency on empirical data. Moreover, the results agree well with the experimental data.

Fouling has a significant negative impact on the lift-drag ratio of blade sections, which would lead to a remarkable reduction of the propulsive efficiency of propellers. The impact gets more severe with increase of the fouling height, however, once the height exceeds a certain limit, the impact becomes almost unchanged. Through detailed examination of the flow structure, such as velocity contours, the disturbance to flow is caused by presence of the model barnacles. For the case of smaller fouling height, the disturbing effect is only restricted in small regions around the barnacles, and the flow restores the normal state below them quickly. However, for case of larger fouling height, the normal state of the flow is destroyed totally and the viscous sublayer almost disappears around the airfoil, thus the flow would be supposed to enter fully rough regime, which could explain that the effect of worse fouling keeps almost constant.

Given the fine mesh employed and the large number of grid elements, the method proposed in this paper is not suitable for three-dimensional airfoils and propellers. In view of the significant effects of fouling and complexity of flow around fouled airfoil surface, however, this study can provide some useful information for industrial application and fluid dynamic academia.

REFERENCES

Abbott, I.H., Von Doenhoff, A.E., 1960. Theory of Wing Sections, including a summary of airfoil data. Dover Publications, New York.

ANSYS, 2011. ANSYS FLUENT Theory Guide. Release 14.0.

Demirel, Y.K., Khorasanchi, M., Turan, O., Incecik, A., et al., 2014. A CFD model for the frictional resistance prediction of antifouling coatings. Ocean Eng. 89, 21–31.

Drela, M., Youngren, H., 2001. XFOIL 6.94 user guide. Aeronautics and Astronautics Engineering, Massachusetts Institute of Technology.

ITTC, 2008. ITTC—Recommended Procedures and Guidelines: Testing and Extrapolation Methods, Propulsion, Performance, Predicting Powering Margins. ITTC.

ITTC, 2011b. ITTC—Recommended Procedures and Guidelines: Practical Guidelines for ship CFD Application. ITTC.

Khor, Y.S., Xiao, Q., 2011. CFD simulations of the effects of fouling and antifouling. Ocean Eng. 38, 1065–1079.

Nikuradse, J., 1933. Laws of flow in rough pipes. NACA Technical Memorandum 1292.

Orme, J.A.C., Masters, I., Griffiths R.T., 2001. Investigation of the effect of biofouling on the efficiency of marine current turbines. In: Proceedings of Marine Renewable Energy Conference (MAREC), 91–99.

Schultz, M.P., 2004. Frictional resistance of antifouling coating systems. J. Fluids Eng. 126(6), 1039–1048.

Schultz, M.P., 2007. Effects of coating roughness and biofouling on ship resistance and powering. Biofouling 23(5), 331–341.

Schultz, M.P., Bendickb, J.A., Holmb, E.R., et al., 2011. Economic impact of biofouling on a naval surface ship. Biofouling 27, 87–98.

Taylan, M., 2010. An overview: effect of roughness and coatings on ship resistance. In: Proceedings of the International Conference on Ship Drag Reduction (Smooth-Ships). Istanbul, Turkey.

Townsin, R.L., 2003. The ship hull fouling penalty. Biofouling 19(Supplement), 9–15.

Walker, M., Atkins, I., 2007. Surface ship hull and propeller fouling management. In: Proceedings of Royal Institution of Naval Architects Conference—Warship: The Affordable Warship.

Woods Hole Oceanographic Institute, 1952. Marine Fouling and its Prevention. George Banta Publishing Co.

The practical application of AHP to select a zone of population evacuation—take Xinzhou District of Wuhan as an example

Jian Qin
WISDRI Engineering and Research Incorporation Limited, Wuhan, China

ABSTRACT: The construction of population evacuation zones is the most important in the construction and its primary task is to select and determine where to construct the evacuation zones. In general, a population evacuation zone should be selected on the basis that it's convenient to shelter and camouflage, easy to maneuver and denfence, and suitable for survival. How to select a population evacuation zone scientifically and appropriately is the primary problem to deal with. To address this issue, this essay lays special emphasis on the application of Analytic Hierarchy Process (AHP) in selecting a population evacuation zone with the theories of *Science of Civil Air Defense*. Besides, it takes Xinzhou District's construction of population evacuation zone as a case to illustrate AHP's practical application.

1 INTRODUCTION

Construction of a population evacuation zone is a core part of the civil air defense system, whose construction quality makes great sense for air-defending of wartime and preventing major disasters of peacetime. Its primary task is to select and determine the appropriate location to construct a population evacuation zone. Thus, how to scientifically and appropriately determine it will directly affect the overall urbanizaiton strategy, the stability of civil air defense system and the preservation for war potential.

2 MAIN REQUIREMENTS FOR THE POPULATION EVACUATION ZONE'S SELECTION

Generally, a population evacuation zone is an area at the township level to house evacuees, store goods and materials and place important facilities. Its selection should meet the requirements of the air defense and anti-disaster integration. Its basic goal is to make the best of it to gurantee and protect the country and its people's lives and property. According to *Science of Civil Air Defense*, the selection of a population evacuation zone must meet the following requirements: First, it should be convenient to shelter and camouflage. Enough space to hold evacuees and necessary facilities, a certain number of underground air defense facilities to shelter people, and a good condition of terrain, landforms and vegetation should be all taken into account. Second, it should be easy to maneuver. Good traffic and road conditions are required so as to gurantee that it is "easy to enter, easy to enter as fast and easy to disperse". Third, it should be applicable for defense. Try to avoid such areas as bases of nuclear power facilities and downstream regions of large-scale water conservancy facilities, etc. that are likely to have air-strike secondary hazard, and avoid major air raid directions, transportation hubs and military facilities, or the key economic targets. Fourth, it should be suitable for survival. Certain life support facilities should be guaranteed, such as food, water, electricity and health care.

3 METHODS OF APPROPRIATELY SELECTING A POPULATION EVACUATION ZONE

3.1 *The method*

The population evacuation zone's selection is actually a complicated multi-objective decision-making matter. It involves many factors. some factors (i.e. the evacuation zone's area, the number of its material reserves, the number of beds for medical treatment, etc.) can be calculated and compared by the quantitative approach, while some others like the degree of transportation convenience, the development of rural supporting infrastructure, and shelter conditions are difficult to calculated or measured by this single approach. To solve this problem, there are many theoretically mature decision-making methods, such as principal component analysis, hierarchical sequence method, ideal point method, utility theory, and AHP. Many factors involved in the process of selecting a population evacuation zone for civil air defense are difficult to quantify. Besides, it's hard to compare different factors by conventional and logical mathematical methods. On account of the facts above, AHP, with a low demand for information and fit for complicated and systematic decision-making problem, is considered as a proper method to help select a population evacuation zone.

3.2 Application of AHP

AHP requires determining an overall objective first according to the nature of the issue. Then create a measure of the total target (the target layer) to achieve the degree of evaluation index system (the index layer). Finally, create a workable solution set according to the index layer (the solution layer) to form a multi-layer structure model of systematic analysis (Fig. 1). With the introduction of an appropriate scale system and the analysis of various goals' (indices') relative importance (expert comparative method), construct a comparative judgment matrix. By mathematical operations, figure out each index's evaluation weight and the relative ranking weight of each solution so as to work out the sequence of each solution's advantages and disadvantages compared with that of the overall objective, for decision makers' reference.

AHP's characteristics is: it uses comparatively less quantitative information to get the decision-making process mathematicized based on in-depth analysis of the nature of the complex decision-making issues, effect factors and their intrinsic relationships, etc., so as to provide simple methods for these complicated decision-making issues with multi-targets, multi-criteria, or no structural characteristics. It is the model and method to make decisions for those complicated systems which are difficult to be fully quantified.

3.3 Comprehensive evaluation index system for a population evacuation zone's selection

The establishment of an evaluation index system is not only the core content of AHP's application,

$$A = \begin{array}{c} & \begin{array}{cccc} a_1 & a_2 & \cdots & a_K \end{array} \\ \begin{array}{c} a_1 \\ a_2 \\ \vdots \\ a_K \end{array} & \left[\begin{array}{cccc} b_{11}^s & b_{12}^s & \cdots & b_{1k}^s \\ b_{21}^s & b_{22}^s & \cdots & b_{2k}^s \\ \vdots & \vdots & \vdots & \vdots \\ b_{k1}^s & b_{k2}^s & \cdots & b_{kk}^s \end{array} \right] \end{array}$$

Figure 1. Sketch of the comparison and judgment matrix.

but also determines the evaluation model's scientificness and soundness. In view of the main requirements for selecting a population evacuation zone in *Science of Civil Air Defense* and the new requirements and ideas of nowadays construction of population evacuation system, the following comprehensive evaluation system for a population evacuation zone is constructed:

The target layer is to find an ideal population evacuation zone; the index layer can be divided into 2 layers according to the needs: the first layer is various reasonable requirements for the population evacuation zone, including regional conditions, distance situations, and security situations; the second layer is the detailed indices that specifically characterize the above conditions. The choice of indices in this layer will directly affect the efficiency and accuracy of the entire evaluation index system. Therefore, we should comprehensively screen the population evacuation zones in view of the relevant theories in *Science of Civil Air Defense* and all the practical situations.

Specifically, the regional conditions' evaluation indicator can be divided into 7 sub-indices:

1. Broad land area: It is often believed that broader the land area is, less crowded the population evacuation zone will be, and more secure people's lives and placement will be. We can use the population density as a quantitative index.
2. Good economic conditions: Better economic conditions mean that the resettlement area is more likely to provide better securities for evacuees (in life and production). Its quantitative indices can be regional GDPs, fiscal income, and per capita disposable income of all alternative evacuation zones.
3. Abundant goods supplies: Abundant goods supplies can gurantee that evacuees have enough daily necessities to consume. Its quantitative indices can be the evacuation zone's crop area, yield, its energy and material consumption, etc.
4. Infrastructure in evacuation zones:. Generally, evacuees are dispersed to various villages, thus it requires that rural infrastructure should have certain gurantee so that it is easy to continue people's life and production and more convenient to correspond with the outside world at the same time. The quantitative indices can be water penetration, electricity, and phone coverage.
5. Good medical conditions: Good medical conditions can provide medical security for evacuees. The quantitative indices can be number of beds per thousand people and number of doctors per thousand.
6. Convenient transportation: A good or bad road condition is directly related to the success or failure of evacuation and its efficiency. The

quantitative indices can be the ratio of villages that have transport service and the degree of available transportation.

7. Good shelter conditions: It requires that there are a certain number of air defense facilities for sheltering or good conditions of terrain, landforms, and vegetation. Its quantitative indices are more vague and generally measured in accordance with the subjective judgment of experts, or can be measured by the terrain height difference and vegetation coverage of various evacuation zones.

The evaluation index of the distance condition can be divided into 3 sub-indices:

1. The evacuation distance for the old and the pregnant: The usual distance is 60 to 80 km from the urban area, mainly on account of the evacuation efficiency and costs.
2. The evacuation distance for workers of research institutions and production units: Its usual distance is no less than 80 km away from the urban area, which is good for both sheltering and convenience for work.
3. The evacuation distance for young adults: Its usual distance is within 50 km of the urban area so as to be able to transfer back and forth in 4 to 6 hours to participate the supportive anti-air strikes activities and production for the urban area.

The evaluation index of the security situation can be divided into 2 sub-indices:

1. The direct influence range for important protection objectives after the attacks: It is mainly the direct influence radius of all important protection objectives attacked.
2. The likely influence range for regions that suffer from secondary disasters: These regions should effectively avoid objectives that tend to produce secondary disasters after being attacked, such as nuclear and chemical facilities, large reservoirs, etc.

The third layer is the solution layer, which is included in the population evacuation zone's evaluation. Generally, in the overall solution level of the civil air defense projects, towns are mainly considered as the evaluation units. Finally, the population evacuation zones' comprehensive evaluation index system is established (see Table 1).

3.4 The hierarchical structure model for population evacuation zones

The construction of the hierarchical structure model for population evacuation zones is mainly based on matrix operations. On account of the method illustrated above, it is assumed that there are k solutions for a city's population evacuation zones, represented by a_1, a_2, ..., a_k. The total target after comparison and election is to obtain the dominance of all the population evacuation zones involved in the evaluation for decision makers to choose from.

3.4.1 Choosing the scale of evaluation value
By comparing the realizations of any one of the indices Cs ($s = 1, 2, ..., 12$), it is found that there are differences among all the solutions. Therefore,

Table 1. The population evacuation zones' comprehensive evaluation index system.

Target layer	Index layer		Solution layer
	First layer	Second level	
Population evacuation zone	Regional conditions (B1)	Regional area (C1)	
		Economic conditions (C2)	
		Material supplies (C3)	
		Infrastructure construction condition (C4)	
		Medical conditions (C5)	Evacuation zone 1
		Convenience condition of transportation (C6)	Evacuation zone 2
		Shelter conditions (C7)	Evacuation zone 3
	Distance situation (B2)	Resettlement distance for the old, weak, sick, disabled and pregnant (C8)	Evacuation zone 4
		Placement distance for workers of research institutions and units of production (C9)	...
		Placement distance for young adults (C10)	Evacuation zone k
	Security situation (B3)	Influence range for important protection objectives (C11)	
		Influence range for regions that may suffer from secondary disasters (C12)	

Table 2. 1–9 scale method.

Scale a_{ij}	Definition
1	Factor i and factor j are of equal importance.
3	Factor i is a little more important than factor j.
5	Factor i is more important than factor j.
7	Factor i is far more important than factor j.
9	Factor i is absolutely more important than factor j.
2, 4, 6, 8	Factor i and factor j comparison value of importance falls in between the 2 adjacent levels above.
Reciprocal 1, $\frac{1}{2},\frac{1}{3},\frac{1}{4},\frac{1}{5},\frac{1}{6},\frac{1}{7},\frac{1}{8},\frac{1}{9}$	The judgement value by comparison of factor j and factor i is the reciprocal number of $a_{ji} = (1/a_{ij})$, $a_{ii} = 1$.

Starry's 1–9 scale method is proposed to display their differences (Table 2).

Starry once used various scale levels for comparison, he concluded that: 1–9 scale is not only the best of all the simple scales, but its comparison result is not worse than these complex scales. A total of 27 comparative scales Starry are: 1 ~ 3, 1 ~ 5, 1 ~ 6, ..., 1 ~ 11; (d+0.1) ~ (d+0.9), where d = 1, 2, 3, 4; 1^P ~ 9^P, where $p = 2, 3, 4, 5...$ etc. He constructed comparative matrix pairs and worked out the weight vectors by comparative judgement of the intensity of light sources placed at different distances. Meanwhile, he contrasted these weight vectors with the actual weight vectors worked out according to physical laws of light intensity and other relevant knowledge. It was found out that the 1–9 scale method is not only simple, but also its effects are comparatively better. Therefore, it is proper to construct elements of comparative matrix pairs with the 1–9 scale method's scales.

3.4.2 Construction of the comparison and judgment matrix

By expert comparison method, any arbitrary index Cs ($s = 1, 2, ..., 12$) can be given a comparison and judgement matrix $A = [b_{ij}^s]_{k \times k}$ ($s = 1, 2, ..., 12$), seen in Figure 1. In A, $b_{ii}^s = 1$, $b_{ij}^s = 1/b_{ji}^s$, $(i, j = 1, 2, ...k)$.

3.4.3 The calculation of maximum eigenvectors

To avoid the judgement inconsistency of the comparison matrixes, their consistency should be checked with the assistance of eigenvectors. By "sum and product" method, the eigenvector $W_s = (w_{1s}, w_{2s}, ..., w_{ks})$ is given. w_{is} is the weight coefficient of the i solution's index s.

$$w_{is} = \bar{w}_{is} \bigg/ \sum_{j=1}^{k} \bar{w}_{is} \tag{1}$$

In this formula, $w_{is} = \sum_{j=1}^{k} \bar{b}_{ij}^s = \sum_{l=1}^{k} \left(b_{ij}^s \bigg/ \sum_{i=1}^{k} b_{ij}^s \right)$, its corresponding maximum eigenvector $\lambda_{max}^s = \sum_{i=1}^{k} \frac{[AWs]_i}{kw_{is}}$.

Table 3. Reference for C_{Rs} values.

s	3	4	5	6	7	8	9	10	11
C_R	0.58	0.90	1.12	1.24	1.32	1.41	1.45	1.49	1.51

Thus the judgement consistency index can be concluded as:

$$\left. \begin{array}{l} C_{Is} = \dfrac{\lambda_{max}^s - k}{k - 1} \\ R_{Cs} = \dfrac{C_{Is}}{C_{Rs}} < 0.1 \end{array} \right\} \tag{2}$$

If formula (2) is workable, it is considered that the judgement is consistent. The value of C_{Rs} can be obtained from Table 3.

3.4.4 Weight sorting according to the total target

The eigenvectors $Ws = (w_{1s}, w_{2s}, ..., w_{ks})$ ($s = 1, 2, ..., 12$) form the eigenvector matrix $W = [w_{ij}]k \times 8$. By analysing weights of the 3 indices (B1~B3)'s comparative importance of the level-1 index layer with the expert comparison method, each index's weight $B = (b_1, b_2, b_3)$ can be given. Calculated by the formula (3), the sort weight vectors P of all the solution layers are obtained, compared with that of the total target. The size of each weight represents each solution's relative importance value or degree of merits.

$$P = B^T W = (p_1, p_2, ..., p_k)^T \tag{3}$$

4 CASE STUDY—XINZHOU DISTRICT OF WUHAN

Take the population evacuation zone's selection of Xinzhou District, Wuhan as an example; we made a comprehensive evaluation of the zone's selection with AHP. The result we have obtained can provide references for Xinzhou District undertaking the construction of the population evacuation zone and its air-defense solutions.

4.1 Background information

Xinzhou District, one of the suburbs of Wuhan, is located in northeastern Wuhan and on the north shore of the middle reaches of the Yangtze River. Its north is the tip of Dabie Mountains. Its east longitude is between 114°30′-115°5′ and, its north latitude is between 30°35′-30°2′. Tuanfeng and Huangpi are respectively on its east and west. It is saparated by the Yangtze River with Qingshan and Ezhou in the south of it. In its north are Hong'an, and Macheng. Its terrain tilts from northeast to southwest. The hills and rivers around it rank like a Chinese character "川", which are commonly known as "one river (Yangtze River), two lakes (Wu Lake, Zhangdu Lake), three canals (Jushui Cannal, Dao Cannal, Shahe Cannal), four hills (Louzhai Hill, Yegu Hill, Changling Hill, Cangyang Hill)".

Xinzhou District has jurisdiction over 9 subdistricts, 3 towns, 2 state-run farms, 1 development area and 1 scenic spot, namely, Zhucheng Street, Yangluo Street, Cangbu Street, Liji Street, Wangji Street, Sandian Street, Jiujie Street, Shuangliu Street, Pantang Street; Phoenix Town, Xugu Town, Xinchong Town; Zhangdu Lake, Longwangzui—the 2 state-run farms; Yangluo Economic Development Area and Daoguanhe Scenic Area. In 2009, the region's total household population was 99.07 million, whose resident population was 85.1 million. Its urbanization reached 46.2% and the population density was 658 people/km².

4.2 Construction of the model

4.2.1 The determination of hierarchical structure

Based on the analysis above, the 9 subdistricts (As Zhucheng and Yangluo are fortified cities in accordance with the superior file, they are not included in the evacuation zones illustrated above.), 3 towns, 2 state-run farms, 1 development area, and 1 scenic spot are taken as alternative solutions for population evacuation zones.

4.3 Comprehensive evaluation of the population evacuation zone with AHP

By experts' scoring and comparison of Xinzhou Statistical Yearbooks' relevant indices, we respectively worked out the ratings between each solution's indices. Finally, according to the laws of AHP, we checked the consistency of each comparative judgment matrix and their results were all smaller than the corresponding random consistent

Table 4. Comprehensive evaluation results of Xinzhou District's population evacuation zone.

Alternatives for evacuation zones	Weight (vector)
Jiujie Street	0.0716
Sandian Street	0.0637
Liji Street	0.0618
Pantang Street	0.0667
Shuangliu Street	0.0656
Xinchong Street	0.0638
Wangji Street	0.0632
Xugu Town	0.0767
Phoenix Town	0.0647
Zhangduhu Farm	0.0652
Longwangzui Farm	0.0689
Daoguanhe Street	0.0769
Zhucheng Street	0.0671
Yangluo Street	0.0589
Cangbu Street	0.0652

indices CR. So it can be concluded that all the judgment matrixes have a comparatively high confidence. On this basis, all solutions' eigenvectors (after normalized), namely, weight vectors can be given.

According to the results above, sort these weight vectors from the large to small and this sort is considered as the basis of a prior choice of the population evacuation zones, namely, Daoguanhe, Xugu Town, Jiujie Street, Longwangzui Farm, Zhucheng Street (the region outside the urban area), Pantang Street, Shuangliu Street Zhangduhu Farm, Cangbu Street, Phoenix Town, Xinchong Street, Sandian Street, Wangji Street, Liji Street, Yangluo Street (the region outside the urban area).

5 CONCLUSION

The selection of the population evacuation zones will directly affect the security of citizens' life and property and social stability in time of wars and disasters. As many factors involved need to be considered for the selection and some of the factors are fuzzy, they cannot be represented by some detailed quantitative indices. Thus, AHP has advantages over this kind of issue. AHP's prominent advantage is that it needs comparatively less data and information so that much time for decision-making can be saved by computer operation. However, it also has its limitations. It requires that all experts involved in the decision-making should be provided with solid professional and extensive socio-economic knowledge and familiar with the project background. To improve its practicality and reliability and to reduce experts' judgment errors, some measures are taken as follows:

1. Appropriately expand the expert advisory scope. Increasing the samples of advisory opinions can improve the probability of obtaining the correct results.
2. Scientifically deal with experts' feedback. Generally speaking, the probability distribution of experts' opinions should be in line with or close to the normal distribution. Mathematical statistics should be adapted to process experts' opinions on it.
3. Select evaluation indices advancing with the times in accordance with the needs and characteristics of the new era's civil air defense construction. At present, the civil air defense should meet both the requirements of air raid protection in wartime and the needs of urban disaster prevention and mitigation in peace time. Therefore, the indices of a population evacuation zone should highlight the relevant content of Two Defense Integration, such as measures of earthquake prevention and flood control.

REFERENCES

Li Bonian, Fuzzy Mathematics and Its Application [M]. Hefei: Hefei University Press, 2007.

National Civil Air Defense Office, Academy of Military Sciences of the People's Liberation Army. *Science of Civil Air Defense* [M]. PLA Publishing House, 2005.

Xinzhou Statistics Bureau of Wuhan. *Xinzhou Statistical Yearbook*. (2008–2009).

Advances in Energy, Environment and Materials Science – Wang & Zhao (Eds)
© *2016 Taylor & Francis Group, London, ISBN 978-1-138-02931-6*

Research on technical reforms of nitrogen and phosphorus removal process in T Sewage Company

Hengxia Shen

College of Environmental Science and Engineering, Ocean University of China, Qingdao, Shandong, China
Rizhao Environmental Protection Bureau, Rizhao, Shandong, China

ABSTRACT: T Sewage Company has three wastewater treatment plants. However, T Company always can't meet the national standard of the total nitrogen and phosphorus emissions and needs to use the nitrogen and phosphorus removal agent to decrease the emissions, making the operating costs increase. This paper compares the aluminum sulfate, ferric chloride, and polymeric aluminum chloride to the sewage PH value, SS, COD, and TP removal effect based on the experiment and obtains the appropriate ratio of gas and water through experiments. We used the experimental results in the T wastewater treatment plant and had good economic and environmental protection results.

1 INTRODUCTION

The composition of pollutants in urban sewage is very complex, which can be divided into three categories. They are physical, chemical, and biological pollutants. Wastewater solids, turbidity, color, temperature characterization of the sewage of physical characteristics; nitrogen, phosphorus, pH, metal salts, all kinds of gas molecules is used to characterize the sewage organic compounds; biochemical oxygen demand BOD, COD, organic carbon and other characterization of the sewage organic compounds; and characterization of bacteria, microbes and toxicity of the sewage biological characteristics. The purpose of urban sewage treatment is through a series of physical, chemical, and biological methods, in order to reduce to acceptable levels of these pollutants in the sewage, to meet emission standards in different waters, or different uses of water resources reuse. T Company is a sewage treatment company, responsible for the surrounding areas of domestic sewage treatment. If the sewage is not treated, the direct discharge can cause the eutrophication of the lake and the river water body. The project of sewage denitrifying phosphorus removal, the effluent quality reach grade a standards and do the phosphorus concentration in the effluent decreased to below 0.5 mg/l; nitrogen removal technology research is needed in the consideration of the various affecting factors, through the field simulation test, determine the biological aerated filter the best gas water ratio and temperature, anti-nitrification in a bio filter appropriate amount of carbon source and the best dosage and make the effluent total nitrogen index down to the following 15 mg/l premise, the implementation of sewage treatment plant operation and energy saving. Through the research on the technology of nitrogen and phosphorus removal in domestic sewage, the sewage treatment plant operation management, which can make the treatment of sewage to achieve the standard discharge, purifies the surrounding water body.

2 PHOSPHORUS REMOVAL TECHNOLOGY RESEARCH

2.1 Current technology

In recent years, T Company has built three sewage treatment plants, which are the west plant, the central and the south sewage treatment plant. Biological aerated filter with it covers an area of small area, good water quality, investment, flexible operation, ability to resist impact load strong advantages by more and more favor, but the phosphorus removal process for biological phosphorus removal, the effect is poorer, and total phosphorus of sewage for 4.270 mg/L, can only remove 10 ~ 15% of the total phosphorus, the effluent total phosphorus cannot achieve stability and compliance, supporting the need for chemical phosphorus removal. Through 3 years of experience in production and operation show that due to the aeration biological filter biological flocculation removal effect, in order to ensure the effluent total phosphorus concentration less than 0.5/1.0 mg/l and total phosphorus of biological aerated filter influent concentration maximum value should be controlled in 0.95/1.45 mg/l.

2.2 Experiment of new technology

Experimental water is collected from the south sewage treatment plant. The water quality is shown in the Table 1.

We need 6 beakers (1 L), 2 beakers (500 mL), electronic balance, spoon, ammonium moly date points spectrophotometric meter experimental drug, aluminum sulfate, ferric chloride and polymeric aluminum chloride purity for industrial grade. Determination of phosphorus: ammonium moly date spectrophotometric method, according to the national standard GB11893-89. For trial use of the medicament of phosphorus removal and the trivalent metal ions and phosphate is in r reaction, so chemical phosphorus removal agent theory based on investment plus the amount should be equal to the amount of phosphate (number of moles). But in view of the possible there are many competing reaction at the same time, domestic sewage in the alkalinity, pH value, trace elements, ligands, etc. and the concentration of SS, varying degrees of impact on the actual reaction, so the actual reaction process is more complicated. Method for the determination of PH was determined by pH meter. Method for the determination of COD: dichromate method, according to GB11914 national standard 89. SS determination method, weight method, with reference to the GB11901-89 national standard of this project examines the three removal agent (i.e., aluminum sulfate, ferric chloride and polymeric aluminum chloride reagent) the best pH value range and aluminum and iron salts of phosphorus removal effect and synergic remove SS and COD, combined with field test to determine the optimum reagent and the best dosage. The six liter beaker, dosage barrel from 1 L, stir the sewage, into coagulant, GJJ-931 six combination of magnetic heating stirrer for rapid mixing slow stirring for 10 min, stirring was stopped after standing for 30 min, the supernatant was analyzed. Study on the dosage and TP removal rate, commonly used dosage coefficient K said. Dosage of coefficient K in here said dosage of molar concentrations of sewage in TP of the molar concentration ratio; theory of K value is 1, but the actual K value usually between 2.3, the typical value is 2.4. According to the three different phosphorus removal agent dosage, study on the relationship between the coefficient and the corresponding removal rate of TP. For aluminum sulfate, ferric chloride and aluminum chloride three chemicals, phosphorus removal rate increases with the increase of the dosage amount coefficient increased. The maximum phosphorus removal rate in the three chemical reagents was 85.2%, 92.1%, and 87.5% respectively. For aluminum sulfate and poly aluminum chloride, K is 2.4, TP removal rate was 85.2% and 84.2%, respectively equal to or less than 1.0 mg/L after treatment with TP and biological gentrification filter section after treatment of biological aerated filter. TP can be less than 0.5 mg/l and reached the urban sewage treatment plant pollutant discharge standard level of a standard. For ferric chloride reagent, K = 2, TP < 1 mg/l, and ultimately make the plant effluent discharge standards. In view of the TP removal effects of three kinds of chemicals, we considered the aluminum sulfate as the best TP removal chemical.

3 NITROGEN REMOVAL TECHNOLOGY RESEARCH

3.1 Current technology

At present, we often use the external carbon sources such as methanol, sodium acetate, and glucose to solve the problem of insufficient carbon source. Methanol is a kind of colorless, transparent, flammable, volatile, toxic liquid, as carbon source, it is expensive and toxic determine, in operation unsafe, easy to employee health harm. Glucose is a kind of simple carbohydrates, is the most widely distributed in nature and most important a monosaccharide. It can also be as a carbon source; sodium acetate is colorless, odorless, crystalline, soluble in water, slightly soluble in ethanol, insoluble ether. The materials such as water hydrolysis, alkali, sodium acetate as carbon source are widely used. De-nitrification process mainly includes aerobic nitrification and anaerobic de-nitrification, in which aerobic nitrification occurred in the two-stage biological aerated filter, effluent total nitrogen for 24~30 mg/L, and the COD value only 50~60 mg/L, cannot meet the normal survival needs of denitrifying bacteria; anaerobic de-nitrification occurred in a denitrifying bio-filter. Therefore, it is necessary to denitrify bio-filter to add carbon source.

3.2 Experiment of new technology

Experimental water is collected from the south sewage treatment plant. The water quality is shown in the Table 2.

Total nitrogen determination method is alkaline potassium sulfate-digestion ultraviolet spectrophotometry, with reference to national standards (GB11894-89). The single pool of field experiment, ratio, aeration monitoring import and the

Table 1. Experimental water quality.

Project	PH	COD/(mg/L)	TP/(mg/L)	SS/(mg/L)
Value	6.7–7.3	324–469	4.2.–7.0	176–264

Table 2. Experimental water quality.

Project	PH	COD/(mg/L)	TP/(mg/L)	SS/(mg/L)
Value	6.7–7.3	324–469	4.2.–7.0	176–264

concentration of ammonia nitrogen in water by different gas water and drawing gas water ratio, ammonia concentration, ammonia nitrogen removal rate curve, and research in the aeration amount and ammonia nitrogen removal rate. In the adjustment process, regular observations of two-stage biological aerated filter of ammonia nitrogen concentration, gas water ratio, in filter effluent ammonia indicators have reached the premise of level of a standard, the gas water ratio is reduced to a minimum, thereby reducing the power consumption. Two-stage biological aerated filter by roots blower aeration, using different gas water ratio, drawing gas water ratio, ammonia concentration, ammonia nitrogen removal rate curve. The results are shown in Figure 1. 1 stands for the water flowing into the plant, 2 stands for the water flowing out of the plant and 3 stands for the nitrogen removal rate.

In the process of biological de-nitrification, electron donor is usually derived from the biodegradable dissolved organic matter in the wastewater, which can be degraded by biological degradation. Although the rational allocation of domestic and foreign researchers on the carbon source and effective utilization has done a lot of research and beneficial exploration, however, due to water restrictions on total carbon source, carbon source in the system cannot be a stable meet the need of de-nitrification. Therefore, the external carbon source is a common way to compensate for the shortage of water source. In the use of different carbon sources, the production of energy is different, and the cell yield is also different. If the metabolic pathway is complex, anti-nitrification rate will be affected; if the cell yield is high, namely, carbon conversion was high proportion of cells, to achieve the same de-nitrification effect when the consumption amount of carbon source corresponding also. Also the different substrates used as carbon source for de-nitrification, microbial on different carbon sources to adapt to the situation may be different, such as a carbon source may within a short time after dosing has obvious removal effect of nitrogen, and other carbon sources need longer time for microbial adaptation. Therefore, the reasonable choice of carbon source has important significance to the de-nitrification process. The through on-site actual debugging, the effects of methanol, glucose, acetate and so on three different carbon source for de-nitrification, in order for

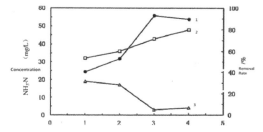

Figure 1. Relationship of sodium acetate input quantity and nitrogen removal rate.

southern and central wastewater treatment plant to select suitable external carbon source. From the experimental data can be obtained when gas water ratio is 3:1, fully able to meet the needs of biochemical reactions, so two sewage treatment plants were by the original a process of wind machine to a two-stage aeration single pool supply change on the grounds of a process fan three two-stage aeration to the single tank of gas, so do the gas water ratio of 3:1 goals, and two sewage treatment plants improved aeration rate since the water good effect, the indicators can achieve standard. The phenomenon of carbon and nitrogen in urban sewage is low in our country, which leads to the problem of insufficient carbon source in biological de-nitrification. For urban sewage with low carbon and nitrogen ratio, it is difficult to reach the national level emission standard after biological treatment.

From Figure 1, we can conclude that the influent of two-stage aeration tank ammonia in the range of 30 mg/L—L, gas water ratio from 1:1 to 4:1 process in, removal rate of ammonia nitrogen by 40.6% up to 93% when gas water ratio is greater than 3:1, effluent ammonia nitrogen removal rate changed little, reach the level of a standard; when gas water ratio of less than 3:1, the effluent water quality is not up to the standard. So when the gas water ratio is 3:1, it can meet the needs of the biochemical reaction, which can ensure the qualified of the ammonia nitrogen in the effluent. It is necessary to do a system reform of the south sewage treatment plant to create the gas water ratio of 3:1.

4 APPLICATION EFFECT OF TECHNICAL REFORMS

We use aluminum sulfate investment amount of empirical formula to T Sewage Company to inspect import water total phosphorus removal effect, and try adjusting dosage to achieve the purpose of saving inhibitors according to the variation of raw water quality. In accordance with the gas and water than 3:1, of southern and central wastewater treatment plant, the aeration system transformation, the reasonable control of the aeration tank aeration and the amount of water, the effluent ammonia the compliance effect of detection to reduce power consumption. Methanol as carbon source to the de-nitrification reaction, the toxicity itself is large, and is easy to produce explosion, security is not high; the removal technology of nitrogen results using sodium acetate as carbon source to the de-nitrification and high safety, and the reaction of high, actual application effect is good, the de-nitrification of effluent total nitrogen can also stable up to standard. Before the study, t sewage treatment plant phosphorus pharmaceutical investment locations for denitrifying bio filter front-end designed dosage 150 mg/L, the effluent total phosphorus cannot be a stable reaches 0.5 mg/l following; after finishing the study. According to the optimal dosage of medicament of phosphorus removal for guiding practical production, adding location change is in the front of the aeration biological filter and the average dosage for 120 mg/L. At this time the phosphorus concentration in the effluent can stable below 0.5 mg/l and meet the national standard of one class A, achieve the expected purpose. The money a year saving can be calculated: $(150–120) \times 0000/1000 \times 365/1000 \times 1800 = 591.3$ thousand Yuan.

Before the research of aeration amount and ammonia nitrogen removal rate, the T Company needs six plant aeration tanks (frequency is 50 Hz), six sets of process of wind machine to six secondary aeration single pool supply (a to form); central wastewater treatment plant secondary aeration tank need 12 Taiwan Craft fan (frequency is 50 Hz) to 12 two-stage aeration single pool supply (one to one). After the research of this project, the connection mode of the aeration pipe is improved. The south wastewater treatment plant first aeration blower pipeline transformation, the specific method is the two-stage aeration tank of each of the three process blower outlet pipeline parallel in together (a total of six blowers, connecting the two groups), in parallel pipeline installed a manual butterfly valve can be achieved by a typhoon, or two Typhoon aircraft to the three seat filter gas. Such two-stage aeration tank change on the grounds of two process fan (but the frequency changed to 35 HZ) to six secondary aeration single tank of gas, so as to achieve gas water ratio of 3:1 goals. Afterwards, the central wastewater treatment plant was transformed as the same method and also has achieved good results.

5 CONCLUSIONS

The drug consumption has reduced more than 15% after using the new technology to guarantee the total nitrogen and total phosphorus meets the standards. Due to the given removal agent investment and consumption is constant, and raw sewage water is constantly changing, resulting in plant effluent total phosphorus, total nitrogen index often substandard; after life sewage removal technology research, through online instrument every three hours monitoring raw water total phosphorus concentration, according to the type of investment, and the volume and raw water total phosphorus concentration timely adjustment of dosage, plant effluent total phosphorus, total nitrogen index of compliance rate increased from 60% to 90%, the drug consumption reduced by more than 15%. The electricity consumption decreased by more than 38%, saving power consumption. By aeration amount and ammonia nitrogen removal rate of research to determine the secondary biological aerated filter economic and efficient gas water ratio, transformation of the blower outlet pipeline connection, reduces the power consumption. An alternative carbon source of the wastewater treatment plant is determined. Considering the health of workers and production operation safety, we find the methanol economy, substitute for sodium acetate. It has good economic and environmental protection results.

REFERENCES

Han Li li. Reconstruction Project for Nitrogen and Phosphorus Removal in Zaozhuang WWTP. China Water and Wastewater [J]. 2011(9): 90–95.

Meng Xin, Wu Kundun, Chen Wei, Zheng Jiandong. Upgrading Reconstruction of Sewage Treatment Plant for Nitrogen and Phosphorus Removal. China Water and Wastewater [J]. 2010(6): 84–86.

Yang Hanxin, Wang Wenxiang, Jiang Xiong, Qiu Zhiping, Liang Zhanxing. Reconstruction Practice of Nitrogen and Phosphorus Removal Process in a Wastewater Treatment Plant of Heshan City. China Water and Wastewater [J]. 2008(9): 83–86.

Zhang Daijun, Lu Peili, Yan Chenmin, Long Tengrui. Application of ASM No. 2 to the study of upgrading WWTP for biological nitrogen and phosphorus removals. Acta Scientiae Circumstantiate [J]. 2013(5): 332–337.

Advances in Energy, Environment and Materials Science – Wang & Zhao (Eds)
© *2016 Taylor & Francis Group, London, ISBN 978-1-138-02931-6*

The development status of medical waste category, management and disposal in China

Zhenbo Bao, Hongjun Teng, Dengchao Jin, Fan Yang, Xinyuan Liu & Yang Li
Tianjin Agricultural University, Tianjin, China

ABSTRACT: Medical waste is classified as number one hazardous waste of *Medical Waste Management Regulations in China*. On the basis of introducing category, common components, characteristics and hazards of medical waste, the laws, regulations and standard, etc. related with Chinese medical waste management and disposal are analyzed and summarized. At last, pointed out that Chinese medical waste management and disposal have gradually legalized and standardized track, to further strengthen the medical waste management and disposal in china and develop new medical waste disposal technology, for the effective and safe disposal of medical waste, the maintenance people healthy and environmental protection, are of great significance.

1 INTRODUCTION

Medical waste refer to the waste with direct or indirect infection, toxic, and other hazardous, which are generated at all levels of medical and health institutions in health care, prevention, health care, and other related activities, the garbage generated by infectious patients treated by medical institutions or suspected patients of infectious diseases should be managed and disposed in accordance with the medical waste. Medical waste may carry bacteria and viruses with the hazards of infectious pathogenic, toxicity, and causticity. Improper disposal of medical waste, not only can cause direct harm to human health, but also can pollute water, air and soil, leading to disease and environment damage. Therefore, to strengthen the awareness of medical waste category, composition, characteristics and harmfulness, and to strengthen the management and disposal of medical waste, have gradually aroused wide attention in the industry (Niu, 2013; Xia, 2013).

2 MEDICAL WASTE CATEGORY, COMMON COMPONENTS, CHARACTERISTICS AND HARMFULNESS

According to the provisions of *The Catalog of Medical Waste Classification* issued by the Ministry of Health and the State Environmental Protection Administration on October 10, 2003, medical waste can be divided into 5 categories of infectious waste, pathological waste, damaging waste, drug waste and chemical waste. Common components,

characteristics and harm of all types of medical waste are shown in Table 1 (Niu, 2013; Xia, 2013; Chen, 2012).

3 THE DEVELOPMENT STATUS OF MEDICAL WASTE MANAGEMENT AND DISPOSAL

Medical waste can pollute environment, spread disease and threat to health, the management and disposal of medical waste has aroused extensive attention of the international community. March 22, 1989, United Nations Environment Program formulated *The Basel Convention on the Control of Trans-boundary Movements of Hazardous Wastes*, medical waste was listed in hazardous waste to be controlled and disposed. May 22, 2001, the international community signed *Stockholm Convention on Persistent Organic Pollutants*, protecting human health and the environment from persistent organic pollutants, controlling and reducing the harm of medical waste. From the mid-1990s, Chinese government gradually increased attention on the safe disposal of medical waste. As Table 2 shows, a series of laws, regulations or standards of medical waste management and disposal were formulated. Before 2003, took the medical waste disposal principle of "who produce, who were harmless", regardless of the size of the hospital medical waste incinerators should be set. However, due to most medical institutions lack professional and technical strength of environmental aspects, the completed incinerator not only can not meet the sterilization requirements, and because there are no good exhaust

Table 1. Category, common components, characteristics and hazards of medical waste.

Category and common components	Characteristics and hazards
1. Infectious waste: ① Material that are polluted by patient's blood, body fluids and excretions. ② Living garbage produced by isolated patients of infectious diseases or suspected patients of infectious diseases. ③ Pathogen media, specimens and bacteria, and viruses storage solution. ④ Various abandoned medical specimens. ⑤ Discarded blood, serum. ⑥ After use of disposable medical supplies and disposable medical devices regarded as infectious waste.	Carry pathogenic microorganisms, contains large amounts of bacteria, viruses, and many toxic chemicals, etc. It is extremely contagious, viral and biological corrosion. Pollute the atmosphere, water, soil and food, have a risk of spreading infectious diseases, resulting in the spread of disease, endangering people's health.
2. Pathological waste: ① Human tissue, organs and other waste generated in the process of surgery and other treatment. ② The tissue and body of medical experiments animal. ③ The abandoned body's tissues and pathological wax block after pathological section.	Carry pathogenic microorganisms, with a certain amount of bacteria and viruses. Pollute the atmosphere, water, soil and food, have a risk of spreading infectious diseases, resulting in the spread of disease, endangering people's health.
3. Damaging waste: ① Used needles and suture needles. ② Used all kinds of medical sharps, comprising: a scalpel, scalpels, knives skin preparation and surgical saws. ③ Used slides, glass tubes and glass ampoules, etc.	The abandoned medical sharps that can stab or cut the body, with a certain amount of bacteria and viruses. Constitute physical damage to the human body, cause bacteria to enter the body, resulting in the spread of disease.
4. Drug waste: ① Abandoned general medicine, such as: antibiotics, non-prescription drugs. ② Abandoned cytotoxic drugs and genotoxic drugs. ③ Abandoned vaccines and blood products.	Expired, eliminate, deteriorated or contaminated waste pharmaceuticals, with a certain toxicity and corrosive. Pollute the atmosphere, water, soil and food, with biological viral and carcinogenic risk, endanger people's health.
5. Chemical waste: ① The waste chemical reagents in medical imaging room and laboratory. ② Abandoned peracetic acid, glutaraldehyde and other chemical disinfectants. ③ Abandoned mercury sphygmomanometer and mercury thermometers.	Toxic, corrosive, flammable and explosive discarded chemicals. Pollute the atmosphere, water, soil and food, incineration is easy to produce dioxins (carcinogenic) and heavy metals and other pollutants.

Table 2. Before 2003 year, the laws, regulations or standards associated with medical waste.

Time (year): laws, regulations or standards	Requirements or purpose
1995: *Law of the People's Republic of China on the Prevention and Control of Environment Pollution by Solid Wastes*	Strengthen the prevention and control of solid waste pollution, and reduce the amount of hazardous solid waste, fully and rationally utilizing, decontamination and disposal of solid waste.
1995: *Graphics Signs for Environmental Protection Solid Waste Storage (Disposal) Site (GB 15562.2-1995)*	Strengthen the solid waste storage and disposal sites supervision and management.
1998: *National List of Hazardous Wastes*	The medical waste NO HW01, as the number one hazardous waste.
2001: *Pollution Control Standard for Hazardous Waste Incineration (GB 18484-2001)*	Provisions for the hazardous waste incineration facility sitting principle, incineration technical performance indicators, the maximum allowable emission limits for incineration emissions air pollutants, incineration residues handling rules emissions and corresponding environmental monitoring.
2001: *Standard for Pollution Control on Hazardous Waste Storage (GB 18597-2001)*	Provisions for hazardous waste packaging, storage facilities siting, design, operation, security and monitoring.
2001: *Standard for Pollution Control on Hazardous Waste Landfill (GB 18598-2001)*	Provisions for the site selection, design, construction, operation, closure and monitoring of landfill.

gas purification system, resulting in emissions of non-compliance, and even produce carcinogen dioxin, serious pollute atmospheric environment, endanger human health (Chen, 2007; Zhao, 2008; Jin, 2006).

In 2003 year, the outbreak of "SARS" epidemic, the urgent need to strengthen the medical waste disposal and management. In 2003–2007 years, a series of regulations, standards, norms, regulations are concentrated formulated, as Table 3 shows.

Table 3. In 2003–2006 years, the laws, regulations or standards associated with medical waste.

Time: laws, regulations or standards	Requirements or purpose
2003: *Medical Waste Management Regulations*	Strengthen safety management of medical waste, prevent the spread of disease and protect the environment.
2003: *Technical Standard for Medical Waste Transport Vehicle (GB 19217-2003)*	Provisions for the special requirements of medical waste transport vehicles, stipulate that the finalized thermal insulation and refrigerated trucks are appropriately transformed for special truck transport of medical waste.
2003: *Technical Standard for Medical Waste Incinerator (GB 19218-2003)*	Prevention and treatment of medical waste incinerator pollution to environment, regulate incinerator design, manufacture, performance and safety use.
2003: *The Classification Catalog of Medical Waste*	Standardize the classification and characteristics of medical waste, identify their common components or specific name.
2003: *Measures for Medical Wastes Management of Medical and Health Institutions*	Provisions for medical waste management of medical and health institutions, effectively prevent and control hazards of medical waste on human health and the environment.
2003: *The Centralized Disposal Technical Specifications of Medical Waste (On Trial)*	Provisions for technical requirements of medical waste temporary storage, transportation and disposal, the training and safety requirements of the relevant personnel, incidents prevention and response measures.
2004: *The Management Administrative Punishment Measures of Medical Waste*	Clear the respective responsibilities of above the county level health administrative departments and environmental protection departments, provisions for the administrative penalties of violating medical waste management regulations.
2004: *Ratified by Adding the Stockholm Convention on Persistent Organic Pollutants (Pops)*	Control emissions of persistent organic pollutants; reduce organic pollutants to the environment.
2005: *Technical Specifications for Medical Waste Centralized Incineration Facility (HJ/T 177-2005)*	Provisions for the construction of medical waste incineration projects, prevent the pollution of medical waste incineration to the environment.
2006: *Technical Specification for Chemical Disinfection Centralized Treatment Engineering on Medical Waste (On Trial) (HJ/T 228-2006)*	Provisions for the practical application of medical waste chemical disinfection treatment technologies, guide the planning, design, construction, inspection and operational management of medical waste chemical disinfection treatment projects.
2006: *Technical Specifications for Microwave Disinfection Centralized Treatment Engineering on Medical Waste (On Trial) (HJ/T 229-2006)*	Provisions for the practical application of medical waste microwave sterilization treatment technology practice, guide the planning, design, construction, operation and management of medical waste microwave disinfection treatment projects.
2006: *Technical Specifications for Steam-Based Centralized Treatment Engineering in Medical Waste (On Trial) (HJ/T 276-2006)*	Provisions for the application of medical waste high-temperature steam treatment technology, to guide the planning, design, construction and operation of medical waste high-temperature steam centralized treatment project.

The medical waste definition, category, collection, temporary storage, transportation and disposal facility construction, operation practices, effectiveness testing and evaluation., are clearly defined, strengthening the management and disposal of medical waste (Yu, 2015; Meng, 2010).

In 2003–2006 years, according to the needs of medical waste effective management and disposal, China has basically established laws and regulations, policies and standard system related with medical waste, a number of medical waste disposal facilities have established, so that medical waste management and disposal have gradually into legalized and standardized track.

As Table 4 shows, in 2007–2014 years, China to further developed a series of laws, regulations or standards related with medical waste disposal, detailed medical waste management and disposal of medical waste, so that medical waste disposal more standardized and institutionalized to ensure the effective and safe disposal of medical waste.

Table 4. After 2007 year, the laws, regulations or standards associated with medical waste.

Time (year): laws, regulations or standards	Requirements or purpose
2007: *The Technology Standard for Hazardous Waste (Including Medical Waste) Incineration Disposal Facilities Dioxin Emission Monitoring (HJ/T 365-2007)*	Provisions for dioxin-like pollutants monitoring in exhaust emissions of hazardous waste incineration facilities and medical waste incineration facilities, reducing dioxin pollution to the environment.
2008: *Standard of Packaging Bags, Containers and Warning Symbols Specific to Medical Waste (HJ 421-2008)*	Provisions for medical waste bags, tool box and containers (barrel) technical requirements, corresponding test methods and inspection rules, and regulations for medical waste warning signs.
2008: Perfectly Revised *National List of Hazardous Wastes*	Again clear medical waste belongs to hazardous waste.
2009: *Technical Specifications for the Supervision and Management to the Operation of Centralized Incineration Disposal Facilities for Hazardous Waste (On trial) (HJ 515-2009)*	Strengthen supervision and management of medical waste incineration facilities operation, to ensure that the standardized operation of medical waste incineration facility.
2010: *Technical Specification of Performance Testing for Facilities of Hazardous Waste (Including Medical Waste) Incineration (HJ 561-2010)*	Provisions for test content, procedures and technical requirements of hazardous waste (including medical waste) incineration facilities involved in performance testing.
2012: *Technical Specifications for Collection, Storage, Transportation of Hazardous Waste (HJ 2025-2012)*	Provisions for technical requirements of hazardous waste collection, storage and transport.
2013: *Technical Specifications for Hospital Sewage Treatment (HJ2029-2013)*	Provisions for Standard design, construction and operation management of hospital sewage treatment works to prevent hospital sewage to pollute the environment.
2013: *Technical Guidelines for Solid Waste Treatment & Disposition Engineering (HJ 2035-2013)*	General technical requirements for solid waste disposal engineering design, construction, inspection, operation and maintenance.
2014: *General Specifications of Engineering and Technology for Hazardous Waste Disposal (HJ 2042-2014)*	Specifies the technical requirements and regulations concerning the application of hazardous waste disposal technology and engineering design, construction, inspection, operation and management process.

4 CONCLUSION

With social and economic development, new diseases, new drugs and medical devices are emerging; medical waste output growth accelerated and medical waste pollution accidents have become frequent. Medical waste treatment results directly related to people's living environment quality, living standards and health must strengthen the safe disposal of medical waste. It needs to have a profound understanding of medical waste classification and hazardous, continue to strengthen the understanding and learning of medical waste disposal technology, standards and codes, actively implement the relevant laws and regulations of medical waste management and disposal, comprehensively promote the construction of medical waste disposal technology system, constantly improve the technological capability and level of medical waste disposal, and to maximize the realization of harmless medical waste, volume reduction and recycling processing target.

ACKNOWLEDGMENTS

The research work was supported by Tianjin SME Technology Innovation Fund No. 13ZXCXSF07000 and Tianjin Science and Technology Special Commissioner Project No. 14 JCTPJC00528.

REFERENCES

Bo. Niu, Feng. Qing. Yu. The Hazards and Disposal of Medical Waste [J]. Guangdong Chemical Industry, 2013, 40(253):152–153,142. (Ch).

Bo. Liu, Hong. Qiang. Liao, Zhong. Wei. Wang, Guang. Wei. Yu, Shi. Qing. Li. Medical waste disposal situation and technical discussion [J]. Chinese Society for Environmental Sciences Annual Conference Proceedings (2010):3504–3506. (Ch).

Deng. Chao. Jin. Steam Sterilizing Treatment Process for Medical Waste and Heat Mass Transfer Models in Moist Sterilization [D]. Tianjin University PhD thesis, 2006, 5:3–4. (Ch).

Hai. Zhao. High Temperature Steam Sterilization Process of Medical Wastes [J]. Environmental Engineering, 2008, S1:209–211. (Ch).

Health Medicine [2003] No. 287. The Classification Catalog of Medical Waste [S]. (Ch).

Hou. Juan. Xia. Study on the Treatment Process of Medical Wastes in Jiangxi Province [D]. Nanchang University Master Thesis, 2013, 5:1–10. (Ch).

Qing. Min. Meng, Xiao. Ping. Chen. A Survey of Medical Waste Pyrolysis and Incineration Treatment [J]. Journal Of Engineering For Thermal Energy And Power, 2010, 25(4):369–373. (Ch).

Ting. Yu. Zhang, Ting. Wang. Harmless Treatment Technology of Medical Wastes In Europe [J]. Shanghai Energy Conservation, 2015(1):35–39. (Ch).

Yang. Chen, Pei. Jun. LI, Chun. Yan. Shao. Analysis of Obstacles Existing in and Countermeasures for Medical Waste Non-incineration Disposal Technology Application [J]. Nonferrous Metals Engineering & Research, 2007(3), 27–29. (Ch).

Yang. Chen, An. Hua. Wu, Qin. Zhong. Feng, Li. Yuan. Liu. Medical waste disposal technique and the source classification strategies [J]. Chinese Journal of Infection Control, 2012, 11(6):401–404. (Ch).

Advances in Energy, Environment and Materials Science – Wang & Zhao (Eds)
© 2016 Taylor & Francis Group, London, ISBN 978-1-138-02931-6

Accessing LUCC and ecosystem service value in Weigan River Basin

Xuning Qiao, Yongju Yang & Hebing Zhang
School of Surveying and Land Information Engineering, Henan Polytechnic University, Jiaozuo, China

ABSTRACT: Land Use/Cover Change (LUCC) makes a decisive role in maintaining ecosystem services. Taking Weigan River Basin as the study area, we construct the transfer matrix of land using overlap analysis function supported by ARCGIS 9.2 software to analyze the dynamics of LUCC. It turns out that: 1) The major conversion types were grass land to cultivated land, unused land to grassland and cultivated land in upstream during 1985–2000. Unused land and cultivated land reduced sharply, while grassland increased rapidly and forest land shrank in size during 2000 and 2007. 2) The land use pattern in the upstream has changed, which led to ecosystem services shifting from food production to water conservation, environment purification, soil formation and protection. Water conservation values of forest land are more than that of cultivated land and grass land. 3) With increase of ecosystem service values in upstream, the land use structure develops towards the direction of protecting water resources.

1 INTRODUCTION

Land use is one of the most effective activities through which human relates to nature. The land use and LUCC have inevitable influence on the ecosystem structure and features. And they affect the chemical composition changes of the atmosphere and lead to global climate change through affecting the biological geochemistry circulation process. Therefore, land use and LUCC make a decisive role in maintaining ecosystem services (Li, 2006). The ecological value of per unit area is different for the land use types, so the conversion among the different land use types will lead to a transmutation to the value of ecosystem services. Costanza published an article by the name of global value of ecosystem services and natural capital in "Nature" magazine in 1997, making the value of ecosystem services estimation principle and method to be clear from the scientific sense (Costanza R., 2007). He is the first to estimate the value of ecosystem services of the global biosphere in the world. This is the most influential research results about the value of ecosystem services. Pimentel etc (1995) summarized and analyzed the research results, which are about natural capital and the value of ecosystem services on the international. And they carried out a comparative study about economic value of the world's biological diversity and United States biological diversity. The results showed that annual economic value of the world's biological diversity are 2928 billion dollars, this estimation results is less than one-tenth of the ecologists in China, Xie Gao-di, etc worked out the equivalent factor table about the value of ecosystem services in China by

analyzing insufficiency and reliable achievements mentioned above (Xie, 2003).

Weigan River Basin is the typical epitome of the arid region in Xinjiang, which can reflect the arid area of mountain—oasis—desert system (Zhang, 2009). By calculating the land use change and values of ecosystem services, how human activities impact on ecological systems and scientific basis for the ecological compensation of river upstream and downstream can be provided. Moreover, a way to solve the problem of ecological security in arid inland river basin can also be given.

2 DATA AND METHODOLOGY

2.1 Data

Data of Land Use/Cover Change (LUCC) in 1985, 2000 and 2007 in Weigan River Basin come from Environmental & Ecological Science Data Center for West China, National Natural Science Foundation of China (http://westdc.westgis.ac.cn). Wheat, corn and cotton are chosen as the major cereal crops in the Weigan River Basin, of which data of area and production come from Xinxiang Statistical Yearbook (1986, 2001, 2008) and Aksu region Statistical Yearbook (1986, 2001, 2008), unit price is from the government network of China cotton website and so on.

2.2 Methodology

Based on the basic assumptions that the value of ecological service function is proportional to biomass in unit area and calculate the output value per unit cultivated land area in the Weigan river basin, this paper ascertains the value of cultivated land ecological service function, as follows (Xiao, 2003).

2.2.1 The determination of the farmland ecological service value per unit area in the study area

We choose wheat, corn, and cotton as the main crops which are maximum planting area in the WeiGan river basin, of which area and production is from XinJiang statistical yearbook and Akesu Region statistical yearbook (2001–2007), unit price is from the Government Network of XinJiang Uighur Autonomous region and the China Cotton Network.

$$E_a = \frac{1}{7}\sum_{i=1}^{n}\frac{m_i p_i q_i}{M} \qquad (1)$$

2.2.2 The determination of the land ecological service value per unit area

The unit price of ecological service function of other land type can be obtained by the equivalent factor table of ecosystem service value in China and the farmland ecological service value per unit area in the study area.

$$E_{ij} = e_{ij}E_a(i = 1, …, 9; j = 1, …, 6) \qquad (2)$$

2.2.3 The determination of ecological service value

Where ESV_a is land ecological service annual value, A_j is jth land area, E_{ij} is the unit price of ecological service function of the ith kind of the jth kind, i is the type of the land ecological service function and j is the land type.

$$ESV_a = \sum_{i=1}^{9}\sum_{j=1}^{6}A_j E_{ij} \qquad (3)$$

3 RESULTS AND ANALYSIS

3.1 Land use/cover change

Unused land and grassland accounts for the largest area of the upstream areas, both of them proportioned 82.2% and 85.11% in 1985 and 2007. Among them, the grass was 38.61% in 1985, 38.52% in 2000, while in 2007 reached 59.81%; unused land is reduced to the 25.27% from the 43.59% in 1985. Changes in various types of land (Fig. 1), it has small changes of various land use area in 15 years from 1985 to 2000. The top two of the changes is the unused land and cultivated land. The former decreased by 0.29%, and the

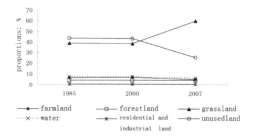

Figure 1. Proportions of land use types in upstream of Weigan River Basin in 1985–2007.

Table 1. Area and proportion of land use in upstream in 1985–2007 unit: hm².

Land use types	1985	2000	2007
Cultivated land	102638.48	107102.48	72521.88
Forest land	64408.42	64552.33	61511.04
Grass land	611710.36	610333.59	930211.85
Water area	110715.12	111563.06	90794.28
Residential and industrial land	4330.7	4793.03	7298.03
Unused land	690614.53	686072.86	393032.57

latter increased by 0.28%. It has dramatic changes in the area of land use in 7 years that from 2000 to 2007. Amplitude of variation is far higher than the first stage. Among them, the grass land increased by 21.29%, the unused land decreased by 18.03%, the Cultivated land decreased by 2.1%, the Residential and industrial and mining land increased by 0.17%,Water area decreased by 1.2%.

3.2 The transfer characteristics of land utilization in upstream area

By the overlaying analysis of land use data of the 1985 and 2000 and with the help of ARCGIS 9.2 software, getting to the transfer matrix of land use types in the upper area of the research from 1990 to 2000. The forest area is 93.98 percent of that in 1985. During 1985 and 2000, in the sequence of the area proportion of the original data, residential and industrial land reduced by 8.86%, the cultivated land reduced by 6.02%, the water area reduced by 1.73%, unused land reduced by 1.71%, the grass land reduced by 1.5% and forest land reduced by 0.01%. The major conversion types are residential and industrial land to cultivated land, cultivated land to grassland and water area to unused land. From 1990 to 2000, 2.9% of cultivated land turned to be grassland, 57.11% of grass land turned to be forest, 1.36% of water area turned to be unused land and 0.19% turned to be cultivated land, 0.52% of unused land turned to be cultivated land and 0.66% turned to be grass land.

In 2000–2007, it made a larger land-use conversion and the cultivated land reduced rapidly. It related to the policy of "Abandoned farmland to grassland". The cultivated land changed from 107102.48 to 72521.88 hectares, the ratio reduced by 32.29%; the grassland changed from 610333.59 to 930211.85 hectares, the ratio increased by 52.41%; the forest changed from 64552.33 to 61511.04 hectares, the ratio reduced by 4.71%; the water land changed from 111563.06 to 90794.28 hectares, the ratio reduced by 18.62%; the residential and industrial land changed from 4793.03 to 7298.03 hectares, the ratio increased by 52.26%; the unused land changed from 686072.86 to 393032.57 hectares, the ratio reduced by 42.71%. The development of unused land is constantly strengthened.

3.3 Calculation of ecological service value

On the basis of studying results of international natural capital and ecosystem service value of Costanza and Pimentel, Xie G.D. provided an evaluation index system of ecological service value that is in national average ecosystem. Ecological service functions of ecosystem are closely related to its biomass of this ecosystem. According to the study of ecosystem biomass in Tarim River Basin by Wang (2007) and Huang (2007), modified indexes of forestland and grassland are obtained from the average of both. Water directly used research results of Huang X. to revise ecological service value per unit area of national ecosystem types. Data of regional wetlands, water and desert ecosystem are difficult to obtain, so values of national average level are used directly. While some scholars research on ecological service value of construction land, but they still are at the exploratory stage, this paper doesn't consider this situation. The calculation result is in Table 2.

3.4 Determination of food production service functions per unit area in farmland

The main food crops are wheat and corn, also a few of rice. The three kinds of crops are main food crops in this river basin, according to methods of calculation for food production service function of farmland ecosystem, draw Ea values of farmland food production service function per unit area of

Table 2. Index of ecosystems in Weigan River Basin.

Ecosystem	Forestland	Grassland	Water
Modified index	0.4622	0.3334	0.94

Weigan River Basin during 1985 and 2007, and the values as follows: 125.7767 yuan/hm^2 in 1985; 515.0298 yuan/hm^2 in 2000; 1519.228 yuan/hm^2 in 2007. Take these as standard per value, and combine ecological service value per unit area of Chinese terrestrial ecosystem and ecosystem corrected parameter of different ecological types in Weigan River Basin, then, calculate ecosystem value of different ecological types in Weigan River Basin.

3.5 Ecological service value

The conversions of land use types caused changes in ecological service values in Weigan River Basin from 1985 to 2007. From 1985 to 2000, the main conversions of land use type were grassland changed into cultivated land, unused land changed into grassland and cultivated land; during 2000 and 2007, large areas of unused land reduced and grassland increased rapidly, while cultivated land also reduced rapidly, forestland began to shrink. But from 1985 to 2007, ecological service economic value had been increasing in the upper reaches of Wegan River Basin. At invariable price of 1990, compared to 1.6239 billion yuan in 1985, it arrived to 5.6517 billion yuan in 2007, which was 3.48 times as 1985, increased 4.0278 billion yuan. Ecological service value of water was in maximum among them, followed by grassland; forestland was closed to cultivated land, and desert (unused land) was in minimum.

Different land use type worked out different ecological service value, the research on ecological service value per unit area of different ecological types in Weigan River Basin from 1985 to 2007 showed that: the main functions of ecological service in farmland ecological system were waste disposal, soil formation and protection, and food production.

Grassland ecosystem played an important role in soil formation and protection, waste treatment and bio-diversity conservation etc ecological service function; the important ecological service functions of forest are soil formation and protection, climate regulation, bio-diversity conservation and water conservation. Get ecological service value per unit area of farmland (cultivated land) ecological system by calculating ecological service value per unit area in Weigan River Basin, and their orders as follows:

Farmland (cultivated land) ecosystem: waste disposal > soil formation and protection > food production > climate regulation > bio-diversity protection > water conservation > climate regulation > raw materials > entertainment culture;
Grassland ecosystem: soil formation and protection > waste disposal > bio-diversity protection > climate regulation > gas regulation > water conservation > food production > raw materials > entertainment culture;

Forest ecosystem (forestland): soil conservation > gas regulation > bio-diversity protection > water conservation > climate regulation > raw materials > waste disposal > entertainment culture > food production;

Water ecosystem (forestland): water conservation > waste disposal > entertainment culture > bio-diversity protection > climate regulation > food production > soil formation and protection > raw materials > gas regulation.

4 CONCLUSIONS

1. The main changes of land use types are grassland change into cultivated land, unused land change into grassland and cultivated land in upper reaches of this research area from 1985 to 2000; large areas of unused land reduced and grassland increased rapidly, while cultivated land also reduced rapidly, forestland began to shrink between 2000 and 2007.
2. The land use pattern in the upstream region of Weigan River Basin has changed, which led to the structure change of ecosystem services, water conservation, environment purification, soil formation and protection instead of former main food production. Water conservation of forestland is far greater than that of cultivated land and grassland, water conservation function of forestland per unit area is close to the study result of Xie G.D., but the value of grass is much lower than that of national average level. Grassland degeneration is serious and biomass is lower. Water conservation function value per unit area is 1.65 times of national average; therefore, water conservation is the most important service function in Basin.
3. The main ecological service functions of farmland were waste treatment, soil formation and protection, and food production; grassland ecosystem played an important role in soil formation and protection, waste treatment and bio-diversity conservation; the important ecological service functions of forestland ecosystem are soil formation and protection, climate regulation, bio-diversity conservation, and water conservation.
4. The ecological service values increased constantly in the upper reaches, and land use will tend to direction that is in favor of protecting water resources of this region. These provide an evidence for compensation that benefit region of downstream to water conservation region of upstream. Forestland and Water area reduction should be paid attention, and reduction of deforestation and water damage become urgent.

ACKNOWLEDGMENT

We are happy to acknowledge the financial support from the National Science & Technology Support Program of China (No. 2012BAJ23B04-2).

REFERENCES

Costanza R., Arge R., Groot R., et al. The value of the world's ecosystem services and natural capital. Nature. Vol. 386 (1997) p. 253–260.

Huang X. Study on ecosystem service function values in Tarim River Basin (Doctorate,. Chinese Academy of Sciences, China 2007).

Li J., Ren Z.Y. The Spatial Analysis of Land Use Ecological Services Value in Loess Plateau in Northern Shaanxi Pronvice. Scientia Agricultura Sinica. Vol. 39 (2006) No. 12, p. 2538–2544.

Li S.D., Mao W.F. Analysis on Surface Water Quality in the Upper Reaches of the Ogan River Basin [J]. Arid Zone Research. Vol. 23 (2006) No. 3, p. 393–398.

Pimentel D., Harvey C., Resosudarmo P., et al. Environmental and economic costs of soil erosion and conservation benefits. Science. Vol. 267 (1995), p. 1117–1123.

Wang H.W. Study on coupling relations between ecosystem services and economic development in arid area (Doctorate,. Chinese Academy of Sciences, China 2007).

Xiao Y., Xie G.D. Economic value of ecosystem services in Mangcuo Lake drainage basin [J]. Chinese Journal of Applied Ecology. Vol. 14 (2003) No. 5, p.676–680.

Xie G.D., Lu C.X., Leng Y.F. Ecological assets valuation of the Tibetan Plateau [J]. Journal of Natural Resources. Vol. 18 (2003) No. 2, p. 189–195.

Zhang X.H., Yang D.G., Liu Y.T. Energy-based sustainability and sensitivity analysis of oasis cropping system: a case study in Weigan River Basin. Acta Ecologica Sinica. Vol. 29 (2009) No. 11, p. 6068–6076.

Advances in Energy, Environment and Materials Science – Wang & Zhao (Eds)
© 2016 Taylor & Francis Group, London, ISBN 978-1-138-02931-6

Comparing different kinds of materials for adsorption of pollutants in wastewater

Pei Wang & Jiangrong Chen
College of Geography and Environmental Sciences, Zhejiang Normal University, Jinhua, China

ABSTRACT: Dirty water is the world's biggest health risk, and continues to threaten both quality of life and public health. Many of our water resources also lack basic protections, making them vulnerable to pollution from factory farms, industrial plants, and activities. Wastewaters are waterborne solids and liquids discharged into sewers that represent the wastes of community life. Primary and secondary treatment removes the majority of BOD and suspended solids found in wastewater. Additional treatment steps have been added to wastewater treatment plants to provide for further organic and solids removals or to provide for removal of nutrients and toxic materials. There have been several new developments in water treatment field in the last years. Graphene-based composite is a new type of nanometer material and shows much better adsorption capacity of pollutants than other materials. This paper summarized and compared the difference between graphene-based composites and other materials in removing various heavy metals and organic pollutants present in wastewater.

1 INTRODUCTION

Recently, the control of water pollution has become increasingly important. Wastewater generally includes domestic wastewater, agricultural wastewater, and industrial sewage. The later, has been found to have an adverse impact on public health and social economy, especially from water pollution caused by heavy metal ions and dyes (Liu et al. 2014). Some potentially toxic heavy metal ions have caused serious diseases, such as cancer, to humans, and have become a great threat to human health (Wang et al. 2014). Thus, there is an urgent demand for the removal of such pollutants. Various methods, such as biological treatment, coagulation, adsorption, chemical oxidation, membrane separation, ion exchange, have been exploited to remove organic dyes and heavy metal ions from water in the past few decades. Among these methods, adsorption is the most versatile and effective method because of its ease of operation, regeneration, and low cost.

In recent years, researchers have invented lots of adsorbents with high selectivity, sensitivity, and good stability for water treatment. These adsorbents include graphene-based composites, biosolid (Liu et al. 2012), activated carbon (Hassan et al. 2014), and many effective adsorbents.

This paper summarized various kinds of adsorbents for water treatment, described the advantages and compared the difference between graphene-based composites and other materials.

2 GRAPHENE-BASED COMPOSITES FOR ADSORPTION OF POLLUTANTS IN WASTEWATER

Graphene-a single atomic layer of graphite has emerged as the "celeb" material of the 21st century. Since its discovery in 2004, graphene has attracted increased attention in a wide range of applications due to its unprecedented electrical, mechanical, thermal, optical, and transport properties. Graphene's high surface-to-volume ratio has resulted in a large number of investigations to study its application as a potential adsorbent for water purification. More recently, graphene-based materials such as graphene oxide, reduced graphene oxide, graphene-inorganic nanostructure composites, and graphene-polymer composites have also emerged as a promising group of adsorbents for the removal of heavy metals and organic pollutants from waste effluents (Huang et al. 2012). Table 1 showed adsorption capacity by various graphene and graphene-based composites.

3 OTHER MATERIALS FOR ADSORPTION OF POLLUTANTS IN WASTEWATER

Water treatment with adsorption processes is an ongoing research area (Erto et al. 2013, Owamah 2013). There are many adsorbing materials except graphene-based composites, such as activated carbon, biosorbents, and inorganic oxides.

Table 1. Adsorption capacity by various graphene and graphene-based composites.

Adsorbent	Pollutant	Adsorption capacity (mg/g)	References
	Adsorption capacity of graphene and graphene-based composites		
Graphene-Fe	Cr(VI)	162	Jabeen et al. 2011
GO	U(VI)	299	Su et al. 2015
GO	Cd(II)	530	Sitko et al. 2013
GO	Zn(II)	345	Sitko et al. 2013
GO	Methylene blue	714	Yang et al. 2011
GO-EDTA	Pb(II)	525	Madadrang et al. 2012
GO-Chitoson	Cu(II)	70	Chen et al. 2013
GO-Chitoson	Methyl blue	468	Yang et al. 2013
GO-Fe$_3$O$_4$	Cu(II)	273	Zhang et al. 2013
GO-FeOOH	As(V)	73.42	Peng et al. 2013
GO-ZrO(OH)$_2$	As(III)	95.15	Luo et al. 2013
G-SO$_3$H/Fe$_3$O$_4$	Safranine	199.3	Chowdhury et al. 2014
G-SO$_3$H/Fe$_3$O$_4$	Neutral red	216.8	Chowdhury et al. 2014
G-SO$_3$H/Fe$_3$O$_4$	Victoria blue	200.6	Chowdhury et al. 2014
Polypyrrole-RGO	Hg(II)	979.54	Chandra & Kim (2011)

Table 2. Adsorption capacity of other materials.

Adsorbent	Pollutant	Adsorption capacity (mg/g)	References
	Adsorption capacity of other materials		
Activated carbon	Methylene blue	1030	Jing et al. 2014
Activated carbon	Acid blue 80	171.5	Wang et al. 2014
WO$_3 \cdot$ H$_2$O	Pb(II)	315	Liu et al. 2014
WO$_3 \cdot$ H$_2$O	Methylene blue	117.8	Liu et al. 2014
BiOBr	Cr(VI)	16.9	Wang et al. 2014
Fe$_3$O$_4$	Pb(II)	43	Wang et al. 2014
Cu$_2$O NCS	Hyaluronic Acid	405.5	Jing et al. 2014
PEI@Mg2SiO4	p-Chiorophenol	286	Chen et al. 2014
Orange peel	Rhodamine B	58.86	Fernandez et al. 2014
Coconut shell	Malachite Green	214	Yagub et al. 2014
Raw clinoptilolite	Fe(III)	98	Abdel-Ghani et al. 2015
Activated carbon	Methylene blue	1030	Hassan et al. 2014
Activated carbon	Acid blue 80	171.5	Luo et al. 2015
WO$_3 \cdot$ H$_2$O	Pb(II)	315	Liu et al. 2014
WO$_3 \cdot$ H$_2$O	Methylene blue	117.8	Liu et al. 2014

Activated carbon is widely used as an adsorbent for water treatment. It is effective to sequester metal ions and dye from water environment. However, activated carbon is expensive (Putra et al. 2014). In recent years, alternative biosorbents have been studied for water clean-up, such as Rice husk and Pine fruit shell-carbon. They are inexpensive, efficient, and practical to be utilized. Inorganic oxides have attracted much attention for the removal of organic dyes and heavy metal ions because of their environmentally benign nature and excellent adsorption properties, such as WO$_3 \cdot$ H$_2$O (Liu et al. 2014), BiOBr (Wang et al. 2014), Cu$_2$O NCS (Jing et al. 2014), PEI@Mg$_2$SiO$_4$ (Chen et al. 2014). Table 2 showed adsorption capacity by various materials.

4 CONCLUSIONS

In summary, there are many materials which could be used for sewage disposal. Compared with other materials, Graphene-based composites with large specific surface area, high selectivity, low-cost, good stability, appear to be very promising for the water treatment. Because of those advantages, graphene-based composites have been considered as materials of future for different other types of applications.

ACKNOWLEDGMENT

This research was supported by National Natural Science Foundation of China (No. 21275131), and Zhejiang Environmental Protection Bureau (No. 2013A025).

REFERENCES

Abdel-Ghani, N.T., G.A. El-Chaghaby, & E.M. Zahran (2015). Cost Effective Adsorption of Aluminium and Iron from Synthetic and Real Wastewater by Rice Hull Activated Carbon. *Am. J. Anal. Chem.* 6, 71–83.

Chandra, V. & K. Kim (2011). Highly selective adsorption of Hg2+ by a polypyrrole-reduced graphene oxide composite. *Chem. Commun.* 47, 3942–3944.

Chen, Y., L. Chen, & H. Bai (2013). Graphene oxide–chitosan composite hydrogels as broad-spectrum adsorbents for water purification. *J. Mater. Chem.* 1, 1992–2001.

Chen, Z., H. Gao, & J. Yang (2014). PEI@Mg2SiO4: and effcient carbon dioxideand nitrophenol compounds adsorbing material. *RSC Adv.* 4, 33866–33873.

Chowdhury, S. & R. Balasubramanian (2014). Recent advances in the use of graphene-family nanoadsorbents for removal of toxic pollutants from wastewater. *Adv. Colloid Interface Sci.* 204, 35–56.

Erto, A., L. Giraldo, A. Lancia, & Moreno-Pirajan (2013). A Comparison between a Low-Cost Sorbent and an Activated Carbon for the Adsorption of Heavy Metals from Water. *Water, Air, Soil Pollut.* 224, 1–10.

Fernandez, M.E., G.V. Nunell, & P.R. Bonelli (2014). Activated carbon developed from orange peels: Batch and dynamic competitive adsorption of basic dyes. *Ind Crop Prod.* 62, 437–445.

Hassan, A.F., A.M. Abdel-Mohsen, & M.M.G. Fouda (2014). Comparative study of calcium alginate, activated carbon, and their composite beads on methylene blue Adsorption. *Carbohydr. Polym.* 102, 192–198.

Huang, X., Qi. X, & F. Boey (2012). Graphene-based composites. *Chem Soc Rev.* 41, 666–686.

Jabeen, H., V. Chandra, & S. Jung (2011). Enhanced Cr(VI) removal using iron nanoparticle decorated graphene. *Nanoscale.* 3, 3583–3585.

Jing, H., T. Wen, & C. Fan (2014). Effcient adsorption/photodegradation of organic pollutants from aqueous systems using Cu2O nanocrystals as a novel integratedphotocatalytic adsorbent. *J. Mater. Chem. A.* 2, 14563–14570.

Liu, B., J. Wang, & J. Wu (2014). Controlled fabrication of hierarchical WO3 hydrates with excellent adsorption performance. *J. Mater. Chem.* 2, 1947–1954.

Liu, T., Y. Li, & Q. Du (2012). Adsorption of methylene blue from aqueous solution by graphene. *Colloid Surface B.* 90, 197–203.

Luo, X., C. Wang, & L. Wang (2013). Nanocomposites of graphene oxide-hydrated zirconium oxide for simultaneous removal of As(III) and As(V) from Water. *Chem. Eng. J.* 220, 98–106.

Luo, X., Z. Zhang, & P. Zhou (2015). Synergic adsorption of acid blue 80 and heavy metal ions (Cu2+/Ni2+) onto activated carbon and its mechanisms. *J. Ind. Eng. Chem.* 2015.

Madadrang, C.J., H.Y. Kim, & G. Gao (2012). Adsorption Behavior of EDTA-Graphene Oxide for Pb (II) Removal. *ACS Appl Mater Interfaces.* 4, 1186–1193.

Owamah, H. (2013). Biosorptive Removal of Pb(II) and Cu(II) from Wastewater Using Activated Carbon from Cas-sava peels. *J. Mater. Cycles Waste.* 1–12.

Peng, F., T. Luo, L. Qiu, & Y. Yuan (2013). An easy method to synthesize graphene oxide-FeOOH composites and their potential application in water Purification. *Mater. Res. Bull.* 48, 2180–2185.

Putra, W., A. Kamari, & S. Yusoff (2014). Biosorption of Cu(II), Pb(II) and Zn(II) Ions from Aqueous Solutions Using Selected Waste Materials: Adsorption and Characterisation Studies. *Journal of Encapsulation and Adsorption Sciences.* 4, 25–35.

Sitko, R., E. Turek, & B. Zawisza (2013). Adsorption of divalent metal ions from aqueous solutions using graphene oxide. *Dalton Trans.* 42, 5682–5689.

Su, Y., S. Yang, & Y. Chen (2015). Adsorption and Desorption of U(VI) on Functionalized Graphene Oxides: A Combined Experimental and Theoretical-Study. *Environ. Sci. Technol.* 49, 4255–4262.

Wang, J., G. Zhao, & Y. Li (2014). One-step fabrication of functionalized magnetic adsorbents with large surface area and their adsorption for dye and heavy metal Ions. *Dalton Trans.* 43, 11637–11645.

Wang, X., W. Liu, & J. Tian (2014). Cr(VI), Pb(II), Cd(II) adsorption properties of nanostructured BiOBr microspheres and their application in a continuous filtering removal device for heavy metal ions. *J. Mater. Chem.* 2, 2599–2608.

Yagub, M.T., T.K. Sen, S. Afroze, & H.M. Ang (2014). Dye and its removal from aqueous solution by adsorption: A review. *Adv Colloid Interfac.* 209, 172–184.

Yang, S., S. Chen, & Y. Chang (2011). Removal of methylene blue from aqueous solution by graphene oxide. *J. Colloid Interface Sci.* 359, 24–29.

Yang, S., J. Luo, & J. Liu (2013). Graphene Oxide/Chitosan Composite for Methylene Blue Adsorption. *Nanoscience and nanotechnology letters.* 5, 372–376.

Zhang, W., X. Shi, & Y. Zhang (2013). Synthesis of water-soluble magnetic graphene nanocomposites for recyclable removal of heavy metal ions. *J. Mater. Chem. A.* 1, 1745–1753.

Advances in Energy, Environment and Materials Science – Wang & Zhao (Eds)
© *2016 Taylor & Francis Group, London, ISBN 978-1-138-02931-6*

The classification of medical waste and its disposal technology

Zhenbo Bao, Hongjun Teng, Dengchao Jin, Jinxing Peng, Nan Wu & Lei Yang
Tianjin Agricultural University, Tianjin, China

ABSTRACT: Medical waste is classified as number one hazardous waste of *Medical Waste Management Regulations in China*. On the basis of introducing the classification and common components of medical waste, the principle and suitability of medical waste treatment technology are analyzed, and the advantages and disadvantages of medical waste treatment technology are summarized. At last, pointed out that it is necessary to adapt suitable medical waste disposal technology in accordance with the advantages and disadvantages of different technology, to maximize the realization of medical waste harmless, volume reduction, and recycling processing target.

1 INTRODUCTION

Medical waste refers to the directly or indirectly infectious, toxic, and other hazardous waste that is produced in medical treatment, prevention, health care, and other related activities of medical and health institutions at various levels. The living garbage produced by patients with infectious diseases or suspected patients with infectious diseases in medical and health institutions should be managed and disposed in accordance with medical waste. Medical waste may carry bacteria and virus, with infected and pathogenic character, biological toxicity, corrosion, and other hazards. The random discharge of medical waste without proper treatment can induce disease, causing direct harm to human health; and can damage the water, air and, soil of environment, causing an indirect harm to human health (Bao, 2013).

2 MEDICAL WASTE CLASSIFICATION AND ITS COMMON COMPONENTS

According to the provisions of *The Catalog of Medical Waste Classification* issued by the Ministry of Health and the State Environmental Protection Administration on October 10, 2003, medical waste can be divided into 5 categories of infectious waste, pathological waste, damaging waste, drug waste, and chemical waste. Common components of all types of medical waste are shown in Table 1 (Xia, 2013; Chen, 2012).

Table 1. Classification and common components of medical waste.

Classification	Common components
1. Infectious waste	① Materials that are polluted by patient's blood, body fluids and excretions. ② Living garbage produced by isolated patients of infectious diseases or suspected patients of infectious diseases. ③ Pathogen media, specimens and bacteria, and viruses storage solution. ④ Various abandoned medical specimens. ⑤ Discarded blood, serum. ⑥ After use of disposable medical supplies and disposable medical devices regarded as infectious waste.
2. Pathological waste	① Human tissue, organs and other waste generated in the process of surgery and other treatment. ② The tissue and body of medical experiments animal. ③ The abandoned body's tissues and pathological wax block after pathological section.
3. Damaging waste	① Used needles and suture needles. ② Used all kinds of medical sharps, comprising: a scalpel, scalpels, knives skin preparation and surgical saws. ③ Used slides, glass tubes and glass ampoules, etc.
4. Drug waste	① Abandoned general medicine, such as: antibiotics, non-prescription drugs. ② Abandoned cytotoxic drugs and genotoxic drugs. ③ Abandoned vaccines and blood products.
5. Chemical waste	① The waste chemical reagents in medical imaging room and laboratory. ② Abandoned peracetic acid, glutaraldehyde and other chemical disinfectants. ③ Abandoned mercury sphygmomanometer and mercury thermometers.

3 MEDICAL WASTE TREATMENT TECHNOLOGY

Medical waste treatment technology can be divided into two major categories of incineration and non-incineration. Incineration technology is mainly high temperature incineration, non-incineration include high temperature steam sterilization, chemical disinfection, electromagnetic sterilization, plasma method, and pyrolysis technology. The principle and suitability of different medical waste treatment technology are shown in Table 2 (Xia, 2013; Chen, 2007; Zhao, 2008; Jin, 2006; Liu, 2010; Zhang, 2015; Meng, 2010).

In the early 1980s, the world's economically developed countries have used incineration technology to treat medical waste. However, since the fluorine-containing substances in medical waste is more, incineration of medical waste is easy to generate a strong carcinogenic dioxins; moreover, with the improvement of people's requirements and awareness of environmental protection, the legal intensity of medical waste treatment to strengthen, most developed countries increasingly recognize that the use of medical waste incineration process security risk. The Stockholm Convention on Persistent Organic Pollutants (Pops) in May 2004 into the implementation phase, the convention requires all signing countries to reduce dioxins and other by-product (Chen, 2007; Zhao, 2008; Jin, 2006; Liu, 2010).

In recent years in the United States, Europe, and other developed countries non-incineration treatment technology has been widely used, and gradually becomes the mainstream of the medical waste treatment technology (Zhao, 2008; Jin, 2006; Liu, 2010). The medical waste incineration treatment technology application is relatively late in China. Before 2006, the domestic related management and technical personnel less contact medical waste incineration technology, the medical waste incineration project running

Table 2. The principle and suitability of different medical waste treatment technology.

Treatment technology	Principle	Suitability
High temperature incineration	Under the condition of 800–1000°C high temperatures and oxygen-rich, the organics in medical waste burn into gas and stable residue, achieve the complete sterilization of medical waste.	Can deal with infectious waste, pathological waste, damaging waste, drug waste and chemical waste.
High temperature steam sterilization	After sorting the medical wastes, under the condition of high temperature and high pressure (gauge pressure 100 kPa, temperature 121°C above, running more than 20 min), ensure that saturated steam can penetrate the interior of medical waste, the microorganism proteins solidify denaturation and inactivated, killing pathogens in medical waste.	Can deal with infectious waste and damaging waste, part of chemical waste, a very small part of pathological waste, can not deal with drug waste.
Chemical disinfection	The crushed medical waste mixing with a concentration of disinfectant (sodium hypochlorite, peracetic acid, glutaraldehyde, ozone, etc.) react, to ensure that medical waste and disinfectant have sufficient contact area and time, decomposing the organic of medical waste and killing or inactivating pathogens.	Can deal with infectious waste and damaging waste, part of chemical waste, a very small part of pathological waste and drug waste.
Electromagnetic sterilization	Medical waste into electromagnetic energy field, using selective energy absorption characteristics of microbial cells. The microbe liquid molecular vibrates according to the frequency of the applied electric field, the microwave energy can be quickly converted into heat energy, heating and killing microorganisms in medical waste.	Can deal with infectious waste and damaging waste, part of chemical waste, a very small part of pathological waste, can not deal with drug waste.
Plasma method	The inert gas in plasma system ionize to glow discharge, instantly generate 1200–3000°C high temperature, so that the organic in medical waste is dehydrated and paralyzed into H_2, CO and other combustible gases, and then after two second burn, kill pathogens in medical waste.	Can deal with infectious waste, pathological waste, damaging waste, drug waste and chemical waste.
Pyrolysis technology	Medical waste is heated in the absence of oxygen or hypoxia, so that medical waste can be decomposed into chemical gas, tar and ash.	Can deal with infectious waste, pathological waste, damaging waste, drug waste and chemical waste.

Table 3. The advantages and disadvantages of different medical waste treatment technology.

Treatment technology	Advantages and disadvantages
High temperature incineration	Advantages: ① Through disinfection sterilization. ② Medical waste volume decreased by 85% to 95%, the volume reduction and decrement effect is obvious. ③ The burning released heat can be reused to generate electricity, heating, etc. ④ The incineration technology is mature, stable and secure. Disadvantages: ① SO_2, NOx, heavy metal gases, Dioxins, polycyclic aromatic hydrocarbons and other toxic substances generated in the combustion must be conducted emissions purification processing. ② A greater amount of fuel gas generates in oxyfuel combustion, the emissions treatment costs of flue gas is high. ③ The combustion system equipment complexity, high investment and operating costs.
High temperature steam sterilization	Advantages: ① Simple operation, low operating and maintenance costs. ② The running media is mainly clean high-temperature steam, safe operation. ③ Prompt and thorough sterilization, which can achieve LOG6 standards. ④ The medical waste after treated and crushing compressed can mix with garbage combustion, to achieve resource recycling. Disadvantages: ① Before disposal medical waste should be strictly sorted. ② The treated medical waste need to be broken and disfigurement treatment. ③ It need be equipped with purification facilities for purifying waste gas and waste liquid generated in processing. ④ The reduction effect of treated medical waste is not obvious, the need to pass living garbage sanitary landfill site or living garbage incinerators for disposal again.
Chemical disinfection	Advantages: ① Process equipment is simple, easy to operate; low processing costs. ② Disinfection process is quick, can achieve high disinfection rates. Disadvantages: ① Before chemical disinfection treatment process to be sorted and crushed process. ② Disinfection process using a large of disinfection liquid impacting on the environment. ③ The high technical requirements for disinfection operator. ④ Sterilization is not complete, but the surface disinfection of waste, rather than deep sterilization.
Electromagnetic sterilization	Advantages: ① High sterilization efficiency. ② Devices can be mobile or fixed. ③ The treated medical waste can be used as solid waste sanitary landfill or incinerated recycling. Disadvantages: ① Waste reduction effect is not obvious. ② Higher construction and operating costs. ③ The radiation protection measures to be taken during disposal.
Plasma method	Advantages: ① High treatment efficiency. ② Small secondary pollution. ③ Medical waste ultimately into glassy solid or slag and other products that can be used directly for final landfill disposal. Disadvantages: ① High technology investment. ② High operating costs.

(*Continued*)

Table 3. (*Continued*)

Treatment technology	Advantages and disadvantages
Pyrolysis technology	Advantages: ① The organics in medical waste convert into the available gas, tar and carbon residue, high degree of resources. ② The secondary smog emissions of NOx, SO_2, HCl, dioxins and soot are small, exhaust gas purification facility investment and operating costs low. ③ Pyrolysis run at the conditions of low temperatures and anoxic, eliminating the generation and diffusion of dioxins, heavy metals and other toxic substances from the principle. ④ Completely harmless treatment, not only completely kill pathogenic microorganisms in medical waste, and can convert chlorine, sulfur, nitrogen and other elements to hydrogen chloride, hydrogen sulfide and ammonia with alkaline liquid washed and absorbed. ⑤ Most heavy metals in medical waste are fixed in the residue. Disadvantages: ① Higher investment and operating costs. ② Technology systems are relatively complex. ③ Require a well-trained operator.

instance less, medical waste incineration facilities construction, and the control parameters of operation process need engineering practice to explore and other factors, which impacted on the application process of non-incineration treatment technology in China. In June 2006, the State Environmental Protection Administration issued Technical Specifications for Steam-based Centralized Treatment Engineering in Medical Waste (On trial) (HJ/T 276-2006), Technical Specification for Chemical Disinfection Centralized Treatment Engineering on Medical Waste (On trial) (HJ/T 228-2006), and Technical Specifications for Microwave Disinfection Centralized Treatment Engineering on Medical Waste (on trial) (HJ/T 229-2006), which established technical and management basis for the planning, design, construction, inspection, and operations management of medical waste non-incineration treatment technologies, medical waste incineration technology gradually showing a rising trend in the construction and application of medical waste centralized disposal facilities (Chen, 2007; Zhao, 2008; Jin, 2006). The advantages and disadvantages of different medical waste treatment technology are shown in Table 3 (Zhao, 2008; Jin, 2006; Liu, 2010; Zhang, 2015; Meng, 2010).

in *China's National List of Hazardous Waste*, its number is the first (Bao, 2013). As social and economic development, new diseases, new drugs, and medical devices are emerging, medical waste output growth accelerated and medical waste pollution accidents have become frequent. Medical waste incineration technology has the advantages of adapting to a wide treatment range of medical waste, the treated medical waste is illegible, thorough disinfection, volume, and mass reduction effect is remarkable, the relevant standard specifications complete, and mature technology, etc., has been used as the main treatment of medical waste, especially for large-scale centralized disposal of medical waste. In high-temperature steam, chemical disinfection, microwave, etc. as the representative of non-incineration technologies, for relatively small-scale centralized disposal of medical waste, the construction and operating costs compared with incineration technology has a large advantage (Jin, 2006; Liu, 2010; Zhang, 2015; Meng, 2010). To maximize the realization of medical waste harmless, volume reduction and recycling processing target, it is necessary to adapt suitable medical waste disposal technology in accordance with the advantages and disadvantages of different technology.

4 CONCLUSION

In *the Basel Convention*, medical waste is classified in Y1 group to be controlled waste categories, its risk characteristics is grade 6.2. Medical waste is the top waste of 47 categories hazardous waste

ACKNOWLEDGMENTS

The research work was supported by Tianjin SME Technology Innovation Fund No. 13ZXCXSF07000 and Tianjin Science and Technology Special Commissioner Project No. 14JCTPJC00528.

REFERENCES

Bo. Liu, Hong. Qiang. Liao, Zhong. Wei. Wang, Guang. Wei. Yu, Shi. Qing. Li. Medical waste disposal situation and technical discussion[J]. Chinese Society for Environmental Sciences Annual Conference Proceedings (2010):3504–3506. (Ch).

Deng. Chao. Jin. Steam Sterilizing Treatment Process for Medical Waste and Heat Mass Transfer Models in Moist Sterilization[D]. Tianjin University PhD thesis, 2006, 5:3–4. (Ch).

Hai. Zhao. High Temperature Steam Sterilization Process of Medical Wastes[J]. Environmental Engineering, 2008, S1:209–211. (Ch).

Health Medicine [2003] No. 287. The Classification Catalog of Medical Waste[S]. (Ch).

Hou. Juan. Xia. Study on the Treatment Process of Medical Wastes in Jiangxi Province[D]. Nanchang University Master Thesis, 2013, 5: 1–10. (Ch).

Qing. Min. Meng, Xiao. Ping. Chen. A Survey of Medical Waste Pyrolysis and Incineration Treatment[J]. Journal of Engineering For Thermal Energy and Power, 2010, 25(4):369–373. (Ch).

Ting. Yu. Zhang, Ting. Wang. Harmless Treatment Technology of Medical Wastes In Europe[J]. Shanghai Energy Conservation, 2015(1):35–39. (Ch).

Yang. Chen, An. Hua. Wu, Qin. Zhong. Feng, Li. Yuan. Liu. Medical waste disposal technique and the source classification strategies[J]. Chinese Journal of Infection Control, 2012, 11(6):401–404. (Ch).

Yang. Chen, Pei. Jun. LI, Chun. Yan. Shao. Analysis of Obstacles Existing in and Countermeasures for Medical Waste Non-incineration Disposal Technology Application[J]. Nonferrous Metals Engineering & Research, 2007(3), 27–29. (Ch).

Zhen. Bo. Bao, Deng. Chao. Jin, Hong. Jun. Teng, Yang. Li. The General Process of Medical Waste High Temperature Steam Sterilization Treatment Technology[J]. Advanced Materials Research, 2013, 807–809:1160–1163.

Advances in Energy, Environment and Materials Science – Wang & Zhao (Eds)
© *2016 Taylor & Francis Group, London, ISBN 978-1-138-02931-6*

Experimental analysis on the mechanical characteristic of single-fractured rock masses under the freeze–thaw condition

Yani Lu
Hubei University of Technology, Xiaogan, Hubei, China

Xinping Li
Wuhan University of Technology, Wuhan, China

Xinghong Wu
Hubei University of Technology, Xiaogan, Hubei, China

ABSTRACT: Physical and mechanical properties of closed cracked rock masses were studied under different freeze–thaw cycles. Seven kinds of specimens with different geometric characteristics were produced using rock-like materials. The influence on different fissure inclinations and fissure lengths on the strength of the rock masses was investigated by the uniaxial compression test. The study found that the uniaxial compressive strength decays badly with the number of freeze–thaw cycles, as well as the elastic modulus. Under the same freeze–thaw cycles and fissure inclination and with the growth of crack length, the uniaxial compressive strength declines dramatically. When other conditions are the same; the inclination has a little effect on the uniaxial compressive strength.

1 INTRODUCTION

In cold regions, the alternating of freeze–thaw cycles causes a series of freeze–thaw disasters, such as the frost heave and thaw collapse, mudslides, landslides, collapses, etc., which have seriously affected the security and stability of cold areas' mining, geotechnical, construction and other projects, as well as devastated cold areas' engineering construction and ecological environment. Therefore, research on rock degradation and deterioration mechanisms and the damage characteristics of rock mass engineering under the freeze-thaw cycle would provide important guidance to the growing number of engineering constructions in cold regions (Li, 2012; Li & Lu, 2013; Lu & Li, 2014).

So far, many scholars have carried out useful research on the freeze–thaw rock. Zhang Ji-Zhou (2008) conducted the freeze–thaw cycle test of silty argillaceous rocks, diabase and dolomitic limestone under two hydration environments, namely distilled water and 1% solution of nitric acid. The test results show that the freeze–thaw damage of the rock is more serious under the acidic condition than under the condition of pure water. Xu Guang Miao (2005) analyzed two different freeze–thaw damage modes of red sandstone and shale, and the two kinds of rock were tested by the

uniaxial compression test under different numbers of freeze–thaw cycles. H. Yavuz (2006) carried out freeze–thaw cycles and mechanical tests on 12 different carbonates, and analyzed the longitudinal wave velocity and the change rules of uniaxial compressive strength and Schmidt intensity's variation under freeze–thaw cycle conditions. The two natural granites with three different moisture contents were subjected to the freeze–thaw cycle test at the temperature range of −20°C to 20°C in the study conducted by Mokhfi Takarli (2008), and investigated the changes in the porosity, permeability, and compressional wave velocity of the granites under different freeze–thaw cycles. Javier Martínez-Martínez (2013) collected rock samples of six different carbonates, and conducted a freeze–thaw cycle experiment (0 times, 12 times, 24 times, 48 times, and 96 times freeze–thaw cycle tests). He also analyzed the bulk diffusion of rocks and the changes in porosity, surface damage characteristics, mechanical properties, and ultrasonic velocity.

The above research results focused on the physical and mechanical properties of the intact rock under the effect of freeze–thaw cycles. However, only a few studies have reported on the physical and mechanical properties of fractured rock masses in nature under the condition of freeze–thaw cycles. Thus, this article uses rock-like

Table 1. Physico-mechanical properties of the material and the prototype.

Materials	Density/ g·cm⁻³	Elastic modulus/Gpa	Poisson's ratio μ	Pressure strength/Mpa	Tensile strength/Mpa	Cohesion/ Mpa	Frictional angle/°
Model	1.93	11.2	0.12	38	8.45	11.4	36
Sandstone	2.31	8.56	0.19	30.24	6.27	8.93	41

materials to produce single-fractured rock masses with different geometric characteristics, explores the physical and mechanical properties and failure characteristics of fractured rock masses through different numbers of freeze–thaw cycles, and provides a reference to the rock mass engineering constructions and the safety of the project operation in cold regions.

2 TESTING PROGRAM

2.1 Materials

Based on a large number of literature, this test selected the model materials similar to rock (brittleness and dilatancy)-cement mortar to produce rock-like model samples. After a lot of trials and comparison, the mixture ratio of the model test was set as cement: quartz sand: silica fume: distilled water: water reducer = 20:16:2.2:5.6:0.2. Table 1 lists the model materials and the main physical and mechanical parameters of sandstone. From Table 1, we can see that the physical and mechanical parameters of both materials are similar, so this model can be classified as a rock-like material.

We selected a Φ 50 mm PVC pipe to cut the rock sample to a cylindrical shape with the length of 100 mm. According to the length and the angle of the design fissures, using the high-speed hacksaw, whose thickness is 0.65 mm, the rock sample was subjected to predetermined cracks, then inserted into a thin plastic paper sheet to produce a finished crack, tauted and fixed with a tape outside the PVC pipe. After pouring cement mortar, vibration, stripping the mold and putting the sample into 20 conservation pools to cure for 28 days, a certain angle and length of closed fracture was finally formed on the rock sample. The geometrical arrangement of fissures is shown in Figure 1, where a is the crack length and α is the fissure inclination.

2.2 Test procedure

The test machine under the condition of constant temperature and humidity of freeze–thaw cycles was set up as follows. First, the sample was frozen at –40°C for 6 h, and then melted at 40°C for 6 h, so that each freeze–thaw cycle lasted for 12 h. Subsequently, the lowest temperature of the

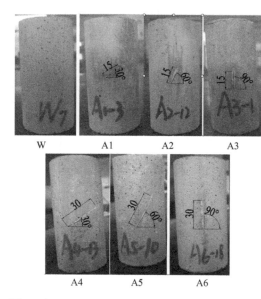

Figure 1. Test model of complete and single-fractured specimens.

freeze–thaw cycle test machine was controlled at –50°C, and humidity was set at 100%.

Seven rock samples were divided into four groups, with 21 samples per group, and then the freeze–thaw cycle was carried out for 0 times, 20 times, 40 times and 60 times. After the test, the quality and volume of the rock samples were determined first, and finally the uniaxial compression test was carried out on the specimens under different freeze–thaw cycles.

The uniaxial compression test was carried out at the Wuhan Geotechnical Engineering Institute by using the RMT-201 rock and concrete mechanical testing machine. The test was conducted with axial displacement control at the displacement rate of 0.005 mm/s.

3 EXPERIMENTAL RESULT ANALYSIS

Complete and single-fractured rock samples were subjected to the uniaxial compression test after 0, 20, 40, and 60 freeze–thaw cycles. The stress–strain curves are shown in Figure 2. For example, in A1-3-20, A1 refers to the rock sample type, 3 represents the number of test items in the group, and 20 indicates 20 freeze–thaw cycles experienced.

From Figure 2, we can summarize the following results:

1. After freeze–thaw cycles, the uniaxial compressive strength of rock samples experiencing 20 times and 40 times of freeze–thaw cycles declines at different degrees, compared with experiencing 0 times of freeze–thaw cycles in which the uniaxial compressive strength decreases slightly, and this decline is not obvious. The results clearly show that the rock samples after the first 40 freeze–thaw cycles experience little injury, but under 60 freeze–thaw cycles, the uniaxial compressive strength decreases more drastically. Thus, it is shown that the rock samples subjected to freezing and thawing have larger injury.

2. The failure mode changes into fragile failure from plastic failure gradually with the increase in the number of freeze–thaw cycles. For the same set of test pieces, the stress–strain curve of no freezing and thawing samples is steeper than that of successive freeze–thaw rock samples. At the same time, the former deformation is relatively smaller than the latter. The peak intensity is suddenly destroyed, which indicates brittle failure characteristics. However, the slope of the curve is gentle after freezing and thawing of the rock samples, and there is some plastic

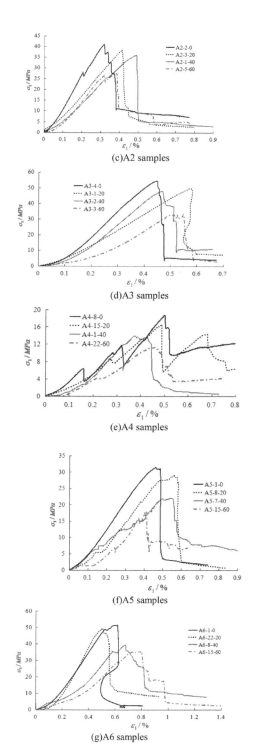

(c)A2 samples

(d)A3 samples

(e)A4 samples

(f)A5 samples

(g)A6 samples

Figure 2. Stress–strain curves of fractured rock specimens subjected to different freeze–thaw cycles.

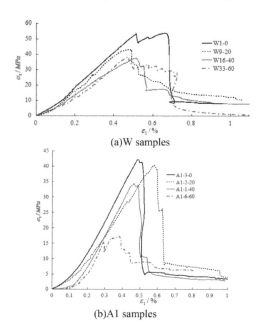

(a)W samples

(b)A1 samples

Figure 2. (*Continued*)

development after the peak intensity is reached, indicating plastic deformation.

3. Two groups with the same fracture length but different fracture inclinations (A1, A2, A3 and A4, A5, A6): under the same number of freeze–thaw cycle conditions, most of the strength of the rock samples is lowest at an inclination of 30°, followed by 60°, when the rock fracture inclination is 90°, the strength is highest, and almost the same as that for the complete fractured rock samples. During fracture inclination change from 30 to 60° and from 60° to 90°, it is shown that the uniaxial compressive strength also increases.

4. The specimens with the same number of freeze–thaw cycles, the same fracture inclination, and different fracture length: the uniaxial compressive strength is significantly reduced with the growth of the fracture length, and the degree of reduction becomes increasingly smaller with the increase in the fissure inclination. For example, for the A1 and A4 specimens with the inclination of 30°, experiencing 20 freeze–thaw cycles, the crack length has a significant effect on the compressive strength, which decreases by 51.2% between long fissure specimen A4 and the short crack specimens A1. However, for two different fracture lengths of the specimen with the same inclination of 90°, the crack length has a little effect on the uniaxial compressive strength, both of them different only by 1.2%.

5. The elastic modulus of the same type of rock samples has varying degrees of reduction with the increase of freeze–thaw cycles, as shown in the stress–strain curve where the slope of the elastic stage becomes increasingly smaller. It is shown that the freezing and thawing cycles cause greater damage to the elastic modulus. The decrease in the elastic modulus of the intact rock samples is smaller than other fractured rock samples. For example, the elastic modulus of the intact rock samples decreased by 39.6% after experiencing 60 freeze–thaw cycles, while the A4 rock samples decreased by 74.6%. This indicates that the presence of cracks has a significant impact on the elastic modulus of the rock mass after freezing and thawing.

4 CONCLUSION

This article investigated the physical and mechanical properties of seven kinds of single-fractured rock samples with different geometrical characteristics through the laboratory experiments. On the basis of the results obtained, the following conclusions can be drawn:

1. First 40 freeze–thaw cycles have a little effect on the uniaxial compressive strength, but the uniaxial compressive strength is significantly reduced after experiencing 60 freeze–thaw cycles.

2. Effect of the fracture geometric characteristics on the uniaxial compressive strength of freeze–thaw rock masses: crack length has a significant effect on the freeze–thaw rock masses and the strength is drastically reduced with the increasing crack length; the impact of fissure inclination on the strength of freeze–thaw rock masses is not very clear; and the uniaxial compressive strength shows a tendency to increase with the increasing inclination of the fractured rock masses under the same conditions.

ACKNOWLEDGMENT

This work was financially supported by the National Natural Science Foundation (51274157) and Hubei Natural Science Foundation (2014CFB575).

REFERENCES

Javier Martínez-Martínez, David Benavente, Miguel Gomez-Heras, et al. Non-linear decay of building stones during freeze–thaw weathering processes. Construction and Building Materials, 38 (2013): 443–454.

Li jielin, Experiment study on deterioration mechanism of rock under the conditions of freezing-thawing cycles in cold regions based on NMR technology [D]. Central South University PhD dissertation, 2012.6.

Li Xinping, Lu Yani, Research on damage model of single jointed rock masses under coupling action of freeze-thaw and loading [J]. Chinese Journal of Rock Mechanics and Engineering, 2013, 32(11): 2307–2315.

Lu Yani, Li Xinping, Wu Xinghong, Fracture coalescence mechanism of single flaw rock specimen due to freeze–thaw under triaxial compression [J]. Rock and Soil Mechanics, 2014, 35(6): 609–615.

Mokhfi Takarli, William Prince, Rafat Siddique, Damage in granite under heating/cooling cycles and water freeze–thaw condition. International Journal of Rock Mechanics & Mining Sciences, 45 (2008): 1164–1175.

Xu Guang-Miao, Liu Quan-Sheng. Overview of developments in three-dimensional fracture research [J]. Chinese Journal of Rock Mechanics and Engineering, 2005, 24(17): 3076–3082.

Yavuz, H., R. Altindag, S. Sarac, et al. Estimating the index properties of deteriorated carbonate rocks due to freeze–thaw and thermal shock weathering. International Journal of Rock Mechanics & Mining Sciences, 43 (2006): 767–775.

Zhang Jizhou, Miao Lin-chang, Yang Zhen-feng, Research on rock degradation and deterioration mechanisms and mechanical characteristics under cyclic freezing-thawing [J]. Chinese Journal of Rock Mechanics and Engineering, 2008, 27(8): 1688–1694.

Effect of irrigation frequency on crop water productivity and economic benefit of oasis jerusalem artichoke (*Helianthus Tuberosus. L*)

Z.Y. Bao, H.J. Zhang & S.X. Chai

Research Center of Engineering Technology on Water Saving Agriculture, Gansu Agricultural University, Lanzhou, China

ABSTRACT: An experiment was conducted to investigate the effect of irrigation frequency on crop productivity and economic benefit of oasis jerusalem artichoke in an arid environment. The results showed that low irrigation frequency significantly ($p < 0.05$) decreased Jerusalem artichoke tuber yield. However, tuber yield of jerusalem artichoke was significantly improved by 38.0%~82.1% in irrigated plots than those not irrigated. Accordingly, net income performance of jerusalem artichoke was similar to tuber yield under various irrigation frequency. In comparison with no irrigation plots, 51.0% to 108.9% of net income was significantly improved in irrigated crops. Besides, the maximum IWUE of jerusalem artichoke was marked in crops irrigated only once during the whole growing season while the minimum marked in those with four times of irrigation. Compared to latter, 31.4%~230.5% of IWUE was significantly increased in other irrigated plots. Therefore, irrigation frequency of twice for crops with the same water supply of 600 $m^3 \cdot ha^{-1}$ respectively at seedling and budding could be applied to effectively improve jerusalem artichoke tuber yield, economic benefit and IWUE in an arid environment.

1 INTRODUCTION

Irrigation frequency has important effect on crop yield formation and life history strategy, but there is seldom research on underground tuber-harvested economic crops (Zhang et al., 2011). Generally, in an arid environment whereas water supply is not always fully guaranteed, the seasonal total water consumption and irrigation water application are not those corresponding to the highest crop yield due to variable climatic conditions and water resource dynamics. Hence, the irrigation strategy should be determined in terms of the relations between crop yield and water consumption and aimed at the optimal economic benefit or water use efficiency (Qian & Y. Li, 2002). Virtually, water could be supplied with reduced frequent irrigation in an arid environment (Kang et al., 2000; Zhang et al., 2007). Theoretically, plant development and photosynthetic assimilation capability could be effectively renewed to a normal growth level through a timely re-watering immediately after light to medium drought (Wang et al., 2007; Liu et al., 2004), and remarkable water-saving, yield increasing and water use efficiency improving could be gained under proper levels of water deficit as well (Cai et al., 2000). At present, researches on jerusalem artichoke aimed at underground tuber harvesting are mainly focused on tuber quality, cultivating techniques involving fertilization, crop ecological adaptation and seawater irrigation (Dai et al., 2009; Shi & Ren, 2008; Wang et al., 2009). However, there is seldom research on effect of irrigation frequency on crop productivity and economic benefit of jerusalem artichoke especially in an arid environment.

In this study we assessed the hypothesis that crop productivity and economic benefit of jerusalem artichoke could be greatly influenced by irrigation frequency in an arid climate. Hence, the objectives of present study were to investigate: 1) Tuber yield and Irrigation Water Use Efficiency (IWUE) of jerusalem artichoke under various irrigation frequencies in an arid environment; and 2) Net income performance of jerusalem artichoke as affected by irrigation frequency in arid areas.

2 MATERIALS AND METHODS

2.1 *Study site description*

The experiment was conducted in March to October 2009 in Zhangye of Gansu Province, PR China (38°56′N latitude, 100°26′E longitude). The site climate was described as 5800 to 6400 MJ m^{-2} total solar radiation, 2932 to 3085 h sunny time, and 128 mm mean precipitation while 2048 mm mean evaporation annually. The soil was loamy with 1.5 $g \cdot cm^{-3}$ in bulk density, 2.7 in specific gravity, 8.4 in pH, and 22.5% in field water capacity within

0 to 100 cm depth. Soil fertility was characterized as 1.37% of organic matter, and 13.4 mg kg⁻¹, 61.8 mg kg⁻¹, 190.4 mg kg⁻¹ in available phosphorus, nitrogen, and potassium within 0 to 20 cm soil profile, respectively.

2.2 Experimental design

The field was divided into plots of 3.5 m × 10.0 m in size, which were arranged in a randomized block design with three replications. Six irrigation treatments and a non-irrigation Check (CK) were established under border irrigation, through which plants were subjected to different irrigation frequency at certain growth stages of seedling, lush foliage, budding, and flowering (Table 1). The irrigation amounts were strictly and precisely controlled using water meters.

2.3 Crop management

Jerusalem artichoke *(Helianthus tuberosus. L)* cultivar "Qingyu 2" was sown on March 28, 2009 and harvested at maturity on October 28, 2009. Spaces were both 45 cm between and within rows. The effective rainfall was 110.7 mm during jerusalem artichoke growing season. Weeds, disease and insects were effectively controlled during the main crop growing stages such as seedling, lush foliage, budding and flowering.

2.4 Measurements and calculations

Soil moisture contents in 0–10 cm, 10–20 cm, 20–40 cm, 40–60 cm, 60–80 cm and 80–100 cm depths were measured gravimetrically every 7 days during crop growing season. At harvest, plants in each plot were harvested for tuber yield of jerusalem artichoke. Crop IWUE was also calculated as tuber yield divided by seasonal total irrigation water applied.

2.5 Statistical analysis

The experimental data were means of three replications and analyzed by SPSS18.0.

3 RESULTS AND DISCUSSION

3.1 Tuber yield

Low irrigation frequency significantly ($p < 0.05$) decreased jerusalem artichoke tuber yield (Table 2). However, the tuber yield of jerusalem artichoke was significantly improved by 38.0%, 40.1%, 45.8%, 65.0%, 82.1% and 66.9% in J1, J2, J3, J4, J5 and J6 than CK, but no significant difference ($p > 0.05$) was found among J1, J2 and J3 as well as between J4 and J6. Compared with J4 and J6, tuber yield in J5 was significantly improved by 10.4% and 9.1%, while that in J1, J2, J3 was significantly reduced by 16.3%, 15.1%, 11.6% and 17.3%, 16.1%, 12.7%, respectively.

3.2 Total irrigation water applied

In comparison with CK, the seasonal total irrigation water was significantly ($p < 0.05$) increased with 600 to 2400 m³ ha⁻¹ in J1, J2, J3, J4, J5 and J6 (Table 2). The maximum irrigation was marked in J6 and the minimum was maintained in J1. Compared to J1 irrigation water was significantly increased with 100%, 100%, 200%, 200% and 300% respectively in J2, J3, J4, J5 and J6. The total irrigation water was significantly decreased with 33.3%, 33.3% in J2 and J3 compared with both J4 and J5, but was significantly increased with 33.3% in J6. However, no significant difference was found ($p > 0.05$) in total irrigation between J2 and J3 as well as between J4 and J5.

3.3 Economic benefits

Net income performance of jerusalem artichoke was similar to tuber yield under various irrigation

Table 1. Irrigation frequency design of jerusalem artichok*.

| Treatment | Irrigation frequency | Irrigation amount in each stage | | | |
		Seedling (m³·ha⁻¹)	Lush foliage (m³·ha⁻¹)	Budding (m³·ha⁻¹)	Flowering (m³·ha⁻¹)
J1	1	600	–	–	–
J2	2	600	600	–	–
J3	2	600	–	600	–
J4	3	600	300	900	–
J5	3	600	900	300	–
J6	4	600	600	300	900
CK	0	–	–	–	–

*Pre-sowing irrigation was not included.

Table 2. Economic benefit of jerusalem artichoke in different treatments**.

Treatments	Tuber yield (t·ha⁻¹)	Irrigation amount (m³·ha⁻¹)	Net income (×10³ RMB·ha⁻¹)	Irrigation water use efficiency (t·m⁻³)
J1	20.83c	600d	8.32c	0.0347a
J2	21.14c	1200c	8.41c	0.0176b
J3	22.00c	1200c	8.84c	0.0183b
J4	24.90b	1800b	10.22b	0.0138d
J5	27.48a	1800b	11.51a	0.0153c
J6	25.19b	2400a	10.30b	0.0105e
CK	15.09d	0e	5.51d	0f

**Letters indicate statistical significance at p < 0.05 level within the same column.

frequency (Table 2). In comparison with CK, net income was significantly (p < 0.05) improved with 51.0%, 52.6%, 60.4%, 85.5%, 108.9% and 86.9% in J1, J2, J3, J4, J5 and J6, respectively. Compared to J4 net income was significantly reduced with 18.6%, 17.7% and 13.5% in J1, J2 and J3 while was significantly improved with 12.6% in J5. However, significant difference didn't occur (p > 0.05) in net income between J4 and J6 as well as among J1, J2 and J3.

3.4 Irrigation water use efficiency

The IWUE of jerusalem artichoke was significantly (p < 0.05) improved by 0.0105 to 0.0347 t·m⁻³ in irrigated plots than those no irrigation, with the maximum marked in J1 plots and minimum in J6 (Table 2). Compared to J6, 230.5%, 67.6%, 74.3%, 31.4% and 45.7% of IWUE was significantly increased in J1, J2, J3, J4 and J5 plots, respectively. However, no significant difference (p > 0.05) was found in IWUE of jerusalem artichoke between J2 and J3.

4 CONCLUSIONS

Low irrigation frequency significantly decreased jerusalem artichoke tuber yield. However, tuber yield of jerusalem artichoke was significantly improved by 38.0%~82.1% in irrigated treatments than those not irrigated, but no significant difference was found among J1, J2 and J3 as well as between J4 and J6. Accordingly, net income performance of jerusalem artichoke was similar to tuber yield under various irrigation frequency. Compared to no irrigation plots, 51.0% to 108.9% of net income was significantly improved in irrigated ones. In addition, the maximum IWUE of jerusalem artichoke was marked in crops irrigated only once during the whole growing season (J1) while the minimum marked in those with four

times of irrigation (J6). In comparison with the latter, 31.4%~230.5% of IWUE was significantly increased in other irrigated plots. Therefore, twice irrigation for crops with the same water supply of 600 m³·ha⁻¹ respectively at seedling and budding could be applied to effectively improve tuber yield, economic benefit and IWUE of jerusalem artichoke in an arid environment.

ACKNOWLEDGEMENTS

This work was mutually supported by Project of Gansu Finance Bureau of China (2012) and Special Scientific Research Fund of Agricultural Public Welfare Profession of China (Grant No. 201303104).

REFERENCES

Cai H., Kang S., Zhang Z., Chai H., X. Hu & Wang J. 2000. Proper growth stages and deficit degree of crop regulated deficit irrigation. *Transactions of the CSAE* 16(3): 24–27.

Dai X., Kang J. & Xu C. 2009. Photosynthetic rate determination of Helianthus tubeuosus under different fertilization. *Sugar Crops of China* 1: 41–42.

Kang S., Shi W. & Zhang J. 2000. An improved water-use efficiency for maize grown under regulated deficit irrigation. *Field Crops Research* 67(3): 207–214.

Liu G., Guo A., Ren S., An S. & Zhao H. 2004. Compensatory effects of rewatering on summer maize threatened by water stress at seedling period. *Chinese Journal of Ecology* 23(3): 24–29.

Qian B. & Li Y. 2002. *Advanced technology research on water-saving agriculture*. Zhengzhou: The Yellow River water conservancy press.

Shi J. & Ren S. 2008. Ecological adaption and its cultivating technology of jerusalem artichoke. *Anhui Agriculture* 8: 33–34.

Wang C., Akihiroz I. & Li M. 2007. Growth and eco-physiological performance of cotton under water stress conditions. *Agricultural Sciences in China* 6(8): 949–955.

Wang J., Liu Z., Long X. & Zhao G. 2009. Effects of growth, photosynthesis and water consumption of Helianthus Tuberosus irrigated with seawater. *Chinese Journal of Soil Science* 40(3): 606–609.

Zhang B., Li F. & Qi G. 2007. Effects of regulated deficit irrigation on grain yield of spring wheat in an arid environment. *Chinese Journal of Eco-Agriculture* 15(1): 58–62.

Zhang H.J., Huang G.B. & Yang B. 2011. Effect of irrigation frequency on life history strategy and yield formation in Jerusalem artichoke (Helianthus tuberosus. L) in oasis of Hexi Corridor. *Acta Ecologica Sinica* 31(9): 2401–2406.

Advances in Energy, Environment and Materials Science – Wang & Zhao (Eds)
© *2016 Taylor & Francis Group, London, ISBN 978-1-138-02931-6*

Numerical study of Zhuanghe coastal water using a two-dimensional finite volume method

Y.Y. Xu, M.L. Zhang & Y. Qiao
School of Ocean Science and Environment, Dalian Ocean University, Dalian, Liaoning, China

ABSTRACT: Based on the finite volume method and unstructured triangular mesh, a depth-averaged 2D explicit scheme mathematical model for dam break and coastal waters is established. This model applies the Roe solver approximate Riemann solution with second-order accuracy to compute the water momentum flux on the grid interface, which is able to calculate the dry-wet moving fronts accurately. The model is firstly verified against two laboratory cases. The calculation result is consistent with the experimental data, which shows that the depth-averaged 2D mathematical model can accurately simulate the evolvement of the dam-break flows. Then, this model is applied to calculate the tide level, flow current and flow direction in Zhuanghe coastal water. By comparing the simulated and measured values, the results show that there has strong agreement between them.

1 INTRODUCTION

With the development of the society, the two-dimensional shallow water equations have been more and more applied to simulate flows in shallow lakes, dam-break, estuaries and coastal zones. A number of numerical methods have been developed to solve these equations. Yoon et al. (2004) applied a second-order upwind finite volume method to solve shallow water equations. The HLL approximates Rieman solver and a limiting technique are used to compute the inviscid flux functions and the slope source term, respectively. Liang et al. (2009) based on a finite volume Godunov-type method presents a well-balanced numerical model for simulating shallow water flows over a wet or dry bed with complex domains.

Along with the fluctuation of the tidal, the water flows has shown up a strong discontinuity in dry-wet boundary on the intertidal zone. So far, there is little research on applied shallow water model to handle those dry-wet boundary questions. The present works concentrate on developing a depth-averaged 2D mathematical model, which can simulate strong discontinuous flow over complex topography, this paper firstly starts with a brief description of modeling method for the hydrodynamic model, and then the examples of applications and discussions are given in next sections.

2 NUMERICAL MODEL

The Shallow Water Equations (SWEs) written in conservation and vector form are (Wang, 2005)

$$\frac{\partial \mathbf{U}}{\partial t} + \frac{\partial \mathbf{F}}{\partial x} + \frac{\partial \mathbf{G}}{\partial y} = \frac{\partial \mathbf{F}_d}{\partial x} + \frac{\partial \mathbf{G}_d}{\partial y} + \mathbf{S} \qquad (1)$$

In which, t is the time, x, y are the horizontal coordinates, U is the vector of conserved variables, F, G, F_d, and G_d are convection fluxes and diffusion fluxes in the x and y directions, respectively, S is source terms which can be defined respectively as follows:

$$\mathbf{U} = \begin{bmatrix} h \\ uh \\ vh \end{bmatrix} \quad \mathbf{F} = \begin{bmatrix} hu \\ hu^2 \\ huv \end{bmatrix} \quad \mathbf{G} = \begin{bmatrix} hv \\ hvu \\ hv^2 \end{bmatrix}$$

$$\mathbf{F}_d = \begin{bmatrix} 0 \\ v_t \dfrac{\partial uh}{\partial x} \\ v_t \dfrac{\partial vh}{\partial x} \end{bmatrix} \mathbf{G}_d = \begin{bmatrix} 0 \\ v_t \dfrac{\partial uh}{\partial y} \\ v_t \dfrac{\partial vh}{\partial y} \end{bmatrix} \mathbf{S} = \begin{bmatrix} 0 \\ -gh\dfrac{\partial \eta}{\partial x} - \tau_{bx} \\ -gh\dfrac{\partial \eta}{\partial y}\, \tau_{by} \end{bmatrix}$$

$$(2)$$

where h is the flow depth, u and v are the depth-averaged velocity in the x and y directions, respectively, η is the water level, τ_{bx} and τ_{by} are the friction slopes in the x and y directions, respectively.

In this study, a cell-centered finite volume method is adopted to sole the SWEs. The computational area is divided into a set of triangular meshes. Integrating Eq. (1) over the area of each control volume, by application of the Green's theorem, a line integral equation can be obtained. After discretizing the line integral equations, the

Roe approximate Riemann solver has been used to calculate the interface fluxes.

3 MODEL CALIBRATION AND APPLICATION

3.1 *Dam break on dry and wet beds*

In this test case a horizontal and frictionless channel of 1200 m long and 1 m wide is considered, this channel is separated into two sections by a dam located at $x = 500$ m (Zhang, 2011). This test case simulated the discontinuous flow and the front wave propagation under wet and dry bed situations. The initial upstream water depth is 10 m and the downstream water depth is 5 m for wet bed and 0 m for dry bed. The computational domain is divided into 9,992 triangular grids, time step is set at 0.001 s. At the initial time, the dam collapsed instantaneously. Figure 1 (a and b) compares the water depth and the velocity for the wet bed case, and Figure 2 (a and b) compares the water depth and the velocity for the dry bed case. The calculated results of water surface profiles and flow velocities follow the analytical values very well, the model has reasonable accuracy for dam-break flows over frictional beds.

3.2 *Partial dam-break test case*

The aim of this test case is to verify the calculation precision and the ability of the mathematical model to capture wet-dry boundary by simulating the flood wave propagation due to a

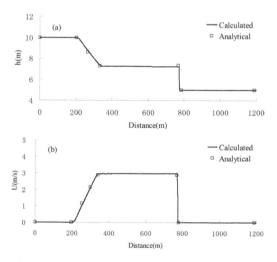

Figure 1. Dam break on wet bed (a): water depth (b): velocity.

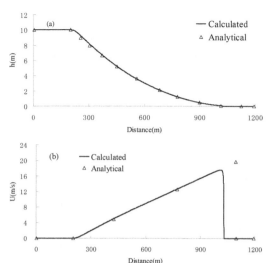

Figure 2. Dam break on dry bed (a): water depth (b): velocity.

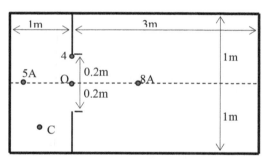

Figure 3. Geometry size and arrangement of measuring point.

partial dam-break. The experiment channel is 4 m long and 2 m wide, consists of a reservoir with water up to 0.6 m contained by a dam at $x = 1.0$ m and a dry bed downstream with three open boundaries (see Fig. 3) (Tseng, 2000). The bottom of the reservoir and floodplain is horizontal. The breach is 0.4 m wide and located at the middle of the dam. The locations of five stations for measuring stage are shown in Figure 3, their coordinates are 5A (0.18 m, 1.0 m), O (1.0 m, 1.0 m), 8 A (1.722 m, 1.0 m), 4 (1.0 m, 1.16 m), and C (0.48 m, 0.4 m).

In this simulation, the computational domain is discretized into 5,018 triangular meshes, the roughness coefficient is set to 0.01 and $\Delta t = 0.001$ s. Figure 4 shows good agreement between the calculated and measured water levels at distinct stations. Inside the reservoir, because of a strong depression

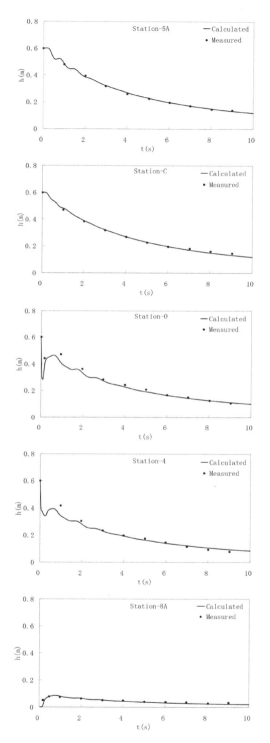

wave that occurs, the water surface continuously descends (see station 5A, C). Due to the effects of boundary reflection, water surface in the reservoir oscillates significantly in the initial stage (see station 5A). At the location of dam break (see station O and 4), the water level falls and rises significantly at the beginning; the numerical model can well reproduce all these details. In general, the mathematic model is capable of reflecting the physical properties of a dam-break flow, and has an advantage in capturing wet-dry boundary.

3.3 *Tidal flow in Zhuanghe coastal water*

The present model is applied to simulate the current situation in Zhuanghe coastal water in the northern Yellow Sea. The computational area is from 122.06° to 123.88° in east longitude and from 122.06° to 123.88° in north latitude, including Changshan channel, Changshan and Shicheng islands (see Fig. 5). This coastal water area having a large number of marine cultivation regions is the important fishery base in the north Yellow sea, so it is important to master the current trend for fishery breeding. In this case, Zhuanghe coastal water area is discretized into 15,746 grid cells, because the domain is complex, a fine grid is used near the coastal line, island area and a coarse grid is used in open sea. At the seaward boundary, the water elevation is specified, with four tidal components being taken into account (M2, S2, K1 and O1). The water elevations are given by the Tidal Model Driver (TMD). The tidal process covers 288 hours from 20 August, 2011 to 1 September, 2011. The Manning's n is set as 0.02. The computational time step is 0.5 s.

Figure 6 shows the comparison of the measured and simulated water levels at Xiao Changshan

Figure 4. Computed and measured water surface levels at five stations.

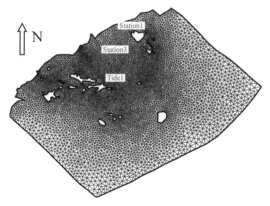

Figure 5. Mesh and measurement stations at computational area.

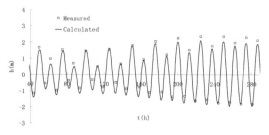

Figure 6. Measured and calculated water level at Xiao Changshan tide station.

tide station. There is strong agreement between the observed and predicted water-surface elevations. The low water levels during neap tide have receivable deviation between the observed and predicted value. The possible reason is that the seaward boundary's water elevations, which were given by the TMD with less precision. Figure 7 and Figure 8 show the comparison of the measured and simulated flow velocities and current direction at two observation stations during neap tide and spring tide, respectively. The (a) and (c) column compares the computed and measured flow

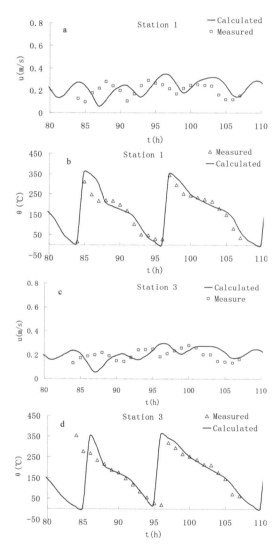

Figure 7. Measured and calculated current velocities and direction during neap tide.

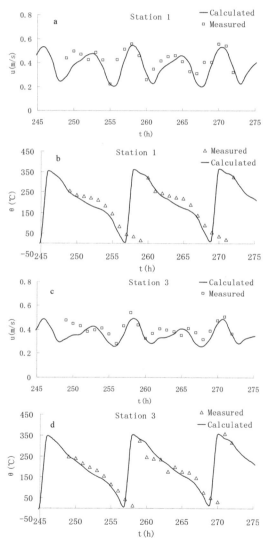

Figure 8. Measured and calculated current velocities and direction during spring tide.

velocities at station 1 and 3, and the (b) and (d) column compares the computed and measured current direction at station 1 and 3. Both the flow velocities and current directions are predicted well during spring tide. The simulated current velocities during neap tide have a slight deviation compared with the measured data (see Fig. 7). The reason for this result: on the one hand, may be associated with the inaccuracy on low tide level calculation, on the other hand, may be the current velocities have some measuring error. In total, one can see that this model can be applied to simulate the flow phenomena of coastal water, and it has practical application value.

4 CONCLUSION

Based on two-dimensional shallow water equations, a depth-averaged 2D mathematical model over fixed beds for dam-break and tide flow has been established by adopting the finite volume method and unstructured triangular mesh. The model's ability to capture the dry-wet boundary has been tested by simulating an idealized and partial dam-break case. And then, the model is applied to calculate the tidal flow in Zhuanghe coastal water to verify its practicability. The advantages of the present model include:

1. The finite volume method is adopted to maintain the conservation of mass strictly in the hydrodynamic model.
2. This model applies to the unstructured triangular grid, has good adaptability to complex boundary.

3. The linear least square reconstruction method has been carried out to simulate the variables on the interface of left and right sides, and a Roe solver approximate Riemann solution is employed to evaluate the water momentum flux on the grid interface. These ensure that the model has second order accuracy.

ACKNOWLEDGMENTS

This work was supported by the Public Science and Technology Research Funds Projects of Ocean (No. 201205023), the Program for Liaoning Excellent Talents in University (LJQ2013077), the Science and Technology foundation of Dalian City (2013J21DW009), the Liaoning Natural Science Foundation (2014020148).

REFERENCES

Liang, Q.T. and Marche, F. 2009. Numerical resolution of well-balanced shallow water equations with complex source terms. *Advances in Water Resources* 32: 873–884.

Tseng, M.H. and Chu, C.R. 2000. Two-dimensional shallow water flows simulation using TVD-MacCormac scheme. *Hydraulic Research* 38: 123–131.

Wang, J.W. and Liu, R.X. 2005. Combined finite volume–finite element method for shallow water equations. *Computers & Fluids* 34: 1199–1222.

Yoon, T.H. and Kang, S.k. 2004. Finite volume model for two-dimensional shallow water flows on unstructured grids. *Journal of hydraulic engineering* 130: 678–688.

Zhang, M.L. and Wu, W.M. 2011. A two dimensional hydrodynamic and sediment transport model for dam break based on finite volume method with quadtree grid. *Applied Ocean Research* 33: 297–308.

Advances in Energy, Environment and Materials Science – Wang & Zhao (Eds)
© 2016 Taylor & Francis Group, London, ISBN 978-1-138-02931-6

Spatial analysis of drought in Haihe River Basin from 1961 to 2010 based on PDECI and SPI

M.J. Yang, Z.Y. Yang, Y.D. Yu, G.Q. Dong & Z.L. Yu
State Key Laboratory of Simulation and Regulation of the River Basin Water Cycle, China Institute of Water Resources and Hydropower Research (IWHR), Beijing, P.R. China
Department of Water Resources, China Institute of Water Resources and Hydropower Research, Beijing, P.R. China

ABSTRACT: Under the background of climate change, the annual precipitation of 47 meteorological stations in Haihe River Basin (HRB) and the surrounding area were analyzed using Precipitation Percentile Index (PDECI) and Standardized Precipitation Index (SPI) to obtain the spatial distribution of different drought levels with the support of GIS. The results show that: (1) from 1961 to 2010, the HRB suffers from different levels of drought. Regions where drought happens most frequently are the northeast and central Hebei province with a PDECI value of 40% and a SPI value of 80%, respectively. (2) The drought concentration area varies with the drought level, from light drought concentrating in northeast HRB to extreme drought concentrating in south HRB, and the drought rate from light to extreme has a decreasing trend. (3) After the comparison of PDECI and SPI on the basis of real drought events, the PDECI shows a better ability in demonstrating the occurrence probability of drought.

1 INTRODUCTION

Under the background of global climate change and the irrational exploitation of water resources, the feature of drought in Haihe River Basin (HRB) is experiencing a new changing trend with four main characters: increasing frequency, expanding drought area, augmenting affected field, and rising loses making the society and economy suffering stronger blows than the normal evolution of regional climate (Weng, et al., 2010). Therefore, to figure out the distribution of different regional drought degrees is becoming increasingly critical to improve the socioeconomic development and to enhance the evolution of ecological environment. In recent decades, many hydrology researchers in China have launched lots of broad and detailed researches to depict and analyze the spatial and temporal variation in HRB using a variety of methods and indexes, and achieved a series of progresses (Yan, et al., 2013, Yang, et al., 2013, Li, et al., 2012). Some profound researches have been done to find out the mechanism of drought in HRB, which indicates that the formation of drought is affected by more than one factor (Zhang, et al., 2011).

This article is based on the daily precipitation data for 50 years and the support of GIS, and the distribution and occurrence rate of different drought degrees was demonstrated and dissected with the methods of Precipitation Percentile Index (PDECI) and Standardized Precipitation Index (SPI) in HRB (Yuan, et al., 2004, Lin, et al., 2012, Zong, et al., 2013, Zhao, et al., 2013). In addition, a comparison was made between the results of the two methods to figure out, which is a better method in representing the drought in HRB.

2 OVERVIEW OF STUDY AREA

Haihe River Basin, between 112°E~120°E and 35°N~43°N, located in the west of Bohai sea, crosses eight provinces including the entire Beijing and Tianjin, the most part of Hebei province, small part of Shanxi province, Henan province, Liaoning province, Shandong province and the Inner Mongolia Autonomous Region. The areas covers a total area of 317,800 km², in which mountain area occupies 189,000 km², 59.5 percent of the total area, and the rest 129,000 km² is the flat area, account for 40.5 percent of the total area. The terrain consists of three types, including plateau, mountain and plain, with elevation rising from southeast to northwest. The western part of HRB is Shanxi plateau and Taihangshan Mountains; the Mongolia Plateau and Yanshan Mountains located in the northern part; the eastern and southeastern part of HRB is the North China Plain (Fig. 1).

The HRB is mainly composed of three rivers: Haihe River, Luanhe River, and Tuhaimajiahe River, the major climate of HRB is the warm and semi-humid continental monsoon climate. The main character of which is that the winter is cold

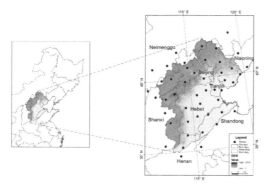

Figure 1. The geological location of the Haihe River Basin (HRB) and DEM (Digital Elevation Model).

and dry while the summer is hot and humid. The annual mean precipitation from 1961 to 2010 is 521 mm, facing a decreasing trend, concentrating in the rainy season (summer), which makes summer flood and inundation frequent. As the social, economic and political center as well as an important granary of China, Haihe River Basin is the core area with high priority of safety insurance. However, the severe and continuous drought makes HRB one of the most calamitous and vulnerable areas in China.

3 DATA AND METHODS

3.1 Data sources

The daily precipitation data from 1961 to 2010 in 47 meteorological stations in and around HRB is obtained from *China Meteorological Data Sharing Service System* and the basic station information are listed in the Table 1 with Annual Mean Precipitation (AMP).

3.2 Data processing

3.2.1 PDECI
The Precipitation Percentile Index is one of the drought indexes that could depict whether the precipitation in a short period is more or less than the average precipitation in a long term. The PDECI is a kind of index that could directly reflect the drought caused by unusual precipitation, which fits for the humid and semi-humid region with annual mean temperature higher than 10°C.

The formula in computing Precipitation Percentile Index (P_a) is as follows:

$$P_a = \frac{P - \bar{P}}{\bar{P}} \times 100\%$$

Table 1. The selected 47 stations in HRB and the AMP (mm).

Code	Station	Long/°	Lat/°	AMP
53898	Anyang	114.400	36.050	564.076
54518	Bazhou	116.383	39.117	506.728
54602	Baoding	115.517	38.850	519.758
54511	Beijing	116.467	39.800	549.606
54423	Chengde	117.950	40.983	520.822
53487	Datong	113.333	40.100	371.904
54714	Dezhou	116.317	37.433	518.439
54736	Dongying	118.667	37.433	578.784
54208	Duolun	116.467	42.183	376.032
54308	Fengning	116.633	41.217	458.316
53564	Hequ	111.150	39.383	462.074
54906	Heze	115.433	35.250	476.141
53463	Huhehaote	111.683	40.817	401.492
53391	Huade	114.000	41.900	314.228
54405	Huailai	115.500	40.400	378.350
54624	Huangye	117.350	38.367	588.688
54725	Huimin	117.533	37.483	561.144
54823	Jinan	116.983	36.600	702.394
54326	Jianpingxian	119.700	41.383	461.522
54539	Leting	118.883	39.433	603.262
54705	Nangong	115.383	37.367	476.334
54449	Qinhuangdao	119.517	39.850	637.688
54436	Qinglong	118.950	40.400	694.734
54606	Raoyang	115.733	38.233	518.172
54808	Shenxian	115.667	36.233	539.922
53698	Shijiazhuang	114.417	38.033	526.232
54454	Suizhong	120.350	40.350	619.368
53772	Taiyuan	112.550	37.783	441.930
54534	Tangshan	118.150	39.667	608.976
54527	Tianjin	117.067	39.083	536.526
54623	Tianjintanggu	117.717	39.050	578.644
54311	Weichang	117.750	41.933	435.102
53593	Weixian	114.567	39.833	400.898
54213	Wengniuteqi	119.017	42.933	349.092
53480	Wulanchabu	113.067	41.033	358.930
53588	Wutaishan	113.517	38.950	747.306
53663	Wuzhai	111.817	38.917	421.052
53986	Xinxiang	113.883	35.317	574.022
53798	Xingtai	114.500	37.067	516.286
53975	Yangcheng	112.400	35.483	597.804
53478	Youyu	112.450	40.000	414.862
53787	Yushe	112.983	37.067	544.278
53673	Yuanping	112.717	38.733	588.688
53399	Zhangbei	114.700	41.150	385.786
54401	Zhangjiakou	114.883	40.783	399.666
57083	Zhengzhou	113.650	34.717	578.554
54429	Zunhua	117.950	40.200	711.414

where P is the precipitation in a short period, mm; $\bar{P}$ is the average precipitation in a long term, mm.
Where:

$$\bar{P} = \frac{1}{n} \sum_{i=1}^{n} P_i$$

where n is the number of years; $i = 1, 2, \ldots, n$.

3.2.2 SPI
The Standardized Precipitation Index is a drought index based only on precipitation that can be used to monitor conditions on a variety of time scales. This temporal flexibility allows the SPI to be useful in both short-term agricultural and long-term hydrological applications.

Computing of the SPI involves fitting a gamma probability density function to a given frequency distribution of precipitation totals for a climate station. The gamma distribution is defined by its frequency or probability density function:

$$g(x) = \frac{1}{\beta^{\alpha}\Gamma(\alpha)} x^{\alpha-1} e^{-x/\beta}$$

where α and β are shaped and scale parameters separately. The maximum likelihood solutions are used to optimally estimate α and β:

$$\alpha = \frac{1}{4A}\left(1 + \sqrt{1 + \frac{4A}{3}}\right)$$

$$\beta = \frac{\bar{x}}{\alpha}$$

where:

$$A = \ln(\bar{x}) - \frac{\sum \ln(x)}{n}$$

where x is the observed precipitation; $\bar{x}$ is the mean precipitation; n is the number of precipitation observations.

The cumulative probability is given by:

$$G(x) = \int_0^x g(x)dx = \frac{1}{\beta^{\alpha}\Gamma(\alpha)} \int_0^x x^{\alpha-1} e^{-x/\beta} dx$$

Because the precipitation might be zero, which is undefined in the gamma function, the cumulative probability becomes:

$$H(x) = q + (1-q)G(x)$$

where q is the probability of a zero. The cumulative probability $H(x)$ is then transformed to the standard normal random variable Z with mean zero and variance of one, which is the value of the SPI.

$$Z = SPI = -\left(t - \frac{c_0 + c_1 t + c_2 t^2}{1 + d_1 t + d_2 t^2 + d_3 t^3}\right)$$
$$\text{for } 0 < H(x) \leqslant 0.5$$

$$Z = SPI = +\left(t - \frac{c_0 + c_1 t + c_2 t^2}{1 + d_1 t + d_2 t^2 + d_3 t^3}\right)$$
$$\text{for } 0.5 < H(x) \leqslant 1.0$$

where:

$$t = \sqrt{\ln\left(\frac{1}{(H(x))^2}\right)} \text{ for } 0 < H(x) \leqslant 0.5$$

Table 2. The annual-scale standard of PDECI and SPI in defining drought levels.

Level	Type	PDECI	SPI
1	No Drought	-15<Pa	-0.5<SPI
2	Light Drought	-30<Pa≤-15	-1.0<SPI≤-0.5
3	Moderate Drought	-40<Pa≤-30	-1.5<SPI≤-1.0
4	Severe Drought	-45<Pa≤-40	-2.0<SPI≤-1.5
5	Extreme Drought	Pa≤-45	SPI≤-2.0

$$t = \sqrt{\ln\left(\frac{1}{(1.0 - H(x))^2}\right)} \quad \text{for } 0.5 < H(x) \leqslant 1.0$$

$c_0 = 2.515517$; $c_1 = 0.802853$; $c_2 = 0.010328$;
$d_1 = 1.432788$; $d_2 = 0.189269$; $d_3 = 0.001308$.

3.2.3 The annual-scale standard

The annual-scale drought level was defined as five types, each standard of whose PDECI and SPI are given in Table 2.

4 RESULTS AND DISCUSSION

4.1 Spatial distribution of AMP

Forty-seven meteorological stations in and around the HRB were chosen and interpolated using IDW in GIS to get the distribution of AMP from 1961 to 2010 statistically (Fig. 2). The AMP decreases from southeast to northwest dramatically from over 700 mm to less than 400 mm. The AMP in northeast Hebei, east Shanxi and west Shandong are relatively high and concentrated while the northwest Hebei, north Shanxi, and a small part of Neimenggu are places where AMP is pretty low. The reason why the distribution of AMP in HRB is so inhomogeneous may mainly be because the mountain that was located in northwest HRB prevented the circumstance of water.

4.2 The spatial distribution characters of different drought levels based on PDECI and SPI

According to the drought levels defined in Table 2, the rate of drought happening in each level in every station during the last 50 years was analyzed statistically and interpolated in HRB using IDW method. Figure 3–7 show the rate distribution of different drought levels calculated in two methods.

Figure 3(a, b) shows a relatively good consistence of the drought happening rate calculated by two methods. The highest drought probability area both are in the middle stripe of HRB from northeast to southwest which was concentrated in northeast and middle west Hebei province. From the comparison of the two methods, the drought

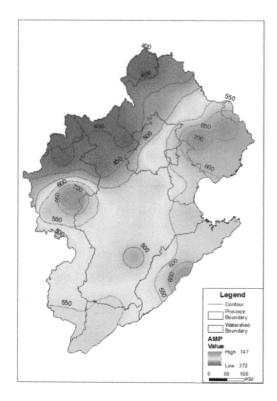

Figure 2.   The distribution of AMP.

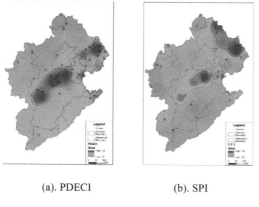

(a). PDECI   (b). SPI

Figure 3.   The rate distribution of drought.

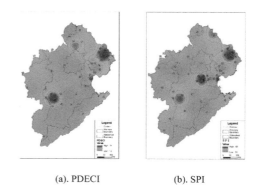

(a). PDECI   (b). SPI

Figure 4.   The rate distribution of light drought.

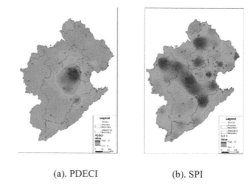

(a). PDECI   (b). SPI

Figure 5.   The rate distribution of moderate drought.

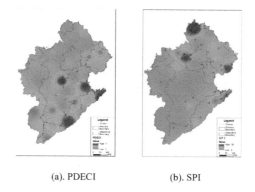

(a). PDECI   (b). SPI

Figure 6.   The rate distribution of severe drought.

happening probability of PDECI is 20%–40%, while the SPI value is 52%–84%, approximately 2 times of PDECI value.

Figure 4(a, b) demonstrates a relatively good consistence of the light drought happening rate calculated by two methods with both highest drought probability areas concentrate in the northeast and southwest of HRB except a high light drought rate in the central Hebei province in SPI. From the comparison of the two methods, the

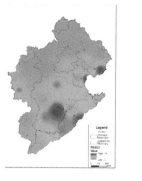 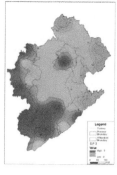

(a). PDECI   (b). SPI

Figure 7. The rate distribution of extreme drought.

light drought happening probability of PDECI is 12%–34%, while the SPI value is higher, from 16% to 60%.

The moderate drought rate illustrated in Figure 5(a, b) varies significantly in two methods. The highest moderate drought probability area in PDECI only concentrates in the center of Hebei province, however, almost the whole HRB suffers a high moderate drought rate on the basis of SPI value. The moderate drought happening rate of PDECI varies from 2% to 22%, while the SPI value is 4%–32%.

The differences between the results of the two methods are even clearly portrayed in Figure 6(a, b). The highest severe drought probability areas calculated by PDECI are in the southeast and center of HRB. When it comes to the SPI, the most probable severe drought areas are the north and east HRB. The drought happening probability of PDECI is 0–7%, however, the SPI value variance is much stronger, from 0 to 24%.

Figure 7(a, b) also depicts different results by two methods. The extreme drought happens most frequently in south and east Hebei province as well as west Shandong province in PDECI. But regions that suffer the most extreme droughts are the whole southwest HRB and the central part of Beijing based on SPI. The drought happening probability of PDECI is 0–10%, similarly, the SPI value is 0–9%.

5 CONCLUSION

This paper investigates the spatial drought variation patterns in HRB based on the PDECI and SPI. The following important conclusions can be drawn from this study:

1. From 1961 to 2010, all area of HRB suffered from different levels of drought, especially in the northeast and middle of HRB where the drought happens most frequently.
2. After the analysis of the spatial distribution of four kinds of drought in HRB, it can be concluded that the areas where the largest drought probability occurs vary with the drought levels. As an association of the results of PDECI and SPI, the light rate happens mostly in northeast and southwest HRB; the moderate drought appears most frequently in the center of HRB; the severe drought occurs in the east part of HRB most likely and the extreme drought happens most often in the south HRB.
3. The drought occurrence rate is much higher as per calculated by SPI than by PDECI except for extreme drought, and the drought concentration areas have some distinctions as well. According to the *Flood and Drought Disasters in Haihe River Basin*, the drought happening rate is approximately 45% in which the light and moderate drought account for 31% and the severe and extreme drought take up 14%, which indicates that the PDECI has a better correspondence with the real drought events (Haihe River Water Conservancy Commission, MWR, et al., 2009).

ACKNOWLEDGMENTS

This study is financially supported by the Project of the Ministry Of Water Resources Public Welfare Industry Special Scientific Research (201401001).

REFERENCES

Haihe River Water Conservancy Commission, MWR., Feng Y., Yao Q.L. et al. 2009. Flood and drought disasters in Haihe river basin. *Tianjin science and technology press.*

Li L.X., Yan D.H., Qin T.L. et al. 2012. Drought variation in Haihe river basin from 1961 to 2010. *Journal of Arid Land Resources and Environment* 26(11): 61–67.

Lin S.J., Xu Y.P., Tian Y. et al. 2012. Spatial and temporal analysis of drought in Qiantang river basin based on Z index and SPI. *Journal of hydroelectric engineering* 31(2): 20–26.

Weng B.S. & Yan D.H. 2010. Integrated strategies for dealing with droughts in changing environment in China. *Resources Science* 32(2): 309–316.

Yan D.H., Yuan Z., Yang Z.Y. et al. 2013. Spatial and temporal changes in drought since 1961 in Haihe River Basin. *Advances in Water Science* 24(01): 34–41.

Yang Z.Y., Yuan Z., Yan D.H. et al. 2013. Study of spatial and temporal distribution and multiple characteristics of drought and flood in Huang-Huai-Hai river basin. *Advances in Water Science* 24(05): 618–625.

Yuan W.P. & Zhou G.S. 2004. Comparison between standardized precipitation index and Z index in China. *Acta Phytoecologica Sinica* 28(4): 523–529.

Zhang Q., Zhang L., Cui X.C. et al. 2011. Progresses and challenges in drought assessment and monitoring. *Advances in Earth Science* 26(7): 763–778.

Zhao X.X. & Wang X.J. 2013. Comparison of different drought index in Kashgar area in Xinjiang, China. *Jilin water resources* 11: 20–23, 26.

Zong Y., Wang Y.J., Zhai J.Q. 2013. Spatial and temporal characteristics of meteorological drought in the Haihe river basin based on standardized precipitation index. *Journal of arid land resources and environment* 27(12): 198–202.

Advances in Energy, Environment and Materials Science – Wang & Zhao (Eds)
© 2016 Taylor & Francis Group, London, ISBN 978-1-138-02931-6

The ground surface displacement of shallow buried circular cavity in a soft layered half-space impacted by SH wave

Yuanbo Zhao, Hui Qi, Xiaohao Ding & Dongdong Zhao
College of Aerospace and Civil Engineering, Harbin Engineering University, Harbin, Heilongjiang, China

ABSTRACT: According to the attenuation characteristic of SH wave scattering, the large-arc assumption method is used to approximate the straight boundary by a circle with a large radius. First, the general forms of wave function are given on the basis of Helmholtz theorem. Then, according to the boundary conditions and the complex Fourier-Hankel series expansion method, the problem is transformed into the infinite linear algebraic equations and the numerical results can be obtained. A numerical example shows that the ground surface displacement amplification coefficient changes with the incident wave number k_1, the wave number parameters combination k^*, and the position transformation of circular hole.

1 INTRODUCTION

With the development of economy, there are more and more population and structures underground in cities, once the earthquake happens, it will cause great loss and damage. Therefore, it has an extremely important practical significance for us to research the influence of underground structures on surface displacement during the seismic analysis. When an earthquake occurs, the vibration spread as seismic wave, which will scatter when encountering an obstacle, affects the surface displacement. There have been a lot of research results for ground surface displacement. MOW, C. C. and PAO, and Y detailed introduced the common method of scattering of elastic waves and dynamic stress concentration (MOW and PAO, 1973). In 1982, the complex function method was used to solve the problem of elastic wave scattering, which extends application of wave function expansion technique (Liu et al., 1982). Ten years later, analytic solution of P-Wave and SH wave scattering around single circular cavity in half-space was obtained by large arc hypothesis method, which turns traditional linear boundary into curved boundary to simplify these questions (Lee and Karl, 1993). Liu S W showed the transient response on the surface under SH wave (Liu and Datta,1991). Trifunac showed analytical solution of surface displacement of semicircle and semi-elliptical alluvial valley under SH wave (Trifunac, 1971, Wong and Trifunac, 1974). Shyu researched ground movement for two valleys under SH wave (Shyu et al., 2014). Aacute studied surface motion of valley in random shape under SH wave (Aacute et al., 2006). The analysis of surface motion of alluvial valley in the overburden for the incident plane SH waves was shown (Liang and Ba, 2007). The closed-form analytical solution of surface motion of a semi-elliptical cylindrical hill for incident plane SH waves was presented in 2011 (Liang and Fu, 2011).

In this paper, the influence of circular cavity shallow buried in a soft layer on surface displacement impacted by SH-wave is analyzed, and numerical results are analyzed by examples.

2 THEORETICAL ANALYSIS

2.1 Problem model

As shown in Figure 1, the problem can be described as a single circular hole that is included in a soft surface overburdened layer of elastic half-space. Where I is the half-space and II is the soft surface overburden layer. G_1 is the shear modulus of the half-space, k_1 is its wave number, ρ_1 is the density of medium; G_2, k_2, ρ_2 and h represents shear modulus, wave number, destiny, and thickness of the surface layer, respectively. T_U and T_D signifies upper and lower boundary of the surface layer, respectively. T_C stands for the boundary of the circular hole, and its radius is r. h_1 and h_2 denotes the distance from the center of the circle to upper and lower boundary, respectively. Use the large arc hypothesis method to approximate the upper and lower boundary with infinite radius of arc so that T_U becomes TT_U and T_D becomes TT_D. The center of circular hole as the origin of coordinates O_2, and the line parallels to T_U as X_2 axis to establish coordinate system $X_2O_2Y_2$. Meanwhile the rectangular coordinate system $X_1O_1Y_1$ is established as the origin O_2 to ensure that the Y_1 axis and the Y_2

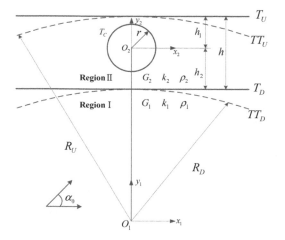

Figure 1. Schematic diagram of the model.

axis are in the same line. In region I, the steady SH-wave incident and the angle of incidence is α_0 which can be obtained by rotating the X_2 axis counter-clockwise. We introduce complex function $z_s = X_s + iY_s$ and $z'_s = X_s - iY_s$ to establish the complex plane z_1 and z_2, where $s = 1, 2$. The transformation relationships of various parameters are as follows:

$$\begin{cases} h = h_1 + h_2 \\ R_U = h + R_D \\ z_2 = z_1 - i(R_D + h_2) \end{cases} \tag{1}$$

2.2 Control equations

This paper studies the out-of-plane shear movement problem of SH-Wave scattering. In the rectangular coordinate system, in X-Y plane, the displacement field of SH-Wave could be expressed as $W(X, Y, t)$, which has nothing to do with Z axis and perpendicular to the X-Y plane. For steady-state problem, the displacement field $W(X, Y, t)$ needs to satisfy the following Helmholtz equation:

$$\frac{\partial^2 W}{\partial X^2} + \frac{\partial^2 W}{\partial Y^2} + k^2 W = 0 \tag{2}$$

where $k = \omega/c$, $c_2 = G/\rho$, ω = the circular frequency of displacement field $W(X, Y, t)$, G = the shear modulus of medium, ρ = the density of medium. The dependencies of displacement field $W(X, Y, t)$ and time t is $\exp(-i\omega t)$, because the problem studied is in a steady state, so $\exp(-i\omega t)$ is left out in the following analysis.

The relationship of stress and strain in the rectangular coordinate system:

$$\tau_{ZX} = G\frac{\partial W}{\partial X}, \; \tau_{ZY} = G\frac{\partial W}{\partial Y} \tag{3}$$

In complex plane, the formula (2) and (3) could be expressed as:

$$4\frac{\partial^2 W}{\partial z \partial \bar{z}} + k^2 W = 0 \tag{4}$$

$$\tau_{ZX} = G\left(\frac{\partial W}{\partial z} + \frac{\partial W}{\partial \bar{z}}\right), \; \tau_{ZY} = iG\left(\frac{\partial W}{\partial z} - \frac{\partial W}{\partial \bar{z}}\right) \tag{5}$$

In the complex plane polar coordinates, the formula (5) could be expressed as:

$$\tau_{Zr} = G\left(\frac{\partial W}{\partial z}\frac{z}{|z|} + \frac{\partial W}{\partial \bar{z}}\frac{|z|}{z}\right),$$
$$\tau_{Z\theta} = iG\left(\frac{\partial W}{\partial z}\frac{z}{|z|} - \frac{\partial W}{\partial \bar{z}}\frac{|z|}{z}\right) \tag{6}$$

2.3 Incident wave field and stress

In the complex plane z_1, the displacement field and stress of incident wave could be expressed as:

$$W^{(I)}_{(z_1, \bar{z}_1)} = W_0 \exp\left[ik_1 \mathrm{Re}\left(z_1 e^{-i\alpha_0}\right)\right] \tag{7}$$

$$\tau^{(I)}_{Zr,(z_1, \bar{z}_1)} = ik_1 G_1 W_0 \exp\left[ik_1 \mathrm{Re}\left(z_1 e^{-i\alpha_0}\right)\right]\mathrm{Re}\left(z_1 e^{-i\alpha_0}/|z_1|\right) \tag{8}$$

$$\tau^{(I)}_{Z\theta,(z_1, \bar{z}_1)} = -ik_1 G_1 W_0 \exp\left[ik_1 \mathrm{Re}\left(z_1 e^{-i\alpha_0}\right)\right]\mathrm{Im}\left(z_1 e^{-i\alpha_0}/|z_1|\right) \tag{9}$$

2.4 Scattered wave field and stress

In the complex plane z_1, the displacement field and stress of scattered wave $W^{(S1)}$ occurred by TT_D in region I could be shown as:

$$W^{(S1)}_{(z_1, \bar{z}_1)} = \sum_{n=-\infty}^{n=+\infty} A_n H_n^{(2)}\left(k_1|z_1|\right)\left(z_1/|z_1|\right)^n \tag{10}$$

$$\tau^{(S1)}_{Zr,(z_1, \bar{z}_1)} = \frac{k_1 G_1}{2}\sum_{n=-\infty}^{n=+\infty} A_n \begin{bmatrix} H_{n-1}^{(2)}\left(k_1|z_1|\right) \\ -H_{n+1}^{(2)}\left(k_1|z_1|\right) \end{bmatrix}\left(\frac{z_1}{|z_1|}\right)^n \tag{11}$$

$$\tau^{(S1)}_{Z\theta,(z_1, \bar{z}_1)} = \frac{ik_1 G_1}{2}\sum_{n=-\infty}^{n=+\infty} A_n \begin{bmatrix} H_{n-1}^{(2)}\left(k_1|z_1|\right) \\ +H_{n-1}^{(2)}\left(k_1|z_1|\right) \end{bmatrix}\left(\frac{z_1}{|z_1|}\right)^n \tag{12}$$

In the complex plane z_1, the displacement field and stress of scattered wave $W^{(S2)}$ occurred in region II by TT_D could be shown as:

$$W^{(S2)}_{(z_1,\bar{z}_1)} = \sum_{n=-\infty}^{n=+\infty} B_n H_n^{(1)}\left(k_2|z_1|\right)\left(z_1/|z_1|\right)^n \tag{13}$$

$$\tau^{(S2)}_{Zr,(z_1,\bar{z}_1)} = \frac{k_2 G_2}{2} \sum_{n=-\infty}^{n=+\infty} B_n \left[\begin{array}{c} H_{n-1}^{(1)}\left(k_2|z_1|\right) \\ -H_{n+1}^{(1)}\left(k_2|z_1|\right) \end{array}\right]\left(\frac{z_1}{|z_1|}\right)^n \tag{14}$$

$$\tau^{(S2)}_{Z\theta,(z_1,\bar{z}_1)} = \frac{ik_2 G_2}{2} \sum_{n=-\infty}^{n=+\infty} B_n \left[\begin{array}{c} H_{n-1}^{(1)}\left(k_2|z_1|\right) \\ +H_{n+1}^{(1)}\left(k_2|z_1|\right) \end{array}\right]\left(\frac{z_1}{|z_1|}\right)^n \tag{15}$$

In the complex plane z_2, the formulas (13)~(15) could be shown as:

$$W^{(S2)}_{(z_2,\bar{z}_2)} = \sum_{n=-\infty}^{n=+\infty} B_n H_n^{(1)}\left(k_2|\otimes|\right)\left(\otimes/|\otimes|\right)^n \tag{16}$$

$$\tau^{(S2)}_{Zr,(z_2,\bar{z}_2)} = \frac{k_2 G_2}{2} \sum_{n=-\infty}^{n=+\infty} B_n \left[\begin{array}{c} H_{n-1}^{(1)}\left(k_2|\otimes|\right)\left(\otimes/|\otimes|\right)^{n-1}\dfrac{z_2}{|z_2|} \\ -H_{n+1}^{(1)}\left(k_2|\otimes|\right)\left(\otimes/|\otimes|\right)^{n+1}\dfrac{\bar{z}_2}{|z_2|} \end{array}\right] \tag{17}$$

$$\tau^{(S2)}_{Z\theta,(z_2,\bar{z}_2)} = \frac{ik_2 G_2}{2} \sum_{n=-\infty}^{n=+\infty} B_n \left[\begin{array}{c} H_{n-1}^{(1)}\left(k_2|\otimes|\right)\left(\otimes/|\otimes|\right)^{n-1}\dfrac{z_2}{|z_2|} \\ +H_{n+1}^{(1)}\left(k_2|\otimes|\right)\left(\otimes/|\otimes|\right)^{n+1}\dfrac{\bar{z}_2}{|z_2|} \end{array}\right] \tag{18}$$

where $\otimes = z_2 + i(R_D + h_2)$.

In the complex plane z_2, the displacement field and stress of scattered wave $W^{(S3)}$ occurred in region II by T_C could be shown as:

$$W^{(S3)}_{(z_2,\bar{z}_2)} = \sum_{n=-\infty}^{n=+\infty} C_n H_n^{(1)}\left(k_2|z_2|\right)\left(\frac{z_2}{|z_2|}\right)^n \tag{19}$$

$$\tau^{(S3)}_{Zr,(z_2,\bar{z}_2)} = \frac{k_2 G_2}{2} \sum_{n=-\infty}^{n=+\infty} C_n \left[\begin{array}{c} H_{n-1}^{(1)}\left(k_2|z_2|\right) \\ -H_{n+1}^{(1)}\left(k_2|z_2|\right) \end{array}\right]\left(\frac{z_2}{|z_2|}\right)^n \tag{20}$$

$$\tau^{(S3)}_{Z\theta,(z_2,\bar{z}_2)} = \frac{ik_2 G_2}{2} \sum_{n=-\infty}^{n=+\infty} C_n \left[\begin{array}{c} H_{n-1}^{(1)}\left(k_2|z_2|\right) \\ +H_{n+1}^{(1)}\left(k_2|z_2|\right) \end{array}\right]\left(\frac{z_2}{|z_2|}\right)^n \tag{21}$$

In the complex plane z_1, the formulas (19)~(21) could be shown as:

$$W^{(S3)}_{(z_1,\bar{z}_1)} = \sum_{n=-\infty}^{n=+\infty} C_n H_n^{(1)}\left(k_2|\oplus|\right)\left(\frac{\oplus}{|\oplus|}\right)^n \tag{22}$$

$$\tau^{(S3)}_{Zr,(z_1,\bar{z}_1)} = \frac{k_2 G_2}{2} \sum_{n=-\infty}^{n=+\infty} C_n \left[\begin{array}{c} H_{n-1}^{(1)}\left(k_2|\oplus|\right)\left(\dfrac{\oplus}{|\oplus|}\right)^{n-1}\dfrac{z_1}{|z_1|} \\ -H_{n+1}^{(1)}\left(k_2|\oplus|\right)\left(\dfrac{\oplus}{|\oplus|}\right)^{n+1}\dfrac{\bar{z}_1}{|z_1|} \end{array}\right] \tag{23}$$

$$\tau^{(S3)}_{Z\theta,(z_1,\bar{z}_1)} = \frac{ik_2 G_2}{2} \sum_{n=-\infty}^{n=+\infty} C_n \left[\begin{array}{c} H_{n-1}^{(1)}\left(k_2|\oplus|\right)\left(\oplus/|\oplus|\right)^{n-1}\dfrac{z_1}{|z_1|} \\ +H_{n+1}^{(1)}\left(k_2|\oplus|\right)\left(\oplus/|\oplus|\right)^{n+1}\dfrac{\bar{z}_1}{|z_1|} \end{array}\right] \tag{24}$$

where $\oplus = z_1 - i(R_D + h_2)$.

In the complex plane z_1, the displacement field and stress of scattered wave $W^{(S4)}$ occurred in region II by TT_D could be shown as:

$$W^{(S4)}_{(z_1,\bar{z}_1)} = \sum_{n=-\infty}^{n=+\infty} D_n H_n^{(2)}\left(k_2|z_1|\right)\left(\frac{z_1}{|z_1|}\right)^n \tag{25}$$

$$\tau^{(S4)}_{Zr,(z_1,\bar{z}_1)} = \frac{k_2 G_2}{2} \sum_{n=-\infty}^{n=+\infty} D_n \left[\begin{array}{c} H_{n-1}^{(2)}\left(k_2|z_1|\right) \\ -H_{n+1}^{(2)}\left(k_2|z_1|\right) \end{array}\right]\left(\frac{z_1}{|z_1|}\right)^n \tag{26}$$

$$\tau^{(S4)}_{Z\theta,(z_1,\bar{z}_1)} = \frac{ik_2 G_2}{2} \sum_{n=-\infty}^{n=+\infty} D_n \left[\begin{array}{c} H_{n-1}^{(2)}\left(k_2|z_1|\right) \\ +H_{n+1}^{(2)}\left(k_2|z_1|\right) \end{array}\right]\left(\frac{z_1}{|z_1|}\right)^n \tag{27}$$

In the complex plane z_2, the formula (25), formula (26), formula (27) could be shown as:

$$W^{(S4)}_{(z_2,\bar{z}_2)} = \sum_{n=-\infty}^{n=+\infty} D_n H_n^{(2)}\left(k_2|\odot|\right)\left(\frac{\odot}{|\odot|}\right)^n \tag{28}$$

$$\tau^{(S4)}_{Zr,(z_2,\bar{z}_2)} = \frac{k_2 G_2}{2} \sum_{n=-\infty}^{n=+\infty} D_n \left[\begin{array}{c} H_{n-1}^{(2)}\left(k_2|\odot|\right)\left(\dfrac{\odot}{|\odot|}\right)^{n-1}\dfrac{z_2}{|z_2|} \\ -H_{n+1}^{(2)}\left(k_2|\odot|\right)\left(\dfrac{\odot}{|\odot|}\right)^{n+1}\dfrac{\bar{z}_2}{|z_2|} \end{array}\right] \tag{29}$$

$$\tau^{(S4)}_{Z\theta,(z_2,\bar{z}_2)} = \frac{ik_2 G_2}{2} \sum_{n=-\infty}^{n=+\infty} D_n \left[\begin{array}{c} H_{n-1}^{(2)}\left(k_2|\odot|\right)\left(\odot/|\odot|\right)^{n-1}\dfrac{z_2}{|z_2|} \\ +H_{n+1}^{(2)}\left(k_2|\odot|\right)\left(\odot/|\odot|\right)^{n+1}\dfrac{\bar{z}_2}{|z_2|} \end{array}\right] \tag{30}$$

where $\odot = z_2 + i(R_D + h_2)$.

255

2.5 Equations set

Because the displacement fields and stress fields expressions of incident and scattering waves have been constructed, according to the boundary conditions that the radial stress of TTU and TC is free and the continuous condition that the displacement and the radial stress of TTD are continuous, the set of equations could be expressed as:

$$
\begin{cases}
(a)\,TT_D\left(|z_1|=R_D\right):\\
\quad W^{(I)}_{(z_1,\bar{z}_1)}+W^{(S1)}_{(z_1,\bar{z}_1)}=W^{(S2)}_{(z_1,\bar{z}_1)}+W^{(S3)}_{(z_1,\bar{z}_1)}+W^{(S4)}_{(z_1,\bar{z}_1)}\\
(b)\,TT_D\left(|z_1|=R_D\right):\\
\quad \tau^{(I)}_{Zr,(z_1,\bar{z}_1)}+\tau^{(S1)}_{Zr,(z_1,\bar{z}_1)}=\tau^{(S2)}_{Zr,(z_1,\bar{z}_1)}+\tau^{(S3)}_{Zr,(z_1,\bar{z}_1)}+\tau^{(S4)}_{Zr,(z_1,\bar{z}_1)}\\
(c)\,T_C\left(|z_2|=R\right):\\
\quad \tau^{(S2)}_{Zr,(z_2,\bar{z}_2)}+\tau^{(S3)}_{Zr,(z_2,\bar{z}_2)}+\tau^{(S4)}_{Zr,(z_2,\bar{z}_2)}=0\\
(d)\,TT_U\left(|z_1|=R_U\right):\\
\quad \tau^{(S2)}_{Zr,(z_1,\bar{z}_1)}+\tau^{(S3)}_{Zr,(z_1,\bar{z}_1)}+\tau^{(S4)}_{Zr,(z_1,\bar{z}_1)}=0
\end{cases}
\tag{31}
$$

Substituting formula (7)~(30) into formula (31), the known quantity to the right hand side of the equals sign, the unknown quantity to the left hand side of the equals sign:

$$
\sum_{n=-\infty}^{n=+\infty}
\begin{bmatrix}
\xi_n^{1,1} & \xi_n^{1,2} & \xi_n^{1,3} & \xi_n^{1,4}\\
\xi_n^{2,1} & \xi_n^{2,2} & \xi_n^{2,3} & \xi_n^{2,4}\\
\xi_n^{3,1} & \xi_n^{3,2} & \xi_n^{3,3} & \xi_n^{3,4}\\
\xi_n^{4,1} & \xi_n^{4,2} & \xi_n^{4,3} & \xi_n^{4,4}
\end{bmatrix}
\begin{bmatrix}
A_n\\ B_n\\ C_n\\ D_n
\end{bmatrix}
=
\begin{bmatrix}
\eta_1\\ \eta_2\\ \eta_3\\ \eta_4
\end{bmatrix}
\begin{matrix}
(a)\\ (b)\\ (c)\\ (d)
\end{matrix}
\tag{32}
$$

For (a), (b), (d), multiply both sides by $\exp(-im\varphi_1)$, for (c), multiply both sides by $\exp(-im\varphi_2)$, integrate them in $(-\pi, \pi)$ respectively, we can get the equation:

$$
\sum_{m=-\infty}^{m=+\infty}\sum_{n=-\infty}^{n=+\infty}
\begin{bmatrix}
\Phi_{nm}^{1,1} & \Phi_{nm}^{1,2} & \Phi_{nm}^{1,3} & \Phi_{nm}^{1,4}\\
\Phi_{nm}^{2,1} & \Phi_{nm}^{2,2} & \Phi_{nm}^{2,3} & \Phi_{nm}^{2,4}\\
\Phi_{nm}^{3,1} & \Phi_{nm}^{3,2} & \Phi_{nm}^{3,3} & \Phi_{nm}^{3,4}\\
\Phi_{nm}^{4,1} & \Phi_{nm}^{4,2} & \Phi_{nm}^{4,3} & \Phi_{nm}^{4,4}
\end{bmatrix}
\begin{bmatrix}
A_n\\ B_n\\ C_n\\ D_n
\end{bmatrix}
=
\sum_{m=-\infty}^{m=+\infty}
\begin{bmatrix}
\Psi_m^{1,1}\\ \Psi_m^{2,1}\\ \Psi_m^{3,1}\\ \Psi_m^{4,1}
\end{bmatrix}
\tag{33}
$$

So we can calculate the coefficient A_n, B_n, C_n, D_n, take them to the formulas and intercept limited items, all the unknown quantities will be found out.

2.6 The ground surface displacement amplification coefficient (W^*)

Define W^* as the displacement amplification coefficient:

$$
W^* = \left\| \left(W^{(S2)}_{(z_1,\bar{z}_1)}+W^{(S3)}_{(z_1,\bar{z}_1)}+W^{(S4)}_{(z_1,\bar{z}_1)} \right)\Big/ W_0 \right\|_{|z_1|=R_U}
\tag{34}
$$

3 NUMERICAL EXAMPLE

In this sample, the results are no-dimensional form so we suppose that the radius of circular cavity is 1. Define the parameters combination $G^* = G_2/G_1$, $k^* = k_2/k_1$. Region I is harder than region II, so k^* is less than 1, that is to say, incident waves enter into the lesser layer from the harder half space.

From Figure 2–4 separately shows the ground surface displacement amplification coefficient W^* with x_1/r when the incident angle is 90°, the circular hole is in the middle of the layer, the thickness of the layer is 3 times of the radius of the circular hole, G^* is 1, the incident wave is in different frequency.

Figure 2 shows that when the incident wave incident in very low frequency, the curve shapes are almost the same, and the difference of amplitude is very small. Figure 3 shows that when the incident wave incident in low and medium frequency band, the curve shapes are different, when k^* is 0.2, the amplitude of curve is least and the curve is approximate to a straight line. When k^* is 0.5, the amplitude of curve is the biggest. Figure 4 shows that when the incident wave incident is in high frequency, the W^* decreases with the increase of k^*. Comprehensive analysis of Figure 2–Figure 4, it can be seen that with the incident wave number increase, the W^* decrease gradually.

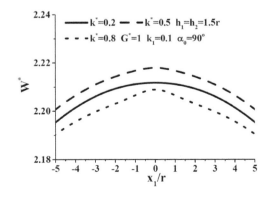

Figure 2. Ground surface displacement amplification coefficient with x_1/r ($k_1 = 0.1$, $G^* = 1$, $h_1 = h_2 = 1.5r$, $\alpha_0 = 90°$).

256

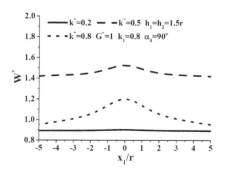

Figure 3. Ground surface displacement amplification coefficient with x_1/r ($k_1 = 0.8$, $G^* = 1$, $h_1 = h_2 = 1.5r$, $\alpha_0 = 90°$).

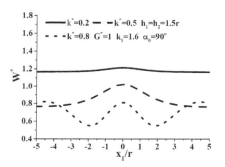

Figure 4. Ground surface displacement amplification coefficient with x_1/r ($k_1 = 1.6$, $G^* = 1$, $h_1 = h_2 = 1.5r$, $\alpha_0 = 90°$).

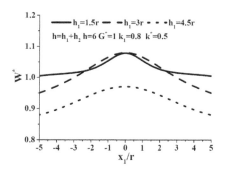

Figure 5. Ground surface displacement amplification coefficient with x_1/r ($k_1 = 0.8$, $k^* = 0.5$, $G^* = 1$, $h = 6$, $\alpha_0 = 90°$).

Figure 5 shows the change of W^* with the location of the circular cavity. When the $h1$ is equal to $1.5r$, the W^* is the largest. When the circular cavity is in the middle of the layer, the maximum of W^* is the same with h_1 is $1.5r$, but the whole curve is

steeper. When h_1 is $4.5r$, the shape of curve is similar with h_1 is $3r$, but the amplitude is reduced by thirty percent.

4 CONCLUSION

Based on the large-arc assumption method, complex function and wave function expansion method, the ground surface displacement of shallow buried circular cavity in a soft layered half-space impacted by SH wave is studied. A numerical example shows that the ground surface displacement amplification coefficient changes with the incident wave number k_1, the wave number parameters combination k^*, the position transformation of circular hole, when the incident wave vertical incidence and shear modulus parameters combination G^* is equal to 1. As a whole, the ground surface displacement amplification coefficient turns to smaller with the increase of the incident wave number. When the parameter h_1 increases and other parameters remain the same, the ground surface displacement amplification coefficient present a decrease trend.

REFERENCES

Aacute, Nchez-Sesma, F.J. & Rosenblueth, E. 2006. Ground Motion At Canyons Of Arbitrary Shape Under Incident Sh Waves. *Earthquake Engineering & Structural Dynamics*, 7.

Lee, V. & Karl, J. 1993. Diffraction of elastic plane P waves by circular, underground unlined tunnels. *European Earthquake Engineering*, 6, 29–36.

Liang, J. & Ba, Z. 2007. Surface motion of an alluvial valley in layered half-space for incident plane SH waves. *Journal of Earthquake Engineering & Engineering Vibration*, 27, 1–9.

Liang, J. & Fu, J. 2011. Surface motion of a semi-elliptical hill for incident plane SH waves. *Earthquake Science*, 24, 447–462.

Liu, S.W. & Datta, S.K. 1991. Transient response of ground surface due to incident SH waves. *Computational Mechanics*, 8, 99–109.

Liu, D., Gai, B. & Tao, G. 1982. Applications of the method of complex functions to dynamic stress concentrations. *Wave Motion*, 4, 293–304.

Mow, C.C. & Pao, Y. 1973. *Diffraction of elastic waves and dynamic stress concentrations*, Crane, Russak.

Shyu, W.S., Teng, T.J. & Yeh, C.S. 2014. Surface Motion of Two Canyons for Incident SH Waves by Hybrid Method ☆. *Procedia Engineering*, 79, 533–539.

Trifunac, M.D. Surface motion of a semi-cylindrical alluvial valley for incident plane SH waves. Bulletin of the Seismological Society of America, 1971. 1770.

Wong, H.L. & Trifunac, M.D. 1974. Surface motion of a semi-elliptical alluvial valley for incident plane SH waves. *Bulletin of the Seismological Society of America*, 64, 1389–1408.

Advances in Energy, Environment and Materials Science – Wang & Zhao (Eds)
© *2016 Taylor & Francis Group, London, ISBN 978-1-138-02931-6*

Discussion on the design of steel skeleton of great span greenhouse in cold region

Lian Zhai & Dong-hui Liu
Jilin Architecture University, Changchun City, Jilin Province, China

ABSTRACT: At present, more and more great span greenhouses were applied to the new products. In view of the rural tourism, it became the bottleneck of mass production in that the ability of anti-wind and anti-snow was weak in cold region. Taking the 9 m span steel skeleton for example, the essay covers the the geometry design parameter, the shape parameter and the material-selection method. At the same time, it also analyzes the bending rigidity and steel consumption between the steel tube and reinforcing bar. Furthermore, it makes use of the finite element software to do the analysis. The results of theoretical analysis on the stress performance and the selection section size of the assigned structural style will provide the reliable reference for structure designers.

1 INTRODUCTION

As the northern rural tourism market was hot in recent years, rural tourism resort model was favored more and more by the investors and tourists. Introducing the ecological agriculture to the rural tourism products can improve the low temperature season rural tourism product development, reduce the comprehensive cost, and improve the return on investment of the enterprise contributing to more powerful vitality for the rural tourism market. At high latitudes in the northeast, the use of large span greenhouse became the core link that was effective utilization of resort space and reduced the cost. Taking Jilin province for an example, reasonable maximum span of general greenhouse being suggested is limited to 6 m according to the local climate conditions, which can't meet the needs of the sightseeing agriculture of the resort. As reception people are much higher than the ordinary greenhouse, reliability of the great span greenhouse designed according to the traditional experience can't be guaranteed. Furthermore, the costs of the foreign large high-grade greenhouse beings introduced are too high. Therefore, the task of independent design of the large span greenhouse is particularly important. The basic functions of the sunlight greenhouse design of the resort must satisfy security, low cost, and large span. The material selection, structure form and spacing of steel skeletons determine the realization of the function of the greenhouse, engineering cost as well as the overall return on investment of the producer to a large extent, so the schedule of the independent design on the steel skeleton should be mentioned as soon as possible. (Zhai 2010).

2 GETTING STARTED ON ARCHITECTURAL SELECTION OF THE STEEL SKELETON

Generally, the larger interior space was needed by sunlight greenhouse. As indicated in Figure 1, steel skeleton without column and masonry wall were a popular form of structure. The daylight effect and heat preservation effect of the structure was good, which was more suitable for the cold area. The steel skeleton whose selections of the parameter were reasonable should permeate the sun as much as possible during the day ensure the appropriate illumination and the temperature of crop growth and enough reserves to heat, while it should have enough insulation performance at night. Usually, it determined the selection of solar greenhouse parameters according to the local geographic latitude, the climate conditions, and the using function, which was considered comprehensively about

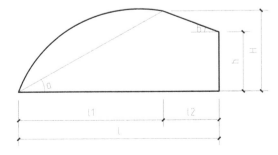

Figure 1. Greenhouse section geometry parameters.

Table 1. The y-coordinate value correspond to horizontal ordinate of each point on the steel skeleton/mm.

Abscissa/mm	0	1000	2000	3000	4000
Span/mm	0	1981	2585	3081	3484

the daylight, heat preservation of the greenhouse crops, fertility and artificial operation factor. Ridge ratio, net span, high slope, the former slope angle of reference, and the height of north wall (or the latter slope angle) are the design parameters which are the control effect on it.

Where L is the indoor net span, L1 is horizontal projection of former slope (southern slope), L2 is the horizontal projection of latter slope (north slope), h is the height of north wall, H is the height of ridge, a is the former slope angle of reference, b is the latter slope angle, and L1/L is ridge ratio.

According to the existing mature experience, the sunlight greenhouse often used 6 m span in the area whose latitude was more than 40°, the ridge ratio was generally selected as 0.8 or so, the height of ridge was about 3 m, the height difference between the roof and the back wall of greenhouse should be above 80 cm, the former slope angle of reference was to ensure more than 20.5°, and the latter slope angle was more than 30°. The shape of the former slope surface was usually adopted bent or curved steel plane truss arch, whose bearing capacity was higher. This design can meet the requirement of crops, and the basic cost is low. For larger span greenhouse, it was generally needed to adjust the height of ridge. Increasing of height can increase the space of the greenhouse, which can be helped in the daylight, but the cost of greenhouse will increase, and it will affect the heat preservation. So the high ridge of the sunlight greenhouse only as planting is unfavorable and exorbitant. It introduces the design content of 9 m span steel skeleton that the purpose is that is conducive to crop growth, facilitated the artificial operation and reduced cost on the premise of guarantee enough structural reliability according to the above parameters to a large span greenhouse. Their shape parameter is listed out in Table 1.

3 DETERMINATION OF LOAD

3.1 Determination method of various kinds of load

Load, which is the external force being effected on the structure, is the basic foundation of structure design, including the natural load such as the wind load, the snow load and the seismic force, and the artificial load such as the heap load, the hoisting load and the maintenance load.

Compared with the traditional structure, the wind load and snow load become the control loads of the greenhouse. Too much load value is wasting material, increasing the cost, and leading to the shadow area increase, and affecting crop growth. Small value can't withstand the wind load and the snow load, causing serious results to life and property security. Therefore, it is an essential prerequisite for the design of sunlight greenhouse steel skeleton that makes sure to confirm the reasonable load determination through the necessary investigation and mathematical statistics, combined with the specification. (GB 50009-2012).

3.1.1 Permanent load

It is mainly including the weight of steel skeleton and all fixed equipment, including insulation shade curtain, lighting and irrigation equipment, etc. which can be selected according to the actual weight.

3.1.2 Variable load

All temporary loads except the wind load and snow load are the variable loads, mainly including straw mattress or insulation quilt and the crop load inside the greenhouse and additional loads. The field test showed that the wet straw mattress weighed about 0.06 KN/m^2; the crop load inside the greenhouse, such as plant hoisted, generally weighed 0.15 KN/m^2; the standard option of roof uniform live load is 0.5 KN/m^2, being referenced on <load code for the design of building structures> (GB50009-2012), the concentrated load of construction and maintenance is generally taken 1.0 KN, which is considered in the most unfavorable position.

3.1.3 Wind load

It is collectively known as wind load expressed as a W that is perpendicular to the surface of the building, and the wind pressure is acted on unit area. It is concerned with wind speed, air density, wind pressure height, the shape and the size of the building, and so on. When it is calculated, the greenhouse steel skeleton and the normal value can be calculated by the following formula:

$$w_K = \mu_S \mu_z \, w_0 \tag{1}$$

where w_0 refers to the reference wind pressure, μ_z refers to change coefficient of height of the wind pressure; μ_S refers to the shape coefficient of wind load.

All items are available on <load code for the design of building structures> (GB50009-2012).

The shape coefficient of wind load is related to the size of the building, whose extreme value generally appears in the area where the wind pressure severe changed, such as the eaves on the gable's top and roof of windward surface when making the structure design of greenhouses. It should be checked on the wind suction withstanding for the plastic membrane roof the greenhouse skeleton.

3.1.4 *Snow load*

Roof snow load standard values on the horizontal plane, should be calculated according to the formula (2)

$$S_k = \mu_r S_0 \tag{2}$$

where μ_r refers to the snow distribution coefficient of roof and S_0 refers to the reference snow pressure (KN/m²). It can be estimated that the design reference period of the greenhouse, according to the number and scale of the construction investment, is made to the design of greenhouses. Then it is determined the reference snow pressure, combining the local meteorological data. The snow distribution coefficient of roof varies depending on the shape of the roof, the assignment is dangerous according to the table 6.2.1 of <load code for the design of building structures> (GB50009-2012), because the snow pressure is easy to make the film deformation prolapsed. μ_r is available: when $\alpha \geq 60°$, $\mu_r = 0$; when $\alpha < 60°$, $\mu_r = 1.0$.

3.2 *Combination of action effects*

Structure may also take a variety of load when the sunlight greenhouse skeleton is installed and used. It is the problem of load combinations that is considered the most unfavorable situation of their overall effect on the structure. In the process of structural calculation, it is usually chosen different combination of the load design value for the following:

Permanent load,
Permanent load + Variable load;
Permanent load + Snow load;
Permanent load + Wind load;
Permanent load + Variable load + Snow load;
Permanent load + Variable load + Wind load.

4 STRUCTURE DESIGN OF THE STEEL FRAME

Carrying capacity of steel skeletons and daylight conditions are the precondition for judging the effect of using of the sunlight greenhouse, so it is particularly important to adopt the reasonable structure of the steel skeleton. The planting industry was developed later in the north in the cold period. Although development speed was faster, the work of technology supporting was relatively lag. With the development needs of planting industry tourism industry, the traditional span greenhouse already cannot satisfy the production needs, so a large span of greenhouse production has become the current trend. But currently, general investment scale is small because the most investors are individual farmers. The structure design of the steel skeleton of the large span greenhouse is not nearly carried on by special design unit. Generally, it is only referenced as the mature experience on the Beijing and Tianjin, which was built through the proper reinforcement with the subcontracted engineering experience, combined with local climate conditions. This kind of engineering reliability was low, which had caused great damage to property with slightly larger snow or wind pressure, although it is usually not easy for the problem to appear. The large span greenhouse must have the high reliability as a sightseeing agriculture. So it must be carried on the professional structure design according to local climate conditions.

The author simulated the current reasonable 9 m span steel structure scheme (shape parameters are shown in Table 1) using the finite element software, and put forward the practical selection of the material cross section (shown in Table 2), which provides the reference for the proposed builders. Figure 2 is the calculation diagram of 9 m span simulation, dot is on behalf of support, and digital is the node number. And truss spacing is 0.9 m, there are tie bars 4 ~ 6 on the longitude, which are welding connections with the truss. The small spacing of the truss and tie bar can decrease the size of bar section, but it impacts the structure light transmittance; the large spacing can increase unit load of bar and improve the overall cost. Select the data after software optimization calculation.

The simulation analysis by finite element software shows that it is the effective way to improve steel skeleton, overall mechanical performance, and increase the comprehensive benefit that steel

Table 2. The materials size of different span steel frames (mm).

Span	Height of ridge	Pipe diameter of upper chord	Equivalent diameter
9000	4200	32×2	15.5

Vertical pipe diameter	Equivalent diameter	Pipe diameter of lower chord	Equivalent diameter
16×1.5	9.3	32×2	15.5

bar being replaced by steel pipe cannot have an obvious increase in the total steel quantity. This is the reason why the moment of inertia of pipe section is larger than steel, bending stiffness is increased significantly, ability to withstand the instability and failure of compression of bottom

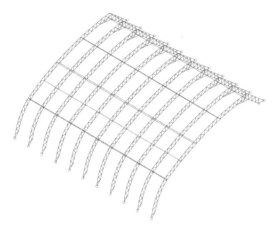

Figure 2.   The calculation sketch plan of steel frame.

chord that frequents the occurrence caused by wind load suction. But because the sunlight greenhouse is in a high humidity environment, it should be payed a special attention to the pipe corrosion problem.

The arch bar diameter is not less than 25 mm, wall thickness is not less than 1.5 mm, which is the requirement of the main materials to the 9 m span of the monomer greenhouse. Such construction steel corrosion resistant ability is stronger. If the hot dip galvanized steel is used, and preservative treatment is welded joint, the durable life of up is to 15 ~ 20 years, which can ensure the long-term benefits of investors. Table 2 shows the dimensions of steel skeleton calculated by software, in which diagonal wall had better choice of 2 mm to improve its corrosion resistance. Table 3–5, respectively, gives the calculation results on the top 10 units of "the stress ratio of Strength", "the stress ratio of stability around the 2-axis stress strength" and "the stress ratio of stability around the 3axis stress strength", the data in the bracket is the combination of number or case number.

The above-mentioned data is shown that it can meet the functional requirements that the intensity of the stress ratio and stress ratio of around 2, 3 axis integral stability is selected. Obviously, units'

Table 3.   The calculation results on the top 10 units of 'the stress ratio of strength'.

No.	Unit no.	Intensity	The overall stability around 2 axis	The overall stability around 3 axis	Resistance to shear stress ratio around 2 axis
1	277	0.61 (13/1)	0.75	0.75	0.00
2	801	0.61 (13/1)	0.75	0.75	0.00
3	539	0.61 (13/1)	0.75	0.75	0.00
4	670	0.61 (13/1)	0.75	0.75	0.00
5	408	0.61 (13/1)	0.75	0.75	0.00
6	932	0.61 (13/1)	0.75	0.75	0.00
7	1063	0.61 (13/1)	0.75	0.75	0.00
8	146	0.61 (13/1)	0.75	0.75	0.00
9	1296	0.61 (13/1)	0.74	0.74	0.00
10	3	0.60 (16/3)	0.72	0.72	0.01

Resistance to shear stress ratio around 3 axis	Slenderness ratio around 2 axis	Slenderness ratio around 3 axis	Around 2 axis W/l	Around 3 axis W/l	Results
0.00	58	58	0	0	Meet
0.00	58	58	0	0	Meet
0.00	58	58	0	0	Meet
0.00	58	58	0	0	Meet
0.00	58	58	0	0	Meet
0.00	58	58	0	0	Meet
0.00	58	58	0	0	Meet
0.00	58	58	0	0	Meet
0.00	58	58	0	0	Meet
0.01	194	194	1/1211	1/1265	Meet

Table 4. The calculation results on the top 10 units of 'the stress ratio of stability around the 2-axis stress strength'.

No.	Unit no.	Intensity	The overall stability around 2 axis	The overall stability around 3 axis	Resistance to shear stress ratio around 2 axis	Resistance to shear stress ratio around 3 axis	Slenderness ratio around 2 axis	Slenderness ratio around 3 axis	Around 2 axis W/l	Around 3 axis W/l	Results
1	215	0.19	0.94 (2/1)	0.94	0.01	0.00	204	30	0	0	Meet
2	1001	0.19	0.94 (2/1)	0.94	0.01	0.00	204	30	0	0	Meet
3	346	0.19	0.94 (2/1)	0.94	0.01	0.00	204	30	0	0	Meet
4	477	0.19	0.94 (2/1)	0.94	0.01	0.00	204	30	0	0	Meet
5	608	0.19	0.94 (2/1)	0.94	0.01	0.00	204	30	0	0	Meet
6	870	0.19	0.94 (2/1)	0.94	0.01	0.00	204	30	0	0	Meet
7	739	0.19	0.94 (2/1)	0.94	0.01	0.00	204	30	0	0	Meet
8	1132	0.19	0.94 (2/1)	0.94	0.01	0.00	204	30	0	0	Meet
9	1227	0.19	0.94 (2/1)	0.94	0.01	0.00	204	30	0	0	Meet
10	84	0.19	0.94 (2/1)	0.94	0.01	0.00	204	30	0	0	Meet

Table 5. The calculation results on the top 10 units of 'the stress ratio of stability around the 3-axis stress strength'.

No.	Unit no.	Intensity	The overall stability around 2 axis	The overall stability around 3 axis	Resistance to shear stress ratio around 2 axis	Resistance to shear stress ratio around 3 axis	Slenderness ratio around 2 axis	Slenderness ratio around 3 axis	Around 2 axis W/l	Around 3 axis W/l	Results
1	215	0.19	0.94	0.94 (2/1)	0.01	0.00	204	30	0	0	Meet
2	1001	0.19	0.94	0.94 (2/1)	0.01	0.00	204	30	0	0	Meet
3	346	0.19	0.94	0.94 (2/1)	0.01	0.00	204	30	0	0	Meet
4	477	0.19	0.94	0.94 (2/1)	0.01	0.00	204	30	0	0	Meet
5	608	0.19	0.94	0.94 (2/1)	0.01	0.00	204	30	0	0	Meet
6	870	0.19	0.94	0.94 (2/1)	0.01	0.00	204	30	0	0	Meet
7	739	0.19	0.94	0.94 (2/1)	0.01	0.00	204	30	0	0	Meet
8	1132	0.19	0.94	0.94 (2/1)	0.01	0.00	204	30	0	0	Meet
9	1227	0.19	0.94	0.94 (2/1)	0.01	0.00	204	30	0	0	Meet
10	84	0.19	0.94	0.94 (2/1)	0.01	0.00	204	30	0	0	Meet

stresses are relatively stable, reasonable force and material applications are more fully for the most dangerous ten bars. This design is carried on in the 50 years of design reference period of Jilin province, so most of the sunlight greenhouse should be safe in the province. It greatly improves the stability of the structure being compared with the steel bar of the equal cross section, and the pipes of the upper chord and the lower chord can ensure the stability of the structure where it can make production safe in the larger wind pressure area.

5 CONCLUSION

On the whole, the basic safeguard of the long-term interests of investors is the comprehensively analyzed, the local load factor, and correctly selection of the materials to build the greenhouse skeleton. The structural design of sunlight greenhouse provides the most direct and basic technical support for the agricultural production. It puts forward further requirements for the engineering technical personnel in the Jilin province that familiarizes with the process of agricultural production and the related finite element calculation software, and applies them to agricultural production potential. It is an effective way to attract more designers and construction personnel to be engaged in an agricultural construction industry, which can fundamentally contribute to the standardization of agricultural construction.

REFERENCES

Lian Zhai, Dong-hui Liu, Shu-yao Song, Yan Wang. Design of Greenhouse in Jilin Province. [J] Journal of Anhui Agricultural Sciences, 2010, 31: 17828–17829.
Load code for the design of building structures GB50009-2012.

Advances in Energy, Environment and Materials Science – Wang & Zhao (Eds)
© 2016 Taylor & Francis Group, London, ISBN 978-1-138-02931-6

Research on spatiotemporal changes of water level corresponding to different flood frequency to ENSO in the Pearl River Delta

Haiyan Qiao, Qiong Jia & Yang Xu
College of Harbor, Coastal and Offshore Engineering, Hohai University, Nanjing, Jiangsu, China

ABSTRACT: Based on Annual Maximum Water Level (AMWL) datasets extracted from 34 gauging stations in the Pearl River Delta (PRD) and the OceanicNino Index (ONI) data in the Niño 3.4 region of 1951–2008, the GEV distribution was used to get the frequency analysis. The water level variation calculated from series of pre-1980 and post-1980 was examined to comprehensively understand flood frequency characteristics. The interdecadal variation in the response of flood level to ENSO in the past fifty years was researched by moving correlation. The spatial maps for flood levels corresponding to different return periods suggest the water level increases gradually from the coast to the riverine system. The flood level increments of different return periods in all sites are obvious. The water levels show a decreasing trend the upper PRD while there is an increasing trend in the middle and lower PRD. The long-term variation of interannual relationship between ENSO and flood level has a significant stage characteristic and it has a great change in the 1980s. The flood level has various responses to various seasonal ENSO.

1 INTRODUCTION

Flooding is an inevitable natural hazard posing a high risk to many places around the world (Shiau, 2003). Low-lying river deltas with extremely high population densities and developed economy are generally more vulnerable to higher risk of flooding. Although it has a small probability of occurrence, catastrophic flooding has historically caused significant economic damages and human loss in many coastal areas. Frequency analysis of extreme water levels across a river system may play a vital role in hazard prevention and mitigation (Al-Futaisi & Stedinger, 1999; Singh & Strupczewski, 2002). Appropriate estimation of the return period of extreme high water level can contribute to better understanding of flood-risk control and then further efforts can be made to mitigate and even take precautions against the heavy consequences caused by flooding. Therefore, substantial attention has been paid to analyze the return period of extreme water levels and numerous studies have been conducted (Pugh & Vassie, 1980; Letetrel et al. 2010).

El Niño-Southern Oscillation (ENSO) is the result of global air-sea couple. It is the strongest signal of interannual climate change of ocean-atmosphere system in the tropical Pacific. Researches have indicated that anomalous temperature and precipitation in many areas are associated with ENSO events closely (Klaus & Michael, 2011). ENSO event is not only a major cause of global climate anomalies, but also leads to the Asian Monsoon anomalies and the occurrence of droughts of China with severe climate disasters and significant economic losses. According to the data provided by National Oceanic and Atmospheric Administration (NOAA) of the United States, there were 18 warm events and 14 cold events from 1951 to 2008.

As the second largest river of China in accordance with streamflow magnitude, the Pearl River, which is comprised of three principal tributaries, namely the West River, the East River and the North River, discharges to the adjoining South Sea through eight outlets (Fig. 1). The Pearl River Delta possesses complex estuarine topography and complicated river networks with the density of channel length per unit area of 0.68–1.07 km/km² (Shi et al. 2012). In recent decades especially since the 1990s, hydrological and geomorphological characteristics of the Pearl River Basin has undergone tremendous variation, leading to alternations of flood return periods due to the superimposed impact of human activities, climate change and other multiple factors. Flooding is gradually increasing and have occurred with "94.6", "98.6", "05.6", "08.6" flood after the 1990s, making the contradiction between economic development and flood disaster risk of the Pearl River Delta region sharper and sharper (Zhang et al. 2009; Jiang et al. 2012). The flood level of the Pearl River Delta is rising in general, the change of the extreme water levels only becomes significant after the 1980s, due to the accumulation of channel changes following the extensive sand excavation in this area.

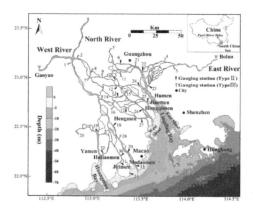

Figure 1. Location of gauging stations in the Pearl River Delta.

2 DATASET AND METHODOLOGY

In order to interpret the past hydrological regimes in terms of future probability of occurrence related to extreme events, instantaneous values of annual maximum water level from 34 gauging stations in the PRD and ONI data in the Niño 3.4 region during 1951–2008 were collected and analyzed in this work. The hydrological data were obtained from the Water Bureau of Guangdong Province, which were strictly controlled before their release to guarantee the quality of data.

Approaches for estimating return period include numerical simulation and frequency analysis, in need of adequate observational data of flood levels at the site under consideration (FEMA, 2004). Fisher and Tippett (1928) proved extreme value distributed asymptotically and summarized three types of extreme value distribution model namely extreme value Type I (Gumbel), Type II (Fréchet) and Type III (Weibull) distribution. The generalized extreme value (GEV) distribution is a reformulation of three probability distribution functions comprising Weibull, Gumbel and Fréchet (Jenkinson, 1955). GEV method is more applicable and has been widely used in hydrology and flood frequency analysis (Xu & Huang, 2011). It can avoid deficiencies of a certain kind of distribution and the results are more reliable than using traditional methods; the corresponding Probability Density Function (PDF) is given by

$$f(x; \mu, \sigma, \xi) = \frac{1}{\sigma} \left[1 + \xi \left(\frac{x - \mu}{\sigma} \right) \right]^{-1/\xi - 1}$$
$$\times \exp \left\{ - \left[1 + \xi \left(\frac{x - \mu}{\sigma} \right) \right]^{-1/\xi} \right\} \quad (1)$$

A GEV model with parameters estimated by the method of maximum-likelihood is recommended for extreme value analysis of annual maxima (FEMA, 2004). By means of estimating the parameters of the model corresponding to the GEV distribution of the annual maxima, the GEV distribution of all 34 gauging stations across the PRD region are sorted. As the results of classification displayed in Figure 1, twenty stations are characterized by the GEV Type II and the other fifteen stations belong to Type III. Using the Kolmogorov-Smirnov method to test the goodness of fit (Chen et al. 2010), the result shows that the fitting results for all sites are accepted at 0.05 level of significance.

The flood level of the PRD is rising in general and the change of it only becomes significant after the 1980s, due to the accumulation of channel changes following the extensive sand excavation in this area. In order to comprehensively understand the water level characteristics in the delta, we also examined water level variation corresponding to different flood frequency in the PRD region calculated from series of pre-1980 and post-1980.

ENSO has a seasonal phase lock. It develops in spring and summer, reaches its peak in autumn and winter and declines in the following summer. Studies suggest the impact of ENSO on climate anomalies is associated with its stage closely (Ni et al. 2000). To reveal the interdecadal variation of correlation between flood level and ENSO at different stages, this study calculated the correlation coefficient between AMWL and ONI from December to February (DJF) of Niño 3.4 region to characterize the influence of ENSO on flood level under peak stage. ONI of DJF is replaced by ONI from June to August (JJA) to characterize the influence of ENSO on flood level under attenuation stage.

Take the moving window n = 11 in moving correlation analysis (Timo & Rasanen) and the moving correlation values are marked in the sixth year. The Pearson's correlation is used to give time-history plot of 11-year moving correlation coefficient.

3 SPATIOTEMPORAL CHANGES OF WATER LEVEL CORRESPONDING TO DIFFERENT FLOOD FREQUENCY

Annual maximum water level in the crisscross river network, serving as one of the most remarkable environment indicators for regional flood risk and water resources management, is a spatially continuous variable. Thus we can quantify AMWL at sites and map the water level with different return periods for the PRD region using the GEV model. Three sites (Daao, Dahengqin and Lezhu) out of

34 have been neglected because the data series are not long enough for analysis.

Figure 2 presents the resulting map of return periods (T = 10, 20, and 50 years) of AMWL. Generally, the water level as shown in Figures 2a–c indicates that flood frequency increases gradually from the coastal areas to the riverine system.

Besides, the water level increment corresponding to different increments of return periods (Figs. 2a–c), revealing the underlying flood risk in the PRD region, can serve as another remarkable indicator in supporting the regional flood risk and water resources management. The bar diagrams (T ranges from 10 to 50 years) in Figure 2a suggest an obvious increment of water level in all 34 sites. The difference between the water level of 50-year and 10-year flood is often larger than 0.5 m in the upper PRD while it is larger than 0.25 m in the middle and lower part. Particularly, the difference at Nanhua station is larger than 1.0 m, which is the largest among 31 sites in the PRD region.

In order to comprehensively understand the water level characteristics in the delta, we also examined water level variation corresponding to different flood frequency in the PRD region calculated from series of pre-1980 and post-1980. Figure 2d shows the change of annual maximum water level from 1958 to 2008. It can be seen that the entire PRD is characterized with a complex spatial pattern in terms of flood level change. Only a few stations have different trend of variation in the water level corresponding to various return periods. Concurrently, it is clear that Lanshi, Makou and Sanshui with significantly positive change trend

in 20-year and 50-year flood also have small negative change trend in water level for 10-year flood. In contrast, Shilong presents a positive change in return periods of 50 years as well as a negative change in return periods of 10 and 20 years.

On one hand, Nanhua and Xinjiapu present a significant downward trend in annual maximum water level, wherein Nanhua shows a negative change with a decrease of −0.75 to −1.50 m and Xinjiapu with a decrease of −1.05 to −0.70 m. On the other hand, upward trend is observed at 26 stations (i.e., 81.25% of all stations). The stations with negative trend are mainly distributed in the upper PRD. Significant increase in AMWL mainly occurs in the middle and lower PRD near the coast.

4 INTERDECADAL VARIATION IN THE RESPONSE OF FLOOD WATER LEVEL TO ENSO

The relationship between winter ENSO and flood level is dominated by positive correlations in the upper PRD region as shown in Figure 3a. The positive and negative correlation appear alternately which the phase of correlation coefficients with temporal change are significant. The correlation from the 1970s to the 1980s is very low. Figure 3b shows that flood level is negatively correlated with summer ENSO before the 1980s. It declines rapidly, even changing inversely from the 1990s to the 2000s. However, the correlation has been maintained at a low level, which means that the difference of interdecadal variation between the flood level and ONI index of Niño 3.4 region both before and after the transition is outstanding.

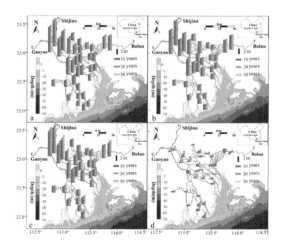

Figure 2. Mapping of water level corresponding to different flood frequency in the Pearl River Delta: (a) toll; (b) pre-1980; (c) post-1980; (d) variations of pre-1980 and post-1980.

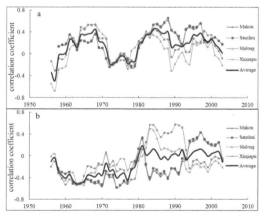

Figure 3. 11-year sliding correlation coefficients between AMWL and ENSO in the upper PRD: (a) winter; (b) summer.

Figure 4a shows that the relationship between winter ENSO and flood level is weak, and the positive and negative correlation appear alternately before the 1970s in the middle PRD. It has maintained a stable negative correlation from the 1970s to the 1980s. The interannual relationship between ENSO and flood level has a great change in the 1980s, and increases rapidly from negative correlation to positive correlation. However, the interannual relationship changes from positive correlation to negative correlation in the early 1990s. The flood level is significantly negatively correlated with summer ENSO (Fig. 4b), and the correlation coefficient has an abrupt change in the 1980s.

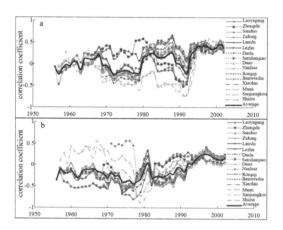

Figure 4. 11-year sliding correlation coefficients between AMWL and ENSO in the middle PRD: (a) winter; (b) summer.

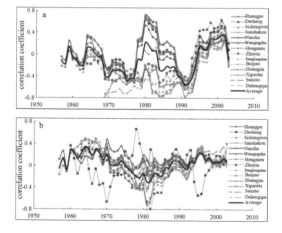

Figure 5. 11-year sliding correlation coefficients between AMWL and ENSO in the lower PRD: (a) winter; (b) summer.

Like the upper and middle PRD, the relationship between winter ENSO and flood level changes greatly in the 1980s in the low PRD (Fig. 5a). The difference is after a long-term stable significant negative correlation, there is a short turning to positive correlation and then becomes another strong negative correlation again. The negative correlation decreases gradually and turns to the positive correlation later. The relationship between summer ENSO and flood level of correlation (Fig. 5b) is stable and weak. The correlation is negative from the early 1970s to the 1990s as the occasion other decades to maintain a positive correlation.

5 CONCLUSION

We quantitatively analyzed the change in water level within the PRD region using GEV methods. Frequency analysis on annual maximum water level is scientifically and practically important in supporting regional water resource management and hazard mitigation planning. Based on annual maximum water level datasets extracted from 34 gauging stations in the PRD and the ONI data in the Niño 3.4 region of 1951–2008, the GEV model was used to get the frequency analysis. Spatiotemporal changes of flood level of pre-1980 and post-1980 was calculated. The interdecadal variation in the response of flood level to ENSO in the past fifty years was researched by using moving correlation. The following conclusions may be draw therefrom:

The resulting maps of annual maximum water level corresponding to different return periods in the PRD region suggest that the water level increases gradually from the tide controlled coastal areas to the riverine system. The most serious risk of flooding is in the coastal region because it is extremely vulnerable to the emerging flood hazards, storm surges and well-evidenced sea level fluctuations. The water level increments corresponding to different increments of flood periods in all 34 sites are obvious. In the upper PRD, the water levels show a decreasing trend while there is an increasing trend in the middle and lower PRD.

In the upper PRD, the transformation of return periods computed for the sub-series is not obvious both at Makou and Sanshui. The flood water level had a significant rise at Shilong and Xinjiapu after the 1980s. A significant long-term change of the flood situation with an increasing tendency of the extreme water level verse return periods occurs in the middle PRD. The alternations of extreme water level verse return periods throughout the estuary are similar on the whole. The return level presents an upward trend especially the four western watercourses of the Pearl River estuary.

The interdecadal variation of correlation between flood water level and ENSO is obvious in the whole PRD. The correlation has a great change in 1980s according to moving correlation analysis. Before and after the abrupt change, the correlation coefficients have obvious differences.

There are significant differences between the response of flood water level to winter ENSO and summer ENSO. The impact of winter ENSO on flood water level is significantly greater than summer ENSO and the fluctuation is stronger, indicating that ENSO in different seasons has different effects on the flood water level. Though the same is El Niño or La Niña event, its impact on the flood water level in the context of different ages has different performance in the Pearl River Delta.

REFERENCES

Al-Futaisi, A. & Stedinger, J.R. 1999. Hydrologic and economic uncertainties and flood-risk project design. *J. Water Resour. Planning Manag.* 125(6): 314–324.

Chen, Z.S, Liu, Z.M, Lu, J.F. 2010. Comparative analysis of parameter estimation methods of generalized extreme value distribution. *Acta Scientiarum Naturalium Universitis Sunyatseni (Natural Sciences).* 49(6): 105–109. (in Chinese).

FEMA (Federal Emergency Management Agency of the United States) 2004. Final draft guidelines for coastal flood hazard analysis and mapping for the Pacific coast of the United States.

Fisher, R.A, & Tippett, L.H. 1928. Limiting forms of the frequency distribution of the largest or smallest member of a sample. *Proc Cambridge Philos Soc.* 24: 180–190.

Jenkinson, A.F. 1955. The frequency distribution of the annual maximum (or minimum) values of meteorological elements. *Quarterly Journal of the Royal Meteorological Society.* 81: 158–171.

Jiang, C.J., Yang, Q.S., Dai, Z.J., et al. 2012. Spatial and temporal characteristics of water level change and its causes in the Zhujiang Delta in recent decades. *Acta Oceanologica Sinica.* 34(1): 46–56. (in Chinese).

Klaus, W. & Michael, S.T. 2011. El Niño/Southern Oscillation behaviour since 1871 as diagnosed in an extended multivariate ENSO index (MEI.ext). *International Journal of Climatology.* 34: 1074–1087.

Letetrel, C., Marcos, M., Martín Míguez, B., Woppelmann, G. 2010. Sea level extremes in Marseille (NW Mediterranean) during 1885–2008. *Continental shelf research.* 30(12): 1267–1274.

Ni, D.H, Sun, Z.B, Zhao, Y.C. 2000. Influence of ENSO cycle at different phases in summer on the East Asian summer monsoon. *Journal of Nanjing Institute of Meteorology.* 23(1): 48–54. (in Chinese).

Pugh, D.T. & Vassie, J.M. 1980. Applications of the joint probability method for extreme sea level computations. *Proceedings of the Institution of Civil Engineers (Part 2).* 69(4): 959–975.

Shi, C., Chen, X.H., Zhang, Q. 2012. Change-points of Water Levels in the Pearl River Delta in January and July for the Last Decades. *Tropical Geography.* 32(3): 233–240. (in Chinese).

Shiau, J.T. 2003. Return period of bivariate distributed hydrological events. *Stoch Environ Res Risk Assess.* 17(1–2): 42–57.

Singh, V.P. & Strupczewski, W.G. 2002. On the status of flood frequency analysis. *Hydrolog. Process.* 16(18): 3737–3740.

Timo, A. & Rasanen, M.K. 2013. Spatiotemporal influences of ENSO on precipitation and flood pulse in the Mekong River Basin. *Journal of Hydrology.* 476: 154–168.

Xu, S.D. & Huang, W.R. 2011. Estimating extreme water levels with long-term data by GEV distribution at Wusong station near Shanghai city in Yangtze Estuary. *Ocean Engineering.* 38(2–3): 468–478.

Zhang, W., Yan, Y.X., Zheng, J.H., et al. 2009. Temporal and spatial variability of annual extreme water level in the Pearl River Delta, China. *Global and Planetary Change.* 6(9): 35–47.

Advances in Energy, Environment and Materials Science – Wang & Zhao (Eds)
© 2016 Taylor & Francis Group, London, ISBN 978-1-138-02931-6

Effect of large flood in 2012 on river planform change in the Inner Mongolia reaches of the Yellow River

S. Yu & K. Wang
Yellow River Institute of Hydraulic Research, Zhengzhou, Henan Province, China

L. Shi
North China University of Water Resources and Electric Power, Zhengzhou, Henan Province, China

ABSTRACT: In 2012 the Yellow River Upper Reaches experienced the largest discharge flood since 1980s, which had significant effect on river planform of the Inner Mongolia river segment. This paper has analyzed feature of the flood and river planform change of Bayagaole~Toudaoguai reached by the use of satellite images, flow and sediment data and cross-sectional data. The results show that the flood in 2012 is characteristic of long duration, large flood volume and low sediment concentration; wandering river segments had no clear change in river regime, but transitional and meandering river segments were smoothened by cut-off.

1 INTRODUCTION

Major changes of river usually only occurs in large floods, which have the main power to convey sediment and organic matter. As a well-known high sediment concentration and wandering river, the Yellow River has been a research focus of all time.

The Inner Monglolia reach is characteristic of a typical wandering although in the Upper Reach. Past research has focused mainly on the effect of aggradation and degradation, water and sediment reduction (Wang, et al., 1996; Hou, et al., 2007; Ta, et al., 2008). Effect of reservoir operation on the reach has been investigated through temporal response processes of Toudaoguai cross-section (Ran et al., 2010). However, little attention has been paid to river planform change, especially by large flood. In 2012, the largest flood occurred in the reaches among the past more than 30 years, which brought about intense effect on the river.

The purpose of this paper is to investigate the river planform response to the large flood in 2012. The large flood may be recognized as a perfect natural experiment for studying the rapid processes of river regime. And this is of importance to practical applications in river management, and particularly to the further improvement of channel regulation in the Upper Yellow River.

2 FIELD SETTING

The Yellow River (Fig. 1) is called the cradle of the Chinese people, which has a total mainstream length of 5464 km, river basin area of 0.75 million km², annual average runoff of 46.4 billion m³, a long-term annual sediment discharge of 1.6 billion, and average sediment concentration of 36.8 kg/m³ measured at Sanmenxia station from 1919 to 1960 (YRCC, 1998; Wu et al., 2004; Wu et al., 2006). The upper reaches is from the headwater to Hekouzhen of Inner Mongolia Autonomous Region, with a length of 3472 km, and supplies about half of the total runoff and only 8% of the sediment flux (Zhao, 1996).

The Inner Monglia reaches are located at the end of the upper reaches, through the Ulan Buh Desert and the Kubuqi Desert, which are the source of sediment by wind. Several reservoirs have been constructed in the Upper Reaches since 1949, among which the Longyangxia reservoir and the Liujiaxia Reservoir have exerted great influence on the Yellow River regime and therefore on fluvial process (Wang et al., 2007). The Liujiaxia reservoir has been put

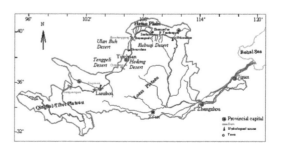

Figure 1. Sketch of the upper Yellow River reaches.

into operation in 1968, located below the Xunhua Gauging Station and with a storage capacity of 57.4×10^8 m³, which is used to hold part of flood water and release it in the dry season. The Longyangxia reservoir between the Tangnaihai Station and the Guide station has been applied in 1986, with the largest storage capacity of 247×10^8 m³ among all the reservoirs, which is capable of multi-year regulating. Besides, there are the Qingtongxia Dam built in 1968 and the Sanshenggong Dam built in 1961 for farmland irrigation in the Hetao Plain, and so on (Ran 2010). By these reservoirs or dams in the mainstream, the Upper Reaches have gradually completely controlled since the 1990s (Ta *et al.*, 2008).

The Inner Monglia Reaches is the northeast of the Yellow River, with a length of 840 km and average width of above 3000 m, which may be divided into two sections: mountainous reaches upstream of the Sanshenggong Dam and alluvial reaches downstream. The alluvial reaches are the research object of the paper. Three types of river pattern occur along the river: wandering from the Bayangaole Station to the Sanhuhekou Station, transitional from the Sanhuhekou Station to the Zhaojunfen Station and bending from the Zhaojunfen Station to Toudaoguai Station.

3 METHODS AND MATERIALS

All the hydrological, cross-section and river planform data adopted in the paper were supplied by the Yellow River Water Conservancy Commission (YRWCC). The measuring methods at all guaging stations and cross sections complied with the technical standards issued by the Ministry of Water Resources of China.

4 RESULTS

4.1 *Characteristics of the large flood in 2012*

The upper reaches of the Yellow River were abundant in rainfall in the July and August of 2012, compared to of previous years. Correspondingly, the main stem at the Lanzhou Station experienced the largest flood with peak flow discharge of 3860 m³/s since 1989. The flood in the inner Mongolia reaches was characteristic of peak flow discharge of more than 2700 m³/s, flood runoff of more than 13 billion m³ based on the hydrological data at the Shizuishan Station, the Bayangaole Station, the Sanhuhekou Station and the Toudaoguai Station.

In addition, the flood had very low sediment concentration of about 3 kg/m³, only with the exception of 5.5 kg/m³ at the Sanhuhekou Station. Sediment delivery was approximate, with the

value between $0.385 \sim 0.416 \times 10^8$t, however, with the exception of 0.756×10^8t at the Sanhuhekou Stations.

Furthermore, the flood had the feature of long duration of high flow discharge and small fluctuation of flow discharge and sediment concentration. For example, at the Bayangaole Station, the flood was in the continuous process of flow discharge more than 2000 m³/s for 44d and sediment concentration mostly between $2 \sim 4$ kg/m³ with average of 3.1 kg/m³.

4.2 *Changes in mainstream shifting scope by the large flood*

Mainstream shifting scopes and channel cut-off ratios are computed based on the satellite remote sensing images from Bayangaole to Toudaoguai in the Inner Mongolia reaches of the Yellow River during May, August, and October in 2012, representing before flood, in flood, after flood, respectively. From Figure 2, mainstream positions at some cross sections had no change with shifting scope of 0. While at other sections they shifted with different values. Especially at sections NO. 21, 27, 48, 56, 65, 83, 97, and 104, mainstream shifting scopes reach local maximum, ranging from 904 to 1962 m. According to water line data, mainstream shifting at the first three sections is caused by river twisting, and at the last five sections (which are marked with ellipses in Fig. 2) is caused by channel cut-off. Apart from the above sections, mainstream shifting has a general trend of increasing until section NO. 43 and then decreasing.

Mean and max mainstream shifting scopes are calculated for the three channel segments. Because channel cut-off is a sudden change of mainstream and therefore leads to great change of mainstream location, mainstream shifting scopes at these sections are excluded. Both mean and max mainstream shifting scopes have a decreasing trend along the river. However, compared with during $2007 \sim 2010$, mean mainstream shifting scopes in 2012 are less

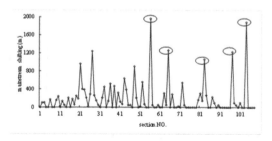

Figure 2. Mainstream shifting scope of the Inner Mongolia reaches of the Yellow River.

in Bayangaole–Sanhuhekou and Zhaojunfen–Toudaoguai, larger in Sanhuhekou–Zhaojunfen.

The great flood leads to channel cut-off at five channel segments, i.e. section NO. 55~57, 64~66, 82~83, 96~97 and 103~104, which are all in the reaches from Sanhuhekou to Toudaoguai. Cut-off ratios are calculated, with the values between 25%~45% and mean of 37%.

5 CONCLUSION

In this paper, the feature of the large flood in 2012 and river planform changes caused by it are analyzed in the Inner Mongolia reaches. The main contribution conclusions from this study are:

1. The flood in 2012 is the largest one among recent several decades, with peak flow discharge of more than 2700 m^3/s, flood runoff of more than 13 billion m^3, and with continuous process of flow discharge more than 2000 m^3/s for 44d and mean sediment concentration of 3.1 kg/m^3, which is very beneficial for the river change.
2. Planform of the wandering reaches from the Bayangaole Station to the Sanhuhekou Station remains, with less mean mainstream shifting scope and more large scale islands, although in the conditions of beneficial flow dynamics. The transitional and the meandering channels mainly experienced cutoffs in local distorted river bends. Some regulation projects are still necessary to maintain relatively stable river planforms.

ACKNOWLEDGMENT

Thanks are expressed to the Yellow River Conservancy Commission for the observation data and satellite remote sensing data.

REFERENCES

Ashmore P.E., 1991. How do gravel-bed rivers braid? Canadian Journal of Earth Sciences 28, 326–341.

Bertoldi W., Zanoni L., Tubino M., 2009. Planform dynamics of braided streams. Earth Surface Processes and Landforms, 34: 547–557.

Chappell A., Heritage G.L., Fuller I.C., et al., 2003. Geostatistical analysis of ground-survey elevation data to elucidate spatial and temporal river channel change. Earth Surface Processes and Landforms 28, 349–370.

Charlton M.E., Large A.R.G., Fuller I.C., 2003. Application of airborne lidar in river environments: the River Coquet, Northumberland, UK. Earth Surface Processes and Landforms 28, 299–306.

Eaton B.C., Lapointe M.F., 2001. Effects of large floods on sediment transport and reach morphology in the cobble-bed SainteMarguerite River. Geomorphology, 40: 291–309.

Feurer D., Bailly J.S., Puech C., Le Coarer Y., Viau A.A., 2008. Very-high-resolution mapping of river-immersed topography by remote sensing. Progress in Physical Geography 32(4), 403–419.

Formann E, Habersack H.M, Schober S., 2007. Morphodynamic river processes and techniques for assessment of channel evolution in Alpine gravel bed rivers. Geomorphology, 90: 340–355.

Ham D.G., Church M., 2000. Bed-material transport estimated from channel morphodynamics: Chilliwack River, British Columbia. Earth Surface Processes and Landforms, 25: 1123–1142.

Hassan M.A., Egozi R., Parker G., 2006. Experiments on the effect of hydrograph characteristics on vertical grain sorting in gravel bed rivers. Water Resources Research, 42 (W09408): 1–15.

Hou S., Chang W., Wang P., et al. Characteristics and cause of formation of channel atrophy at Inner Mongolia section of the Yellow River [J]. Yellow River, 2007, 29(1): 25–26, 29. (in Chinese).

Hou S., Chang W., Wang P., et al. Fluvial processes in Inner Mongolia reach of the Yellow River [J]. Journal of Sediment Research, 2010, (6): 44–50. (in Chinese).

Hou S., Wang P., Chang W., 2007, Evaluation on the volume of scour and fill of Inner Mongolia section of the Yellow River. Yellow River, 29(1): 21–23. (in Chinese).

Kiss T., Sipos G., 2007. Braid-scale channel geometry changes in a sand-bedded river: significance of low stages. Geomorphology, 84: 209–221.

Lane S.N., Westaway R.M., Hicks D.M., 2003. Estimation of erosion and deposition volume in a large, gravel-bed, braided river using synoptic remote sensing. Earth Surface Processes and Landforms 28, 249–271.

Li S.S., Millar R.G., Islam S., 2008. Modelling gravel transport and morphology for the Fraser River gravel reach, British Columbia. Geomorphology, 95: 206–222.

Long H., Du Y., Wu H., et al., 2007. Channel aggradation shrinkage and its impact on ice flood in Ningxia-Inner Mongolia reach of the Yellow River, 29(3): 25–26. (in Chinese).

Marcus W.A., Fonstad M.A., 2008. Optical remote mapping of rivers at sub-meter resolutions and watershed extents. Earth Surface Processes and Landforms 33(1), 4–24.

Mosley M.P., 1982. Analysis of the effect of changing discharge on channel morphology and instream uses in a braided river, Ohau River, New Zealand. Water Resources Research 18(4), 800–812.

Murray B., Paola C., 1994. A cellular model of braided rivers. Nature, 371: 54–57.

Rajaguru S.N., Gupta A., Kale V.S., et al., 1995. Channel form and processes of the flood-dominated Narmada River, India. Earth Surface Processes and Landforms, 20: 407–421.

Ran L., Wang S., Fan X., 2010. Channel change at Toudaoguai station and its responses to the operation of upstream reservoirs in the upper Yellow River. J. Geogr. Sci., 20(2): 231–247.

Rodrigues S., Breheret J., Macaire J., et al., 2006. Flow and sediment dynamics in the vegetated secondary channels of an anabranching river: the Loire River (France). Sedimentary Geology, 186: 89–109.

Ta W.Q., Xiao H.L., Dong Z.B., 2008. Long-term morphodynamic changes of a desert reach of the Yellow River following upstream large reservoirs' operation. Geomorphology, 97(3/4): 249–259.

Wang Y., Feng X., Wang L., et al., 1996. Influence of upstream reservoirs on the Inner Mongolia reach of the Yellow River. Yellow River, 1: 5–10. (in Chinese).

Wang S., Hassan M.A., Xie X., 2006a. Relationship between suspended sediment load, channel geometry and land area increment in the Yellow River delta. Catena 65, 302–314.

Wang Y., Li J., Liang H., 2006b. Granularity change in lower Yellow River channel. Soil and Water Conservation in China 8, 14–23.

Wang H., Yang Z., Saito Y., et al., 2007. Stepwise decreases of the Huanghe (Yellow River) sediment load (1950–2005): Impacts of climate change and human activities. Global and Planetary Change, 57: 331–354.

Westaway R.M., Lane S.N., Hicks D.M., 2003. Remote survey of large-scale braided, gravel-bed rivers using digital photogrammetry and image analysis. International Journal of Remote Sensing 24(4), 795–815.

Wu B.S., Wang Z.Y., Li C.Z., 2004. Yellow River Basin management and current issues [J]. J. Geographical Sci., 14, 24–37.

Yellow River Conservancy Commission (YRCC), 1998. Summary of the Yellow River Basin. The Yellow River Annuals, Vol. II, Henan People's Press, Zhengzhou, China (in Chinese).

Zhao Wenlin, 1996. Water Resources Science and Technology Series of the Yellow River: Sediment. Zhengzhou: Yellow River Water Resources Publishing House, 1–807. (in Chinese).

Advances in Energy, Environment and Materials Science – Wang & Zhao (Eds)
© *2016 Taylor & Francis Group, London, ISBN 978-1-138-02931-6*

Test of hysteretic freezing characteristics of clay during freezing and thawing

Dongyang Li & Bo Liu

School of Mechanics and Civil Engineering, China University of Mining and Technology, Beijing, China

ABSTRACT: In order to study the hysteretic freezing characteristics of clay, a test method is proposed to measure the relationship between Unfrozen Water Content (UWC) and temperature in clay. Based on the lump system analysis, the interior temperature of sample remains essentially uniform all the time during a heat transfer process if its volume is sufficiently small. Four small-volume clay samples were frozen and thawed by natural convection in isothermal air, and their temperature-time curves were recorded. Based on the principle of energy conservation, a thermal model was proposed to predict the characteristic curves of UWC through analyzing the relationship between temperature change and heat energy in each stage of freezing and thawing cycle. This method was applied to test clay with moisture content of 35%. Four clay samples were frozen to −27°C in −30°C isothermal air and then thawed by natural convection in 28°C isothermal air. The results showed that the UWC in freezing process is always greater than that in thawing process, and the maximum difference is about 4.3%~4.6% for the temperature in the range of −4~−5°C. Thus, the effect of hysteresis should be considered when measuring and analyzing the thermal and mechanical properties of clay.

1   INTRODUCTION

When soil freezes and thaws, the Unfrozen Water Content (UWC) is dependent not only on the temperature but also on the change trend of the temperature, i.e. the unfrozen water characteristic curves during freezing and thawing are not identical. Usually, the UWC is greater in the freezing than in thawing, this phenomenon is known as freezing hysteresis. Many soil exhibit hysteresis, which, in a systems sense, is the memory of past states of the system (Guymon and Luthin, 1974). Williams (1962) observed the hysteresis in the curves of specific heat change in a soil during freezing and thawing. Hysteresis of unfrozen water in soil was also found and reported when researchers measured the UWC using Nuclear Magnetic Resonance (NMR) (Tice, et.al., 1981) and Time Domain Reflectometry (TDR) (Stähli and Stadler, 1997). Brian D. Smerdon and C.A. Mendoza (2010) investigated the hysteretic freezing characteristic of riparian peatlands in the western boreal forest in Canada using TDR. Oleg Petrov and István Furó (2010) explored the freezing and melting temperature hysteresis of water in Vycor porous glass and controlled pore glass using NMR cryoporometry and NMR relaxometry techniques.

During the last decades, some empirical methods have been developed for measuring UWC in soils. Tsytovich (1975) suggested an equation for determination of UWC in soils by testing its plasticity index, plastic limit and negative temperature. Anderson and Tice (1973) had summarized a simple power function containing 2 parameters. Typical values of the characteristic parameters for several soils were studied based on the specific surface by many researchers in 1973~1988 (Akagawa, 1988; Anderson and Tice, 1972; Anderson and Tice, 1973; Kay, Tice and Sheppard, 1981; Oliphant, Tice and Nakano, 1983; Patterson and Smith, 1981; Simth, 1984; Smith, 1985). Michalowski (1993) described the UWC in frozen soils using a 3-parameter function which is flexible enough to accommodate many soils. Based on the results of 141 DSC experiments on 6 mono-mineral soils, Kozlowski (2007) presented a semi-empirical model describing the variation of the UWC in soil-water systems as a function of temperature.

However, all of these empirical methods predict UWC in soil without making a clear distinction between freezing and thawing. Actually, hysteresis of unfrozen water in soils plays an important role in many heat transfer problems, such as engineering, geophysics, meteorology and agriculture.

So, understanding of the hysteresis of UWC in frozen soils will help one quantitatively analyze the values of soil's thermal properties in these problems. In this article, we will develop a new method to explore the hysteretic freezing characteristics of clay.

2 TEST PROCEDURES AND RESULTS

The samples are undisturbed clay soils extracted from Shandong province (China). The height of thin-wall sampler is 47 mm and inner diameter is 32 mm. The specific density of soil particle is 2.75 and specific heat of dry soil is 0.769 kJ/(kg·K) at −10°C and 0.845 kJ/(kg·K) at 10°C. The test procedures are as follows:

1. Weigh the net weight of thin-wall steel samplers.
2. Wrap soil samples with fresh-keeping film. Weigh soil and sampler, and then calculate the mass of undisturbed soil m (Table 1), as shown in Figure 1.
3. Put temperature sensors (Pt100, 3 mm diameter) in the center of samples to monitor the temperature, as shown in Figure 2.

2.1 *Testing clay samples during freezing*

Put all samples into freezer and freeze them in air at constant temperature of −30°C from initial temperature of 16°C to a negative temperature by natural convection. Record the time history of temperature, as shown in Figure 3. (Due to space limitation, only 4 cases are presented in this article).

As shown in Figure 3, the time history curve of temperature can be roughly divided into 3 stages.

1. First stage, the temperature decreases sharply with time when temperature of soil exceeds 0°C. It means that the heat in soil samples quickly dissipates to the outside cold air and the water in soils has not transformed into ice yet. This stage is cooling process of unfrozen soil. (Fig. 4).
2. Third stage, the temperature obviously decreases again when temperature is lower than −1.6°C, but it is not as fast as the first stage because certain amount of liquid water gradually changes into ice with the temperature decreasing in soil. This stage is the cooling process of frozen soil. (Fig. 4).

2.2 *Testing clay samples during thawing*

Freeze the soil samples up to T_{0t} (about −27°C) and then heat the samples in isothermal air at constant temperature of 28°C by natural convection. Record the time history of temperature until the sample temperature approaches air temperature, T_a (about 15°C) (Fig. 5). Then, dry the sample and weigh it. Lastly, evaluate the mass of soil particle

Table 1. Samples data of clay.

No.	m (g)	w (%)	T_{0f} (freezing) (°C)	T_{af} (freezing) (°C)	T_{0t} (thawing) (°C)	T_{at} (thawing) (°C)
Clay-A1	63.272	35.058	16.0	−30.0	−27.0	28.0
Clay-A2	63.199	35.353	16.0	−30.0	−27.0	28.0
Clay-A3	63.015	35.813	16.0	−30.0	−27.0	28.0
Clay-A4	63.747	35.010	16.0	−30.0	−27.0	28.0

Figure 1. Clay samples.

Figure 2. Fix up temperature sensors.

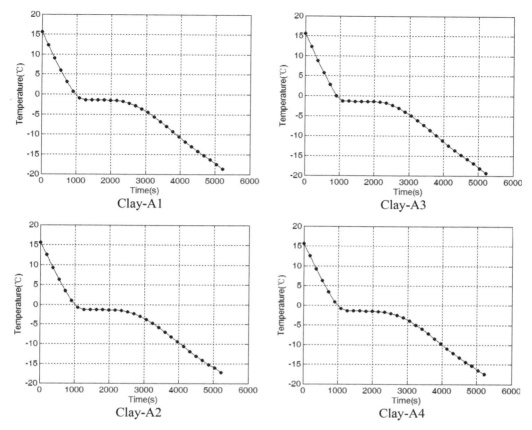

Figure 3. Recorded time history of temperature during freezing at core center.

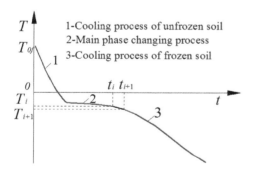

Figure 4. The time-history of temperature of soil sample during freezing.

and total water content w. All test data are listed in Table 1.

Similarly, the time history of temperature during thawing can also be roughly divided into 3 stages (see Fig. 5):

1. First stage, the temperature increases sharply with time till soil temperature reaches $-8°C$. This implies that the phase composition remains constant or undergoes unnoticeable changes when temperature is below $-8°C$. This stage is the warming process of frozen soil (Fig. 6).

2. Second stage, the temperature slowly increases from $-5°C$ to $-1.6°C$ and then stays within a steady range of $-1.6\sim-1.2°C$ because latent heat of fusion is required when ice transforms to water. So, the latent heat of fusion prevents the temperature from increasing until most ice transforms into water. This stage is the major phase changing process. The soil consists of both frozen and melted components. (Fig. 6).

3. Third stage, the temperature obviously increases again when temperature is above $-1.2°C$. It indicates that the phase change is completed and all ice in soil has been melted. This stage is the warming process of melted soil. (Fig. 6).

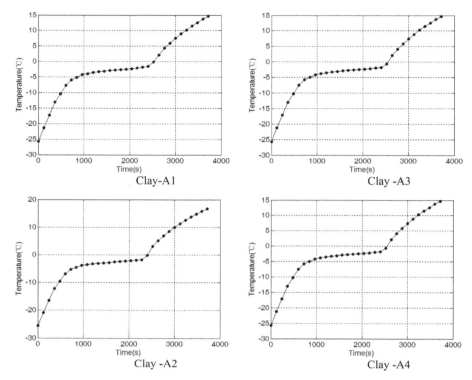

Figure 5. Recorded time history of temperature during thawing at core center.

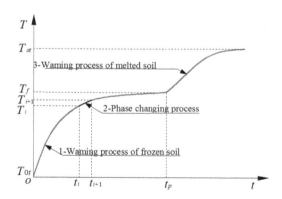

Figure 6. The time-temperature history of soil sample thawing.

3 MODELING UNFROZEN WATER CHARACTERISTIC CURVE

3.1 *Temperature change without phase transformation*

During cooling process of unfrozen soil (Fig. 4), the temperature decreases sharply with time because the heat in soil samples quickly dissipates to the outside cold air and the water in soils has not transformed into ice yet. This stage is the cooling process of unfrozen soil. In this process, the change of soil's specific heat is very small thus can be neglected. According to lumped parameter method, the cooling process of unfrozen soil without phase transformation can be described by:

$$T(t) = T_{af} + (T_{0f} - T_{af}) \exp\left(-\frac{t}{k_u}\right) \qquad (1)$$

where k_u is characteristic time of unfrozen soil, s.

$$k_u = \frac{mC_u}{h_f A} \qquad (2)$$

where h_f is convective transfer coefficient between unfrozen soil and cold air; A is the outside surface area of soil sample; and C_u is mass specific heat of unfrozen soil.

$$C_u = \frac{C_s + wC_w}{1 + w} \qquad (3)$$

278

where C_s, C_w is the specific heat of soil particle and water respectively; w is the total water content. Combining Eqs. (1), (2) and (3), convective heat transfer coefficient between soil and air during freezing can be obtained, as listed in Table 2.

The same method can be used for the analysis of the warming process of melted soil. During the warming process of melted soil, the temperature rises as the heat transfer by convection between soil and warm air and the ice in soil has totally melted. The change of soil's specific heat may be neglected. According to lumped parameter method, the warming process of melted soil without phase transformation can be described as:

$$T(t) = T_{at} + (T_f - T_{at})\exp\left(-\frac{t - t_p}{k_m}\right) \tag{4}$$

where t_p is the time that ice phase change takes to complete, s.

$$k_m = \frac{mC_u}{h_m A} \tag{5}$$

The characteristic time k_m can be obtained by fitting the warming process curve of melted soil with Eq. (4). The convective transfer coefficient (h_m) can be solved from Eqs. (3) and (5). The parameters resulted from best-fitting are listed in Table 2.

3.2 Phase changing process

In the main phase change process during freezing, the heat of soil continuously dissipates into air by the means of convection, but the soil's temperature always stay within a steady range, as shown in curve 2 in Figure 4. The reason is that, the latent heat of fusion holds the temperature until most liquid water transforms into ice. The soil temperature stays in a stable range due to the interaction between air cooling convection and latent heat of water.

After all free water changes into ice at the freezing point, part of loosely bound water gradually

transforms into ice with the temperature decrease, as shown in curve 3 in Figure 4. However, the amount of loosely bound water is very small, so its latent heat cannot be balanced by the heat loss in the cooling air convection, as a result, the temperature of soil decreases gradually.

Assuming that the temperature decreases from T_i to T_{i+1}, and the increment of temperature is extremely small and changes linearly with time (Fig. 4).

$$\beta_i = \frac{T_{i+1} - T_i}{t_{i+1} - t_i} = \frac{T - T_i}{t - t_i} \tag{6}$$

The total energy leaving the sample during this process is

$$Q_s = \int_{t_i}^{t_{i+1}} hA\,(T_a - T)\,dt \tag{7}$$

Substitution of Eq. (6) into Eq. (7) leads to

$$Q_s = hA\int_{t_i}^{t_{i+1}} [T_a - T_i - \beta_i(t - t_i)]\,dt \tag{8}$$

When the water content keeps constant, the change in the total energy of the sample during this time interval is

$$Q_c = mC_f(T_{i+1} - T_i) \tag{9}$$

where C_f is specific heat of frozen soil,

$$C_f = \frac{C_s + w_u(T_i)C_w}{1 + w} + \frac{w - w_u(T_i)}{1 + w}C_i \tag{10}$$

where $w_u(T_i)$ is unfrozen water content at temperature T_i. Note that the Eq. (9) is an approximate calculations, it is only appropriate if the difference between T_{i+1} and T_i is small.

Based on energy conservation principle, the decrease in total energy of the sample during freezing Q_c is equal to the difference between total energy leaving the sample Q_s and the latent heat of water changed into ice. That is,

$$Q_c = Q_s - \Delta m_i L \tag{11}$$

where Δm_i is the mass of water changed into ice, g, and L is latent heat, 334 kJ/g.

The unfrozen water content at T_{i+1} can be determined from

$$w_u(T_{i+1}) = w_u(T_i) + \frac{\Delta m_i}{m_s} \tag{12}$$

Table 2. Fitted data of clay.

No.	C_u kJ/(kg·K)	k_u (s)	h_f W/(m²·K)	k_m (s)	h_m W/(m²·K)
A1	1.7075	2272	8.0739	1657	11.0653
A2	1.7128	2342	7.8426	1430	12.8423
A3	1.7212	2160	8.5220	1620	11.3562
A4	1.7066	2334	7.9095	1621	11.3908
Mean	1.7210	2277	8.0870	1582	11.6637
Std	0.0067	84	0.3059	102	0.7992

where m_s is the mass of soil particles, g. The unfrozen water content is equal to total water content at freezing point T_f, i.e.,

$$w_u(T_f) = w \qquad (13)$$

The UWC at each temperature point can be derived and estimated from Eq. (12). The unfrozen water content characteristic curve during thawing can be obtained in the similar manner. Based on the principle of energy conservation, the total heat energy absorbed by frozen soil consists of two parts, one is used for soil temperature rising, the other is for ice changing into water. Thus, the analysis and solution are similar to those in thawing (Liu Bo and Li Dongyang, 2012).

4 RESULTS AND DISCUSSION

From the analysis above, the convective transfer coefficient between soil and cold air h is obtained by fitting measured value during cooling process of unfrozen soil or warming process of melted soil with Eq. (1) or Eq. (4). Based on the principle of energy conservation, the amount of ice that changes phase can be solved by the amount of latent heat absorbed or released during phase change. The UWC characteristic curve during freezing or thawing can be deduced with Eqs.(12) and (13), as shown in Figure 6.

In Figure 6, w_{uf} and w_{ut} are unfrozen water content versus temperature during freezing and thawing, respectively. $w_{uf} - w_{ut}$ is the difference between two processes. The following experimental phenomena were observed:

1. The significant transitional phase occurs at $0 \sim -2°C$ in both freezing and thawing processes. The UWC is sensitive to the temperature in this temperature range, but the difference is not obvious between two processes. The phase change of free water is independent of freezing and thawing process. In this case, most ice comes from free water.
2. When the temperature rises from $-17°C$ to $-8°C$ in the thawing process, the change of UCW is negligible. When the temperature reaches $-8°C$ in thawing process, the change becomes noticeable. During freezing, a lot of loosely bound water is frozen after the temperature drops from $-8°C$ to $-17°C$, and the UCW is greater in freezing than in thawing. So, the phase change of loosely bound water is strongly dependent of freezing and thawing process.

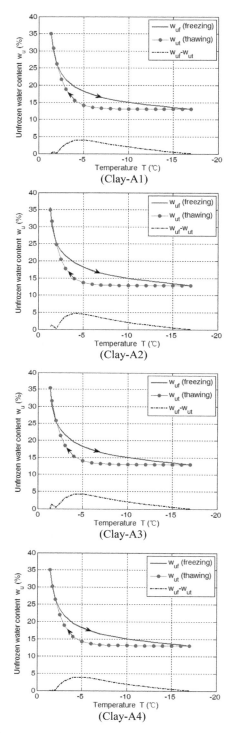

Figure 7. Characteristic curves of unfrozen-water content during freezing and thawing.

280

3. Comparison of clay's UWC during freezing and thawing indicates that, the largest difference of UWC w_{uf}-w_{ut} is about 4.3%~4.6% when the temperature range is −4°C ~−5°C which nearly reaches to 1/3 of unfrozen water content in practically frozen state (−17°C). For this reason, it is necessary to make a distinction between freezing and thawing process when measuring and analyzing the thermal and mechanical properties of frozen clays.

4. In the test method proposed in this article, the hysteretic freezing characteristic of clay-water system can be successfully observed.

5 CONCLUSION

The results show that, the phase change of free water is independent of freezing and thawing process when the significant transitional phase occurs at 0~−2°C, in this scenario, the difference of UCW is indistinguishable between two processes. However, in the temperature range of −8°C to −17°C, the phase change of loosely bound water is strongly dependent of freezing and thawing process, and the UCW is greater during freezing than during thawing.

Based on the comparison of the clay's UWC curves during freezing and thawing, it is found that the largest difference is about 4.3%~4.6% in the temperature range of −4°C~−5°C, which nearly reaches to 1/3 of UWC in practically frozen state (−17°C). For this reason, it is necessary to make a distinction between freezing and thawing process when measuring and analyzing the thermal and mechanical properties of frozen clay.

REFERENCES

Akagawa, S. (1998). Experimental study of frozen fringe characteristics. Cold Reg. Sci. Technol. 15(3): 209–23.

Allen, R. Tice, Duwayne, M. Anderson (1981). Unfrozen water contents of submarine permafrost determined by nuclear magnetic resonance, Engineering Geology Volume 18, Issues 1–4, December 1981, Pages 135–146.

Brian, D. Smerdon, C.A. Mendoza (2010). Hysteretic freezing characteristics of riparian peatlands in the Western Boreal Forest of Canada, HydrologicAL Processes, 24, 1027–1038 (2010).

Guymon, G.L. and J.N. Luthin (1974). A coupled heat and moisture transport model for Arctic soils, Water Resour. Res., 10(5), 995–1001.

Kay, B., M. Fukuda, H. Izuta, and M. Sheppard (1981). The importance of water migration in the measurement of the thermal conductivity of unsaturated frozen soils. Cold Reg. Sci. Technol. 5: 95–106.

Kozlowski. (2007). A semi-empirical model for phase composition of water in clay–water systems. Cold Regions Science and Technology Vol. 49, 226–236.

Liu, Bo, Li, Dongyang, A simple test method to measure unfrozen water content in clay-water systems, Cold Regions Science and Technology Volume 78, July 2012, Pages 97–106.

Manfied Stähli, Daniel Stadler (1997). Measurement of water and solute dynamics in freezing soil columns with time domain reflectometry, Journal of Hydrology, Volume 195, Issues 1–4, August 1997, Pages 352–369.

Oleg Petrov and István Furó (2011). A study of freezing-melting hysteresis of water in different porous materials. Part I: Porous silica glasses, Microporous and Mesoporous Materials 138(2011) 221–227.

Oliphant, J.L., A.R. Tice, and Y. Nakano (1983). Water migration due to a temperature gradient in frozen soil. In Proc. 4th Int. Conf. on Permafrost, Fairbanks, Alaska. Washington, D.C.: National Academy Press, vol. 1, pp. 951–956.

Patterson, D.E., and M.W. Smith (1981). The measurement of unfrozen water content by time-domain reflectometry: Results from laboratory tests. Can. Geotech. J. 18(1): 131–44.

Radoslaw L. Michalowskia (1993). A constitutive model of saturated soils for frost heave simulations. Cold Regions Science and Technology Vol.22, No. 1. 47–63.

Smith, M.W. (1984). Soil freezing and frost heaving at the Caen experiment. In Proc. Seminar on Pipelines and Frost Heave, Caen, France. Ottawa: Carleton University, pp. 19–22.

Smith, M.W. (1985). Observation of soil freezing and frost heave at Inuvik, Northwest Territories, Canada. Can. J. Earth Sci. 22(2): 283–90.

Tsytovich, N.A. (1975). The mechanics of frozen ground, McGraw-Hill, New York. 43–45.

Wiiliams, P.J. (1962) Specific heats and unfrozen water content of frozen soils. Proceedings of first Canadian conference on permafrost, April 1962. National research council of Canada, Technical memorandum No. 76, 109–126.

Advances in Energy, Environment and Materials Science – Wang & Zhao (Eds)
© 2016 Taylor & Francis Group, London, ISBN 978-1-138-02931-6

Color/Turbidity detection of water supply pipeline cleaning and its application

Xue Li Xu

Daqing Oil Field Co., Water Service Company Pipeline Company, Daqing, Heilongjiang, China

ABSTRACT: This paper introduced the log normal distribution model of color (pseudo color) and turbidity, as well as the application of this model in tap-water pipe line cleaning. Color and turbidity are commonly used technical indicators to evaluate the cleaning effect of tap water pipeline, and is an important monitoring method to understand the status of water supply network. The lognormal distribution equation of color and turbidity of the monitoring water pipeline cleaning was given to point out the relationship between parameter combinations of the internal state of pipeline, the significance of parameters was given as following: (1) The shape parameter indicates the degree of concentration of sediment cleaned out; (2) Time shift parameter indicates clean degree of tap water without cleaning, which is compared to the clean degree of pure water; (3) The scale parameter indicates the cleaning speed of sedimentation. The examples show that the model parameters are of a certain practical significance on the internal situation analysis of pipeline.

1 INTRODUCTION

Urban water supply pipeline safety and hygiene cleaning which is related to the important projects of the health of urban residents attract home and abroad general concern (Alexander, 2005; Liu, 2009). It is also one of the important prevention measures of pipeline corrosion and fouling adhesion, which cause drinking water secondary pollution (Vlassis, 2010). The scientificalness of tap water pipe cleaning effect evaluation method and reliability of indicators, and urban water resources monitoring (Ashok, 2011; Nikoo, 2011) are not only one of the critical links in supply of clean water, but also the hot topic issues in studying environment and water resources (Ngwenya, 2013; Wu, 2011; Han, 2013). According to GB5749-2006 standard, in the 4 major categories of conventional indicators in drinking water, water turbidity and color are sensory traits and general chemical indicators ranked in the top two common and important scientific evaluation indicators. According to the physical meaning of turbidity and color, they carry a lot of information in pipeline. They are not only used for water quality analysis but also for the diagnosis of internal conditions of pipelines. The color and turbidity indicators of sensory traits and common chemical indicators for water supply pipe cleaning in this paper are to establish the modified logarithmic normal distribution equation model, with the discussion of the application of the model parameters in the secondary pollution caused by pipeline aging corrosion and fouling.

This method in this paper provides a feasible way for the evaluation of pipeline cleaning effect and the prejudgment of changes in the pipeline.

2 COLOR/TURBIDITY OBSERVATION MODEL

Micro particle distribution and geometry identifying have an important influence in many fields of natural science, material science, environmental science, and military science. Related to distribution type, parameter explanation research has always been paid the main attention in related industry (Lemb, 1993; Alexander, 2005; Tomáš, 2011; Corrochano, 2015). Kolmogorov points out in literature that the size of crushing product obeys normal distribution. Georges Fournier and Miroslaw Jonasz point out in literature (Miroslaw, 1996) that the particle size of the particles in the sea water satisfies lognormal distribution. Before Kolmogorov, Rosin P. and Rammler E. studied the distribution law of coal particles according to Weibull distribution, and raised the Rosin Rammler distribution of the grain size of coal particles (Rosin, 1993). Litterature (Taya, 2012) points out, in some cases, the Nuliyama–Tanasawa distribution is more suitable for the particle size distribution of micro bubbles. And this indicates that the grain size distribution curve shapes possibly change a lot with the matters and the different conditions of particle produced (Tomáš, 2011; Rosin, 1933; Taya, 2012). Whether the color (According to the national standard, uses the Platinum cobalt colorimetric method to measure the chromaticity, false color) and turbidity distribution of the tap water pipelines obey the known distribution needing to be judged according to the sample data. Based on the character of the actual sample turbidity and color variation curve shape, which is similar to the logarithmic normal distribution, and the model of literature, we make the following assumptions: 1)the three-dimensional polyhedron with the suspended particles as its vertex in the water, no longer contains suspended particles influencing water color and turbidity; the polyhedron is called "water particles"; 2) with the increase of suspended particles, "water particles" is "crushed" into smaller size "water particles", the process meets the gridding process in litterature of different size of particles grinding; 3) the cleaning process is seen as smashing "water particles"; and the diameter of "water particles" is particle size is proportional to the cleaning time. According to these 3 hypotheses, the distribution of water particles of the different diameters in this process of cleaning meets the lognormal distribution model. The same type of suspended particle's flux is proportional to the suspended particle density. According to the definition of water color and turbidity, the suspended particle distribution density (the numbers in unit volume) is proportional to the color and turbidity. According to the definition of "water particles", the distribution density of suspended particle is proportional to the density of "water particles". Thus, the color/turbidity value is proportional to the density distribution of "water particles". Log normal distribution function is $g(x)$:

$$g(x) = \begin{cases} \dfrac{e^{-0.5\left(\frac{\ln(x)-\mu}{\sigma}\right)^2}}{x\sigma\sqrt{2\pi}}, & x > 0 \quad \int_{-\infty}^{+\infty} g(x)dx = 1 \\ 0, & x \leq 0 \end{cases} \quad (1)$$

Under the assumption that the $g(x)$ is proportional to the $f(x)$, $f(x)$ is the same with the logarithm normal distribution $g(x)$ in shape in formula (1). In order to reflect the initial suspended solid condition in the model, the initial water suspended particle quality is assumed to be the suspended matter produced in the initial stage of cleaning, which caused the initial deviation value x_0. The ratio coefficient of $g(x)$ and $f(x)$ is N, then the color/turbidity model is $f(x) = N*g(x - x_0)$. The equation of color/turbidity variation with time with initial time deviation is

$$f(x) = \begin{cases} N\dfrac{e^{-0.5\left(\frac{\ln(x-x_0)-\mu}{\sigma}\right)^2}}{(x-x_0)\sigma\sqrt{2\pi}}, & x > x_0, \quad N = \int_{x_0}^{+\infty} f(x)dx \\ 0, & x \leq x_0 \end{cases}$$

$$(2)$$

In formula (2), x is cleaning time, σ is curve shape parameter in lognormal distribution, and μ is the logarithm average value of sample taking time point. Through the substitution as follow $z_\sigma = (x - x_0)$, $\lambda = \exp(-\mu/\sigma)$, formula (2) is deduced into the following color/turbidity model:

$$f(x) = Ng(x-x_0) = \begin{cases} N\dfrac{e^{-0.5\left(\frac{\ln(x-x_0)-\mu}{\sigma}\right)^2}}{(x-x_0)\sigma\sqrt{2\pi}}, & x > x_0, \\ 0, & x \leq x_0 \end{cases}$$

$$z = (x-x_0)^{\frac{1}{\sigma}}, \quad \lambda = e^{-\frac{\mu}{\sigma}}, \quad N = \int_{x_0}^{+\infty} f(x)dx \quad (3)$$

Among them, $f(z)$ indicates color/turbidity with time changing; x is the time; ratio coefficient N indicates the observation accumulation value (total turbidity integral value and total color integral value); x_0 indicates observation time's backward deviation of initial water quality compared with pure water. And initial water quality can be comprehended as "pure water" of no pollution by pipeline cleaning; x_0 indicates pipeline cleaning time for pure water reaches the degree of initial water; Initial model color/turbidity value show the clean degree of initial water quality compared to pure water. $\lambda = \exp(-\mu/\sigma)$ is scale parameter, mode is $x_{max} = x_0 + \exp(\mu-\sigma^2)$, the peak of the model is $f(x_{max}) = N(2\pi)^{-0.5}\sigma^{-1}$ $\exp(0.5\sigma^2-\mu)$. Scale parameter $\lambda = \exp(-\mu/\sigma)$ and noise-signal ratio (NSR) (mathematical expectation/variance. In the literature (Rafael, 2007) it is indicated with square form) are very similar, the latter is commonly used in image communication signal quality judgment (Rafael, 2007; Schroeder, 1999). In this paper, μ is the logarithm average value of the time, and is not the expectation value. The λz relationship in this paper is based on formula (3) is called a scale parameter to the time of λ. The product λz of the formula (3) indicates the meaning of expansion of independent variable for λ. This paper explains 4 parameters according to the explanations of log normal distribution and the expansions of relevant particle parameter distribution, combining with the real color and turbidity data in pipeline washing. For more definition and meaning for parameters of formula (3), you can refer to relevant litteratures for log normal distribution and the particle distribution (Lemb, 1993; Miroslaw, 1996). According to the observation of color and turbidity data, the initial water quality results in the observation time backward deviation of color and turbidity. In order to solve the problem of observation time backward deviation, formula (3) is added with time deviation value x_0 on the log normal distribution model. In order to keep the consistency with the observation values, an expansion parameter N is needed. If no parameters change except x_0, the curve will deviate with x_0. Different σ's peak point is in $(x_0, x_0 + \exp(\mu))$ range. With the smaller of σ, the peak point is closer to $x_{max} = x_0 + \exp(\mu)$ on the right, and at around the peak the whole curve is pulled up with waist shrinking. With the smaller of λ the curve is lengthened and the peak is down; x_{max} moves to the right infinitely and the integral value is more and more dispersed.

Different characters of the model will be presented according to the different conditions caused by internal sedimentation or secondary pollution or other reasons. According to the characteristics of sediment shedding and drifting, the meanings of parameters in typical cases are listed as following:

a. If the sediment in the tube wall is easy to fall off and be washed, shape parameter and scale parameter are relatively big;
b. If the sediment on the wall is not easy to fall off or be washed, shape parameter and the scale parameter are relatively small;
c. If there are both (a) and (b) cases on the tube wall, the shape parameter is big and the scale parameter is small;
d. If there is a continuous secondary pollution source, the translation parameter will be naturally big;
e. If the size of the particle is large or the particle is heavy or the adhesion force is strong, the scale parameter becomes smaller and the cleaning time is relatively longer;
f. If the situation (a) to (c) and (d) exist simultaneously, the parameters' conditions are as stated in previous.

In the color and turbidity fitting on model, considering the instability of the initial observation data, the nonlinear least squares fitting algorithm has been added the fading character. Nonlinear least square model is

Objective function: $\min\limits_{x_0,\sigma,\mu,N} F(x_0,\sigma,\mu,N)$;
Nonlinear least square method

$$F(x_0,\sigma,\mu,N) = \sum_{i=1}^{m} w_i (f(x_i) - y_i)^2,$$

The fading factor $w_i = \frac{1}{(m-i+1)^2}$, the sampling value of the chroma and turbidity;

Constraint condition: $-\infty < x_0 < \min\limits_{1 \le i \le m}\{x_i\}$, $0 < \sigma < +\infty, 0 \le \mu < +\infty, 0 < N < +\infty$. Among them, m is the sample number. Calculate the parameters of x_0, σ, μ, and N through the MATLAB trust region method.

3 PIPELINE FLUSHING EFFECT ANALYSIS

This paper is based on the data of Daqing Qianjin trunk, Xingxi trunk. The network flushing uses a strong oxidizing agent of sodium hypochlorite. The water soluble sodium hypochlorite reacts in the water: $NaClO + H_2O \rightarrow HClO + NaOH$. Sodium hypochlorite's dissociation rate is 99.99%, so it can be identified as sodium hypochlorite that is all dissociated as HClO. Before the turbidity and color peaks, there is always a minimum valley whose water quality is near the qualified water. The minimum value is the static water quality after dosing. This value is not seen as the tube wall dosing cost. In accordance with the regulations of the GB5749-2006, drinking water turbidity limit is 1NTU and when it is to the water resource limitation and water purification technology conditions limitations, it is 3 NTU. Drinking water color (platinum cobalt colorimetric method) value's maximum limitation is 15 degrees. These indicators are the termination conditions of flushing. The air compressor pumps according to the actual situation, the pressure value verifies. And the outlet of the tube is spraying. The dosing pump's flow is 8 m³/h. In order to prevent the pollution of aq sterile, all emissions are not let out but be sewered into pipe canal. By such closed management, the discharged sewage is isolated from the external environment. It is an important technical parameter to control the air pressure and the gas flow in the actual cleaning project. If the pulse flow in pipelines is in the formation of turbulence, it will form the best spray type cleaning effect. Control the flow of water, on one hand to prevent advection resulted from too large a flow, on the other hand to prevent the air water separation of upper air and below water resulted by too small flow. The air volume and the flushing effect commonly present a positive correlation. It is needed to prevent tube rupture caused by too heavy pressure. So water spraying in the terminal of tube is needed to be seen in actual operation. If there is no water spraying in the terminal, it suggests the pressure is low, and if there is straight fast water flow in the end, it suggests the pressure maybe too high. So aiming at different flushing region, different pressure, and air flow volume are adjusted to. The following experiment is actually measured, and color is measured through Platinum cobalt colorimetric method.

Example 1: Qianjin trunk is setted two air pumping points and one sewage let-outs at the end. As for its flushing length is relatively long, series washing method is taken considering the time limit to guarantee the flushing quality. Color and turbidity model fitting curves are shown in Figure 1 and Figure 2. Figure 1 is color model parameters

Table 1. Daqing Qianjin water supply pipeline basic data and flushing water quality change model coefficients and errors.

Number		Channel name		Pipeline material	Diameter (mm)		Flushing distance (km)		Flushing time (h)	
1		Qian jin trunk		Cast iron	600		10.7		24	
Time	9:30	11:30	13:00	14:30	18:00	20:30	22:00	1:30	3:40	7:00	8:30
Color	116	306	290	550	495	305	217	102	41	14	14
Turbidity	5.26	26.9	21.6	106	94.7	52	19.2	15.61	2.17	0.96	0.91

Model parameter	σ	N	λ	x_0	Mean square error
Color model	0.2443	5618.40	0.000008	−10.40	42.39
Turbidity model	0.3749	830.48	0.004719	0.00	9.74

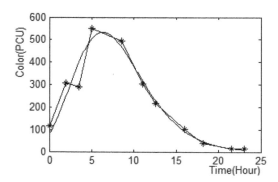

Figure 1. 24-hour cleaning chroma changes of Qianjin water supply pipeline.

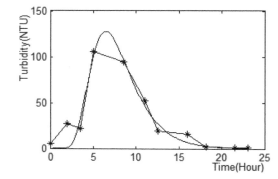

Figure 2. 24-hour cleaning turbidity changes of Qianjin water supply pipeline.

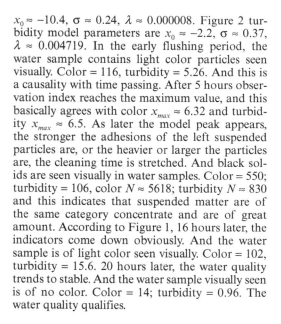

$x_0 \approx -10.4$, $\sigma \approx 0.24$, $\lambda \approx 0.000008$. Figure 2 turbidity model parameters are $x_0 \approx -2.2$, $\sigma \approx 0.37$, $\lambda \approx 0.004719$. In the early flushing period, the water sample contains light color particles seen visually. Color = 116, turbidity = 5.26. And this is a causality with time passing. After 5 hours observation index reaches the maximum value, and this basically agrees with color $x_{max} \approx 6.32$ and turbidity $x_{max} \approx 6.5$. As later the model peak appears, the stronger the adhesions of the left suspended particles are, or the heavier or larger the particles are, the cleaning time is stretched. And black solids are seen visually in water samples. Color = 550; turbidity = 106, color $N \approx 5618$; turbidity $N \approx 830$ and this indicates that suspended matter are of the same category concentrate and are of great amount. According to Figure 1, 16 hours later, the indicators come down obviously. And the water sample is of light color seen visually. Color = 102, turbidity = 15.6. 20 hours later, the water quality trends to stable. And the water sample visually seen is of no color. Color = 14; turbidity = 0.96. The water quality qualifies.

Example 2: Xingxi channel is of large diameters and complicated materials. Respectively, it is made from cast iron, screw steel pipe, and glass steel pipe. Two air pumping points and one sewage let-out points are set. Air is pumped in from the points at the two terminals, and sewage is let out in the middle with the water spraying out from the emission point. Color and turbidity model curve fitting is shown as Figure 3 and Figure 4. Color $x_0 \approx -2.6$, $\sigma \approx 0.38$, $\lambda \approx 0.005297$; turbidity $x_0 \approx -2.2$, $\sigma \approx 0.54$, $\lambda \approx 0.062356$. It can be seen in Figure 2, in the early flushing period, the color is 111, turbidity is 10.7. And the water sample takes a look of grey and black. After 4-hour flushing, the color is 497 and the turbidity is 37.6. And the indicators of the water quality reach the maximum. And this time, black precipitate can be seen in the water sample. And this is basically agreed with color $x_{max} \approx 3.7$ and turbidity $x_{max} \approx 3.3$. Color $N \approx 2605$, turbidity $N \approx 162$. This indicates suspended particles of the same category are flushed out gradually, and the cleaning quantity is normal. After 11-hour flushing, the color is 41, and turbidity is 2.17, and the

Table 2. Daqing Xingxi water supply pipeline basic data and flushing water quality change model coefficients and errors.

Number	Channel name	Pipeline material	Diameter (mm)	Flushing distance (km)			Flushing time (h)		
1	Xingxi channel	Cast iron/steel iron	700	Cast iron 5.57, steel iron 4.59			15		
Time	9	10	11	12	13	14	15	16	17	18	19
Color	111	197	127	497	395	373	291	212	113	107	96
Turbidity	10.7	13.2	10.1	37.6	29.8	26.7	10.6	11.2	7.51	6.87	2.37
Time	20	21	22	22:30	23	23:30	0
Color	41	27	15	13	14	13	13
Turbidity	2.17	1.13	1.01	0.99	1	0.91	0.92

Model parameter	σ	N	λ	x_0	Mean square error
Color model	0.3782	2604.99	0.005297	−2.60	53.62
Turbidity model	0.5383	161.99	0.062356	0.00	4.90

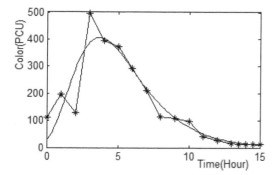

Figure 3. 15-hour cleaning chroma changes of Xingxi water supply pipeline.

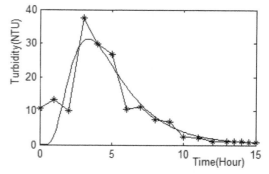

Figure 4. 15-hour cleaning turbidity changes of Xingxi water supply pipeline.

water indicators decrease. The water sample contains several black impurities. After flushing 14.5 hours, the color is 13 and the turbidity is 0.91. The water sample is of no color and it's qualified for the requirement.

Example 2 Channel diameter is 600 mm, example 2 channel diameter is 700 mm, the sewage in emission terminal are all sprayings. The volume of example 2 is larger than example 1. Although their diameter and flushing distance is different, their water flow speed and water pulse impact are of little difference. Actually including example 1 and example 2, water flow speed and pulse impact in major cleaning projects are commonly the same. According to the shape and scale parameter in the two examples, the shape parameters of color and turbidity of example 1 are smaller than example 2. This indicates example 1 appear suspended solid more concentrated than example 2. The scale parameter of example 1 is smaller than example 2. This indicates

example 1 stretches flushing time in compare with example 2. The channel length of example 1 is longer than example 2. The suspended load cleaned out along the channel in example 1 are more than in example 2. And this emerges as suspended load released out at the terminals. According to the fact that the time shift parameters of example 1 is bigger than example 2 and the fact the turbidity time shift parameter of example 1 is smaller than example 2, the example 1 has color pollution resource and the suspended matter have little to do with color pollutions. Example 1 has light color suspended particles in the early flushing, well the example 2 has black precipitation in the middle flushing. Searching the history material of pipelines, Example 1 is produced in 1983, and its material is cast iron, which has 42 valves. Example 2 is produced in 1980, half of which is made of cast iron and half is steel. Its diameter is 700 mm, and has 23 valves. The character of example 1 is having many valves, and the material is cast iron with a 600 mm diameter. This is indicated in example 1, there is high probability of secondary pollution caused by valve aging. In both examples the pipelines are old. In example 1 the shedding particles indicates that there is secondary pollution in some part of channel. The diameter of example 2 is 100 mm bigger than example 1. And the incidence of bacterial corrosion and the formation of shedding particles of example 2 is higher than that of example 1. The black particles and black washing emissions indicate some part of the tube appear more severe anaerobic reaction (Niu, 2006). In the water supply pipeline, sulfate reducing bacteria in the formation of oxidized iron bacteria anaerobic environment for pipe wall corrosion produce black rust sulfide (Huang, 2011). According to the color value of the two examples are all relatively bigger, it can be inferred that there are heavy black sulfide pollution by anaerobic sulfur reaction on the tube wall of both examples. According to the cumulative amount $N \approx 830$ of example 1 is bigger than that $N \approx 162$ of example 2, it can be inferred that suspended adhesion degree of example 1 is higher than example 2. The color of suspended solid and dissolved substances, which is tone and described of saturation, express the components of matter. In example 1 colorN/turbidityN ≈ 6.8, in example 2 color N/turbidityN ≈ 16.1, this indicates that the substances cause color change in 1 turbidity unit in example 2 is bigger than example 1. Of the material in the case of 1 unit is more than 1. This also reflects the facts that in Example 2 there are mainly anaerobic corrosion reaction inside pipes while in Example 1 there are more iron rust pollution caused at pipe valves.

4 CONCLUSION

Based on the log normal distribution, the color/turbidity model can be used to fit the actual sampling data of the color and turbidity. According to the comparison between the model parameters and the actual situation, the curve shape, time scale and time shift parameter have some practical significance. The time shift parameter implies the initial water quality state, which includes the possibility of the occurrence of the secondary pollution. Time scale reflects the information of the length of the pipeline and the intensity of the adhesions on the tube walls, and the shape of the curve reflects the concentration degree of suspended solid emission. To solve the problem of suspended particle sizes determination, space distribution of sedimentary and the problem of the estimate of left sedimentary inside tubes, there will be further experiments and researches combining model in this paper and water quality model in the future, so as to provide basis for pipeline cleaning, condition evaluation and secondary pollution source decision.

REFERENCES

Alexander Omelchenko, Alexander A. Pivovarov, W. Jim Swindall. Modern Tools and Methods of Water Treatment for Improving Living Standards [M]. NATO Science Series IV, Earth and Environmental Series– Vol. 48. Published by Springer (formerly Kluwer Academic Publishers), 2005, part I: 3–8.

Ashok Lumb, T.C. Sharma, Jean-François Bibeault. A Review of Genesis and Evolution of Water Quality Index and Some Future Directions [J].Water Qual Expo Health, 2011, 3:11–24.

Corrochano B.R., J.R. Melrose, A.C. Bentley, P.J. Fryer, S. Bakalis. A new methodology to estimate the steady-state permeability of roast and ground coffee in packed beds [J]. Journal of Food Engineering, 2015, 150:106–116.

Elaine Irving, Megan Cooley, Karoliina Munter, et al. Water for life: healthy aquatic ecosystem information synthesis and initial assessment of the status and health of aquatic ecosystems in Alberta: surface water quality, sediment quality and non-fish biota [R]. Technical Report #278/279-01, Alberta Environment, Edmonton, AB, Canada.

Han Mei, Fu Qing, Zhao Xing-ru, et al. Study on the Limit of Potassium Permanganate Index in Centralized Drinking Water Sources [J]. Research of Environmental Sciences, 2013, 26(10):1126–1131.

Huang Chunkai, Zhao Peng, Zhang Hongwei, et al. Analysis of the characteristics of corrosion scale in water supply pipe [J]. Water & Wastewater Engineering, 2011, 37(9):155–159.

Lemb M., T.H. Oei, H. Eifert, et al. Technegas: A Study of Particle Structure, Size and Distribution [J]. European Journal of Nuclear Medicine, 1993, 20(7):576–579.

Liu Yan, Zheng Bing-hui, Wang Jun, et al. Risk Assessment for Drinking Water Sources of C City: A Case Study [J]. Research of Environmental Sciences, 2009, 22(1):52–59.

Miroslaw Jonasz, Georges Fournier. Approximation of the size distribution of marine particles by a sum of log-normal functions [J]. Limnology and Oceanography, 1996, 41(4):744–754.

National standard of the people's Republic of China, GB5749-2006, Standards for drinking water quality.

Ngwenya N., E.J. Ncube, J. Parsons. Recent advances in drinking water disinfection: successes and challenges [J]. Rev Environ Contam Toxicol, 2013, 222:111–170.

Nikoo M.R., R. Kerachian, S. Malakpour-Estalaki, et al. A probabilistic water quality index for river water quality assessment: a case study [J]. Environmental Monitoring and Assessment, 2011, 181(1–4):465–478.

Niu Zhang-bin, Wang Yang, Zhang Xiao-jian, et al. Analysis of the Characteristics of Corrosion Scale in Drinking Water Distribution Systems [J]. Research of Environmental Sciences, 2006, 27(6): 1150–1154.

Rafael C. Gonzalez, Richard E. Woods, Steven L. Eddins. Digital image processing using MATLAB [M]. Beijing: Publishing House of Electronics Industry, China, 2007:334.

Rosin, P., Rammler, E. The Laws Governing the Fineness of Powdered Coal [J]. Journal of the Institute of Fuel, 1933, 7:29–36.

Schroeder D.J. Astronomical optics [M]. Academic Press, (2nd ed.), 1999:433.

Taya C., Y. Maeda, S. Hosokawa, et al. Size Distributions of Micro-bubbles Generated by a Pressurized Dissolution Method. The 7th International Symposium on Measurement Techniques for Multiphose Flows 2011, AIP Conf. Proc., 2012, 1428:199–206.

Tomáš Vítěz, Petr Trávníček. Particle size distribution of a waste sand from a waste water treatment plant with use of Rosin–Rammler and Gates–Gaudin–Schumann mathematical model [J]. 2011, 59(3):197–202.

Vlassis Likodimos, Dionysios D. Dionysiou, Falaras Polycarpos. CLEAN WATER: water detoxification using innovative photocatalysts [J]. Rev Environ Sci Biotechnol, 2010, 9:87–94.

Wu Chunde, Chen Lu, Xu Yabin, et al. Application of Sodium Hypochlorite for Disinfection of Secondary Water Supply [J]. Journal of South China Normal University (Natural Science Edition), 2011, No. 4: 89–97.

Advances in Energy, Environment and Materials Science – Wang & Zhao (Eds)
© *2016 Taylor & Francis Group, London, ISBN 978-1-138-02931-6*

Study on the influence of ecological protection measures on water conservation capacity in the Sanjiangyuan Region

Jiaqi Zhai, Yong Zhao, Haihong Li, Qingming Wang & Kangning Chen
State Key Laboratory of Simulation and Regulation of Water Cycle in River Basin, China Institute of Water Resources and Hydropower Research, Beijing, China

ABSTRACT: The water conservation capacity of ecosystem is calculated and its temporal and spatial variation characteristics are analyzed based on the distributed hydrological model. According to the ecological protection measures, such 7 scenarios as conversion of grazing land to grassland, conversion of farmland to forest, treatment of black-soil beach grassland, rodent pest control and wetland protection are set in this study. The results show that the water conservation capacity is in a significant increase. Compared with the former protective measures, the water conservation capacity comes to 620 million cubic meters under the protection planning scenarios, increasing by 0.6%.

1 INTRODUCTION

The Sanjiangyuan Region is the headstream of Yangtze River, the Yellow River and the Lancang River, and it is known as the Chinese Water Tower. It is not only an important freshwater-supply region but also a water conservation region in China. Meanwhile, many ecological protection projects are constructed in this important area. It is significant for revealing service function of ecosystem and evaluating the effect of ecological protection and construction projects by studying the change of water conservation capacity of ecosystem in the Sanjiangyuan Region.

Although the study on the Sanjiangyuan Region becomes a hotspot at present, Some differences still exist in the cognition to the study results of water conservation function in the Sanjiangyuan Region. Liu (2005a, 2005b, 2006) has conducted diverse studies in terms of ecological function, ecological service value, and water conservation function and value of ecosystem in the Sanjiangyuan Region; Shi (2012) and Chen (2012a, 2012b) have evaluated the ecosystem service value of the grassland in the Sanjiangyuan Region, and performed a quantitative analysis on the change of ecosystem service value of the grassland in the Sanjiangyuan Region in four different periods, including 1985, 1996, 2000 and 2008. They believed that the main service function of Sanjiangyuan Region is to adjust climate and conserve water. Lai (2013a, 2013b) had some quantitative analyses on the change of ecosystem services before and after the first phase of ecological protection in the Sanjiangyuan Region, and he

held that the water conservation function of the Sanjiangyuan Region plays an important role in adjusting regional water cycle, balancing water quantity and improving hydrology and ecology. Based on the previous study results, some scholars carried out fruitful work in terms of service function and water conservation in the Sanjiangyuan Region. Nevertheless, there are still some common problems not to be resolved. For example, no study on the matter and value amount of water conservation service function in the Sanjiangyuan Region; most of studies only focus on the subsection of ecosystem service function, but ignoring of the results obtained by conceptualized and simple algorithm, and relatively-rough calculation is done for key procedures including precipitation, transpiration and runoff. Moreover, large-scale regional averages of time and space data are used as input ones, so that it is hard to present the difference between the different river basins and areas in the Sanjiangyuan Region. This difference is bigger than the one between complex terrain, climate and hydrological status in the Region. Some analyses lack on the key influential factors and dynamic evolution process for the matter and value amount of water conservation function of the Sanjiangyuan Region, so that the study is hard to be done on the evolution mechanism of water conservation. Furthermore, current studies focus on the spatial scope, including 16 counties and 1 town planned in the first phase of ecological protection of the Sanjiangyuan Region, whereas the time analysis is only limited to the changes after 2000. So, relevant results lack on the scope planned in the second phase and long-term time sequence. It is hard to support the scientific and technological demand on implementing the second phase of planning and construction of the ecological compensation system in the Sanjiangyuan Region. Therefore, an urgent study should be done on the function of ecosystem water conservation in the plan in the second phase in the Sanjiangyuan Region at national and provincial levels. Only by this way, scientific and technological support can be available to protect the Sanjiangyuan Region, and the ecological civilization can be constructed nationally.

A new perspective is proposed in this paper based on the principle of water cycle by integrating and analyzing concept, connotation and algorithm of water conservation in the ecosystem. The evolution process of water cycle is simulated by the water allocation and cycle model in consideration of characteristics of hydrological connection between water cycle elements in the frigid high-altitude area in the Sanjiangyuan Region. At the same time, basic data are provided for calculating of regional water conservation capacity and time and space evolution characteristics. The multiple ecological protection scenarios and ecological protection measures, that have been taken or are being taken, exert the influences on water conservation capacity of the ecosystem in Sanjiangyuan Region, and this influences are is simulated and analyzed in this paper.

2 STUDY AREA AND DATA

The Sanjiangyuan Region is located in the south of Qinghai Province (Fig. 1), whose geological place is 89°24′~102°27′ east longitude and 31°39′~37°10′ north latitude. It covers a total area of 395,000 square kilometers, accounting for 54.6% of total

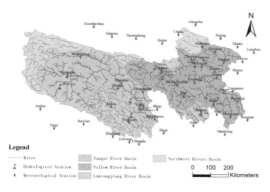

Figure 1. Location of the study area in the Sanjiangyuan Region.

area in Qinghai Province. It involves 21 counties of 4 Tibetan Autonomous Prefectures (Yushu, Guoluo, Hainan and Huangnan), and Tangulashan Town of Golmud City.

Meteorological data in this study come from the daily process data from 1980 to 2012 in 43 meteorological stations in the Sanjiangyuan Region, and were released by the National Meteorological Information Center of China Meteorological Administration; DEM data are 30-meter resolutions issued by the NASA in 2009; land use and soil information data are sourced from the Chinese land use data in 2005, and soil information data are issued by the Data Center for the Resources and Environmental Science of CAS; the hydrological data are from the monthly runoff process in 10 hydrological stations monitored by the Hydrological Administration of Qinghai Province from 1980 to 2012.

3 METHODS

3.1 Basic concept of water conservation

Water conservation is one of the basic ecosystem functions, as well as one of the important elements included in the indirect use value of the ecosystem service function. The function of the water conservation has been always one of the priorities for ecology and the hydrology. In many study results, however, no consensus is reached for the basic definition of water conservation in different cognitions and demands. Generally, water conservation capacity refers to the functions of intercepting precipitation, conserving water in the soil, increasing the precipitation, slow-releasing surface runoff, supplementing underground water and adjusting river runoff. The regional ecosystem has such functions in various vegetation canopy, litter and soil interception, infiltrated and stored precipitation and water stored in the glacier. So, water conservation is also defined as the function of intercepting precipitation or adjusting runoff in the ecosystem. The interception of precipitation refers to interception and storage, rainfall amount including interception by canopy, defoliation and soil. The function of adjusting runoff refers to delay of surface runoff, increase of soil and underground runoffs, decrease of abundant conditions and addition to litter, control of soil erosion and improvement of river water quality by influencing the hydrological process, promoting redistribution of precipitation, affecting the movement of water in soil and changing water-converging conditions. In addition, there is a bigger gap between water conservation capacities of the different ecosystems due to the biodiversity and diverse ecosystems. Besides, climate, soil depth, texture, terrain and landscape may affect water conservation capacity of ecosystem. In general, the water conservation can be divided into three categories as follows:

Category I: It is characterized by water conservation volume with its water conservation capacity. It focuses on the space of water conservation that the woodland has. Based on the considerations, it can be divided into two sub-categories—general water conservation capacity and soil water conservation capacity; the former considers water conservation volume intercepted by canopy, defoliation and soil layer, while the latter only applies the water conservation volume intercepted by the soil layer. It is believed that the water conservation volume intercepted by the canopy and the defoliation layer will be vaporized and return to atmosphere in the short time, and only the water volume conserved by soil is in a final conservation.

Category II: it refers to the water conservation with the precipitation reserve. The precipitation is mainly conserved by woodland. It is believed that the decreased volume is the water volume conserved by woodland by comparing the decrease of the runoff between the woodland and naked land. In addition, the remaining part in percentage of vaporization and spread volume of canopy and trunks is the water conservation volume by calculating of annual statistical data for special regions.

Category III: it is characterized by water conservation volume based on the principle of water volume equilibrium. In Category I, the difference between annual average precipitation and vaporization is considered as the water conservation volume, while the volume of the runoff in Category I is regarded as the water conservation capacity in Category II. In addition, the growth of underground runoff in Categories I and II is regarded as the water volume conserved by forest in a comparison with woodland and naked land.

Because there are different understandings in terms of concept and connotation of water conservation, various algorithms and diverse results are available for water conservation capacity. They are different in number, even in an opposite conclusion. Based on the conclusion of the cognition and understanding on the water conservation, the water conservation of the ecosystem can be explained in two aspects: one is water conservation capacity of the ecosystem, that is, water conservation capacity is reflected by the maximum water volume conserved in the defoliation of soil surface and the soil pore, which is similar to the maximum capacity of the reservoir; the other aspect is water conservation flux of the ecosystem, that is, besides rapid runoff on the surface, there is also the water volume conserved in the soil for short term or long term after the rainfall. It will be vaporized and infiltrated to supply underground water and adjust the water source of river runoff. This is similar to the leak of a reservoir and water conservation.

3.2 Methods for water conservation capacity

Due to different understanding to the water conservation concept, many algorithms have been derived, e.g. method of general water conservation capacity (Deng Kunmei, 2002; Li Jing, 2008; Qin Jiali, 2009), method of precipitation reserve (Zhang Sanhuan, 2001), method of annual runoff volume (Jiang Haiyan, 2005; Zhang Wenguang, 2007) and method of water volume equilibrium (Xiao Han, 2000; Cheng Genwei, 2004). The calculation principle and process are as follows:

① Method of general water conservation: Total water is intercepted by vegetation canopy, surface defoliation and soil layer, and the result reflects the maximum water conservation capacity or static conservation capacity of the ecosystem. The details of the algorithm are as follows:

$$Q = Q_1 + Q_2 + Q_3 \qquad (1)$$

where, Q, Q_1, Q_2 and Q_3 represent water conservation volume, water volume intercepted by the vegetation canopy; water volume is intercepted by the defoliation on the ground surface and water volume is intercepted by the soil layer of the ecosystem, respectively. The calculating unit goes by cubic meter. The volume intercepted by the canopy can be estimated by the interception rate of canopy and precipitation; the water volume conserved by the defoliation can be calculated by the volume conserved by defoliation and maximum water conservation capacity; the volume conserved by the soil layer can be calculated by non-capillary porosity and soil thickness. This method considers the interception to the precipitation more comprehensively by different layers generally, and it is helpful for comparing the size of the function on intercepting the precipitation from different layers. However, the calculation that needs a great many actually measured data is relatively complex.

② By comparing the runoff of vegetation on the surface like woodland and grassland as well as decrease of runoff on naked land, the water conservation volume decreases, and its calculation method is as follows:

$$Q = A \cdot (J_0 \cdot K) \cdot (R_0 - R_g) \qquad (2)$$

where, A stands for area (m^2); J_0 is total volume of annual average precipitation (mm); K

is percent of calculating regional runoff generation precipitation; R_0 and R_g are runoff generation coefficients of naked land and some ecosystem. The meanings of other symbols are the same as above.

③ The method of water volume equilibrium regards the ecosystem as one black box, starting from the input/output of the water volume. The remaining water volume is derived by the water volume equilibrium, i.e. annual precipitation of the ecosystem deducting the water volume flowed out directly. The method of water volume equilibrium is the foundation to study the mechanism of the water conservation. It can reflect actual water conservation volume of the ecosystem in relatively accurate manner, and easy to be operated. So this method is regarded as the most perfect one for calculating water conservation volume, as well as the algorithm used most frequently. There are many understandings about the water volume flowed out directly. For example, evapotranspiration volume or runoff volume on the surface applies the direct volume flowed out. By analyzing the mechanism of water conservation and the concept of water conservation flux, it is believed that the water conservation volume is derived by the precipitation deducting the runoff on the surface. The details of the algorithm are as follows:

$$Q = A \cdot (P - R_s)/1000 \qquad (3)$$

where: A is the area (m²); P is annual precipitation of the ecosystem (mm); R_s is the runoff on the surface of the ecosystem (mm); the meanings of other signs are the same as above.

It is indicated that every method has its respective merits and limits, and the method of general water conservation capacity needs a great deal of actually measured data as the base, and ignores the impact of vegetation consumption, whose calculation value only reflects the maximum water conservation volume theoretically. The method of the precipitation reserve adopts the empirical value of the vaporization volume, but ignores the impact of the surface runoff. The method of annual runoff volume is based on the assumption that the water conservation volume is equal to annual runoff volume. But it is much different from the actual status, with a bigger error in calculation result. The method of the water volume equilibrium considers the factors of the evapotranspiration volume and ground surface runoff. This method is the basis for studying the water conservation volume of the ecosystem, as well as the most perfect one for calculating the water

conservation volume theoretically. It is used most widely, however, this method ignores the time and space difference in terms of the vaporization and the runoff, so that the serious error occurs in calculation.

3.3 Water cycle element and flux simulation

Water cycle variables including precipitation, runoff and evapotranspiration and soil water volume l need to be acquired to calculate the water conservation capacity. The Water Allocation and Cycle Model (WACM) can provide the information on the change of the element flux in many scales of time and space, and new technical means for calculating the water conservation capacity. The WACM developed independently is planned to be used for simulating the water cycle elements in the Sanjiangyuan Region and the change process of their flux, and it can provide data information needed in calculating and evaluating water conservation capacity of the ecosystem in the Sanjiangyuan Region. WACM model is a platform of water allocation and cycle model based on water allocation, cycle and change and energy change process of their accompanying matters (C, N). Its core is the water cycle module that can provide the means for analysis of water resources allocation, natural-artificial composite water cycle simulation, matter cycle simulation, drought evaluation, climate change and impact brought by human activities (Zhao Yong, 2007; Zhai Jiaqi, 2012). The WACM water cycle module mainly includes the processes like evaporation, transpiration, snow cover and snow melt, runoff generation on the surface, water infiltration in soil, converging river network, underground water and artificial water use. The processes of evaporation and transpiration are calculated by Penman formula, Noilhan-Planton formula, Penman-Monteith and their revised formulas according to water area, naked land, vegetation and residential area. The simulation of runoff generation flux considers the backwater effect of the reservoir on the surface. The penetration of the water in soil is calculated by Horton formula, and the shallow and deep soil layers are calculated by Richards equation. The confluence on the slope and river confluence is calculated by the simple one-dimension Saint-Venant equation. The shallow and deep layers of underground water in the plain are simulated with the plane two-dimension underground water model, and underground water in the mountain area is simulated with the vertical equilibrium model. The artificial water used includes the simulation of drainage for agricultural irrigation and drainage for industrial and living. The former can simulate the process of water conversion of various irrigation and drainage channels

Code	Scenario	Description of the scenario
G10	Benchmark scenario	The protection measure not taken in current status.
G11	Returning grazing land to grassland	The protection measure of returning grazing land to grassland taken in 64.4×10^4 km².
G12	Rodent pest control	The protection measure of rodent pest control taken in 11.6×10^4 km².
G13	Treating black soil beach	The protection measure of treating black soil beach taken in 0.35×10^4 km².
G14	Woodland protection	The protection measure of returning the farm to the forest and cultivating the forest by closing the mountain taken in 0.31×10^4 km².
G15	Wetland protection	The protection measure of wetland protection taken in 0.11×10^4 km².
G16	General protection scenario	In general scenario, all the measures taken in the scenarios of G11-G15 as above.

based on the water amount according to different irrigation types (channel irrigation, well irrigation, combination of the channel and well irrigation, and the agriculture raised by rainwater); the latter can simulate general nodes of the industrial water, living water, water consumption and drainage in the area involved. The detailed organization, basic principle and model parameters can be seen from the references (Zhao Yong, 2007; Zhai Jiaqi, 2012).

3.4 Ecological protection measures and scenarios in Sanjiangyuan Region

The main measures for the first phase of protection plan in the Sanjiangyuan Region include conversion of grazing land to grassland, rodent pest control, treatment of black soil beach grassland, woodland protection and wetland protection, and 7 scenarios are set in this study as shown in Table 1.

4 RESULTS AND DISCUSSION

4.1 Model calibration and verification

The data information of monthly runoff process actually were measured from 3 hydrological stations in Tuotuohe, Xinzhai and Zhimenda of Changjiangyuan region, 2 hydrological stations in Xiangda, Xilaxiu of Lancangjiang region and 5 hydrological stations in Tangnaihe, Shangcun, Damitan, Jungong and Jimai of Yellow River region from 1980 to 2012. Its Calibration Period (CP) is from 1980 to 1999, and the Verification Period (VP) is from 2000 to 2012. The verified results can be seen in Table 2. It is found that the Relative Error (RE) of the calibration is controlled at 19%, the relevant coefficient (R^2) is above 0.84, and the efficiency coefficient (Ens) is above 0.68; the relative error during verification period is 14%,

the relevant coefficient is above 0.77, and the efficiency coefficient is above 0.62. They all meet the precision requirement.

4.2 Time and space change characteristics of the water conservation capacity of the Ecosystem in Sanjiangyuan Region

Figure 2 and Table 3 show the change of water conservation of the ecosystem in the Sanjiangyuan region from 1980 to 2012. It is found that average water conservation volume in Sanjiangyuan region for years is 111.9 billion cubic meters, and the water conservation volume in nearly 3 decades has been increasing, with bigger fluctuation over years. The water conservation status in 1980s is worst with a capacity of only 104.99 billion cubic meters, and the water conservation volume in 1980s maintained at a multi-year average level, with a capacity of about 109.93 billion cubic meters. The water conservation status since 2005 has been improved significantly at a capacity of 124.9 billion cubic meters.

According to the characteristics of annual and inter-annual changes of water conservation volume in various regions, the water conservation volume of the ecosystem in the Yangtze River regions in 1984 and 1990 is lower than 40 billion cubic meters, which is lower than the normal year level significantly. Comparing with 1980s, the status of the water conservation in Yangtze River regions is in a change process from violent decrease and increase, and it deceases by nearly 10% comparing with multi-year average. This is because the water conservation decreases most seriously in this region. Since 2005, the deterioration of the ecosystem has been better controlled with precipitation after the protection of ecosystem is strengthened, and the water conservation of the ecosystem has increased rapidly by 14% comparing with multi-year average. In the Yellow River regions, the water conservation volume in 1990, 2000 and 2002 is lower than 40 billion cubic meters, which were lower than

Table 2. Model calibration and verification results.

Num	Station	RE CP	RE VP	R² CP	R² VP	Ens CP	Ens VP
1	Zhimenda	6%	1%	0.944	0.936	0.881	0.826
2	Xinzhai	3%	−1%	0.840	0.773	0.681	0.590
3	Tuotuohe	18%	13%	0.862	0.837	0.706	0.624
4	Xiangda	5%	2%	0.872	0.843	0.693	0.655
5	Xialaxiu	6%	1%	0.891	0.856	0.783	0.717
6	Tangnaihai	7%	−5%	0.926	0.916	0.834	0.803
7	Shangcun	−8%	−12%	0.862	0.883	0.708	0.686
8	Damitan	19%	14%	0.859	0.888	0.703	0.753
9	Jungong	5%	−8%	0.914	0.915	0.808	0.808
10	Jimai	6%	−7%	0.877	0.883	0.697	0.707

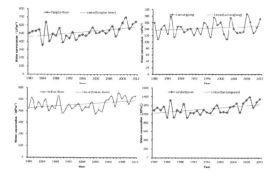

Figure 2. Dynamic change of water conservation in Sanjiangyuan region (1980–2012).

the normal year level significantly. General change of the water conservation increases, but the fluctuation of inter-annual change is relatively significant. From inter-annual characteristics, the water conservation volume in 1990s is lowest, decreasing by 6% comparing with multi-year average. It is recovered in 2000, and the function of the water conservation in 2005 is improved significantly, which increases by 11% comparing with multi-year average. General change of the water conservation in Lancangjiang regions goes slow, and its increasing tendency was not so significant as Yangtze River regions and Yellow River regions. This indicates that the change of the ecosystem in this area is relatively stable, with steady and normal function of water conservation. During the process of change, it decreases a little in 1990s, with a little changing range. However, the water conservation increases by less than 5% by 2000.

According to space change, the water conservation increases from the northwest to southeast, e.g. the water conservation of Yalong river in Yangtze River drainage and Daxia river and Yaohe river in Yellow River region; the water conservation increases from the upper reaches to lower reaches of Yangtze River. It is believed that the lower reaches have relatively rich precipitation, with high vegetation coverage and good ecosystem. Conversely, the water conservation decreases from the upper reaches to lower reaches in Yellow River drainage. This mainly because that the upper reaches have more precipitation, with vast area rarely populated and less impact of human activities. However, the impact of excessive grazing and water and soil loss on ecosystem is common with the intensity of human activities in the lower reaches, so that the ecosystem is damaged, and water conservation capacity declines.

4.3 Analysis on the impact of ecological protection measures on water conservation capacity

Table 4 shows the change of water conservation of the ecosystem in the Sanjiangyuan region under different scenarios of ecological protection measures. It is found that the water conservation volume increases significantly after adopting the ecological protection measures. Compared with the scenarios without protection measures, the water conservation volume under the first phase of planning scenario increases by 0.6% with an increase of 620 million cubic meters; the water conservation in Yangtze River region increases most, with the most significant protection effect.

According to water conservation effect in the different protection measures, the result of returning the grazing to the grassland is the most significant under the first phase of planning scenario, and the water conservation volume increases by 0.5%, with an increase of around 580 million cubic meters. The woodland protection scenario presents an increase of water conservation volume by 0.3%, with an increase of around 330 million

Table 3. Comparison and analysis of water conservation in Sanjiangyuan region in different periods (unit: 10^8 m^3).

Year	Yangtze river	Lancang river	Yellow river	Northwestern	Sanjiangyuan
1980s	503.2	140.1	456.1	187.6	1099.3
1990s	475.5	141.3	433.2	178.8	1049.9
2000s	551.7	144.7	474.1	200.2	1170.5
2010–2012	604.7	147.6	501.3	211.3	1253.6
1980–2004	495.1	140.9	442.6	184.4	1078.6
2005–2012	592.4	147.6	509.0	211.2	1249.0
1980–2012	518.7	142.5	458.7	190.9	1119.9

Table 4. The impact of the first phase of ecological protection measure on the water conservation of the ecosystem in Sanjiangyuan region (unit: 10^8 m^3).

Scenario	Region			
	Yangtze river	Lancang river	Yellow river	Sanjiangyuan
G10				
WCC	518.7	142.5	458.7	1119.9
G11				
WCC	522.2	143.4	460.1	1125.8
ΔW	3.5	0.9	1.4	5.8
R (%)	0.7%	0.6%	0.3%	0.5%
G12				
WCC	519.1	142.6	458.9	1120.6
ΔW	0.3	0.1	0.2	0.7
R (%)	0.1%	0.1%	0.1%	0.1%
G13				
WCC	518.8	142.5	458.8	1120.1
ΔW	0.1	0.0	0.1	0.1
R (%)	0.0%	0.0%	0.0%	0.0%
G14				
WCC	522.0	142.5	458.7	1123.3
ΔW	3.3	0.0	0.0	3.3
R (%)	0.6%	0.0%	0.0%	0.3%
G15				
WCC	518.7	142.5	458.7	1119.9
ΔW	0.0	0.0	0.0	0.0
R (%)	0.0%	0.0%	0.0%	0.0%
G16				
WCC	522.3	143.5	460.3	1126.2
ΔW	3.6	1.0	1.6	6.2
R (%)	0.7%	0.7%	0.4%	0.6%

*WCC: Water conservation capacity.

cubic meters. But the result of water conservation volume under wetland protection is not significant yet at a less protection range.

5 CONCLUSIONS

The ecosystem in the Sanjiangyuan region is very fragile, and the ecological protection measures play a positive role in improving regional water conservation capacity. Based on the water allocation and cycle model, i the time and space characteristics can be simulated and analyzed for the water conservation capacity in the Sanjiangyuan region from 1980 to 2012; the water resource effect can be compared and analyzed in different protection scenarios including conversion of grazing to grassland, farmland to the forest, treatment of black soil beach, prevention and treatment of rodent pest and wetland protection in the Sanjiangyuan region. The results indicate that the water conservation volume in Sanjiangyuan region increases significantly. Compared with the scenario without protection measures, the water conservation volume under the first phase of planning scenario increases by 0.6%, with a growth of 620 million cubic meters, while the water conservation volume under the second phase of planning scenario increases by 1.4%, with an increase of 1.56 billion cubic meters.

ACKNOWLEDGEMENTS

This research has been financed and supported by the National Natural Science Foundation of China (51309249, 51379216), Special Scientific Study of Public Welfare of the Ministry of Water Resources (201401041), National Program on Key Basic Research Project (2010CB951100) and National Science and Technology Support Program during the 12th Five-year Plan (2012BAC19B03), and it is hereby acknowledged.

REFERENCES

Chen Chunyang, Tao Zexing, Wang Huanjiong and Dai Junhu. Evaluation on the Service Value of the Ecosystem in the Grassland of Sanjiangyuan Region. Progress of Geography, 2012a, 31(7): 147–153.

Chen Chunyang, Dai Junhu, Wang Huanjiong and Liu Yachen. The Change of Service Value of the Ecosystem in Sanjiangyuan Region Based on Land Use Data, 2012b, 31(7): 970–977.

Cheng Genwei and Shi Peili. The Water Conservation Benefit and Its Value Evaluation of the Forest in the Upper Reaches of Yangtze River. Science on Chinese Water and Soil, 2004, 2(4): 17–20.

Deng Kunmei, Shi Peili and Xie Gaodi. Study on the Water Conservation Capacity and Value of the Forest Ecology System in the Upper Reaches of Yangtze River. Resources Science, 2002, 24(6): 68–73.

Jiang Haiyan, Jiang Chunying, Xu Dongyan et al. Value Estimation of the Ecological Service Function of the Water Conservation in the Forest of Mountain Area at the East of Liaoning.

Lai Min, Wu Shaohong, Dai Erfu, Yin Yunhe, Pan Tao and Zhao Dongsheng. The Change of Service Value of the Ecosystem in Sanjiangyuan Natural Reserve under the Background of Ecological Construction. Progress of Geography, 2013a, 31(1): 8–17.

Lai Min, Wu Shaohong, Dai Erfu, Yin Yunhe and Zhao Dongsheng. Evaluation on the Indirect Use Value of the Ecosystem in Sanjiangyuan Region. Journal of Natural Resources, 2013b, 28(1): 38–50.

Li Jing & Ren Zhiyuan. The Change of Time and Space on the Value of Water Conservation of the Ecosystem in Yellow Soil Plateau at the North of Sha'aXi. Journal of the Ecology, 2008, 27(2): 240–244.

Liu Minchao, Li Diqiang, Wen Yanmao and Luan Xiaofeng. Ecological Function Analysis and Value Assessment on the Ecosystem in Sanjiangyuan Region. Journal of Plan Resources and Environment, 2005a, 25(9): 1280–1286.

Liu Minchao, Li Diqiang, Luan Xiaofeng and Wen Yanmao. Service Function and Value Assessment of Ecosystem in Sanjiangyuan Region. Journal of Plan Resources and Environment, 2005b, 25(1): 40–43.

Liu Minchao, Li Diqiang, Wen Yanmao and Luan Xiaofeng. Functional Analysis and Value Assessment on the Water Conservation Capacity of the Ecosystem in Sanjiangyuan Region. Resources and Environment in Yangtze River Drainage, 2006, 15(3): 405–408.

Qin Jiali, Yang Wanqin and Zhang Jian. Water Conservation Capacity and Value Evaluation of the Typical Ecosystem in the Upper Reaches of Minjiang. Journal of Application and Environment Biology, 2009, 15(4): 453–458.

Shi Fantao & Ma Renping. Functional Analysis on the Ecosystem of the Grassland in Sanjiangyuan Region. Grass Industry and Pasturage.

Xiao Han, Ouyang Zhiyun and Zhao Jingzhu. Initial Exploration on the Service Function of the Forest Ecology System and Its Ecological Value Evaluation—An example of the Tropical Forest in Jianfengling Mountain of Hainan Island. Journal of Applied Ecology, 2000, 11(4): 481–484.

Zhai Jiaqi. Theory on Water-Nitrogen-Carbon System and Its Application Study in the River basin. China Institute of Water Resources and Hydropower Research, 2012.

Zhang Sanhuan, Zhao Guozhu and Tian Yunzhe. Study on the Evaluating the Ecological Environment Value of the Forest Resources in Hunchun Forest Area, Changbaishan Mountain. Journal of Yanbian University: Natural Science Version, 2001, 27(2): 126–134.

Zhang Wenguang, Hu Yuanmai, Zhang Jing et al. Water Conservation and Value Change of the Forest Ecology System in the Upper Reaches of Minjiang River. Journal of the Ecology, 2007, 26(7): 1063–1067.

Zhao Yong, Lu Chuiyu and Xiao Weihua. The Study on the Rational Allocation of General Water Resources (II)-Model. Journal of Hydraulic Engineering, 2007, 38(2): 43–49.

Advances in Energy, Environment and Materials Science – Wang & Zhao (Eds)
© *2016 Taylor & Francis Group, London, ISBN 978-1-138-02931-6*

Study on the ground subsidence caused by pipe jacking

Xiankai Bao, Chunhui Huang & Huanhuan Liu
College of Architecture and Civil Engineering, Inner Mongolia University of Science and Technology, Baotou, Inner Mongolia, China

ABSTRACT: The surface deformation is a prominent problem in the pipe-jacking construction. It has been the focus of attention of geotechnical industry. This paper describes the mechanism of surface deformation caused by pipe jacking, and the brief factors affecting this discussion; in addition to the engineering example, the law of surface deformation is analyzed to study if the surface deformation can cause the pipe jacking for the future to provide some reference.

1   INTRODUCTION

As a trenchless technology, pipe jacking method has been widely used in underground engineering construction because of its unique technical advantages. Moreover, pipe jacking which requires no excavation formation surface is possible to cross the existing buildings, such as roads, railways, rivers, ground, etc. It has small disturbance to the surrounding environment, which saves space compared to the open-cut construction method and has a certain economy compared to the shield in the same construction conditions (Ren, 2004). However, there is an important problem about pipe jacking that is how to effectively control surface subsidence caused by construction (Qiao, 2000). This paper firstly discusses the surface deformation of the mechanism and influencing factors, then the practical engineering of the surface deformation has been analyzed.

2   ON THE GROUND SUBSIDENCE CAUSED BY FACTORS

Disturbance of the soil caused by Pipe jacking cannot be ignored. If it is not treated promptly, it will lead to construction delay, ground subsidences, and other construction accident casualties (Zhao, 2008). The key factors for the surface deformation are following.

2.1   *Soil pressure*

Soil Pressure plays an important role to maintain the body balance of the excavation soil. Pipe jacking is in the top process; underbreak will occur when the head of pipe jacking is faster and the one that is unearthed is slower, which leads to soil in the soil warehouse which cannot be discharged in time and soil pressure is greater than earth pressure at rest of soil; overbreak will occur when the head of pipe jacking is into slower, soil pressure is less than earth pressure at rest of soil and soil mass of excavation face is excavated excessively. Underbreak and overbreak will cause the soil mass of excavation to lose their balance and make the ground deformate. About the calculation of warehouse pressure of pipe jacking, there is no fixed method. Generally, it refers to the local geology, which is determined by the experience and makes adjustments based on field monitoring at any time.

2.2   *Soil properties*

Effects of soil properties on soil displacement are mainly on soil deformation modulus, when other conditions remain unchanged, the soil deformation modulus is larger and surface disturbance caused by pipe jacking is smaller. Conversely, the soil deformation modulus is smaller and surface disturbance is greater.

2.3   *Jacking speed*

Jacking speed has a more important control value in the earth pressure balance pipe jacking construction. The size jacking speed will have a significant impact on stabilization of soil pressure of the excavation face, balance of soil amount in and out warehouse and injection of the slurry. Therefore, a reasonable jacking speed value is also particularly important for the construction.

As we can see, there are many factors which can influence the surface subsidence. Although with the deepening of the theoretical research it has obtained certain research results. Considering factors still exist such as weakness; in view of the ground settlement caused by construction, there is only the macroscopic observation, research about the subtle changes in the internal conduct is little. Then, there is the theoretical study, most of them only considers the individual factors and ignores the interactional relationship between various factors, for example, grouting. Attention is often payed to the individual aspects of configuration, grouting amount, grouting location, and interaction relationship between pipe jacking and soil is ignored, which leads to the gap between research results and practical.

3 MECHANISM ANALYSIS OF THE SUBSIDENCE

Disturbance on the surrounding soil in the process of pipe jacking construction is the root of the surface subsidence (Yu, 1997). Main parts are as follows.

3.1 Surface subsidence before pipe jacking arriving at monitoring

When the distance between jacking position and monitoring stations is far, it can produce disturbance to soil under the influence of the vibration of the blade wheel. As soil particles have certain porosity, pores contain a certain amount of water and air. So, under the top into disturbance, the water and air are squeezed out from pore, which makes the soil particles move relatively, soil will get compressed, and there will be a small of the ground at this time. As the pipe jacking moves forward continually, the distance to the monitoring points shortens, the extrusion of soil is more intense, the ground begins to swell, and with the closer of jacking machine, the soil deformation is also faster and greater.

3.2 Surface subsidence when pipe jacking arriving at monitoring

When the distance between jacking position and monitoring stations is close, the soil mass in front of the pipe jacking which suffers extrusion of jack, vibration and stirring of the cutter is in a state of a complex. The soil excavation, propulsion, balance of mud or gas pressure can make the volume of soil change and cause ground settlement or upliftment.

3.3 Surface subsidence when pipe jacking crossing monitoring

When the distance between jacking position and monitoring stations is close, the soil mass in front of the pipe jacking which suffers extrusion of jack,

vibration and stirring of the cutter is in a state of a complex. On the one hand, under the influence of excavation, soil horizontal stress decreases because of the stress relaxation of soil. On the other hand, horizontal stress increases due to the propulsion, balance of mud or gas pressure. If the reduced value of the horizontal stress is greater than active earth pressure, soil mass of excavation face will collapse. If increment value of the horizontal stress is greater than passive earth pressure, soil mass will be made outward, which causes the upliftment and deformation of the ground. These two factors would bring some changes to the volume of the soil and cause the ground settlement or uplift.

3.4 Surface subsidence after pipe jacking crossing monitoring

In order to reduce the frictional resistance during construction, the diameter design of the subsequent socket is 2–5 cm smaller than the diameter of the jacking machine. So, after the tail of the jacking machine crossing monitoring, soil around pipeline will move to the tube wall to fill the following socket, which could make the soil produce movement. In order to keep the soil stable and reduce friction in the jacking process, grouting processes must be disposed at the periphery of the pipe joint in the construction process, and the grouting pressure should be in a certain range.

4 AN ENGINEERING EXAMPLE PROJECT SUMMARY

4.1 Project outline

Underground tunnel construction project of AErDing Street of Baotou city (Inner Mongolia university of science and technology) is located in Baotou city, main span of the underground crossing construction is 64.8 m, thickness of covering soil is about 10.95 m, maximum burial depth of gallery is 10.95 m, and the surface was covered with artificial filling soil widely. Modern pluvial sediments of the quaternary Holocene (Q4) is consisted of powder sand, gravel, and round gravel layer; the quaternary Pleistocene (Q3) constitutes of green-yellow lacustrine deposits, charcoal grey silty clay and silt interbed.

A, B, and C are the three line sections along A Er Ding Street which are 18 m, 33 m, and 48 m, respectively, from the starting well. Each line includes five points; each point distance of line is different. The extending distance on both sides of A Er Ding Street by the middle of the channel monitoring are 5 m and 7 m, respectively. The surface settlement monitoring during the pipe jacking construction layout is shown in Figure 1.

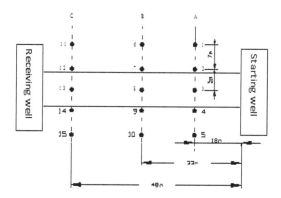

Figure 1.   Diagram of monitoring arrangement.

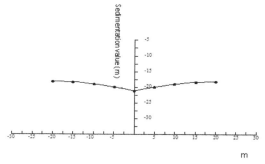

Figure 2.   Diagram of monitoring section settlement.

Table 1.   Subsidence of monitoring point 3.

Rearview (m)	Foresight (m)	Dispersion (m)	Cumulative subsidence (mm)
1.64800	1.73703	−0.08903	0.05
1.64845	1.73799	−0.08954	−0.46
1.36982	1.46118	−0.09136	−2.28
1.28525	1.27833	−0.09837	−2.37
1.33782	1.43468	−0.09686	−2.41
1.33787	1.43450	−0.10583	−9.53
1.34753	1.45926	−0.14173	−17.52
1.29580	1.40752	−0.11172	−22.64

4.2   The monitoring results and analysis

As the soil excavation, the cumulative settlement of the monitoring of axis is increasing, such as monitoring point 3 (shown in Table 1); the rest of the monitoring is not here.

As time goes on, the displacement of monitoring point 3 is constantly increasing, after the pipe jacking through monitoring is about 18 m, the settlement which reaches about 22.64 mm is achieving the construction warning value, others are all within the specified value. Maximum subsidence of the remaining two axis points, 3 and 13, are 20.64 mm and 21.37 mm, respectively. Settlement is reaching warning value mainly because the top speed is too fast (4 cm/min), which causes the disturbance of soil largely, and the grouting pressure is small (0.6 Mpa), which is unable to maintain the stability of soil and support the surface. The measures through lowering the top speed, keeping it at 2 cm/min and increasing the grouting pressure 0.9 Mpa control the settlement within the prescribed scope successfully (Pang, 2013). The soil settlement volume of the monitoring lateral reduces gradually by jacking the axis center to both sides, and with jacking axis as the center, settlement of both sides is distributed symmetrically. Taking A as the monitoring section for instance, the concrete is shown in Figure 2.

When monitoring point 3 of the axis reaches the maximum value of 22.74 mm, sedimentation value of point 2 and 4 are 18.71 mm and 18.71 mm, sedimentation value of point 1 and 5 are 16.54 mm and 16.33 mm, sedimentation value along the axis is distributed symmetrically.

When the pipe jacking through a certain range is centered with monitoring, soil settlement of monitoring will change dramatically, monitoring soil will produce a significant sedimentation rapidly. And when it exceeds a certain range, the settlement changes flatten out gradually. Sedimentation value of monitoring point 3 is analyzed, because monitoring point 3 is close to starting well, point 3 has a tiny upliftment that is about 0.05 mm. Then as the top continues, sedimentation value increases continuously. When it passes monitoring location, sedimentation hasn't reached maximum settlement and settlement value of 9.53 mm. When, from the pipe jacking to monitoring is 18 m, sedimentation reaches a maximum of 22.64 mm, then settlement verges to be gentle and no large fluctuations occur.

To sum up, with the entering of jacking machine, settlement monitoring points are changing constantly, and the change of value is different. It has its own trends, including Transverse, Longitudinal and each of its points. When the pipe jacking machine is in a different position, the settlement value is related to jacking machine position. Subsidence value of each point is different, which is in accord with deformation mechanism.

5   CONCLUSIONS

There are many factors that affect the surface deformation of pipe jacking, both macro factors and the micro factors. In the theoretical study of the future, the goal is not only to discuss the existing research results and to improve, but also to explore the new domain, more attention should be payed to the microcosmic influencing factors, in particular, and theoretical research work should be obtained more carefully. At the same time, there are some differences between the surface subsidence rules under different construction technologies and geological conditions. So a better understanding of the surface subsidence laws of pipe jacking will have an important role in guiding.

REFERENCES

Pang Chenjun, Bao Xiankai. Numerical simulation of rectangular pipe jacking [J]. Construction Technology, 2013(42):410–412.

Qiao Hongwei, Deng Zhihui. Ground subsidence prediction of pipe jacking [J]. West-china explore engneering, 2000(6):113–114.

Ren Jianwen. Trenchless Pipe Jacking Construction Technology [J]. West-China explore engneering, 4(1):33–34.

Yu Bingquan. Pipe Jacking Construction Technology [M]. BeiJing: China Communications Press, 1997.

Zhao Xiangjun, Chen Jikun. Design of pipe jacking technology [J]. Hydraulic construction, 2008(5):26–27.

Advances in Energy, Environment and Materials Science – Wang & Zhao (Eds)
© *2016 Taylor & Francis Group, London, ISBN 978-1-138-02931-6*

Experimental study of cyclic loading history influence on small strain shear modulus of saturated clays in Wenzhou

Xiaobing Li
Research Center of Coastal and Urban Geotechnical Engineering, Zhejiang University, Hangzhou, P.R. China

Chuan Gu, Xiuqing Hu & Guoqun Fu
College of Civil Engineering and Architecture, Wenzhou University, Wenzhou, P.R. China

ABSTRACT: Small strain shear modulus G_{max} is a key parameter in soil dynamics. Although the cyclic loading on saturated soils induces a decrease of effective stress and a rearrangement of soil skeleton which may both lead to a degradation in undrained stiffness, only the contribution of effective stress reduction to the G_{max} degradation is considered in Hardin-Richart equation, and that of soil fabric change is neglected. In this paper, undrained cyclic triaxial tests in combination with bender element tests were conducted on saturated clays in Wenzhou. The influences of cyclic loading history on the small strain shear modulus were studied. The test results show that: in the beginning of tests, G_{max} of saturated clays under cyclic loading history is moderately lower than the corresponding value under non-cyclic loading history at the same effective stress, however, when the effective stress reduces to some extent, G_{max} decreases suddenly and reflects the failure of soil skeleton.

1 INTRODUCTION

The small strain shear modulus G_{max} (when shear strain $\gamma < 10^{-5}$) is an essential parameter in soil dynamics, especially in the analysis of soil response under earthquakes. Many researches (Harding and Richart 1963; Hardin and Black 1968; Tokimmatsu et al 1986; Finn et al 1976) have been conducted to establish the equations of G_{max} related to the mean effective principal stress, void ratio, overconsolidation ratio and other factors. The most well known model is proposed by Hardin and Richart, as:

$$G_{max} = AF(e)(\sigma'_m)^n OCR^k \qquad (1)$$

where A is an empirical constant related to the influences of soil fabric; n is an empirical determined exponent reflecting the effects of mean effective principal stress σ'_m; σ'_m is calculated as $\sigma'_m = (\sigma'_v + 2\sigma'_h)/3$, in which σ'_v is vertical effective consolidation stress and σ'_h is horizontal effective consolidation stress; k is the exponent of OCR that depends on plasticity index; e is void ratio and $F(e)$ is void ratio function, which is usually given by (for saturated clays):

$$F(e) = 1/(0.3 + 0.7e^2) \qquad (2)$$

Although Equation (1) has been widely used and proved in practical practice, however, many researchers (Drnevich et al 1976; Wichtmann and Triantafyllidis 2004; Zhou and Chen 2005) have indicated that it cannot consider the influence of cyclic loading history. It is indicated that not only the effective stress (due to the development of excess pore water pressure) or void ratio (in the condition of dry sand, etc.) is changed after relative large cyclic loading, but also the soil fabric has been different from the initial state. Equation (1) does not consider the influences of fabric changes on the small strain response of soils. Many laboratory investigations (Shirley and Hampton 1978; Dyvik and Madshus 1985) have been carried out to study the influences of cyclic loading history on G_{max} for sands or other granular materials, however, according to the authors' knowledge, few researches have been conducted for saturated clays. In this paper, two series of cyclic triaxial tests were carried out, one with cyclic loading history and another without cyclic loading history, which serves as a comparison. The bender elements are used associated with the cyclic triaxial tests to measure the shear wave velocity and then to calculate the small strain shear modulus.

2 TEST PROGRAM

In this paper, the small strain shear modulus is measured based on the bender element tests.

In comparison with the traditional methods such as resonant column tests or improved cyclic triaxial tests, the bender element tests not only can increase the precision of measurement, but also can work in combination with the cyclic triaxial tests and thus can apply to the condition of cyclic loading history. The bender elements used in this study is shown in Figure 1. The shear wave velocity is calculated by:

$$V_s = \frac{L}{\Delta t} \qquad (3)$$

In which L is travel distance and Δt is travel time. The small strain shear modulus can be calculated as:

$$G_{\max} = \rho V_s^2 \qquad (4)$$

In the process of bender element tests, the most important and difficult point is the definition of travel time Δt. In this paper, the travel time is determined based the method proposed by Zhou. The sketch is shown in Figure 2, in which we can see that the travel time is the interval between the initial point of transmitting wave and the first arrival point of the receiving wave.

Two series of tests were conducted. The first is the bender element tests without cyclic loading history. After consolidation under certain confining pressure, the shear wave velocity was measured. For one of the specimens, 50, 100, 150, 200, 250, 300 kPa was applied continuously; and for the other three specimens, 100, 200, 300 kPa were applied respectively. This series of tests is mainly serving as the comparison with the

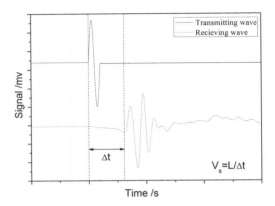

Figure 2. Sketch of determining the travel time.

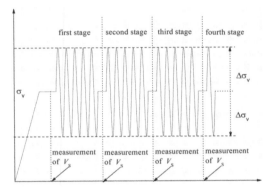

Figure 3. Sketch of the test with cyclic loading history.

second series. The second series of tests is combination of cyclic triaxial tests and bender element tests with cyclic loading history. Three specimens were consolidated under confining pressure of 100, 200, 300 kPa, respectively. Then cyclic tests were conducted in a staged process, every stage contained about 10~50 loading cycles. At the internal of every stage, bender element tests were carried out. The cyclic loading was applied until the failure of specimens. The sketch of the test procedure is shown in Figure 3. The detailed test program is shown in Table 1, in which CSR is the cyclic stress ratio, defined as:

$$CSR = \frac{q^{ampl}}{2\sigma'_m} \qquad (5)$$

where q^{ampl} is amplitude of cyclic deviatoric stress.

Figure 1. The bender elements used in this study.

Table 1. Test program.

Test name	Test number	Confining pressure (kPa)	Frequency (Hz)	CSR
Bender element tests without loading history	NO. 1-1	50, 100, 150 200, 250, 300	–	–
	NO. 1-2	100	–	–
	NO. 1-3	200	–	–
	NO. 1-4	300	–	–
Bender element tests with loading history	NO. 2-1	100	0.1	0.192
	NO. 2-2	100	0.1	0.219
	NO. 2-3	200	0.1	0.186
	NO. 2-4	200	0.1	0.226
	NO. 2-5	300	0.1	0.201
	NO. 2-6	300	0.1	0.246

3 TEST RESULT

3.1 Tests without cyclic loading history

Relationship between small strain shear modulus and effective confining pressure is shown in Figure 4. The results indicate that the two parameters are related well. With the increasing of effective confining pressure, the small strain shear modulus increases continuously.

Relationship between the small strain shear modulus normalized by F(e) and effective confining pressure is shown in Figure 5. Because the soil used in this study is remolded saturated clay and normally consolidated, thus OCR is not considered. The fitted equation based on Equation (1) is:

$$G_{max}/F(e) = 4.504 \cdot (\sigma'_m)^{0.594} \qquad (6)$$

3.2 Tests with cyclic loading history

Based the effective stress theory, in dynamic tests, with the development of excess pore water pressure, the effective stress will decrease, i.e.,

$$\sigma'_m = \sigma'_{m0} - \Delta u \qquad (7)$$

In which, σ'_{m0} is the initial effective confining pressure, and Δu is excess pore water pressure.

Therefore, in the tests with cyclic loading history, the residual excess pore water pressure (Δu_r), residual axial strain ($\Delta \varepsilon$) and travel time (Δt) were measured at the internal of every stage. Therefore, the mean effective principal stress and shear wave velocity can be calculated as:

$$\sigma'_m = \sigma'_{m0} - \Delta u_r \qquad (8)$$

$$V_s = (1 - \Delta \varepsilon)L_0/\Delta t \qquad (9)$$

In which L_0 is initial travel distance.

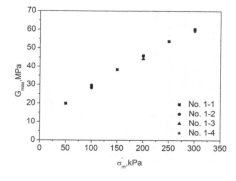

Figure 4. Relationship between small strain shear modulus and effective confining pressure.

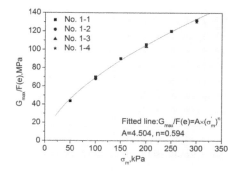

Figure 5. Relationship between normalized small strain shear modulus and effective confining pressure.

Based on Equations (8) and (9), the relationship between small strain shear modulus with cyclic loading history and mean effective principal stress can be presented, which is shown in Figure 6. The initial effective confining pressures in Figure 6 (a),

305

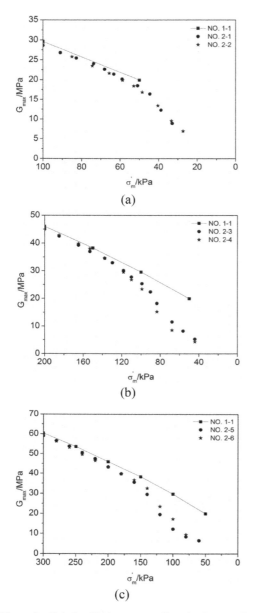

(a)

(b)

(c)

Figure 6. Relationship between small strain shear modulus with loading history and mean effective principal stress.

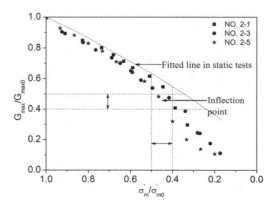

Figure 7. Relationship between normalized small strain shear modulus with loading history and normalized mean effective principal stress.

reaches some degree, the small strain shear modulus decreases suddenly, and result in an inflection point. After the appearance of inflection point, the excess pore water pressure develops continuously and the small strain shear modulus decreases quickly. At the end of tests, the small strain shear modulus with cyclic loading history degrade to only about 25% of the corresponding small strain shear modulus without cyclic loading history.

Relationship between normalized small strain shear modulus G_{max}/G_{max0} and normalized mean effective principal stress σ'_m/σ'_{m0} is shown in Figure 7. Where G_{max0} and σ'_{m0} are the initial small strain shear modulus and mean effective principal stress in the beginning of tests, respectively. It can be seen that the relationship between G_{max}/G_{max0} and σ'_m/σ'_{m0} is well related without the influence of initial effective confining pressure. Moreover, we can see that the inflection point is at the range of 0.4~0.5 normalized small strain shear modulus and 0.4–0.5 normalized mean effective principal stress.

The test results show that the shear modulus with loading history is very different from the one without loading history. The changes of soil fabric induced by pre-cyclic loading make great influences on the small strain response of saturated clays in Wenzhou. It is also found that different with the results of cyclic loading history on sands, the influence of cyclic loading history on small strain shear modulus for saturated clays can not only reflect the degradation of soil fabric but also reflect the failure of soil fabric.

4 CONCLUSIONS

The influences of cyclic loading history on the small strain shear modulus of saturated clays are

(b), and (c) are 100, 200 and 300 kPa, respectively. It can be seen that in comparison with the small strain shear modulus without cyclic loading history, the small strain shear modulus with loading history is moderately lower in the beginning of tests. The reduction is about 5%~8%. However, as the tests progress, when the excess pore water pressure

studied in this paper. The following conclusions can be drawn from the test results:

1. For the remolded clay used in this study, relationship between small strain shear modulus without cyclic loading history and effective confining pressure can well be fitted by the Hardin Equation.
2. The influences of cyclic loading on the small strain shear modulus of saturated clays are great. At the beginning of tests, G_{max} is moderately lower than the corresponding G_{max} without cyclic loading history; however, when the effective stress reduces to some extent, G_{max} decreases suddenly and reflects the failure of soil skeleton.

ACKNOWLEDGMENTS

The work presented in this paper is supported by the National Natural Science Foundation of China (51478364), the the Provincial Natural Science Foundation of Zhejiang (S20090034) and the Provincial College Students' Science and Technology Plan of Zhejiang (2014R424042).

REFERENCES

Drnevich V.P., Hall JR, Richart Fe JR. 1976. Effects of amplitude of vibration of the shear modulus of sand[C]//*Proceeding of International Symposium on Wave Propagation and Dynamic Properties of Earth Materials*, Albuquerque, NM.

Finn W.D.L., Byrne P.M., Martin G.R. 1976. Seismic response and liquefaction of sands[J]. *Journal of Geotechnical Engineering Division, ASCE* 102(8): 841–56.

Finn W.D.L., Lee K.W., Martin G.R. 1977. An effective stress model for liquefaction[J]. *Journal of Geotechnical Engineering Division, ASCE* 103(6): 517–33.

Hardin B.O., Black W.L. 1968. Vibration modulus of normally consolidated clay[J]. *Journal of Soil Mechanics and Foundations Division, ASCE* 94(2): 453–69.

Hardin B.O., Richart Jr F.E. 1963. Elastic wave velocities in granular soils[J]. *Journal of Soil Mechanics and Foundations Division, ASCE* 89(1): 33–65.

Shirley D., Hampton L.D. 1978. Shear-wave measurements in laboratory sediments[J]. *Journal of the Acoustical Society of America* 63(2): 607–613.

Tokimatsu K., Yamazaki T., Yoshimi Y. 1986. Soil liquefaction evaluation by elastic shear moduli[J]. *Soils and Foundations* 26(1): 25–35.

Wichtmann T., Triantafyllidis Th. 2004. Influence of a cyclic and dynamic loading history on dynamic properties of dry sand, Part 1: cyclic and dynamic torsional prestraining[J]. *Soil Dynamics and Earthquake Engineering* 24(2), 127–147.

Advances in Energy, Environment and Materials Science – Wang & Zhao (Eds)
© *2016 Taylor & Francis Group, London, ISBN 978-1-138-02931-6*

Application on geophysical monitoring technology of geological structure detection in Jinfeng coal mine

Z.H. Lu, X.P. Lai, Y.Z. Zhang & P.F. Shan
Energy School, Xi'an University of Science and Technology, Xi'an, China

X.J. Yue
Shenhua-NingXia Coal Ltd., Yin chuan, China

ABSTRACT: In order to find out roof aquifer space distribution laws and geological abnormal body, such as internal concealed faults, small structures in No. 18 coal seam of Jinfeng mine are found. Through the detection of concealed geological conditions in working face, combined with field data of mine geological exploration and development, using the radio wave tunnel perspective method, the research mark out 7 tectonic anomalies in the area of this working face; by using audio frequency electric perspective and three figure—double predictive evaluation method, it forecasts water inrush disaster of water bearing stratum and detects two abnormal stripes whose apparent conductivity is greater than 6 s/m. Combined with electrical prospecting, the research deduces the southern relatively rich water distribution area is larger than the northern area in geophysical prospecting area, and the rich-water distribution area has a certain connection with water channels like faults. Thus it is concluded that the two stripes' position: the conductivity fracture zone height is about 59.67 m to 63.12 m, the caving zone height is between 30.91 m to 31.83 m, and coal seam roof water filling water containing No. 18 coal is to loose bed ripples induced by the roof at the crack between paragraphs sandstone aquifer.

1 INTRODUCTION

The unknown status of mine water disaster, geological structure, gas, and other disaster occurrences are the main reason for the frequent coal mine accidents. Using advanced detection technology and equipment to do advanced detection and prediction is crucial. The research foundation of Geophysical work mainly based on differences in electric, density, magnetic, and elastic, radioactive and other physical properties of various media; using different physical methods and geophysical instruments to detect natural or artificial geophysical field change; by the analysis and research of the geophysical data to infer and explain the geological and mineral distribution situation (Qu et al. 2006, Zhong et al. 2007). The detection of groundwater mainly includes the technology of direct current electric prospecting, radio wave penetration, audio frequency, coal transient electromagnetic and infrared detection, etc. Radio wave penetration method can distinguish the normal and abnormal area of the working face, and can detect the geological structures that cause the electrical parameter changes. Audio frequency electric perspective can detect the water conductive structures of the roof and floor within the working face.

Transient electromagnetic method has a unique advantage in the response of yield water geological body as a non-contact electromagnetic method in time domain (Zhao & Wu 2012).

2 GEOLOGICAL STRUCTURAL CHARACTERISTICS OF MINE AREA

Jinfeng coal mine is located in the middle of Majiatan mine area. The main exploited coal seam is No. 18 coal. According to mine exploration data and actual roadway revealed (Wang et al. 2011, Hu et al. 2011), 18 coal's belongs to the stable coal seam, whose thickness is between 2.49 m~5.5 m. The coal seam roof and floor sandstone fissure and bedding is developed and thickness of thin bedding fine less than 3 cm, interlayer is filled with plant debris, which occurred poor connection; along the tendency, coal seam becomes thinner from west to east; along the strike, it changes from thicker to the South northward. According to the safety production principle, "forecast at first, excavation after exploration and taken measures", so the main aquifer spatial distribution, rich water zoning characteristics and working face concealing faults, and small structures such as geological anomalous body in 18 coal roof should have been identified previously.

3 APPLICATION OF INTEGRATED GEOPHYSICAL MONITORING TECHNOLOGY

3.1 Radio wave tunnel perspective

3.1.1 Detection principle

The radio wave X-ray technology is the kind of instrument and data processing system which is based on the characteristics of electromagnetic wave propagation in coal seam. If the electromagnetic wave emitted from the source is transmitted through the coal seam, where there is concerned fault, the collapse column, rich in water, water and concentration of roof caving goaf geological anomalous body, the electromagnetic wave received energy will be significantly weakened, and form the perspective shadow (abnormal area). Electromagnetic wave penetration in the coal (rock) layer of the two roadways or two drillings, assumes that the radiation source (antenna axis) is the midpoint of the O as the origin, in approximate homogeneous isotropic coal (rock) layer, distance from the observation point P to point o r, on electromagnetic wave field strength HP can be represented by:

$$H_P = H_0 \frac{e^{-\beta r}}{r} f(\theta) \qquad (1)$$

Type: H_0 is the antenna around the coal (rock) layer initial intensity, under certain transmission power; β for coal (rock) layer of electromagnetic wave absorption coefficient; R points to the straight line distance point O; $f(\theta)$ is the direction factor, $f(\theta) = \sin(\theta)$. H_0 is a constant when the radiation conditions do not change with time. The absorption coefficient is the main parameter which affects the amplitude of the electric field. The bigger the value is, the bigger the field intensity changes.

3.1.2 Detection application and results analysis

In drift way transport and launch entry, emitting point and the receiving point were layout at distance between points 10 m, launch point spacing 50 m, each firing points corresponding to multiple received measurement points. When detecting the launch point, each launch time is 3 min, receiving time 3 min, and renewal time 2 min. Using WKT-E type radio wave perspective, using high resolution fixed point scanning method was used to detect (frequency 0.5 MHz), the radio wave penetration 7 delineated more concentrated anomaly area, as shown in Figure 1. The two regions of the larger range and the attenuation range of the electric field are located near the stop line and the cut hole. The three abnormal zones are mainly caused by the large faults or buried faults in the working surface of the working face. Three anomaly areas in Cut hole area, in roadway excavation, unrevealed fault of any size, also did not find any geological structure; but it does not rule out in the working face that conceals the larger fault or other geologic anomalous body (variation of coal seam, broken roof, roof fissure development). Abnormal areas are inferred to have been revealed by $JL_{12\text{-}7}$ fault caused by; No. 5 abnormal area inferred to have been revealed by $JL_{12\text{-}2}$ fault, caused from the return airflow roadway extending to the region; 6 abnormal area inferred for the working face of the hidden fault fracture zone caused by; 7 abnormal area, combined with regional geological situation, inference for two lanes have been revealed by $JL_{12\text{-}1}$ fault caused by. From cut hole at 500 m, it has been revealed by no faults obvious decay reaction, infer the region revealed $JL_{12\text{-}3}$, $JL_{12\text{-}4}$, $JL_{12\text{-}5}$, $JL_{12\text{-}16}$ fault, in the face of extending length is not long, little impact of mining working face.

3.2 Mine audio electrical penetration

3.2.1 Detection principle

Audio electrical penetration as one of important branches in the mine directs the current method. The method is based on the theory of the electric field distribution of the point source, according to the detection data, calculate the working face roof and floor depth range of apparent conductivity value, then this parameter, for mapping, analysis and interpretation of geophysical exploration method. Because of the absorption of the low resistance to

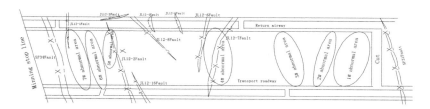

Figure 1. Anomalies area by radio wave.

the current, the current density will decrease and the apparent conductivity will be increased in the corresponding receiving position of the tunnel. Coal seam and roof and floor (sandstone and mudstone) normally have a significant difference, and relative to the coal seam, the top and bottom plate is the high resistance layer, so it can be used by a three layer model to simulate electrical layer combination to simulate electrical layer combination for full at any point in the space of potential formula:

$$U_{i,j} = \frac{\rho_i I}{4\pi}\left[\frac{1}{L} + \sum_{n=1}^{k} k_n(i,j)\right] \qquad (2)$$

$U_{i,j}$ is the i layer of point source in the jth layer potential; ρ_i for the ith layer resistivity value; I for the current intensity; L is the distance from the supply point and observation point; $k_n(i,j)$ is a function of reflection coefficient, $k_n(i,j) = F(L,d,\theta,\rho m)$; d for point source from where coal seam roof distance; and θ is dipole axis observation point and direction angle. Amplitude and width of $U_{i,j}$, depending on the scale of the anomalous body occurrence, abnormal body and country rock electrical difference of anomalous body, from the baking distance and other factors.

3.2.2 *Detection application and results analysis*

By the full range of detector detection, in haulage roadway floor along with the cut hole layout launch a pole (cumulative layout 7 root steel electrode, the numbers for A1~A7); in return to the airflow roadway roof and vertical roadway arranged to receive very M, N, 10 m spacing, arranged in two sides of roadway corner. When the probe is detected, the emitter is arranged well in the first pitch 50 m, then the M_0, N_0 are arranged along the cutting hole, and then the emitter is turned on the emitter A to record the corresponding current and potential difference. After the measurement, move M, N with the distance of 10 m, and successively measure till the data is collected. Fully ranged exploration construction, single phase detection length is 300 m, after each phase measurement, steel electrodes are sequentially continuing with detection until detected mining stop line, as shown in Figure 2.

Under normal circumstances, the resistivity of rock (coal) will not change under the same condition, however, the resistivity will change when the rock (coal) fissure is developed or filled with water. From low to high, the electrical resistivity increases from cold to warm color. The higher apparent conductivity value and the better the formation conductivity is, the higher possibility of water abundance is (Wang 2011, Yong et al. 2011). Survey results are show in Figure 3, the upper section of the floor of working face. The apparent conductivity of the rock layer is 1~26 s/m, the average value is 4.2 s/m, and the standard deviation is 1.8 s/m, which has 2 apparent conductivities, which is greater than 6 s/m of the abnormal band. "No. 1 abnormal section" is mainly developed in the rock group of Ueda which is sandstone dominated, while exception is large and has good continuity. It shows that in the abnormal range, the lateral connectivity is better, and the water is rich and relatively strong. "No. 2 abnormal" is mainly developed in group sandstone in the field, the anomaly amplitude is small, the number of regions is in the shape of a circle, and the continuity is poor, shows that the water is relatively weak in the range of anomaly. From the structural point of view, "No. 1 abnormal" is in the SF_{24} fault extension section, "No. 2 abnormal" mainly in the SF_{33} around the fault. According to the contrast and analysis of the structural characteristics of the abnormal shape and the area, the apparent electrical conductivity anomaly is caused by water or fracture development of the corresponding part of rock stratum and relatively rich water. In the test area, "No. 1 and No. 2 abnormal" are mainly developed in the 0~40 m layer and their performance characteristics in different layers are shown in Table 1.

3.3 *Evaluation and analysis of "Three pictures and double forecasting"*

"Three pictures and double forecasting" evaluation method is here to solve the three big problems of seam roof filling water, charge of water channel and water filling strength of roof water in rush evaluation method. "Three pictures" refers to the roof caving safety zoning map, roof water filling containing water rich water zoning map and roof

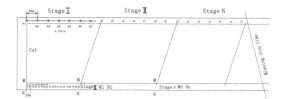

Figure 2. Probe arrangement of the full length phase.

Figure 3. The results of the audio frequency electric perspective of the floor rock stratum.

Table 1. The abnormal distribution of the 0~40 m inner layer of coal seam floor.

Abnormal number	Return-air roadway/m	Haulage roadway/m	0~40 m layer Abnormal threshold	Abnormal range	Remarks
No. 1	Secure channel (south) 0~200	Secure channel (south) 0~130	6	6~12	Stronger abnormal continuity
No. 2	Secure channel (south) 635~840	Secure channel (south) 725~900	6	6~12	Abnormal discontinuity

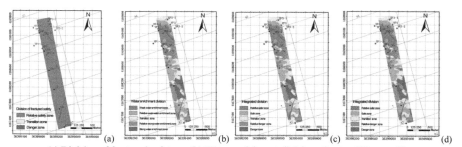

(a) Division of fractured safety   (b) Water enrichment division of water filling aquifer
(c) Integrated division of water bursting conditions   (d) Evaluated result of "Three pictures and double forecasting"

Figure 4. Prospected picture of roof coal.

Table 2. The outcomes of two belts' altitude based on actual data.

Methods	Water-flowing fractured zone Climax location/m	Water-flowing fractured zone Development height/m	The multiple of excavation thickness	Caving zone Climax location/m	Caving zone Development height/m	The multiple of mining thickness
Simple hydrology	101.11	63.12	13.72	132.40	31.83	6.92
Color TV	104.56	59.67	12.97	133.32	30.91	6.72

Chung (water bursting condition comprehensive zoning map; "double forecasting" refers to under the natural and man-made transformation state of mining work face segment and the whole engineering prediction of water inflow. As shown in Figure 4, combined with electrical prospecting, inferred from geophysical exploration area (Quaternary, ancient near and bedrock weathering zone), relative water rich area, distribution area of southern than northern, the distribution of relative water rich area and fault such as water conducting channel with contact. The additional exploration drilling (dg8, hole DG9) simple hydrological observation results and borehole color TV and measuring results comparison. It is concluded that the two belts vertex position; and through calculation, we get the two belts development height, see Table 4.

From Table 2, the difference between two results is not significant. The excavation height was determined by the results of the two methods of value. The development height of the water flowing fractured zone is 59.67~63.12, which is 12.97~13.72 m times of the coal seam, and the height of caving zone is 30.91~31.83 m, which is 6.72~6.92 times of the coal seam. This means that during the working face excavation, the coal seam roof water filling sources includes No. 18 coal seam to loose layer, roof caving zone of each segment of the sandstone aquifer.

4 CONCLUSION

1. Using wireless electrical technology perspective to delineate 7 structural anomaly zone of

the working face; through audio frequency electric perspective method and "Three pictures and double forecasting" evaluation methods, successfully forecast rich water aquifer water inrush disaster and detect two apparent conductivity is greater than 6 s/m anomalies. Finally conclude the height of conducted fissure water is 59.67 m~63.12 m and caving zone is 30.91 m~31.83 m.

2. In order to verify the rich district and weak watery zones curtained by electricity, underground drilling drainage technique was applied here, which shows that electrical prospecting results are reliable. According to the rich district range and distribution rules by exploration, the cumulative drainage reached to be 58 million m³.

REFERENCES

Hu W.Y., Li D.B., Cao H.D. (2011) Shenhua Ningxia Coal Group Jinfeng mine hydro geological exploration report of the first mining area. R Institute of coal science and Industry Group.

Qu H.F., Liu Z.G., Zhu H.H. (2006) Technique of synthetic geologic prediction ahead in tunnel informational construction. J Chinese Journal of Rock Mechanics and Engineering 25(6):1241–1251.

Wang P., Guo T.H., Qiao W. (2011) Shenhua Ningxia Coal Group Jinfeng first coalface roof water drainage design and research. R Ningxia coal exploration engineering company.

Wang R. (2011) Study on geological mechanical condition and risk of water inrush from coal seam roof and floor. D China University of Mining and Technology.

Yong Z.C., Song H.J., Meng X.L. (2011) Application of the mine transient electromagnetic method in the exploration of the working face roof. J Information technology 27:95–96.

Zhao X.Y., Wu C.Y. (2012) Mine transient electromagnetic method is applied to the shaft into the water. J Mine Construction Technology 33(1):22–25.

Zhong S.H., Sun H.Z., Wang R. (2007) The present situation of the geological prediction in front of tunnel face and its development. J Chinese Journal of Engineering Geophysics 4(3):180–185.

Architectural environment and equipment engineering

Advances in Energy, Environment and Materials Science – Wang & Zhao (Eds)
© 2016 Taylor & Francis Group, London, ISBN 978-1-138-02931-6

Stress monitoring in the ARMA model of Changchun subway construction applications

Xu-Qing Zhang, Ji-Kai Zhang, Min-Shui Wang & Guo-Dong Yang
College of Geoexploration Science and Technology, Jilin University, Changchun, China

ABSTRACT: This paper presents the application of the ARMA time-series analysis model for the analysis and forecast of stress data in Changchun subway construction. The results show that the model has an ideal accuracy in the forecast after the smooth process of the data. As the step size increases, the accuracy drops. Thus, the results indicate that the time-series analysis model is effective in the short-term forecast.

1 INTRODUCTION

The emphasis and difficulty of deformation monitoring are the analysis methods. After processing the deformation monitoring data by establishing an appropriate function model, we can get the accurate forecast, avoid accidents as well as construct scientifically (Zhang, 2010; Huang, 2010; Sun, 2013). There are many methods for analyzing the deformation monitoring data. The time-series analysis model analyzes a series of time-variant, dynamic and interrelate data series (Pan, 2001; Wang, 2006). Thus, this model is commonly applied for analyzing the deformation monitoring data of building settlement. This paper preliminarily discusses the construction stress analysis by using the time-series analysis model.

2 TIME-SERIES ANALYSIS PRINCIPLES

The basic idea of the time-series analysis model is as follows: in the smooth, zero-mean, normal time series {Xt}, Xt is related to each of its former n step values Xt − 1, Xt − 2, … Xt − n, as well as to each of its former m steps interferences a_{t-1}, a_{t-2}, ..., a_{t-m} (m, n = 1, 2, ..., n).

ARMA (p, q) is a p-order autoregressive q-order moving average model, which is one of the time-series analysis models.

The establishment of the ARMA time-series analysis model includes the model order determination, parameters estimation, and model test.

Model order determination often uses the partial correlation, AIC criterion, F-test, white noise test, and criterion function. In the precondition of determining the model order, the model parameters can be estimated preliminarily; after establishing, the quality of the model needs to be tested. It can be achieved by calculating the error

a_t between the primary data and the model, i.e. to test whether a_t is white noise or not. If the error sequence is random, the established model contains the trend term of original time series and can be used to predict; if not, the established model is not perfect and needs to be established again.

2.1 AIC criteria to order

AIC criterion, namely the minimum information criterion, was proposed by the Japanese statistician Akaike in 1974. This criterion is suitable for less number of sample data time series; it can also give the best estimate of ARMA model parameters and model order of the ARMA model. When having a large enough sample size, sample autocorrelation function will be close to the observed data of time-series autocorrelation function. Within the prescribed scope order from low to high, the model calculates the AIC value in turn, and finally determines to use its minimum order number as the optimal order number (Sun, 2013).

Maximum likelihood estimate model parameters can be expressed as follows:

$$AIC = (n-d) \log \sigma^2 + 2(p+q+2).$$

The least-squares estimate model parameters are given as follows:

$$AIC = (n-d) \log \sigma^2 + (p+q+2) \log n.$$

where n is the number of samples, as the fitting residual sum of squares, and d, p, q are parameters.

The practical value of p and q is generally not more than 2, otherwise the function will be volatile.

ARMA (p, q) fixed order sequence AIC criterion is as follows: selecting the parameters p, q makes true that $AIC = \min$.

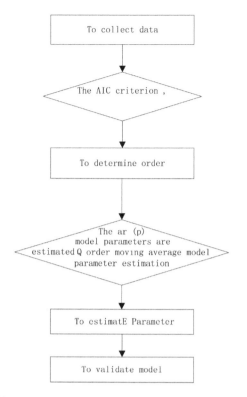

Figure 1. The establishment of the time-series analysis model.

2.2 Parameter estimation

After the model identification, we determine the time sequence function model, to determine the order number based on preliminary estimate model parameters.

2.3 Model validation

The model is validated to test the quality of the established model, by calculating the error a_t between the original data and the model data, to determine whether inspection a_t is white noise or not. If the sequence a_t is random, then the trend of the model contains the original time series, so the model can be used to predict. If the error sequence a_t is not random, then the model could not be perfect enough, and need to perform modeling again.

3 SUBWAY STRESS MONITORING USING THE TIME-SERIES ANALYSIS MODEL

This paper takes the concrete supporting reinforcement stress monitoring point GJYL-0201 that is in Wei Xing subway station of Changchun as an example, and analyzes the monitoring data by using the ARMA (p, q) model.

The observation time of the monitoring point was from the beginning of stress monitoring to the capping of the open foundation pit, and the foundation pit deformation monitoring lasted 161 days, with a total of 37 phases.

3.1 Data preprocessing

Data preprocessing includes the elimination and interpolation of the singular values, the construction of equidistant data sequence, and a smooth process. According to the 3δ criterion, the observation values of 36th, 43th, and 101th are singular values, which are needed to be removed, and then the observation values of pretest-posttest to interpolate are averaged.

The observation period of original observation data is not stable and the time interval of every observation is not the same. However, a certain amount of sample observations and evenly distributed sample data is needed on the analysis and forecast using the method of time-series analysis. So, the data are needed to be interpolated before the time-series analysis. In order to keep the consistency between the original observations and the new ones, as well as to avoid the distortion of interpolation, this paper uses linear interpolation to interpolate the observations into 41 phases at a 4-day interval. At the same time, with the aim of testing the feasibility and

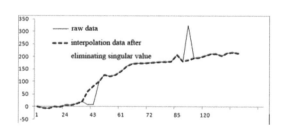

Figure 2. The original and preprocessing data.

Table 1. The result of the stationary test.

	Statistic T	Spearman's correlation coefficient	Critical value of T, ta
Original data	29.679212	0.978571	2.022691
4-order stationary data	0.301722	0.048258	2.022691

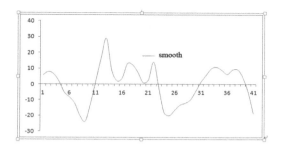

Figure 3.   Stationary data.

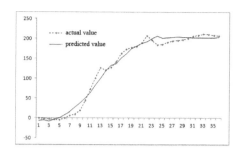

Figure 4.   The contrast of one step predictive data and measured data.

Table 2.   The contrast of the 5-phase prospective results and the measured data.

Phase number	Interpolation data/KN	Prediction data/KN	Prediction error	Relative error	Steps
37	211.8429	206.8973	4.9456	2.33%	1
38	215.8342	209.4606	6.3736	2.95%	2
39	216.8676	214.7177	2.1499	0.99%	3
40	216.4199	222.9044	−6.4845	−3.00%	4
41	214.4911	234.3534	−19.8623	−9.26%	5

accuracy of time-series analysis for the monitoring and analysis of reinforcement stress, this paper uses the first 36 phase data to set up the time–series analysis model, and forecast by the last 5 phase data.

3.2   Data stationary test

The time series is stable in the time-series analysis model. This paper uses the Daniel test to test the stationarity of data. The Daniel test is based on Spearman's correlation coefficients. From the test results summarized in Table 1, the original data are proved to be non-stationary. Combining the high volatility of stress observation data, we try to choose the higher-order fit. So, this paper chooses a 4-order carve fit to perform the stabilized treatment.

The 4-order curve-fitting equation is given as follows:

$$\hat{X}_t = 0.0000041082t^4 - 0.0014t^3 + 0.1334t^2$$
$$- 1.6332t - 4.431$$

3.3   Order determination and parameter estimation of the model

Using the AIC criteria to determine the order, the fluctuation of simulated values is found to be strong when p, q > 3 in the test. So, in this paper, the selection of p,q is within 3.

The calculating results show that the model is the ARMA (1,1) model. The expression of the model is as follows:

$$x_t = \phi_1 x_{t-1} - \theta_1 a_{t-1} + a_t$$

The parameters are as follows: $\phi_1 = 0.1485$ $\theta_1 = 0.1843$.

3.4   Forecast

According to the established function model, the one step simulation is based on the established function model, whose biggest fitting error is 21.04, fitting square error is 8.36, and forecast relative error is within 10% relative to the reading interval [0,240], so the established model has a higher prediction accuracy.

The trends of the predictive data and measured data are basically the same, so the fitting effect of the ARMA (1,1) model is good.

As the prediction phases increase, the accuracy will decrease, but the short-term prediction is excellent. In contrast to other data of stress monitoring points, the conclusion is basically the same.

4   CONCLUSION

According to the above analysis and calculating results, the application of the ARMA model for the stress monitoring data analysis is feasible and the forecast effect is obvious.

1. The application of the ARMA model for the analysis of construction stress monitoring data has a higher prediction accuracy, and there is a certain reliability and accuracy.
2. As the prediction steps of the ARMA model increase, the prediction accuracy decreases. So, the ARMA model is better for short-term prediction.

REFERENCES

Huang Sheng Xiang, Yin Hun, Jiang Zheng. Building Deformation Monitor Data Processing (Second Edition) [M]. Wuhan University Press. 2010.10

Lin Xun. Application of Time Series Analysis in Buildings Deformation Monitoring [D], Jilin University, 2005.

Pan Guo rong, Wang Sui hui, Liu Da jie. Dynamic Model and Forecasting of Ground Deformation in Subway Construction [J]. Journal of Tongji University, 2001, 29(3):294–298.

Sun Tong He. An Application of Time Series Analysis and Its Application in Measurement [J]. Eomatics & Spatial Information Technology. 2013, 36(3):12–13.

Wang Changfeng, Chen Xingchong. Stress Monitoring and Processing Method for the Construction of Prestressed Concrete Girder [J]. Journal of Lanzhou Jiaotong University, 2006, 25(4):24–17.

Wang Yan, Applied Time Series Analysis [M]. China Renmin University Press, 2005.7.

Zhang Jian, A Case Study of Time Series Analysis Applied for Construction Monitoring of Long-span Bridge [D]. South China University of Technology, 2010.

Zhang Shan Wen, Lei Jie Ying, Feng You Qian. MATLAB in the application of time series analysis [M]. Xian university of electronic science and technology press. 2007.

Advances in Energy, Environment and Materials Science – Wang & Zhao (Eds)
© 2016 Taylor & Francis Group, London, ISBN 978-1-138-02931-6

Simulating analysis and research on transformer winding deformation fault

Y. Chen, Q. Peng & J. Tang
Sichuan Electric Power Research Institute, Chengdu, China

Y.H. Yin
Chengdu Brainpower Digital Technology Co. Ltd., China

ABSTRACT: Focusing on transformer winding deformation fault issue, this paper proposes an analog analysis circuit of transformer winding deformation fault by combining equivalent circuit model of transformer with frequency response method. It studies parametric variation of fault analog circuit and frequency-response variation corresponding to different deformation faults. Based on this circuit, this paper puts forward a simulating device for transformer winding deformation fault, which is equipped with frequency response tester. This device can display graphically waveform of transformer's frequency response, which can achieve a quick and simple way to identify the problem of transformer winding deformation, so it can be used for experimental teaching on winding deformation frequency response analysis and self-checking for winding deformation frequency response tester.

1 INTRODUCTION

Transformer is one of the important power system equipment, the transformer suffered accidental collision and impact during transport and the impact of current fault state, resulting in a certain degree of deformation mechanical structure of the transformer winding which is prone to accidents.

Since the failure of the transformer winding deformation and power system operation will cause serious harm, and the conventional test methods cannot effectively detect such defects, it can only be verified by crane inspection. This not only takes a lot of manpower and resources, but also causes certain hazards to the transformer itself. Moreover, under the current power system operation, conducting a prolonged power outage detection on large transformers is very difficult. The relevant provisions of the national grid put the prevention of near area short-circuiting in the transformer winding and the deformation failure detection of transformer winding on a very important position. Right now, many countries and even put the detection in the first place of the transformer preventive pilot project. So the focus is that it can quickly detect deformation fault in internal winding deformation transformer winding in the field without hanging the hood. There are two methods for winding deformation test at home and abroad: "short-circuit impedance method"[1] and the "frequency response analysis."[2] And "frequency

response analysis" is the most common and most effective means for detecting winding deformation. The accuracy of output sine wave signal frequency in "frequency response analysis" should be less than 0.01%[3,4], more sensitive and more accurate compared to the short-circuit impedance method.

The principle of "frequency response analysis" is based on the equivalent circuit of the transformer, which can be seen as a common ground of the two-port network. The frequency characteristics of the two-port network can be used to describe the transfer function $H(j\omega) = U_0(j\omega)/U_i(j\omega)$[5,6]. This method is characterized by the transfer function description of the network known as the "frequency response analysis." As each transformer has its own corresponding response curve, the winding deformation, the internal parameter changes will lead to changes in the transfer function[7,8]. The transformer windings can be found whether it is changed or not by analyzing and comparing the frequency response curve transformer[9–11].

2 PRINCIPLE OF FREQUENCY RESPONSE METHOD AND EQUIVALENT MODEL OF TRANSFORMER WINDING

Under the effect of high frequency voltage, each winding of the transformer can be considered as a two-port network formed by linear passive linear resistors, inductance (mutual inductance), distributed parameter capacitance, internal characteristics of which can be described by transfer function H(jω). If the winding deformation failure, the internal windings distributed inductance, capacitance and other parameters will inevitably change, resulting in an equivalent network transfer function H(jω) of the poles and zeros change, so that the frequency response characteristics of the network change.

The test of transformer winding deformation through the "frequency response analysis" is done by detecting the frequency response characteristic of each winding transformer, and the test results compare longitudinal or transverse direction, based on differences in amplitude frequency response determines the transformer winding changes that may occur. After comparing the results, the transformer winding changes that may occur can be determined by the differences in amplitude frequency response.

Transformer coils are generally designed for cake-type structure, its purpose is to consider the insulation and pressure, but there are gaps between the cake which is easy to heat, the coils cake has a close capacitance, and inductance coil to the ground and to the other phase, and to other voltage coil. In addition, the casing includes ground capacitance, and capacitor in joints, all those have structure parameters according to their location in these structures, so according to their structure, it can constitute the equivalent circuit of a transformer winding during the test. When the frequency exceeds 1 kHz, the basic core of the transformer has no effect. Each winding can be regarded as a passive linear two-port network consists of resistors, capacitors, inductors and other distribution parameters configured, as shown in Figure 1, The amplitude frequency response characteristics of the transformer windings is obtained by the manner of frequency scanning. Continuously change the frequency f (angular frequency ω = 2f) of the applied sinusoidal excitation source Vs, measure the ratio of the terminal voltage V_o and the voltage excitation signal amplitude of V_i at different frequencies.

At last, we can enact the winding amplitude frequency response curve of the excitation and responder ports. In the figure, L_s represents the coil inductance, C_g represents the winding-to-ground capacitance, C_s represents the capacitance between the cake. V_i, V_o represent the equivalent terminal voltage excitation and response terminal voltage in the network, Vs is sinusoidal excitation signal source voltage, Rs is signal source output impedance, R is matched resistors.

The frequency response curve measured through "frequency response analysis" is represented by a common logarithm formula, as shown in the voltage amplitude of the process of formula (1).

$$H(f) = 20Log[V_o(f)/V_i(f)] \qquad (1)$$

where: H(f) is the value of transfer function |H(jω)| when the frequency is f; $V_o(f)$ is the peak or RMS |$V_o(jω)$| of response voltage when the frequency is f, V_i (f) is the peak or RMS |$V_i(jω)$| when the terminal voltage excitation frequency f, $V_o(f)$ and $V_i(F)$ should use both Peak or RMS at the same time, and should be the same kind value.

V_i is the input sweeping signal, V_o is the output response signal, which actually represents a current flowing through Ro, the ratio V_o/V_i would represent a reactance change. If the deformation phenomenon of axial, radial dimension change occur in winding, Ls, Cs, Cg and other distribution parameters in the network will inevitably change, leading to changes of the distribution of zeros and poles in its transfer function H(jω). Therefore, the deformation of transformer windings can be diagnosed by comparing the frequency response of transformer windings.

3 ANALOG CIRCUIT AND EXPERIMENTAL RESULTS

The analog circuit simulating the winding consists of the third order inductance (mutual inductance), the capacitor circuit. There are six relay switches Cg "winding capacitance to ground", Cs "capacitance between cake," L "inductance coil" to achieve the value of different fault types of action. The analog part of the circuit is shown in Figure 2.

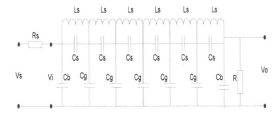

Figure 1.　The equivalent circuit of transformer winding.

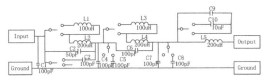

Figure 2.　The analog circuit parts of winding.

The transformer winding analog circuit controls these three parameters: the Cg "winding capacitance to ground", Cs "capacitance between the cakes" to increase and decrease L "coil inductance" through six relays. The default is that relay 1 is closed; L1 is not connected to the circuit. Relay 2 is closed; C3 is not connected to the circuit. Relay 3 is open, L3 is connected to circuit. Relay 4 is closed; C5 is not connected to the circuit; relay 5 is open, C8 is connected to circuit. Relay 6 is open, C10 is connected to circuit.

The default waveform of analog circuit parts of winding is shown in Figure 3.

When simulating "local compression" fault, increase the Cs (capacitance between cake), open relay 2, C3 and circuit phase, increasing the capacitance between the analog circuit Cs cake, winding through the frequency response waveform as Figure 4.

By comparing two waveforms as shown in Figure 5, it can be clearly seen that the "resonance point frequency" of "local compression" waveform distortion state moved to the left, its resonance frequency point becomes smaller; the resonance point of the peak is reduced accordingly.

When simulating the fault deformation category of "coil off shares", the waveform is shown in Figure 6 by increasing the inductance of the coil Ls, opening relay 1, connecting L1 to the circuit, increasing the coil inductance Ls of the analog circuit.

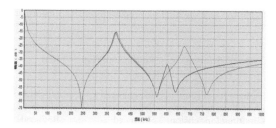

Figure 5. The waveform comparison between normal state and "local compression".

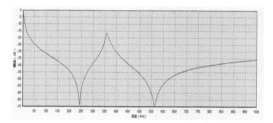

Figure 6. The "coil broken" waveform of transformer winding.

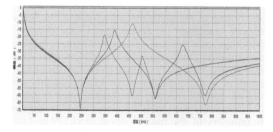

Figure 7. The waveform comparison of different deformation failures.

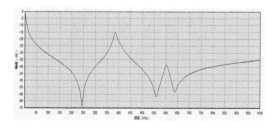

Figure 3. The normal waveform of transformer winding equivalent circuit.

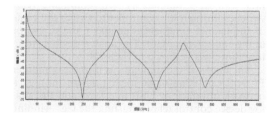

Figure 4. The "local compression" waveform of transformer winding.

As it can be seen, frequency resonance point of waveform in deformed state "coil off shares" moved to the left, the resonance point reduced from four to three, the peak resonance point unchanged, which is in line with the theoretical waveform distortion.

By simulating different fault conditions, the change of the resonance frequency and the peak point of the resonance point can be clearly seen from the waveform as shown in Figure 7.

By changing Ls, Cs, Cg and other distribution parameters of the mathematical model, the common type of deformation can simulate the changes of the resonance frequency and the resonance point peak of the transformer windings, which is in the Table 1.

Table 1. Deformation parameters table of transformer winding.

Types of deformation	Inductance (Ls)	Cake capacitances (Cs)	Capacitance to ground (Cg)	Frequency of resonance point	Peak of resonance point
Overall deformation					
Transport collision	Unchanged	Unchanged	Reduced	Right (all)	Unchanged
Overall compression	Increased	Increased	Unchanged	Left (1)	Increased (1)
Compression tension	Reduced	Reduced	Unchanged	Right (1)	Reduced (1)
Partial deformation					
Partial compression and tension	Unchanged	Increased Reduced	Unchanged	Left Right	Reduced Increased
Interturn short circuit	Reduced	Increased	Unchanged	Right (L)	Increased (L)
Coil strand breakage	Increased	Unchanged	Unchanged	Left (L)	Unchanged
Metal foreign object	Unchanged	Increased	Unchanged	Left (L)	Increased (M, H)
Wire displacement	Unchanged	Unchanged	Increased (close to the shell) Reduced (close to the coil)	Left (H) Right (H)	Reduced (H) Increased (H)
Axial distortion	Unchanged	Reduced	Reduced	Right	Reduced (L) Increased (M)
Width (diameter) deform	Inside reduced Outside increased	Unchanged	Inside increased Outside reduced	Left (1) Right (M, H)	Unchanged

4 STRUCTURE OF THE SIMULATION DEVICE AND WORKFLOW

According to the mathematical model of the transformer windings, a device simulating distorted windings can created. The transformer winding deformation fault simulation device is characterized by the base of frequency characteristics mathematical model with the use of analog circuits and digital control circuit, the transformer winding deformation symptom is simulated by changing the parameters of the mathematical model. The fault simulation device which can simulate faults such as the common transformer local twist, metallic foreign body, the overall displacement transformer fault may provide the normal operating mode and fault mode. It uses the transformer winding deformation frequency tester to do sweep test on simulation apparatus, providing the normal operating mode and the frequency response curve of different failure modes. In the premise that the transformer winding deformation tester functions well, we can very clearly found the frequency response curve of transformer winding deformation when the parameters changed. This fault simulator can be used for teaching method of the winding deformation response, and can also be used for on-site rapid testing for transformer winding deformation tester frequency response function, and to exclude test failure caused by the failure of the instrument itself.

Since each winding of the transformer can be seen as a distributed passive linear two-port network parameters consisting of linear resistance, inductance (mutual inductance), capacitors and the like. The actual work of the transformer windings can be simulated by transformer winding equivalent circuit consisting of resistance, inductance, capacitance. Different types of deformation fault can be simulated by controlling the relay switches through the controller to switch the value of inductance (mutual inductance), the capacitor in the equivalent circuit. The analog devices structure flowchart is shown in Figure 8.

The mode of the simulator is that the computer PC software connect an analog device through a network interface or USB interface, and the analog device control relay adjusting the Cg "winding capacitance to ground", Cs "capacitance between the cake", Ls "coil inductance" variation these three parameters of the analog portion to simulate

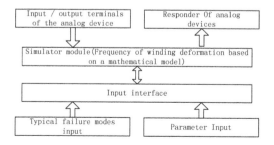

Figure 8. The structure flow chart of the simulator.

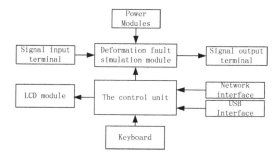

Figure 9. The working flow chart of the simulator.

different deformation fault in the transformer windings. The corresponding fault waveform to determine and confirm the type of winding deformation failure can be obtained through the transformer winding deformation test device. Work flow is shown in Figure 9.

5 CONCLUSION

This paper proposes an analog analysis circuit of transformer winding deformation fault by combining equivalent circuit model of transformer with frequency response method. It studies parametric variation of fault analog circuit and frequency-response variation corresponding to different deformation faults. Based on this circuit, this paper puts forward a simulating device for transformer winding deformation fault, which is equipped with frequency response tester. So the resonance frequency and the resonance point of the peak point of the waveform changes can be easily seen, which provides a quick and simple way to identify the problem of transformer winding deformation. In addition, it can be furthermore used for experimental teaching on winding deformation frequency response analysis and

self-checking for winding deformation frequency response tester.

REFERENCES

[1] Zeng Gang-yuan. Measuring Short-circuit Reactance Is an Effective Method of Judging Transformer Winding Deformation [J]. Transformer, 1998, 8(35): 15–19.

[2] Huang Hua, Zhou Jian-guo, Jiang Yi-ming, et al. Diagnosis of Winding Deformation of Transformer by Impedance Method and FRA [J]. High Voltage Engineering, 1999, 2(25): 70–73.

[3] [GB1094.5-85] National Bureau of Standards. Power transformers—Part V (the ability to withstand short circuit) [GB]. 1985, 11, 22.

[4] [589] State Power Corporation. Power to prevent major accidents Twenty-five key requirements [Z]. 2000, 9, 28.

[5] Cheng Wenfeng, La yuan. Transformer winding frequency response test method related problems [J]. Electrical Measurement & Instrumentation, 2013.7.10.

[6] He Wei, Liu Yigang, Hu Guohui based on short-circuit reactance method of distribution transformer winding deformation online diagnosis [J]. Electrical Measurement & Instrumentation, 2014.8.27.

[7] Liu Lipeng, Qiao Yuliang, Telecommunications for detected by the current source transformer winding deformation. [J]. Electrical Measurement & Instrumentation 2014, 51 (8).

[8] Yan Yu-ling, Jiang Jian-wu. The Theoretical Analysis and Experimental Study on Transformers Winding Deformation [J]. High Voltage Apparatus, 2010, 45(5): 44–59.

[9] Chen Sheng, Chen San-yun. Case Analysis of Overstandard of Low Voltage Short Circuit Impendance in Transformertest [J]. High Voltage Apparatus, 2009, 45(2): 108–111.

[10] He Wen-lin, Chen Jin-fa. A Study on Detecting Winding Deformation in Transformers with FRA Method [J]. Electric Power. 2000, 33(12): 39–42.

[11] Yao Sen-jing. Application of Lateral Comparison Approach in Testing Transformer Winding Deformation [J]. Guangdong Electric Power, 2000, 13(4): 11–14.

Advances in Energy, Environment and Materials Science – Wang & Zhao (Eds)
© *2016 Taylor & Francis Group, London, ISBN 978-1-138-02931-6*

Study on the equipment maintenance support efficiency evaluation based on the gray correlation analysis method

Xiang Zhao, Ming Guo & Yongjun Ruan
6th Department of Ordinance Engineering College, Shijiazhuang, Hebei Province, P.R. China

ABSTRACT: The scientific evaluation of equipment maintenance support efficiency is very important for the unit equipment management. This paper studied the multi-objective maintenance support efficiency evaluation problem. First, an index system of equipment maintenance support efficiency was built. Second, the theory of the gray correlation analysis method was described. Third, a multi-objective evaluation model of equipment maintenance support efficiency was built based on the gray correlation analysis method. Finally, the model was validated. The evaluation model was identified as a rationality and feasible model. The results of the evaluation model show a good reference value for finding weakness in the equipment maintenance support system efficiency.

1 INTRODUCTION

Equipment maintenance support is very important in keeping and resuming the tactical and technological performance of the equipment. Equipment maintenance support evaluation is a multi-objective decision-making problem. It is complex and hardly evaluated. There are no unique and distinct standard applied to the evaluation process. Gray correlation analysis is suitable for a multi-objective decision-making problem. So, the gray correlation analysis method is adopted to evaluate the equipment maintenance support efficiency. A hierarchical structure model was constructed for the evaluation of equipment maintenance support efficiency. The relative index weights are calculated by the Analytic Hierarchy Process (AHP) method. The equipment maintenance support efficiency is evaluated synthetically by the gray correlation analysis method. Finally, the model combines the qualitative analysis and the quantitative calculation.

2 THE BUILDING OF THE INDEX SYSTEM OF THE EQUIPMENT MAINTENANCE SUPPORT EFFICIENCY

The system of the equipment maintenance support is a complex system. The efficiency evaluation contains multiple objects and rules. Therefore, a scientific evaluation index system should be built. According to the analysis of the characteristics of the equipment maintenance support process, the index system of the equipment maintenance support efficiency is built by the top-down approach. Its constitution is based on the principles of

systematic, independence, hierarchy, and testability. The index system of the equipment maintenance support efficiency has three levels. The first one is the global index; the second level is the sub-index; the third level is the attribute index. After analyzing the process of equipment maintenance

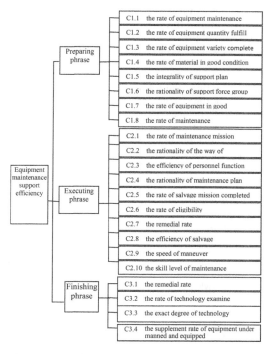

Figure 1. Equipment maintenance support efficiency evaluation index system.

support in the task, there exist three sub-indices, including the preparing phrase index, executing phrase index, and finishing phrase index of maintenance support task. As the node index, the attribute index is the quantification index about the maintenance support efficiency. For example, it needs to evaluate the equipment repair efficiency at the preparing phrase. It is equal to the quotient of the repaired equipment quantity and total broken-down equipment quantity*100%. The index system of the equipment maintenance support efficiency is shown in Figure 1.

3 EQUIPMENT MAINTENANCE EFFICIENCY EVALUATION MODEL BASED ON THE GRAY CORRELATION ANALYSIS METHOD

3.1 Idea of the gray correlation analysis method

Professor Julong Deng, a famous scholar in Huazhong University of Science and Technology, founded the gray system theory in 1982. It is used to study the incomplete information system. The aim of the gray correlation analysis method is to seek the quantificational method of measuring the correlation degree among the factors, so that the key influence infectors of the system situation can be found. The comparative method of the system situation quantification is to obtain the correlation coefficient and the correlation degree between the reference sequence and the comparing sequences (sub-sequence). It has a basic idea that the method involves a correlation sorting analysis in forecasting evaluation. The gray correlation degree analysis method has the following characteristics: first, regarding the system as a series of attributes; second, choosing one reference sequence (the standard sequence); third, computing the correlation degree between the reference serial and example serial; finally, obtaining the sorted correlation degree. As a simplification matter of the multiple objective decision method, the gray correlation analysis

method is suitable for the equipment maintenance support evaluation.

Because the equipment maintenance support efficiency evaluation is a complex matter, affected by stochastic factors, the gray correlation degree analysis method could resolve it effectually in a simplified and straight manner.

3.2 The construing process of the model

The process of modeling equipment maintenance support efficiency by the gray correlation degree analysis method is as follows:

1. Calculating the evaluation factor, getting the quantitative evaluation of evaluating factors according to the index definition. We can get the mark about the qualitative system from the experts. An example of the five-grade system is presented in Table 1.
   According to the standard, we can get all the remarks of the evaluating factors from experts and experienced commanders quantitatively. The data are given in Table 2.
   In Table 2, F_0 represents the reference ideal system (all of the items have full marks), and thus is the optimal system. It can also be established with practical definition. F_i represents the system to compare and $F_i(j)$ represents the evaluating mark of the jth factor in the ith system, with $i = 1, 2, \ldots\ldots m, j = 1, 2, \ldots\ldots n$.
2. Normalization processing. Every numerical value of factors obtained from the mark is divided by the full marks and the quantificational evaluating table is made up of all of the result.
3. Computing the absolute deviation, minimum difference and maximum difference

Absolute deviation: $\delta_{0i} = |F_0(j) - F_i(j)|$ (1)

Minimum difference:
$\delta_{min} = \text{mini}\,\text{minj}|F_0(j) - F_i(j)|$ (2)

Maximum difference:
$\delta_{max} = \text{maxi}\,\text{maxj}|F_0(j) - F_i(j)|$ (3)

Table 1. An example of the five-grade system.

Evaluating factors	Values				
	5	4	3	2	1
The battlefield maintenance fulfillment rate	Fastest	Faster	Middling	Slower	Slowest
The qualification rate of maintenance	Highest	Higher	Middling	Lower	Lowest
The speed of maneuver	Fastest	Faster	Middling	Slower	Slowest
...	...	...	...	...	...
Factor n		...	...	...	...

Table 2. The statistic of the result.

System	The factor of judge						
	The battlefield maintenance fulfillment rate	The qualification rate of maintenance	The speed of maneuver	...	Factor j	...	Factor N
F_0	$F_0(1)$	$F_0(2)$	$F_0(3)$	...	$F_0(j)$	...	$F_0(n)$
F_1	$F_1(1)$	$F_1(2)$	$F_1(3)$	...	$F_1(j)$	...	$F_1(n)$
...	...	...	...	...	...	...	...
F_i	$F_i(1)$	$F_i(2)$	$F_i(3)$	...	$F_i(j)$	...	$F_i(n)$
...	...	...	...	...	...	...	...
F_m	$F_m(1)$	$F_m(2)$	$F_m(3)$	...	$F_m(j)$	...	$F_m(n)$

4. Computing correlation coefficient and the correlation degree

The correlation coefficient is marked by $\zeta_{0i}(j)$, whose mathematical expression can be represented as follows:

$$\zeta_{0i}(j) = \frac{\delta\min + p\delta\max}{\delta oi(j) + p\delta\max} \qquad (4)$$

The density p refers to the distinguishing correlation coefficient, which is usually $p = 0.5$. The correlation coefficient ranges from 0 to 1. The higher the correlation coefficient, the better the correlation is.

The correlation degree refers to the correlation between two factors. From Formula (2)–(4), we can achieve many correlation coefficients during the comparing process. So, we regard the average of the correlation coefficient as the system correlation degree.

The relational degree is given as follows:

$$U_{0i} = \frac{1}{n}\sum_{1}^{n} \zeta_{0i}(j)$$

where n is the total number of factors.

5. Sorting of the correlation degree. We can get the system sequence according to the correlation degree. Then, the system efficient evaluation order can be obtained. Finally, the optimal system among the evaluation system can be chosen.

4 THE APPLICATION EXAMPLE OF THE MODEL

For the sake of simplicity, we chose three evaluating factors (the battlefield maintenance fulfillment rate, the qualification rate of maintenance, and the speed of maneuver) from the index system in order to verify the model. It showed how the gray correlation analysis method can be used in the equipment maintenance support efficiency evaluation process.

4.1 Evaluation model definition

Let us suppose that there are six systems to be evaluated, namely F_1, F_2, F_3, F_4, F_5, and F_6. The evaluating marks can be obtained from experts by using the evaluating method.

The reference system is marked as F_0. We define the factor value as (5, 5, 5). We then ascertain the system's equipment maintenance efficiency sequence by the gray correlation analysis method.

4.2 The process of model evaluation

1. Quantificational evaluation of the factors, as shown in Table 3.
2. Normalization processing, as shown in Table 4.
3. Computing the absolute difference, minimum difference, and maximum difference

The absolute difference is given as follows:

$$\delta_{01} = \{0.6, 0.4, 0.2\}$$

$$\delta_{02} = \{0.2, 0, 0.2\}$$

$$\delta_{03} = \{0.4, 0.4, 0.2\}$$

$$\delta_{04} = \{0, 0.2, 0.2\}$$

$$\delta_{05} = \{0.2, 0.4, 0\}$$

$$\delta_{06} = \{0.2, 0.2, 0.4\}$$

From the above result, we see that the minimum difference is $\delta_{min} = 0$ and the maximum difference is $\delta_{max} = 0.6$.

4. Computing correlation coefficient and correlation degree

The correlation coefficient is $p = 0.5$:

$$\zeta_{01} = \{0.33, 0.43, 0.6\}$$

Table 3. Quantificational evaluation of the factors.

| System | Factor | | |
	The battlefield maintenance fulfillment rate	The qualification rate of maintenance	The speed of maneuver
F_0	5	5	5
F_1	2	3	4
F_2	4	5	4
F_3	3	3	4
F_4	5	4	4
F_5	4	3	5
F_6	4	4	3

Table 4. Evaluation matrix after normalization processing.

| System | Factor | | |
	The battlefield maintenance fulfillment rate	The qualification rate of maintenance	The speed of maneuver
F_0	1	1	1
F_1	0.4	0.6	0.8
F_2	0.8	1	0.8
F_3	0.6	0.6	0.8
F_4	1	0.8	0.8
F_5	0.8	0.6	1
F_6	0.8	0.8	0.6

$$\zeta_{02} = \{0.6, 1, 0.6\}$$

$$\zeta_{03} = \{0.43, 0.43, 0.6\}$$

$$\zeta_{04} = \{1, 0.6, 0.6\}$$

$$\zeta_{05} = \{0.6, 0.43, 1\}$$

$$\zeta_{06} = \{0.6, 0.6, 0.43\}$$

The correlation degree is given as follows:

$$U_{01} = \frac{1}{3}\{0.33 + 0.43 + 0.6\} = 0.4533$$

$$U_{02} = \frac{1}{3}\{0.6 + 1 + 0.6\} = 0.7333$$

$$U_{03} = \frac{1}{3}\{0.43 + 0.43 + 0.6\} = 0.4867$$

$$U_{04} = \frac{1}{3}\{0.6 + 0.6 + 1\} = 0.7333$$

$$U_{05} = \frac{1}{3}\{0.6 + 0.43 + 1\} = 0.6767$$

$$U_{06} = \frac{1}{3}\{0.6 + 0.6 + 0.43\} = 0.5433$$

4.3 The analysis of the evaluation result

The result of the system of maintenance support efficiency sorting is as follows: F2 = F4 > F5 > F6 > F3 > F1. It shows that System 2's efficiency and System 4's efficiency are better than the efficiency of other systems. The comparison of their efficiency by adding evaluation factors is necessary.

5 CONCLUSIONS

According to the analysis and application, there are both advantages and shortcomings in the evaluation of equipment maintenance support efficiency by the gray correlation analysis method. When we obtained the evaluation data, the evaluation result can be computed by the formula manual calculation, which is simple and convenient. Through the efficient value sorting and the mission requisition, the equipment support personnel can make a more reliable choice. In order to increase the accuracy of the evaluation, we need to decrease the influence of the subjective factor and ensure the objectivity of the original data and the quantitative evaluation.

REFERENCES

Deng julong. 2002. Foundation of Grey Theory, press of Huazhong University of Science and Technology.
Gan maozhi etc. 2005. Military equipment repair engineering, National Defense Industrial Press.
IP W.C. 2009. Applications of grey correlation analysis method to river environment quality evaluation in China, correlation analysis method. J Hydrology. 33:284–290.
Liu sifeng. 2008. The theory and application of Grey system theory, Beijing: Science Press.
Lv feng. The comparisons study of seven grey correlation analysis methods. 2000. Journal of University of Wuhan industry University, 22(2):41–43.

Advances in Energy, Environment and Materials Science – Wang & Zhao (Eds)
© 2016 Taylor & Francis Group, London, ISBN 978-1-138-02931-6

Effect of polyurea reinforced masonry walls for blast loads

J.G. Wang

School of Engineering Science, University of Science and Technology of China, Hefei, Anhui, China

H.Q. Ren, X.Y. Wu & C.L. Cai

Luoyang Water Engineering and Technology Research Institute, Luoyang, Henan, China

ABSTRACT: This paper presents recent efforts that used polyurea to reinforce clay brick masonry walls in dynamic event. In order to investigate the peak pressure of damage, failure modes and failure mechanisms of spray-on polyurea reinforced clay brick masonry walls subjected to blast, six tests were performed. The results of tests suggest that the primary damage of the clay brick masonry wall are the fracture of the brick and the mortar crack that extended from top to bottom at the center of the wall. The deformation of clay brick wall is minimal. The polyurea layer can significantly improve blast resistance, and turn the collapse of unreinforced wall into local mortar joint separation or the development of flexure in the walls. The presence of polyurea coat approach offers the potential advantage of more-efficiently absorbing strain energy of host structure and prevents the collapse and structural failure of the wall, and minimizes producing deadly fragments of the reinforced walls.

1 INTRODUCTION

Most existing buildings were not designed to withstand blast loading (Hamoush et al., 2001). Most casualties and injuries sustained during external explosion are not caused by the bomb detonation, but rather by the disintegration and fragmentation of wall that can be propelled at high velocities by the blast (Knox et al., 2000). Therefore, a crucial tactic to defeating this threat is to improve the resistance of wall without breaking apart and contributing to the fragment problem. The resistance of a wall to blast loads can be enhance by increasing the mass and ductility of the wall with additional reinforcement materials, including concrete, steel, carbon, and glass fiber-composites (Barbero et al., 1997, Slawson et al., 1999, Crawford et al., 1997a, Crawford et al., 1997b, Corbi, 2013, Faella et al., 2011, Bui and Limam, 2014, Mosallam, 2007, Buchan and Chen, 2007). However, the feasibility of widespread application is challenged by difficulties in developing cost- and time-efficient methods of applying the material to the structure.

The Air Force Research Laboratory (AFRL) at Tyndall Air Force Base, Fla., has conducted research to evaluate the potential of polymer in the phases of the project (Connell, 2002, Davidson et al., 2004, Davidson et al., 2005, Knox et al., 2000). The full-scale explosive tests suggested that the elastomeric polymer coating effective improved the blast resistance of unreinforced masonry unit walls (Connell, 2002, Davidson et al., 2004, Buchan

and Chen, 2007). The observations and conclusions from both the testing and finite element modeling indicate that the elongation capacity of polymer and the strong bond between the polymer and masonry are critical for the polymer-reinforced masonry unit walls (Davidson et al., 2004, Davidson et al., 2005, Knox et al., 2000). The elastomeric coating on the interior of the wall effectively minimizes the deadly secondary fragmentation and potential for collapse of unreinforced concrete masonry unit walls. But material properties of concrete masonry unit block used in the tests of AFRL evidently is different from clay brick.

Most existing building, such as office buildings, residential buildings, and restaurants, are constructed by clay brick or autoclaved aerated concrete block which are the most common construction material utilized throughout China. While the material capability is different from the one of concrete blocks which were used in the tests of AFRL. The masonry, which was constructed by clay brick or aerated block provides adequate strength for conventional design loads, but it does not meet the minimum design standards mandated for blast protection of new and renovated facilities. So the peak pressure of damage, failure modes, and failure mechanisms of spray-on polyurea reinforced masonry constructed by clay brick and aerated block were investigated in this research to determine the effectiveness of the polyurea to improve the blast resistance of unreinforced masonry unit walls. This paper summarizes the results from a series of explosive tests that involved a spray-on polyurea. Test methodology, mechanisms of effectiveness, and dynamic response of reinforced walls are discussed. This work is ongoing, and other papers and reports are being developed that will present the results of subsequent tests and provide greater technical detail.

2 TEST PROCEDURES

2.1 Wall specimens

In the six explosive tests, two wall sizes were used in the experimental series. As can be seen in Figure 1, the first four test walls, nominally 3.6 m wide by 2.8 m tall, were constructed by clay bricks with 42-courses tall and approximately 17 blocks wide. The last two test walls, two unreinforced masonry wall panels were constructed inside the reaction structure and constructed by clay brick. The masonry walls were approximately 1.20 m wide, 2.88 m high and 0.24 m thick with a 0.24 × 0.24 × 2.88 m steel reinforced concrete column on both sides of the wall panels. At the interval location around the masonry wall, grout was filled that typically used to connect the top of a wall to the roof and column on both sides. Figure 1 illustrates the wall specimens. The lower bottom edges of the masonry walls were embedded in foundation soils in all explosive tests. In general, after one month, the wall specimens were used in the explosive test.

2.2 Mortar and bricks

A standard mortar was used in the construction of the masonry panels. The mortar was mixed to the proportion specification of JGJ 137-2001 and J129-2001 standard. The design value of the strength grade about mortar used in the tests was M5. The average compressive strength and the uniaxial tensile strength obtained by testing were 8.9 MPa and 0.297 MPa, respectively (Guiqiu et al., 2008, Chunyi et al., 2011).

Common red clay bricks and autoclaved aerated concrete blocks readily available from building suppliers were used. The nominal dimensions clay brick were 24 cm × 11.5 cm × 5.3 cm about clay brick. The test results which were conducted on the clay bricks showed that the average compressive strength and the Young modulus are 19.7 MPa and 1 3000 MPa, respectively (Weizhong and Hao, 2013, Chunyi et al., 2011). The key mechanical properties of the blocks are reported in Table 1.

2.3 Composite material

The polyurea composite material used in the explosive test was modified. Through to adjusting proportion of isocyanate-group and amino group, it could achieve the performance requirements of explosive test. The results showed that NCO content in prepolymer was increased to

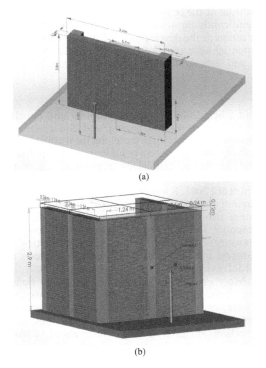

(a)

(b)

Figure 1. Specimens textures, dimensions in m: (a) clay masonry unit wall; (b) reaction structure.

Table 1. Material properties of blocks and mortar (Guiqiu et al., 2008, Weizhong and Hao, 2013, Jiang et al., 2006, Chunyi et al., 2011).

Material	Density/ g·cm⁻³	Young modulus/MPa	Poisson ratio	Compressive strength/MPa	Tension strength/MPa
Clay brick	1.8	15400	0.16	19.4	0.513
Mortar	2.1	4450	0.21	8.9	0.297

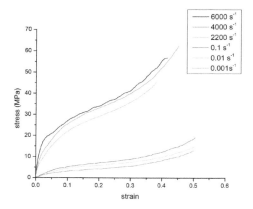

Figure 2. Engineering stress vs. strain curves for various strain rates.

Table 2. Material properties of polyurea.

Property	Value
Modulus of elasticity/MPa	23
Elongation at Rupture	510%
Breaking strength/N·m	77
Adhesion (concrete)/MPa	4.2
Density/g·cm⁻³	1.02

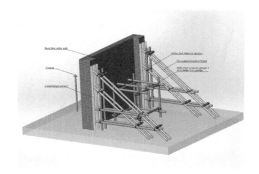

Figure 3. Explosive test setup and support system illustrating of masonry unit wall.

specific proportion, so the composite material that satisfy the request of performance could be prepared, it's tensile break strength is 23 MPa and breaking elongation ratio is 510%. Experimental studies of polyurea behavior under varying strain rates observe that the material's response is highly nonlinear and strongly strain rate dependent. Figure 2 presents the high strain rate compressible test data of polyurea in this work. The mechanical properties of the composite material mentioned in Table 2.

2.4 Test setup

Six explosive tests were conducted to evaluate the effectiveness of the polyurea. Each wall specimens was vertically constructed two brick columns with 0.36 m wide by 0.36 thick on the back surface as edge support. As can be seen in Figure 3 on the back faces of two columns, steel pipes were mounted onto a frame constructed to support the wall from the back side. The steel pipes and U-steel were Q235 normal steel, which followed the current national standard of China of GB/T3091. The Q235 normal steel is carbon structural steel which compresses strength is 235 MPa. Partially reinforced meant that a layer of polyurea was applied only to the front of masonry unit wall and fully reinforced meant that it was applied to the front and back of some masonry unit walls.

2.5 Explosive charge and instrumentation

The walls were subjected to a certain amount of TNT. Each of the tests involved an explosive charge positioned away from unreinforced or reinforced masonry unit walls. Explosive charge sizes and distances from the front face of test wall were provided here and shown in Table 3. Pressure and deflection experienced by the walls were measured using pressure gauges and deflection gauges as illustrated in Figure 1. Two reflected pressure gauges were mounted in pipe in the test walls, and located at the center and one fifth point along the mid-height of the walls, respectively. The deflection gauge was mounted at the center of test walls.

Table 3. Experimental condition of six tests.

Test	Charge weight (kg)	Standoff (m)	Burst height (m)	Thickness (mm)	Support system
Test 1 (Y2)	2.0	1.0	1.4	0	No
Test 2 (TQ-Z-J-D-5)	5.0	1.0	1.4	3 (Partially)	Yes
Test 3 (TQ-Z-J-2-4)	8.0	1.0	1.4	3 (Fully)	Yes
Test 4 (TQ-Z-J-1)	15.0	1.3	1.4	3 (Both faces)	–
Test 5 (TQ-Q-W-2)	20.0	1.3	1.4	3 (Both faces)	–

The pressure gauges and deflection gauges used in the explosive test were PVDF pressure sensor and LVDT displacement sensor, respectively. The former has large measuring rang (10 MPa–300 MPa) and broaden adaption environment; the latter was SMW-WYDC-100 L LVDT. But the deflection data of test wall were not obtained due to the severe explosive test except that in Test 4.

3 TEST RESULTS

3.1 Clay brick unit wall

3.1.1 Unreinforced and no support
In order to evaluate the blast resistance of unreinforced clay brick wall, the control wall was tested using 2 kg of TNT, detonated 1 m away in Test 1. This charge produced the measured pressure of 10.5 MPa at the center and 5.51 MPa at the one fifth point of wall, respectively. The back face of the control wall did not support system. As can be seen in Figure 4, the unreinforced wall underwent large deflections and significant damage. The control wall collapse and therefore was completely destroyed due to the block separated at mortar layers above the midheight. The wall failed at midheight and the top half folded over the bottom half, rotating into the back ground. The fold-over section could be seen near the center of the length of the wall. All of the spalled portions stayed near the back of the wall and impacted the back ground. The bottom half of test wall stayed without separating at the mortar joints and fracture on the front face. The results of the control wall suggested that some support system should be placed on the back of brick column to enforce one-way flexure and to reduce the collapse potential of test wall.

3.1.2 Partially reinforced and support
The Test 2 was performed to investigate the effectiveness of polyurea layer and the dynamic response of partially reinforced wall (polyurea on the front face) subjected to blast. The thickness of polyurea layer on the front face (impact face) was 3 mm. Based on the result of Test 2, the support system was used in this explosive test. The explosive charge was increased and the standoff distance was the same as Test 1. The values of the peak pressure for the two gauges were 36.37 MPa and 28.18 MPa, respectively. Figure 5 shows that the partially reinforced wall underwent large deflections and remained the integrity of wall without collapse occurring. The primary damage was a vertical crack about 0.15 m wide that formed over the half-height of the wall above midheight. At failure, the initial vertical crack propagated through the thickness of the wall from the center of the top to the center of the wall, and came into being three bifurcated cracks which were diagonal cracks from the center to the edge of the wall. But the polyurea layer remained intact with some tensile strain marks along the cracks. No sign of damage and fracture was observed in the reinforced layer. The other vertical crack propagated approximately the height of the wall in the left joint between the wall and the column. The mortar joints of the test wall at the bottom became separated as a result

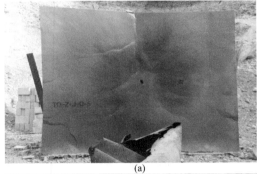

(a)

(b)

Figure 5. Partially reinforced wall in Test 2: (a) impact face; (b) back face.

Figure 4. Response of Test 1: impact face of control wall.

of the deflection of the wall. As expected, the results from the test suggested that the blast resistance of the partially reinforced wall exceed that of unreinforced.

3.1.3 *Fully reinforced and support*

In the Test 3, the test wall was spray-on polyurea layer with 3 mm on the front face (impact face) and the back face (opposite to the impact face) to investigate the failure modes and the pressure peak of the blast resistance limit. There was the support system in this test. The explosive charge was increased and the standoff distance was not changed compared to Test 2. The peak pressures obtained from the gauges on the front face of test wall were 54.73 MPa and 46.66 MPa, respectively. The Figure 6 shows that a major vertical crack with 0.7 m height occurred in the back side at the top center of the wall. The crack, however, did not propagate completely across the face or through the thickness to the front side of the wall. No other sign of crack and fracture was observed on the

surfaces of the reinforced wall. Two cracks were occurred at the left column about height 1.2 m and 1.9 m, respectively. Only one crack was observed at the right column near the bottom. The deflections were small at the crack of the column and hence the effect of the bending was ignored. As expected, the polyurea materials significantly increased the stiffness of the fully reinforced wall compared to the partially reinforced one. This increase in stiffness translated into a significant increase in pressure and a decrease in displacement obtained at the ultimate flexural resistance.

Since the retrofitted wall of Test 3 did not reach a capacity limit, the Test 4 explosive charge was increased (2.5 times) while the standoff distance was same, which resulted in much greater energy imparted onto the wall. The support system, instrumentation, and test wall were essentially the same as Test 3. The pressure measurements at the wall surface were 155.21 MPa and 123.55 MPa, respectively, and approximately triple that of Test 3. The data collected from the deflection gauge was 2.99 cm.

As seen in the Figure 7, the wall rotated about the bottom of the wall, falling onto the back ground with the front of the wall facing up. The primary damage was the separating of most mortar joints and the vertical crack propagated completely over the wall. The test wall underwent severe deformation and collapsed into many spalled portions, which remained the frame of wall. The polyurea layer separated completely from the whole front face of the wall and tore into some pieces due to the tensile strain along the cracks. No sign of some debris which reached the ground far away the back of wall was observed. The results showed that although the wall completely disintegrated because of the extreme energy imparted by the blast, the polymer held much of the spalled portions together. The polyurea material contained the debris and offered protection for occupants against an explosive charge at a relatively close distance.

(a)

(b)

Figure 6.   Fully reinforced wall in Test 3: impact face (left); back face (right).

Figure 7.   Collapse result of fully reinforced wall in Test 4.

3.2 Clay brick wall

In order to investigate the effect of polyurea reinforced framed structure constructed by clay brick, the reduced-scale dynamic experiments were conducted which evaluate the effectiveness of the polyurea to improve the blast resistance.

3.2.1 Fully reinforced

In order to investigate the peak pressure of damage and failure mode of test wall panels, Test 5 employed 15.00 kg of TNT to produce the structure loading. This charge produced a measured pressure of 41.13 MPa at the column and 26.80 MPa at the center of wall panel. The thickness of polyurea layer on the interior and exterior of right panel was 3 mm, respectively. Figure 8 shows that the exterior cement plaster of control wall fracture entire wall width along the central eight courses due to the direct shock load pressure. Accordingly, approximately twenty courses blocks incurred interior cement plaster fracture along the width and mid-height of control wall. The vertical crack propagated completely from the top to the bottom of the wall along the junction. The horizontal crack propagated across the face while not through the thickness to the back side of the wall. But on the side wall of control wall, two cracks were observed and propagated to the back side of the wall. No deflection was observed at the top and side edges of the control wall. The posttest inspection showed that no sign of fracture and tensile failure of polyurea layer were observed in this wall. The wall panels in the reaction structure sustained significant damage. It suggested that polyurea retrofit helped the wall to remain intact and prevented debris from entering the test structures.

Since the retrofitted wall of Test 1 did not reach a capacity limit, the Test 6 explosive charge was increased to 20.00 kg and the standoff distance was not changed, which resulted in much greater

Figure 9. Posttest configuration of Test 2: impact face (left); back face (right).

energy imparted to the structure. The reaction structure, instrumentation, and test panels were essentially the same as Test 5.

The peak pressure measurements at the wall were approximately 56.85 MPa at the column and 37.05 MPa at the center of wall panel. Figure 9 showed that control wall was destroyed. A circular gaping hole with size of 1.1×1.25 m occurred at the control wall due to the larger energy imparted by the blast. Most of the debris which fell apart from brick rupture stayed near the interior of the structure. The wall failed at the connection at the top and both sides. Two cracks propagated completely from the top to the bottom of the wall and to the back side of the wall. Approximately, two-thirds of the exterior cement plaster incurred fracture subjected to blast. Two major cracks occurred at the side wall of control wall, did not propagate completely across the face but through the thickness to the back side of the wall. No sign of flexure or face fracture was observed from the reinforced wall, but the polyurea layer presented some tensile strain marks due to exterior cement plaster compressive strain under blast loading. No fracture or strain marks occurred at the interior face of reinforced wall. It suggested that the blast resistance of test wall far exceeded the impact pressure in Test 6.

4 ANALYSIS AND DISCUSSION

Failure description of the system under blast loading is highly sensitive to the peak pressure, impulse, and support conditions (Davidson et al., 2005, Drysdale et al., 1994). When the standoff distance is relatively small, the peak pressure or impulse of the wall would be the failing criterion. In the explosive tests, the peak pressure varied due to a series of interference factors, however, the shape and duration of the load curve did not vary substantially. So the impulse was

Figure 8. Response of Test 1: side wall (top-left); back face (bottom-left); impact face of test wall (right).

the judgment standard of wall failing in this paper. Methods of computing the impulse from explosive detonations is available in J. Henrych. Table 4 represents the measurements captured by the pressure gauges and the impulses computed by the positive pressure impulse formula of J. Henrych.

4.1 Effect of polurea retrofit

It was apparent that both the partially and fully reinforced walls performed much better than the unreinforced wall subjected to blast because of the much delaying of cracking and final failure and the significantly improving of the blast resistance. Compared to the unreinforced, the reinforced wall appeared to be much stronger and more rigid than the unreinforced wall because (1) the initial sign of cracking for the reinforced wall occurred without collapsing; (2) the impact force in the reinforced wall was higher with increased more 4 fold at clay brick wall; (3) no sign of fracture was observed on the face. This can be attributed to the existence of the additional polyurea layer in the reinforced wall, which can effectively increase the blast resistance of the wall compared with the control wall.

4.2 Effect of reinforced patterns of polyurea

As respected, the test walls with different reinforced patterns could resist different impact force. The partially reinforced wall was sprayed polyurea on the front directly behind the blast load, and as a result, it experienced more deflections and had lower impact forces than the fully reinforced wall behind the blast load. It should be noted that fully

reinforced clay brick wall could resist the impulse of 2.85 MPa·ms, but completely collapsed at the impulse of 6.99 MPa·ms. So the blast resistance of clay brick wall by fully reinforced with 3 mm thickness polyurea was between the two impulse values, which was more than 2 fold of the blast resistance limit of partially reinforced wall.

4.3 Failure modes of test walls

The primary failure mode of unreinforced wall was the separating between block and mortar joint which cause the loss of structural integrity under blast loads. But in the reinforced test wall, the primary failure mode was the vertical and diagonal cracks which occurred at the mortar joints and completely traverse the thickness of the blocks. The primary crack propagated completely from the top to the bottom of the wall, and developed furcation at the center of the wall due to the one-way flexure mode of text wall and the shear strain on the bottom half. In the explosive test of clay brick wall, the presence of polyurea layer improved the flexure resistance of test wall due to the increase of the effective tangent modulus of reinforced wall (Urgessa and Maji, 2009, Blazynski, 1987). Furthermore, the polyurea increased the blast resistance of test wall due to the highly pressure sensitive and the stiffness increasing remarkably with increasing pressure (Amini et al., 2010a, Amini et al., 2010b). Consequently, a part of impact energy was transferred to the polyurea and dissipated in the deformation of the polyurea layer. The fully reinforced wall appeared to be much stronger and more rigid than partially reinforced.

Table 4. Summary of six explosive tests.

Test number	Gauge 1 (MPa)	Impulse (MPa·ms)	Gauge 2 (MPa)	Impulse (MPa·ms)	Failure characterization
Test 1 (Y2)	10.51	0.63	5.51	0.54	Control wall severely collapsed above burst height without front face fracture.
Test 1 (TQ-Z-J-D-5)	36.37	1.57	28.18	1.02	Wide crack propagated completely the thickness of the wall with large deformation. The polyurea layer intact with some tensile strain marks.
Test 2 (TQ-Z-J-2-4)	54.73	2.85	46.66	1.63	The initial crack occurred at the center of the top without deformation.
Test 3 (TQ-Z-J-1)	155.21	6.99	123.55	5.05	The wall rotated about bottom and severely collapsed due to overload. Polyurea torn and separated completely from front face.
Test 4 (TQ-Z-J-2)	41.13	3.05	26.8	2.12	Crack propagated across the face of control wall with fracture of cement plaster. Reinforced wall remain intactness.
Test 5 (TQ-Z-J-1)	56.85	4.34	37.05	3.06	A circular gaping hole occurred at the control wall due to severely overload. the polyurea layer presented some tensile strain marks.

5 CONCLUSIONS

A series of impact tests have been conducted to investigate the dynamic impact response of walls strengthened with polyurea layer. This paper summarizes the major findings from this experimental program. Based on the data, the polyurea reinforced is found to improve significant blast resistance and flexure resistance of the wall under impact loading. The polyurea layer could effectively contain the splintered wall components, remained intact of the wall, and could prevent serious injury to person inside a building subjected to blast. For reinforced clay brick walls, typical failure mode is vertical cracks which propagated completely through the thickness of the wall with the rupture of block. The difference of both failure modes is the result of difference between bond strength of mortar joint and material properties of brick.

Several other parameters can be included in the future tests, such as thickness of polyurea layer, the relative position of polyurea layer, and different types of test wall. The materials and application procedures would be optimized to better protect building and facility occupants against the effects of blast event.

ACKNOWLEDGMENTS

The tests described herein were conducted by the Luoyang Water Engineering and Technology Research Institute at Luoyang, China. The writers would like to thank Ren Huiqi and Wu Xiangyun for their support on this project at the Luoyang Water Engineering and Technology Research Institute.

REFERENCES

2001. JGJ 137-2001 J129-2001, Technical code for perforated brick masonry structures.

Amini, M. & Isaacs, J. (eds) 2010a. Experimental investigation of response of monolithic and bilayer plates to impulsive loads. *International Journal of Impact Engineering*, 37, 82–89.

Amini, M. & Isaacs, J. (eds) 2010b. Investigation of effect of polyurea on response of steel plates to impulsive loads in direct pressure-pulse experiments. *Mechanics of Materials*, 42, 628–639.

Barbero, E.J. & Davalos, J.F. (eds). Year. Reinforcement with advanced composite materials for blast loads. *In:* Building to Last, 1997. ASCE, 663–667.

Blazynski, T.Z. 1987. *Materials at high strain rates*, Springer.

Buchan, P. & Chen, J. 2007. Blast resistance of FRP composites and polymer strengthened concrete and masonry structures–A state-of-the-art review. *Composites Part B: Engineering*, 38, 509–522.

Bui, T.T. & Limam, A. 2014. Out-of-plane behaviour of hollow concrete block masonry walls unstrengthened and strengthened with CFRP composite. *Composites Part B: Engineering*, 67, 527–542.

Chunyi, X. & Ming, L. (eds) 2011. Experiment and Numerical Simulation on Axial Compressive Performance of Autoclaved Fly Ash Solid Brick Masonry Columns. *Trans. Tianjin Univ.*, 17, 454–460.

Connell, J.D. 2002. *Evaluation of elastomeric polymers for retrofit of unreinforced masonry walls subjected to blast*. University of Alabama at Birmingham.

Corbi, I. 2013. FRP reinforcement of masonry panels by means of c-fiber strips. *Composites Part B: Engineering*, 47, 348–356.

Crawford, J. & Bogosian, D. (eds). 1997a. Evaluation of the effects of explosive loads on masonry walls and an assessment of retrofit techniques for increasing their strength. *In:* 8th International Symposium on Interaction of the Effects of Munitions with Structures.

Crawford, J.E. & Malvar, L.J. (eds) 1997b. Retrofit of reinforced concrete structures to resist blast effects. *ACI Structural Journal*, 94.

Davidson, J.S. & Fisher, J.W. (eds) 2005. Failure mechanisms of polymer-reinforced concrete masonry walls subjected to blast. *Journal of Structural Engineering*, 131, 1194–1205.

Davidson, J.S. & Porter, J.R. (eds) 2004. Explosive testing of polymer retrofit masonry walls. *Journal of Performance of Constructed Facilities*, 18, 100–106.

Drysdale, R.G. & Hamid, A.A. (eds) 1994. Masonry structures, behaviour and design. Canadian Masonry Design Center, Mississauga, Ontario.

Faella, C. & Martinelli, E. (eds) 2011. Masonry columns confined by composite materials: experimental investigation. *Composites Part B: Engineering*, 42, 692–704.

Guiqiu, L. & Chuxian, S. (eds) 2008. Analyses of the Elastic Modulus Values of Masonry. *Journal of Hunan University (Natural Sciences)*, 35, 29–32.

Hamoush, S.A. & Mcginley, M.W. (eds) 2001. Out-of-plane strengthening of masonry walls with reinforced composites. *Journal of Composites for Construction*, 5, 139–145.

Jiang, X. & Xuefu, X. (eds) 2006. Experimental Study on Rock Deformation Characteristics Under Cycling Loading and Unloading Conditions. *Chinese Journal of Rock Mechanics and Engineering*, 25, 3040–3045.

Knox, K.J. & Hammons, M.I. (eds) 2000. Polymer materials for structural retrofit. *Force Protection Branch, Air Expeditionary Forces Technology Division, Air Force Research Laboratory, Tyndall AFB, Florida*.

Mosallam, A.S. 2007. Out-of-plane flexural behavior of unreinforced red brick walls strengthened with FRP composites. *Composites Part B: Engineering*, 38, 559–574.

Slawson, T. & Coltharp, D. (eds). 1999. Evaluation of anchored fabric retrofits for reducing masonry wall debris hazard. *In:* Proc., 9th Int. Symp. on Interaction of the Effects of Munitions with Structures.

Urgessa, G.S. & Maji, A.K. 2009. Dynamic response of retrofitted masonry walls for blast loading. *Journal of Engineering Mechanics*, 136, 858–864.

Weizhong, F. & Hao, Y. 2013. Research on the Compression Strength Test of Common Clay Brick By Rebound Measures. *Research & Explore*, 31, 35–37.

Advances in Energy, Environment and Materials Science – Wang & Zhao (Eds)
© 2016 Taylor & Francis Group, London, ISBN 978-1-138-02931-6

Development and optimization of an improved vacuum assisted conventional extraction process of epigoitrin from *Radix Isatdis* using orthogonal test design

Y.Q. Wang, Z.F. Wu, J.P. Lan, X. Wang & M. Yang
*Key Laboratory of Modern Preparation of Traditional Chinese Medicine, Ministry of Education,
Jiangxi University of Traditional Chinese Medicine, Nanchang, China*

ABSTRACT: An improved Vacuum Assisted Heat Reflux Extraction (VAHRE) technique was proposed and applied for the extraction of epigoitrin from *R. Isatidis*. The extraction condition was carefully optimized with the aid of single factor experiment and orthogonal array including the boiling temperature, ethanol concentration, extraction time and extraction cycles. Compared with conventional reference extraction methods, the VAHRE technique gave higher extraction yield due to the less thermal degradation of epigoitrin and more efficient release from plant matrix with the aid of vacuum degree. Epigoitrin was optimally extracted from *R. Isatidis* by using 40% ethanol, at 60°C, for 2 h, 1 cycle. Under these conditions, the yield value of epigoitrin reached 0.599 mg/g. The results indicated that VAHRE was a simple and efficient technique for extracting epigoitrin from *R. Isatidis*, which might shows great potential for becoming an alternative technique for industrial scale-up applications.

1 INTRODUCTION

Radix Isatidis, the dried rhizomes of *Isatis indigotica* Fort., is officially recorded in Chinese Pharmacopoeia in 2010 edition under the name "Ban-Lan-Gen", which is one of the most important crude herbs and has been commonly used in traditional Chinese medicine for more than 2000 years. *R. Isatidis* is responsible for many therapeutic actions, e.g., antibacterial, antiviral, antiendotoxin, antitumor, antiinflammatory and immune-regulatory activities (Zheng, 2003; Hsuan, 2009; Fang, 2004; Chung, 2011). Especially, during Severe Acute Respiratory Syndrome (SARS) outbreak in 2002 to mid-2003, "Isatis root granules" was one of the most popular medicines for prevention and treatment of the disease in China. This is an example of the long-standing popularity of Chinese medicine for trustworthy antiviral therapy.

Many reports in the literature put indigo, organic acids as quantitative markers which are not the antiviral compounds in *R. Isatidis* and very difficult to extract by the orthodox extraction method using water (Zhou, 2010). Recently, it was reported that a thione compound, epigoitrin, which possesses significant activity of virus-resistance and has been recorded as an official quantitative marker of *R. Isatidis* in Chinese Pharmacopoeia in 2010 edition (Xu, 2005).

Where as, there have been few reports on the extraction of epigoitrin from *R. Isatidis*, Heat Reflux Extraction (HRE) is the main extraction method in recent research. It usually requires long extraction time, high temperature, and extraction efficiency is low due to oxidation and hydrolysis. With the development of modern analytical techniques, many novel extraction techniques appeared, such as microwave assisted extraction and ultrasonic assisted extraction, but complex equipment construction, high equipment expenditure and low material throughput have made them difficult in industrial scale-up applications. Therefore, it is essential and desirable to improve conventional extraction technique and establish an economical and high efficient extraction method of epigoitrin from *R. Isatidis*.

Vacuum Assisted Heat Reflux Extraction (VAHRE) is an economic, simple and efficient method (ZL200420024138.7, PCT/CN2005/000023) which can be upgraded by adding a vacuum controller device on the basis of the conventional equipment. Preliminary tests showed the VAHRE could accelerate the release of solute from the plant matrix by the vacuum assisted breakdown of cell components, and facilitate the solid-liquid mass transfer between the extraction solvent and matrix. The purpose of this work was to optimize the VAHRE procedure for extracting epigoitrin from *R. Isatidis*, which has not been reported before. An orthogonal array was employed to analyze the interaction among the VAHRE operating factors.

2 MATERIALS AND METHODS

2.1 Material and reagents

The roots of *Isatis Indigotica* was purchased and identified from Jiangxi Gexuan Medicinal Plant Co. Ltd., Zhangshu, China by Prof. Ming Yang (Jiangxi University of Traditional Chinese Medicine). Reference substance of epigoitrin was supplied by the National Institute for the Control of Pharmaceuticals and Biological products (Beijing, China). Analytical grade ethanol was purchased from Xirong Chemical Reagent Co., Ltd. (Guangdong, China). Chromatographic grade acetonitrile was purchased from Tedia Company Inc. (USA).

2.2 Apparatus

Agilent 1200 HPLC equipped with a Variable-Wavelength Ultraviolet Detector (VWD) (Agilent Technologies, USA) was used for HPLC analysis. A Phenomenex reversed-phase Gemini C_{18} column (250×4.6 mm, 5 μm) and a Phenomenex C_{18} guard column were used for all chromatographic analysis. VAHRE experiments were carried out with a V850 Vacuum Controller (BUCHI Labortechnik AG, Switzerland).

2.3 Determination of epigoitrin by HPLC

Determination of epigoitrin extracted from *R. Isatidis* by various extraction methods was analyzed by the HPLC method. Acetonitrile and water were used as the mobile phase. The elution program was as follows: 0–10 min (2% acetonitrile), 10–20 min (2–15% acetonitrile), 20–25 min (15–100% acetonitrile), and 25–35 min (100% acetonitrile). The flow rate was 1.0 ml/min and the injection volume was 5 μl. The detection wavelength was set at 245 nm. All analyses were carried out under isothermal conditions at 25°C. The calibration curve for the determination of epigoitrin was constructed under the optimal conditions. The linear regression equations and correlation coefficients for epigoitrin were $A = 3514.4 \, m - 0.2741$ ($R^2 = 0.9996$), where A was the peak area of epigoitrin (mAUS), m was the amount of injected epigoitrin (μg). Linear range of epigoitrin was 0.0274–0.2 μg. The results implied the HPLC method was reliable for quantitative analysis of epigoitrin.

2.4 Vacuum Assisted Heat Reflux Extraction (VAHRE)

The extraction process of epigoitrin from *R. Isatidis* was performed in a VAHRE system with different vacuum degree and extraction time settings. The roots of *Isatis Indigotica* were

milled into powder using a XY-200 high-speed grinder (Xiangzhu Songqing Machine Co., Ltd., Zhejiang, China). The powder was sifted through No. 4 (65-mesh) sieve for the extraction experiments. The powder (15 g) was placed in a round-bottom flack and mixed with twenty times the volume of ethanol solvent (varying ethanol levels of 40–80%, v/v) for special time at room temperature, and then extracted using VAHRE at different boiling temperature for special time according to the experimental design. After being extracted, the mixture was filtered under vacuum through Whatman No.1 paper (Whatman-Xinhua Filter Paper Co., Zhejiang, China), then the supernatants were collected and filtered through a 0.45 μm nylon filter before being analyzed by HPLC. The extraction yield of epigoitrin was calculated as follows: $Y = A_e/A_m$, where Y was the amount of epigoitrin extracted from *R. Isatidis* (mg/g), A_e was the amount of epigoitrin extracted from *R. Isatidis* (mg), and A_m was the amount of plant material used in extraction (g).

2.5 Optimization of epigoitrin extraction

Optimization of the extraction condition for epigoitrin from *R. Isatidis* has not been reported. After carefully studying the results obtained from the previous experiments, the following factors were investigated: ethanol concentration (%), extraction time (h), boiling temperature (°C) and extraction cycles. Then three influential factors were screened for further investigations by an orthogonal array (L_9 (3^4)) after the single factor test. All the experiments were performed in triplicate. The data were analyzed by SPSS statistical software.

2.6 Conventional extraction techniques

Comparison experiments were performed in order to illuminate the advantages and disadvantages of VAHRE. Heat Reflux Extraction (HRE) is the traditional extraction process, thus the reference extraction method was performed using a classical reflux apparatus with 15 g of the drug powder for 2 h. Extraction was operated using 300 mL of 40% ethanol aqueous as extraction solvent, under the optimized VAHRE conditions except vacuum assist.

3 RESULTS AND DISCUSSION

3.1 Effect of ethanol concentration on the extraction yield

In general, solvent is considered an important parameter for extraction process because it may affect the solubility of the target ingredients. Owing to the polarity of epigoitrin and the toxicity of some solvents, ethanol water solution was chosen as extraction solvent. A series of extractions were carried out with a different ethanol concentration (20, 40, 60, 80 and 100%) to evaluate the effect of ethanol concentration. The extraction conditions were as follows: Sample: 15 g, extractant volume: 300 mL, boiling temperature: 60°C, extraction time: 2 h, extraction 1 cycle. As seen in Figure 2, the yield value of epigoitrin reached the peak when the ethanol concentration was 40%, at higher or lower concentrations, the yield was suppressed. *R. Isatidis* included a large variety of chemical components such as alkaloids, glycosides and polysaccharides. There into, glycosides have relatively high affinity with water due to their polarity. Therefore, the possible reasons is that 40% ethanol as extraction solvent could enhance the solubility of glycosides, and then improve the mass transfer of target alkaloids (epigoitrin) due to the solubilization of glycoside. A similar result has been reported by Guo (Guo, 2013).

3.2 Effect of boiling temperature on the extraction yield

In order to evaluate the performance of boiling temperature in VAHRE process, several boiling temperature (from 40 to 80°C) of the solvent were tested by adjusting the system vacuum degree using vacuum controller. Figure 2 shows that the extraction yield increased obviously before boiling temperature reached 60°C, and then decreased with the further increase of boiling temperature.

One possible reason was that VAHRE could accelerate the release of solute from the plant matrix by the vacuum assisted breakdown of cell components. Besides, an increase in the working

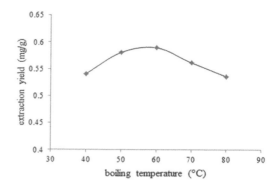

Figure 2. Effect of boiling temperature on the extraction yield of epigoitrin from *R. Isatidis*.

temperature favored extraction, enhancing both the solubility of the solute and the diffusion coefficient (Zhao, 2007). However, as a result of the interference caused by chemical and thermal degradation, or reaction with other components, solute can be degraded at higher temperature (Durling, 2007). In this case, the working temperature at 60°C got the highest yield of epigoitrin.

3.3 Effect of extraction time on the extraction yield

To some extent, the extraction time plays another predominant role in the VAHRE. Figure 3 revealed the extraction yield of epigoitrin increased with improving extraction time from 1 to 2 h. However, when the extraction time continue to extend from 2 to 4 h, the yield of epigoitrin decreased slowly. Therefore, 2 h was chosen as the maximum range of extraction time in subsequent experiments.

3.4 Effect of extraction cycles on the extraction yield

As shown in Figure 4, the extraction yield of epigoitrin levitated with extraction cycles extended from 1 to 4 cycles. There was only a 0.018 mg/g increase in extraction yield from 1 to 4 cycles. Hence, taking into account the saving of extraction time and energy, extraction 1 cycle was considered to be enough for the extraction procedure.

3.5 Orthogonal test design

On the basis of the above single factor experiments, an orthogonal test was designed to further optimization extraction parameters. The following factors were selected: ethanol level (*A*): 40, 60 and 80%, boiling temperature (*B*): 40, 60 and 80°C and extraction time (*C*): 1, 2 and 3 h. The extraction

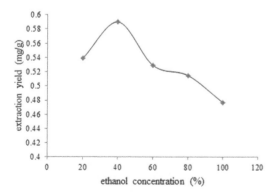

Figure 1. Effect of ethanol concentration on the extraction yield of epigoitrin from *R. Isatidis*.

341

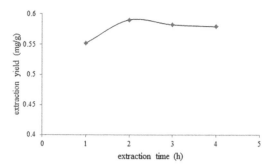

Figure 3. Effect of extraction time on the extraction yield of epigoitrin from *R. Isatidis*.

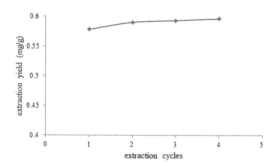

Figure 4. Effect of extraction cycles on the extraction yield of epigoitrin from *R. Isatidis*.

Table 1. Orthogonal test design and results.

No.	A	B	C	Yield (mg/g)
1	40	40	1	0.541
2	60	60	1	0.509
3	80	80	1	0.456
4	40	60	2	0.599
5	60	80	2	0.491
6	80	40	2	0.438
7	40	80	3	0.546
8	60	40	3	0.487
9	80	60	3	0.521
K_1	0.502	0.562	0.489	
K_2	0.509	0.496	0.543	
K_3	0.518	0.472	0.498	
R	0.090	0.054	0.016	
F	19.312	7.480	0.566	
Significant[a]	*			

a*, Significant ($p < 0.05$).

results performed under orthogonal design conditions were shown in Table 1. All experiments aimed at optimizing the conditions of extraction for greatest efficiency. Each experiment was conducted in triplicate and the average yield value of epigoitrin was used for statistical analysis. All the experiments were randomly carried out to avoid any kind of bias.

K_1–K_3 was the average yield value of epigoitrin under various investigated conditions, and the maximum value was the optimum value. In addition, according to the largest donating rule, the factor with the largest range value has the greatest effect on the extraction efficiency of epigoitrin. Extreme difference analysis of orthogonal test revealed that the influential order of three factors on the extraction yield of epigoitrin from *R. Isatidis* was $A > B > C$, which was in compliance with the order based on *F* values in variance analysis (Table 1).

To evaluate which factors had the greatest influence on the extraction efficiency of epigoitrin, variance analysis (ANOVA) was applied to assess the results. The results indicated that ethanol concentration (*A*) had significant effect on the yield value of epigoitrin ($p < 0.05$). Taking into

account the saving of extraction time and energy, the result of the single factor test and extreme difference analysis, the optimum condition for VAHRE was $A_1B_2C_2$ namely ethanol concentration 40% (v/v), boiling temperature 60°C and extraction time 2 h. Under the optimized conditions, the optimal yield value of epigoitrin reached 0.599 mg/g, which was higher than the yield value obtained from conventional HRE (0.516 mg/g). That is to say, there are 16.1% increases in extraction yield as a result of using VAHRE procedure. The findings are consistent with those reported in the literature by our groups (Wu, 2014; Han, 2013; Huang, 2013), which have shown that applying vacuum technique to the extraction of bioactive components from plants can significantly improve extraction yield compared to conventional extraction methods.

4 CONCLUSIONS

Through this optimization study, an improved VAHRE method was developed to extract epigoitrin from *R. Isatidis*. The optimum extraction conditions were set up using single factor test and L_9 (3^4) orthogonal experimental design. Compared with conventional reference extraction methods, the present technique gave a significant increase in extraction yield, which seemingly ascribed to enhanced mass transfer while reducing the working temperature and preventing the interference caused by chemical and thermal degradation due to the higher temperature. These results indicated that VAHRE was a simple and efficient technique for extracting epigoitrin from *R. Isatidis*, which might shows great potential for becoming an alternative technique for industrial scale-up applications.

ACKNOWLEDGMENTS

This work was supported by National Natural Science Foundation of China (No. 81173565), Key discipline of Jiangxi University of Traditional Medicine Young Teachers Cultivation Plan (No. 2013jzzdxk048), and Training Programs of Innovation and Entrepreneurship for Undergraduates (No. 201310412045).

REFERENCES

Chung, Y.C. Tang, F.Y. Liao, J.W. Chung, C.H. Jong, T.T. Chen, S.S. Tsai, C.H. & Chiang, E.P. 2011, *Isatis indigotica* induces hepatocellular cancer cell death via caspase-independent apoptosis-inducing factor translocation apoptotic pathway in vitro and in vivo. *Integrative Cancer Therapies* 10(2): 201–214.

Durling, N.E. Catchpole, O.J. Grey, J.B. Webby, R.F. Mitchell, K.A. Foo, L.Y. & Perry, N.B. 2007, Extraction of phenolics and essential oil from dried sage (*Salvia officinalis*) using ethanol-water mixtures. *Food Chemistry* 101(4): 1417–1124.

Fang, J.G. Shi, C.Y. Tang, J. Wang, W.Q. & Liu, Y.H. 2004, Screening of active fraction of antiendotoxin from *Folium Isatidis*. *Chinese Traditional and Herbal Drugs* 35(1): 60–62.

Guo, C.Y. Wang, J. Hou, Y. Zhao, Y.M. Shen, L.X. & Zhang, D.S. 2013, Orthogonal test design for optimizing the extraction of total flavonoids from *Inula helenium*. *Pharmacognosy Magazine* 9(35): 192–195.

Han, L. Huang, J. Yang, X.M. Wu, Z.F. & Yang, M. 2013, Study on process of vacuum extraction for *Salvia miltiorrhiza* water-soluble components. *China Journal of Traditional Chinese Medicine and Pharmacy* 28(11): 3201–3203.

Hsuan, S.L. Chang, S.C. Wang, S.Y. Liao, T.L. Jong, T.T. & Chien, M.S. 2009, The cytotoxicity to leukemia cells and antiviral effects of *Isatis indigotica* extracts on pseudorabies virus. *Journal of Ethnopharmacology* 123(1): 61–67.

Huang, J. Yang, X.M. Zhang, D.K. Han, L. Wu, Z.F. & Yang M. 2013, Investigation of influencing factors for vacuum in water vacuum extraction process. *Chinese Journal of Experimental Traditional Medical Formulae* 19(9): 9–12.

The State Pharmacopoeia Commission of the People's Republic of China, Pharmacopoeia of the People's Republic of China, 2010, I. 191–191, China Medical Science Press, Beijing.

Wu, Z.F. Wang, Y.Q. Yang, M. Wang, F. Huang, J. & Han, L. 2014, Investigation of stability and optimization of vacuum extraction technology for active constituents in *Gardeniae Fructus*. *Chinese Journal of Experimental Traditional Medical Formulae* 20(6): 10–14.

Xu, L.H. Huang, F. Chen, T. & Wu, J. 2005, Antivirus constituents of Radix of *Isatis indigotica*. *Chinese Journal of Natural Medicines* 3(6): 359–361.

Zhao, S. Kwok, K.C. & Liang, H. 2007, Investigation on ultrasound assisted extraction of saikosaponins from *Radix bupleuri*. *Separation and Purification Technology* 55(3): 307–312.

Zheng, J.L. Wang, M.H. Yang, X.Z. & Wu, L.J. 2003, Study on bacteriostasis of *Isatis indigotic* Fort. *Chinese Journal of Microecology* 15(1): 18–19.

Zhou, W. Zhang, X.Y. Xie, M.F. Chen, Y.L. Li, Y. & Duan, G.L. 2010, Infrared-assisted extraction of adenosine from *Radix Isatidis* using orthogonal experimental design and LC. *Chromatographia* 72(7): 719–724.

Advances in Energy, Environment and Materials Science – Wang & Zhao (Eds)
© 2016 Taylor & Francis Group, London, ISBN 978-1-138-02931-6

Calculation of key parameters on laser penetrating projectile steel

Guifei Song, Liangchun Li, Shaoguang Wang & Fujun Xia
Storage and Supply Room Ordnance Engineering Institute, Shijiazhuang, China

ABSTRACT: With the theory of laser radiating metal materials, based on the background of small-bore ammunition disposal, correlative parameters on laser penetrating projectile steel far away 30 m target ammunition. This provides a simple and quick calculation method for laser penetrating ammunition. Calculation results are helpful to the laser choice and development.

1 INTRODUCTION

Using a laser to destroy unexploded ordnance, mines, and other dangerous explosives is a new model, which not only has a high destruction efficiency, easy operation, safety and low risk, and easy to implement rapid destruction of long-range mobility, to meet the complex geographical conditions dangerous explosives destroyed on the spot needs. United States from the mid-1980s began a study of the laser to destroy mines. After nearly 20 years of efforts, developed a laser ammunition disposal system—Zeus, successfully applied to Afghanistan, counter-terrorism under battlefield conditions by a roadside bomb in Iraq, superficial mines, improvised explosive devices, and other terrorist explosives exclusion and destruction (Ren, 2009). China Ordnance Equipment Research Institute and the Institute of Engineering Corps, the first in July 2007 were applied to the destruction of landmine vehicle systems engineering research, carried out a series of laser destruction of landmines test (Li, 2008). Compared with the mines, ammunitions have harder projectile material, more superior performance, greater thickness, and they are more difficult to be penetrated by current laser drilling. Therefore, to carry out the relevant parameters of the laser penetrating projectile body is one of the key technologies of laser destruction of ammunition for the laser to destroy munitions systems; engineering has great practical significance. In this paper, the interaction between the laser and the theory of metallic materials, in order to destroy a small-caliber ammunition for the background, calculated at 30 m distance to the target ammunition laser penetration of a projectile penetrating steel-related parameters for the analysis of laser penetration destruction of ammunition provided a simple and efficient method of calculation, but also for the selection and development of the laser provides a theoretical basis for reference.

2 LASER PROJECTILE PENETRATION PROCESS

Laser penetration projectile basic process is (Dong, 2007): laser beam generated from the laser into the atmosphere through the first beam control systems, atmospheric transmission after a certain distance, and ultimately transferred to the target at a distance of ammunition, and bombs by laser material interaction, laser penetrating projectile body to complete the process. Capacity and effectiveness of laser damage depends on the laser target ammunition and ammunition deposited on the target power density, and low-level laser distance transmission attenuation is an important factor affecting the laser power density. Laser projectile penetration process, shown in Figure 1.

Figure 1. Course of laser penetrating projectile.

3 CALCULATION

3.1 Description of the problem

A small-caliber ammunition has a maximum wall thickness of 3 mm. Now a 3 mm diameter hole is wanted to be penetrated at its body of 30 meters. Seek laser power density, penetration time and the laser power at the target ammunition. The model of elastomer steel chemical composition is shown in Table 1; parameters are shown in Table 2.

3.2 Basic assumptions

① Do not consider the effects of the curvature of the cylindrical portion of the cartridge. Laser penetrating projectile body is assumed to be vertical penetration of laser flatbed models;

② Assuming the laser beam in space was fundamental Gaussian distribution;

③ Laser penetrating projectile body during melting and vaporization only consider the phenomenon;

④ Laser distance attenuation at low altitude, only consider the atmospheric absorption;

⑤ Assuming there is no attenuation of the beam transformation.

3.3 The laser parameters at the target cartridge (Li, 1998)

Elastomer penetration laser process, if appropriate adjustment of laser intensity, a melt layer is excluded outset, the penetration rate can be accelerated, the molten layer is formed equal to the axial velocity. And if the material is formed on the steam incident almost transparent, then there is the array surface evaporation power balance equation as follows:

$$I_{\alpha} = \frac{L_v dh(t)}{dt} + \frac{KT_v}{2\sqrt{\pi/\alpha t}[1 + 2h(t)/r]} \quad (1)$$

In the formula,
I_{α}-radiation intensity absorbed by material surface, W/cm^2;
L_v-evaporation energy per unit of material, J/cm^3;
T_v-evaporation temperature, K;
K-thermal conductivity, $W/cm \cdot K$;
α-temperature conductivity, cm^2/s;
r-radius of the hole, cm;
h-depth of the hole, cm;
t-time, s.

According to Equation (1) can obtain the contribution rate of the evaporation effect in penetration speed:

Table 1. Chemistry components of the projectile steel.

C	S_i	Mn	Cr (≤)	Ni (≤)	P (≤)	S (≤)	Cu (≤)
0.55~ 0.65	0.17~ 0.40	0.50~ 0.80	0.30	0.30	0.050	0.050	0.20

Table 2. Correlative parameters of the projectile steel.

Density (g/cm³)	Specific heat capacity (J/g·K)	Thermal conductivity (W/cm·K)	Evaporation energy per unit (J/cm³)
7.84	0.136	0.3475	48944

$$\left(\frac{dh}{dt}\right)_v = \frac{I_{\alpha}}{L_v} - \frac{KT_v}{2L_v\sqrt{\pi/\alpha t}(1 + 2h/r)}$$

Melting effect on the penetration of the contribution (when the melting temperature is equal to half the evaporation temperature) as follows: $0.35\sqrt{\alpha/t}$.

So the whole penetration rate is the sum of both:

$$\frac{dh}{dt} = \frac{I_{\alpha}}{L_v} - \frac{KT_v}{2L_v\sqrt{\pi/\alpha t}(1 + 2h/r)} + 0.35\sqrt{\alpha/t}$$

For most metals,

$$\frac{L_v}{T_v c\rho} \approx 5, \text{ and } \alpha = K/c\rho$$

In the formula,
c-the specific heat capacity of the material, $J/g \cdot K$;
ρ- the density of the material, g/cm^3.
It can have laser drilling speed:

$$\frac{dh}{dt} = \frac{I_{\alpha}}{L_v} - \left\{0.17\sqrt{\alpha/t}\left[1 - \frac{2h}{r}\right]\right\}$$

If the melting temperature was half of the evaporation temperature, the aperture value of about:

$$d = d_f + 0.7\sqrt{\alpha/t}$$

In the formula, d_f is the beam diameter on the surface of the shells.

According to the above relation, absorption intensity I_{α}, energy E_a and pulse duration time τ needed in forming a hole of "h" and "d" at the cartridge by laser penetration can be calculated

$$I_a = \frac{L_v h(d - 0.7\sqrt{\alpha t})}{\tau(d - 1.2\sqrt{\alpha t})}$$

$$E_a = \frac{\pi I_a \tau d_f^2}{4}$$

$$\tau = \frac{L_v h}{I_a}$$

In fact the laser intensity increases with hole depth should be increased, so as to effectively remove the material, so that condensation does not pore walls can also choose a more moderate further parameters:

$$I_a = \frac{5.5 L_v h\alpha}{d^2}$$

$$E_a = 0.5 L_v h d^2$$

$$\tau = \frac{0.33 d^2}{\alpha}$$

$$\tau = \frac{0.33 d^2}{\alpha}$$

$$d_f = 0.6d$$

According to the above empirical formula, you can calculate the parameters of the laser target ammunition, as shown in Table 3.

3.4 The Laser Long Distance Attenuation under the Condition of a Low Attitude (Yang, 2007)

Depending on the temperature, composition, ionization state of the vertical distribution characteristics, the atmosphere can be divided into three layers: the troposphere (<12 km), the stratosphere (12~60 km), ionosphere (60~2000 km). Since the troposphere above little impact on the laser transmission, the transmission of the laser affect the actual atmospheric effects of troposphere mainly refers to the laser transmission. Laser distance projectile penetration, is generally about 1.5 m transfer from the ground, which is typical of low altitude transmission. Troposphere water vapor condensation of gas molecules and with the absorption of the laser.

Table 3. Correlative parameters of laser.

Power density (W/cm²)	Energy (J)	Pulse duration (s)	Beam diameter (cm)
292442	660.7	0.09	0.18

Atmospheric attenuation is usually caused by the laser by the following formula:

$$I(R) = I_0 \exp\left[-\int_0^R \mu(r, \lambda) \mathrm{d}\lambda\right]$$

In the formula,

$I(R)$-the intensity of laser of a λ wavelength after an "R" distance propagation in the atmosphere;

I_0-beam intensity launched by laser;

$\mu(r, \lambda)$-atmospheric attenuation coefficient in an "r" distance.

Atmospheric haze is the most common natural phenomenon. For laser transmission loss due to haze and other aerosol particles can be applied Mie theory or empirical formulas to calculate and predict, often based on the concentration of atmospheric aerosols reflect estimated visibility. Formulas commonly used empirical models to predict the haze caused by the laser attenuation coefficient is:

$$\mu = \frac{3.912}{V_b}\left(\frac{0.55}{\lambda}\right)^a$$

From the above formula for sunny weather to haze, etc. are available. Where V_b Atmospheric visibility; a Wavelength correction factor, and with visibility about; under different visibility conditions, a Values are:

$$a = \begin{cases} 0.585 V_b^{1/3} & V_b \leq 6 \text{ km} \\ 1.3 & 10 \leq V_b \leq 12 \text{ km} \\ 1.6 & V_b = 23 \text{ km} \end{cases}$$

3.5 Laser power density

According to "paragraph D", it could be calculated that the laser power density should be at least 296565 in a distance of 30 meters to the target.

4 EPILOGUE

Consolidating results, the following conclusions could be achieved:

1. Engineering calculations can quickly calculate and analyze the key parameters of projectile steel using laser penetration, which could be used as the basis of theoretical analysis penetrating munitions scrap.
2. The results of the calculation are low, mainly because the laser penetrating missile and laser attenuation model has been simplified, but as a basis to carry out trials to select the laser is feasible.

ACKNOWLEDGMENT

I would like to express my gratitude to all those who helped me during the writing of this thesis.

My deepest gratitude goes first and foremost to Professor Li Liangchun, for his constant encouragement and guidance. He has walked me through all the stages of the writing of this thesis.

Second, I would like to express my heartfelt gratitude to other professors and senior engineers, without them I would have not been able to write this paper. Similarly, I thank my friends who helped me both in work and daily life.

Last my thanks would go to the authors in my reference. Based on their papers I know the development of Third party logistics. At the same time, their papers also bring me many points.

REFERENCES

Dong Hai-yan, Li Wei, Dai Ming et al. Research of high power fiber laser atmosphere propagation [J]. Optical Technique, 2007, 33(6): 830–831.

Li Wei, Zhao Yong, Chen Xi, et al. High Power Fiber Laser Used to Destroy Ammunition by Buring [J]. Laser & Optoelectronics Progress, 2008, (7): 39–42.

Li You-sheng. Study on laser deep-hole drilling [J]. Laser Technology, 1998, 22(2): 99–100.

Ren Guo-guang. Laser Weapons for Anti-unexploded Ordnance and Anti-improvised Explosive Devices [J]. Laser & Infrared, 2009, 39(3): 233–238.

Yang Rui-ke, Ma Chun-lin, Han Xiang-e, et al. Study of the attenuation characteristics of laser propagation in the atmosphere [J]. Infrared and Laser Engineering 2007, 36(9): 415–416.

Advances in Energy, Environment and Materials Science – Wang & Zhao (Eds)
© 2016 Taylor & Francis Group, London, ISBN 978-1-138-02931-6

The research on temperature control of mass concrete

Jun Deng, Yangbo Li & Shengliang Wu
College of Hydraulic and Environmental Engineering, Three Gorges University, Yichang, Hubei, China

ABSTRACT: The excessive temperature gradients are generated in the mass concrete construction of the most common and serious quality common fault. So the temperature control technology to ensure regular normal construction under the condition of using conventional materials has become an important issue. The graduation design on the mass concrete crack control by temperature control during construction period are studied, and combined with concrete dam during the construction that has been carried on the quantitative pouring piece temperature field simulation calculation and analysis, the design of hot areas is analyzed and the specific content is as follows: from the mechanism of concrete crack, temperature control measures and so on, qualitative analysis was carried out on the mass concrete temperature control; by comparing the temperature variation of different conditions it has been focused on the temperature field simulation analysis of the construction of concrete dam in the cold regions so that it has a more in-depth understanding of temperature changes of cold region on the impact of the dam.

1 INTRODUCTION

Cracks which are caused by the temperature deformation of mass concrete are the most common and serious quality defects in the process of construction, so studying measures of temperature control and crack prevention to ensure the normal construction that workers use conventional materials to construct something under normal conditions has become an important issue. The cement hydration heat is the main temperature factor in mass concrete. Due to the cement hydration and the poor thermal conductivity of concrete it will generate a lot of hydration heat in the first few days so that it can lead to the temperature rise and volume expansion of concrete. Also the changes of outside temperature of the concrete will cause the changes of internal temperature of the concrete. In continental climate or cold regions the changes of outside temperature often become the main factors affecting the concrete temperature deformation. And the main feature of temperature cracks which are different from other cracks is the concrete of expansion or closure along with the change of temperature (Yu, 1992). In order to prevent cracks in the concrete it needs to take effective temperature control measures. The initial temperature difference between inside and outside and the late temperature difference of the foundation are considered to be the main reason for the formation of cracks, and thus, it will achieve better results to take crack control measures which are holding the concrete surface and embedding internal cooling pipes (Wang, 2007).

2 MEASURES OF TEMPERATURE CONTROL AND CRACK PREVENTION AND EVALUATION OF PROGRAMS OF MASS CONCRETE

To prevent the cracks of concrete the key is to reduce the temperature difference of the structure of concrete between inside and outside and to reduce the temperature difference of the concrete foundation. The measure to reduce the temperature difference inside and outside is the surface heat preservation and internal cooling. The surface heat preservation can prevent the temperature difference of the structure of concrete between inside and outside from distributing excessive heat on the surface of the concrete, and cooling water can bring the early hydration heat through the cooling water pipe to reduce the temperature difference between inside and outside and the magnitude of the temperature drop (Wang, 2010). Technologies, such as the arrangement of the cooling water pipes inside new concrete, the use of hanging space template, setting post-poured strips, and so on can achieve reducing the temperature difference of the concrete foundation. After reaching the maximum temperature, in order to carry out joint grouting taking artificial cooling measures to shorten the cooling-off period of the reduction in temperature so that it can drop the temperature concrete as quickly as possible to the grouting temperature. Creating an evaluation system to evaluate the temperature state of mass concrete it needs in general from the aspects which are concrete placing temperature, pouring temperature, the maximum temperature, vertical temperature distribution in the axial temperature distribution and so on. So it can determine the current weaknesses of temperature control measures and then provide the basis for adjusting the temperature control measures (Zhu, 1997). According to the different characteristics of the construction of the concrete from pouring to joint grouting, it analyzes mainly from the three aspects which are the casting process, the temperature duration curve, the temperature distribution.

3 CALCULATION PRINCIPLE OF CONCRETE TEMPERATURE FIELD

Because it is usually stratified pouring and its temperature parameters also changes with time for mass concrete, so it is necessary to consider various factors in the calculation. Here are theories and methods, which are simulation of unsteady temperature field described:

Unsteady temperature field must satisfy Heat conduction equation in internal concrete:

$$\frac{\partial T}{\partial \tau} = a\left(\frac{\partial^2 T}{\partial x^2} + \frac{\partial^2 T}{\partial y^2} + \frac{\partial^2 T}{\partial z^2}\right) + \frac{\partial \theta}{\partial \tau} \quad (1)$$

In this formulation: a—temperature conductivity; θ—adiabatic temperature rise of concrete.

It can not give each temperature field in the presence of thermal temperature with the heat conduction differential equation in the last equation (Liu, 2002). Only if it has the conditions that there are thermal differential equations and boundary conditions, the temperature field can be uniquely determined. Boundary conditions of Thermal temperature typically include initial and boundary conditions.

Initial and boundary conditions:

$$T(x,y,z,t) = T_0(x,y,z) \quad (2)$$

Boundary conditions for heat conduction (Jiang, 2005):

1. The first boundary condition: the surface temperature of the concrete is a known function of time, that is:

$$T(t) = f_1(t) \quad (3)$$

2. The second boundary conditions: the heat flow of the concrete surface is a known function of time, that is:

$$\lambda \frac{\partial T}{\partial n} = f_2(t) \quad (4)$$

3. The third boundary conditions: Heat transfer condition on concrete boundary is known, that is:

$$\lambda \frac{\partial T}{\partial n} = \beta(T - T_a) \quad (5)$$

In this formulation: β—equivalent thermal exchange coefficients on concrete surface; T_a—ambient air temperature; λ—thermal conductivity of concrete. The third boundary condition can be transformed into the first boundary condition and the second boundary condition under certain conditions.

The third type of boundary condition can be transformed into the first type and the second type of boundary under certain conditions. As $\beta \to 0$, the third type of boundary condition can be converted to the second type of boundary condition; if $\beta \to \infty$, the third type of boundary condition can be converted to the first type of boundary condition.

The derived formula considers the cooling water pipe as a negative heat source, considering effects on water pipe cooling from the average sense. The equivalent heat conduction equation is as follows:

$$\frac{\partial T}{\partial t} = a\nabla^2 T + (T_0 - T_w)\frac{\partial \phi}{\partial t} + \theta_0 \frac{\partial \Psi}{\partial t} \quad (6)$$

$$\phi = e^{-pt} \quad (7)$$

In this formulation: $k = 2.09 - 1.35\xi + 0.320\xi^2$, $\xi = \lambda L/c_w \rho_w q_w$; D is the diameter and T is the time.

Combined with formula (7) it can get an average temperature of concrete under cooled water.

$$T(t) = \sum e^{-p(t-\tau)}\Delta\theta(\tau) = \theta_0\Psi(t) \quad (8)$$

Since the temperature field problems relates to many aspects of the design. And the content of heat transfer analysis is very complex. Therefore, this article will select MSC. Marc software is a finite element analysis software to handle highly nonlinear problems. It offers a wide range of heat transfer analysis capabilities to support the above types of heat transfer analysis. Simulation analysis of data obtained are more accurate and the actual match. The software simulation analysis of data obtained is accurate.

4 THE CALCULATION OF TEMPERATURE FIELD OF MASS CONCRETE DURING THE CONSTRUCTION PERIOD IN COLD REGIONS

4.1 Calculation parameters

1. Working condition 11: Concrete which is put in storage (pouring temperature is taken monthly mean temperature) is started pouring on May

Table 1. The basis of the temperature difference at the second bunker.

The basis of the temperature difference at the second bunker	Working condition 11	Working condition 12	Working condition 13
	20.72°C	19.35°C	18.66°C

Table 2. The temperature difference between inside and outside at the second bunker.

The temperature difference between inside and outside at the second bunker	Working condition 11	Working condition 12	Working condition 13
	27.92°C	26.55°C	25.86°C

1. The intermittent is 7 days without thermal insulation. Surface coefficient of heat release β is 1005 kJ/(m·d·°C), temperature conductivity of concrete is 0.08074 m²/d, Concrete thermal conductivity of concrete is 184.9 kJ/(m·d·°C), the density of concrete is 2663 kg/m³, specific heat is 860 J/((kg·°C), temperature conductivity of bedrock is 0.08074 m²/d, Concrete thermal conductivity of bedrock is 184.9 kJ/(m·d·°C), matrix density is 2663 kg/m³, specific heat is 860 J/((kg·°C), adiabatic temperature rise of bedrock is 0. In order to consider the impact of sunshine environmental temperature is on the basis of monthly average temperature plus 2°C.

2. Working condition 12:Concrete which is put in storage (pouring temperature is taken monthly mean temperature) is started pouring on May 1. The intermittent is 10 days without thermal insulation. The intermittent is 7 days without thermal insulation. Surface coefficient of heat release β is 1005 kJ/(m·d·°C), temperature conductivity of concrete is 0.08074 m²/d, Concrete thermal conductivity of concrete is 184.9 kJ/(m·d·°C), the density of concrete is 2663 kg/m³, specific heat is 860 J/((kg·°C), temperature conductivity of bedrock is 0.08074 m²/d, Concrete thermal conductivity of bedrock is 184.9 kJ/(m·d·°C), matrix density is 2663 kg/m³, specific heat is 860 J/((kg·°C), adiabatic temperature rise of bedrock is 0. In order to consider the impact of sunshine environmental temperature is on the basis of monthly average temperature plus 2°C.

3. Working condition 13: Concrete which is put in storage (pouring temperature is taken monthly mean temperature) is started pouring on May 1. The intermittent is 14 days without thermal insulation.

4.2 Calculation of MSC. Marc Software

The Research shows that the ambient temperature is 2°C, the annual average temperature is 9.2°C in cold. The formula respectively has two: The maximum temperature-the ambient temperature = the temperature difference between inside and outside; the maximum temperature-the annual average temperature = the basis of the temperature difference (Song, 2009).

The basis of the temperature difference at the second bunker (Table 1).

The temperature difference between inside and outside at the second bunker (Table 2).

By contrast *Hydraulic Structures*, *Concrete Gravity design specifications SL319-2005* and other norms it will choose the smaller hydration heat of the cement of concrete and the smaller temperature difference for the best.

The scope of the basis of the temperature difference of this article is in 19°C~ 22°C in cold. After comparing three working conditions the basis of the temperature difference of the condition 11 and the condition 12 are the best to meet the requirements and then finds that working condition 12 about the temperature difference between inside and outside at the second bunker is smaller than working condition 11 about the temperature difference between inside and outside at the second bunker. Considering all things it finds that working condition 12 is the best to meet the requirements of this article, so the optimum condition is working condition 12.

5 CONCLUSION

The sector in dam construction has been plagued by the temperature control of concrete on dam for a few decades. The main reason is that the temperature field and the thermal creep stress of computational theory is not yet perfect. There are many influential factors (Wang, 2003), so there are still a lot of works to be done. The authors suggest the following observations:

1. Solar radiation and diurnal temperature difference have a big influence on the temperature

of concrete surface of pouring- cube when it pours concretes in hot season. It is proposed to strengthen the water conservation of the warehouse and the water conservation of the continuous spray to improve the casting quality of concrete (Zhu, 2006);

2. In order to reduce the temperature rise of the hydration heat it mainly uses moderate portland cements and Low thermal cements and selects the optimum aggregate gradation, the amount of fly ash with high-quality, admixtures, and uses reasonable thickness of concrete, the interim period and initial water cooling measures;

3. In general, it often chooses the pre-cooling method of coarse aggregates to significantly reduce the temperature of concrete of pouring-cube through the design of temperature control of concrete on dam. Currently, it often takes measures about water cooling, air cooling, etc. Taking pre-cooling measures of coarse aggregates could lead the temperature dropped to below freezing.

4. Although the calculation of 3D temperature field of concrete cooling pipe is theoretically more rigorous, its computational scale for three-dimensional temperature field simulation of the entire dam and creep stress field on the high RCC dam is sizable. Now the method may not be realized by computer.

REFERENCES

Bofang Zhu, Effect of cooling by water flowing in non-metal pipes embedded in concrete dams, Journal of construction Eng. ASCE. Vol. 125, No. 1, Jan, 1997.

Haiqing Liu and Wanqing Zhou, Finite Difference Method Analysis of 3D Unsteady Temperature Field in Mass Concrete, Proceedings of the Seventh International Symposium on Structural Engineering for Young Experts, Science Press, 2002.

Jiang Z.B., F. Jin, J. Sheng, Numerical simulations of water waves due to landslides, J. Yangtze River Research Institute, (2005).

SL319-2005. Concrete Gravity Dam Design Specification.

Song X.Y. and A.G. Xing, Two-dimensional numerical simulation of landslide waves by Flunt, Hydrogeol. Eng. Geol. 3 (2009) 90–94.

Wang Y., K.L. Yin, Analysis of movement process of landslide in reservoir and calculation of its initial surge height, Earth Sci. J. China University of Geosci. 28 (5) (2003) 579–582.

Wang Zhen-hong, Zhang Guo-xin, Liu Yi, Liu You-zhi, Study on Temperature Control and Crack Prevention of High-performance Slag Concrete Structure during Construction. IEEE. 2010.

Wang Zhen-hong, Zhu Yue-ming, Temperature Control and Anti study of thin-walled concrete structures during construction [J]: Architecture & Technology of Xi'an University, 2007,39(6): 773–778.

Yu Xiao-zhong, Ma Shu-fang, Hazard analysis and repair of Concrete fissure(a), dam and safety, 1992, (3): 27–38.

Zhu Bo-fang, Situation and prospects of temperature control and prevention cracks on concrete dam [J]: Journal of Hydraulic Engineering. 2006 37(12).

Advances in Energy, Environment and Materials Science – Wang & Zhao (Eds)
© 2016 Taylor & Francis Group, London, ISBN 978-1-138-02931-6

The heat-insulating property evaluation and application of high permeability-high strength concrete materials

Yi-zhong Tan, Yuan-xue Liu & Pei-yong Wang
Chongqing Key Laboratory of Geomechanics and Geoenvironmental Protection, Department of Civil Engineering, Logistical Engineering University, Chongqing, China

ABSTRACT: The high permeability-high strength concrete belongs to the typical of porous materials. It is mainly used in underground engineering for cold area, it can act the role of heat preservation, also to be the bailing and buffer layer. When the high permeability-high strength concrete materials was used in the cavity for separate lining to a reality engineering in northern Tibet, it was shown that the method could protect the around permafrost well than unfilled engineering. It can achieve the requirements to be the heat insulation material.

1 INTRODUCTION

There are widely distributed permafrost and seasonal frozen soil at high altitude. A study has discussed that the Qinghai Tibet Plateau in China has a most widely and thickness permafrost distributing, also the temperature is the lowest all of China. The annual average temperature in these areas was −2°C to −6°C (Luo, 2012). In all kinds of geotechnical engineering, either in constructing or operating period, the temperature must be strictly controlled, especially for the residency underground engineering. Unless to do so, the internal temperature can meet to the requirements for inhabiting, also the frozen soils (rock) can be ensured in a station of stabilization when the temperature was changed. Jae-Sung Kwon (2009), Wei (2011) and Chen (2014) considered that if the high permeability-high strength concrete material was chosen to be the filling material which was played a role of bailing and absorption of vibrating in underground space, the key parameter to measure the heat-insulating property was the effective coefficient of heat conductivity. So it is important to choose a suitable predicting model which can meet its own character to calculate the thermal conductivity. So far as to get the direct relationship type, which can reflect the mixture and the effective coefficient of heat conductivity directly, it must to be a significance trying in theorizing and practicing.

At present, colleagues have developed the effective thermal conductivity of porous materials deeply. They have also got so much excellent results. Such as Bozomolov (1941) were built a theory equation to describe the thermal conductivity for porous materials. This model was built from the bulk factor for solid phase and fluid phase. Daizo

Kunni (1960) had a mathematical model to predict the thermal conductivity of porous rocks through the total parameters theory. But the model had lost the sight of the impact with porosity, then it was limited to be using in practical application.

2 THE STRUCTURE COMPOSITION OF HIGH PERMEABILITY-HIGH STRENGTH CONCRETE

The structure of high permeability-high strength concrete material is cemented the uniform size aggregates by gelatinization materials, which are composed of cement, polyacrylate, and water. The porous material is a structure of frame-ventage work, which is named as skeleton pore structure in engineering. When the material is cut by the cutting machine, the specimen slice maps have been obtained, later the real slices of high permeability-high strength concrete are shown in Figure 1. The effective thermal conductivity of porous materials are contacted with porosity closely (Pabst, 2014;

Figure 1. The real slices of high permeability-high strength concrete.

Li, 2001). According to prior experiment, a phenomenon was found that the porosity of the high permeability-high strength concrete was determined by size of aggregate and thickness of polymer slurry which were packed around the uniform size aggregates. When the material was produced by the way of non-vibration or plane—vibration, the strength was mainly supplied by the force of friction and bond strength.

3 APPLICATION OF HIGH PERMEABILITY-HIGH STRENGTH CONCRETE MATERIALS

3.1 Calculation models

From the predicting model and experiment measuring data, it was known that the cavity for separate

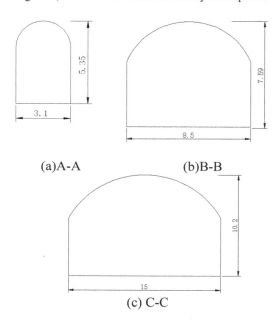

(a)A-A   (b)B-B

(c) C-C

Figure 2. The abridged general view for the three representing sections.

Table 1. The calculation parameters.

Materials	Permafrost	Reinforced concrete lining	Preliminary bracing
Thermal conductivity [J/(m·s·K)]	2.04	2.56	2.56
Density (kg/m³)	2456	2500	2200
Specific heat (kg·K)	1524	880	880

lining which was built in high-cold and high-altitude area was filled with high permeability-high strength concrete materials could reduce the heat delivery well. But the effectiveness should be proved by computing as well. So in this section, from the characteristics of underground engineering in high-cold and high-altitude area a reality engineering in northern Tibet was chosen. In the reality engineering, three representing sections were selected, there were entry section A-A, standard section B-B and synthesis section C-C. The entry section A-A had a width of 3.1 m and height of 5.35 m. The entry section B-B had a width of 8.5 m and height of 7.5 m. The entry section C-C had a width of 15 m and height of 10.2 m. Then the abridged general view for the three representing sections was listed in Figure 2.

3.2 Calculation parameters and meshing model

In the construction, it was required that the inner temperature should be kept at 20°C, while the adjacent rock was permafrost had the temperature below –4°C. Then the parameters of the used materials were listed in Table 1, where the thermal conductivity of the high permeability-high strength concrete materials was obtained by theory calculating and experiment measuring in prophase.

In this segment, the heat-transfer was calculated by the software of COMSOL Multiphysics which was used widely. First, it could be built a hemihedral symmetry calculating model of entry section A-A, standard section B-B and synthesis section C-C. In order to get an accurate result, the computing range was far away from the sections. Then the parameters of the materials was taken into these calculated models and got the meshing graphs, which were exploded in Figure 3.

3.3 The comparison and appreciation for calculating results

From the calculating of the two-dimension models, the isallothermic charts which were extracted in the computing results of filled and unfilled structure. Then the isallothermic charts were shown in Figure 4. It was found that the isallothermic charts for different sections was distributed by a character of concentric circles distally. It was seen that the heat-transfer in filled and unfilled structures had large differences in different sections. Under the same condition, the cavity for separate lining was filled by the high permeability-high strength concrete materials could stop the heat-transfer effectively. Then on the basis of the outer surface for initial lining, ten fixed points were selected which were distributed along the direction of frozen soil in 50 m range. So the temperature distributing of each fixed points were counted in Figure 5.

From upper Figure 5, the temperature on the initial lining would rise at 288 K (15°C) immediately by the unfilled structure under the temperature head. Then the temperature of the structure was gradually stabilized at 290K (17°C). But when the cavity

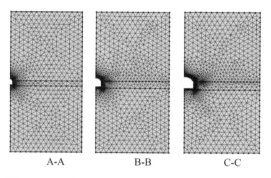

A-A   B-B   C-C

Figure 3. The meshing graphs of the calculating models.

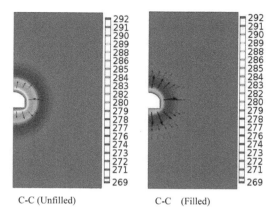

C-C (Unfilled)   C-C (Filled)

Figure 4. The isallothermic charts in different sections.

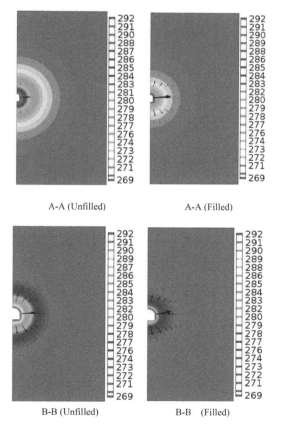

A-A (Unfilled)   A-A (Filled)

B-B (Unfilled)   B-B (Filled)

Figure 4. (Continued)

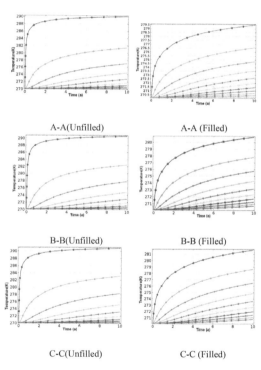

A-A(Unfilled)   A-A (Filled)

B-B(Unfilled)   B-B (Filled)

C-C(Unfilled)   C-C (Filled)

Figure 5. The temperature change curves with time in different sections of surrounding rock.

of the separate lining was filled in high permeability-high strength concrete materials, the heat-transfer rate of speed might slow down. It is also found that the changing of the temperature was related by the style of sections. When the time was gone to the tenth year, the temperature of section A-A would drive to 279K (16°C), of section B-B would drive to 282K (9°C), of section C-C would drive to 284K (11°C). From the tendency of the curves, the 0°C loop around the section A-A of unfilled structure could elongate to 15m in 10 years, the section B-B was 8 years, the section C-C was 7 years, while the filled structure of the separate lining was little. It was proved that the high permeability-high strength concrete materials could protect the around permafrost well, it also had a great ability to be the heat insulator.

4 CONCLUSION

From the structure of high permeability-high strength concrete, it belongs to the porous materials, also, according to the experimental and calculating data, this material is equivalence to the insulation material which is commonly used in underground engineering. It can achieve the basic requirements to be the heat insulation material. When the high permeability-high strength concrete materials was used in the cavity for separate lining to reality engineering in northern Tibet, it was shown that the method could protect the surrounded permafrost well than unfilled engineering. So the high permeability-high strength concrete to be the material of heat-insulating material had high evaluating of property.

ACKNOWLEDGMENTS

The financial supports provided by the National Natural Foundation of China (No. 50979112) and the National Natural Foundation of Chongqing China (No, CSTC, 2014jcyjA30011).

REFERENCES

Bozomolov V.Z.; Chudnovsky A.F.,: Agrophysical, Trans, 20(3):12–20(1941).

Chen, Y.F.; Zhou, S.; Hu, R., et al.: Estimating effective thermal conductivity of unsaturated bentonites with consideration of coupled thermo-hydro-mechanical effects, International Journal of Heat and Mass Transfer, 72 (18):656–667(2014).

Daizo Kunni.; J. M. Smith.,: Heat Transfer Characteristic of Porous Rocks, A.I.Ch.E. Journal, 40(6):71–78(1960).

Gaosheng Wei; Yusong Liu; Xinxin Zhang et al.: Thermal conductivities study on silica aerogel and its composite insulation materials, International Journal of Heat and Mass Transfer, 54 (10):2355–2366. (2011).

Jae-Sung Kwon.; Choong Hyo Jang.; Haeyong Jung., et al.,: Effective thermal conductivity of various filling materials for vacuum insulation panels, International Journal of Heat and Mass Transfer, 52 (12):5525–5532 (2009).

Li, M.W.; Zhu, J.C.; Yin, Z.D,.: Analysis of effective thermal conductivity of particle dispersive composites, Gong Neng Cai Liao, 20(4):397–398(2001).

Luo, J.; Niu, F.J.; Lin, Z.J.,: Permafrost features around a representative thermokarst lake in Beiluhe on the Tibetan Plateau", Journal of Glaciology and Geocryology, 34(5):1110–1118(2012).

Pabst, W.; Gregorová, E.,: Conductivity of porous materials with spheroidal pores, Journal of the European Ceramic Society, 34 (4): 2757–2766(2014).

Advances in Energy, Environment and Materials Science – Wang & Zhao (Eds)
© 2016 Taylor & Francis Group, London, ISBN 978-1-138-02931-6

Research on the injection characteristics of biogas engine

Z.H. Fang, K. Wang, Y. Sun & G.J. Guo
Shanghai Normal University, Shanghai, China

ABSTRACT: The biogas injection characteristics of biogas engine are the basic data to control biogas engine, accurately. In the biogas engine, the biogas injection flow is easily affected by some factors like injection pressure and injection time; the injection characteristic is also a direct impact on the overall performance of the engine. This paper chooses the C8051F340 microcontroller as the controlling chip, designs a rapid response injection driving circuit and studies the biogas injection characteristic on the different conditions of injection pressure, injection time, the sizes of injector, which can be used to determine the injector's type, size and the controlling range of the parameters in the biogas engine, which also laid the foundation of the precise control of biogas engine.

1 INTRODUCTION

With the increasing emphasis on environmental protection, and the shortage of oil, coal and other fossil energy, looking for a new, clean, and high-effective energy has become the most critical issue in the world (Feng 2011). As biogas is burning, the heat value is high, but the emission is very low, it has a great development prospect using biogas as the fuel of engine.

The controlling of biogas engine is related with a lot of subsystems like the fuel supply, mixing, ignition and others, and the engine has different running conditions in the real time, which means the injection amounts of gas change a lot. While the precise gas supply is controlled by the injection timing and the injection pulse width of the electronic-control injector of engine, the pulse width determines the amount of gasper injection, which has directly affected the overall performance of the engine (Zhang 2011, Guo & Mao 2011). Therefore, the study of characteristics of the gaseous fuel injector appears to be very important. At present, people has paid a lot of attention on the studies of injection characteristics of liquid fuel of engine, but less on injection characteristics of gaseous fuel engine, considering about the different characteristics between the mechanical response and electromagnetic response of gaseous fuel engine, the analysis and research on the injection characteristics of gaseous fuel engine is needed, which can also be used to determine the type of injector and the controlling range of the parameters of the injector. Since the flow characteristics of the injector is vulnerably affected by the injection pressure, the injection time, the flow diameter of injector and other factors; we need to do some research on

the biogas injection characteristics under different running conditions and external conditions.

2 BIOGAS INJECTION CHARACTERISTICS TEST DEVICE

Biogas injection characteristics test device is shown as Figure 1.

This paper uses a biogas pump to pressure 3 bar, then uses a pressure regulation valve (SNS) to regulate gas pressure to the required value, controls the opening and closing of the injector by the MCU and drive circuit, and measures the gas flow in the real time through the gas flow meter. The gas flow meter we use is the type of LZB-10, with the range of 0–2000 L/h, which is referenced by the 465 engine. The real displacement of 465 engine is 0.998L, with the bore/stroke are 65.5/74 mm, then the displacement of a signal cylinder is about

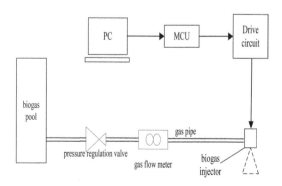

Figure 1. Biogas injection characteristics test device.

0.25L. Assumed the inflation efficiency of the engine is 80%, then the maximum intake flow of a signal cylinder ($Q_{air-max}$) is 300 L/min, as shown.

$$Q_{air-max} = 0.25 \text{ L/2r}*3000 \text{ r/min}*80\% = 300 \text{ L/min}$$

Assumed the minimum intake flow takes 5% of the maximum flow, then the minimum intake flow of a signal cylinder ($Q_{air-min}$) is as shown.

$$Q_{air-min} = 300 \text{ L/min}*5\% = 15 \text{ L/min}$$

Then we can calculate the range of intake flow of a signal cylinder (Q_{gas}) by assuming the desired air-fuel ratio is 10, which is shown in Formula 3.

$$Q_{gas} = (15\sim300) \text{ L/min}*1/10$$
$$= 1.5\sim30 \text{ L/min} = 90\sim1800 \text{ L/h}$$

The above results have indicated that the chosen gas flow meter meets the requirement. This experiment uses C8051F340 MCU, which can output multiple PWM waves with variable duty cycle and can change the injection time. The C8051F340 MCU is a fully integrated, on-chip system MCU, which uses the core of CIP-51 microcontroller to develop software with the standard 805x assembler and compiler. The kernel consists of four counters/timers with 16 bites, two full duplex UART which has extended the baud rate configuration, one enhanced SPI port, the internal RAM about 4352 bytes and up to 40 I/O pins. In addition, the MCU has several key improvements in the core and peripherals of CIP-51, which has improved the overall performance a lot and can be easier to use in the final application, one of the most important is that there is an on-chip programmable counter named PCA is integrated in the C8051F340 MCU, which could just satisfies the requirements of outputting multiple PWM waves with variable duty

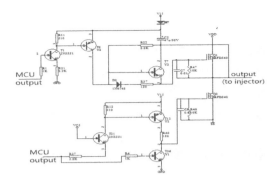

Figure 3. Drive circuit of the injector.

Figure 4. The drive circuit board.

cycle. This experiment uses a solenoid valve as biogas injector. The drive voltage of biogas injector is 12V and the rated power of biogas injectors 5W, the diameter of the injector is 1.4 or 2.2 mm. The biogas injector is shown as Figure 2.

When designing the drive circuit of the injector, it is needed to make sure that the opening and closing of the injector is faster and the entire power consumption is as small as possible. Figure 3 is a drive circuit, we can see that when the input is 0, the output of drive circuit is 1, and the biogas injector is open; while when the input is 1, the output of drive circuit is o, and the biogas injector is closed.

Figure 4 is the drive circuit board.

3 EXPERIMENT AND RESULTS ANALYSIS

3.1 Experiment conditions

In the experiment, the variable factors are injection pressure, injection time, and injector diameter, so the above variable factors can be assumed as follows:

The actual biogas pressure in the engine cylinder is generally about 1–3 bar, we tests the biogas

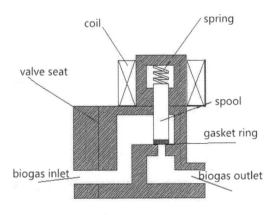

Figure 2. Biogas injector.

injection characteristics with the pressure of 1.2, 2, 2.8 bar.

We test the biogas injection characteristics with the injector diameter of 1.4, 2.2 mm.

3.2 The biogas injection characteristics of different injector diameter

Set the injection cycle is 40 ms (engine speed is 3000 r/min), observe the biogas injection characteristics with different injector diameter and injection pressure, as shown in Figures 5, 6. Compared to Figures 5 and 6, we can find:

The injection flow is proportional to the injection time, but the injection flow tends to be stable with the continuous increasing.

a. When the injection time is 0–2 ms, there is almost no flow passing through the injector.

Because of the electromagnetic and mechanical response, the solenoid doesn't open, therefore no flow, compare to the gasoline solenoid which usually has no flow in 1–1.5 ms, the gas solenoid has longer stroke and slower response speed;

b. When the injection time is 2–5 ms, the injector has just started to work, the flow is very small and the flow curve is unstable, the electromagnetic and mechanical response is still very slow among this period, so the flow is unstable;

c. When the injection time is 5–20 ms, the injector is working stable and the injection flow is almost proportional to the injection time in linear segment, so when keeping the injection pressure constant, the flow can be improved by increasing the injection time according to the linear phase of 5–20 ms;

d. When the injection time is above 20 ms, the flow is stable and the solenoid is totally open, even if increasing the injection time, the flow will not change a little.

3.3 The biogas injection characteristics of different injection pressure

As shown in Figures 5–6, the injection flow is proportional to the injection pressure, if it wants to shorten the injection time and increase the injection flow, just increase the injection pressure. In the fuel injection subsystem of actual gas engine, in order to increase the flow per unit, the gas can be pressurized according to the characteristics.

3.4 The biogas injection characteristics of different injector cycles

Set the injection cycle is 60, 40, 30 ms, the injection pressure is 2.8 bar and the injector diameter is 2.2 mm. Observe the biogas injection flow, as shown in Figures 7–9.

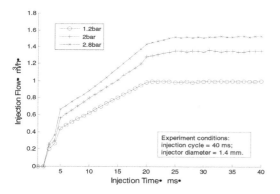

Figure 5.   Diameter = 1.4 mm gas flow characteristics.

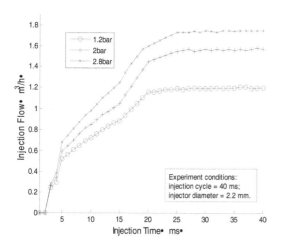

Figure 6.   Diameter = 2.2 mm gas flow characteristics.

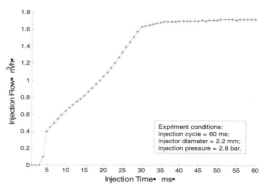

Figure 7.   Injection cycle = 60 ms, biogas injection characteristics.

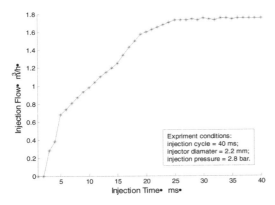

Figure 8. Injection cycle = 40 ms, biogas injection characteristics.

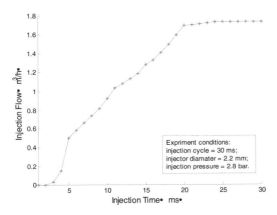

Figure 9. Injection cycle = 30 ms, biogas injection characteristics.

Compared with Figures 6–8, we can find that the maximum injection flow of the injector isn't related with injection cycle (engine speed), and when the injection pressure is 2.8 bar and the diameter of injector is 2.2 mm, the maximum injection flow is 1.73 m³/h, which matches the intake gas quantity indicator of engine cylinder, therefore, the injector used in this experiment meets the actual situation.

4 CONCLUSIONS

Based on the variable factors in the gas injection process, this paper study the gas flow characteristics in different conditions, sums up the rules of gas flow characteristics and determine the ejector's type and the controlling range of the parameters in the biogas power engine. The flow features are as follows:

1. The injection flow is proportional to the injection time and the injector diameter, its maximum flow isn't related with injection cycle, and the injection flow tends to be stable with the continuous increasing of the injection time;
2. When the injection time is 5–20 ms, the injection flow is almost proportional to the injection time in linear segment, so when keeping the injection pressure constant, the flow can be increased by longer the injection time according to the linear phase of 5–20 ms;
3. The injection flow is proportional to the injection pressure, if it wants to shorten the injection time and increase the injection flow, just larger the injection pressure. In the fuel injection subsystem of actual gas engine, in order to increase the flow, the gas can be pressurized according to the characteristics;
4. The above summary of gas flow characteristics has laid the foundation of the precise control of biogas power engine.

ACKNOWLEDGMENT

This work was financially supported by the Research Program of Science and technology Committee of Shanghai (12160503000).

REFERENCES

Fen L. et al. the Modeling of Gas Injector and the Simulation of Dynamic Characteristics [J]. Shanghai Gas, 2011:17–19.

Guo B. The Experiment Study on the Performance of Natural Gas Injection Used in Vehicle [J]. Modern Vehicle Power, 2011(3):15–18.

Tian L.Y. The Experiment Research of the Fuel Injection Characteristics of the Internal Combustion Engine [D]. Shanghai Jiaotong University, 2009.

Xing G.J. The Effective Ways to Improve the Efficiency of the Gas Injection [J]. Metal School, 2010, 36(4):445–448.

Zhang L.F. The Mathematical Modeling of the gas injector of common-rail and the Study of dynamic characteristics [D]. Tongji University, 2011.

Advances in Energy, Environment and Materials Science – Wang & Zhao (Eds)
© *2016 Taylor & Francis Group, London, ISBN 978-1-138-02931-6*

Study on relationship of land use & landscape pattern and water quality in Dahuofang reservoir watershed

Ruichao Guo & Xuhui Li
Key Laboratory of Environment Change and Water-Land Pollution Control (University of Henan Province), College of Environment and Planning, Henan University, Kaifeng, China

ABSTRACT: Inappropriate land use is considered a main factor in the deterioration of water quality. To investigate the relationship between land use and water quality (RLWs), 5 physicochemical parameters (TN, NO_3-N, NH_3-N, NO_2-N and TP) of water from 77 sampling sites of Dahuofang Reservoir watersheds were monitored during 2012-2013. Based on "source-sink" landscapes, the sampling sites were classified into six groups. RLWs in three periods were analyzed via stepwise multiple regression models. The landscapes of forest and farmland cover types played significant roles in determining the surface water quality during the low-flow, high-flow, and mean-flow periods based on the results of a stepwise linear regression. These results may provide incentive for the local government to consider sustainable land use practices for water conservation.

1 INTRODUCTION

Surface waters can be contaminated by natural processes, such as precipitation inputs and erosion, and by anthropogenic disturbances through agricultural, industrial, and urbanization activities that increase the consumption of water resources (Shrestha & Kazama, 2007; Singh et al., 2004). Many landscape factors can influence surface water quality of a river, such as channel slope, vegetation on the banks and riparian zone conditions (Townsend et al., 1997). However, one of the most important factors is land use and land cover (Griffith et al., 2002).

Effective analytical tools, such as geographical information systems (GIS) and multivariate statistics, are able to deal with spatial data and complex interactions, and are coming into common usage in the relationship between land use and the water physicochemical characteristic (Allan & Johnson, 1997; Tong & Chen, 2002; Mehaffey et al., 2005; Brion et al. 2011; Wang et al., 2013; Ye et al., 2014). Tong & Chen (2002) found a significant relationship between land use and in-stream water quality. They showed much higher concentrations of nitrogen and phosphorus in agricultural and impervious urban land than that in forest land surfaces. Mehaffey et al. (2005) suggested that total nitrogen (TN) and total phosphorous (TP) are significantly positively related to agricultural land. Although the study on linking of LULC and water quality have become more common in the past 20 years (Ahearn et al., 2005), there is still large potential to improve the recognition of the relationship between land use and water quality owing to the advances in the technology of GIS and

multivariate statistics. It is recognized that both land use and landscape pattern should be considered in analyzing stream conditions. Some researchers argue that the analyses of LULC and water quality should be conducted during the storm reason, because the landscape has a most intimate influence on the watercourse (Johnson et al., 1997; Arheimer & Liden, 2000). Others hold that to get reliable results, water quality should be analyzed during different seasons (Arheimer & Liden, 2000).

As an important source of drinking water, the Dahuofang Reservoir sustains 23 million residents in Liaoning Province, Northeast China. However, there is great concern about the surface water quality in Dahuofang Reservoir watershed because of rapidly occurring land-use changes, including deforestation, increased agriculture and mining, and urbanization. Therefore, it is an urgent task to study the effects of land use & landscape pattern on surface water quality. This study was performed to explore the effects of land use types land use & landscape pattern on the physicochemical characteristics of the water in Dahuofang Reservoir watershed. The aims of this study were as follows: (a) to analyze the spatial and temporal distribution characteristics of the surface water quality; (b) to clarify the effect of land use on water quality in Dahuofang Reservoir watershed.

2 MATERIALS AND METHODS

2.1 Study area

Dahuofang Reservoir is situated in the east of Liaoning Province in China, and provides drinking water for Shenyang and neighboring cities. Dahuofang Reservoir watershed (124°48′30″–125°21′30″E, and 41°47′50″–42°08′40″N) includes the Hun River and the Suzi River, which are located in Qingyuan County and Xinbin County, respectively (Figure 1). The main corridor of the Hun River is 110 km long, and it drains an area of 2332 km² which is equivalent to 59.47% of the total area of Qingyuan County. The main corridor of the Suzi River is 119 km long and it drains an area of 2087 km² which is equivalent to 47.0% of the total area of Xinbin County.

The study area is an extension of the southwest range of the Changbai Mountains, and is characterized by low mountains and foothills. The southeast is relatively high; the northwest is low-lying; and the central area is hilly with a maximum elevation of 1116 m. It belongs to the temperate continental monsoon climate region, with hot summer and cold winter. The average annual precipitation ranges from 700 to 900 mm, and as much as 70–80% of annual precipitation falls from June to August. The predominant land use type in this region is forest. With the development of economy, however, the land use has a growing conversion into agriculture and housing. As a consequence, deterioration of the aquatic

environment in Dahuofang Reservoir watershed is becoming serious.

2.2 Samples collecting and analysis

According to the monthly mean precipitation, three distinctive periods were determined: low-flow (April and May), high-flow (June–August), and mean-flow (September and October). In this study, Dahuofang Reservoir watershed was divided into 77 subwatersheds, as shown in Figure 1. Water samples were collected from outlets of the 77 watersheds in the May (low-flow), August (high-flow), and October (mean-flow) respectively. To quantify the chemical analyses, 1000 mL water was collected four times at each sampling site. Then the sampled water was mixed well and 500 mL water was stored in clean polythene bottles. Sulfuric acid was added into each sample to adjust pH value between 1 and 2, water samples were stored at 4°C and transported to the laboratory for advanced analysis. All the sampling sites were located by GPS and the surrounding landscapes were recorded too. Water pollution in this region is mainly caused by indiscriminate discharge of domestic sewage, application of chemical fertilizer, industrial pollution, etc. Thus, these parameters included nitrate-nitrogen (NO_3-N), ammonium nitrogen (NH_3-N), nitrite- nitrogen (NO_2-N), total nitrogen (TN), and total phosphorus (TP) were used to characterize water pollution. All the parameters were performed following standard methods (State Environment Protection Bureau of China 2002).

2.3 Land use analysis

The boundaries of the whole watershed and 77 sub-watersheds were produced from 1:250,000 Digital Elevation Model (DEM) by using ArcGIS 9.3 Desktop GIS software. Due to the rough scale of DEM,

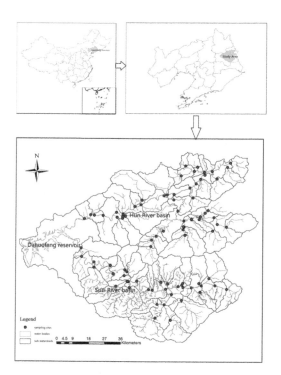

Figure 1.   Study catchments and sampling site locations.

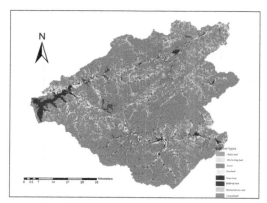

Figure 2.   Map of the land use types.

then, the boundaries were modified by hand based on the 2.5 m resolution image of advanced earth observation satellite (ALOS. 2009). The ALOS image was also used to draw the land use map of the study area by hand. In these processes, 4, 3, 2 bands were combined and the projection of both images and vector formats were Universal Transverse Mercator (UTM), zone 51. In order to make the results as wildly usable as possible, the land uses were classified into eight types (Figure 2): Paddy land, Dry farming land, Forest, Grassland, Water body, residential land, Mining/industry land and Unused land.

Different land uses have different influences on the ecological environment (Li et al., 2008). The "source" and "sink" landscape identification is performed at the annual scale. The dominant "source" landscape types were paddy land, dry farming land, residential land, mining/industry land and unused land. The "sink" landscape types were forests, grasslands, and water bodies.

2.4 Statistical analysis

Data analyses (e.g., mean, standard deviation, maximum and minimum concentrations) were performed by SPSS 18.0 for Windows in this study. Spatial autocorrelation statistics were used to examine the spatial variations of water quality. Pearson correlation analysis was implemented to test the relationship between land use types and physicochemical parameters of the water. Stepwise linear regression were applied to extract the most important physicochemical parameters of the water during the different rainfall periods using the SPSS 18.0 software package.

3 RESULTS AND DISCUSSION

3.1 Land use composition and sub-watersheds similarity

Analysis showed that three types of landscapes, including forest land, dry land, and paddy land,

dominated at all of watershed, with the sum of proportions more than 80%. Conversely, proportion of other types of landscapes, such as grassland, mine, and unused land, was less than 5%. An obvious feature was that landscape, were dominant at all of sub-watersheds, with the sum of proportions varying from 50.77% to 92.88%.

Based on the analysis of the "source" and "sink" landscape (Table 1, Figure 3), these 77 subwatershed were divided again into six groups. The landscape types and area statistics are shown in Table 1. Based on the analysis of the proportion of "source" and "sink" landscapes (Table 1, Figure 3), these 77 sub-watershed were divided again into six groups. The landscape types and area statistics are shown in Table 1. The proportion of "source" landscapes occupied the largest areas at group I – group III, which were located in the broad river valleys, most of which were flat and shallow. These areas, because of fertile soil, available irrigation system and abundant natural resources, were economic-developed and populous area in watershed. Water quality deterioration is serious in these area. Group IV-group VI, the proportion of "sink" landscapes was greater than that of other land-use types, were located in the headwater region.

3.2 Temporal variation of water quality

Water quality indicators vary greatly according to season. On the whole, TN concentration ranges from 1.673 to 12.151 mg/L, exceeded surface water Class III water standard (GB3838-2002). In the 77 water quality monitoring sections, only one sample of TN concentration was 1.673 mg/L, others are at Grade V and worse than Grade V, serious pollution. TP concentration ranges from 0.017 to 0.818 mg/L, 88% of samples did not exceeded surface water Class III water standard. NO_3-N and NH_4-N concentration ranges from 0.998 to 7.071 mg/L and 0.123 to 1.192 mg/L, respectively. Most water quality indicators show higher concentration during the high-flow compared with

Table 1. The classification of source-sink landscape.

Groups	"Source" (%)	"Sink" (%)	Sub-watersheds
I	≥30	≤70	W1, W3, W9, W12, W19, W21, W24, W41, W47, W48, W49, W50, W59, W69
II	25–30	70–75	W4, W13, W14, W15, W31, W34, W52, W61, W64, W72, W73, W74, W77
III	20–25	75–80	W2, W5, W6, W20, W35, W39, W43, W53, W76
IV	15–20	80–85	W7, W10, W16, W17, W18, W22, W23, W25, W26, W29, W30, W32, W37, W38, W41, W45, W46, W51, W57, W75
V	10–15	85–90	W8, W28, W36, W40, W42, W54, W55, W60, W65, W68, W70, W71
VI	≤10	≥90	W11, W27, W33, W56, W58, W62, W63, W66, W67

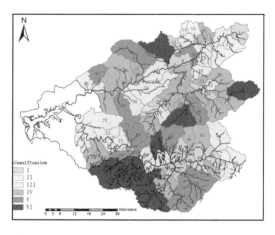

Figure 3. The classification map of source-sink landscape.

Table 2. Stepwise multiple regression models for LULC within variant contributing zones and non-point pollutant.

Parameters	Regression models	R^2	P
Group I			
NH$_3$-N[c]	NH$_3$-N = 0.226 + 0.13X$_1$	0.351	0.026
Group II			
NH$_3$-N[a]	NH$_3$-N = 7.232 − 0.093X$_3$	0.786	<0.001
NO$_3$-N[c]	NO$_3$-N = 3.317 − 2.005X$_4$	0.391	0.022
Group III			
TP[a]	TP = 0.032 + 0.036X$_6$ − 0.039X$_7$	0.876	0.002
TP[c]	TP = 0.048X$_4$ − 0.034	0.679	0.006
Group V			
NH$_3$-N[a]	NH$_3$-N = 0.053X$_2$ − 0.288	0.429	0.009
NH$_3$-N[c]	NH$_3$-N = 0.479 − 0.164X$_4$	0.203	0.046
Group VI			
TN[a]	TN = 0.561 + 4.967X$_6$	0.675	0.012
NO$_3$-N[c]	NO$_3$-N = 22.548 − 0.228X$_3$	0.597	0.025

a, b, and c represented for low-flow, high-flow and mean-flow, respectively. X$_1$, X$_2$, X$_3$ X$_4$, and X$_6$ represented for paddy land, dry land, forest land, grassland, and residential land.

the other two periods. During the high-flow, the average values of TN, TP, and NO$_3$-N are 6.470, 0.153, and 3.954 mg/L, respectively. It is interesting that the concentration of NH$_4$-N is highest in low-flow, and that the average content is 0.551 mg/L. This finding is interpreted as follows: River inputs take the form of rainwater, ice-melt and snow-melt, lake water, swamp and groundwater recharge, with rainfall serving as the main source. Given the limited urbanization and lack of sewage treatment measures in the watershed, industrial sewage and most domestic sewage and household waste are discharged directly into the river system. Additionally, there are many livestock and poultry factories within the watershed, most of which have non-standardized feces pools, so that liquid dung is also discharged directly to ditches and flows into rivers. In particular, due to the low rainfall prior to the rainy season, the domestic and industrial sewage is the main source of river feeding, which results in a high ammonia-nitrogen concentration.

3.3 Statistical relationships between land use and water quality

In general, there is a significant negative correlation between forest coverage and NH$_3$-N indicators, suggesting that forest land is an important predictor of water quality change, and that increased forest cover can mitigate the deterioration of water quality to a certain degree. Farmland shows a highly positive correlation with nitrogen-containing compounds. As one of the essential components of "Source" landscapes, farmland is usually considered an important pollution source that degrades water quality.

Numerous studies have shown significant relationships between the percentages of agricultural or unban land cover and nutrient concentrations in streams and rivers (Wang et al., 2013; Ye et al., 2014).

4 CONCLUSIONS

The results of this study show that forest played an important role in maintaining clean water, whereas farmland deteriorated the surface water quality due to non-point agricultural pollutants. It is fundamental to preserve sufficient forest land area and to control agriculture to maintain good water quality in in Dahuofang reservoir watershed. However, previous studies did not consider the influence of landscape scales on water quality, and so might have underestimated the threshold values. Finally, sustainable land use practices and scientific environmental regulations are required for water conservation and management in Dahuofang reservoir watershed.

ACKNOWLEDGMENTS

This work was funded by the National Natural Science Foundation of China (41201494, 41201520, and 41430637).

REFERENCES

Ahearn DS, Sheibey RW, Dahlgren RA, et al. 2005. Land use and land cover influence on water quality in the last free-flowing river draining the western Sierra Nevada, California. Journal of Hydrology, (313):234–247.

Allan JD & Johnson L. 1997. Catchment-scale analysis of aquatic ecosystems. Freshwater Biology, 37(1): 107–111.

Arheimer B & Liden R. 2000. Nitrogen and phosphorus concentrations from agricultural catchments—influence of spatial and temporal variables. Journal of Hydrology, 227(1):140–159.

Brion G, Brye K, Haggard B, et al. 2011. Land-use effects on water quality of a first-order stream in the Ozark Highlands, Mid-Southern United States. River Research and Applications, 27(6):772–790.

Chinese State Environment Protection Bureau (CSEPB). 2002. Water and Wastewater Monitoring Analysis Methods, 4th ed., Chinese Environment Science Press, Beijing, China.

Griffith JA, Martinko EA, Whistle JL, et al. 2002. Interrelationships among landscapes, NDVI, and stream water quality in the US central plains. Ecological Applications. 12:1702–1718.

Johnson L, Richards C, Host GE, et al. 1997. Landscape influences on water chemistry in Midwestern stream ecosystems. Freshwater Biology, 37(1):193–208.

Li Y, Wang ZG, Peng SL, et al. 2008. Impact of land use patterns on eco-environment in pearl River Estuary. Progress Geography 27 (3), 55–59 (in Chinese).

Mehaffey MH, Nash MS, Wade TG, et al. 2005. Linking land cover and water quality in New York City's water supply watersheds. Environmental Monitoring and Assessment, 107(1–3):29–44.

Shrestha S & Kazama F. 2007. Assessment of surface water quality using multivariate statistical techniques: a case study of the Fuji river basin, Japan. Environmental Modelling & Software, 22(4):464–475.

Singh KP, Malik A, Mohan D, et al. 2004. Multivariate statistical techniques for the evaluation of spatial and temporal variations in water quality of Gomti River (India)—a case study. Water Research, 38(18): 3980–3992.

Tong ST & Chen W. 2002. Modeling the relationship between land use and surface water quality. Journal of Environmental Management, 66(4):377–393.

Townsend CR, Arbuckle CJ, Crow TA, et al. 1997. The relationship between land use and physicochemistry, food resources and macroinvertebrate communities in tributaries of the Taieri River, New Zealand: A hierarchically scaled approach. Freshwater Biology, 37:177–191.

Wang RZ, Xu TL, Yu LZ, et al. 2013. Effects of land use types on surface water quality across an anthropogenic disturbance gradient in the upper reach of the Hun River, Northeast China. Environmental Monitoring and Assessment, 185:4141–4151.

Ye Y, He XY, Chen W, et al. 2014. Seasonal Water Quality Upstream of Dahuofang Reservoir, China—the Effects of Land Use Type at Various Spatial Scales. Clean- Soil, Air, Water, 42 (10):1423–1432.

Advances in Energy, Environment and Materials Science – Wang & Zhao (Eds)
© *2016 Taylor & Francis Group, London, ISBN 978-1-138-02931-6*

Theoretical and experimental study of volumetric heat transfer coefficient for Direct-Contact Heat Exchanger

Junwei Huang
Faculty of Mechanical and Electrical Engineering, Yunnan Agricultural University, Yunnan, P.R. China

Shibo Wang & Jianhang Hu
State Key Laboratory of Complex Nonferrous Metal Resources Clean Utilization, Kunming University of Science and Technology, Yunnan, P.R. China

ABSTRACT: This paper proposes a novel model of volumetric heat transfer coefficient for Direct-Contact Heat Exchanger (DCHE). This model is based on an integration of single bubble heat-transfer characteristics and the drift-flux model. Bases on experiment, experimental volumetric heat transfer coefficient is obtained at variations initial heat transfer temperature difference, refrigerant flow rate, heat transfer oil flow rate, heat transfer oil height, respectively. These experimental values are used to validate the theoretical model of this paper. The result show that there is a good agreement between the theoretical and experimental values, and the errors are almost within 20%. The model may improve our insight into the dependencies of the total heat transfer performance of DCHE, indirectly, on the exchanger design.

1 INTRODUCTION

Direct-contact heat transfer occurs when immiscible fluids are brought into contact at different temperature, resulting in the evaporation of the fluid that has a lower boiling point (Mahood 2008). Direct-Contact Heat Exchanger (DCHE) use the liquid-vapor phase change of the working fluid inside the heat exchanger. On the other hand, DCHE between two fluids without any heat transfer wall. It has lots of advantages over indirect heat exchanger because of simpler of structure, more rapidly exchanges heat, and higher thermal storage density (Kar et al. 2007; Nomura et al. 2013). A DCHE have been recommended for water desalination, crystallization, energy recovery from industrial waste, thermal energy storage, ice-slurry production, etc. (Lemenand et al. 2010; Nomura et al. 2013; Hyun et al. 2005; Hawlader & Wahed 2009).

Because of the complexity of the multiphase flow in exchangers, the process is better described in terms of volumetric heat transfer coefficient. It is a common engineering practice to use a volumetric heat transfer coefficient to describe the DCHE thermal performance (Kulkarni & Ranade 2013). Core and Mulligan used a population balance model to predict the volumetric heat transfer coefficient (Core & Mulligan 1990). Mori proposed a model that assumes no nucleation delay in initially monodispersed drops and a heat transfer to each of the drops (Mori 1991). Due to simultaneous

evaporation, that can be approximated by an empirical correlation for heat transfer to an isolated drop evaporating in a quiescent. Song et al. used a population balance model to predict the volumetric heat transfer coefficient for direct-contact evaporation in a bubble column (Song et al. 1999). And this model is based mainly on the energy balance and the population balance. Zhang et al. established a models that take into account the evaporation of continuous phase water into the dispersed phase and the two-phase droplets break-up (Zhang et al. 2001). Which calculated results showed good agreement with the experimental values.

This paper presents an attempt to derive a model of volumetric heat transfer coefficient. To establish the model, the analysis is divided into the first and second stage on the basis of fundamental assumption in the process of direct contact heat transfer. And the purpose of experimental setup is to validate the model of volumetric heat transfer coefficient. The model may improve our insight into the dependencies of the total heat transfer performance of DCHE, indirectly, on the exchanger design.

2 THEORETICAL ANALYSIS

2.1 Physical system

Our research was carried out in a counterflow spray column in which the dispersed phase is injected in the form of saturated liquid drops of a uniform size, resulting in an immediate start of evaporation. The dispersed phase is extract heat from the continuous phase, therefore the continuous phase has a temperature higher than the boiling point of the dispersed phase. The model is intended to yield the relationship between the rate of evaporation of droplets and their displacement from their source and site of nucleation, the process is better described in terms of volumetric heat transfer coefficients. Schematically shown in Figure 1.

2.2 Fundamental assumptions

To establish the model, the most fundamental assumption is that there are two stages in the process of direct contact heat transfer. In the first stage the individual droplets behave independently of one another, while in the second stage it is recognized that the droplets affect each other and begin to coalesce or fragment. In other words the number density of the droplets varies with the height in the second stage. Accordingly, the analysis is divided into two steps—the first stage and the second stage. At the same time, before deriving the volumetric heat transfer coefficients, the following assumptions should be introduced for convenience.

1. The multidroplet direct-contact evaporation can be described with single droplet correlations for the heat-transfer rate of individual droplets (Shimizu & Mori 1988).
2. During both the two stage it is assumed that the evaporating two-phase droplets (bubble) are spherical.
3. During both stages it is assumed that the relative velocity of the two-phase flow can be represented by a simple drift-flux formulation.

2.3 Mathematical description

The model of Smith et al. offered a fundamental mathematical framework to describe the

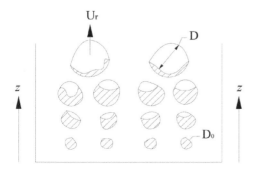

Figure 1. Illustration of two-phase bubble size dependence on height (Smith et al. 1982).

relationship between the rate of evaporation of droplets and their displacement from their source and site of nucleation (Smith et al. 1982). Figure 3 schematically shows this phenomena. Such the equation of volumetric heat-transfer coefficient could be written as:

$$h_v(z) = \frac{1}{z} \int_0^z A_b(z') n_b(z') h_b(z') dz' \tag{1}$$

where z is the axial coordinate, A_b is the droplet surface area, n_b is the number density of two-phase droplets, h_b denotes the surface heat transfer coefficient per unit surface area for a single droplet.

To simplify subsequent calculations, a new terms of r is defined as the ratio of the instantaneous equivalent spherical diameter D to its (initial) value D_0 preceding nucleation. Hence, Eq. (1) becomes:

$$h_v\left[z(r)\right] = \frac{1}{z(r)} \int_{r_0}^r A_b(r') n_b(r') h_b(r') \frac{dz'}{dr} dr' \tag{2}$$

During both the first stage and the second stage it is assumed that the droplets do depart from sphericity so that the surface area can be approximated by

$$A_b(r) = \pi D^2 = \pi D_0^2 r^2 \tag{3}$$

The relative velocity between the droplets and continuous phase in the both stages, U_r, can be regarded as the drift-flux model (Wallis 1969)

$$U_r = (1-\alpha)^{n-1} U \tag{4}$$

where the single droplet velocity is given by:

$$U = U_0 r^y \tag{5}$$

According to the Eq. (2) given above, we use the correlations of the surface area heat transfer rate for a single two-phase droplet. Huang derived a analytical formula for the Nusselt number (Huang 2013):

$$Nu = \frac{\left[\frac{k_f}{k_c} \frac{R_v}{R_d}\left(1+\sin\frac{\beta}{2}\right) + 0.0447 \mathrm{Re}_0^{0.78} \mathrm{Pr}^{1/3} \frac{(1-\cos\beta)}{2} \right]}{\left[1 + \frac{3JaFr_0\tau}{\mathrm{Re}_0\mathrm{Pr}} \left[\frac{2k_f}{k_c}\frac{R_v}{R_d}\left(1+\sin\frac{\beta}{2}\right) + 0.0447\mathrm{Re}_0^{0.78}\mathrm{Pr}^{1/3}(1-\cos\beta) \right] \right]^{1/3}} \tag{6}$$

U
R
Bubble 1
Continous
Liquid 3
Drop 2

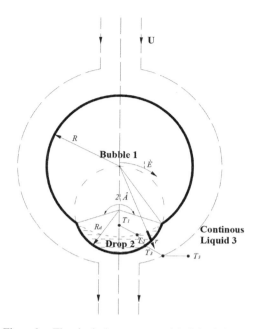

Figure 2. The physical geometry model of single bubble.

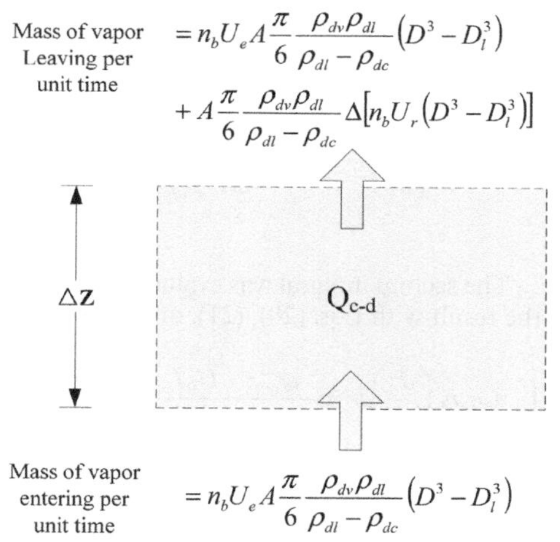

Mass of vapor
Leaving per
unit time
$= n_b U_e A \frac{\pi}{6} \frac{\rho_{dv}\rho_{dl}}{\rho_{dl}-\rho_{dc}} (D^3 - D_l^3)$
$+ A \frac{\pi}{6} \frac{\rho_{dv}\rho_{dl}}{\rho_{dl}-\rho_{dc}} \Delta[n_b U_r (D^3 - D_l^3)]$
ΔZ
Q c-d
Mass of vapor
entering per
unit time
$= n_b U_e A \frac{\pi}{6} \frac{\rho_{dv}\rho_{dl}}{\rho_{dl}-\rho_{dc}} (D^3 - D_l^3)$

Figure 3. Diagram of energy balance calculation in the second stage (Smith et al. 1982).

where β is the half-evaporation open angle, shown in Figure 2.

In addition, the define of Nusselt number is:

$$Nu = \frac{h_b D}{k_c} = \frac{C}{r} \tag{7}$$

where

$$C = \frac{k_f}{k_c}\frac{R_v}{R_d}\left(1+\sin\frac{\beta}{2}\right) + 0.447\,\mathrm{Re}_0^{0.78}\,\mathrm{Pr}^{1/3}\frac{(1-\cos\beta)}{2} \tag{8}$$

1. Volumetric heat transfer coefficients in the first stage
During the first stage it is assumed that the individual droplets behave independently and do not coalesce or fragment each other, therefore, $n = 1$ and $y = 0$ in Eqs. (4) and (5) during this stage.
At the same time, the number density of the droplets do not varies with the height in the first stage.
Hence,

$$n_b(r) = n_{b0} \tag{9}$$

The dispersed phase volume fraction is obtained from:

$$\alpha(r) = \frac{\pi}{6} D_0^3 n_b(r) r^3 \tag{10}$$

The relationship between r and z in Eq. (2) can be determined by solving for the single droplet evaporation rate:

$$\frac{d}{dz}(\rho_{dv} V_{dv}) = \frac{1}{U_0}\frac{q}{L} \tag{11}$$

The dispersed phase vapor volume per droplet is given by:

$$V_{dv} = (\pi/6)\frac{\rho_{dl} D_0^3}{\rho_{dl} - \rho_{dv}}(r^3 - 1) \tag{12}$$

While the heat-transfer rate per droplet is given by:

$$q = h_b(r) A_b(r) \Delta T \tag{13}$$

Substituting Eqs. (12) and (13) into Eq. (11) yields:

$$\frac{\pi}{2}\frac{\rho_{dl}\rho_{dv}}{\rho_{dl} - \rho_{dv}} D_0^3 r^2 \frac{dr}{dz} = \frac{h_b A_b \Delta T}{U_0 L} \tag{14}$$

When Eqs. (9) and (10), and (14) are combined with Eq. (2), the result is:

$$h_v(r) = \frac{\alpha_0}{z(r)} U_0 \frac{L}{\Delta T}\frac{\rho_{dl}\rho_{dv}}{\rho_{dl} - \rho_{dv}}(r^3 - 1) \tag{15}$$

The distance z is expressed as a function of r by integrating Eq. (14). Substituting Eqs. (3) and (7) into Eq. (14) and integrating the resultant equation yields:

$$\frac{r^3 - 1}{3} = Bz \qquad (16)$$

Where

$$B = \frac{2h_{b0}\Delta T}{U_0 D_0 L} \frac{\rho_{dl} - \rho_{dv}}{\rho_{dl}\rho_{dv}} \qquad (17)$$

Hence, Eq. (15) becomes:

$$h_v(r) = \frac{6\alpha_0 h_{b0}}{D_0} \qquad (18)$$

Eq. (18) can be also be expressed as a function of z:

$$h_v(z) = 3B\alpha_0 U_0 \frac{L}{\Delta T} \frac{\rho_{dl}\rho_{dv}}{\rho_{dl} - \rho_{dv}} \qquad (19)$$

2. Volumetric heat transfer coefficients in the second stage

During the second stage it is assumed that the individual droplets coalescence or fragmentation each other, while the larger cap-shaped droplets that result from evaporation have velocities approximately proportional to the square root of their diameter, therefore, $n = 4$ and $y = 1/2$ in Eqs. (4) and (5) during this stage.

In this stage the droplet number density, n_b, can be calculated by using:

$$n_b(r) = n_{b0}\left(\frac{r_a}{r}\right)^3, \quad r > r_a \qquad (20)$$

where the subscript a indicates that the quantity is evaluated at the onset of agglomeration.

The energy balance calculation is shown schematically in Figure 2, where Q_{c-d} is the rate of heat transfer from the continuous phase to the dispersed phase.

A heat balance yields in the limit:

$$\frac{d}{dz}\left[U_r\left(1 - \left(\frac{r_l}{r}\right)^3\right)\right] = \frac{6h_b\Delta T}{rLD_0} \frac{\rho_{dl} - \rho_{dv}}{\rho_{dl} - \rho_{dv}} \qquad (21)$$

The Eq. (4) in the second stage becomes:

$$U_r = (1 - \alpha_{max})^{-1} U_0\left(\frac{r}{r_a}\right)^{1/2} \qquad (22)$$

The relationship is derived by invoking the principle of conservation of dispersed mass flux which, in conjunction with Eq. (20), yields:

$$r_l^3 = \frac{U_0}{U_r(r)}\left(\frac{r}{r_a}\right)^3 \qquad (23)$$

Substituting Eq. (23) into Eq. (21), and combining the result with Eqs. (7) and (22), yields:

$$\frac{2}{7r_a^{1/2}}\left(r^{7/2} - r_a^{7/2}\right) = \frac{12(1 - \alpha_{max})h_{b0}\Delta T}{U_0 D_0 L}$$
$$\times \frac{\rho_{dl} - \rho_{dv}}{\rho_{dl}\rho_{dv}}(z - z_a) \qquad (24)$$

During the second stage, Eq. (2) assumes the form:

$$h_v = \frac{1}{z_a + (z - z_a)}\left[\int_1^{r_a} A_b n_b h_b\left(\frac{dz}{dr}\right)dr \right.$$
$$\left. + \int_{r_a}^r A_b n_b h_b\left(\frac{dz}{dr}\right)d \right] \qquad (25)$$

The first integral in Eq. (25) was evaluated in the first stage

$$\int_1^{r_a} A_b n_b h_b\left(\frac{dz}{dr}\right)dr = \alpha_0 U_0 \frac{L}{\Delta T} \frac{\rho_{dl}\rho_{dv}}{\rho_{dl} - \rho_{dv}}\left(r_a^3 - 1\right) \qquad (26)$$

The second integral was evaluated combining the result with Eqs. (20), (21), and (22)

$$\int_{r_a}^r A_b n_b h_b\left(\frac{dz}{dr}\right)dr = \frac{\alpha_{max}}{1 - \alpha_{max}} \frac{U_0 L}{\Delta T}$$
$$\times \frac{\rho_{dl}\rho_{dv}}{\rho_{dl} - \rho_{dv}}\left(r^{1/2} - r_a^{1/2}\right) \qquad (27)$$

Finally, Substituting Eq. (16) for z and Eq. (24) for $z - z_a$ into Eq. (25), and integrating the resultant equation yields:

$$h_v(r) = 2\frac{h_{b0}}{D_0} \frac{\alpha_0\left(r_a^3 - 1\right) + \dfrac{\alpha_{max}}{1 - \alpha_{max}}\left(r^{1/2} - r_a^{1/2}\right)}{\dfrac{r_a^3 - 1}{3} + \dfrac{1}{21r_a^{1/2}} \dfrac{r^{7/2} - r_a^{7/2}}{1 - \alpha_{max}}} \qquad (28)$$

As a function of z, Eq. (28), the volumetric heat transfer coefficient becomes:

$$h_v(z) = \frac{2h_{b0}}{D_0 Bz}\left[3\alpha_0 Bz_a + \frac{\alpha_{max}}{1-\alpha_{max}}\right.$$
$$\times\left(\left(1+\frac{7}{2}Bz_a + 21(1-\alpha_{max})B(z-z_a)\right)^{1/7}\right.$$
$$\left.\left.-\left(1+\frac{7}{2}Bz_a\right)^{1/7}\right)\right]$$

(29)

Where:

$$h_{b0} = \frac{k_f R_v}{2R_d^2}\left(1+\sin\frac{\beta}{2}\right)$$
$$+ 0.447 R_0^{0.78}\, Pr^{1/3}\frac{k_c(1-\cos\beta)}{2D_0}$$

(30)

3 EXPERIMENTS

3.1 *Experimental apparatus*

Figure 4 shows the schematic of experimental setup. It consists of a DCHE (1), a electric heater (2), a storage vessel (4), a plate condenser (5), a centrifugal pump (7) and a gear oil pump (8). The DCHE is a spray column of 480 mm inside diameter, 1500 mm height with four temperature-measuring holes and three viewing windows (see Fig. 5). The Heat Transfer Oil (HTO) is employed as the continuous phase in all runs. The *R245fa* (1,1,1,3,3 pentafluoropropane) is the dispersed phase in all runs.

Figure 4. Schematic of experimental equipment of direct-contact heat transfer.
(1) The DCHE (2) electric heater (3) temperature control device (4) storage vessel (5) plate condenser (6) centrifugal pump (7) gear oil pump (8) frequency conversion control cabinet (9) highspeed shutter video camera (10) level gauge (11) K-type thermocouple (12) liquid mass flowmeter (13) gas mass flowmeter (14) pressure gauge (15) thermometer.

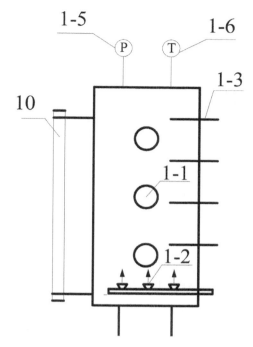

Figure 5. The DCHE.
(1-1) viewing windows (the second) (1-2) nozzle (1-3) K-type thermocouple (1-4) level gauge (1-5) pressure gauge (1-6) thermometer.

Table 1. Conditions of experimental operations.

$\Delta T(°C)$	$U_0(m\cdot s^{-1})$	$U_c(m\cdot s^{-1})$	$Z(m)$
80~120	0.04~0.12	0~0.72	0.4~0.6

3.2 *Performance evaluation of DCHE model*

The experimental volumetric heat transfer coefficient, h_{exp}, which is given by (Thongwik et al. 2008):

$$h_{exp} = \frac{Q}{V \times LMTD}$$

(31)

where V is the volume of continuous phase in DCHE and Q is the rate of heat transfer from the continuous phase to the dispersed phase, given by

$$Q = \dot{m}(h_{do} - h_{di})$$

(32)

where $\dot{m}$ is mass flow-rate of dispersed phase steam, and h is enthalpy of dispersed phase. The *LMTD* in Eq. (31) is the logarithmic mean temperature difference, which is defined as

$$LMTD = \frac{(T_{ci} - T_{do}) - (T_{co} - T_{di})}{\ln\frac{(T_{ci} - T_{do})}{(T_{co} - T_{di})}} \qquad (33)$$

where T is temperature. In all above of equations, the subscript c refers to continuous phase, d refers to the dispersed phase, i refers to the inlet, and o refers to the outlet.

3.3 Operational conditions of the experiment

The experiment is repeated by varying the initial heat transfer temperature difference ΔT, refrigerant flow rate U_0, HTO flow rate U_c, and HTO height Z to study their effects on the volumetric heat transfer coefficient. The conditions of experimental operations are shown in Table 1.

4 COMPARISON OF THEORY AND EXPERIMENT

Figure 6~Figure 9 show the comparison of theoretical and experimental values of the volumetric heat transfer coefficient at variations initial heat transfer temperature difference, refrigerant flow rate, HTO flow rate, HTO height, respectively. The theoretical values is calculated by Eq. (29). The experimental values is calculated by Eq. (31). The Eq. (29) has a good ability to predict the volumetric heat transfer coefficient. The predicted results are in good agreement with experimental data.

The Figure 10 shows that errors between the theoretical and experimental values are almost within 20% for the different experimental conditions. It indicates that the theoretical function of volumetric heat transfer coefficient is reliable. The errors are mainly caused by the following reasons. One is the rate of heat released to the environment is not considered

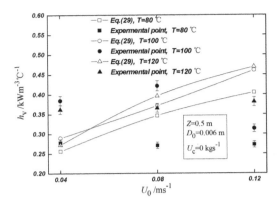

Figure 7. Comparison of theoretical and experimental values of the volumetric heat transfer coefficient at refrigerant flow rate.

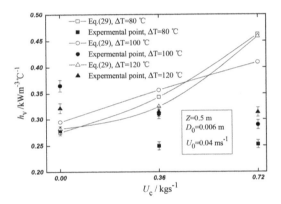

Figure 8. Comparison of theoretical and experimental values of the volumetric heat transfer coefficient at HTO flow rate.

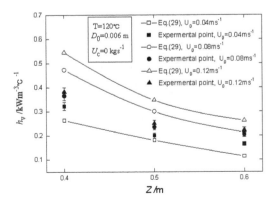

Figure 9. Comparison of theoretical and experimental values of the volumetric heat transfer coefficient at HTO height.

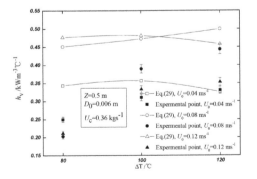

Figure 6. Comparison of theoretical and experimental values of the volumetric heat transfer coefficient at initial heat transfer temperature difference.

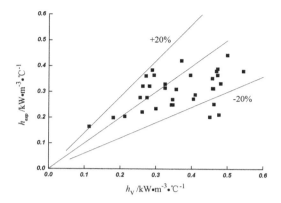

Figure 10. The deviations between tested values and the results predicted by Eq. (29).

in calculating the theoretical values. The other is the uncertainties in the experimental measurements. Therefore, we should not overlook some experimental findings that cannot be interpreted on the basis of the present function of volumetric heat transfer coefficient. These observations, contradictory to the present function, are left for further experimental and theoretical examinations in the future.

5 CONCLUSIONS

In this paper, the process of direct contact heat transfer of the DCHE is divided into two stages, the first and second stage, where a novel model of volumetric heat transfer coefficients were developed, respectively. This model is based on an integration of single bubble heat-transfer characteristics and the drift-flux model. The experimental setup was established used to validate the function of volumetric heat transfer coefficient in this paper. The comparison of theoretical and experimental values of the volumetric heat transfer coefficient at variations initial heat transfer temperature difference, refrigerant flow rate, HTO flow rate, HTO height, respectively. In general, there is a good agreement between the theoretical and experimental values, and the errors are almost within 20% for the different experimental conditions. Hence, further confidence can only be gained by present function to more successful experimental and theoretical examinations in the future.

ACKNOWLEDGMENTS

This work is supported by the Special Preliminary Research of 973 Plan, China (2014CB460605), the National Natural Science Foundation of China (Project Nos. 51406071), and the Natural Science Foundation of Yunnan Province (2013FB020).

REFERENCES

Core K.L., Mulligan J.C., Heat transfer and population characteristics of dispersed evaporating droplets, AIChE J. 36 (1990).

Hawlader M.N.A., Wahed M.A., Analyses of ice slurry formation using direct contact heat transfer, Applied Energy, 86(7C8) (2009) 1170–1178.

Huang J. W., The heat transfer performance and optimization study on ORC direct-contact evaporator, Ph. D. Thesis, 2013.

Hyun Y.J., Chun W.G., Kang Y.H., An experimental investigation into the operation of a direct contact heat exchanger for solar exploitation, International Communications in Heat and Mass Transfer, 32(3C4) (2005) 425–434.

Kar S., Chen X.D., Nelson M.I., Direct-Contact Heat Transfer Coefficient for Condensing Vapour Bubble in Stagnant Liquid Pool, Chemical Engineering Research and Design, 85(3) (2007) 320–328.

Kulkarni A A., Ranade VV., Direct Contact Heat Transfer via Injecting Volatile Liquid in a Hot Liquid Pool: Generation and Motion of Bubbles. Chemical Engineering Science, (0) (2013).

Lemenand T., Durandal C., Della Valle D., Peerhossaini H., Turbulent direct-contact heat transfer between two immiscible fluids, International Journal of Thermal Sciences, 49(10) (2010) 1886–1898.

Mahood H.B., Direct-contact heat transfer of a single volatile liquid drop evaporation in an immiscible liquid, Desalination, 222(1C3) (2008) 656–665.

Mori Y.H., An analytic model of direct-contact heat transfer in spray-column evaporators, AIChE J. 37 (1991) 539–546.

Nomura T., Tsubota M., Oya T., Okinaka N., Akiyama T., Heat Release Performance of Direct-Contact Heat Exchanger With erythritol as Phase Change Material, Applied Thermal Engineering, 61(2), (2013) 28–35.

Nomura T., Tsubota M., Oya T., Okinaka N., Akiyama T., Heat storage in direct-contact heat exchanger with phase change material, Applied Thermal Engineering, 50(1) (2013) 26–34.

Shimizu Y., Mori Y.H., Evaporation of single liquid drops in an immiscible liquid at elevated pressures: experimental study with n-pentane and R113 drops in water, Int. J. Heat Mass Transfer 31 (1988) 1843–1851.

Smith R.C., Rohsenow W.M., Kazimi M.S., Volumetric heat transfer coefficient for direct-contact evaporation, Trans. ASME J. Heat Transfer 104 (1982) 264–270.

Song M, Steiff A, Weinspach PM., Direct-contact heat transfer with change of phase: a population balance model. Chemical Engineering Science, 54(17)(1999) 861–3871.

Thongwik S., Vorayos N., Kiatsiriroat T., Nuntaphan A., Thermal analysis of slurry ice production system using direct contact heat transfer of carbon dioxide and water mixture, International Communications in Heat and Mass Transfer, 35(6) (2008) 756–761.

Wallis, G.B. One-Dimensional Two-Phase Flow, McGraw-Hill, New York, 1969.

Zhang P., Wang Y.P., Guo C.L., Wang K. Heat transfer in gas-liquid-liquid three-phase direct-contact exchanger [J]. Chemical Engineering Journal, 84(3) (2001) 381–388.

Advances in Energy, Environment and Materials Science – Wang & Zhao (Eds)
© 2016 Taylor & Francis Group, London, ISBN 978-1-138-02931-6

Research on dynamic analysis of pod vibration based on ANSYS/LS-DYNA

Jinyu Jiang & Shiming Li
Mechanical and Electrical Engineering College, Hubei Polytechnic University, Huangshi, China

ABSTRACT: Finite element method model of the pod vibration was established by using ANSYS/LS-DYNA software. Finite element dynamic response analysis was carried out. The stress distribution of pod under harmonic loads was obtained in vibration process. And the displacement, velocity, and acceleration with the time variation were also got. It was more accurate simulated the whole impact process of pupa and silkworm shell. The method can be used to analyze the vibration system of pod.

1 INTRODUCTION

Vibration detection method was used during the research of non-destructive pod testing. According to the characteristics of cyclic movement of the vibration device of pod, the dynamic characteristics of the vibration device of pod was analyzed. The results of the modal analysis can be used to analyzing the signal and selecting the characteristic values of signal. This research of dynamics analysis has formed a deep theory and practice for the research of non-destructive testing instrumentation of pod.

2 A FINITE ELEMENT MODEL OF POD VIBRATION

Finite element model is used to simulate the physical system with unknown quantities, and different physical systems correspond to different quantities. The main elements of the finite element model are node, element, real constant, material properties, boundary conditions, and loads. Creating a finite element model is the model of the process by the geometric entities Meshing. The element attributes are required before the grid partitioned, which included element type, real constant material model, etc. These properties are very important for finite element analysis, which not only affected the grid division, but also greatly affected the accuracy of the solution. The LS-DYNA model of pod was created by pre-processor. The general process includes.

The establishment of pod' finite element model was shown in Figure 1. It contained two parts: the finite element model establishment of silkworm shell that was the white reticulated shell and the finite element model establishment of pupa that was the Purple solid sphere in Figure 1. According to the actual shape of the shell and the pupa

chrysalis, the shell of pod was similar to the shell of ellipsoid. Long diameter and short diameter of ellipsoid were 36 mm and 22 mm which was the average size of the shell. Ellipsoid shell thickness was 0.8 mm. It was formed around the major axis of the ellipsoid rotating housing. The pupa was approximately spherical of diameter 10 mm.

2.1 The finite element model establishment of silkworm shell

In ANSYS/LS-DYNA explicit dynamic analysis, the geometry model of silkworm shell was shown in Figure 2. The establish steps were as follows:

① According silkworm shell shape features, select unit type shell SHELL163, using Lagrange algorithm, thickness of the shell of 0.0008 m, shell shear factor 5/6.

② Set of solid models silkworm shell material type and attributes, density silkworm shells, Young's modulus of elasticity, Poisson's ratio, 0.3.

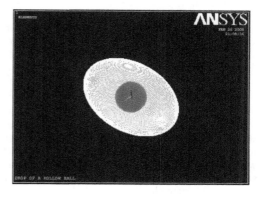

Figure 1. Finite element modeling of pod.

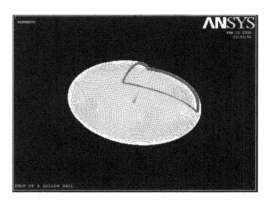

Figure 2. The finite element model of silkworm shell.

③ First elliptical coordinates, define the key points, generated line to give the same both inside and outside the geometric center of the ellipse, generate elliptical torus, the cross-sectional shape as the silkworm shell; the resulting elliptical torus around the long axis of the revolution, the formation of elliptical ball housing, silkworm shell entity model is set up.

④ Divided grid, a finite element model of the process is the solid model geometry mesh, finite element model and solid model is different, solid modeling mathematically expresses the geometry of the prototype, actually Finite element analysis do not participate, you need to be converted to a node entity or unit, all applied to the geometric entity boundary loads, constraints must ultimately be transferred to the finite element model of the node or element. So meshing is particularly important, you can use the free time division to be a reasonable choice of grid size, silkworm shell length direction of about 0.036 m, mesh size selected in the 0.003 m~0.001 m, can be further refined based on local needs.

⑤ The definition of silkworm shell node group element, to convert the coordinate system to spherical coordinates, select a radius of 0.01 m and 0.04 m radius of two spheres (to ensure that all nodes on the silkworm shells can choose to) to form a spherical shell of all nodes, nodes as silkworm shell component. Silkworm shell component was a total of 4800 nodes when the mesh size was 0.001 m.

As diagram that is shown in Figure 2 silkworm shell finite element model, similar to the entity model and real, create the finite element model with modeling requirements, shown in purple surrounded is the entire body section to a part of the site, whichcan clearly see the hull of silkworm and silkworm shell thickness.

2.2 The finite element model establishment of pupa

The geometrical model establishment of pupa was similar with silkworm shell. The pupa geometric model was shown in Figure 3. It was divided in the following steps:

① According to the pupa shape and its characteristics, element type for shell elements shell 163 was selected by using a Lagrange element integration algorithm and the diameter of pupa chrysalis as 0.01 M.

② Set pupa solid model material types and attributes, pupa density of 100 kg/m³, elastic modulus of 30 GPa, Poisson's ratio of 0.2.

③ The pupa modeling method was relatively simple. First in the spherical coordinate system, set the center location as the origin of coordinates, pupa chrysalis radius was 0.005 M, and then generated the silkworm solid sphere model.

④ The diameter of the pupa chrysalis was 0.01 M. So, the mesh size should be set below in 0.002 m in order to ensure that the pupa model in the mesh did not appear in distortion. Under the premise of satisfying precision and distortion, grid size was improved as much as possible in order to simplify the calculation. Sometimes it needed to be repeated for the model to be properly modified, reasonable grid division. This was a very tedious work.

⑤ Defined the pupa chrysalis node group element, the coordinate system transformed to the global coordinate system. The radius of the sphere was 0.006 m. It contained all the nodes, as the pupa node group element. When the mesh size was 0.001 M, pupa chrysalis group element was a total of 272 nodes. The establishment of element was fully prepared for the contact collision surface selection and definition.

The finite element model of pupa was shown in Figure 3. The intermediate blue sphere was the entity model of silkworm chrysalis. The node of silkworm shell element was displayed around the pupa model. For the convenience of modeling and solving, the pod under the action of the

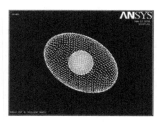

Figure 3. The finite element model of pupa.

exciting force was analyzed when pupa obtained a certain speed and separated with silkworm shell.

The relative position was shown in Figure 3. Silkworm shell was due to resonant force at this time. The separated pupa did free fall at certain initial velocity. The separated pupa was only by gravity at this point. The load and constraints were realized in subsequent analysis.

3 DYNAMIC RESPONSE ANALYSIS OF POD VIBRATION

The various parts of the actual motion state were very complex in the process of the pod vibration. The finite element dynamic response analysis was very difficult to predict the vibration.

The pod vibration device was greatly simplified in this study. The finite element model was established. Finite element dynamic response analysis was carried out. The stress distribution of pupa and silkworm shell and time response of each node displacement were obtained in vibration process.

3.1 Imposed load and constraint

The main purpose of finite element analysis was the structure inspection or the components response under a certain load conditions. Therefore, it was a key step to develop appropriate load conditions in the analysis.

The size and class of load was determined after analyzing the force of the vibration device./ In test device of pod vibration, the silkworm shell and pupa had two states when the excitation to reach steady state: separation state and the collision state of pupa and silkworm shell.

In separation state, silkworm shell by fixture passed a cosine harmonic thrust and pupa did free fall at a certain speed only by gravity. These two forces can be approximated to uniform distribution. In the collision state, pupa and silkworm shell increased the impact of the opposite direction and applied to the concentrated load of model node. The acceleration diagram of pupa vibration and silkworm shell load time history curve were shown in Figures 4 and 5. The components on the vibration of constraint and dynamic loading were realized.

3.2 The dynamic response analysis of pod vibration

The stress of pod at different time was under the harmonic load by using ANSYS dynamic response analysis for the finite element model. And the displacement, velocity, and acceleration with the time variation were also observed. Stress

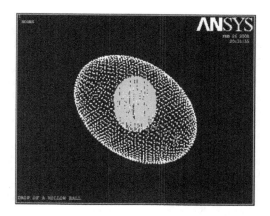

Figure 4. Acceleration diagram of pupa vibration.

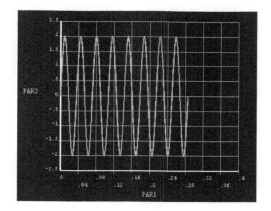

Figure 5. Silkworm shell load time history curve.

analysis of silkworm shell and pupa in different time was shown in Figures 6 and 7. The left was the pupa before the collision. And the right was the pupa at collision in Figure 6. The left was the silkworm shell before the collision and the right was the silkworm shell at collision in Figure 7. They were screenshot images of pod stress variation process. The figure more intuitive displayed the stress distribution of silkworm shell and pupa before and after the collision. It was in agreement with the actual vibration. And it indirectly reflected the reliability of finite element analysis program.

Displacement of pupa and silkworm shell with time was analyzed.

Under the action of sinusoidal excitation force, the node displacement-time curve of silkworm shell was shown in Figure 8 (left). System displacement hysteresis can be seen and the displacement gradually presented the change of sine law. Local turning was due to the collision of silkworm shell and pupa, which caused small changes. The overall trend of curve was consistent with the actual situation.

Under the action of sinusoidal excitation force, the node displacement-time curve of pupa was shown in Figure 8 (right). There was no obvious regular pattern in the displacement change. It had no obvious rules from the figure and needed further analysis.

The two curves in Figure 8 were compared and analyzed. A preliminary conclusion was found. Collision frequency of pupa and silkworm shell is higher than excitation frequency, which is consistent with power spectral band of pod. So, pod vibration dynamic response could be deeply analyzed to perfect frequency range selection analysis of pod vibration signal.

4 CONCLUSION

The finite element model of the pod vibration was established by using ANSYS/LS-DYNA software. And the dynamic response was analyzed. It was more accurately simulated the whole impact process of pupa and silkworm shell. It was consistent with the actual vibration in calculating the vibration displacement of the silkworm shell. The method can be used to analyze the vibration system of pod.

ACKNOWLEDGMENT

This research was financially supported by the scientific research project of Hubei Provincial Education Department (2015): Research on intelligent control system based on the opening degree of the steam turbine switching valve.

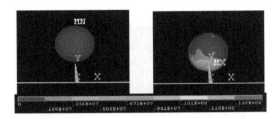

Figure 6. The stress distribution of pupa before collision (left) and collision (right).

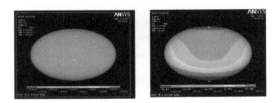

Figure 7. The stress distribution of silkworm shell before collision (left) and collision (right).

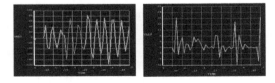

Figure 8. Node displacement-time curve of silkworm shell (left) and pupa (right).

REFERENCES

Ju R., Hsiao B. 2004, Drop Simulation for portable Electronic Products. In: The 8th International LS-DYNA Users Conference [J]. Michigan., 5(14):1–6.

National Instruments Corporation. 2004, Measurement and Automation Catalog.

Drucker D.C., Prager W. Soil mechanics and plastic analysis or limit design [J]. Apply Math, 1952, 10(2):157–165.

Livermore, 2006, Software Technology Corporation. LS-DY-NA Theoretical Manual [M]. USA: Livermore Soft-ware Technology Corporation,.

Kobbelt L.P., Bischoff S., Botsch M., et al. 2006, Geometric Modeling Based on Polygonal Meshes [EB/OL]. http://www-i8.informatik.rwth-aachen.de/uploads/media/tutorial.Pdf.

Sherry L., Jim D., John G. Super L U, 2006, Version 3.0 [CP/OL]. Lawrence Berkeley National Laboratory, University of California.

http://crd.lbl.gov/~xiaoye/SuperLU/.

Advances in Energy, Environment and Materials Science – Wang & Zhao (Eds)
© 2016 Taylor & Francis Group, London, ISBN 978-1-138-02931-6

The Influence of different testing feed velocity on the dynamic modulus of elasticity of lumber

Pengfei Zhang
Beijing Forestry Machinery Research Institute of the State Forestry Administration, Beijing, China

Wei Zhang
Beijing Forestry Machinery Research Institute of the State Forestry Administration, Beijing, China
Forestry New Technology Research Institute, Chinese Academy of Forestry, Beijing, China

Qianwei Zhang, Hailong He & Zheng Jin
Beijing Forestry Machinery Research Institute of the State Forestry Administration, Beijing, China

ABSTRACT: In this paper, FD1146 Lumber Stress Grading Machine and Mechanical Testing Machine were used respectively to measure the dynamic and static modulus of elasticity of the same batch of lumber. The relationship between dynamic and static modulus of elasticity was analyzed, and how the test values of dynamic modulus of elasticity changed under different feed velocity of the machine was also studied. The results indicate that: the dynamic and static modulus of elasticity of lumber are highly correlated, and the correlation coefficient is 0.921; with the method of mechanical stress grading, the average test value of dynamic modulus of elasticity is 3.53% higher than the average one of static modulus of elasticity; as the feed velocity of lumber gradually increases, the test values of dynamic modulus of elasticity decrease accordingly, but the deviation is not significant; FD1146 Lumber Stress Grading Machine could provide a good assessment of the static modulus of elasticity of lumber.

1   INTRODUCTION

Lumber (all the word "lumber" in this paper refers to structural lumber) is the major material of modern wood-structure buildings. In recent years, lumber structure buildings have been more and more widely used in China. To make sure the material can fulfill the requirements of architecture structures on indicators like physical mechanic properties and appearance quality, lumber must be graded before putting into use.

There are two major lumber grading methods: visual grading and machine stress grading. Mechanical stress grading method is to use mechanical methods to exert constant deformation (or load) on the sample to test the corresponding load (or deformation), then calculate the strength of the tested lumber and classify them. This method could assess the mechanical strength of lumber fast and accurately. The development and industrial application of this method is relatively mature, for example, the Model 7200 HCLT Grading Machine of Metriguard, America, and also the MK5A of Plessey Computermatics Company, England.

At present, the main method used by researchers at home and aboard is to test the dynamic modulus of elasticity and static modulus of elasticity of lumber, so as to analyze the relationship between dynamic and static modulus of elasticity, also to assess different grading methods and the influence of lumber defection on lumber grading. Erikson (2000) and others used visual grading and Mechanical Stress Grading (MSR) method to classify lumber respectively, proving that MSR could replace visual grading method in the grading of dimension lumber. Concu et al (2013) used visual grading and MSR to grade lumber, and studied how the properties of timber could affect its value. Hassan (2013), Rohanová (2010) and others tested the dynamic modulus of elasticity of lumber based on Bending Vibration Method and Longitudinal Vibration Method, and analyzed the correlation between dynamic and static modulus of elasticity. Viguier (2014) adopted Mechanical Bending Grading method and X-ray Radiography technique to test the dynamic modulus of elasticity of lumber, and analyzed the influence of grain angle on the strength of lumber. Jiang (2008) and others used Longitudinal Fundamental Frequency Vibration method to test the dynamic modulus of elasticity and static modulus of elasticity of lumber, and discussed the correlation between dynamic and static modulus of elasticity. Zhou (2009) and others used ultrasonic wave, longitudinal vibration, transversal vibration, and bending methods to grade lumber, and analyzed the influence of different grading methods on the grades of dimension lumber. Zhang (2011), Shen (2011) and others from Beijing Forestry University summarized the current situation of dimension lumber grading in China, also studied the modulus of elasticity of dimension lumber and board based on vibration stress method.

There are many factors affecting the test results of the dynamic modulus of elasticity of lumber. Mainly, based on FD1146 Lumber Stress Grading Machine developed by Beijing Forestry Machinery Research Institute of the State Forestry Administration, this paper studied the change in tested values of the dynamic modulus of elasticity of lumber under different feed velocities. This study has certain instructional meaning for practical testing of dynamic modulus of elasticity of lumber.

2 MATERIALS AND METHOD

2.1 Materials

The test materials used in this paper were two kinds of lumber, spruce and Douglas fir, provided by Suzhou Crownhomes Co. Ltd. For each kind of lumber, three sizes, namely, 2 × 4, 2 × 6, and 2 × 8, was selected respectively, and 5 pieces were taken for each size, as shown in Table 1. Spruce and Douglas fir are two kinds of typical structural lumber. Spruce lumber has light and soft texture, straight, well-distributed veins and few knots, its fine structure making it easy to process; Douglas fir, on the other hand, is heavy and durable, has high compressive and bending strength and high decay resistance. Using these two kinds of lumber for experiment, is quite representative.

2.2 Methods

The test of dynamic modulus of elasticity of lumber employed FD1146 Lumber Stress Grading Machine (as shown in Fig. 1) developed by Beijing

Figure 1. FD1146 Lumber Stress Grading Machine.

Forestry Machinery Research Institute of the State Forestry Administration. This machine could adopt different feed velocities for lumber testing according to the demands of practical production. Its operating principle is as follows: use two up and down loading rollers to exert constant deformation on the lumber, collect the corresponding loads of the loading rollers in the motion process of the lumber, process the data of the modulus of elasticity in real time, calculate the dynamic modulus of elasticity (E_{MSR}) of lumber, then grade and classify the lumber and spray codes on them according to certain standards.

The INSTRON mechanical testing machine is used to measure the static modulus of elasticity (Es) of lumber, and three-points bending test is defined and implemented according to the GB/T28987-2012 Standard. The deflection of the neutral axis of the beam at the center of the span is measured with respect to a straight line joining two reference points equidistant from the reactions and on the neutral axis of the beam. The distance between the supporting points is 18 times the height of the sample, the loading speed is 10 mm/min, and the deflection data is collected in the loading range of 1000–2000 N. Then, according to the relationship between load and bending deflection value, the static modulus of elasticity (Es) is calculated using Formula (1).

$$E_S = 23\Delta FL^3 / 108\Delta eb^3 d \qquad (1)$$

where L is the span length (mm); b is the width of sample (mm); d is the thickness of the sample (mm); ΔF is the difference between two levels of the load below proportional limit (N); Δe is the deflection caused by ΔF (mm).

Before all the samples put into experiment, their real sizes were measured, and lumber moisture contents were measured by the moisture tester (TESTO 606-2) to make sure their moisture

Table 1. Sawn lumber for experiment.

Lumber number	Species	Number of samples	Sizes (inches)
No. 1 ~ No. 5	Spruce	5	2 × 4
No. 6 ~ No. 10	Spruce	5	2 × 6
No. 11 ~ No. 15	Spruce	5	2 × 8
No. 16 ~ No. 20	Douglas fir	5	2 × 4
No. 21 ~ No. 25	Douglas fir	5	2 × 6
No. 26 ~ No. 30	Douglas fir	5	2 × 8

contents were about 12%. For each group of dimension lumber, the test for dynamic modulus of elasticity was conducted first. Three different feed velocities of FD1146 Lumber Stress Grading Machine, namely, 10 Hz (0.5 m/s), 15 Hz (1.0 m/s) and 20 Hz (1.5 m/s), were selected for the experiment, and each lumber was tested 5 times under each feed velocity, using the average of the five test values as the dynamic modulus of elasticity (E_{MSR}) of that lumber. Then, mechanical testing machine was used to test the static modulus of elasticity (E_S) of each lumber, and the relationship between dynamic and static modulus of elasticity was compared and analyzed.

3 RESULT AND DISCUSSION

3.1 *The test results of dynamic and static modulus of elasticity*

After using mechanical stress grading method and static three-points bending method to test the same batch of lumber, the test values of dynamic modulus of elasticity (E_{MSR}) and static modulus of elasticity (E_S) are summarized in Table 2.

Table 2. The statistics of Lumber dynamic and static modulus of elasticity test values.

Elastic property	E_{MSR}	E_S
Min/GPa	8.93	8.52
Max/GPa	18.03	17.3
Average/GPa	12.73	12.28
Standard deviation	2.45	2.64
Variance	6.00	6.97
Variation coefficient/%	19.25	21.5

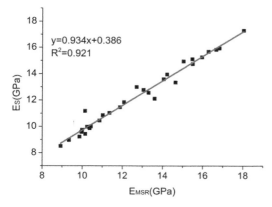

Figure 2. The correlation between dynamic and static modulus of elasticity.

When dynamic and static modulus of elasticity is compared, the average value of E_{MSR} has been found higher than the average of E_S, and the average deviation between E_{MSR} and E_S is 3.53%. One possible reason is that lumber is a kind of viscoelastic body instead of elastomer. Static modulus of elasticity, obtained from static three-points bending methods which takes a longer testing time, usually contains the factor of viscosity strain. Dynamic modulus of elasticity, on the other hand, could ignore the viscosity strain as the testing time

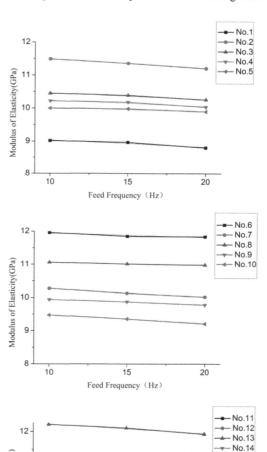

Figure 3. (*Continued*)

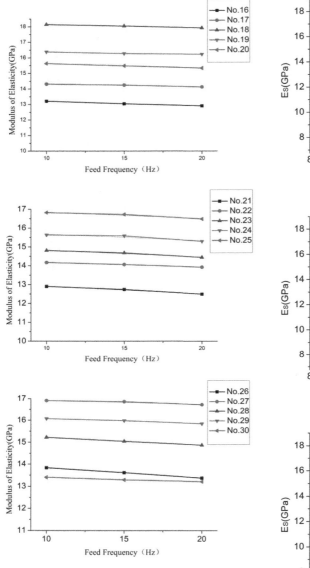

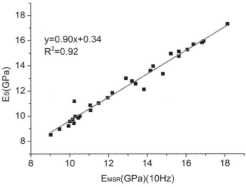

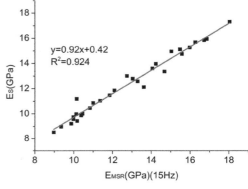

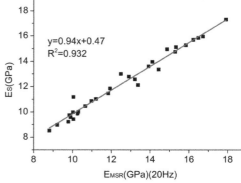

Figure 3. The variation of dynamic modulus elasticity tested under different feed velocities of different dimension lumber.

Figure 4. The correlation between dynamic and static modulus of elasticity tested under different feed velocities.

is quite short, and elasticity phenomenon is pure. As a result, the values of E_{MSR} are relatively higher than the values of E_S.

The correlation between dynamic modulus of elasticity E_{MSR} and static anti-bending modulus of elasticity E_S is shown in Figure 2. It can be seen that the dynamic modulus of elasticity E_{MSR} obtained from mechanical stress grading method and the static modulus of elasticity E_S are highly correlated, and their correlation coefficient (R^2) was 0.921, which is significantly correlated at 0.01 level. This shows that FD1146 Lumber Stress Grading Machine could provide a good assessment of the static modulus of elasticity of lumbers.

3.2 The influence of lumber feed velocity on the test results of dynamic modulus of elasticity

Under three different feed velocities, the dynamic modulus of elasticity of lumber for the three sizes of spruce and Douglas fir were tested respectively, and the variation relationship of each lumber's dynamic modulus of elasticity tested under different feed velocities was established, as shown in Figure 3.

It can be seen from Figure 3 that:

1. For spruce and Douglas fir of different sizes, when their dynamic modulus of elasticity were tested by using mechanical bending grading method, the test values of dynamic modulus of elasticity of lumber showed a decreasing trend as the lumber feed velocities gradually increased.
2. For the two kinds of lumber, spruce and Douglas fir, their tested values of dynamic modulus of elasticity under different feed velocities were all quite close, among which the No. 15 lumber had the most significant deviation, namely, 3.87% (0.42 GPa), and the testing process of the machine was quite stable on the whole.

3.3 The comparison of test results of static modulus of elasticity under different feed velocities

Under different lumber feed velocities, the correlation between static modulus of elasticity and dynamic modulus of elasticity of all sizes of lumber were studied. It can be learned that when the feed velocity was 20 Hz (1.5 m/s), the correlation coefficient was relatively high ($R^2 = 0.932$), and the values were relatively close to the static modulus of elasticity, as shown in Figure 4.

4 CONCLUSION

1. For the two groups of dimension lumber used for the test, namely, spruce and Douglas fir, the values of dynamic modulus of elasticity obtained from mechanical stress grading method were relatively higher than that of static modulus of elasticity in general. However, the values of the two modulus were quite close, and the variation rate was no bigger than 4.1%. This indicates that mechanical stress grading machine could provide a good assessment to the static modulus of elasticity of lumber.
2. When different sizes of lumber was tested on FD1146 Lumber Stress Grading Machine, the dynamic modulus of elasticity of the same lumber gradually decreased as the lumber feed velocities increased, but the variation rate was not significant, with the highest deviation rate being 3.87%, and the stability of the machine's testing process is relatively high.
3. When the test values of dynamic modulus of elasticity and static modulus of elasticity were compared under different feed velocities, it can be seen that the correlation between dynamic modulus of elasticity and static modulus of elasticity was the highest when the feed velocity was 20 Hz (1.5 m/s). Suitable lumber feed velocity could be selected according to practical production.

Mechanical stress grading method can well predict the static anti-bending elastic properties of lumber. And with high degree of automation, it is very suitable for industrialized application. In the future, we would continue to analyze the influence mechanism of lumber feed speed to testing the dynamic modulus of elastic and the influence factors of variation trend to be detected, and realize the accurate evaluation of the mechanical properties of the lumber.

ACKNOWLEDGMENT

Fund: Special Fund for Forest Scientific Research in the Public Welfare, Item Number: 201104064; Fundamental Research Funds for Central Public Welfare Research Institutes project, Item Number: CAFINT2 014C17.

REFERENCES

Alena Rohanová & Rastislav Lagaňa & Vladimír Vacek. 2010. Static and dynamic modulus of spruce structural timber. *Forestry and Wood Technology*, 72: 229–232.

Concu G. & de Nicolo B. & Trulli N. et al. 2013. Strength Class Prediction of Sardinia Grown Timber by Means of Non Destructive Parameters. *Advanced Materials Research*, 778: 191–198.

Erikson R.G. & Gorman T.M. & Green D.W. et al. 2000. Mechanical grading of lumber sawn from small-diameter lodgepole pine, ponderosa pine, and grand fir trees from Northern Idaho. *Forest Products Journal* 50(7/8): 59–65.

GB50005-2003. Code for design of timber structures[S].

Hassan K.T.S. & Horáček P. & Tippner J. 2013. Evaluation of Stiffness and Strength of Scots Pine Wood Using Resonance Frequency and Ultrasonic Techniques. *BioResources* 8(2): 1634–1645.

Jiang Jinghui & Lü Jianxiong & Ren Haiqing et al. 2008. Assessment of different grade dimension lumber by dynamic modulus of elasticity. *Journal of Nanjing Forestry University(Natural Sciences Edition)* 32(2): 63–66.

Jiang Jinghui & Lü Jianxiong & Ren Haiqing et al. 2008. Evaluation of modulus of elasticity for dimension lumber by three nondestructive techniques. *Journal of Zhejiang Forestry College*, 25(3): 277–281.

Lou Wanli. 2013. Grading of Structural Larch Dimension Lumber. *Journal of Building Materials* 16(4): 734–738.

Ni Jun & Yang Chunmei. 2011. Capability and Application of Lightweight Wooden Structure. *Jiangsu Construction* 5: 95–99.

Shen Shijie & Yang Yang & Chen Yong et al. 2011. Evaluation of Mechanical Properties of Structural Larch Sawn Lumber by Machine Stress Rating. *Science & Technology Review* 29(6): 54–56.

Sun Yan-liang & Zhang Hou-jiang & Zhu Lei et al. 2011. Non destructive Inspection Methods for Elastic Modulus of Wood and Relevant Research Status. *Forestry Machinery & Woodworking Equipment* 39(7): 9–11.

Viguier J. & Jehl A. & Collet R. et al. 2014. Improving strength grading of timber by grain angle measurement and mechanical modeling. *Wood Material Science & Engineering* (ahead-of-print): 1–12.

Zhang Wei & Wang Jian-gong & Zhou Hai-bin et al. 2013. Development and Application of FD1146 Timber Stress Grading Machine. *China Forest Products Industry* 6:41–43.

Zhou Haibin & Ren Haiqing & Lü Jianxiong et al. 2009. Effects of Grading Methods on Dimension Lumber Grades for Chinese Fir Plantation. *Journal of Building Materials* 12(3): 296–301.

Advances in Energy, Environment and Materials Science – Wang & Zhao (Eds)
© 2016 Taylor & Francis Group, London, ISBN 978-1-138-02931-6

The mechanism of residents' participation of architectural heritage conservation

Xiaoyu Ma
Xi'an Jiaotong University (XJTU), Xi'an, Shaanxi Province, China

ABSTRACT: The protection and use of architectural heritage are closely related with the lives of ordinary citizens. As important stakeholder, they have the right and obligation to participate. However, China's current architectural heritage protection and utilization mode makes ordinary citizens have very few opportunities to participate. We intend to find a complementary combination of the "top-down" and "bottom-up" protecting approach, to build a new, more effective, and multi-participating mechanism of action to ensure the participation of the residents.

1 INTRODUCTION

In October 2005, "Xi'an Declaration" was adopted at the ancient city of Xi'an. The Declaration stresses the importance of the environment for the protection of architectural heritage. Architectural heritage itself is of course important, but the surrounding environment also cannot be ignored. Architectural heritage surrounding environment includes material and non-material environment. When we make protection plan, we should take real surrounding environment into consideration. This environment includes the structures, dimensions, shape of the heritage, the surrounding buildings scenery pattern, spatial pattern, etc. It is the transition media of the architectural heritage toward surrounding urban space environment. It is exactly because of such space buffer, architectural heritage is able to receive adequate protection, and to avoid direct contact with the city and the disruption and destruction. In addition, interaction between people and the environment can bring architectural heritage new vigor to maximize its effect.

The protection and use of architectural heritage, an inheritance of culture, is not only the responsibility of the government, but is also closely related with the daily life of citizens, especially the life of residents who live near the architectural heritage. The general public, as stakeholders of the protection and utilization of architectural heritage, has the right and obligation to participate. Meanwhile, as compared with the government stakeholders are more vulnerable, public opinion is most likely to be ignored, and therefore, their right to participate needs to be protected most.

2 SITUATION OF ARCHITECTURAL HERITAGE CONSERVATION PLAN

With the completion of the legal system, democratic awareness of the public growing, more and more people are concerned about the protection of architectural heritage, and hope to be able to participate. And the final beneficiaries of the planning results are ordinary citizens, especially those living near the heritage. So the public needs and views on the surrounding architectural heritage has become an important factor in planning.

However, the current protection and use pattern of architectural heritage is basically a "top-down" approach to planning. More specifically, this means that mainly under the auspices of the relevant experts and government-led investment funds, relying on technical support from professional planners, a protection mechanism is formed.

In this mechanism during operation, the participation mechanism is incomplete, information is not open, participatory process is imperfect, and the government has monopoly of factors such as the decision is right. As a result, the general public, especially those living in the vicinity of the architectural heritage, is difficult to really get involved therein. In a word, the current situation shows that China lacks an effective mechanism to ensure the participation of the public.

We intend to build a new mechanism of action on the basis of the current heritage protection mechanism to ensure the participation of ordinary citizens in the architectural heritage conservation and utilization planning process. Finally, we want to find a complementary combination of the "top-down" and "bottom-up" protecting approach, to build a new, more effective, and multi-participating mechanism.

3 ACTION MECHANISM TO GUARANTEE THE PARTICIPATION OF RESIDENTS

On the basis of legislation, a complete architectural heritage conservation and utilization planning procedures can ensure maximum public participation. If there is no established procedure, because of the different situations of different project, significant randomness will exist in the planning. Hence we should adopt a set of established procedures to minimize the impact of a variety of man-made or accidental factors, and therefore, to guarantee enough participation of citizens.

According to a large number of references and case studies, we believe that such a new mechanism should include two elements of human and systems. Protection of architectural heritage needs cooperation of many different parties, such as government and residents. Because the interest of ordinary residents is more vulnerable, to guarantee their participation, stakeholders must be studied as build elements. Apart from man-made factors, different measure to promote the participation of residents has little relevance to each other, and each measure cannot have significant effect if operated alone. So we include such measures as sub-mechanisms into the new mechanism.

3.1 Construction element 1: Stakeholders

Here, stakeholders refer to the people finally affected by the protection and planning process. According to decision theory, if different stakeholders are allowed to participate in the decision-making process and to express their needs, it will have positive impact on the decision results and reduce the potential negative factor. The general public is the most direct stakeholders, and therefore, has the right and obligation to participate. In this process, transparency, openness, and involvement of participation, need planning department, oversight bodies and GNO to guarantee cooperatively.

In the planning process, the main stakeholders are: government departments (construction administrative departments, Cultural Relics Bureau, the Environmental Protection Agency, etc.), planners, experts, NGO and the general public. Where the general public can be divided into residents living near the architectural heritage and the general public. Evidently, residents living near heritage are impacted most significantly by the protection and use of architectural heritage. However, other residents are much less influenced.

These stakeholders in accordance with the attitude can be divided into: Decision Type, such as some government department; Support Type, which have low possibility of objection, such as

planners and designers; Neutral Type, such as experts and scholars; Against Type, which have larger possibility of objections, NGO community is the typical example; Stochastic Type, the possibility is not fixed, such as ordinary residents.

The Table 1 shows the relationship between different stakeholders.

If the final completion of the program is a product, then the final user is general public, so planners and designers should serve the public first.

3.2 Construction element 2: Sub-mechanism

After extensive literature review and case studies, we summarize a variety of policies and measures to protect and promote participation of the residents into four mechanisms, which are the four fundamental sub-mechanism of the final mechanism. The four sub-mechanisms are education system, legal system, supervision mechanism and NGO mechanisms, which cover the whole process of residents' participation in protection of architectural heritage, including the early stage, the feedback and the final stage.

Table 1. Nature of stakeholders.

Stakeholders	Requirement	Character
Government	Lead conservation planning, maximize the social, economic and cultural benefits	Decision
Planners	Rational planning	Support
Experts	Provide professional support	Neutral
NGO	Convey the views of the general public	Oppose
Citizens	Improve their living environment	Stochastic

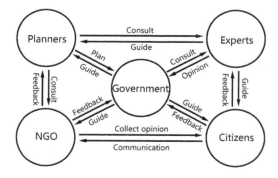

Figure 1. The relationship between stakeholders.

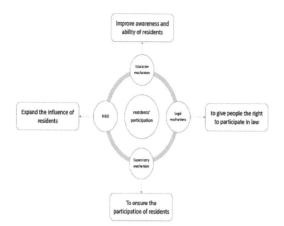

Figure 2.　The relationship between stakeholders.

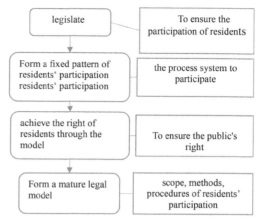

Figure 3.　Operation process of legal mechanisms.

The role of these four mechanisms in ensuring the participation of the residents is illustrated shown in Figure 2.

3.2.1　*Legal mechanisms*

Sound legal mechanism can guarantee public participation in the planning process. First, this can be done by legislation. Second, an established and clear pattern for residents to participate will help. The law must ensure the public's right to know, to express. Ultimately, the law must state clearly citizen's rights, extent, methods, and procedures to participate. Before the plan is designed, public opinion should be adopted. When the plan is formulated, it must be publicly displayed. If it turns out that the public is not satisfactory enough, then the plan needs to be adjusted until the public satisfy the plan.

3.2.2　*Supervisory mechanism*

In order to prevent the decision-maker's monopoly of planning authority, independent oversight bodies consisting of NGO, ordinary residents and some officers of government should be set up. This oversight body is not under the control of government, and is able to keep balance between the government and the general public. This approach refers to the British plan monitoring system, and is characterized by a certain degree of improvement.

Supervisory mechanism has always been throughout the entire process to ensure that every aspect of the operation is transparent and open.

3.2.3　*Education mechanism*

The goal of education mechanism is to enhance the citizens' awareness and ability to participate in the protection of architectural heritage. The expertise of the general public is relatively scarce. So when

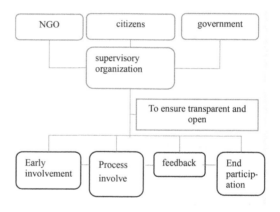

Figure 4.　Operation process of supervisory mechanism.

they participate in the process of planning, their idea may be contrary to the scientific one, and cannot be clearly and professionally expressed. Or they cannot understand the plan displayed publicly well.

The solutions to these problems depend on the improvement of residents' relevant ability. But this is a slow process. Therefore, the government needs to provide opportunities for residents to learn relevant professional knowledge and encourage more residents to participate in the process. Specifically, the government could utilize the internet and traditional media to advertise and give lectures to residents.

3.2.4　*NGO mechanism*

NGO called the Non-Governmental Organizations. NGO is a social force that can connect to different

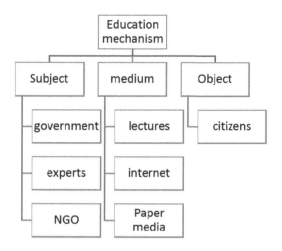

Figure 5. Elements of supervisory mechanism.

Figure 6. Operation process of NGO.

academic areas, interest groups, and government departments. Through NGO, the public interest may be able to be expressed and reflected better. At the same time, NGO and government can cooperate and promote each other. The current development of China's civil society organizations is not sufficient because the government has not delegated enough authority. With reasonable NGO management and operation system, NGO and government can promote each other so that we can guarantee the participation of the residents better.

During planning, there is information asymmetry between the government planners and the general public. Meanwhile, they have different cognition and demand for the planning, which gives rise to a lot of divergence and deviation. However, the NGO is independent of the Government and it usually attracts more ordinary people and convey their voices. As a result, NGO has the advantage to collect various opinions and integrate them. In this way, it has become a bridge between citizens and government to communicate. Such communication enables ordinary citizens more positively in the participation of the protection of architectural heritage.

After a long period of development, the NGO in Western countries have much larger size, influence on government and more financial resources. China still lacks NGO organization with extensive social influence. So, more attention to this aspect should be paid.

3.3 Integrated action mechanism

If the legislative mechanisms, feedback mechanisms, supervisory mechanisms, education system, and NGO are integrated, they can play their respective maximum effectiveness. We reconstruct these four sub-mechanisms and form a new mechanism of action, in order to balance the various stakeholders better. At the same time, let four sub-mechanisms play a role in every stage of the participation of residents in order to guarantee their participation. The mode of this action can be illustrated by the Figure 7 Operation process of Integrated action mechanism (see the appendix).

REFERENCES

He Fang, Tang Long. Public participation strategy in the transformation of the old district [J]. China real estate, 2008.
Wang Hui. Absent Participation [D]. Lanzhou University, 2006.
Yin Hongli. Public participation in the transformation of the old urban areas [D]. Fudan University, 2010.

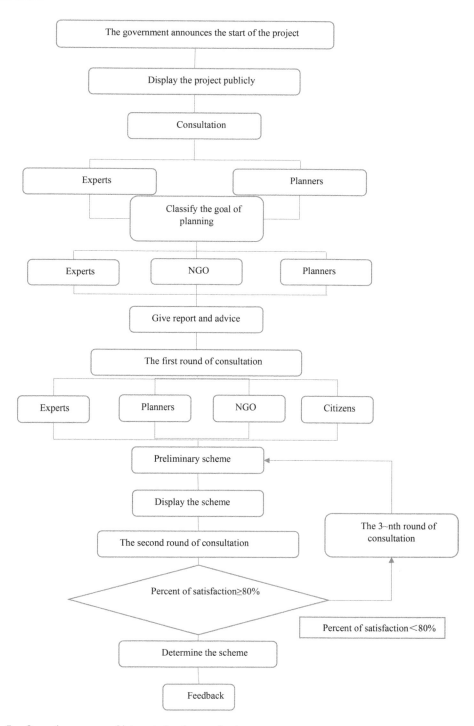

Figure 7.   Operation process of integrated action mechanism.

Advances in Energy, Environment and Materials Science – Wang & Zhao (Eds)
© 2016 Taylor & Francis Group, London, ISBN 978-1-138-02931-6

Analysis of displacements and temperature increments of the half space subjected to a flat cylindrical heat source

J.C.-C. Lu & W.-C. Lin
Department of Civil Engineering, Chung Hua University, Hsinchu, Taiwan, R.O.C.

F.-T. Lin
Department of Naval Architecture and Ocean Engineering, National Kaohsiung Marine University, Kaohsiung, Taiwan, R.O.C.

ABSTRACT: This paper presents the closed-form solutions of displacements and temperature increments of a half space subjected to a flat cylindrical heat source. The strata are modeled as a homogeneous isotropic linear elastic half space. The governing equations of the mathematical model are based on the theory of thermoelasticity. In the formulation of mathematical model, the thermal stresses of the half space obey Newton's second law and Hooke's law. Besides, the energy conservation and heat conduction law are introduced to formulate the basic governing equations of thermal flow. The software Mathematica is used for symbolic computations to present the analytic solutions. The solutions can be used to test numerical models and the detailed numerical simulations of the thermal-elastic processes near the heat sources.

1 INTRODUCTION

The heat source such as a canister of radioactive waste buried in the strata lead to thermomechanical responses. Radioactive wastes are usually deposited deep to isolate them from the living environment of human beings. A linear model was adequate for a repository design based on technical conservatism.

Lu and Lin (2006) displayed the transient ground surface displacement produced by a point heat source/sink through analog quantities between poroelasticity and thermoelasticity. Ai and Wang (2015) employed Laplace-Hankel transform to derive the axisymmetric thermal consolidation of multilayered thermoelastic porous media due to a heat source. Laplace and Fourier transform techniques were used to generate the distribution of thermal stresses and temperature for a half space with heat sources and body forces by El-Maghraby (2010). Lotfy (2011) aimed to investigate the transient disturbances created by an internal line heat source that suddenly starts moving uniformly inside a visco-elastic half space. Rosati and Marmo (2014) presented the closed-form expressions of the displacements, stress and temperature fields induced by a uniform heat source acting over an isotropic half space. Sarkar and Lahiri (2012) displayed a three-dimensional problem for a homogeneous, isotropic and thermoelastic half-space subjected to a time-dependent heat source without

energy dissipation. Beside, Lu, Lin and Lin (2010) solved the problem of the homogeneous isotropic elastic half space subjected to a circular plane heat source and presented the closed-form solutions.

The ground surface is predominantly modeled as a flat surface (Nowacki, 1986). The sedimentary soils or rocks usually have laminated structures. The original primary layers of strata are always laid out horizontally on mathematical formulation, and abundant groundwater can be found from the layers. The groundwater moving velocity is generally very slow, and thermal conduction generated by nuclear waste or geothermal can occur in the laminated strata. The flat cylindrical heat source is introduced in this paper on the modeling of thermal conduction through the layered strata.

In this paper, attention is focused on the closed-form solutions of the displacements and temperature increments for an isotropic stratum subject to a flat cylindrical heat source. The homogeneous isotropic soil mass is modeled as a linear elastic isothermal half space. The governing equations of the mathematical model are based on the theory of thermoelasticity (Nowacki, 1986). Based on the fundamental solutions of the half space due to a point heat source, the software Mathematica is used to derive the desired closed-form solutions in this paper. The solutions can be used to test numerical models and the detailed numerical simulations of the thermoelastic processes near the heat sources.

2 MATHEMATICAL MODELS

Figure 1 presents a point heat source of constant strength buried in an elastic half space at a depth h. The stratum is considered as a homogeneous isotropic medium with a vertical axis of symmetry. The fundamental equations of thermoelasticity consist of the displacement equations of motion and the equation of heat conduction. Considering a point heat of constant strength located at point $(0, h)$, the basic governing equations of the elastic stratum for linear axially symmetric deformation can be expressed in terms of displacements and temperature increments of the stratum in the cylindrical coordinates (r, z) as follows (Lu & Lin, 2006):

$$G\nabla^2 u_r + \frac{G}{1-2\nu}\frac{\partial \varepsilon}{\partial r} - G\frac{u_r}{r^2} - \beta\frac{\partial \vartheta}{\partial r} = 0, \quad (1a)$$

$$G\nabla^2 u_z + \frac{G}{1-2\nu}\frac{\partial \varepsilon}{\partial z} - \beta\frac{\partial \vartheta}{\partial z} = 0, \quad (1b)$$

$$\lambda_t \nabla^2 \vartheta + \frac{Q}{2\pi r}\delta(r)\delta(z-h) = 0, \quad (1c)$$

where $\nabla^2 = \partial^2/\partial r^2 + 1/r\,\partial/\partial r + \partial^2/\partial z^2$ is defined as the Laplacian operator. The displacements u_r and u_z are in the radial and axial directions, respectively. The symbol ϑ is the temperature increment measured from the reference state, $\varepsilon = \partial u_r/\partial r + u_r/r + \partial u_z/\partial z$ is the volume strain of the stratum, and the thermal expansion factor $\beta = 2G(1+\nu)\alpha_s/(1-2\nu)$. The constants ν, G and α_s are Poisson's ratio, shear modulus, and linear thermal expansion coefficient of the skeletal materials of the stratum, respectively. The constant λ_t defines the thermal conductivity of the thermal-elastic medium, and the symbol $\delta(x)$ is the Dirac delta function.

We assume that the half space surface at $z = 0$ is a traction-free and isothermal boundary for all times. The rational basis of the isothermal assumption on the flat surface is that the nuclear waste or

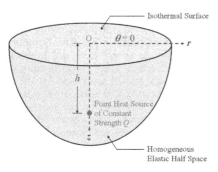

Figure 1. Point heat source buried in a thermoelastic half space.

geothermal is deposited or generated at a great depth on the mathematical modeling. Besides, the ground surface is in contact with atmosphere, and it is kept at isothermal condition at each specific time. Hence

$$\sigma_{rz}(r,0) = 0,\ \sigma_{zz}(r,0) = 0,\ \text{and}\ \vartheta(r,0) = 0. \quad (2a)$$

here σ_{ij} are the incremental stress components.

It is reasonable to assume that the point heat source has no effect at the boundary $z \to \infty$ for all times. Hence

$$\lim_{z\to\infty}\{u_r(r,z),u_z(r,z),\vartheta(r,z)\} = \{0,0,0\}. \quad (2b)$$

The fundamental solutions of displacements and temperature increments of the stratum can be derived from the differential equations (1a) to (1c) corresponding with the boundary conditions at $z = 0$ and $z \to \infty$.

3 FUNDAMENTAL SOLUTIONS

Using Hankel integral transformation with respect to the variable r, the fundamental solutions of the long-term horizontal displacement $u_r(r,z)$, vertical displacement $u_z(r,z)$, and temperature increment $\vartheta(r,z)$ of the half space due to a point heat source are obtained as follows:

$$u_r(r,z) = \frac{(1+\nu)\alpha_s Q}{8\pi(1-\nu)\lambda_t}\left[\frac{r}{\tilde{R}_1} - \frac{r}{\tilde{R}_2^*} + (3-4\nu)\frac{rh}{\tilde{R}_2\tilde{R}_2^*}\right.$$
$$\left. - \frac{rz}{\tilde{R}_2\tilde{R}_2^*} - \frac{2hrz}{\tilde{R}_2^3}\right], \quad (3a)$$

$$u_z(r,z) = \frac{(1+\nu)\alpha_s Q}{8\pi(1-\nu)\lambda_t}\left[\frac{z-h}{\tilde{R}_1} - (3-4\nu)\frac{h}{\tilde{R}_2}\right.$$
$$\left. - \frac{z}{\tilde{R}_2} - \frac{2hz(z+h)}{\tilde{R}_2^3}\right], \quad (3b)$$

$$\vartheta(r,z) = \frac{Q}{4\pi\lambda_t}\left(\frac{1}{\tilde{R}_1} - \frac{1}{\tilde{R}_2}\right), \quad (3c)$$

where $\tilde{R}_1 = \sqrt{r^2 + (z-h)^2}$, $\tilde{R}_2 = \sqrt{r^2 + (z+h)^2}$ and $\tilde{R}_2^* = \sqrt{r^2 + (z+h)^2} + z + h$. The equations (3a) to (3c) are the fundamental solutions of the thermoelastic half space due to a point heat source.

4 CLOSED-FORM SOLUTIONS

The closed-form solutions of the horizontal displacement $u_r(r,z)$, vertical displacement $u_z(r,z)$ and temperature increment $\vartheta(r,z)$ due to a flat cylindrical heat source with radius b and thickness t buried at a depth h, as shown in Figure 2, can be derived from equations (3a) to (3c). Considering a unit volume dV located at a distance s from the center of flat cylindrical heat source. The heat strength of this unit area is qdV, and it can be approximated as a point heat source. The increment of displacements u_r, u_z, and temperature increment ϑ due to the elementary flat cylindrical heat source can be obtained by substituting $r - s$ for r, $z + u$ for z, and $qsd\theta dsdu$ for Q in equations (3a) to (3c). Thus, the total increment of displacements and temperature rise of the stratum can be determined by the integration with radial limits of $s = 0$ to $s = b$, vertical limits of $u = h$ to $u = h + t$, and circumferential limits of $\theta = 0$ to $\theta = 2\pi$. Using Mathematica to complete the symbolic calculations, the closed-form solutions are given as below:

$$u_r = \frac{(1+v)\alpha_s q}{8\pi(1-v)\lambda_t}[J_1 - J_2 + (3+4v)J_3 - J_4 - J_5], \quad (4a)$$

$$u_z = \frac{(1+v)\alpha_s q}{8\pi(1-v)\lambda_t}[J_6 - (3+4v)J_7 - J_8 - J_9], \quad (4b)$$

$$\vartheta = \frac{q}{4\pi\lambda_t}(J_{10} - J_{11}), \quad (4c)$$

where

$$J_1 = \frac{\pi}{3}\Big[\big(2r(z-t) - b(z+t)\big)R_{r-b,z+t}$$
$$+ 2r(z+t)R_{r,z+t} + (2r+b)zR_{r-b,z} - 2rzR_{r,z}$$
$$+ \big(z^3 + 3tz^2 + 3t^2z - t^3\big)\ln\big(r - b + R_{r-b,z+t}\big)$$
$$+ (z+t)^3 \ln\big(r + R_{r,z+t}\big) + r^3\ln\big(z + t + R_{r,z+t}\big)$$
$$+ (r - 2b)(r+b)^2\ln\big(z + t + R_{r-b,z+t}\big)$$
$$+ z^3\ln\big(r - b + R_{r-b,z}\big)$$
$$+ (r-b)^2(r+2b)\ln\big(z + R_{r-b,z}\big)$$
$$- z^3\ln\big(r + R_{r,z}\big) - r^3\ln\big(z + R_{r,z}\big)\Big], \quad (5a)$$

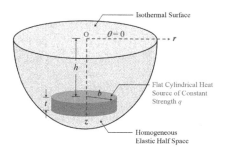

Figure 2. Flat cylindrical heat source in a thermoelastic half space.

$$J_2 = \frac{\pi}{6}\Big[6bt(2z + 4h + t) + (5r + 4b)(z + 2h)R_{r-b,z+2h}$$
$$+ 5r(z - 2h)R_{r,z+2h} + 5r(z + 2h + t)R_{r,z+t+2h}$$
$$+ \big(4b(z - 2h + t) + 5r(z + 2h + t)\big)R_{r-b,z+t+2h}$$
$$+ 6rt(2z + 4h + t)\ln\frac{r-b}{r}$$
$$+ 2\big(z^3 + 6hz^2 + 12h^2z - 8h^3\big)\ln\big(r - b + R_{r-b,z+2h}\big)$$
$$+ (r-b)^2(r+2b)\ln\big(z + 2h + R_{r-b,z+2h}\big)$$
$$+ 6r\big(z^2 + 4hz - 4h^2\big)\ln\frac{2(z + 2h + R_{r-b,z+2h})}{r(r-b)(z+2h)^2}$$
$$+ 2(z + 2h)^3\ln\big(r + R_{r,z+2h}\big) - r^3\ln\big(z + 2h + R_{r,z+2h}\big)$$
$$+ 6r(z + 2h)^2\ln\frac{2(z + 2h + R_{r,z+2h})}{r^2(z+2h)^2}$$
$$+ 2(z + 2h + t)^3\ln\big(r - b + R_{r-b,z+2h+t}\big)$$
$$+ (r - 2b)(r+b)^2\ln\big(z + 2h + t + R_{r-b,z+2h+t}\big)$$
$$+ 6r(z + 2h + t)^2\ln\frac{2(z + t + 2h + R_{r-b,z+2h+t})}{r(r-b)(z+2h+t)^2}$$
$$+ 2\big((z+t)^3 + 6h(z+t)^2 + 12h^2(z+t) - 8h^3\big)$$
$$\cdot\ln\big(r + R_{r,z+2h+t}\big) + r^3\ln\big(z + 2h + t + R_{r,z+2h+t}\big)$$
$$+ 6r\big((z+t)^2 + 4h(z+t) - 4h^2\big)$$
$$\cdot\ln\frac{2(z + 2h + t + R_{r,z+2h+t})}{r^2(z+2h+t)^2}\Big], \quad (5b)$$

$$J_3 = \pi h\Big[-2bt + (r - b)R_{r-b,z+2h} + rR_{r,z+2h}$$
$$+ (r + b)R_{r-b,z+2h+t} - rR_{r,z+2h+t} + 2rt\ln\frac{r}{r-b}$$
$$+ (z + 2h)^2\ln\big(r - b + R_{r-b,z+2h}\big)$$
$$+ 2r(z + 2h)\ln\left(-\frac{2(z + 2h + R_{r-b,z+2h})}{r(r-b)(z+2h)}\right)$$
$$+ \big(z^2 + 4hz - 4h^2\big)\ln\big(r + R_{r,z+2h}\big)$$
$$+ 2r(z - 2h)\ln\left(-\frac{2(z + 2h + R_{r,z+2h})}{r^2(z+2h)}\right)$$
$$+ \big((z+t)^2 + 4h(z+t) - 4h^2\big)\ln\big(r - b + R_{r-b,z+2h+t}\big)$$
$$+ 2r(z - 2h + t)\ln\left(-\frac{2(z + 2h + t + R_{r-b,z+2h+t})}{r(r-b)(z+2h+t)}\right)$$
$$+ (z + 2h + t)^2\ln\big(r + R_{r,z+2h+t}\big)$$
$$+ 2r(z + 2h + t)\ln\left(-\frac{2(z + 2h + t + R_{r,z+2h+t})}{r^2(z+2h+t)}\right)\Big], \quad (5c)$$

393

$$J_4 = \frac{\pi}{6}\Big[-6bt(2z+2h+t)-r^3\ln\left(z+2h+R_{r,z+2h}\right)$$
$$+(2bh+4hr-2bz-rz)R_{r-b,z+2h}-r(z+4h)R_{r,z+2h}$$
$$-\left(r(z-4h+t)+2b(z+h+t)\right)R_{r-b,z+2h+t}$$
$$+r(-z+4h-t)R_{r,z+2h+t}+6rt(2z+2h+t)\ln\frac{r}{r-b}$$
$$+2(z+2h)^2(2z+h)\ln\left(r-b+R_{r-b,z+2h}\right)$$
$$+6rz(z+2h)\ln\left(-\frac{2(z+2h+R_{r-b,z+2h})}{r(r-b)(z+2h)}\right)$$
$$+(r-b)^2(r+2b)\ln\left(z+2h+R_{r-b,z+2h}\right)$$
$$+2\left(2z^3+9hz^2+12h^2z-4h^3\right)\ln\left(r+R_{r,z+2h}\right)$$
$$+6rz(z-2h)\ln\left(-\frac{2(z+2h+R_{r,z+2h})}{r^2(z+2h)}\right)$$
$$+2\left(2(z+t)^3+9h(z+t)^2+12h^2(z+t)-4h^3\right)$$
$$\cdot\ln\left(r-b+R_{r-b,z+t+2h}\right)+(r-2b)(r+b)^2$$
$$\ln\left(z+2h+t+R_{r-b,z+2h+t}\right)+6r\left((z+t)^2+2h(z-t)\right)$$
$$\cdot\ln\left(-\frac{2(z+2h+t+R_{r-b,z+2h+t})}{r(r-b)(z+2h+t)}\right)$$
$$+2(z+2h+t)^2(2(z+t)+h)\ln\left(r+R_{r,z+2h+t}\right)$$
$$+r^3\ln\left(z+2h+t+R_{r,z+2h+t}\right)+6r(z+t)(z+2h+t)$$
$$\cdot\ln\left(-\frac{2(z+2h+t+R_{r,z+2h+t})}{r^2(z+2h+t)}\right)\Big],\tag{5d}$$

$$J_5 = 2h\pi\Big[-(r+b)R_{r-b,z+2h}+rR_{r,z+2h}-rR_{r,z+2h+t}$$
$$+(b+r)R_{r-b,z+2h+t}+h(2z-h)\ln\left(r-b+R_{r-b,z+2h}\right)$$
$$-\left(z^2+h^2\right)\ln\frac{4(r-b-R_{r-b,z+2h})}{(r-b)(z^2-h^2)(z+2h)}$$
$$+2hr\ln\left(z+2h+R_{r-b,z+2h}\right)+h(2z+h)\ln\left(r+R_{r,z+2h}\right)$$
$$-\left(z^2+h^2\right)\ln\frac{4(r+R_{r,z+2h})}{r(z^2-h^2)(z+2h)}-2hr\ln\left(z+2h+R_{r,z+2h}\right)$$
$$-\left(z^2-h^2\right)\ln\frac{4(r-b-R_{r-b,z+2h+t})}{(r-b)(z^2-h^2)(z+2h+t)}$$
$$+(h+t)(2z+h+t)\ln\left(r-b+R_{r-b,z+2h+t}\right)$$
$$-2hr\ln\left(z+2h+t+R_{r-b,z+2h+t}\right)$$
$$+\left(2h(t+z)+t(t+2z)-h^2\right)\ln\left(r+R_{r,z+2h+t}\right)$$
$$-\left(z^2-h^2\right)\ln\frac{4(r+R_{r,z+2h+t})}{r(z^2-h^2)(z+2h+t)}$$
$$+2hr\ln\left(z+2h+t+R_{r,z+2h+t}\right)\Big],\tag{5e}$$

$$J_6 = \frac{\pi}{3}\Big[\left(2b^2-br-r^2+2(t+z)^2\right)R_{r-b,z+t}$$
$$+\left(r^2-2(t+z)^2\right)R_{r,z+t}-\left(r^2+2z^2\right)R_{r,z}$$
$$+3r\left(z^2+2tz-t^2\right)\ln\left(r-b+R_{r-b,z+t}\right)$$
$$+3r(z+t)^2\ln\left(r+R_{r,z+t}\right)+3rz^2\ln\left(r-b+R_{r-b,z}\right)$$
$$+\left(2z^2-2b^2-br-r^2\right)R_{r-b,z}-3rz^2\ln\left(r+R_{r,z}\right)\Big],\tag{5f}$$

$$J_7 = h\pi\Big[-(z+2h)R_{r-b,z+2h}+(z+2h)R_{r,z+2h}$$
$$+(z+2h+t)R_{r-b,z+2h+t}+(z-2h+t)R_{r,z+2h+t}$$
$$+2r(z+2h)\ln\left(r-b+R_{r-b,z+2h}\right)$$
$$-\left(r^2+b^2\right)\ln\left(z+2h+R_{r-b,z+2h}\right)$$
$$+2r(z-2h)\ln\left(r+R_{r,z+2h}\right)-r^2\ln\left(z+2h+R_{r,z+2h}\right)$$
$$+2r(z-2h+t)\ln\left(r-b+R_{r-b,z+2h+t}\right)$$
$$-\left(r^2-b^2\right)\ln\left(z+2h+t+R_{r-b,z+2h+t}\right)$$
$$+2r(z+2h+t)\ln\left(r+R_{r,z+2h+t}\right)$$
$$+r^2\ln\left(z+2h+t+R_{r,z+2h+t}\right)\Big],\tag{5g}$$

$$J_8 = \frac{\pi}{3}\Big[\left(-2b^2-2h^2+br+r^2-5hz-2z^2\right)R_{r-b,z+2h}$$
$$+\left(2h^2-r^2+5hz+2z^2\right)R_{r,z+2h}$$
$$+\left(2b^2+2h^2-br-r^2+2(t+z)^2+5h(t+z)\right)R_{r-b,z+2h+t}$$
$$+\left(2(z+t)^2+5h(z+t)-r^2-2h^2\right)R_{r,z+2h+t}$$
$$+3rz(z+2h)\ln\left(r-b+R_{r-b,z+2h}\right)$$
$$-3h\left(r^2-b^2\right)\ln\left(z+2h+R_{r-b,z+2h}\right)$$
$$+3rz(z-2h)\ln\left(r+R_{r,z+2h}\right)+3hr^2\ln\left(z+2h+R_{r,z+2h}\right)$$
$$+3r\left((z+t)^2+2h(z-t)\right)\ln\left(r-b+R_{r-b,z+2h+t}\right)$$
$$-3h\left(r^2+b^2\right)\ln\left(z+2h+t+R_{r-b,z+2h+t}\right)$$
$$+3r(z+t)(z+2h+t)\ln\left(r+R_{r,z+2h+t}\right)$$
$$-3hr^2\ln\left(z+2h+t+R_{r,z+2h+t}\right)\Big],\tag{5h}$$

$$J_9 = 2h\pi\Big[zR_{r-b,z+2h}-zR_{r,z+2h}+(z-t)R_{r-b,z+2h+t}$$
$$+(z+t)R_{r,z+2h+t}-2hr\ln\frac{2(-r+b+R_{r-b,z+2h})}{hr(r-b)(z+2h)}$$
$$-\left(r^2+b^2\right)\ln\left(z+2h+R_{r-b,z+2h}\right)-2hr\ln\left(-\frac{2(r+R_{r,z+2h})}{hr^2(z+2h)}\right)$$
$$-r^2\ln\left(z+2h+R_{r,z+2h}\right)+2hr\ln\left(-\frac{2(r-b-R_{r-b,z+2h+t})}{hr(r-b)(z+2h+t)}\right)$$
$$-\left(r^2-b^2\right)\ln\left(z+2h+t+R_{r-b,z+2h+t}\right)$$
$$+2hr\ln\left(-\frac{2(r+R_{r,z+2h+t})}{hr^2(z+2h+t)}\right)+r^2\ln\left(z+2h+t+R_{r,z+2h+t}\right)\Big],\tag{5i}$$

$$J_{10} = \pi\Big[(z+t)R_{r-b,z+t} + (z-t)R_{r,z+t} - zR_{r-b,z}$$
$$+zR_{r,z} + 2r(z-t)\ln(r-b+R_{r-b,z+t})$$
$$+2r(z+t)\ln(r+R_{r,z+t})$$
$$-(r^2-b^2)\ln(z+t+R_{r-b,z+t})$$
$$+r^2\ln(z+t+R_{r,z+t}) + 2rz\ln(r-b+R_{r-b,z})$$
$$-(r^2+b^2)\ln(z+R_{r-b,z})$$
$$-2rz\ln(r+R_{r,z}) - r^2\ln(z+R_{r,z})\Big], \qquad (5j)$$

$$J_{11} = \pi\Big[-(z+2h)R_{r-b,z+2h} + (z+2h)R_{r,z+2h}$$
$$+(z+2h+t)R_{r-b,z+2h+t} - (2h-t-z)R_{r,z+t+2h}$$
$$+2r(z+2h)\ln(r-b+R_{r-b,z+2h})$$
$$-(r^2+b^2)\ln(z+2h+R_{r-b,z+2h})$$
$$+2r(z-2h)\ln(r+R_{r,z+2h})$$
$$-r^2\ln(z+2h+R_{r,z+2h})$$
$$+2r(z-2h+t)\ln(r-b+R_{r-b,z+2h+t})$$
$$-(r^2-b^2)\ln(z+2h+t+R_{r-b,z+2h+t})$$
$$+2r(z+2h+t)\ln(r+R_{r,z+2h+t})$$
$$+r^2\ln(z+2h+t+R_{r,z+2h+t})\Big],$$
$$\qquad (5k)$$

and

$$R_{i,j} = \sqrt{i^2 + j^2}. \qquad (6)$$

5 CONCLUSIONS

Based on the half space fundamental solutions of themoelasticity shown in equations (3a) to (3c), the closed-form solutions of the homogeneous elastic half space for axially symmetric deformation of the horizontal displacement, vertical displacement and temperature increment subjected to a flat cylindrical heat source are presented by equations (4a) to (4c). The examination of the closed-form solutions (4a) to (4c) are completed by the numerical results, not presented in this manuscript. The limit of the flat cylindrical heat source induced closed-form solutions converge to point heat source induced fundamental solutions as the radius and thickness of the flat cylindrical heat source tend to approach zero.

ACKNOWLEDGEMENTS

This work is supported by the National Science Council of Republic of China through grants NSC102-2221-E-216-022 and NSC85-2211-E-216-001.

REFERENCES

Ai, Z.Y. & Wang, L.J. 2015. Axisymmetric thermal consolidation of multilayered porous thermoelastic media due to a heat source, *International Journal for Numerical and Analytical Methods in Geomechanics*: doi: 10.1002/nag.2381.

El-Maghraby, N.M. 2010. A generalized thermoelasticity problem for a half-space with heat sources and body forces, *International Journal of Thermophysics* 31(3): 648–662.

Lotfy, Kh. 2011. Transient thermo-elastic disturbances in a visco-elastic semi-space due to moving internal heat source, *International Journal of Structural Integrity* 2(3): 264–280.

Lu, J. C.-C. & Lin, F.-T. 2006. The transient ground surface displacements due to a point sink/heat source in an elastic half-space, *Geotechnical Special Publication No. 148*: 210–218.

Lu, J. C.-C., Lin, W.-C. & Lin, F.-T. 2010. Closed-form solutions of the homogeneous isotropic elastic half space subjected to a circular plane heat source. *Geotechnical Special Publication No. 204*: 79–86.

Nowacki, W. 1986. *Thermoelasticity*, Polish Scientific Publishers, Warszawa: Pergamon Press.

Rosati, L. & Marmo, F. 2014. Closed-form expressions of the thermo-mechanical fields induced by a uniform heat source acting over an isotropic half-space, *International Journal of Heat and Mass Transfer* 75: 272–283.

Sarkar, N. & Lahiri, A. 2012. A three-dimensional thermoelastic problem for a half-space without energy dissipation, *International Journal of Engineering Science* 51: 310–325.

Advances in Energy, Environment and Materials Science – Wang & Zhao (Eds)
© *2016 Taylor & Francis Group, London, ISBN 978-1-138-02931-6*

Research on high-rise residential district environment landscape problems and countermeasures

Y. Sun & G.L. Gao

College of Civil Engineering and Urban Construction, Jiujiang University, China

ABSTRACT: High-rise residential district has become the trend of urban resident development, and high-rise residential district has higher and higher requirements on environment landscape with rapid development of social economy in China and accelerated urbanization process. Environment landscape plays an important role in high-rise residential district. In the paper, environment landscape in high-rise residential district and countermeasures are studied. President situation and problems in China high-rise environment landscape are analyzed, thereby proposing countermeasures to enhance environment landscape of high-rise residential district. In the paper, it is believed that green area should be enhanced, spatial scale should be improved, traffic jam problems should be reasonably solved, construction of the neighborhood should be strengthened, thereby optimizing environment landscape of high-rise residential district, and improving resident living quality of residents in high-rise residential area.

1 INTRODUCTION

People have higher and higher requirements on residential environment with progress of science and continuous development of society. Environment landscape has become an important problem of modern residential district design. However, China's urbanization process is accelerated, and urban land tension has larger contradiction with high demand on environment landscape quality. Therefore, environment landscape of high-rise residential district has become an important problem that should be solved by real estate industry in China currently. Scientific and rational combination of habitable and ecological landscape with high-rise residential space is the important problem that should be considered at present.

2 ANALYSIS ON PRESENT CONDITION OF ENVIRONMENT LANDSCAPE IN HIGH-RISE RESIDENTIAL DISTRICT

2.1 Improvement of environment landscape consciousness

It is very important to improve quality of residential environment in residential district due to high volume rate in high-rise residential district. In high-rise residential district, ground environment landscape has become an important standard to measure quality of high-rise residential district.

Environment consciousness is gradually improved, and therefore developers pay more and more attention to design and development of landscape environment. Cost for landscape environment is gradually improved. The garden-type model house has become a selling point of high-rise residential district. Garden-type high-rise residential district has become a pronoun of upscale landscape, which obtains attention of purchasers.

2.2 Floor space expansion and green rate increase

High-rise residential district has higher demand for ventilation, sunshine, etc. Therefore, the spacing of high-rise residential district is larger that of ordinary multi-storey residential district. Corresponding field area and green area are also increased. Improvement of green land rate is an important indicator to improve residential quality in residential district. Actual condition of current high-rise residential district in China shows that floor spacing in China high-rise residential district is expanded, green land rate is also gradually improved, which has become main advertisement for many developers to propagandize own residential district. Land can be utilized more intensively and effectively through expanding floor space and improving green land rate, thereby avoiding previous environment and social problems due to high-rise high density residential districts. It is an important mode of sustainable development in modern society.

2.3 *Improvement of underground space utilization efficiency and ground landscape increase*

There are more and more vehicles in modern society. Raito of parking lots in high-rise residential district is increased, and parking lot problems can not be solved on the ground. Utilization ratio of underground space should be improved for increasing parking lots. In addition, high-rise residential district has higher land price usually, and land is relative tense, thereby ground resources must be saved for sorting and developing underground space. Land can become landscape garden. In China most high-rise residential district is mostly used as landscape space except those for fire-fighting channel.

3 PROBLEMS IN ENVIRONMENT LANDSCAPE OF THE HIGH-RISE RESIDENTIAL DISTRICT

3.1 *Smaller activity space and per capita green area*

Activity space and per capita green area are generally smaller in high-rise residential district. Positive side and lateral side spacing among all buildings in the high-rise residential district are larger due to sunshine, fire-fighting and reasons in other aspects. Some garden green land seems to have larger area, however high-rise residential district has larger population density, total population is high in the residential district, and therefore average per capita green space is small. Residents in high-rise residential district may ignore average per capita green space under general condition. There are more shrubs, pool and other landscape in environment landscape of high-rise residential district. The landscape has better landscape effect, which can not meet activity demand of residents in high-rise residential district with higher population density. Visual aesthetics of environment beauty in environment landscape in high-rise residential district has conflict with activity demand of residents sometimes. The developer should pay attention to choose. In addition, Shadow area in back and side parts is larger in high-rise residential district. Citizen activities are few in shadow area. Few people have activities in the north residential district especially in winter. Resident utilization rate is lower.

3.2 *Relatively depressed spatial scale*

Plot ratio refers to the ratio between total building area and construction land area.

Plot ratio calculation formula is shown as follows: plot ratio = total construction area/ total land area.

Plot ratio reflects architecture development strength in limited scope of land. Plot ratio is closely related to building type and environmental quality. Currently, China urban land resource is tense, population is gradually increased, thereby leading to shortage of place to live in the urban area. Residential building is gradually developed upwards, namely land plot ratio is increased for solving the problem of shortage for residential space. High-rise living district is characterized by larger population density, and residents can easily feel depressive feeling if architecture layout is not rational enough. Currently, most high-rise residential district space is in '#' shape. The central garden is surrounded by high buildings. People suffer from the sense of being observed during activity in central garden. In addition, there are more building shadow areas in the south of many high-rise residential districts in China, and therefore utilization rate of residents is lower. Some architecture designers arrange square pavement in environment landscape of high-rise residential area in order to obtain better plane effect. Some pavement patterns are more exquisite in medium and small space dimensions in multi-story residential district. However, if square pavement is set in spatial scale in high-rise residential district, affinity of residential space is lost, and it is difficult to provide people with more livable living environment.

Relevant research shows that the most ideal landscape environment land plot ratio of high-rise residential district is between 2.5 and 3.5. It is not necessary to achieve higher plot ratio. Plot ratio of high-rise residential district should be controlled within more rational scope. In addition, China issued 'Code for Urban Residential District Planning and Design'. It regulates that plot ratio of development and construction in residential district should be less than 3.5. However, plot ratio of high-rise residential district is mostly higher than 3.5 in China. Table 1 shows that the plot ratio of high-rise residential district is between 3.5 and 6.0. There is larger gap with ideal landscape environment. Excessive development leads to lower environment quality in high-rise residential district, thereby leading to larger space pressure, traffic jam and other problems.

3.3 *Complex traffic and insufficient neighborhood relation*

High-rise residential districts are mostly open. Cross relationship between buildings and roads is simpler. Since high-rise residential district has higher population density, higher requirements on fire control, function increase and other reasons, high-rise residential districts suffer from higher traffic pressure, and it is easy to produce traffic

Table 1. Relationship between construction category and volume rate.

Category type	Plot ratio
High-grade single-family villas	FAR ≤ 0.3
Ordinary single-family villas	0.3 < FAR ≤ 0.5
Double and townhouse	0.5 < FAR ≤ 0.8
Full multi-storey residential building or mixed lower villa	0.8 < FAR ≤ 1.2
Ordinary multi-storey residential building or mixed small high-rise residential building	1.2 < FAR ≤ 1.5
Multi-storey and small high-rise residential building	1.5 < FAR ≤ 2.0
Small high-rise residential building	2.0 < FAR ≤ 2.5
Small high-rise building and high-rise residential building (less than 18 floors)	2.5 < FAR ≤ 3.5
High-rise residential building (higher than 18 floors and building height less than 100 m);	3.5 < FAR ≤ 6.0
Residential and commercial complex or skyscrapers	FAR > 6.0

jam condition. Private cars are increasing rapidly due to improvement of living standards. Parking lots for most high-rise residential buildings are not sufficient in China. Many private cars are parked in landscape area, thereby leading to complex and confused traffic in high-rise residential district. In addition, public space of all floors is not clearly regulated for residents in high-rise residential district, thereby leading to unlimited activity scope area. They meet unfamiliar people during outward activity. In addition, modern residents spend less time in outdoor activities, and they are not much willing to take part in activities outdoors. There are few spontaneous group activities in high-rise residential district. Neighbors do not contact frequently at mean time, thereby leading to vaguer neighborhood in high-rise residential district, relatively indifferent neighborhood relationship and less communication.

4 COUNTERMEASURES TO STRENGTHEN LANDSCAPE ENVIRONMENT CONSTRUCTION IN HIGH-RISE RESIDENTIAL DISTRICT

4.1 *Increase of green area*

High-rise residential districts should adopt a variety of means to increase the green area of the high-rise residential district. Landscape environment should be arranged according to construction space during design of green system. Road greening, residential area greening, public space greening, etc. should be focused, thereby forming multilevel greening system composed of point, face and line. The developer should make full use of terrain for vertical greening development during transverse greening, such as balcony greening, roof greening, metope greening, etc. Three-dimensional greening system should be constructed, three-dimensional greening not only can increase the green area and

diffuse environment of high-rise residential district for improving environment landscape of high-rise residential district, but also can reduce radiation heat, glare, etc. of some roofs and son at low floor, heat at rainwater penetration isolation roof can be reduced, etc. In addition, green land of high-rise residential district should be closer and closer to residents, which should be set in place where residents frequently pass and can be reached naturally, and therefore residents can view them conveniently. The landscape should have suitable scale, and it should be livable. Open layout should be adopted, and therefore residents can approach to the green land and enjoy green land actually.

4.2 *Improvement of space scale*

High-rise residential district has generally higher plot ratio in China. Larger space pressure can be caused due to improved plot ratio. The space scale should be continuously improved due to depression caused by high-rise residential district. Gradation of high-rise residential district can be increased. It should be vertical, three-dimensional and abundant with diversified level, thereby increasing comfort of high-rise residential district. Spatial crowded feeling can be solved by underlying overhead mode in high-rise residential district. Complete overhead or local overhead can be selected for releasing space crowded sense according to actual condition in high-rise residential district. Underlying overhead mode of high-rise residential district can increase space of citizen public activities, thereby enriching space mode of high-rise residential area. People can achieve more relaxed and opener vision. Buildings in high-rise residential districts are higher and higher in China at present. Bottom overhead mode can be adopted for improving spatial scale of high-rise residential district, thereby providing more public space for residents in high-rise residential district, thereby constructing more landscape environment, and improving living environment of high-rise residential district effectively. In addition, environment of high-rise residential districts are mostly used by old people and children. Activity space problem should be regarded as focus for environment landscape design when high-rise residential district has tense land, and livable space environment landscape can be constructed.

4.3 Reasonable solution to heavy traffic problem

High-rise residential district has higher population density and higher demands on service facilities, especially demand of modern residents on parking lots. Therefore, underground space should be fully utilized. Parking lots should be constructed in underground space for meeting demand of residents in high-rise residential district on parking lots. Supporting auxiliary facilities can be arranged in underground space as far as possible, thereby saving ground space and constructing more landscape. In addition, roads at different levels belong to corresponding space levels, road at different levels, especially motor vehicle road, should be connected level by level as far as possible. Layout of road network, proportion of road land, width and section form of various roads can be planned rationally according to resident demand and resident behavior. Traffic flows should be rationally organized, and motor vehicle and pedestrians can be separated under the precondition of reducing mutual interference between motor vehicles and pedestrians.

4.4 Reinforcement of neighborhood construction

Neighbor contact should be based on the precondition of guiding neighbors to participate in outdoor activities, such as improvement of participation in chat, playing cards, fitness, chess and other activities. Neighborhood relationship construction in high-rise buildings should be strengthened through external environment construction firstly. External environment construction should be highly targeted. Activity characteristics in the space should be increased, thereby promoting neighborhood in high-rise residential district For example, pattern tiles with rich texture, curved shape as well as novel and lively form should be paved in walkways to buildings. The walkway is fit with green land construction. Rest, fitness facilities and landscape sketch are equipped besides the road, which can become better place for leisure, fitness and exchange in the living district. In outdoor site layout, neighborhood level reaching use environment of residents should be designed through activity path and space integration. For example, activity space suitable for the level should be set in corresponding group scope according to certain service radius. Certain public green land can be integrated. Ring roads can

be jointly used between two neighboring buildings according to visual distance. One land for children activity can be shared, and similar marks, colors, etc. can be adopted, thereby creating public space at hierarchical levels, arousing sense of identity and sense of field, and setting up neighborhood aggregation consciousness at corresponding scope.

5 CONCLUSION

China residents have higher and higher requirements on living environment landscape with economic development and living standard improvement in China. High-rise residential district has higher population density. Residential environment landscape construction should be reinforced in order to improve resident comfort and sense of happiness. In the paper, it is believed that comfort of high-rise residential district can be improved through analyzing living requirements of high-rise residential district, development requirements of resident environment landscape, and constantly optimizing environment landscape of high-rise residential district. The paper believes that green area should be increased in high-rise residential building aiming at landscape condition and problems in China high-rise residential district, space scale can be improved, traffic jam problems should be rationally solved, neighborhood construction should be enhanced for optimizing environment landscape in high-rise residential building, and improving habitability of high-rise residential district.

REFERENCES

Guo Hongbo. Analysis and application of ecological garden landscape design in modern residential district inhabitation environment design. Chinese and Foreign Construction. 2010 (04):78–80.

Han Lei. Development and evolution of residential district environment landscape design. Beauty and Time (in). 2012 (7):112–114.

Su Shaoquan. Residential district ecological landscape planning and design. Building Materials in Jiangxi Province. 2012 (3):64–65.

Wang Ruifang. On environmental landscape design in residential district. Chinese Horticulture Abstract. 2010 (02):43–45.

Wu Weifeng. Existing problems and countermeasures in garden landscape design. Anhui Agriculture Bulletin (semimonthly). 2012 (16):55–56.

Xiong Yijun. On landscape environment design of urban residential district. Chemical Industry in Jiangxi Province. 2009 (3):12–14.

Yu Rutao. On urban environment, high-rise residential building and natural ecological impact. Guide of Being Rich by Science and Technology. 2012 (05):31–33.

Zhang Guixiu. Study of residential district garden landscape design. Jilin Agriculture. 2012 (04):90–92.

Advances in Energy, Environment and Materials Science – Wang & Zhao (Eds)
© 2016 Taylor & Francis Group, London, ISBN 978-1-138-02931-6

Regulation mode analysis and selection of power plant boiler fan

Yufu Li, Jun Zhao, Shougen Hu & Hailiang Jia
University of Shanghai for Science and Technology, Shanghai, China

ABSTRACT: Nowadays the large power plants mostly operate by adopting the method of peak-shaving. Since boiler load changes with electricity consumption, to keep the variable load units or large units of powder feeding system running stably and efficiently, air volume and air pressure of a primary air fan has to be regulated timely and effectively. In this paper, fan regulating modes of high efficiency can be selected through testing the fan operation efficiency under baffle adjustment and variable speed control respectively in variable conditions, thus providing a basis for fan regulation strategies.

1 INTRODUCTION

As one of the main boiler auxiliary equipments in power plants, a fan consumes about 2%–3% of the total generating capacity of a plant. With energy issues emerging and the growing use of electricity, people pay more and more attention to the economy of primary air fan. To reduce electrical energy consumption of the fan, we need to design a high efficient fan type and adopt the best regulating mode. According to the characteristics of primary air fan operation and the situation of power plant operation, the current focus should be placed on research and adaptation of the best adjustment method. For one thing, currently the vast majority of fans used in power plants are already high efficient, with workpiece ratio above 80%–85%, thus it is hard to greatly improve the efficiency of a fan itself; for another thing, the design output of a fan is always 20%–30% higher than its practical output, or even greater, which leads the fan operational efficiency to be below its maximum, i.e., the fan. For variable load units, the fan needs to operate in a long time at a very low load, so even with a high efficient type, the actual workpiece ratio is still low. Operation surveys showed that: about 50% fans run with efficiency below 70% and 12% with efficiency below 50%. In order to reduce the power consumption of a primary air fan, we primarily need to improve its efficiency at a low load, the only way of which is to adjust the fan by adopting the best.

2 SELECTION OF THE BOILER MILLING SYSTEM, FAN CAPACITY AND MARGIN PRESSURE

The type of pulverizing system and coal mill should be determined after the comparison based on the scope of possible changes in coal types, coal quality, coal mill conditions and the load nature, with the burner structure and boiler furnace structure considered additionally. Basic standards are provided in DLT 5145-2002 "Technical Requirements for Design and Calculation of Coal Pulverizing System" and DL5000-2000 "Technical Regulations of Power Plant Design". For large capacity units with appropriate coal types, medium-speed coal mills are more preferred. And in this situation, if the air preheater can meet requirements, we should adopt a cooling fan milling system of direct blow pressure, and a single-speed centrifugal fan is appropriate for a trisector preheater. The basic air volume of a fan should be calculated according to the design coal type, including the primary air volume of a boiler required at maximum continuous evaporation, the air leakage amount of air preheater that manufacturers assure on the primary air side one year later, as well as the loss of coal mill seal air volume provided by the fan. As for the air volume margin, it should not be less than 35%, also a temperature margin should be added, which is determined by "outdoor design temperature for summer ventilation". Air volume is calculated according to formula (1) and (2) and appropriate pressure head margin of the fan should be 30%. But if the fan is in series operation with a blower, its pressure head margin can be increased to 35%.

$$Q_{cal} = \frac{B_B g_1 r_{ha} \times 10^3}{Z_{Fan} \cdot \rho_{la} \left(1 - \varphi_{le,AH}\right)} \tag{1}$$

$$p_0 = p \frac{\rho_0}{\rho} \tag{2}$$

In formula:

Q_{cal}—fan calculation ventilation, m³/h

B_B—boiler coal burning at rated load, t/h

g_1—the amount of desiccant obtained by thermodynamic calculation in the first test, kg/kg

r_{ha}—share of a hot air dryer in the first test

Z_{Fan}—number of hot primary air fans

ρ_{la}—density of cold air, kg/m³

$\varphi_{le,AH}$—leakage rate of primary air in air preheater, i.e., percentage of air leakage in primary air flow in preheater entrance

p_0—pressure head on the fan performance curve, Pa

p—design pressure, Pa

ρ_0—gas density corresponding to performance curves, kg/m³

ρ—gas density under design conditions, kg/m³.

3 CHOICES OF REGULATING MODES FOR PRIMARY AIR FANS

There are two main air volume control modes of primary air fans. One adjustment is to install inlet guide vanes at the entrance of the fan air duct. By varying the angle of inlet guide vanes, we can change the characteristic curve of a fan and generate a "pre-swirl air flow" at the entrance of fan impeller, thus achieving the purpose of adjusting air volume. The other mode is the variable speed adjustment of fan drives, which regulates air amount by changing speed, using the proportional relationship between fan flow and fan. The regulating mode of a primary fan should be chosen according to its flow variation. For a primary air fan of variable load unit, hydro-viscous speed clutches, fluid couplings and inverters all can be used to adjust the variable speed, but which device to adopt and its corresponding adjustment range must be determined after a detailed comparison in technic and economic evaluation. For instance, if we choose frequency converters, we still need to configure inlet guides for fans. Also, the capacity of converters must be determined according to the shaft power, which corresponds to 90% (or even lower) of fan flow at the fan's TB point (maximum operating point). In this way, not only can we save investments by adopting a smaller capacity inverter, but also we can achieve the highest regulating efficiency of a fan. However, when fan flow is variable above 90% of rated flow, we generally do not adopt frequency control modes but use inlet guide vane adjustments. Since the inverter itself has power loss, it will not produce a great energy saving effect with efficiency at about 90%. Thus, regulating efficiency of fan inlet guide vanes would be higher than that of frequency control mode under the condition that fan flow is above 90% of its rated.

4 THE EFFECT OF CONSTANT SPEED REGULATION AND VARIABLE FREQUENCY ADJUSTMENT ON THE FAN

In this part, we will look into a case of RJ29-DW2620F primary air fans in Limited 350 MW Unit of Guo Hua Ningdong Power Company (located in Ningxia, Gansu province, China). Take these centrifugal double-suctioned and double-feet-upholding as an example. The fans before transformation, were in low working efficiency with baffle adjustment. Their performance parameters are shown in Table 1.

And their actual operating parameters are shown in Table 2, measured by the fan performance test under economic operating conditions of the unit. Results show that under selection conditions, an average of flow allowance is about 38% and the head margin is about 40%. So, there exists a great loss because of a large selection flow allowance of the primary air fan.

In the boiler operation regulation, inlet baffle opening of a fan should be adjusted according to unit loads. Figure 1 shows the control curve of inlet baffle opening. In a high load (330 MW) condition, the opening degree of fan inlet baffle is about 90%. And the lower the unit load, the smaller the baffle opening. In a low load (200 MW) condition, a fan inlet baffle opening is less than 60%. In this case, the resistance losses are so great that the fan efficiency is seriously effected. But when we adopt variable frequency adjustment, fan speed changes with the unit load, as is shown in Figure 2.

Now we analyze the fan efficiency under baffle adjustment and frequency adjustment respectively in the same load conditions. When the unit load is low, fan efficiency under baffle adjustment gradually increases with the load. But when we adopt frequency regulations, fan efficiency is relatively

Table 1. Performance parameters of primary air fans.

Name	Calculation conditions	Selection conditions
Volume flow in fan inlet (m³·h⁻¹)	310475	372438
Mass flow in fan inlet (kg·h⁻¹)	310475	392040
Total pressure of the fan (kPa)	19.21	23.05
Shaft power of the fan (kW)	2036.5	2713.3
Total pressure efficiency of the fan (%)	85.2	85
Fan speed (r·min⁻¹)	1455	1455

Table 2. Actual operating parameters of fans.

Working conditions	Entrance volume flow (m³·s⁻¹)	Selection flow allowance %	Full indenter kPa	Selection pressure head margin %
Selecting condition	103.46	–	23.05	–
Counting condition	86.24	20	19.21	20
Operating condition	74.81	38.3	16.43	40.3

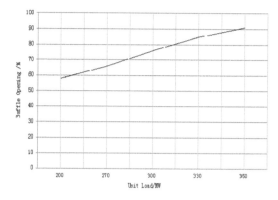

Figure 1. Opening curve of fan inlet baffle under constant speed adjustment.

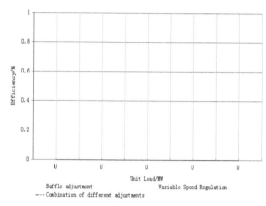

Figure 3. Efficiency curves under different adjustments.

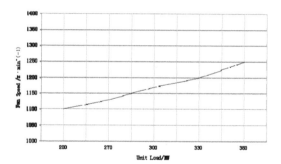

Figure 2. Variation curve of fan speed under variable frequency adjustment.

stable, remaining above 80% during variable load operation. However, if the boiler is running above 95% of the rated load, power loss of frequency converters will be greater than the throttling loss of baffle adjustment. In this circumstance, fan efficiency under baffle adjustment is higher than that under frequency regulation. When baffle adjustment, variable speed regulation and inlet guide vane adjustment are combined together, the fan can maintain a high efficiency no matter the load is high or low. As shown in Figure 3, regulating the fan with a combination of various adjustment modes can not only ensure it run efficiently in a low-load condition, but also avoid large power losses produced by variable speed device consumption with boiler high load running. Thus, it is of great significance to improve fan efficiency by combining a variety of regulations in power plants.

5 SUMMARY

Optimized running of centrifugal fans is an important direction of boiler energy conservation in large-scale power plants. In the above case with 350 MW unit primary air fans, after analysis and comparison of fan efficiency under baffle adjustment and frequency regulation, we now can come the following conclusions:

1. The fan margin should be selected according to actual operating conditions of power plants.
2. Under inlet baffle adjustment, great throttling losses are produced in low load conditions, leading to a low fan efficiency, which is greatly influenced by the baffle opening. But if the baffle opening is greater than 90%, fans can be high efficient.
3. Frequency adjustments ensure fans maintain high operating efficiency, but when the load reaches 95% of the rated load, fan efficiency is lower under frequency regulation than that under baffle adjustment.

4. Fan regulation modes should be selected based on the actual boiler operations. Since variable load units change with time, adopting just one single adjustment is not appropriate. Contrarily, we should combine a variety of regulating modes such as baffle adjustment and frequency control to ensure the efficient operation of the fan.

ACKNOWLEDGEMENTS

Fund project: project of Shanghai Commission in science and technology. Project number: 14110502400.

REFERENCES

Deng Xiandong, The research and practice of energy-saving modification on primary fan of Zhang Jiakou power plant 300 MW unit boiler [D]. North China Electric Power University. 2003, 12: 6–9.

Dou Huike, Yang Fusheng, Hu Yu, Optimization on inlet baffle plate operation of air fans of large CFB boiler [J]. North China Electric Power. 2014, (5). DOI:10.3969/j.issn.1003-9171.2014.05.015.

Han Zhicheng, Cai Guangyu, Zeng Yanfeng, Song Hongmei, Optimal operation research of primary air in a 600 MW boiler [J]. Electric power, 2011.

Li Shuangjiang, Lv Ming, Jia Shaoguang, Economic analysis on single row configuration of forced, induced, and primary air fans for a 600 MW boiler [J]. Thermal Power Generation. 2012, 41(8). DOI:10.3969/j.issn.1002-3364.2012.08.006.

Liu Jiayu, Qi Chunsong, Choice of centrifugal fan regulating in power plant and economic evaluation [J]. Electric power, 1990, (9).

Zhang Yongming, Hao Wenyi, Li Zhendong, Qiao Yanxiong, Wu Chenguang, Energy-saving modification of boiler centrifugal primary fan [J]. Inner Mongolia Electric Power. 2014, 6.

Advances in Energy, Environment and Materials Science – Wang & Zhao (Eds)
© *2016 Taylor & Francis Group, London, ISBN 978-1-138-02931-6*

Optimization of the matching performances between pump and valve in the boom converging loop of a PFC excavator

Wenhua Jia & Dasheng Zhu
School of Mechanical Engineering, Nanjing Institute of Technology, Nanjing, China

Chenbo Yin & Hui Liu
School of Mechanical and Power Engineering, Nanjing University of Technology, Nanjing, China

ABSTRACT: In the present study, with the adoption of a hydraulic excavator (SY215C8M, Sany Heavy Industry, China) as the object, the matching characteristics between the variable pump and the multi-directional control valve in the boom converging loop were analyzed. An optimization scheme was proposed to improve the matching performances, in which the throttling characteristics of the U-type groove were optimized using a Particle Swarm Optimization (PSO) algorithm, and thus, the structural parameters with preferable throttling characteristics could be obtained. Moreover, in order to verify the feasibility of the proposed optimization scheme, an experimental platform based on the hydraulic system of the excavator was built. The results indicate that the pressure fluctuation and loss can be effective suppressed after optimization.

1 INTRODUCTION

The performance of a hydraulic excavator depends on whether the match among various subsystems is reasonable or not. In recent years, the researchers from many universities and institutions at home and abroad have conducted a great deal of research on the static and dynamic behaviors of main pump, the matching performances of engine power, as well as the orifice area of the throttling groove in the multiple directional control valve spool, and a lot of achievements were gained [1–4]. In a Positive Flow Control (PFC) system, the mismatch between pump and valve will produce pressure impact and heat near the valve port, i.e., leading to a lot of energy waste. Even more seriously, the whole hydraulic system may fail. As a consequence, for a PFC hydraulic system in the excavator, the studies on the matching characteristic between pump and valve will be of great practical significance.

2 PRINCIPLE OF AN ELECTRONIC PFC HYDRAULIC SYSTEM

2.1 *Indices reflecting the throttling characteristics of the U-type groove*

In a PFC hydraulic system [5], the pilot signals generated by a pilot-operated joystick are transmitted to the variable pump and the 3-way 6-port multiple directional control valve simultaneously. The pilot-operated oil pump generates the pressure signals, which are used for controlling the displacement of multiple directional control valve spool. At the same time, the pilot signals collected by the controller are transmitted to the flow control device of the variable pump for controlling its displacement. Since the time difference exists during the transmission of pilot signals, and moreover, the variable pump and multiple directional control valve present different response characteristics to pilot signals, the mismatch between pump and control valve occurs frequently.

When the displacement of the variable pump increases too quickly, and simultaneously the openings of the multiple directional control valve spool

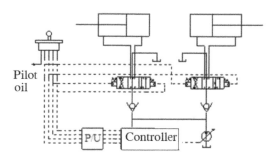

Figure 1. Schematic diagram of the PFC hydraulic system.

are too small, the pressure loss and fluctuation are easily generated. To solve these problems, we proposed a solution for optimizing the structural dimension of the throttling groove in the multiple directional control valve.

2.2 Indices reflecting the throttling characteristics of the U-type groove

From the distribution structure of the throttling groove in the boom multiple valve spool, we can observe that, when the spool is opened, the U-type groove plays a throttle governing role. As shown in Figure 2, by optimizing the structural dimension of the U-type groove, the gradient of orifice area when the spool is opened can be enhanced, and thus, the pressure impact and fluctuations can be weakened.

By analyzing the orifice area of the valve port of the U-type groove, we can find that the overall pressure drop of the valve port is allocated to two cross sections (A_1 and A_2) according to a certain proportion. To be specific, the principle of allocation is that, on each section (A_1 and A_2), the pressure drop is inversely proportional to its orifice area, i.e., the pressure drops on two sections exhibit the following relation:

$$\frac{\Delta p_1}{\Delta p_2} = \left(\frac{C_{d2}A_2}{C_{d1}A_1}\right)^2 = k^2 \qquad (1)$$

in which k is referred to as the allocation coefficient of the pressure drops at two valve ports in the U-type groove [8, 9]. For a flow section, a smaller concentrated pressure drop is indicative of less serious cavitation corrosion. Therefore, when designing the structure of a U-type groove, we should consider how to adjust the structural dimension so that the pressure drops on two sections are equal, i.e., the allocation coefficient of the pressure drops at two valve ports, k, should approach 1. Accordingly, the average value of k can be taken as an optimization objective.

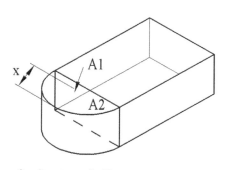

Figure 2. Structure of a U-type groove.

For the valve port of a U-type groove, in order to achieve a favorable cavitation characteristic with different openings, the value of k should be close to the average value or fluctuate around the average value. In accordance with the lease-square fitting principle, the mean square deviation value, $D(k)$, can be adopted as another optimization objective in measuring the allocation of pressure drops at the valve ports.

3 OPTIMIZATION OF THE THROTTLING CHARACTERISTICS OF THE GROOVE

3.1 Multi-objective optimization function for the throttling characteristics

Generally, the hydraulic radius, R_h, should be as large as possible in designing the size of a valve spool, i.e., the larger the better. Comparatively, the average value and the variance value of the pressure drop allocation coefficient should be as small as possible.

The optimization function can be written as:

$$Y_1 = \min(y_1) = \min\left(\frac{1}{R_{h2}}\right) = \min\left(\frac{2R+H}{2RH}\right)$$

$$Y_2 = \min(E(k)) = \min\left(\frac{\int_0^L \frac{C_{d1}A_1(x)}{C_{d2}A_2(x)}dx}{L}\right) \qquad (2)$$

$$Y_3 = \min\left(\sum_{i=1}^n (k_i - E(k))^2\right)$$

in which C_{d1} and C_{d2} denote the flow coefficients on two flow sections, A_1 and A_2, respectively. In the present work, C_{d1} and C_{d2} were set as 0.68 and 0.69, respectively.

To comprehensively evaluate the throttling characteristics of the U-type groove, we made a weighted summation on Eq. (2). After transformation, the optimization function can be rewritten as:

$$Y = W_1Y_1^* + W_2Y_2 + W_3Y_3 \qquad (3)$$

in which W_1, W_2 and W_3 denote the weight coefficients, respectively, and moreover, $W_1 + W_2 + W_3 = 1$.

Since the hydraulic radius R_h is a unit with dimension, we then should make it dimensionless through appropriate operations. With regard to the objective function Y_1, we should perform the following processing:

$$Y_1^* = \frac{y_1 - y_{\min}}{y_{\max} - y_{\min}} \qquad (4)$$

in which y_{min} and y_{max} denote the maximum and minimum values of R_h, respectively.

The parameters used for representing the geometric dimension of a U-type groove mainly include the radius R, the depth H, and the length L, which were then adopted as the optimization objectives in the present work. The ranges of these three parameters, namely, the constraint condition can be written as:

$$\begin{cases} 2 \leq R \leq 3 \\ 1.5 \leq H \leq 3.5 \\ 7.5 \leq L \leq 8 \end{cases}$$

We then compiled a Standard Particle Swarm Optimization (SPSO) program, in which the number of particle swarms was set as 20, the dimension of particles was set as 10, the number of loops was set as 500 and the three weight coefficients, W_1, W_2 and W_3, were all set as 1/3. Subsequently, the objective function representing the throttling characteristics of the U-type groove was solved. After optimization, the geometric dimensions are listed as follows, $R = 2.9$ mm, $H = 3.5$ mm and $L = 8$ mm, respectively, and the results are displayed in Figure 3.

Using two different structural parameters, $R_{2.5}H_{2.8}L_8$ and $R_{2.9}H_{3.5}L_8$, namely, the parameter in initial design and the parameter after optimization, the indices reflecting the throttling characteristics of the U-type groove were calculated and compared, with the results listed in Table 1.

As shown in Table 1, the U-type groove with the optimized structural parameters exhibits a preferable through-flow performance. After optimization, both average and variance of K decrease significantly, which is conductive to suppressing the pressure drop and fluctuation at the valve port. Accordingly, the matching performance between variable pump and multiple directional control valve can be improved. Compared with the average

Table 1. Comparison of the indices reflecting the throttling characteristics of the U-type groove with two different structural parameters.

Structural parameters	R_{h2}/mm	$E(k)$	$D(k)$
Before optimization			
$R_{2.5}H_{2.8}L_8$	1.7949	1.2414	49.0393
After optimization			
$R_{2.9}H_{3.5}L_8$	2.1828	0.9752	30.6420
Throttling characteristics	+21.61%	−21.45%	−37.52%

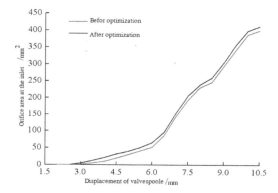

Figure 4. Comparison of the orifice areas at the inlet opening before and after optimization when the boom was lifted.

value, the variance value exhibits a steeper decline, suggesting that the values of K are always close to the average value within the range of the opening. In other words, when oil flows, the generated pressure drops on two flow sections are basically equal, and therefore, the impacts of oil on the valve spool and the induced cavitation corrosions can be effectively suppressed.

3.2 Analysis of the orifice area of valve spool after optimization

Figure 4 displays the orifice areas at the inlet spool before and after optimization, from which we can observe that the orifice area after optimization is larger and varies more early. During the movement of valve spool, the orifice area after optimization varies more smoothly. It means that, after optimization, the pressure difference at the valve port decreases and the oil flows more smoothly. Accordingly, both the action accuracy and operation stability of the working device can be enhanced.

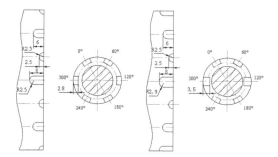

Figure 3. Comparison of the structural dimension of the throttling groove at the inlet opening when the boom was lifted.

4 EXPERIMENTS

In the present work, the excavator (SY215C8M, Sany Heavy Industry, China), as shown in Figure 5, was selected as the experimental prototype. The rated working pressure of the main pump is 343 bar and the maximum pump discharge is 2*102.5 cc/rev. The displacement of the rotary motor is 210 L/min and the rate flow of the multiple directional control valve is 456 L/min, respectively. The excavator has boom and stick cylinder retaining valves, which can achieve a series of advanced functions such as boom raise and swing priority.

4.1 Comparison of pressure responses of the boom valve spool before and after optimization

The improved boom spool was then installed at the multiple directional control valve, and the hydraulic pipelines of the experimental prototype were connected. After checking, the signal sensor and date acquisition equipment were then installed at the pipelines. Figure 5 displays the experimental results of pressure at the oil inlet of the boom valve spool.

By analyzing the single-action experimental data of the boom valve spool before and after optimization, we can obtain the following measured parameters, as listed in Table 2.

The experimental curves and measured data, as shown in Table 2, indicate that the pressure peak at the inlet of boom valve spool after optimization is reduced by approximately 24 bar, and the maximum overshoot decreases by approximately 22.5%. After optimization, the pressure impact on the boom valve also decreases and its distribution tends to be more uniform. The pressure response time at the inlet of boom valve spool is shortened by 0.259 s,

Table 2. Measured parameters.

Single action	P_s(bar)	P_w(bar)	M_p	t_s/s	t_m/s	t_{bs}/s
Head port						
Before	159.42	118.535	34.5	1.183	0.2531	4.45
After	135.18	120.792	12	0.924	0.1375	4.365
Rod port						
Before	5.3	4.963	6.7	1.355	0.372	4.45
After	4.4	3.994	10.2	1.103	0.347	4.365
Head port						
Before	155.36	115.912	34	1.475	0.3372	4.404
After	134.04	112.516	19.2	1.0355	0.1934	4.32
Rod port						
Before	4.48	4.226	6	1.54	0.483	4.404
After	4.12	3.807	8.2	1.15	0.426	4.32

which is beneficial to improving the stationarity of the valve spool's movement. Additionally, the time during which the oil cylinder is expanded to the maximum is reduced by approximately 0.08 s.

5 CONCLUSIONS

In order to solve the mismatch between variable pump and reversing valve, an optimization scheme was proposed in this article. First, the optimization on the throttling characteristics of the U-type groove based on particle swarm optimization was conducted, and then the structural parameters of the groove with preferable throttling characteristics were obtained. Finally, the feasibility of the optimization scheme was proved by the experiments.

After optimization, when the spool is opened, the pressure impact on the valve port and the overshoot decrease effectively, and the response of the outlet pressure to the pressure at the outlet of variable pump exhibits a remarkable rise. Accordingly, the problems such as pressure loss and fluctuation induced by too fast increase of the variable pump displacement and too small openings of multiple directional control valve can be satisfactorily solved. Moreover, the matching performances between variable pump and multiple directional control valve can be effectively improved, and the operating performances of the excavator, as well as the service life of the hydraulic components can also be enhanced.

ACKNOWLEDGMENT

This work was supported in part by The National Natural Science Fund, Jiangsu Natural Science

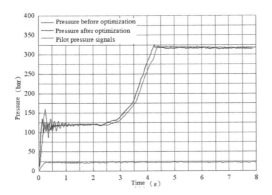

Figure 5. Comparison of the pressure responses at the oil inlet of the boom valve spool before and after optimization (during the lifting process).

Foundation, University of Jiangsu Natural Science Foundation, Nanjing Institute of Technology and SANY Co., Ltd in Jiangsu. Lecturer, Support Fund No. 11302097, BK20130741, 13KJB460009, QKJA201201.

REFERENCES

Zhu Jiangong, Study on constant power control of variable pump, Chinese Hydraulics & Pneumatics, 2005, 7, 61–62.

Jia Wenhua, Yin Chenbo, Gao Donghui and Chen Kelei, Analysis on positive flow pump control system of hydraulic excavator, Journal of Nanjing University of Technology (Natural Science Edition), 2011, 11(6):98–101.

Jia Wenhua, Yin Chenbo and Cao Donghui, Coordination analysis of movable boom preferential lifting of excavator, Coal Mine Machinery, 2013, 34 (5).

Ji Hong, Wang Dongsheng, Shohei Ryu and Fu Xin, Flow control characteristic of the orifice in spool valve with notches, Transactions of the Chinese Society for Agricultural Machinery, 2009, 40 (1): 198–202.

Yuan Shihao, Yin Chenbo and Liu Shihao, Two-step throttle properties of hydraulic valve ports, Journal of Drainage and Irrigation Machinery Engineering, 2012, 30 (6): 713–720.

Jia Wenhua, Yin Chenbo and Cao Donghui, Analysis and renovation of proportional throttle loading valve port matching characteristics, Chinese Hydraulics & Pneumatics, 2013, 1: 83–85. The above material should be with the editor before the deadline for submission. Any material received too late will not be published.

Advances in Energy, Environment and Materials Science – Wang & Zhao (Eds)
© *2016 Taylor & Francis Group, London, ISBN 978-1-138-02931-6*

The role of micro/nano-structure in the complex wettability of butterfly wing

Gang Sun & Yan Fang

School of Life Science, Changchun Normal University, Changchun, Jilin, China

ABSTRACT: The micro-morphology of the butterfly wing surface was characterized by a Scanning Electron Microscope (SEM). The Contact Angle (CA) and Sliding Angle (SA) of water droplet on the wing surface were measured by an optical CA meter. The role of multi-dimensional micro/nano-structure in the complex wettability and self-cleaning performance of the wing was investigated. The wetting mechanism was discussed from the perspective of biological coupling. The butterfly wing is of superhydrophobicity (CA 150~157°) and low adhesion (SA 1~4°), and exhibits hierarchical rough micro-morphology including primary structure (scales), secondary structure (longitudinal ridges and latitudinal links) and tertiary structure (stripes). The scales play a crucial role in the complex wettability of the wing surface. In micro-dimension, the smaller the width is or the bigger the spacing of the scale is, the stronger the hydrophobicity of the wing is. In nano-dimension, the smaller the height is or the smaller the width is or the bigger the spacing of the longitudinal ridge is, the stronger the hydrophobicity of the wing is. On the wing surface without scales, the CA decreases by 23.4%~46.3%, the SA increases above 65°. The average rate of $CaCO_3$ pollution removal from the wing surface is as high as 88.0%. There is a positive correlation ($R^2 = 0.8934$) between pollution removal rate and roughness index of the wing. The cooperation of chemical composition and micro/nano-structure leads to the complex wettability and excellent self-cleaning function of the wing. The butterfly wing can be used as a biomimetic template for design and fabrication of multi-functional interfacial materials.

1 INTRODUCTION

Wetting of solid substrates by liquids is one of the most important properties widely used by technological applications (Wei et al. 2009). Surface wettability depends basically on the surface roughness and the free energy (Wang & Jiang 2007). In the recent years, the interfacial materials with desirable characteristics have attracted more and more attention due to the wide applications in industrial, military, engineering and domestic fields. After long-term natural selection, many creatures have evolved peculiar (superhydrophobic, self-cleaning, anti-adhesive, anti-icing, anti-wearing, etc.) body surfaces to adapt to the environment (Zheng et al. 2007). Insect wing, one of the most complicated three-dimensional periodical substrate in nature, has become a popular template for biomimetic preparation of functional materials (Fang et al. 2007). The authors have conducted some work on the anisotropism and wettability of butterfly wing (Fang et al. 2008). In the current work, the role of multi-dimensional micro/nano-structure in the complex wettability of butterfly wing was investigated. The results may bring insight for design and fabrication of novel intelligent materials.

2 MATERIALS AND METHODS

2.1 *Materials*

The specimens of ten butterfly species were collected in Ji'an City, Jilin Province of northeast China. The wings were cleaned, desiccated and flattened, then cut into 5 mm × 5 mm pieces from the discal cell (Fig. 1). The distilled water for measurements of CA and SA was purchased from Tianjin Pharmaceuticals Group Co. Ltd., China. The volume of water droplets was 5 μl. The $CaCO_3$ particle was purchased from Shanghai Aibi Chemistry Preparation Co. Ltd., China.

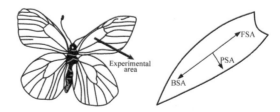

Figure 1. The experimental area and the SAs in different directions on the butterfly wing.

The size distribution of $CaCO_3$ particle was 5~10 μm.

2.2 Characterization of micro/nano-structure

After gold coating by an ion sputter coater (Hitachi E-1045, Japan), the wing pieces were observed and photographed by a SEM (Hitachi SU8010, Japan). Using Photoshop software, the microstructural parameters of the wing surfaces were measured in the SEM images.

2.3 Measurements of CA and SA

Using an optical CA measuring system (DataPhysics OCA20, Germany), the CA of water droplet on the wing surface was measured by sessile drop method at room conditions of $(25±1)$ °C and relative humidity of approximately 80%. The SA of water droplets was measured along three different directions, including forward SA (FSA, the SA of droplet from wing base to wing terminal end), backward SA (BSA, the SA of droplet from wing terminal end to wing base) and perpendicular SA (PSA, the SA of droplet perpendicular to the major axis of wing) (Fig. 1). The water droplet was dripped on the sample table in a horizontal position, then the inclination degree of the table was raised 1° each time until the droplet rolled off freely. The inclination degree of the table was recorded as the SA value.

2.4 FT-IR measurement

After grinding finely, 5~8 mg of wing samples were mixed homogeneously with 200 mg of KBr and pressed into a thin slice. The absorbance was measured by means of FT-IR (Nicolet FT-IR200, USA). The chemical composition of the wing surface was analyzed by the FT-IR spectra.

2.5 Removal of $CaCO_3$ particles

The wing pieces were affixed to glass slides with double-sided adhesive tape, and put on the sample table of OCA20. Five mg of $CaCO_3$ particles were evenly spread on the discal cell of the wing. A water droplet from an injector fell on the $CaCO_3$ area. The sample table was inclined 3°, and the droplet flowed through the contaminated area. A stereo microscope (Zeiss SteREO Discovery V12, Germany) was used to observe the removal of $CaCO_3$ particles. An electronic analytical balance (Shimadzu AUX-120, Japan) was used to measure the mass of $CaCO_3$ residual on the wing surface. The removal rate was calculated.

3 RESULTS AND DISCUSSION

3.1 The effect of scale on hydrophobicity and adhesion of the wing surface

The wing surface exhibits hierarchical rough structures composed of primary structure (the micrometric scales) [Fig. 2(a)], secondary structure (the submicro longitudinal ridges and latitudinal links on the scales) [Fig. 2(b)] and tertiary structure (the nano stripes on the longitudinal ridges and latitudinal links) [Fig. 2(b)]. The cross-section of the longitudinal ridge is triangular. The micro-morphological parameters of the butterfly wing surface are shown in Table 1.

The wing surfaces are superhydrophobic (CA 150~157°) (Table 2). Meanwhile, the water SAs on the wing surfaces are extremely small (FSA 1~4°, BSA 6~12°, PSA 8~14°) (Table 2). There are significant differences between the SAs in various directions ($p < 0.01$). The asymmetrical sliding behavior of water droplet on the wing surface ascribes to the anisotropic micro-morphology and the different energy barriers. The butterfly wing surface, just like the lotus leaf and the rice leaf (Yao et al. 2011), displays superhydrophobicity and low adhesion; while some other natural surfaces like the peanut leaf and the rose petal exhibit high hydrophobicity and high adhesion (Bhushan & Her 2010). The distinct complex wettability results from the different micro-morphologies. The scales play a crucial role in the wetting behavior of droplet on the butterfly wing. In a contrast test, the scales were removed from the wing surfaces. The CA decreases by 23.4%~46.3%, all the SAs (FSA, BSA, PSA) increase above 65° (the maximum inclination angle of the sample table is 65°).

3.2 The relationship between hydrophobicity and micro/nano-structure of the wing surface

The wing surface is relatively rough with superhydrophobicity and heterogeneity. A composite

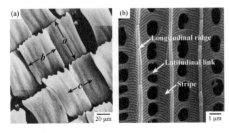

Figure 2. The multiple-dimensional rough micromorphology of the butterfly wing surface (SEM images). *(a) Primary structure (scales); (b) Secondary structure (longitudinal ridges and latitudinal links) and tertiary structure (stripes).

Table 1. Micro-morphological parameters of the butterfly wing surface.

Species	Scale (μm)			Longitudinal ridge (μm)		
	Length (a)	Width (b)	Spacing (c)	Height (d)	Width (e)	Spacing (f)
Aglais urticae	175	92	89	0.75	0.81	1.84
Apatura iris	214	74	132	0.63	0.93	1.73
Argynnis paphia	368	88	94	0.82	0.82	1.65
Damora sagana	252	84	103	0.67	0.90	1.82
Erynnis montana	179	96	115	0.78	0.77	1.71
Libythea celtis	186	85	96	0.85	0.85	1.63
Lycaena dispar	325	84	110	0.94	0.74	2.14
Pyrgus maculatus	282	67	129	1.15	0.68	1.90
Thecla betulae	286	81	95	0.96	0.84	1.58
Vanessa cardui	324	83	93	1.02	0.95	2.07
Average	259	83	106	0.86	0.83	1.81

Table 2. Wetting property and self-cleaning performance of the butterfly wing surfaces.

Species	CA(°)		Measured SA(°)			CaCO₃ removal rate (%)	RI
	Measured	Predicted	FSA	BSA	PSA		
Aglais urticae	152	146	3	8	11	89.2	2.7
Apatura iris	154	149	2	9	9	86.5	2.3
Argynnis paphia	150	145	4	6	14	88.0	2.7
Damora sagana	151	149	1	12	13	90.3	3.2
Erynnis montana	151	147	2	7	9	81.6	1.9
Libythea celtis	154	148	3	10	12	85.7	2.6
Lycaena dispar	157	151	2	9	8	90.4	3.1
Pyrgus maculatus	153	150	4	7	10	88.5	2.8
Thecla betulae	155	152	1	10	11	87.1	2.6
Vanessa cardui	152	151	2	8	12	92.2	3.6
Average	153	149	2	9	11	88.0	2.8

*RI, the roughness index of wing surface, is denoted by $\sqrt{\frac{4d^2}{e^2}+1}$, where d and e represent height and width of the longitudinal ridge, respectively.

contact is formed between the droplet and the surface. The contact behavior of a water droplet can be described by the Cassie-Baxter equation:

$$\cos\theta_c = \phi_s\cos\theta_e + \phi_s - 1 \qquad (1)$$

where θ_c is the apparent CA of a droplet on a composite surface, ϕ_s is the area fraction of solid ($0 < \phi_s < 1$, calculated from micro-morphological parameters of the wing surface), θ_e is the intrinsic CA of water on an ideal flat surface (approximately 95° for the butterfly wing). The contact state of a water droplet on the micro/nano structure of the wing surface is shown in Figure 3.

In this case, Eq. (1) can be modified for the theoretical (predicted) CA (θ_t) as follows:

$$\cos\theta_t = \sqrt{\frac{4d^2}{e^2}+1} * \frac{be}{cf} * \cos\theta_e + \frac{be}{cf} - 1 \qquad (2)$$

Based on Eq. (2), the predicted CAs were calculated (Table 2). Taking predicted CAs as independent variable y^*, measured CAs as dependent variable y, the degree of fitting was judged by:

$$Q = \Sigma(y - y^*)^2 \qquad (3)$$
$$R_{New} = 1 - (Q/\Sigma y^2)^{1/2} \qquad (4)$$

where Q is sum of square of deviations, R_{New} is the coefficient of determination in nonlinear regression equation. The calculated R_{New} values are 0.922~0.964 for the ten butterfly species. There is no significant difference between the measured CAs and the predicted CAs, demonstrating the

413

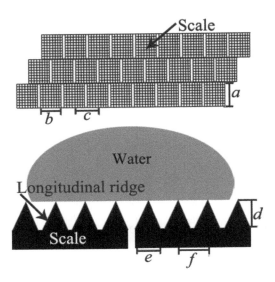

Figure 3. The contact state of a water droplet on the micro/nano structure of the wing surface.

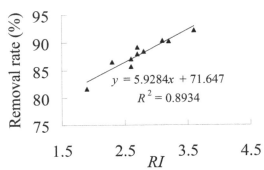

Figure 4. The relationship between the removal rate of $CaCO_3$ particle and Roughness Index (RI) of the wing surface.

hydrophobicity model based on micro-morphology is in good accord with the Cassie-Baxter equation. In the light of Eq. (2), the smaller the values of $\sqrt{\frac{4d^2}{e^2}+1}$ and $\frac{be}{cf}$ are, the bigger the theoretical CAs are. Namely, in micro-dimension, the smaller the width is or the bigger the spacing of the scale is, the stronger the hydrophobicity of the wing surface is; in nano-dimension, the smaller the height is or the smaller the width is or the bigger the spacing of the longitudinal ridge is, the stronger the hydrophobicity of the wing surface is. Naturally hydrophobic material such as chitin is the chemical foundation for the high hydrophobicity on the butterfly wing. Much higher hydrophobicity, however, cannot be induced by the chemical composition alone. Superhydrophobicity of the wing surfaces is attributed to a cooperation of hydrophobic material and rough micro-morphology. Here, only the primary and secondary microstructures of the wing surfaces were incorporated in the hydrophobicity models. In fact, the more subtle tertiary microstructure (nano stripes) also contributes to the surface roughness and hydrophobicity. The hydrophobicity model involving tertiary microstructure is likely to offer more accurate CA predictions.

3.3 The relationship between self-cleaning property and micro/nano-structure of the wing surface

Due to the rough structures on the wing surface, most contaminating particles settle on the tips of microtextures, the actual contact area between particle and microstructure is very small. The adhesive force between particles and water droplet is much larger than that between particles and microstructure, thus the particles can be "trapped" and taken away easily by the rolling droplet. The removal rate of contaminating particle has no significant correlation ($R^2 < 0.1$) with the scale parameters (length, width, spacing) or the longitudinal ridge parameters (height, width, spacing), but has significant correlation ($R^2 = 0.8934$) with Roughness Index (RI) of the wing surface (Fig. 4). RI, the magnitude of surface roughness, is the ratio of the real area to the geometry projection area. The superhydrophobicity and self-cleaning characteristic of the wing surface ascribes to the coupling effect of hydrophobic material and rough structure. The self-cleaning function endows the butterfly with the ability to lighten body burden, increase flight efficiency and optimize energy budget. Thus, the butterfly can get more opportunities to survive.

4 CONCLUSIONS

The butterfly wing surface displays hierarchical micro-morphology including primary structure (scales), secondary structure (longitudinal ridges and latitudinal links) and tertiary structure (stripes). The wing surface is of low adhesion (SA 1~4°) and superhydrophobicity (CA 150~157°). The micro-morphology, dynamic behavior of water droplets, and self-cleaning performance of the wing surfaces show remarkable anisotropism. The scales play a crucial role in the complex wettability of the butterfly wing. In micro-dimension, the smaller the width is or the bigger the spacing of the scale is, the stronger the hydrophobicity of the wing surface is. In nano-dimension, the smaller the height is or the smaller the width is or the bigger the spacing of the longitudinal ridge is, the stronger the hydrophobicity of the wing surface is. On the wing surface without scales, the CA decreases by 23.4%~46.3%, the SA increases above 65°. The average rate of $CaCO_3$ particle removal from the wing surface is as high as 88.0%. There is a good positive correlation ($R^2 = 0.8934$) between pollution removal rate and roughness index of the wing surface. The coupling effect of hydrophobic material and rough microstructure contributes to the special wettability and excellent self-cleaning characteristic of the wing. The butterfly wing can serve as a template for biomimetic design and preparation of micro-controllable superhydrophobic surface and nano dust-free coatings. This work not only promotes our understanding of wetting mechanism of natural surfaces, but may offer inspirations for development of novel interfacial material with multi-functions.

ACKNOWLEDGEMENTS

This work was financially supported by the National Natural Science Foundation of China (50875108), the Natural Science Foundation of Jilin Province, China (201115162), Science and Technology Project of Educational Department of Jilin Province, China (2009210, 2010373, 2011186). Dr. Prof. Yan Fang is the corresponding author of this paper.

REFERENCES

Bhushan, B. & Her, E.K. 2010. Fabrication of superhydrophobic surfaces with high and low adhesion inspired from rose petal. *Langmuir* 26(11): 8207–8217.

Fang, Y., Sun, G., Wang, T.Q. & Cong, Q. 2007. Hydrophobicity mechanism of non-smooth pattern on surface of butterfly wing. *Chinese Science Bulletin* 52(5): 711–716.

Fang, Y., Sun, G. & Cong, Q. 2008. Effects of methanol on wettability of the non-smooth surface on butterfly wing. *Journal of Bionic Engineering* 5(2): 127–133.

Wang, S.T. & Jiang, L. 2007. Definition of superhydrophobic states. *Advanced Materials* 19(21): 3423–3424.

Wei, P.J., Chen, S.C. & Lin, J.F. 2009. Adhesion forces and contact angles of water strider legs. *Langmuir* 25(3): 1526–1528.

Yao, X., Song, Y.L. & Jiang, L. 2011. Applications of bio-inspired special wettable surfaces. *Advanced Materials* 23(6): 719–734.

Zheng, Y.M., Gao, X.F. & Jiang, L. 2007. Directional adhesion of superhydrophobic butterfly wings. *Soft Matter* 3(2): 178–182.

Advances in Energy, Environment and Materials Science – Wang & Zhao (Eds)
© *2016 Taylor & Francis Group, London, ISBN 978-1-138-02931-6*

Reinforcement simulation for ramp bridge

Yu Li & Biao Pan
Key Laboratory of Ministry of Communications for Bridge Detection and Reinforcement Technology,
School of Highway, Chang'an University, Xi'an, Shanxi, China

ABSTRACT: Highway ramp bridge is taken to be object of study by using two reinforcement design. And the FEA model is established to do the numerical simulation for reinforced highway ramp bridge. Then checking calculation is done to study the flexural load-bearing minimal capacity of normal section, shear load-bearing maximal capacity of inclined section, crack width of roof of box girder in service stage. The result shows that the load-bearing capacity of reinforced highway ramp bridge can satisfy the code requirement. The reinforcement measure, FEA model and calculation are suggested to reinforce the highway ramp bridge. So, some meaningful references are provided for the further research on reinforcement design of highway ramp bridge.

1 INTRODUCTION

Due to the influence of various factors (repeated action of long-term load, the shortage of steel consumption, construction irregularities, etc.), the bridge structure may cause crack by the redistribution of the internal stresses in service stage. And this leads to the bridge's retirement in advance before unreached design working life. So it is necessary to study the practical experience of bridge reinforcement technology (Housner, 1956; Xu, 1997; Ingham, 1997; Kareem, 1995; Larose, 1995; Igusa, 1990).

In this paper, highway ramp bridge is taken to be object of study by using two reinforcement schemes (steel plate-concrete composite technique, bonding steel plate to outside of web plate and tensioning vertical pre-stressed near the fulcrum). And the FEA model is established to do the numerical simulation for reinforced highway ramp bridge.

2 PROJECT OVERVIEW

Qiaosi-Yuhang-Tanghe highway is an important part of Hangzhou circle city road, located in the north of Hangzhou. It's 29.297 km long, opened in 2001. The route crosses with state highway 104 near Gouzhuang, designed with interchange. Gouzhuang interchange C ramp bridge with 6 spans; vertical layout as follows: $16\,m + 4 \times 20\,m + 16\,m$, 115.10 m long; ramp bridge horizontal layout as follows: 0.5 m (barriers) + 7.5 m (roadway) + 0.5 m (barriers). The upper structure is cast-in-place reinforced concrete continuous box girder, box girder height is 1.2 m, single box single room,

roof width is 8.48 m, base plate width is 4.1 m, cantilever length is 2.19 m; box girder web plate thickness is 30~45 cm, roof thickness of middle section is 20 cm, base plate thickness is 15 cm.

Maintenance and management department found that box girder web plate and base plate of C ramp bridge all had different degrees cracks in the daily maintenance and regular inspection, the maximum crack width was 0.8 mm. By identifying for the original bridge found: (1) design according to the old specification, original design load is automobile—beyond 20, trailer—120; (2) the upper box girder appears larger web plate crack, from visual inspection the maximum crack width is about 0.8 mm; (3) it is found that the shear capacity of upper box girder exceeds specification limit through checking calculation.

3 REINFORCEMENT SCHEMES

C ramp bridge reinforcement aims to: (1) through reinforcement measures to improve load-bearing capacity and meet the requirements of load rating of highway grade I; (2) treating bridge structure crack and repairing and reinforcing the structural damage parts to ensure durability. According to the bridge current situation and combining with the structure calculation, reinforcement schemes can be divided into two parts: regular crack repair and main structure reinforcement.

3.1 *Regular crack repair*

For the width of structural stressed cracks is greater than or equal to 0.15 mm, repairing through pressure grouting; for non-structural stressed crack and its width is less than 0.15 mm cracks, adopting sealed method.

3.2 *Main structure reinforcement*

Through checking calculation, when the shear load-bearing capacity of C ramp bridge can't meet the requirements of highway first grade load, 20 meters mid-span flexural capacity also does not meet the requirements. Therefore, adopt the following two kinds of reinforcement schemes to C ramp bridge:

1. Plan 1: Using steel plate-concrete composite technique to improve the flexural (shear) capacity of structure;
2. Plan 2: Bonding steel plate to outside of web plate and tensioning vertical pre-stressed near the fulcrum, improving structure's shear load-bearing capacity. Within 0.5 L (L per span length) in the base plate of box girder, bond steel plate to reinforce in the same direction as bridge, improving the flexural load-bearing capacity of structure.

4 ESTABLISHMENT OF FEA MODEL

The FEA model of Gouzhuang interchange C ramp bridge was established by using discrete structure FEA method. According to the structure characteristics of ramp bridge, it was simplified on the premise of ensuring that its load and rigidity were consistent with the actual structure. The specific modeling steps are as below: (1) using beam element to simulate concrete beam; (2) main reinforcement uses HRB335, stirrup uses R235; (3) the steel plate of strengthening scheme uses Q235 and Q345. The FEA model of Gouzhuang interchange C ramp bridge and its cross-section as shown below, constraint conditions of the model are shown in Table 1.

5 THE CALCULATION RESULTS OF REINFORCEMENT DESIGN

5.1 *Checking calculation to flexural load-bearing capacity of normal section*

Under the action of load combination 3 and 4, flexural load-bearing maximal capacity of normal section is as shown below. Among them, Mn is absolute value of structural allowable flexural load-bearing capacity; rMu is design value of structural combination of action effects flexural load-bearing capacity.

Under the action of load combination 3 and 4, flexural load-bearing minimal capacity of normal section as below.

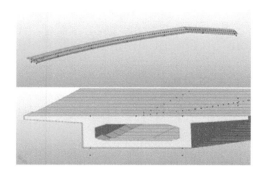

Figure 1. FEA mode.

Table 1. Load combination condition.

Combination	The state of checking	Load combination
1	Load-bearing capacity	Fundamental combination: 1.2D + 1.4T[1]
2	Load-bearing capacity	Fundamental combination: 1.2D + 1.4T[2]
3	Load-bearing capacity	Fundamental combination: 1.2D + 1.4M + 1.12T[1]
4	Load-bearing capacity	Fundamental combination: 1.2D + 1.4M + 1.12T[2]

*Note: D is for Constant load effect; M is for Highway-I lane loading effect; T[1] indicates overall warming effect; T[2] said the overall cooling effect.

5.2 *Checking calculation to shear load-bearing capacity of inclined section*

Under the action of load combination 3 and 4, shear load-bearing maximal capacity of inclined section is show as below. Among them, Vn is absolute value of structural allowable shear load-bearing capacity; rVd is design value of structural combination of action effects shear load-bearing capacity.

Under the action of load combination 3 and 4, shear load-bearing minimal capacity of inclined section is shown as below.

Conclusion of checking calculation: Checking calculation is done to study the shear load-bearing capacity of inclined section, according to mandatory provisions, design formula: $\gamma S \leq R$ of article 5.1.5 and article 5.2 in "Code for Design on Reinforced and Pre-stressed Concrete Structures of Highway Bridge and Culvert" (JTG D62-2004). The results of design value of structural shear load-bearing capacity in service stage were greater than design value of structural combination of action effects in service stage, so the design of shear load-bearing capacity of inclined section meets the requirements.

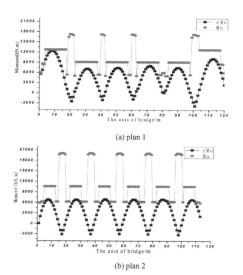

(a) plan 1

(b) plan 2

Figure 2. Result for flexural load-bearing maximal capacity of normal section.

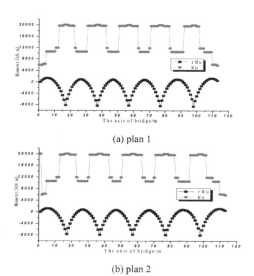

(a) plan 1

(b) plan 2

Figure 3. Result for flexural load-bearing minimal capacity of normal section.

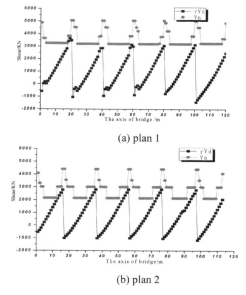

(a) plan 1

(b) plan 2

Figure 4. Result for shear load-bearing maximal capacity of inclined section.

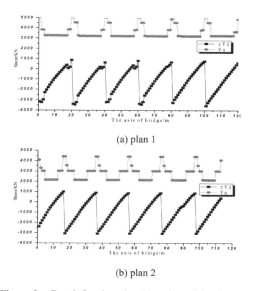

(a) plan 1

(b) plan 2

Figure 5. Result for shear load-bearing minimal capacity of inclined section.

5.3 *Checking calculation to crack width in service stage*

Checking calculation is done to study if the crack width in service stage and tension reinforced stress under short-time effect meet the corresponding permissible value in code, according to mandatory provisions of article 6.4.3 and article 6.4.4 in "Code for Design on Reinforced and Pre-stressed Concrete Structures of Highway Bridge and Culvert" (JTG D62-2004). Among them, the W_AC is permissible value of crack width in service stage. W_tk is design value of combination of action effects. Checking calculation conclusion: crack width in service stage meet the corresponding permissible value in code.

419

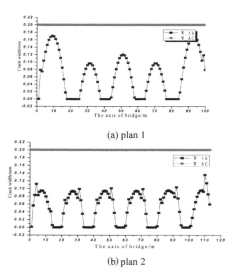

(a) plan 1

(b) plan 2

Figure 6. Results for crack width of floor of box girder in service stage.

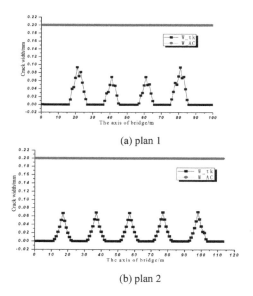

(a) plan 1

(b) plan 2

Figure 7. Results for crack width of roof of box girder in service stage.

6 CONCLUSION

This report establishes the FEA model of "Reinforcement design of Gouzhuang interchange C ramp bridge of Hangzhou circle city road", according to the design drawings of ramp bridge and reinforcement schemes provided by others. Draw conclusions as below:

1. Checking calculation is done to study the flexural load-bearing capacity of normal section of reinforced concrete, according to mandatory provisions, design formula of article 5.1.5 and article 5.2 in "Code for Design on Reinforced and Pre-stressed Concrete Structures of Highway Bridge and Culvert" (JTG D62-2004). Checking calculation conclusion: the results of design value of structural flexural load-bearing capacity and shear load-bearing capacity of reinforcement C ramp bridge were greater than design value of combination of action effects in service stage, so the design of flexural load-bearing capacity shear load-bearing capacity of normal section meet the requirements.

2. Checking calculation is done to study if the crack width in service stage and tension reinforced stress under short-time effect meet the corresponding permissible value in code, according to mandatory provisions of article 6.4.3 and article 6.4.4 in "Code for Design on Reinforced and Pre-stressed Concrete Structures of Highway Bridge and Culvert" (JTG D62-2004). Among them, the W_AC is permissible value of crack width in service stage. W_tk is design value of combination of action effects. Checking calculation conclusion: crack width of reinforcement C ramp bridge in service stage meet the corresponding permissible value in code.

ACKNOWLEDGMENTS

This work was financially supported by the National Natural Science Foundation of China (Grant No. 51408042) and Natural Science Foundation of Shaanxi Province (2014JQ7253).

REFERENCES

Housner G.W. Limit Design of Structures to Resist Earthquake [C]. Proceedings of 1st Conference on Earthquake Engineering, Berkeley, 1956.

Ingham T.J., Rodriguez S., Nader M. Nonlinear Analysis of the Vincent Thomas Bridge for Seismic Retrofit [J]. Computers & Structures, 1997, 64(3):1221–1238.

Igusa T., Xu K. Wide band-response characteristics of multiple subsystems with high modal density [C]// Proceedings of the Second International Conference on Stochastic Structural Dynamics. USA, 1990.

Kareem A., Kline S. Performance of multiple mass dampers under random loading [J]. Journal of Structural Engineering, 1995, 121(2):348–361.

Larose G.L., Falco M., Cigada A. Aeroelastic Response of the Towers for the Proposed Bridge over Stretto di Messina [J]. Journal of Wind Engineering and Industrial Aerodynamics, 1995, 57:363–373.

Xu Y.L., Ko J.M., Yu Z. Modal Analysis of Tower-Cable System of Tsing Ma Long Suspension Bridge [J]. Engineering structures, 1997, 19(10):857–867.

Advances in Energy, Environment and Materials Science – Wang & Zhao (Eds)
© *2016 Taylor & Francis Group, London, ISBN 978-1-138-02931-6*

Single-variable model for a dynamic soil structure

Xi Chen, Yingxue Yan & Xiaojie Hou
School of Civil Engineering and Architecture, Shaanxi University of Technology, Hanzhong, Shaanxi, China

ABSTRACT: On the basis of dynamic shear strain and stress, the dynamic parameters of soil structures can be derived by generalizing the parameters of soil structures under static conditions. During their analysis, a unified expression has been obtained, which can be used to employ the dynamic parameters of soil structures by expressing their dynamic shear moduli. In this article, the model for the parameters of soil structures has been built, by describing a single-variable in its active role, i.e., the soil dynamic parameter model based on the same dynamic shear stress method and the same dynamic shear strain method.

1 INTRODUCTION

The soil structure is an internal intrinsic factor of its strength and deformation (Tian, 2010). In Engineering practice, the importance of soil structure has been considered unanimously. In the present century, the soil structure has been regarded as one of the core issues for understanding the soil mechanics (Shen, 1996).

There are the three basic ways to study soil structure: (1) microstructure morphology method, (2) solid mechanics method, and (3) soil mechanics method[3]. The soil mechanics method was favored by many researchers owing to its great potential and good prospects for further development; so far, several useful results have been achieved. Xie Dingyi (1999) and Qi Jilin (1999) have proposed the parameter-reflecting soil structure in quantitative terms. They have introduced the quantitative parameters in framing the soil constitutive relation due to its deformation and strength, which made the study of soil structure a new insight. Shao Shengjun (2004) proposed another method of structural parameters by taking the ratio between the differences of principal stress of undisturbed soil and disturbed soil and the saturated soil. They further developed the structural parameters of the stress and the strain of an undisturbed loess. Chen Cunli (2006 & 2007) had defined structural quantitative parameters in accordance with the void ratio between an undisturbed soil and a remolded soil, the saturated soil under the same pressure. Their research further explored the change laws with press and the water content of the soil structure of an undisturbed loess by experiment. Luo Yasheng (2004 & 2006) had proposed the parameters of the strain-CSP reflecting soil structure and the parameters of an established soil dynamic structure, which developed a model under complex conditions of stress by extending the parameter of strain-CSP to a dynamic state. On the basis of

dynamic strength conditions, Tian Kanliang (2010) had established the dynamic structural parameters. They had further explored the development of laws for the dynamic structures of soil.

The careful investigation of the parameters for soil dynamic structures reveals that they are not unified in terms of their formation and owing to their multivariable expression. The nonuniform formation is not convenient to determine the nature of the soil structure due to the reason that it is very hard to distinguish the degree of influence for different variables of the soil structure. It can cause hindrance while estimating the parameters of dynamic structures. On the basis of such issues, this article has: (1) formulated the parameters of a unified dynamic structure, and (2) developed a single-variable model for the dynamic parameters of soil structure and evaluated its rationality.

2 SINGLE-VARIABLE MODEL FOR THE PARAMETERS OF A DYNAMIC STRUCTURE

2.1 *A unified form for the parameters of a dynamic soil structure*

The dynamic parameters of soil structure can be obtained by extending the parameters of soil structures under the static conditions. On the basis of dynamic shear strain and stress, the mathematical relation for the dynamic parameters of a soil structure can be written in terms of Eqs. (1) and (2) as follows:

$$m_{\tau d} = \frac{\gamma_{dr}\gamma_{ds}}{\gamma_{do}^2} \tag{1}$$

$$m_{\gamma d} = \frac{\tau_{do}^2}{\tau_{dr}\tau_{ds}} \tag{2}$$

where, $m_{\tau d}$ and $m_{\gamma d}$ are the dynamic parameters of soil structure on the basis of dynamic shear strain and stress respectively; γ_{do} is the dynamic shear strain of an undisturbed loess at a dynamic shear stress of τ_d; γ_{ds} is the dynamic shear strain of a saturated loess at a dynamic shear stress of τ_d; γ_{dr} is the dynamic shear strain of a remolded loess at a dynamic shear stress of τ_d; τ_{do} is the dynamic shear stress of an undisturbed loess at a dynamic shear strain of γ_d; τ_{ds} is the dynamic shear stress of a saturated loess at a dynamic shear strain of γ_d; τ_{dr} is the dynamic shear stress of a remolded loess at a dynamic shear strain of γ_d.

By assuming that the undisturbed loess at a dynamic shear strain of γ_{do}, the saturated loess at a dynamic shear strain of γ_{ds} and the remolded loess at a dynamic shear strain of γ_{dr} in the relation of Eq. (1) are based on the dynamic shear strain of $m_{\gamma d}$. The dynamic shear stress (τ_d) is a required condition to be determined, therefore, Eq. (1) can be simplified as follows:

$$m_{\tau d} = \frac{\gamma_{dr} \gamma_{ds}}{\gamma_{do}^2} = \frac{\dfrac{\tau_d}{G_{dr}} \dfrac{\tau_d}{G_{ds}}}{\left(\dfrac{\tau_d}{G_{do}}\right)^2} = \frac{G_{do}^2}{G_{dr} G_{ds}} \qquad (3)$$

where, G_{do} is an undisturbed loess of the dynamic shear modulus at a dynamic shear stress of τ_d; G_{ds} is a saturated loess of the dynamic shear modulus at a dynamic shear stress of τ_d; G_{dr} is a remolded loess of the dynamic shear modulus at a dynamic shear stress of τ_d.

Similarly, by assuming that the undisturbed loess at a dynamic shear stress of τ_{do}, the saturated loess at a dynamic shear stress of τ_{ds} and the remolded loess at a dynamic shear stress of τ_{dr} in the relation of Eq. (2) are based on the dynamic shear stress of $m_{\gamma d}$. The dynamic shear strain (γ_d) is a required condition to be determined, therefore, Eq. (2) can be simplified as follows:

$$m_{\gamma d} = \frac{\tau_{do}^2}{\tau_{dr} \tau_{ds}} = \frac{(G_{do} \gamma_d)^2}{G_{dr} \gamma_d G_{ds} \gamma_d} = \frac{G_{do}^2}{G_{dr} G_{ds}} \qquad (4)$$

where, G_{do} is an undisturbed loess of the dynamic shear modulus at a dynamic shear strain of γ_d; G_{dr} is a saturated loess of the dynamic shear modulus at a dynamic shear strain of γ_d; G_{dr} is a remolded loess of the dynamic shear modulus at a dynamic shear strain of γ_d.

Through the above process of simplification, we have found that the parameters of soil dynamic structure are based on the dynamic shear stress and strain, which have the same formation as of a dynamic soil structure and these can be described with a unified form as follows:

$$m_d = \frac{G_{do}^2}{G_{dr} G_{ds}} \qquad (5)$$

where, m_d can be described as the required parameter of a dynamic soil structure.

On the basis of Eq. (5), it can be possible to deduce: (1) $m_d = m_{\tau d}$, when the dynamic shear moduli of G_{do}, G_{ds} and G_{dr} have been obtained under the dynamic shear stress, and (2) $m_d = m_{\gamma d}$, when the dynamic shear moduli of G_{do}, G_{ds} and G_{dr} have been obtained under the dynamic shear strain. Since the relationship between the soil dynamic shear stress and the soil dynamic shear strain is not linear and elastic, therefore, the dynamic parameters of soil structure on the basis of methods for the dynamic shear stress and the dynamic shear strain are different. Keeping this in view, it is always necessary to indicate the relevant method, which is being used to determine the parameters of a dynamic soil structure. The parameters for the dynamic soil structure are theoretically equal to the starting point of its active role (refer to the dynamic shear stress and/or shear strain) by excluding the special point. Owing to the reason that the dynamic shear stress and shear strain from the starting point of their active role are zero, the magnitude of dynamic shear modulus represents its maximum extent at the point and it has nothing to do with either the soil dynamic shear stress or the dynamic shear strain. Therefore, it is concluded that $m_d = m_{\tau d} = m_{\gamma d}$ at this point.

Next, the parameters of dynamic soil structure in a unified form are based on the dynamic shear modulus. These have a positive significance to avoid the parameters of soil structure, which are unable to solve the initial states of dynamic soil shear stress and the shear strain that can be taken equal to zero.

2.2 Single-variable model of a dynamic soil structure on the basis of Hardin-Drnevich hyperbola model

On the basis of testing on huge datasets to determine the parameters of dynamic soil structures, it has been inferred that the Hardin-Drnevich hyperbolic model can be used to describe the hyperbolic relationship between the soil stress and the soil strain under dynamic loading, i.e.,

$$\tau_d = \frac{\gamma_d}{a + b\gamma_d} \qquad (6)$$

where, a and b are the test parameters.

The dynamic shear modulus can mathematically be expressed as follows:

$$G_d = \frac{1 - b\tau_d}{a} \tag{7}$$

or,

$$G_d = \frac{1}{a + b\gamma_d} \tag{8}$$

where, $m_d = m_{\tau d}$, when the dynamic shear moduli of G_{do}, G_{ds} and G_{dr} are obtained under the same dynamic shear stress.

Next,

$$\tau_{do} = \tau_{dr} = \tau_{ds} \tag{9}$$

Thereafter,

$$G_{do} = \frac{1 - b_1 \tau_{do}}{a_1} \tag{10}$$

$$G_{dr} = \frac{1 - b_2 \tau_{dr}}{a_2} = \frac{1 - b_2 \tau_{do}}{a_2} \tag{11}$$

$$G_{ds} = \frac{1 - b_3 \tau_{ds}}{a_3} = \frac{1 - b_3 \tau_{do}}{a_3} \tag{12}$$

where, a_1 and b_1 are the test parameters of an undisturbed loess, a_2 and b_2 are the test parameters of a remolded loess, and a_3 and b_3 are the test parameters of a saturated loess.

On the basis of the method for dynamic shear stress, a single-variable model of a dynamic soil structure can be obtained by substituting the Eqs. (10)–(12) into the Eq. (5) as follows:

$$m_d = m_{\tau d} = \frac{G_{do}^2}{G_{dr} G_{ds}}$$

$$= \frac{a_2 a_3 (1 - b_1 \tau_{do})^2}{a_1^2 (1 - b_2 \tau_{do})(1 - b_3 \tau_{do})} \tag{13}$$

where, $m_d = m_{\gamma d}$, when the dynamic shear moduli of G_{do}, G_{ds} and G_{dr} are obtained under the same dynamic shear strain.

and,

$$\gamma_{d0} = \gamma_{dr} = \gamma_{ds} \tag{14}$$

Therefore,

$$G_{do} = \frac{1}{a_1 + b_1 \gamma_{do}} \tag{15}$$

$$G_{dr} = \frac{1}{a_2 + b_2 \gamma_{dr}} = \frac{1}{a_2 + b_2 \gamma_{do}} \tag{16}$$

$$G_{ds} = \frac{1}{a_3 + b_3 \gamma_{ds}} = \frac{1}{a_3 + b_3 \gamma_{do}} \tag{17}$$

On the basis of the method for a dynamic shear strain, a single-variable model for the parameters of a dynamic soil structure can be obtained by substituting the Eqs. (15)–(17) into the Eq. (5) as follows:

$$m_d = m_{\gamma d} = \frac{G_{do}^2}{G_{dr} G_{ds}}$$

$$= \frac{(a_2 + b_2 \gamma_{do})(a_3 + b_3 \gamma_{do})}{(a_1 + b_1 \gamma_{do})^2} \tag{18}$$

It is observed that the above-derived Eq. (18) is consistent with the mathematical relations, as available in the existing literature (Wang, 2010).

On the basis of Eqs. (13) and (18), it can be concluded that the parameters of dynamic soil structure along with its active role as single variable can be obtained by using the methods of the dynamic shear stress and/or the dynamic shear strain, which may calculate their dynamic shear moduli of G_{do}, G_{ds} and G_{dr}. The following are their positive and key points: (1) the parameters of dynamic soil structure can reflect intuitively the relationship between its active role in single variables and its dynamic structure, and (2) the parameters of a dynamic soil structure are expressed in a single variable that can simplify the process of estimation by realizing the numerical analysis.

3 THE MODEL VERIFICATION

3.1 Test method and data

The two different groups of dynamic torsional shear experiments were used to test the rationality of parameters for dynamic soil structures, which were expressed by a single variable model in this article. Each set of soil samples had included a natural undisturbed sample with dry density (ρ), a remolded sample with the same moisture content and dry density, and a saturated undisturbed sample. The specific properties of soil samples are as follows: Luochuan loess was taken from a soil at a depth of 3 m; its natural moisture content was measured as 12%, which was obtained by using the drying method, whereas, the dry density and the saturated moisture content were approximately equal to 1.32 g/cm³ and 38% respectively.

The soil sample is the hollow cylinder with an outer diameter of 70 mm, an inner diameter of 30 mm and the height of 100 mm. The soil samples include both undisturbed and remolded samples. The undisturbed samples were extracted directly from the natural loess, the remolded samples were pressed with uniform velocity by using the dynamic torsional shear three-axis pressure injector. After the collection of soil samples, we have used the method of air-drying and water-film transfer to control the moisture content of the samples. The saturated samples were extracted directly from the bottom of samples by employing the technique of water penetration, and, in this regard, the dynamic torsional shear three-axis apparatus was used.

The tests were conducted by using the instrument, i.e., DTC-199 type of electro-hydraulic periodic torsional loading three-axis apparatus. These tests followed the vibration triaxial tests, which fulfilled the specifications of "regulations of soil test [SL237-1999]". The tests, adopted in this article, have exhausted the consolidation for drainage with confining pressures of 100 KPa and 200; and, in this regard, they used the consolidation ratio of 1.5.

During testing, the frequency of a periodic horizontal shear was pressed in accordance to the equivalent form of a sine wave on the top of the sample and that was 1 Hz. When drainage was not exhausted, we have imposed a fixed vibration times (10) load until the destruction of sample. The relationship curve between the dynamic shear stress and the dynamic shear strain, as obtained during the test, has been shown in Figure 1; the parameters of curve fitting have been shown in Table 1.

3.2 Model test

The dynamic parameters of soil structures (i.e., $m_{\tau d}$ and m_d) were obtained as follows: (1) the $m_{\tau d}$ was based on the dynamic shear strain, which was obtained by reading directly the test data, therefore, it can be treated as the "measured" values; and (2) the m_d was based on the method of dynamic shear stress, which was obtained by directly calculating the Eq. (13), therefore, it can be treated as "calculated" values. Similarly, the dynamic parameter of soil structures, i.e., $m_{\gamma d}$ was based on the dynamic shear stress, which was taken as the "measured" values; whereas, the dynamic parameter of soil structure, i.e., m_d was based on the dynamic shear strain, which was obtained by directly calculating the Eq. (18), therefore, it can be treated as the "calculated" values. The rationality of the mode was judged by comparing the two types of numerical relationship between the measured values and the calculated values. Their respective numerical relationships had been shown in Figures 2 and 3, and Tables 2 and 3.

During the inspection of models for the parameters of soil structures, which were based on the stress method, the "calculated" values were assessed by using the Eq. (1); whereas, the "measured"

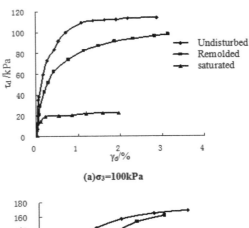

(a)σ_3=100kPa

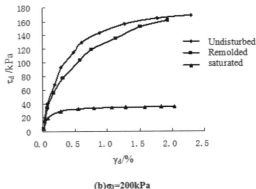

(b)σ_3=200kPa

Figure 1. The curves of dynamic shear stress and strain for Luochuan loess.

Table 1. Optimized parameters, as obtained from the curve fitting.

Confining pressure σ_3/kPa	Specimen type	$a/10^{-4}$	$b/10^{-4}$	R^2
100	Undisturbed loess	0.15	80.59	0.99
	Remolded loess	0.28	95.47	1.00
	Saturated loess	0.37	419.24	0.96
200	Undisturbed loess	0.16	52.07	1.00
	Remolded loess	0.24	50.97	1.00
	Saturated loess	0.22	268.27	1.00

values were calculated by using the Eq. (13). When the confining pressure was 100 KPa, the dynamic shear stresses of controlling conditions were taken as: 0, 5, 10, 15, 20 KPa; whereas, when the confining pressure was 200 KPa, the dynamic shear stresses of controlling conditions were taken as: 0, 5, 10, 15, 20, 25, 30 KPa. The numerical relationship, so obtained, has been shown in Figure 2 and Table 2.

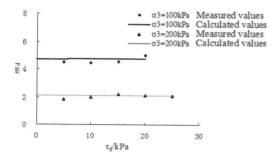

Figure 2. The measured point for the parameters of dynamic soil structures on the basis of dynamic shear strain and their calculation curve, which is based on the method of dynamic shear stress.

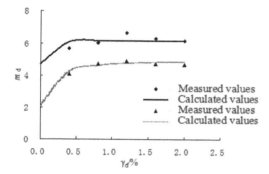

Figure 3. The measured point for the parameters of dynamic soil structures on the basis of dynamic shear stress and their calculation curve, which is based on the method of dynamic strain.

Table 2. The comparison of parameters for dynamic soil structures between their measured value on the basis of method for dynamic shear strain and the calculated value on the basis of method for dynamic shear stress.

Confining pressure σ_3/kPa	Dynamic shear stress τ_d/kPa	Measured value $m_{\tau d}$	Calculated value m_d	Errors
100	0	–	4.71	–
	5	4.48	4.72	5.36%
	10	4.47	4.73	5.74%
	15	4.52	4.74	4.80%
	20	4.99	4.75	−4.92%
200	0	–	2.09	–
	5	1.84	2.10	13.91%
	10	1.98	2.10	5.88%
	15	2.21	2.10	−4.83%
	20	2.13	2.10	−1.45%
	25	2.09	2.11	0.90%

Note: Check the rate of estimation error as: $(m_d - m_{\tau d})/m_{\tau d} \times 100\%$.

Table 3. The comparison of parameters for dynamic soil structures between their measured value on the basis of method for dynamic shear stress and the calculated value on the basis of method for dynamic shear strain.

Confining pressure σ_3/kPa	Dynamic shear strain γ_d/%	Measured value $m_{\gamma d}$	calculated value m_d	Errors
100	0.0	–	4.71	–
	0.4	5.69	6.10	7.21%
	0.8	6.01	6.17	2.56%
	1.2	6.65	6.18	−7.14%
	1.6	6.29	6.18	−1.80%
	2.0	6.15	6.18	0.47%
200	0.0	–	2.09	–
	0.4	4.10	4.25	3.74%
	0.8	4.74	4.63	−2.41%
	1.2	4.90	4.77	−2.76%
	1.6	4.75	4.84	1.77%
	2.0	4.72	4.88	3.42%

Note: Check the rate of estimation error as: $(m_d - m_{\tau d})/m_{\tau d} \times 100\%$.

During the inspection of models for the parameters of soil structures, which were based on the strain method, the "calculated" values were assessed by using the Eq. (2); whereas, the "measured" values were calculated by using the Eq. (18). When the confining pressures were 100 and 200 KPa, the dynamic shear strain of controlling conditions were as: 0, 0.4%, 0.8%, 1.2%, 1.6%, 2.0%. The numerical relationship, so obtained, has been shown in Figure 3 and Table 3.

On comparison of Figure 2 and Table 2 with Figure 3 and Table 3, we have found that the parameters of soil structures expressed by a single variable are consistent with their measured values; therefore, the model can be regarded as a "reasonable". Of course, there are still some deviations between the measured values and calculated values; their main reasons may be the numerical accuracy, the dynamic shear stress and the dynamic shear strain. These discrete points, as obtained during the testing, are often not required during the numerical analysis, therefore, the fitting of those discrete points into a curve that may be read with the numerical points cannot only increase the time involved in the calculation, but also treated as inaccurate. Keeping this in view, we have adopted a unique technique in the proposed model of this article, i.e., the dynamic parameters of soil structures can be achieved by substituting the parameters of fitting tests and variables into the model. This unique technique led to an output, which was high in accuracy and efficient in calculation.

On comparison, the calculated values when their active role were zero (refer to Tables 2 and 3), we have found that the dynamic parameters of soil structures, which were based on the methods of dynamic shear stress and dynamic shear strain, were numerically equal to each other. This finding has verified our inference that the dynamic parameters of soil structures at starting points of their active role can theoretically be treated equal (refer to Section 2.1).

4 CONCLUSIONS

On the basis of dynamic shear strain and stress, the dynamic parameters of soil structures can be derived by generalizing the parameters of soil structures under the static conditions. During their analysis, a unified expression has been obtained, which can be used to employ the dynamic parameters of soil structures by expressing the dynamic shear modulus. In this article, a model on the basis of parameters for soil structure has been built, by describing a single-variable in its active role.

The proposed model has actually been examined on the basis of the methods for dynamic shear stress and the dynamic shear strain; their beneficial results have been shown in this article, of which, some key findings can be summarized as follows:

The parameters of soil structures on the basis of the dynamic shear strain and the dynamic shear stress can be regarded as "unified", i.e., the dynamic parameters of soil structures can be expressed along with their dynamic shear moduli. The dynamic parameters of soil structures along with a unified form on the basis of their dynamic shear modulus have a positive consequence to avoid the parameters of soil structures, which were unable to solve the initial state of dynamic soil shear stress and the dynamic soil shear strain of "zero" values.

In this article, the model for the parameters of soil structures has been developed, which is based on the methods for the dynamic shear stress and the dynamic shear strain. It can actually examine its active role for a single-variable description. The positive consequence for the dynamic parameters of the soil structures with its active role as a single variable is that it can reflect intuitively the influence of its active role on the dynamic parameters of soil structures and it can simplify the process of calculation by realizing the speed of its numerical calculation.

Since the relationship between the soil dynamic shear stress and the soil dynamic shear strain is not linear and elastic, therefore, the dynamic parameters of soil structure on the basis of methods for the dynamic shear stress and the dynamic shear strain are different. Keeping this in view, it is always necessary to indicate the relevant method, which is being used to determine the parameters of a dynamic soil structure. The parameters for the dynamic soil structure are theoretically equal to the starting point of its active role. Owing to the reason that the dynamic shear stress and shear strain from the starting point of their active role are zero, the magnitude of dynamic shear modulus represents its maximum extent and it has nothing to do with either the soil dynamic shear stress or the dynamic shear strain.

REFERENCES

Tian Kan-liang, Zhang Hui-li. Dynamic test study of loess structural property based on strength conditions [J]. Rock Mechanics and Engineering, 2010, 29(11):2356–2361.

Shen Zhu-jiang. Mathematics model of soils structure—the key issue of 21th century soil mechanics [J]. Chinese Journal of Geotechnical Engineering, 1996, 18(1):95–97.

Xie Ding-yi, QI ji-lin, Soil structure characteristics and new approach in research on its quantitative parameter [J], Chinese Journal of Geotechnical Engineering, 1999, 21(6):651–656.

Qi Ji-lin, Soil structure characteristics and study of its quantitative parameter [D], Xi'an: Xi'an University of Technology Doctoral Dissertation, 1999.

Shao Sheng-jun, Zhou Fei-fei, Long Ji-yong. Study of the structural and quantitative parameters of undisturbed loess [J]. Chinese Journal of Geotechnical Engineering, 2004, 26(4):531–536.

Chen Cun-li, Gao Peng, He Jun-fang. Undisturbed loess taking into account the structural effects of the equivalent linear model [J]. Chinese Journal of Geotechnical Engineering, 2007, 29(9):1330–1336.

Chen Cun-li, Hu Zai-qiang, Gao Peng. Structural and deformation characteristics of undisturbed loess [J]. Rock and Soil Mechanics, 2006, 27(11):1891–1896.

Chen Cun-li, Gao Peng, Hu Zai-qiang. Moistening deformation characteristics and its structural relationship of the loess [J]. Rock Mechanics and Engineering, 2006, 25(7): 1352–1360.

Luo Ya-sheng, Xie Ding-yi, Shao Sheng-jun. Soil structural parameters under complex stress conditions [J]. Rock Mechanics and Engineering, 2004, 23(24):4248–4251.

Luo Ya-sheng, Xie Ding-yi. Structural constitutive relation of soils under complex stress conditions [J]. Journal of Sichuan University (Engineering Science), 2005, 37(5):14–18.

Luo Ya-sheng. Variation characteristics of soil structure and structural constitutive relation of unsaturated loess [J]. Xi'an: Xi'an University of Technology Doctoral Dissertation, 2003.

Wang Zhijie, Luo Ya-sheng, Yang Yong-jun. Study of dynamic structural characteristics of unsaturated loesses in different regions [J]. Rock and Soil Mechanics, 2010, 31(8):2459–2464.

Advances in Energy, Environment and Materials Science – Wang & Zhao (Eds)
© 2016 Taylor & Francis Group, London, ISBN 978-1-138-02931-6

Study of influencing factors on the strain localization formation of saturated clay under the plane strain condition

Xiangyu Gu & Taiquan Zhou
School of Environment and Civil Engineering, Jiangnan University, Wuxi, Jiangsu, China

ABSTRACT: Production condition and formation mechanism of normally consolidated clay and lightly over-consolidated clay under the plane strain condition were analyzed by the modified Cam clay model using the finite element software ABAQUS. Production condition and formation mechanism of the shear band were studied by influencing factors on strain localization formation. The influencing factors including loading speed, size of the soil specimen, normally consolidated condition and lightly over-consolidated condition have an effect on the distribution of pore water pressure, variation of the effective stress path of soil in shear band and out of shear band and the angle between the shear band and horizontal line in the formation process of the strain localization band. The results show that the size of soil specimen can influence the number of shear bands. Loading speed can influence the distribution of the void ratio. Consolidation state can influence the stress path of nodes, which are inside and outside the shear band. The influencing factors will affect the angle between the shear band and the horizontal line, the width of the macroscopic shear band, and the starting position of the shear band. The formation mechanism of the shear band is that during the loading process, the void ratio of elements on some shear surface increases significantly, but the effective stress decreases. Elements on this section reach the critical bearing capacity ahead of others. At the same time, the strain energy of other elements will be released rapidly, which leads to the deformation of this section that increases rapidly and then to the formation of the macroscopic shear band.

1 INTRODUCTION

Compared with concrete and rock-like materials, strain localization of clay soil generally occurs by water–soil coupling. Because formation and development of shear bands are affected by the deformation properties of soil, boundary constraint conditions and loading speed, it is certainly difficult to study the strain localization of shear bands of clay soil. So far, studies of shear bands are mainly concentrated on the aspects of production condition and influencing factors of shear bands, width of shear bands, angle between the shear bands and horizontal line and numerical simulation of formation and development of the shear band. In addition, researchers at home and abroad have so far studied the characteristics of the shear band in three aspects including theoretical analysis, soil test and numerical analysis. Work has been carried out on numerical simulation as follows. Xuebin Wang (2005) analyzed the phenomenon of refraction and reflection of the shear band of rocks by using the finite difference method. Zhengyin Cai (2008) suggested that there is a close link between the strain localization and characteristics of strain softening of materials. Lianmin Xue (2005 & 2006) studied the characteristics of the shear band of

over-consolidated clay, including the boundary effect and the effect of the loading rate on the basis of Nakai's sub-loading constitutive model. Castelli (2009) simulated the propagation law of the shear band in a soil specimen with initial imperfection using the boundary element method, and found that the initial imperfection of materials is one of the factors for the formation of the shear band. The above research has mostly studied the influencing factors and propagation law of the shear band; meanwhile, researchers have studied the relationship between unstable failure and formation mechanism of the shear band from the perspective of the constitutive model and the bifurcation theory.

In this paper, the production condition and formation mechanism of normally consolidated soils and lightly over-consolidated clay under the plane strain condition were analyzed by the modified Cam clay model using the finite element software ABAQUS.

2 CALCULATION MODEL OF ABAQUS AND RELATED PARAMETERS

Under the plane strain condition, the specimen size, whose thickness is the unit one, is shown in Figure 1 and Table 1. Specimen A was divided into

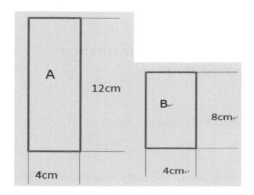

Figure 1. Specimen size.

Table 1. Specimen size.

Numbering of model	Strain rate (%)	Model size (cm)
A_N		
1	1	4×12
2	0.1	
3	0.01	
$B_O\,B_N$		
1	1	4×8
2	0.1	
3	0.01	

Table 2. Parameters of the calculation.

Parameters	Normally consolidated	Lightly over-consolidated
Index of isotropic compression λ	0.191	0.191
Isotropic rebound coefficient κ	0.0043	0.0043
Initial void ratio	1.10	1.12
Compressed yield stress	200 (kpa)	200 (kpa)
Initial average effective stress	200 (kpa)	120 (kpa)
Stress ratio on critical state M	1.14	1.14
Permeability coefficient k	$1.63*10^{-9}$	$2.70*10^{-9}$
Poisson's ratio ν	0.3	0.3

300 (10*30 = 300) elements and Specimen B was divided into 128 (8*13 = 128) elements; meanwhile, the specimens were longitudinally and bilaterally symmetric. First, confining pressure was applied to simulate the initial consolidation condition. Then, the simulation process was divided into two steps. First, isotropic consolidation pressure was applied, at the pressure of 200 kpa under the normally consolidated condition and at the pressure of 120 kpa under the lightly over-consolidated condition. Second, displacement loading was applied, in which the vertical displacement was 20 percent of the height of the corresponding specimen. We ensured the flow of the pore water and the uniformity of pore pressure in the specimens. What's more, the rate of vertical loading was far less than the permeability coefficient. The initial condition of the calculation and material parameters of the modified Cam clay model are summarized in Table 2, where N represents normally consolidated condition and O represents lightly over-consolidated condition.

The upper end and the lower end of specimen are assumed as the drainage boundary. Boundary conditions, whose upper end and lower end are displacement boundary conditions, are shown in Figure 2a, b. We ensured that the longitudinal strain of points in

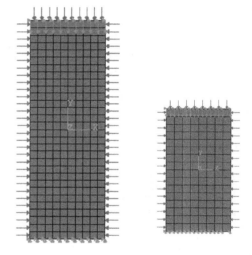

Figure 2. a) Model A and boundary conditions b) Model B and boundary conditions.

the upper end was always equal in the loading process and the lower end of the specimens was fixed.

Formation mechanism of shear band was analyzed in three aspects of specimen size, loading speed and consolidation state as follows.

3 RESEARCH ON INFLUENCING FACTORS

3.1 *Effect of model size on the shear band*

When the vertical compressive displacement of the top of specimen B_{N-1} was 16 mm (20 percent of the height of specimen B_{N-1}), under the conditions of simulating drainage consolidation, the calculation results showed that the specimen was structurally symmetric and the top of the specimen developed two intercrossed shear bands. We then kept the calculation condition unchanged and changed only the specimen size. When the vertical compressive displacement of the top of specimen A_{N-1} was 24 mm (20 percent of the height of specimen A_{N-1}), the specimen was divided into two parts, namely the upper part and the lower part. Every part developed two intercrossed shear bands and the distance between the end of the shear band and the top of the specimen decreased. Meanwhile, the angle between the shear band and the horizontal line decreased and the width of the shear band decreased.

3.2 Effect of loading speed on the shear band

Given the calculation models of specimens B_N, A_N, and B_O under the conditions of axial strain reaching 20% in the loading process, the results showed that the inclination angle of the shear band under the high strain rate condition was larger than the angle under the low strain rate condition, but the width of the shear band under the low strain rate condition was larger than that under the high strain rate condition. The distribution of pore water pressure of normally consolidated clay and lightly over-consolidated clay was analyzed. The distribution of pore water pressure of the specimen, in the conditions of strain rate reaching 0.1%/min, was relatively uniform. The difference between the maximum and minimum values was relatively small; meanwhile, it increased under the conditions of the strain rate reaching 1%/min.

3.3 Effect of consolidation state on the shear band

Given the calculation models of specimens B_{N-1} and B_{O-1} under the conditions of axial strain reaching 20%, the results showed that the inclination angle of the shear band under the lightly over-consolidated condition was larger than the angle under the normally consolidated condition and so was the width of the shear band. The log strain of the models showed that, in the same position, the log strain of the lightly over-consolidation specimen was larger than that of the normally consolidation specimen. What's more, pore water pressure of the lightly over-consolidated specimen reached the peak ahead of others with the increasing strain. In addition, the growth rate of pore water pressure of the lightly over-consolidated specimen was lower than that of pore water pressure of the normally consolidated specimen.

3.4 Mechanism analysis on the shear band

The positions of three picked nodes of specimen B are shown in Figure 3. Node 77 is located in the center of the shear band and nodes 74 and 23 are located outside the shear band. Nodes 77 and 74 are located in the same horizontal line; meanwhile, nodes 77 and 23 are located in the same vertical line.

Pore water pressure–axial strain curve of three nodes of specimen B_{N-1} was analyzed. The results showed that the pore water pressure in the shear band was in full accord with that outside the shear band.

As shown in Figure 4, the void ratio of soil near the shear band increased significantly. The void ratio

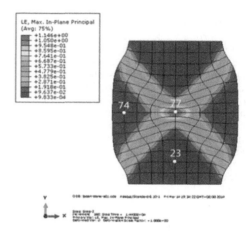

Figure 3. Positions of the nodes of specimens B.

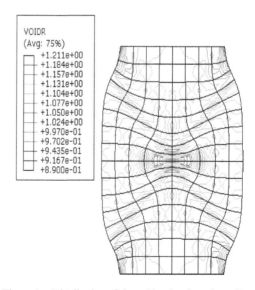

Figure 4. Distribution of the void ratio of specimen B_{N-1}.

of soil near the shear band increased and the pore water pressure decreased, which led to the decrease in the shear strength of the region, thus resulting in the formation of the shear deformation zone.

Stress paths of three nodes of specimen B_{N-1} were analyzed. The results showed that stress paths in the shear band were in accord with those outside the shear band before the peak shear stress appeared. When the shear band appeared, there were differences between stress paths in the shear band and those outside the shear bands. Furthermore, stress paths outside the shear band started to unload, but those in the shear band presented loading morphology until they intersected with the critical state line. According to the frontal analysis, the void ratio of node 77 in the shear band increased, so the bearing capacity of the nodes decreased rapidly to form a weak element, which led to the formation of the macroscopic shear band.

E-ln p curves of three nodes of specimen B_{N-1} were analyzed. The results showed that nodes including 77, 74, and 23 started to move from the isotropic normal consolidation line to the critical state line before the shear band appeared. Node 74 (outside the shear band) and node 23 (in the shear band) were in a significantly unloading expanded state after the shear band appeared.

In the same way, the positions of three picked nodes of specimen A_{N-1} are shown in Figure 5. The three picked nodes are 248, 244, and 171. The pore water pressure–axial strain curve of three nodes of specimen A_{N-1} was analyzed. The results showed that the pore water pressure in the shear band was in full accord with those outside the shear band. In Figure 6, the distribution of the void ratio of specimen A_{N-1} was analyzed. The results showed that the void ratio of soil in the shear band increased and the effective stress decreased. The stress paths

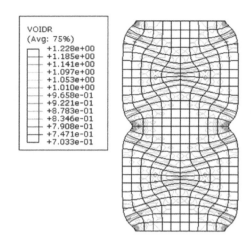

Figure 6.　Distribution of the void ratio of specimen A_{N-1}.

of three nodes of specimen A_{N-1} were analyzed. Furthermore, the E-ln p curves of three nodes of specimen A_{N-1} were analyzed. The results showed that the three nodes started to move from the isotropic normal consolidation line to the critical state line before the shear band appeared. Node 244 (outside the shear band) and node 171 (in the shear band) were in a significantly unloading expanded state after the shear band appeared. The results were in full accord with the results of specimen B_{N-1}.

The lightly over-consolidated specimen B_{O-1} was analyzed. The positions of three picked nodes of specimen B_{O-1} are shown in Figure 3. The pore water pressure–axial strain curve of three nodes of specimen B_{O-1} was analyzed. In addition, the distribution of the void ratio of specimen B_{O-1} was analyzed. The results showed that the pore water pressure in the shear band was in full accord with that outside the shear band. What's more, the void ratio of soil near the shear band increased more significantly.

The stress paths of three nodes of specimen B_{O-1} were analyzed. The results showed that the stress path of node 77 in the shear band was in accord with that of node 74 outside the shear band, but the stress of node 23 below the shear band had a small increase before the peak shear stress appeared. When the shear band appeared, stress paths in the shear band were not in accord with those outside the shear band, and stress paths outside the shear band started to unload when node 23 was ahead of node 74. Furthermore, stress paths in the shear band were in the loading state until they intersected with the critical state line.

The E-ln p curves of three nodes of specimen B_{O-1} were analyzed. The results showed that the three nodes started to move to the critical state line before the shear band appeared. However, node 74

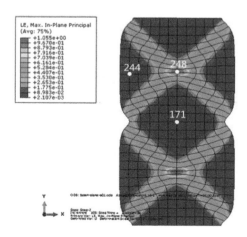

Figure 5.　Positions of the nodes of specimen A_{N-1}.

and node 23 were in a significantly unloading the expanded state after the shear band appeared.

According to the previous analysis, in the process of forming the macroscopic shear band, the changes of pore water pressure were consistent in the specimen. The void ratio near the shear band increased significantly, but the effective stress and bearing capacity decreased. When loading reached the peak stress, elements outside the shear band released the strain energy rapidly. Moreover, elements in the shear band could not bear large deformation, which led to the formation of the macroscopic shear band.

4 CONCLUSION

Model size: specimen size is the factor that affects the characteristics of the shear band. When the height-to-width ratio is small, the specimen develops two intercrossed shear bands. When the height-to-width ratio is large, the specimen develops two or more pairs of intercrossed shear bands; meanwhile, the inclination angle of the shear band and the width of the shear band are decreased.

Loading speed: loading speed is the factor that affects the characteristics of the shear band. The larger the loading speed is, the larger the inclination angle of the shear band is and the more uniform the distribution of the void ratio is. Furthermore, the width of the shear band decreases.

Consolidation state: consolidation state is the factor that affects the characteristics of the shear band. The inclination angle of the shear band under the lightly over-consolidated condition was larger than the angle under the normally consolidated condition and so was the width of the shear band. This is because the lightly over-consolidated specimen whose void ratio is larger is destroyed ahead of the normally consolidated specimen.

Explanation on the formation mechanism of the shear band: shear band has a number of appearances. During the loading process, the void ratio of elements on some shear surface increases significantly, but the effective stress decreases.

Elements on this section reach the critical bearing capacity ahead of others. At the same time, the strain energy of other elements will be released rapidly, which leads to the deformation of this section that increases rapidly and then to the formation of the macroscopic shear band.

5 DISADVANTAGES AND FUTURE PROSPECT

The research objective of this article involves normally consolidated clay and lightly over-consolidated clay. In fact, heavy over-consolidated clay often has some characteristics that are different from those of normally consolidated clay and lightly over-consolidated clay. The simulation method proposed in this article can be applied to the research of heavy over-consolidated clay to develop the research objectives and research range of this article.

Specimen size, loading speed and consolidation state were considered in this article. However, initial imperfection and restrained condition also have an effect on the shear band in practical engineering. Therefore, this is also important to simulate other influencing factors by using the method proposed in this article.

REFERENCES

Castelli M., Allodi A., Scavia C. A numerical method for the study of shear band propagation in soft rocks. *International Journal for Numerical and Analytical Methods in Geomechanic* 2009, 33(13): 1561–1587.

Lianmin Xue, Hehua Zhu, NAKAI Teruo. Numerical simulation of shear band in overconsolidated clay. *Rock and soil Mechanics* 2006, 27(1): 61–66.

Lianmin Xue. Effects of boundary condition and loading speed on shear band localization. *Journal of Hydraulic Engineering* 2005, 36(1): 9–15.

Xuebin Wang. Complexity of shear band patterns and discreteness of stress-strain curves. *Rock and soil Mechanics*, 2005, 26(1): 25–30.

Zhengyin Cai. Progressive failure of sand and its numerical simulation. *Rock and soil Mechanics* 2008, 29(3): 580–585.

The experiment and study of concrete inorganic protective preparation in cold regions

Shaozhen Wang

Tianjin College, University of Science and Technology Beijing, Tianjin, China

ABSTRACT: We examined the cold region of the concrete inorganic protective agent through the experimental study of the concrete 30 flexural, the carbonization and the wear-resisting influence. The experimental results indicated that this inorganic protective agent produced on the concrete 30 can increase its flexural strength by about 10%, and the carbonization and the wear resistance can be, respectively, improved by 36 and 45%. This product was used in the Ningan Wolong River Reservoir in Heilongjiang Province for strengthening purposes, which produced an ideal effect.

1 INTRODUCTION

Impermeability of concrete, frost resistance and compressive strength are important indicators of the durability of concrete. Similarly, anti-carbonation properties of concrete, abrasion resistance, and flexural strength are the durability of concrete evaluation criteria. To further test the effect of the inorganic protective agent of concrete in cold regions, we performed bending, carbonation and abrasion tests.

2 TEST ANALYSIS

2.1 Analysis of flexural strength

Experiment with universal hydraulic tester flexural strength of concrete specimens, methods of operation and the results referring to JTG E30-2005, using a 150 mm × 150 mm × 550 mm specimen of each of group 3 were conducted. The specimen was subjected to a standard curing condition for 3 days, 7 days, 14 days, and 28 days. According to the preparation of six surfaces evenly brushing, the samples were placed on a hydraulic universal testing machine to measure the flexural strength of concrete bl°Ck. The test result is shown in Table 1 and Figure 1.

As can be seen from Table 1 and Figure 1, after applying the inorganic protective agent of concrete in cold regions, the flexural strength of concrete increases significantly. For example, C30 concrete's flexural strength increases by more than 10% when it is brushed.

2.2 Analysis of carbonation resistance

On the basis of "hydraulic concrete testing procedures" SL352-2006, anti-carbonation specimen

Table 1. C30 concrete test results.

Name category	Concrete 30	
	Original specimen	Test specimen
Carbonation depth/mm	18.7	11.9
Abrasion quality/g	12.3	6.7
Flexural strength/MPa		
3 days	24.2	26.9
7 days	31.5	34.8
14 days	39.2	43.7
28 days	42.1	49.2

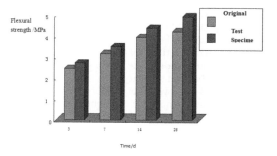

Figure 1. Flexural strength test result.

using 100 mm × 100 mm × 100 mm cube concrete specimens, each of group 3, the reference member and the brushing member of a group were performed. Under the standard curing condition for 28 d, according to the preparation method of treatment even for brushing the sides, on the carbonization chamber after the standard acceleration test

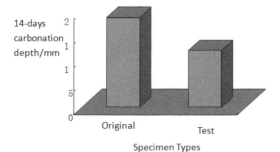

Figure 2. Anti-carbonation test results.

lasting 14 days, the carbonation depth of concrete in the cold area was obtained before brushing the inorganic protective agent of concrete specimens. The result is shown in Table 1 and Chart 2.

As can be seen from Table 1 and Figure 2, after applying the concrete inorganic repellant in the cold region, the smear specimen's carbonation depth for 14 days is lower than the reference member. For example, C30 concrete specimens, after brushing the concrete in cold regions of the inorganic protective agent, under standard conditions after carbonization for 14 days, its carbonation depth is reduced by 36%. Visible, concrete inorganic repellant brushing in cold regions can improve the carbonation resistance of concrete.

2.3 Analysis of abrasion

Evaluation of abrasion resistance of concrete pavement does not currently have clear test criteria to follow, and not easy to simulate in the laboratory scene abrasion conditions. We used the abrasion scrubbing method (DL/T 5150-2001) proposed by the American scholar T.C. Liu, and inorganic abrasion tests were carried out to improve the greatest degree of simulation of concrete structures under the actual abrasion situation.

Figure 3 shows the underwater abrasion test machine. When we scrubbed the size of φ30 cm × 10 cm cylindrical concrete specimens, placed in an inner diameter of 31.1 cm, 45.0 cm high in the steel drums, with the speed of 14010 rad/min the bench driven submerged paddle stirring rod that rotated the concrete surface water so that the ball went back and forth jumping and rolling (ball specifications: diameter of 25 mm, 10, 35 to 19 mm, 12.5 mm of 25). Thus, the surface of concrete specimens has an impact abrasion.

The value method is given in Equation (1), with mass loss of concrete specimen's representative value for the amount of abrasion being accurate to 0.1 g:

$$\Delta g = g_0 - g_1 \qquad (1)$$

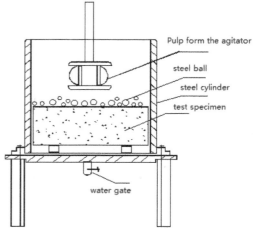

Figure 3. Underwater abrasion test machine.

where
Δg-test bl°Ck abrasion mass loss;
g_0-saturated surface dry before scrubbing, the total mass of the specimen;
g_1-after scrubbing surface dry after saturation, the total mass of the specimen.

The test selects to characterize the quality loss abrasion resistance. Underwater punch used in the test specimen grinding the test machine will grind for 12 h. Mass loss of concrete is shown in Table 1 and Figure 4.

Brushing cold area concrete inorganic repellant, abrasion resistance and abrasion resistance of concrete reference concrete specimens are compared to a different extent. From Table 1 and Chart 4, it can be seen that the use of concrete in the cold area of the specimen 12 h inorganic repellant abrasion mass is smaller than the reference quality abrasion parts, suggesting that the abrasion resistance of the test piece out-performs the benchmark piece.

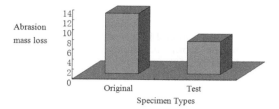

Figure 4. Abrasion test result.

Under the same circumstances, the mass loss of C30 concrete specimens is at least 45% less than the reference members, suggesting that the concrete inorganic protective agent in cold regions can improve the abrasion resistance of concrete.

3 MECHANISM

Cold regions' concrete inorganic repellant improves and enhances the durability of concrete based on the following mechanisms.

1. Inorganic protective material particles fill the pores of the concrete structure concrete base densely;
2. Inorganic repellant brushing the concrete in cold regions: inorganic protestant hydration reaction generates a new product so that the original loose-compacting concrete structures and concrete cement hydration products are formed;
3. Inorganic repellant brushing the concrete in cold regions: inorganic repellant layer pore structure and pore structure are compared with the original concrete compacting;
4. Cold regions' concrete inorganic repellant brushing: brushing the concrete base material interface layer that is dense and stable.

4 PROJECTS

Wolong River Reservoir is located in Ning'an, Heilongjiang Province, Shiyan Nan Toad River tributary, and it is 17 km from the control drainage area 120 m², whose total volume is 13.8 million m³, and belongs to third-class engineering, according to the 50 years of flood design, thousand years of flood check design with 1.52 acres of irrigated paddy fields. The reservoir was built in the 1980s. After decades of its operation, the presence of dam abutment leakage and reservoir spillway water hole resulted in disrepair, preventing the normal use of the reservoir. If these problems are not solved and early reinforcement measures are not taken, it will affect the safe operation of the reservoir. To ensure

Table 2. Concrete site testing data.

Name	Correction rebound value	Intensity conversion value of the surveyed area of concrete (MPa)
Origin place	35.2	28.3
Test place	34.0	30.1

(a) Not used and not over water (b) Used and Not over water

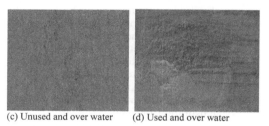

(c) Unused and over water (d) Used and over water

Photo 1. Comparison of the concrete surface effect.

the normal use of the reservoir, in May 2005, Wolong River Reservoir reinforcement project was started. The project budget and the total investment was 33.23 million yuan, the total project amount was 177 500 m³, and the construction period was 2 years. In the project reinforcing the Wolong River Reservoir; we applied the cold area's concrete inorganic repellant to the actual project and, by contrast, obtained good results.

First, after a 3-day investigation and analysis of the construction site, we selected a cross section, where the Wolong River Reservoir section was considered as a practical application test sample of the new concrete inorganic protective agent. The cross section evaluated in a year's time, after the effects of 7, 8, and 9 months of Wolong River flooding of the water, and the Northeast cold winter, showed a large temperature difference between day and night tests. It is well known that Heilongjiang Province is the country's coldest province. The average temperature range in the month of January is −30.9°C–14.7°C, and there is a relatively large temperature difference between day and night in summer and the average summer temperature is about 18°C, with the extreme maximum temperature being 41.6°C. Concrete structure in such a harsh climate condition is vulnerable to damage. After a year of practice, in order to detect the practical application of the new concrete inorganic protective agent, the actual site of the concrete cross section was analyzed by means of a field observation and non-destructive testing with inorganic repellant new concrete cross section compared with that not using the repellant cross section, which is obviously preserved.

5 CONCLUSION

1. Concrete specimens coated with concrete inorganic repellant in the cold area: its flexural strength is increased especially in the early strength of the specimen, from about 10% to 15%. This is due to the penetration of inorganic repellant surface material that reacts with the concrete compound to plug the gap, a phenomenon that particularly increases the early strength of concrete producing a great influence. The early strength of concrete with increasing conservation, micro-cracks, and voids caulking concrete can not only reduce the stress concentration area, but also make concrete inside the water not to dissipate, promote cement hydration and increase its early strength. Its late strength increase is weakened. This may be due to the cold area of the inorganic protective agent, which is primarily the result of concrete action on the concrete surface after completion of the reaction to the weakening of the role of the late strength of concrete.

2. After brushing the cold area's inorganic protective agent carbonation resistance of concrete is improved after 14 days of C30 concrete carbonation depth, the reference member can be reduced by 3.6%.

3. The test showed that the use of concrete in cold regions of inorganic protestant specimen 12 h abrasion mass is smaller than the reference quality abrasion parts. Under the same circumstances, the mass loss of C30 concrete specimens is at least 45% less than the reference member.

4. By using the actual project, it was proved that the cold region of the inorganic protective agent can be used to create concrete and concrete work can be built, which effectively improves the performance of concrete, development and application of concrete in the cold area of the inorganic protective agent, and can greatly reduce the resources waste, reduce disease concrete, and improve the efficiency of concrete use.

REFERENCES

Jielong Cai, Concrete surface protection materials Application. Guangdong Water Resources and Hydropower 2014, (11).

JingLi. New protective surface coating of concrete progress and development. Shanghai Coatings, 2013, (6).

Shengli Zhang, Bridge concrete structure of the surface protective coating applied research concrete [J], 2006, (5): 91–93.

Advances in Energy, Environment and Materials Science – Wang & Zhao (Eds)
© 2016 Taylor & Francis Group, London, ISBN 978-1-138-02931-6

Research on the effect of the durability of arch foot concrete of the Xixihe bridge as the reinforcement corrosion

Kai Li

The 2nd Engineering Co. Ltd., China Railway Construction Bureau, Hebei Tangshan, China

ABSTRACT: This article establishes a nonlinear finite element model based on the general finite element method, and simulates the stress of the reinforced rust caused by the reinforced concrete in the arch foot of the Xixihe bridge. Based on the finite element numerical analysis, the calculation method of the corrosion rate of reinforced concrete corroded reinforcement is presented. Numerical calculation and analysis, including the concrete component angle of reinforcement corrosion and central reinforcement corrosion of uniform corrosion expansion force of concrete member, based on previous reinforcement corrosion distribution theory analysis results, combines with the results of the numerical analysis of the corrosion expansion force, to calculate the arch foot concrete cover cracking moment of steel corrosion rate, predict the Xixihe bridge arch foot concrete protective layer cracking time, and provide the reference for the design.

1 ENGINEERING BACKGROUND

With the increasing cost of engineering maintenance, the durability has become a serious problem in construction engineering. The corrosion cracking of reinforced concrete is often considered as one of the marks of the normal limit state of the structural members of concrete structures. The numerical analysis method can solve many problems that cannot be solved or are difficult to solve. It is a powerful tool for analyzing and studying the structure of concrete. With a deeper understanding of material properties, material constitutive relations correctly reveal, and using computer software and hardware equipment, the numerical analysis method that has become a concrete structure analysis is an important method.

For the arch foot of the Xixihe Bridge, the numerical analysis method can be used to determine the concrete members with corroded expansion cracking moment of corroded expansion pressure, and then according to the concrete cover cracking moment corrosion expansion pressure and corrosion rate of the relationship, the concrete cover cracking moment reinforcement corrosion rate calculation method can be put forward, which is very meaningful to study the durability of the concrete structure.

The starting mileage of the new railway Xixihe bridge in DK416 + 910.55 include: the length of DK417 + 404.15, the main span as 240 m, the main span as 493.6 m, the bearing type as the CFST arch bridge, and the main girder section form as the box girder. The main arch ring of the CFST structure and the arch span center site is 240 m. The main arch is in the inclined plane of the arch axis for catenary, with the arch axis coefficient (m) of 2.2 and the span ratio of 1/4.364. The arch rib height is 5.7 m, 3.0 m wide, each rib by 4 limb Phi 1100×20 mm steel pipe, the upper and lower chord all by steel pipe and two limbs during two 20 mm-thick steel plate connections

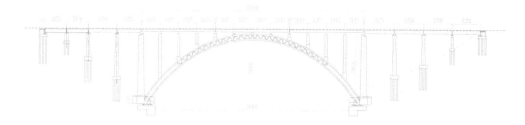

Figure 1. Xixihe bridge layout plan.

is dumbbell-shaped, and the length of the arch rib is uniform; from arch foot arch rib ends all about 53.0 m range between the lower chord each by two 16 mm-thick steel plate connections, which constitute the solid web section, the arch rib section is in a box. The overall structure diagram and the arch cross section diagram of the bridge are shown in Figure 1.

2 FINITE ELEMENT ANALYSIS

2.1 Model establishment

In this paper, the analysis of the steel corrosion pressure of the concrete members is carried out by using ANSYS finite element software. In the basic unit of the ANSYS finite element program, the element SOLID65 can describe the crack of concrete unit. Therefore, SOLID65 is often applied in the analysis of concrete stress and cracking. Because this research focuses on the phenomenon that the concrete is cracked by the rust pressure, in order to simulate the crack of the crack of the concrete, the SOLID65 element is adopted. According to the characteristics of SOLID65 element, the analysis model of constitutive relation and failure criterion was applied to the Prager model and Willam & Warnke five-parameter failure criterion, crack dispersed fixed crack model (smeared fixed crack model), and the displacement convergence criterion. The concrete size of the finite element analysis model is 2350 mm*3000*650 mm, the thickness of the reinforcement layer is 40 mm, the diameter of the reinforcement is 30 mm, and the concrete is C45. The diagonal part and the middle bar are, respectively, analyzed and simulated. Figure 2 shows the finite element model, and the position of the reinforcement is 1~5 number, respectively. At the bar position, the ring force is used to simulate the steel bulge, as shown in Figure 3.

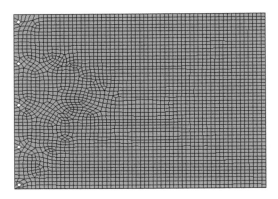

Figure 2. Finite element model.

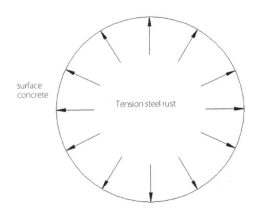

Figure 3. Schematic diagram of the reinforced rust expanding force.

2.2 Result analysis

According to the finite element calculation analysis, when the corner corrosion of uniform corrosion expansion pressure reaches 1.45 MPa and concrete member is subjected to the corrosion expansion pressure that reaches the limit value, the ring to the corrosion expansion pressure is added to the maximum value and cannot continue loading. Department of steel corrosion produced uniform corrosion expansion pressure that reaches 1.52 MPa, and concrete member is subjected to the corrosion expansion pressure that reaches the limit value, as presented in Table 1. The results of the calculation of the corrosion of the central bar are more accurate than the calculation results of the corner. This is because the corrosion of the central bar has more peripheral concrete constraints. The corrosion of reinforcement in the diagonal part and the middle part is analyzed, respectively, and the method and results are significant for determining the corrosion pressure of the reinforced concrete at the crack moment. Figure 4 shows a concrete cracking moment of corroded expansion force diagram; the figure can be clearly seen at different positions of reinforced in the corroded expansion force that leads to the distribution of the size of the concrete cracking moment of the corroded expansion force, from which it can be obviously seen that the middle position of the concrete is not easy to crack, and the corners of the concrete are easy to crack.

3 Determination of steel corrosion

For the uniform corrosion of the rebar, the corrosion of the steel bar will produce a bigger pressure of rust. According to the earlier research results of Zhao Yuxi et al., the corrosion of rebar and the

Table 1.	Different positions of reinforced corrosion at different positions of concrete.				
Steel bar	1	2	3	4	5
Rust swelling pressure/MPa	1.45	1.48	1.52	1.46	1.42

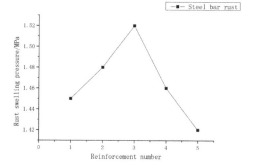

Figure 4. Schematic diagram of reinforced rust expansion force of concrete cracking.

Table 2.	The corrosion depth and corrosion depth of the reinforced bar at different positions.	
Reinforcement number	$q(\theta)$/MPa	x/μm
1	1.45	2.11
2	1.48	2.36
3	1.52	2.49
4	1.46	2.19
5	1.42	2.01

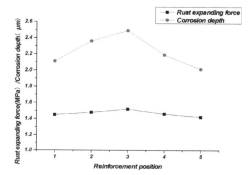

Figure 5. Relationship between the corrosion depth and the corrosion depth of the rebar.

corrosion of the reinforced pressure are satisfied by formulas (1) and (2) as follows.

Where $q(\theta)$ is the corrosion expansion pressure; x is the depth of steel corrosion; n is the volume expansion of corrosion rate; R is the bar radius c concrete protective layer thickness; E_c are the concrete elastic modulus and Poisson's ratio; and E_r are the rust of the elastic modulus and Poisson's ratio. The relationship between the different reinforcement corrosion pressures and the depth x of the rebar rust layer can be determined by the following formula:

$$q(\theta) = \frac{\left(\sqrt{(n-1)\cdot\rho(\theta)+1}-1\right)R}{\dfrac{(1+v_c)R(R+c)^2 + (1-v_c)(R+c)^3}{E_c(2R\cdot c + c^2)}} + \dfrac{n(1-v_r^2)R\cdot\sqrt{(n-1)\cdot\rho(\theta)+1}}{E_r\left\{\left[(1+v_r)n-2\right]+2/\rho(\theta)\right\}} \quad (1)$$

$$x = R - \sqrt{1-\rho(\theta)}\cdot R \quad (2)$$

According to formulas (1) and (2), the corrosion rate of the reinforcement of the concrete protective layer can be determined, as shown in Table 2. From Table 2, it can be seen that the corrosion expansion force of Steel No. 3 is 1.52, and the maximum corrosion depth of Steel No. 3 is 1.52, indicating that the reinforced durability of the number 3 position is the best, and not susceptible to corrosion.

The rust of the number 5 position is the smallest (only 2.01), and the corrosion depth of the rebar is 1.42; therefore, the durability of the 5 position is the worst, and most susceptible to corrosion. According to this method, the failure of the concrete structure reinforcement corrosion depth can be estimated effectively, which has a great significance for the study of the durability of the concrete structure such as the Xixihe bridge arch foot concrete structure design to provide a reference.

4 CONCLUSION

1. Through the establishment of the finite element model, the Xixihe bridge arch foot concrete uniform corrosion expansion pressure is simulated and analyzed, including the angle steel corrosion and the central reinforcement corrosion. The finite element model can better simulate the concrete protective layer by the corrosion expansion pressure cracking, and calculate the concrete protective layer failure corresponding to the corrosion expansion pressure.

2. According to the results of the theoretical analysis and the finite element analysis, the concrete cover cracking moment reinforcement corrosion rate calculation method is put forward.

With the known concrete conditions and structural erosion, this method can estimate the concrete protective layer cracking moment of the steel bar corrosion rate, which is beneficial for the study of the durability of the concrete structure.

REFERENCES

Alonso C., Andrade C., Rodriguez J., et al. Factors controlling cracking of concrete affected by reinforcement corrosion [J]. Materials and Structures, 1998, 31:435–441.

Liu Y.P., Weyers R.E., Modeling the time-to-corrosion cracking in chloride contaminated reinforce concrete structures [J]. ACI Materials Journal, 2012, 95:675–681.

Molina F.J., Alonse C., Andrade C. Covercracking as a function of bar corrosion: part 2-Numberical model [J]. Materials and Structures, 2007, 26:532–548.

Yan Fei. Atmospheric environment [D]. Research on several problems of concrete durability. Hangzhou: Zhejiang University, 1999.

Zhang Weiping. Damage prediction of reinforced concrete structures and its durability assessment [D]. Shanghai: Tongji University, 1999.

Zhao Yu, Jin Weiliang. Corrosion cracking of reinforced concrete members during corrosion cracking of steel bars [J]. Journal of water conservancy, 2004, 35 (11): 97–101.

Zheng Jianjun, Zhou Xinzhu. Analytical solution for corrosion damage of reinforced concrete structures [J]. Journal of hydraulic engineering, 2004, 35 (12):62–68.

Advances in Energy, Environment and Materials Science – Wang & Zhao (Eds)
© 2016 Taylor & Francis Group, London, ISBN 978-1-138-02931-6

Factors affecting the rapid excavation speed of rock roadways based on the blasting method

Guo-hui Li, Ji Li, Xin Wu & Jie-ping Wu
School of Engineering, Sichuan Normal University, Chengdu, Sichuan, China

ABSTRACT: Based on the current background of China's coal mines, major constraint factors restricting the excavation speed of blasting methods are analyzed: These include: (1) geological structures, especially in complex geological conditions; (2) equipment configurations, especially limited capacities of transportation equipment; (3) construction technologies; stale blasting technologies; (4) organizational processes, wasted time and non-parallelization in key processes; and (5) construction managements, unscientific construction organizations. Finally, some improvement measures to overcome the aforementioned factors are proposed.

1 INTRODUCTION

Most of the rock roadways are constructed by traditional blasting methods, and many coal mines achieve a high level in the rapid excavation speed of roadways. Due to the limitations of geological, economic, equipment, technology, and other conditions, the level of mechanizations and process on the excavations are still lagging behind the developed countries, and there is also a weak link of coal mining (Qin et al. 2012). The current performance indicators in the excavation speed and quality are embodied in the following aspects: low speed has only an average monthly footage of 60~70 m, and low efficiency has an average of less than 1.0 m/(work-day).

The excavation speed of roadways directly affects the starting node of coal mining, and the economic efficiency of enterprises. It is one of the key issues faced by China's coal mine construction at present. To improve the speed, this paper takes the excavation of domestic roadways as the research object and blasting methods as the research content, and analyzes and researches some factors influencing the speed, and puts forward the methods to overcome these constraint factors.

2 FACTORS AFFECTING THE SPEED OF RAPID EXCAVATION

2.1 Geological structures

Geological structures, reflecting rock hardness, roof and proof stability condition, and gas emission and water inflow and other aspects, are the primary factors affecting the rapid excavation speed of rock roadways.

Better geological structures provide a better foundation, but complex structures become a bottleneck for improving the excavation speed. Taking rock hardness as an example, the low rock hardness coefficient leads to a higher speed in blasting roadways, but results in increased time and difficulty in roadway supports. In contrast, the high rock hardness coefficient leads to the lower time and difficulty in roadway supports, but results in the decreased efficiency of boreholes and the increased time in blasting roadways. Taking the water inflow as the next example, if the inflow increases in a roadway of some water-rich area, such as the river, wastewater deposition often has a negative impact on constructions; as a result, some measures are required to take to control the wastewater deposition. The increased inflow increases the construction cost, and thereby also influences the excavation speed.

2.2 Equipment configurations

1. *Low-efficient drill equipment.* Drilling is the main process in a roadway excavation. The rational selection of the drill equipment and the method will directly determine the excavation speed. Air leg drills (e.g. 7655, YT-28, YTP26) of the small output power and the low drilling speed are commonly used in roadways. The use of new drilling bits and multi-machine operation for the air leg drills, or the hydraulic and electric drill equipment, will improve excavation efficiency (Fan 2009).

2. *Low-efficient transportation equipment.* The process of loading and transporting coal gangue employing large amounts of labor force is the most time-consuming process, accounting for about 35%~50% cycle time, and also is the key

factor to restrict the excavation speed in blasting methods (Yuan et al. 2009; Song & Gang 2007).

1. *Unmatched capacity between loading and transportation equipment.* Usually, the speed of loading equipment is faster than that of transportation equipment. So, the former has to wait for the latter, which cannot transport gangue as quickly as possible. A bottleneck appears in the transportation system rather than in the loading system because the transportation system has disadvantages of much links, weak abilities and high labor intensities. These disadvantages raise a question of capacity mismatch between loading and transportation (strong loading capacity). For a reasonable match between loading and transportation equipment, improving the capacity and efficiency of the transportation equipment becomes an important research in the excavation and transportation equipment development (Qin et al. 2004; Fan & Zhang 2007; Yuan et al. 2009).

2. *Frequent equipment failures.* In the Qidong coal mine, Qin (Qin et al. 2012) found that with the continuous extension of excavated sections, if the equipment were not strictly maintained in accordance with the maintenance system, the gradually increased equipment would lead to some machinery equipment failure, which would have a great impact on the improvement of the excavation speed.

3. *Low-efficient supporting equipment.* The pneumatic anchor drilling machines are widely used in the supporting process. The machines have some common features such as low degree mechanization, high labor intensity, poor security, and slow speed.

4. *Many failures of auxiliary equipment.* Pipelines that transport air and water are easy susceptible to serious leakage because of many union joints. Furthermore, it is easy to see the pipeline failures with a greater and farther excavation distance. The failures of the water transportation waste occur on average 10~30 min every shift. The air pipeline failures affect the speed of the ventilation and smoke exhaust, which leads to a long time of smoke exhaust after blasting (Qin et al. 2004). These kinds of factors seriously affect the operation of the next process usually, which is one of the factors affecting the excavation speed.

2.3 *Construction technologies*

1. Unscientific cutting mode causes the low-efficient cyclic footage. At present, the shallow hole single wedge-(shaped) cut mode is used in most excavation projects, in which the hole depth is 1.4~1.5 m and the cyclic footage is 1.2~1.4 m.

2. Unscientific blasting processing causes a poor-quality roadway. The phenomenon of imprecise drilling, contour irregularities, less drilling, and multi-charge leads to a low efficiency of borehole, serious over-break and under-break, thus causing a poor-quality roadway, in which inconveniences for late repairs affect the speed and quality of excavation (Qin et al. 2012).

3. Unscientific support processing hinders the safety quality and the speed of excavation. The current common bolting and shotcreting method is better than the previous shed shoring and arching supporting method in terms of the construction speed, safety quality, labor intensity and production cost (Qin et al. 2004). But there are still some details that need to be improved. Over-break and poor excavation sections seriously result in the poor installation quality, increasing of actual thickness and shotcrete workload of hanging nets and anchor bolts. The sequence operation of the mixing and spraying process leads to the low utilization ratio of time. The number of anchor bolts is too much to increase the time costs of drilling and installation. These backward supporting technologies not only have the hidden danger of security with poor surrounding rock sections, but also cause the low speed of supporting and restrict the increasing speed of excavation.

2.4 *Process organizations*

Linear, sequential and unparallel organization among processes is also a factor influencing the speed of excavation. Moreover, unscientific and unimplemented cycle schedules, on the whole, a complex process itself, and the handover of details between processes consumes a lot of time in improving the speed of excavation.

2.5 *Construction managements*

1. *Low-quality employees.* The quality of constructors and technicians for the higher mechanization and information equipment, and the quality of management personals for scientific and advanced management concepts, are highly required in rapid excavation projects. However, because of arduous industry, not only the first-line constructors, but also technical staff and management personnel, have a relatively low cultural level. The constructors are not in accordance with the technical specification for the operation, who are not familiar with the production processes and do not understand

the production equipment performance, and perform unskilled operations. The low-quality technical staff choose unreasonable designs and imperfect production processes for roadway excavation. The grass-root management personnel could not manage and thus do not perform effectively well on on-site management, and conscientiously execute the formulation of the management system. The top management personnel lack field experience and technical abilities.

2. *Unreasonable operating system.* The working efficiency of the traditional "three-eight system" is generally low after half of a regular cycle, due to physical decline after the long time work of workers. In order to realize the rapid excavation, "four-six system" can be employed.

3. *Ways for improving the speed of rapid excavation.* The basic principle of the thoughts: based on the suitable "method" for enterprises, the coordination and unity, performance matching of the four factors among the equipment configurations, construction technologies, process organizations and construction managements are ensured.

2.6 Reasonable selection and configuration of equipment

1. *Drilling equipment.* New type of drills, multi-machine operation and other measures are used to improve the drilling speed and tap their potential time. But due to the limited equipment performance, its development level has reached its high point, in which noise and pollution are difficult to overcome. In the future, the medium and large coal mines, relatively abundant in technology and capital, can consider hydraulic drill carriages to improve drilling efficiency. The devices have the advantages of high power, high efficiency, fast speed of drill eye, small noise, and low air pollution. The strength of the enterprise can also explore the application of digging machines.

2. *Loading and transportation equipment.* The shuttle mine cars have outstanding performances in the construction of traffic tunnels and water conservancies. The loading and transportation efficiency can be improved exponentially by using and optimizing it.

3. *Support equipment.* Reasonable selection of the anchor and its equipment is used to improve the quality of installation. On the one hand, encouraging the adoption of high strength, reliability and pre-tightening force anchor bolts reduces the quantity of the bolts. On the other hand, the introduction of the development of new anchor machineries, such as hydraulic single-type anchor drilling equipment or rotary roof bolter, and electric anchor drilling machines, improves the work efficiency.

4. *Auxiliary equipment.* Fixing the people for a fixed position and continuing to act conscientiously, the site management system on auxiliary equipment, carries out the equipment cares and maintenances.

2.7 Continuous improvement of construction technology

According to the geological conditions, such as rock properties, and engineering conditions, such as construction equipment, the suitable blasting method and support technology are selected.

1. Using composite cutting methods and middle or deep-hole smooth blasting technologies, the cyclic footages and blasting effect can be improved. The cyclic footage is affected mainly by the blasting effect, which is decided by the cutting methods and blasting parameters such as the eye-deep, charge form. The development trend in the future is a composite cutting method combined with a variety of cutting methods. Xinxing Coal Mine in the Hegang Mining Group using a "bottom double wedge cutting method" (Liu 2010), Weishan Zhaoyang Coal Mine in the Xinguang Group a "throwing funnel wedge cutting method", Qidong Coal Mine presenting an improved wedge cutting method combination of the above two advantages (Qin et al. 2012) have achieved good results, which has a very high reference value. The key technology of the smooth blasting method is to control the peripheral eye distance and charge quantity. In order to improve the cyclic footages and explosion effect, starting from increasing the depth and diameter of holes, we can use middle or deep-hole smooth blasting technologies to enhance the excavation effect, instead of the shallow-hole blasting technology in the current roadway excavation mostly.

2. *Reasonable choice to improve the support process.* To ensure the safety support, we could do the advance timbering or the initial support, and the strengthening support in the adverse geological sections under the condition of the surrounding rock. To improve the support process, techniques such as the wet spraying process and the suspension roof support with the capacity of self-support of the surrounding rock can reduce the labor intensity of workers, and also improve the construction environment.

2.8 Optimization management of process organization

To shorten the time of each process as far as possible, a whole cycle time should thus be shortened. The time of each process and cycle is determined, and then a reasonable and efficient cycle chart is compiled.

The use of a parallel and cross-operation system is regarded as the second measure of the optimization of process organizations, such as the parallel operation between a shift and a quality and safe inspection of excavated sections using multi-drilling equipment in a drilling process at the same time.

2.9 *Scientific management in construction organization*

Improving the qualities of personnel engaged in excavations can promote their work efficiency. Improving the systems of operations can maximize the capacities of equipment and staff. The application of fine managements can ensure continuities and effectiveness of each construction cycle. The establishment of security systems can strengthen excavation preparation work and field managements.

3 CONCLUSIONS AND PROSPECTS

The excavations of rock roadways are still based on blasting methods in China's coal mines. Although the excavation technologies had made a great progress, the key common problems faced by coal mining enterprises are still the conflict between the excavation and mining. This paper analyzes five factors that restrict the rapid excavation speed of rock roadways, such as geological structures, equipment configurations, construction technologies, process organizations, and construction managements, and puts forward some methods to overcome the constraint factors. The next work should further study rapid excavation theories and construction technologies, which are suitable for various rocks, combining these factors with the suggested methods.

ACKNOWLEDGMENTS

This project was supported by the 251 Key Talent Training Project of Sichuan Normal University and the Scientific Research Fund of Sichuan Provincial Education Department (14ZB0030).

REFERENCES

Fan Yi-ning, Zhang Jun-wen 2007. Our country high production highly effective minepit construction the present situation and countermeasure. Coal Technology 26 (7): 5–7.

Fan Yong 2009. Analysis on the influence factors to roadway quick excavation. Shanxi Coal 29: 20–21, 30.

Liu Yan-jun 2010. Fast Driving Technology of Cutting at Bottom in Rock Roadway. Coal Technology 29 (2): 75–76.

Qin Qing-ju, Zhang Liang & Wang Hua 2012. Analysis of the application of fast excavation in rock roadway technology in Qidong coal mine. China Mining Magazine 21 (7): 87–89.

Qin Zhong-cheng, Wang Tong-xu, Li Zheng-long, Liu Hong-ru & Gao Hong-liang 2004. Test research of integrated technology and process of rock roadway drivage with high-efficiency and high-speed. Ground Pressure and Strata Control 2: 34–36.

Song Hong-wei, Liu Gang 2007. Sinking and driving engineering. Beijing: China Coal Industry Publishing House.

Yuan Wen-hua, Ma Qin-yong, Liu Han-xi & Peng Ji-yong 2009. Reconstruction of gangue transportation system and its application to rapid excavation in rock roadway. Coal Engineering 12: 40–42.

Advances in Energy, Environment and Materials Science – Wang & Zhao (Eds)
© *2016 Taylor & Francis Group, London, ISBN 978-1-138-02931-6*

Effect of high strength reinforced steel and axial compression ratio on the hysteretic behavior of rectangular bridge piers

Xian Rong, Dandan Xu & Ping Liu
School of Civil Engineering, Hebei University of Technology, Tianjin, China
Civil Engineering Technology Research Center of Hebei Province, Tianjin, China

ABSTRACT: Four reinforced concrete rectangular bridge piers were tested under a low cycle loading to study the seismic behavior of HRB500 high strength reinforced steel. The effect of axial compression ratio, the longitudinal reinforcement and stirrup strength grade on specimens' bearing capacity, deformability, ductility, and hysteretic behavior were analyzed. It shows that the bearing capacity, deformability and hysteretic behavior of a high strength reinforced concrete rectangular bridge piers are improved as compared to the ordinarily ones. The longitudinal reinforcement and the stirrup strength grade have a different effect on deformability. For axial compression ratio, the small one can significantly improve the bearing capacity, deformability and hysteretic behavior of specimens, but it has a negative effect on the bearing capacity.

1 INTRODUCTION

HRB500 steel, as a kind of high strength hot rolled ribbed bar with excellent comprehensive properties, is much more commonly applied in structural engineering and bridge engineering. HRB500 steel possesses many characteristics such as high tensile strength, good ductility and excellent welding ability. (Fan et al, 2013; Dan Dubina et al, 2014; Tavallali Hooman et al, 2014) Promoting the application of HRB500 steel in the concrete structure can significantly reduce reinforcement amount, and achieve good economic and social benefits. Therefore, in order to improve the bridge piers' seismic behavior under the action of strong earthquake, it has very important theoretical and practical significance by using the longitudinal reinforcement and stirrup with high-strength steel. (Li, 2010; Zhang et al, 2010; Li et al, 2014). The hysteretic behavior of four concrete bridge piers with HRB500 steel were studied by pseudo-static test in order to study the application of high strength reinforced bar in bridge engineering.

2 TEST SURVEY

2.1 Specimen design

In this experiment, four rectangular reinforced concrete bridge piers were designed, including one common reinforced concrete bridge pier specimen and three specimens with HRB500 steel. The effect of axial compression ratio, the longitudinal reinforcement and the stirrup strength grade on specimens' hysteretic behavior were analyzed.

Four bridge pier specimens had the same stirrup spacing with 80 mm, the same stirrup diameter with 10 mm and the same longitudinal reinforcement diameter with 16 mm. The specimens were composed of pier shaft and base. Model size parameters were same. The pier shaft height was 1280 mm. The base height was 500 mm. We enhanced steel in the pier top to prevent pier top local failure. The specimens design parameters are shown in Table 1. The specific dimensions and reinforcements are shown in Figure 1. Figure 1, (1) represents stirrup which has two kinds of HPB300 and HRB500, (2) represents longitudinal reinforcement with two kinds of HRB335 and HRB500.

2.2 Material mechanical properties

Before loading, we conducted steel tensile test and concrete mechanical property test according to related standards. The tensile properties of steel are showed in Table 2. The concrete grade was C50. Through actual measurement, the average of

Table 1. Design parameters of specimens.

Specimen	Axial compression ratio	Longitudinal reinforcement strength grade (MPa)	Stirrup strength grade (MPa)
SJ1	0.07	335	300
SJ2	0.07	335	500
SJ3	0.07	500	500
SJ4	0.14	500	500

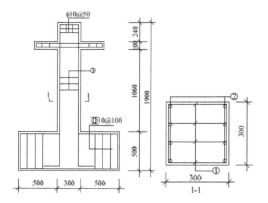

Figure 1. Dimensions and reinforcements of specimens (mm).

Table 2. Tensile properties of steel.

Steel type	Diameter (mm)	Yield strength (MPa)	Ultimate strength (MPa)	Elastic modulus (× 10⁵ MPa)
HPB300	10	352	543	1.95
HRB335	16	383	546	1.95
HRB500	10	585	750	1.92
HRB300	16	600	756	1.92

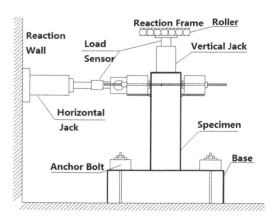

Figure 2. Diagram of the test apparatus.

concrete cubic compressive strength was 58.1 MPa and the average of axial compressive strength was 38.9 MPa.

2.3 Test loading scheme

The low cycle load was applied by hydraulic servo actuator during the pseudo-static test. The test used the loading method of load-displacement control. The axial load was applied on the pier top with hydraulic jack. Horizontal cycle load was applied by the hydraulic jack on the back wall. At the initial test, the load control method was used by the way of hierarchical loading. Each load control grade cycled once. When the load reached ultimate load, the loading method was replaced by the displacement control. The yield displacement adopted the displacement value at this point. And each load control grade cycled three times. When the specimen's bearing capacity was down to 85% of the ultimate bearing capacity, the specimen was a failure. Then we stopped the test. Diagram of the test apparatus is shown in Figure 2.

3 TEST RESULT ANALYSIS

3.1 Bearing capacity, deformability and ductility

Ductility is an important measurement index in structural seismic analysis. (Jiang et al, 2013; Liu et al, 2012; Du et al, 2011). The ductility of the member means that the member is able to withstand large deformation before failure when its bearing capacity significantly reduces. Ductility is usually evaluated with displacement ductility ratio.

Mathematical expression of displacement ductility ratio:

$$\mu_u = \frac{\Delta_u}{\Delta_y} \tag{1}$$

In Equation (1), μ_u = displacement ductility ratio; Δ_u = failure displacement, which is corresponded with failure load; and Δ_y = yield displacement. The unit of Δ_u and Δ_y are millimeter (mm).

Mathematical expression of failure load:

$$F_u = 85\% F_{max} \tag{2}$$

In Equation (2), F_u = failure load and F_{max} = ultimate load. The unit of F_u and F_{max} are kN.

The bearing capacity, displacements and the displacement ductility ratio of the specimens are shown in Table 3. These numbers are the average value of loading and reversal loading. The unit of load is kN. The unit of displacement is mm.

1. Bearing capacity analysis:
The yield load and ultimate load of SJ2 are increased by 4.6% and 2.8%, respectively, than SJ1. It means that the stirrup with HRB500-class high tensile reinforcement makes the bearing capacity of bridge pier specimens improved, but the impact is not obvious. Compared to SJ2 with SJ3, the yield load and ultimate load of SJ3

Table 3. Bearing capacity, displacements and ductility coefficient of specimens.

Steel type	SJ1	SJ2	SJ3	SJ4
Cracking load	33.17	31.17	32.50	45.17
Ultimate load	78.35	81.94	102.08	107.01
Yield load	93.84	96.50	116.00	125.67
Cracking displacement	1.43	1.58	2.23	2.37
Yield displacement	12.08	10.38	14.16	12.75
Ultimate displacement	23.68	34.31	38.56	28.10
Failure displacement	42.38	46.04	56.87	37.47
Displacement ductility ratios	3.52	4.74	4.04	2.96

are increased by 24.6% and 20.2%, respectively, than SJ2. We can observe the bearing capacity of bridge pier specimens remarkably improved by using HRB500 longitudinal reinforcement. Compared to SJ3 with SJ4, the yield load and ultimate load of SJ4 are increased by 4.8% and 8.3%, respectively, than SJ3, which shows the big axial compression ratio can enhance the specimens' bearing capacity.

2. Deformability analysis:

The cracking displacement, yield displacement, ultimate displacement, and failure displacement of SJ2 are, respectively, increased by 10.5%, −14%, 44.9%, and 8.3% than SJ1. Overall, HRB500-class stirrup makes the deformability of bridge pier specimens improved. Compared to SJ2 with SJ3, the four displacements of SJ3 are increased a lot more than SJ2, which represents that the deformability is heightened by using HRB500-class longitudinal reinforcement. However, the displacements of SJ4 are lower than SJ3. The research result confirms that the bigger the axial compression ratio, the more bad the pier specimens' deformability.

3. Ductility analysis:

The ductility ratio of SJ2 is 34.7% higher than SJ1. So, stirrups of HRB500-class high strength reinforced bar can significantly improve the ductility of the specimens. But, the ductility ratio of SJ3 is decreased by 14.8% when compared with SJ2, which means HRB500 longitudinal reinforcement will reduce the specimens' ductility. Contrast SJ3 with SJ4, SJ4's ductility ratio is 26.7% lower than SJ3. It can be inferred that small axial compression ratio is favorable to the ductility of specimens.

3.2 Hysteretic behavior analysis

The hysteretic curve, as one of the important indexes to study the seismic behavior of the bridge pier, is the load-displacement curve of specimen under low cyclic loading. The shape of hysteretic curve can reflect the seismic behavior of specimens, such as the bearing capacity, deformability, energy dissipation capacity, etc (Xiao, 2010; Deng, 2010). The area surrounded by the hysteretic curve reflects the specimens' ability of energy dissipation. So, the area is larger, the energy dissipation capacity is stronger, and the seismic performance is better. The hysteretic curves of the specimens are shown from Figure 3 to Figure 6.

According to the shape of the hysteresis loop, the analysis is as follows:

The hysteresis loop shape of specimens changes from spindle into bow shape gradually. The curve slope gradually decreases with the increase of the load during the initial stage of loading. After several times of cyclic loading, the inflection point is formed on the curve. In the initial stage of unloading, the hysteretic curve is steeper, and the curve tends to be gentle with the decrease of the load. The influence of reinforcement strength and axial compression ratio on the hysteretic behavior of the specimens is mainly analyzed.

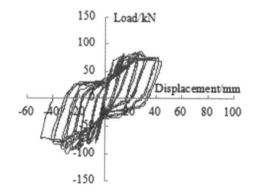

Figure 3. Hysteretic curves of SJ1.

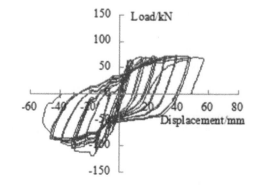

Figure 4. Hysteretic curves of SJ2.

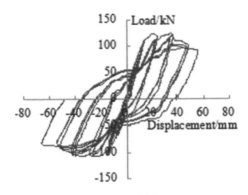

Figure 5. Hysteretic curves of SJ3.

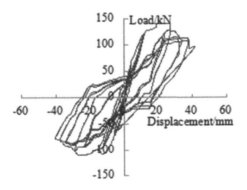

Figure 6. Hysteretic curves of SJ4.

1. Effect of reinforcement strength:
   Comparison between SJ1 and SJ2 hysteresis curve can be found, SJ2's curve area and ultimate displacement is much larger, and the curve is much plumper. It means that the hysteretic behavior of bridge pier with HRB500-class stirrup is better than the common one. SJ3 has a larger curve area, plump curve and bigger ultimate displacement and load when compared with SJ2. So, the bearing capacity and hysteretic behavior of bridge piers with HRB500-class longitudinal reinforcement is better than the common ones.
2. Effect of axial compression ratio:
   The longitudinal reinforcement and stirrup of SJ3 and SJ4, which use HRB500-class high strength reinforced bar, have different axial compression ratio. Comparison between SJ3 and SJ4 hysteresis curve can be found, SJ3 has larger curve area, plumper curve and bigger ultimate displacement and load. The research result confirms that small axial compression ratio can improve the hysteretic behavior of specimens.

In summary, high strength stirrup and longitudinal reinforcement can improve the hysteretic behavior of bridge piers. Reducing the axial compression ratio has a good effect on the hysteretic behavior of the bridge piers.

4 CONCLUSION

1. According to the analysis of longitudinal reinforcement strength's influence, HRB500-class steel can significantly improve the bearing capacity, deformability and hysteretic behavior of the specimens, but the ductility will be reduced. On the whole, high strength longitudinal reinforcement is advantageous to the seismic behavior of bridge piers.
2. According to the analysis of stirrup strength's influence, HRB500-class high-strength stirrup can significantly improve the specimens' deformability, ductility and hysteretic behavior. It makes the bearing capacity improved, but the effect is not obvious.
3. According to the analysis of stirrup strength's influence, reducing the axial compression ratio can improve the deformability, ductility and hysteretic behavior. But, the larger one has great effect on the bearing capacity. Therefore, designing appropriate axial compression ratio is important in the practical engineering.

SUPPORTED PROJECTS

This work is supported by the Tianjin Natural Science Foundation (12JCYBJC14100), Science and Technology Plan Projects of Hebei Province Transportation Hall (Y-2011052, Y-2012041), Key Project for Science and Technology Research of Hebei Province Higher Education (ZD2010123), and the Research Plan for Construction Science and Technology of Hebei Province (2014-122, 2014-123).

REFERENCES

Bingnan Li, Hang Dai, Jiwen Zhang. *Research on realization of seismic ductility by high strength bar for railway piers with low reinforcement ratios* [J]. Industrial Construction, 2014, 44(8): 88–91+97.

Dan Dubina, Aurel Stratan, Adrian Ciutina. *High strength steel in seismic resistant building frames* [J]. Steel construction, 2014, 7(3): 173–177.

Guiqian Li. *Experiment study and numerical analysis on seismic performance of reinforced concrete bridge columns* [D]. Chongqing: Chongqing Jiaotong University, 2010.

Lizhong Jiang, Guangqiang Shao, Jingjing Jiang. *Experimental study on seismic performance of solid piers with round ended cross-section in high-speed railway* [J]. China Civil Engineering Journal, 2013, 46(03): 86–95.

Qin Zhang, Jinxin Gong, *Load-deformation relations of reinforced concrete columns under flexural-shear failure* [J]. Journal of Architecture and Civil Engineering, 2010, 27(3): 78–84.

Tavallali Hooman, Lepage Aadres, Rautenberg Jeffrey M. *Concrete beams reinforced with high-strength steel subjected to displacement reversals* [J]. ACI Structural Journal, 2014, 111(5): 1037–1047.

Xiuli Du, Mingqi Chen, Qiang Han. *Experimental evaluation of seismic performance of reinforced concrete hollow bridge columns* [J]. Journal of Vibration and Shock, 2011, 30(11): 254–259.

Yanhui Liu, Ping Tan, Fulin Zhou. *Experimental study of seismic performance for the double-chambers thin-wall concrete pier* [J]. China Civil Engineering Journal, 2012, 45(S1): 90–95.

Yanqing Deng. *Experimental studies on seismic behavior of HRB500 RC columns* [D]. Chongqing: Chongqing University, 2010.

Yi Xiao. *Research on hysteretic behavior and resilience model of the damaged member* [D]. Changsha: Central South University, 2010.

Zhong Fan, Lin Xu, Yuan Fang, et al. *Discussion on the application of high-strength reinforcement* [J]. Structural Engineers, 2013, 29(6): 171–177.

Advances in Energy, Environment and Materials Science – Wang & Zhao (Eds)
© *2016 Taylor & Francis Group, London, ISBN 978-1-138-02931-6*

The application on fuzzy neural control in boiler liquid level control system

Huanxin Cheng, Guoqing Zhang & Jing Li
Qingdao University of Science and Technology, Qingdao, Shandong, China

Li Cheng
Xinjiang Technical Institute of Physics and Chemistry, Chinese Academy of Sciences, Urumqi, Xinjiang, China

Lingling Kong
Research Institute of Physical and Chemical Engineering of Nuclear Industry, Tianjin, China

ABSTRACT: For the character of nonlinear and the time-varying in industrial boiler liquid level control process, which has been difficult to meet the requirements of control precision of nonlinear system based on the conventional PID cascade control, the paper puts forward the boiler liquid level control based on fuzzy neural network. It aims to put the self-learning ability of neural network into fuzzy control, which does not need to establish specific mathematical model to achieve the level control. Through the MATLAB design and simulation of the boiler liquid level system, the results are compared with the conventional PID control and fuzzy PID control, then it shows that the method is effective.

1 INTRODUCTION

In industrial production, the boiler is widely used in various industries as the main controlled object. The main parameters of its outputs are the Liquid level. For these nonlinear systems with the characteristics that are delayed and time-varying, it is difficult to build a mathematical model. Using the conventional PID control it is difficult to meet the optimal control of a variety of conditions (Ding 2009; Lou & Liu 2010). In this study, a new method of controlling boiler liquid level is presented for the operation variable (opening of the valve) and the controlled variables (boiler liquid level) in Boiler Level Control process. This method uses the neural network to the training samples, and by fuzzy reasoning and genetic algorithm to determine membership function of the system. Finally, network weights and fuzzy rules need to be trained and adjusted by the BP neural network to meet performance requirements of the system.

2 THE DESIGN OF NEURAL FUZZY CONTROLLER

2.1 The establishment of neural fuzzy controller

This fuzzy neural network is applied Mamdani reasoning. We designed a two-dimensional neuro fuzzy controller with two input variables and one output variable. Its structure diagram is shown in Figure 1, which is composed of the input layer, fuzzy layer, the fuzzy inference layer, and the defuzzification layer (Zhao et al. 1996, Fu et al. 2010). The two input variables of this system are liquid level deviation e and liquid level deviation variation ec and the corresponding fuzzy variables are E and EC, respectively. The output value u is related to the fuzzy variable U. Each layer of network concrete structure is defined as follows (Liu, 2013).

The first layer is the input layer. This design uses liquid level deviation e and liquid level deviation change ec as two input variables of controller.

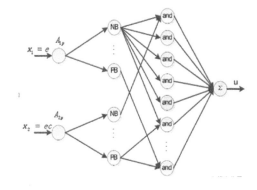

Figure 1. Neural fuzzy control structure diagram.

Each of them has 7 fuzzy subsets, and the variables of the input layer are passed to the next layer through the activation function.

$$\begin{cases} I_i^{(1)} = x_i \\ O_i^{(1)} = I_i^{(1)} \end{cases} \quad i = 1, 2 \tag{1}$$

I_i^j and O_i^j represent the output of the j layer of I neurons.

The second layer is fuzzy layer. At this level there are 14 nodes, representing E and EC fuzzy subset membership of each element. The fuzzy subsets of these two fuzzy inputs are:

E = {PB, PM, PS, ZO, NS, NM, NB}

EC = {PB, PM, PS, ZO, NS, NM, NB}

$$I_{ij}^{(2)} = \begin{cases} 0 & x \le a_{ij} \\ (x_i - a_{ij})/(b_{ij} - a_{ij}) & a_{ij} \le x \le b_{ij} \\ (c_{ij} - x_i)/(c_{ij} - b_{ij}) & x \le c_{ij} \end{cases}$$

$$j = p = 1, 2, \dots 7$$

$$O_{ij}^{(2)} = A_{ij} = \exp(I_{ij}^{(2)}) \tag{2}$$

The membership function of the fuzzy input variable is used as the membership function of the fuzzy input variables.

A_{ij} represents a subset of the membership of triangular membership function of the i-th input corresponding to the j-th fuzzy sets. a_{ij}, b_{ij}, and c_{ij} represents the i-th input corresponding to the three parameters of triangular membership functions of j-th fuzzy sets. The weights of the connecting layer and the fuzzy layer are 1.

The third layer is the fuzzy inference layer. The main function of this layer is to realize the judgment function of fuzzy logic. The number of neurons in this layer represents the number of control rules. There are 49 nodes in the liquid level control, which means there are 49 fuzzy rules. The reasoning rule of this fuzzy layer can be expressed as: if x_1 is A_{1p} and x_2 is A_{2p} then U is W_k. Among them, x_2 and x_1 are the input variables E and EC of the system. A_{1p} and A_{2p} are the corresponding membership subsets. U is output. W_k is the corresponding membership fuzzy subset. Set network weights of this layer are 1.

$$\begin{cases} I_K^{(3)} = A_{1p}(x_1) \cdot A_{2p}(x_2) & p = 1, 2 \dots 7 \\ O_K^{(3)} = I_K^{(3)} & k = 1, 2 \dots 49 \end{cases} \tag{3}$$

The fourth layer is the defuzzification layer. There is only one output layer in this level. Its function is clarity calculation. In this paper, we use weighted averaging method for defuzzification.

$$\begin{cases} I^{(4)} = \sum_{k=1}^{49} O_k^{(3)} \cdot W_k \\ O^{(4)} = U = \dfrac{I^{(4)}}{\sum_{k=1}^{49} O_k^{(3)}} \end{cases} \tag{4}$$

U is the output of the fuzzy neural network. W_k represents the connection weights of the third and fourth layers.

2.2 Parameter adjustment based on BP network

After the neuro-fuzzy controller is established, we use BP neural network training to adjust parameter values and fuzzy rules. Assuming the expected output value of the liquid level control system is $r(k)$ and the actual output value is $y(k)$. Then, using gradient descent method, get the center value and width value of membership function and the weight of system:

$$\begin{aligned} \frac{\partial J_0}{\partial w} &= \frac{\partial J_0}{\partial y(k)} \cdot \frac{\partial y(k)}{\partial u(k)} \cdot \frac{\partial u(k)}{\partial w} \\ &= -(y_t(k) - y(k)) \cdot \frac{\partial y(k)}{\partial u(k)} \cdot \frac{\partial u(k)}{\partial w} \end{aligned} \tag{5}$$

$$w(k+1) = w(k) + \eta \cdot (y_t(k) - y(k)) \\ \cdot \frac{\partial y(k)}{\partial u(k)} \cdot \frac{O_k^{(3)}}{\sum_{k=1}^{49} O_k^{(3)}} + \lambda \Delta w(k) \tag{6}$$

Do the following approximation:

$$\frac{\partial y(k)}{\partial u(k)} \approx \frac{y(k) - y(k-1)}{u(k) - u(k-1)} \tag{7}$$

The weights of the fuzzy neural network are:

$$w(k+1) = w(k) - \eta \cdot \frac{\partial J_0}{\partial w} + \lambda \Delta w(k) \tag{8}$$

Similarly, the parameter correction value of the fuzzy neural network is:

$$a_{ij}(k+1) = a_{ij}(k) - \eta \cdot \frac{\partial J_0}{\partial a_{ij}} + \lambda \Delta a_{ij}(k)$$

$$b_{ij}(k+1) = b_{ij}(k) - \eta \cdot \frac{\partial J_0}{\partial b_{ij}} + \lambda \Delta b_{ij}(k)$$

$$c_{ij}(k+1) = c_{ij}(k) - \eta \cdot \frac{\partial J_0}{\partial c_{ij}} + \lambda \Delta c_{ij}(k)$$

η is the learning rate and λ is the momentum coefficient.

3 THE LOOP CONTROL OF BOILER LIQUID LEVEL CONTROL

We use the double impulse control loop. The concrete control loop is shown in Figure 2.

As we can see from Figure 2, the steam flow as the feed forward signal is introduced into the control loop. Its main function is to prevent liquid level adjustment process from the influence of spurious signals and to make sure level control changing is in the right direction in the beginning. In this way, we can reduce the impact of the water supply and the water level to the control. It can also reduce the over time. Ensure good static characteristics and quality control.

4 SIMULATION OF BOILER LIQUID LEVEL CONTROL SYSTEM

By the relationship between the material imbalance and the thermal equilibrium, we can get the dynamic specific equation of boiler bubble water level adjustment object:

$$T_1T_2\frac{d^2h}{dt^2} + T_1\frac{dh}{dt} = \left(T_w\frac{du_w}{dt} + K_wu_w\right) - \left(T_D\frac{du}{dt} + K_Du_D\right) \quad (9)$$

In this formula: h is the height of the bubble liquid level. T_W is time constant of water supply. T_D is time constant of steam flow. K_w is magnification of water supply. K_D is a magnification of steam flow. D is steam flow. W is water supply of boiler. T_1 and T_2 are Time constants(s). In the formula (9), the water supply and steam flow disturbance are considered. The main consideration is the effect of water flow when we establish the mathematical model and the steam flux is the compensation of actual control.

In Simulink of MALTLAB, according to the structure of the designed controller and membership function of the individual variables and Fuzzy

rule, we establish controller needed by systems. Make some assumptions for boiler liquid level control system. The structure of conventional PID control is basically the same as the structure of fuzzy control and structure of neuro fuzzy control system. We adopt step signal as the input signal to do some equivalent simplification for the mathematical model of controlled subject. Simulation in the same load, even if there are disturbances, we assume that the time and the size of the disturbances are exactly the same.

When we only use PID control, the parameters of the controller are $K_p = 0.03$, $K_d = 1$, and $K_i = 0.004$. Step response curve of a conventional PID is shown in Figure 3. We can see from the figure: Overshoot is larger. Excessive time is longer. Control effect is not good.

When we use fuzzy PID control, the parameters of the controller are finally selected by continuously adjusting as $K_p = 0.093$, $K_d = 0.008$. Figure 4 is step response curve of fuzzy PID. We can see from this figure that overshoot and over time is relatively large and the control effect is not ideal.

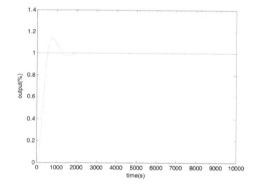

Figure 3.   Conventional PID step response curve.

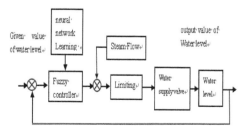

Figure 2.   Schematic diagram of boiler water level control loop.

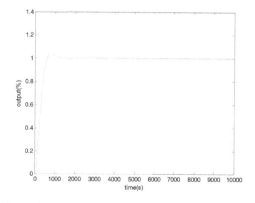

Figure 4.   Fuzzy PID step response curve.

In making fuzzy rules of neuro-fuzzy network, we follow the following principles. When the error of the boiler liquid level is large, the controller increment mainly eliminates error. On the contrary, the controller increment mainly makes sure of the stability of the system (Zhao et al. 2006). To ensure the overshoot meet the requirements of the system, Fuzzy rules are developed combined with relevant industry experience. System scale factors are $K_e = 1.2$ and $K_{ec} = 0.3$. Learning rate is $\eta = 0.2$, momentum coefficient $\lambda = 0.02$. The initial value of the weight coefficient is the random variable of [−0.5, 0.5]. The simulation is shown in Figure 5.

From the above simulation figure we can see that when we apply the fuzzy neural network control system, the overshoot is small and the system can reach stability in short time. The time of adjustment is relatively short. So the control effect of neural network control system is relatively stable. At the same time static and dynamic performance is good, it can be expected to achieve good control effect.

5 PRACTICAL APPLICATION

Put the control parameters into boiler control system parameter module and adjust according to the actual situation. The real time control chart of steam boiler bubble water level is shown in Figure 6.

It can be seen from the figure that the deviation of the measurement value and the setting value is not big and meet the deviation range. The effect is good.

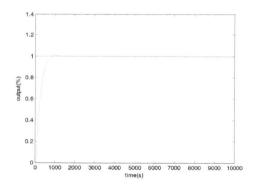

Figure 5. Step response curve of the fuzzy neural network.

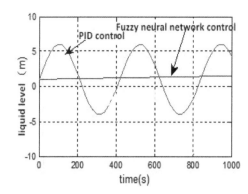

Figure 6. Real time control chart of boiler liquid level.

6 CONCLUSION

For the character of nonlinear and the time-varying in industrial boiler liquid level control process, we proposed a neuro-fuzzy control method for liquid level control system. The simulation results show that the design can realize the liquid level control and guarantee the static and dynamic performance of the system. So this method is significant for boiler liquid level control.

REFERENCES

Chunyuan Zhao, et al. Neural fuzzy control of boiler drum level [J]. Journal of ShenYang Ligong University, 2006, 26(4): 9–12.

Guhong Ding Design of liquid level fuzzy control system based on Matlab [J]. 2009, 9(4): 258–260.

Huixuan Fu, Hong Zhao, et al. Matlab neural network application design [M]. Beijing: Press of machinery industry, 2010.

Rui Liu. Application of fuzzy Neural Network in Power Plant Main Steam Temperature Control System [D]. Chang'an University, 2013.

Wei Lou, Xiangdong Liu. Application of fuzzy control in the boiler drum water level control system [J]. Boiler Technology. 2010, 41(1): 27–30.

Zhenyu Zhao, et al. Fundamentals and Applications of fuzzy theory and neural networks [M]. Tsinghua University Press, 1996.

Advances in Energy, Environment and Materials Science – Wang & Zhao (Eds)
© *2016 Taylor & Francis Group, London, ISBN 978-1-138-02931-6*

Simulation on hydraulic fracturing propagation using extended finite element method

Dongxue Wang, Bo Zhou & Shifeng Xue
College of Pipeline and Civil Engineering, China University of Petroleum, Qingdao, China

ABSTRACT: As a kind of effective simulation methods, extended finite element method was developed rapidly after being proposed. In this article, this method was used to studying fracturing propagation cases, and the model built in article was simulated after simplifying the boundary conditions, and finally this method was successfully carried out in this hydraulic example. A curve of the displacement versus stress of the point on the crack tip during the whole propagation process, which was gotten by analyzing the simulation result, and it was corresponding to the reality. In this article, the simulation process and result were largely optimized.

1 INTRODUCTION

Since 1947, when hydraulic fracturing was successfully used to increasing production in USA, this technology was developed rapidly to the mean method for the production in most oil fields. However, as yet, the study on the explicit principle is not very good. The traditional study method was to do experiments or simulate on the bases of finite element method, and the simulation was the main method. But, the fact is that, the influence of the meshing was huge to the traditional element method, which meant, different meshing numbers or method made a difference to the result so that the simulation would produce big tolerance (Rethore, 2005). Therefore, the simulation based on extended finite element method would be an effective one to study the hydraulic fracturing technology (Nagashima, 2004).

In this article, the basis theorem of the extended finite element method is introduced in details. And the element displacement mode, the level set method, the stress intensity factor, and the energy fracture criterion are introduced in the followed part. Then in the third part, the simulation model was built according to reality, and boundary conditions were simplified effectively, and the calculation based on the extended finite element method was put into effect. After the simulation job, an analytic job was done in detail. In the article, a group of curves of the vertical and horizontal displacement in pace with stress of the point on the crack tip is given by two figures, and the tendency of the displacement and the stress is analyzed in words. As a result of the analysis, a big advantage can be got, that is the conclusion part, which is that, the result conformed to the

theoretical analysis, and the simulation process was optimized.

2 BASIS THEOREM

The extended finite element method was put forward as an effective simulation method for discontinuity geometries, and met with great favor of many researchers; so, it developed rapidly (Chopp D., 2002). As a new method to solve the discontinuity geometries, it was developed in the framework of the traditional finite element method, to put enrichment functions, which could also optimize the calculation process and result in traditional functions to illustrate the discontinuity (Wagner, 2001). The enrichment functions included: the Heaviside function and the Crack tip displacement function. The enrichment functions made the XFEM one great superiority that did not exist in traditional method (Belytschko, 1999): the meshing was independent outside of the discontinuity, which avoided the disadvantage of re-meshing caused by crack tip singularity, to increase the computational efficiency largely and make the result much more accurate (Sukumar, Moes, Moran, Belytschko, 2000).

2.1 Element displacement mode

The displacement mode of traditional finite element method can be expressed as:

$$u = \sum_i N_i u_i \tag{1}$$

where N_i is shape function, and u_i is the displacement of the nodes, and i is the nodes set (Stolarska, Chopp, Moes, Beltyschko, 2003).

In XFEM, to describe the discontinuity of the crack, the mode is expressed as:

$$u = \sum_i N_i \left(u_i + H(x)a_i + \sum_{k=1}^{4} \phi_k b_{ki} \right) \quad (2)$$

where a_i, b_{ki} are the freedom of enrichment nodes, and just because of these two enrichment functions' addition, the extended finite element method owns its unique advantages. The Heaviside function $H(x)$ is expressed as:

$$H(x) = \begin{cases} 1 & (x > 0) \\ -1 & (x < 0) \end{cases} \quad (3)$$

Which shows the discontinuity of the interface displacement. And ϕ_k shows the approximation of the crack tip in local coordinate system:

$$\phi_k = \begin{cases} \sqrt{r} \sin\dfrac{\theta}{2} \\ \sqrt{r} \cos\dfrac{\theta}{2} \\ \sqrt{r} \sin\theta\sin\dfrac{\theta}{2} \\ \sqrt{r} \sin\theta\cos\dfrac{\theta}{2} \end{cases} \quad (4)$$

where r and θ are two parameters of the local coordinates of the crack tip.

2.2 The level set method

The level set method is a numerical technique of track the motion of interfaces. It describes the Interface shape by zero level set function, and gets to changing the interface shape though analyzing the level set function, especially avoiding re-meshing at the time when the level set upgrades, which just conforms to the extended finite element method (Ventura G., 2006). So, in the calculation on the motion of the interface, the fixed mesh can be applied. As is showed in Figure 1, the position needed in calculation can be described by two level set functions φ and ϕ, which are perpendicular to each other. The whole field is divided into three parts: general elements, penetrated elements and crack tip elements. And those kinds of elements can be expressed by the level set functions (Stolarska, 2001). In detailed, general elements can be described as:

$$\varphi_{max}\varphi_{min} \geq 0 \quad (5)$$

And Penetrated elements:

$$\phi_{max} < 0, \phi_{min} < 0, \varphi_{max}\varphi_{min} \geq 0 \quad (6)$$

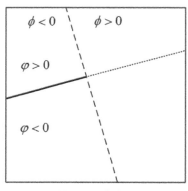

(a)One crack tip

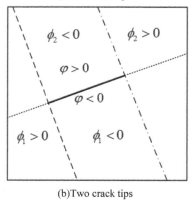

(b)Two crack tips

Figure 1.   The level set functions.

And Crack tip elements:

$$\phi_{max}\phi_{min} < 0, \varphi_{max}\varphi_{min} \leq 0 \quad (7)$$

2.3 Stress intensity factor

Stress intensity factor can be used to express the stress field of the crack tip and the size of the displacement, in fracture mechanics (Karihaloo, 2006). It shows the amount of deformation, the loads applied to the model, and also can be used to express the tendency or the motion of the crack propagation (Chahine E., 2006).

The calculation of the stress intensity factor mainly includes numerical method and analytic method. But analytic method only applied to some simple models, so most cases need the numerical method. And according to the definition, the stress intensity factor of the mode I and mode II crack can be expressed as:

$$\begin{aligned} K_I &= \sigma\sqrt{\pi a} \\ K_{II} &= \tau\sqrt{\pi a} \end{aligned} \quad (8)$$

where the σ and τ are the nominal stress. For mode *I* crack, σ is the biaxial tensile stress of an infinite tablet with a central crack, the length of which is $2a$; and for mode *II*, τ is shearing stress at infinity of the same tablet (Erdogan, 1963).

2.4 *Fracture criterion*

For most cracks, given a fixed pressure, when K reaches to critical value K_c, the crack will become unstable, and propagate (Peters, 2005). There are many fracture criterions, and in this article, the energy release rate G is used. There are two assumptions in the fracture criterion. The first is that the crack will propagate along the direction of the minimum energy release rate; and the other one is that when the energy release rate comes to the critical value G_c, the crack begins to propagate (Chopp D.L., 2003).

In cases of plane strain, energy release rate of the crack, which propagates in the direction of the crack line can be expressed as:

$$G = G_I + G_{II} = \frac{1-\mu^2}{E}\left(K_I^2 + K_{II}^2\right) \tag{9}$$

So, when the model material is given, the energy release rate is totally depended on the stress near the crack and the path of the crack propagation.

3 NUMERICAL EXAMPLE

Considering a reservoir, a model can be built like Figure 2, and be analyzed as a strain problem with unit thickness. The boundary conditions are assumed as: *AD* is considered as a symmetrical boundary, and the other three *AB*, *BC*, and *CD* are put as zero displacement boundaries. The horizontal ground stress is applied on these three boundaries, while an injection pressure is applied on the crack. When the horizontal ground stress is considered, it would be divided into two parts: the maximum horizontal stress and the minimum one.

The geometry size and loads factors are given in Table 1. As showed in Table 1, "*L*" is the geometry boundary size, "*a*" is the crack length, "σ_x" is the minimum horizontal ground stress, while the maximum horizontal ground stress is σ_y, and the injection pressures is σ.

As showed in Table 2, there are some relevant parameters of the model material. In this model, linear elasticity materials were adopted. In the table,

Table 1. Geometry size and loads factors.

L/m	a/m	σ_x/MPa	σ_y/MPa	σ/MPa
3	0.5	16	11	11

Table 2. Material parameters.

E/GPa	μ	σ_t/MPa	G_I/(N/M)	G_{II}/(N/M)	G_{III}/(N/M)
3.82	0.3	6	40200	40200	0

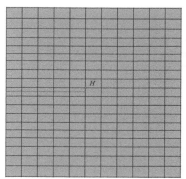

(a) Original model

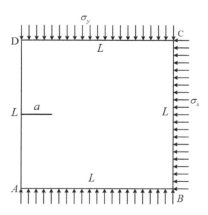

Figure 2. Numerical model.

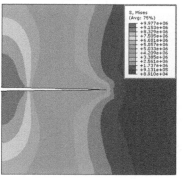

(b) Stress nephogram

Figure 3. Original model and stress nephogram.

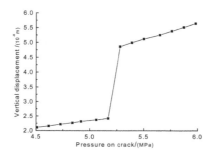

Figure 4. Curve of vertical displacement versus pressure.

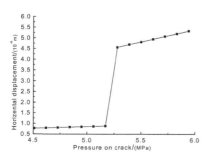

Figure 5. Curve of horizontal displacement versus pressure.

"E" is the elastic modulus, "μ" is the poisson ratio, "σ_t" is critical tensile strength, and "G_I, G_I, G_{III}" are three energy release rates. And the initial crack and propagation model are showed in Figure 3.

As is showed in Figure 3, point H adjoins to the crack tip so that the displacement can be easily monitored. Figure 4 and Figure 5 are shown its two different displacements. In Figure 4, "v" represents the vertical displacement, while in Figure 5, u stands for the horizontal displacement, and s is the corresponding stress.

As is showed in Figure 4 and Figure 5, when the stress comes to 5.17 Mpa, the vertical displacement increases to a big value of 4.86×10^{-6} m, along with the horizontal displacement that increases to 4.56×10^{-6} m, which means the crack propagates. After this sudden change, the displacement grows in a constant speed again. In the simulation result, it can be seen, that the crack propagates along horizontal direction. And this is because the model's boundary condition is symmetrical about the crack line. So, in the two stress ranges (before and after), changes of the displacement of the point H is linear, which is conformed to the beginning hypothesis.

4 CONCLUSION

In this article, extended finite element method was applied to simulating hydraulic fracturing, which intends to study the possibility of its feasibility in reality. The simulation on the reality model shows that it's much more convenient to use the extended finite element method for the hydraulic fracturing propagation problems. And the analysis of the result shows that the application of the XFEM made the calculation process much more convenient, and the results more accurate. So, it is feasible to apply the extended finite element in actual operating mode.

REFERENCES

Belytschko T., Black T. Elastic crack growth in finite elements with minimal remeshing [J]. *International Journal for Numerical Method in Engineering*, 1999, 45:601–620.

Chahine E., Laborde P., Renard Y. A quasi-optimal convergence result for fracture mechanics with XFEM [J]. Academie des Sciences, Paris, Ser. I, 2006, 342:527–532.

Chopp D., Dolbow J.E. A hybrid extended finite element/level set method for modeling phase transformations [J]. *International Journal for Numerical Methods in Engineering*, 2002, 54(8):1209–1233.

Chopp D.L., Sukumar N. Fatigue crack propagation of multiple coplanar cracks with the coupled extended finite element and fast marching method [J]. International Journal of Engineering Science, 2003, 41(8):845–869.

Erdogan. On crack extension in plates under plane loading and transverse shear [J]. *Trans. ASME. Journal of Basic Engng*. 1963.

Karihaloo. Improving the accuracy of XFEM crack tip field using higher rderquadrature and statically admissible stress recovery [J]. International Journal for Numerical Methods in Engineering, 2006, 66(9):1378–1410.

Nagashima, Suemasu. Application of extended finite element method to fracture of composite materials. *European Congress on Computational Methods in Applied Sciences and Engineering (ECCOMAS)*, Finland, 2004.

Peters. Numerical aspects of the extended finite element method [J]. Proceedings of Applied Mathematics and Mechanics, 2005, 5:355–356.

Rethore J., Gravouil A., Combescure A. A combined space-time extend finite element method [J]. *International Journal for Numerical Methods in Engineering*, 2005, 64:260–284.

Stolarska M., Chopp, Moes et al. Modelling crack growth by level sets in the extended finite element method [J]. International Journal for Numerical Methods in Engineering, 2001, 5l(8):943–960.

Stolarska, Chopp, Moes, Beltyschko. Modelling crack growth by level sets. 2003.

Sukumar, Moes, Moran, Belytschko. Extended finite element method for three dimensional crack modeling [J]. *International Journal for Numerical Methods in Engineering*, 2000, 48 (11):1549–1570.

Ventura G. On the elimination of quadrature subcells for discontinuous functions in the extended finite-element method [J]. International Journal for Numerical method in Engineering, 2006, 66:761–795.

Wagner G.J., Moes N., Liu W.K, Belytschko T. The extended finite element method for rigid particels in stokes flow [J]. *International Journal for Numerical Methods in Engineering*, 2001, (51):293–313.

Advances in Energy, Environment and Materials Science – Wang & Zhao (Eds)
© *2016 Taylor & Francis Group, London, ISBN 978-1-138-02931-6*

Experimental study of the pre-split crack effect on the stress wave propagation caused by explosive

Jianjun Shi, Xing Wei & Huaming An
School of Civil and Environmental Engineering, University of Science and Technology Beijing, Beijing, China

Haili Meng
China Academy of Railway Sciences, Beijing, China

ABSTRACT: Presplitting blasting produces a pre-splitting crack between blasting area and surrounding rock, which can prevent stress wave from travelling through surrounding rock, reduce the damage or destruction of the rock outside the excavation scope, and maintain maximally the original strength and stability of rock. In order to study the attenuation of explosion stress wave through the pre-splitting crack, a larger number of small concrete model specimens have been systematically studied, and its regularity of vibration reduction was obtained according to different crack width, different charge and different filling medium. The results of research show that: crack width is an important factor affecting vibration reduction rate; the vibration reduction rate of pre-splitting crack without fillings is higher than that with fillings; the denser is the filling material, the better is the vibration reduction effect; the wider is the pre-splitting crack, the higher is the strain wave attenuation rate. This study reflects the formation process of pre-splitting crack, which is helpful to guide the pre-splitting blasting in practical engineering.

1 INTRODUCTION

Presplitting blasting is always adopted in the railway or highway stone cutting excavation. In pre-splitting blasting engineering, a pre-splitting crack can be produced between blasting area and surrounding rock in order to prevent blasting stress wave from spreading towards surrounding rocks. It can reduce the damage or destruction of the rock out of the excavation area, and maximally maintain the original intensity and stability of rock [Whang, 2013]. However, its interaction process and attenuation law is not clear.

At present, researchers have studied the mechanism and effect of vibration reduction based on theoretical analysis and numerical simulation [Hu, 2011 & Chen, 2011]. Li Xibing has studied the influence of weak structural plane on stress wave propagation according to interface strength described with Coulomb Friction Law [Li, 1993]. Myer has studied the interaction of explosive stress wave with pre-splitting fissure according to interface model described with joint stiffness. Wang Mingyang has put forward the simplified calculation model of explosion seismic wave passing through weakening layer and obtained the attenuation equation based on Viscoelastic Wave Theory. In the actual projects, rock mass is not the ideal elastic medium. Thus, it is very difficult

to obtain the analytical solution of interaction between explosive stress wave and pre-splitting crack. Some results have been achieved by some researchers in the study of pre-splitting crack by the method of numerical simulation. For example, Liang Kaishui used LS-DYNA analysis software to simulate the vibration reduction of pre-splitting cracks, which have different width, length, and depth in open-pit mine [Liang, 2006 & Dai, 2012]. At present, there are only a few experimental studies about the effect of pre-splitting crack on stress wave propagation [Guo, 2010]. In this paper, we have studied systematically the attenuation of seismic wave passing through pre-splitting crack and drawn vibration reduction law in the conditions of different crack width, different charge mass and different medium.

2 THE IMPACT OF PRE-SPLITTING CRACK ON SEISMIC WAVE PROPAGATION

Blasting seismic wave, a kind of elastic-plastic stress wave [Xu, 2006] is produced when shock wave attenuates to far off areas after explosion. Generally, the formation scope of seismic wave is 150 times greater than the radius of cartridge. Sound wave is also a kind of elastic wave and its frequency is close to the frequency of seismic wave. In blasting engineering, attenuation law or propagation velocity of sound waves in rock mass before and after blasting can be used for analysis of the index such as internal structural state of rock mass, mechanics parameter, damage degree of blasting vibration, and so forth. Therefore, we can get the attenuation law of seismic wave transiting through pre-splitting crack by studying the behavior of sound wave on the same condition.

2.1 Model specimen

The model specimens (Fig. 1) are cement mortar cube, which is produced by casting of 425 # silicate cement and fine sand. There are two kinds of specimens. The first is that one specimen (40 × 40 × 28 cm) is divided into two pieces (30 × 40 × 28 cm and 10 × 40 × 28 cm respectively); the second is that there is a 14 cm deep fracture in the line 30 cm away from the long side of specimen (40 × 40 × 28 cm), and five kinds of different crack width is 0, 1, 2, 3, and 4 cm respectively. Besides, we made the standard cube for measuring physical and mechanical parameters of experimental materials. Its size is 10 × 10 × 10 cm. And it is conserved on the same conditions with model specimens. Physical and mechanical parameters of model specimens are shown in Table 1.

2.2 The impact of width of pre-splitting crack with filling on vibration reduction

Using the first model specimens, we change crack width by moving the position of the small block. The crack is filled with fine sand (particle size is less

than 1.25 mm). The propagation velocity of sound wave is tested by Acoustic Tester. Comparing with the propagation velocity of sound wave, which travels through rock mass without fissure, we can obtain the effect of pre- pre-splitting crack on wave propagation [Lu, 1997]. According to the test results, the relationship between vibration reduction rate and crack width is fitted into a curve. It is shown in Figure 2. The pictures in Figure 3 are in sequence the waveform, which sound wave travels through the rock with crack width of 0, 2, 4, and 8 mm.

It indicates in Figure 2 that wave velocity declines and vibration reduction rate increases gradually following with the increase of crack width.

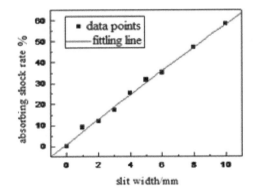

Figure 2.   Fitted curve of vibration reduction rate and crack width.

Figure 1.   Model specimens.

Table 1.   Physical and mechanical parameters of specimens.

Density, g/cm³	Elastic wave velocity, m/s	Modulus of elasticity, GPa	Compressive strength, MPa
2.108	3606	19.8	34.2

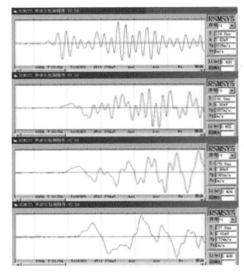

Figure 3.   Testing results of sound waves of different crack width (0, 2, 4, 8 mm).

Figure 3 indicates that the medium in pre-splitting crack has the property of filtering high frequency wave. Furthermore, the wider the fissure is, this property is more obvious.

2.3 The impact of different filling medium on vibration reduction

Using the first model specimens, we fill up pre-splitting cracks with different medium, and then test the propagation velocity of sound wave in each specimen with different crack width. Table 2 is the propagation velocity of sound wave when the filling is medium coarse sand (particle size, 1.25 ~ 2.5 mm). Table 3 is propagation velocity of sound wave when the filling is medium fine soil (particle size, less 0.63 mm). These experimental results show that wave velocity declines and vibration reduction rate increases gradually along with the increase of crack width. The fitted curve of crack width and vibration reduction rate of different filling medium (course sand, fine sand and soil) is shown in Figure 4.

It is shown in Figure 4 that if crack width of specimens is the same, the vibration reduction rate

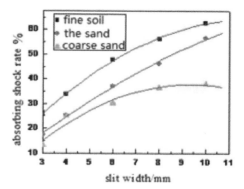

Figure 4. Fitted curve of different medium.

Table 4. Propagation velocity of sound waves in pre-splitting crack without filling medium.

Crack width, mm	Specimen 1	Specimen 2	Average velocity m/s	Vibration reduction rate, %
1				
Not through	3571	3636	3604	20.09
Through	2893	2866	2880	
2				
Not through	3588	3534	3561	24.01
Through	2717	2695	2706	
3				
Not through	3571	3540	3556	31.38
Through	2469	2410	2440	
4				
Not through	3509	3571	3540	40.03
Through	2109	2136	2123	

Table 2. Propagation velocity of sound waves (coarse sand).

Crack width/mm	Specimen			Average velocity m/s	Vibration reduction rate %
	1	2	3		
0	3226	3195	3242	3221	0
3	2778	2803	2817	2799	13.10
4	2439	2381	2428	2416	24.99
6	2222	2273	2258	2251	30.11
8	2041	2062	2038	2047	36.45
10	2000	2020	1985	2002	37.85

Table 3. Propagation velocity of sound waves (soil).

Crack Width/mm	Specimen			Average velocity m/s	Vibration reduction rate %
	1	2	3		
0	3226	3195	3242	3221	0
1	2857	2899	2863	2873	10.80
2	2740	2728	2703	2824	15.43
3	2410	2353	2372	2378	26.17
4	2151	2127	2118	2132	33.81
6	1667	1681	1693	1680	47.84
8	1429	1389	1406	1408	56.29
10	1205	1183	1220	1203	62.65

of fine soil is higher than that of fine sand and the vibration reduction rate of fine sand is higher than that of course sand. It means that the smaller is the particle size of filling medium, the higher is the rate of vibration reduction. The main reason is that filling material with higher density can absorb more wave energy.

2.4 The impact of width of pre-splitting crack without filling on vibration reduction

The second model specimens are used in this part of experiment. There is nothing but air in the pre-splitting crack of all specimens. The crack width of specimens is 1, 2, 3, and 4 mm, respectively. Two propagation velocity of elastic wave are tested: one is that it travels through pre-splitting crack, the other is not. If elastic wave travels through pre-splitting crack, it will diffract.

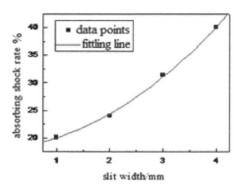

Figure 5. Fitted curve of vibration reduction rate and crack width of pre-splitting crack of no fillings.

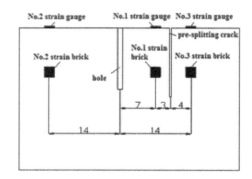

Figure 6. Relative position of the strain brick and strain gauge.

Its propagation distance can be calculated by the triangle side relationship. Measured elastic wave velocity is shown in Table 4 and the same two specimens are noted as specimen 1 and specimen 2, respectively.

Table 4 shows that as elastic wave transits through pre-splitting crack without fillings, vibration reduction rate gradually increase along with the increasing width of pre-splitting crack. Comparing the data of Table 5 with Table 2 and Table 3, we can arrive at a conclusion that the vibration reduction rate of pre-splitting crack without fillings is higher than that of pre-splitting crack with fillings. Therefore, the phenomenon of filtering wave is not present in these experiments and the sound waves travel through crack by diffraction.

3 THE IMPACT OF PRE-SPLITTING CRACK ON STRAIN WAVE PROPAGATION

3.1 *Experiment design*

The second model specimens are used in this part of experiment. Three strain bricks must be embedded inside the specimens and three strain gages must be attached on the specimen surface ahead of time. The relative position of strain brick, strain gauge and pre-splitting crack is shown in Figure 6. Burial depth of Strain brick is 10 cm. Blasthole (diameter is 1 cm and depth is 12 cm) is drilled in the center of specimen by percussion drilling. Centering on blasthole, the strain bricks and strain gages of No. 2 and 3 are distributed symmetrically. By testing and comparing the strain of all points where strain bricks and strain gauges locate, we can obtain the impact of pre-splitting fissure on wave propagation. In this experiment, we choose RDX explosives, put them into glass tubes (external diameter, 1 cm; inner diameter, 0.8 cm) in order

for full coupling of explosives and blasthole wall, insert electric detonator into explosive top, and plug up the porthole with absorbent cotton.

3.2 *Strain test system*

After explosion, shock wave spreads outward along the radiation direction, and rock mass is deformed caused by the impact of shock wave. The deformation has close relationship with shock wave energy and dielectric properties. Blasting dynamic strain measurement can indicate the characteristics of internal stress wave parameters of medium and the stress state of components' location. At present, in domestic, dynamic strain is generally measured by blasting resistance strain testing-it transform the mechanical quantity (strain) change into resistance change depend on the resistance strain gauge installed on the specimens, and then transform resistance change into the output of voltage or current. Strain test system is made up of the strain gauge, dynamic strain gauge, TST3406 dynamic testing analyzer and computer.

3.3 *Test result and analysis*

The result of dynamic strain measurement is an oscillogram that represents the stain of each measuring point. Maximum strain value of the instantaneous strain state of can be obtained by comparing the waveform of tested strain wave with and that of the known amplitude. In general, the method of standard waveform acquisition is called calibration. Calibration of dynamic strain amplitudes can be obtained from the calibration device in deformeter. After calibration, strain value of the measuring point can be calculated by the following equation:

$$\varepsilon_t = \frac{h}{H} \varepsilon_b$$

In this equation: ε_i, instantaneous strain value of measuring points; h, panel height of strain curve in a moment; H, the distance from Strain zero line to Standard strain line that has similar height with; h, ε_b standard strain values corresponding with H.

1. Strain test of same pre-splitting crack at condition of different dosage

   Test 10 specimens (crack width, 1 mm) of different doses. The dose is 1.0, 1.5, 2.0, 2.5, 3.0 g, respectively. The testing results indicates that the strain value of all points in specimens enlarges along with the increase of dosage and the nearer to blasthole the testing point, the greater the strain values. When the dosage increased to 3.0 g, the strain value of No.1 strain brick has been increased to 17518 $\mu\varepsilon$. The same as the hole distance strain brick and strain gauge, the strain value of strain brick 3 and strain gauge 3 is smaller than that of strain brick 2 and strain gauge 2. It means that strain wave attenuates when it travels through pre-splitting crack. Thus surrounding rock is protected for the reason that strain value of surrounding rock is effectively reduced. As strain wave passes through pre-splitting crack, strain value reduces on average by 17.17%. However, dose has little effect on the amount of reduction of strain value.

2. Strain test of different pre-splitting crack at condition of same dosage

   In this experiment, the crack width of specimens is 0, 1, 2, 3, and 4 mm. After explosion of 1.5 g RDX, the strain value of all strain bricks and strain gages is tested. According to the result, strain wave attenuation rate and crack width is fitted into a curve (Fig. 7). It indicates that the wider is the pre-splitting crack, the greater is the attenuation rate and energy decrement of strain wave. When crack width increases to 3.0 g, strain wave attenuation rate increases to 37.75%. In conclusion, pre-splitting crack must have enough width so that it can be capable of reducing vibration effectively.

3. Strain test of pre-splitting crack filled with medium

   When the crack is filled up with fine sand, the wilder the crack is, the greater the strain wave attenuation rate is. On the same conditions of dose and crack width, the strain wave attenuation rate of pre-splitting crack filled with medium is higher than that filled without medium. It means that inhibition of strain wave is declined when crack is filled up with medium.

3.4 Analysis of blasting effect

According to the observation of blasting effect of specimens after explosion, blasting effect found pre-splitting crack has a great influence on fragment effect. Mainly manifested in the following aspects:

1. Fragment effect of both sides is significantly different-the concrete before crack is damaged seriously, but the concrete after crack is relatively complete which indicates that this part of concrete is fully protected.

2. The shape of blasting crater has changed. The funnel opening is approximately circular when there is no pre-splitting crack. However, when there is pre-splitting crack in concrete, the funnel opening is no longer a complete circle and truncated by pre-splitting crack.

3. The development of blasting fracture is hindered by pre-splitting crack. The blasting results show that blasting fracture is blocked by pre-splitting crack when it extends outwards. There is no fracture in protected concrete.

4. Filling medium in pre-splitting crack will lower protection effect. The protected concrete of specimens with fillings, all other factors being equal, is damaged more seriously than that of specimens without fillings for the reason that some of explosive stress wave pass through crack and damage the protected section.

4   CONCLUSIONS

1. Pre-splitting crack has hindering function for seismic wave propagation. Whether there is filling medium in pre-splitting crack or not, it can prevent the spread of seismic wave.

2. Under the condition of the same crack width, vibration reduction effect of pre-splitting crack without fillings is much better than that with fillings. If there is no filling medium in the crack, seismic wave will diffract through crack bottom.

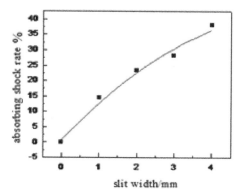

Figure 7.   Fitted curve of strain wave attenuation rate and crack width.

If there is filling medium in the crack, seismic wave can pass through filling medium directly, but the medium in pre-splitting crack has the property of filtering high frequency wave. Besides, the smaller is its particle size, the higher is the rate of vibration reduction.

3. Strain experiment proves that strain wave passing through pre-splitting crack will have certain attenuation, when the crack is filled up with fillings, the attenuation will decrease; crack width is concerned with strain wave attenuation rate; when crack width increases from 1 mm to 4 mm, strain wave attenuation has increased from 17.17% to 37.75%. The wider is the pre-splitting crack, the greater is the attenuation rate and energy decrement of strain wave.

ACKNOWLEDGMENTS

This work was financially supported by the National Natural Science Foundation of China (51208036) and "the Fundamental Research Funds for the Central Universities (FRF-TP-14-075A2)".

REFERENCES

Chen Qing-kai, Zhu Wan-cheng. Mechanism of the crack formation induced by pre-split blasting and design method for the pre-split blasting hole space [J]. Journal of Northeastern University (Natural Science), 2011, 32(7):1024–1027.

Dai Bing, Zhao Guo-yan. Numerical Simulation Analysis of Blast Parameters Matching based on Cracking of Pre-split Blasting [J]. Blasting, 2012, 29(1):10–14.

Fangtian Wang, Shihao Tu, Yuan Yong, Yufeng Feng, Fang Chen, Hongsheng Tu. Deep-hole pre-split blasting mechanism and its application for controlled roof caving in shallow depth seams [J]. International Journal of Rock Mechanics and Mining Sciences, 2013, 64:112–121.

Guo Yao, Meng Haili, Qi Yanjuan, Xue Li. Study on Mechanism of the transmission of blasting vibration in the pre-split crack [J]. Chinese Journal of Scientific Instrument, 2010, 31(4):17–20.

Hu Jianhua, Lei Tao, Zhou Keping. Chen Qingfa [J]. Effect of blasting vibration on pre-splitting crack in filling-environment. Journal of Central South University (Science and Technology), 2011, 42(6):1704–1709.

Li Xi-bing. Influence of the structural weakness planes in rock mass on propagation of stress waves [J]. Explosion and Shock Waves, 1993, 13(4):334–342.

Liang Kai-shui, Chen Tian-zhu, Yi Chang-ping. Numerical simulation for vibration-isolating effect of vibration-isolating slot [J]. Blasting, 2006, 23(3):18.

Lu Wenbo, Lai Shixiang, Dong Zhenhua. Analysis of vibration isolating effect of pre-splitting crack in rock excavation by blasting [J]. Explosion and Shock Waves, 1997, 17(3):193–193.

Myer L.R. Effects of Single Fracture on Seismic Wave Propagation [C]//Barton & Stephansson, eds Rock Joints. Rotterdam: Bakema, 1990:458–467.

Advances in Energy, Environment and Materials Science – Wang & Zhao (Eds)
© 2016 Taylor & Francis Group, London, ISBN 978-1-138-02931-6

Research on ultimate strength of confined concrete with circular stirrups

MengYuan Lu & Dongsheng Gu
College of Environment and Civil Engineering, Jiangnan University, Wuxi, Jiangsu, China

ABSTRACT: With external confinement, the transverse deformation of concrete is restricted, thus the compress behavior of core concrete is improved and the concrete compressive strength and deformation capacity can be greatly enhanced. According to confining materials and confining models, confined concrete is classified into stirrup confined concrete, steel tube confined concrete, FRP confined concrete, and other new material confined concrete. There are many common stirrup confined concrete constitutive models, but their forms are complex and not easy for calculating. This article first makes a brief introduction to the stirrup restraint concrete. And then inspired by Richart linear model based on the hydrostatic pressure, through the analysis of experimental data, the author puts forward an improved simple model for calculating the axial compressive strength of stirrup confined concrete, and concluded the liner relationship between the ultimate strength of confined concrete with the lateral confining stress on concrete from transverse reinforcement. And this article then verifies this theory through the existing Mander's constitutive model; two models coincided basically in calculating the result. So the meaning of this article is mainly to put forward a simplified model for calculating the ultimate bearing capacity of confined concrete with stirrups and its research results can be used as a reference for engineers.

1 RESEARCH BACKGROUND

Confined concrete refers to concrete using external confinement to improve its own original compression feature, in order to improve its compressive strength and deformation performance. The bigger lateral pressure concrete under, the greater the ability to restrict or restrain the transverse deformation. Confined concrete is a quite classic subject of structure, and is also one of the most basic research field of modern structural engineering. This has maintained a great interest in research at home and abroad, and also has established a systematic theory of confined concrete.

At present, there is a lot of research that has been carried out about the stirrup confined concrete and FRP confined concrete at home and abroad. All kinds of finite element method and the development of computer technology have created favorable conditions for the nonlinear analysis of reinforced concrete structure and components. However, researchers seldom focus on the linear simplification that can be applied in practical engineering.

By constraining the core concrete, the compressive strength and ductility of the concrete can be improved, the concrete structure is particularly important for seismic region. Appropriately increasing the stirrups and improving the structure form is one of the most simple, economic and effective measures to improve the seismic performance of structure.

Stirrup enhancement effect of the confined concrete mainly depends on its ability to exert binding force on the surface of the core concrete. The greater the lateral confining stress is, the greater the enhancement of concrete can be. Binding force is mainly affected by the following factors.

Stirrup restraint of concrete depends on two aspects, one is the relative size of the effective confined concrete area, and the second is the magnitude of the lateral confining stress applied to effective area of confined concrete. During the stirrup restraint, the two aspects are interrelated. In general, the bigger the effective relative area of confined concrete is, the more significant the role of the binding force plays, the greater the stirrup restraint towards the concrete. Effective area of confined concrete and the magnitude of the stirrup binding force depend on the amount of stirrup, the stirrup spacing, the mechanical properties (mainly refers to the yield strength of the stirrup) and the decorated form.

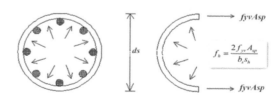

Figure 1. Stirrup confined concrete.

2 INSPIRED BY RECHART MODEL

As early as 1928, Richart had proposed a confined concrete model. The strength of the concrete model is mainly based on the experimental results of the hydrostatic pressure. Here the model is:

$$\frac{f_{cc'}}{f_{co'}} = 2.05\lambda_h + 1 \tag{1}$$

The model has clearly put forward the linear relations between the ultimate strain of confined concrete and the characteristic value of the stirrup.

Therefore the author of this article considered that under the effect of stirrups, the ultimate strain of confined concrete and the lateral confining stress on concrete from transverse reinforcement may exist like the similar linear relationship, which can be used to greatly simplify the calculation.

On the basis of the analysis of the collected data, the following simplified model is put forward as following.

3 SIMPLIFICATION OF CONFINED CONCRETE MODEL

Use the mathematical tools to turn the statistics into equation, the function image and the function expression are as follows:

$$y = 0.185x \tag{2}$$

That is:

$$\rho_c f_{yv} = 0.185\left(f_{cc'} - f_{co'}\right) \tag{3}$$

$$\frac{1}{2}\rho_v \frac{f_{yv}}{f_{co'}} = 0.185\left(\frac{f_{cc'}}{f_{co'}} - 1\right) \tag{4}$$

In the above formula, ρ_v refers to the volumetric stirrup ratio, f_{yv} refers to the yield strength of

Table 1. The collected statistics of confined concrete.

Setting	$\rho_c f_{yv}$ (Mpa)	$f_{cc'} - f_{co}$ (Mpa)	$\rho_c f_{yv}$ (Mpa)	$f_{cc'} - f_{co}$ (Mpa)
	6.94	0.95	16.90	3.02
	11.00	1.68	17.90	3.38
	12.50	1.85	17.90	3.08
	10.90	2.66	17.80	3.90
	13.90	3.03	18.60	3.93
	15.40	1.77	18.90	3.38
	16.10	1.78	19.90	3.34
	14.00	3.51	21.90	4.27
	15.00	3.31	21.30	5.33
	16.80	2.59	24.00	3.38
	17.30	2.61		

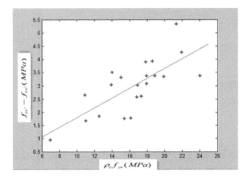

Figure 2. Simplification of calculating model.

stirrup, $f_{cc'}$ refers to the ultimate strength (peak stress) of confined concrete, and $f_{co'}$ refers to the compressive strength of unconfined concrete.

In addition, by introducing a new variate of stirrup characteristic value of λ_h to control the amount of stirrup in each volumetric unit:

$$\lambda_h = \rho_v \frac{f_{yv}}{f_{co'}} = \frac{4A_{sp}}{d_s s_h} \cdot \frac{f_{yv}}{f_{co'}} \tag{5}$$

Among the above, A_{sp} refers to the area of spiral bar, d_s refers to diameter of spiral, s_h is the vertical stirrup spacing.

And the link between the lateral confining stress on concrete from transverse reinforcement f_h and the stirrup characteristic value is λ_h:

$$\lambda_h = 2 \times \frac{f_h}{f_{co'}} \tag{6}$$

That is to say, the liner link between the ultimate strain of confined concrete and the lateral confining stress on concrete from transverse reinforcement is:

$$f_{cc'} = 5.4 f_h + f_{co'} \tag{7}$$

Therefore, according to the analysis, in the range of the commonly used amount of stirrups, the relationship between the ultimate strain of confined concrete and the the lateral confining stress on concrete from transverse reinforcement can be simplified into the formula above. The verification of this model through Mander's model is as follows.

4 THE VALIDATION OF MANDER'S MODEL

In the finite element analysis of concrete, the stirrup restraint of core concrete of the compression

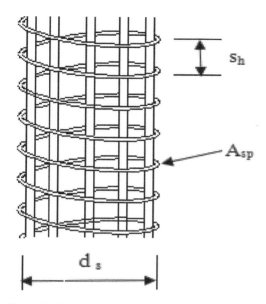

Figure 3. Parameter of stirrups.

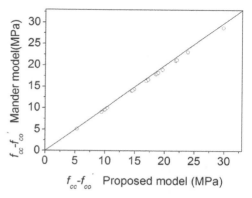

Figure 4. Comparison of Mander's model and proposed model.

member section can be responded through the equation of stress-strain full curves of concrete, it is to use the constitutive model of confined concrete. There are kinds of stress and strain models of confined concrete that has been put forward, in which Mander's model has been the most widely used. It applies to the circular stirrup as well as the rectangular stirrup. The following first makes a brief introduction towards Mander's model, and then verify the above linear relationship through this theory.

Mander etc. made 31 full-scale axial compression specimens, including three kinds of section forms of round, square, rectangular, and multiple reinforcement methods.

The Mander's model shows that when the effective binding force of the cross section in two directions are the same, it is on the premise of cylinder of this article studies, the link between the maximum of confined concrete compressive strength $f_{cc}{'}$ and unconstrained concrete compressive strength $f_{co}{'}$ is as follows:

$$\frac{f_{cc}{'}}{f_{co}{'}} = -1.254 + 2.254\sqrt{1+\frac{7.94 f_h{'}}{f'_{co}}} - \frac{2 f_h{'}}{f'_{co}} \quad (8)$$

The f_h values for the effective restraint stress when the transverse reinforcement get yield, for circular cross section:

$$f_h{'} = 2 f_{yv} A_{sp} /(d_s s_h) \quad (9)$$

In the formula above, f_h is the effective restraint stress of stirrup, A_{sp} refers to the area of spiral bar, d_s refers to diameter of spiral, s_h is the vertical stirrup spacing.

Compare the result of this simplified calculating model with Mander's model, the result just as the following picture shows. From which it can be concluded that the result of this two models matches each other well.

5 CONCLUSION

This paper proposed an the improved simple model for calculating the axial compressive strength of stirrup confined concrete, and concluded the liner relationship between the ultimate strain of confined concrete and the lateral confining stress on concrete from transverse reinforcement. This has largely simplified the calculating and proved that it also has a good agreement with the existing models. Above all, this paper provides a reference for the engineers and has certain practical value.

With the large development of national construction, reinforced concrete structure has developed rapidly in our country. However, there are a series of problems come to exist. Concrete structure's weakness lies in the heavy weight of itself, the large seismic effect and the brittle failure. Through the effective restraint of concrete, brittleness of concrete can be improved and the structures could have a good seismic performance.

Stirrup confined concrete has been most widely used for its unique advantages and has a long history of development. In addition, the stirrup confined concrete and the application of some new materials such as FRP confined concrete is still the direction of development in structure engineering in the future.

ACKNOWLEDGMENT

The authors would like to acknowledge financial support from the Natural Science Foundation of Jiangsu province (No. BK20131105), Six talent peaks project in Jiangsu Province (No. JZ-013), Key Technology Support Program of Shu qian (No. S201406).

REFERENCES

Cusson D., Paultre P. (1995) "Stress- strain model for confined high strength concrete". Journal of Structural Engineering. pp 468–477.

Mander J.B. M.J.N. Priestley & R. Park (1988) "Theoretical Strain-Stress Model For Confined Concrete" ASCE. pp 1804–1826.

Murat Saatcioglu & Salim R. Razvi (2002) "Displacement-Based Design of Reinforced Concrete Columns for Confinement" ACI Structural Journal/January-February. pp 10.

Sheikh S.A., Uzumeri S.M. (1982). "Analytical model for concrete confinement in tied columns" Journal of the Structural Division, ASCE. pp 2703–2722.

Shi Qingxuan et al (2009). "Development and prospects of hoop reinforcement confined concrete" Journal of Building Structures (Supplementary Issue 2) pp 109–114. In Chinese.

Advances in Energy, Environment and Materials Science – Wang & Zhao (Eds)
© 2016 Taylor & Francis Group, London, ISBN 978-1-138-02931-6

Ground motion of half space with ellipse inclusion and interfacial crack for SH waves

Xiaohao Ding, Hui Qi, Yuanbo Zhao & Dongdong Zhao
College of Aerospace and Civil Engineering, Harbin Engineering University, Harbin, Heilongjiang, China

ABSTRACT: Based on elastodynamics, the scattering problem of SH-wave by an shallowly buried ellipse inclusion in bi-material half plane is analyzed by using complex function method and Green's function method. Firstly, the conformal mapping method and image method are employed in area I to construct expression of the scattering wave field. Secondly, the interface crack is constructed using crack dividing technique. With the aid of 'conjunction' technique, unknown forces system is adding on the uncracked section of the interface to satisfy the displacement and stress continuity condition. Through Fourier series expansion technology, a series of algebraic equations that determinate the unknown forces are set up. Finally, the displacement on the horizontal surface and the scattering wave field in the research model can be obtained.

1 INTRODUCTION

The scattering of elastic waves by complex terrain is attractive to many researchers and engineers because of the importance for civil engineering and earthquake engineering applications. Consequently, many researches and studies focusing in this field have been carried out over the past decades. Lee and Trifunac (1979) investigated the response of tunnels to incident SH-waves using image method. Chen (2004) investigated the scattering of SH-waves by an arbitrary shape cavity in half space. QI (2011) solved the scattering problem of SH-waves by a cylindrical inclusion in right-angle plane with beeline crack. Lee (2013) studied the scattering of SH-waves by semi-elliptical hills in half space. Luo (2010) presented an analytical solution to diffraction of SH-waves by an underground semi-circular cavity using wave function expansion method.

The focus of this paper is to develop a semi-analytical method for solving the scattering problem of elastic waves by ellipse inclusion shallowly embedded in a bi-material half plane with interface crack. Several common methods or techniques in elastic research, such as conformal mapping method (Chen & Liu 2004), image method (Lee et al. 1979), crack dividing technique (Qi et al. 2012), interface 'conjunction' technique (Qi & Yang 2012) are used in this paper.

2 MODELING AND ANALYSIS

As Figure 1 shows, there are an ellipse inclusion of half-macroaxis a and half-brachyaxis b and an interfacial crack of a length 2A in a bi-material half space. The distances from the center of the inclusion to the vertical interface and horizontal surface are d and h, respectively. The depth of the crack center point o'' is l, the incident angle is α_0.

The scattering problem of SH waves has only anti-plane displacement G which satisfy the Helmholtz equation. The governing equation of this paper with omitted the time harmonic factor exp $(-i\omega t)$ can be written as:

$$\frac{\partial^2 G}{\partial z \partial \bar{z}} + \frac{1}{4} k^2 G = 0 \qquad (1)$$

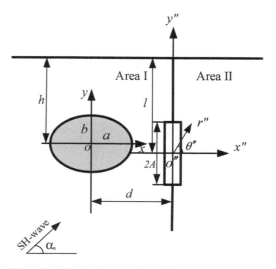

Figure 1. Sketch of research model.

In which, z and $\bar{z}$ are the complex variables, $k = \omega/c_s$ is wave number, ω and $c_s = \sqrt{\mu/\rho}$ are the circular frequency of steady state response and the shear velocity of the media, ρ and μ are the mass body density and the shear modulus of the media, respectively.

For wave scattering problems involving ellipse inclusion in the complex $(z, \bar{z})$ plane, we need use conformal mapping method. The mapping function we use in this paper has the following form:

$$Z = \omega(\eta) = R\left(\eta + \frac{m}{\eta}\right), \eta = Re^{i\theta} \tag{2}$$

The mapping function map the outside of the inclusion in the $(z, \bar{z})$ plane into the region $|\eta| > 1$. Consequently, the corresponding governing Eq. (1) in $(\eta, \bar{\eta})$ plane takes on the following form:

$$\frac{1}{\omega'(\eta)\overline{\omega'(\eta)}}\frac{\partial^2 G}{\partial \eta \partial \bar{\eta}} + \frac{1}{4}k^2 G = 0 \tag{3}$$

The relative stresses can be written as:

$$\begin{cases} t_{rz} = \dfrac{\mu}{R|\omega'(\eta)|}\left(\eta\dfrac{\partial G}{\partial \eta} + \bar{\eta}\dfrac{\overline{\partial G}}{\partial \eta}\right) \\ t_{\theta z} = \dfrac{i\mu}{R|\omega'(\eta)|}\left(\eta\dfrac{\partial G}{\partial \eta} - \bar{\eta}\dfrac{\overline{\partial G}}{\partial \eta}\right) \end{cases} \tag{4}$$

3 THEORETICAL FORMULATIONS

3.1 Green's function

As Figure 2 shows, with the aid of image method, the displacement field produced by the line source force $\delta(z - z_0)$ loading on the vertical surface in a complete elastic quarter-plane can be written as the follow in the $(\eta, \bar{\eta})$ plane:

$$G^{(i)}(\eta, \bar{\eta}) = \frac{i}{2\mu}\Big[H_0^{(1)}\big(k_1|\omega(\eta) - \omega(\eta_0)|\big) \\ + H_0^{(1)}(k_1|\omega(\eta) - \overline{\omega'(\eta_0)}|) \Big] \tag{5}$$

where, $H_0^{(1)}(\bullet)$ is the first kind Hankel function of 0 order, $\omega(\eta_0) = d + i(h - y_0)$, $\omega'(\eta_0) = d + i(h + y_0)$.

The scattering wave in area I can be written as:

$$G^{(s)}(\eta, \bar{\eta}) = \sum_{n=-\infty}^{\infty} A_n \sum_{j=1}^{4} S_n^{(j)} \tag{6}$$

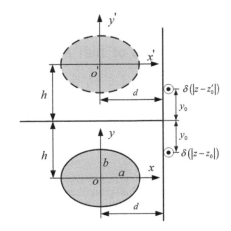

Figure 2. Images model of area I.

in which,

$$\begin{cases} S_n^{(1)} = H_n^{(1)}\big[k|\omega(\eta)|\big]\left[\dfrac{\omega(\eta)}{|\omega(\eta)|}\right]^n \\[2mm] S_n^{(2)} = H_n^{(1)}\big[k|\omega(\eta) - 2ih|\big]\left[\dfrac{\omega(\eta) - 2ih}{|\omega(\eta) - 2ih|}\right]^{-n} \\[2mm] S_n^{(3)} = (-1)^n H_n^{(1)}\big[k|\omega(\eta) - 2d|\big]\left[\dfrac{\omega(\eta) - 2d}{|\omega(\eta) - 2d|}\right]^{-n} \\[2mm] S_n^{(4)} = (-1)^n H_n^{(1)}\big[k|\omega(\eta) - 2ih - 2d|\big] \\[2mm] \qquad \times\left[\dfrac{\omega(\eta) - 2ih - 2d}{|\omega(\eta) - 2ih - 2d|}\right]^n \end{cases}$$

and A_n is an unknown coefficient determined by the boundary conditions.

The standing wave in ellipse inclusion can be written as:

$$G^{(t)}(\eta, \bar{\eta}) = \sum_{n=-\infty}^{\infty} B_n J_n\big(k_3|\omega(\eta)|\big)\left[\dfrac{\omega(\eta)}{|\omega(\eta)|}\right]^n \tag{7}$$

and B_n is an unknown coefficient determined by the boundary conditions.

In the $(\eta, \bar{\eta})$ plane, the boundary conditions around the ellipse inclusion $(|\eta| = 1)$ can be expressed as follows:

$$\begin{cases} G^{(i)} + G^{(s)} = G^{(t)} \\ \tau_{rz}^{(i)} + \tau_{rz}^{(s)} = \tau_{rz}^{(t)} \end{cases} \tag{8}$$

Eq. (8) can be solved by using Fourier series expansion and effective truncation.

The Green's function needed in area I and area II can be written as:

$$\begin{cases} G_1(\eta,\overline{\eta}) = G^{(i)} + G^{(s)} \\ G_2(\eta,\overline{\eta}) = G^{(i)} \end{cases} \qquad (9)$$

3.2 Plane wave

As the Figure 3 shows, with the aid of image method, the incident wave in area I:

$$W^{(i,e)} = W_0 \exp\left\{\frac{ik_1}{2}\left[(\omega(\eta)-ih)e^{-i\alpha_0} + (\overline{\omega(\eta)}+ih)e^{i\alpha_0} \right.\right.$$
$$\left.\left. +(\omega(\eta)-ih-2d)e^{-i\gamma_0} + (\overline{\omega(\eta)}+ih-2d)e^{i\gamma_0}\right]\right\} \qquad (10)$$

where, $\gamma_0 = \pi - \alpha_0$, α_0 and W_0 are incident angle and the amplitude of the incident wave respectively.

Similarly, the equivalent reflected wave and refracted wave are:

$$W^{(r,e)} = W_1 \exp\left\{\frac{ik_1}{2}\left[(\omega(\eta)-ih)e^{-i\alpha_1} + (\overline{\omega(\eta)}+ih)e^{i\alpha_1} \right.\right.$$
$$\left.\left. +(\omega(\eta)-ih-2d)e^{-i\gamma_1} + (\overline{\omega(\eta)}+ih-2d)e^{i\gamma_1}\right]\right\} \qquad (11)$$

$$W^{(f,e)} = W_2 \exp\left\{\frac{ik_2}{2}\left[(\omega(\eta)-ih)e^{-i\alpha_2} + (\overline{\omega(\eta)}+ih)e^{i\alpha_2} \right.\right.$$
$$\left.\left. +(\omega(\eta)-ih-2d)e^{-i\gamma_2} + (\overline{\omega(\eta)}+ih-2d)e^{i\gamma_2}\right]\right\} \qquad (12)$$

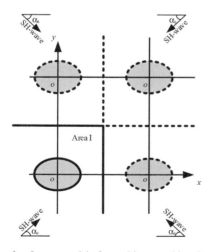

Figure 3. Image model of area I impacted by plane SH waves.

The scattering waves and standing waves have the same form with the Green's when the model is impacted by SH waves.

The Green's function we need are:

$$\begin{cases} G_1 = G^{(i)} + G^{(s)} \\ G_2 = G^{(i)} \end{cases} \qquad (13)$$

3.3 Conjunction function

As shown in Figure 4, a pair of opposite forces with the multitude $[-\tau_{\theta''z''}^{(I)}]$ and $[-\tau_{\theta''z''}^{(II)}]$ are applied to the left and right side of the section of the region where the crack will appear, resulting in a stress-free section as interfacial crack. Meanwhile, unknown force systems f_1 and f_2 are loaded on the sections outside the crack-section to satisfy the continuity conditions of stress and displacement on the interface (Qi & Yang 2012). According to the stresses and displacement continuity condition at the linking section and the Green's function we have obtained, the integral equations with unknown anti-plane forces can be expressed as:

$$\int_A^l f_1(r_0'',\beta_2)[G_1(r'',\beta_1;r_0'',\beta_2) + G_2(r'',\beta_1;r_0'',\beta_2)]dr_0''$$
$$+ \int_A^\infty f_1(r_0'',\beta_1)[G_1(r'',\beta_1;r_0'',\beta_1) + G_2(r'',\beta_1;r_0'',\beta_1)]dr_0''$$
$$= [-W^{(S)}]_{\theta_0''=\beta_1} + \int_0^A \tau_{\theta''z''}^{(I)}(r_0'',\beta_2)G_1(r'',\beta_3;r_0'',\beta_2)dr_0''$$
$$- \int_0^A \tau_{\theta''z''}^{(I)}(r_0'',\beta_3)G_1(r'',\beta_3;r_0'',\beta_2)dr_0''$$
$$+ \int_0^A \tau_{\theta''z''}^{(II)}(r_0'',\beta_2)G_2(r'',\beta_3;r_0'',\beta_2)dr_0''$$
$$- \int_0^A \tau_{\theta''z''}^{(II)}(r_0'',\beta_3)G_2(r'',\beta_3;r_0'',\beta_2)dr_0''$$

$$(14)$$

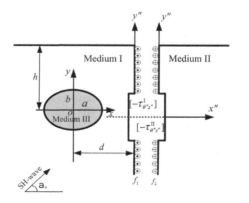

Figure 4. Conjunction model.

$$\int_A^l f_1(r_0'',\beta_2)[G_1(r'',\beta_2;r_0'',\beta_2)+G_2(r'',\beta_2;r_0'',\beta_2)]dr_0''$$

$$+\int_A^\infty f_1(r_0'',\beta_1)[G_1(r'',\beta_2;r_0'',\beta_1)+G_2(r'',\beta_2;r_0'',\beta_1)]dr_0''$$

$$=[-W^{(S)}]_{\theta''=\beta_2}+\int_0^A \tau_{\theta''z''}^{(I)}(r_0'',\beta_2)G_1(r'',\beta_2;r_0'',\beta_2)dr_0''$$

$$-\int_0^A \tau_{\theta''z''}^{(I)}(r_0'',\beta_3)G_1(r'',\beta_2;r_0'',\beta_3)dr_0''$$

$$+\int_0^A \tau_{\theta''z''}^{(II)}(r_0'',\beta_2)G_2(r'',\beta_2;r_0'',\beta_2)dr_0''$$

$$-\int_0^A \tau_{\theta''z''}^{(II)}(r_0'',\beta_3)G_2(r'',\beta_2;r_0'',\beta_3)dr_0''$$

$$(15)$$

In which, G_1 and G_2 are the Green's functions in area I and II respectively. The equations above can be solved by using Fourier series expansion and effective truncation. Then, all the unknown parameters in the process of solving the surface displacement have been obtained.

3.4 Surface displacement

In the area $x''< 0$:

$$W^{(1)} = W^{(I)} + \int_A^l f_1(r_0'',\beta_2)G_1(r',\theta';r_0'',\beta_2)dr_0''$$

$$+\int_A^\infty f_1(r_0'',\beta_1)G_1(r',\theta';r_0'',\beta_1)dr_0'' \qquad (16)$$

where, $W^{(I)} = W^{(i,e)} + W^{(r,e)} + W^{(s)}$.

In the area $x''< 0$:

$$W^{(2)} = W^{(II)} - \int_A^l f_2(r_0'',\beta_2)G_2(r',\theta';r_0'',\beta_2)dr_0''$$

$$-\int_A^\infty f_2(r_0'',\beta_1)G_2(r',\theta';r_0'',\beta_1)dr_0'' \qquad (17)$$

where, $W^{(II)} = W^{(f,e)}$.

4 RESULTS

For the numerical calculation, set the ratio $b/a = 0.8$, term number of series $n = 7$. Here the dimensionless parameters $\mu_1^* = \mu_2/\mu_1 = 1.0$, $\mu_2^* = \mu_3/\mu_1$, $k_1^* = k_2/k_1$, $k_2^* = k_3/k_1$, are the ratio related to the parameters of medium I, II, III, and $l = h = 5a$, $d = 1.5a$, $2A = a$, k_1a is the incident wave number, α_0 is incident angel.

Figure 5(a) presents the distribution of surface displacement with the incident angel. As we can see, the surface displacement has the maximum when SH waves disturb the model horizontally. Figure 5(b)–(c) show the ratios related to the parameters of medium I, II, III have effects on the distribution of surface displacement.

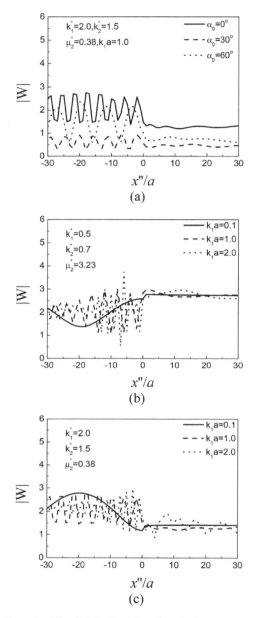

Figure 5. The distribution of surface displacement.

5 CONCLUSIONS

In this paper, the analytical solution of surface displacement of the bi-material half space with interface cracks and ellipse inclusion impacted by SH wave is given by using the complex function method and Green's function method. The distribution of the displacement is influenced by the incident angel and the vertical interface. As we can see, the interface will reduce the amplitude of the displacement in area II when medium I is softer than medium II, because the soft medium absorbs more energy. There are some theoretical and applied value on construction of underground structure near the interface between different soil and fundamental science.

REFERENCES

Chen Zhi-gang, Liu Dian-kui. Dynamic response on a shallowly buried cavity of arbitrary shape impacted by vertical SH-wave [J]. Earthquake Engineering Vibration, 2004, 24(4): 32–36.

Chen J.T., Chen P.Y., Chen C.T. Surface motion of multiple alluvial valleys for incident plane SH-waves by using a semi-analytical approach [J]. Soil Dynamic and Earthquake Engineering, 2008, 28: 58–72.

LEE V.W., Trifunac M.D., ASCE A.M. Response of tunnels to incident SH-waves [J]. Journal of the engineering mechanics division, 1979, 8(EM4): 643–659.

Lee V.W., Manoogian M.E. Surface motion above an arbitrary shape underground cavity for incident SH wave [J]. European Earthquake Engineering, 1995, 8(1): 3–11.

Lee V.W., Amornwongpaibun A. Scattering of anti-plane (SH) waves by a semi-elliptical hill: I—Shallow hill [J]. Soil Dynamics and Earthquake Engineering, 2013, 53: 116–125.

Luo Hao, V.W. Lee, Liang Jian-wen. Anti-plane (SH) waves diffraction by an underground semi-circular cavity: analytical solution [J]. Earthquake engineering and engineering vibration, 2010, 9(3): 385–396.

Liu Gang, Han Feng, Liu Dian-kui, et al. Conformal mapping for the Helmholtz equation: Acoustic wave scattering by a two dimensional inclusion with irregular shape in an ideal fluid [J]. The Journal of the Acoustical Society of America, 2012, 131(2): 1055–1065.

Qi Hui, Yang Jie, Li Hong-liang, Yang Zai-lin. Scattering of SH-wave by a cylindrical inclusion in right-angle plane with arbitrary beeline crack [J]. Journal of Vibration and Shock, 2011, 30(5): 208–212. (in Chinese).

Qi Hui, Yang Jie. Dynamic analysis for circular inclusion of arbitrary positions near interfacial crack impacted by SH-wave in half-space [J]. European Journal of Mechanics/A Solids, 2012, 36: 18–24.

Ren Yun-yan, Zhang Li, Han Feng. Dynamic load analysis of underground structure under effect of blast wave [J]. Applied Mathematics and Mechanics (English edition), 2006, 27(9): 1281–1288.

Advances in Energy, Environment and Materials Science – Wang & Zhao (Eds)
© *2016 Taylor & Francis Group, London, ISBN 978-1-138-02931-6*

The performance study of notch-stud connections of Timber-Concrete Composite beam

Guojing He, Hongzhi Xiao, Liping Chen & Li Li
College of Civil Engineering and Mechanics, Central South University of Forestry and Technology Changsha, Hunan, China

ABSTRACT: The notch-stud connections for timber-concrete composite beams is obtained by cutting a notch from the timber beam and reinforcing it with stud in the notch, and filling it with concrete during the pouring of the concrete slab. Compared to mechanical fasteners, this type of connection has the merit of high stiffness, strength and slip resistance. In this paper, the results of nine variations sizes of rectangular notch of the connections and one notched connection without studs are presented. And these specimens were loaded to failure under shear force. Geometrical variations are the depth, width and length of the notch, use or not of a stud, diameter of the stud. The purpose of this study is to identify the factors affecting the mechanical properties and to optimize the connection detail. Through this study, the shear bearing capacity of the studs were researched and the load-slip curve was obtained. And a quarter of the specimens were modeled by ABAQUS. Finally, the experimental results were agreed with numerical results well.

1 INTRODUCTION

In today's low carbon economy era, timber as environmentally friendly material is paid a high degree of attention by people. But for the mechanical properties of timber, which in compression, span and structure form is limited. In order to make up for the inadequacy of mechanical properties of timber, the composite structure is usually considered, and the Timber-Concrete Composite (TCC) structure is one of the most common composite structures. This kind of composite structure makes full use of the respective advantages of timber and concrete. Timber has strong toughness, good tensile strength and best anti-seismic property. And the concrete has excellent stiffness and higher compressive strength and stability (Ceccotti, 1995). The performance of the TCC beam is significantly influenced by the behavior of the shear connection. There is the longitudinal shear force in the contact surface of a TCC beams, and the main purpose of the shear connector is the transfer of the longitudinal shear between timber and concrete and effectively prevent the wave action between concrete slab and timber, and also improve the portfolio performance (Wang, 2012). Some ductility is desirable since both timber and concrete exhibit quite a brittle behavior in tension and compression, respectively, and the plasticization of the connection is the only source of ductility for the TCC

system (Ceccotti, 2006; Seibold, 2006). However, the connection system needs to be inexpensive to manufacture and install and it also needs to be installed conveniently.

Shear key connection is the timber beam with a notch, then pouring concrete and the notch fill with concrete slab to form the whole. And with a bolt or lag screw to strengthen the connection with concrete. Such a connection is also called notch-stud connection. Its stiffness, strength, and slip resistance is strong (Deam, 2008). As a result of the existence of a small amount of slip and shear key, the horizontal shear falls into a small area of the pressure, and stud in tension resistance the vertical load. Shear key itself, therefore, are not under shear, and the stress distribution is showed in Figure 1.

For this kind of connection, the length and depth of the notch and different types of studs are the main factors influencing the effect of the combination. This paper reports the results of experimental tests recently performed on different notch size connection systems. The purpose was to identify the factors affecting the mechanical properties and, ultimately, to optimize the connection detail. Geometrical variations include width, depth, and length of the notch, use or not of a stud, and different diameters of the studs. And the result of a quarter of the specimens was modeled by ABAQUS compared with the experimental results.

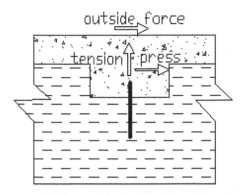

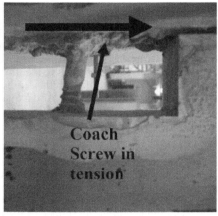

Figure 1. Detail of the shear connector.

Table 1. Example table caption.

Con-nection type	Length × Width × Depth (mm)	Stud diameter (mm)	F_{max} (KN)		Slide (mm)
			Exp.	Anal.	
A1	150 × 60 × 50	16	46.43	50.17	3.14
A2	150 × 60 × 40	16	42.63	49.17	4.17
A3	150 × 60 × 20	16	42.05	48.66	4.61
A4	100 × 60 × 50	16	38.96	41.18	4.33
A5	50 × 60 × 50	16	28.01	33.44	3.12
A6	150 × 40 × 50	16	34.00	39.14	5.00
A7	150 × 80 × 50	16	49.35	54.56	4.46
B1	150 × 60 × 50	13	34.42	37.74	5.35
B2	150 × 60 × 50	10	30.37	34.55	6.02
C1	150 × 60 × 50	–	29.83	38.37	8.17

2 PUSH-OUT TEST OF NOTCH-STUD CONNECTIONS

An experimental parametric study is essential for the optimization of the notch shape so that the best compromise between labor cost and structural efficiency is achieved. The performance of different connector shapes listed in Table 1 and they were evaluated through push-out shear tests performed on timber-concrete composite blocks (see Fig. 2). Variations of the size of notch included the length, depth, and width of the notch. Diameter of 10 mm and 13 mm of studs were also inserted in the centre of the notches, while in other cases no stud was used. A total of 10 different types of connection were selected. Three push-out specimens were then constructed for each connection type, for a total of 30 specimens. The push-out tests were performed with the universal testing machine (see Fig. 3). The size of the timber is $160 \times 300 \times 400$ (mm), and concrete is C30, the size is $60 \times 300 \times 400$ (mm). Detail component parameters are shown in Table 1.

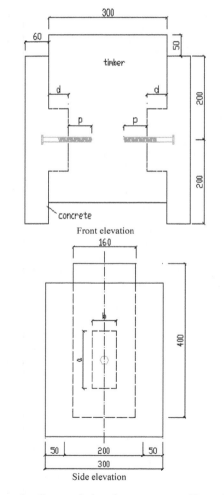

Front elevation

Side elevation

Figure 2. Symmetrical push-out test setup (dimensions in mm).

476

Figure 3.   Push-out test.

2.1   *Results and discussion*

The results in terms of shear strength (Fmax), slip and Damage form are summarized in Table 1. The strength Fmax is defined as the largest value of shear force per connector monitored during the test for slips not larger than 15 mm (Cen, 1991). When the slip of push-out specimens is greater than 15 mm, the corresponding load when you pick up the slip is 15 mm for the ultimate load. Launch failure modes of the specimens are ductile fracture.

The most important factors affecting the connection performances were found to be the length of the notch (compare Fmax for specimens A1, 46.43kN, A4,38.96 and A5,28.01kN) and the width of notch (compare Fmax for specimens A6, = 34kN, and A7, 49.35kN). Generally, all of the specimens failed by shear in the concrete (see photo in Fig. 4), hence a longer length of notch is

Figure 4.   Destructional forms.

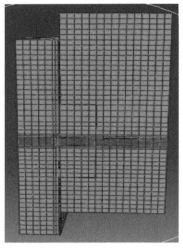

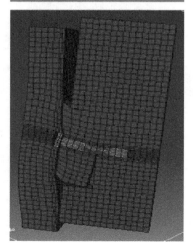

Figure 5. Numerical simulation.

necessary to improve the shear strength. The only source of ductility was provided by the stud, which also significantly increased the resistance. The presence of a coach screw significantly enhanced the stiffness of the connection (compare Fmax for specimens A1,46.43kN, and C1, 29.83kN).

3 NUMERICAL MODELING OF NOTCH-STUD CONNECTIONS

Numerical analysis of the connections were carried out using the finite element program ABAQUS (Zhuang, 2008). In order to simplify the calculation, only a quarter of the real geometry of the push-out specimens were modeled. The process of loading and the mesh are represented in Figure 5. A three-dimensional eight-node element with reduction of integral functions (C3D8R) was selected to model the solid parts of the specimen (concrete, timber, and stud). And the method of calculation is full Newton iteration.

In this paper, three-dimensional nonlinear finite element model of timber-concrete composite specimen was established by ABAQUS and analyzed by the mechanical properties of the shear connector of timber-concrete composite structure. The plastic damage constitutive stress-strain relationship of concrete and the bilinear model of the stud and timber were adopted. The reasonable formula and constitutive relation to calculate the stress-strain relationship of concrete, timber, and steel is chosen. For the interface between the concrete and the timber and for the stud-concrete and timber-concrete surfaces, 3D contact elements were used to simulate the friction among these materials. The non-linear numerical analysis was carried out by increasing the displacement on the top of specimen.

The development of the numerical model is not yet concluded. At the present time, the numerical results are not fully to be satisfied. For the convergence problems, possible reasons could be a mesh not adequate or the utilized material models, and also could be the contact elements. Some refinement and further investigation is, therefore, needed for the calibration of the 3D model on the experimental results.

4 CONCLUSION

Given to the results of the shear push-out tests of timber-concrete composite structure performed, it can be concluded that rectangular notches cut from of the timber beam and reinforced with stud are an excellent connection system compared to the other connection systems. High shear strength

478

can be achieved, along with acceptable post-peak behaviour characterized by gradual decrease in strength. The most important factors affecting the connection performance were found to be the length and width of the notch and the presence of a stud. Lastly, the experimental results were compared with numerical results carried out using a 3D finite element model implemented in ABAQUS software package. Predicted and experimental failure mechanisms agreed reasonably well, although some refinement and further investigation are needed to fully calibrate the model on the experimental results in terms of both strength and stiffness.

ACKNOWLEDGMENTS

This work was financially supported by the China National Natural Science project (NSFC Project No. 51478485) and the State Bureau of Forestry 948 projects (2014-4-51).

REFERENCES

Ceccotti, A. "Timber-concrete composite structures." Timber Engineering STEP 2. 1st Edition Centrum Hout. The Netherlands. 1995; pp. E13/1–12.

Ceccotti, A., Fragiacomo, M. & Giordano, S. "Long-term and collapse tests on a timber-concrete composite beam with glued-in connection." Materials and Structures, RILEM, Special Volume "Research for Reliable Timber Structures", 40(1), 2006; pp. 15–25.

Cen Comite European de Normalisation. "Timber structures—Joints made with mechanical fasteners—General principles for the determination of strength and deformation characteristics." EN 26891. Brussels, Belgium.1991.

Deam B.L., Fragiacomo M., and Buchanan A.H. Connections for composite concrete slab and LVL flooring systems. Materials and Structures, 2008, 41(3):495–507.

Seibold, E. "Feasibility study for composite concrete-timber floor systems using laminated veneer lumber in NZ." Dissertation Thesis. University of Karlsruhe, Germany. 2004.

Wang Bin, Xiao Fei. Timber-concrete composite beams research status review [J]. Jiangsu building, 2012148 (3).

Yeoh D., Fragiacomo M. Performance of Notched Coach Screw Connection for Timber-Concrete Composite Floor System.

Zhuang Zhuo, Based on ABAQUS finite element analysis and application. Tsinghua University Press, 2008, 12, 31.

Advances in Energy, Environment and Materials Science – Wang & Zhao (Eds)
© 2016 Taylor & Francis Group, London, ISBN 978-1-138-02931-6

Vibration test and finite element analysis of a timber construction with curbwall

Fangfang Qian
Huaiyin Institute of Technology, College of Architecture and Civil Engineering, Jiangsu, China

Weidong Lu, Weiqing Liu, Xiaowu Cheng, Depeng Lv & Xingxing Liu
Nanjing Tech University, College of Civil Engineering, Jiangsu, China

ABSTRACT: For inheritance and development of Chinese culture, the historical cultural districts have been reconstructed in most parts of China. As the same time, existing timber constructions have been repaired and lots of antique timber constructions have been built. In this paper, a timber construction with curbwall was made a vibration test. Two models of timber constructions based on the test actual construction were simulated by the finite element software. This paper made a comparative analysis of curbwall's impact on the dynamic performance and structural stiffness. Finite element analysis result was compared with vibration test result. It turned out that curbwall has a significant impact on the dynamic performance of timber constructions. It can improve the lateral stiffness of the structure. Finally, connecting measures and ring beams are suggested to strengthen the curbwall and the structure. It provides the reference and basis for practical design.

1 INTRODUCTION

1.1 Introduction

The ancient Chinese timber constructions have been proven by history with the superior seismic performance. However, there are few studies on the dynamic characteristics of the actual timber structures. (Wu, 2010) Through the summary of the earthquake damage survey and the field investigation, the earthquake damage losses and casualties of timber constructions were mainly from the curbwall's destruction and collapse.

1.2 Engineering overview

The project was a new-built antique timber construction with two stories, as shown in Figure 1

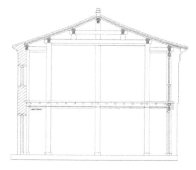

Figure 2. Section.

and Figure 2. It was a wooden structure with post and lintel construction and three sides of curbwalls. Half section of the timber column was surrounded by brick wall. Connecting measures were between the brick wall and the timber columns. The size was 14.4m long, 6.5m wide, and 3.3m high in the first layer, 3.4m high in the second layer, and 8.4m high on the top of the roof.

2 FINITE ELEMENT ANALYSIS

2.1 Model building

In order to obtain the dynamic characteristics of the similar constructions, the above engineering

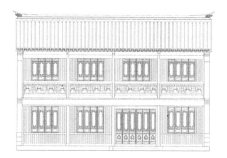

Figure 1. Elevation.

example was modeled and analyzed with the finite element analysis software. (Yu, 2010) The timber structure Model 1 (without curbwalls) and Model 2 (with curbwalls) were simulated, respectively, as shown in Figure 3 and Figure 4.

Beams and columns adopted general beam elements, curbwall took wall elements, and roof boards used plane elements with 4 nodes. (Gao et al. 2011; Zhao et al. 2000) The actual construction's curbwalls were only on three sides, the other side existed wooden doors and windows. The lateral stiffness of wooden doors and windows could be ignored, compared with the lateral stiffness of brick curbwalls. So the curbwalls were arranged only on three sides during modeling analysis. The mortise and tenon joints of the wooden structure simulated the semi-rigid feature in the elastic stage by the method of beam and release.

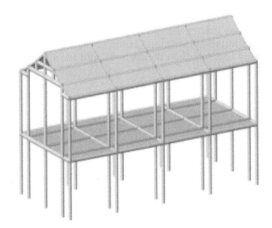

Figure 3. Model 1 (without curbwalls).

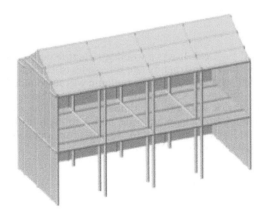

Figure 4. Model 2 (with curbwalls).

Purline end and rafter end were set to the hinge, column foot was the same. Mortise and tenon joint nodes belong to semi-rigid nodes on the mechanical properties, and can absorb a certain amount of bending moment. Timber structure is in the elastic stage under the effect of more severe earthquake, so the semi-rigid feature of the nodes was simulated with the initial rotational stiffness of mortise and tenon joint nodes. It was simulated with the release of the beam end in the finite element software. The release coefficient was obtained by testing the ratio of the initial rotational stiffness and the beam end line stiffness.

2.2 Modal analysis

Model 1 and Model 2 were all carried out modal analysis in order to compare the modes of vibration between the model without curbwalls and the one with curbwalls. The first order mode shapes of the two models all vibrated along the longitudinal direction, the second ones all vibrated along the lateral direction, the third ones twisted in the XY plane. It was in conformity with the vibration mode law of construction. Lateral stiffness was greater than the longitudinal stiffness because horizontal timber frame existed in the model. The longitudinal vibration occurred on the side without walls in Model 2 in consideration of the clamping effect of curbwalls. Because of the constraint function of the cross wall on both ends, larger vibration amplitude occurred in the middle of the timber roof truss.

2.3 Curbwalls' impact on the dynamic characteristics of the timber structure

In order to discuss curbwalls' impact on the dynamic characteristics of the timber structure, the results of the first three periods and frequencies of Model 1 and Model 2 were made a contrastive analysis, as shown in Table 1.

From the results available in the table, it can be concluded curbwall has a significant impact on the dynamic characteristics of the timber structure. The natural frequency of the timber structure with

Table 1. Calculation results of periods and frequencies between two models.

Modal order	Model 1		Model 2	
	Frequency (Hz)	Period (s)	Frequency (Hz)	Period (s)
1	0.48	2.11	3.05	0.32
2	0.50	1.99	3.37	0.30
3	0.66	1.51	5.26	0.19

curbwalls is 6 times of the one without curbwalls. Thus it can be seen that lateral stiffness of the structure is improved significantly and the elastic deformation of timber frame under horizontal load is reduced significantly because of the clamping effect of curbwalls.

3 SITE DYNAMIC TESTS

3.1 *Pulsating method test*

Pulsating method analyzed the dynamic characteristics of constructions through the measurement of the environmental random vibration. A large number of tests show that pulsation of constructions and bridges has an important characteristic that can clearly reflect its inherent frequency and other vibration characteristics.

3.2 *Site test*

The test was made by DH5922 dynamic signal testing system and DH610H sensors. In this test, the sampling frequency was 50 Hz, and the sampling time was 30 minutes. (Wang et al. 2005; Lu et al. 2010) The test content was the construction's response under ambient excitation. Positions of measure points were shown in Figure 5. The vibration along X direction (lateral direction) and Y direction (longitudinal direction) on each point was tested, respectively. Measure points were arranged on both the first and second floor.

3.3 *Test results*

According to the process and analysis of the collected data, the acceleration time-history curves and cross-power spectrum from the site test are shown in Figure 6 and Figure 7. Based on the analysis of time-history curves, the dynamic characteristics of this construction were gained, as shown in Table 2.

3.4 *Analysis and comparison on the dynamic test results*

The periods and frequencies of Model 2 (with curbwalls) under the single direction vibration were calculated by the finite element software, and then compared with the tested results. The calculation results were shown in Table 3.

Under the single direction vibration, the dynamic characteristics results of the tested structure and the finite element model-Model 2 were shown in Table 4.

From the comparison of the finite element calculation and test results, it is concluded that the

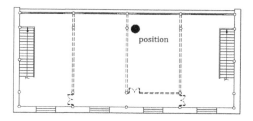

Figure 5. Layout of measuring points.

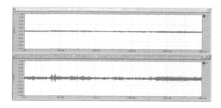

(a) At EW (longitudinal direction)

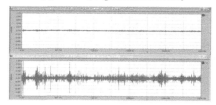

(b) At NS (lateral direction)

Figure 6. Measured time-history curves.

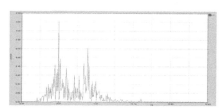

(a)At EW (longitudinal direction)

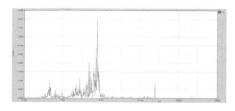

(b) At NS (lateral direction)

Figure 7. The cross-power spectrums and coherence functions between the reference points and the test points.

Table 2. Dynamic characteristics test results.

Modal order	Longitudinal direction			Lateral direction		
	Frequency (Hz)	Period (s)	Damping ratio (%)	Frequency (Hz)	Period (s)	Damping ratio (%)
1	2.61	0.38	2.75	3.91	0.26	2.51
2	5.64	0.18	2.08	6.93	0.14	1.76

Table 3. The single direction vibration results of Model 2.

Modal order	Longitudinal direction		Lateral direction	
	Frequency (Hz)	Period (s)	Frequency (Hz)	Period (s)
1	3.05	0.32	3.38	0.29
2	4.93	0.20	4.38	0.23

Table 4. The frequency comparison of the results.

Modal order	Longitudinal direction			Lateral direction		
	Test (Hz)	Calculation (Hz)	Error (%)	Test (Hz)	Calculation (Hz)	Error (%)
1	2.61	3.05	16.8	3.91	3.38	13.6
2	5.64	4.93	12.6	6.93	4.38	36.8

frequency of the finite element calculation is more consistent with the tested one under the longitudinal effect, thus the second order tested result is obviously higher than the finite element calculated under the lateral effect. Analyzing the reason, it may be associated with the larger brick wall stiffness in practical engineering. Mortar strength, masonry techniques and block quality may all affect the brick wall stiffness. While modeling, the curbwall was simulated by wall elements, therefore it's not enough fine.

4 SEISMIC FORTIFICATION MEASURES OF WALLS

In the frequent earthquake disasters, wall's destruction and collapse is one of the main causes of economic losses and casualties. To reduce damage from the wall, reducing weight of wall, strengthening the integrity of wall, strengthening connections between walls and the timber frame are feasible. And they also make a contribution to improve the seismic performance.

The connecting methods between the wall and the timber frame are as follows: connecting steels or steel wire meshes are set at the joint of

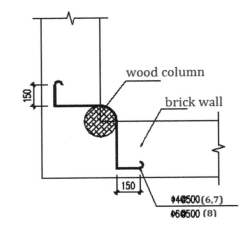

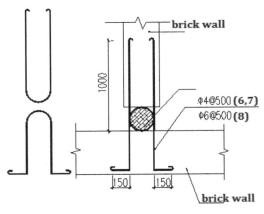

Figure 8. Connecting measures between wood column and brick retaining wall.

longitudinal and lateral wall; connecting details are needed between ring beams and timber columns; wall nails can be used to connect the timber frame. The connecting measures between wood columns and brick curbwalls can refer to Figure 8. The wall nails are shown in Figure 9.

Setting ring beams is an effective measure to enhance the integrity and ability to resist collapse. And the effect is very obvious.

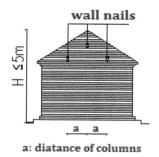

a: diatance of columns

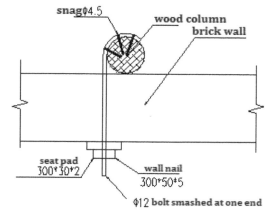

Figure 9. Wall nails.

5 CONCLUSION

1. Curbwall in the traditional timber constructions can significantly improve the lateral stiffness of the structure. The natural frequency of the structure in consideration of the curbwall's lateral effect is 6 times of the frequency of the timber frame only. Therefore the curbwalls' impact on the lateral stiffness of the structure should be considered during the anti-seismic and anti-wind design of similar timber constructions.
2. The test results and the calculated results of the dynamic characteristics are similar during the finite element analysis of timber constructions with curbwalls, with the modeling method given in this paper. The test damping ratio of this timber construction with curbwalls is about 2.75%.
3. In order to reduce casualties and economic losses caused by the wall collapse during the earthquake, connecting measures between curbwall and the timber frame, and ring beams are suggested.

REFERENCES

Gao Dafeng, Cao Pengnan, Ding Xinjian, 2011. Simplified analytical research on ancient Chinese timber structures. *Journal of earthquake engineering and engineering vibration*, 31(2):175–181.

Li Guoqiang, Li Jie, 2002. Dynamic test theory and application of engineering structures. *Beijing science press*.

Lu Weidong, Lan Zongjian, Liu Weiqing, 2010. Vibration test of a high-rise building and seismic performance Analysis. *World earthquake engineering*, 26(1): 169–174.

Peng Yong, 2010. Seismic performance evaluation of technical analysis of masonry-timber structure. *Bubiness China*, 196:236–237.

Sui Yunkang, Chang Jingya, Ye Hongling, 2011. Numerical simulation of semi-rigid joint nodes of timber structures. *Journal of Beijing university of technology*, 37(9):1298–1303.

Wang Wenbo, Luo Shizhong, Liu Junjie, 2005. Reliability test of building structures with the theory of the madal identification of environmental incentive. *Science and technology of overseas building*, 26(3):116–118.

Wu Ti, 2010. The application about measurements of dynamic properties during the protection of excellent historical buildings. *Sichuan building science*, 36(6):60–64.

Xu Yafeng, Bai Shouman, 2003. Vibration test and analysis of constructions. *West-China exploration engineering*, 11(1):151–152.

Yu Zhixiang, Zhao Shichun, Wu Hao, 2010. Numerical simulation of seismic behavior of retrofitted masonry-timber structure of Jushi Building on Qingcheng Mountain. *Journal of southwest jiaotong university*, 45(2):179–184.

Zhao Junhai, Yu Maohong, Yang Yansong, Sun Jiaju, 2000. Finite element analysis of ancient Chinese timber structures. *China civil engineering journal*, 33(1):32–35.

Advances in Energy, Environment and Materials Science – Wang & Zhao (Eds)
© *2016 Taylor & Francis Group, London, ISBN 978-1-138-02931-6*

Studies on internal forces of shield lining segments with different design models

Kewei Ding & Yaguang Wang

School of Civil Engineering, Anhui Jianzhu University, Hefei, Anhui, China

ABSTRACT: In the process of analyzing the internal force of shield lining segments, it is difficult to obtain satisfactory calculation accuracy with general load structure method because of the particularity of the structure. With the finite element software, this paper analyzes the stress of the segment by analyzing the structure characteristics and stress characteristics of shield lining segments and using beam-spring model and modified routine model of layer structure method. Then we compared the internal force of the segment with different design models and analyzed the characteristics of the two models, which would be beneficial to the segment structure design later.

1 INTRODUCTION

With the development of economy, more and more subways have been constructed in the city. The segment of lining structure is the most important component of the subway. At present load structure method has been used mostly for the design of underground structure, which considers the generated load as the role of ground on structure aimed to calculate the internal force and deformation of lining under the load based on this assumption. However, owing to the specificity of the underground structure, the surrounding rock deformation would cause elastic resistance, which might have impacted on the stress of underground structure. With this background, the structure method is a more rational substitute for the load structure method. This method regards the lay and lining as a whole, and calculates the internal force and deformation of the lining and the surrounding layer on the basis of the continuum medium mechanics.

There are different calculation models available to the lay structure method. This paper uses a beam-spring model and modified routine model for simulation and calculation with finite element software and compares the differences between the above two models by analyzing the mechanical characteristics, processes, and simulation results.

2 MODIFIED ROUTINE MODEL

Modified routine model is the most widely used in the analysis of internal force of the segment. A ring of lining segment was assembled in staggered pattern in the shield tail of tunneling, the joints which had waterproof filling materials and connection bolt cause lower rigidity. And the mechanical characteristics of the joints were also related to the structure, the thickness of the liner, the pre-tightening force of the bolt present a certain nonlinear (Zeng, 2005) that was complicated to calculate. The bending rigidity of segment of modified routine model introduced the ηEI (η is effective ratio of the bending rigidity) to consider the simulation of the joints and loss of the rigidity of a ring of segments.

3 BEAM-SPRING MODEL

This model simplified the stress of the segment to plain strain. Considering the impact of segments and joints, the segment was simulated by 1D beam element, and elastic connection was used to simulate radial and tangential joints of the segments.

3.1 Beam element

There exited axial force, shearing force, and bending moment acting on the nodes at the ends of the beam element and elastic connection linked with the nodes of the beam element. There are some parameters such as radial spring stiffness coefficient k_n, tangential spring stiffness k_s and rotation spring stiffness coefficient k_θ. Because of the liner elasticity of the beam element, we could work out the node displacement and node force's matrix expression by using Castigliano's second theorem when we worked out the strain complementary energy.

$$\{F\} = [k]\{\delta\} \tag{1}$$

Using the static balance condition, we could find out the relational expression between the node displacement and node force of the beam's nodes.

$$\begin{Bmatrix} F_1 \\ F_2 \end{Bmatrix} = \begin{bmatrix} k_{11} & k_{12} \\ k_{21} & k_{22} \end{bmatrix} \begin{Bmatrix} \delta_1 \\ \delta_2 \end{Bmatrix} \qquad (2)$$

3.2 Axial joints of segment

The axial joints of segment were connected by bolts with waterproof materials filling, which resisted the effect of axial force N, shearing force Q, and bending moment M in the bolts and generated the tangential relative displacement Δv and radial relative angle $\Delta\theta$. So radial, tangential, and rotation spring stiffness coefficient should be defined when joints were simulated by elastic connection.

According to the paper (Zhu, 2000), we got the relational expression of segment spring's the elastic connection in a whole ring.

$$\begin{Bmatrix} N \\ Q \\ M \end{Bmatrix} = \begin{bmatrix} k_n & & \\ & k_s & \\ & & k_\theta \end{bmatrix} \begin{Bmatrix} \Delta u \\ \Delta v \\ \Delta\theta \end{Bmatrix} \qquad (3)$$

3.3 Longitudinal joints of rings

The segments were assembled under staggered joint erection, the longitudinal bolts of which connected the rings. We considered that there was the tension between the segments at the side of tensile and the pressure at the compress when the tunnel lining structure produced the bending deformation in the longitudinal direction. Similarly, the longitudinal joints of rings also exited the shearing force which was radial direction and orthogonal to the tension and pressure. So we simulated longitudinal joints of rings with elastic connection by setting up longitudinal spring affected by compression and radial spring affected by shear, and rotation spring that wasn't needed any more (Su, 2007).

The longitudinal joints' relational expression between force and displacement is:

$$\begin{Bmatrix} N_q \\ Q_q \end{Bmatrix} = \begin{bmatrix} k_{nq} & \\ & k_{qq} \end{bmatrix} \begin{Bmatrix} \Delta u' \\ \Delta v' \end{Bmatrix} \qquad (4)$$

In this expression, N_q is the longitudinal axial and Q_q is the radial shearing force. k_{nq}, k_{qq} refers to the spring stiffness coefficients and $\Delta u'$, $\Delta v'$ means the longitudinal and radial displacement.

4 GROUND SPRING

The lay structure method should consider the effect of the lining's deformation on the surrounding rock. The deformation caused by the change of structure internal force could cause the elastic resistance of surrounding rock. We established the ground spring element. The force of ground spring is proportional to the layer's deformation, and we called the ratio foundation resistance coefficient (Zhu, 2006) that could obtain the coefficient by experiments. The elastic resistance is assumed by the linear Winkle elastic foundation Theory (Yuan, 2008) and uses a kind of model of peripheral spring. The ground spring is simulated by curved spring, which is compressed in normal direction only.

5 BOUNDARY CONDITIONS AND LOAD

As the plane model we built, we set up longitudinal displacement constraint on the beam element as boundary conditions. With the geology and design drawings for the engineering as the reference, we could calculated the load of the vertical earth pressure, lateral earth pressure, and counter force of arch bottom by using the soil column principle (Ding, 2001). Due to the ground spring that has been established for simulating the soil's elastic resistance, the load acting on the outermost edge of the segment cannot be considered.

6 AN ENGINEERING EXAMPLE

We used Hefei Metro Project as an example in this paper and calculate the internal forces of segment by the two models. Tunnel segments were assembled by using the domestic common method as 5+1 pieces assembled. It divides into 6 pieces and staggered joint assembled with 45°. The buried depth of tunnel was set to 12.7m. External diameter of tunnel: 6.0m, interior diameter of tunnel: 5.4m. The layer parameters are shown Table 1.

1. The section size of the segment's pieces:thickness of lining segment: H = 0.3m, width of lining segment: B = 1.5m, the section's moment of inertia: I = 0.003375m⁴, the material is elastic concrete, E = 34500000KN/m², the bending rigidity of the segment is EI.
2. In the modified routine model, effective ratio of the bending rigidity is η = 0.8. We achieved the reduction of the segment's bending rigidity through reducing the section's moment of inertia.
3. The spring stiffness coefficients of joints' elastic connection: k_n = 1149000KN/m, k_q = 786000KN/m, k_θ = 120000KN*m/rad (positive direction) and k_θ = 120000KN*m/ rad (Negative direction). k_{nq} = 1149000KN/m, k_{qq} = 786000KN/m.

Table 1. Layer parameters.

| Layer | Density/kg·m⁻³ | Consolidated quick shear | | Coefficient of subgrade reaction/ MPa/m | |
		Cohesion/kPa	Fric/(°)	Vertical	Level
Clay (1)	19.5	47	12	45	45
Clay (2)	19.6	50	13	50	55
Clay (3)	19.6	52	13	50	55

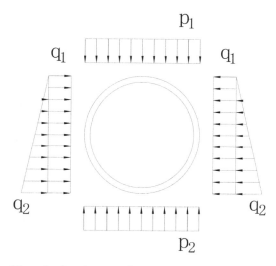

Figure 1. Load system of structure.

4. The value of clay's foundation resistance coefficient is $1*10^4$KN/m³. The characteristic is compressed and be in normal direction only.
5. Due to the plane model, we should multiply the load we calculated by the width of the segment, then we got the result imposed on the beam element as the load of continuous beam. Specific values are as follows: $p_1 = 377.25$KN/m, $p_2 = 300.45$KN/m, $q_1 = 119.4$KN/m, $q_2 = 106.05$KN/m. The load distribution is in the following Figure 1.

7 ANALYSIS OF COMPUTING RESULTS

As shown in the segment's deformation pattern, we realized the deformation law of the segment in these two models was consistent, which is in accordance with the features of segment stress. As the internal force diagram (Fig. 2,3,4,5) indicated the maximum moment occurs at the bottom or crown of the tunnel, and the maximum axial force occurs at the outermost edge at around. The computing results of two models had the same variation trend and

Figure 2. The axial force of segment with beam-spring model.

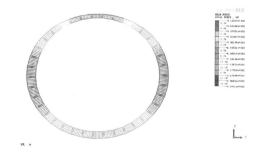

Figure 3. The axial force of segment with modified routine model.

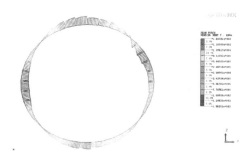

Figure 4. The bending moment of segment with beam-spring model.

489

the numerical value difference was not significant. But we could see the sudden change obviously in the internal force diagram which was calculated in the beam spring model.

For comparing the feature of segment's internal force under these two models, we selected four points (the crown, the bottom, and the outermost edge of right side and left side) of the tunnel as Figure 6.

Figure 5. The bending moment of segment with modified routine model.

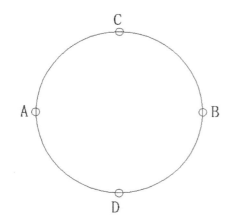

Figure 6. Observation point on the segment.

Table 2. The internal force of the four points.

	Beam-spring model		Modified routine model	
	Axial force (KN)	Bending moment (KN*m)	Axial force (KN)	Bending moment (KN*m)
A	871.19	−142.55	946.11	−195.51
B	522.98	−79.87	950.72	−194.51
C	323.81	157.68	418.92	206.51
D	727.31	94.91	725.47	156.43

8 CONCLUSION

This paper used the finite element method and thought of the design model of layer structure method. Considering the stress feature of the segment's joints and the interaction with the layer, we built the numerical model and calculation. Then we got the more accurate and more realistic results, which is consistent with the actual situation.

1. As shown in the deformation pattern, the segments had an outward displacement at the outermost edge at the round, it would be elastic resistance in this position. Compared with the elastic resistance we calculated in formula method, the simulation result was more accurate and reliable, which is considering that the joint of the ground spring and the compatible deformation of the layer.
2. As the beam-spring model is a discontinuous medium model, we could see the obvious sudden changes in the internal force diagram either the axial force or the bending moment, which suggested the segment's joints had the nonlinear characteristic and a complicated mechanical feature. We should pay more attention to the process of design and construction.
3. The modified routine model is a homogeneous medium, so the internal force distribution is symmetrical. As the load and the structure is symmetric in the vertical axis, the results showed that the modified routine model has the characteristics of the elastic center.

The modified routine model and the beam spring model were plane models simulating with the condition of plane strain. It suggested that 3D model should be built for complicated simulation and precise analysis. However, these two kinds of design models could be more convenient to simulate, and the calculating results be accurate enough to meet the requirement of the engineering.

ACKNOWLEDGMENTS

This project 11472005 was supported by National Natural Science Foundation of China and Anhui Provincial Science and Technology Research Project Funding through grant No. 1501041133.

REFERENCES

Ding Chun lin et al., 2001. Comparison of Calculating Methods for Internal Force of Segment Lining for Metro Shield-driven Tunnel [J]. *Underground Space*, 2001, 03:208–214+239–240.

Su Zong-xian, He Chuan. 2007. Shell-spring-contact model for shield tunnel segmental lining analysis and its application [J]. *Engineering Mechanics*, 10:131–136.

Yuan Jian-yi, Zhou Shun-hua, Gong Quan-mei. 2008. Study on Design Method of Lining Internal Forces of Shield Tunnel under-across Railway [J]. *Chinese Journal of Underground Space and Engineering*, 2008, 02:290–294.

Zeng Dong yang, He Chuan. 2005. Study on Factors Influential in Metro Shield Tunnel Segment Joint Bending Stiffness [J]. *Journal of The China Railway Society*, 04:90–95.

Zhu Hehua, Cui Maoyu, Yang Jinsong. 2000. Design model for shield lining segments and distribution of load [J]. *Chinese Journal of Geotechnical Engineering*, 2000, 02:190–194.

Zhu Wei, Huang Zheng-rong, Liang Jing-hua. 2006. Studies on shell-spring design model for segment of shield tunnels [J]. *Chinese Journal of Geotechnical Engineering*, 2006, 08:940–947.

Advances in Energy, Environment and Materials Science – Wang & Zhao (Eds)
© 2016 Taylor & Francis Group, London, ISBN 978-1-138-02931-6

Constitutive relationship of super high-strength concrete filled steel box

Xiaoming Chen, Jin Duan & Yungui Li
China State Construction Technical Center, Beijing, China

ABSTRACT: The seismic performance of ultra-tall buildings should be estimated by elastic-plastic time-history analysis via general FEA software. In order to increase the bearing capacity of concrete, super high-strength concrete may be used and the confinement of steel is necessary to improve its ductility and bearing capacity. A concise constitutive relationship of this mechanical model is developed by introducing both the concrete strength defined by European code and a parameter of confinement effect into skeletons presented in nation code. The numerical results show that it can simulate the mechanical behavior reasonably.

1 INTRODUCTION

To ensure the structural safety enough under expected rare seismic wave, elastic-plastic deformation should be checked for those structures which are more than 150 meters high or those most important buildings as regulated in the "Code for Seismic Design of Buildings" (2010a). In the other national code of "Technical specification for concrete structures of tall building" (2010b), the principle of height was detailed as "for those structures more than 200 meters high, deformation should be checked by elastic-plastic time-history analysis."

Till now, most of the professional software has presented relative nonlinear modules for both static analysis and dynamic analysis. But more or less, these professional softwares are not completely qualified for so many complex jobs, especially for simulating the material nonlinearity together with geometrical nonlinearity. On the other hand, to consider the accuracy and efficiency at the same time is nearly impossible too. Therefore, almost all of these kinds of analysis have to depend on the general FEA software. In this general FEA software, ABAQUS may be used most widely for its powerful nonlinear solver.

Both implicit method and explicit method in ABAQUS can simulate material nonlinearity and geometrical nonlinearity at the same time very well by using abundant beam elements and shell elements. For simulating the nonlinearity of concrete, three kinds of constitutive model are available, including smeared cracking model, cracking model, and plastic damage model (2006). It also can simulate the reinforcement easily by the function REBAR or REBAR LAYER for frame and shell, respectively. So it has been used for analyzing many ultra-tall buildings, such as the Pingan Tower and Shanghai World Finance Center. Compared with those normal high-rise buildings, there are still some unconventional factors of elemental model and constitutive model that would confront users in the analysis of these ultra-tall buildings.

By using an example of ultra-tall building, the constitutive model proposed by Chinese code was combined with the definition of super high-strength concrete in European code for ABAQUS to formulate concise constitutive model. In this formulation, the effect of brittleness was taken into account by a discount factor and the confinement of steel tube and steel plate to concrete is also considered by a simple parameter.

2 STRUCTURAL MODEL

The ultra-tall building that was analyzed has 200 floors and more than 1000 meters high. It has four towers, and these towers are connected together at the interval of each 20 floors as shown in Figure 1. Its seismic performance should be

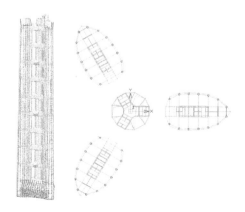

Figure 1. Structural model.

analyzed under seismic fortification intensity 7, in which, the peak acceleration is 220gal.

For the incredible structural volume, it will not only make the efficiency and reliability of analysis much more difficult but also will introduce strict requirements for strength and ductility of materials. Thus, the grade of concrete being used in vertical members is even up to C120. Except for the concrete filled steel tube, shear walls with double layer of steel-plate is used nearly in all the layers except the top part. For these series problems, the FEA model was researched for ABAQUS (Fig. 1).

3 CONSTITUTIVE MODEL

The constitutive relationships of both concrete and steel are presented in "Code for design of concrete structures" (2010c). For the concrete from C30 to C80, the tension skeleton curve of concrete is defined as follows:

$$\sigma = (1 - d_t)E_c\varepsilon \tag{1}$$

$$d_t = \begin{cases} 1 - \rho_t[1.2 - 0.2x^5] & x \le 1 \\ 1 - \dfrac{\rho_t}{\alpha_t(x-1)^{1.7} + x} & x > 1 \end{cases} \tag{2}$$

And the compression skeleton curve of concrete is:

$$\sigma = (1 - d_c)E_c\varepsilon \tag{3}$$

$$d_c = \begin{cases} 1 - \dfrac{\rho_c n}{n - 1 + x^n} & x \le 1 \\ 1 - \dfrac{\rho_c}{\alpha_c(x-1)^2 + x} & x > 1 \end{cases} \tag{4}$$

For super high-strength concrete which exceeds C80, the mechanical properties have already been researched by Pu XinCheng (2002) and Yu Zhiwu (2003) through experimental methods, but these achievements have not been introduced into the national code till now. Otherwise, in European code, super high-strength concrete was described systemically. So it can be used to define the mechanical behavior from C90~C120 for this example.

The strength of cylinder can be expressed with the strength of cubic as follows:

$$f_{ck,cube} = 0.85f_{cu,k} \tag{5}$$

Reference to European code, the brittleness should be considered by introducing a parameter which is taken as 0.74 to discount the compression strength of super high-strength concrete as follows:

$$f_{ck} = 0.74f_{ck,cube} \tag{6}$$

And the tensile strength for C90~C120 can be written as:

$$f_{ck} = 0.7f_{ctm} \tag{7}$$

Where:

$$f_{ctm} = 2.12\ln(1 + f_{cm}/10) \tag{8}$$

And:

$$f_{cm} = f_{ck,cube} + 8 \tag{9}$$

The Young's modulus is as follows:

$$E_c = \dfrac{10^5}{2.2 + \dfrac{34.7}{f_{cu,k}}} \tag{10}$$

Based on these definitions, the skeleton curves of C90~C120 are shown in Figure 2 together with C50~C80.

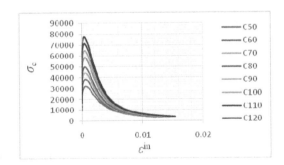

Figure 2. Skeletons of compression.

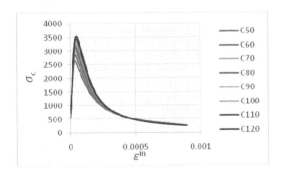

Figure 3. Skeletons of tension.

494

4 CONFINEMENT EFFECT OF STEEL

Concrete filled steel box can improve both the bearing capacity and ductility of concrete significantly, so it is used widely in high buildings. But this idea cannot be introduced into shear walls easily because of the buckling behavior of steel plates, so only a unique layer of steel plate is used at the neutral plane of shear wall for increasing bearing capacity. As shown in Figure 4, steel box walls are used in the building shown in Figure 1 to achieve high performance under expected rare earthquake.

The shear wall shown in Figure 4 has a section behavior which is more like concrete filled steel tube and is usually the effect of buckling, which may be negligible. The stiffness of the diaphragm also can be neglected, then this section can be simulated with sandwich shell element through the definition for each layer of material, thickness, integral point and so on. As to the confinement effect of steel plates to concrete, it can be estimated as concrete filled steel tube. Susantha (2001) has formulated this constitutive model through the diameter-thickness ratio. In this paper, confinement effect parameter, which is formulated by Han Linhai (2007), is used together with the constitutive model of Eq. 1 to Eq. 4. The confinement effect parameter can be written as:

$$\xi = \frac{A_s f_y}{A_c f_{ck}} \tag{11}$$

where
f_y is the yield strength of steel;
A_s is the section area of steel plates;
f_{ck} is the characteristic value of compression strength of concrete;
A_c is the section area of concrete.

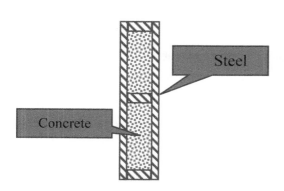

Figure 4. Section of steel-box wall.

5 NUMERICAL RESULTS

Based on the mechanical model mentioned above, the damage and equivalent plastic strain results of elastic-plastic time-history analysis under seismic wave for the structure shown in Figure 1 are presented in Figure 5 to Figure 8, and the relative time-history of energy dissipation are also given in Figure 9 and Figure 10.

It can be seen that the compression damage of concrete in both columns and walls is controlled in a low level by the steel box. Compared with the steel-plate shear walls, the walls at the top of the building are damaged much more seriously without the confinement of steel box, and the equivalent plastic of rebar layer is also more significant than the others.

Usually compression damage is the main reason which will decrease the bearing capacity or

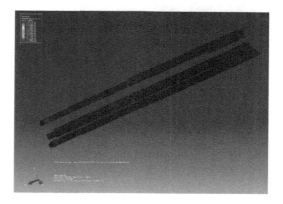

Figure 5. Compression damage of walls.

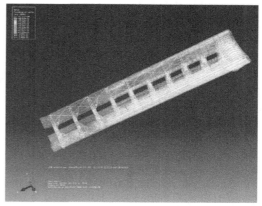

Figure 6. Compression damage of frames.

Figure 7. Equivalent plastic strain of walls.

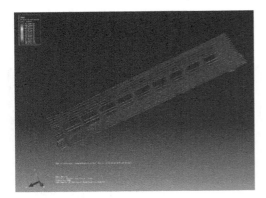

Figure 8. Equivalent plastic strain of frames.

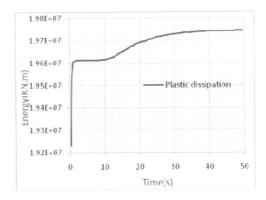

Figure 9. Time-history of plastic dissipation.

even collapse under the action of seismic wave. But compared with the plastic dissipation in this model, only significant energy is dissipated by concrete damage profited from concrete filled steel box for both columns and walls.

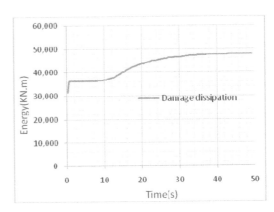

Figure 10. Time-history of damage dissipation.

6 CONCLUSION

Super high-strength concrete, concrete filled box steel and double layers steel-plate walls may be absolutely necessary for ultra-tall buildings. Reasonable and concise mechanical model for elastic-plastic analysis is the key point through general FEA software. The parameter of confinement effect, mechanical behavior of concrete proposed by European code and the skeletons of concrete presented by Chinese codes are used together to formulate a concise constitutive relationship. The numerical example shows that it can be used easily for simulating the behavior of confined super high-strength concrete under seismic wave.

REFERENCES

ABAQUS Inc (2006). ABAQUS User Manual, V6. 5. 5.
China architecture & building press (2010a). Code for Seismic Design of Buildings, Beijing, China (2010a).
China architecture & building press (2010b). Technical specification for concrete structures of tall building, Beijing, China.
China architecture & building press (2010c). Code for design of concrete structures, Beijing, China.
Han Linhai (2007). Concrete filled steel tubular structure-theory and application [M]. 2nd ed. Beijing: Science Press.
Pu Xincheng, Wang Chong, Wu Zhijun, et al (2002). The study on strengths and Deformability of C100~C150 Super High Strength & High performance Concrete. Concrete, 10:3~7.
Susantha K., Ge H., Usami T. (2001). Uniaxialstress-strainrelationship of concrete by various shaped steel tubes [J]. Engineering Structures, 23(10):1331–1347.
Yu Zhiwu, Ding Faxing (2003). Unified calculation method of compressive mechanical properties of concrete. Journal of Building Structures, vol 24(4):41~46.

Advances in Energy, Environment and Materials Science – Wang & Zhao (Eds)
© 2016 Taylor & Francis Group, London, ISBN 978-1-138-02931-6

Offshore wind turbine aerodynamic damping analysis and apply in semi-integrated analysis method

B. Wang
Powerchina Huadong Engineering Corporation Limited, Hangzhou, China

W.H. Wang & X. Li
State Key Laboratory of Coastal and Offshore Engineering, Dalian University of Technology, Dalian, China

Y. Li
Chinese-Deutsch Institute for Applied Engineering, Zhejiang University of Science and Technology, Hangzhou, China

ABSTRACT: Aerodynamic damping has significant effects on the dynamic responses of OWT sub-structure. In recent years aerodynamic damping analysis methods were derived on the basis of blade element momentum theory. In this paper, an aerodynamic damping numerical model is suggested on the base of fluid mechanics theory. This model takes the effects of air viscosity and Reynolds number into account. Using Fortran language, an aerodynamic analysis module considering the effect of aerodynamic damping was developed. Comparison of OWT analysis results by the semi-integrated model and the fully coupled model is carried out.

1 INTRODUCTION

1.1 Introduction

The semi-integrated analysis method was widely used in OWT sub-structure design, but the aerodynamic damping was neglected. Some researcher have presented that the aerodynamic damping should be considered in the OWT sub-structure optimization design and seismic analysis. Thus, for fixed bottom OWT analysis, the aerodynamic damping should be considered in the semi-integrated method.

Valamanesh V. (2014) derived a closed-form solution for the aerodynamic damping of HAWTs responding dynamically in the for-aft and side-to-side directions on the basis of blade element momentum theory of integration form, and further discussed the aerodynamic damping influence in OWT seismic analysis. Salzmann D. (2005) performed an evaluation with different aerodynamic damping analysis methods found that the aerodynamic damping analysis method with constant rotor speed wind turbine becomes inaccurate when applied to variable speed turbines, and the control system has significant effects on the damping results. Van P. (2013) derived an aerodynamic damping matrix that can be used to the wind turbine foundation design, and proved that the reduction of the structural response due to aerodynamic damping

is significant. Hansen M. (2006) compared the free decay method and stochastic subspace method that used in aerodynamic damping analysis, and found that the second method can handle the deterministic excitation signals. Shirzadeh R. (2013) adopted the OMA method to determine the damping value of the fundamental F-A mode of an OWT, considering the effects of the aerodynamic hydrodynamic and soil loads. Devriendt C. (2013) presented comparative study between different techniques that were used to identify the damping values of an OWT on a monopile foundation, and found that the damping ratios can directly be obtained from vibrations of the tower under ambient excitation from wave and wind loading.

1.2 Aerodynamic damping analysis methods

Valamanesh V. (2014) derived a closed-form solution for the aerodynamic damping in the for-aft and side-to-side directions based on BEM.

$$\xi_{AD,x} = \frac{c_{AD}}{2\sqrt{km}} = \frac{N_b(A+B)}{2\sqrt{km}} \tag{1}$$

$$\xi_{AD,y} = \frac{c_{AD}}{2\sqrt{km}} = \frac{N_b(B'+A')}{4\sqrt{km}} \tag{2}$$

$$A = \rho \int V_w(1-a)[C_L\cos(\phi) + C_D\sin(\phi)]c(r)dr \tag{3}$$

$$B = \frac{1}{2}\rho \int \Omega r(1+a')f(C_L,C_D,\alpha)c(r)dr \qquad (4)$$

$$f(C_L,C_D,\alpha) = \left(\frac{\partial C_L}{\partial \alpha}+C_D\right)\cos(\phi)$$
$$+ \left(\frac{\partial C_D}{\partial \alpha}-C_L\right)\sin(\phi) \qquad (5)$$

$$A' = \frac{1}{2}\rho \int [V_w(1-a)]f'(C_L,C_D,\alpha)c(r)dr \qquad (6)$$

$$f'(C_L,C_D,\alpha) = \left(\frac{\partial C_L}{\partial \alpha}+C_D\right)\sin(\phi)$$
$$+ \left(C_L-\frac{\partial C_D}{\partial \alpha}\right)\cos(\phi) \qquad (7)$$

$$B' = \rho \int \Omega r(1+a')[C_L\sin(\phi)-C_D\cos(\phi)]c(r)dr \qquad (8)$$

where $\xi_{AD,x}$ is the aerodynamic damping in the F-A direction, $\xi_{AD,y}$ is the aerodynamic damping in the S-S direction, ρ is the air density, C_L and C_D are the coefficients of lift and drag force respectively, ϕ is the angle of inflow, c is the chord length of the blade, N_b is the number of blades, Ω is the rotor speed.

Based on assumptions of small angle of inflow, high tip speed ratio, and unstalling, Garrad derived aerodynamic damping formula with constant speed wind turbine as the following.

$$c_{AD} = \frac{1}{2}\rho V_{rot}cC_{L\alpha} \qquad (9)$$

$$\xi_{AD} = \frac{c_{AD}}{4m\omega_n} = \frac{\rho V_{rot}cC_{L\alpha}}{4m\omega_n} \qquad (10)$$

where V_{rot} is the tangential fluid velocity in rotor plane.

1.3 Aerodynamic damping derivation using fluid mechanics theory

This derivation hypothesized the blade as a panel, and neglected the airfoil effects on flow distribution. Based on Fluid Mechanics, the local frictional drag of a panel regardless of thickness can be calculated by Equation (11).

$$\tau = 0.332\,\mu U\sqrt{\frac{U}{\upsilon x}} \qquad (11)$$

where τ is the panel local drag, μ is the fluid dynamic viscosity, υ is the kinematic viscosity.

Based on the former assumptions, Equation (12) can be rewritten as Equation (13).

$$\tau = 0.332\,\mu U\sqrt{\frac{U}{\upsilon c}} \qquad (12)$$

So the resultant drag of a single blade is

$$W = -2\int_0^R\int_0^C 0.332\,\mu \dot{x}\sqrt{\frac{\dot{x}}{\upsilon c}}dcdr$$
$$= -2\int_0^R\int_0^C 0.332\,\mu \dot{x}c^{-1}\sqrt{\frac{\dot{x}c}{\upsilon}}dcdr$$
$$= -2\int_0^R\int_0^C 0.332\mu \dot{x}c^{-1}\sqrt{\mathrm{Re}}dcdr$$
$$= -0.664\int_0^R\int_0^C \dot{x}\frac{\rho\upsilon\sqrt{\mathrm{Re}}}{c}dcdr \qquad (13)$$

For the derivation in this section, the whole OWT structure consider as a single DOF mass-spring system, which all masses are concentrated at the hub with tower lateral stiffness. So the equation of motion in F-A direction simplified as:

$$m\ddot{x}+kx = N_b * W \qquad (14)$$

Equation (13) can be substituted into equation (14), so equation (14) can be taken as:

$$m\ddot{x}+kx = N_b * -0.664\int_0^R\int_0^C \dot{x}\frac{\rho\upsilon\sqrt{\mathrm{Re}}}{c}dcdr \qquad (15)$$

$$m\ddot{x}+N_b * 0.664\int_0^R\int_0^C \dot{x}\frac{\rho\upsilon\sqrt{\mathrm{Re}}}{c}dcdr+kx=0 \qquad (16)$$

The damping in equation (16) caused by fluid viscosity is defined as:

$$c_\upsilon = N_b * 0.664\int_0^R\int_0^C \frac{\rho\upsilon\sqrt{\mathrm{Re}}}{c}dcdr \qquad (17)$$

If the blade simplified as finite panel, Equation (17) can be rewritten as:

$$c_\upsilon = 3 * 0.664\int_0^R\int_0^C \frac{\rho\upsilon\sqrt{\frac{\dot{x}c}{\upsilon}}}{c}dcdr$$
$$= 1.992\int_0^R \rho\upsilon\sqrt{\frac{\dot{x}}{\upsilon}}dr\int_0^C \frac{1}{\sqrt{c}}dc$$
$$= 3.984\int_0^R \rho\upsilon\sqrt{\mathrm{Re}}dr$$
$$= 3.984\sum_{i=1}^n \rho\upsilon\Delta r\sqrt{\mathrm{Re}} \qquad (18)$$

So the numerical model of aerodynamic damping caused by fluid is defined as Equation (19).

$$c_\nu = 3.984 \sum_{i=1}^{n} \rho u \Delta r \sqrt{\text{Re}} \qquad (19)$$

where n is the number of blade elements, Re is the Reynolds number of a blade element.

2 AERODYNAMIC DAMPING ANALYSIS OF HW 5 MW OWT

2.1 Basic parameters of HW 5 MW OWT

Combined the NREL 5 MW baseline wind turbine and its tower with the pentapod sub-structure and foundation of a practical OWT, an integrated structure system of OWT, named HW 5 MW OWT is suggested. The upper structure configuration of HW 5 MW is same with the NREL 5 MW. The basic parameters of upper structure are listed in Table 1. Table 2 lists the pitch control strategy of NREL 5 MW wind turbine. Figure 1 illustrates pentapod dimensions of HW 5 MW OWT.

Table 1. Upper structure basic parameters.

Parameters	Values
Rating	5 MW
Rotor, hub diameter	126 m, 3 m
Hub height	90 m
Cut-in, rated, cut-out wind speed	3 m/s, 11.4 m/s, 25 m/s
Cut-in, rated rotor speed	6.9 rpm, 12.1 rpm
Overhang, shaft tilt, precone	5 m, 5°, 2.5°
Rotor mass	110,000 kg
Nacelle mass	240,000 kg

Table 2. Blade pitch control of NREL 5 MW wind turbine.

Wind speed (m/s)	Rotor speed (rpm)	Pitch angle (°)
11.4	12.1	0.00
12.0	12.1	3.83
13.0	12.1	6.60
14.0	12.1	8.70
15.0	12.1	10.45
16.0	12.1	12.06
17.0	12.1	13.54
18.0	12.1	14.92
19.0	12.1	16.23
20.0	12.1	17.47
21.0	12.1	18.70
22.0	12.1	19.94
23.0	12.1	21.18
24.0	12.1	22.35
25.0	12.1	23.47

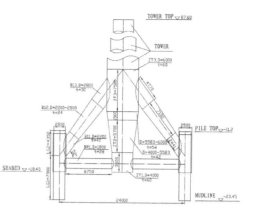

Figure 1. Pentapod structure of HW 5 MW OWT (unit: elevation, m; dimension, mm).

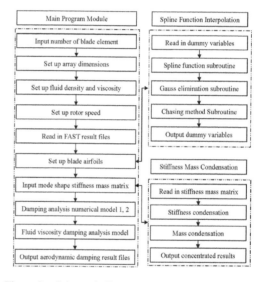

Figure 2. Schematic flow chart of aerodynamic damping analysis module.

2.2 Aerodynamic damping analysis module

In order to research the wind turbine aerodynamic damping during operation, the Aerodynamic Damping Analysis Module on the basis of FAST V7.0 was developed by Fortran language.

The damping analysis module is composed of three sub-modules such as main program control module, spline function interpolation module, structure stiffness and mass condensation module. The aerodynamic damping analysis module involves the new fluid viscosity damping numerical model. Figure 2 illustrates the analysis procedure of this module.

Figure 3. Partial FE model of HW 5 MW OWT in FAST.

2.3 *Numerical model of HW 5 MW OWT in FAST*

In order to perform aero-elastic analysis in the range of cut-in and cut-out wind speed, a partial model of HW 5 MW shown in Figure 3 was established in FAST V7.0. The aero-elastic analysis results files were post processed by the damping module.

3 RESULTS OF HW 5 MW AERODYNAMIC DAMPING ANALYSIS

3.1 *Aerodynamic damping time history*

3.1.1 *Results of Valamanesh method in F-A direction*

Figure 4 displays the aerodynamic damping ratio results at cut-in, rated and cut-out wind speeds. AD-FA1 represents the aerodynamic damping ratio calculated by Valamanesh method in F-A direction. In the range of cut-in and rated wind speeds, the steady values of aerodynamic damping ratio obviously increase with the increasing wind speed, due to the effects of pitch control strategy. However, the aerodynamic damping ratio decreases when the wind speed exceeds cut-out wind speed. Because of the rotor speed, aerodynamic damping at cut-out speed is higher than that at the cut-in wind speed.

3.1.2 *Results of Garrad method in F-A direction*

Figure 5 shows the aerodynamic damping ratio results by Garrad method. AD-FA2 represents the

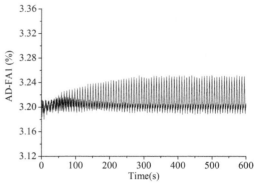

(a) Cut-in wind speed – 3m/s

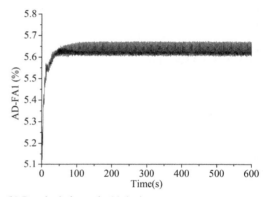

(b) Rated wind speed - 11.4m/s

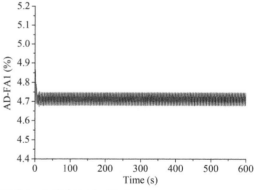

(c) Cut-out wind speed – 25m/s

Figure 4. Results of Valamanesh method.

aerodynamic damping ratio calculated by Garrad method in F-A direction. In the range of cut-in and rated wind speed, because of neglecting the effects of rotor speed, the steady values of aerodynamic ratio increase slightly. Due to blade pitch control, the aerodynamic damping ratio decreases when the

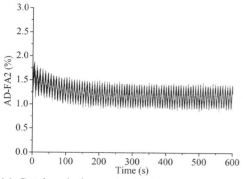

(a) Cut-in wind speed – 3m/s

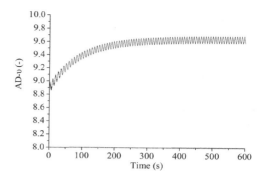

(a) Cut-in wind speed – 3m/s

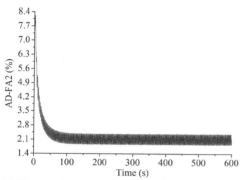

(b) Rated wind speed - 11.4m/s

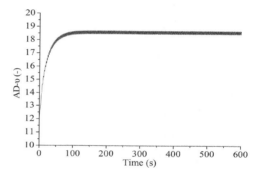

(b) Rated wind speed - 11.4m/s

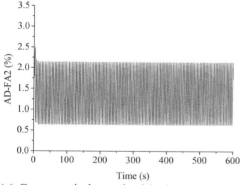

(c) Cut-out wind speed – 25m/s

Figure 5.   Results of Garrad method.

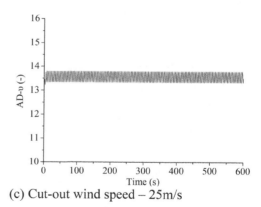

(c) Cut-out wind speed – 25m/s

Figure 6.   Results of fluid viscosity damping model.

wind speed is larger than the rated wind speed. For the Garrad method, not all the blade elements satisfy the assumption of small angle of inflow very well, so this method may give inaccurate results.

3.1.3 *Results of fluid viscosity damping model*

Figure 6 displays the aerodynamic damping results at cut-in, rated and cut-out wind speeds. AD-υ represents the aerodynamic damping calculated by fluid viscosity damping model. In the range of cut-in and rated wind speeds, the steady value of aerodynamic damping increase with the increasing wind speed, due to the effects of fluid viscosity. However, the aerodynamic damping decreases out of the range of cut-out wind speed. Due to the effects of the relative fluid velocity and blade element Reynolds, the aerodynamic damping at cut-out speed is higher than that at the cut-in wind speed.

3.2 *Aerodynamic damping results summarization*

The derivation of aerodynamic damping was performed by using the above mentioned methods on the basis of the partial model of HW 5 WM OWT.

Figure 7 illustrates the aerodynamic damping ratio variation with respect to the wind speed in the F-A direction derived by the Valamanesh method.

From Figure 7, in the scope of cut-in and cut-out wind speed, the aerodynamic damping increase significantly. The aerodynamic damping ratio mean value has a maximum at the wind speed of 12 m/s, close to the rated wind speed 11.4 m/s. In the range of 11.4 m/s and 16 m/s, the aerodynamic damping ratio variation becomes more complex, due to the effects of control strategy and local vibration, but above the rated speed, the aerodynamic damping ratio decreases. Because of the influence of rotor speed, the statistic values at the cut-out wind speed are higher than the cut-in wind speed.

Figure 8 displays the aerodynamic damping ratio variation with respect to the wind speed in the F-A direction derived by the Garrad method.

From Figure 8, the statistical values have a maximum value at the rated wind speed. In the range of cut-in and cut-out wind speed, the aerodynamic damping values increase slightly relative to the Valamanesh method, due to the neglects of rotor speed effects. Due to the influence of blade pitch control, the aerodynamic damping ratio decrease obviously above the rated wind speed. The statistical values at the cue-out wind speed are nearly equal to the cut-in wind speed. That is different from the Valamanesh method. Not all the blade elements satisfy the assumptions of Garrad method at the same time, so this method may give inaccurate results at some operation status.

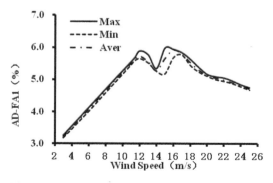

Figure 7. AD-FA1 statistic values variation with respect to wind speed.

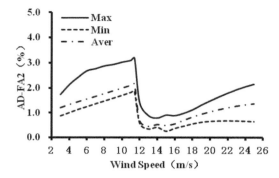

Figure 8. AD-FA2 statistic values variation with respect to wind speed.

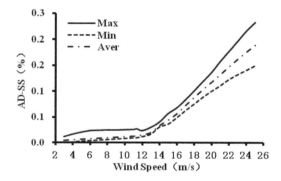

Figure 9. AD-SS statistic values variation with respect to wind speed.

Figure 9 shows the aerodynamic damping results in the S-S direction derived by the Valamanesh method.

From Figure 9, the aerodynamic damping ratio results in the S-S direction are smaller than the F-A direction results, but the values increase obviously above the rated wind speed. So the control strategy has significant influence on the aerodynamic damping in the S-S direction, but their values are small enough to be neglected in the analysis.

Based on the comparisons of the two methods, it can conclude that the Valamanesh method is better than the Garrad method. Since the former method is based on less assumptions and proves that the aerodynamic damping depends on directions, the aerodynamic damping analysis should use different numerical models in different directions.

3.3 *Aerodynamic damping application in semi-integrated analysis method*

The semi-integrated method is widely used in OWT sub-structure design. In recent years researchers

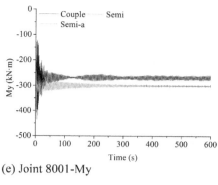

(a) Joint 8001-Fx

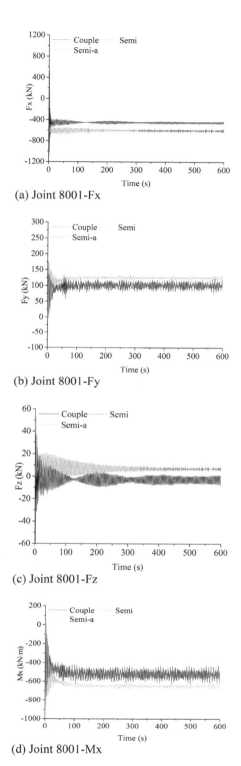

(b) Joint 8001-Fy

(c) Joint 8001-Fz

(d) Joint 8001-Mx

Figure 10. (Continued)

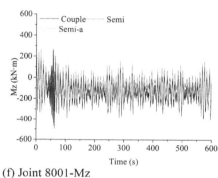

(e) Joint 8001-My

(f) Joint 8001-Mz

Figure 10. Response at joint 8001 of Member 8001–9001.

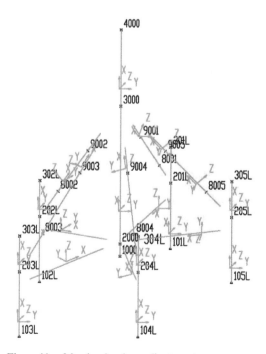

Figure 11. Member local coordinate system.

503

found that the semi-integrated method may get higher structure responses than the fully couple method. Thus, to structure optimization design, the aerodynamic damping should be considered in semi-integrated method.

Considering of aerodynamic damping, a semi-integrated analysis of the HW 5 MW OWT is performed and compared with the fully coupled analysis.

Figure 10 shows the internal forces of sub-structure members by different analysis models. The Couple represents the fully coupled model, the Semi represents the semi-integrated model, the Semi-a represents the semi-integrated model with the aerodynamic damping. The internal force coordinate system can refer to Figure 11.

Figure 10 display the differences of the member internal forces by these three models. The Semi results have greater mean value that the Couple results, but it nearly equal to the Semi-a results. The Semi results have wider amplitude range than the Semi-a results, due to the effects of aerodynamic damping, such as Figure 10 (b) (c) (d).

4 CONCLUSIONS

The aerodynamic damping analysis module was developed to research the effects of aerodynamic damping on the wind turbine. Considering Reynolds number, the fluid viscosity damping was derived based on fluid mechanics. By the comparison, the Valamanesh method was better than the Garrad method. Finally, the differences of the semi-integrated model, coupled model, and semi-integrated model with aerodynamic damping effects were studied.

ACKNOWLEDGEMENT

This work is funded by the National Natural Science Foundation of China (Grant No. 51121005), and supported by the Science Foundations of Powerchina Huadong Engineering Corporation Limited (No. KY120228-03-07 and KY2014-02-41) and the Open Fund Project of State Key Lab of Coastal and Offshore Engineering in Dalian University of Technology (No. LP1413). Their financial supports are gratefully acknowledged.

REFERENCES

Christof Devriendt, Pieter Jan Jordaens, Gert De Sitter, Patrick Guillaume. Damping estimation of an offshore wind turbine on a monopile foundation. IET Renewable Power Generation, 2013, 7(4): 401–412.

David Laino, A. Craig Hansen. User's guide to the wind turbine aerodynamics coputer software AeroDyn. Windward Engineering, LC, Salt Lake City, UT:2002.

Jason M. Jonkman, Marshall L. Buhl. FAST user's guide. National Renewable Energy Laboratory, CO, Technical Report No. NRRL/EL-500-38230, 2005.

Jonkman B.J., J.M. Jonkman. Addendum to the user's guides for FAST A2AD, and Aerodyn released March 2010-February 2013. Tech. rep., National Renewable Energy Laboratory, Golden, Colorado, 2013.

Martin O.L. Hansen. Aerodynamics of wind turbines. Routledge, 2015.

Morten. Hansen, Kenneth Thomsen, Peter Fuglsang, Torben Knudsen. Two methods for estimating aeroelastic damping of operational wind turbine modes for experiments. Wind Energy, 2066, 9(1–2): 179–191.

Salzmann D., J. van der Tempel. Aerodynamic damping in the design of support structures for offshore wind turbines. Paper of Copenhagen Offshore Conference, 2005.

Sandy Butterfield, Walter Musial, George Scott. Definition of a 5-MW reference wind turbine for offshore system development. Golden, CO: National Renewable Energy Laboratory, 2009.

Shirzadeh R., C. Devriendt, M.A. Bidakhvidi, P. Guillaume. Experimental and computational damping estimation of an offshore wind turbine on a monopole foundation. Journal of Wind Engineering and Industrial Aerodynamics, 2013, 120: 96–106.

Valamanesh V., A.T. Myers. Aerodynamic damping and seismic response of horizontal axis wind turbine towers. Journal of Structure Engineering, 2014.

van der Male P., K.N. van Dalen, A.V. Metrikine. Aerodynamic damping of nonlinearily wind–excited wind turbine blades. Proceedings of the EAWE 9th PhD Seminar on Wind Energy in Europe, September 18–20, 2013, Uppsala University Campus Gotland, Sweden.

Advances in Energy, Environment and Materials Science – Wang & Zhao (Eds)
© 2016 Taylor & Francis Group, London, ISBN 978-1-138-02931-6

Interaction gesture recognition for highway 3D space alignment design

L.D. Long, X.S. Fu & H.L. Zhu
School of Civil Engineering and Transportation, South China University of Technology, Guangzhou, Guangdong, P.R. China

Y.Q. Wang
Guizhou Expressway Group Co. Ltd., Guiyang, Guizhou, P.R. China

ABSTRACT: It's the human-computer interaction prerequisite for true 3D highway alignment design to efficiently navigate and interactively design highway 3D center curve in 3D virtual environment by making full use of the powerful capabilities of Kinect to identify human body joints and track their space motion and position coordinates. According to the features and function requirements to design highway 3D center curve in 3D virtual environment, this paper is designed with main interaction gestures, and has a method that defines a gesture by the combination of several typical pose snapshots, and then to calculate the matching degree between the snapshots and skeleton data frames to recognize a specific gesture after analyzing the characteristics of gestures. Finally, the test proved the efficiency and accuracy of the recognition method proposed by this paper is able to meet well the requirements of highway 3D space curve design on gestures recognition.

1 INTRODUCTION

One of the key issues of true 3D highway design is the automatic drawing and dynamic interactive edit of highway 3D space centerline in Virtual Environment (VE) to get the optimal alignment (Fu et al., 2014). The weakness of traditional interaction tools like a 2D mouse or 1D keyboard used to interact with 3D VE are two-fold: user could not efficiently modify the 3D viewpoint to flexibly navigate (Kang et al., 2011), and achieve the direct manipulation to highway alignment object in VE and dynamically move a spatial entity on X, Y and Z-direction. With the launch of motion sensors represented by Kinect, the outstanding capabilities to detect the depth information of objects in its visual field and to recognize and locate more than 20 joints of human body (Shotton et al., 2013) make it an easy target to interact with 3D VE with gesture for the navigation of 3D viewpoint (Kang et al., 2011) and manipulation 3D highway alignment objects.

After logic interaction gesture designing, to efficiently and accurately recognize them is the basic requirements for the gesture-based human-computer interaction. With the launch of Kinect, the outstanding joints recognizing and tracking capabilities, strong anti interference performance and low price of which lead to a worldwide researches on 3D gesture-based user interface. For Kinect-based gesture recognition, current studies mainly concentrated in the analyses of the static space position of multiple joints to recognize body pose or the movement track of single joint to recognize some simple motion, and recognition methods of the motion described by the combination of the space movement of multiple joints are rare. Furthermore, most of studies focus on the improvement of recognition accuracy and are out of the real demands of the specific application. Moreover, greater accuracy usually means more complex recognition algorithm, which will consume more computing capacity and decrease the recognition efficiency, and lead to the great difficulties for the real-time gesture recognition of interaction gestures, which is the determinant of a gesture-based 3D user interface.

This paper designs the main interaction gestures based on the function requirements of highway 3D alignment design, then proposes a gesture recognition method based on pose snapshots of multiple joints: through analyzing the space movement tracks or coordinates of key joints involving in a gesture, several typical segments could be picked up to define the gesture using the combination of them, then doing the matching calculation between gesture definition and the skeleton data frames from Kinect for accurate and efficient recognition. Finally, a test is carried out, and the results verify that the recognition method proposed in the paper met well with the recognition accuracy and efficiency requirements of highway 3D alignment design.

2 RELATED WORK

The feature analysis of gesture recognition has experienced three stages: 2D human body contour, 3D surface depth and space movement track of joint (Mihail et al., 2012). In the first stage, RGB video is taken as original data: Bobick et al. (Bobick et al., 2001) extracted the human profiles from video images and stacked them to track the changes of human contour so that motion history and energy images were formed. With the development of depth camera, gesture recognition based on depth data got some studies. Li (Li et al., 2010) propose the bag of 3D point model, which samples 3D feature points of human body 3D surface and matching with motion description graph which can reduce error rate greatly. Owing to the release of Kinect, it was easier to acquire the accurate 3D coordinates of body joint. Sait C. et al (Celebi et al., 2013) used DTW to conduct template matching aiming at kinetic characteristic of 6 joints of human body and the accuracy reach 96.7%. Youwen W. et al (Wang et al., 2012) used HMM to achieve accurate recognition of motion of single hand joint. Xiaodong Y. et al (Yang et al., 2012) used Naïve Bayes Nearest Neighbor to classify the gestures and further improve the accuracy.

Whatever the feature of is analyzed, gesture recognition methods mainly include 3 types: algorithm matching, template matching and machine learning. Algorithm matching tracks spatial position of human body joints and achieves fast matching calculation on gesture based on triangle geometry. Which is fast and simple to define and recognize gesture, and easy to code and modify, hence is applied well to recognize those gestures with a small number and a big difference. Template matching like DTW, computes the similarities between gesture movement and template data to classify some systematic gestures. These methods are very accurate and reliable, but usually complicated and time-consuming to calculate (Wang et al., 2012). Machine learning methods are widely used for gesture recognition, including Bays Maximum Likelihood Estimate, Artificial Neural Network, Support Vector Machine and Hidden Markov Model, and so on. These kinds of methods are stable and robust, and can be used to recognize all kinds of gestures, even some very subtle motion through appropriate training. However, the accuracy depends on the gesture features selected and the quantity and quality of training data set, and the lots of good training data often is difficult to obtain (Wu et al., 2012).

3 INTERACTION GESTURE DESIGN

Gesture includes static pose and dynamic motion, and is the main input way of Natural User Interface (NUI). On the basis of the general gesture design principles of NUI (Yu, 2012), interaction gesture should be natural, in line with human nature and habits, consistent with the knowledge of target users, meet the functional requirements etc. In addition, being easy to understand and meaning clear, and a minimal numbers are also followed. In accordance with different control objects when designing interactively 3D highway alignment in VE, this paper divided all gestures into 2 parts: 3D object manipulation and 3D viewpoint navigation gesture.

3.1 3D viewpoint navigation

Design and optimization of 3D highway alignment in VE with realistic terrain and landscape requires user to be allowed to review the alignment designing in a flexible multiple perspectives to evaluate the environment harmonization, and analyze the continuity, consistency and 3D sight distance to find where it is unreasonable, and then revise to ensure the design quality. Therefore, to modify the coordinate, height, azimuth, and tilt of viewpoint optionally by gesture is necessary to meet these requirements.

According to different control objects, 3D navigation includes 4 gestures: panning, zooming, rotation, and tilt (KamelBoulos et al., 2011). To get the natural and intuitive gesture design, the experiences obtaining in using paper drawing and touchscreen devices are applied. As shown in Figure 1, suppose a virtual touch plane ($z = z_{head} - \Delta z$) existing in front of user, when hand passes through it and carries out a gesture, the navigation operation will be triggered. What's more, the movement speed value is used as the control parameter to modify the map window.

3.2 3D highway alignment object manipulation

Highway spatial alignment design classically involves a two-stage: the optimal space centerline design and cross section alignment design. Owing to cross section alignment could be designed automatically by the intersection and union operation of space centerline and digital terrain model, the interaction gestures in this paper only focus on the needs of interactively 3D centerline modeling. First, to create a set of control points, and then input the related design parameter values, is the basic method of 3D spatial curve modeling (Arangarasan et al., 2000; Bourdot et al., 2010). Like the cubic spline curve and pH curve, highway 3D centerline has a set of space control points, therefore, manipulations of which could be summed up in 6 gestures, namely pointing, design parameter value input, and creation, selection, translation and deletion of control point.

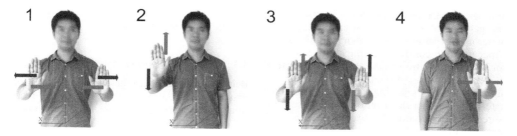

Figure 1. Gestures for 3D viewpoint navigation. (1) Zooming. (2) Tilt. (3) Rotation. (4) Panning.

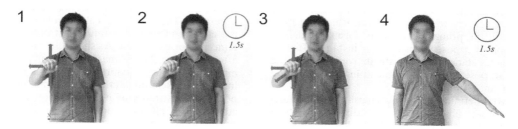

Figure 2. Typical gestures for highway design. (1) Pointing. (2) Creation and Selection of control point, On-screen keyboard tap. (3) Translation of control point. (4) Deletion of control point.

Highway alignment object manipulation gestures mainly include creation, selection, translation, and deletion of control point, pointing and on-screen keyboard tap. As shown in Figure 2, gesture design imitates the interaction gesture in real life, for instance, right hand is used more frequently, a fist means grab and drag but a palm means release. Such design not only can reduce the learning load, but allows designer to immerse completely in the design work to make full use of initiative and creativity.

4 GESTURE RECOGNITION

4.1 *Analysis of interaction gesture characteristic*

Accuracy and efficiency of gesture recognition are essential indicators to verify the naturalness of a natural user interface, and is the basic factors to prove design efficiency too. But they also conflict with each other: greater accuracy usually means more complex recognition algorithm, which will consume more computing capacity and decrease the recognition efficiency ultimately. Moreover, considering the following characteristics of interaction gestures design and recognition in this paper: (1) Including both static pose and dynamic motion; (2) Few in number, only 8 gestures; (3) Can be described by the simple movement of one or multiple joints; (4) Real time recognition is demanded to reduce delay; (5) The same or similar

gesture is used for different operation goals; (6) Be easy to change the gesture design according to the user experience.

According to the requirements of highway 3D alignment design, the gesture recognition proposed by this paper takes into account the efficiency rather than accuracy. After using the primary constraint, such as control point selecting status or cursor position, to determine which operation will be trigger for a gesture, the features of each gesture are analyzed to pick up the typical segments which inevitably appear, and then define the gesture with the combination of these segments. Finally, match calculation between gesture definition and the skeleton data frames from Kinect to recognition the gesture.

4.2 *Gesture definition*

Gesture recognition method based on pose snapshots of multiple joints (Yu, 2012) defines, respectively, the spatial constraint and time constraint through extracting and combining pose snapshots. Such methods achieves the efficient recognition of dynamic spatiotemporal movement track of multiple joints only through the matching calculation of the static spatial coordinates of them, and will greatly simplified the recognition.

1. Primary constraint. Through reusing the gestures, the same or similar gestures could realize multiple manipulation goals, and reduce the

number of gesture and learning load of user, and get a more user-friendly user interface and better user experience. As shown in Table 1, this paper uses the control selecting status and cursor hovering position as the primary constraint to decide which gesture the same motion is. Similarly, control selecting status is also used to determine whether the deletion gesture needs to be recognized to reduce the waste of computing resource and improve recognition efficiency.

2. Spatial constraint. Pose snapshots are several must happen segments of a gesture. In the Kinect coordinate system, the spatial coordinate constraint can be set in X, Y and Z-direction, respectively, for each snapshot, which, together with the primary constraint, forms the definition of a pose snapshot. For instance, Table 2 shows static space coordinate constraint of the two pose snapshots included in both Zooming In and Zooming Out.

3. After defining each pose snapshot of a gesture, combine all snapshots according to the time sequence and confirm the cycle index of the combination to complete the gesture defining ultimately. For dynamic motion, gesture definition usually includes several snapshots, and the cycle index depends on the specific gesture, but for static pose, it usually is the multiple cycles of single snapshot, and the cycle index depends on the duration of the pose and frame rate.

The definition of Zooming In and Zooming Out are shown in Table 3.

4.3 Gesture recognition process

Being consistent with gesture defining process, matching calculation of gesture recognition also includes pose snapshot recognition and the combination recognition:

1. Matching calculation of pose snapshot. Pose snapshot reflects the space position of multiple key joint participating in movement in a certain moment of the occurrence process of a gesture. The matching calculation of spatial constraint uses skeleton data frame from Kinect as original data. As shown in Figure 3, after judging the primary constraints, the matching of the 3D coordinate data of joints included in skeleton data frame and the spatial position constraints of pose snapshot will be computed. If all constraints are met, the pose snapshot happened.

2. Matching calculation of snapshot combination. The movement track of a gesture in physical space is described by the combination of pose snapshots and the cycle index of combination. Accordingly, matching calculation of temporal constraints is used to judge whether the time sequence of snapshots is met (See Fig. 4).

Table 1. Primary constraint.

Situation	Primary constraint	Yes/No	Gesture
Same pose	Cursor hovers over a control point	Yes	Creation
		No	Selection
Similar motion	One control point is selected	Yes	Translation
		No	Pointing

Table 2. Spatial constraint.

Diresction	1st pose snapshot	2nd pose snapshot
Constraint in Z-direction*	$Z_{HL}<Z_{EL}$ & $Z_{HR}<Z_{ER}$	$Z_{HL}<Z_{EL}$ & $Z_{HR}<Z_{ER}$
Constraint in Y-direction*	$Y_{HipC}<Y_{HL}<Y_{SL}$ & $Y_{HipC}<Y_{HR}<Y_{SR}$	$Y_{HipC}<Y_{HL}<Y_{SL}$ & $Y_{HipC}<Y_{HR}<Y_{SR}$
Constraint in X-direction*	$X_{HL}>X_{EL}$ & $X_{HR}<X_{ER}$	$X_{HL}<X_{EL}$ & $X_{HR}>X_{ER}$

*H-Hand, E-Elbow, S-Shoulder, L-Left, R-Right, C-Center, *Hip*-Hip.

Table 3. Temporal constraint.

Gesture	Combination of pose snapshots	Cycle index
Zooming In	1st Pose snapshot → 2nd Pose snapshot	1
Zooming Out	2nd Pose snapshot → 1st Pose snapshot	1

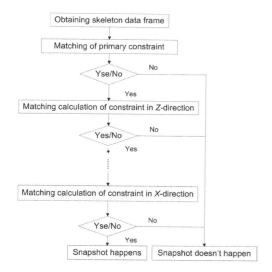

Figure 3. Matching calculation process of a pose snapshot.

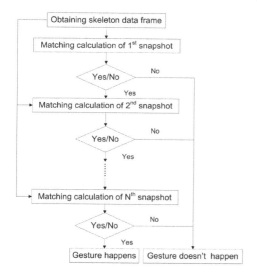

Figure 4. Matching calculation process of snapshots combination.

When each snapshot happens in turn, the result of gesture detection is positive.

5 TEST AND ANALYSIS

5.1 Test design

To verify the practicability of the gesture rec-ognition above, the efficiency and accuracy of recognition needs to be quantitatively measured. This paper invited 13 designers to join the test. They have good experience with computer but NUI like motion sensing game. After detailing the operation of system prototype, test items, proce-dures to all testers, and practiced 10 min training, the test began. In order to directly indicate the effi-ciency and accuracy, a basic test and integrative test is carried out.

1. Basic test. Every tester carries out each test gesture 10 times, namely each test gesture will be tested 130 times in total. The practicability index of recognition includes efficiency and accuracy. Efficiency will be presented by the time from starting to detect the 1st snapshot to confirm the gesture, and the accuracy is the per-centage of the number of correct recognition.
2. Integrative test. The efficiency and accuracy is usually expressed as user operation efficiency, therefore, the test is designed to measure the time consumption of the process navigating from the initial windows of 3D-map to the tar-get window. In order to directly indicate the efficiency and accuracy, all test results are com-pared with those of a conventional mouse and keyboard.

5.2 System prototype

According to the test design, this paper used the Microsoft Kinect for Windows SDK as the driver software of Kinect sensor. With the support of DotNet Framework, this paper used the Extensi-ble Application Markup Language (XAML) for GUI design. Furthermore, to simplify the develop-ment of system prototype and quickly to realize the navigation of 3D viewpoint and manipulation of 3D highway object, this paper used Google Earth (GE) as 3D VE and the Placemark of which as 3D control point, then carried on the second-ary development of GE with Google Earth KML and COM API technology to develop a workable system prototype.

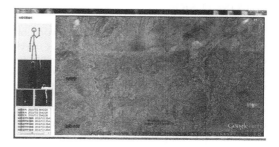

Figure 5. Screenshot of system prototype interface.

Table 4. Result of basic test.

Test item	Zooming in	Rotation clockwise	Tilt forward	Creation	Deletion
Efficiency (ms)	209	222	267	2097	1872
Accuracy (%)	98.2	94.5	93	90.8	96.4

Table 5. Result of integrative test.

Input method	Mean time (s)	Ratio
Gesture	78.23	1.459
Mouse/Keyboard	53.61	

5.3 Test result and analysis

After test completing, test data is collated to indicate the practicability of the recognition method proposed by this paper to highway 3D alignment design.

1. Basic test. As shown in Table 4, the average time consumption of gesture recognition is used to indicate the efficiency: the more average time consumption, the poorer efficiency. Efficiency moves in concert with the number of the snapshots included in a gesture: dynamic motion usually has less snapshots and higher efficiency. In addition, Accuracy of every test gesture is more than 90%. Such result meets largely the requirements of highway 3D alignment design.
2. Integrative test. After calculating the average navigation time of all testers, the test result is shown in Table 5. Just looking at the mean time, result is negative: gesture spent more time than mouse/keyboard. However, considering that all testers have grown accustomed to mouse/keyboard but only practice the gesture-based navigation for 10 minutes, and the navigation time depends on not only the efficiency and accuracy of recognition method but the user experiences, this disparity is acceptable accordingly.

6 CONCLUSION

To interact with 3D VE by gesture is a workable solution to break through the obstacles for developing true 3D highway design caused by conventional 1D and 2D input devices like mouse and keyboard. According to the function, accuracy and efficiency requirements of highway 3D alignment design on gesture recognition, this paper designs the main gestures and analyzes their characteristics and proposes the recognition method based on pose snapshots of multiple joints, which is very simple and reliable, and is not sensitive to the changes of environment and light conditions. Furthermore, this method works well in recognizing both static pose and dynamic motion, and the recognition accuracy and efficiency meet well the requirements of highway 3D alignment design. When testing, recognition errors came up occasionally due to the low location accuracy of Kinect and the similarity between gestures. When advanced motion sensors, like Leap Motion, are applied, and stricter constraints are used to define a gesture, the accuracy and efficiency of recognition will be improved greatly.

ACKNOWLEDGMENT

This research was financially supported by the National Natural Science Foundation of China (NSFC, 51278202).

REFERENCES

Arangarasan, R. & Gadh, R. 2000. Geometric modeling and collaborative design in a multi-modal multi-sensory virtual environment. *Proceeding of the ASME 2000 Design Engineering Technical Conferences and Computers and Information in Engineering Conference* 2000: 10–13.

Bobick, A.F. & Davis, J.W. 2001. The recognition of human movement using temporal templates. *Pattern Analysis and Machine Intelligence, IEEE Transactions on* 23(3): 257–267.

Bourdot, P. & Convard, T. & Picon, F. et al. 2010. VR–CAD integration: Multimodal immersive interaction and advanced haptic paradigms for implicit edition of CAD models. *Computer-Aided Design* 42(5): 445–461.

Celebi, S. & Aydin, A.S. & Temiz, T.T. et al. 2013. Gesture Recognition Using Skeleton Data with Weighted Dynamic Time Warping. *Computer Vision Theory and Applications.*

Fu, X.S. & Long, L.D. & Li, H.F. & Ge, T. 2014. Motion-based true three-dimensional highway alignment design method and system framework. *Journal of South China University of Technology* 42(8): 91–96.

KamelBoulos, M.N. & Blanchard, B.J. & Walker, C. et al. 2011. Web GIS in practice X: a Microsoft Kinect natural user interface for Google Earth navigation. *International journal of Health Geographics* 10(1): 45–50.

Kang, J. & Seo, D. & Jung, D.A. 2011. Study on the control method of 3-dimensional space application using Kinect system. *International Journal of Computer Science and Network Security* 11(9): 55–59.

Li, W. & Zhang, Z. & Liu, Z. 2010. Action recognition based on a bag of 3d points. *Computer Vision and Pattern Recognition Workshops (CVPRW), 2010 IEEE Computer Society Conference on.* IEEE: 9–14.

Mihail, R. & Jacobs, N. & Goldsmith, J. 2012. Static Hand Gesture Recognition with 2 Kinect Sensors. *Proc. Int. Conf. on Image Processing, Computer Vision, and Pattern Recognition.*

Shotton, J. & Sharp, T. & Kipman, A. et al. 2013. Real-time human pose recognition in parts from single depth images. *Communications of the ACM* 56(1): 116–124.

Wang, Y. & Yang, C. & Wu, X. et al. 2012. Kinect Based Dynamic Hand Gesture Recognition Algorithm Research. *Intelligent Human-Machine Systems and Cybernetics (IHMSC), 2012 4th International Conference on. IEEE* 2012.1: 274–279.

Wu, D. & Zhu, F. & Shao, L. et al. 2012. One shot learning gesture recognition with Kinect sensor. *Proceedings of the 20th ACM international conference on Multimedia. ACM* 2012: 1303–1304.

Yang, X. & Tian, Y.L. 2012. Eigenjoints-based action recognition using naive-bayes-nearest-neighbor. *Computer Vision and Pattern Recognition Workshops (CVPRW), 2012 IEEE Computer Society Conference on. IEEE,* 2012: 14–19.

Yu, T. 2012. *Practices of kinect application development.* Beijing: Publishing House of Machinery Industry.

Advances in Energy, Environment and Materials Science – Wang & Zhao (Eds)
© *2016 Taylor & Francis Group, London, ISBN 978-1-138-02931-6*

Research on the united tension formulas of short bridge suspenders for two typical kinds of boundary conditions

Jingbo Liao, Huailin Liu & Guangwu Tang
State Key Laboratory of Bridge Engineering Structural Dynamics, Chongqing Communications Research and Design Institute, Chongqing, China

ABSTRACT: The tension formulas of bridge suspenders of the clamped-clamped boundary condition or the clamped-hinged boundary condition, aren't explicitly expressed as the classical formulas of hinged-hinged boundary condition, this is because the frequency functions are the transcendental function of the bridge suspender, which contains the hyperbolic and trigonometric function that is needed for the nonlinear iteration and may cause the data overflow. In the paper, based on the nonlinear iteration method and the numerical fitting theory, the tension formulas of bridge suspenders are deduced and the expressions with the units are constructed under two typical kinds of boundary conditions by a new non-dimensional parameter η which is combined with the natural frequency formulas of the hinged-hinged beam. Finally, the numerical comparative study shows both of the relative error of the fitting formulas of the two kinds of boundary conditions are under ±5.0%. The fitting formulas provide a fast identification method for the suspender tension estimation.

1 INTRODUCTION

In order to grasp the structural characteristics of cable-stayed bridge in the process of the construction and operation, it is very necessary to know well the load conditions of the bearing suspenders. At present, the vibration method is the common technology to estimate the tension of the suspenders, which owns lots of superiorities, such as more economical, more effort, more time-saving than other methods, and is very suitable for the tension identification in the operation process; this is because the frequencies of the vibration method are so easily obtained.

First, the short suspenders should be defined. In china, Hu Hai-chang uses the energy method to study the local effect of tension on the convergence rate of the deflection solution of the beam (Hu, 1979&1987); however, the dynamical definition of the short suspenders is not given. In the reference (Li, 2014), the definition of the short suspenders are preliminarily descripted, that is "when the effect of the geometrical stiffness due to the tension is more obvious than physical stiffness, the element is considered as string, conversely as beam" but the quantitative definition isn't still provided.

Additionally, some practical factor should be considered, for example, the practical bridge suspenders may be wrapped up by the concrete tube or anchoring end, which have an influence on the vibrating properties. There are few researches to focus on this area. These problems are preliminarily studied by the parameter identification based on the

optimization theory in the reference (Tang, 2011; Liao, 2007), it should be noted that the essential problem of the vibration isn't deeply analyzed, even sometimes there is some confusion about its understanding; the typical case is that the bending stiffness EI, the line density ρA, and the cable length l and the tension T in the finite element model are all chosen as the identification parameter. The reason of that is the finite element model explicitly containing these parameters. In fact, the four parameters are also expressed in the vibration frequency functions, but it can be transferred only to two independent parameters, such as α and β, namely, the frequency functions are only related to the independent two parameters α and β (Zui, 1996). Therefore, the number of the identification parameters and the change interval of that are determined. It obviously can improve the efficiency of the identification method. But it is inevitably the identifying process that still needs the repeating trial and error.

Recently, the researchers at home and abroad focus mainly on how to express the tension estimation process of the suspenders as practical formulas for two typical kinds of boundary conditions, namely, the clamped-clamped and clamped-hinged boundary condition, similar to the hinged-hinged boundary condition. This is really the parameter fitting question. Hiroshi, et al., launched a creative study by which the natural frequencies of the suspenders are united by the frequencies of the classical strings, and a new nondimensional parameters η are introduced, and then the solution of the frequency functions are approximated by the piecewise fitting function (Zui, 1996). The implicit expression is given in the paper (He, 2012), the elastic bearings and the additional mass and the tension are considered. Zhang Wei studies the practical formulas, which are suitable with the tension calculation for the small tension and the large sag of the clined cable (Zhang, 2012). Fang Zhi, et al. studied the fitting relation between the tension and the bending stiffness, length, density, and frequency by the regulation of the solution of the frequency functions under the clamped-clamped boundary condition (Zhi, 2012; Fang, 2007). Liu Zhijun, et al. (2011) deduces the numerical fitting formulas of the tension, containing the bending stiffness, length, density, and frequency of the suspenders under the clamped-hinged boundary condition, which still needs to be solved by the nonlinear function. Li Guojie, et al. (2009) provides the non-dimensional relation between the tension and the frequency by the least square method. All of the above formulas need the repeating trial and error. This is because that the segmented point is closely related with the tension. First, the tension is predicted, and the interval is determined based on the predicting tension, and then a new estimating tension is obtained; further, the tension is also used to update the predicting tension, repeated to do this process until the error satisfies the meet. The whole process has way more intricacy and not fit for practical engineering. It is interesting that the repeated trial and error is avoidable by exchanging the tension and the frequency (Liao, 2014).

In this paper, the shear effect, inclination and the sag-to-span ratio of the suspenders aren't considered; the effect of the bending stiffness on the vibration is preserved, and a new nondimensional parameters η are introduced by the natural frequency of the Euler–Bernoulli beam under the hinged-hinged boundary conditions, and the approximate tension formulas of suspenders are deduced based on the nonlinear iteration method and the piecewise fitting function, further the united tension formulas of the suspenders are constructed for the two typical kinds of boundary conditions. Finally, the precision of the fitting formulas is studied by comparing the fitting formulas with the direct solution of iterative method for the two typical boundary conditions.

2 THE FREQUENCY FUNCTION OF THE SUSPENDER FOR TWO TYPICAL KINDS OF BOUNDARY CONDITIONS

The vibration differential function of the suspender is:

$$EI\left[\frac{\partial^4 y(x,t)}{\partial x^4}\right] - T\frac{\partial^2 y(x,t)}{\partial x^2} + \rho A\frac{\partial^2 y(x,t)}{\partial t^2} = 0 \quad (1)$$

In which, EI, T, and ρA are, respectively, the bending stiffness, the tension, the line density of the suspender, $y(x, t)$ represents the displacement of the suspender with the time and position.

$Y(x, t)$ can be translated to a product in which the dependence of y on x, and t is separated by the method of separation of variables, that is:

$$y(x,t) = Y(x)\sin(\omega t + \varphi) \quad (2)$$

The function (2) is substituted into the function (1) with $k^4 = \omega^2 \rho A/EI$, $\alpha^2 = Tl^2/EI$, we can get:

$$\frac{d^4Y}{dx^4} - \alpha^2\frac{\partial^2 Y}{\partial x^2} - k^4 Y = 0 \quad (3)$$

The general solution of the function (3) is supposed to be:

$$Y(x) = c_1 ch\lambda_2 x + c_2 sh\lambda_2 x + c_3\cos\lambda_1 x + c_4\sin\lambda_1 x \quad (4)$$

In which c_1, c_2, c_3, c_4 are the undetermined constants by the boundary conditions, and λ_1, λ_2, respectively, that is:

$$\lambda_1 = \sqrt{-\frac{\alpha^2}{2} + \sqrt{\frac{\alpha^4}{4} + k^4}} \quad (5)$$

$$\lambda_2 = \sqrt{\frac{\alpha^2}{2} + \sqrt{\frac{\alpha^4}{4} + k^4}} \quad (6)$$

(5) function and (6) function can be equally translated to:

$$2\lambda_1\lambda_2 = 2k^2 \quad (7)$$

$$\lambda_2^2 - \lambda_1^2 = \alpha^2 \quad (8)$$

Under the clamped-clamped boundary condition and the clamped-hinged boundary condition, the frequency function of the suspender are as follows:

$$ch\lambda_2\sin\lambda_1/\lambda_1 - sh\lambda_2\cos\lambda_1/\lambda_2 = 0 \quad (9)$$

$$2\lambda_1\lambda_2(1 - ch\lambda_2\cos\lambda_1) + (\lambda_2^2 - \lambda_1^2)sh\lambda_2\sin\lambda_1 = 0 \quad (10)$$

3 THE UNITED TENSION FORMULAS OF THE SUSPENDER UNDER TWO TYPICAL KINDS OF BOUNDARY CONDITIONS

The function (8) can be equally translated to

$$\lambda_2^2 = \alpha^2 + \lambda_1^2 \tag{11}$$

The function (11) is substituted into the function (9) and the function (10), which contain λ_1 and λ_2. Then the functions (12) and (13) become

$$ch\sqrt{\alpha^2 + \lambda_1^2}\sin\lambda_1/\lambda_1 - sh\sqrt{\alpha^2 + \lambda_1^2}\cos\lambda_1/\lambda_2 = 0 \tag{12}$$

$$2\lambda_1\sqrt{\alpha^2 + \lambda_1^2}\left(1 - ch\sqrt{\alpha^2 + \lambda_1^2}\cos\lambda_1\right)$$
$$+ \alpha^2 sh\sqrt{\alpha^2 + \lambda_1^2}\sin\lambda_1 = 0 \tag{13}$$

If the parameters α^2 and the frequency order i are provided, the variable quantity λ_1 can be solved by the functions (12) and (13). According to the reference (Qin, 2014 & Liao, 2013) for the different frequency order i, the variable quantity λ_1 can be expressed as follows:

$$\lambda_{1,i} = (i + \mu_i)\pi \tag{14}$$

The function (12) and (13) are related with μ_i and α^2. Therefore, for a given α^2 and a frequency order number i and μ_i can be determined by solving the above functions. That is to say μ_i is the function which contains frequency order i and α^2. For two typical kinds of boundary conditions, the range of μ_i is obtained by the reference (Qin, 2014 & Liao, 2013), as follow:

$$0 \le \mu_i \le 1/4 \tag{15}$$

$$0 \le \mu_i \le 1/2 \tag{16}$$

And then the effect of the α^2 on μ_i is analyzed in the low order frequency range. If the contribution of the tension to the vibration effect is far less than the bending stiffness under two typical kinds of boundary conditions, the suspender is regarded as Euler–Bernoulli beam, and the value of μ_i approaches to the upper bound value when $i = 1$. The need to point out is the upper bound value of (16) is actually 0.50561 (Fang, 2010), when the frequency order of Euler–Bernoulli beam equals to 1 under the clamped-clamped boundary condition, however, for the actual suspender, it is receivable that the upper bound value takes 1/2.

$$\lim_{\alpha^2 \to 0} \mu_i = 1/4 \tag{17}$$

$$\lim_{\alpha^2 \to 0} \mu_i = 1/2 \tag{18}$$

Under the low order frequency, if the contribution of the bending stiffness to the vibration effect is far less than the tension under two typical kinds

of boundary conditions, the suspender is regarded as string, and then μ_i approaches to the lower bound value, that is:

$$\lim_{\alpha^2 \to \infty} \mu_i = 0 \tag{19}$$

Further, μ_i is not only related to α^2, but also the frequency order i. With the increase of the frequency order i, the effect of the bending stiffness on the vibration becomes more and more obvious, and the suspender is also regarded as Euler–Bernoulli beam, μ_i approaches to 1/4 and 1/2 for two typical kinds of boundary conditions, as follows:

$$\lim_{i \to \infty} \mu_i = 1/4 \tag{20}$$

$$\lim_{i \to \infty} \mu_i = 1/2 \tag{21}$$

And expression (5)'s identity is transformed as follows:

$$\lambda_1^4 + \alpha^2\lambda_1^2 = k^4 \tag{22}$$

The expression (5), which contains the circular frequency ω_i, is expressed as the natural frequency f_i (Hz), as follows:

$$k^4 = 4\pi^2 f_i^2 \rho Al^4/EI \tag{23}$$

And then a new non-dimensional parameter is got from (23), that is:

$$\eta_i = 4f_i^2 \rho Al^4/EI \tag{24}$$

The non-dimensional parameter η_i is got through the natural frequencies of the suspenders that are non-dimensionally handled by the natural frequencies of the Euler–Bernoulli beam under the hinged-hinged boundary condition. But the normalized parameters are the natural frequencies of the string in the reference [6].

Substitute (14) and (23) into (22), the following functions with the new nondimensional parameter η_i is obtained:

$$(i + \mu_i)^4 \pi^2 + (i + \mu_i)^2 \alpha^2 = \eta_i \tag{25}$$

Based on $\alpha^2 = Tl^2/EI$ and (24), (25) is expressed as the formulas which contain the tension of the suspender:

$$T = \frac{4f_i^2 \rho Al^2}{(i + \mu_i)^2} - \frac{EI}{l^2}(i + \mu_i)^2 \pi^2 \tag{26}$$

515

For the clamped-hinged boundary condition, $\mu(\eta_i) \leq 1/4$, the clamped-clamped boundary condition is $\mu(\eta_i) \leq 1/2$.

Infact (16) also holds for the hinged-hinged boundary condition when $\mu(\eta_i) = 0$, further, the classical tension formulas are naturally obtained:

$$T_{JJ} = \frac{4 f_i^2 \rho A l^2}{i^2} - \frac{EI}{l^2} i^2 \pi^2 \tag{27}$$

Therefore, (26) is the united tension formula of the suspender under two typical kinds of boundary conditions.

4 DETAILED EXPRESS OF μ_i FOR TWO TYPICAL KINDS OF BOUNDARY CONDITIONS

The transcendental function (12) and (13), which contains the hyperbolic function and trigonometric function, are solved by the nonlinear iteration, and then substituted μ_i into (25), η_i is the detailed expression, and the concrete operations are similar to the paper [6], as follows:

1. Given are a series of α^2, where μ_i is solved by the nonlinear iteration, such as Newton-Raphson method, based on the transcendental function (12) and (13).
2. Substitute μ_i into the function (25), η_i is obtained.
3. A series of data (η_i, μ_i) are produced by the nonlinear iteration method.
4. The piecewise function is used to fit (η_i, μ_i), the solutions of the (12) and (13) are explicitly expressed by the piecewise function.

Using the above method, η_i is selected as the variable or the segmented point which can avoid the repeated trial and error in the tension estimation, which takes precedence over the tension as the segmented point of [6] in the process of the tension estimation; further, the tension identification efficiency is improved.

Figure 1 shows that the new non-dimensional parameter η_1, varies with μ_1 when the frequency order i equals to 1, and the black line and the blue line, respectively, represent the clamped-clamped and clamped-hinged boundary conditions, after the same. It is concluded that the range of μ_1 is [0, 0.50561] for the clamped-clamped boundary conditions, and [0, 1/4] for the clamped-hinged boundary conditions, which further verifies (15) to (21).

In order to explicitly express the relation between η_1 and μ_1, it is important to choose the appropriate segmented point and the fitting function, such as

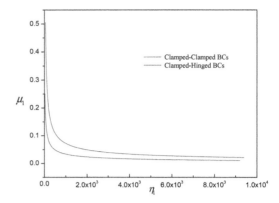

Figure 1. The relationship between η_1 and μ_1. Under two typical kinds of boundary conditions ($i = 1$).

Table 1. The relationship between η_1 and μ_1. Under the clamped-hinged boundary condition ($i = 1$).

μ_1	η_1
$1.34e-1 + 3.10e-2 \times \exp(-\eta_1/25.32)$ $- 3.27e-3 \times \exp(-\eta_1/11.77)$	[24.087, 4.00e1)
$6.43e-2 + 1.86e-1 \times \exp(-\eta_1/21.30)$ $+ 1.62e-1 \times \exp(-\eta_1/89.90)$	[4.00e1, 1.40e2)
$1.62e-2 + 1.02e-1 \times \exp(-\eta_1/148.25)$ $+ 4.66e-2 \times \exp(-\eta_1/1029.81)$	[1.40e2, 3.50e3)
$3.02e-3 + 1.52e-2 \times \exp(-\eta_1/4557.168)$ $+ 7.61e-3 \times \exp(-\eta_1/30478.64)$	[3.50e3, 9.00e4)
0.00	[9.00e4, ∞)

Table 2. The relationship between η_1 and μ_1. Under the clamped-clamped boundary condition ($i = 1$).

μ_1	η_1
$2.15e-1 + 8.3453\,e-1 \times \exp(-\eta_1/48.07075)$	[50.72, 1.10e2)
$1.4489e-1 + 5.7276\,e-1 \times \exp(-\eta_1/82.92)$	[1.10e2, 2.00e2)
$8.379e-2 + 2.9622e-1 \times \exp(-\eta_1/201.33)$	[2.00e2, 6.00e2)
$1.945e-2 + 1.1988e-1 \times \exp(-\eta_1/474.54)$ $+ 5.309e-2 \times \exp(-\eta_1/3077.71)$	[6.00e2, 1.00e4)
$5.47e-3 + 2.165e-2 \times \exp(-\eta_1/7109.58)$ $+ 1.289e-2 \times \exp(-qq/39417.23)$	[1.00e4, 1.00e5)
0.00	[1.00e5, ∞)

the polynomial function, the exponential function, the power function and so on, and then the least square method is used to come closer to the fitting error to approach the minimum as soon as possible.

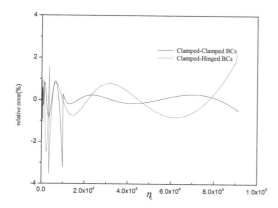

Figure 2. The relative error between the formula and iterative method under two typical kinds of boundary conditions ($i = 1$).

In the Figure 1, with the increase of η_1, the value of μ_1 shows the declining tendency. First, η_1 rapidly falls, and then the change trend of μ_1 becomes steady and slow; therefore, it is expressed by the piecewise exponential function. In this paper, two fitting formulas are, respectively, given in the Table 1 and Table 2 for two typical boundary conditions.

The fitting formulas are easy to be obtained, similar to the Table 1 and Table 2 for two typical boundary conditions when the frequency order i is equal to or greater than 2.

Figure 2 shows that the comparison between the fitting formulas and the direct solution of iterative method for the two typical boundary conditions by the relative error method. With the increase of η_1, both of the relative errors for the two typical boundary conditions are lower than ±5%, which shows that the precision of the fitting formulas meets the engineering requirements.

5 CONCLUSION

In this paper, a new non-dimensional parameter η is introduced by the natural frequency of the Euler–Bernoulli beam under the hinged-hinged boundary conditions, and the approximate tension formulas of suspenders are deduced based on the nonlinear iteration method and the piecewise fitting function, further the united tension formulas of the suspenders are constructed for the two typical kinds of boundary conditions. Finally, the precision of the fitting formulas are studied by comparing the fitting formulas with the direct solution of iterative method for the two typical boundary conditions, the numerical study shows that both of the relative error for the two typical boundary conditions are lower than 5%. The above fitting formulas are used to identify the tension of the suspender without repeating trial and error, suitable for the practical operation and improve the precision and efficiency, which provides a fast identification method for the suspender tension.

ACKNOWLEDGEMENTS

This paper is supported by Chongqing Training Program for the Science and Technology Talents (Grant No. cstc2014kjrc-qnrc30003) and West China Communications Construction Science & Technology project of the PRC (Grant No. 2013 319 740 080 and Grant No. 2013 364 740 600). We gratefully appreciate them.

REFERENCES

Fang Tong, Xue Pu. Theory of vibration and application [M]. Xi'an: Northwestern Polytechnical University Press, 2010.

Fang Zhi, Wang Jian-qun, Yan Jiang-ping. The tension measurement of cables and suspenders with frequency method [J]. Journal of vibration and shock, 2007, 26(9):78–82.

He Wei, Chen Huai, Wang Bo. Study of suspender tension measurement based on frequency method with complex boundary conditions [J]. China civil engineering journal, 2012, 45(3):93–98.

Hiroshi Zui, Torhru Shinke, Yoshio Namita. Practical formulas for estimation cable tension by vibration method [J]. ASCE Journal of structural engineering, 1996, 122(6):651–656.

Hu Hai-chang. Energy method and local effects [J]. Chinese Journal of theoretical and applied mechanics, 1979, 4:353–359.

Hu Hai-chang. Natural Vibration Theory of Multi-degree Freedom Structure [M]. Science Press, Beijing, China, 1987.

Li Guo-qiang, Gu Ming, Sun Li-ming. Theory of Cable Vibration, Dynamic, Detection and Vibration Control [M]. Science Press, Beijing, China, 2014.

Li Guo-qiang, Wei Jin-bao, Zhang Kai-ying. Theoretical and experimental study on cable tension estimation by vibration method accounting for rotational end restraints [J]. Journal of Building Structures, 2009, 30(5):220–226.

Liao Jing-bo, Tang Guang-wu, Meng Li-bo, et al. A system for cable wireless automation measurement and identification of bridge suspender [P]. China 201410659522.2, 2015.

Liao Jing-bo, Tang Guang-wu, Meng Li-bo, Zhen Gang. An approximate frequency formula of clamped cable [J]. Journal of Vibration and Shock, 2013, 32(6): 149–151.

Liao Wei-yang. Vibration-based test validation and program design of tension estimation and parameter identification of bridge short cables [D]. Chongqing: Chongqing Jiaotong University, 2007.

517

Liu Zhi-jun, Rui Xiao-ting. Research on tension measurement of cables and suspenders by vibration method [C]//. The 10th national vibration theory and application. Nan Jing, 2011. 753–759.

Qin Xiao-ping, Liao Jing-bo, Meng Li-bo. An approximate frequency formula of clamped-hinged cable and its solving condition [C]// China civil engineering society. Beijing: People's Communication Press, 2014. 969–973.

Tang Xiao-bing, Yuan Zhen-hua, Zhang Jian-chao, DI Hong-yan. Genetic algorithm applied in tension measurement of uniform sectional boom fixed at both ends [J]. Journal of highway and transportation research and development, 2011, 28(8):79–84, 94.

Zhang Wei, Wang Guang-zhen, Sun Yong-ming. Analysis on Frequency and Force of Stay Cable Considering Flexural Rigidity [J]. Journal of highway and transportation research and development, 2012, 29(7):64–69, 75.

Zhi Fang, Wang Jian-qun. Practical formula for cable tension estimation by vibration method [J]. ASCE Journal of bridge engineering, 2012, 17(1): 161–164.

Advances in Energy, Environment and Materials Science – Wang & Zhao (Eds)
© 2016 Taylor & Francis Group, London, ISBN 978-1-138-02931-6

Application of Probabilistic Fracture Mechanics in fatigue life evaluation of crane beam

Gang Xue & Yutong Meng

The school of Architecture and Civil Engineering, Baotou, Inner Mongolia Autonomous Region, China

ABSTRACT: In order to improve the precision of the evaluation on life of crane beam, the probability statistics are introduced in evaluation. The principle of Probabilistic Fracture Mechanics (PFM) is introduced, and the calculation process of PFM is discussed. The corresponding formula that is derived from this is the relationship between life and reliability, which is given. Meanwhile, the fatigue life of the in-service crane beam is evaluated using PFM. The residual fatigue life of the crane beam under the alternating load is thus obtained.

1 INTRODUCTION

Nominal stress method and local stress and strain method are contained by traditional fatigue life evaluation method. The fatigue life of the members is then calculated by using Miner's rule and the SN curve of the members. Varying degrees of defects or limitations are located in these traditional fatigue life evaluation methods, and the fatigue life evaluation method based on fracture mechanics has been turned to be a mature, accurate method in recent years. It can be introduced to evaluate the remaining fatigue life of steel crane girders in service. But the parameters C and m in the Paris formula are considered constant in the current theory of fracture mechanics, in fact C and m are random variables, and are subjected to a statistical distribution. Among the structural members, crane beam environment of metallurgy factory building is extremely complex. In addition to the alternating load of the crane, the external force is caused by near mechanical equipment, initial imperfection in the member, the level of development of lots of invisible crack during its service, and the uncertainty of material properties, et al. These random parameters are shown a certain probability distribution situation. The accuracy of the evaluation (Wang, 2006) will be improved if the probability statistics is put into evaluating the remaining fatigue life of steel crane girders in service.

2 MECHANISM AND METHODS OF PROBABILISTIC FRACTURE MECHANICS

2.1 The formula of crack propagation life

The structure cannot be fractured immediately after the crack appears, but the crack will continue to expand until it reaches the critical instability value, this phase of experience is called remaining fatigue life. The relationship among da/dN, a, $\Delta\sigma$, and material properties have been established by expressing the crack's growth rate. Many kinds of expressions are contained by it, but among them, the simplest and most widely used is the formula Pairs. It's expressed as follows:

$$\frac{da}{dN} = C(\Delta K)^m \tag{1}$$

In the formula:
C, m—the material constant (temperature, environment, and frequency) about the test condition;
ΔK—the stress intensity factor range.

The crack of crane beam is caused by fatigue failure, it is perpendicular to the crack plane, and it is I crack, so it is the open crack. Stress intensity factor $K = \alpha\sigma\sqrt{\pi a}$, stress intensity factor ranges as follow:

$$\Delta K = \alpha\Delta\sigma\sqrt{\pi a} \tag{2}$$

In the formula:
α—form factor coefficients, crack size, the location;
σ—the nominal stress is located in crack location by calculating in no crack;
a—crack size.

The formula of crack propagation life is obtained by putting formula (2) into formula (1).

$$N_c = \frac{2}{C(m-2)(\alpha\Delta\sigma\sqrt{\pi})^m}\left(a_0^{1-\frac{m}{2}} - a_c^{1-\frac{m}{2}}\right) \tag{3}$$

In the formula:
N_c—crack propagation life;

α_0—initial crack length;

α_c—length of crack after extend.

In order to simplify the process of actual projects, m and α will be taken as constant, so the number of independent variable in the N_c is four.

2.2 The relationship between fatigue life and reliability

In the actual project, the reliability is connected with the fatigue life; the residual life in reliability is obtained by taking the logarithm on the both side of formula (3) to calculate:

$$\lg N_c = \lg \frac{2}{(m-2)(\alpha\sqrt{\pi})^m} - \lg C - m\lg\Delta\sigma$$
$$+ \lg\left(a_0^{1-\frac{m}{2}} - a_c^{1-\frac{m}{2}}\right). \tag{4}$$

Set : $B = \dfrac{2}{(m-2)(\alpha\sqrt{\pi})^m}; \ a = a_0^{1-\frac{m}{2}} - a_c^{1-\frac{m}{2}}.$

The formula (5) turns into $lgN_c = lgB - lgC - mlg\Delta\sigma - lga$.

Where C, $\Delta\sigma$, and a are the random variables. If C, $\Delta\sigma$, and a follows lognormal distribution, life N_c will follow lognormal distribution. But only C follows a lognormal distribution and compares it with other parameters, where C plays a leading position. In order to make life and reliability together, set $\Delta\sigma$ and a as distributed normally, so N_c will follow a lognormal distribution.

If N_c is distributed normally, $\lg N_c$ will be distributed normally:

$$\lg N_c \sim N\left(\mu_{\lg N_c}, \sigma^2_{\lg N_c}\right)$$

Mean of $\lg N_c$: $\mu_{\lg N_c} = \mu_{\lg B} - \mu_{\lg C} - m\mu_{\lg\Delta\sigma} + \mu_{\lg a}$

Variance : $\sigma^2_{\lg N_c} = \sigma^2_{\lg C} + m^2\sigma^2_{\lg\Delta\sigma} + \sigma^2_{\lg a}$

Reliability:

$$P_r = P(N_c > \overline{N}) = P(\lg N_c > \lg\overline{N})$$
$$= P\left(\frac{\lg N_c - \mu_{\lg N_c}}{\sigma_{\lg N_c}} > \frac{\lg\overline{N} - \mu_{\lg N_c}}{\sigma_{\lg N_c}}\right)$$

A standard normal distribution up quantile μ_{p_r} is introduced, $\mu_{p_r} = \frac{\lg N_c - \mu_{\lg N_c}}{\sigma_{\lg N_c}}$, reliability P_r is obtained by checking the standard normal distribution table through μ_{p_r}.

$$\lg N_c = \mu_{\lg N_c} + \mu_{p_r}\sigma_{\lg N_c} \tag{5}$$

$$N_c = 10^{\mu_{\lg N_c} + \mu_{p_r}\sigma_{\lg N_c}} \tag{6}$$

The quantile μ_{pr} can be obtained when the Fatigue life N_c is known, and reliability P_r is obtained by checking table or calculating. The quantile μ_{pr} can be obtained by inverse calculation when reliability P_r is known. Through the formula (6), the fatigue life can be known in this reliability.

3 THE FINITE ELEMENT ANALYSIS

3.1 Selection of the element type and model

Eight nodes (I, J, K, L, M, N, O, P) are contained by solid185, three direction (X, Y, Z) and three displacements of space are contained by it, so it is a 3-d solid element. The trait of plastic, creep, expansion, stress hardening, large deformation, and large strain are included in this element type. In addition, the solid185 based on solid45 has been optimized, so the characteristic of small calculation scale and high calculation accuracy are contained by it. From what has been discussed above, the solid185 is selected, and it is used in building model of crane beam.

The influence of the upper part of the crane on the upper part of crane beam is studied during the process of the reciprocating motion in this paper. The material of crane beam is Q345 steel, so the density of steel $\rho = 7800\text{kg/m}^3$, elastic modulus E = $2.1\times10^{11}\text{N/m}^2$, poisson ratio $v = 0.3$. The thickness of stiffener is $t_s = 12\text{mm}$. The thickness of end-stiffeners is 30mm, width is 300mm; according to the construction requirement of variable beam, the right angle mutation bearing is designed. The thickness of inserted plate is 50mm, width is 840mm, and length is 1800mm. The length of insertion is 800mm, the thickness of end-plate is 50mm, width is 8400, and height is 1000mm. The structure shown in Figure 1.

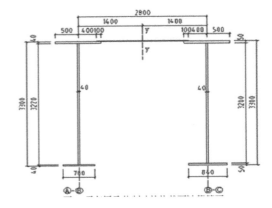

Figure 1. Diagram of structure.

520

3.2 The simulation analysis of dynamic load

In fact, the crane beam of metallurgy workshop is impacted by alternating dynamic load, so in order to simulate a real forced state of crane beam, load should be divided into different sub-step and applied to the corresponding position of crane beam. It is similar to using the promise that the remaining fatigue life of steel crane beam is calculated by fatigue module. In the specification stipulated, the normal load value is selected and the load of crane isn't multiplied by dynamic coefficient. So the load is 766kN, and the role of a single crane is bared by crane beam.

After the dynamic load is applied, the POST26 is dealt with after entering the time history, so the change curve of nodes that can be observed is obtained during the dynamic load moving, the deflection change curve of 84144 node is shown as Figure 2.

3.3 The simulation analysis of fatigue

The powerful module of fatigue is contained by ANSYS, the computation basis of static stress is contained by it, and the parameters of fatigue are calculated; the load and stress cycle number is ensured, and the stress of this location is reserved; the stress concentration factor of target position is ensured, and the parameters of every stress cycle is defined, so ANSYS is used to evaluate the remaining fatigue life and to calculated damage. As shown as in Figure 3, the position of the most deflected region is located in time near the load in 10.3s; the maximum deformation position is similar to standard, and the fatigue crack caused by this position; the node of analysis of fatigue can be obtained after the position of maximum deformation and maximum stress are magnified, as shown as in Figure 3.

As shown as in Figure 9, the 84144 node is located between the web and the flange, the most

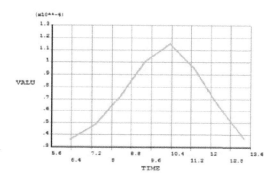

Figure 2. The deflection change curve of 84144 node.

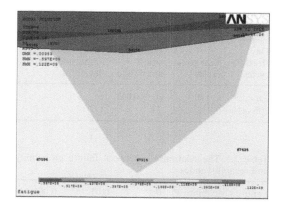

Figure 3. The distribution map of nodes.

stress seen is in this position, which is the fouling position of fatigue damage. So this node is defined as a reference point of fatigue analysis.

4 THE FATIGUE LIFE EVALUATION OF STEEL CRANE GIRDERS IN SERVICE

The objective crane beam belongs to an iron company's factory; the crane beam is broken seriously in the test, so the evaluation is done. The material of the crane beam is Q345 steel. The fatigue property of Q345 steels is similar to 16Mn steels (Yang, 2006), so that the fatigue property of 16Mn steels is used in the test. The statistical result of logarithm of crack growth parameter C is listed by related documents, $\lg C \sim N (-11.8117, 0.256662)$.

The dynamic strain value that is transferred by strain gauge of the lower flange at the maximum bending moment of the crane beam is recorded by parameter acquisition instrument, and the equivalent stress range is obtained after the dynamic strain value is multiplied by elastic modulus. The operating frequency of the crane and deadweight are not constant every time as observed by the field, so the test is done for consecutive number of days. Limited by long time test, the stress amplitude of crane and the measured results of the running crane is statistically analyzed in paper (Xing, 2002), and the stress cycle number and the stress amplitudes are obtained as shown in Table 1.

Normal distribution is set into $Y \sim N (\mu, \sigma^2)$. Set $Y = \lg X$, so X is shown as a normal distribution. Conversion relation between statistical parameters as follows:

$$\mu_{\lg X} = \lg \frac{\mu_X}{\sqrt{1 + \delta^2_X}} \tag{8}$$

521

Table 1. The measured results of dynamic stress in the maximum bending moment.

Day stress cycles	Year stress cycles	Equivalent stress amplitude	Equivalent stress amplitude standard deviation
Σn_i	$n_Y (10^5)$	$\mu_{\Delta\sigma}$ (MPa)	$\sigma_{\Delta\sigma}$
206~418	0.752~1.53	73	1.77

Table 2. The relationship between failure probability and reliability index.

Failure probability P_f	0.1	0.01	0.001
Reliability index β	1.28	2.33	3.09
Failure probability P_f	1.0×10^{-4}	3.2×10^{-5}	1.0×10^{-6}
Reliability index β	3.72	4.00	4.75

$$\sigma^2_{\lg X} = \frac{\lg\left(1 + \delta^2 X\right)}{2.303} \tag{9}$$

Y is a coefficient of variation of the variable X in the formula (Chen, 2005). It's the ratio between the standard deviation and the mean. As shown in the date, the initial crack size is between 0.05mm~0.25mm. Setting $\mu_{a_0} = 0.20$mm and coefficient of variation as $\delta_{a_0} = 0.4$.

According to the formula (8) and formula (6), $\mu_{\lg a_0} = -0.7312$, $\sigma^2_{\lg a_0} = 0.0280$, so $\lg a_0 \sim N$ (−0.7312, 0.0280).

According to the formula (4) and formula (5), the statistical parameter $\lg N_c$ is as follow:

$$\mu_{\lg N_c} = \lg\frac{2}{\pi^{3/2}} - \mu_{\lg C} - 3\mu_{\lg\Delta\sigma} + \frac{1}{2}\mu_{\lg a_0} \tag{10}$$

$$\sigma^2_{\lg N_c} = \sigma^2_{\lg C} + 9\sigma^2_{\lg\Delta\sigma} + \frac{1}{4}\sigma^2_{\lg a_0} \tag{11}$$

$\lg\Delta\sigma, \lg a_0$, and $\lg C$ are put into the formula (10) and formula (11), so $\mu_{\lg N_c} = 6.1436$, $\sigma^2_{\lg N_c} = 0.0739$.

According to the formula (7), N_c can be obtained by reliability P_r. The relationship between failure probability and reliability index (Xin, 2004) is as shown in Table 2.

As shown as Table 2, the first three results of reliability index have engineering values, so the result in the maximum number of stress cycles is as follows:

When failure probability $P_f = 0.001$, so $P_r = 0.999$, and the up quantile is $\mu_{p_r} = -\beta = -3.09$,

$$\lg N_c = \mu_{\lg N_c} + \mu_{P_r}\sigma_{\lg N_c} = 6.1436 - 3.09$$
$$\times \sqrt{0.0739} = 5.3036$$

So $N_c = 10^{5.3036}$, $N_c = \frac{10^{5.3036}}{1.53 \times 10^5} = 1.31$ years in the maximum number of stress cycles.

Similarly available:
When $P_f = 0.01$, $N_c = 2.11$ years;
When $P_f = 0.1$, $N_c = 4.08$ years.
So according to the result, the remaining fatigue life of steel crane girders in service is between 1.31~4.08 years.

5 CONCLUSION

The crane girder is affected by many uncertain factors and bearing dynamic loads, so the force is extremely complex, the fatigue life evaluation method that is based on fracture mechanics is put forward in this paper. The uncertainty parameter can be seen as the random variable in this method, the impact of uncertainties on the structural integrity assessment is reduced, and high practical engineering value is contained in this way.

The simulation of dynamic load and calculations are done by the finite element analysis software ANSYS, and the fatigue life and damage are calculated. As shown as in the result of calculation, the most stress amplitude of the crane beam, the fatigue life and the level of damage are changed as the position of stress. The most stress amplitude, the smallest fatigue life, and the most level of damage are located between the web and the flange. But when it's far away the position that is located between the web and the flange, the stress amplitude and the level of damage is decreased, whereas the fatigue life is increased. The initial imperfection of material is located in the structure, for instance, the micro-crack, the void, the impurity, and the low strong section, et al. The position of the initial imperfection is influenced by the alternating load, and gets damaged by stress concentration. The level of damage is increased as the stress amplitude increases. The position is located between the web and the flange, as it is one of the weakest areas of the upper part of the crane beam, and the stress concentration is the main cause of the damage that is caused by the fatigue crack that appeared early.

As shown as the example of the life evaluation of steel crane girders, the fatigue life evaluation method that is based on fracture mechanics accords with actual boundary condition of the crane beam. It can be applied to evaluate the remaining fatigue life of steel crane girders in service. The remaining fatigue life of the crane beam can be obtained in this way; this method has high analysis precision, and as reference has been improved, this approach has a broader applicability.

ACKNOWLEDGMENT

The author thanks the financial support of Inner Mongolia Science Foundation, grant 2015MS0552.

REFERENCES

Chen Chuan-rao. Fatigue and Fracture [M]. Wuhan: Huazhong University of Science & Technology Press, 2005.

Wang Chun-sheng, Chen Ai-rong, Nie Jian-guo, et al. Probabilistic Assessment of Remaining Fatigue Life for Steel Bridge [A]. Computational Mechanics, 2006, 23(4): 408–413.

Xin Kun-tao, Yue Qing-rui, Liu Hong-bing. Fatigue Dynamic Reliability Analysis for Steel Crane Girders [J]. China Civil Engineering Journal, 2004, 37(08): 38–40.

Xing Kun-tao. Study on Fatigue Reliability and Safety Control of Steel Structure Crane in Service [D]. Dalian: Dalian University Of Technology, 2002.

Yang Xiao-hong. Remaining Fatigue Life Assessment of Steel Bridges Based on Reliability Theory [D]. Zhenjiang: Jiangsu University, 2006.

Advances in Energy, Environment and Materials Science – Wang & Zhao (Eds)
© 2016 Taylor & Francis Group, London, ISBN 978-1-138-02931-6

Failure analysis of large, low head water pump units

Baoyun Qiu, Jinyu Cao & Yizhou Yang
School of Hydraulic, Energy and Power Engineering, Yangzhou University, Yangzhou, China

ABSTRACT: In order to improve reliability and durability of large, low head water pump units, investigation of many typical pumping stations was performed, and failure causes of critical components of the pump units were analyzed. The results show that: poor material properties, large sediment concentration of the lubricating water and poor installation quality of the pump units will aggravate wear of the water-lubricating metalloid guide bearing; Running at off-design condition of the pumps and the sediment of the pumped water will speed up cavitation and erosion of the impeller blades; Hot operation, operating voltage, mechanical effect of startup and shutdown and the moisture and dirt in the environment are the main reasons which result in insulation failure of electric machines mating the pumps. The results of the paper are significant for design of reliability, durability and maintenance management of the large water pump units.

1 INTRODUCTION

Large, low head water pump units in China are mainly large vertical axial-flow pump units and vertical guide vane mixed-flow pump units. Due to the large volume, large numbers of assembly units and complexity of the structure, there are many factors affecting reliability with complex mechanisms. Research and analysis of the failure reasons of the critical components of the pump units are conducive to improve the reliability and durability of the units by taking measures from the aspects of design and manufacturing, operation management.

Through investigation and survey of many large pumping stations, the main failure modes of large low-lift water pump units are as follows: wear of water-lubricating guiding bearing of the pump, water seal failure of oil-lubricated guiding bearing of the pump, impeller cavitation and fracture of blades of the pump (Cao (2008) & Wang (1998)), insulation failure of the motor stator (Cao (2008)), thrust bearing burnout of the electric machine (Qiu (2000)) etc.

2 ANALYSIS OF PUMP FAILURE

2.1 Wear of water-lubricating guiding bearing

Water-lubricating metalloid guide bearing has found its wide application in large water pumps because of its simple structure, low cost, and simple and convenient maintenance (Qiu (2005)). Metalloid guiding bearing of large pumps mainly consists of polyurethane guide bearing, rubber guide bearing, etc.

Due to small load for vertical unit pumps, water lubrication can be used in the metalloid bearing.

If the river water is of good quality with less sediment, the water can be used directly for the bearing lubrication, otherwise, additional pressure clear water will be required for water lubrication.

Due to the low carrying capacity and poor wear-resisting performance of metalloid guide bearing, and poor lubrication performance using water as lubrication medium, generally, the polyurethane guide bearing of the large vertical pump will have a wear of 0.5–1.5 mm through 20000 h operation.

If sediment charge of lubricating water is large, the material of the guide bearing and the shaft neck will wear quickly. Excessive wear of the guide bearing will cause large throw and intensive vibration of the impeller and the pump shaft, even accident of blades impacting the shell in severe case. It is the main failure form of metalloid guide bearing that the wear of the bearing and journal exceeds the limit value.

The wear reasons of water lubricated metalloid guide bearing are:

1. The poor wear resistance and durability of the bearing material. Abrasion resistance and durability of the bearing materials are related to the composition and proportion of the material and processing quality, and even guide bearing materials of different batches have considerably differences in wear-resistant and durable performances.

2. The poor quality of lubricating water. Water pump guide bearing is lubricated directly by river water, when the sediment concentration is high, sediment will enter into bearing clearance, which will accelerate the wear of bearing materials and shaft neck, shorten the service life of the bearing. Taking a pumping station as

an example in Jiangsu province installed large vertical direct drive axial-flow pump units with impeller diameter of 2.0 m, the pump initially used the oil lubricated Babbitt guide bearing. Since the water seal is easy to fail and the leakage water will flood and damage the Babbitt guide bearing, bakelite guide bearing of P23 is used instead, with additional pressure clear water used as lubricating medium. As a result, the application in guide bearings of two water pumps is successful, the pumps had been running 50000 h, and the wear is only about 1 mm. But the other pump guide bearing which was directly lubricated by river water has a service life of only 100–200 h, and then was changed back to oil lubricated Babbitt bearing.

3. The heavy bearing load. The same polyurethane guide bearing is respectively used for vertical pumps, inclined and horizontal pumps. Since the guide bearing of vertical pump only bears unbalanced force of rotational parts, the bearing has a small load and a long running life, the running time generally could reach 10000–20000 h. For inclined, horizontal pumps, guide bearing bears the weight of the impeller, the pump shaft and other rotating components, the bearing has a load of about 10 times or larger of the guide bearing load of same size vertical pump, the excessive wear of the guide bearing and even the blade impacting shell will happen only after running 100–250 h.

4. The poor quality of installation. Large concentric deviation of the pump guide bearing nest as a benchmark of the pump unit centre and oversize throw of the pump shaft will cause pump shaft pressure on the bearing. The unbalanced magnetic pull because of uneven air gap between stator and rotor of the motor and non-axisymmetric incoming flow of the impeller will increase the guide bearing load; the vibration of rotating parts will be caused by unequal installation angles of blades and other factors, which is not conducive to formation of lubricating liquid membrane of the pump guide bearing. All those problems of installation quality, will affect the service life of the guide bearing.

2.2 The soaking of oil lubricated guide bearing

As known in Figure 1, the oil lubricated guide bearing of the large vertical water pump is located in the hub of guide blade, a water seal is equipped between the hub of guide blade and the hub of wheels, and the upside of the guide blade is equipped with the protection pipe, to prevent water from entering the bearing. Leakage water of seals entering from the bottom passes though the drain pipe embedded in guide vane and is discharged.

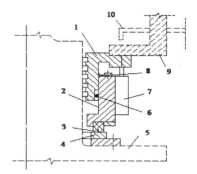

1-Stationary seat 2-Stationary ring seat 3-Stationary ring 4-Rotating seal ring 5-Impeller hub 6-O-ring seal 7-Spring and spring box 8-Sealing rubber and press plate 9-Guide blade hub 10-Drain pipe

Figure 1.   The structure of mechanical end face seal.

The oil lubricated Babbit guide bearing has great bearing capacity, good abrasion resistance, good durability, without aging problem. The main fault and failure form is that leakage of large quantities of water because of the failure of water seal device causes soaking of oil lubricated guide bearing because leakage water can not be drained in time.

1. The main reasons for the leakage as the failure of end-face water seal devices of oil lubricated guide bearing are as follows:

   a. The stationary ring seat can't pressure in the rotating seal ring by its own gravity and spring pressure after lifting the unit and vibration because the sealing rubber rope between the end-face seal of stationary ring seat and fixed seat is assembled too tight.

   b. The uneven contact surfaces of rotating ring and stationary ring can not mesh well. The leakage is usually large at initial period of operation, and gets much smaller after a period of operation time (Lin & Qiu (2006)).

   c. With the increase of sealing surface wear of the end-face sealing device, the spring pressure decreases, the pressure in the sealing surface decreases, and seal leakage water increases when running. Seal leakage water will increase rapidly when the wear of the end face seal exceeds the limit value.

2. The main cause that leakage water of end face seal device can not be drained off is that the deposition of sediment concentration of the water blocks the drainage pipes (Huang (2012)).

2.3 Failure of pump cavitation

Normally, the pump impeller blade and shell made of cast steel or cast iron material will inevitably produce different degree of cavitation after running for a period of time. Cavitation pits are formed in blade surface or even the parts of the blades drop out, the hydraulic performances of the pump deteriorate; the clearance between outer edge of blades and shell increases, leakage loss increases, which will lead to sharply decline of pump efficiency.

It is commonly believed that the water pump cavitation will occur when the *NPSH*a of the pump device in pumping station is smaller than the NPSHr, through which we can confirm the suction height of the pump installation while design the pumping station:

$$H_s = \frac{p_a}{\rho g} - \frac{p_v}{\rho g} - h_s - \Delta h_r \qquad (1)$$

where ρ for water density (kg/m³); g for gravity acceleration (m/s²); p_a for atmospheric pressure (Pa); p_v for vaporization pressure of water at that temperature (Pa); h_s for head loss of the pump suction pipe (m); Δh_r for *NPSH*r of the pump (m). But for many practical pumping stations determined the pump installation height according to Eq. (1), even the water pump being installed much lower, serious cavitation will also occur in a short period of 3–4 years.

Usually, cavitation performance of pumps is got through performance test of water pump in laboratory using clear water, and the cavitation mechanism of actual pump stations is much more complicated.

1. The root reason of water pump cavitation is that cavitation bubbles, which generate due to the drops of flow pressure, collapse and perish in the surface of flow passage components. The smaller the flow surface pressure is, the bigger cavitation intensity is. Cavitation performance and inlet flow pattern of water pump have an effect on the situation of pump cavitation.
2. When the water pump operate in off-design operation condition, the water flow will have a large angle with the blade tangent direction, and flow impact will occur at inlet of the blade, which will produce eddies or flow separation on the back side of the blade (small flow rate condition) or the front side (large flow rate condition) and thus cause larger pressure drop.
3. Natural water drawn from rivers and reservoirs by prototype pump stations is different from the water used in laboratory experiment, the former contains plenty of humus, microorganisms, which can increase initial cavitation pressure of water and thus cavitation will be more serious.
4. Natural water contains large amount of sediment and other solid particles, which pass through the surfaces of flow passage components, wear the surfaces, cause surfaces rough and uneven,

worsen flow regime near the surfaces, promote and strengthen water cavitation; abrasion and cavitation promotes each other.

The cavitation speed of water pumps of the third station is more than three or four times of the one of the fourth station, the reason is that water pumps of the third station had poor cavitation performance, what's more, the poor inflow in inlet passage worsened the water pump cavitation performance. The pump impeller had serious blade surface cavitation and clearance cavitation obviously which affected the energy performance of the water pump before periodic overhaul of the third station and the cavitation repair welding must be carried out when overhaul. The pump cavitation status of the fourth station is much better, generally, after two cycles of operation, only simple cavitation repair welding was required for the impeller blade.

2.4 The fracture of pump blade

Through investigation on pumping stations, it was found that impeller blades of many large and medium-sized axial-flow pumps and guide vane mixed-flow pumps had crack and even fracture phenomenon, mostly occurred in blade root.

A pump station originally used cast steel impeller and impeller shell, whose cavitation is serious. After transformation and renovation, the impeller and the shell were made of stainless steel. There is no cavitation occurred for operation of 15–17 years, which avoid cavitation problem, but in impeller blade root of a few pumps the cracks were found, even the blade fracture for some pumps.

After blade fracture, not only the pump hydraulic performance deteriorates seriously, but also it may cause damage of other parts of the unit.

Blade fracture is mainly due to the fatigue failure caused by exciting forces and resonance when the water pump is in operation. Due to stress concentration, especially the manufacturing quality problems (if any casting sand holes or pores) of blade root, firstly crack generates, and then crack propagates, and finally fractures.

Main reasons of pump blade crack and fracture are as follows:

1. The stress concentration. Crack, fracture firstly occurs on the edge of the blade root where there is stress concentration. Lack of blade thickness and strength, unreasonable shape, all will increase stress concentration.
2. The poor blade material performance.
3. Quality defects of the blade manufacturing. For example, there is casting holes in the blade stress concentration place.
4. Instability flows passing through the impeller. Unreasonable shape of inlet passage and

seriously sedimentation at inlet of the pumping station will easily deteriorate the inlet flow patterns of water pumps.

5. The pump operation under over rated head. The differential pressure between pressure side and vacuum side of the blade increases, and the pump may operate in the "saddle-shaped" unstable region; blades bear more than allowable bending moment and aggravate vibration, fatigue fracture is easily occurred.

6. When the pressure change frequency of the unstable flow through the impeller blade and inherent frequency of the blade is the same or close to each other, blade resonance will occur and fatigue fracture of the blade will speed up.

7. Foreign bodies enter into the water pump with flow, which block up the rotating impeller and cause blade vane mechanical fracture.

2.5 The seal failure of pump shaft

The packing seal installed at the pump shaft stretching out pump body belongs to motive seal, bad treatment will cause a lot of water emitting, which even endanger the above motor.

A large amount of leakage of packing seal is mainly due to loose packing, the large axis throw, or the eccentricity of packing box cause excessive wear of stuffing. Therefore, we should pay attention to installation and maintenance.

When packing seal has a large amount of water leakage, pressing fitting packing again is the common method. The tightness of packing is suitable only when there is drops of leakage. But to get a radical treatment, throw of the pump shaft should

be decreased and rotation center of the pump shaft should be well concentric with the stuffing box seat.

3 ANALYSIS OF FAILURE CAUSE OF MOTOR

3.1 Insulation failure

After the large pump unit running for a long time, the insulation of the electric machine will be gradually ageing under the influence of various factors; electrical insulation of stator and rotor winding of the motor and between the core lamination stacks will be damaged. Once insulation failure occurs during the operation, local or most coils will be burnt, causing significant economic losses and safety accidents.

The causes of the motor insulation aging are shown in Table 1 which mainly include: (1) Thermal aging. The insulating materials produce a variety of physical and chemical changes due to heat for a long time; (2) Electrical aging. The insulation aging is caused by partial discharge, leakage and electrocorrosion; (3) Mechanical aging. Major symptoms are fatigue crack, loose, wear of insulation material and so on; (4) Environmental aging. That is insulation erosion of dust, moisture, oil, salt and other corrosive substances of pollution. For example, air duct ventilation of some pumping stations is not reasonable. Although taking the heat, it brings the water vapor, speeds up the aging of stator winding insulation because of humid environment.

Table 1. Factors of motor insulation aging and degradation phenomenon.

Degradation factors	Patterns of manifestation	Deterioration symptoms
Heat	Succession	Shrinkage, chemical metamorphism, lower mechanical strength, worse heat dispersion
	Thermocycling	Separation layer, chap, deformation
Electricity	Running voltage	Partial discharge corrosion, Surface leakage fulgurize
	Surge voltage	Arborescence discharge
Machinery	Vibration	Wear
	Shock	Separation layer, chap
	Curve	Separation layer, chap
Environment	Moisture absorption Dirt Moisture condensation Soaking Conductive material stained	Leakage current increase, formation of a surface leakage passage and carbonization burning marks
	Oil, medicine stained	Erosion and chemical modification

3.2 The burnout failure of sliding thrust bearing

The sliding thrust bearing is the key damageable part of a large vertical motor. Burnout of thrust pads because of high temperature is common failure of the motor.

The main causes of thrust pad burnout are as follows:

1. Too small design load of thrust bearing or thrust bearing overload. When the load factor of thrust bearing (the ratio of working load and design load) $\varepsilon > 0.9$, the bearing is easy to burnout (Qiu (2000)).
2. Poor installation quality of thrust bearing. Each piece of thrust pad has different installation height, unevenly stress, if the stress of some thrust pad exceeds limit value, bearing alloy will burn out and damage firstly, and then the stress of other pieces of pads increases and burns out one by one, which will cause a set of thrust pads consecutively burning out and damaged in an instant.
3. Other reasons of thrust pads burnout are: (a) Oil metamorphism because used for a long time and bearing alloy separate out; (b) The design cooling capacity of oil cooler in oil cylinder is not enough; (c) Blockage in cooling water pipe and misoperation cause cooling capacity decreasing or lost.

4 CONCLUSIONS

The main failure reasons of large low-lift pumping units are as follows:

1. The damage of key and easy to grind components with relative motion. Including water-lubricating metalloid guide bearing and shaft neck of the pump, rotating seal ring and stationary ring seat of water sealing device of oil lubricated Babbit guide bearing, packing and shaft neck of pump shaft packing seal, motor sliding thrust bearing, etc. Main reasons of wear aggravation and even large peeling for this kind of parts are: material defects of the parts, overload, poor lubricating medium (such as lubricating water containing amount of sediments) and poor quality of installation.
2. The failure of pump impeller blade. Water pump blade has two failure modes: cavitation and fracture. Primary cavitation is earlier in

internal pumps of actual pumping station, cavitation occurs more easily. When the pump operates diverging design conditions, NPSHr will increase and cavitation will occur easily. Blade surface damage is usually the results of cavitation and abrasion of sediments in the water combined action. Fracture usually occurs in the root of impeller blade for stress concentration; both flow vortex shock excitation and resonance of the blade cause the root material fatigue, which is relate to pump impeller blade structure, working condition and flow stability.
3. Insulation failure of motor. Main reasons are: thermal, electrical, mechanical and environmental factors. High temperature, thermocycling and mechanical action can cause insulation deterioration, strength reduction, separation layer, chap and deformation. Operation voltage and its impact will cause corrosion of partial discharge, burning marks; moisture and dirt in the environment can form surface leakage passage, which will cause leakage increase and form carbide burning marks.

ACKNOWLEDGMENTS

This work was supported by the National Natural Science Foundation of China (No. 51379182) and the Author Foundation of the National Excellent Doctorial Dissertations of China (No. 2007B41).

REFERENCES

Cao, Haihong 2008. *Study on Durability of Large Vertical Pump Units*. Yangzhou University.

Huang, Gen 2012. *Study on Reliability of Large Vertical Pump Units*. Yangzhou University.

Lin, Haijiang and Qiu, Baoyun 2006. Study on large pump guide bearing end face sealing. *Lubrication and Sealing* (10): 161–164.

Qiu, Baoyun 2000. Analysis on burnout of Babbit Thrust Pad of Motor in Pumping station. *Journal of Yangzhou University (Natural science edition)* (1): 62–65.

Qiu, Baoyun 2005. *Theories and key technology of Large Water Pump Devices*. Beijing: China water power press.

Wang, Qinshi 1998. Analysis and Process of 2000 ZLQ-13. 4–8 Pump Blade Fracture of Dongjiang-shenzhen Water Supply Cause. *Drainage and irrigation Machinery*, (2): 8–10.

Advances in Energy, Environment and Materials Science – Wang & Zhao (Eds)
© 2016 Taylor & Francis Group, London, ISBN 978-1-138-02931-6

Study for seismic behavior of dovetail mortise-tenon joints with gap damaged

Kai Feng, Kun Kang, Xianjie Meng, Jing Hou & Tieying Li
Taiyuan University of Technology, Taiyuan, Shanxi, China

ABSTRACT: Effect of dovetail joints damage for the overall seismic performance of the component is very large; there are many forms of damage. With reference to the Yingzao Fashi, two frame nodes model have been designed in accordance with the original size, including a node intact frame column model and a dovetail mortise tenon joints of band gap of damage model. Extraction of the friction contact surface of the dovetail joints can be done by setting a macro gap of 5 mm, and making the separation of tenon and mortise. The component applied is low cyclic loading and by studying dovetail joint destruction influence on seismic performances. Hysteretic curve of two models show the "pinch" phenomenon, but the damaged one is worse, stiffness degradation is obvious and energy dissipation capacity is weaker. Therefore, the loss of the friction contact surface will greatly affect the seismic performance of the structure.

1 INTRODUCTION

In Chinese ancient historic building, the most representative one is that of Yingxian Wooden Pagoda, Feiyun Floor, the Forbidden City, and so on which stands for thousands of years with a high historical significance and research value. Though there are mortise and tenon structures of ancient buildings without nails; the structure is still able to put members closely linked. The tenon node is a semi-rigid joint, which can bear a certain moment and has a good deformation capacity and energy absorption. The tenon is divided into straight tenon, tenon and dovetail. The dovetail is named after its shape which is like swallow tail, end wide roots are narrow. The dovetail can resist a certain tension, though the shear capacity is poor. The dovetail is a common ancient structure connecting member. Therefore, study on the seismic perform-ance of the dovetail is very necessary for the pro-tection of ancient buildings.

At present, the domestic and foreign scholars have made some research on the seismic perform-ance of ancient buildings wood structure dovetail mortise tenon joints. But most studies have concen-trated in intact dovetail joints, such as Gao Dafeng analyzed mortise tenon joints' working mechanism, and determined the rotational stiffness of mortise tenon joint through research on simulation of wood frame model. He also put forward the calculation model of this type of structure under horizontal earthquake action. Yao Kan and Xu Minggang studied the seismic performance of ancient timber structure dovetail tenon joints. Mortise and tenon

connection failure patterns, hysteretic curve, energy consumption and other characteristics have been summed up. Besides, some scholars considered the damage of the mortise and tenon joint, such as King studied several damaged forms of mor-tise tenon joints common through artificial simu-lation, made comparisons and analyzed seismic performances of degradation between damaged tenon joints and intact ones. Sui Gong carried out the shaking table test of ancient wooden buildings dovetail model. Based on the previous research this paper makes use of finite element modeling and simulation experimental method and investigates the effect of dovetail node damage on the seismic performance of ancient buildings.

2 EXPERIMENTAL STUDY ON SEISMIC PERFORMANCE OF DOVETAIL TENON DESIGN

2.1 Specimen design

To study the damaged dovetail node seismic per-formance for a large wooden structure effect, this paper designs the original size experiment model which is referred to by Yingzao Fashi. Parameters are as follows.

2.2 Design of damaged specimen

To dovetail, for example, through the thousands of years, there have been changes in the wooden mate-rial, the ancient structure itself and other external force. Its overall structure and the parts have been

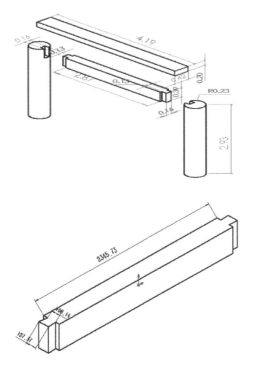

Figure 1. Sketch of the model.

Table 1. Sketch of the model.

Member name	Dimension name	Original dimension	Model size/mm
Column	Column diameter	36	468
	Column height	225	2925
Fang	Length	322	4186
	Width	32	416
	Height	15	195
Dovetail	Tenon width	12	156
	Tenon neck width	10	130
	Length	10	130
Lan E	Height	30	390
	Width	20	260
	Length	221	2873

damaged in varying degrees. Under different loads, this paper compares and analyzes the seismic performance between the intact dovetail and the damaged one with a gap. In this way, this paper studies the seismic performance while its lack of friction is caused by weakening.

2.2.1 Damaging mechanism of dovetail

Wood organisms belonging to a porous material, the breeding of various organisms, lets wood to suffer long from an irreversible damage, due to the shrinkage, insects, fungi, corrosion, surface carbonized cellulose degradation and other reasons, its appearance and mechanical properties are largely changed. The connection between the components will be gradually increased. Besides, mortise and tenon suffered more cracking and damaging. Both sides of dovetail joints have great changes and loss will be more serious.

At the same time, ancient building suffered in a long period of complex stress environment. In the long-term pressure of the upper structure, and the majority of members is in yield compression and repeatedly greater loads. Damaged tenon mouth due to friction missing, damaging of the wood surface is very obvious. Dovetail damage, although more subtle, withstands compression and friction, mortise and tenon wear more serious.

In general, under the combined effect of organic and environmental stress, isolation mortise and tenon mouth is more extensive common destruction. And the construction timber loss rate of is 5–10%.

2.2.2 Design of defect

There are many forms of dovetail damaging, including internal holes and macro cracks. This paper is based on the practical investigation, and the gap between the components is 5 mm and the gap is accounting for 6. 41% section of specimens, which is in reasonable range.

2.3 Experimental material

Wood is a porous bio-material, and it is typical of the various anisotropic materials. It shows different property on significant axial, tangential and radial. This experiment will be assumed to be orthotropic

Figure 2. Actual damaged.

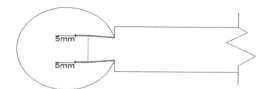

Figure 3. Cross-section design.

Table 2. Mechanical of the wood.

EX/MPa	14190	PRX	0.522	GXY/MPa	710
EY/MPa	710	PRYZ	0.483	GYZ/MPa	38
EZ/MPa	710	PRXZ	0.522	GXZ/MPa	710

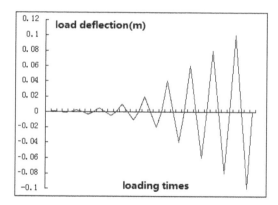

Figure 4. Loading program.

wood, ignoring the difference between the radial and tangential. For the elastic constants of wood referred to reference, where unit of E, G is MPa. As is shown, parameters of North China larch are below.

2.4 Experimental loading scheme

The vertical load is divided into 10 KN and 20 KN. Before applying horizontal load, the vertical load is applied in the Fang.

Maintain constant vertical load, and apply the horizontal displacement progressively. According to the experimental standard ISO-16670, the level of load with variable amplitude displacement mode load, while the required displacement curve is experimentally determined ultimate displacement. In this test, limit of the displacement is 100 mm. As a result, the experiment was terminated after 1%, 3%, 5%, 10%, 20%, 40%, 60%, 80%, and 100% of the peak displacement with two triangular wave cycles sequentially.

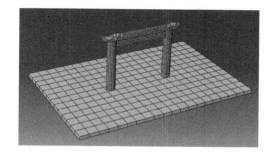

Figure 5. Meshing.

3 ABAQUS FINITE ELEMENT MODEL

3.1 Modeling

The study is based on ABAQUS finite element software simulation. To simulate the real situation, the test simulates the floor as the ground, limiting the movement and deformation of the floor and the provisions of floor stiffness is great as it does not deform and slip. After the experiment, the same friction setting member properties including the timber and the ground were considered consistent friction between the components which were not considered the different static and sliding frictions. All the friction between the members is hard, allowing members' separation and the coefficient of friction is set to 0.4.

3.2 Load

In practice, the load from the upper member of the brackets will be applied on the column. To simulate the real situation, inertial force is applied on both the Fang respectively, after that horizontal displacement changes as the scheme.

3.3 Meshing

This experiment is meshed in structured meshing technique and columns are swept by meshing technique in advanced algorithms. Unit is shaped as hexahedral and the unit type is set to C3D8R. The result is as follows.

4 ABAQUS NUMERICAL SIMULATION RESULTS AND ANALYSIS

4.1 Experimental results

This experiment is to study the earthquake response of structure. As a result, with increase of the horizontal displacement, larger slip structure and larger slip deformation took place. Meanwhile, the vertical load is large, second-order effects of gravity are relatively clear, these are all role models to produce large deformation. As shown for the seamless structure under experimental phenomena, 20 KN loads that framework undergone significant lateral shift, but not unstable failure. Other experiments are similar.

4.2 Hysteretic curve

Hysteretic curve structure under the cyclic loading between force and inelastic strain curve. It is the comprehensive expression of the seismic performance of the structure. The area represents capacity of energy dissipation. Therefore, hysteretic loops more fully, indicating that the seismic performance of the structure is better. After the original analysis, the hysteretic curves are obtained as follows:

1. Due to the special structure of dovetail joint, hysteretic curve appears obviously as asymmetric phenomenon. At the same time, hysteretic curves of four kinds of models show obviously pinched "shrink" phenomenon and shape the anti "Z", which indicates that the four kinds of models produced are of a larger slip and with increasing lateral load and slip it increases. Although the whole structure of the larger slip, but the overall structure remains stable and did not collapse.
2. In the early stage of loading, as shown in the figure (a), hysteresis loops of well-developed and nodes are generally in the elastic stage. After unloading, the structure can be recovered rapidly. And with the increase of load level, hysteretic curve becomes steeper and slope increases, which means mortise and tenon bites close; when the load continues to increase, the slope of the curve reduces. It can be seen from the figure, the larger the horizontal displacement, the

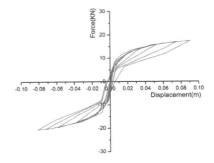

(a) 10KN Intact

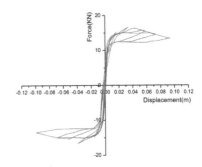

(b) 10KN Damaged

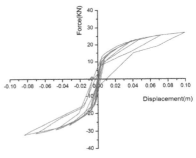

(c) 20KN Intact

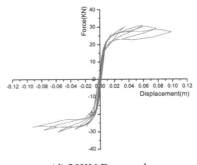

(d) 20KN Damaged

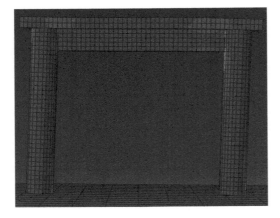

Figure 6. Experiment process.

Figure 7. Hysteretic curve.

curve is slower. The reference experiment, found between the structures of increasing extrusion shows the "pull tenon" phenomenon. With the increasing displacement amplitude, cycle delay curves in a rising stage almost before similar, along the development of the previous cycle. On the contrary, when the unloading displacement, curves do not coincide and show a slope gradually becoming small or even zero. This phenomenon is due to the gradual loading after each other between the mortise and tenon loose.

3. Compare chart (a) and (b). It is found that the damaged zone of fracture and the slope of the component develops soon, the damaged components are with certain initial stiffness, but the structure to withstand the reaction has reached a high degree with the increase of load level. Hysteretic curve slope reaches its peak after rapidly reducing and hysteresis loops renders back to shrink the state. Compared with the figure (a), damaged component of hysteresis loop is narrower and forward back-ward difference and joint slips are more. Because there is a gap between the mouths of the tenon and mortise, lateral friction binding deletion, between the column and the tenon mainly by a positive pressure transfer force, tenon and mortise and tenon mouth between the compression stroke variable length. Figure (b) in hysteresis loop area of the envelope is small, dovetail joint energy dissipation function is not fully realized, component energy dissipation capacity of the weak.

4. Compare chart (a) and (c), when the vertical load is applied to a different structure, it can also have an impact on the structure of the whole antibody. Structure vertical load larger, as more full hysteresis loop dissipate more energy. As shown in (c), the tenon plastic deformation nodes are more obvious. In fact, at this time, between the tenon and mortise compression more closely, friction greater displacement under the action of the same level of compression deformation under development more structure more seismic energy dissipation.

5. Compare chart (b) and (d), although increasing 10 KN, seismic capacity of structures still remain narrow hysteresis loop, hysteresis loop slope deterioration has slowed and the overall structure energy dissipation capacity has not changed much. The overall structure of energy dissipation capacity has not been much improved, but due to the increase of vertical load. Due to lack of structure seismic performance of friction between the tenon and Mao export will greatly, structure to withstand the force. Comparison of four hysteresis curve, when the lateral displacement of the load is small, displacement consistent reaction force

changes the range, second-order effects the performance curve of gravity is not obvious. When lateral displacement load increases, the upper portion of the lateral displacement due to the gravity of loads causes an additional additive effect, namely gravity P-Δ effect, hysteresis curve inflection point, accelerate the reduction of the stiffness, and increase the level of slippage increases.

4.3 Skeleton curve

Skeleton curve reflects the member stiffness changes. It is another important indicator of seismic behavior of structures. The experiments compare the skeleton curve four cases and draw conclusion as follows:

1. Due to the unique structure dovetail, four skeleton curves are to maintain the overall trend that is similar, but the pros and cons are asymmetrical.
2. In this experiment, the dovetail tenon frame node force turn can be divided into two stages, i.e., elastic stage, yield hardening stage. The vertical load is applied, between the components of the space which is compressed and components are connected closely. So in the beginning, skeleton curves have a certain slope which means that the structure can bear a certain load. With the increase of lateral displacement loads, structure slippage increases coupled with gravity second order effect. The slope of the curve becomes smaller, the joint stiffness decreases and the structure gradually yields enhancement, energy dissipation capacity to become weak. If the lateral load continues to increase, the structure has a tendency to collapse.
3. Compared with damaged joints, the intact joints' stiffness has a good growth and degradation even slowly, with mortise and tenon mouth gradually, the structure still has a

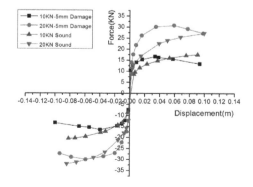

Figure 8. Skeleton curve.

certain carrying capacity and the risk does not occur. With the sharp changes of stiffness, the overall seismic performance of the structure is relatively low and sensitive to the load. If the lateral displacement continues to in-crease, the structure at any time may occur instable.

5 CONCLUSION

1. In the larger lateral loads, complete and damaged dovetail tenons are similarly modified. Both of them will have some degrees of pulling the tongue. Because of friction, good dovetail appears late and the tongue pulls to a smaller extent. On the contrary, damaged dovetail joints shows obvious structural stiffness degradation and poor seismic energy dissipation capacity. The seismic performance of the whole structure is weakened obviously.
2. Within a certain range, the greater vertical load on the structure can have a better stiffness and seismic performance. Because of the vertical load increasing, the static friction force structure is not easy to occur while limiting the deformation of the structure. But when subjected to large lateral loads, gravity second-order effects are more pronounced. The larger vertical load will have a greater moment, affordability over structure and easy instability failure.
3. Lack of tenon and mortise mouth friction will directly result in an overall energy dissipation capability which means less prone to pull the tongue phenomenon. Besides, it is significantly weakening the rigid structure and reducing the seismic performance of the structure. Dovetail member present, most of the ancient building due to natural and human causes, mortise and tenon mouth mostly there is a big gap, which is unfavorable for long-term storage member.

ACKNOWLEDGMENT

This work was financially supported by the National Natural Science Foundation of China (No. 5127832-4). The instructor is Li Tieying, and we are very grateful to all reviewers for their constructive comment on this study.

REFERENCES

Chen Zhiyong, Zhu Enchun, Pan Jinglong. Numerical simulation of the mechanical properties of wood under complex stress state [J]. Journal of Computational Mechanics, 2011, 28 (04): 629–634.

Fan Chengmou, Zhang Shengdong, Chen Song. Basic principles of wood structure [M]. Beijing: China Building Industry Press, 2008.

Fang Dongping, Yu Maohong, Yu Miyamoto. Experimental study of the structural characteristics of ancient wooden structure building engineering mechanics [J], 2000 (02): 75–83.

ISO-16670. Timber structure-Joints made with mechanical fasters Quasistatic reversed cyclic test method [s]. International Organization for Standard, 2003.

King W.S. Yen J.Y. Yen Y.N. Joints characteristics of traditional Chinese wooden frames [J]. Engineering Structure, 1996, 18 (8): 635–644.

Li Tieying. Yingxian Tower Status Structure and Mechanism of damaging elements [D]. Taiyuan University of Technology, 2004.

Liu Dunzhen. Chinese architectural history [M]. Nanjing: Nanjing University Institute of Technology Press, 1953.

Pan Jinglong, Zhu Enchun. Wood design principle [M]. Beijing: China Building Industry Press, 2009.

Sui Gong, Zhao Hongtie, Xue Jianyang, etc. Shaking table test of the ancient palacestyle wooden architecture model [J]. Journal of Building Structure, 2010, 31 (2).

Xu Minggang, Qiu Hongxing. Tenon ancient wooden buildings on seismic [J]. Building Science, 2011, 27 (7): 56–68.

Yang Xuejian, Zhang Pengcheng, Zhao Hongtie. Study of the ancient wooden structure seismic mechanism [J]. Xi'an University of Architecture & Technology, 2000 (01): 8–11.

Yao Kan, ZHAO Hongtie, Ge Hongpeng. Experimental study of ancient wooden structure tenon connection properties [J]. Engineering Mechanics, 2006, 23(10): 168–173.

Zhang Pengcheng, Zhao Hongtie. Chinese ancient wooden frame deformation and stress behavior [J]. World Earthquake Engineering, 2003 (30): 9–14.

Zhao Hongtie, Zhang Haiyan, Xue Jianyang. Analysis of ancient buildings wood structure dovetail joint stiffness [J]. Xi'an University of Architecture & Technology (Natural Science), 2009 (04): 450–454.

Advances in Energy, Environment and Materials Science – Wang & Zhao (Eds)
© *2016 Taylor & Francis Group, London, ISBN 978-1-138-02931-6*

Finite element analysis of the flange sealing of pipes under the condition of vibration

Zhirong Yang
China Special Equipment Inspection and Research Institute, Beijing, China

Dayong Zhang
Dalian University of Technology Panjin, Liaoning, China
Dalian Ocean University, Dalian, Liaoning, China

Liang Sun
China Special Equipment Inspection and Research Institute, Beijing, China

Guojun Wang
Dalian Ocean University, Dalian, Liaoning, China

ABSTRACT: Under the conditions of vibration, the pipeline structure can produce relatively severe vibration. The strong vibration can cause the fatigue failure of the pipeline and the leaks of the flange. Thus, it is necessary to analyze the flange strength and pipe stress under the conditions of vibration by the finite element method. The flanged piping system is not only an important component, but also an error-prone component in the design and use. In this paper, based on the data monitored on the offshore platform in Bohai Bay, the deck acceleration induced by ice vibration serves as excitation of pipelines. A mechanical model of structural vibration of pipeline system is built. With finite element modeling, the dynamic responses of the main pipeline system and the local structure of pipeline induced by deck vibration are calculated. It accurately calculates the dynamic responses of the pipe and flange connection under vibration conditions. The conclusions of this paper have practical and guiding significance on the design and failure analysis of the connection system of the pipe and flange under the conditions of vibration.

1 INTRODUCTION

Sea ice is the dominated environmental load of marginal structures in Bohai Bay. The design of ice-resistance platforms has usually been considered the maximum bearing capacity under the limited ice load. However, the effects of offshore platforms and facilities induced by the dynamic ice load are not been considered. Based on the data monitored on the platforms, sea ice can induce the periodic load apparently, and make the offshore jacket structures to vibrate with major acceleration (Yue & Bi 2000). Recently, there has been analogous accident hidden danger in Bohai Bay induced by the dynamic ice load. Current design practice of pipeline is based on linear analysis where stress due to design loads are calculated and compared with allowable stress. Owing to this and much other overlapping conservatisms, piping systems at present tend to employ a large number of supports and are usually rather rigid. These rigid and overly constrained systems may experience high dynamic responses under ice load. So it is necessary to

analyze dynamic response by ice-induced vibration for gas pipeline system on the jacket platforms.

For pipeline system, the widespread vibrations are being induced by reciprocating compressors and pumps, two phase flow, wind, earthquake, water hammer, etc. In terms of action form of excitation, pipeline vibration can be divided into two kinds separately induced by internal media and external loads. It was found that the current design code of pipe deals with vibration problems using simplified methods or experience. At present, vibration research of pipeline system can be divided into two aspects: one is pulsatile flow in pipe under time-varying conditions due to pump and valve operation inducing vibrations; the other is dynamic characteristics and responses of pipe. Based on the different mechanism of pulsatile flow, using fluid mechanics theory, the former analyses the regulation of pulsatile flow, and makes it within control range. By means of vibration analysis of pipeline, the latter researches vibration regulation, and takes effective measures to promote the aseismatic capacity of pipeline system. Many researchers focused particularly on solving vibration problems of pipe as follows: the interaction between flow or gas and pipe (Ziade & Sperling 2001, Usik & Hyuckjin 2003, Li & W. 2000), current and submarine pipe (X & B 1999, Gong & Lu 2000), and the nonlinear response of piping to seismic loads (Luciano 1995, Soliman & Datta 1995). However, there remains a need for detailed systematic study of the dynamic response by ice-induced vibration for complex gas pipeline system on the jacket platforms. The main reasons for it are that piping systems including many supports, pipe elements (e.g. bends, tees, flanges, valves, etc.), equipments, steel structures, are so complicated that building model is very difficult; compared with earthquake and wind load, the research of ice load is not so sophisticated that the excitation of pipeline on the jacket platforms can be determined difficultly. Based on monitoring, the phenomenon of pipeline vibration induced by dynamic ice load is similar to the effect of over ground pipelines to random ground motion. Ice-structure interaction induces the vibration of platform deck (the vibration mechanism and analysis of offshore platforms induced by dynamic ice is not explained in this paper), and then its acceleration response is used as the input to the piping supports. Based on the data monitored on the platforms of JZ202 in Bohai Bay, the typical history curve of deck acceleration serves as excitation of pipelines on the jacket platform. Then a mechanical model of structural vibration of pipeline system is built. With finite element modeling, considering the influence of the complex supports, devices and elbows of pipelines, the dynamic responses of the main pipeline system and the local structure of pipeline induced by deck vibration are calculated. The results show ice-induced vibration and the form of pipeline have big effect on the vibration property of the pipeline system. Lastly, some suggestions are proposed to degrade the risk induced by sea ice. Dynamic analysis of natural gas pipeline system exposure to ice-induced vibration provides theory reliance for producing safely of platforms in ice zone.

2 RESPONSE OF OFFSHORE PLATFORM BY ICE-INDUCED VIBRATION

In Bohai Bay, the offshore jacket structures are designed flexibly, as sea ice is not as thick as in the polar area. At the same time, the ice velocity is very fast because of the strong current and seasonal wind. In the last few years, the data monitored on platforms reveals that significant ice-induced vibrations and resonant vibration took place on these jacket platforms regardless of whether or not ice-breaking cones are installed in Bohai Bay.

JZ202 jacket platforms, flexible ice-resistance structures, are located in the north part of the Bohai Bay. In order to analyze dynamic response by ice-induced vibration for gas pipeline system on the jacket platforms, the writes applied accelerometers on JZ202 platform decks to measure the responses induced by ice load and recorded data the whole winter without stopping.

The data monitored shows that resonance can be induced at a certain speed corresponding to a given ice thickness. The amplitude and frequency of the ice force depend mainly on the ice feature itself. This causes the amplitude of structure response to be random and like that induced by wave or wind. Figure 1 shows the typical time-series of deck acceleration response. Then by means of spectral analysis, the frequency spectrum of the acceleration response is given in Figure 2. Based on monitoring, the phenomenon of pipeline vibration induced by dynamic ice load is similar to the response of pipelines subjected to random ground motion excitation. Ice-structure interaction induces the vibration of platform deck, and then its acceleration response is used as the input to the piping supports. The difference from ground motion excitation is that durational time of platform vibration by ice load is longer and the characteristic of response is a typical narrow-band random process.

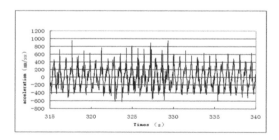

Figure 1. Typical time-series of deck acceleration response of JZ202 platforms.

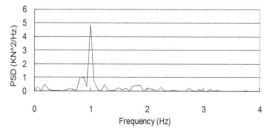

Figure 2. Typical acceleration response dominant frequency.

3 MECHANICAL MODEL OF STRUCTURAL VIBRATION OF PIPELINE SYSTEM BY ICE-INDUCED VIBRATION

In strict sense, structural vibration of pipeline system by ice-induced vibration belongs to random vibration of limitless freedom system, so that piping systems must be analyzed involving a large amount of computation. For this reason, the complex pipeline structural dynamics problem is formulated for structures discretized as systems with a finite number of degrees of freedom. We now write the equations of an MDF system subjected to ice-induced vibration a (t). The dynamic response of the structure to this excitation is defined by the displacement x, velocity $\dot{x}$, and acceleration $\ddot{x}$. Dynamic equations of natural gas pipeline system exposure to ice-induced vibration:

$$[M]_P\{\ddot{x}\} + [C]\{\dot{x}\} + [K]\{x\} = -[M][I]a(t) \qquad (1)$$

where $[M]$ is the mass matrix, $[C]$ is the damping matrix, $[K]$ is the stiffness matrix of the pipeline system, $\{I\}$ is the unit matrix.

Usually, the vibration equation of multi-degree-of-freedom system can be solved by means of immediate integration or modal superposition. Immediate integration method is enforced integral calculus to dynamic equation directly. Modal superposition method determines the natural frequencies ω_n and modes Φ_n of undamped free vibration system firstly, which is a matric characteristic value problem. Then using orthogonality of modes, original dynamic equation can be inverted a new equation, whose freedoms are not hookup mutually. Lastly dynamic equations of respective freedom can be had integration and superposition. From this analysis, we can derive the vibration equation of pipeline system by dominical coordinate description:

$$[M]_P\{\ddot{x}\}_P + [C]_P\{\dot{x}\}_P + [K]_P\{x\}_P$$
$$= -[M]_P[I]A(t) \qquad (2)$$

where $[M]_P = [\Phi]^T[M][\Phi]$; $[C]_P = [\Phi]^T[C][\Phi]$; $[K]_P = [\Phi]^T[K][\Phi]$ $\{x\} = [\Phi]\{x_P\}$.

The matrix $[\Phi]$ is called the modal matrix for the eigenvalue problem. From these, we can see $[M]_P$, $[K]_P$ are both diagonal matrices. If the square matrix $[C]$ is diagonal (else, suppose there is a proportional relation between $[C]$ and $[M]$, $[K]$, $[C] = \alpha[M] + \beta[K]$, where α, β are proportional constants), Eq. 2 represents N uncoupled differential equation, such as Eq. 3, in modal coordinate $\{x\}_P$, and the system is said to have classical damping classical modal analysis is applicable to such system.

$$M_P\ddot{x}_{Pi} + C_{Pi}\dot{x}_{Pi} + K_{Pi}x_{Pi} = -M_{Pi}A(t) \qquad (3)$$

Which in standard form is:

$$\ddot{x}_{Pi} + 2\zeta_i\omega_i\dot{x}_{Pi} + \omega_i^2 x_{Pi} = -A(t) \qquad (4)$$

The random vibration Eq. 4 may be solved by means of the response spectrum analysis or response history analysis. The former needs for the platform deck acceleration response spectrum, and derives the displacement, velocity and acceleration response spectrum of pipeline. The later uses response history analysis procedure, by means of number integration, solves the Eq. 4, and gets dynamic response history of pipe. Based on different hypothesis, number integration includes linear acceleration method, Wilson-θ, Newmark's method, etc. This paper uses the second method.

Through analysis, we can see that the amplitude and stress value are very high when excitation frequency of pipe (deck acceleration response frequency induced by ice vibration) is close to self-sustained oscillation of pipeline system (resonance of vibration system happens in $\omega = \sqrt{1-2\zeta^2}\,\omega_p$, not in $\omega = \omega_P$).

4 NUMERICAL ANALYSIS OF PIPELINE VIBRATION BY ICE INDUCED VIBRATION

Reaching any definitive dynamic response of piping system is a difficult task, because piping systems are notoriously irregular in their configurations, so that a great many piping systems must be analyzed involving a large amount of computation. To partially alleviate this computational burden, finite element numerical simulation of pipeline induced by ice vibration is adopted. Using the CAESAR II computer program (CAESAR II is a PC-based pipe stress analysis software program developed, it is an engineering tool used in the mechanical design and analysis of piping system.), the main pipeline system and the local structure on the JZ202MSW platform are modeled by simple beam elements. It is supposed that gas pressure is constant, i.e. the vibration of pipe induced by gas is very small, corresponding induced by ice vibration.

5 PRELIMINARIES OF EXISTING SYSTEM

The existing piping system is located on the JZ202MSW platform and is of early design. Since it is not readily accessible for maintenance, it is over designed, has many supports and is very rigid. Figure 3 is the main pipeline system, including production, testing and quantification pipes.

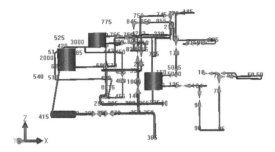

Figure 3. The main pipeline systems of JZ202 MSW platform.

Figure 4. The local structure of pipeline.

It starts from the brow and ends at the seabed where it is anchored. The pipe consists of two heaters, one separator, one emitter, 56 bends, 34 valves (including flanges) and 8 tees. The second model is the local structure of pipeline, whose fixed form is cantilever beam structure, and connects with the platform as shown in Figure 4.

6 NUMERICAL RESULTS

Based on the data monitored on the platform of JZ202MSW, the typical history curve of deck acceleration (such as Fig. 1) serves as the excitation of pipelines on the jacket platform, and is used as input to piping supports. This paper uses modal superposition and Wilson-θ method to derive the dynamic responses of the two models induced by ice vibration. The fundamental natural frequencies of the main gas pipeline system and the local pipeline structure are respectively 0.77 Hz and 1.48 Hz. The maximum values of dynamic responses of the main part of gas pipeline are shown in Table 1. The forces on the flanges of the pipeline are shown in Table 2.

Table 1. The maximum values of dynamic responses of the quantification pipeline.

Node	Dynamic stress (Mpa)	Translations (cm)			Rotations (deg.)		
		X	Y	Z	X	Y	Z
420	10.624	0.321	3.790	0.420	0.302	0.049	0.352
425	74.336	0.321	2.844	0.199	0.269	0.105	0.334
430	21.145	0.314	1.854	0.013	0.201	0.050	0.294
435	42.470	0.371	2.829	0.260	0.260	0.209	0.425
440	70.866	0.459	2.813	0.345	0.244	0.303	0.593

Table 2. The forces on the flanges of the main pipelines.

Flange number	FX (N)	FY (N)	FZ (N)	MX (N·m)	MY (N·m)	MZ (N·m)
315~320	609	1330	787	998	437	382
455~460	89	578	627	1186	477	259
675~680	116	156	80	151	100	175

From the numerical results of pipelines dynamic analysis, we can easily see:

1. The dynamic responses of the main part of gas pipelines are very complex induced by ice vibration, and the response magnitude of quantification pipelining is higher than other pipelines;
2. Because the fundamental frequency of deck acceleration response (i.e. the fundamental frequency of pipelines excitation) is about 1.3 Hz (Fig. 2), and the natural frequency of JZ202MSW platform is 1.3 Hz, the natural frequency of local pipeline structure (1.481 Hz) is close to the above two values, the second resonance comes into being easily;
3. The forces on the flanges of the pipelines induced by platform vibration are complicated and not neglected, which is the main reason of the looseness of the flanges.

Consequently, in order to prevent fatigue fracture of pipelines and leakage of flanges, vibration control measures should be considered. Vibration control may be achieved mainly by controlling the piping (including the main pipeline systems and the local pipe structure) natural frequency and/or stress control, when it is not practical or the vibration source cannot be eliminated.

7 CONCLUSIONS

Based on the data monitored on the platforms of JZ202 in Bohai Bay, a mechanical model of structural vibration of pipeline system is built.

With finite element modeling, the dynamic responses of the main pipeline system and the local structure of pipeline induced by deck vibration are calculated. The results show ice-induced vibration and the form of pipeline have big effects on the vibration property of the pipeline system. Dynamic response analysis of natural gas pipeline systems exposed to ice-induced vibration on offshore platform provides theory reliance for safe producing of existing offshore platforms and design of new platforms in ice zone.

The supports and restraints of pipeline systems are the main components received the level vibration of platform deck, and served as border condition determine the modal of pipeline systems. So in order to improve the ant seismic capability of pipelines, staff should dispose the supports and restraints reasonably, adjust the supports stiffness to prevent whiplash effect, and use flexibility pipe joint to improve buffer function. Based on dynamic property of pipeline systems, effective vibration control measure should be used, such as selecting hydraulic damper, tuner, etc.

ACKNOWLEDGEMENTS

This material is based on work funded by the National quality inspection of public welfare scientific research project under Award No. 201310161 and General Administration of Quality Supervision, Inspection and Quarantine of the People's Republic of China technology project under Award No. 2013QK019. This financial support is gratefully acknowledged.

REFERENCES

Gong, S.W, and Lu, C. (2000). "Structural Analysis of a Submarine Pipeline Subjected to Underwater Shock," *J Pressure Vessels and Piping*, Vol 77, pp 417–423.

Li, C.J., and Wang, Y.C. (2000). "Analysis of the Vibration in Gas Pipeline System," *J Natural Gas Industry*, Vol 20, pp 80–83.

Luciano Lazzeri (1995). "On the Nonlinear Response of Piping to Seismic Loads," *J Pressure Vessel Technology*, Vol 123, pp 324–331.

Soliman, H.O., and Datta, T.K. (1995). "The Seismic Response of a Piping System to Non-stationary Random Ground Motion," *J Sound and Vibration*, Vol 180, pp 459–473.

Usik Lee, and Hyuckjin, Oh. (2003). "The Spectral Piping System due to Flow in a Spherical Elbow," *J Fluids and Structures*, Vol 15, pp 751–767.

Xu, T., and Bai, Y. (1999). "Wave-induced Fatigue of Multi-span Pipelines," *J Marine Structures*, Vol 12, pp 83–106.

Yue, Q.J., and Bi, X.J. (2000). "Ice-induced Jacket Structure Vibrations in Bohai Sea" *J Cold Regions Engineering*, Vol 14, pp 81–92.

Ziade, S., and Sperling, H. (2001). "Vibration of a High-pressure Element Model for Pipeline Conveying Internal Steady Flow," *J Engineering Structures*, Vol 25, pp 1045–1055.

Advances in Energy, Environment and Materials Science – Wang & Zhao (Eds)
© *2016 Taylor & Francis Group, London, ISBN 978-1-138-02931-6*

A method for parameter optimization of locking dowel base on the orthogonal experiment

Shi-tong Chen
College of Architecture and Civil Engineering, Beijing University of Technology, Beijing, China
Hebei Engineering Research Center for Traffic Emergency and Guarantee, Shijiazhuang Tiedao University,
Shijiazhuang, Hebei, China

Wen-xue Zhang & Xiu-li Du
College of Architecture and Civil Engineering, Beijing University of Technology, Beijing, China

Yao-hui Zhang
Hebei Engineering Research Center for Traffic Emergency and Guarantee, Shijiazhuang Tiedao University,
Shijiazhuang, Hebei, China

ABSTRACT: In order to improve the seismic performance of continuous girder bridge by fully utilizing the aseismic potential of sliding-pier, locking dowel device with acceleration activation was proposed. The structural form and operating mechanism of the aseismic device was explained and the mechanical model was established. In order to achieve the optimum aseismic effect, a parameter optimization method of locking dowel based on an orthogonal experimental design was proposed. The problem of solving the numerical simulation was transformed to the optimum combination of the level of each factor, which affects the aseismic effect, and the objective of acquiring the optimum combination of parameters of locking dowel using a relatively small number of tests was achieved. Results of the analysis showed that in the prerequisite, the site condition and bridge structures are determined, the connecting stiffness is the primary factor affecting the aseismic effect. Orthogonal experimental design provides new ideas for the parameter optimization of aseismic devices.

1 INTRODUCTION

A continuous girder bridge is commonly used in bridge engineering. To meet the requirement of deformation due to temperature load, fixed support is setup in one bridge pier for each span while movable supports are setup on the rest of bridge piers. The horizontal load due to longitudinal seismic action produced by the upper structure is mainly sustained by the fixed pier. Past earthquake incidents showed that the longitudinal seismic displacement response of the girder of continuous girder bridge is larger, which is likely to result in the damage of expansion joints and support, and in worst cases, the collapse of girders. Aseismic design is the most economic and effective approach to improve the seismic performance of bridges. Study (Wanc, 2003) suggested that the aseismic effect is affected primarily by the constitution of earth vibration frequency, the parameters of support as well as the natural frequency of vibration of the bridge.

Study (Liu, 2012) showed that the aseismic effect of double hyperboloid spherical support is significantly affected by the coefficient of friction as well as the distance to the centre of the sphere. Study showed that aseismic support with reasonable parameters is capable of effectively reducing the structural displacement and internal force response. Study (Chen, 2008) showed that the improper application of aseismic support could lead to serious consequences.

Existing studies on the optimum combination of parameters of aseismic devices by comprehensive analysis on impact factors of aseismic support is rare. Based on existing research and recently developed locking dowel device, the orthogonal experimental design principle is introduced and the optimum combination of parameters of locking dowel for continuous girder bridge is evaluated in accordance of a selected 7-span continuous girder bridge project. The analytical approach of the multiple optimum combinations of parameters of aseismic support is evaluated.

2 OPERATING MECHANISM OF LOCKING DOWEL AND CONSTITUTION OF ELEMENT

Locking dowel consists of a locking ball, a locking dowel bracket, a boxed cofferdam and a locating base, as shown in Figure 1. The mechanical model is shown in Figure 2. The locking dowel is installed between the girders of continuous girder bridge. During normal service, the locking ball is located in the curved slot on the locking dowel bracket. When the earthquake occurs, the locking ball moves within the curved slot before the pier-top acceleration reaches the activation threshold value. The locking dowel is not activated and longitudinal free sliding state exists between the sliding-pier and the girder. When the pier-top acceleration first reaches the activation threshold value, the locking dowel is activated, the locking ball rolls falls into the slot of the locating base. Only when the relative displacement between the girder and sliding-pier is larger than the gap of the locking dowel g_p, the collision occurs between the locking ball and the rubber bearing of upper cofferdam and lower locating base. The relative displacement of the girder and sliding-pier is restricted in order for the sliding-pier to resist horizontal seismic load together with fixed piers. Rubber bearing is setup on the bottom opening part of the boxed cofferdam as well as the slot of the locating base in order to prevent rigid collision when the locking ball falls into the slot of the locating base and to adjust the connecting stiffness of the locking dowel.

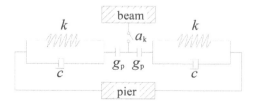

Figure 1.   Configuration of locking dowel.

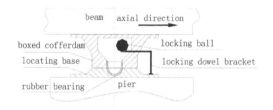

Figure 2.   Dynamical model.
where a_k—the activation threshold value of acceleration; k—the connecting stiffness of the locking dowel; c—the unit damping coefficient; and g_p—the spacing of locking.

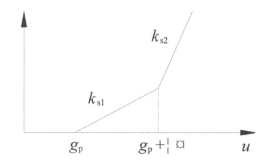

Figure 3.   The force-displacement relationship of locking dowel.

For ease of demonstration, let:

$$a(t^*) = a_k \tag{1}$$

where $a(t^*)$ is the pier-top acceleration at t^*; t^* is the time when pier-top acceleration first reaches the activation threshold value of acceleration.

Under the effect of seismic force, the rubber bearing is compressed until the maximum compression Δ is reached and the connecting stiffness of the locking dowel varies. The force-displacement relationship of the locking dowel is demonstrated in Figure 3.

The constitutional equation of the locking dowel unit is given as follows according to the operating mechanism of the locking dowel, where d_i is the relative displacement between the girder and pier:

$$F(t) = \begin{cases} 0 & ; & t < t^* \\ 0 & ; & t \geq t^*, |d_i| < g_p \\ k_{s1} \cdot (|d_i| - g_p) & ; t \geq t^*, g_p + \Delta > |d_i| \geq g_p \\ k_{s2} \cdot (|d_i| - g_p - \Delta); & t \geq t^*, |d_i| \geq g_p + \Delta \end{cases} \tag{2}$$

3   CALCULATING MODEL

The analysis of the aseismic application of locking dowel was conducted based on a seven-span equal-height continuous girder beam with a dimension of 62.5 m + 5 × 96 m + 62.5 m. As shown in Figure 4, the weight of the main girder was 36,300 t, the height of bridge pier was 20 m with a vertical bending moment of inertia of 30 m⁴, cross-sectional area of 15 m² and the elastic modulus of concrete was taken as 3.45×10^{10} N/m². The finite element model of the complete bridge was established by ANSYS. The girder and pier were simulated by beam element. The nonlinear connection of the

Figure 4. Model for a large-span continuous bridge.

locking dowel was incorporated by the locking dowel unit shown in Figure 2. It was assumed that the bridge pier remained linear elastic during the analytical process and the bridge pier was fixed onto the ground.

For ease of analysis, two conditions were incorporated in this study: condition 1 is the original bridge design model, i.e. bridge pier 4# was hinged to the main girder and girder on other bridge piers was able to slide freely; condition 2 is the locking dowel model where the dowel was activated by acceleration, i.e. bridge pier 4# was hinged to the main girder and acceleration activated locking dowels were setup between the girder and piers 2#, 3#, 5# and 7#. During the analysis, the energy loss due to collision was ignored, i.e. $c = 0$.

Aseismic ratio λ_1 is used to represent the aseismic effect of locking dowel aseismic system for continuous bridge and load increment ratio λ_2 is used to represent the increment of seismic load sustained by sliding-pier, which are respectively calculated as:

$$\lambda_1 = \frac{R_{max} - R_{c,max}}{R_{max}} \times 100\%,$$
$$\lambda_2 = \frac{R_{c,max} - R_{max}}{R_{max}} \times 100\% \qquad (3)$$

In which R_{max} is the analyzed maximum seismic response in condition 1 while, $R_{c,max}$ is the analyzed maximum seismic response in condition 2.

4 ORTHOGONAL EXPERIMENTAL DESIGN

Typical points can be selected by orthogonal experimental design from the complete experimental program to conduct selected experiment.

4.1 Parameter optimization of locking dowel

The aseismic mechanism of locking dowel of continuous girder bridge is to utilize the aseismic potential of sliding-pier by adding the amount of seismic load sustained by the sliding-pier in order to reduce the

seismic response of the fixed piers and improve the seismic performance of the bridge. Therefore, the primary control target is the aseismic ratio, followed by the load increment ratio of each sliding-pier.

In order to fully evaluate the aseismic effect of locking dowel, based on the prerequisite that the site condition and bridge structure is identical, and with reference to the constitutional relationship of the locking dowel, three important factors, namely activation threshold value, locking gap and connecting stiffness were selected in the analysis in considering the damping coefficient has little effect on the aseismic performance of the support. A total of five levels were selected for each factor, as shown in Table 1.

When the impact factors and the respective levels are determined, the arrangement of orthogonal experimental plan, i.e. the selection of orthogonal table, becomes the key part of the process. According to the three determined impact factors and respective five levels, a $L_{25}(5^6)$ orthogonal table was selected to conduct orthogonal experimental design. The designation of "25" suggests the number of rows in the orthogonal table, that is to say, the original number of numerical simulation 5^3 had been reduced to 25.

4.2 Orthogonal experimental analysis

According to the experimental plan of $L_{25}(5^6)$ orthogonal table, the aseismic ratio of locking dowel aseismic system and the load increment ratio of the sliding-pier of continuous girder bridge were analyzed. The dynamic seismic input included: El-Centro Wave and TAR_TARZANA Wave in type-II site condition, Lanzhou Wave and CPC_TOPANGA Wave in type-III site condition. During the calculation, the peak value of acceleration was adjusted to 0.4 g. Energy loss during the collision process was ignored, i.e. $c = 0$. Limited by space, Table 2 shows only the aseismic ratio and load increment ratio of sliding-pier in each condition under the effect of El-Centro Wave (the load increment ratio was represented by the maximum shear force on the bottom of 3# sliding-pier).

Table 1. Factors for orthogonal table and respective levels.

Factors	$a_k/(m/s^2)$	g_p/m	$k/(kN/m)$
Factor no.	A	B	C
Level 1	0.1	0.0040	1×10^5
Level 2	0.3	0.0050	1×10^6
Level 3	0.6	0.0075	1×10^7
Level 4	1.0	0.0090	1×10^8
Level 5	1.5	0.0100	1×10^9

Table 2. Results based on orthogonal experiment.

Exp. no.	Factors A/(m/s²)	B/10⁻³m	C/(kN/m)	λ_1/%	λ_2/%
1	0.1	4	1×10^5	34.8	442.0
2	0.1	5	1×10^6	49.3	803.3
3	0.1	7.5	1×10^7	52.0	777.2
4	0.1	9	1×10^8	50.6	733.1
5	0.1	10	1×10^9	49.3	750.6
6	0.3	4	1×10^6	45.1	566.3
7	0.3	5	1×10^7	45.1	516.4
8	0.3	7.5	1×10^8	45.1	581.4
9	0.3	9	1×10^9	45.1	624.7
10	0.3	10	1×10^5	30.9	379.4
11	0.6	4	1×10^7	45.1	529.6
12	0.6	5	1×10^8	45.1	542.1
13	0.6	7.5	1×10^9	45.1	580.7
14	0.6	9	1×10^5	30.6	396.5
15	0.6	10	1×10^6	45.1	591.1
16	1	4	1×10^8	45.1	544.2
17	1	5	1×10^9	45.1	549.1
18	1	7.5	1×10^5	30.5	416.1
19	1	9	1×10^6	45.1	604.4
20	1	10	1×10^7	45.1	616.1
21	1.5	4	1×10^9	44.8	520.4
22	1.5	5	1×10^5	44.8	320.3
23	1.5	7.5	1×10^6	45.1	636.6
24	1.5	9	1×10^7	45.1	643.5
25	1.5	10	1×10^8	45.1	658.9

It is evident from Table 2 that: 1) among the typical 25 simulating analysis, the aseismic ratio had a minimum value of 30.52% (Experimental plan 18) and a maximum value of 51.95% (Experimental plan 3), suggesting that the use of locking dowel on continuous girder bridge achieved outstanding aseismic effect; 2) At the same time when aseismic effect was achieved, the seismic load carried by the sliding-pier increased by a significantly varied amount. For example in experimental plans 21 and 22, although a 44.80% aseismic ratio was achieved, the load increment ratio in Plan 21 was 520.42% compared with 320.26% in Plan 22. The finding shows that in order for the utilization of locking dowel on continuous girder bridge to achieve optimum effect, not only should the aseismic ratio be considered but also the load increment of sliding-pier.

4.3 Range analysis

The range R refers to the difference of the maximum value and minimum value of a group of data. It represents the degree of discreteness of a group of data and can be used as a parameter to evaluate the significance of a factor. The value of R represents the degree of impact of the variation of level on the test result. The larger the range, the higher influence the variation of that factor has on the test results. Factors with the most range are the primary factors. The results of range analysis of aseismic ratio and load increment ratio are given in Tables 3 and 4, respectively. Figure 5 show the variation of aseismic ratio and load increment ratio when the level of each factor changes.

In which Σk_i is the statistical parameter of the respective factor at level i; $\overline{\Sigma k_i}$ is the average value of Σk_i.

It is evident from Table 3 that connecting stiffness is the most significant factor of the aseismic ratio of locking dowel, followed by the acceleration activating value. Locking gap has little effect on the aseismic ratio of the locking dowel. It can be determined by the aseismic ratio that the optimum combination is A1B 2C3. Analysis from Table 4 shows the connecting stiffness is the most significant factor on the load increment of sliding-pier, followed by the acceleration activating value. Locking gap has little effect on the load increment ratio of the sliding-pier. It can be determined by the load increment ratio that the optimum combination is A3B 1C1 (in range analysis of load increment ratio, the smaller the load increment ratio the better).

The analysis of Figure 5(a) shows that: 1) with the increase of acceleration activating value, the aseismic ratio exhibited a tendency of initial reduction and followed up increase with a stable zone in between; 2) with increase of locking gap, the aseismic ratio first increased then reduced. The variation of aseismic ratio is smooth when last three levels of the locking gap varies; 3) when

Table 3. Range analysis of aseismic ratio.

Index	Aseismic ratio Factor A	Factor B	Factor C
Σk_1	235.97	214.80	171.64
Σk_2	211.07	229.29	229.55
Σk_3	210.86	217.63	232.17
Σk_4	210.74	216.42	230.83
Σk_5	224.77	215.27	229.22
$\overline{\Sigma k_1}$	47.19	42.96	34.33
$\overline{\Sigma k_2}$	42.22	45.86	45.91
$\overline{\Sigma k_3}$	42.17	43.53	46.43
$\overline{\Sigma k_4}$	42.15	43.28	46.17
$\overline{\Sigma k_5}$	44.95	43.06	45.84
R	5.05	2.90	12.11
Sequence	2	3	1
Optimum plan	A1	B2	C3

Table 4. Range analysis of load increment ratio.

Index	Load increment ratio		
	Factor A	Factor B	Factor C
Σk_1	3506.26	2602.46	1954.24
Σk_2	2668.08	2731.10	3201.60
Σk_3	2640.03	2991.97	3082.87
Σk_4	2729.82	3002.26	3059.71
Σk_5	2779.72	2996.12	3025.48
$\overline{\Sigma k_1}$	701.25	520.49	390.85
$\overline{\Sigma k_2}$	533.62	546.22	640.32
$\overline{\Sigma k_3}$	528.01	598.39	616.57
$\overline{\Sigma k_4}$	545.96	600.45	611.94
$\overline{\Sigma k_5}$	555.95	599.22	605.10
R	173.25	79.96	249.47
Sequence	2	3	1
Optimum plan	A3	B1	C1

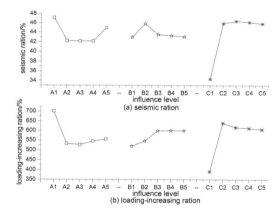

Figure 5. Changing trend with the influence level of the factor.

the connecting stiffness changes from level C1 to level C2, the aseismic ratio increases significantly. When the connecting stiffness varies from level C2 to level C4, the sensitivity of aseismic ratio on the variation of connecting stiffness was reduced significantly.

It can be known from Figure 5 (b) that: 1) the load increment ratio of sliding-pier exhibited a trend of first reduce then increase with the increase of acceleration activating value, and the rate of reduction between A1 level to A2 level was larger than the rate of variation between A2 level to A5 level; 2) During the increase of locking dowel gap, when it changed from B1 level to B3 level, the load increment ratio exhibited significant increase and then became stable; 3) with the increase of connecting stiffness, the load increment ratio exhibited a trend of first increase then decrease. When the connecting stiffness varied from C1 level to C2 level, the load increment of sliding-pier was significant.

5 CONCLUSIONS

1. Analysis showed that the utilization of locking dowel for continuous girder bridge could achieve outstanding aseismic effect.
2. Range analysis showed that connecting stiffness is the most significant factor on the aseismic effect of locking dowel, the acceleration activating value is the significant factor while the locking gap is the least significant factor.

The orthogonal experimental method for locking dowel and the analytical approach for the optimization of impact factors are valuable for the popularization and application of aseismic analysis and parameter optimization of aseismic support for continuous girder bridge.

REFERENCES

Aashto Guide Specifications for Seismic isolation Design (3rd Edition) [S].
Chen Yong-qi, Wang Jing, Liu lin. Failure Analysis of Overseas Seismic Isolated Bridges [J]. Earthquake Resistant Engineering and Retrofitting, 2008, 30(5): 41–47. in Chinese.
Liu Jun, Wang He-xi. Application of Double Spherical Seismic Isolation Bearing in a Rigid Frame-Continuous Girder Bridge [J]. Journal of Railway Science and Engineering, 2012, 9(3): 117–123. in Chinese.
Wanc Li, Yan Gui-ping, Sun Li. Analysis of Seismic Response of Isolated Bridges with LRB [J]. Engineering Mechanics, 2003, 20(5): 124–129. in Chinese.

Advances in Energy, Environment and Materials Science – Wang & Zhao (Eds)
© 2016 Taylor & Francis Group, London, ISBN 978-1-138-02931-6

Research on crack propagation process of concrete using photoelastic coatings

Hongbo Gao, Dongsheng Song & Sujuan Ouyang
College of Civil Engineering and Architecture, Hainan University, Haikou, China

ABSTRACT: An advanced, accurate visual research method for photoelastic coatings is presented, in comprehending fracture properties of concrete. Using this method, whole crack propagating processes that appeared as interference fringes on wedge-splitting specimens were recorded utilizing a still camera and a digital video camera. The experimental results show that the crack propagation of concrete includes different stages: the crack initiation, the stable propagation, and the unstable propagation.

1 INTRODUCTION

Concrete is one kind of heterogeneous composite material, and its failure and fracture mechanism is very complicated. In the world, there is not enough understanding on the failure mechanism of concrete, especially its brittle fracture mechanism and basic mechanical properties.

In order to comprehend fracture mechanism of concrete and describe the crack propagation in concrete structures, it is necessary to correctly determine fracture parameters of concrete and comprehend crack propagation process before the unstable fracture, special physical property and mechanical behavior of the area near the crack tip. So, in the research on fracture mechanics of concrete, it is of important academic and practical significance on study of fracture mechanism to utilize photoelastic coatings (Zandman et al. 1977) to directly investigate the stable crack propagation process of concrete specimens before their failure. The work on this field performed by Niu (1991) shows that the processing procedure of the photoelastic coatings is simpler than that of Moire interferometry and does not need a reference grating. However, the further study is necessary.

The focal point of the experimental study in this paper is to present and directly use photoelastic coatings in fracture analysis of concrete specimens. Through this advanced visual method, the whole procedure of crack initiation, stable propagation and unstable propagation, i.e. the final failure, in all 24 wedge splitting specimens was observed and detected. In the experiments, reflection polariscopes were used to measure fringe-orders of isochromatic-fringes and propagation lengths of fringes, of which the shape looks like "V", in each specimen under different loads. Cameras were utilized to take photographs of the isochromatic-fringe patterns,

and digital video camera wholly recorded the crack propagating processes that appeared on photoelastic coatings. Then, the visual observation results of the crack initiation, the stable propagation and the unstable fracture under different loading stages were obtained.

2 EXPERIMENTS

2.1 Specimens

The specimen dimensions as indicated in Figure 1 are given in Table 1. All wedge splitting specimens were cast according to the same concrete mix from several batches within two weeks due to great amount of the concrete used. The concrete mix proportion is 1: 1.94: 3.17: 0.56 (cement: sand: aggregate: water by wt.). The maximum size of coarse aggregates, which are crushed limestone

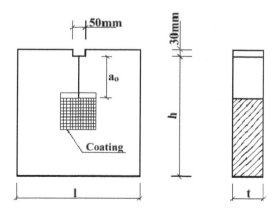

Figure 1. Configuration and geometry of wedge splitting specimen.

Table 1. Dimensions of specimens.

Specimens	l × h × t (mm)	a_0/h	Number of specimens
WS20	200 × 200 × 200	0.4	4
WS40	200 × 200 × 200	0.4	4
WS60	200 × 200 × 200	0.4	4
WS80	200 × 200 × 200	0.4	4
WS100	200 × 200 × 200	0.4	4
WS120	200 × 200 × 200	0.4	4

produced in the Dalian Area, is 20 mm and that of river sand is 5 mm. The initiation notches were made using greased steel plates with thickness of 2 mm. The notch-to-depth ratio (a_0/h) of all specimens is 0.4. A minimum of four specimens is required for each set. After casting, all specimens were maintained by watering for 28 days. The experimental age of all specimens is 70 days. The concrete cube compressive strength with 70 days experimental age is 29.5 MPa, the tensile strength 2.33 MPa, the modulus of elasticity 30.5 GPa and the Poisson's ratio 0.21.

2.2 Photoelastic coatings

An epoxy resin was used in the experiments as the coating matrix material. The manufacture of each coating is exactly performed to meet the following requirements: (1) high strain-optic sensitivity; (2) low modulus of elasticity to minimize reinforcing effects; (3) low initial stress; (4) high strain-stress proportion limit and strain-optics proportion limit; (5) low creep; and (6) good machinability. The modulus of elasticity of the epoxy resin coating E_c is 4.63 GPa, strain-optic sensitivity K is 0.104 and material-fringe value (strain) f_e is 755×10^{-6} cm per fringe.

2.3 Apparatus and testing procedure

A 5000 kN testing machine was used to achieve a stable test. Figure 2 shows the testing ground and the loading arrangement using a wedge splitting testing method. The wedge splitting testing method was proposed by Hillemeier & Hilsdorf (1977) to measure fracture energy of concrete, then further developed by Bruehwiler and Wittmann (1990) for performing stable fracture mechanics tests to study fracture behavior of concrete. In this study, two round steel bars under the bottom of a wedge splitting specimen simply supported the self weight of the tested specimen. The supports and the loading arrangements were such that the applied forces acting on the wedge splitting specimens are statically determined.

A clip gauge was utilized to measure the crack opening displacement COD. The COD was measured at the center of the notch to minimize possible errors caused by eccentricity. The crack mouth opening displacement COD and the applied load P were recorded continuously by computer data collecting system in the experiments. Normally, the load was monotonously applied to the tested specimen and a completed P-COD curve was recorded. For some specimens, the load was monotonously applied to the maximum load, and then after the maximum load was exceeded, unloading was done at the maximum load or at the 95% of maximum load so that unloading compliance can be determined, and then reloading procedure began again until the load became near to zero. The typical curves of the load P versus COD are shown in Figure 3.

During the tests, a reflection polar scope, accompanied with a portable digital video camera and a scientific camera were utilized to observe and record interference fringes appearing on each coating. The optical system of V-type utilized in the

Figure 2. Testing ground and wedge splitting loading arrangements.

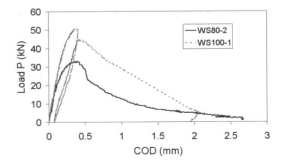

Figure 3. The plots of load versus crack opening displacement measured from specimens WS80-2 and WS100-1.

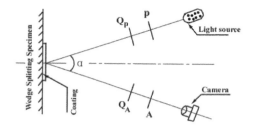

Figure 4. The illustration of an optical system of V—type used in the reflection polariscope. (P = Polarizer; A = Analyzer; Q = Quarter-wave plate).

investigations is illustrated in Figure 4. Normally, about 10 photographs of the interference fringes appeared on a photoelastic coating during the testing procedure for a specimen could be recorded.

3 RESULTS AND DISCUSSION

In the experiments, all isochromatic-fringe patterns that appeared on their photoelastic coatings under different loads were recorded through cameras. Figure 5 show the whole processes of crack propagation from crack initiation to unstable fracture under different loads recorded from the two specimens WS80–2. Some crack propagation traces, after peak load was exceeded, are given, too. According to the investigation of Zandman et al., 1977, the non-uniform fringes, of which the shape likes "V", called V-type fringes, indicate the presence of a crack. This means that the V-type fringes shown in Figures 5 disclose the crack propagating processes in the two specimens.

The preformed crack on the coating coincides with that on the specimen. Comparing the crack propagating processes in Figures 5 with the correspondingly real crack propagated traces, after unstable fracture occurred, i.e. the peak load was achieved; it is not difficult to find that each tip of the V-type fringe is completely consistent with a corresponding realistic fracture trace. So, the crack propagating processes appeared on the photoelastic coatings are worth of confidence.

According to the research achievement, 0.25% can be served as the tensile strain ultimate of concrete. The fringe-order value N is calculated using the formula (1), (Zandman et al. 1977), which gotten through the stress-optic law substituting all relevant parameters' values.

$$N = \frac{2\left(1+v_c\right)\cdot t_e \varepsilon_c}{\left[1+\dfrac{t_e}{t_c}\cdot\dfrac{E_e}{E_c}\cdot\dfrac{1+v_c}{1+v_e}\right]\cdot f_\varepsilon} \quad (1)$$

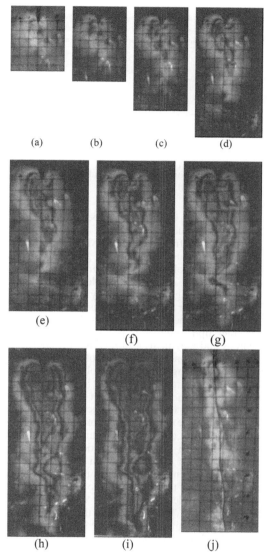

Figure 5. Crack propagation process on the photoelastic coating of specimen WS80-2 (Unit: cm).
(a) $P = 0.582\,P_{max}$; (b) $P = 0.638\,P_{max}$; (c) $P = 0.706\,P_{max}$;
(d) $P = 0.780\,P_{max}$; (e) $P = 0.836\,P_{max}$; (f) $P = 0.881\,P_{max}$;
(g) $P = 0.921\,P_{max}$; (h) $P = 0.972\,P_{max}$; (i) $P = 0.983\,P_{max}$;
(j) Crack propagation trace after peak load.

In formula (1), t_c represents the thicknesses of concrete specimens; t_e is the thickness average value of all epoxy coatings, where $t_e = 0.26$ cm; E_c and E_e are the modulus of elasticity of concrete and epoxy coating materials, respectively; v_c and v_e are the possion's ratios of concrete and epoxy coating materials, respectively; ε_{tu} is the ultimate tensile strain value of concrete material, 0.25%; f_ε is material-fringe value

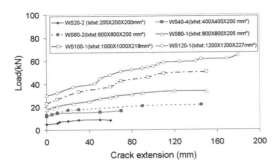

Figure 6. Relationship between load and crack extension (P-Δa curves).

(strain), where $f_\varepsilon = 755 \times 10^{-6}$ cm per fringe. So, The fringe-order value N is 0.21.

According to fringe alignment, when isochromatic fringe-order value $N = 0.21$, the relevant color is between black color and gray color and approaches to the gray color. As an example, please see the crack tip detected using the photography fringe alignment shown in Figure 5 (g). This means that the tension strain ε_{tu} of concrete materials used in the tests at the crack tip is assumed to be 0.0025. Through this analysis and judgment, it means that the crack initiation on the concrete when the color of photoelastic coatings changes from black color to gray color. So, the tip of gray V-type fringe is just the tip of specimen's extending crack. With the observation results recorded through the reflection polariscope and cameras, the crack propagation lengths of each specimen under different loads can be determined easily. Figure 6 shows the relationship between load and crack extension length ($P - \Delta a$ curve). In the measurements, the measuring distance is from the tip of the preformed notch to the point where the tensile strain capacity of concrete was firstly exceeded. This means that the "crack extension length" termed in this paper already included a length of the fracture process zone conventionally termed in literature. Therefore, in this paper, the critical crack extension length Δa_c is defined as such a distance that is from the tip of the performed notch to the corresponding point where the tensile strain value ε is larger than ε_{tu}, the ultimate of tensile strain of concrete at the peak load.

4 CONCLUSION

Through the study in this paper, we could conclude that in the research on the fracture mechanism of concrete, especially the observation on the crack propagating processes in the prototype concrete specimens or practical structures, the photoelastic coating is one kind of effective method, which possesses many advantages compared with other methods of experimental strain and stress analysis. It is a full-field, precise, concise, visual method that is convenient to be kept by taking photographs or films, so it is worth of applying and spreading. Using this method in our experiments, we have investigated that a clear stable propagating process of macro-crack exists before unstable fracture occurred. With the increase of specimen size, the crack propagation becomes more and more obvious and can be seen by naked eyes. The fracture process of the concrete structure is a complicated and relatively long-term one that includes the crack initiation, the stable propagation and the unstable propagation. Before the unstable fracture happened, the observable macro-cracks have already appeared on the concrete specimens. It was found that the critical crack extension length Δa_c increases with the increase of specimen size within the range of specimen sizes used in the experiments.

Furthermore, it is found that with the same initial notch-to-depth ratio (a_0 / h), the larger the size of the specimen is, the earlier the crack initiation will occur. On the contrary, under the same pre-requirement, the smaller the size of the specimen is, the later the crack initiation will be. So, for small size specimens, the crack initiation of their macro-cracks is relatively late, and the failure will occur soon after the crack initiation. It could be expected that it would be comparatively difficult to observe the stable propagation of the main cracks in fracture experiments on small size concrete specimens.

ACKNOWLEDGMENTS

This work was financially supported by the Natural Science Foundation of China (No. 51268010) and Natural Science Foundation of Hainan province (No. 512116).

REFERENCES

Brühwiler, E. and Wittmann, F.H., The wedge splitting test, a method of performing stable fracture mechanics tests. Engineering Fracture Mechanics, 35(1–3) (1990), 117–125.

Hillemeier, B., and Hilsdorf, H.K., Fracture mechanics studies on concrete compounds. Cement and Concrete Research, 7(5) (1977), 523–535.

Niu, Yanzhou, and Tu, Chuanlin, The Investigation on the Crack Propagation Process of Big Aggregate Concrete Using Photoelastic Coatings. Journal of Hydroelectric Engineering, Beijing, 4(1991), 38–44.

Zandman, F., Render, S. and Dally, J.W., Photoelastic Coatings, Published Jointly by the Iowa State University Press and Society for Experimental Stress Analysis (SESA), First Edition, 1977.

Environmental materials

Advances in Energy, Environment and Materials Science – Wang & Zhao (Eds)
© *2016 Taylor & Francis Group, London, ISBN 978-1-138-02931-6*

Synthesis and adsorption of OMMT/AHL grafted PAA superabsorbent composite

Yifan Xu, Hongxing Zhao, Guopeng Chen, Enxiao Lian & Yanli Ma
College of Material Science and Engineering, Northeast Forestry University, Harbin, China

ABSTRACT: Organic montmorillonite/acid hydrolysis lignin graft poly(acrylic acid) superabsorbent composite (OMMT/AHL-g-PAA) was prepared by intercalation polymerization of montmorillonite, acrylic acid and acid hydrolysis lignin using ammonium persulphate as an initiator and N,N′-ethylene bisacrylamide (MBA) as a crosslinker. Adsorption behavior of water and Pb(II) ion on OMMT/AHL-g-PAA were investigated. The maximum adsorption capacity of Pb(II) reached $239.18 \, mg \cdot g^{-1}$. The adsorption kinetics of OMMT/AHL grafted PAA fitted a pseudo-second-order kinetic model, indicating that chemisorption contributes mainly to Pb(II) adsorption. The solution pH values had a major impact on Pb(II) adsorption with optimal removal observed around pH 3–5 and 8–9.

1 INTRODUCTION

With the rapid development of metal plating equipment, mining, tanning, pesticide industry, heavy metals and organic compounds in wastewater is directly or indirectly discharged into the environment. Different from the organic pollutants (Hao Cui et al. 2012), heavy metals are not biodegradable. Many techniques have been used for the removal of toxic heavy metals from the water, such as chemical precipitation, ion exchange, membrane filtration, adsorption, coagulation, flocculation, flotation and electrochemical treatment (Suksabye P. & Thiravetyan P. 1987). The adsorption process, for convenient, cheap and effective, is often considered as the most appropriate the method of removing inorganic and organic pollutants (Aihua Sun et al. 2006).

The adsorption efficiency mainly depends on the type of sorbent. The adsorption capacity of polyacrylic acid hydrogel can reach 1000 g/g, with structure of three-dimensional hydrophilic network in water, was widely used in agriculture, tissue engineering, drug delivery, water treatment and sensor research (Panzavolta, S. et al. 2014). Effect of carboxyl electrostatic can influence metal cation adsorption capacity to the composite polymer. But due to the low mechanical and dynamics properties of polyacrylic acid hydrogel hinders its practical application in water treatment.

The lignin from paper-making, textile and other cellulose preparation industry is a resource of non-toxic, renewable and abundant, which is a connection with more hydrophobic degradable natural polymer (Satvinder S. et al. 2013) by C—C bonds and ether bond. The flocculation of lignin is very beneficial for improving the adsorption capacity of hydrogels for heavy metal ions (Uzochukwu C. Ugochukwu et al. 2014), and the hydroxyl, carbonyl, ether and benzene methyl in lignin molecule can form hydrogen bonds or π—hydrogen bonds, the hydroxyl and carbonyl of lignin can interact with metal ions. In summary, lignin containing graft hydrogel may be applied for the wastewater treatment fields of heavy metal and dye.

In a previous work of ours a composite based on OMMT and PAA was characterized. The results confirmed the formation of well dispersed ordered intercalated assemble layers of MMT/PAA matrix. The presence of acid hydrolysis lignin in intercalated three-dimensional composite internal structures was confirmed by uniform -dispersion. The amount of Pb(II) was calculated by atomic spectrophotometer. The purpose is preparation of montmorillonite/acid hydrolysis lignin grafted polyacrylic acid compound to improve the biodegradability and strength of gel material using the method of graft copolymerization. We study the sensitivity and expansion of the adsorption capacity of Pb(II) for pH, temperature and ionic strength to determine the dynamic mechanism of compound.

2 EXPERIMENTS

2.1 *Material and instrument*

Commercial alkali lignin was supplied by Tralin Paper Co., Ltd. (Shandong, China). All reagents were of analytical grade and were used without further purification. All pH measurements were made with a pHS-3 digital pH-meter (Shanghai Lei Ci Device Works, Shanghai, China) and Atomic Absorption Spectrometry (AAS) (TAS-990, P general, China).

2.2 *Preparation of OMMT/AHL-g-PAA*

Preparation of acid hydrolysis lignin: 0.54 g crude alkali lignin was poured into 400 mL HCl (1 mol/L) and keep pH as 5 at 80°C for 4 h, and then filtration, drying to get acid hydrolysis lignin.

Preparation of OMMT: 8 g Na-montmorillonite and 200 mL distilled water were added into the three-neck flask and adjusted the pH to 7, stirred and added 60 mL 0.14 mol/L hexadecyl trimethyl ammonium bromide solution, raised the temperature to 70°C and reacted 2 h. After the reaction, the slurry was filtered, washed with water, dried and shattered to get a white powder organic montmorillonite (OMMT).

Preparation of montmorillonite/acid hydrolysis lignin graft poly(acrylic acid): 1.56 mL acrylic acid that degree of neutralization is 60%, 3 wt% OMMT, 0.7 wt% N,N—methylene-bis-acrylamide and 1 wt% ammonium persulfate were mixed and stirred for 5 min, and then heated to 80° and kept 2 h to get montmorillonite/poly(acrylic acid) composite. At the same time, 0.2 g, 0.4 g, 0.6 g, 0.8 g, 1.0 g acid hydrolysis lignin were separately added 8 mL of 1 M NaOH solution, 0.05 g FeSO4 and 1 mL of 30% H_2O_2 and kept temperature at 60°C for 30 min, mixed it with above the composite at 60°C and kept for 2 h, the product was washed with anhydrous ethanol for 3 times, vacuumed for drying at 50°C to prepare montmorillonite/Acid Hydrolysis Lignin (AHL) graft poly (acrylic acid). Then, montmorillonite/acid hydrolysis lignin graft poly (acrylic acid) was grinding into 20 to 40 mesh particles for testing and characterization.

And frames, copy these texts paragraph by paragraph without including the first word (which includes the old tag). It is best to first retype the first words manually and then to paste the correct text behind. When the new file contains all the text, the old tags in the text should be replaced by the new Balkema tags (see section 3). Before doing this apply automatic formatting (AutoFormat in Format menu).

2.3 *Characterization*

2.3.1 *Determination of environmental sensitivity*

The pH, temperature and concentration of NaCl swelling properties of OMMT/AHL grafted PAA were determined by measuring their hydrogels swelling ratio. To determine their pH-responsive properties, 0.1 g freeze-dried hydrogel samples were immersed in buffer solutions with different pH values (pH 3–11). The solutions were kept at 25°C without NaCl. To measure their temperature-responsive properties, 0.1 g freeze-dried hydrogel samples are immersed in different temperature buffer solutions (4–60°C). The pH values were kept at 6 without NaCl. To determine their ionic strength-responsive properties, 0.1 g freeze-dried hydrogel samples were immersed in buffer solutions with different concentration of NaCl (1–20 mg/L). The solutions were kept pH values at 6 in 25°C temperature. After a fixed period of time, the hydrogels were taken out from the buffer solutions. After being wiped off the excess solutions on the surfaces weighed it.

2.3.2 *Pb(II) adsorption experiments*

The experiments were conducted by containing the OMMT/AHL grafted PAA and 100 ml of heavy metal wastewater (Pb ionic concentration = 375 µg/mL) in a 250 mL sealed grinding mouth erlenmeyer flask and then shaking at 150 rpm. The system pH, ionic strength (adjusted by adding NaCl) and temperature were varied from pH 3 to 9, 1 to 20 mg/L and 15 to 45°C, with particle size 20 to 40 mesh. The Pb(II) that remained in the supernatant was measured using an atomic absorption spectrophotometer and calculate sample adsorption capacity of lead ions.

3 RESULTS AND DISCUSSION

3.1 *Environmental sensitivity of OMMT/AHL grafted PAA hydrogels*

To determine pH value and time effects on OMMT/AHL grafted PAA, a batch experiment was performed on the adsorption of water by OMMT/AHL graft poly(acrylic acid) with different ratios of AHL grafted PAA (w/w,0.2:1,0.4:1,0.6:1,0.8: 1,1:1). And OMMT/AHL grafted PAA with the ratio of AHL grafted PAA (w/w, 0.2:1) was chosen to measure sensitivity of the equilibrium swelling ratio by different ionic strength and temperature. From Figure 1, a series of OMMT/AHL grafted PAA have analogous swelling curves. With concentration of H^+ in the solution increased, the swelling rate of composite hydrogel increased at pH range of 3 to 5, that reason is H^+ repulsion between H^+ in the carboxylic acid of compound gel and H^+ in the solution. When the solution was adjusted to pH 6, due to the H^+ concentration decreased and carboxy groups were growed into carboxylate anions, the swelling rate of composite hydrogel was decline. Phenol is weakly acidic and at high pH gives the phenolate anion $C_6H_5O^-$, the carboxylate anion of OH^- in solution also form a new repulsion system. Since then, with the OH^- concentration increased,

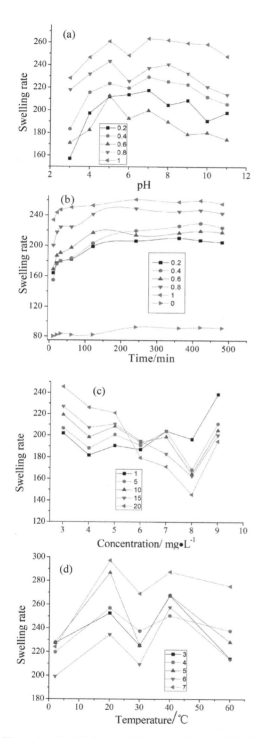

Figure 1. Equilibrium swelling ratio of several kinds of acid hydrolysis lignin content hydrogels on effect of (a)pH, (b) time, (c) ionic strength, (d) temperature.

hydrogel swelled again (see Fig. 1 a). Different ratios of AHL grafted PAA's kinetics curves of composite obeyed Higuchi equation (see Fig. 1b). Compare Figure 1(a) with Figure 1(b), with the ratios of AHL grafted PAA increased, the swelling ratio of composite increased too. However the ratio was 0.6:1, its corresponding swelling ratio of composite falls sharply. Density of composite hydrogel increased led to penetration resistance of water. That due to acid hydrolysis lignin grafted polypropylene making the system neutral.

The 0.2:1 ratio (w/w) of OMMT/AHL grafted PAA was chose to determined sensitivity of ionic strength and temperature. With the increasing of concentration of NaCl composite hydrogel expansion was increasing at pH range of 3 to 8. At this time concentration of NaCl have the negative effect on the composite hydrogel. After concentration of NaCl was higher than 8 mg·L^{-1}, the carboxy turns into carboxylate, Na$^+$ ions made swelling rate of composite to decline sharply. Carboxyl and OH$^-$ in solution formed a new repulsion system.

3.2 *Effect of pH value on Pb(II) adsorption of OMMT/AHL grafted PAA*

The influence of pH of the solution on adsorption capacity of Pb(II) by OMMT/AHL grafted PAA was studied under a fixed initial Pb(II) concentrations (375 µg/mL) within the pH range of 3–9 (Fig. 2). The lead removal capacity increased abruptly with increasing the pH and reached almost maximum adsorption around an initial pH = 4.0 for 0.2:1, 0.4:1 and 0.8:1 (w/w) of AHL to PAA investigated. Increasing initial pH of solution favors Pb-OH bond formation between the reactive groups on the OMMT/AHL grafted PAA and Pb(II). The adsorption capacity was then

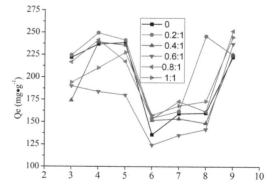

Figure 2. Effect of pH of equilibrium solution on adsorption of Pb(II) by OMMT/AHL grafted PAA that swelling ratios of AHL to PAA were 0.2:1, 0.4:1, 0.6:1, 0.8:1 and 1:1 (w/w).

sharply fall within the pH range of 5–6. When pH increased to 7, Pb(OH)$^+$ and Pb(OH)$_2$ species could form and become the dominant species as pH increases. Pb(OH)$^+$ and Pb(OH)$_2$ species has higher affinity to oxygen-containing binding sites. Meanwhile, with increasing pH, protons are released from the carboxyl group groups, leaving more binding sites available for Pb adsorption. Therefore, the minimum adsorption capacity stayed high at pH = 6–8, possibly due to the fact that Pb(OH)$^+$ species have a higher formation constant and can steadily stay in the equilibrium solutions. Further increasing the pH above 9.0 increased the adsorption, possibly due to the fact that Pb(OH)$_2$ species formed precipitation at pH > 9 have a lower solubility and cannot steadily stay in the equilibrium solutions.

3.3 Pb(II) adsorption kinetic studies of OMMT/AHL grafted PAA

The influence of contact time was investigated using 375 mg L^{-1} initial Pb(II) concentration at different samples (PAA, OMMT/PAA and OMMT/AHL grafted PAA). Figure 3A shows the adsorption capacity of Pb(II) by those samples as a function of contact time. It can be seen that the adsorption rate was very high during the first 4 h of the process. The removal of Pb(II) maintained at a constant level after 8 h contact time. In order to evaluate the controlling mechanism of adsorption processes, pseudo-first-order and pseudo-second-order kinetic equations were used to test the experimental data. The pseudo-first order kinetic model (Hao Cui et al. 2012) was suggested and its linear form can be formulated as:

$$\ln(Q_e - Q_t) = \ln Q_e - k_1 t$$

Pseudo-second order kinetic model (Hao Cui et al. 2012) can be expressed as:

$$t/Q_t = 1/k_2 Q_e^2 + t/Q_e$$

where Q_e and Q_t are the amount of Pb(II) adsorbed at equilibrium and time t (mg·g^{-1}), espectively. k_1 is the equilibrium rate constant of the pseudo-first order adsorption (min^{-1}) which is determined from the slope of plot of $\ln(Q_e - Q_t)$ versus t (Fig. 3a). k_2 is the equilibrium rate constant of the pseudo-second order adsorption (g mg^{-1} min^{-1}) which is similarly determined from the slope of plot of t/Q_t versus t (Fig. 3b).

The parameters including rate constants k_1, k_2 and correlation coefficients of PAA, OMMT/PAA and OMMT/AHL grafted PAA at T = 298 K was calculated and the results were listed in Table 1. The

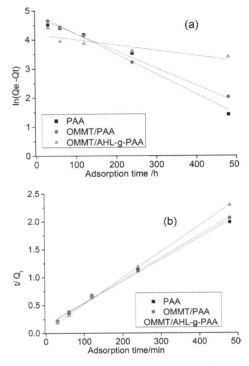

Figure 3. Adsorption kinetics (a) pseudo-first-order and (b) pseudo-second-order of Pb(II) OMMT/AHL grafted PAA.

Table 1. Parameters for kinetic models of Pb(II) adsorption onto OMMT/AHL grafted PAA.

Sample	PAA	OMMT/ PAA	OMMT/ AHL grafted PAA
Q_e (mg·L^{-1})	245.8	240.55	239.18
Pseudo-first-order model in 480 min			
k_1 (min^{-1})	0.0069	0.0059	0.0019
$Q_{e\,calc}$ (mg·g^{-1})	133.62	120.54	65.5
R^2	0.9863	0.9904	0.7
Pseudo-second-order model in 480 min			
k_2 (g·mg^{-1}·min^{-1})	1.19 × 10^{-4}	1.12 × 10^{-4}	4.01 × 10^{-4}
$Q_{e\,calc}$ (mg·g^{-1})	256.4103	250	217.3913
R^2	0.993	0.9971	0.9997

correlation coefficient values of the pseudo-first-order model are low and the calculated equilibrium adsorption capacity do not agree with experimental values, indicating that the pseudo-first order kinetic model is poor fit for the adsorption processes of those samples for Pb(II). However, for pseudo-second order kinetic equation, the correlation coefficients are found to be higher than 0.99, and its calculated equilibrium adsorption capacities fit well with the experimental data. These suggest that the adsorption data are well represented by pseudo-second-order kinetics, which indicates that the adsorption process was controlled by chemical complexation.

4 CONCLUSIONS

In this study, OMMT/AHL grafted PAA composite was prepared and the adsorption of Pb(II) ions by this composite was investigated in batch systems. The maximum adsorption capacity of Pb(II) reached 239.18 mg·g^{-1}. The adsorption kinetics of OMMT/AHL grafted PAA fitted a pseudo- second-order kinetic model, indicating that chemisorption contributes mainly to Pb(II) adsorption. The solution pH values had a major impact on Pb(II) adsorption with optimal removal observed around pH 3–5 and 8–9. Based on these results, a highly effective Pb(II) removal from aqueous solutions is anticipated for OMMT/AHL grafted PAA.

ACKNOWLEDGEMENT

This work was supported by the Fundamental Research Funds for the Central Universities (No. DL13CB09), the Youth Science Foundation of Heilongjiang Province of China (No. QC2014C011) and the Chinese "Twelfth Five Year" National Science and Technology Plan Project in Rural Areas (No. 2012BAD24B0403).

REFERENCES

Aihua Sun, Zhigang Xiong, Yiming Xu, Adsorption and photosensitized oxidation of sulfide ions on aluminum tetrasulfophthalocyanine-loaded anionic resin, Journal of Molecular Catalysis A: Chemical, 2006, Vol(259):1–6.

Fenfen Guo, Wenjing Shi, Wan Sun, Xuezhi Li, Feifei Wang, Jian Zhao and Yinbo Qu, Differences in the adsorption of enzymes onto lignins from diverse types of lignocellulosic biomass and the underlying mechanism, Biotechnology for Biofuels, 2014, 7:38.

Giuseppina Sandri, Maria Cristina Bonferoni, Franca Ferrari, Silvia Rossi, Carola Aguzzi, Michela Mori, Pietro Grisoli, Pilar Cerezo, Marika Tenci, Cesar Viseras, Carla Caramella, Montmorillonite–chitosan–silver sulfadiazine nanocomposites for topical treatment of chronic skin lesions: In vitro biocompatibility, antibacterial efficacy and gap closure cell motility properties, Carbohydrate Polymers, 2014, Vol (102): 970–977.

Hao Cui, Yan Qian, Qin Li, Qiu Zhang, Jianping Zhai, Adsorption of aqueous Hg(II) by a polyaniline/attapulgite composite, Chemical Engineering Journal, 2012, Vol(211–212): 216–223.

Panzavolta, S., Gioffrè, M., Bracci, B., Rubini, K. and Bigi, A. Montmorillonite reinforced type A gelatin nanocomposites. J. Appl. Polym. Sci. 2014, 131, 40301(1–6).

Satvinder S. Panesar, Sinto Jacob Manjusri Misra, Amar K. Mohanty, Functionalization of lignin: Fundamental studies on aqueous graft copolymerization with vinyl acetate, Industrial Crops and Products, 2013, 46: 191–196.

Suksabye P., Thiravetyan P. Cr(VI) adsorption from electroplating plating wastewater by chemically modified coir pith. J Environ Manage. 2012, vol (102):1–8.

Uzochukwu C. Ugochukwu, Martin D. Jones, Ian M. Head, David A.C. Manning, Claire I. Fialips, Biodegradation and adsorption of crude oil hydrocarbons supported on "homoionic" montmorillonite clay minerals, Applied Clay Science, 2014, Vol (87): 81–86.

Advances in Energy, Environment and Materials Science – Wang & Zhao (Eds)
© 2016 Taylor & Francis Group, London, ISBN 978-1-138-02931-6

Distribution characteristics of saturated hydrocarbon and UCM, organic matter sources and oil-gas indicative significance of core P7327 in the Chukchi Sea

Qingying Zhao, Ronghua Chen, Haisheng Zhang & Bing Lu
Second Institute of Oceanography, SOA, Hangzhou, China

ABSTRACT: Research molecular component characteristics and characteristic ratios of n-alkanes, isoprenoid, Unresolved Complex Mixtures (UCM) and sterane of sediment core P7327 in the Chukchi Sea. Results show that component characteristics of n-alkanes at most sections are similar; bimodal pattern shows dual contribution of terrigenous and marine sources. High-carbon hydrocarbon/low-carbon hydrocarbon ratio (H/L) of n-alkanes in sediment core (1–121 cm) is 0.299~1.326, C_{27}/C_{19} ratio is 1.392~7.565, lipoid ratio TAR and ACL are 1.015~5.531 and 0.930~7.610 respectively. These 4 parameters further show that, sources from terrigenous higher plants and marine sources (lower algae, fungi and submerged plants) coexist in modern sediments of the Chukchi Sea. Researches also show that Carbon Preference Index (CPI) of sediment core (7–184 cm) is 1.208~1.664, close to 1; pristane/ phytane ratio (pr/ph) is low at most sections; a lot of Unresolved Complex Mixtures (UCM) prevail in the samples; U/R ratio is 2.2–4.3. These 3 parameters show that petroleum hydrocarbon with high maturity is input, and subject to biodegradation later. According to intact n-alkanes series and obvious UCM coexistence in sediment core P7327 of the Chukchi Sea (in particular, carbon chain length of n-alkanes in 182–184 cm sediment is obviously shorter than that of other sections), sterane isomerization parameter $C_{29}20S/(20S + 20R)$ and $C_{29}\beta\beta/(\beta\beta+\alpha\alpha)$ is 0.37–0.47 and 0.45–0.53 respectively, which reach or close to isomerization balance end point and belong to a mature range. Different from contemporary sediments, maturity of organic matters perhaps mean that, contemporary sediments of the Chukchi Sea may be mixed with petroleum source. These petroleum hydrocarbons may come from leaked oil and gas reservoirs in the deep sea (including pollution from human activities), and may be affected by microbial action of bacteria.

1 INTRODUCTION

Saturated hydrocarbon in marine bottom sediments mainly consists of n-alkanes, isoprenoids, steranes, terpanes and biomarkers. Their refined molecular components and distribution characteristics directly indicate primary organic matter source, bacterial degradation, sedimentary environment, seabed oil and gas shows (Brassell et al, 1986; Wolff et al, 1986; Wang Tieguan et al, 1995). Unresolved Complex Mixture (UCM) in saturated hydrocarbon is formed by bacterial biodegradation of petroleum; in the extract from modern sediments, fraction chromatogram of saturated hydrocarbon has UCM, which is regarded to be an existence evidence of petroleum organic matter. UCM in crude oil contains 250,000 compounds (Sutton et al, 2005), with obvious upheaval in crude oil chromatogram subject to biodegradation (Peters et al, 2005). These almost indistinguishable compounds may contain a lot of undiscovered organic geochemistry information (Gough & Rowland, 1991).

With 58.7×104 Km² area, Chukchi Sea is a marginal sea of the Arctic Ocean located in Chukchi Peninsula, between Alaska Peninsula and Wrangel Island. As a shallow shelf sea in the south of the Arctic Ocean, it plays an important role in matter and energy exchange between the Arctic Ocean and the Pacific (Walsh et al, 1989; Zhao Jinping et al, 2003). In addition, after the world's first ocean drilling in the North Pole has obtained late Paleocene-middle Miocene sediments with rich organic matters, people guess that the North Pole boasts rich oil and gas resources; there may be oil under all ice sheets (Revkin et al, 2004), which attract extensive attention from global scholars and international institutions (USGS, 2008). However, due to awful weather and difficulty in sea exploration (floating ice and large-scale neogene basalt), full exploration hasn't been carried out so far. As the North Pole has been increasingly warming and ice sheets of the Arctic Ocean have been melting quickly in recent years, natural barrier to oil and gas resources in the North Pole is gradually collapsing. So Russia is launching a strategic plan for exploration and development of oil and gas in continental shelf of the North Pole (Han et al, 2011; 2012). By referring to literatures, it is found that so far, Chinese scholars' researches about the North Pole mostly focus on natural science, and seldom involve exploration of seabed oil and gas resources in the North Pole and the Arctic Ocean.

With organic geochemical methods, this paper researches biomarker compound components of saturated hydrocarbon (n-alkanes, isoprenoid, sterane) and morphological characteristics of UCM in sediment core of the Chukchi Sea, obtains organic matter sources, sedimentary condition, bacterial degradation and recognition of possible oil and gas leakage in the deep Chukchi Sea, and thus provides geochemical evidence for researches about ecological environment, oil and gas resources in the Chukchi Sea.

2 GEOLOGICAL OVERVIEW OF RESEARCH AREA

Like other seas in the North Pole, Chukchi Sea has accumulated all kinds of terrigenous sediments, but the function of the Arctic Ocean is weakened due to the Pacific water input through Bering Strait. In bottom deposits in open area of Chukchi Sea, organic matter source is vital movement of plankton, organisms (plant debris) and dissolved humus (essential part) that inflow with the Pacific water. Moreover, formation, movement and storage of oil and gas resources are inseparable from landmass evolution. Present tectonic morphology of the North Pole is formed after complicated tectonic evolutions of many landmasses (Lawrence et al, 2008). Researches show that Canada Basin (Canada Basin may be the earliest basin of the Arctic Ocean) is sunken and reestablishes structure outline, which is the reason for depression of North Chukchi. Sedimentary basins on the continental shelf of Chukchi Sea have been developed due to strong sedimentation of the depression and accumulation of sediments. As the thickness of settled layer increases, there are favorable conditions for formation and accumulation of oil and gas. It is thought that the Cretaceous—Cenozoic continental shelves and continental slope basins have huge oil and gas potentials; oil and gas gather in sedimentary covers and basements of basins (Thurston & Theiss, 1987).

2.1 Sample collection

Sediment core P7327 was collected by "Snow Dragon" icebreaker during the First Chinese National Arctic Research Expedition in Chukchi Sea of the Arctic Ocean (73°25′N, 164°57′W,

81 m water depth), near the margin of Canada Basin. Location of sediment core P7327 is shown in Figure 1; this gravity-piston type sample is 184 cm long. It mainly consists of ash black sandy clay and argillaceous silt. Seen from distribution of fine particle sediment sample, Chukchi Sea mainly has a stable sedimentary environment with relatively weak hydrodynamic condition.

2.2 Pretreatment and analysis before test

The sample was dried at a temperature lower than 50°C, smashed to 100 mesh (powder) and extracted for 72 h with dichloromethane and methyl alcohol at 3:1 volume ratio. Extracting solution of every layer owned obvious shade deviation; through rotary evaporation, the extract was dried. Then, with normal hexane, above concentrated solution deposited and got rid of asphaltene, and went through chromatographic separation with silica gel column; with n-hexane, benzene and dichloromethane/methyl alcohol (1:1), saturated hydrocarbon, aromatic hydrocarbon and impurities were leached and separated. Saturated hydrocarbon component was used for gas chromatography and chromatography—mass spectrometry.

n-alkanes: it used HP5890 Gas Chromatograph (GC) and elastic quartz capillary column (DB-5, 30 m × 0.25 mm inner diameter and 0.17 μm coating thickness). Temperature rise program: initial temperature was 80°C, temperature rise rate was 5°C/min and final temperature was 280°C which was kept for 30 min.

Sterane: Gas Chromatograph-Mass Spectrograph (GC-MS) was used. Vaporizing chamber temperature: 310°C, transmission line temperature: 310°C, temperature rise program: initial temperature of column: 100°C, temperature rise rate: 4°C/min, final temperature of column: 3154°C, constant temperature was kept for 22 min. Chromatographic column DB5-MS 60 m × 0.25 mm × 0.25 um, column pressure: 170 kPa, split ratio: 20:1, carrier gas: helium, line speed: 27 cm/s, collection mode: circle

Figure 1.   Map of sampling stations of P7327 hydrocarbon source rocks in Chukchi Sea.

scan (SCAN)/Selected Ion Monitoring (SIM), ion source temperature: 250°C, scanning speed: 0.46 scan/s, ionization mode: Electron Impact (EI).

3 RESULT AND CONCLUSION

3.1 Recognition of molecular assembly characteristics and sources of saturated hydrocarbon

3.1.1 N-alkanes and source

As a widespread biomarker compound in marine environment, n-alkanes mainly have artificial source, biological source and natural source. In general, n-alkanes from petroleum hydrocarbon input own high maturity and low main peak carbon number, without obvious odd-even predominance (Hostettler et al, 1999). Generally speaking, in n-alkanes chromatogram, if pre-peak n-alkanes with low carbon number are predominant, n-alkanes mainly come from marine source; if post-peak n-alkanes with high carbon number are predominant, n-alkanes mainly come from terrestrial source; bimodal group means mixed marine and terrestrial source (Zhu Chun et al, 2005; Yang Dan et al, 2006). C_{17} and C_{19} in low carbon chain distribution mainly come from algae, while n-alkanes with even number carbon of C_{16}~C_{22} are generally bacterial sources (Guo Zhigang et al, 2006; Gao et al, 2008). n-alkanes with medium carbon chain length (C_{23}~C_{27}) come from large fresh water and marine plants; large submerged plants are mainly C_{23} and C_{25} (Ficken et al, 2000; Mead et al, 2005). n-alkanes with long carbon chain come from surface wax, leaves, pollen, spores and fruits of terrigenous higher plants (Goni et al, 1997), with the richest C_{27}, C_{29} and C_{31}; there is obvious odd-even predominance in C_{25}~C_{33} (Silliman et al, 2003); C_{27}/C_{19} reflects relative proportion of external source

and endogenous hydrocarbon. Moreover, average content of carbon atoms in Average Carbon Chain Length (ACL) of long-chain n-alkanes (C_{25}~C_{33}) from terrigenous higher plants indicates input extent of petroleum hydrocarbon in the sediment. When the sediment has petroleum hydrocarbon input, ACL of the sediment drops to some extent (Jeng et al, 2006).

Among 7 sections of sediment core P7327 in the Chukchi Sea, carbon number distribution range of n-alkanes is C_{15}~C_{33}, mainly with bimodal pattern. Main peak carbons of low and medium carbon peaks are C_{17}, C_{19} or C_{25}, while main peak carbon of high carbon peaks is C_{27}, without obvious odd-even predominance. It means that organic matters come from marine fungi, algae and terrigenous plants. Composition structures of n-alkanes among all sections have significant changes: $\Sigma C_{27+29+31}$ of high carbon chain is 0.123~28.149, $\Sigma C_{21+23+25}$ of medium carbon chain is 1.023~16.282, $\Sigma C_{15+17+19}$ of low carbon chain is 1.495~5.491, and TAR is 1.015~ 5.531, showing that organic matters in the sediment have more terrigenous sources than marine sources.

3.1.2 Isotrenoid and sedimentary environment (oxidation/reduction)

Phytane is predominant in reducing environment, whereas pristine is predominant in oxidation or weak oxidation environment. Therefore, high Pr/Ph reflects oxidation palaeo environment, and low Pr/Ph shows reducing environment. Environmental pH value also affects phytol conversion: acid environment is good for pristane generation and alkaline environment is good for phytane, so Pr/Ph relative content in sedimentary organic matter marks conversion environment of original organic matter diagenesis. It is generally acknowledged that Pr/Ph<1 indicates anoxic reducing sedimentary

Table 1. Geochemical parameters of n-alkanes in recent sediments from the Chukchi Sea.

Sediment core P7327 (cm)	Carbon number range	$\Sigma C_{27+29+31}$	$\Sigma C_{21+23+25}$	$\Sigma C_{15+17+19}$	ACL	TAR	nC_{27}/nC_{19}	L/H	CPI	Pr/Ph	Pr/nC_{17}	Ph/nC_{18}
1~2	nC16~C33	28.149	16.282	5.089	7.610	5.531	7.565	0.299	1.264	3.017	1.079	0.370
7~8	nC15~C31	3.496	5.233	3.443	1.040	1.015	1.648	1.205	1.664	1.183	0.930	0.821
40~41	nC16~C31	2.939	6.424	2.413	0.930	1.218	1.392	1.326	1.650	0.784	0.936	1.041
86~87	nC16~C33	3.944	7.615	2.257	1.177	1.747	1.577	0.968	1.651	1.243	0.730	0.465
107~108	nC15~C33	14.626	11.452	5.491	3.899	2.664	2.760	0.543	1.341	1.866	0.914	0.409
120~121	nC16~C33	5.328	4.389	1.495	1.418	3.564	3.136	0.616	1.331	0.687	0.942	0.796
182~184	nC15~C27	4.123	1.023	1.524	0.068	2.705	0.044	8.653	1.208	0.774	0.445	1.030

Note: ACL: average content of carbon atoms in long-chain n-alkanes (C_{25-33}); L/H: low/high carbon molecular ratio [ΣC_{15-23}/ΣC_{24-33}]; CPI: carbon preference index [C_{21-33}]; pr/ph: pristane/phytane; TAR: $\Sigma C_{27+29+31}$ (total content of n-alkanes with terrigenous source predominance)/$\Sigma C_{15+17+19}$(total content of n-alkanes with terrigenous source predominance).

environment, while Pr/Ph>1 is oxidizing condition (Volkman et al, 1988). In addition, Pr/Ph ratio is related with petroleum hydrocarbon input and maturity of organic matter to a large extent. Researches show that when petroleum hydrocarbon is input into sediments, Pr/Ph ratio is generally close to or less than 1.0 (Volkman et al, 1992; Zaghden et al, 2007).

Pristane (2,6,10,14—tetramethyl pentadecane) and phytane (2,6,10,14—tetramethyl hexadecane) come from some special organisms. As a pioneer of pristine and phytane, chlorophyllous phytol widely exists in the nature—it derives from phytol side chain, it is relatively stable, its resistance to microbial attack is stronger than n-alkanes, so Pr/Ph ratio is widely used as a palaeoenvironment index. Phytol is reduced under reducing conditions, and converted to phytane after hydrogenation and dehydration. Under oxidizing conditions, phytol is oxidized, decarboxylated and converted to pristine (Volkman et al, 1986). Regarding Pr/Ph ratio at all sections of P7327, it is 3.017 at the top (1~2 cm), much larger than 1. It means that overlying deposits are in strongly oxidizing sedimentary environment; alkane molecular structure changes a lot, showing that organic matters in sediment core have evolved. As the depth increases, Pr/Ph ratio tends to reduce. In early sections (120~121 cm, 182~184 cm), Pr/Ph ratio reduces. It means that sediment core is under reducing conditions; phytane has appeared before pristine. As the depth increases, long-chain hydrocarbon converts to short-chain hydrocarbon. At 182~184 cm sections, with the input of bacterial hydrocarbon, increase of buried depth and enhancement of bacterial degradation, abundance of molecules with low carbon number tends to rise significantly.

3.1.3 Sterane composition and source
Sterane is formed by original sterol through reduction action during sedimentation. Cholestane C_{27}, ergostane C_{28} and sitostane C_{29} ternary combination figure can distinguish combination features of matter sources. The basis is that C_{27} sterane (whose predecessor is C_{27} sterol) mainly has marine source, C_{29} sterane (whose predecessor is C_{29} sterol) mainly has higher plant source, while C_{28} sterane has dual source: it comes from higher plants and phytoplankton (Li et al, 2009). However at present, indicative significance of regular sterane source is controversial. This research mainly identifies steranes with m/z 217 as base peak. At all sections of sediment core P7327, steroid mainly includes 20R, $5\alpha(H)14\alpha(H)17\alpha(H)$-20S, $5\alpha(H)14\beta(H)17\beta(H)$-20S and regular steranes series with 20S configuration, as well as $13\beta(H)17\alpha(H)$-20S, 20R, $13\alpha(H)17\alpha(H)$-20S and rearranged steranes series

with 20S configuration. In most cases, 24-methyl-$5\alpha(H)14\alpha(H)17\alpha(H)$-C29-cholestane (20R) has the highest content.

Regular steranes boast absolute predominance among steroid; main difference between rearranged steranes and regular steranes is that C-10 and C-13 methyl are rearranged onto C-5 and C-14. In sedimentary strata, $\Sigma(C_{27}+C_{28})/\Sigma C_{29}$ value of regular steranes is distributed in 0.31–0.49, showing that considerable quantity of algae and aquatic organisms are input. At most sections of sediment core, content of sterane is $C_{27} > C_{28} < C_{29}$; sterane C_{29} is predominant, meaning that organic matter source is mixed (TenHaven et al, 1988). Considerable quantity of 4- methyl steranes exist at all sections; they mainly come from methylotrophic bacteria or other bacteria with 4- methyl steranes. Undoubtedly, pregnane and high pregnane are direct biodegradation products of animals.

3.2 Existence evidence and source of petroleum hydrocarbon

In most cases, organic matters in modern sediments represent the latest sedimentary organic matters. Without geological thermal evolution, their molecular spatial configuration keeps features of biological configuration. When organic matters with thermal maturity, such as organic matters from petroleum, are mixed into modern sediments, they are displayed by geochemical indexes and molecular markers such as CPI value, isoprenoid, UCM and sterane organic matter maturity. Odd-even predominance of n-alkanes drops; CPI value is on the low side, with low molecular weight and high relative content of Pr and Ph. There is also geological configuration biomarker of modern sediments after thermal maturation.

3.2.1 Evidence from geochemical parameters of n-alkanes and isoprenoid
Locate tables close to the first reference to them in After cyclic organic matters, such as mature organic matters and petroleum hydrocarbon, are mixed into modern sediments, ordinary geochemical parameters show relatively high maturity (Bence et al, 1996). For example, odd-even predominance of n-alkanes drops; CPI value is on the low side; isoprenoid alkanes with low molecular weight C_{13}, C_{16}, C_{17}, Pr and Ph have high relative contents. At all sections of sediment core P7327, ACL value is low and changes in 0.068~7.610, which is obviously affected by petroleum hydrocarbon input, especially 182~184 cm section. CPI is usually used to evaluate n-alkanes source of marine sediments. n-alkanes from different sources have different CPI values: special CPI of terrestrial plant epicuticular wax is 4~10, while CPI from crude oil is close to 1

(Tolosa et al, 1996). In sedimentary strata of this research, CPI is 1.208~1.664, obviously lower than modern (or contemporary) normal marine sediments, and far less than special CPI of terrestrial plant epicuticular wax. It is close to 1, perhaps because of mixing effect of crude oil input, some terrigenous higher plants and marine organisms.

The ratio of pristane, phytane to adjacent n-alkanes—Pr/nC$_{17}$ and Ph/nC$_{18}$ can serve as comparison indexes and micro-biological degradation marker of compounds (Wang et al, 1988; Atlas et al, 1991). At all sections of P7327, Pr/nC$_{17}$ ratio is 0.445~1.079 and Ph/nC18 ratio is 0.370~1.041. Generally in petroleum or slightly degraded petroleum, contents of C$_{17}$ and C$_{18}$ are higher than Pr and Ph (Blumer et al., 1972). C$_{17}$/Pr and C$_{18}$/Ph are greater than 1. However compared to adjacent isoprene components Pr and Ph, n-alkanes components C$_{17}$ and C$_{18}$ are much easily degraded in marine environment, so this ratio in marine sediments is smaller than previous value. As degradation increases, this ratio may be lower than 1 (Gao et al., 2008). Therefore, when crude oil is seriously degraded, it is necessary to use more stable biomarkers to measure petroleum degradation.

3.2.2 *Evidence for unresolved mixture (UCM)*
Appearance of UCM (unresolved complex mixture) in the sediments is closely related with petroleum input, and bacterial microbial action contributes to UCM to some extent (Bouloubassi et al., 2001). It is widely accepted that UCM is a product of petroleum degradation and weathering. Compared to n-alkanes, UCM is more resistant to weathering and degradation, so it easily keeps and accumulates in marine environment. Although ordinary one-dimensional capillary gas chromatography owns very high resolution, it still cannot separate the compounds completely. A series of combined compounds appear on the chromatogram, represented by continuous baseline lifting (Fig. 2), which mainly consists of branched chain alkanes and cycloparaffin (Li Zeli et al., 2011). Predecessors have realized long ago that biodegradation oil from different sources and after different secondary changes has different UCM forms and carbon number distribution ranges (Killops

and Al-Juboori, 1990; Gough et al., 1992; Ventura et al, 2008). Generally, UCM content in unpolluted marine sediments is less than 10 μg/g^{-1} (Tolosa et al., 1996). If ratio of U (indistinguishable complex hydrocarbon on chromatogram)/R (distinguishable hydrocarbon on chromatogram) is more than 2 and less than 4, it means that the sediments have been polluted by petroleum; if U/R ratio is more than 4, it means that the sediments have been seriously polluted by petroleum. UCM value is calculated with planimetry (Boehm and et al., 1978).

UCM is widespread in all sections of sediment core P7327, and is represented by obvious baseline upheaval of low carbon peaks. Based on spatial distribution features of UCM, it is shown that the sediments have petroleum hydrocarbon input to varying degrees; bacterial microbial degradation contributes to UCM in sediments to some extent (Seki et al, 2006). UCM in short-chain n-alkanes attributes to micro-biological degradation of natural organic matters, and may be affected by petroleum hydrocarbon input. UCM is distributed in C$_{22}$~C$_{33}$, showing petroleum component input (Simoneit et al., 1986). Generally speaking, original organic matters in contemporary seabed sediments are immature organic matters without thermal evolution. Morphologic change feature of obvious UCM bulge in the samples shows the input of mature organic matters, as well as strong micro-biological degradation and oxidation.

3.2.3 *Evidence from sterane isomerization parameter*
In organic matters of sediments, sterane biomarker composition feature is closely related with sedimentary environment, organic parent material source and thermal evolution feature. They exist in geologic body stably, always keep specific form and distribution during conversion from living body to geologic body, and own good "heredity". Therefore, they provide accurate dimension of fingerprint information about parent material source and maturity of organic matters. In crude oil with high thermal maturity, distribution pattern of C$_{27}$, C$_{28}$ and C$_{29}$ regular sterane shows obvious V-shaped feature (Huang et al., 1979). Thermal stability of rearranged sterane is better than that of regular

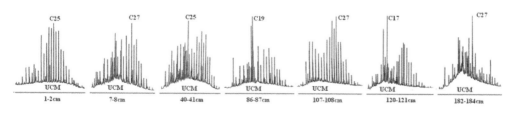

Figure 2. Gas chromatogram of n-alkanes in recent sediments from the Chukchi Sea.

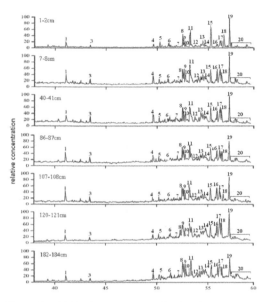

Figure 3. Mass chromatogram of m/z = 217 (steranes) from different sediment sections of Core P7327.
[Peak number: 1: 5α-C21, pregnane; 3: 5α-C22, risepregnane; 4: 13β, 17α-C27, rearranged cholestane (20S); 5: 13β, 17α-C27, rearranged cholestane (20R); 6: 13α, 17β-C27, rearranged cholestane (20S); 7: 13α, 17β-C27, rearranged cholestane (20R); 8: 5α, 14α, 17α-C27, cholestane (20S); 9: 5α, 14β, 17β-C27, cholestane (20R); 10: 5α, 14β, 17β-C27, cholestane (20S); 11: 5α, 14α, 17α-C27, cholestane (20R); 12: 24-methyl, 5α, 14α, 17α-C28, cholestane (20S); 13: 24-methyl, 5α, 14β, 17β-C28, cholestane (20R);14: 24-methyl, 5α, 14β, 17β-C28, cholestane (20S); 15: 24-methyl, 5α, 14α, 17α-C28, cholestane (20R); 16: 24-ethyl, 5α, 14α, 17α-C29, cholestane (20S); 17: 24-ethyl, 5α, 14β, 17β-C29, cholestane (20R); 18: 24-ethyl, 5α, 14β, 17β-C29, cholestane (20S); 19: 24-ethyl, 5α, 14α, 17α-C29, cholestane (20R); 20: 4α-methyl, 24-ethyl, 5α, 14α, 17α-C30, cholestane (20S); 21: 4α-methyl, 24-ethyl, 5α, 14β, 17β-C30, cholestane (20R); 22: 4α-methyl, 24-ethyl, 5α, 14β, 17β-C30, cholestane (20S)].

sterane; the ratio of rearranged sterane to regular sterane can serve as maturity parameter, and sterane isomerization parameters C29ββ/(αα + ββ) and C2920R/20(R + S) are used to judge oil and gas maturity. As maturity increases during thermal evolution, the ratio increases gradually (Aboul-Kassin et al, 1996).

Sterane isomerization parameters C_{29}20S/(20S + 20R) and C_{29}ββ/(ββ + αα) in this research is 0.37–0.47 and 0.45–0.53, which reach or close to isomerization balance end point and belong to a mature range. Generally speaking, original organic matters in contemporary seabed sediments are immature organic matters without thermal evolution. Sterane

isomerization parameters in sediment core P7327 show maturity or low maturity feature, meaning this sediment core has accepted leakage input of mature organic matters from deep oil and gas reservoirs.

4. CONCLUSION

1. In saturated hydrocarbon of sediment core P7327, n-alkanes have bimodal pattern, pristane/phytane ratio is low, H/L is 0.299~1.326, C27/C19 ratio is 1.392~7.565, lipoid ratio TAR and ACL are 1.015~5.531 and 0.930~7.610 respectively. Above distribution features of n-alkanes show that terrigenous higher plant source and marine source (lower fungi, algae and submerged plants) coexist in contemporary sediments in the Chukchi Sea.
2. UCM content is relatively high, CPI value is close to 1, sterane isomerization parameters C2920S/(20S + 20R) and C29ββ/(ββ + αα) is 0.37–0.47 and 0.45–0.53, showing that the sediments have high maturity. Different from contemporary sediments, maturity of organic matters in sediment core P7327 perhaps mean that, contemporary sediments of the Chukchi Sea may be mixed with petroleum source. These petroleum hydrocarbons may come from pollution from human activities, leaked oil and gas reservoirs in the deep sea, and may be affected by microbial action of bacteria.

ACKNOWLEDGEMENT

Funded by National Natural Science Foundation of China (No. 41276199). Sample collection has been vigorously supported by the crew of "Snow Dragon" and personnel of relevant units, as well as Chinese Arctic and Antarctic Administration. We'd like to thank them.

The above material should be with the editor before the deadline for submission. Any material received too late will not be published.

REFERENCES

Aboul-Kassin T.A.T., Simoneit B.R.T. Lipid geochemistry of surficial from the coastal environment of Egypt I. Aliphatic hydrocarbons-characterization and sources[J]. Marine Chemistry, 1996, 54:135–158.

Atlas B.A. Petroleum Microorganism[M]. Translated by Huang Difan, Tan Shi, Yang Wenkuan et al. Beijing: Petroleum Industry Press, 1991, 1~37.

Bence A.E., Kvenvolden K.A., Kennicutt M.C., et al. Organic geochemistry applied to environmental assessments of Prince Willian Sound, Alaska, after the Exxon Valdez oil spill—a review[J]. Org Geochem, 1996, 24:7–42.

Blumer M., Sass J. Indigenous and Petroleum—derived hydrocarbons in polluted sediment[J]. Mar pollut Bull, 1972, 6:92–94.

Boehm P. and Quinn J. Benthic hydrocarbons of rhode island sound[J]. Estuarine and Coastal Marine Science, 1978, 6:471–494.

Bouloubassi I., Fillaux J., Saliot A. Hydrocarbons in surface sediments from the Changjiang (Yangtze River) Estuary[J]. East China Sea Marine Pollution Bulletin, 2001, 42:1335–1346.

Brassell S.C., Eglinton G., Mo F.J. Biological marker compounds as indicational history of the Maoming oil shale[J]. Org Geochem, 1986, 10:927–941.

Ficken K.J., Li B., Swain D.L. An n-alkane Proxy for the sedimentary inputs of submerged/floating freshwater aquatic macrophytes[J]. Organic Geochemistry, 2000, 31:745–749.

Gao X., Chen S. Petroleum pollution in surface sediments of Daya Bay, South China, revealed by chemical fingerprinting of aliphatic and alicyclic hydrocarbons[J]. Estuarine, Coastal and Shelf Science, 2008, 80:95–102.

Goni M.A., ruttenberg K.C., Eglinton T.L. Sources and contribution of terrigenous organic carbon to surface sediments in the Gulf of Mexico[J]. Nature, 1997, 389:275–278.

Gough M.A. & Rowland S.J. Characterization of unresolved complex mixtures of hydrocarbons from lubricating oil feedstocks[J]. Energy and Fuels, 1991, 5:869–874.

Gough M.A., Rhead M.M., Rowland S.J. Biogegradation studies of unresolved complex mixtures of hydrocarbons: model UCM hydrocarbons and the aliphatic UCM[J]. Organic Geochemistry,1992,18:17–22.

Guo Zhigang, Yang Zuosheng, Lin Tian. Carbon Isotope Composition and Provenance Analysis of Monomer N-alkanes in Argillaceous Area of the East China Sea[J]. Quaternary Sciences, 2006, 26:384–390.

Han Xueqiang. Russian Exploration and Development Strategy of Oil and Gas Resources[J]. Petroleum Science Forum, 2011, 30:47–55.

Han Xueqiang. Russian Exploration and Development Status Quo of Oil and Gas Resources in Continental Shelf Sea Area[J]. Petroleum Science Forum, 2012, 31:45–49.

Harji R.R., Yvenat A., Bhosle N.B. Sources of hydrocarbons in sediments of the Mandove estuary and the Marmugoa harbor, west coast of India[J]. Environ Int, 2008, 34:959–965.

Hostettler F.D., Perera W.E., Kvenvolden K.A., et al. A record of hydrocarbon input to San Francisco Bay as traced by biomarker profiles in surface sediment and sediment cores[J]. Marine Chemistry, 1999, 64:115–127.

Huang, Wengen, Meinschein W.G. Sterols as ecological indicators[J]. Geochim Cosmochim Acta, 1979, 43:739–744.

Jeng, W.L. Higher plant n-alkane average chain length as an indicator of petrogenic hydrocarbon contamination in marine sediments[J]. Marine Chemistry, 2006, 102:242–251.

Killops S.D. and Al-Juboori M.A.H.A. Characterisation of the Unresolved Complex Mixture (UCM) in the gas chromatograms of biodegraded petroleums[J]. Org. Geochem, 1990, 15:147–160.

Li Shuanglin, Zhang Shengyin, Zhao Qingfang et al. Geochemical Characteristics and Indicative Significance of Saturated Hydrocarbon in Bottom Sediments of South Yellow Sea[J]. Marine Geology Letters, 2009, 25:1–7.

Li Zeli, Ma Qimin, Cheng Hai'ou et al. Research on Characteristic Parameters of N-alkanes in Overlaying Deposits at Jinzhou Bay[J]. Environmental Science, 2011, 11:3300–3304.

Mead R., Xu Y., Chong J. Sediment and Soil organic matter source assessment as revealed by the molecular distribution and carbon isotopic composition of n-alkanes[J]. Organic Geochemistry, 2005, 36:363–370.

Moritz. R.E. and Perovich, D.D. 1996. Surface Heat Budget of the Aretic Ocean (SHEBA)[J]. Science.

Peters K.E., Walters C.C., Moldowan J.M. The biomarker guide: Biomarkers and Isotopes in Petroleum Systems and Earth History[J]. Cambridge: Cambridge University Press, 2005.

Revkin A.C. Under all that ice, maybe oil[J]. New Yoek Times, 2004, 30 November.

Seki O., Yoshikawa C., Nakatsuka T., et al. Fluxes, source and transport of organic matter in the westerm Sea of Okhotsk: stable carbon isotopic ratios of n-alkanes and total organic carbon[J]. Deep-Sea Research I, 2006, 53:253–270.

Silliman J.E., Schelske C.L. Saturated hydrocarbons in the sediments of Lake Apopka, Florida[J]. Organic Geochemistry, 2003, 34:253–260.

Simoneit B.R.T. Characterization of organic constituents in aerosols in relation to their origin and transport: a review[J]. International Journal of Environmental Analytical chemistry, 1986, 23:207–237.

Sutton P.A., Lewis C.A., Row land S.J. Isolated of individual hydrocarbons from the unresolved complex hydrocarbon mixture of a biodegraded crude oil using preparative capillary gas chromatography[J]. Organic Geochemistry, 2005, 36:960–970.

TenHaven H.L., De Leeuw J.W., Sinninghe Damaste J.S., et al. Application of biological markers in the recognition of palaeopersaline environments. In: Lacustrine Petrolum Source Rocks[J]. Geological Society Special Publication, 1988, 40:123–130.

Tolosa I., Bayona J.M., Albaie J. Aliphaatic and Polycyclic aromatic hydrocarbons and sulfur/oxygen derivatives in northwesteen Mediterranean sediments: spatial and temporal variability, fluxes, and budgets[J]. Environmental Science and Technology, 1996, 30:2495–2503.

USGS. Circum-Arctic Resouece Appraisal: Estimates of Undiscivered Oil and Gas North of the Arctic Circle[J]. USGS Fact Sheet 2008–3049, 2008, 1–3.

Ventura G.T., Kening F., Reddy, C.M., et al. Analysis of unresolved complex mixtures of hydrocarbons extracted from Late A rchean sediments by comprehensive two—dimensional gas chromatography (GCXGC)[J]. Organic Geochemistry, 2008, 39:846–867.

Volkman J.K., Maxwell J.R. Acyclic isoprenoids as biological makers[A]. Jonts R.B. Biological makers in the sedimentary record[J]. New York, Elsevier Publishers, 1986, 1~42.

Volkman J.K., Burton H.R., Everitt D.A., et al. Pigment and liquid compositions of algal and bacterial communities in Ace Lake. Vestfold Hills, Antarctica[J]. Antarctics. Hydrobiologica, 1988, 165:41~57.

Volkman J.K., Holdsworth D.G., Neill G.P., et al. Identification of natural, anthropogenic and petroleum hydrocarbons in aquatic sediments[J]. Science of The Total Environment, 1992, 112:203–219.

Walsh J.J. Arctic carbon sinks: Present and future, Global Biogeochem[J]. Cycles, 1989, 3(4), 393–411, doi:10.1029/GB003i004p00393

Wang Qijun, Chen Jianyu. Geochemistry of Oil and Gas[M]. Wuhan: China University of Geosciences Press, 1988, 99~108, 283.

Wang Tieguan, Zhong Ningning, Hou Dujie et al. Formation Mechanism and Distribution of Low-maturity Oil and Gas[M]. Beijing: Petroleum Industry Press, 1995, 56–58.

Wolff G.A., Lamb N.A., Maxwell J.R. The origin and fate of 4-methyl steroid hydrocarbons I. 4-methyl steranes[J]. Geochem Cosmchim Acta, 1986, 50:335–342.

Yang Dan, Yao Longkui, Wang Fangguo et al. Carbon Molecular Assembly Features of N-alkanes in Modern Sediments at the South China Sea, and its Indicative Significance[J]. Journal of Marine Sciences, 2006, 4:29–39.

Zaghden H., Kallel M., Elleuch B., et al. Sources and distribution of aliphatic and polyaromatic hydrocarbons in sediments of Sfax, Tunisia, Mediterranean Sea[J]. Marine Chemistry, 2007, 105:70–89.

Zhao Jinping, Zhu Dayong, Shi Jiuxin. Annual Variation Feature and Main Connected Factors of Sea Ice in the Chukchi Sea[J]. Advances in Marine Science, 2003, 21:13–131.

Zhu Chun, Pan Jianming, Lu Bing et al. N-alkanes Molecular Assembly in Modern Sediments at Yangtze Estuary and Nearby Sea Areas, and Its Indication of Organic Carbon Migration and Distribution[J]. Acta Oceanologica Sinica, 2005, 4:59–67.

Advances in Energy, Environment and Materials Science – Wang & Zhao (Eds)
© 2016 Taylor & Francis Group, London, ISBN 978-1-138-02931-6

Absorption kinetics of CO_2 in MEA promoted K_2CO_3 aqueous solutions

Wenhua Si, Chenlu Mi & Dong Fu

School of Environmental Science and Engineering, North China Electric Power University, Baoding, Hebei, China

ABSTRACT: The solubility and absorption rate of CO_2 in MEA promoted potassium carbonate (K_2CO_3) aqueous solution were measured at normal pressure with temperatures ranging from 303.2–323.2 K. The temperature and the mass fractions dependences of the solubility and saturated CO_2 loading were determined. The influence of the mass fraction of MEA on the absorption rate of CO_2 was illustrated.

1 INTRODUCTION

The greenhouse effect caused by the emissions of CO_2 from industrial processes and coal-fired boilers seriously impacted the sustainable development of the economy in recent decades. The limitations of CO_2 emissions are urgently required.

There are various technologies available for the separation of CO_2 from the flue gas of conventional fossil fuel fired power plants, e.g., chemical absorption, physical absorption, cryogenic methods, membrane separation, and biological fixation (Um et al., 2003). Chemical absorption process is generally recognized as the most effective technology (Rao and Rubin, 2002). Many solvents have gained widespread acceptance as viable solvents for pre and post combustion capture of CO_2, but the most effective solvents are generally considered to be potassium carbonate solvents or aqueous alkanolamines, including Monoethanolamine (MEA), Diethanolamine (DEA), N-Methyldiethanolamine (MDEA), Di-2-Propanolamine (DIPA) and so on (Choi et al., 2009; Ghosh, Kentish and Stevens, 2009).

Previous investigators have explored the solubility and reaction rate of CO_2 in aqueous potassium carbonate (Benson, Field and Jimeson, 1954; Benson, Field and Haynes, 1956; Tosh et al., 1959). The main advantages of potassium carbonate solution for CO_2 removal are the high chemical solubility of CO_2 in the carbonate/bicarbonate system, low solvent costs and the low energy requirement for regeneration (Ghosh, Kentish and Stevens, 2009). However, due to the poor reaction kinetics, a rate promoter is considered essential to improve the rate of CO_2 absorption in K_2CO_3 aqueous solution (Danckwerts and McNeil, 1967; Shrier and Danckwerts, 1969; Astarita, Savage and Longo, 1981; Savage and Sartori, 1984; Versteeg et al., 1996; Cullinane and Rochelle, 2005; Thee et al., 2012).

As a most frequently used alkanolamines absorbent, MEA is considered to be an good promoter for K_2CO_3 aqueous solution reaction rates, due to its high reactivity with CO_2 and low solvent cost (Rao and Rubin, 2002; Mandal et al., 2003; Abu-Zahra et al., 2007; Anderson et al., 2011; Qader, 2011; Thee et al., 2012).

The main purpose of this work is to determine an appropriate addition of MEA to K_2CO_3 aqueous solution, so that the high absorption rate and large CO_2 loading can be simultaneously achieved. To this end, we measured the solubility and absorption rate of CO_2 in K_2CO_3 aqueous solution promoted by small amount of MEA at 1 atm and series of temperatures, and demonstrated the influence of the mass fraction of MEA on the solubility, CO_2 loading and absorption rate.

2 EXPERIMENTAL SECTION

K_2CO_3 and MEA were purchased from Huaxin chemical Co., with mass purity ≥ 99%. They were used without further purification. Aqueous solutions of MEA-K_2CO_3 were prepared by adding doubly distilled water. The uncertainty of the electronic balance (FA1604A) is ± 0.1 mg.

To avoid the vast crystallization of $KHCO_3$ (the reaction product of K_2CO_3 and CO_2), in each aqueous solution, the mass fraction of K_2CO_3 is less than 0.25, and the mass fraction of MEA is no greater than 0.075.

2.1 Apparatus and procedure for absorption rate of CO_2

The absorption performance of aqueous solutions was measured by the equipment composed of one high-pressure CO_2 tank, two Mass Flow

Controllers (MFC), one absorption bottle, one constant temperature water bath, one desiccator and one CO_2 analyzer (Advanced Gasmitter by Germany Sensors Europe GmbH, the accuracy is ± 2%). The flask was immersed into the thermostatic bath and the temperature of the solution can be regulated within 0.1 K. During the experiment, CO_2 from a high-pressure tank was inlet into the first MFC to maintain a constant flow rate and then into the absorption bottle and absorbed by the solution. The residual and unabsorbed gas firstly flowed into the desiccator and then into the CO_2 analyzer. The gas concentration was measured by the CO_2 analyzer, and the flow rate was measured by the second MFC. Both the data of gas concentration and flow rate were recorded by the computer. In all the experiments, the rotational speed of the magnetic stirrer is fixed as 1000 rpm. The schematic diagram for the measurements of saturated solubility and the absorption rate of CO_2 is shown in Figure 1.

2.2 Results and discussion

The solubility and the absorption rate of CO_2 in MEA-K_2CO_3 aqueous solution were measured at normal pressure with temperatures ranging from 303.2–323.2 K. Time dependence of dissolution of CO_2 under different temperatures is shown in Figure 2, indicating that at higher temperatures, the saturated absorption may be achieved with shorter time and the absorption rate of CO_2 increases with the increases of temperature. The solubilities n and CO_2 loading α under different temperatures and different mass fractions of amines are shown in Table 1.

From Table 1, one may find that at given w_1 and w_2, the solubility of CO_2 decreases with increasing temperature. At given temperature and w_1, it's

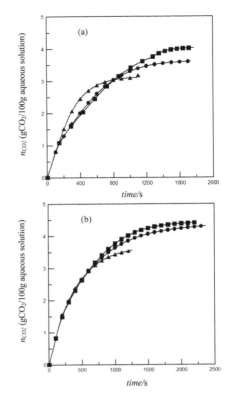

Figure 2. The influence of temperature on the absorption rate of CO_2 in MEA (2)-K_2CO_3(1) aqueous solutions. Symbols: ■ T = 303.2 K ● T = 313.2 K ▲ T = 323.2 K (a): $w_2/w_1 = 0.025/0.10$ (b): $w_2/w_1 = 0.05/0.10$.

Table 1. Solubilities n (gCO$_2$/100 g aqueous solution) and CO_2 loading α (molCO$_2$/[molMEA + molK$_2$CO$_3$]) in MEA (2)-K_2CO_3 (1) aqueous solutions at 1 atm.

		303.2 K		313.2 K		323.2 K	
w_1	w_2	n	α	n	α	n	α
0.1	0	3.12	0.98	3.12	0.97	3.04	0.95
	0.025	4.03	0.81	3.60	0.73	3.24	0.65
	0.050	4.41	0.65	4.30	0.63	3.56	0.52
	0.075	5.59	0.65	5.05	0.58	4.14	0.48
0.2	0	6.24	0.98	6.15	0.97	5.75	0.90
	0.025	6.65	0.81	6.25	0.76	5.84	0.71
	0.050	6.84	0.69	6.27	0.63	6.08	0.61
	0.075	6.87	0.58	6.51	0.55	6.07	0.51

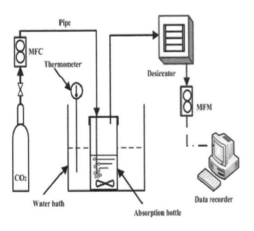

Figure 1. The schematic diagram.

worth noting that although the absolute solubility of CO_2 (defined as gCO$_2$/100 g aqueous solution) increases with the increase of w_2, the CO_2 loading (defined as molCO$_2$/[molMEA + molK$_2$CO$_3$]) decreases with the increase of w_2.

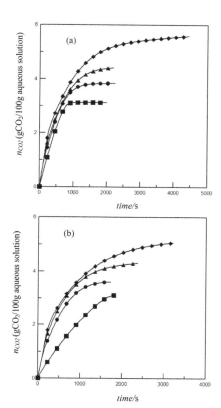

Figure 3. The influence of mass fraction of MEA on the absorption rate of CO_2 in MEA (2)-K_2CO_3 (1) aqueous solutions. Symbols: ■ $w_2/w_1 = 0.00/0.10$; ● $w_2/w_1 = 0.025/0.10$; ▲ $w_2/w_1 = 0.05/0.10$; ◆ $w_2/w_1 = 0.075/0.10$ (a): 303.2 K (b): 313.2 K.

Figure 3 shows the influence of the mass fraction of MEA on the absorption rate. Comparison shows that without MEA promotion, CO_2 is relatively slowly absorbed by K_2CO_3 aqueous solution, however, the absorption rate increases very significantly when the K_2CO_3 aqueous solution is promoted by little amount of MEA. Taking into account the fact that the CO_2 loading decreases with the increase of w_2, $w_2 = 0.075$ is close to the optimal addition of MEA, under which both high CO_2 absorption rate and large CO_2 loading can be simultaneously achieved.

3 CONCLUSIONS

In this study, the solubility and absorption rate of CO_2 in MEA promoted K_2CO_3 aqueous solution were measured at normal pressure with temperatures ranging from 303.2–323.2 K. The influence of the addition of MEA on the solubility and absorption rate of CO_2 was illustrated. Our results showed that:

1. When the mass fraction is of very small value, MEA promotes the absorption rate of CO_2 very significantly. Moreover, the absorption rate of CO_2 increases with the increase of temperature;
2. Both the temperature and the mass fraction have influences on the solubility of CO_2 and CO_2 loading. The solubility of CO_2 respectively increases and decreases with increasing mass fraction of MEA and temperature. However, CO_2 loading decreases with the increase of both mass fraction of MEA and temperature.

ACKNOWLEDGEMENT

The authors appreciate the financial support from the National Natural Science Foundation of China (Nos. 21276072 and 21076070), the Natural Science Funds for Distinguished Young Scholar of Hebei Province (No. B2012502076), the Fundamental Research Funds for the Central Universities (No. 11ZG10), and the 111 project (B12034).

REFERENCES

Abu-Zahra, M.R., Schneiders, L.H., Niederer, J.P., Feron, P.H., & Versteeg, G.F. (2007). CO_2 capture from power plants: Part I. A parametric study of the technical performance based on monoethanolamine. *International Journal of Greenhouse gas control, 1*(1), 37–46.

Anderson, C., Scholes, C., Lee, A., Smith, K., Kentish, S., Stevens, G., & Hooper, B. (2011). Novel pre-combustion capture technologies in action–results of the CO2CRC/HRL Mulgrave capture project. *Energy Procedia, 4,* 1192–1198.

Astarita, G., Savage, D.W., & Longo, J.M. (1981). Promotion of CO_2 mass transfer in carbonate solutions. *Chemical Engineering Science, 36*(3), 581–588.

Benson, H.E., Field, J.H., & Haynes, W.P. (1956). Improved process for CO_2 absorption uses hot carbonate solutions. *Chem. Eng. Prog, 52*(10), 433–438.

Benson, H.E., Field, J.H., & Jimeson, R.M. (1954). CO/sub 2/absorption: employing hot potassium carbonate solutions. *Chem. Eng. Prog.;(United States), 50*(7).

Choi, W.J., Seo, J.B., Jang, S.Y., Jung, J.H., & Oh, K.J. (2009). Removal characteristics of CO_2 using aqueous MEA/AMP solutions in the absorption and regeneration process. *Journal of Environmental Sciences, 21*(7), 907–913.

Cullinane, J.T., & Rochelle, G.T. (2006). Kinetics of carbon dioxide absorption into aqueous potassium carbonate and piperazine. *Industrial & engineering chemistry research, 45*(8), 2531–2545.

Danckwerts, P.V., & McNeil, K.M. (1967). The effects of catalysis on rates of absorption of CO_2 into aqueous amine—potash solutions. *Chemical Engineering Science, 22*(7), 925–930.

Ghosh, U.K., Kentish, S.E., & Stevens, G.W. (2009). Absorption of carbon dioxide into aqueous potassium carbonate promoted by boric acid. *Energy Procedia*, *1*(1), 1075–1081.

Mandal, B.P., Biswas, A.K., & Bandyopadhyay, S.S. (2003). Absorption of carbon dioxide into aqueous blends of 2-amino-2-methyl-1-propanol and diethanolamine. *Chemical Engineering Science*, *58*(18), 4137–4144.

Qader, A., Hooper, B., Innocenzi, T., Stevens, G., Kentish, S., Scholes, C., ... & Zhang, J. (2011). Novel post-combustion capture technologies on a lignite fired power plant-results of the CO2CRC/H3 capture project. *Energy Procedia*, *4*, 1668–1675.

Rao, A.B., & Rubin, E.S. (2002). A technical, economic, and environmental assessment of amine-based CO_2 capture technology for power plant greenhouse gas control. *Environmental science & technology*, *36*(20), 4467–4475.

Savage, D.W., Sartori, G., & Astarita, G. (1984). Amines as rate promoters for carbon dioxide hydrolysis. *Faraday Discussions of the Chemical Society*, *77*, 17–31.

Shrier, A.L., & Danckwerts, P.V. (1969). Carbon dioxide absorption into amine-promoted potash solutions. *Industrial & Engineering Chemistry Fundamentals*, *8*(3), 415–423.

Thee, H., Suryaputradinata, Y.A., Mumford, K.A., Smith, K.H., da Silva, G., Kentish, S.E., & Stevens, G.W. (2012). A kinetic and process modeling study of CO_2 capture with MEA-promoted potassium carbonate solutions. *Chemical Engineering Journal*, *210*, 271–279.

Tosh, J.S., Field, J.H., Benson, H.E., & Haynes, W.P. (1959). *Equilibrium study of the system potassium carbonate, potassium bicarbonate, carbon dioxide, and water* (No. BM-RI-5484). Bureau of Mines, Pittsburgh, Pa. (USA).

Um, H.M. (2000). The Study on the Development of Demo Plant Scale Carbon Dioxide Separation and Conversion Technologies in Power Station. *Korea Energy Management Co-operation R&D, Korea, Rep.*

Versteeg, G.F., Van Dijck, L.A.J., & Van Swaaij, W.P.M. (1996). On the kinetics between CO_2 and alkanolamines both in aqueous and non-aqueous solutions. An overview. *Chemical Engineering Communications*, *144*(1), 113–158.

Advances in Energy, Environment and Materials Science – Wang & Zhao (Eds)
© 2016 Taylor & Francis Group, London, ISBN 978-1-138-02931-6

Kinetics in Mn (II) ion catalytic ozonation of Papermaking Tobacco Slice wastewater

Saiyan Chen, Youming Li & Lirong Lei
State Key Laboratory of Pulp and Paper Engineering, South China University of Technology, Guangzhou, Guangdong, China

ABSTRACT: This study elucidated the Chemical Oxygen Demand (COD) degradation efficiency of the Papermaking Tobacco Slice (PTS) wastewater with Mn (II) ion catalyzed ozonation. The effect of catalyst dosage and initial pH on COD removal was investigated. The results reveal that COD removal was effected by initial pH value and Mn (II) ion can promote COD removal rate. In Mn (II) ion catalytic ozonation system, three types of oxidants, including hydroxyl radicals, manganese ions and ozone, contribute to the oxidation reactions of pollutants in the PTS wastewater and both the manganese ions and hydroxyl radicals are the more important oxidants for COD removal than ozone. Moreover, manganese ions play a more important role for the COD removal than hydroxyl radicals under neutral and alkaline condition, especially at pH 7, at which the contribution ratio of molecular ozone, hydroxyl radicals and manganese ions is 1:2.32:4.96.

1 INTRODUCTION

Papermaking Tobacco Slice (PTS) is produced using tobacco wastes like tobacco stalks, fragments by paper-making technology and about 70 m³ wastewater was generated when 1 ton of PTS is produced. Various pollutants including cellulose, hemicelluloses, lignin, organic acids, alkaloids, nicotine and tannins had been found in PTS wastewater (Meher et al. 1995; Sponza 2002; Wang et al. 2011), which contributed to high concentration of COD and color in wastewater. Conventional physicochemical processes and biological processes have been used for treatment of the PTS wastewater and effectively decrease the pollution level of the wastewater. However, due to the presence of recalcitrant pollutants and toxic substances to microorganisms in the PTS wastewater (Saunders & Blume 1981; Munari 1986), PTS wastewater contains various organics in terms of residual COD of above 500 mg·L⁻¹ and developing an effective advanced treatment process has been an urgent problem for the PTS production industry.

Ozone processes, which can react with multiple pollutants, have been used for treatment of pulp and paper effluents and effective removal of COD and color had been reported (Meza et al. 2011; Lei & Li 2014). Mn (II)/O₃ progress has been studied for treatment of various pollutants and better treatment efficiency for pollutants degradation than single ozonation had been achieved (Gracia et al. 1998; Ma & Graham 1999; Andreozzi et al. 2001).

However, there are only a few literatures about single ozonation or Mn (II)/O₃ progress for treatment of PTS wastewater. In this paper, treatment efficiency of the PTS wastewater in Mn catalytic ozonation was investigated. The contribution ratio of different oxidations in the Mn (II) ion catalytic ozonation process was researched and analyzed.

2 METHODS

2.1 Wastewater sampling

Wastewater sample was taken from a PTS plant located in southern area of China. The plant produces tobacco slice by papermaking process utilizing tobacco waste. After collection, wastewater sample was immediately stored at 4°C until use. The COD and color of the actual effluent were 500 mg·L⁻¹ and 2200 C.U., and pH value was 8.0–8.1.

2.2 Experimental procedure

The glass ozone bubble reactor with a 500 mL flask was loaded with 400 mL of wastewater sample and placed in thermostatic water bath. The water bath was adjusted to the desired temperature during reaction process. A certain amount of $MnSO_4 \cdot H_2O$ and Na_2CO_3 was added into the reactor. Then, a 1 L·min⁻¹ flow of O_3 (19.98 mg·L⁻¹) oxygen-ozone mixture was produced by a laboratory ozone generator (Chuanghuan, model CH-ZTW) from pure oxygen and bubbled to the reactor through a porous gas diffuser situated at its bottom. The flow rate of the mixture-gas was continuously monitored with a Rota meter incorporated. Ozone in the gas phase was measured by the standard potassium iodide absorption method (APHA 1992). The un-reacted ozone leaving the reactor was scrubbed in 2% KI solution. Each group experiments were carried three times. Results were average values.

Before each experiment, pH value was adjusted to desired level using dilute sodium hydroxide or sulphuric acid solution. Samples were withdrawn at regular time intervals. The determination of the COD and color at 465 nm were carried out on a Hach spectrophotometer (DR2800, Hach, Loveland, CO, USA) according to the standard methods (APHA 1992). The pH was measured with a Sartorious PB-10 pH meter.

3 RESULTS AND DISCUSSION

3.1 Effect of Mn (II) ion dosage on ozonation efficiency

To evaluate the effect of CO_3^{2-} dosage on COD removal, experiments were conducted at CO_3^{2-} dosage of 0, 0.01, 0.015, 0.02, 0.025 mM, pH 8 and a temperature of 25°C. Variations of COD removal for different CO_3^{2-} dosage with time by ozonation are shown in Figure 1. It can be seen that COD removal rate during ozonation obviously decreased with the increasing of CO_3^{2-} dosage. However, it has no significant increase when the CO_3^{2-} dosage was above 0.02 mM and the minimum COD removal of 25.0% is observed at 0.025 mM in 90 min. The results indicate the optimal CO_3^{2-} dosage is 0.02 mM. During the single ozonation progress without CO_3^{2-}, color removal rate was all above 90%.

R^2 value nearly 0.99 (Table 1) indicates single ozonation process with CO_3^{2-} is first order kinetics. In this system, oxidations specify is nearly only ozone. During single ozonation process without CO_3^{2-}, the oxidation specify was ozone and hydroxyl radicals and R^2 value is nearly 0.99, too. This oxidation process is first order kinetics, too. Then it can be considered the total reaction rate constant was constituted from molecular ozone and hydroxyl radicals. First order kinetics is a simplification of pseudo first order due to huge excess concentration of COD in the system, which is not limiting in the reactions. But the research was conceived model on the base of HO• was scavenged completely.

To evaluate the effect of Mn (II) ion dosage on COD removal, experiments were conducted at 0, 0.1, 0.2, or 0.3 mM, a temperature of 25°C, and pH 8. Variations of COD removal for different Mn (II) ion dosage are shown in Figure 2(a).

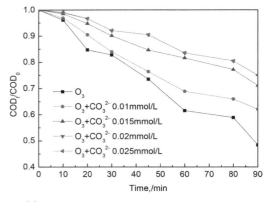

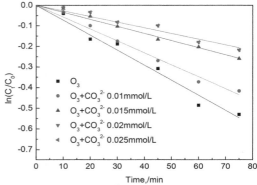

Figure 1. Effect of CO_3^{2-} dosage on COD removal and reaction kinetics during single ozonation process.

Table 1. Analysis of kinetics adding CO_3^{2-} in ozonation system.

Reaction system	CO_3^{2-} (mM)	k (×10^{-3})	R^2
S.O	0	7.28	0.99066
S.O.C	0.01	5.80	0.99471
S.O.C	0.015	3.43	0.99446
S.O.C	0.020	2.76	0.98112
S.O.C	0.025	2.76	0.9799

It can be seen that COD removal during ozonation obviously enhances with increasing of Mn (II) ion dosage from 0 to 0.2 mM. However, it has no significant increase when Mn (II) ion dosage increases 0.3 mM and maximum COD removal of 69.6% is reached at Mn (II) ion dosage of 0.3 mM in 90 min. The result is similar to the previous study (EI-Raady et al. 2005), which reported that in Mn (II)/O_3 progress of citric acid, removal of citric acid increased with increasing of Mn (II) concentration in the range 0.05–0.5 mM, but

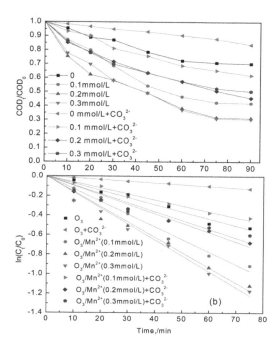

Figure 2. Effect of Mn (II) ion dosage on (a) COD removal and (b) reaction kinetics during ozonation process (CO_3^{2-} = 0.02 mM).

further increasing Mn (II) dosge had no significant effect on removal of citric acid.

Ozonation is more effective by addition of Mn (II) ion for the involvement of hydroxyl radicals generated by reaction of ozone and manganese ions (Eq. (1)). However, reactions also happen between manganese ions and hydroxyl radicals in the reaction system (Eq. (2)) (Wu et al. 2008), which result in the decrease of hydroxyl radicals concentration and reduce the degradation efficiency of organics during ozonation process. This may partially explain why COD removal does not obviously increase when Mn (II) ion dosage increases from 0.2 to 0.3 mM.

$$Mn^{(n-1)+} + O_3 + H^+ \rightarrow Mn^{n+} + O_3 + \bullet OH \qquad (1)$$

$$Mn^{(n-1)+} + \bullet OH \rightarrow HO^- + Mn^{n+} \qquad (2)$$

At the same time, the manganese ions in the reaction system can participate in the degradation reaction of organic pollutants to enhance the treatment efficiency of the PTS wastewater. Authors suggested that both the Mn (IV) and Mn (VII) ions were involved in the oxidation reaction of pollutants in the Mn (II)/O_3 system. Some organic acids (like pyruvic acid, oxalic acid) can be degradation in Mn (IV)/O_3 the homogeneous

system (Andreozzi et al. 1998; Andreozzi et al. 2001). The elimination of acids (acetic acid) for the reaction between Mn (IV) or Mn (VII) enhance the alkaline of system, which will generate more hydroxyl free radicals to enhance the oxidation efficiency of PTS wastewater. Some pollutants will be Manganese complexes with Mn (II) to form intermediate products. It may be easily oxidized by ozone (Legube & Leitner 1999). Some authors suggested that Mn (II) could react with O_3 to generate hydrous manganese oxide, which attacks pollutants in the wastewater and accelerates the degradation efficiency (Ma & Graham 1997). The possible reaction pathways proposed can be described as follows:

$$Mn^{2+} + O_3 + 2H^+ \rightarrow Mn^{4+} + O_2 + H_2O \qquad (3)$$

$$Mn^{4+} + 1.5O_3 + 3H^+ \rightarrow Mn^{7+} + 1.5O_2 + 1.5H_2O \qquad (4)$$

$$Mn^{2+} + Mn^{4+} \rightarrow 2Mn^{3+} \qquad (5)$$

$$Mn^{7+} + 1.5C_2O_4^{2-} \rightarrow Mn^{4+} + 3CO_2 \qquad (6)$$

$$Mn^{7+} + 1.5CH_3COCOOH + 1.5H_2O$$
$$\rightarrow 1.5CH_3COOH + 1.5CO_2 + Mn^{4+} + 3H^+ \qquad (7)$$

$$Mn^{4+} + CH_3COOH \rightarrow Mn^{2+} + products \qquad (8)$$

$$Mn^{3+} + O_3 + (RR2)^{2-} + H^+$$
$$\rightarrow Mn^{2+} + O_2 + HO \bullet + products \qquad (9)$$

In these reactions, Mn (IV) and Mn (VII) have stronger oxidation than other low charge ions, which can decompose the pollutants fast. From scientific aspects, manganese is not an oxidizer, but in the reaction by catalysis and by giving it some selectivity or preference on the path way of reaction, manganese can be considered as promoter of free radical reaction and simply called oxidizer.

Consequently, while the ozone possibly remains as the main oxidizer in ozonation process, three oxidants-hydroxyl radicals, manganese ions and ozone could be important oxidants to oxidize organics in PTS wastewater in Mn (II)/O_3 progress. The hydroxyl radicals have higher oxidizing potential and are less selective than ozone, resulting in better mineralization efficiency of organics in PTS wastewater. Because carbonate is inorganic radical, it cannot generate organic intermediates to influence COD value in next reactions like organic radical scavenger, as 2-propanol, et al.

To evaluate the contribution ratio of three single oxidants' degradation efficiency in PTS wastewater in Mn (II)/O_3 progress, experiments were conducted with CO_3^{2-}, which reacts with hydroxyl radicals. For comparison the contribution ratio of three oxidants, it can be considered hydroxyl radicals can react with radical scavenger completely. This is hypothesis and a conceived substrate. Thus, 0.02 mM Na_2CO_3 was enough to react with hydroxyl radicals produced in the system. It was added into ozonation process and the results were shown in Figure 2 (a).

It was seen from Figure 2(a) that COD removal declines from 54.5% to 38.4% during Mn (II)/O_3 progress with CO_3^{2-}. That indicated hydroxyl radical was important oxidant to degradation organics in PTS wastewater. Even in no-catalytic ozonation, COD removal decreases from 29.9% to 16.4% in 90 min after 0.02 mM CO_3^{2-} was added into the reaction system, indicating significant effect of hydroxyl radicals on pollutants removal in single ozonation process. This is because hydroxyl radicals can be produced by reaction of molecular ozone and hydroxyl ion at alkaline condition during single ozonation process.

A first order kinetic model was proposed for COD removal of PTS wastewater by ozonation. The experimental data fulfill first order kinetic equation. The natural logarithm of COD concentration versus reaction time yields straight lines for both no-catalytic ozonation and Mn (II)/O_3 process, It can be seen from Figure 1 (b) that the reaction rate constant increased with increasing the Mn (II) ion dosage, and significantly decreased with 0.02 mM CO_3^{2-}.

In ozonation process, COD removal was resulted by direct oxidation of ozone and indirect oxidation of hydroxyl radicals. The total reaction rate constant can be described as the following equation (Eq. (10)).

$$k = k_{[O3]} + k_{[\bullet OH]} \qquad (10)$$

In single ozonation process, COD removal was from ozone and hydroxyl radicals at pH of 8.0, temperature of 30 °C, and ozone concentration of 14.76 mg·L^{-1}. Reaction rate constant (k) was constituted by $k_{[O3]}$ and $k_{[\bullet OH]}$. After CO_3^{2-} was added, it was molecular ozone to react with organics because hydroxyl radicals reacted with CO_3^{2-} in the system. The total COD removal rate can be described as the following equation (Eq. (11)).

$$r_a = r_{O3} + r_{\bullet OH} \qquad (11)$$

In Table 2, k value was the natural logarithm of COD removal rate. After calculation, the reaction of ozone and hydroxyl radicals with organics was conformed to the first order kinetics. In the next experiments, the dosage of CO_3^{2-} was 0.02 mM.

Contribution of Mn (II) ion can be calculated by the COD removal rate by Mn (II) ion and ozone (in group ⑤) did away with that by ozone (in group ②). In group ② the dosages of CO_3^{2-} was

Table 2. Non-catalytic ozonation and inhibition by carbonates.

No.	Reaction system	k ($\times 10^{-3}$)	R^2	r_a
1	S.O.	7.28	0.99069	r_{O3}, r_{OH}
2	S.O.C (0.01 mM)	3.03	0.99642	r_{O3}
	S.O.C (0.015 mM)	1.71	0.99941	r_{O3}
	S.O.C (0.02 mM)	1.66	0.99875	r_{O3}
	S.O.C (0.025 mM)	1.52	0.99715	r_{O3}
3	S.O.C (0.01 mM)	3.45	0.9517	r_{OH}
	S.O.C (0.015 mM)	4.92	0.98298	r_{O3}
	S.O.C (0.02 mM)	4.99	0.98576	r_{O3}
	S.O.C (0.025 mM)	5.15	0.98634	r_{O3}

0.01, 0.015, 0.02 and 0.025 mM, but in the previous experiments the optimal radical scavenging was 0.02 mM. So in the next experiments of catalytic ozonation system, the dosage of CO_3^{2-} was 0.02 mM, too.

The COD removal rate from Mn (II) ion was shown in group ⑦. It is calculated by that in later three groups of group ⑤ subtracted it in group ①. For calculation COD removal rate kinetics parameter of hydroxyl radicals, the radical scavenger (CO_3^{2-} of 0.02 mM) was added. The data in group ⑥ was the COD removal rate from hydroxyl radicals, which was reduced that in group ⑤ by that in group ④. The amount of hydroxyl radicals in group ⑥ was more than that in ozonation process. So oxidation was stronger and reaction was first order kinetics. In experiments, the kinetics parameter (k) was largest and the COD removal rate was fastest when Mn (II) ion dosage was 0.2 mM. So in the next experiments, the dosage of Mn (II) ion and CO_3^{2-} were 0.2 mM, 0.02 mM, respectively.

$$r_a = r_{O3} + r_{OH} + r_{Mn} \qquad (12)$$

During single ozonation process, single ozonation with radical scavenger (CO_3^{2-}) process, Mn (II)/O_3 process and Mn (II)/O_3 with CO_3^2 process, all the reactions accorded with the first order kinetics. During Mn (II)/O_3 process, COD removal can be contributed by ozone, hydroxyl radicals and manganese ions. The total reaction rate constant can be described as the following equation (Eq. (13)).

$$k = k_{[O3]} + k_{[\bullet OH]} + k_{[rest]} \qquad (13)$$

Because ozone availability is fixed in the three systems, the ozone concentration in water and the direct oxidation of ozone can be considered constant. When CO_3^{2-} was added in the single ozonation system, COD removal was from only direct oxidation of molecular ozone, and the total reaction rate constant k equals to $k_{[O3]}$. During the

Table 3. Catalytic ozonation and inhibition by carbonates.

No.	Reaction system	k ($\times10^{-3}$)	R^2	r_a
1	S.O.	7.28	0.99069	r_{O3}, r_{OH}
4	S.O.M. (0.1 mM)	12.94	0.995	r_{O3}, r_{OH}, r_{Mn}
	S.O.M. (0.2 mM)	15.61	0.98853	r_{O3}, r_{OH}, r_{Mn}
	S.O.M. (0.3 mM)	16.28	0.99949	r_{O3}, r_{OH}, r_{Mn}
2	S.O.C	1.71	0.99941	r_{O3}
5	S.O.C.M. (0.1 mM)	6.06	0.99425	r_{O3}, r_{Mn}
	S.O.C.M. (0.2 mM)	9.61	0.98845	r_{O3}, r_{Mn}
	S.O.C.M. (0.3 mM)	9.47	0.9818	r_{O3}, r_{Mn}
2	S.O.C	4.92	0.98298	r_{OH}
6	S.O.C.M. (0.1 mM)	4.49	0.97446	r_{OH}
	S.O.C.M. (0.2 mM)	3.2	0.90202	r_{OH}
	S.O.C.M. (0.3 mM)	3.85	0.97772	r_{OH}
7	S.O.C.M. (0.1 mM)	3.87	0.98145	r_{Mn}
	S.O.C.M. (0.2 mM)	6.96	0.96853	r_{Mn}
	S.O.C.M. (0.3 mM)	6.51	0.93799	r_{Mn}

Table 4. Effect of initial pH on ozonation efficiency.

No.	Reaction system	k ($\times10^{-3}$)	R^2	r_a
8	pH 5, S.O.M	9.7	0.91467	r_{O3}, r_{OH}, r_{Mn}
	pH 7, S.O.M	14.16	0.99708	r_{O3}, r_{OH}, r_{Mn}
	pH 8, S.O.M	15.61	0.98853	r_{O3}, r_{OH}, r_{Mn}
	pH 10, S.O.M	15.67	0.97139	r_{O3}, r_{Mn}
9	pH 5, S.O.M.C	5.31	0.96556	r_{O3}, r_{Mn}
	pH 7, S.O.M.C	10.19	0.99415	r_{O3}, r_{Mn}
	pH 8, S.O.M.C	9.61	0.98845	r_{O3}, r_{Mn}
	pH 10, S.O.M.C	9.6	0.98932	r_{O3}, r_{Mn}
10	pH 5, S.O.M.C	1.45	0.96174	r_{OH}
	pH 7, S.O.M.C	1.63	0.9806	r_{OH}
	pH 8, S.O.M.C	3.2	0.90202	r_{OH}
	pH 10, S.O.M.C	1.38	0.98815	r_{OH}
11	pH 5, S.O.M.C	2.35	0.96425	r_{Mn}
	pH 7, S.O.M.C	3.57	0.95363	r_{Mn}
	pH 8, S.O.M.C	6.96	0.96853	r_{Mn}
	pH 10, S.O.M.C	7.08	0.97714	r_{Mn}

Note: in Table 2~Table 4, the data of groups ①, ②, ④, ⑤, ⑧ and ⑨ were from experiments, the others were calculated.

Mn (II)/O_3 process, when CO_3^{2-} was added in the ozonation system, the reaction kinetics of COD removal can be interpreted by the direct oxidation of molecular ozone and degradation reaction from manganese ions, and the total reaction rate constant k can be described as follows:

$$k = k_{[O3]} + k_{[rest]} \qquad (14)$$

The contribution ratio of three types of oxidation reactions can be calculated using respective kinetic constants in the Mn (II) ion catalytic ozonation process, which was shown in Table 1. It can be seen from Table 1 that in the single ozonation system, the contribution ratio of molecular ozone and hydroxyl radical is 1:3.26, indicating that hydroxyl radical is a more important oxidant for the COD removal of the PTS wastewater than molecular ozone. Meanwhile, it can be found from Table 1 that in the Mn (II) ion catalytic ozonation process, both hydroxyl radicals and manganese ions are the dominant oxidants and play more important roles for COD removal of the PTS wastewater than molecular ozone during ozonation.

3.2 Effect of initial pH on ozonation efficiency

Initial pH had obvious effect on degradation efficiency of organics during Mn (II)/O_3 progress according to literatures (Andreozzi et al. 2001; Wu et al. 2008). Under the different pH value, the number of hydroxyl radical generated by ozone molecular was various with the increasing of pH value of the reaction system. Mn/O_3 had good efficiency of degradation of oxalic acid in the acid system according to literatures (Zang et al. 2009). The role of OH^- will influence Mn ion stability in solution. In the previous study (Chen et al., 2014), the optimal degradation efficiency was at pH 8.0. To ensure the systematically of study, experiments were conducted at pH 5, 7, 8 or 10, Mn (II) ion dosage of 0.2 mM, and a temperature of 25°C. Furthermore, to investigate the effect of hydroxyl radicals

Table 5. Analysis of contribution ratio of three types of oxidations for the COD removal in reaction systems with CO_3^{2-}.

Reaction system	CO_3^{2-} (mM)	k ($\times10^{-3}$)	$k_{[O3]}$ ($\times10^{-3}$)	$k_{[.OH]}$ ($\times10^{-3}$)	$k_{[rest]}$	$k_{[O3]}$:$k_{[.OH]}$:$k_{[rest]}$	R^2
S.O.	0	7.28	1.71	5.57	0	1:3.26:0	0.99066
S.O.	0.02	1.71	1.71	0	0	———	0.99941
S.O.M. (0.1 mM)	0	12.49	1.71	6.43	4.35	1:3.76:2.54	0.995
S.O.M. (0.1 mM)	0.02	6.06	1.71	0	4.35	———	0.99245
S.O.M. (0.2 mM)	0	15.61	1.71	6	7.9	1:3.751:4.62	0.98853
S.O.M. (0.2 mM)	0.02	9.61	1.71	0	7.9	———	0.98853
S.O.M. (0.3 mM)	0	16.98	1.71	7.22	7.35	1:3.76:2.54	0.99499
S.O.M. (0.3 mM)	0.02	9.06	1.71	0	7.35	———	0.97247

577

Table 6. Analysis of contribution ratio of three oxidants to the COD removal in ozonation system at different initial pH.

pH	Mn (mM)	CO_3^{2-} (mM)	k (×10⁻³)	$k_{[O3]}$ (×10⁻³)	$k_{[.OH]}$ (×10⁻³)	$k_{[rest]}$	$k_{[O3]}$:$k_{[.OH]}$:$k_{[rest]}$	R^2
5.0	0	0	6.06	1.71	4.35	——	1:2.54:0	0.94328
5.0	0.2	0	9.7	1.71	4.39	3.6	1:2.57:2.11	0.91467
5.0	0.2	0.02	5.31	1.71	——	3.6		0.96556
7.0	0	0	6.99	1.71	5.28	——	1:3.09:0	0.99522
7.0	0.2	0	14.16	1.71	3.97	8.48	1:2.32:4.96	0.99708
7.0	0.2	0.02	10.19	1.71	——	8.48		0.99414
8.0	0	0	7.28	1.71	5.57	——	1:3.26:0	0.99066
8.0	0.2	0	15.61	1.71	6	7.9	1:3.51:4.62	0.98853
8.0	0.2	0.02	9.61	1.71	——	7.9		0.98845
10.0	0	0	5.78	1.71	4.07	——	1:2.38:0	0.9915
10.0	0.2	0	15.67	1.71	6.07	7.89	1:3.55:4.61	0.97142
10.0	0.2	0.02	9.6	1.71	——	7.89		0.98932

on COD removal, 0.02 mM CO_3^{2-} was added in the ozonation system as hydroxyl radicals' scavenger. Variations of the COD removal at different initial pH with time are shown in Figure 2 and analysis of contribution ratio of the three types of oxidation reactions to the COD removal is indicated in Table 6.

It can be seen from Figure 3 that in Mn (II)/O₃ system, COD removal obviously enhances with the increase of initial pH and COD removal efficiency varies from 43.9% to 70.5% in 90 min. At the same time, the COD removal obviously decreases after addition of CO_3^{2-}, indicating important effect of hydroxyl radicals on degradation efficiency of pollutants in the PTS wastewater during the Mn (II)/O₃ process.

In Table 6, it can be seen that the variation tendency of contribution ratio of different oxidation reactions is different between single ozonation and Mn (II) ion catalytic ozonation system. In ozonation system, the contribution ratio of ozone and hydroxyl radical enhances from 1:2.54 to 1:3.26 when initial pH increases from 5 to 8 and then decreases to 1:2.38 when the initial pH further increases to 10, indicating that hydroxyl radicals fulfill the more important role for the COD removal at pH 8.

In Mn (II)/O₃ system, it is found from Table 6 both manganese ions and hydroxyl radicals are the more important oxidants than ozone and the contribution ratio of three oxidizers varies from 1:2.57:2.11 to 1:3.55:4.61 when the initial pH increases from 5 to 10. On the other hand, manganese ions play a more important role than hydroxyl radicals at pH 7, 8 and 10, especially at pH 7, at which the contribution ratio of the three types of oxidation reactions is 1:2.32:4.96. Moreover, hydroxyl radicals play a more important role at higher pH of 8 and 10 in the Mn (II) ion catalytic ozonation process.

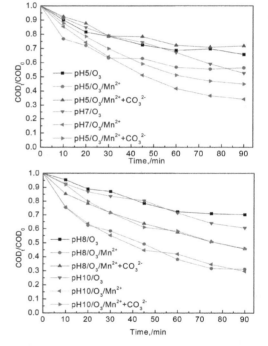

Figure 3. Effect of initial pH on COD removal during catalytic ozonation process (Mn (II) = 0.2 mM; CO_3^{2-} = 0.02 mM).

4 CONCLUSIONS

The experimental results reveal that Mn (II) ion can increase the COD removal of the PTS wastewater during ozonation process. The COD removal enhances with the increase of initial pH and varies from 43.9% up to 70.5% in 90 min during ozonation catalyzed by Mn (II). In the single ozonation system, both hydroxyl radicals and molecular ozone contribute to the oxidation reactions of pollutants, but hydroxyl radicals fulfill the more important role for the COD removal than molecular ozone. However, in the Mn (II) ion catalytic ozonation system, three types of oxidants including hydroxyl radicals, manganese ions and molecular ozone contribute to the oxidation reactions of pollutants in the PTS wastewater and both manganese ions and hydroxyl radicals are the more important oxidant for the COD removal than molecular ozone. Moreover, manganese ions play a more important role for the COD removal than hydroxyl radicals at neutral and alkaline condition, especially at pH 7, at which the contribution ratio of molecular ozone, hydroxyl radicals and manganese ions is 1:2.32:4.96. Hydroxyl radicals play a more important role for the COD removal at higher pH of 8 and 10 in the Mn (II) ion catalytic ozonation process.

SYMBOLS

S.O.—single ozonation
C.O.—catalytic ozoonation
S.O.C.—S.O. with carbonates
C.O.C.—C.O. with carbonates
S.O.M.—S.O. with Manganese
S.O.C.M.—S.O.C. with Manganese
r_a—total COD removal rate
r_{O3}—COD removal rate from ozone
r_{OH}—COD removal rate from hydroxyl radicals
r_{Mn}—COD removal rate from manganese ions
k—total reaction rate constant, $mg \cdot L^{-1} \cdot min^{-1}$
$k_{[O3]}$—reaction rate constant ascribed to direct oxidation of ozone, $mg \cdot L^{-1} \cdot min^{-1}$
$k_{[\cdot OH]}$—reaction rate constant ascribed to indirect oxidation of hydroxyl radicals, $mg \cdot L^{-1} \cdot min^{-1}$
$k_{[rest]}$—reaction rate constant ascribed to oxidation of manganese ions, $mg \cdot L^{-1} \cdot min^{-1}$

REFERENCES

APHA (American Public Health Association). 1992. Standard Methods for the Examination of Water and Wastewater, 19th ed., USA.

Andreozzi, R. et al. 1998. The ozonation of pyruvic acid in aqueous solutions catalyzed by suspended and dissolved manganese. Water Res. 32 (5): 1492–1496.

Andreozzi, R. et al. 2001. Kinetic modeling of pyruvic acid ozonation in aqueous solutions catalyzed by Mn (II) and Mn (IV) ion. Water Res. 35(1): 109–120.

Chen Saiyan, et al. 2014. Ozone Oxidation Pretreatment of Wastewater from Paper-making Process Reconstituted Tobacco Production. Tobacco Chemistry, 12: 27–31.

EI-Raady, A. et al. 2005. Catalytic ozonation of citric acid by metallic ions in aqueous solution. Ozone Sci. Eng. 27(6): 495–498.

Gracia, R. et al. 1998. Mn (II)-catalysed ozonation of raw EBRO river water and its ozonation byproducts. Water Res. 32(1): 57–62.

Legube, B. & Leitner, N.K.V. 1999. Catalytic ozonation: a promising advanced oxidation technology for water treatment. Catal. Today 53(1): 61–72.

Lei, L., & Li, Y. 2014. Effect of ozonation on recalcitrant Chemical Oxygen Demand (COD), color and biodegradability of Hardwood Kraft Pulp (KP) Bleaching Effluent. BioResources 9(1): 1236–1245.

Ma, J., & Graham, N.J.D. 1997. Preliminary investigation of manganese-catalyzed ozonation for the destruction of atrazine. Ozone Sci. Eng. 19(3): 227–240.

Ma, J., & Graham, N.J.D. 1999. Degradation of atrazine by manganese-catalysed ozonation for the destruction of atrazine-effect of humic substances. Water Res. 33(3): 785–793.

Meher, K.K. et al. 1995. Biomethan-ation of tobacco waste, Environ. Pollut. 90(2): 199–202.

Meza, P.R. et al. 2011. Reduction of the recalcitrant COD of high yield pulp mills effluents by AOP. Part 1. Combination of ozone and activated sludge. Bio Resources 6(2): 1053–1068.

Munari, M. 1986. Quantitative determination of nicotine content in protein extracted from tobacco. Tobacco Journal International 2: 128–132.

Saunders, J.A. & Blume, D.E. 1981. Quantitation of major tobacco alkaloids by high performance liquid chromatography. Journal of Chromatography 205, 147–154.

Sponza, D.T. 2002. Toxicity studies in a tobacco industry biological treatment plant. Water Air Soil Poll. 134(1–4): 137–164.

Wang, M. et al. 2011. Optimization of Fenton process for decoloration and COD removal in tobacco wastewater and toxicological evaluation of the effluent. Water Sci. Technol. 63(11): 2471–2477.

Zang X.J. et al. 2009. Quantitative study on mechanism of Mn (II) catalytic ozonation of oxalic acid. CIESC Journal. 60(2): 428–434.

Advances in Energy, Environment and Materials Science – Wang & Zhao (Eds)
© *2016 Taylor & Francis Group, London, ISBN 978-1-138-02931-6*

Formula optimization of dust coagulation agent for application in open-pit iron ore mine blasting

E.W. Ma
School of Mines Engineering, University of Science and Technology Liaoning, Anshan, China

Q.S. Wu
Mining Company of Anshan Iron and Steel Group Corporation, Anshan, China

X.Y. Wu & Q.C. Lang
Tianjin Recyclable Resources Institution, All China Federation of Supply and Marketing Cooperatives, Tianjin, China

Z.J. Cui
School of Environmental Science and Engineering, Shandong University, Jinan, China

ABSTRACT: A new kind of dust coagulation agent for application in open-pit iron ore mine blasting was prepared, which composed of surfactant factor, moisture factor, and coagulation factor. The optimum formula of the dust coagulation agent was obtained according to the results of the single-factor tests and the orthogonal tests. The reagents of span-80, Na_2SiO_3 and soluble starch were chosen as the surfactant factor, moisture factor, and coagulation factor, with the concentration of 0.05%, 20%, and 0.20%, respectively. The highest dust suppression efficiency of the agent was 97.50%. The results show that the raw material of the dust coagulation agent has wide resource, low cost and outstanding dust suppression effect. It has broad application prospects in blasting dust suppression in the open-pit iron ore mine.

1 INTRODUCTION

A large amount of dust is produced in the process of blasting. The dust from the blasting process of open-pit iron ore mine has the characteristics of the large instantaneous dust, high concentration, and small particle size. According to the relevant statistical data, the ratio of PM10 (particulate matter 10) was nearly 90% after blasting in open-pit iron ore mine. After blasting, the dusts with large particle sizes are settled in a short time. While, the settlement rate of the particles with small size (below 10 µm) are low, and the airborne dust are distributed in the pit, which not only degrades the quality of surrounding environment such as atmosphere and water, but also leads to serious physiological damage to the workers in the workplace, results in pneumoconiosis to different extent, and endangers the lives and health of workers.

So far, there was little research on the dust suppression measures during blasting process. For PM10, the regular dust suppression measures were not applicable in this condition, such as machinery dust removal, and electric dust precipitation. Dust coagulating technique was the emerging fine dust control technique in recent years. Coagulating is the process of fine particles contact with each other and forming large particles by physical or chemical methods. At present, the coagulating techniques include electric agglomeration, acoustic agglomeration, vapor phase coagulation, thermal coagulation, magnetic coagulation, photocoagulation, and chemical coagulation et al.

In view of the work environment of open-pit mining, chemical coagulation technique has the feasibility in dust suppression, especially in the suppression of PM10. Therefore, the study of high efficiency dust coagulation agent has a great significance for dust suppression.

This study focuses on the dust suppression of open-pit iron ore mine blasting in Anshan area (in China). The high efficiency dust coagulation agent with good effect of water conservation, dust coagulation and moisture sorption was prepared by the single-factor tests and orthogonal tests, and dust suppression efficiency was evaluated in this study.

2 EXPERIMENTAL

2.1 Sample preprocessing

The open-pit iron mine blasting dust sample was taken from Anshan area (in China). In order to study the dust suppression effect of fine particles, the sample needs to be preprocessed. Figure 1 shows the process of blasting dust pretreatment. The final sample was obtained by broken-grinding and gravity settling for 48 h. The size distribution of the sample was tested by laser particle analyzer (BT-2003) and the D90 of the sample was below 10 μm.

2.2 Materials

The dust coagulation agent was composed of the surfactant factor, moisture factor, and coagulation factor. Analytical reagents were used in the experiment. The surfactant factor, such as Sodium Dodecyl Benzene Sulfonate (SDBS), Sodium Dodecyl Sulfate (SDS), span-80 and tween-80, was selected by wet ability test equipment. The water loss rate of the solutions of $CaCl_2$, $MgCl_2$, glycerol ($C_3H_8O_3$) and sodium silicate (Na_2SiO_3) were tested for selecting the moisture factor. The coagulation factor was chosen from soluble starch, NaCMC and PAM.

2.3 Experimental methods

Single-factor tests and orthogonal tests were taken in this study. The surfactant factor, moisture factor, and coagulation factor were selected by the single-factor tests. Then, the formula of the agent was optimized by orthogonal tests.

2.3.1 Wet ability test

Wetting velocity was used to represent the wet ability and it was measured by the method in GB/T16913-2008. The wetting velocities of the solutions of SDBS, SDS, span-80 and tween-80 were measured. The concentrations of surfactants were 0.05%, 0.10%, 0.15%, and 0.20%, respectively. Surfactant type and optimum concentrations could be determined.

2.3.2 The evaporation resistance test

The evaporation resistance properties of the solutions of $CaCl_2$, $MgCl_2$, glycerol and sodium silicate were tested. The concentrations of the solutions were 0.05%, 0.10%, 0.15%, and 0.20%, respectively. The dust sample (10 g) was taken in a dish, the

solutions were sprayed on dish surface, and then dried the dish in the drying oven at 75 °C, recorded the quality every 1 h and calculated the water loss rate. The formula of water loss rate was calculated as follows:

$$\eta_1 = \frac{w_0 - w_i}{w_1} \times 100\% \tag{1}$$

where, η_1 is the water loss rate of the dust, %; w_0 is the initial quality of the dish, g; w_i is the quality after drying i h, g; w_1 is the quality of the solution, g. The type of the moisture absorbent and its optimum concentration could be determined.

2.3.3 The selection of the coagulation factor

The coagulation factor was chosen from NaCMC, soluble starch and PAM. The solutions were prepared with the concentrations of 0.05%, 0.10%, 0.15%, and 0.20%, respectively. The solution viscosity was tested by viscosity meter (NDJ-1). The dust sample of 30 g was taken in the dish. The solution of 10 ml sprayed on its surface and then dried the sample at 105 °C for 4 h. After that, sieved the sample with 100 mesh sieve and shocked 100 times. The qualities of oversize and undersize were measured, and then the mass ratio (R) was calculated. Considering the solution viscosity and the mass ratio (R), the coagulating effect of the solution was evaluated.

2.3.4 Orthogonal tests

According to the results of single-factor tests, the orthogonal tests selected surfactant factor, moisture factor, and coagulation factor as the influencing factors. Each factor set three levels. The orthogonal tests were carried out according to the orthogonal table of L9 (3^4). Dust suppression efficiency as evaluation index was detected under the condition of laboratory simulation. Figure 2

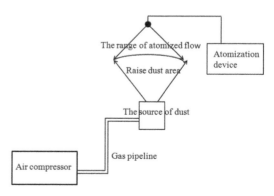

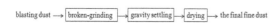

Figure 1. The pretreatment process of the blasting dust.

Figure 2. The simulation device of dust formation.

shows the simulation device of the dust formation in laboratory.

A dish (Φ12 cm) was used to collect the dust in the test. The collecting time was 2 min. The initial quality of dust fall (M_0) was obtained after naturally settling for 24 h. The quality of the dust sample was 100 g in the test. The operating pressure of the air compressor was 6 bar. The coagulation agent sprayed on the dust area about 10 second and the dish was collected after 2 min. The dish was dried at 120 °C for 2 h. The quality of dust fall was named M. The dust suppression efficiency (η) was calculated by the formula as follows:

$$\eta = \frac{M - M_0}{M} \times 100\% \qquad (2)$$

where: η is the dust suppression efficiency, %; M is the quality of the falling dust after spraying the coagulation agent, g; M_0 is the initial quality of the falling dust without spraying the coagulation agent, g.

3 RESULTS AND DISCUSSION

3.1 Analysis of wet ability

Due to the different hydrophobic group, different surfactants have different adsorption capacities for dust sample. The adsorption capacity for the dust sample can be reflected by the wetting velocity. The maximum wetting velocity can be reached by adding certain surfactant. Figure 3 shows the wetting velocity of dust sample with different surfactants. As can be seen from Figure 3, the solution of span-80 (0.15%) has the higher wetting velocity. Therefore, 0.05%–0.15% solutions of span-80 were chosen as the surfactant factors in the orthogonal tests.

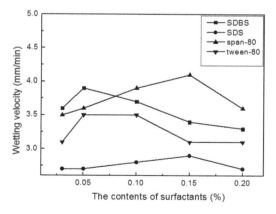

Figure 3. The wetting velocity of dust sample with different surfactants.

3.2 Analysis of water loss rate

Figure 4 shows the variation of water loss rate with different absorbents (CaCl$_2$, MgCl$_2$, C$_3$H$_8$O$_3$, and Na$_2$SiO$_3$). With the increasing of the absorbent concentrations, the rate of water loss decreased obviously. As can be seen from Figure 4 (D), the lowest water loss rate was obtained as the concentration of Na$_2$SiO$_3$ was 25%. Therefore, 0.05%–0.15% solutions of Na$_2$SiO$_3$ were chosen as the moisture factors in the orthogonal tests.

3.3 Analysis of the coagulation factor

Table 1 shows the viscosity and the mass ratio of oversize to undersize (R) of coagulation factors with different concentrations. With the increase of their concentrations, all the viscosity of the coagulation factors increased. The solution of 25% NaCMC has the highest viscosity, which was 21.15 MPa·s. At the same concentration, the solution of NaCMC has the maximum viscosity and the solution of PAM took second place. The viscosity of the soluble starch had a little change as the concentration increased from 0.05% to 0.25%. According to previous study, the solution sprayed would be difficult when the viscosity was higher than 4 MPa·s. Although NaCMC solutions have the higher viscosity, but it didn't meet the requirement as the coagulation factor.

In Table 1, the value of R increased with increasing the concentration of coagulation factors. The maximum of R was 86.26% when the concentration of PAM was 0.25%. It indicated that the effects of coagulation and adhesion were enhanced by increasing the concentration of solution. The fine dust after spraying the agent was adhered easily. Considering the concentration, viscosity and the R value of the solution, although the PAM solutions with concentration above 1.15% have a good adhesion effect, their viscosities were above 4 MPa·s, which was not good for spraying. Therefore, the solution of PAM was not suitable as the coagulation factor for dust suppression.

For soluble starch solution, with increasing the concentration from 0.05% to 0.25%, the viscosity of the solution increased from 1.07 MPa·s to 1.18 MPa·s, the value of R increased from 5.23% to 43.99%. Especially, as the concentration of starch solution was between 0.15% and 0.25%, the R value was between 43.05% and 43.99%. Thus, it can be seen that the starch solutions, in certain range of concentrations, both meet the requirements of spray on solution viscosity and have good adhesion effect. Therefore, 0.15%–0.25% starch solutions were chosen as the coagulation factors in the orthogonal tests.

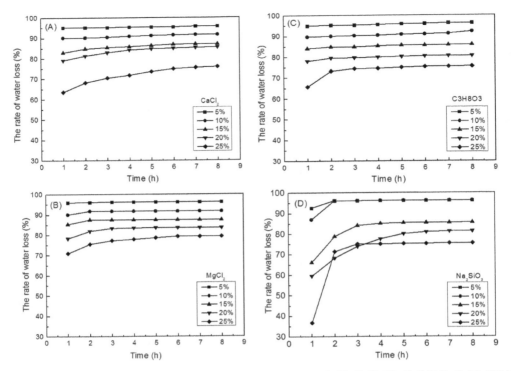

Figure 4. The variation of water loss rate with different absorbents. (A: $CaCl_2$; B: $MgCl_2$; C: $C_3H_8O_3$; D: Na_2SiO_3).

Table 1. The results of the coagulation factor test.

Coagulation factors	Concentration (%)	Viscosity (MPa·s)	R (%)
Soluble starch	0.05	1.07	5.23
	0.10	1.08	7.18
	0.15	1.10	43.05
	0.20	1.18	43.99
	0.25	1.18	43.70
NaCMC	0.05	5.46	19.58
	0.10	7.78	44.55
	0.15	10.34	49.44
	0.20	13.42	60.93
	0.25	21.15	62.20
PAM	0.05	2.07	8.91
	0.10	2.63	15.29
	0.15	5.36	44.12
	0.20	5.78	51.93
	0.25	5.97	86.26

Table 2. The three factors three levels orthogonal table.

Levels	Span-80 (A)/%	Na_2SiO_3 (B)/%	Soluble starch (C)/%
1	0.05 (A1)	15 (B1)	0.15 (C1)
2	0.10 (A2)	20 (B2)	0.20 (C2)
3	0.15 (A3)	25 (B3)	0.25 (C3)

efficiency as index, orthogonal design of three factors three levels was taken to optimize the formula and verified test. Table 2 shows the design of the three factors three levels orthogonal table. Table 3 shows the orthogonal tests results and the analysis of the results.

As can be seen from Table 3, in the three factors, the maximum of range was caused by the factor A, the minimum was caused by the factor C. The order of importance degree for the three factors was A > B > C.

For the factor of A, the maximum of dust suppression efficiency was obtained in level 1. For the factors of B and C, the maximum of dust suppression efficiency were all obtained in level 2. The results show that the optimum formula of the dust suppression agent was A1B2C2, which is span-80

3.4 Analysis of the orthogonal tests

According to the single-factor tests, the orthogonal tests selected span-80 as the surfactant factor, Na_2SiO_3 as the moisture factor and soluble starch as the coagulation factor. Taking dust suppression

Table 3. Results analysis of orthogonal tests.

No	Factors (%)			Dust suppression efficiency (%)
	A (span-80)	B (Na$_2$SiO$_3$)	C (soluble starch)	
1	0.05 (1)	20 (2)	0.15 (1)	79.89
2	0.05	20	0.25 (3)	83.60
3	0.05	25 (3)	0.20 (2)	96.30
4	0.10 (2)	15 (1)	0.25	37.57
5	0.10	25	0.15	42.86
6	0.10	25	0.20	22.75
7	0.15 (3)	15	0.20	88.36
8	0.15	15	0.25	45.50
9	0.15	20	0.15	46.56
K1	259.79	171.43	169.31	
K2	103.18	210.05	207.41	
K3	180.42	161.91	166.67	
$k1$	86.60	57.14	56.44	
$k2$	34.39	70.02	69.14	
$k3$	60.14	53.97	55.56	
Range (R)	52.21	16.05	13.58	
Factors order	A > B > C			
Optimization levels	A1	B2	C2	
Optimal combination	A1B2C2			

with the concentration of 0.05%, Na$_2$SiO$_3$ with the concentration of 20% and soluble starch with the concentration of 0.20%.

The optimum formula (A1B2C2) was not existed in the orthogonal tests, so the additional test was added. The dust suppression efficiency was 97.50% in the additional test (A1B2C2), which was higher than the maximum (96.30%) in the orthogonal tests (A1B3C2).

In this study, the obvious dust suppression effect can be obtained by spraying the coagulation agent, which was composed of span-80 (0.05%), Na$_2$SiO$_3$ (20%) and soluble starch (0.20%). The results show that the raw material of the dust suppression agent has wide resource low cost and outstanding dust suppression effect. It has broad application prospects in blasting dust suppression in the open-pit mine.

4 CONCLUSIONS

A new kind of dust coagulation agent was prepared, which composed of surfactant factor, moisture factor and coagulation factor. Through the single-factor tests and the orthogonal tests, span-80 was chosen as the surfactant factor, it improved the wet ability to the dust. The moisture factor selected Na$_2$SiO$_3$, which has the lowest water loss rate in the test. Soluble starch was used as the coagulation factor, which enhanced the coagulation effect for the dust. Taking dust suppression efficiency as index, orthogonal design of three factors three levels was

taken to optimize the formula and verified test. The optimum formula of the dust coagulation agent was span-80 (0.05%), Na$_2$SiO$_3$ (20%) and soluble starch (0.20%). The highest dust suppression efficiency of the dust coagulation agent was 97.50%. The dust suppression agent has the advantages of wide raw material sources, simple production, low cost, and outstanding dust control effect, which has broad application prospects in blasting dust suppression in the open-pit iron ore mine.

REFERENCES

Copeland, C.R. & Kawatra, S.K. 2011. Design of a dust tower for suppression of airborne particulates for iron making. *Minerals Engineering* 24(13):1459–1466.

Ding, C. & Nie, B.S. 2011. Experimental research on optimization and coal dust suppression performance of magnetized surfactant solution. *Procedia Engineering* 26:1314–1321.

Ji, J.H. & Hwang J. 2004. Particle charging and agglomeration in DC and AC electric fields. *Journal of Electroatatices* 61(1):27–68.

Lu, X.X. & Wang, D.M. 2015. Experimintal investigation of the pressure gradient of a new spiral mesh foam generator. *Process Saftey and Environmental Protection* 94:44–54.

Markauakas, D. & Maknickas A. 2015. Simulation of acoustic particle agglomeration in poly-dispersed aerosols. *Procedia Engineering* 102:1218–1225.

Zhou, D. & Luo, Z.Y. 2015. Preliminary experimental study of acoustic agglomeration of coal-fired particles. *Procedia Engineering* 102:1261–1270.

Advances in Energy, Environment and Materials Science – Wang & Zhao (Eds)
© *2016 Taylor & Francis Group, London, ISBN 978-1-138-02931-6*

Properties of hemihydrate phosphogypsum and application to solidified material

Ciqi Liu, Qinglin Zhao & Kongjin Zhou
State Key Laboratory of Silicate Materials for Architecture, Wuhan University of Technology, Wuhan, Hubei, China
School of Materials Science and Engineering, Wuhan University of Technology, Wuhan, Hubei, China

Jianqiu Li
Guizhou Chuanheng Chemical Engineering Limited Liability Company, Guizhou, China

ABSTRACT: In this paper, the chemical and mineralogical properties of hemihydrate phosphogypsum, such as, chemical composition, crystal water content, mineralogical phase, and microstructure have been obtained by comprehensive analysis including chemical analysis, X-Ray diffraction analysis, differential thermal analysis, and environmental scanning electron microscope analysis. Based on this, the lime-fly ash stabilized hemihydrate phosphogypsum solidified materials are prepared. The feasibility is discussed from mechanical and durability performances and then some suggestions are proposed to practical application. It is shown that the hemihydrate phosphogypsum for this research is a air hardening binder, whose crystal water content is 16.42%, main mineralogical phases are 18.88% $CaSO_4 \cdot 1/2H_2O$, 72.85% $CaSO_4 \cdot 2H_2O$ and a small content of quartz. The unconfined compressive strength of lime-fly ash solidified material is 3.26MPa and 5.99MPa for 7 days and 28 days, respectively. Thus, this study is important to guide the comprehensive utilization of industrial solid waste of hemihydrate phosphogypsum.

1 INTRODUCTION

In accordance with different wet process of phosphoric acid production, phosphogypsum by-product has different property. The chemical formula of crystal water in phosphogypsum—n value depends on the phosphoric acid process that uses 2.0 for the dihydrate process and 0.5 for the hemihydrate process (a high effective process for phosphoric acid production). At present, dihydrate process is popular in most phosphorus chemical enterprises, and hemihydrate process are used only in several factories including Guizhou Chuanheng Chemical Co., Ltd which has a hemihydrate process phosphoric acid device of 150,000 tons per year with the hemihydrate phosphorus gypsum output of 750,000 tons.

However, it will bring in serious pollution and big resource-consuming if the hemihydrate phosphorus gypsum is discharged without waste disposal. The by-product of phosphogypsum created by dihydrate process has been used in solidified materials and it has shown good properties (Shen, 2008; Shen, 2009; Hua, 2010; Shen, 2003; Zhou, 2007). It is reported that the hemihydrate phosphogypsum and lime-Fly Ash (FA) system show good gypsum-based binder characteristics (Zhou, 2006; Mei, 2012; Shen, 2007; Qi, 2009). In view of this, hemihydrate phosphogypsum's characteristics have been studied in order to improve the utilization of hemihydrate phosphogypsum and study the feasibility in the solidified materials.

2 RAW MATERIALS AND TEST METHODS

2.1 Raw materials

1. Hemihydrate phosphogypsum: coming from the phosphogypsum yard of Guizhou Chuanheng Chemical Limited Liability Company. The 0.3 mm and 0.075 mm sieve passing rate is 50.61% and 20.03% respectively, and the density is 2.358 g/cm^3. The corresponding chemical composition analysis is shown in Table 1.
2. FA: coming from a power plant in Guizhou, whose appearance color is brown, the corresponding chemical composition analysis is also shown in Table 1. The 0.3 mm and 0.075 mm sieve passing rate is 98.98% and 86.50% respectively, and the density is 2.023 g/cm^3.
3. Lime: coming from a lime plant in Guizhou. The effective calcium- and magnesium oxide content is 77.97%. The lime is magnesium lime with 5.27% MgO content, which belongs to the first class in accordance with the Chinese Industrial Norm for Building Material JC/T 480-92.

Table 1. Chemical compositions of raw materials (%).

Raw materials	Hemihydrate phosphogypsum	Lime	Fly ash
CaO	33.04	72.70	36.87
SiO_2	2.09	5.01	17.87
Al_2O_3	0.27	0.54	7.34
Fe_2O_3	0.10	1.05	4.24
MgO	0.051	5.27	3.05
P_2O_5	1.05	0.017	–
F	0.46	Undetected	–
TiO_2	Undetected	0.14	1.16
SO_3	44.30	1.15	2.93
Crystal water	16.42	–	–
Loss	18.09	13.80	25.50

2.2 Test methods

1. The chemical compositions of hemihydrate phosphogypsum

   To analyze the chemical compositions of hemihydrate phosphogypsum, raw materials have been dried at the temperature of 40°C with some wind and then directly ground in a ceramic mortar and all samples have been filtered by a 0.075 mm sieve and the powder is used for chemical analysis testing. The pressing power pellets method is developed by XRF for chemical analyzing and testing.

2. X-Ray Diffraction (XRD) analysis

   The mineralogical phase composition of the hemihydrate phosphogypsum is analyzed by D8 Advance XRD machine (made in German) in the experiments. The rated power of X-ray diffraction is 12 kW, the scanning range is from 5° to 60° (2θ) and the moving target material is Cu.

3. Environmental Scanning Electron Microscope (ESEM) analyzing method

   In order to observe the microstructure of hemihydrate phosphogypsum and its hydrated production, Philips environmental scanning electron microscope, XL-30 ESEM with integral energy dispersive X-ray micro-analysis system have been applied. The application of ESEM can realize the direct observation of hemihydrate phosphogypsum without conductive film. The samples need to be dispersed by absolute ethyl alcohol before the testing. The acceleration voltage in the normal observation is 12.0 kV. With integral energy, dispersive X-ray micro-analysis could be applied to determine some certain element's content.

4. Differential Thermal Analysis (DTA)

   In order to figure out the phase changing characterization of the hemihydrate phosphogypsum in the heating process, DTA test is applied. 50 to 150 mg samples have been ground to less than 0.075 mm and dried. And the temperature range of analysis is from 20 to 1000°C.

5. Mechanical property test of solidified materials

   The test is carried out in accordance with the Standard Norm JTG E51-2009, Test Methods of Materials Stabilized with Inorganic Binders for Highway Engineering. In order to determine the optimum water content and maximum dry density of the inorganic binders, the heavy compaction method is applied in accordance with the Compaction Test Standards of Ministry of Communications T0804-94. Unconfined compressive strength test can be tested on the basis of the Standard of Ministry of Communications T0805-94 with the sample size being Φ 50 mm × 50 mm.

3 RESULTS AND ANALYSIS

3.1 Characteristics of hemihydrate phosphogypsum

The characteristics of hemihydrate phosphogypsum are significantly different from that of dihydrate phosphogypsum by hemihydrate phosphoric acid process and that of dihydrate phosphoric acid process, respectively.

1. Air hardening

   The hemihydrate phosphogypsum is an air-hardening binder and can gain some certain mechanical strength though $CaSO_4 \cdot 1/2H_2O$ is transformed into $CaSO_4 \cdot 2H_2O$. The air hardening of hemihydrate phosphogypsum can be verified by the hardened hemihydrate phosphogypsum after being deposited for a certain time, shown in Figure 1.

2. Chemical characteristics

   The main raw materials in this research are hemihydrate phosphogypsum, lime and brown

Figure 1. The hardened hemihydrate phosphogypsum in the disposal yard.

FA, and the chemical compositions of the raw materials are shown in Table 1.

Judging from Table 1, the main composition of hemihydrate phosphogypsum are CaO and SO_3, the total phosphorus in P_2O_5 format and fluorine is 1.05% and 0.46% respectively and the loss is 18.09%. The content of $CaSO_4$ in hemihydrate phosphogypsum is 75.31% if 44.30% SO_3 is all transformed into $CaSO_4$. Taking into account crystal water content, the content of $CaSO_4 \cdot 1/2H_2O$ and $CaSO_4 \cdot 2H_2O$ is 18.88% and 72.85% respectively. Actually, the fresh hemihydrate phosphogypsum is all nearly in $CaSO_4 \cdot 1/2H_2O$ format; however, the hemihydrate phosphogypsum can be transformed into dihydrate phosphogypsum due to the over 20% free water in the fresh hemihydrate phosphogypsum, and as time goes on, the hemihydrate phosphogypsum is continuously transformed into dihydrate phosphogypsum. In this research, the hemihydrate phosphogypsum in the yard is used as the major raw material for the experiment.

As a matter of fact, the hemihydrate phosphogypsum should be dried timely or the crystal water content should be paid closer attention because the mechanical performance of solidified materials is greatly affected by the change of crystal water content.

Besides, the FA is considered as substandard grade by the Standard Norm GB/T 1596-2005 (Fly ash used for cement and concrete) and as the type of high calcium FA due to the high content of CaO, which is as high as 36.87%, and the low content of SiO_2 and Al_2O_3. The color of this FA is brown due to its high content of iron. The FA, stored near Guizhou Chuanheng Chemical Industry Limited Liability Company, is chosen as the raw material for the optimum unitization.

3. Mineralogical composition characteristics

In order to determine the mineralogical phase characteristics of hemihydrate phosphogypsum, XRD test is carried out and the results are shown in Figure 2.

The XRD pattern of hemihydrate phosphogypsum is consistent with the results of chemical analysis. The main mineralogical phases of hemihydrate phosphogypsum are $CaSO_4 \cdot 2H_2O$ and $CaSO_4 \cdot 1/2H_2O$. A large amount of dihydrate phosphogypsum is contained in hemihydrate phosphogypsum after being deposited and the content of $CaSO_4 \cdot 2H_2O$ is larger than that of $CaSO_4 \cdot 1/2H_2O$ by the intensity of XRD diffraction peaks, meanwhile, a little quartz is contained in hemihydrate phosphogypsum.

4. Thermal stability

To understand clearly the phase transformation of hemihydrate phosphogypsum in practical application, which is very important to temperature control, comprehensive thermal analysis test is carried out and the thermal patterns are shown in Figure 3. It shows that the crystal water of dihydrate phosphogypsum is released at the temperature of 40°C, so the temperature should be controlled lower than 40°C when the test of water content is carried out, otherwise, the water content will be overestimated.

If the drying temperature of hemihydrate phosphogypsum is above 80°C, the optimum water content of lime-FA stabilized material is larger than the normal, which will cause that the water content of lime FA stabilized is overlarge and then the strength formation and improvement are affected. The crystal water of hemihydrate phosphogypsum is lost at 153.3°C, meanwhile, the hemihydrate phosphogypsum is transformed into solubility anhydrite indicated

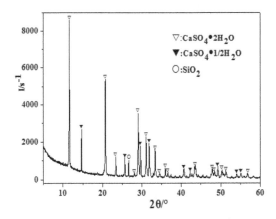

Figure 2. XRD pattern of hemihydrate phosphogypsum.

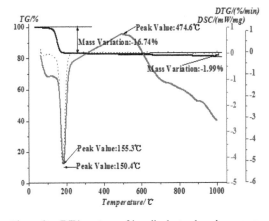

Figure 3. DTA pattern of hemihydrate phosphogypsum.

by the weightlessness. At 474.6°C, the solubility anhydrite is transformed into insolubility anhydrite, as an exothermic peak shown in Figure 3, but the mass does not change.

The loss of hemihydrate phosphogypsum is 18.09% by the chemical analysis which is consistent with the result of differential thermal analysis, 18.73%. But the two methods are different in testing method, equipment, sampling, and weight, so there is 3.5% relative error. The crystal water is calculated as 16.74% by the thermal analysis, which is consistent with the result of the chemical analysis, about 16.42%.

In general, the pavement temperature will not be over 80°C, so the removal of crystal water at the high temperature doesn't need to be considered when the fresh hemihydrate phosphogypsum is used as solidified materials. But hemihydrate phosphogypsum has been deposited for a certain time, which contained a certain dihydrate phosphogypsum, crystal water loss in dihydrate phosphogypsum which should be considered when the temperature is higher than 40°C in summer. However, the mechanical performances of lime-FA and hemihydrate phosphogypsum solidified materials won't be affected, even the strength is increased, but the water should be increased properly in construction accordingly.

5. Microstructure of hemihydrate phosphogypsum

The microstructure of hemihydrate phosphogypsum is studied by environmental scanning electron microscope shown in Figure 4. In Figure 4, the crystal morphology of hemihydrate phosphogypsum is mostly granular and some cross-grown polymeric crystals are on the surface which is more than 95% of the total due to that hemihydrate phosphogypsum is partly transformed into dihydrate phosphogypsum and the particle sizes range from 5 μm to 20 μm, mainly about 10 μm. At the same time, a little scattered growth of single crystals exists. The hemihydrate phosphogypsum is beneficial to the compaction in solidified materials thanks to that the particle is relatively small, with more uniform particle size distribution and clear edge of particle crystal. The FA is simulated of sulfate activity owing to the large contact area of the hemihydrate phosphogypsum and FA, which is proved by performance testing result of solidified materials of hemihydrate phosphogypsum.

3.2 Mechanical performance of hemihydrate phosphogypsum-lime-FA system

The compaction test of hemihydrate phosphogypsum-lime-FA is carried out in accordance with the Standard of Test Methods of Materials Stabilized

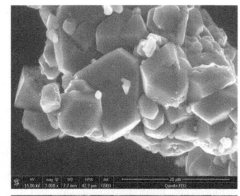

Figure 4. Environmental scanning electron microscope of hemihydrate phosphogypsum.

Table 2. 7d Unconfined compressive strength for hemihydrate phosphogypsum-lime-fly ash system.

Name	Solidified materials of hemihydrate phosphogypsum-lime-fly ash
Maximum dry density (g/cm³)	1.42
Optimum water content (%)	23
Degree compaction (%)	95
Unconfined compressive (MPa)	
R_{7d}/Rir	3.26/3.00
Cv (%)	4.8
R_{28d}/Rir	5.99/4.54
Cv (%)	14.7

with Inorganic Binders for Highway Engineering (JTG E51-2009), the maximum dry density and optimum water content of mixture have been achieved by the compaction test, the samples size of Φ 50 mm × 50 mm are formed as 95% compaction and the unconfined compressive strength is carried out after being cured for 7 days, and the results are shown in Table 2.

In Table 2, the unconfined compressive strength of hemihydrate phosphogypsum-lime-FA system (3.26 MPa) after 7-days' curing is much larger than that of the normal lime-FA solidified materials (about 0.8 MPa). With the increasing of curing age, the unconfined compressive strength of hemihydrate phosphogypsum and lime-FA solidified materials increases gradually, which achieves as high as 5.99 MPa after 28-days' curing.

4 CONCLUSIONS

1. Characteristics of hemihydrate phosphogypsum
   Hemihydrate phosphogypsum is a air-hardening binder and the main chemical compositions are CaO and SO_3, with little harmful ingredients, so it can play a very good role in binder and activator. The crystal water content of hemihydrate phosphogypsum is 16.42% and the main mineralogical phases are $CaSO_4 \cdot 1/2H_2O$ and $CaSO_4 \cdot 2H_2O$ with the content being 18.88% and 72.85% respectively. The hemihydrate phosphogypsum are mainly granular crystals and cross section sheet growth polymer crystals.
2. Feasibility of hemihydrate phosphogypsum for the solidified materials
   The unconfined compressive strength of hemihydrate phosphogypsum and lime-FA solidified material is 3.26 MPa and 5.99 MPa for 7-days' curing and 28-days'curing respectively.
3. Suggestions for hemihydrate phosphogypsum lime-FA solidified materials
4. The hemihydrate phosphogypsum can be transformed into dihydrate phosphogypsum due to that the hemihydrate phosphogypsum contains a mount of free water. If not dried, the binder characteristic is lost and the mechanical properties of solidified material are affected.
5. The dihydrate phosphogypsum can be transformed into hemihydrate phosphogypsum at the temperature of 40°C and the hemihydrate phosphogypsum can be transformed into anhydrite at the temperature of 80°C. So the temperature should be controlled when the water content of hemihydrate phosphogypsum-lime-FA solidified materials is tested, otherwise the addition of water will be affected in construction.
6. Due to that the hemihydrate phosphogypsum-lime-FA solidified materials is a gypsum-based binder system, though the water stability and freezing resistance are improved by the addition of lime and FA, the durability should be considered in application to the construction. At the same time, the hemihydrate phosphogypsum-lime-FA solidified materials should be avoided being prolonged immersion in water and freeze-thaw damage.

ACKNOWLEDGMENTS

The authors are grateful to the National Natural Science Foundation of China for supporting the research reported in this paper, under grant number 51202173.

REFERENCES

Hua, M.J., Wang, B.T., Chen, L.M., Wang, Y.H., Quynh, V.M., He, B., Li, X.F. 2010. Verification of lime and water glass stabilized FGD gypsum as road sub-base. Fuel 89: 1812–1817.
Mei, F.D., Hou, J.J., Liu, Z.D. 2012. Research on activity characteristics on composite cementitious materials based on phosphogypsum. Procedia Engineering 43: 9–15.
Qi, Z.H., Xu, X.Y., Zhu, M.L., Hu, Y. 2009. Unconfined compressive strength of mixture of phosphogypsum-fly ash-lime-clay, Proc. of Int. Symp. on Geoenvironmental Eng., ISGE2009, Hangzhou, 745–748.
Shen, W.G., Jiang, J., Zhang L., Wang, K., Zhou, M.K. 2008. A Study on Properties of Phosphogypsum Modified Lime-Fly Ash Road Base Materials. HIGHWAY (01): 141–145. (in Chinese)
Shen, W.G., Zhou, M.K., Ma, W., Hu, J.Q., Cai, Z. 2009. Investigation on the application of steel slag-fly ash–phosphogypsum solidifie d material as road base material. Journal of Hazardous Materials 164: 99–104.
Shen, W.G., Zhou, M.K., Yu, C.J., Wu, S.P., Peng, L. 2003. Study on the properties of phosphogypsum modificated lime-fly ash road base coures materials. Journal of Wuhan University of Technology 25(10): 34–37. (in Chinese)
Shen, W.G., Zhou, M.K., Zhao, Q.L. 2007. Study on lime–fly ash–phosphog ypsum binder. Construction and Building Materials 21: 1480–1485.
Zhou, M.K., Zha, J., Shen, W.G. 2007. Design, preparation and property of phosphorous slag road base materials. Acta Scientiarum Naturalium Universitatis Sunyatseni 46(6): 159–160.
Zhou, M.K., Zhao, Q.L., Zha, J., Shen, W.G. 2006. The proportion, performances and mechanism of phosphogypsum-lime-fly ash binder, 16. Internationale Baustofftagung: Tagungsbericht, in Weimar, Germany, 10837–10844.

Study on chemical oxidation of acetone gas

Wenxia Zhao, Hui Kang & Ailing Ren

School of Environmental Science and Engineering, Hebei University of Science and Technology, Shijiazhuang, China

ABSTRACT: VOCs (Volatile Organic Compounds) pollution problem causes more and more attention of people. Chemical oxidation is concerned because of its oxidation, high speed, and high efficiency. In this paper, sodium hypochlorite (NaClO) solution was used to purify acetone gas, and the effect factors were studied. The results showed that the removal efficiency of acetone gas was highest, up to 98%, when pH value was about 11, oxidation reduction potential was 680 mV. As the middle product, trace amount of acetic acid in the exhaust gas was detected by GC-MS.

1 INTRODUCTION

Volatile Organic Compounds (VOCs) refers to the organic compounds whose saturated vapor pressure is greater than 133 kPa or whose boiling point is between 50°C and 260°C under the normal temperature (Marco, 2014). VOCs mainly comes from the petroleum and chemical industry, the pharmaceutical industry, the printing industry, transportation and so on, which including a variety of hydrocarbons, halogenated hydrocarbons, alcohols, aldehydes, ketones, ethers, acids, and so on (An, 2014). These pollutants emission causes not only great resources waste, but also serious environmental problems. VOCs temporarily or permanently, damage the breathing, blood, liver and skin, for example, benzene and Benzopyrene can cause human cancer (Wu, 2012). Therefore, the treatment and recovery of VOCs were paid more attentions in the world.

The control technologies of VOCs are usually divided into two categories, one is destructive method, and the other is recovery method (Abdullahi, 2014). The former includes oxidation method (Benjamin, 2012) and biological method (Susant, 2014), the latter mainly includes absorption method (Hsu, 2012), adsorption method (Jibril, 2015), combustion method (Wu, 2012), and condensation method (Lin, 2005) and so on. Because VOCs have the characteristics of wide sources and complex components, so in the practical application, single technology is usually difficult to achieve VOCs effective purification, so the combined technology has become the current developing direction. Chemical oxidation is concerned because of its oxidation, high speed, and high efficiency (Wu, 2014). In this paper, acetone gas was selected as objective VOCs pollutant, sodium hypochlorite (NaClO) solution was used to oxidate acetone; the combination method of the absorption with chemical oxidation was studied.

2 EXPERIMENTAL

2.1 Materials and reagents

Acetone (CH_3COCH_3); Sodium hypochlorite (NaClO, 10% of available chlorine); Hydrochloric acid (HCl); Sodium hydroxide (NaOH) reagent grades are AR; they were purchased Shijiazhuang Reagent Factory.

2.2 Experimental process

The experimental system includes two parts of gas distribution system and absorption system, the absorption system comprises an absorption tower and a liquid recirculation system, the experimental setup is showed in Figure 1.

Here the small flow air (0–5 mL/min) is leaded into the acetone solution, the large flow air (0–500 mL/min) as dilution gas is leaded into the mixture tube, which is used to ensure the mixed acetone concentration stable. The absorption tower is a cylindrical chamber, whose diameter is 5 cm and height is 45 cm, in which the filler is ceramic rings. When NaClO solution is passed into the absorption tower by the circulating water pump, the oxidation reaction is performed on the surface of the ceramic rings. The absorption liquid is returned into the circulating water box after the reaction. When the system is stable, acetone concentration before and after reaction is measured by gas chromatography, the purifying efficiency of acetone gas is studied.

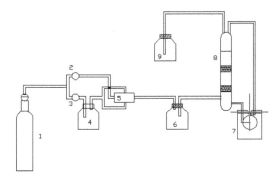

Figure 1.   Experimental setup.
1. Air tank; 2. Flow-meter (0–500 mL/min); 3. Flow-meter (0–5 mL/min); 4. Acetone; 5. Mixture tube; 6. Buffer bottle; 7. Circulating water box; 8. Absorption tower; 9. Exhaust gas treatment device.

2.3   Methods

2.3.1   Determination of the pH value
As chemical absorption liquid, the NaClO solution was adjusted by hydrochloric acid solution and sodium hydroxide solution. After the system was stable, pH value of the circulation absorption liquid was measured by pH meter with pH composite electrode.

2.3.2   Determination of the redox potential
The redox potential of the NaClO solution was measured by pH meter with 501 ORP composite electrodes.

2.3.3   Determination of acetone gas concentration
Acetone gas concentration was determinated by Gas chromatograph (GC-14C, Shimadzu Corporation) with a 50 m capillary column and a flame ionization detector. The analysis conditions were 90°C of column temperature, 150°C of inlet temperature and 200°C of detector temperature. The peak time of acetone was about 1.77 min.

Removal efficiency of acetone gas was calculated as follows:

$$\eta(\%) = \frac{C_1 - C_2}{C_1} \times 100\% = \frac{S_1 - S_2}{S_1} \times 100\% \quad (1)$$

where C_1 and C_2 are the inlet concentration and export concentration, mg/m^3, S_1 and S_2 are the import and export of peak area.

2.3.4   Determination of gas phase products
Gas phase products in the exhaust gas was analyzed by gas chromatography mass spectrometry (GC-MS, QP2010, Shimadzu Corporation).

3   RESULTS AND DISCUSSION

3.1   Effect of pH on acetone removal rate
When the acetone air flow was 305 mL/min and the NaClO solution flow was 70 mL/min, acetone gas was absorbed and chemically oxidized by NaClO solution under different pH as shown in Figure 2.

The removal rate of acetone initially increased and then decreased with pH, when pH was about 11, the removal rate was the highest, about 98.7%. The reason was that when pH value was too low, it was not easy to generate ClO$^-$ ion, so the oxidation rate of acetone was decreased, while when pH value was too high, it was not conducive to the reaction to the forward movement, so the removal effect of acetone was also reduced.

3.2   Effect of oxidation-reduction potential on acetone removal rate
For a solution system, the oxidation reduction potential is a composite result of a variety of redox reaction of oxidant with reducing substances. Although it is not as the index of oxidant and reducing substance concentration, but it helps to understand the electrochemical characteristics and analyse the properties of the solution, so it is a comprehensive index. When the acetone air flow was 305 mL/min and the NaClO solution flow was 70 mL/min, acetone gas was absorbed and chemically oxidized by NaClO solution under different oxidation-reduction potential as shown in Figure 3.

From Figure 3, when the oxidation-reduction potential was about 680 mV, the removal rate of acetone gas was the highest; the removal efficiency can reach above 98%. Obviously, the potential was too high or too low was not conducive to acetone removal.

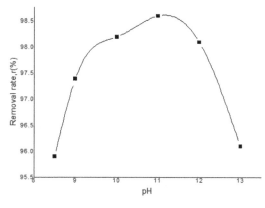

Figure 2.   Relation curve of acetone removal rate with pH.

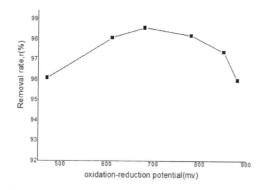

Figure 3. Relation curve of acetone removal rate with oxidation-reduction potential.

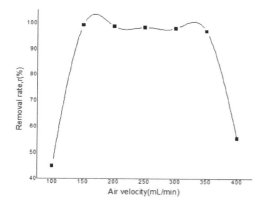

Figure 4. Relation curve of acetone removal rate with air flow.

3.3 Effect of gas flow on acetone removal rate

When pH value of the sodium hypochlorite was 11 and the NaClO solution flow was 70 ml/min, acetone was purified by NaClO solution under different gas flow as shown in Figure 4.

It can be seen from Figure 4, at initial stage, the removal rate of acetone gas steadily increased with the gas flow, and then it remained basically unchanged from 150 mL/min to 350 mL/min, next it sharply decreased after 350 mL/min.

3.4 Effect of NaClO solution concentration on acetone removal rate

The removal rate of acetone gas changed with the effective chlorine in NaClO solution as shown in Figure 5.

It can be seen from Figure 5, the removal rate of acetone gas increased with the effective chlorine in NaClO solution concentration before 12.5%, then

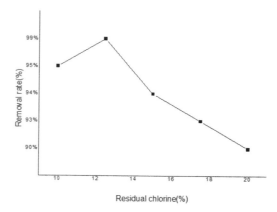

Figure 5. Effect curve of effective chlorine on removal rate.

it decreased, the highest removal efficiency can reach 98.8%.

3.5 Determination of gas phase products

The components of the exhaust gas were analyzed by GC-MS. Except for trace amount of acetone and acetic acid, no other substances were found in the exhaust gas. Acetone attributed to the uncompleted purification, while acetic acid was the middle product from degraded acetone. The specific reaction was not clear, further study needs to be done.

4 CONCLUSIONS

This study showed that the NaClO solution could effectively purify acetone gas. In the Influence factors, pH value and the oxidation-reduction potential had important effects on the removal of acetone gas. When pH of NaClO solution was about 11 the oxidation-reduction potential was maintained at 680 mV and the residual chlorine is 12.5%.

ACKNOWLEDGMENT

This work was financially supported by Hebei Province Science and Technology Support Program (14273712D-2), Higher School Science and Technology Research Projects of Hebei Province (ZD2014101), National Science and technology support program (2014BAC23B04-03), Hebei Province Science and Technology Support Program (14273712D), Dr. Scientific Research Start-up Fund of Hebei University of Science and Technology (QD201015), and Five Big Platform

Open Fund of Hebei University of Science and Technology (2014-11).

REFERENCES

Abdullahi, E.M. et al. Temperature and air-water ratio influence on the air stripping of benzene, toluene and xylene [J]. Desalin. Water Treat. 2014, 1–8.

Benjamin Solsona, et al. Total oxidation of VOCs on Au nanoparticles anchored on Co doped mesoporous UVM-7 silica [J]. Chemical Engineering Journal, 2012:391–400.

Ching-Yi Wu. Oxidative scrubbing of DMS-containing waste gases by hypochlorite solution [J]. 2014:596–602.

Jibril Mohammed, Noor Shawal Nasri. Adsorption of benzene and toluene onto KOH activated coconut shell based carbon treated with NH3 [J]. International Biodeterioration & Biodegradation, 2015, 102:245–255.

Ling-Jung Hsu, Chia-Chang Lin. Binary VOCs absorption in a rotating packed bed with blade packings [J]. Journal of Environmental Management, 2012, 98:175–182.

Marco Ragazzi, et al. Effluents from MBT plants: Plasma techniques for the treatment of VOCs [J]. Waste Management, 2014, 34:2400–2406.

Susant Kumar Padhi, Sharad Gokhale. Biological oxidation of gaseous VOCs—rotating biological contactor a promising and eco-friendly technique [J]. Journal of Environmental Chemical Engineering, 2014, 4: 2085–2102.

Taicheng An, et al. Pollution profiles and health risk assessment of VOCs emitted during e-waste dismantling processes associated with different dismantling methods [J] Environment International, 2014,73:186–194.

Wu M, et al. Catalytic combustion of chlorinated over VOx/TiO2 catalysts [J]. Catalysis Communications, 2012, 18:72–75.

Xiangmei (May) Wu, et al. Exposures to Volatile Organic Compounds (VOCs) and associated health risks of socio-economically disadvantaged population in a "hot spot" in Camden, New Jersey [J]. Atmospheric Environment, 2012, 57:72–79.

Yu-Chih Lin, et al. Control of VOCs emissions by condenser pre-treatment in a semiconductor fab [J]. Journal of Hazardous Materials, 2005, 120:9–14.

The research on the non-isothermal crystallization process of POM fiber

Hua Tan
Guangxi Key Laboratory of Road Structure and Materials, Guangxi Transportation Research Institute, Nanning, Guangxi, China

Yun Wang & Xiaoxi Hu
Guangxi Key Laboratory of Road Structure and Materials, Nanning, Guangxi, China
College of Petroleum and Chemical Engineering, Qinzhou University, Qinzhou, Guangxi, China

Jiaqi Huang & Chun Li
Guangxi Key Laboratory of Road Structure and Materials, Guangxi Transportation Research Institute, Nanning, Guangxi, China

ABSTRACT: Polyoxymethylene (POM) fiber is one of the synthetic fiber, which has the best comprehensive properties and good prospect in application. In this paper, the non-isothermal crystallization process of POM fiber was studied using Mo Zhishen method with different cooling rate. The results showed that it was ideal to study on the non-isothermal crystallization process of POM fiber by the means of Mo Zhishen method. The crystallization peak temperature of POM decreases with the increase of cooling rate, semicrystallization time is shortened, and the crystallization rate is accelerated. The crystallization rate slows down, as the crystallinity increases.

1 INTRODUCTION

Polyoxymethylene (POM) is a kind of linear polymer with high density and high crystallization (Cao et al. 2008). The POM fiber prepared by polyoxymethylene can inherit the most advantages of polyoxymethylene and give play to its potential advantage. It has high strength, high modulus, excellent dimensional stability and thermal stability, high alkali resistance, chemical corrosion resistance, light resistance, weather resistance, and abrasion resistance. When POM fiber is applied to the concrete matrix, the researchers found that the high strength and high modulus performance could improve the concrete strength, excellent alkali resistance could resist strong alkaline corrosion, and the excellent dispersion of POM fibers in the concrete matrix could help POM fibers uniformly dispersing in the concrete mixing process (Liu et al. 2013; Hou et al. 2013). The POM fiber is one of the synthetic fiber which has the best comprehensive properties and has good application prospect (Li et al. 2008). In this paper, the non-isothermal crystallization process of POM fiber was studied using Mo Zhishen method with different cooling rate.

2 MATERIALS AND METHODS

2.1 Materials

POM fiber was produced by Yuntianhua Co. LTD (Yunnan, China) and the melt index was 9.34 g/min under 190 °C and 2160 g load.

2.2 The DSC and TG testing of POM fiber

The POM fiber was cut into pieces and the quality of the sample was 5~15 mg. Then the sample was heated using TGA/DSC 1 simultaneous thermal analyzer (METTLER-TOLEDO, Switzerland) from room temperature to 500 °C with 10 °C/min. The all process was using nitrogen protection and nitrogen flow was 20 mL/min.

2.3 The testing of non-isothermal crystallization kinetics of POM fiber

The POM fiber was divided into five copies, and each quality was 5~15 mg. The automatic baseline system was used for heating and cooling curves. The POM fiber was heated using TGA/DSC 1 simultaneous thermal analyzer (METTLER-TOLEDO, Switzerland) from room temperature to 220 °C with 5 °C/min, 10 °C/min, 15 °C/min, 20 °C/min, and 25 °C/min, respectively. The temperature that was 220 °C was maintained for 3 min, and then the temperature was reduced to room temperature correspondingly with 5 °C/min, 10 °C/min, 15 °C/min, 20 °C/min, 25 °C/min, respectively. The all process was using nitrogen protection and nitrogen flow was 20 mL/min.

3 RESULTS AND DISCUSSION

3.1 The results of DSC and TG testing of POM fiber

Figure 1 and Figure 2 is TG curve and DSC curve of POM fiber from room temperature to 500 °C, respectively. Based on the TG curve, it can be seen that POM fiber is broken down completely when the sample was heated to 500 °C. Figure 2 shows that POM fiber has an endothermic peak at about 170 °C. The melting point of POM is 164.3 °C. So this endothermic peak is produced by the melting of POM. At about 360 °C, there is an endothermic

peak that is caused by thermal decomposition of POM which produces the formaldehyde gas. There is an endothermic peak at about 410 °C which is caused by complete decomposition of POM.

3.2 The influence of different cooling rate on the crystallization behavior of POM fiber

Figure 3 is DSC cooling curves of POM fiber with different cooling rate. Figure 4 is the relationship between the relative crystallinity of POM fiber and crystallization time. Based on Figure 3 and Figure 4, a series of useful parameters could be obtained, such as the initial crystallization temperature (T_0), the end crystallization temperature (T_{end}), crystallization peak temperature (T_p), the enthalpy (ΔH) in the process of crystallization, and half crystallization temperature ($t_{1/2}$), etc., which are shown in Table 1.

From Figure 3 and Table 1, it can conclude that the larger the cooling rate are, the crystallization peak of POM is nearer to the low temperature

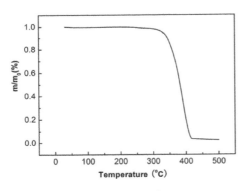

Figure 1. The TG curve of POM fiber.

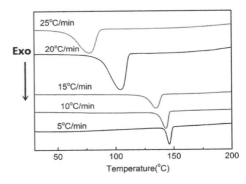

Figure 3. DSC cooling curves of POM fiber with different cooling rate.

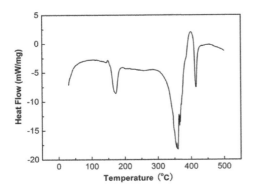

Figure 2. The DSC curve of POM fiber.

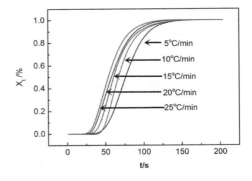

Figure 4. The relationship between the relative crystallinity of POM and crystallization time.

Table 1. The crystallization parameters POM under different cooling rate.

Cooling rate/ (°C/min)	5	10	15	20	25
T_0/°C	149.9	146.5	146	112.33	77.75
T_{end}/°C	137.7	127.8	104.5	71	37.92
T_p/°C	146.1	143.1	134.7	104.33	77.92
$t_{1/2}$/s	76	67	61	63	54
ΔH/(J/g)	8.71	15.82	31.7	39.13	46.7

direction and the wider the crystallization peak shape is. The reason is that when the cooling rate is low, the high temperature sustains for a long time. POM fiber has a plenty of time to form the nucleus, the molecular chains have enough time to arrange and pile neatly, and the higher temperature area is the area that the largest crystallization rate appears. With the increase of cooling rate, on the one hand more parts of macromolecular chain diffuse into the crystal lattice at lower temperatures. On the other hand at low temperature the ability of the molecular chain activity is poorer, and the larger degree of supersaturation makes the nucleation rate quicker, crystal growth faster, which causes more imperfect crystallization and crystallization peak width changes obviously. Figure 4 shows that the cooling crystallization process of POM fiber has experienced three stages. The first stage is the crystallization induction phase which is mainly crystal nucleus formation stage. Due to the first stage at higher temperature, the formation and damage of crystal nucleus are happening under the action of heat, so obvious crystallization exotherm on the DSC curve does not appear and the relative crystallinity is very low. The second stage is given priority to crystal growth. Crystal grows rapidly after crystal nucleus formation and the relative crystallinity is nearly 100%. The third stage is the crystallization equilibrium phase. With the increase of time, the relative crystallinity is 100%, which basically reaches the equilibrium state. Figure 4 shows that, the length of the second stage of the non-isothermal crystallization of POM fiber are greatly influenced by the cooling rate. The crystallization rate is proportional to the cooling rate. Shorter time is needed to reach the crystallization equilibrium, so semic-rystallization time is shortened.

3.3 The non-isothermal crystallization kinetics study of POM fiber

This experiment adopts Mo Zhishen method to predict and analyze the non-isothermal crystallization process of POM fiber (Ren et al. 2003). First the problem of non isothermal crystallization

kinetics is solved by the Avrami and Ozawa method (Avrami, 1939; Avrami, 1939; Ozawa, 1971). The process is as follows.

The dynamics relationship of Avrami equation is Equation (1)

$$1-X_t = \exp(-Z_t t^n) \tag{1}$$

Take logarithm on both sides of Equation (1), Equation (2) could be obtained.

$$\lg[-\ln(1-X_t)] = \lg Z_t + n\lg t \tag{2}$$

In Equation (2), n is Avrami index. Z_t is the crystallization rate constant which related to the nucleation and growth rate.

The dynamics relationship of Ozawa equation is Equation (3).

$$1-X_t = \exp[-K(T)/\Phi^m] \tag{3}$$

Take logarithm on both sides of Equation (3), Equation (4) could be obtained.

$$\lg[-\ln(1-X_t)] = \lg K(T) - m\lg\Phi \tag{4}$$

In Equation (4), K(T) is the cooling function in the non-isothermal crystallization process when the temperature is T. m is Ozawa index. Φ is cooling rate, °C/min.

The Mo method puts forward the point that based on Equation (2) and Equation (4). Equation (5) could be obtained. Further more, it can get Equation (6) and Equation (7).

$$\lg Z_t + n\lg t = \lg K(T) - m\lg\Phi \tag{5}$$
$$\lg\Phi = (1/m)\lg[K(T)/Z_t] - (n/m)\lg t \tag{6}$$
$$\lg\Phi = \lg F(T) - a\lg t \tag{7}$$

Here, $F(T) = [K(T)/Z_t]^{1/m}$, a = n/m.

According to Equation (7), with the same relative crystallinity, the plot of $\lg\Phi - \lg t$ is obtained. It gets a straight line with the intercept of $\lg F(T)$, and the slope of a. The physical meaning of F(T) is the cooling rate when achieving a given relative crystallinity at a unit time. To achieve a given relative crystallinity, the larger F(T) is, the lower the crystallization rate is.

According to the Mo Zhishen method, the crystallization kinetics data of POM fiber under different cooling rate are processed from Equation (7) and the results are shown in Figure 5 and Table 2. As can be seen in Figure 5, there is a good linear correlation between $\lg\Phi$ and $\lg t$, which indicates that the Mo Zhishen method can successfully describe the non-isothermal crystallization process of POM fiber. As can be seen from Figure 5 and Table 2, F(T) increases with the increasing of relative crystallinity, which illustrates that larger

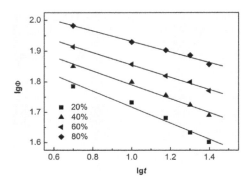

Figure 5. The relationship between lgΦ and lgt.

Table 2. The crystallization kinetics data of POM fiber.

X_t/%	20	40	60	80
a	0.26	0.23	0.2	0.17
F(T)	1.98	2.02	2.06	2.1

cooling rate in unit time leads to larger crystallinity. Crystallization rate decreased with the increase of relative crystallinity. This is because that with the increase of crystal content in the melt, the crystallinity increased, the resistance of the melt macromolecular migrating to spherulite surfaces increases, and the crystallization rate reduces.

4 CONCLUSIONS

The TG curve shows that POM fiber is broken down completely at 500 °C. The DSC curve shows that POM fiber has endothermic peaks at about 170 °C, 360 °C, and 410 °C. The shapes of Polyoxymethylene (POM) crystallization curves under the different cooling rates are similar. The crystallization peak temperature of

POM decreases with the increase of cooling rate, semicrystallization time is shortened, and the crystallization rate is accelerated. It is ideal to study on the non-isothermal crystallization process of POM by the means of Mo Zhishen method. The crystallization rate slows down as the crystallinity increases.

ACKNOWLEDGMENTS

This work is supported by the open project of Guangxi Key Lab of Road structure and materials (grant number 2012 gxjgclkf-005) and the Science and Technology Research Project for Universities of Guangxi (grant number 2013YB258).

REFERENCES

Avrami M. Kinetics of phase change I: General theory [J]. Journal of Chemical Physics, 1939, 7(12): 1103–1112.

Avrami M. Kinetics of phase change I: Transformation-time relationship for random Distribution of Nuclei [J]. Journal of Chemical Physics, 1939, 8(2): 212–224.

Cao Lingling, Wang Yong, Wang Yimin. The development and application of polyoxymethylene fiber [J]. Synthetic Technology and Application, 2008, 23(1): 38–41.

Hou Shuai, Wang Wennian, Zeng Xiansen, Li Xiaofei. Study on splitting tensile strength of polyoxymethylene fiber reinforced concrete [J]. Journal of Wuhan Textile University, 26(3): 39–42.

Li Xiangming, Qin Jun, He Yong, Wang Hua. The research and application of polyoxymethylene fiber [J]. Synthetic Fiber in China, 2008, 37(12): 5–9.

Liu Lu, Hou Shuai, Wu Jing, Wang Luoxing. Study on the flexural performance of polyoxymethylene fiber reinforced concrete [J]. Journal of Wuhan Textile University, 26(3): 35–38.

Ozawa T. Kinetics of non-isothermal crystallization [J]. Polymer, 1971, 12(3): 150–158.

Ren Minqiao, Zhang Zhiying, Mo Zhishen, Zhang Hongfang. The development of research on the later stage kinetics in polymer crystallization [J]. Polymer Bulletin, 2003 (3): 15–22.

Advances in Energy, Environment and Materials Science – Wang & Zhao (Eds)
© *2016 Taylor & Francis Group, London, ISBN 978-1-138-02931-6*

Preparation of composite flocculant and its application in urban recycled water treatment

Changgang Xue & Fengtao Wang
State Grid Henan Electric Power Research Institute, Zhengzhou, China

Chunyi Tong
College of Biological Sciences, Hunan University, Changsha, China

ABSTRACT: Chitosan (CS) microsphere (CSM) was prepared by using water-soluble CS with emulsification and cross-linking process. The composite flocculant (PFS-CSM) was prepared by Polymeric Ferric Sulfate (PFS) and CSM. Antibacterial property was tested while coliform and PFS-CSM were co-cultured. The results showed that compared with the traditional flocculant PFS, various water quality indicators obviously increase, and the consumption reduces by 40%. Under the optimum reaction condition, the turbidity, chromaticity and COD removal rate reach 96.3%, 78.1% and 96.3% respectively. PFS-CSM performs strong inhibition effect on the microorganism which can significantly reduce the existing chlorine type fungicide dosage when treating recycled water. PFS-CSM is a potential new type water treatment agent, and it is worth further promotion and application.

1 INTRODUCTION

With the shortage of water resources in China, power plants' raw water pretreatment system begins to adopt urban recycled water as the source of recycling water. But recycled water contains high dissolved salts, microorganism, hardness, and COD, which easily makes the condenser heat exchange pipe and circulating water pipeline to have the following phenomena—corrosion, scaling and slime. There is no doubt that it will affect the unit operating safety and economy. So reclaimed water can only be recycled and used in the production after a certain processing by adding flocculant and fungicide.

Inorganic polymer flocculant is cheap with good water treatment effect. But the water contains aluminum ions and iron ions. Specifically, aluminum ions are harmful to people's health and iron ions have some effect on the color of water. Meanwhile, as inorganic flocculant dosage is usually big, it will create sludge, make it more difficult to dispose, and bring secondary pollution problem as well as high processing cost (Liu, 2007; Carlos, 2006).

Chitosan is rare natural macromolecule material with positive charge in the nature. It has the advantage of wide sources and biodegradable, which has obvious advantages as water treatment cationic flocculant. Compared with those commonly used inorganic flocculants, its dosage is less, processing speed is faster, sludge is easier to be handled without any secondary pollution, etc

(Wang, 2011; Krishnamoorthi, 2006). However, the cost of organic flocculant is high, which makes it hard to use in large-scale. To solve the defects of single flocculant, some scholars combined ordinary chitosan with inorganic polymer flocculant to prepare composite flocculant in recent years, which has obtained good application. But there is no report of composite flocculant to be prepared by chitosan microspheres or nanoparticles and inorganic flocculant.

Chitosan has good broad-spectrum antibacterial effect. If it is made into nanoparticles, its bacteriostatic effect strengthens (Ding, 2005). Compared with the ordinary chitosan, water soluble chitosan is easier to degrade, and has better antibacterial property (Yang, 2005).

Water soluble chitosan is used to prepare chitosan microspheres, and combined with polymeric ferric sulfate to obtain composite flocculant. On this basis, it conducts antibacterial performance test, and applies to recycled water treatment. Results show that compared with traditional PFS flocculant, the dosage of composite flocculant decreases by 40%, and water treatment effect is better. At the same time, because of the addition of CSM, it has excellent antibacterial effect, which can reduce the use of fungicides in the late water treatment process. It indicates that PFS-CSM is a kind of potential compound flocculant, and it can be applied as new agent to deal with circulating water (urban recycled water) for power plants.

2 EXPERIMENTS

2.1 *Reagent and instrument*

Medicine and reagent: the molecular weight of water soluble chitosan is 5000; other medicine and reagents such as polymeric ferric sulfate, paraffin oil and glutaraldehyde are domestic analytical pure; Colibacillus is provided by this laboratory.

Instrument: scanning electron microscope (JEOL Company, Japan); Zetasizer potential size analyzer (Malvern Company, UK); 19 HW-1 stirring hot plate (Jiangsu, China), ultrasonic generator (Shanghai Xinzhi Institute), Orion AQ4500 turbidity meter (Thermo, America), UV-1600 ultraviolet and visible spectrophotometer (Beijing Rayleigh Company); water purification system.

2.2 *Methods*

2.2.1 *CSM preparation*

Measure paraffin oil 150 mL, add Span80 making up 3% of the total volume, high-speed mixing and blending, form stable microemulsion as oil phase. Measure a certain amount of chitosan, dissolve in 15 mL ultrapure water to disperse and let stand. After 30 min, both chitosan solution and 0.5 mL glutaraldehyde are added in the oil phase to keep stirring reaction for 2 hours. Emulsion is washed by acetone and ethanol solution three times, followed with precipitation and lyophilization, and then powder CSM with surface positively charged is obtained.

2.2.2 *CSM additive solution preparation*

Weight 1 g CSM powder, ultrasonic is dispersed in 200 mL double distilled water, let stand at room temperature, mixture of 1 g/L CSM suspension is prepared.

2.2.3 *PFS additive solution preparation*

PFS solution whose mass fraction is 10% (density is 1.5 g/cm³) is diluted into 1 g/L add liquid.

2.2.4 *Test water quality*

Circulating water in the power plant adopts urban recycled water from Xinxiang Jiatai Lake sewage treatment plant, water quality indicator is: pH = 7.49; turbidity is 49 NTU; electrical conductivity is 2300 µs/cm; COD = 79.6 mg/L; chromaticity is 40; E. coli number is 10000/L.

2.2.5 *Composite flocculant preparation and water treatment test*

It investigates the influence of different mass ratios between PFS and CSM on urban recycled water turbidity, COD and chroma removing rate and chooses the best quality ratio; it also examines the influence of urban recycled water pH value, composite flocculant dosing quantity and stirring speed on various quality indexes after treatment.

2.2.6 *Composite flocculant effect on the growth of Colibacillus*

Culture in LB liquid culture medium e. coli bacteria liquid after activation, then add a certain amount of sterilized PFS-CSM solution to co-culture, take 100 L co-culture bacterium solution every 30 min. Ultraviolet spectrophotometer is used to inspect OD600 value, according to the proportion of the synthesis of two kinds of flocculants, take the same amount of PFS and CSM as contrast, calculate the flocculant inhibition rate.

3 RESULTS AND DISCUSSIONS

3.1 *CSM representation*

Figure 1 is the potential distribution of CSM dispersion in acid solution, CSM is positively charged, and the potential is 12 mV, which can improve the strength of PFS in water with the positive charge, raise the adsorption ability of negative charge suspended material in the water, and provides a theoretical basis for enhancing the flocculation effect of PFS. In addition, by reducing the dosage of PFS, it can reduce the concentration of iron ions residue in water. Figure 2 is SEM microgragh of CSM. It is clear that particle size distribution is 2~4 µm with the average at about 3 µm. Microspheres show neat morphology and good dispersivity.

3.2 *Water treatment effect of PFS-CSM prepared by different mass ratio*

Figure 3 shows water treatment effect of composite flocculant prepared by different mass ratio PFS-CSM. As can be seen from the figure, with

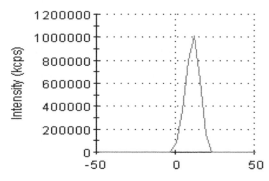

Figure 1. Image of potential of CSM in weak acid solution.

Figure 2. SEM micrograph of CSM.

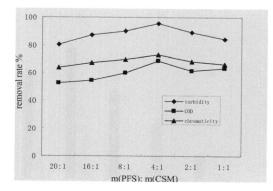

Figure 3. Effect of m (PFS): m (CSM) on the performance of flocculant.

the increase of CSM ratio in composite flocculant, turbidity, COD and chroma removing rate generally increase. When m (PFS): m (CSM) is 4:1, these three indexes reach the highest level, which is 95.5%, 68.8% and 73.5%, respectively. However, if the proportion of CSM continues to increase, the effect of treatment is worse. This is because the flocculation performance of inorganic flocculant is mainly related to electrostatic adsorption of colloid particles, while organic flocculant is mainly through the chelation and middle bridging action of colloidal particles. Because CSM surface has positive charge, adding the flocculant can neutralize colloidal particles on the surface of the negative charge. Meanwhile, through the bridging action, the colloidal particles become dense and easier to sedimente (Zeng, 2009). But when the proportion of chitosan microspheres is big and that of PFS is small, the potential of PFS in the reaction liquid is reduced, the formation of large colloidal particles are not able to completely polymerization settled by the adsorption effect of PFS, part of the

colloid particles left in the water, which causes composite flocculant turbidity, COD and chromaticity removal rate significantly decrease.

3.3 Water treatment effect of PFS-CSM with different dosage

Figure 4 shows water treatment effect of composite flocculant with different PFS-CSM dosage and compares with PFS turbidity removal rate. With the increase of PFS-CSM dosage, its turbidity, COD and chromaticity removal rate significantly reduce generally increase. When the dosage is 15 mg/L, these three indexes reach the highest value, which is 95.4%, 72.5% and 78.3%. The turbidity removal rate gets to the maximize number 90% when PFS dosage is 25 mg/L. It shows that composite flocculant not only has better treatment effect, but also less dosage, which is only 60% of the dosage of PFS. The adding of CSM enhances the adsorption ability of PFS. But when the PFS-CSM dosage is too large, it comes with excessive colloid flocculant adsorption. Due to its electrical repulsion, colloid particles suspend again and unable to settle down, which reduces the effect of water treatment (Wang, 2011). Thus, it is important to add appropriate flocculant dosage because too high or too low cannot reach the best effect.

3.4 Water treatment effect of PFS-CSM on test water quality under different pH value

Figure 5 shows water treatment effect of composite flocculant under different pH value and compares with PFS turbidity removal rate. From the figure, it is clear that turbidity and COD removal rate reach the highest value, which is 94.8% and 73.2% when pH is 8.6. This is because H^+ can be combined with

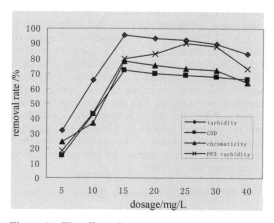

Figure 4. The effect of dosage on the performance of flocculant.

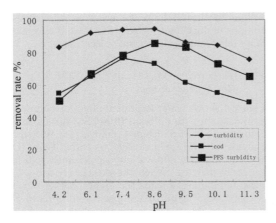

Figure 5. The effect of pH on the performance of flocculant.

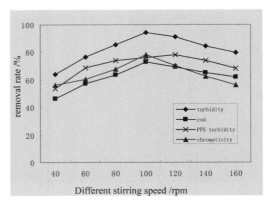

Figure 6. The effect of RPM on the performance of flocculant.

a large number of amino in composite flocculant under low pH environment to form positively charged protons amine-NH_3^+, decrease the chelating ability of NH_2, and cause the total adsorption amount of flocculant on colloid particles to decrease; When the pH is too high, most of the metal ions are easy to conduct hydrolysis reaction to form hydroxyl complexes, thereby weaken the effect of flocculant and NH_2 and reduce the adsorption capacity. So, when the reaction liquid is in pH 6 ~ 9.5, the water treatment effect can be maintained at a high level.

3.5 Water treatment effect of PFS-CSM under different stirring speed

Figure 6 shows water treatment effect of PFS-CSM under different stirring speed and compares with PFS turbidity removal rate. As can be seen from the figure, all the three indexes reach the highest level when rotational speed is 100 rpm, which is 94.3%, 73.1% and 78.1%, respectively. Compared with the same amount of PFS turbidity removal rate, it is nearly 20% higher. Under the condition of low speed, flocculating agent cannot fully react with colloid particles in the solution, drug combination is not uniform, and the condensation effect is poor. When the speed is too high, floc formed will be scattered, broken, and unable to coagulate. Therefore, the control of the appropriate reaction speed is one of the important factors to achieve good coagulation effect.

3.6 PFS-CSM bacteriostatic efficacy detection

Figure 7 is the inhibition of PFS-CSM on bacteria growth. As can be seen from the figure, the bacteria inhibition rate of PFS has maintained at about 2%

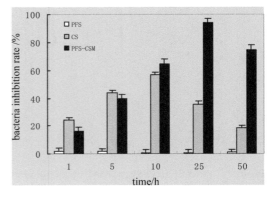

Figure 7. Growth inhibition ratios to E. coli of PFS, CS and PFS-CSM.

in 50 h test process. Taking the factor of test error into account, it can be basically believed that there is little inhibiting effect on bacteria. In the first 10 hours, PFS-CSM bacteriostatic effect is similar to normal CS; at about 25 hours, PFS-CSM bacteria inhibition rate reaches 94%, which basically inhibits the growth of bacteria. The results also confirms that CSM shows stronger bacteriostasis effect than CS, and indicates that PFS-CSM can not only be used as a flocculating agent, but also has the potential to dramatically reduce the existing chlorine fungicide dosage to conduct the effect of a dose of dual-use.

3.7 Water treatment effect of PFS-CSM on recycled water under the optimum reaction condition

Table 1 shows the effect of power plant treat urban recycled water under the condition of Composite

Table 1. The results of flocculant on the performance.

Title	Turbidity removal rate %	Chromaticity removal rate %	COD removal rate %	Bacteria inhibition rate %
PFS	80.2	52.7	58.4	2.1
PFS-CSM	96.3	78.1	73.1	86.2

Flocculant m (PFS): m (CSM) is 4:1, dosage is 15 mg/L, pH is 8.6 and speed of agitator is 100 rpm. Compared with single inorganic flocculant PFS, PFS-CSM has obvious advantages on the water treatment indexes, which is because CSM can neutralize and reduce colloid particle surface charge. By bridging action, flocculation particle size is larger, becomes dense, and much easier to sediment. In addition, as a result of the chitosan antibacterial ability, prepared PFS-CSM has good effect on microorganism inhibition.

4 CONCLUSIONS

a. Neat morphology and better uniformity of CSM is prepared by reverse microemulsion method, which then compound with PFS to obtain composite flocculant PFS-CSM.
b. Under different reaction conditions, PFS-CSM shows good turbidity, chrominance reduced and COD removal effect on the recycled water sample. Compared with the traditional flocculant PFS, various water quality indicators obviously increase, and the consumption reduces by 40%. Under the optimum reaction condition, the turbidity, chromaticity and COD removal rate reach 96.3% 78.1% and 96.3% respectively.

c. PFS-CSM performs strong inhibition effect on the microorganism because of the effect of CSM antibacterial function, which can significantly reduce the existing chlorine type fungicide dosage when treating recycled water. The realization of the aim of a dose of dual-use indicates that PFS-CSM is a potential new type water treatment agent, and it is worth further promotion and application.

REFERENCES

Carlos N., Luis M.S., Elena F., et al. *Polyacrylamide induced flocculation of a cement suspension [J]*. Chemical Engineering Science, 2006, 61(8):2522–2532.

Ding Derun, Shen Yong. *Antibacterial finishing with chitosan derivatives and their nano particles[J]*. 2005, 14: 12–14.

Krishnamoorthi S., Singh R.P. *Characterization of Graft Copolymer Based on Polyacryamide and Dextran [J]*. J Appl Polym Sci, 2006, 101(5):2109–2111.

Liu Bingtao, Lou Yuanzhi, Xu Fei. *Effect of polymeric aluminum chloride/chitosan composite flocculant on activated sludge [J]*. Environmental Chemistry, 2007, 26(1):42–45.

Wang Jiusi, He Zhaozhao, Guo Lixin, et al. *Preparation of composite flocculant (polyferric silicate sulfate and chitosan) and its application in wastewater treatment [J]*. Chinese Journal of Applied Chemistry, 2011, 28(1): 27–31.

Yang Yumin, Ma Zhenxiang, Yin Jicheng, et al. *Study on antimicrobial activity of chitosan water soluble derivatives [J]*. Chin J Public Health, 2005, 21(9), 1080–1081.

Zeng Defang, Cheng Jie, Xiao Jianguo. *Experimental study on chitosan composite flocculant for treating paper-making wastewater [J]*. Chinese Journal of Environmental Engineering, 2009, 3(10), 1808–1811.

Advances in Energy, Environment and Materials Science – Wang & Zhao (Eds)
© 2016 Taylor & Francis Group, London, ISBN 978-1-138-02931-6

Effects of cadmium on calcium and manganese uptake, and the activity of tonoplast proton pumps in pakchoi roots

G.W. Sun
College of Horticulture, South China Agricultural University, Guangzhou, China

Z.J. Zhu
Department of Horticulture, College of Agriculture and Biotechnology, Zhejiang University, Hangzhou, China
Department of Horticulture, School of Agricultural and Food Science, Zhejiang A & F University, Lin'an, China

X.Z. Fang
Research Institute of Subtropical Forestry, CAF, Fuyang, China

ABSTRACT: Pakchoi plants were grown in a hydroponic solution supplemented with cadmium to investigate the toxicant's effects on calcium and manganese uptake, and on the activity of tonoplast proton pumps in the roots. Cadmium was found to inhibit pakchoi growth, and the concentrations of both calcium and manganese decreased with increasing cadmium concentrations in the growth medium. Cadmium also significantly reduced the H^+-ATPase (EC 3.6.1.3) (H^+-transporting adenosine triphosphates) activity of tonoplasts. Conversely, H^+-PPase (EC 3.6.1.1) (H^+-pyrophosphatase) activity of tonoplasts significantly increased at high cadmium concentrations, which implied that H^+-PPase might play an important role in the response to cadmium-induced stress.

1 INTRODUCTION

Cadmium (Cd^{2+}) is one of the most toxic environmental pollutants for plants. It can inhibit plant growth (Hayat et al. 2011); and interfere with metabolic processes such as photosynthesis (Wan et al. 2011), nitrogen metabolism (Shama et al. 2010), and active oxygen metabolism (Podazza et al. 2012). Moreover, several studies have suggested that mineral nutrients might be affected by the presence of Cd^{2+}, and that the interactions between Cd^{2+} and other nutrients may lead to changes in nutrient availability in plant and physiological disorders, as well as a reduction in growth and yield (Erdem et al. 2012).

Calcium (Ca^{2+}) is an essential macronutrient for plant growth and a well-known component in signal transduction pathways. It can activate and regulate many processes in eukaryotic cells. Small intracellular changes in Ca^{2+} concentration can regulate processes such as elongation and division of cells and metabolism (Horvath et al. 1996). Ca^{2+} influences the uptake of Cd in plants (Ismai, 2008) and alleviates metal toxicity stress by restoring plant metabolism and chlorophyll (Zhang et al. 1998). Exogenous Ca^{2+} application for could alleviate Cd toxicity stress (Wan et al. 2011). Manganese (Mn^{2+}), another important micronutrient for plant growth, is essential for several important metabolic processes (Mukhopadhyay & Sharma, 1991), including the photolysis of H_2O by photosystem II and the assimilation of NO_2^- in the chloroplast. Therefore, any cadmium-induced deficiencies in calcium or manganese might lead to impairment of these processes.

In general, vacuole is considered as the main storage place for heavy metals in plant cells. Sequestration of metal ions inside the vacuole involves active transport systems operating in the tonoplast. It is well known that pumping of heavy metals across the tonoplast is energized directly by ATP hydrolysis (primary active transporters: ABC-type proteins and P_{1B}-ATPases) or by the transmembrane pH gradient (secondary active antiporters) (Kramer et al. 2007). The vacuolar membrane possesses 2 proton pumping enzymes, vacuolar H^+ transporting ATPase (V-ATPase) and vacuolar H^+ transporting pyrophosphatase (V-PPase), generating proton motive force which energizes secondary active transport systems (Serrano et al. 2007). H^+-ATPase employs ATP, while H^+-PPase uses Pyrophosphate (PPi) as energy to pump protons into the vacuole, establishing a proton electrochemical potential that is employed to energize a wide range of secondary transport processes (Blarkla & Pantoja, 1996). Migocka et al. (2011) reported that the tonoplast and vacuole have a major function in Cd efflux from the cytosol in the roots of cucumber subjected to Cd stress. Tonoplast proton pumps with a different energy source allow plants to maintain transport processes into the vacuole even under stressful conditions. However, their functional relation and relative contributions to ion storage and detoxification remain unclear (Krebs et al. 2010). The purpose of this study was to investigate the effect of Cd^{2+} on the uptake of Ca^{2+} and Mn^{2+}, and on the activity of proton pumps in the root tonoplasts of pakchoi, in order to understand the mechanisms of Cd^{2+} toxicity in plants.

2 MATERIALS AND METHODS

2.1 Plant materials

Experiments were carried out at the Institute of Vegetable Science, Zhejiang University. Seeds of pakchoi (*Brassica campestris ssp. Chinensis* (L.) Makino *cv.* Hangzhouyoudong) were germinated on moist vermiculite. When the seedlings grew to the 4-leaf stage, they were selected for uniformity and transplanted to 4 L plastic containers filled with the nutrient solution. Each container held 4 plants. The nutrient solution (pH 6.0) contained $Ca(NO_3)_2$ 3.0 mM, KNO_3 4.0 mM, KH_2PO_4 1.0 mM, $MgSO_4$ 1.0 mM, $MnCl_2$ 3.6×10^{-3} mM, H_3BO_3 4.5×10^{-2} mM, $CuSO_4$ 8×10^{-4} mM, $ZnSO_4$ 1.5×10^{-3} mM, H_2MoO_4 9.1×10^{-5} mM, and EDTA-Fe 9.0×10^{-2} mM. The nutrient solution was aerated with a pump continuously and refreshed twice weekly. Cd^{2+} was added as $CdSO_4$ at concentrations of 0, 1, and 10 mg·L^{-1}. The plants were harvested 10 days after Cd treatment, and separated into shoots and roots. The roots were washed 3 times with deionized water, and the fresh weights of shoots and roots were measured. Young roots (10 g) were sampled for determining the activities of H^+-ATPase and H^+-PPase, while the remaining roots and shoots were weighed after oven drying (65°C) to a constant weight.

2.2 Experimental methods

Tonoplast vesicles were prepared according to a modified method by Ballesteros et al. (1998). The activities of H^+-ATPase and H^+-PPase were determined by modification of a method reported by Wang & Sze (1985).

The dried shoot and root materials were ground to a fine powder and digested with a mixture of HNO_3 and $HClO_4$ (2:1). Concentrations of Cd^{2+}, Ca^{2+} and Mn^{2+} were determined simultaneously by inductively coupled plasma atomic emission spectroscopy (ICP-AES, Optima 5300 DV, Perkin-Elmer American).

3 RESULTS AND DISCUSSION

3.1 Plant growth

The addition of 1 mg·L^{-1} Cd^{2+} did not significantly affect the shoot dry weight, but 10 mg·L^{-1} Cd^{2+} significantly reduced the shoot dry weight of pakchoi (Table 1). Moreover, the root dry weight of pakchoi decreased with an increase in Cd^{2+} concentrations. The addition of 10 mg·L^{-1} Cd^{2+} reduced the shoot dry weight and root dry weight by 49% and 40%, respectively. While the growth inhibition produced by Cd^{2+} could be due to nutritional imbalance as has been described in different plant species (Zhang et al. 2000; Sandalio et al. 2001; Ammar et al. 2008), the results of the present study verified a dose effect, with increasing inhibition concomitant with the increase in Cd^{2+} concentrations in the culture medium. Moreover, general effects of toxicity were observed, including the presence of smaller leaves, chlorosis and brown roots in pakchoi.

3.2 Cd²⁺ concentrations in the shoots and roots of pakchoi

Cd^{2+} concentrations in shoots and roots increased significantly with increasing Cd^{2+} concentration in the culture solution (Table 2). Most of the Cd^{2+} taken up accumulated in the root, and only a relatively small amount was transported from root to the shoot. The concentrations of Cd^{2+} in the tissues of treated plants were similar to that observed by other authors under similar growing conditions (Zhang et al. 2000; Wu et al. 2003),

Table 1. Effects of Cd on the growth of pakchoi.

Cd treatment	0 mg·L^{-1}	1 mg·L^{-1}	10 mg·L^{-1}
Shoot dry weight	1.38 ± 0.07	1.44 ± 0.05	0.84 ± 0.16
Root dry weight	0.1 ± 0.01	0.09 ± 0.01	0.06 ± 0.01

Table 2. The contents of Cd and Ca, Mn in shoots and roots of pakchoi.

Cd treatment	0 mg·L^{-1}	1 mg·L^{-1}	10 mg·L^{-1}
Shoot Cd	/	108.6 ± 0.8	601.9 ± 20.4
Root Cd	/	675.2 ± 93.0	3226.4 ± 129.2
Shoot Ca	31.5 ± 0.6	30.3 ± 2.1	25.1 ± 1.4
Root Ca	6.2 ± 0.2	5.8 ± 0.3	3.4 ± 0.1
Shoot Mn	94.5 ± 0.9	68.4 ± 1.7	52.7 ± 1.6
Root Mn	175.7 ± 9.9	115.4 ± 2.5	49.2 ± 0.4

and the concentrations of Cd^{2+} in both shoot and root were proportional to their concentrations in the growth medium, while the increased sensitivity to Cd^{2+} in the roots was in agreement with results from our previous study (Zhu et al. 2004).

3.3 Effect of Cd^{2+} on Ca^{2+} and Mn^{2+} uptake

The concentrations of Ca^{2+} and Mn^{2+} in both shoots and roots decreased with increasing Cd^{2+} concentration in the nutrient solution (Table 2). The presence of ions with similar physical properties (i.e., valence and ionic radius) in the medium can cause disturbances in their uptake and distribution in plants. In the present study, Ca^{2+} concentration in plant tissues decreased in the presence of Cd^{2+}, which may be due to transport of Ca^{2+} being competitively impeded or displaced by Cd^{2+} ions, which are toxic to plants. Another possibility is that Cd^{2+} is taken up by plants using the Ca^{2+} transporter (Skórzyń ska-Polita et al. 1998). Examining the correlation of calcium levels with some photosynthetic parameters, Ramalho et al. (1995) confirmed that Ca^{2+} plays a significant role in the stabilization of chlorophyll and in the maintenance of high photochemical efficiency at PSII. Greger & Lindberg (1987) also reported that effects of toxic Cd^{2+} on sugar beetroots resembled symptoms characteristic of Ca^{2+} deficit. Wan et al. (2011) reported that both the plant growth and the activity of diurnal photosynthetic system remain the least altered under Cd-induced toxicity stress, which suggests that Ca^{2+} in the proximity of plasma membrane is proficient in alleviating Cd toxicity by reducing the cell-surface negativity and competing for Cd^{2+} ion influx. Cd-induced photosynthetic system impairment was rehabilitated by exogenous Ca fortification possibly through countering the uptake of Cd^{2+} (Ismai, 2008).

Manganese is a micronutrient essential for several important metabolic processes and is a co-factor for several Mn-metalloproteins, including the mitochondrial Mn-superoxide dismutase (Bowler et al. 1992), and plant peroxidases essential to H_2O_2 scavenging (Rodríguez-Marañón & van Hystee, 1994). A few studies have been conducted on the relationship between Cd and Mn, but their results are inconsistent. Some studies have demonstrated a reduction in the content of Mn^{2+} in the presence of Cd^{2+}, while conversely, tissue accumulation of Cd^{2+} declined when the concentration of Mn^{2+} was increased (Cataldo et al. 1983; Thys et al. 1991). However, a modest increment in Mn^{2+} concentration in lettuce shoots was observed under several Cd^{2+} treatments, which was associated with a marked reduction in root Mn^{2+} content (Gárate et al. 1992). Treatment with Cd at 0.5 $\mu g\ m \cdot L^{-1}$ increased the shoot concentration of Mn, but decreased the root concentrations of

Mn (Zhu et al. 2004). In this study, the concentrations of Mn^{2+} in both shoots and roots decreased in the presence of Cd^{2+}, and it is possible that the decrease of Mn^{2+} concentrations in the plant tissues resulting from Cd^{2+} influence metabolic processes related to Mn^{2+}, and further consequently affected plant growth. Our previous study confirmed that 10 $mg \cdot L^{-1}$ Cd^{2+} could affect photosynthesis of pakchoi, with a significant decrease in the net photosynthesis rate (Sun et al. 2005).

3.4 H^+-ATPase activity

Compared with the control, H^+-ATPase activity was not significantly changed by the addition of Cd^{2+} at 1 $mg \cdot L^{-1}$, but 10 $mg \cdot L^{-1}$ Cd^{2+} significantly reduced the activity of H^+-ATPase (Fig. 1). The change of H^+-ATPase activity was dependent on the heavy metal, concentration of the heavy metal, and the time of treatment. Kabata et al. (2008) have demonstrated that Cd, Cu, and Ni had no significant effect on the activity of vacuolar ATPase in cucumber roots when applied at 10 μM concentration for 2 h. However, on treatment with Cd and Cu at 10 or 100 μM for 2 or 24 h, the presence of Cu ions in the nutrient solution significantly stimulated V-ATPase activity, measured as both ATP-dependent H^+ transport and ATP hydrolysis, whereas Cd ions diminished it. Cu-induced stimulation of enzymes depended on metal concentration and time of exposure, and reached the highest level in roots treated with 100 μM $CuCl_2$ for 24 h-more than 200% in comparison to the control. Cadmium inhibited the hydrolytic and transporting activity of V-ATPase to a similar extent under all treatment conditions by 25 and 35% (approximately), respectively (Kabata et al. 2010). Vacuolar H^+-ATPase, or V-ATPase, is a primary active pump located at the vacuolar membrane (tonoplast) of the plant cell (Sze, 1985; Sze et al. 1992). H^+-ATPase pumps protons from the cytoplasm to the lumen of the

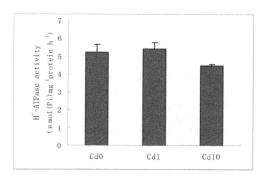

Figure 1. The H^+-ATPase activity of tonoplasts in pakchoi root.

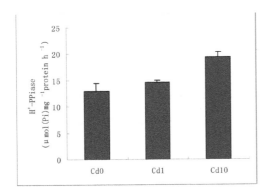

Figure 2. H⁺-PPase activity of tonoplasts in pakchoi root.

vacuole using the energy released by ATP hydrolysis, thereby creating an electrochemical H⁺-gradient that is the driving force for a variety of transport events of ions and metabolites (Ballesteros et al. 1997). Vacuolar compartmentalization plays a very significant role in Cd^{2+} detoxification and tolerance by preventing the free circulation of Cd^{2+} ions in the cytosol (di Toppi & Gabbrieli, 1999). Additionally, in the presence of Mg-ATP, Cd-phytochelatin complexes (as well as apo-phytoxhelatins) are transported against the concentration gradient across the tonoplast by specific carriers, and they accumulate inside the tonoplast vesicles at a concentration up to 38 times higher than that in the external solution (Salt & Rauser, 1995).

Because of the importance of the V-ATPase in the plant vacuole, it is expected that the activity of the H⁺-ATPase can be modulated to cope with environmental and metabolic changes (Dietz et al. 2001). Another Cd^{2+} detoxification mechanism might involve immobilization of the ions by interaction with cell wall components or by chelating to phytochelatins in the cell cytosol, which could be considered to be a protective mechanism to avoid the accumulation of Cd^{2+} in photosynthetic and reproductive tissues (Leita et al. 1992). Disturbances in the supply of mineral elements and generation of reactive oxygen species in the presence of Cd^{2+} were reported in our previous studies (Sun et al. 2004a, b; 2008; Zhu et al. 2004). In this experiment, Cd^{2+} reduced H⁺-ATPase activity of the tonoplast, suggesting that it affected Cd-PC transport at high concentrations and that Cd-PC transport appears to be stimulated by both nutrient deficiencies and Cd^{2+} stress. Moreover, Cd^{2+} detoxification and tolerance by vacuolar compartmentalization was affected, allowing for increased concentration of free Cd^{2+} in the cytosol, and subsequent injury to the plant cells of pakchoi.

3.5 H⁺-PPase activity

The addition of $1\ mg \cdot L^{-1}\ Cd^{2+}$ did not affect the activity of H⁺-PPase, but $10\ mg \cdot L^{-1}$ Cd significantly enhanced it. Wang (1990) reported that the properties of H⁺-PPase are different from those of H⁺-ATPase. Qian & Liu (1995) confirmed that H⁺-ATPase plays a key role in maintaining an electrochemical H⁺-gradient across membranes under salt stress, and H⁺-PPase performs only a supplementary role in pumping protons. They further suggested that H⁺-PPase might be more sensitive to Na⁺. Functioning of vacuolar pyrophosphatase (V-PPase) was not modified by Cu and Cd as clearly as the functioning of V-ATPase. PPi hydrolysis catalyzed by V-PPase seemed to be relatively insensitive to both metals. The only exception was 24-h exposure of roots to $100\ \mu M\ Cu^{2+}$, which caused an increase of approximately 40% in the hydrolytic enzyme activity (Kabata et al. 2010). In the present study, H⁺-PPase activity increased significantly at high Cd^{2+} concentrations. Therefore, we hypothesized that the response of H⁺-PPase to Cd^{2+} stress might be a protective mechanism involving Cd^{2+} detoxification, and it may further play a primary role in proton pumping under high Cd^{2+} stress.

4 CONCLUSIONS

The present study indicated that Cd^{2+} can inhibit the growth of pakchoi and reduce both Ca^{2+} and Mn^{2+} uptake, which causes disturbance of physiological metabolisms related to Ca^{2+} and Mn^{2+}. The reduction of H⁺-ATPase activity at high Cd^{2+} concentrations might affect Cd-PC transport, therefore allowing for more free circulation of Cd^{2+} ions in the cytosol, causing injury to plant cells and inhibiting growth. Moreover, H⁺-PPase activity increased significantly at high Cd^{2+} concentrations, implying that it played a primary role in pumping protons under high Cd^{2+} stress.

ACKNOWLEDGEMENT

This work is financially supported by the 3rd phase "211 Project" of Guangdong Province, and the Project of China Agriculture Research System (CARS-25-C-04).

REFERENCES

Ammar, W.B. Nouairi, I. Zarrouk, M. Ghorbel, M.H. & Jemal, F. 2008. Antioxidative response to cadmium in roots and leaves of tomato plants. *Biologia Plantarum*, 52: 727–731.

Ballesteros, E. Blumwald, E. Donarie, J.P. & Blever A. 1997. Na+/H+ antiport activity in tonoplast vesicles isolated from sunflower roots induced by NaCl stress. *Physiologia plantarum*, 99: 328–334.

Ballesteros, E. Kerkeb, B. Donaire, J.P. & Belver, A. 1998. Effects of salt stress on H+-ATPase activity of plasma membrane-enriched vesicles isolated from sunflower roots. Plant Science, 134: 181–190.

Blarkla, B.J. & Pantoja, O. 1996. Physiology of ion transport across the tonoplast of higher plants. *Annual Review of Plant Physiology and Plant Molecular Biology*, 47: 159–184.

Bowler, C. van Montagu, M. & Inze, D. 1992. Superoxide dismutase and stress tolerance. *Plant Molecular Biology*, 43: 83–116.

Cataldo, D.A. Garland, T.R. & Wildung, R.E. 1983. Cadmium uptake kinetics in intact soybean plants. *Plant Physiology*, 73: 844–848.

di Toppi, L.S. & Gabbrieli, R. 1999. Response to cadmium in higher plants. *Environmental and Experimental Botany*, 41: 105–130.

Dietz, K.J. Tavakoli, N. Kluge, C. Mimura, T. Sharma, S.S. Harris, G.C. Chardonnens, A.N. & Golldack, D. 2001. Significance of the V-type ATPase for the adaptation to stressful growth conditions and its regulation on the molecular and biochemical level. Journal of Experimental Botany, 52: 1969–1980.

Erdem, H. Kinay, A. Ozturk, M. & Tutus, Y. 2012. Effect of cadmium stress on growth and mineral composition of two tobacco cultivars, *Journal of Food, Agriculture and Environment*, 10: 965–969.

Gárate, A. Ramos, I. & Lucena, J.J. 1992. Efecto del cadmio sobre la absorción y distribución de manganeso de distintas variedades de *Lactuca. Suelo Planta*, 2: 581–591.

Greger, M. & Lindberg, S. 1987. Effects of Cd²⁺ and EDTA on young suger beets (*Beta vulgaris*). II. Net uptake and distribution of Mg, Ca and Fe(II)/Fe(III). *Physiologia plantarum*, 69: 81–86.

Hayat, S. Hasan, S.A. & Ahmad, A. 2011. Growth, nitrate reductase activity and antioxidant system in cadmium stressed tomato (*Lycopersicon esculentum* Mill.) cultivars, *Biotechnologie agronomie societe et environment*, 15: 401–413.

Horvath, G. Droppa, M. Oravecz, A. Raskin, V.I. & Marder, J.B. 1996. Formation of the photosynthetic apparatus during greening of cadmium-poisoned barley leaves. *Planta*, 199: 2380–243.

Ismail, M.A. 2008. Involvement of Ca²⁺ in alleviation of Cd²⁺ toxicity in common bean (*Phaseolas vulgaris* L.) plants. *Research Journal of Agriculture and Biological Sciences*, 4: 203–209.

Kabała, K. Janicka-Russak, M. Burzyn'ski, M. & Kłobus, G. 2008. Comparison of heavy metal effect on the proton pumps of plasma membrane and tonoplast in cucumber root cells. *Journal of Plant Physiology*, 165: 278–88.

Kabała, K. Janicka-Russak, M. & Kłobus, G. 2010. Different responses of tonoplast proton pumps in cucumber roots to cadmium and copper, J Plant Physiol, 167: 1328–1335.

Kramer, U. Talke, I.N. & Hanikenne, M. 2007. Transition metal transport, *FEBS Letters*, 581: 2263–2272.

Krebs, M. Beyhl, D. Gorlich, E. Al-Rasheid, K.A.S. Marten, I. Stierhof, Y.D. Hedrich, R. & Schumacher, K. 2010. Arabidopsis V-ATPase activity at the tonoplast is required for efficient nutrient storage but not for sodium accumulation, *Proceedings of the National Academy of Sciences of the United States of America* 107: 3251–3256.

Leita, L. Baca-García, M.T. & Maggioni, A. 1992. Cadmium uptake by *Pisum sativum:* Accumulation and defense mechanism. *Agrochimia*, 36: 253–259.

Migocka, M. Papierniak, A. Kosatka, E. & Klobus, G. 2011. Comparative study of the active cadmium efflux systems operating at the plasma membrane and tonoplast of cucumber root cells. *Journal of Experimental Botany*, 62: 4903–4916.

Mukaopadhyay, M.J. & Sharma, A. 1991. Manganese in cell metabolism of higher plants. *The Botanical review*, 51: 117–149.

Podazzaa, G. Ariasb, M. & Pradoc, F.E. 2012. Cadmium accumulation and strategies to avoid its toxicity in roots of the citrus rootstock *Citrumelo, Journal of Hazardous Materials*, 215–216: 83–89.

Qian, H. & Liu, Y.L. 1995. Relationship between Na⁺/H⁺ antiport of tonoplast vesicles isolated from barley roots and salt compartmentation in plants. *Journal of Nanjing Agricultural University*, 18(2): 16–20 (in Chinese).

Ramalho, J.C. Rebelo, M.C. Santos, E.M. Antunes, L.M. & Nunes, A.M. 1995. Effects of calcium deficiency on *Coffea arabica.* Nutrient changes and correlation of calcium levels with some photosynthetic parameters. *Plant and Soil*, 172: 87–96.

Rodríguez-Marañón, M.J. & van Hystee, R.B. 1994. Plant peroxidases: interaction between their prosthetic groups. *Phytochemistry*, 37: 1217–1225.

Salt, D.E. & Rauser, W.E. 1995. Mg ATP-dependent transport of phytochelatins across the tonoplast of oat roots. *Plant Physiology*, 107: 1293–1301.

Sandalio, L.M. Dalurzo, H.C. Gómez, M. Romero-Puertas, M.C. & del Río, L.A. 2001. Cadmium-induced changes in the growth and oxidative metabolism of pea plants. *Journal of Experimental Botany*, 52: 2115–2126.

Serrano, A. Perez-Castineira, J.R. Baltscheffsky, M. & Baltscheffsky, H. 2007. H⁺-PPases: yesterday, today and tomorrow. *IUBMB Life*, 59: 76–83.

Sharma, A. Sainger, M. Dwivedi, S. Srivastava, S. Tripathi, R.D. & Singh, R.P. 2010. Genotypic variation in *Brassica juncea* (L.) Czern. cultivars in growth, nitrate assimilation, antioxidant responses and phytoremediation potential during cadmium stress, *Journal of Environmental Biology*, 31: 773–780.

Skórzyńska-Polita, E. Tukendorfa, A. Selstamb, E. & Baszyńskia, T. 1998. Calcium modifies Cd effect on runner bean plants. *Environmental and Experimental Botany*, 40: 275–286.

Sun, G.W. Zhu, Z.J. & Fang, X.Z. 2004a. Effects of different cadmium levels on active oxygen metabolism and H₂O₂-scavenging system in *Brassica campestris* L. *ssp. chinensis. Agricultural Sciences in China*, 3(4): 305–309 (in Chinese).

Sun, G.W. Zhu, Z.J. & Fang, X.Z. 2004b. Effects of different cadmium levels on the growth and antioxidant enzymes in *Brassica campestris* L. *ssp. Chinensis* (L.) *Makino. Acta Horticulturae Sinica*, 31(3): 378–380 (in Chinese).

Sun, G.W. Zhu, Z.J. Chen, R.Y. & Liu, H.C. 2008. Effect of cadmium on nitrogen accumulation and activities of nitrogen assimilation enzymes in pakchoi. *Acta Horticulturae*, 768: 545–550.

Sun, G.W. Zhu, Z.J. Fang, X.Z. Chen, R.Y. & Liu, H.C. 2005. Effect of cadmium on photosynthesis and chlorophyll fluorescence of pakchoi. *Plant Nutrition and Fertilizing Science*, 11(5): 700–703 (in Chinese).

Sze, H. 1985. H⁺-translocating ATPase: advances using membrane vesicles, *Annual Review of Plant Physiology*, 36: 175–208.

Sze, H. Ward, M.W. & Lai, S. 1992. Vacuolar H⁺translocating ATPases from plants: structure, function, and isoforms, *Journal of Bioenergetics and Biomembranes*, 24: 371–382.

Thys, C. Vanthomme, C.P. Schrevens, E. & De Proft, M. 1991. Interactions of Cd with Zn, Cu, Mn and Fe for lettuce (*Lactuca sativa* L.) in hydroponic culture. *Plant, Cell & Environment*, 14: 713–717.

Wan, G.L. Najeeb, U. Jilani, G. Naeem, M.S. & Zhou, W.J. 2011. Calcium invigorates the cadmium-stressed *Brassica napus* L. plants by strengthening their photosynthetic system, *Environmental Science and Pollution Research*, 18: 1478–1486.

Wang, Y.Z. & Sze, H. 1985. Similarities and differences between the tonoplast-type and mitochondrial H⁺-ATPase of oat roots. *The Journal of Biological Chemistry*, 260: 10434–10443.

Wang, Y.Z. 1990. Pyrophosphatase on tonoplast in plant. *Plant Physiology Communications*, 4: 73–78 (in Chinese).

Wu, F.B. Zhang, G.P. & Dominy, P. 2003. Four barley genotypes respond differently to cadmium: lipid peroxidation and activities of antioxidant capacity. *Environmental and Experimental Botany*, 50: 67–78.

Zhang, G.P. Fukami, M. & Sekimoto, H. 2000. Genotypic differences in effects of cadmium on growth and nutrient compositions in wheat. *Journal of Plant Nutrition*, 9: 1337–1350.

Zhang, S.G. Gao, J.Y. Song, J.Z. & Weng, Y.J. 1998. Salt stress on the seedling growth of wheat and its alleviation through exogenous Ca(NO₃)₂, *Triticale Crops*, 18: 60–64.

Zhu, Z.J. Sun, G.W. Fang, X.Z. Qian, Q.Q. & Yang, X.E. 2004. Genotypic differences in effects of cadmium exposure on plant growth and contents of cadmium and elements in 14 cultivars of Bai Cai. *Journal of Environmental Science and Health. Part. B*, 39(4): 675–687.

Advances in Energy, Environment and Materials Science – Wang & Zhao (Eds)
© *2016 Taylor & Francis Group, London, ISBN 978-1-138-02931-6*

ABA's antioxidant protection role in salt-stressed cells of bloom-forming cyanobacteria *Microcystis aeruginosa*

Erjuan Chen, Hongfen Xue, Yongjuan He, Xiaoqian Zhang & Xiaolan Chen
School of Life Sciences, Yunnan University, Kunming, Yunnan Province, China

Guobin Deng
Yunnan Academy of Biodiversity, Southwest Forestry University, Kunming, Yunnan Province, China

ABSTRACT: The effect of exogenous Abscisic Acid on salt stress response of bloom-forming cyanobacteria *Microcystis aeruginosa* was studied in this paper. The results showed that 20 mg · l⁻¹ exogenous ABA can effectively enhance the resistance ability of *M. aeruginosa* cells to NaCl stress at the concentration of 0.25 M. Exogenous ABA can restore the growth inhibition in salt-stressed cells. By promoting the Superoxide Dismutase and Glutathione Peroxidase activities and inducing cellular polysaccharides biosynthesis. Exogenous, ABA can effectively help scavenge the excess superoxide radicals in the cells caused by salt stress. Also, NaCl treatment can increase the ABA synthesis in *M. aeruginasa* cells and secretion into the extrocellular environment. These results indicated that ABA might also play important role in salt-stress response and resistance of bloom-forming cyanobacteria *M. aeruginosa*.

1   INTRODUCTION

1.1   *Microcystis aeruginosa bloom and its effect on environment*

Cyanobacterium *Microcystis aeruginosa* is a bloom-forming species distributed globally in fresh waters, which has been recognized as an environmental problem since it significantly increases the water turbidity and causes serious taste and odor problems. It can also produce various biologically active substance, some of them highly toxic and lethal to a variety of organisms, including humans (Codd et al., 1997; Zhang et al., 2011).The blooming mechanisms of cyanobacteria have been extensively studied (Kaplan et al., 1988; Burkert et al., 2001; Yang et al., 2008), and the highly adaptation capability to various stress is essential for its dominate numerically and frequently in natural water bodies. However, the mechanism under its stress adaptation remains poorly understood.

1.2   *ABA's function in cyanobacteria*

Abscisic Acid (ABA) is an isoprenoid plant hormone that regulates physiological adaptation to various environmental stresses, including water deficiency, osmotic stress, and low temperature (Hauser et al., 2011). In cyanobacteria, ABA can change the ratio of the heterocyst to vegetative cell in *Nostoc* 6720, led to an increase in Ca²⁺ uptake and caused an increase in heterocyst frequency (Huddart et al., 1986). ABA also led to an increase in nitrogenase activity (Marsalek et al., 1992a; Hartung et al., 2010). The discovery that cyanobacteria produce ABA (Zahradnickova et al., 1991; Marsalek et al., 1992b) also indicates that ABA may be an important physiological factor in cyanobacteria. But little is known about the effect of ABA on stress-resistance in the cyanobacteria. We previously reported that ABA can increase the resistance of *M. aeruginosa* cells to metal stress (Li et al, 2012), here we report that ABA also functions in its salt stress response.

2   MATERIAL AND METHODS

2.1   *Cyanobacteria culture and treatment*

A xenic culture of the planktonic cyanobacterium *M. aeruginosa* (FACHB905) was used in the experiments. The strain was obtained from the Culture Collections of the Freshwater Algae of the Institute Hydrobiology (FACHB-Collection; Wuhan, China). The strain was grown under constant cool-fluorescent light intensity of 20 umol photons PAR m⁻² s⁻¹, 12:12 LD cycle and temperature of 25 ± 1°C in BG11 medium (Rippka et al., 1979). After 5 days, the strains were harvested by centrifugation, washed 3 times by BG11 medium, and the pellets were inoculated into 250 ml Erlenmeyer flasks (cell concentration about 1 mg l⁻¹ chl a). Then the flasks are separated into three groups: 1) Control group: BG11 medium was used as control; 2) Salt-stress group: BG11 medium was enriched with NaCl to yield total final concentration of 0.25 M; 3) NaCl-stress + ABA treatment group: add ABA with a final concentration of 20 mg l^{-1} to group 2). Each treatment was repeated 3 times. NaCl and ABA concentrations and the exposure times in this experiment were based on previous toxicity bio-assays (unpublished data).

2.2 Extracts of M. aeruginosa

Cell extracts were prepared from 50 ml of cells harvested after each treatment by centrifugation at 7000 rpm for 10 min. The pellets were resuspended in 5 ml of 50 mM sodium phosphate buffer, pH 7.0, and extracted by ultrasonic and liquid nitrogen. Debris was removed from the extract by centrifugation at 7000 rpm for 10 min at 4°C. The supernatants were used as crude extract for biochemical analysis.

2.3 Measurements of physiological parameters

The optical density and chlorophyll a concentration were used to track cyanobacterial growth and calculate cell numbers. Chlorophyll a was extracted in 80% acetone and measured by absorption measurements at 680 and 750 nm. Total Superoxide Dismutase (SOD) and Glutathione Peroxidase (GPX) activity were determined by the method described by Wu et al., 2007. The method for measurement of superoxide radicals O_2^- is described by Bossuyt and Janssen, 2004.

2.4 Measurement of ABA content

Cells of control and NaCl-stressed group were harvested after each treatment by centrifugation at 7000 rpm for 10 min. The supernatant and pellets were resuspended in 5 ml of methanol for ABA extraction. HPLC method was employed using Agilent 1100 series with a chromatogram column of C18 (4.6 mm.i.d. × 250 mm), and the mobile phase was composed of CH3CH:0.02M HAc buffer solution (30:70), and was detected at 254 nm under 25°C of column temperature. Also, Dianchi Lake water which contains large quantity of M. aeruginosa cells was used for same ABA extraction and measurement.

2.5 Statistical analysis

All experiments were performed in 3 replicates. Data presented in this study are presented in means-standard deviation (SD).

3 RESULTS

3.1 Effect of ABA on the growth of M. aeruginosa under salt-stressed conditions

NaCl treatment induced dramatic decrease of cell number of M. aeruginosa within 24 h, while ABA, added at concentration of 20 mg l^{-1} to medium, and increased M. aeruginosa cell number after 48 h's exposure. Figure 1 presents the results of experiments on the influence of ABA on the cell growth of M. aeruginosa under both normal and NaCl-stressed conditions. Exogenous ABA increased survival cell number by 125% compared to the NaCl-stressed group after 48 h of exposure.

3.2 Effect of ABA on the physiological parameters of M. aeruginosa under NaCl stress

Salt stress can induce oxidative stress by generating ROS (Reactive oxygen species) which has been reported in various algae (Pinto, 2003). Oxidative stress directly damages proteins, amino acids, nucleic acids, porphyrin,s and phenolic substances, which attributes the decrease of cell number after stress in Figure 1.

In this study, contents of superoxide anion radicals content in cells increased sharply after NaCl treatment (0.278 vs. 0.046 ug per 10^6 cells), while it is consistent with previous reports in other organisms. Exogenous ABA can effectively dampen the promotion induced by salt and keep the harmful ROS at certain level. As can be seen from Figure 2, ROS content in the stressed cells with ABA is kept the same as in the control cells, which are 0.052 in former and 0.046 ug per 10^6 cells in latter.

Enzymatic and non-enzymatic mechanisms exist in organisms for ROS scavenge. SOD (Superoxide

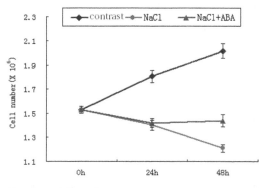

Figure 1. Effect of exogerous ABA on the growth of M. aeruginosa under salt stress.

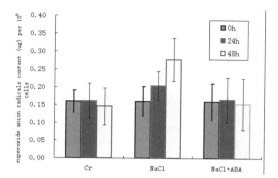

Figure 2. Effect of exogerous ABA on the superoxide anion radicals content of *M. aeruginosa* under salt stress.

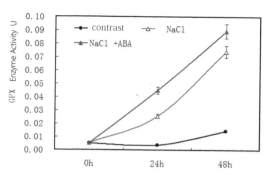

Figure 3. Effect of ABA on GPX activity of *M. aeruginosa* cell under salt stress.

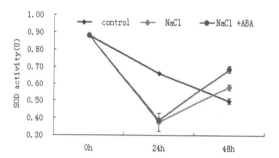

Figure 4. Effect of ABA on SOD activity of *M. aeruginosa* cell under salt stress.

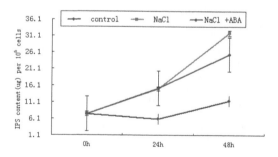

Figure 5. Effect of ABA on IPS content of *M. aeruginosa* cell under salt stress.

Dismutase) and GPX (Glutathione Peroxidase) are two main components of those scavenge enzymes. As shown by Figures 2 and 3, GPX and SOD activities in stress cells are induced by NaCl, while ABA can promote the stimuli of the enzyme activity. While GPX activity maintains at certain level in control cells throughout the whole culture period, it constantly increases in NaCl-adding cells. And in cells with additional ABA as well as NaCl, GPX activity climbs faster than cells only under salt stress. Unlike GPX, SOD activity in cells under salt stress increases after 48 h and preceded by a decrease stage. ABA treatment doubles the SOD activity in cells than salt-stress only cells () as shown in Figure 3. Induction of these antioxidant enzymes efficiently increased the survival rate and regeneration of cells under stress (Fig. 1).

Other than enzymatic mechanism which catalyzes the ROS into nontoxic products, non-enzymatic chemicals always bind and consume the ROS. Polysaccharides including intracellular and extracellular (IPS and EPS) were reported involving in ROS scavenge. IPS synthesis induction was observed in salt stressed cells in our study as shown by Figure 4. It increases from 11 in control to 33 and 25 (ug per 10^6 cells) in NaCl-stress cells and ABA + NaCl cells.

3.3 NaCl treatment can induce ABA synthesis in M. aeruginosa cells

Then we detected the ABA contents in supernatant and cells in different group. Using HPLC, the contents of samples were measured and listed in Table 1.

In pure culture of *M. aeruginosa* cells, ABA level is under the detectable region by the method adopted in this study. While ABA contents in salt stress cells are 5.88 mg/L, ABA content reaches a

Table 1. ABA contents in supernatant and cells in different group of *M. aeruginosa*.

ABA content (mg/L)	Supernatant	Cells
Control group	/	/
NaCl group	5.88 ± 0.10	0.45 ± 0.02
Dichi Lake water	4.19 ± 0.27	2.12 ± 0.08

concentration of 0.45 mg/L in supernatant, which is 13 times of those in the cells. For the sample from Dianchi blooming water, ABA content is 2.12 mg/L in the mix cells, while 4.19 mg/L in the supernatant.

4 DISCUSSION

ABA can react against oxidative damage caused by salt and osmotic stress. Catalase and ascorbate peroxidase activities were shown to be significantly higher in ABA treated cells (Joshua et al., 2010). ABA may be important in mitigation of oxidative damage in stressed algae. In this study, we found ABA also function in oxidative scavenge and thus enhance the salt resistance of *M. aeruginosa* cells, which may contributes to the dominance in natural water bodies and form bloom.

ABA can be induced by NaCl stress in *Nostoc* (Zahardnickova, 1991; Marsalek, 1992). Our results also showed ABA can be induced by salt stress in *M. aeruginosa* cells, which may decrease active oxygen species caused by salt and other stress via induction of GPX and SOD, also by stimulating the synthesis of IPS. Based on these results, we suggest that induction of f ABA in cyanobacteria may also be an essential mechanism involving in salt stress. It might also function as harm restoring in adaptation to salt stress since all the parameter measures in this experiment were changed after 48 h of ABA addition as shown in Figures 2, 3, 4, and 5. Whether the salt-stress adaptation is only based on the universal's mechanism such as ROS-scavenge or has its special signal transduction pathway remains unknown and worthy of further research.

As universal harmony existed in all kingdom of organism, ABA plays important ecological role in natural habitat including the bloom-forming water bodies (Hauser, 2011). Our experiment also proved that in blooming Dianchi Lake, both the water and the mix cells contain high levels of ABA. Although the resource of ABA keeps unknown, its function on bloom-forming cyanobacteria can not be ignored. As proved by this study, ABA, in spite of it origin (form cyanobacteria itself or other organisms like bacteria, plants and animals), can effectively enhance the resistance of bloom-forming species, *M. aeruginosa* to salt stress at least by increasing the cells' antioxidant capability.

Most interestingly, the content of ABA in supernatant of salt stressed sample is much higher than that of cells, which indicate that ABA was induced to synthesis and secrete out of cells and into the environment. Secretion of ABA into surrounding water may be partial of mechanism which cyanobacteria adopted to compete with other organisms.

To study the role of ABA in the stress-resistance of bloom-forming cyanobacteria and its inter-species competition will help us to find new methods to control the overproduction of cyanobacteria and the formation of harmful bloom.

5 CONCLUSION

This study proved that ABA can efficiently enhance the salt stress of bloom-forming cyanobacteria *M. aeruginosa*. After treatment with high concentration of NaCl, Cell number decreased sharply because of the oxidation damage induced by salt stress. While exogenous ABA can decrease the level of ROS by inducing the GPX and SOD activity and synthesis of protective substance such as IPS. Also the synthesis of ABA can be induced by NaCl in pure cultured *M. aeruginosa* cells. And most importantly, high level of ABA exists in the natural bloom water body like Dianchi Lake in which *M. aeruginosa* is the dominant species. All these results indicate the importance of ABA on resistance and dominance of cynaobacteria, which may also contribute to inter-specific competition of cyanobacteria in various environments.

ACKNOWLEDGMENTS

This research was supported by NSFC (NO. 30960036 and 31260283) and YNUY (NO. 201421).

REFERENCES

Bossuyt, B.T.A. & Janssen, C.R. 2004. Long-term acclimation of pseudokirchneriella subcapitata (Korshikov) Hindak to different copper concentrations: changes in tolerance and physiology. *Aquat Toxicol*. 68(5):61–74.

Burkert, U., Hyenstrand, P., Drakare, S., et al. 2001. Effects of the mixotrophic flagellate Ochromonas sp. on colony formation in Microcystis aeruginosa. *Aquatic Ecology* 35:9–17.

Codd, G.A., Ward, C.J., Bell, S.G. 1997. Cyanobacterial toxins: occurrence, modes of action, health effects and exposure routes. *Arch. Toxicol*. 19:399–411.

Hartung, W. 2010. The evolution of abscisic acid (ABA) and ABA function in lower plants, fungi and lichen. *Funct Plant Biol*. 37:806–812.

Hauser, F., Waadt, R. & Schroeder, J.I. 2011. Evolution of abscisic acid synthesis and signaling mechanisms. *Current Biology* 21:R346–R355.

Huddart, H., Smith R.J., Langton, P.D., Hetherington, A.M., Mansfield, T.A. 1986. Is abscisic acid a universally active calcium agonist? *New Phytol*. 104: 161–173.

Kaplan, D., Christiaen, D., Arad. S. 1988. Binding of heavy metals by algal carbohydrates. *Algal Biotechnology* 179–187.

Li, S.B., Lv, J., Xue, H.F., et al. 2012. Enhanced Cu^{2+}-stress resistance of Microcystis aeruginosa by ABA. *Advanced Material Research*. 343–344, 1229–1235.

Marsalek, B., Simek M. 1992. The effect of abscisic and gibberellic acids on nitrogenase activity and growth of the cyanobacterium Nostoc muscorum Agardh. *Archiv Hydrobiologie Supplementband* 94:119–127.

Wu, Z.X., Gan, N.Q., Huang, Q., Song, L.R. 2007. Response of Microcystis to copper stress-Do phenotypes of Microcystis make a difference in stress tolerance? *Environmental Pollution* 147:324–330.

Zahradnickova, H., Marsalek, B., Polisenska, M. 1991. High-performance thin-layer chromatographic and high-performance liquid chromatographic determination of abscisic acid produced by cyanobacteria. *Chromatography*. A555:239–245.

Zhang, X.J., Chen, C., Lin, P.F., Hou, A.X., Niu, Z.B., Wang, J. 2011. Emergency drinking water treatment during source water pollution accidents in China: origin analysis, framework and technologies. Environ. *Sci. Technol.* 45:161–167.

Advances in Energy, Environment and Materials Science – Wang & Zhao (Eds)
© *2016 Taylor & Francis Group, London, ISBN 978-1-138-02931-6*

Synthesis of highly monodisperse Polystyrene microspheres and assembly of the polystyrene colloidal crystals

C. Yang, L.M. Tang, Q.S. Li, L.T. Feng, A.L. Bai, H. Song & Y.M. Yu
State Key Laboratory of Heavy Oil Processing, College of Chemical Engineering, China University of Petroleum (East China), Qingdao, Shandong, P.R. China

ABSTRACT: Highly monodisperse Polystyrene (PS) colloidal spheres were successfully synthesized through a facile emulsifier-free emulsion polymerization approach. The size and distribution of the PS colloidal spheres were systematically investigated in terms of the polymerization reaction time, reaction temperature, monomer dosage, and initiator concentration. The results showed that the size of monodisperse PS colloidal spheres tended to increase with an increase in the polymerization reaction time, and the size of monodisperse PS colloidal spheres could be easily controlled in a wide range from 180–500 nm. The particles grew quickly during the initial reaction stage (1–5 h), and the growth rate reached steady-state after 6 h. The particle sizes were reduced with an increase in the reaction temperature. An increase in monomer concentration led to an increase in the size of monodisperse PS colloidal spheres, and an increase in the initiator concentration also slightly increased the size of PS colloidal particles. In a simple "floating self-assembly" method, the PS colloidal spheres floated on the air–water interface and self-assembled into large area three-dimensional, ordered colloidal crystals, and their structures were investigated by scanning electron microscopy.

1 INTRODUCTION

Colloidal crystals are structures with a periodic arrangement that forms spontaneously when monodisperse colloidal spheres are in aqueous solution. The periodic arrangement is similar to that of atoms in a crystal. Because the sizes of colloidal spheres are generally 0.1–1.0 µm, which is on the sub-micro scale, the colloidal crystals enable the fabrication of all-optical integrated circuits, which play an important role in computing systems and optical communications. The preparative process of colloidal crystals is inexpensive and relatively simple, and therefore, the use of colloidal crystals as templates for other periodic structures ranging from nanometers to micrometers in size has received significant attention in recent years. In addition, colloidal crystals have the diffractive optical properties of photonic crystals, which make them potentially useful as new types of optical sensors and filters, optical switches, and high-density magnetic data storage devices (Zhang et al. 2014; Guo et al. 2012; Park & Xia 1999; Siwich et al. 2001).

At present, the most familiar colloidal crystals are SiO$_2$ colloidal crystals and polymer colloidal crystals. Among them, Polystyrene (PS) nano- and microspheres are of particular interest, and those with controlled particle morphology are the basis for many of today's advanced high-performance polymer materials, which are extensively used in various fields including medical science and biology and as highly ordered nanostructured material templates (Pietrovito et al. 2015; François et al. 2015). Although many methods are available for preparing PS spheres, such as miniemulsion polymerization, seed emulsion polymerization, dispersion polymerization, suspension polymerization, and flash nanoprecipitation, these polymerization methods have apparent drawbacks given that they require multistage polymerization or the use of hazardous solvents and a considerable amount of surfactant. Moreover, the polymerization reaction time is normally greater than 10 h. The emulsifier-free emulsion polymerization in water, which is called soap-free emulsion polymerization, is an environmentally desirable choice for the preparation of monodisperse polymer particles at low impurity levels, and colloidal spheres prepared by this method versus other methods feature better monodispersity and cleaner surfaces (Li & Salovey 2000; Shibuya et al. 2014). Therefore, this method has been widely used for preparing monodisperse polymer colloidal spheres.

For the further self-assembly of monodisperse polymer microspheres to fabricate three-dimensional (3D) ordered colloidal crystals, several approaches, such as gravitational sedimentation, centrifugal sedimentation, vertical deposition, physical confinement, spin-coating, and an emulsion crystallization method, have been proposed. However, the task of rapidly preparing a large area of colloidal crystals without defects remains a challenge. In the "floating self-assembly" method, the spherical particles self-assemble into a 3D ordered structure at the air–water interface due to the strong attractive forces at a meniscus between a substrate and colloidal particles, and this method was reported to be effective for the rapid preparation of a large area of colloidal crystals without defect (Im et al. 2002; Zhang et al. 2011).

In the present study, we synthesized highly monodisperse PS colloidal spheres via an emulsifier-free emulsion polymerization approach. To better understand the mechanism by which PS microspheres were formed, parameters such as the polymerization reaction time, reaction temperature, monomer concentration, and initiator concentration were varied and the corresponding changes in PS microscpheres were examined in detail. Additionally, rapid self-assembly of colloidal crystals was induced using the facile "floating self-assembly" method and the structure of the obtained colloidal crystals was also investigated.

2 EXPERIMENTAL

2.1 *Chemicals*

Styrene (St, 99.5%) purchased from Sinopharm Chemical Reagent Co., Ltd. (China) was purified briefly by washing three times with aqueous NaOH (5 wt%), followed by washing with double-distilled water until neutral pH to remove the polymerization inhibitor. Potassium persulfate (KPS, 99.5%) and sodium hydroxide (NaOH, 96.0%) of analytical grade were purchased from Sinopharm Chemical Reagent Co., Ltd. Nitrogen was obtained from Qingdao Tianyuan Gas Co., Ltd. Other regents were of analytical grade and utilized without further purification.

2.2 *Synthesis of PS colloidal spheres*

The PS colloidal spheres were synthesized via an emulsifier-free emulsion polymerization method. Typically, double-distilled water (500 mL) was poured into a 1-L glass reaction kettle with four necks. After purging the vessel with nitrogen for about 1 h with magnetic stirring at 400 rpm, the purified styrene was added to the glass reaction kettle. Then the kettle was placed in an 80 °C water bath. Polymerization was initiated by adding a specific amount of KPS. During the polymerization, samples were collected at specific time intervals in order to characterize the PS particles formed

and to calculate the corresponding monomer conversion. The polymerization reaction lasted 10 h. The mixture was finally allowed to cool to room temperature, and the obtained PS spheres were kept suspended in the mother liquor until use.

2.3 *Assembly of PS colloidal crystals*

Glass substrates were cleaned ultrasonically with acetone followed by double-distilled water. They were then treated in freshly prepared Piranha solution (a 3:1 mixture of concentrated sulfuric acid with 30% hydrogen peroxide) to obtain a hydrophilic surface. After being rinsed copiously first with double-distilled water and then with ethanol, the glass substrates were dried under a nitrogen gas flow before use. The assembly of the colloidal crystals from PS spheres in suspension was conducted via the floating self-assembly technique. A PS colloidal suspension (15 mL) with a typical concentration of approximately 1.00 wt% was poured into a cylindrical clear glass beaker (50 mL). The glass beaker containing the PS colloidal suspension was sonicated for 30 min and then placed in a convection furnace at 75 °C. The 3D ordered colloidal crystals were obtained as the liquid evaporated.

2.4 *Measurement and characterization*

The particle size and distribution were investigated by dynamic light scattering (DLS, Nano-S, Malvern Instruments Ltd). All measurements in this study were taken at a temperature of 25 °C and measured three times to assess result repeatability. Images of the surface and cross sections of the colloidal crystals were taken via scanning electron microscopy (SEM, Japan Hitachiltd, S4800), and gold spray processing was applied to the sample surface before SEM. The monomer conversion of styrene was determined by taking an aliquot of PS (about 15 g) from the reaction kettle. The conversion was then determined from the solid content after drying the PS sample at 105 °C.

3 RESULTS AND DISCUSSION

3.1 *Effect of reaction conditions on PS particle size*

3.1.1 *Effects of polymerization reaction time*
To explore the possible size range of PS particles prepared via soap-free emulsion polymerization, polymerization reactions were carried for different reaction time with a monomer concentration of 0.61 M and initiator concentration of 2.29 mM. Figure 1 shows the growth of the PS particles according to the polymerization

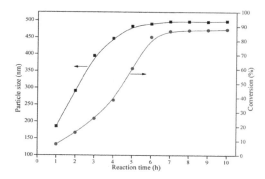

Figure 1. Dependence of the mean particle size of PS spheres (■) and monomer conversion (●) on the polymerization reaction time.

reaction time. Clearly, the particle size grew as the polymerization reaction time increased, and the size of as-synthesized PS colloidal spheres could be controlled within a wide range from 180–500 nm. The PS particles grew quickly during the early stages of reaction (1–5 h), and the growth rate reached steady-state after 6 h. During styrene polymerization, no surfactant was present in the system, and the radicals generated in the aqueous phase in the absence of micelles began reacting with the monomer dissolved in the aqueous phase. The chains formed were colloidally unstable and kept collapsing on each other to attain stability. The mass of collapsed chains was then swollen with the monomer and monomer continued to polymerize within these particles, such that they gradually came together to form PS particles of larger size (Mittal et al. 2007). As the reaction time increased, the monomer concentration in the system continually decreased, and the growth of the PS sphere size slowed and finally stopped. It was also clearly observed that monomer conversion increased with increasing polymerization time. It has been reported that rapid increases in monomer conversions higher than 40% are due to the Trommsdorff effect (gel effect) after the disappearance of monomer droplets (Friis & Hamielec 1976, Chiu & Shih 1986). These results demonstrate that the reaction time is important in controlling the PS sphere size. Because the monomer conversion eventually reached approximately 88% after 8 h of polymerization, the polymerization time in our subsequent experiments was fixed at 8 h.

3.1.2 Effects of polymerization reaction temperature

The effects of reaction temperature on the PS particle size and size distribution are illustrated in

Table 1. As the reaction temperature increased, the sizes of the PS particles decreased proportionally. As the temperature varied from 80–95°C, the average size of particles decreased from 499 to 318 nm while particles remained monodisperse. There may be two factors responsible for this observation. One is the factor of chain initiation, by which the particles sizes might be affected by the degree of concentration of primary free radicals produced by the KPS initiator, which is a function of the reaction temperature. At higher temperatures, many oligomer free radicals were generated in a short time compared to the amount formed at lower temperatures, resulting in more primary nuclei and smaller PS spheres. The other potential factor is chain growth, by which the diameter of the polymer particles represents the degree of polymerization to some extent. Chain growth is well known to be an exothermic reaction, and thus, increasing the reaction temperature can accelerate the rate of chain growth. However, increasing the reaction temperature also causes the chain growth reaction equilibrium to shift to the left, inhibiting chain growth and resulting in a decrease in the sizes of PS particles.

3.1.3 Effects of monomer concentration

The average sizes and size distributions of PS spheres formed at different monomer concentrations are shown in Table 2, and the results showed that the average size of PS microspheres could be increased but the low polydispersity index maintained by increasing the monomer concentration. The negative charges (i.e., sulphate ions) from the KPS initiator moieties at the end of the polymer chains make these ends hydrophilic, and as a result, these groups are present on the surface of the particles, which gives colloidal stability to the

Table 1. Size of PS microspheres with different polymerization reaction temperature.

Sample number	I	II	III	IV
T/°C	80	85	90	95
D/nm	499	411	362	318
PdI	0.050	0.027	0.025	0.026
Conversion/%	87.22	87.69	92.87	85.95

Table 2. Size of PS microspheres with different monomer concentration.

Sample number	V	VI	III	VII
[St]/M	0.17	0.35	0.61	0.87
D/nm	210	321	362	439
PdI	0.020	0.028	0.025	0.030
Conversion/%	85.01	95.79	92.87	94.20

particles. However, as the amount of styrene monomer increased for a given amount of initiator, the surface charge density of each colloidal particle decreased and the electrostatic repulsion weakened between PS particles. Therefore, the colloidal particles were thermodynamically unstable system and tended to aggregate to reduce free energy, resulting in an increase in particle size. In addition, increasing monomer concentration caused the chain growth reaction equilibrium to shift to the right, promoting chain growth and increasing the sizes of the PS particles. Clearly, the increase in monomer concentration resulted in a 2 × increase in PS sphere size but did not, however, affect their size distribution.

3.1.4 *Effects of initiator concentration*
The average particle sizes and polydispersity indices for PS particles formed at various KPS initiator concentrations are shown in Table 3, and the results showed that the average particle size could be successfully increased and the low polydispersity index retained by increasing the initiator concentration. The PS particle size could be controlled from 340–396 nm by changing the initiator concentration from 1.54–3.75 mM. The dependence of the final particle size on the initiator concentration is known to be based on two cases: the case in which the particle size decreases with increasing initiator concentration (Wang et al. 2002; Xu et al. 1999), and the case in which the particle size shows the reverse trend (Chern & Lin 2000). The former mechanism is easily understandable, because as the concentration of initiator increases, the concentrations of primary free radicals and sulphate ions both increase, and a large number of nuclei are formed at the initial stage. Thus, the final particle size becomes small for a given amount of monomer. In addition, hydrophilic sulphate ions play a role in maintaining the stability of PS particles by preventing the agglomeration of PS colloidal particles. Thus, the number of PS particles of smaller sizes increases. On the other hand, the mechanism for the latter case is based on the fact that the ionic strength of the system increases with an increase in initiator concentration. When the static electricity force of the PS particles is greater than the coagulation force between the particles, the system will become stable. However, an increase in the ionic strength of the system with increasing initiator concentration might weaken electrostatic repulsion between the PS particles. Therefore, the system becomes increasingly unstable, which makes aggregation of the initial particles especially easy, and then the PS particles form polymer spheres of larger size. In our experiment, the KPS mass as a weight percentage with respect to styrene monomer was 0.71~1.78%. In addition, the molar concentration of KPS was very low. Therefore, the mechanism for the latter case corresponds to our observations. These experimental results indicate that higher initiator concentrations resulted in PS colloidal particles larger than those prepared with lower concentrations of initiator. Clearly, this work provides an easy and general approach to the preparation of monodisperse polymer spheres.

3.2 *Assembly and morphology of PS colloidal crystals*

Colloidal crystals were assembled from PS colloidal spheres via floating self-assembly. Colloidal crystals, also known as colloidal arrays, are usually face-centered cubic (fcc) close-packed and hexagonal close-packed (hcp). Theoretical analysis indicates that the face-centered cubic close-packed structure is more stable thermodynamically (Woodock 1997). It is possible to make such a structure dominant in colloidal crystals by carefully controlling the experimental conditions (Ding et al. 2004). To confirm the structure of the crystals formed at the air–water interface, we observed single crystals of PS colloidal particles with a diameter of 230 nm via SEM. Figure 2a clearly demonstrates that the as-prepared colloidal particles basically had a face-centered cubic close-packed structure corresponding to a (111) crystalline plane over the sample area. This finding provided a satisfactory solution and fit well with established theory and phenomena. Although there may be some defects in the structure in one plane, the structural defects may be caused by differences in the particle sizes of the colloidal spheres, which can be minimized by strictly controlling the

Table 3. Size of PS microspheres with different initiator concentration.

Sample number	VIII	III	IX	X
[KPS]/mM	1.54	2.29	3.03	3.75
D/nm	340	362	380	396
PdI	0.036	0.025	0.026	0.032
Conversion/%	75.43	92.87	89.16	95.10

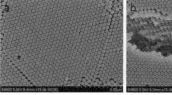

Figure 2. Colloidal crystals self-assembled from PS colloidal microspheres with a diameter of 230 nm at 75°C: (a) top-view SEM image and (b) cross-sectional SEM image.

synthesis conditions to obtain spheres with a uniform particle size. Figure 2b shows an SEM image of the cross-section of a colloidal crystal prepared using the floating self-assembly approach. The obtained colloidal crystal had a 3D ordered structure and exhibited a regular arrangement from the top surface to the bottom surface.

4 CONCLUSION

In summary, we have successfully synthesized monodisperse PS colloidal spheres through emulsifier-free emulsion polymerization and studied the effects of varying the polymerization reaction time, reaction temperature, monomer dosage, and initiation concentration. We have demonstrated the production of particles ranging in size from 180–500 nm and believe it should be possible to form particles beyond this size range. The experimental results indicate that the particle size increases with increasing polymerization reaction time, but decreases with increasing reaction temperature. An increase in monomer dosage led to an increase in the size of monodisperse PS colloidal spheres. When the initiator concentration increased, the sizes of PS colloidal particles increased slightly. Additionally, we have devised a facile and effective approach to fabricate a large area of 3D ordered colloidal crystals by the "floating self-assembly" approach. The obtained colloidal crystals had a high quality 3D ordered structure, and the microspheres were in a face-centered cubic arrangement. Importantly, our method can be used to assemble colloidal particles into a 3D ordered structure that can provide an ideal template for other ordered 3D porous materials.

ACKNOWLEDGMENTS

This work is supported by the Foundation of the National Natural Science Foundation of China (Grant No. 51172284), the Fundamental Research Funds for the Central Universities (No. 15CX06048A), and the Foundation for Outstanding Young Scientist in Shandong Province (No. BS2011HZ024).

REFERENCES

Chern, C.S. & Lin, C.H. 2000. Particle nucleation loci in emulsion polymerization of methyl methacrylate. *Polymer* 41: 4473–4481.

Chiu, W. & Shih, C. 1986. A study of the soap-free emulsion polymerization of styrene. *Journal of Applied Polymer Science* 31: 2117–2128.

Ding, J. Gao, J.N. Tang, F.Q. 2004. Fabrication of colloidal crystal array by self-assembly method. *Progress in Chemistry* 16: 321–326.

François, A. Riesen, N. Ji, H. Afshar, V.S. Monro, T.M. 2015. Polymer based whispering gallery mode laser for biosensing applications. *Applied Physics Letters* 106: 031104.

Friis, N. & Hamielec, A.E. 1976. Gel-effect in emulsion polymerization of vinyl monomer. *ACS symposium series*, Washington, DC 24: 82–91.

Guo, C. Zhou, C. Sai, N. Ning, B.A. Liu, M. Chen, H. Gao, Z. 2012. Detection of bisphenol A using an opal photonic crystal sensor. *Sensors and Actuators B-Chemical* 166–167: 17–23.

Im, S.H. Lim, Y.T. Suh, D.J. Park, O.O. 2002. Three-dimensional self-assembly of colloidals at a water-air interface: A novel technique for the fabrication of photonic bandgap crystals. *Advanced Materials* 19: 1367–1369.

Li, J.Q. & Salovey, R. 2000. "Continuous" emulsifier-free emulsion polymerization for the synthesis of monodisperse polymeric latex particles. *Journal of Polymer Science Part A: Polymer Chemistry* 38: 3181–3187.

Mittal, V. Matsko, N.B. Butté, A. Morbidelli, M. 2007. Functionalized polystyrene latex particles as substrates for ATRP: Surface and colloidal characterization. *Polymer* 48: 2806–2817.

Park, S.H. & Xia, Y. 1999. Assembly of mesoscale particles over large areas and its application in fabricating tunable optical filters. *Langmuir* 15: 266–273.

Pietrovito, L. Cano-Cortés, V. Gamberi, T. Magherini, F. Bianchi, L. Bini, L. Sánchez-Martín, R.M. Fasano, M. Modesti, A. 2015. Cellular response to empty and palladium-conjugated amino-polystyrene nanospheres uptake: A proteomic study. *Proteomics* 15: 34–43.

Shibuya, K. Nagao, D. Ishii, H. Konno, M. 2014. Advanced soap-free emulsion polymerization for highly pure, micron-sized, monodisperse polymer particles. *Polymer* 55: 535–539.

Siwick, B.J. Kalinina, O. Kumacheva, E. Miller, R.J.D. 2001. Polymeric nanostructured material for high-density three-dimensional optical memory storage. *Journal of Applied Physics* 90: 5328–5334.

Wang, D. Dimonie, V.L. Sudol, E.D. Elaasser, M.S. 2002. Dispersion polymerization of n-butyl acrylate. *Journal of Applied Polymer Science* 84: 2692–2709.

Woodcock, L.V. 1997. Entropy difference between the face-centered cubic and hexagonal close-packed crystal structures. *Nature* 385: 141–149.

Xu, J.J. Li, P. Wu, C. 1999. Formation of highly monodispersed emulsifier-free cationic poly(methylstyrene) latex particles. *Journal of Polymer Science Part A: Polymer Chemistry* 37: 2069–2074.

Zhang, J. Wang, M. Ge, X. Wu, M. Wu, Q. Yang, J. Wang, M. Jin, Z. Liu, N. 2011. Facile fabrication of free-standing colloidal-crystal films by interfacial self-assembly. *Journal of Colloid and Interface Science* 353: 16–21.

Zhang, J. Tian, Y. Ling, L.T. Yin, S.N. Wang, C.F. Chen, S. 2014. Versatile hydrogel-based nanocrystal microreactors towards uniform fluorescent photonic crystal supraballs. Journal of Nanoparticle Research 16: 2769–2774.

Advances in Energy, Environment and Materials Science – Wang & Zhao (Eds)
© *2016 Taylor & Francis Group, London, ISBN 978-1-138-02931-6*

Thermogravimetric analysis of Polyoxymethylene fiber used for cement concrete modification

Jianghong Huang
Guangxi Key Laboratory of Road Structure and Materials, Guangxi Transportation Research Institute, Nanning, Guangxi, China

Yun Wang & Xiaoxi Hu
Guangxi Key Laboratory of Road Structure and Materials, Nanning, Guangxi, China
College of Petroleum and Chemical Engineering, Qinzhou University, Qinzhou, Guangxi, China

Chun Li
Guangxi Key Laboratory of Road Structure and Materials, Guangxi Transportation Research Institute, Nanning, Guangxi, China

ABSTRACT: Isothermal and non-isothermal thermogravimetric analysis of polyoxymethylene fiber used for cement concrete modification using a thermogravimetric analyzer, and the thermal decomposition kinetics of polyformaldehyde fiber was studied. The results showed that POM fiber started to degrade at about 250 °C, and finished degration at 500 °C. The non-isothermal degration and the isothermal degration of POM fiber followed the first order reaction. When in the isothermal degration progress, the higher the holding temperature was, the shorter the starting degration time of POM fiber was, and the higher the weightlessness was. The processing temperature of POM fiber should not exceed 220 °C.

1 INTRODUCTION

With the continuous development of national economy, highway transportation presents the characteristics of "big flow, heavy trend". The performance requirements of pavement layer material is higher and higher. Compared with ordinary concrete, fiber concrete not only can improve the crack resistance of pavement structure, permeability resistance, frost resistance and withstand the impact toughness of wheel load (Deng 2003), but also reduce the thickness of pavement, increase the plate plane size, reduce the number of joint, to a certain extent, can improve the driving comfort, and prolong the service life of concrete pavement, reduce maintenance cost, under the same traffic conditions (Yang 2008). Fiber reinforced concrete pavement materials, therefore, is increasingly applied in airport runways, highways, bridges and other important project, and exhibited excellent (Shen et al. 2004). At present, the fibers used in fiber reinforced concrete with PP, PAN, PET, PVA, PA, Kevlar, carbon fiber, and more and more high performance fibers were used in concrete (Corinaldesi et al. 2012; Habel et al. 2008).

Polyoxymethylene (POM) is an engineering thermoplastic with excellent physical and processing properties, and good chemical resistance, which is a crystalline polymer, insoluble in water and acetone, extremely difficult to soluble in dilute acid solution and alkali solution. It has the following advantages: good comprehensive performance and coloration; the best fatigue resistance of all thermoplastics; strength, rigidity, impact strength and creep resistance and other excellent performance; high creep and stress relaxation ability; long-term use of dimension stability; excellent friction and wear resistance; using with wider range of temperature; excellent resistance to light (Cao et al. 2008; Li et al. 2008). The researchers introduced the POM fiber in cement concrete, and found 0.9% POM fiber had good dispersion; POM fiber reinforced concrete had good performance and resistance to bending splitting tensile performance; POM fiber could effectively prevent the cracks of concrete plastic shrinkage and expansion (Liu et al. 2013; Hou et al. 2013). There is a huge market in the fiber reinforced concrete material for POM fiber.

In this paper, isothermal and non-isothermal thermogravimetric analysis of polyoxymethylene fiber used for cement concrete modification using the thermogravimetric analyzer, and the thermal decomposition kinetics of polyoxymethylene fiber was studied. It is expected that, data support would be provided for the manufacturing process of fiber used in concrete engineering.

2 MATERIALS AND METHODS

2.1 *Materials*

POM fiber was produced by Yuntianhua Co. LTD (Yunnan, China) and the melt index was 9.34 g/min under 190 °C and 2160 g load.

2.2 *The non-isothermal thermogravimetric testing of POM fiber*

The POM fiber was cut into pieces and the quality of each sample was 5~15 mg. Then the sample was heated using TGA-50 thermogravimetric analyzer (Shimadzu, Japan) from room temperature to 500 °C with 5 °C/min, 10 °C/min, 15 °C/min, 20 °C/min, and 25 °C/min, respectively. The all process was using nitrogen protection and nitrogen flow was 20 mL/min.

2.3 *The isothermal thermogravimetric testing of POM fiber*

The POM fiber was divided into five copies, and the quality of each sample was 5~15 mg. The POM fiber was heated using TGA-50 thermogravimetric analyzer (Shimadzu, Japan) from room temperature to 180 °C, 200 °C, 220 °C, and 240 °C with 10 °C/min, respectively. The above temperature was maintained for 2.5 hour. The all process was using nitrogen protection and nitrogen flow was 20 mL/min.

3 RESULTS AND DISCUSSION

3.1 *The non-isothermal thermogravimetric analysis of POM fiber*

The non-isothermal thermogravimetric curves of POM fiber are shown in Figure 1. As seen in Figure 1, POM fiber starts to decompose at about 250 °C and finishes decomposition at 500 °C. The slower heating rate is, the lower the initial decomposition temperature is.

The reaction order can be obtained following the Equation (1) by means of Coats-Redgern method (Li et al. 2014).

$$\lg\frac{-ln(1-c)}{T^2} = \lg\frac{AR}{\beta E} - \frac{E}{2.303RT} \tag{1}$$

Plotting $\lg[-ln(1-c)/T^2]$ to $1/T$. If a straight line is got, it shows for the first order reaction. If it is not the first order reaction, the bottom of straight line occurs deviation. Based on the thermogravimetric data with heating rate as β = 5 K/min, 10 K/min, 15 K/min, 20 K/min, 25 K/min and different weightlessness rate C (C = 0.10, 0.15, 0.20, 0.15, 0.30), the Coats-Redfern figure of POM fiber is obtained, which shows in Figure 2.

Figure 2 shows that $lg[-ln(1-c)/T^2]$ has good linear relationship with $1/T$. This illustrates that the thermal decomposition reaction of POM fiber is the first order reaction.

A differential treatment is carried out to the thermogravimetric curves in Figure 1, and the biggest weightlessness rate temperature T_{max} is obtained under different heating rate β. β are 5 K/min, 10 K/min, 15 K/min, 20 K/min, 25 K/min, and the corresponding T_{max} are 552 K, 584 K, 598 K, 605 K, 615 K, respectively. Kissinger maximum weight loss rate method is used to analyze the thermal decomposition activation energy as shown in Equation (2) (Kissinger 1957).

$$\ln\frac{\beta}{T_{max}^2} = \ln\left[-\left(\frac{AR}{E}\right)(1-c)_{max}\right] - \frac{E}{RT_{max}} \tag{2}$$

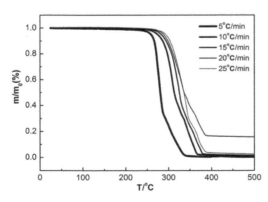

Figure 1. The non-isothermal thermogravimetric curves of POM fiber.

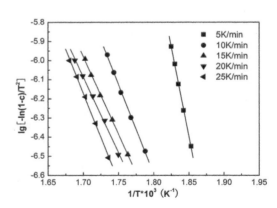

Figure 2. The relationship of $lg[-ln(1-c)/T^2]$ and $1/T$ (The linear correlation coefficient: 5 K/min, –0.99939; 10 K/min, –0.99922; 15 K/min, –0.99939; 20 K/min, –0.99846; 25 K/min, –0.9992).

Plotting $\ln\frac{\beta}{T_{max}^2}$ to $1/T_{max}$, and the fitting line is in Figure 3. According to Figure 3, POM fiber decomposition activation energy is 62.77 KJ/mol.

3.2 The isothermal thermogravimetric analysis of POM fiber

Figure 4 shows the thermogravimetric analysis curves of POM fiber in the same holding time under different temperature. The starting degration time and the last weightlessness of POM fiber in the same holding time under various temperature are listed in Table 1.

As seen in Figure 4 and Table 1, the higher the holding temperature is, the shorter the starting degration time of POM fiber is, and the higher the weightlessness is. There is no obvious degration when the holding temperature and the holding time of POM fiber are 180 °C and

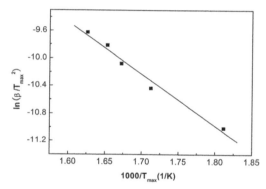

Figure 3. The relationship of $\ln\frac{\beta}{T_{max}^2}$ and $1/T_{max}$ (The linear correlation coefficient: −0.9896).

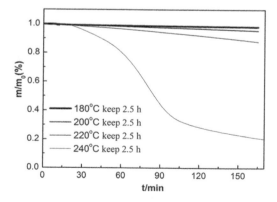

Figure 4. The thermogravimetric analysis curves of POM in the same holding time under different temperature.

Table 1. The starting degration time and the last weightlessness of POM fiber in the same holding time under various temperature.

	Starting degration time (min)	Last weightlessness (%)
180 °C	No obvious degration	1.94
200 °C	95	4.55
220 °C	40	12.9
240 °C	29	79.8

2.5 h, and the last weightlessness is 1.94% under this condition. When the holding temperature is at 240 °C, POM fiber starts to degrade at the holding time of 29 min, and the last weightlessness reaches 79.8%, which demonstrates that the processing temperature of POM fiber should not exceed 220 °C.

Further isothermal degration rule study can be carried out by the initial degration stage on the thermogravimetric curve of POM fiber. The basic formula of thermal analysis kinetics is as follow:

$$\frac{dc}{dt} = k(1-c)^n \tag{3}$$

K is the reaction rate constant, n is reaction order.

The thermal degration reaction of POM fiber is first order reaction by the result of non-isothermal thermogravimetric analysis. So, when $n = 1$, the integration equation of Equation 3 is obtained:

$$-In(1-c) = kt \tag{4}$$

Plotting $In(1-c)$ to t, four good linear straight lines are obtained, as shown in Figure 5. The slope is the degradation rate constant k. The resulting degradation rate constant and the corresponding linear regression coefficient are listed in Table 2.

As seen in Figure 5 and Table 2, the isothermal degration of POM fiber follows the first order reaction. With the increase of temperature, the degration rate increased. The degration rate at 240 °C is 45 times higher than that at 180 °C.

According to Arrhenius equation:

$$k = A\exp(-E/RT), \tag{5}$$

$$\log k = -E/2.303RT + \log A, \tag{6}$$

enter the Equation 6 with k under different temperature. Plotting $\log k$ to $1/T$, and the linear flitting result is shown in Figure 6.

According to Figure 6, slope is calculated for 6.20414, intercept is 9.52573. Therefore, it can

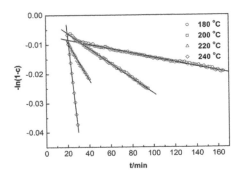

Figure 5. The relationship of $-ln(1-c)-t$ at different temperature.

Table 2. The degradation rate constant and the corresponding linear regression coefficient of POM fiber at different temperature.

	Degradation rate constant k (min^{-1})	Corresponding linear regression coefficient R
180 °C	7.48×10^{-5}	6.31×10^{-4}
200 °C	2.56×10^{-4}	0.99926
220 °C	6.31×10^{-4}	0.99813
240 °C	3.39×10^{-3}	0.99945

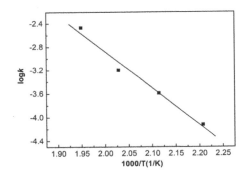

Figure 6. The relationship of log$k$ and $1/T$ (corresponding linear regression coefficient R: −0.98913).

obtain activation energy $E = 118.79$ kJ/mol, and frequency factor $A = 3.36 \times 10^{12}$ min^{-1}.

4 CONCLUSIONS

POM fiber starts to degrade at about 250 °C, and finishes degration at 500 °C. The non-isothermal degration of POM fiber follows the first order reaction. According to Kissinger's maximum weightlessness method, POM fiber decomposition activation energy is 62.77 KJ/mol. The higher the holding temperature is, the shorter the starting degration time of POM fiber is, and the higher the weightlessness is. The processing temperature of POM fiber should not exceed 220 °C. The isothermal degration of POM fiber follows the first order reaction. With the increase of temperature, the degration rate increased the activation energy of this progress is 118.79 kJ/mol.

ACKNOWLEDGEMENTS

This work is supported by the open project of Guangxi Key Lab of Road structure and materials (grant number 2012 gxjgclkf-005) and the Science and Technology Research Project for Universities of Guangxi (grant number 2013YB258).

REFERENCES

Cao Lingling, Wang Yong, Wang Yimin. The development and application of polyoxymethylene fiber [J]. Synthetic Technology and Application, 2008, 23(1): 38–41.
Corinaldesi V., Moriconi G. Mechanical and thermal evaluation of ultra high performance fiber reinforced concretes for engineering applications [J]. Construction and Building Materials, 2012, 26(1): 289–294.
Deng Zongcai. High performance synthetic fiber concrete [M]. Beijing: science press, 2003.
Habel K., Charron J., Braike S., Hooton R.D., Gauvreau P., and Massicotte B. Ultra-high performance fiber reinforced concrete mix design in central Canada [J]. Canadian Journal of Civil Engineering, 2008, 35(2): 217–224.
Hou Shuai, Wang Wennian, Zeng Xiansen, Li Xiaofei. Study on splitting tensile strength of polyoxymethylene fiber reinforced concrete [J]. Journal of Wuhan Textile University, 26(3): 39–42.
Kissinger H.E. Reaction kinetics in differential thermal analysis [J]. Analytical Chemistry, 1957, 29(11): 1702–1706.
Li Hongshan, Guo Jianwei, Liao Yanming, Zheng Yirong, Li Shanji. Investigation of thermal properties of Eu$_2$O$_3$/HDPE composite [J]. Chinese Journal of Colloid & Polymer, 2014, 32(2): 55–57.
Li Xiangming, Qin Jun, He Yong, Wang Hua. The research and application of polyoxymethylene fiber [J]. Synthetic Fiber in China, 2008, 37(12): 5–9.
Liu Lu, Hou Shuai, Wu Jing, Wang Luoxing. Study on the flexural performance of polyoxymethylene fiber reinforced concrete [J]. Journal of Wuhan Textile University, 26(3): 35–38.
Shen Rongxi, Cui Qi, Li Qinghai. New type of fiber reinforced cement matrix composites [M]. Beijing: China building industry press, 2004.
Yan Qibin. Mechanical properties of synthetic fiber-reinforced high-strength concrete on the pavement project. Fujian Building Materials, 2008, (3): 6–8.

Advances in Energy, Environment and Materials Science – Wang & Zhao (Eds)
© *2016 Taylor & Francis Group, London, ISBN 978-1-138-02931-6*

Synthesis of uniform zirconium oxide colloidal particles by hydrothermal method and the influence factor analysis

A.L. Bai, H. Song, Q.S. Li, Y.Q. Wang, C. Yang & Y.M. Yu
State Key Laboratory of Heavy Oil Processing, College of Chemical Engineering, China University of Petroleum (East China), Qingdao, China

ABSTRACT: The stable zirconium oxide (ZrO_2) colloidal particles with a narrow size distribution were synthesized via hydrothermal approach. The size and distribution of the as-synthesized ZrO_2 nanoparticles were investigated in terms of reaction time, precursor concentration, and reaction temperature. Dynamic Light Scattering (DLS) were applied to monitor the evolution of particle size and distribution in synthesis process. The structure and morphology of the samples were studied using X-Ray Diffractometer (XRD), TEM and HRTEM. HRTEM images displayed that the ZrO_2 nanospheres were actually composed of tiny ZrO_2 nanocrystalline. XRD revealed that the colloidal particles had a monoclinic structure and the crystallinity depended strongly on the temperature and reaction time. The growth mechanism of the colloidal spheres has been elucidated in detail. The whole growth process can be summarized as a reversible nucleation-aggregation-dissolution process. The process parameters affected the size and size distribution by the connection of the supersaturation.

1 INTRODUCTION

In the past decades, colloidal materials have aroused widespread concern in many fields. Among the colloidal materials, those particles with a diameter in the submicrometer scale are of particular interest because they can be used as building blocks for self-assembly into 2D or 3D photonic crystals (Huang et al. 2006). Highly monodisperse ZrO_2 colloidal spheres in micrometer and submicrometer size ranges should be ideal building blocks for formation of photonic crystals because of its relatively high refractive index and unique physicochemical properties. In addition, ZrO_2 nano-materials can also be applied in various areas, such as photocatalyst (Maeda et al. 2008), ceramic biomaterials, nanocomposite materials (Zhang, et al. 2000; Pouretedal et al. 2012). It is particularly urgent for the production of ZrO_2 particles with controlled size and morphology. Different synthesis methods, including hydrothermal method (Zhu et al. 2007), microemulsion method (Tai et al. 2004), sol–gel method (Widoniak et al. 2005), or electrolysis of inorganic salts (He et al. 2001), have been successfully devised. In contrast, hydrothermal methods can obtain easily the ZrO_2 NPs with the narrow size distribution at a relatively lower reaction temperature, facilitates a decrease in agglomeration between particles, offers a uniform composition, can maintain the purity of product and can control particle morphology (Byrappa et al. 2007). More importantly, forced hydrolysis of inorganic

salt solutions provides a promising chemical route for large-scale production of ZrO_2 sols because this approach ultilizes inexpensive inorganic salts as starting materials and has relatively simple, low energy heating requirements.

In the present study, we devised a facile tuning strategy for the synthesis ZrO_2 colloidal spheres, that is, specific reaction parameters was selected to work with the reaction system in a coordinate way. By utilization of this strategy, we could delicately tune the size from the nanometer to submicrometer scale. Meanwhile, the formation of ZrO_2 and the evolution of particle size were investigated in detail which can offer some reference for high industrial product and academic research.

2 EXPERIMENTAL SECTION

2.1 Chemicals

Zirconyl chloride octahydrate ($ZrOCl_2 \cdot 8H_2O$, 99% pure) and absolute ethanol (99.7% purity) of analytical grade were purchased from Sinopharm Chemical Reagent Co., Ltd. All employed chemicals, were used as received without further purification. Deionized water was prepared in the experiment for use.

2.2 Synthesis

In our present experiment, zirconium oxychloride hydrate ($ZrOCl_2 \cdot H_2O$) was used as the starting material. In a typical procedure, a certain amount of $ZrOCl_2 \cdot H_2O$ was dissolved in 40 mL of the alcohol-water mixture (the volume ratio of water to alcohol was 5:1) and stirred for 0.5 h to attain transparent solution. Then, the solution was transferred into a 50 ml Teflon-lined stainless steel autoclave and sealed. The autoclave was placed in a homogeneous reactor with a temperature control to heat the sample from 110°C to 150°C and held at this temperature for a given time. No additional surfactant was used. In the meantime, the autoclave was rotated at a 10 rpm to mix well during heat treatment. After the hydrothermal treatment, the autoclave cooled naturally to room temperature, a milky ZrO_2 colloidal suspension was obtained.

3 RESULTS AND DISCUSSIONS

3.1 *The analysis of influence factor*

Figure 1 showed the XRD patterns of ZrO_2 NPs. Several peaks at 2θ degrees of 24.09°, 27.95°, 31.36°, 34.06°, 35.16°, 49.21°, and 55.29° can be seen, which implied that the as-synthesized the ZrO_2 NPs has a monoclinic structure. The strongest diffraction peak at around 27.95° corresponds to the (−111) plane. XRD patterns of sample a, b, and c repeated all the diffraction peaks with similar characteristics, whereas sample c showed less intensity than sample a and b. When the temperature increased from 110 to 120 °C and reaction time was extended from 24 h to 48 h, the peak intensity increased which showed that the crystallinity depends strongly on the temperature and reaction time used. According to the Scherrer's equation, $D = 0.89 \lambda / \beta \cos\theta$, where λ is the wavelength for CuKα (0.15406 nm) radiation and β is the Full Width at Half Maximum (FWHM) of the (−111) and (220) peaks. The crystallite diameters are about 20 nm.

In this study, ZrO_2 NPs could be selectively obtained by varying the synthetic parameters, such as precursor concentration, reaction time and the hydrothermal temperature. Figure 2(a–f) showed the TEM images of representative samples of ZrO_2 NPs with average diameters of 60 nm (sample b), 120 nm (Sample c), 105 nm (Sample d),180 nm (Sample e), and 200 nm (Sample f) respectively.

To analyze the detailed structure of the ZrO_2 NPs, HRTEM were performed. Figure 3 showed a closer view of one ZrO_2 sphere with average size of 200 nm (Figure 2f). The insets in Figure 3 are their corresponding enlarge figure of HRTEM images. The lower resolution image displayed the surface is smooth, but the high resolution TEM image revealed the surface of the ZrO_2 is coarse and the individual nanocrystals were stacked loosely inside the sphere. The size of a nanocrystal is about 3.3 nm. This result was the evidence that the ZrO_2 spheres are composed of numerous ZrO_2 nanocrystals. Moreover, a lattice spacing of 0.27 nm was observed in the ZrO_2 structure and indicated the high crystallinity ZrO_2. The HRTEM results showed the primary particle sizes were relatively smaller than that estimated from the XRD results. This deviation could be attributed to the fact that XRD reflected crystallinity instead of the actual morphology of particles. In HRTEM images, the small particles are counted, but large particles are picked in XRD pattern.

By suitably adjusting these parameters, the evolution from cubic to sphere morphology can occur, and the diameter will also change simultaneously. The DLS were applied to monitor the evolution of particle size and distribution in synthesis process. Figure 4 showed the DLS results for the-synthesized ZrO_2 NPs with various concentrations. The results

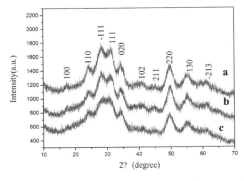

Figure 1. XRD patterns of ZrO_2 nanocrystals prepared at $[Zr^{4+}] = 0.4$ M for different reaction parameters: (a) at 110°C, for 48 h, (b) at 120°C, for 24 h, (c) at 110°C, for 24 h.

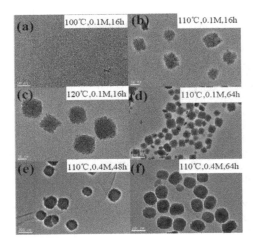

Figure 2. Typical TEM images of ZrO_2 NPs prepared under various reaction conditions.

Figure 3. High-magnification TEM image of a single ZrO₂ nanosphere. The insets are the corresponding enlarge Figure of HRTEM images.

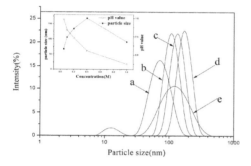

Figure 4. The size distribution curves of ZrO₂ nanoparticles prepared by different concentration (at 120°C for 16 h: (a) 0.05 M, (b) 0.1 M (c) 0.2 M (d) 0.5 M (e) 1.0 M.

showed that with the concentration of ZrOCl₂ increased, the particle size also increased continuously in the range of 0.1–0.5 M. From Figure 2d and 2f, it can be seen that the change of the particle shape from cube to sphere occurred as the monomer concentration increased. This can be explained that the nanocrystals particle can dissolve and reprecipitate on the crystal surface reversibly in the solution. When the concentration was further increased to 1 M, we found that the product became more transparent and very small amounts of the product with smaller size could be obtained. Meanwhile, the size decreased clearly and the size distribution became slightly broadened. The hydrogen ion concentration in solution increased with ZrOCl₂ concentration which was shown in the inset of Figure 4. In the previous literature, it is generally reported that the solubility of zirconia in acid increases with decreasing pH (Baes et al. 1976). The hydrogen ion concentration has an important effect on the dissolution of the primary particles and probably affects the forming the secondary aggregated particles.

Figure 5 showed the DLS curve of the sample prepared at different reaction time. When the reaction time was just 2 h, no product was detected from DLS measurement. When the reaction time is 4 h, the average size is very small and the size distribution was bimodal. When the reaction time was further extended, the particle size increases linearly with the reaction times. Meanwhile, the pH value was decreased with the increase of reaction time, which indicated the hydrolysis level is on a continuous increase. The reaction time has no obvious effect on the morphology of ZrO₂ NPs, as shown in Figure 2b and Figure 2d. However, we can find the ZrO₂ NPs composed of loosely aggregated nanocrystalline when reaction time was 16 h, but the aggregation of ZrO₂ NPs became more and more compact at the reaction period of 64 h. Besides, we find an interesting and counter-intuitive phenomenon in Figure 6. The results were different especially to high concentration (0.4 M) when solutions were treated over a relatively long time, the average diameter decreased with the prolonging of reaction

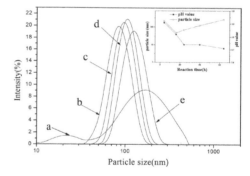

Figure 5. The size distribution curves of ZrO₂ nanoparticles prepared by different reaction time at 0.1 M: (a) 4 h, (b)16 h, (c) 24 h, (d) 40 h, (e) 64 h.

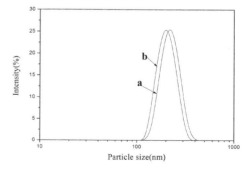

Figure 6. The DLS curves curves of ZrO₂ nanoparticles prepared by different reaction time at 0.4 M: (a) 64 h (b) 72 h.

time. The possible reason is that the change of the growth rate is proportional to the solubility change in highly concentrated solutions.

Figure 7 showed the size and size distributions curve of the sample prepared at 110, 120, 130, 140, and 150°C. When the reaction temperature increased, the particle size also increased. The growth mechanism and final structure of these ZrO_2 particles can be interpreted by the research of LaMer (LaMer et al. 1950). According to their literature, nuclei particles that are less than 20 nm are formed during an initial stage of the hydrolysis reaction. As the reaction temperature increases, the hydrolysis rate increased so that more monomer particles are formed. Thus, the particles became larger by aggregating of monomers. However, the size distribution turned wider gradually with further increase of temperature. Higher temperatures meaned that the agglomeration of particles happened easily and resulted in the generation of boarder size distribution. Moreover, some of the changes in the morphology can be seen from the Figure 2b and 2c, When the temperature reached 120°C, the products are stacked more compact than in the 110°C. Therefore, the monodisperse NPs can be obtained by selecting reaction parameters in a coordinate way combining with the characters of effect factors.

3.2 *The formation mechanism of ZrO_2 NPs*

The reaction route for the formation of ZrO_2 nanospheres can be summarized as follows:

$$Zr^{4+} + H^+ = Zr(OH)_4 \text{ (s)} + H^+ \tag{1}$$

$$Zr(OH)_4 \text{ (s)} = ZrO_2 \text{ (particle)} + H_2O \tag{2}$$

$$ZrO_2 \text{ (particle)} \rightarrow ZrO_2 \text{ (sphere)} \tag{3}$$

The present preparation of ZrO_2 microspheres can be considered as a reversible nucleation-aggregation-dissolution process. In the process, the continuous, but slow formation of the small nanocrystals, the attachment of these nanocrystals to form "cores", and a surface growth of the"cores" to form uniform microspheres through aggregation of additional nanocrystals, further dissolution of the microspheres that has been formed in the acidic solution, constitutes the whole growth process (Fig. 8) of the ZrO_2 NPs.

In the previous literature (Talapin et al. 2001), the growth rate of a particle is formulated as follows:

$$\frac{dr}{dt} = V_m D[Z]_\infty \left[\frac{S - \exp\left[\frac{2\gamma V_m}{rRT}\right]}{r + \frac{D}{k_p}\exp\left[\alpha \frac{2\gamma V_m}{rRT}\right]} \right] \tag{4}$$

where $[Z]_\infty$ is the monomer concentration, r is the radius of the particle, t is reaction time, γ is the surface free energy per unit area, V_m is the molar volume of crystal, k_B is Boltzmann constant, T is temperature, D is the diffusion coefficient of the monomer, k_p is the precipitation reaction constant of the monomer on the surface of bulk crystal, and R is the transfer coefficient for the dissolution reaction, S is supersaturation of the system. If introduced a parameter, $K = (RT/2\gamma V_m)(D/k_p)$, when $S \gg 1$ and/or $K \ll 1$, the smaller particles grow faster than the larger ones, leading to the "focusing" of the size distribution. On the other hand, when $S \ll 1$ and/or $K \gg 1$, the size distribution becomes broader. Obviously, supersaturation has a great effect on the size distribution.

This is the focus of our discussions in this section, and we attempted to show that the size distribution control mechanism in synthetic process. Our experimental data matched well with the kinetics analysis. In the synthesis of ZrO_2 NPs,

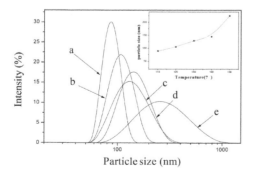

Figure 7. The size distribution curves of ZrO_2 nanoparticles prepared by different temperature: (a) 110°C (b) 120°C (c) 130°C (d)140°C (e) 150°C.

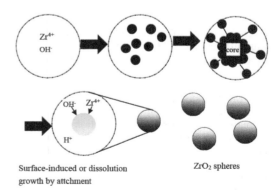

Surface-induced or dissolution growth by attchment

ZrO_2 spheres

Figure 8. Schematic representation of the growth process for ZrO_2 particles.

the supersaturation increases with increasing temperature and precursor concentration in synthetic process. The nucleation process occurs only when the supersaturation level is kept high enough to overcome the interfacial free energy. The monomer supply rate should be equal to or higher than the reducing of the monomer caused by the nucleation and growth of the nanocrystals.

In Figure 2a, when the temperature is too low (100°C), no obvious crystal particle could be observed. Similar phenomenon was observed when the reaction time is shorter or in highly concentrated solutions. Meanwhile, in Figure 4 and 5, we observed that the size distribution were bimodal. In high H^+ ions concentration solution, the accumulation of the monomers will be slower or more difficult due to increasing of solubility. When shorter reaction time is provided, the monomer accumulation is also little and the particles are smaller. The low supersaturation leads to the broader size distribution according to equation (4). And at the same time, based on Gibbs-Thomson effect, the smaller particles have higher surface free energy, the monomers moved easily from smaller to larger particles, redistribution of the monomers occured, the size distribution broadened due to the Ostwald ripening process. In Figure 4 and Figure 5, we observed that the size distribution turned from bimodal to unimodal with the variations of time and the concentration. When the concentration was suitable or the reaction time was further prolonged, the size distribution turned unimodal, and the width of the peak gets much narrower, mainly as a result of supersaturation increasing. The result may be that the "cores"are formed through attachment, the nucleation process was terminated, and only the growth process proceeded, they grow larger with reaction time, no additional nucleation, the size begin to focus, the size is also increasing continually with the time.

Through the analysis above, the precursor concentration and temperature play an active role in the accumulation of the monomer, but H^+ ions concentration was just opposite. We believe that our current theoretical research would be helpful to understand the size and size distribution control mechanism of the synthetic procedures.

4 CONCLUSION

In summary, a facile hydrothermal process was devised to prepare monodisperse ZrO_2 NPs by using the alcohol-water mixed solution as solvent. A mechanism for the formation of ZrO_2 NPs has been studied. The size and size distribution of particles was strongly dependent on the critical

reaction parameters including the hydrothermal temperature and time, the precursor concentration. The formation of the ZrO_2 NPs is attributed to the aggregation and dissolution. When the $ZrOCl_2$ concentration is very high or the reaction time is short, the size distribution was bimodal. In order to obtain uniform size ZrO_2 NPs, the temperature, $ZrOCl_2$ concentration and reaction time must be adjusted to improve supersaturation, as far as possible combining the formation kinetics.

ACKNOWLEDGMENTS

This work is supported by the Foundation of the National Natural Science Foundation of China (Grant No. 51172284).

REFERENCES

Baes, C.F. & Mesmer, R.E. 1976. Wiley, New York. 152–159.

Byrappa, K. & Adschiri, T. 2007. Hydrothermal technology for nanotechnology, Prog. Cryst. Growth Ch. 53:117–166.

He, T. Jiao, X., Chen, D. 2001. Synthesis of zirconium sols and fibers by electrolysis of zirconium oxychloride, J. Non-Cryst Solids. 283: 56–62.

Huang, J. Tao, A.R. Connor, S. He, R. 2006. A General Method for Assembling Single Colloidal Particle Lines, Nano Lett. 6: 524–529.

LaMer, V.K. & Dinegar, R.H. 1950. J. Am. Chem. Soc. 72:4847.

Maeda, K. Terashima, H. Kase, K. Higash, M. 2008. Surface Modification of TaON with Monoclinic ZrO_2 to Produce a Composite Photocatalyst with Enhanced Hydrogen Evolution Activity under Visible Light, Bulletin of the Chemical Society of Japan 81: 927–937.

Pouretedal, H.R. Tofangsazi, Z. Keshavarz, M.H. 2012. Photocatalyticactivity of mixture of ZrO_2/SnO_2, ZrO_2/CeO_2 and SnO_2/CeO_2 nanoparticles, J. Alloys Compd. 513: 359–364.

Tai, C.Y. & Hsiao B.Y. 2004. Preparation of spherical hydrous-zirconia nanoparticles by low temperature hydrolysis in a reverse microemulsion, Colloid Surface A. 237: 105–111.

Talapin, D.V. Rogach, A.L. Haase, M. Weller, H.J. 2001. Evolution of an Ensemble of Nanoparticles in a Colloidal Solution: Theoretical Study, J. Phys. Chem. B. 105: 12278–12285.

Widoniak, J. & Eiden-Assmann, S. 2005. Synthesis and Characterisation of Monodisperse Zirconia Particles, Eur. J. Inorg. Chem, 3149–3155.

Zhang, Q. Shen, J. Wang, J. Wu, G. 2000. Sol-gel derived ZrO_2-SiO_2 highly reflective coatings, Int. J. Inorg. Mater. 2: 319–323.

Zhu H, Yang D, Zhu L. 2007. Hydrothermal Synthesis and Characterization of Zirconia Nanocrystallites, Am. Ceram. Soc. 90: 1334–1338.

Advances in Energy, Environment and Materials Science – Wang & Zhao (Eds)
© *2016 Taylor & Francis Group, London, ISBN 978-1-138-02931-6*

An electro-thermal model for Li-ion battery under different temperatures

P. Chen, F. Sun, H. He & W. Huo

National Engineering Laboratory for Electric Vehicles, School of Mechanical Engineering, Beijing Institute of Technology, Beijing, China

ABSTRACT: Lithium ion battery receives a great concern for the researchers all over the world. An electro-thermal model that can simultaneously predict the voltage and temperature of the lithium ion battery is critical for a good battery system design. In this paper, an electro-thermal model has been built using the famous Dualfoil model with addition of thermal effects. Multiple experiments were conducted to validate the capability of the electro-thermal model at variant temperatures and different current rates. The built electro-thermal model can predict the voltage response well under variant working conditions. However, the temperature prediction of the electro-thermal model needs further improvement.

1 INTRODUCTION

The Li-ion power cell occupies much of the electric vehicle market. It is free from memory effect and pollution, which makes it more competitive than other kinds of cells, such as lead-acid cell (Song et al. 2011). Safety is one of the most important properties of Li-ion cell, and it is the main reason for undermining the development of lithium ion battery (Wu et al. 2011). An electro-thermal coupled model can simulate and predict the behaviors of voltage and temperature of lithium ion batteries (Gu & Wang 2000). Such an electro-thermal coupled model can help design of a battery system (Lee et al. 2013) and reduce the possibility of safety problems (Song et al. 2013).

A physical based Dualfoil model, that can capture the electrochemical behavior of lithium ion battery, has been proposed (Chaturvedi et al. 2010) and improved with thermal effects (Guo & White 2013). And an electro-thermal coupled Dualfoil model has been founded (Ye et al. 2012). And there is a must for the electro-thermal model to simulate the electro-thermal behavior under variant temperatures.

However, the electro-thermal coupled Dualfoil model has high computational complexity, thus the coding of the model may be very complex. Multiphysics software COMSOL® is very famous for multi-physics simulation. It can be used for building an electro-thermal coupled model for lithium ion battery through user friendly interface with less coding work (Cai & White 2011). However, they did not extend their work on the battery model under different temperature and different working conditions. In this paper, we developed an electrochemical lithium ion battery model in COMSOL Multi-physics considering different temperatures and different working conditions. Correlated experiments were conducted and the model is verified by the results of the experiments.

2 MODEL

2.1 Electrochemical model

The equations related to electrochemical model are Equation 1–10, as listed in Table 1. Correlated variables and their physical meanings in Table 1 have been listed in Table 2. Six variables exist in the electrochemical model including $\{\Phi_s, \Phi_e, c_s, c_e, j, i_s\}$.

Except the six variables, there are many fixed physical parameters in the model. Those physical parameters have been listed in Table 3, with their descriptions and values. The equilibrium potentials varying with the stoichiometric coefficient for both the cathode and the anode have been shown in Figure 1.

2.2 Thermal model

The equations related to thermal model are Equation 11–14, as listed in Table 4.

Compared to variables in the electrochemical model, there is only one more variable in the thermal model, the temperature T. Thus there is no need to list the variables in the thermal model. And the name and values of the physical parameters involved in the thermal model are listed in Table 5.

Table 1. Equations for the electrochemical model.

Name of equation	Expression of equation	
Ohm's law	$-\sigma_s^{\text{eff}}\dfrac{\partial \Phi_s}{\partial x} = i_s$	(1)
Charge conservation, solid	$-\dfrac{\partial i_s}{\partial x} = \sigma_s^{\text{eff}}\dfrac{\partial^2 \Phi_s}{\partial x^2} = aFj$	(2)
Charge conservation, liquid	$\dfrac{\partial}{\partial x}\left(\sigma_e^{\text{eff}}\dfrac{\partial \Phi_e}{\partial x}\right) = -aFj$ $+\dfrac{2RT(1-t_+^0)}{F}\dfrac{\partial}{\partial x}\left(\sigma_e^{\text{eff}}\dfrac{\partial \ln c_e}{\partial x}\right)$	(3)
Substance conservation, solid	$\dfrac{\partial c_s}{\partial t} = \dfrac{1}{r^2}\dfrac{\partial}{\partial r}\left(D_s r^2 \dfrac{\partial c_s}{\partial r}\right)$	(4)
Substance conservation, liquid	$\varepsilon_e \dfrac{\partial c_e}{\partial t} = \dfrac{\partial}{\partial x}\left(D_e^{\text{eff}}\dfrac{\partial c_e}{\partial x}\right) + (1-t_+^0)aj$	(5)
Butler-Volmer equation	$j = i_0 \cdot \left[\exp\left(\dfrac{\alpha_n F}{RT}\eta_s\right)\right.$ $\left. -\exp\left(-\dfrac{\alpha_p F}{RT}\eta_s\right)\right]$	(6)
Reaction over potential	$\eta_s = \Phi_s - \Phi_e - U - j \cdot R_{SEI}$	(7)
Exchange current density	$i_0 = k \cdot (c_e)^{\alpha_n}\left(c_{s,\max} - c_{s,e}\right)^{\alpha_n}(c_{s,e})^{\alpha_p}$	(8)
SOC in the particle	$y = \dfrac{3}{R_s^3}\int_0^{R_s} r^2 \dfrac{c_s}{c_{s,\max}}\, dr$	(9)
Potential changes with temperature	$U = U_{\text{ref}}(y) - (T - T_{\text{ref}})\left[\dfrac{dU}{dT}\right]$	(10)

Table 2. The variables in the electrochemical model.

Symbol	Description	Unit
Φ_s	Electrical potential in solid phase	V
Φ_e	Electrical potential in electrolyte phase	V
c_s	Particle concentration in solid phase	mol/m^3
c_e	Salt concentration in electrolyte phase	mol/m^3
j	Molar flux	mol/m$^2\cdot$s
i_s	Electrode current density	A/m^2

Table 3. The physical parameters in the electrochemical model.

Symbol	Unit	Value		
i_{1C}	A/m^2	15		
F	A·s/mol	96485		
t_+^0	1	0.363		

Symbol	Unit	Negative	Separator	Positive
L	m	120×10^{-6}	25×10^{-6}	110×10^{-6}
σ_s^{eff}	S/m	$\varepsilon_s^{1.5}\sigma_s$		$\varepsilon_s^{1.5}\sigma_s$
σ_s	S/m	$100\times g(T)$		$100\times g(T)$
ε_s	1	0.471		0.297
σ_e^{eff}	S/m	$\varepsilon_e^{1.5}\sigma_e$	σ_e	$\varepsilon_e^{1.5}\sigma_e$
σ_e	S/m	$\sigma_e = f(c_e)$	$\sigma_e = f(c_e)$	$\sigma_e = f(c_e)$
ε_e	1	0.357		0.444
D_s	m^2/s	1.5×10^{-13} $\times h(T)$		5×10^{-13} $\times g(T)$
D_e^{eff}	m^2/s	$\varepsilon_e^{1.5}D_e$	D_e	$\varepsilon_e^{1.5}D_e$
D_e	m^2/s	7.5×10^{-11}	7.5×10^{-11}	7.5×10^{-11}
R_s	m	4×10^{-6}		2×10^{-6}
A	1/m	$3\varepsilon_s/R_s$		$3\varepsilon_s/R_s$
α_n, α_p	1	0.5		0.5
R_{SEI}	$\Omega\cdot$m^2	0.001		0.001
K	m/s	2×10^{-11}		2×10^{-11}
$c_{s,max}$	mol/m^3	21,000		20,000
y	1	0.80		0.09
U_{ref}	V	Figure 1		Figure 1
dU/dT	V/K	-0.00005		-0.0001

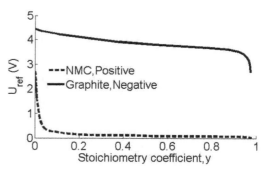

Figure 1. The equilibrium potential $U_{\text{ref}}(y)$ in Equation 10 in Table 1 of the cathode and anode used in the model.

2.3 Revision terms considering temperature variations

The existing Li-ion battery model in COMSOL Multiphysics is only available at room temperature (298 K). Big changes would come to some properties of the cell when the ambient temperature leaves far away from 298 K. As we are building a model considering variant temperatures, some parameters, i.e. D_s and σ_s, has been revised by revision terms $g(T)$ and $h(T)$, as listed in Table 3, to help the model to fit with experimental data at variant temperatures. The revision term $g(T)$

Table 4. Equations for the thermal model.

Equation name	Equation expressions
Heat generation	$Q = Q_{rev} + Q_{irr} + Q_{ohm} + Q_{short}$ (11)
Reversible heat	$Q_{rev} = FajT \dfrac{\partial U}{\partial T}$ (12)
Irreversible heat	$Q_{irr} = Faj(\Phi_s - \Phi_e - U - j \cdot R_{SEI})$ (13)
Ohmic heat	$Q_{ohm} = \sigma_s^{eff}\left(\dfrac{\partial \Phi_s}{\partial x}\right)^2 + \sigma_e^{eff}\left(\dfrac{\partial \Phi_e}{\partial x}\right)^2$ $+ \dfrac{2\sigma_e^{eff}RT}{F}(1 - t_+^0)\dfrac{\partial \ln c_e}{\partial x}\dfrac{\partial \Phi_e}{\partial x}$ (14)

Table 5. The physical parameters in the thermal model.

Symbol	Parameter	Unit	Value
ρ	The density	kg/m³	2100
C_p	The specific thermal capacity	J/kg·K	1100
h	Convection coefficient	W/m²·K	5
T_∞	Ambient temperature	K	283/298/308

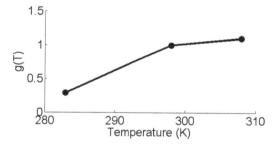

Figure 2. The revision term $g(T)$ used to revised the D_s and σ_s, to fit the experimental data at variant temperatures.

varying with the temperature has been shown in Figure 2, whereas the revision term $h(T)$ is very complex and has been embedded inside the COMSOL software.

3 EXPERIMENTS

3.1 *Experimental settings*

8 Ah lithium ion battery with the code of SPIM08HP bought from CITIC GUOAN MGL was employed in the battery test to validate the built electro-thermal model. The battery was charged/discharged using a HT-V5C200D200-4

Figure 3. The instruments used in the experiments.

Table 6. Matrix of the test design.

Temperature	Current rate		
	1/3C	1C	2C
283 K	√	√	×
298 K	√	√	√
308 K	√	√	×

tester manufactured by KINTE, and the temperature was set at fixed values using a YINHE thermal chamber, as shown in Figure 3.

3.2 *Test design*

The test includes two parts: 1) Reference Performance Test (RPT), which is used to calibrate the maximum available capacity of the cell; 2) charge/discharge experiments of the cell using different current rates under different ambient temperatures.

The RPT process to test the maximum available capacity of the cell is conducted under a fixed temperature of 298 K. First, fully charge the cells using 1/3C constant current followed by a constant voltage charging profile. Second, remain static for 30 minutes. Third, discharge the cell using 1C current until the cell voltage reaches the cutoff limit. After that, repeat the previous three steps for three times. If the capacity deviation of the three test results is within 2%, then average the results to get the maximum available capacity of the cell.

The charge-discharge experiments of the cell using different current rates under different ambient temperatures are more complex than the RPT. The charge is conducted using 1/3C constant current followed by a constant voltage charging at 298 K, the same as that in RPT. However, the discharge process is conducted under variant temperatures (283 K, 298 K, 308 K) using different current rates (1/3C, 1C. 2C), as listed in Table 6. Table 6 is also the test matrix describing all the tests that are used for validating the built electro-thermal model. The conducted tests are marked as √, whereas those not conducted marked as ×.

4 MODEL VALIDATION AND DISCUSSION

The voltage prediction of the model can fit well with the experimental data at different temperatures under different current rates, as shown in Figure 4. The dotted lines in Figure 4 denote the data acquired from the experiments, whereas the solid lines represent the simulated data from the model.

It can be seen from Figure 4(a) that the voltage prediction of the model for 1C discharge slightly deviates from the experimental data, whereas the voltage prediction for 1/3C discharge fits well with the experiment. That may be caused by the high polarization and low reactivity within the battery at low temperature, especially at high current rate. The model can predict the voltage under variant current rates at 298 K, because the COMSOL

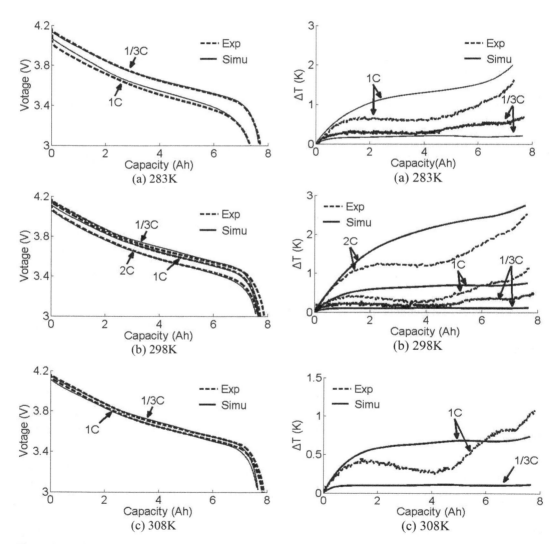

Figure 4. Validation of the electro-thermal model: the voltage curves at different ambient temperatures under different current rates.

Figure 5. Validation of the electro-thermal model: the temperature curves at different ambient temperatures under different current rates.

model is specially applied to a temperature around 298 K. And the voltage prediction for battery working at 308 K is also good, except for those differences a low State Of Charge (SOC), because low SOC always accompany with high polarization, which was not included in current electro-thermal model.

The temperature rise, ΔT, was collected to evaluate the capability of the electro-thermal battery model. It can be seen that the predicted temperature of the model did not fit that well with the experimental data comparing with the prediction result for the voltage curves. The reason for the obvious deviations might be 1) the temperature of the thermal chamber may fluctuate although fixed temperature was set; 2) the temperature sensor may not be pasted tightly on the surface of the battery; 3) the parameter dU/dT in Table 3 changes with battery SOC, however, in our model dU/dT has a lumped value.

5 CONCLUSION

In this paper, an electro-thermal model was built to simulate the voltage and temperature behaviors of lithium ion battery under different working conditions. For the prediction of the voltage, the model can fit well with the experimental data, except for high current rate at low temperature. For the prediction of the temperature, the model gets a better fit at higher discharge rate. The possible reason is that the temperature derivative of equilibrium potential (dU/dT) is set as a constant in the model, whereas dU/dT may varies with SOC in reality. Future work will focus on adjusting dU/dT to acquire a better temperature prediction of the electro-thermal model.

ACKNOWLEDGMENT

This paper is funded by National Natural Science Foundation of China under the contract number of 51276022. And it is also funded by Science and technology support program of Ministry of Science and Technology of China under the contract number of 2013BAG10B01.

REFERENCES

Cai, L. & White, R.E. 2011. Mathematical modeling of a lithium ion battery with thermal effects in COMSOL Inc. Multiphysics (MP) software. *Journal of Power Sources* 196: 5985–5989.

Chaturvedi, N.A. et al. 2010. Algorithms for advanced battery management systems: modeling, estimation, and control challenges for Lithium-ion batteries. *IEEE Control Systems Magazine* 6: 49–68.

Gu, W.B. & Wang, C.Y. 2000. Thermal-electrochemical modeling of battery systems. *Journal of the Electrochemical Society* 147(8): 2910–2922.

Guo, M. & White, R.E. 2013. A distributed thermal model for a Li-ion electrode plate pair. *Journal of Power Sources* 221: 334–344.

Lee, K.J. et al. 2013. Three dimensional thermal-, electrical-, and electrochemical-coupled model for cyclindrical wound large format lithium-ion batteries. *Journal of Power Sources* 241: 20–32.

Song, L. et al. 2013. A review on the research of thermal models for Lithium ion battery cell. *Automotive Engineering* 35(3): 285–291.

Song, Y.H. et al. 2011. Present status and development trend of batteries for electric vehicles. *Power System Technology* 35(4): 1–7.

Wu, K. et al. 2011. Safety performance of Lithium-ion battery. *Progress in Chemistry* 23(2/3): 401–408.

Ye, Y. et al. 2012. Electro-thermal modeling and experimental validation for lithium ion battery. *Journal of Power Sources* 199: 227–238.

Capacity degradation of commercial LiFePO$_4$ cell at elevated temperature

Wenpeng Cao

School of Materials Science and Engineering, Central South University, Changsha, Hunan, China
Shenzhen Institute of Advanced Technology, Chinese Academy of Sciences, Shenzhen, Guangdong, China

Juan Li & Zhengbin Wu

Shenzhen Institute of Advanced Technology, Chinese Academy of Sciences, Shenzhen, Guangdong, China

ABSTRACT: The cyclelife test of commercial 22650-type lithium iron phosphate/graphite batteries were performed at room and elevated temperatures in order to deduce the aging mechanisms for capacity fading. A number of non-destructive electrochemical techniques, such as capacity recovery test with small current, electrochemical impedance spectroscopy, differential-voltage and differential—capacity analysis, and destructive analysis of electrode are discussed in details. Compared with the cell cycled at 55°C, the lower 25°C reached its end of life faster, and a higher discharge current of 3C should be responsible for the power loss of cylinder batteries and difference of capacity fading between 25 and 55°C because a great of proportion of capacities can be regained through a small current. The lithium loss mechanism was considered to be the dominated reason for all degradation of batteries and persistent damage and regeneration of SEI layer was the leading cause of active lithium loss.

1 INTRODUCTION

Since the pioneering work of Padhi et al. (Padhi et al., 1997), olivine-type lithium iron phosphate (LiFePO$_4$) electrode has attracted wide attention with its well-known advantage of high theoretical capacity (ca. 170 mAhg^{-1}), low cost, low toxicity, and suitable potential of lithium-ion insertion/extraction (ca. 3.45 V vs. Li$^+$/Li) and so on. At present, the well combination of the graphite negative electrode and the LiFePO$_4$ positive electrode in a li-ion battery is one of the most promising candidates for vehicle's power source.

Investigation of the failure mechanism of LFP batteries is of vital importance to either durable batteries design, or service life extension. Evaluation of the permanent and temporary capacity or power fading is difficult for cells were influenced by various factors and the underlying mechanism is very complicated (Omar et al., 2015, Vetter et al., 2005, Dupré et al., 2010, Dubarry et al., 2014, Sarasketa-Zabala et al., 2015, Zhang et al., 2011). Over the past few years, substantial efforts have been donated for the study of capacity fading at elevated temperature (Liu et al., 2010, Amine et al., 2005, Chang et al., 2008, Koltypin et al., 2007). In 2005, J. Vetter et al. elaborated the aging mechanisms in lithium-ion battery and showed that the higher temperature not only enhances the kinetics of lithium insertion/removal process into/from the host lattice, but also changes the morphology and composition of SEI layers (Vetter et al., 2005). Amine et al. showed a distinction of cycle performance when experienced 100 cycles with a degradation of 70% at 55°C, 41% at 37°C and nearly no fading at room temperature. Postmortem analysis revealed the dissolution of Fe^{2+} from the LiFePO$_4$ electrode and subsequent deposition on the carbon anode, which act ascatalyst accelerates the formation of SEI layer and constantly consumes the active lithium and imposes high surface resistance as a consequence (Amine et al., 2005). Zaghib et al. focused on the relationship between iron dissolution and Fe impurities in the LiFePO$_4$, and showed a good capacity retention without iron dissolution (Zaghib et al., 2006).

In this work, capacity fading of commercial lithium-ion batteries was studied with a uniformed changed charge and discharge regime at different temperatures, and various techniques which contain non-destructive electrochemical methods and destructive analyses of both electrochemical and physical were applied to diagnose the aging mechanisms of LiFePO$_4$ batteries. Electrochemical techniques contain the diagnosis of capacity recovery with a small current, electrochemical impedance spectroscopy, differential-voltage and differential-capacity analysis. While destructive methods include half-cell test and posttest XRD and SEM analysis were performed for further elucidation of capacity fading. And it was confirmed that the cell-failure modes were similar and mainly caused by loss of lithium inventory.

2 EXPERIMENT

22650-type cylindrical Lithium-ion cell of 2 Ah nominal capacity at ambient temperature were obtained from Optimum Nano, Shenzhen. Cycling condition covered temperatures of room condition (ca. 25°C) and higher level (55°C), Cycling test were performed on Neware BTS-5V10A system (Neware Electronics Co. Ltd, Shenzhen), the route were fixed at 1C constant current charging to a maximum voltage of 3.65 V, and a voltage hold was applied until the current was less than C/20 or 100 mA, and a constant 3C discharge current was applied until the voltage was down to 2.5 V, there was a rest time of 5 minutes when charging/discharging transformation happens. Periodically, cells were taken off the cycling test and were subjected to several health diagnosis measurements, including capacity recovery test with a small current (0.2C), Electrochemical Impedance Spectroscopy (EIS) analysis, and differential-voltage and differential-capacity analysis. For EIS testing, 50% SOC was adjusted through small current, and the open-circuit potential was perturbed by an AC sinusoidal potential of 5 mV amplitude and a frequency ranging from 10 kHz to 0.1 Hz. The EIS were acquired with a Gmary reference 600.

Selected cells were disassembled after being discharged to 2.5 V or fully charged in a dry glove box filled with argon. The fully charged negative tapes and separators were taken pictures, and both electrodes from fully discharged cells were sampled for electrochemical half-cell analysis after being immersed in Dimethyl Carbonate (DMC) for ten minutes. Each sample was prepared by scraping the electrode material from one side of the current collector and being punched into a disk with a diameter of 14 mm. Capacities of positive and negative electrode were measured in half-cells, where lithium metals were used as counter electrodes. The half-cells were cycled at a 1/10C rate on a Land battery tester.

3 RESULTS AND DISCUSSION

3.1 Non-destructive electrochemical analysis

Figure 1 shows the discharge curves of the cells after different numbers of cycles at 25 and 55°C. There is a visible decrease of voltage plateau possibly caused by increased inner resistance accompanied nonstop growth of SEI layer. At 25°C, the capacity of cell decreases to 76% of its initial capacity after 425 cycles. However, when cycling temperature

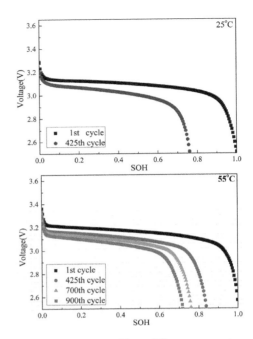

Figure 1. Discharge profiles at different temperatures.

increased to 55°C, the cell shows a slower capacity loss with 84% capacity retention at the same 425 cycles and an equal 76% up to 700 cycles, 72% to 900 cycles. These results are in contrast to those reports of capacity fading at different temperatures (Song et al., 2013, Amine et al., 2005), and it shows that the cell cycled at a higher temperature may not always lose its capacity quickly.

The capacity recovery when cycled with a small current (nearly 0.2C) was shown in Figure 2. And it illustrates the degradation of lithium-ion batteries can be classified into power loss and capacity loss. Power loss is mainly influenced by polarization and can be regained after a long relax time or cycling with tiny current while capacity loss is caused by fatal damage of battery components and can't recovered permanently. Figure 2 shows that after 300 cycles, the recovered capacities are 4.8% at 25°C and 5% at 55°C, which are approximately equal. But after 600 cycles, they are 47.6% and 8.7% respectively, and a higher proportion of 203% when cycled more than 900 times at 25°C. The results of capacity recovery test indicates a larger amount of power loss exists at lower 25°C than 55°C because terrible ion transmission arises electrochemical polarization without enough time for lithium ion shuttling back and forth between positive and negative. In a word, lithium ion couldn't diffuse to a far distance lead to polarization should be responsible for the distinction of power loss between 25 and

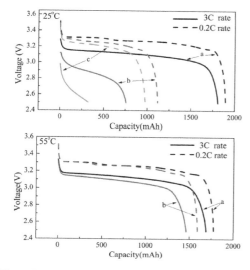

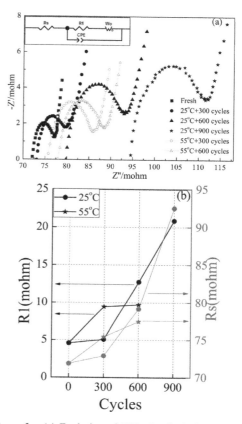

Figure 2. Voltage plots with a small discharge current. Left: 25°C, (a) 300th, (b) 600th, (c) 900th. Right: 55°C, (a) 300th; (b) 600th.

Figure 3. (a) Evolution of EIS of cells during cycling at 25°C and 55°C, and the used equivalent circuit in this study. (b) The variation of resistance Rs and R1 with cycling number at temperature 25 and 55°C.

55°C, while the real reasons of capacity loss or their relevance still needs further analysis.

The voltage plots over the same current showed visible discrepancy of voltage platform and represented significant impedance increase exists, and the difference was larger at 25°C revealed the rapid growth of impedance than 55°C, and this can be demonstrated by following EIS.

EIS is a powerful in situ characterization technique that allows probe specific processes in electrochemical systems from the characteristic impedance behaviors along certain frequency ranges, more details about EIS in general are given in Refs. (Suresh et al., 2002, Reichert et al., 2013). The evolution of impedance spectra in the nyquist form of the Li-ion cells at SOC value of 50% during cycling over different temperatures and a Randles circuit initially proposed by John Edward Brough Randles that consist of an active electrolyte resistance Rs in series with the parallel combination of the Constant Phase Element (CPE) and an active charge transfer resistance R1 and a specific electrochemical element of Warburg element (Wo) were illustrated in Figure 3(a) (Randles, 1947). Characteristic parameters are selected by the program Zview and compared for different operating conditions. And the variation of resistance Rs and R1 with cycling number at temperature 25 and 55°C were shown in Figure3 (b).

As labeled in Figure 3(b), the value of resistances increased with cycle number at both 25 and 55°C. After 300 cycles all resistances of the cell at 25°C increased significantly when compared with 55°C. These results were consistent with the

tendency of discharge capacity and potential difference of small current over the whole life, indicating there is relevance between impedance increase and degradation of cells especially at a lower 25°C.

The capacity-voltage curve of the graphite/LiFePO₄ cell contains lots of information about both electrodes and it could be visible when treated with differential. As the constant single plateau of LiFePO₄ potential profile, the curve of cylinder battery represents the characteristic of anode (Liu et al., 2010). The peak in the curve of differential-voltage (dV/dQ) vs. Q mainly represents the transformation of carbides, namely, distinct x value in Li_xC_6 (Liu et al., 2010, Safari and Delacourt, 2011), and the distance between any two peaks represents the amount of active graphite materials. Whereas the peak in the curves of differential-capacity (dQ/dV) vs. V represents the different equilibrium, and the area under every peak represents the content of the corresponding equilibrium phase (Liu et al., 2010, Safari and

Delacourt, 2011). Figure 4 shows the differentiation curves over the whole life at 25 and 55°C.

In the plots of differential-voltage (Figure 4a and 4b), as the cells ages, all peaks move toward a lower capacity, and especially, the extreme right peak, meaning that the contents of active lithium available for discharge gradually decreases with cycling, and after 300 cycles, the magnitude of migration 25°C is heavier than 55°C, this is in vary with the former EIS analysis. The distance between any two peaks is proportional to the amount of active negative materials. And the reduction of distance also reveals the loss of an active graphite material. But this decrease in graphite content does not directly lead to capacity loss, because of sufficient space for active lithium embedding in the anode, which is consistent with results of following half-cell experiment.

The plots of differential—capacity are widely used as sensitive to cell performance decay that clearly shows the potential regions where phase transition happens. General characteristics are

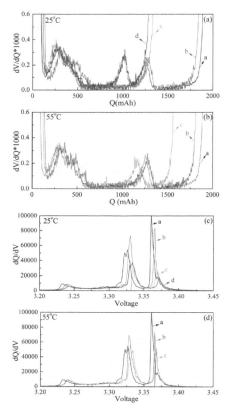

Figure 4. Evolution of the (a), (b) differential-voltage and (c), (d) differential-capacity plots of the graphite/LiFePO₄ cells. a, Fresh; b, 300th cycle; c, 600th cycle; d, 900th cycle.

consistent with previous observation with graphite anode. The area under the peak at 3.36 V, which corresponds to the abscissa of the right peak in differential-voltage curve gradually decreases with cycling, which represents the loss of recyclable lithium.

To summarize, the analysis of differential curves shows significant loss of recycled lithium and slight degradation of graphite, and the consumption of active lithium rather than the depletion of graphite should be the main reason for capacity fading.

3.2 Destructive integrated analysis

To further explore the origin of capacity degradation, the fully charged cylindrical batteries were disassembled. Figure 5 shows the images of graphite tapes with fully lithium embedded. The surface of fresh graphite was golden without any defects, but after 300 cycles at 25°C, positions with poor lithium insertion in the middle of tape can be observed and elsewhere remains well. When the batteries reached their end of life, the surface of battery cycle at 25°C appears to be smooth, and no visible structural damage can be detected compared with a pristine cell. In contrast, anode tape cycled at 55°C shows a wide range of bad lithium insertion. Nevertheless, the cell capacity aged at 55°C remains higher and resistance remains lower than 25°C.

The batteries was also dismantled when discharged to a cut-off voltage of 2.5 V in order to examine the capacity recession of both electrodes, and sections were punched out and tested in coin cells. Figure 6 shows the first and second charge/discharge profiles of positive and negative electrode with a slow 0.1C rate. On the first charge of the LiFePO₄ cathode, only part of the expected Li was recovered, with a range of 84% from fresh cathode to 61% from

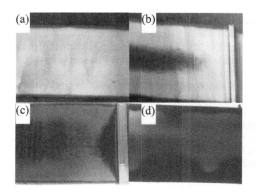

Figure 5. Images of negative tapes with fully lithium intercalation. (a) Fresh; (b) 25°C + before aged; (c) 25°C + aged; (d)55°C + aged.

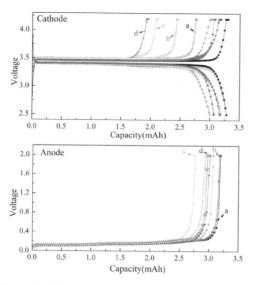

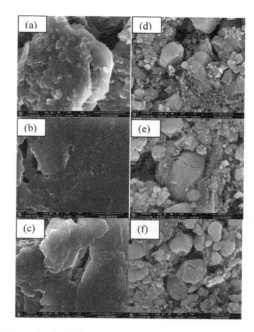

Figure 6. Voltage curves of fresh and aged electrode in half-cell test. Solid symbols:1st cycle; Hollow symbol: 2nd cycles. (a) Fresh; (b) 25°C + before aged; (c) 25°C + aged; (d) 55°C + aged.

Figure 7. SEM images of positive electrodes (a–c) and negative (e–f) electrodes (a, d), Fresh; (b, e), 25°C + aged; (c, f), 55°C + aged.

aged at 55°C. The first charge of the cycled cathode reveals how much active lithium remains in the cell after a number of cycling (Safari and Delacourt, 2011, Shim and Striebel, 2003), and the maximum capacity of the second charge process with abundant lithium source reflects the degree of cathode damage. Comparison of the first charge capacity with the maximum capacity shows that only a small amount of damage happens in cathode especially at 55°C while the anode experienced more, but it remains eligible when compared with battery degradation. The main reason of battery failure is the consumption of lithium, which is usually associated with the continual formation and dissolution of SEI (Safari and Delacourt, 2011, Shim and Striebel, 2003).

Neither electrode showed apparent capacity degradation in the half-cell test; almost all electrodes can exhibit the very similar voltage profiles to the fresh electrode in the case of sufficient lithium source. We can be confident that the capacity fading in the full cell is not due to the capacity degradation of electrode, but the consumption of lithium through a side reaction caused by instability of SEI layer on the anode should be the main reason.

Figure 7 compares the SEM morphologies of positive and negative electrode surface of fresh and after testing at 25 and 55°C. Clear boundaries can be distinguished in fresh graphite electrode, and the electrode cycled at 25°C shows more SEI deposition when comes to end of life. Its surface is covered with a thick SEI layer and the boundary of graphite

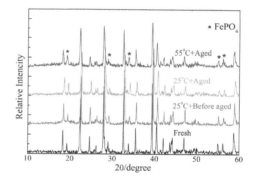

Figure 8. XRD patterns of LiFePO$_4$ positive electrode.

is ambiguous, while graphite aged at 55°C can be partially distinguished. SEM analysis of carbon electrodes further confirms the aging of LiFePO$_4$/graphite batteries is mainly caused by continuous growth of SEI films, which consumes active lithium and increases the inner resistance. On the contrary, there is no obvious change in the morphology of the cathode, but some tiny cracks and little evidence of capacity depression caused by positive electrode.

Figure 8 shows the XRD patterns of the LiFePO$_4$ electrodes, which sampled from fresh cell, aged and before aged cell with cycling at 25 or 55°C. There is no apparent change in the olivine structure after cycling. Most Bragg lines maintain the same angle and relative intensity with the pristine material except very slight shift in diffraction peaks, which means no damage happens on olivine-type structure during the whole lifetime. A series of diffraction peaks are detected because of insufficient active lithium.

4 SUMMARY

Capacity degradation of LiFePO$_4$/graphite cells cycled at 25 and 55°C were studied. Capacity decreased to 76% of its initial capacity after 425 cycles at 25°C but 84% at 55°C, greater drop of voltage platform was detected at 25°C than 55°C after 300 cycles, which was consistent with EIS conclusion. Destructive analysis revealed only slight damage happens on negative electrode which has no influence on cells' capacity degradation, and no observable recession on cathode. A higher 3C discharge current should be responsible for the power loss of cylinder batteries and difference of capacity fading between 25 and 55°C for a great of proportion of capacities can be regained through a small current. The lithium loss mechanism was considered to be the dominated reason for all degradation of batteries and persistent damage and regeneration of SEI layer was the leading cause of active lithium loss.

REFERENCES

Amine, K., Liu, J. & Belharouak, I. 2005. High-temperature storage and cycling of C-LiFePO4/graphite Li-ion cells. *Electrochemistry Communications*, 7, 669–673.

Chang, H.-H., Wu, H.-C. & Wu, N.-L. 2008. Enhanced high-temperature cycle performance of LiFePO4/carbon batteries by an ion-sieving metal coating on negative electrode. *Electrochemistry Communications*, 10, 1823–1826.

Dubarry, M., Truchot, C. & Liaw, B.Y. 2014. Cell degradation in commercial LiFePO4 cells with high-power and high-energy designs. *Journal Of Power Sources*, 258, 408–419.

Dupr, N., Martin, J.-F., Degryse, J., Fernandez, V., Soudan, P. & Guyomard, D. 2010. Aging of the LiFePO4 positive electrode interface in electrolyte. *Journal of Power Sources*, 195, 7415–7425.

Koltypin, M., Aurbach, D., Nazar, L. & Ellis, B. 2007. On the stability of LiFePO4 olivine cathodes under various conditions (electrolyte solutions, temperatures). *Electrochemical and Solid-State Letters*, 10, A40–A44.

Liu, P., Wang, J., Hicks-Garner, J., Sherman, E., Soukiazian, S., Verbrugge, M., Tataria, H., Musser, J. & Finamore, P. 2010. Aging Mechanisms of LiFePO4 Batteries Deduced by Electrochemical and Structural Analyses. *Journal of the Electrochemical Society*, 157, A499–A507.

Omar, N., Firouz, Y., Gualous, H., Salminen, J., Kallio, T., Timmermans, J.M., Coosemans, T., Van Den Bossche, P. & Van Mierlo, J. 2015. 9—Aging and degradation of lithium-ion batteries. *In:* Franco, A.A. (ed.) *Rechargeable Lithium Batteries*. Woodhead Publishing.

Padhi, A.K., Nanjundaswamy, K. & Goodenough, J. 1997. Phospho—olivines as positive—electrode materials for rechargeable lithium batteries. *Journal of the electrochemical society*, 144, 1188–1194.

Randles, J.E.B. 1947. Kinetics of rapid electrode reactions. *Discussions of the faraday society*, 1, 11–19.

Reichert, M., Andre, D., R Smann, A., Janssen, P., Bremes, H.G., Sauer, D.U., Passerini, S. & Winter, M. 2013. Influence of relaxation time on the lifetime of commercial lithium-ion cells. *Journal of Power Sources*, 239, 45–53.

Safari, M. & Delacourt, C. 2011. Aging of a Commercial Graphite/LiFePO4 Cell. *Journal of the Electrochemical Society*, 158, A1123–A1135.

Sarasketa-Zabala, E., Gandiaga, I., Martinez-Laserna, E., Rodriguez-Martinez, L.M. & Villarreal, I. 2015. Cycle ageing analysis of a LiFePO4/graphite cell with dynamic model validations: Towards realistic lifetime predictions. *Journal of Power Sources*, 275, 573–587.

Shim, J. & Striebel, K.A. 2003. Cycling performance of low-cost lithium ion batteries with natural graphite and LiFePO4. *Journal of Power Sources*, 119–121, 955–958.

Song, H., Cao, Z., Chen, X., Lu, H., Jia, M., Zhang, Z., Lai, Y., Li, J. & Liu, Y. 2013. Capacity fade of LiFePO4/graphite cell at elevated temperature. *Journal of Solid State Electrochemistry*, 17, 599–605.

Suresh, P., Shukla, A. & Munichandraiah, N. 2002. Temperature dependence studies of ac impedance of lithium-ion cells. *Journal of applied electrochemistry*, 32, 267–273.

Vetter, J., Novak, P., Wagner, M.R., Veit, C., Moller, K.C., Besenhard, J.O., Winter, M., Wohlfahrt-Mehrens, M., Vogler, C. & Hammouche, A. 2005. Ageing mechanisms in lithium-ion batteries. *Journal of Power Sources*, 147, 269–281.

Zaghib, K., Ravet, N., Gauthier, M., Gendron, F., Mauger, A., Goodenough, J.B. & Julien, C.M. 2006. Optimized electrochemical performance of LiFePO4 at 60°C with purity controlled by SQUID magnetometry. *Journal of Power Sources*, 163, 560–566.

Zhang, Y., Wang, C.-Y. & Tang, X. 2011. Cycling degradation of an automotive LiFePO4 lithium-ion battery. *Journal of Power Sources*, 196, 1513–1520.

Advances in Energy, Environment and Materials Science – Wang & Zhao (Eds)
© *2016 Taylor & Francis Group, London, ISBN 978-1-138-02931-6*

Oxygen Reduction Reaction on sulfur doped graphene by density functional study

J.P. Sun & X.D. Liang
School of Electrical and Electronic Engineering, North China Electric Power University, Beijing, P.R. China

ABSTRACT: The mechanism of sulfur doped graphene catalyzing Oxygen Reduction Reaction (ORR) is studied using Density Functional Theory (DFT). The active site of the catalyst was analyzed, and the ORR reaction path was simulated. The results show that, contrary to N-doped graphene, ORR reaction intermediates absorbed on the dopant sulfur atom. The doped graphene has evidently catalytic effect on absorbed OOH, and the reaction follows the four-electronic transfer path.

1 INTRODUCTION

Low temperature Proton Exchange Membrane Fuel Cell (PEMFC) can convert chemical energy through the electrode reaction directly into electrical energy. Because the reaction process does not involve burning and the product is water, it is regarded as an ideal power source replacement of traditional internal combustion engine. However, one of the main factor affecting the industrialization of fuel cells, is Oxygen Reduction Reaction (ORR) occurring on the cathode need precious metal Platinum (Pt) as catalyst, which has limited reserves and is costly. So developing cheaper, stable non-Pt catalyst with high performance becomes an important topic. (Sun, 2013; Liu, 2014).

Graphene is a two-dimensional carbon material composed of SP2 hybrid carbon atoms, with unique physical and chemical properties. Because of its good electrical conductivity, and catalytic activity after doping, graphene doped with nonmetallic atoms as Oxygen Reduction Reaction (ORR) catalytic electrode materials of fuel cells, is attracting more and more interests. (Kong, 2014) It was found that carbon nanotubes (Gong, 2009; Yu, 2010) and graphene (Qu, 2010; Sheng, 2011) doped with N atoms possess good catalytic activity for ORR, which is equal or even higher than commercial Pt/C catalyst. Dai, et al. (Gong, 2009) argued that this kind of high catalytic activity is due to the larger electronegativity of N (3.04) compared to C (2.55). In addition, smaller electronegativity elements than C atom are introduced into graphene, such as B (Panchokarla, 2009; Sheng, 2012; Wang, 2011; Wang, 2012) and P (Liu, 2011). The results showed that doping atoms can change the electrical properties and chemical activity of graphene system, and have very good catalytic

effect on oxygen reduction reaction. Recently, S-doped graphene has been synthetized through various methods, and showed catalytic activity comparable to N-doped graphene system. (Yang, 2012; Park, 2014; Chen, 2014)

Oxygen reduction reaction mechanisms were studied on various nonmetallic element doped graphene based on density functional theory. The results showed that N-doped graphene (Kim, 2011; Zhang, 2011; Yu, 2011) and S-doped graphene (Zhang, 2014), are efficient catalysis for oxygen reduction reaction on the electrode. Yu, et al. (Yu, 2011) studied the impact of solvent and coverage on catalytic reaction in detail. Xia, et al studied influence of the dopant position of N-doped graphene on oxygen reduction-oxygen precipitation reaction (Zhang, 2011), analyzed four configurations of S-doped graphene by using clusters model, and the path of the oxygen reduction reaction (Zhang, 2014). Being first found, much research works both experimental and theoretical focused on N-doped graphene. In this paper, using first principle calculation method based on density functional theory, S-doped graphene is studied as the catalysis of oxygen reduction reaction in acid environment, and the catalytic mechanism is discussed.

2 METHODS

In this paper, calculation about electronic properties of graphene and its adsorption system based on spin-polarized density functional theory of the Vienna Ab-inito Simulation Package (VASP) was carried out (Kresse, 2012), with Projector-Augmented Wave (PAW) method describing the interaction between the ion core and electron. Electron exchange-correlation functional was represented with the Generalized Gradient Approximation (GGA), and the model of Perdew-Burke-Ernzerhof (PBE) was used for the nonlocal corrections. A kinetic energy cutoff of 500 eV was used with a plane-wave basis set. Graphene cell uses 4 * 4 * 1, a total of 32 atoms, with a vacuum space of 20 Å. The integration of Brillouin zone was conducted using 11 * 11 * 1 Monkhorst-Pack grid with the Γ-point included, and the error of the energy was not greater than 0.01 eV. Atoms were relaxed and convergence precision was −0.02. H, C, O, S valence electron configuration is respectively, H 1 s^1, C 2 $s^2 2p^2$, O 2 $s^2 2p^4$, and S 3 $s^2 3p^4$, Spin was considered in the calculation.

Adsorption energy: $E_{ad} = E_{gra} + M − (E_{gra} + E_M)$ $E_{gra} + M$ refers to the graphene system energy after doping atoms and adsorbing molecules, E_{gra} refers to the graphene system energy after doping atoms, E_M is the energy of molecules.

3 RESULTS AND DISCUSSION

3.1 Sulfur-doped graphene

As shown in Figure 1 (a), 4 * 4 * 1 graphene cell, a total of 32 C atoms, are selected as the super cell. Replaced one C with S atom, S-doped graphene system model was built. In the figure, the yellow balls denote C atoms, and the blue ball is S. It is generally accepted that dopant atoms and the nearby C atoms form bonds in doped graphene, resulting in electronic charge density redistributes. The charged atoms could be the catalytic active positions. Therefore, we first discussed ORR catalytic activity position of S-doped graphene by analysis of charge distribution and density of states.

After optimizing the structure of above model, Bader charge analysis is used for each atom of S-doped graphene system to obtain quantity of charge. The charge distribution of S and nearby C atoms are shown in Figure 1 (b). The charge on the C atoms in intrinsic graphene is very small, and evenly distributed. In contrast, charge redistribution happened after doping. That is, net charge distributed on the dopant and other atoms. In the figure, the red balls denote positively charged atoms, and the green ones are negatively charged atoms. Positively charged S atom gains 0.39e, while the negatively charged adja cent C atoms lose 0.11e.

Zhang et al. (Zhang, 2011; Zhang, 2012) studied ORR catalytic mechanism of N-doped graphene in the fuel cell, showing that ORR catalytic active sites are closely related to unevenly distributed charge. Therefore, the C atoms of intrinsic graphene do not have catalytic activity. After doping, the N atom was negatively charged, while the C atoms adjacent to the N atom lost electrons and became positively charged. These C atoms act as the catalytic active sites of ORR. Like N-doped graphene, in S-doped graphene, the results show that the S atom lose electrons, positively charged. The C atoms adjacent to the S became negatively charged, showing that charge transferred between S atom and the nearby C atoms.

Analysis of Partial Density of States (PDOS) variation before and after doping is a common method to study interaction between atoms and the charge transfer. Figure 2 shows the calculated Total Density of States (TDOS) of S-doped graphene, and Partial Density Of States (PDOS) of S and neighbor C atoms. Among them, (a) shows total density of states of S-doped graphene, (b) shows partial density of states of sulfur atom in sulfur-doped graphene (c) shows partial density of state of neighboring C atoms. For comparing, the figure also shows the density of state of an isolated S atom and a C atom in intrinsic graphene. Here, (d) shows isolate S atom, (e) shows C atom in intrinsic graphene; In addition, in terms of different types of

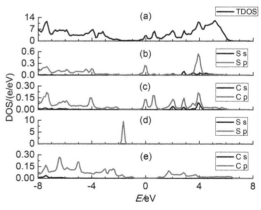

Figure 2. (a) Sulfur-doped graphene; (b) Sulfur in S-doped graphene; (c) carbon atom in sulfur-doped graphene; (d) isolated sulfur atom; (e) carbon atom in intrinsic graphene.

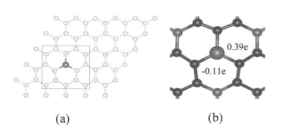

(a)   (b)

Figure 1. (a) Sulfur-doped graphene structure; (b) the atoms charges, S and neighboring C.

valence electrons involved in bonding, in (b), (c), (d) and (e), the black lines denote s state partial density of state, the red line denotes p state partial density of state.

From Figure 2 (b), (c), it can be seen that S and C atoms in graphene strongly interacted after doping. There are evident overlaps between the electronic states of two atoms. The main interaction occurs between p electronic states of S and C, as shown in (b). Two resonance peaks in S atomic partial density of states appeared. One is located around the Fermi level, and another one is located 4 eV above Fermi level. There also formed a weak energy band between −4 eV and −8 eV under the Fermi level. Compared to isolated atom, after being doped into graphene, the energy levels of S atom split and extended into bands. Because the density of states 4 eV above Fermi level corresponds to the empty state, or with no electrons filling, Figure 2 (b) shows that S atom loses electrons after doping. Similarly, comparing Figure 2 (c) and (e) to analysis of electronic gain and loss on C atoms in graphene. The above results are consistent with the results obtained by Bader charge analysis.

It is generally accepted that positively charged atoms are conducive to adsorption of intermediates or atom groups in ORR, becoming the catalytic activity sites. Therefore, the active sites of N-doped graphene are C atoms adjacent to N atoms. Contrary to N-doped graphene, S atom lost electron became positively charged, while carbon atoms adjacent to dopant S atom obtained negative charges after doping. As a reasonable inference, the active sites of S-doped graphene are the S atom. The following ORR reaction process optimization showed that O_2 and OOH tended to adsorbed on S atoms, which confirmed this conclusion.

3.2 ORR reaction paths

The research shows that ORR on the fuel cell cathode has two paths (Yeager, 1984; Tsuda, 2007), one is four-electron transfer path, $O_2 + 4H+ + 4e- \rightarrow 2H_2O$, and O_2 is directly reduced into H_2O, without the formation of intermediate H_2O_2. Another one is two-electron transfer way, $O_2 + 2H+ + 2e- \rightarrow H_2O_2$, producing intermediate H_2O_2. The former path is more efficient. Zhang et al. [17] have found that four-electron transfer path dominated in the ORR reaction of doped graphene, therefore, we are more interested in four—electron path and will discuss it.

ORR four-electron transfer steps simulation is shown in Figure 3. First, OOH adsorbs on the surface of S-doped graphene, forming G-OOH after optimization, as shown in Figure 3 (a). Then, H is added to the structure and made close to the S atom far from the graphene plane in OOH. After optimization, we find that O—O bond is broken,

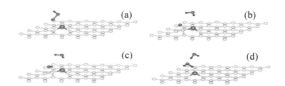

Figure 3. (a) adsorbing OOH; (b) first H atom added (c) second H atom added; (d) third H atom added.

forming a H_2O molecule and O atom adsorbed on graphene (G-O), as shown in Figure 3 (b). After that, adds the second H to the structure, and make it closing to the G-O, forming G-OH after optimization, as shown in Figure 3 (c). In the end, adds the third H into the system, resulting in C—O bond breaking and the second H_2O forming, as shown in Figure 3 (d). The above simulation shows that ORR catalyzed by O-doped graphene follows four—electron transfer paths indeed.

In the four-electron reaction, adsorption of OOH or O_2, is the first step of ORR, and is also a key step. This is because that O—O bond rupture is the prerequisite of four-electron transfer path occurring. Dissociation energy of molecular state O—O bond can be as high as 494 KJ/mol. However, after adsorbing on graphene, because S and O atoms have stronger interaction, O—O bond energy can be significantly reduced, making subsequent O—O bond rupture reaction possible. Therefore, we first focused our study on the states of doped graphene after adsorption of OOH and O_2.

Figure 4 shows the structure of S-doped graphene adsorbing OOH before and after optimization. Here, a, a′ stand for the structures before optimization, and b, b′ stand for the optimized structure. a, b stands for the side view before and after optimization, and a′ b′ stands for vertical view before and after optimization. The yellow balls denote C atoms, blue ball S atom, pink balls O atoms, and brown balls H atoms. O atom initially is placed on the top of C atom, and OOH perpendicular to the surface of graphene. After optimization, C atom close to dopant S sticks out of graphene plane of 1.01 Å. As shown in Figure 1, forming a tetrahedron structure, as shown in Figure 4(b). S and C atoms getting close to each other, the length of C—S bond is 2.54 Å, while the length of O—O bond is 1.44 Å. The energy of graphene adsorbing OOH is −3.06 eV. After optimization, O—O bond length in OOH increases from the isolated state of 1.320 Å to 1.44 Å. It shows that, OOH and graphene form strong chemical bond. At the same time, the adsorption weakens O—O bond, causing increase of the bond length.

In order to understand the catalytic effect deeply, we further studied the related mechanism

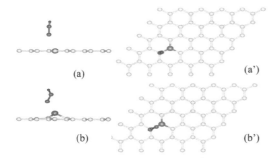

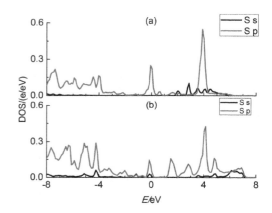

Figure 4. The structure of sulfur-doped graphene adsorption OOH.

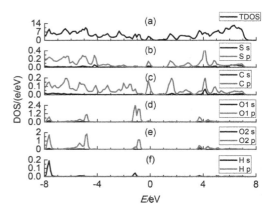

Figure 5. (a) total density of states; PDOS (b) S atom; (c) C adjacent to S; (d), (e) two O atoms in OOH; (f) H atom.

Figure 6. Partial density of states of S atom in S-doped graphene (a) before adsorption of OOH; (b) after adsorption of OOH.

by the density of state analysis. Figure 5 shows density of states of S-doped graphene adsorbed with OOH. Here, figure (a) represent the total density of states of the system, and (b) for the partial density of state of S atom, (c) for partial density of state of C atom near the OOH, (d) for partial density of state of O atom in OOH near to dopant S atom, (e) for partial density of state of O atom far away from S atom, (f) for partial density of state of H atom. The black lines denote s state partial state density, and red lines denote p state partial state density. Figure 6 shows partial density of states of the dopant S atom before and after adsorption of OOH, including (a) before adsorbing OOH, (b) after adsorbing OOH.

Comparing Figure 5 with Figure 2, we can see that, before and after adsorbing OOH, the density of state of the system changed, especially, the dopant S atom. From Figure 6 (b), we can clearly see that, compared with figure (a) that before adsorption, in 0~4 eV energy range above Fermi level, the p electronic state density peak broadened and lowered noticeably. As the states above Fermi energy level are not occupied by electrons, we can infer that dopant S lost electrons after adsorption. Similarly, from analysis of O in OOH, it shows that O atom gets electrons after adsorption. Because the HOMO orbital in OOH is antibonding orbital, electrons filling will weaken the bond strength of O$-$O in OOH, further resulting in the bond broken and the H_2O generation.

Bader analysis results show that, OOH gains overall 0.325e electron charge after adsorption. After further analysis, we find that the dopant S atom loses 0.2e electron charges after adsorption. Therefore, the gained charge of OOH should be partly derived from the C atoms, and partly from the dopant S atom, which plays the role of a bridge. From Figure 6, we can see that, in addition to p electron states, s electron states also evidently changed, and simultaneously filling of the states, which could be attributed to the change of S

atomic hybrid states after adsorption. It is important to further analyze and understand catalysis mechanism of S-doped graphene.

Adsorption process of O_2 on the graphene is also optimized, and the results are similar to OOH adsorption. After S-doped graphene adsorbing O_2, the C atom next to the dopant O atom protrudes graphene plane with 0.96 Å forming a tetrahedron structure. The length of S$-$O bond is 2.64 Å, and the length of O$-$O is 1.323 Å. The length of O$-$O bond changes from the isolated state of 1.320 Å to 1.323 Å. Compared with the adsorption of OOH, S-doped graphene has less impact on O$-$O bond interaction. Besides, the adsorption energy of O_2 is -2.48 eV, and the adsorption energy of OOH -3.06 eV. It suggests that, G-OOH is more stable than G-O_2 in energy, which could infer that OOH adsorption in the first electron transfer step reaction may be a more possible way to happen in ORR.

4 CONCLUSIONS

By using Density Functional Theory (DFT), we studied the ORR catalytic reaction process occurring on the Sulfur-doped graphene. The results showed that, the catalytic active site of ORR was the dopant S atom. Through analysis of the charge distribution and the density of states, we found that S atom loses electrons positively charged and C atoms adjacent to S atom gain electrons, and negatively charged. The optimization of various intermediates adsorption process showed that the ORR reaction occurring on S-doped graphene follows the way of four electrons path, in which OOH adsorption is more likely to happen than O_2 adsorption. And further study is needed to understand the ORR mechanism of S-doped graphene, which has a significance of guiding experimental works.

REFERENCES

Chen, Y.; Li, J.; Mei, T.; Hu, X.G.; Liu, D.W.; Wang, J.C.; Hao, M.; Li, J.H.; Wang, J.Y.; Wang, X.B., "Low-temperature and one-pot synthesis of sulfurized graphene nanosheets via in situ doping and their superior electrocatalytic activity for oxygen reduction reaction," J. Mater. Chem. A., 2, 2014, pp. 20714–20722.

Gong, K.P.; Du, F.; Xia, Z.H.; Durstock, M.; Dai, L.M., "Nitrogen-Doped Carbon Nanotube Arrays with High Electrocatalytic Activity for Oxygen Reduction," Science., 323, 2009, pp. 760–764.

Kim, H.; Lee, K; Woo, S.L.; Jung, Y.S., "On the mechanism of enhanced oxygen reduction reaction in nitrogen-doped graphene nanoribbons," Phys. Chem. Chem. Phys., 13, 2011, pp. 17505–17510.

Kong, X.K.; Chen, C.L.; Chen, Q.W., "Doped graphene for metal-free catalysis," Chem. Soc. Rev., 43, 2014, pp. 2841–2857.

Kresse, G.; Furthmuller, "Efficient iterative schemes for ab initio total-energy calculations using a plane-wave basis set," J. Phys. Rev. B., 54, 1996, pp. 11169.

Liu, M.M.; Zhang, R.Z.; Chen, W. "Graphene-Supported Nanoelectrocatalysts for Fuel Cells: Synthesis, Properties, and Applications," Chem. Rev., 114(10), 2014, pp. 5117–5160.

Liu, Z.W.; Peng, F.; Wang, H.J.; Yu, H.; Zheng, W.X.; Yang, "Novel Phosphorus-Doped Graphite Layers with High Electrocatalytic Activity for O2 Reduction in Alkaline Medium," J. Chem. Int. Ed., 50, 2011, pp. 3257–3261.

Panchokarla, L.S.; K.S. Subrahmanyam; S.K. Saha; A. Govindaraj; H.R. Krishnamurthy; U.V. Waghmare and C.N.R. Rao, "Synthesis, structure, and properties of boron- and nitrogen-doped graphene," Adv. Mater., 21, 2009, pp. 4726–4730.

Park, J.A.; Jang, Y.J.; Kim, Y.J.; Song, M.S.; Yoon, S.; Kim, D.H.; Kim, S.J., "Sulfur-doped graphene as a potential alternative metal-free electrocatalyst and Pt-catalyst supporting material for oxygen reduction reaction," Phys. Chem. Chem. Phys., 16(1), 2014, pp. 103–9.

Qu, L.T.; Liu, Y.; Baek, J.B.; Dai, L.M., "Nitrogen-doped graphene as efficient metal-free electrocatalyst for oxygen reduction in fuel cells," ACS Nano, 4(3), 2010, pp. 1321–1326.

Sheng, Z.H.; Shao, L.; Chen, J.J.; Bao, W.J.; Wang, F.B.; Xia, X.H., "Catalyst-Free Synthesis of Nitrogen-Doped Graphene via Thermal Annealing Graphite Oxide with Melamine and Its Excellent Electrocatalysis," ACS Nano., 5(6), 2011, pp. 4350–4358.

Sheng, Z.H.; Gao, H.L.; Bao, W.J.; Wang, F.B.; Xia, X.H., "Synthesis of boron doped graphene for oxygen reduction reaction in fuel cells," J. Mater. Chem., 22, 2012, pp. 390–395.

Sun, S.G.; Chen S.L., Electrocatalysis. Chemical industry press. 1st ed.; Beijing, 2013.

Tsuda, M.; Kasai, H., "Proton Transfer to Oxygen Adsorbed on Pt: How to Initiate Oxygen Reduction Reaction," J. Phys. Soc. Jpn., 76, 2007, pp. 024801.

Wang, S.; Iyyamperumal, E.; Roy, A.; Xue, Y.; Yu, D.; Dai, L., "Vertically Aligned BCN Nanotubes as Efficient Metal-Free Electrocatalysts for the Oxygen Reduction Reaction: A Synergetic Effect by Co-Doping with Boron and Nitrogen," Angew. Chem., 50(49), 2011, pp. 11756–11960.

Wang, S.; Zhang, L.; Xia, Z.H.; Roy, A.; Chang, D.W.; Baek, J.B.; Dai, L., "BCN Graphene as Efficient Metal-Free Electrocatalyst for the Oxygen Reduction Reaction," Angew. Chem., 51(17), 2012, pp. 4209–4212.

Yeager, "Electrocatalysts for O2 reduction," E. Electrochim. Acta., 29, 1984, pp. 1527–1537.

Yang, Z.; Yao, Z.; Li, G.F.; Fang, G.Y.; Nie, H.G.; Liu, Z.; Zhou, X.M.; Chen, X.A.; Huang, S.M., "Sulfur-Doped Graphene as an Efficient Metal-free Cathode Catalyst for Oxygen Reduction," ACS Nano., 6(1), 2012, pp. 205–211.

Yu, D.S.; Zhang, Q.; Dai, L.M., "Highly Efficient Metal-Free Growth of Nitrogen-Doped Single-Walled Carbon Nanotubes on Plasma-Etched Substrates for Oxygen Reduction," J. Am. Chem. Soc., 132(43), 2010, pp. 15127–15129.

Yu, L.; Pan, X.L.; Cao, X.M.; Hu, P.; Bao, X.H., "Oxygen reduction reaction mechanism on nitrogen-doped graphene: A density functional theory study," J. Catal., 282, 2011, pp. 183–190.

Zhang, L.P.; Niu, J.B.; Dai, L.M.; Xia, Z.H., "Effect of Microstructure of Nitrogen-Doped Graphene on Oxygen Reduction Activity in Fuel Cells," Langmuir, 28(19), 2012, pp. 7542–7550.

Zhang, L.P.; Niu, J.B.; Li, M.T.; Xia, Z.H., "Catalytic Mechanisms of Sulfur-Doped Graphene as Efficient Oxygen Reduction Reaction Catalysts for Fuel Cells," J. Phys.Chem.C., 118 (7), 2014, pp. 3545–3553.

Zhang, L.P.; Xia, Z.H.; "Mechanisms of Oxygen Reduction Reaction on Nitrogen-Doped Graphene for Fuel Cells," J. Phys. Chem. C., 2011, 115(22), 2011, pp. 11170–11176.

Advances in Energy, Environment and Materials Science – Wang & Zhao (Eds)
© *2016 Taylor & Francis Group, London, ISBN 978-1-138-02931-6*

Study of thermodynamics on the formation of tetrahydrofuran hydrate by React Calorimeter (RC1e)

Xiaodong Dai
Shengli College China University of Petroleum, Dongying, Shandong, China

Xiaoming Hu
Shandong Shida Shenghua Chemical Company, Dongying, Shandong, China

Kun Liu, Hongyan Li & Yue Liang
Shengli College China University of Petroleum, Dongying, Shandong, China

Wei Xing
China University of Petroleum (East China), Qingdao, Shangdong, China

ABSTRACT: Kinetics and thermodynamics during the formation of tetrahydrofuran hydrate was studied by React Calorimeter (RC1e). The influence of THF concentration to hydrate formation was investigated. During hydrate formation, the two components system consisted by THF and water would go through nuclear induction, nuclear formation and hydrate growth. The nuclear formation point could be determined by temperature and thermal changes. While, through analysis to nuclear formation temperature, energy in induction period, and heat release, it could be found that THF concentration had influence to solid composition in solution. With lower 20 ml THF, it preferred to form more ice, but with more than 20 ml, it preferred to form more THF hydrate.

1 INTRODUCTION

As the demand of energy consumption is growing, the exploration and application of natural gas hydrate resources, which has large reserves and high energy density, is becoming more and more attractive. For reasonable utilization of natural gas hydrate resources, scientists demand a detailed knowledge about their physical and chemical properties. With the development of advanced equipments and analysis methods, researchers have widely investigated the structure, formation process and dissociation process of hydrates, which benefited to understand the characteristics of hydrates well. Because the formation conditions for natural gas hydrate are so harsh for simulation, tetrahydrofuran hydrate is commonly used as a model compound to study the formation and dissociation processes of natural gas hydrate for experimental research.

Tetrahydrofuran is a hydrophilic compound, which can form II hydrate at atmospheric pressure and cooling conditions. Its structure and properties, as well as kinetic and thermodynamic process of its hydrate are very typical (Bai, 2005; Ma, 2006). RC1e (Mettler-Toledo), can accurately measure the process heat online with a control resolution of 0.2 mK. Meanwhile, it has torque measurement and control system, which can monitor the torque change semi-quantitatively (Dai, 2011). In this study, RC1e was used to investigate the process parameter variations with setting programs. Under different concentrations researches on hydrate formation process and different concentrations, could help us to explain the mechanism of tetrahydrofuran hydrate formation from the perspective of thermodynamics and kinetics.

2 EXPERIMENTAL

Reaction Calorimeter (RC1e, Mettler-Toledo, Swiss) with control temperature range of RC1e is −70~300°C, having the isothermal, adiabatic and crystallization temperature modes. It can measure the process heat changes online and select the appropriate baseline for integration. Setting program for cooling procedures by iControl software, could measure the process parameters of tetrahydrofuran hydrate formation. RC1e together with ReactIR on-line analysis system can be used to measure the infrared absorption using infrared

(FTIR) technology for the determination of system composition (Jia, 2012).

Added 200 ml distilled water to the reactor, and mixed with different dosages of tetrahydrofuran so that to obtain solutions with different concentrations. When injection port plug screwed well, solution was stirred for 5 min to mix well. Program was set for the process of tetrahydrofuran hydrate formation.

3 RESULTS AND DISCUSSIONS

3.1 *Temperature and heat during the hydrate formation*

A mixture of 10 ml THF in 200 ml water was used as a sample to elaborate the heat and temperature changes during the hydrate formation.

As shown in Figure 1, when Tj was decreased from 5°C with the cooling rate of 1°C/min, the solution system energy was taken away by the cooling medium, which led to the decreasing of Tr. Once Tj and Tr decreased to −10°C and −7.4°C, respectively, Tr suddenly increased to −1.5°C. This phenomenon might be resulted from the formation of hydrate nucleation, which released heat through the phase change. No crystal was observed in the reactor at this moment. Consequently, Tj was maintained at −10°C, but Tr was still at −1.5°C, which indicated that the two components system reached the phase equilibrium. The system temperature would not change as long as the present of liquid phase. Later, crystalline substances appeared within the reactor solution. Thus, Tj started to increase artificially for the protection of stirring system and the experiment was ended.

RC1e can be used to measure the heat changes during the process of the hydrate formation and analyze thermodynamic process, where qaccu (power of system heat storing), qflow (heat flux

power of reactor), qr (power of system heat generating), and qr = qflow + qaccu. According to the curve of qaccu, qaccu was negative as Tj in the process of decreasing, which indicated that the solution system was dissipating heat through the vessel wall (Fig. 2). At this moment, qr was zero, meaning that the solution itself did not have any physical process with heat generating. As Tr increased from −7.4°C to −1.5°C, qaccu and qr increased simultaneously. This phenomenon suggested that internal changes happened in the solution system during this process and phase change happened in the two component system accompanying the release of considerable heat. Because the produced heat from hydrate formation can be removed by vessel wall, the temperature of the solution in the reactor kept constant with qaccu = 0. A constant positive qr suggested that the two component system had a sustained heat release during the formation of hydrate with phase transition process (Qi, 2008). For the protection of stirring system, Tj was increased artificially, when qaccu and qr became meaningless.

3.2 *THF concentration on hydrate formation*

Certain dosages (5 ml, 10 ml, 12.5 ml, 15 ml, 17.5 ml, 20 ml, 22.5 ml, 25 ml, 30 ml, 50 ml) of THF were added into 200 ml water, respectively. The formation process of hydrate was quite similar to crystallization, including nucleation and growth. The process of formation of the stable hydrate nucleation with a larger size than a critical dimension produced a new phase, which was relatively slow and named induction period (Chen, 2008).

As shown in Figure 2, Qaccu was negative accompanying by the energy release before the formation of hydrate in the two component system.

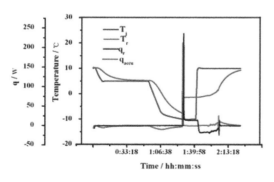

Figure 1. Temperature and heat changes of 10 ml (THF)/200 ml (water) during the hydrate formation.

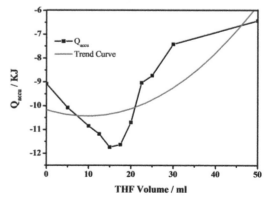

Figure 2. Influence of THF concentration on accumulation heat of the induction period.

With the increase dosages of THF from 5 ml to 20 ml, the released energy increased gradually to a maximum value. However, the released energy decreased if the THF dosages were larger than 20 ml. The temperature Tr for hydrate nucleation formation had a similar variation process during the addition of increasing amount of THF.

Tetrahydrofuran can dissolve in water with any ratios and form a two component system. According to the phase rule, solidification point is the temperature for solid-liquid phase equilibrium, when the steam pressures for solid phase and liquid phase are equal. Due to the existence of THF in the THF solution, the vapor pressures of the solution and water only can be equal when the temperature is below the water freezing point. The larger the THF concentration is, the lower the vapor pressure it has. Therefore, with the increase of THF concentration, the released energy during the hydrate nucleation induction period also increased and nucleation temperature dropped (Fig. 3). When the additional content of THF was higher than 20 ml, the number of nucleation increased and the hydrate formation became much easier as the THF concentration increased. Additionally, hydrate nucleation temperature increased and the released energy decreased.

The Qr was compared during the first 5 min from the beginning of heat releasing during the nucleation formation. In Figure 4, Qr increased as the additional volume of THF increased to 20 ml, which meant that more and more water existed in the system. When the additional volume of THF was larger than 20 ml, it would take 15 min for the system to begin the stable heating releasing period of hydrate. Thus, the first 15 min of integrated heat was compared. Overall, the released heat still increased with the increasing concentrations

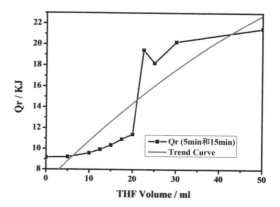

Figure 4. Influence of THF concentration on the peak heat release of hydrate formation.

of THF. Considering the influence of THF concentrations on the storage heat and nucleation temperature of the hydrate induction period, it was concluded that the concentration of THF influenced the composition of the solid formed in the solution. When THF amount was lower than 20 ml, it was easy to form ice; if it was greater than 20 ml, hydrate formed.

4 CONCLUSIONS

Tetrahydrofuran hydrate formation process was studied with the application of RC1e. The kinetics of hydrate nucleation induction, nucleation and growth can be determined by analyzing the temperature and heat changes. When the solution cooled to the super cooling state, the two components systems composed of tetrahydrofuran and water kept releasing heat slowly to the outside. Once the solution reached a certain temperature (nucleation temperature), the two component system release considerable heat by transformation with the occurrence of exothermic peak. Later, the system temperature remained essentially unchanged, and a stable hydrate growth stage occurred.

The concentration of THF influences the composition of the formed solid in solution. According to the analysis of the nucleation induction period of temperature, energy and heat, it is concluded that: 1) a small amount of THF can reduce the freezing point of a solution. This solution is consistent with the two component immiscible system phase rule, where cooling is easy to form ice; 2) with large amount of THF, the two component system can form hydrate. As the THF percentage increases, hydrate forms more easily.

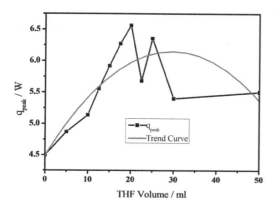

Figure 3. Influence of THF concentration on heat release during hydrate formation.

REFERENCES

Bai, X.D., Zhang, L., Li, L., Huang, J.J., Wang, Zh.Y., Yan, Y., Xu, F.H., A new method to evaluate the performance of hydrate inhibitors. *Chemical Engineering of Oil and Gas,* 2005, 34(6), 445–447.

Chen, G.J., Sun, C.Y., Ma, Q.L., Gas hydrate science and technology. Science and Technology of Gas Hydrate, Beijing: *Chemical Industry Press*, 2008, 124–159.

Dai, X.D., Jia, Z.Q., Sun, L., Gao, L.B., Zhang, Zh. H., Guo, H.F., The study of heat treatment mechanism for wax crude oil using RC1e calorimetric reactor. *Oil & Gas storage and transportation*, 2011, 30(5), 359–361.

Jia, Z.Q., Dong, J., Li, Q.R., Li, J., Xiong, Y.M., X.D. Dai, Liu, Ch., Hydrate formation process of tetrahydro-furan tested by RC1e and ReactIR. *Oil & Gas storage and transportation*, 2012, 31(11), 865–867.

Ma, Y.H., Gou, L.T., He, X.X., Liu, F.R., Study on the formation kinetics of the hydrate above 0°C. *Natural gas Geoscinece*, 2006, 17(2), 244–248.

Qi, X.Y., Fan, Sh.Sh., Liang, D.Q., Li, D.L., Chen, J.G., 3A molecular sieve of tetrahydrofuran hydrate formation process. *Science in China Series B Chemistry*, 2008, 38(3), 170–176.

Advances in Energy, Environment and Materials Science – Wang & Zhao (Eds)
© 2016 Taylor & Francis Group, London, ISBN 978-1-138-02931-6

In-situ synthesis of LSCF-GDC composite cathode materials for Solid Oxide Fuel Cells

J. Li
Electric Vehicle Research and Development Center, Shenzhen Institutes of Advanced Technology,
Chinese Academy of Sciences, Shenzhen, P.R. China
Academy of Fundamental and Interdisciplinary Sciences, Harbin Institute of Technology, Harbin, P.R. China

Z. Wu
Electric Vehicle Research and Development Center, Shenzhen Institutes of Advanced Technology, Chinese
Academy of Sciences, Shenzhen, P.R. China

N. Zhang, D. Ni & K. Sun
Academy of Fundamental and Interdisciplinary Sciences, Harbin Institute of Technology, Harbin, P.R. China

ABSTRACT: $La_{0.6}Sr_{0.4}Co_{0.2}Fe_{0.8}O_{3-\delta}$-$Gd_{0.2}Ce_{0.8}O_{2-\delta}$ (LSCF-GDC) composite material is co-synthesized by the in-situ sol-gel method for Solid Oxide Fuel Cell (SOFC) cathodes. For the in-situ sol-gel method, the composite precursors are mixed at the sol state, which makes more uniform mixture than the solid state. Both the co-synthesized and physically mixed LSCF-GDC composites have been investigated for phase homogeneity by TEM energy-dispersive X-ray spectroscopy mapping and backscatter electron image. The results indicate that the dispersion of LSCF and GDC at the nanometer and micrometer scale is more homogeneous in the co-synthesized composites. Electrochemical impedance spectroscopy measurements confirm that the in-situ synthesized composites show lower polarization resistance than the physically mixed composites. An anode-supported fuel cell consisting of the in-situ synthesized LFCS-GDC composite cathode demonstrates a maximum power density of 1068 mW cm^{-2} at 800°C, which is promising for intermediate temperature SOFC systems.

1 INTRODUCTION

In recent years, there has been considerable interests in intermediate temperature (600~800°C) solid oxide fuel cells (IT-SOFCs) (Zhou et al., 2009, Fu et al., 2007, Park et al., 2011), as they allow the use of lower-cost interconnect and balance-of-plant materials than high temperature (≈1000°C) SOFCs. Reduced temperature also has reduced problems with sealing and thermal degradation. While improvements in IT-SOFCs have been achieved, their performance is still limited by cathode overpotential. Thus, it is of great importance to identify improved cathodes for IT-SOFCs.

For SOFC cathodes, the oxygen reduction reaction takes place at the Triple-Phase Boundary (TPB) where gas, electronic conductive and ionic conductive materials are in contact with each other. Therefore, sufficient electronic and ionic conductivity, as well as appropriate porosity are required for SOFC cathodes. It is well known that combination of electrode material with electrolyte dramatically improve the electrochemical activity of the electrode (Kim et al., 2007, Leng et al., 2008, Brant et al., 2001). The improved electrochemical performance of the composite cathode is due to the extended TPB area from cathode/electrolyte interface to the whole cathode. Furthermore, the composite cathode shows better thermal and chemical compatibility with electrolyte than the single phase cathode. At present, mixed-conductive perovskite $La_{0.6}Sr_{0.4}Co_{0.2}Fe_{0.8}O_{3-\delta}$ is one of the most promising cathode materials for IT-SOFCs. Addition of an ionic conductive phase, such as $Zr_{0.84}Y_{0.16}O_2$ (YSZ) (Chen et al., 2010) or $Gd_{0.2}Ce_{0.8}O_{2-\delta}$ (GDC)/ $Sm_{0.2}Ce_{0.8}O_{2-\delta}$ (SDC) (Murray et al., 2002, Wang et al., 2005, Xu et al., 2006) into the LSCF electrode could significantly enhance its performance at intermediate temperatures. It is confirmed by Murray et al. (Murray et al., 2002) that addition of 50 wt% GDC to $La_{0.6}Sr_{0.4}Co_{0.2}Fe_{0.8}O_3$ electrodes resulted in a factor of about ten times reduction in the polarization resistance.

Generally, LSCF and GDC powders are separately fabricated and physically mixed to prepare LSCF-GDC composites. However, this method usually suffers from the disadvantages related to tedious technology, time-consuming, and especially that the resultant composites are coarse particles after high temperature sintering due to inhomogeneous dispersion. In order to solve the problem, in-situ synthesis of LSFC-GDC composites which have pure phase constituents and homogeneous distribution has been proposed in the present work. For the in-situ sol-gel method, LSCF and GDC precursors are mixed at the sol state, which makes more uniform mixture than the solid state. Then, LSCF-GDC composites are obtained by co-sintering which simplifies the fabrication process. As a result, the in-situ synthesized composites with uniform dispersion are expected to increase the TPB area and enhance the electrochemical performance of LSCF-GDC composite cathodes for IT-SOFCs.

2 EXPERIMENTAL

LSCF-GDC composites were co-synthesized by in-situ sol-gel method according to the stoichiometry of $La_{0.6}Sr_{0.4}Co_{0.2}Fe_{0.8}O_{3-\delta}$ and $Gd_{0.2}Ce_{0.8}O_{2-\delta}$. The raw materials were all analytical grade and used as received without further purification. Stoichiometric amount of $La(NO_3)_3 \cdot 6H_2O$, $Co(NO_3)_2 \cdot 6H_2O$, $Fe(NO_3)_3 \cdot 9H_2O$ and $Sr(NO_3)_2$ were dissolved in distilled water. Subsequently, citric acid was added and the mole ratio of citric acid to the total metal cations was 1.8. Then, the pH was adjusted to 9.0 by ammonia and the solution was kept at 80°C under stirring until a transparent sol was formed. The experimental procedure to prepare GDC sol was similar, using $Gd(NO_3)_2 \cdot 6H_2O$ and $Ce(NO_3)_2 \cdot 6H_2O$ as raw materials and the mole ratio of citric acid to total metal cations was 1.5. Then, the precursors were mixed under stirring and dehydrated at 120°C yielding the composite xerogel. Except for the section discussing the effect of composite composition on the performance of LSCF-GDC composite cathode, LSCF and GDC in the composites were all 50:50 in weight ratio. The resulting xerogel was held at 450°C for 5 h to ensure the complete decomposition of organics and further heated to 850~1000°C to form the composite powders.

Thermogravimetric analysis of the LSCF-GDC xerogel was carried out at the temperature range between room temperature and 1000°C in flowing air, using a Netzsch STA4497 TG-DSC (Germany). Formation of LSCF and GDC phases were confirmed by Rigaku D/max-ΠB X-Ray Diffractometer (XRD) using Cu Kα radiation. The specific surface area of the as prepared powders was measured using nitrogen adsorption measurements with a surface area and porosity analyzer (ASAP 2020, Micromeritics, USA). Morphology and microstructure of the samples

were examined by Scanning Electron Microscope (SEM, FEI Quanta 200f, Netherland) equipped with X-ray Energy Dispersive Spectroscope (EDS, LeicaS440). Transmission Electron Microscopy (TEM) imaging and energy-dispersive X-ray spectroscopy (EDX) were performed using Tecnai G2 F30 TEM (FEI, Netherland) at the accelerating voltage of 200 kV. The samples were ball-milled for 24 h and then dispersed on a copper grid.

YSZ electrolytes were prepared according to Fu (Fu et al, 2007). Electrochemical Impedance Spectroscopy (EIS) measurements were performed on a potentiostat/galvanostat (model PARSTAT® 2273, Princeton applied research) in a frequency range of 100 kHz~10 mHz at an amplitude of 10 mV. A three-electrode setup was used with the screen-printed LSCF-GDC composite cathode (firing at 1000°C for 2 h) as working electrode. Measurements were taken in air over a temperature range of 600~800°C, at zero bias. The spectra were fitted using analysis software (ZSimpWin, PerkinElmer Instruments).

Anode-supported fuel cells were prepared according to Fu (Fu et al, 2007). The LSCF-GDC composites were screen-printed on the YSZ electrolyte and fired at 1000°C for 2 h. For the cell tests, 100 sccm H_2 (3 vol% H_2O) and air are used as fuel and oxidant. The cell performance was measured with Arbin equipment (USA).

3 RESULTS AND DISCUSSION

3.1 Crystalline phase and microstructure characterization

The TG-DSC curves of the in-situ synthesized LSCF-GDC xerogel is shown in Figure 1. The analysis shows that most of the organics are removed at temperatures <400°C. The 47.6% weight loss between 161°C and 322°C and the exothermic peak at 231°C are due to the decarboxylation reaction of citric acid and decomposition of free NH_4NO_3. In addition, the 23.8% weight loss between 322°C and 397°C which corresponds to the exothermic peak at 355°C is attributed to decomposition reactions

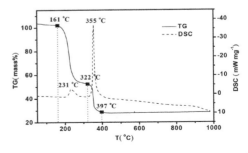

Figure 1. TG-DSC curves for the LSFC-GDC xerogel.

of metal nitrates. From temperatures above 400°C, there is almost no weight loss in the TG curve, indicating that the organic decomposition reactions have been completed <400°C. The unapparent endothermic peaks at temperatures >400°C indicate that LSCF and GDC phases are thoroughly formed.

Figure 2 shows the XRD patterns for the pure LSCF and GDC, as well as the in-situ synthesized composites firing at different temperatures. The LSCF-GDC composites calcined at 400°C displays broad and weak diffraction peaks, showing a non-crystalline phase. After firing at 600°C, broad diffraction peaks emerge corresponding to fluorite GDC and perovskite LSCF phases. In addition, it is obvious from Figure 2e that after firing at 850°C, diffraction peaks of the perovskite LSCF phase show narrow full width at half maximum, while those of the fluorite GDC phase are relatively broad. The result indicates that LSCF grows faster than GDC and exhibits higher crystallinity after firing at 850°C. The fluorite GDC phase also shows narrow FWHM when the firing temperature further increases to 950°C, as shown in Figure 2g. Therefore, it is indicated that LSCF and GDC phases have been successfully co-synthesized by in-situ sol-gel method. The XRD data suggest that there are no observable chemical reactions between LSCF and GDC.

Figure 3a and b show the SEM micrographs of the pure phase LSCF and GDC calcined at 900°C for 2 h. There are agglomerations of LSCF consisting of 100~200 nm small particles. The GDC powders are sheet shaped agglomerations which is typical morphology of the powders prepared by citric sol-gel method (Natile et al., 2008). The SEM micrographs of the in-situ synthesized LSCF-GDC composites calcined at 850~1000°C are shown in Figure 3c~f. As shown in Figure 3c, the composite powder fired at 850°C consists of sheet shaped agglomerations, which is similar to

Figure 3. SEM micrographs of (a) LSCF and (b) GDC prepared by sol-gel method calcined at 900°C for 2 h, and co-synthesized LSCF-GDC composites calcined at (c) 850°C, (d) 900°C, (e) 950°C, (f) 1000°C for 2 h.

that of GDC powder. As the calcination temperature increases, the morphology of the composite powders gradually changes into agglomerations of spherical small particles, due to the grain growth during the high temperature thermal treatment.

Figure 3a and b show the SEM micrographs of the pure phase LSCF and GDC calcined at 900°C for 2 h. There are agglomerations of LSCF consisting of 100~200 nm small particles. The GDC powders are sheet shaped agglomerations which is typical morphology of the powders prepared by citric sol-gel method (Natile et al., 2008). The SEM micrographs of the in-situ synthesized LSCF-GDC composites calcined at 850~1000°C are shown in Figure 3c~f. As shown in Figure 3c, the composite powder fired at 850°C consists of sheet shaped agglomerations, which is similar to that of GDC powder. As the calcination temperature increases, the morphology of the composite powders gradually changes into agglomerations of spherical small particles, due to the grain growth during the high temperature thermal treatment.

Figure 4a and b show the TEM micrographs of the in-situ synthesized and physically mixed LSCF-GDC composites. The images suggest that the in-situ synthesized composite consists of homogeneously dispersed nanoparticles of around 50 nm in diameter, but the physically mixed composite shows a wide particle size distribution. Figure 4c and d show the EDX mappings of the in-situ synthesized and physically mixed LSCF-GDC composites. The in-situ synthesized composites show a uniformly high concentration of La, Sr, Fe and Ce (the doped Co and Gd are not analyzed) throughout the sample, indicating that LSCF and GDC are homogeneously dispersed in the composites. For the physically mixed composites, La, Sr and Fe coexist in hundred domains, whereas Ce exists in sub-hundred domains. It is indicated from Figure 4d that LSCF and GDC phases are obviously distinguished in the physically mixed

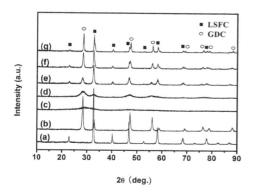

Figure 2. XRD patterns for (a) LSCF, (b) GDC (calcined at 900°C for 2 h) and co-synthesized LSCF-GDC composites calcined at (c) 400°C, (d) 600°C, (e) 850°C, (f) 900°C and (g) 950°C for 2 h.

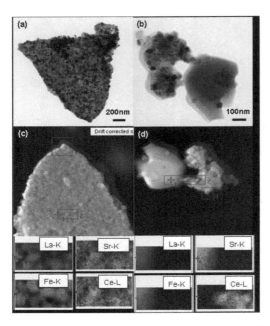

Figure 4. TEM micrographs of (a) co-synthesized and (b) physically mixed LSCF-GDC composites; EDX maps (La Kα, Sr Kα, Fe Kα, Ce Lα) of (c) co-synthesized and (d) physically mixed LSCF-GDC composites.

composites. Furthermore, as shown in Figure 4a and b, the particles size of LSCF is much smaller in the in-situ synthesized composites than that of physically mixed samples, since LSCF and GDC hinder the grain growth of each other during high temperature thermal treatment.

Since the EDX mappings show the phase dispersion at the nanometer scale, the distribution of LSCF and GDC phases in the composites at the micrometer scale are analyzed by the backscatter electron images, shown in Figure 5a and b. For clear images, the composite powders were pressed into pellets followed by sintering at 1200°C for 5 h. The EDS patterns shown in Figure 5c and d correspond to the selected regions in Figure 5a. It is obvious from the images that LSCF (dark part) and GDC (bright part) are clearly distinguished. In addition, compared with the physically mixed composites, LSCF and GDC phases disperse homogeneously in the in-situ synthesized composites, without severe agglomerations of each phase, as demonstrated by Kim et al. (Kim et al., 2007). Furthermore, LSCF and GDC form much larger domains in the physically mixed composites than the in-situ synthesized sample. Thus, based on the above results, it is indicated that LSCF-GDC composites with smaller particle size and homogeneous dispersion are obtained by in-situ sol-gel method.

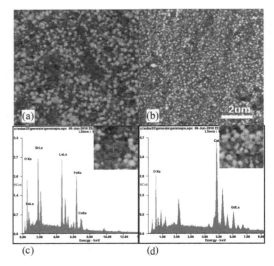

(c)   (d)

Figure 5. Backscatter electron images of (a) co-synthesized and (b) physically mixed LSCF-GDC composites sintered at 1200°C for 5 h; (c) and (d) EDS patterns of LSCF-GDC composites, the inset shows the corresponding selected area.

The in-situ synthesized LSCF-GDC composites are potential to extend the TPB area and enhance the electrochemical performance of the LSCF-GDC composite cathode.

3.2 Electrochemical characterization

The Nyquist plots of the in-situ synthesized LSCF-GDC composite cathode as functions of cathode composition and powder calcination temperature are shown in Figure 6. The complex impedance curves presented typically two depressed arcs and the spectra were evaluated by using the equivalent circuit $LR_{el}(QR_{ct})(QR_d)$. Here, the inductance L is attributed to high frequency artifacts arising from the apparatus. R_{el} is ohmic resistance, and $(R_{ct}Q)$, (R_dQ) represent the High Frequency (HF) and Low Frequency (LF) arc, respectively. The HF arc R_{ct} is probably associated with charge-transfer processes which include oxide ion diffusion in the bulk of cathode and incorporation of oxygen ions from TPB into YSZ lattice. The LF arc R_d is attributed to the diffusion processes which include adsorption-desorption of oxygen, oxygen diffusion at the gas-solid interface and surface diffusion of intermediate oxygen species (Jørgensen et al., 1999, Jiang et al., 2003, Chen et al., 2003).

Shown in Figure 7 are the polarization resistances between the in-situ synthesized LSCF-GDC composite cathodes and the YSZ electrolytes, as a function of cathode composition. It is indicated that the LSCF-GDC composite cathode

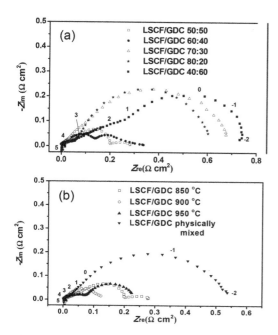

Figure 6. (a) Nyquist plots of co-synthesized LSCF-GDC composite cathodes as a function of LSCF-GDC composite composition; (b) Nyquist plots of physically mixed and co-synthesized LSCF-GDC composite cathodes as a function of powder calcination temperature.

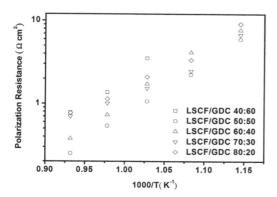

Figure 7. Polarization resistances between LSCF-GDC composite cathode and YSZ electrolyte as determined from impedance spectroscopy under open circuit conditions, as a function of LSCF-GDC cathode composition.

with the weight ratio of 50:50 shows the lowest polarization resistance. At 800 and 750°C, the polarization resistance is estimated to be about 0.25 and 0.59 Ω cm^{-2}, respectively. In addition, it is implied that when the weight ratio of LSCF in the composites is too low or too high (i.e., LSCF/

GDC 40:60 or 80:20), the composite cathodes exhibit large polarization resistances. It is well known that for a cathodic material that possesses significant mixed ionic and electronic conductivity, the oxygen electrode reaction can proceed via the following steps (Adler et al., 1998, Einguedé et al., 2001): (a) oxygen diffusion in the gas phase, (b) dissociative adsorption reaction, (c) charge transfer reaction, (d) oxide vacancy diffusion in the bulk electrode, and (e) ion transfer at the electrode-electrolyte interface. Each of these individual reaction steps contributes to the overall electrode polarization. However, there is usually one step with rate constants remarkably smaller than the others, termed the rate-limiting step. Therefore, for the quick response of the successive steps, sufficient electronic and ionic conductivity is simultaneously required. Thus, when the weight ratio of LSCF in the composite cathodes is too low or too high, the insufficient electronic or ionic transportation results in the poor performance.

Except for the LSCF-GDC cathode composition, the powder calcination temperature also has significant influence on the cathode polarization resistances, due to the great change of the powder morphology. Figure 8 shows the polarization resistances between the in-situ synthesized LSCF-GDC composite cathode and the YSZ electrolyte, as a function of powder calcination temperature. The results suggest that the composite cathode prepared by the in-situ synthesized LSCF-GDC powder fired at 900°C shows the lowest polarization resistance. At 800 and 750°C, the polarization resistance is estimated to be about 0.15 and 0.32 Ω cm^{-2}, respectively. The Nyquist plots of the optimized LSCF-GDC composite cathode at different operation temperatures is shown in Figure 9.

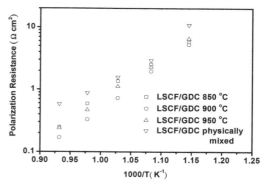

Figure 8. Polarization resistances between LSCF-GDC cathode and YSZ electrolyte as determined from impedance spectroscopy under open circuit conditions, as a function of LSCF-GDC powder calcination temperature and fabrication technique.

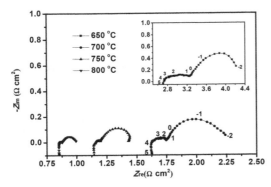

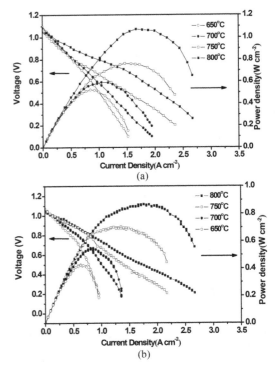

Figure 9. Nyquist plots of the optimized co-synthesized LSCF-GDC composite cathode at test temperatures of 650–800°C. The numbers indicate the frequency logarithm.

Figure 10. Cell voltage and power density as a function of current density for the cells consisting of Ni-YSZ anode, 15 μm thick YSZ electrolyte, and (a) co-synthesized and (b) physically mixed LSCF-GDC composite cathode fired at 1000°C for 2 h.

The cathode using the powder fired at 850°C shows the highest polarization resistance. It is indicated from Figure 3c that the composite powder fired at 850°C consists of sheet shaped agglomerations of small particles, so it is difficult to get a homogeneously dispersed connective network and uniformly porous structure based on the powder. Thus, the poor performance can reasonably be explained. For the in-situ synthesized powder fired at 950°C, severe agglomerations are observed as shown in Figure 3e. The specific surface area of the in-situ synthesized powder fired at 850°C is 11.3 m² g⁻¹, and the value decreases to 7.8 m² g⁻¹ for the powder fired at 950°C. Therefore, the reduced specific surface area results in the decrease of the TPB area and thereby the poor electrochemical performance.

The polarization resistances of the physically mixed LSCF-GDC composite cathode are also provided in Figure 8. As expected, the composite cathode prepared by the physically mixed powders shows higher polarization resistance than the entire in-situ synthesized composite cathodes. The result confirms that the in-situ synthesized LSCF-GDC composites could enhance the electrochemical performance. Based on the analysis above, the main reason is that the uniform dispersion in the in-situ synthesized composites provides sufficient and fast transportation of electrons and ions for the oxygen reduction reaction. Furthermore, the in-situ synthesized LSCF-GDC powders show smaller particle size with larger surface area which is favorable for providing extended TPB area where the oxygen reduction reaction takes place.

The cell voltage and power density of the fuel cell as a function of current density are shown in Figure 10a and b, under various fuel cell operating conditions. The fuel cells used YSZ electrolyte membranes of about 10 μm thick, together with

1.2 mm Ni-YSZ anodes. The Open Circuit Voltages (OCVs) are all above 1.05 V, indicating that the YSZ electrolyte is sufficiently dense and exhibits negligible electronic conductivity. The peak power density of the fuel cell consisting of the in-situ synthesized LSCF-GDC composite cathode reaches 1068, 762, 592, 524 mW cm⁻² at 800, 750, 700 and 650°C, respectively, while the value for the fuel cell consisting of the physically mixed LSCF-GDC composite cathode is correspondingly 871, 692, 548, 425 mW cm⁻². Thus, the fuel cell using the in-situ synthesized composite cathode exhibits higher power output which is very encouraging, especially for operating at intermediate temperatures to dramatically reduce the costs of SOFC systems.

4 CONCLUSIONS

In this work, LSCF-GDC composites for SOFC cathodes have been prepared by in-situ sol-gel method. The EDX mappings and backscatter electron images indicate that LSCF and GDC phases are uniformly dispersed in the in-situ synthesized composites, compared with the physically mixed composites. The influence of experimental conditions on the polarization resistance between the LSCF-GDC composite cathode and the YSZ electrolyte has been systematically studied. The in-situ synthesized LSCF-GDC composite with a weight ratio of 50:50 calcined at 900°C for 2 h shows the lowest polarization resistance of 0.15 and 0.32 Ω cm^{-2} at 800°C and 750°C, respectively, much lower than the values of the physically mixed composites. The peak power density of the fuel cell using in-situ synthesized LSCF-GDC composites is encouraging. Furthermore, the in-situ synthesized method proposed herein shows the potential to be developed as a facile and promising route to prepare many composite materials for various applications.

REFERENCES

Adler S.B. 1998. Mechanism and kinetics of oxygen reduction on porous $La_{1-x}Sr_xCoO_{3-\delta}$ electrodes. Solid State Ionics, 111, 125–134.

Brant M.C., Matencio T., Dessemond L. & Domingues R.Z. 2001. Electrical and Microstructural Aging of Porous Lanthanum Strontium Manganite/Yttria-Doped Cubic Zirconia Electrodes. Chemistry of Materials, 13, 3954–3961.

Chen J., Liang F., Yan D., Pu J., Chi B., Jiang S.P. & Li J. 2010. Performance of large-scale anode-supported solid oxide fuel cells with impregnated $La_{0.6}Sr_{0.4}Co_{0.2}Fe_{0.8}O_{3-delta}$ + Y_2O_3 stabilized ZrO_2 composite cathodes. Journal of Power Sources, 195, 5201–5205.

Chen X.J., Khor K.A. & Chan S.H. 2003. Identification of O_2 reduction processes at yttria stabilized zirconia|doped lanthanum manganite interface. Journal of Power Sources, 123, 17–25.

Fu C., Sun K., Zhang N., Chen X. & Zhu D. 2007. Electrochemical characteristics of LSCF–SDC composite cathode for intermediate temperature SOFC. Electrochimica Acta, 52, 4589–4594.

Jiang S.P. 2003. Issues on development of $(La,Sr)MnO_3$ cathode for solid oxide fuel cells. Journal of Power Sources, 124, 390–402.

Jørgensen M.J., Primdahl S. & Mogensen M. 1999. Characterisation of composite SOFC cathodes using electrochemical impedance spectroscopy. Electrochimica Acta, 44, 4195–4201.

Kim J., Song R., Kim J., Lim T., Sun Y. & D. Shin. 2007. Co-synthesis of nano-sized LSM–YSZ composites with enhanced electrochemical property. Journal of Solid State Electrochemistry, 11, 1385–1390.

Leng Y., Chan S.H. & Liu Q. 2008. Development of LSCF–GDC composite cathodes for low-temperature solid oxide fuel cells with thin film GDC electrolyte. International Journal of Hydrogen Energy, 33, 3808–3817.

Murray E.P., Sever M.J. & Barnett S.A. 2002. Electrochemical performance of $(La,Sr)(Co,Fe)O_3$-$(Ce,Gd)O_3$ composite cathodes. Solid State Ionics, 148, 27–34.

Natile M.M., Poletto F., Galenda A., Glisenti A., Montini T., Rogatis L.D. & Fornasiero P. 2008. $La_{0.6}Sr_{0.4}Co_{1-y}Fe_yO_{3-\delta}$ Perovskites: Influence of the Co/Fe Atomic Ratio on Properties and Catalytic Activity toward Alcohol Steam-Reforming. Chemistry of Materials, 20, 2314–2327.

Park Y.M., Kim J.H. & Kim H. 2011. In situ sinterable cathode with nanocrystalline $La_{0.6}Sr_{0.4}Co_{0.2}Fe_{0.8}O_{3-\delta}$ for solid oxide fuel cells. International Journal of Hydrogen Energy, 36, 5617–5623.

Ringuedé A. & Fouletier J. 2001. Oxygen reaction on strontium-doped lanthanum cobaltite dense electrodes at intermediate temperatures. Solid State Ionics, 139, 167–177.

Wang W.G. & Mogensen M. 2005. High-performance lanthanum-ferrite-based cathode for SOFC. Solid State Ionics, 176, 457–462.

Xu, X. Jing Z., Fan X. & Xia C. 2006. LSM–SDC electrodes fabricated with an ion-impregnating process for SOFCs with doped ceria electrolytes. Solid State Ionics, 177, 2113–2117.

Zhou X., Sun K., Gao J., Le S., Zhang N. & Wang P. 2009. Microstructure and electrochemical characterization of solid oxide fuel cells fabricated by co-tape casting. Journal of Power Sources, 191, 528–533.

Advances in Energy, Environment and Materials Science – Wang & Zhao (Eds)
© *2016 Taylor & Francis Group, London, ISBN 978-1-138-02931-6*

Ni-Co/AlMgO$_x$ catalyzed biodiesel production from Waste Cooking Oil in supercritical CO$_2$

Muyao Xi
China–Japan Union Hospital, Jilin University, Changchun, P.R. China

Wenyu Zhang
Department of Food Science and Engineering, Jilin University, Changchun, P.R. China

Mengni Cui, Xuebin Chu & Chunyu Xi
China–Japan Union Hospital, Jilin University, Changchun, P.R. China

ABSTRACT: Ni-Co/AlMgO$_x$ bimetallic catalysts were evaluated for the production of biodiesel from low quality oil such as Waste Cooking Oil (WCO) containing 15 wt.% free fatty acids. The effects of catalyst preparation conditions such as pH of precipitating solution was the most effective catalyst in simultaneously catalyzing the transesterification of triglycerides and esterification of Free Fatty Acid (FFA) present in WCO to methyl esters. The variation of the CO$_2$ pressure exhibited a large impact on the reaction rate and product distribution. The optimization of reaction parameters with the most active Ni-Co/AlMgO catalyst showed that at 180°C, 1:10 oil to alcohol molar ratio and CO$_2$ pressure 8 MPa, a maximum ester yield of 96 wt.% could be obtained. The catalysts were recycled and reused many times without any loss in activity.

1 INTRODUCTION

Biodiesel fuel from vegetal oil attracts attention as a promising one to be substituted for conventional diesel fuel (Goya, Seal, Saxena 2008; Ma, Hanna 1999; Ranganathan, Narasimhan, Muthukumar 2008); however, the use of biofuels have become controvertial. Biodiesel is usually produced by alkaline transesterification of triglyceride to methyl esters (Gui, Lee, Bhatia 2008; Pinzi, Garcia, Lopez-Gimenez, Luque de Castro, Dorado, Dorado 2009; Achten, Mathijs, Verchot, Singh, Aerts, Muys 2007). The use of cheap low quality feed stocks such as Waste Cooking Oil (WCO), animal fat and tall oil instead of refined vegetable oil will help in improving the economical feasibility of biodiesel. Currently, the inexpensive and large quantity of WCO from households and restaurants are collected and used as either animal feed or disposed causing environmental pollution. Thus, WCO offers signifycant potential as an alternative low–cost biodiesel feedstock which could partly decrease the dependency on petroleum–based fuel (Becker, Makkar 2008; Makkar, Becker 2009; Srivastava, Prasad 2004).

Conventionally, biodiesel is produced using homogeneous mineral acid or alkali catalysts. However, these catalysts have major disadvantages, where the cost of production increases due to

catalyst being consumed and a series of purification steps. When FFAs are present, they react with the homogeneous alkali catalysts, form unwanted soap by-products and deactivate the catalyst. Thus, heterogeneous Catalysis has been attracting much attention these years (Srivastava, Prasad 2004; Helwani, Othman, Aziz, kim, Fernando 2009). A range of catalysts including Mg/Al hydrotalcites, alkali nitrate and alkali carbonate-loaded Al$_2$O$_3$, polymer resins, sulfated-tin and zirconia oxides and tungstated-zirconia have also been reported. Most of these catalysts lose their activity on recycle and/or require pretreatment of the feedstock to remove the FFAs and water.

ScCO$_2$ is an environmentally benign reaction medium, which is cheap, nontoxic, and non-flammable, its mild critical point (Tc = 304 K, Pc = 7.29 MPa) and variable physical properties are attractive for its application in the chemical reactions as a replacement of the conventional organic solvents.

In this work, in an attempt to develop a high activity catalyst with long cycle life, the double-metal catalysts, Ni-Co/AlMgO$_x$ are synthesized and evaluated for biodiesel preparation from waste cooking oil. Also, Influence of various reaction parameters such as molar ratio of WCO to alcohol and CO$_2$ pressure was studied in the present investigation. The preparation parameter of the catalyst was optimized. It indicates that the precipitation process is sensitive to the pH value of the precipitating solution. The following research is to determine the optimized pH that leads to the best catalyst performance.

2 EXPERIMENT

2.1 Preparation of the catalyst

$Ni\text{-}Co/AlMgO_x$ bimetallic catalysts were prepared by using conventional coprecipitation method with aqueous ammonia solution as precipitating agent. The pH value of the precipitating solution was controlled in the desired rang. The precipitation was carried out at room temperature (22–24°C). The precipitate was washed with de-ionized water, dried overnight in the air at 120°C, and then calcined in the air at 800°C for 6 h.

2.2 Characterization

The composition of the catalysts was analysed at the Saskatchewan Research Council Analytical Laboratory using Inductively Coupled Plasma Mass Spectrometry (ICP-MS). The BET surface area, porous volume and average pore diameter were measured by N_2 adsorption at the temperature of −196°C using Micromeritics ASAP 2010. Approximately 0.2 g catalyst was used for each analysis. The degassing temperature was 200°C to remove the moisture and other adsorbed gases from the catalyst surface. X-Ray powder Diffraction (XRD) analysis was conducted using a Rigaku/Rotaflex Cu rotating anode X-ray diffraction instrument under the ambient drying condition.

2.3 Synthesis of biodiesel

The simultaneous transesterification and esterification of WCO containing 15 wt.% free fatty acids in a 500 cm^3 Teflon-lined high-pressure stainless steel batch reactor. A certain amount of WCO (100 g), Anhydrous methanol and catalyst were then loaded to the reactor, and the reactor was sealed and flushed three times with 2 MPa CO_2 to remove the air. After flushing, the reactor was heated up to 180°C, stirring speed of 600 rpm optimized previously. Then liquid CO_2 was compressed into the reactor using a high-pressure liquid pump to the desired pressure. The reaction mixture was stirred continuously with a magnetic stirrer for a certain time. The reactions were carried out for a period of 8 h unless otherwise stated. The products were analyzed using Gel Permeation Chromatography (GPC) equipped with a RI detector and two 300 mm −7.8 mm phenogel columns connected

in series. Tetrahydrofuran (THF) was used as a mobile phase and the triglycerides, diglycerides, monoglycerides and methyl esters in the product were quantified by comparing the peak areas of their corresponding standards. The percentage free fatty acid content was determined from the acid value following the AOCS method.

2.4 Catalysts reusability tests

The catalysts separated from the reaction mixture by filtration were initially washed with hexane to remove non-polar compounds such as methyl esters on the surface. Further, the catalysts were washed with methanol to remove polar compounds such as glycerol and finally dried at 85°C overnight.

3 RESULTS AND DISCUSSION

3.1 pH effect

It has been known that the pH value of the solution significantly affect phase states of metal hydroxide using coprecipitation method. We research the effect of precipitating solution pH on catalyst activity. When a metal salt is dissolved in water, the metal ions are solvated by water molecules. The coordinated water molecules can be deprotonated to produce hydroxide or even oxide species. The deprotonation processed can be expressed by the following equilibria which depend on the charge of the metal and the pH of the solution:

$$[M\text{-}OH_2]^{n+}\text{-------}[M\text{-}OH]^{(n-1)+}\text{-----------}[M\text{=}O]^{(n-2)+}$$

The pH values significantly affect the chemical compositions and states of the catalysts. It is interesting to see that precipitation only occurred in the pH range of 8.0–9.0 when the pH value of the precipitating solution was adjusted in a broader range. Table 1 shows the properties of bimetal catalysts synthesized in various pH value of the precipitating solution. Beyond that range, sol-gel was obtained. 5 samples were prepared at 5 pH levels, #1–3 at pH 8.6, 8.3 and 8.8 respectively; and #4

Table 1. Properties of synthesized catalysts.

Catalyst	Phase state	BET surface (m^2/g)	Pore volume (ml/g)
1# (pH = 8.6)	Precipitation	155.84	0.328
2# (pH = 8.3)	Precipitation	126.15	0.307
3# (pH = 8.8)	Precipitation	118.46	0.319
4# (pH = 9.5)	Sol-gel	75.85	0.245
5# (pH = 7.5)	Sol-gel	68.89	0.224

and 5 at pH = 9.5 and 7.5, respectively. As shown in Figure 1, the precipitated catalyst had similar catalytic performance, highly active and stable. But catalysts made out of the pH range did not.

The BET surface area, porous volume and average pore diameter were measured (Table 1). The result shows that the precipitated catalysts possess higher surface area, larger pore volume and smaller pore size, compared with the sol-gel catalysts, which would affect the catalyst performance. It have been found in our previous work that the superior performance of Ni-Co catalysts comes from high BET surface area, small pore diameter, relative easy reducibility, good metal dispersion, high metallic surface.

3.2 Molar ratio of oil to alcohol

The ratio of methanol to oil molar is one of the most important parameters that affects the yield of methyl esters. Theoretically, the transesterification of vegetable oil requires three moles of methanol per mole of triglyceride. Since the transesterification of triglyceride is a reversible reaction, the excess of methanol shifts the equilibrium towards the direction of ester formation. It is reported that in order to shift the equilibrium towards forward direction, use of high molar ratios of oil to alcohol such as 1:40 and even 1:275. In the present work, an increase in oil to alcohol molar ratios from 1:6 to 1:12 resulted in a significant effect on the ester yield (Fig. 2). The yield of methyl esters increased from 78 to 96 wt.% corresponding to a FFA content of about 1 wt.% after 8 h reaction time. The excess methanol used in the reaction can be collected by distillation and reused.

3.3 Influence of CO_2 pressure

Figure 3 shows the results of the ester yield at different pressures of CO_2. The variation of the CO_2

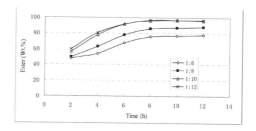

Figure 2. Effect of oil to alcohol molar ratio on ester yield. Reaction conditions: reaction temperature 180°C, stirring speed 600 rpm, CO_2 pressure 8 MPa.

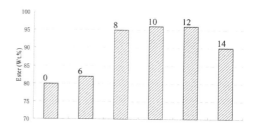

Figure 3. Influence of CO_2 pressure. Reaction conditions: reaction temperature 180°C, stirring speed 600 rpm, molar ratio of oil to alcohol 1:10.

pressure exhibited a large impact on the reaction rate and product distribution, the yield to ester was 82% in the absence of solvent or at a lower CO_2 pressure of 6 MPa (subcritical region), while it increased up to 96% at CO_2 pressures above 8 MPa (supercritical region). The yield to the ester changed very slightly at a pressure range of 8–12 MPa. When CO_2 pressure was raised up to 14 MPa or higher, both the yield to the ester and reaction rate decreased. These results may be explained with the dilution effect of $ScCO_2$.

4 CONCLUSIONS

In a word, a new and cheap Ni-Co/AlMgO$_x$ bimetallic catalysts has been successfully used in the biodiesel from low quality oil such as Waste Cooking Oil (WCO) containing 15 wt.% free fatty acids in scCO$_2$. The effects of catalyst preparation conditions such as pH of precipitating solution was the most effective catalyst. The catalysts were recycled and reused many times without any loss in activity. The optimization of reaction parameters with the most active Ni-Co/AlMgO catalyst showed that at 180°C, 1:10 oil to alcohol molar ratio and and CO_2 pressure 8 MPa a maximumester yield of 98 wt.% could be obtained. The findings have potential for wide spread applications in academic and

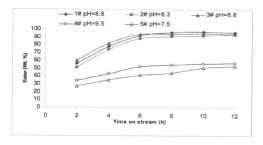

Figure 1. Influence of pH value of the precipitating solution on catalytic. Reaction conditions: reaction temperature 180°C, molar ratio of oil to alcohol 1:10, stirring speed 600 rpm, CO_2 pressure 8 MPa.

industrial scale production of biodiesel from low quality feedstock's containing high free fatty acid.

REFERENCES

Achten W.M.J., Mathijs E., Verchot L., Singh V.P., Aerts R., Muys B. 2007. Jatropha biodiesel fueling sustainability? *A perspective. Biofuels Bioprod Biorefin*; 1:283e91.

Becker K., Makkar H.P.S. 2008. Jatropha curcas: a potential source for tomorrow's oil and biodiesel. *Lipid Technol*; 20:104e7.

Goyal H.B., Seal D., Saxena R.C. 2008. Bio-fuels from thermochemical conversion of renewable resources: a review. *Renew Sust Energy Rev*; 12:504–17.

Gui M.M., Lee K.T., Bhatia S. 2008. Feasibility of edible oil vs. nonedible oil vs. *waste edible oil as biodiesel feedstock. Energy*; 33:1646–53.

Helwani Z., Othman M.R., Aziz N., kim J., Fernando W.J.N. 2009. Solid catalysis for transesterification of triglycerides with methanol. Appl Catal A; 363:1e10.

Ma F., Hanna M.A. 1999. Biodiesel production: a review. *Bioresour Technol*; 70:1–15.

Makkar H.P.S., Becker K. 2009. Jatropha curcas, a promising crop for the generation of biodiesel and value-added copro-ducts. *Eur J Lipid Sci Technol*; 111:773e87.

Pinzi S., Garcia I.L., Lopez-Gimenez F.J., Luque de Castro M.D., Dorado G., Dorado M.P. 2009. The ideal vegetable oil-based biodiesel composition: a review of social, economical and technical implications. *Energy Fuels*; 23:2325–41.

Ranganathan S.V., Narasimhan S.L., Muthukumar K. 2008. An overview of enzymatic production of biodiesel. *Bioresour Technol*; 99:3975–81.

Srivastava A., Prasad R. 2000. Triglycerides-based diesel fuels. *Renew Sust Energy Rev*; 4:111e33.

Advances in Energy, Environment and Materials Science – Wang & Zhao (Eds)
© 2016 Taylor & Francis Group, London, ISBN 978-1-138-02931-6

Special wettability of locust wing and preparation of biomimetic polymer film by soft lithography

Gang Sun & Yan Fang
School of Life Science, Changchun Normal University, Changchun, Jilin, China

ABSTRACT: The special wettability and micro/nano-structure of locust wing were investigated by a video-based Contact Angle (CA) meter and a Scanning Electron Microscope (SEM). The wetting mechanism was discussed from the perspective of biological coupling. Locust wings were used as biomimetic templates to fabricate multi-functional polymer (PDMS, Polydimethylsiloxane) films by soft lithography. The natural wing surface exhibits hierarchical micro/nano-structures and high adhesive superhydrophobicity, the water CA is 154°, the water Sliding Angle (SA) is higher than 180°. The prepared polymer film reproduces faithfully the surface microstructures of the bio-template, and displays a good hydrophobicity and high adhesion (CA 146°, SA > 180°). The complex wettability of the natural and artificial locust wing surfaces ascribes to the coupling effect of material element (hydrophobic composition) and structural element (rough micro-morphology). Locust wing can be employed as a biomimetic template for design and fabrication of special functional surface.

1 INTRODUCTION

The interfacial material with special properties and functions is attracting more and more attention due to valuable theoretical importance and application potential in industrial and domestic fields (Wang & Jiang 2007). In recent years, many researches on superhydrophobic and self-cleaning surfaces have been focused on lotus leaf, on which a droplet displays a special Cassie state called "lotus state" (low adhesive superhydrophobicity) (Feng et al. 2002). However, little has been focused on another special contact state of a droplet on hydrophobic surface called "Gecko state" (high adhesive superhydrophobicity). The superhydrophobic material with high adhesion can be used as a "mechanical hand" in no-loss microfluidic transport (Wei et al. 2009). After millions of years of evolution, many animals and plants have possessed distinctive body surfaces which are superhydrophobic, self-cleaning, anti-adhesive, anti-corrosive, drag reducing and anti-wearing (Yao et al. 2011). Insect wing surface, one of the most complicated three-dimensional periodical substrates in nature, has become a popular biomimetic fabrication template because of its excellent characteristics. Inspired by the complex hierarchical morphologies in biological systems, some biomimetic materials have been prepared by photolithography, vapor deposition, electrospin, interference lithography, plasma treatment and sol-gel technique (Bhushan & Her 2010). We have done some work on microstructure and wettability of butterfly wing surface (Fang et al. 2014, Fang et al. 2015, Fang et al. 2008, Fang et al. 2007, Sun et al. 2009), in this work it was found that the locust wing is of high adhesive superhydrophobicity. The locust wing was used as a template to fabricate biomimetic polymer films by soft lithography, which faithfully reproduced the wing micro-morphology and showed high adhesive hydrophobicity. This work may bring interesting insights into biomimetic design of micro-controllable superhydrophobic surface and fabrication of novel self-cleaning material.

2 MATERIALS AND METHODS

2.1 *Materials and reagents*

Locust specimens (*Atractomorpha lata*) were collected in Changchun City, Jilin Province of northeast China. The wings were cleaned, desiccated and flattened, then cut into pieces of 8 mm × 8 mm from the hind wing. The PVA (polyvinyl alcohol) used in the first soft transfer was purchased from Sinopharm Chemical Reagent Co., Ltd, China. The PDMS (Sylgard 184 Silicone Elastomer Kit) used in the second soft transfer was purchased from Dow Corning, USA.

2.2 *Preparation of biomimetic polymer film*

In the first soft transfer, the locust wing piece was affixed to a glass slide with double-sided adhesive tape as the primary template. PVA solution (mass concentration 10%) was dropped homogeneously on the wing surface. After 24 h under the ambient temperature of $(25 \pm 1)°C$, the PVA film with the inverse structure of the wing surface was peeled carefully off the primary template. In the second soft transfer, the prepared PVA film was affixed to a glass slide with double-sided adhesive tape as the secondary template. The prepolymer PDMS and curing agent were mixed in a volume ratio of 10:1. The PDMS mixture was dropped homogeneously on the surface of PVA film, then degassed in a vacuum chamber. Having been baked at 120°C for 1.5 h, the PDMS was solidified and could be peeled off the secondary template with tweezers. The surface structures on the prepared PDMS film were similar to those on the natural locust wing.

2.3 Measurement of wetting angles

Using an optical CA measuring system (DataPhysics OCA20, Germany), the CA of water droplet on the wing surface was measured by sessile drop method at room conditions of $(25 \pm 1)°C$ and relative humidity of approximately 80%. The water droplet was dripped on the sample table in a horizontal position, then the inclination degree of the table was raised 1° each time until the droplet rolled off freely. The inclination degree of the table was recorded as the SA value.

2.4 Characterization of surface microstructure

After gold coating (approx. 15 nm) by an ion splash instrument (Hitachi E-1045, Japan), the natural and artificial wing surfaces were observed, photographed and characterized under a SEM (Hitachi SU8010, Japan).

3 RESULTS AND DISCUSSION

3.1 Micro-morphology of the natural locust wing surface

The locust wing surface possesses complicated hierarchical microstructures. The wing veins are very clear as grids, constituting the primary microstructure of wing surface [Fig. 1(a)]. The micrometric pillar gibbosities with uniform size distribute regularly and densely in the vein grids, constituting the secondary microstructure of wing surface [Fig. 1(b)]. The diameter of gibbosity is 4.2~9.8 μm, the height is 4.7~7.5 μm, the spacing is 9.8~14.2 μm. The numerous nano corrugations between pillar gibbosities constitute the tertiary microstructure of wing surface [Fig. 1(c)].

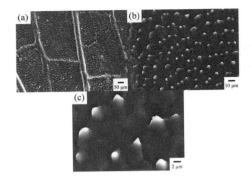

Figure 1. Hierarchical rough microstructure of the natural locust wing surface (SEM images). *(a) Primary microstructure (wing veins); (b) secondary microstructure (pillar gibbosities); (c) tertiary microstructure (nano corrugations).

3.2 Micro-morphology of the artificial biomimetic surface

After the first soft transfer, the prepared PVA film exhibits the inverse structure of the wing surface. The reverse hollow arrays [Fig. 2(a)] and pillar gibbosities [Fig. 2(b), 2(c)] can be found on the PVA film, including macrogrooves and porous microstructures. The width of the macrogrooves is 19.7 μm. The diameter of the porous microstructures is 6.8 μm, the spacing is 11.5 μm.

After the second soft transfer, the PDMS films were prepared. The macro/micro/nano structures on the natural wing surface are retained on the PDMS film. The wing vein is very clear [Fig. 3(a)]. The length of the rectangular vein grid ranges from 535 nm to 720 nm, the width from 165 nm to 230 nm. The size and distribution of primary and secondary structures are the same as those on the natural locust wing surface [Fig. 3(b)]. The diameter and spacing of the gibbosities are 6.2 μm and 11.6 μm, respectively. Some nanostructures can also be observed [Fig. 3(c)]. The artificial wing surface duplicates the structures on the natural locust wing surface very well.

3.3 Complex wettability of the natural and artificial locust wing surfaces

The water droplet stands on locust wing surface as a sphere. The CA of water droplet on the locust wing is 154°, showing that locust wing surface is a natural bio-surface with superhydrophobicity. The diameter of a water droplet is about 2.1 mm, far outweighing the spacing (9.8~14.2 μm) of pillar gibbosity. The droplet can neither enter totally the grooves between gibbosities, nor contact fully with wing surface. A composite contact occurs.

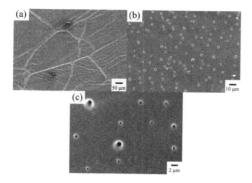

Figure 2. Micro-morphology of the prepared PVA films (SEM images). *(a) Inverse structure of wing veins on the PVA film; (b), (c) inverse structures of pillar gibbosities on the PVA film.

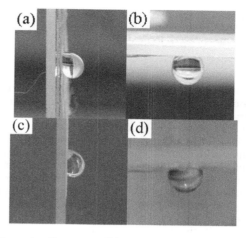

Figure 4. Contact conditions of a water droplet on the natural locust wing and artificial PDMS film. *(a), (b) On the vertical and inverted locust wing; (c), (d) on the vertical and inverted PDMS film.

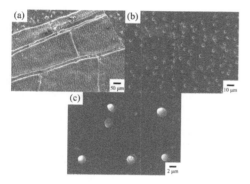

Figure 3. Micro-morphology of the prepared PDMS films (SEM images). *(a) Artificial wing veins on the PDMS film; (b), (c) artificial pillar gibbosities on the PDMS film.

Owing to the chemical composition (waxy layer), the wing surface can achieve hydrophobicity (CA 105°); owing to the multiple-dimensional rough microstructure, locust wing surface can achieve superhydrophobicity (CA over 150°). The combination of hydrophobic composition and multidimensional rough microstructures contributes to the superhydrophobicity of the wing surface.

Meanwhile, the water droplet displays high adhesion on locust wing surface, and appears as "Gecko state". The droplet does not leave wing surface at any angle of inclination, even verticalized or inverted [Fig. 4(a), (b)]. This property resembles that of peanut leaf and rose petal (Feng et al. 2002). Apart from superhydrophobicity, a "Gecko state" droplet exhibits high adhesion and can be pinned on the substrate effectively; whereas a "Lotus state" droplet exhibits low adhesion and extremely small CA hysteresis. The wing surfaces of locust

and butterfly (both belonging to Class Insecta) show high adhesive superhydrophobicity and low adhesive superhydrophobicity, respectively, which results from their different surface microstructures. The spacing between micrometric gibbosities on locust wing surface (averagely 13.46 μm) is 7.7 times of that on butterfly wing surface (averagely 1.74 μm), the density of micrometric gibbosity on locust wing surface is far smaller than that on butterfly wing surface (Fang et al. 2007). On a locust wing, the micrometric structures can be partially wetted by water droplets. Relatively less air is trapped and sealed between water droplet and wing surface. As a water droplet is removed from the locust wing surface, negative pressure is produced due to the exchange of confined air, so high adhesive force is induced. While on a butterfly wing, a water droplet stands on the tips of micrometric gibbosities, much air remains under the droplet. The solid-liquid-gas Triple Contact Lines (TCL) are expected to be contorted and extremely unstable, leading to the low adhesion between the water droplet and the butterfly wing surface.

The prepared rough PDMS film displays a high hydrophobicity, the water CA on it is 146°. Whereas, the water CA on the smooth PDMS film is just 116°. The artificial PDMS film also has high adhesion [Fig. 4(c), (d)]. The prepared PDMS film does not achieve the superhydrophobicity of the natural locust wing because of the partial reproduction of locust wing nanostructure on it. It can be concluded that the superhydrophobicity of the locust wing is determined cooperatively by micro- and nano—structures. Up to now the reproduction of nano-structure is one of the methodological difficulties in the field of biomimetic materials.

4 CONCLUSIONS

Locust wing surface possesses multi-dimensional rough microstructures (primary structure, secondary structure, and tertiary structure) and high adhesive superhydrophobicity (CA 154°, SA > 180°). Using locust wing as template, polymer (PDMS) films with multifunction were prepared successfully by a two-step soft imprint method. The prepared PDMS films reproduce faithfully the macro/microstructures of the bio-template and exhibit a good hydrophobicity and a high adhesion (CA 146°, SA > 180°). The fabrication method in this paper is simple, efficient and reliable, without the need for costly apparatus and processing. The cooperation of the multi-dimensional surface structures and hydrophobic composition is the origin for the special wettability of natural locust wing surfaces and artificial polymer film. It is believed that locust wing can serve as a biomimetic template contributing to the next generation of bio-inspired multi-functional devices. This work may bring interesting insights for preparation of micro-controllable superhydrophobic surface and no-loss microfluidic transport channels.

ACKNOWLEDGEMENTS

This work was financially supported by the National Natural Science Foundation of China (50875108), the Natural Science Foundation of Jilin Province, China (201115162), Science and Technology Project of Educational Department of Jilin Province, China (2009210, 2010373, 2011186). Dr. Prof. Yan Fang is the corresponding author of this paper.

REFERENCES

Bhushan, B. & Her, E.K. 2010. Fabrication of superhydrophobic surfaces with high and low adhesion inspired from rose petal. *Langmuir* 26(11): 8207–8217.

Fang, Y., Sun, G. & Bi, Y.H. 2014. Preparation and characterization of hydrophobic nano silver film on butterfly wings as bio-template. *Chemical Research in Chinese Universities* 30(5): 817–820.

Fang, Y., Sun, G. & Cong, Q. 2008. Effects of methanol on wettability of the non-smooth surface on butterfly wing. *Journal of Bionic Engineering* 5(2): 127–133.

Fang, Y., Sun, G., Bi, Y.H. & Zhi, H. 2015. Multiple-dimensional micro/nano structural models for hydrophobicity of butterfly wing surfaces and coupling mechanism. *Science Bulletin* 60(2): 256–263.

Fang, Y., Sun, G., Wang, T.Q. & Cong, Q. 2007. Hydrophobicity mechanism of non-smooth pattern on surface of butterfly wing. *Chinese Science Bulletin* 52(5): 711–716.

Feng, L., Li, S.H., Li, Y.S., Li, H.J., Zhang, L.J., Zhai, J., Song, Y.L., Liu, B.Q., Jiang, L. & Zhu, D.B. 2002. Super-hydrophobic surfaces: from natural to artificial. *Advanced Materials* 14(24): 1857–1860.

Sun, G., Fang, Y., Cong, Q. & Ren, L.Q. 2009. Anisotropism of the non-smooth surface of butterfly wing. *Journal of Bionic Engineering* 6(1): 71–76.

Wang, S.T. & Jiang, L. 2007. Definition of superhydrophobic states. *Advanced Materials* 19(21): 3423–3424.

Wei, P.J., Chen, S.C. & Lin, J.F. 2009. Adhesion forces and contact angles of water strider legs. *Langmuir* 25(3): 1526–1528.

Yao, X., Song, Y.L. & Jiang, L. 2011. Applications of bio-inspired special wettable surfaces. *Advanced Materials* 23(6): 719–734.

Advances in Energy, Environment and Materials Science – Wang & Zhao (Eds)
© *2016 Taylor & Francis Group, London, ISBN 978-1-138-02931-6*

Novel graphene-based composites for adsorption of organic pollutants in wastewater treatment

Mengyou Zhou
College of Chemistry and Life Sciences, Zhejiang Normal University, Jinhua, China

Jianrong Chen
College of Geography and Environmental Sciences, Zhejiang Normal University, Jinhua, China

ABSTRACT: Removal of various pollutants, including both inorganic and organic compounds from the environment, especially in the wastewater, is a big challenge. These pollutants terribly threaten both our environment and human being. With the development of science and technology, there are a number of physical, chemical, and biological methods to control the water pollution successfully. And adsorption technique is usually simple and can work effectively. The adsorption capacity of materials depends on its porous structure and surface property. Graphene-based composite is a kind of new type of nanometer material, a novel material. Graphene shows a high adsorption capacity. And in this paper, recent research achievements were reviewed on the application of graphene-based composites in wastewater treatment. It presented a high adsorption capacity to some kinds of organic pollutants.

1 INTRODUCTION

Since the industrial revolution and agricultural activities, rapid developments have largely contributed to the severe pollution in our environment. Because there are a lot of organic pollutants discharging from industry and households, these pollutants terribly threaten both the eco-environment and human health. So, we have to pay more attention to the ecological balance and human life, especially the wastewater. Generally speaking, wastewater includes domestic wastewater, agricultural wastewater, and industrial sewage (Shi et al. 2014). In wastewater, organic pollutants, especially dyes, pesticides, antibiotics, phenols, aromatic amines, oils, and so on, are needed to be removed. They are much harmful to our water system and human being.

With the development of science and technology, there are a number of physical, chemical, and biological methods to control the water pollution successfully. Among these various ones, adsorption is widely used and considered as a simple and easy way. It has high efficiency to remove different types of pollutants, including organic and inorganic pollutants from wastewater. In addition, adsorption does not result in secondary pollution by producing harmful substances during the process (Wang et al. 2013).

Graphene is a two-dimensional nanomaterial that has between one and ten layers of sp2-hybridized carbon atoms arranged in six-membered rings (Lü et al. 2012). It has unique morphology, chemical, thermal, electronic, and mechanical properties. Lots of efficient adsorbents are based on nanomaterials especially graphene, currently. So, these graphene-based adsorbents are developed quickly to meet the increasing requirement for high performance materials for organic pollutants removal, because they show the high adsorption capacities.

There have been many reports on the adsorption of organic pollutants in wastewater treatment. The purpose of this paper was to summarize the applications of various kinds of graphene-based adsorbents for organic pollutants.

2 APPLICATIONS OF GRAPHENE-BASED COMPOSITES FOR ADSORPTION OF ORGANIC POLLUTANTS

Adsorption is the surface phenomenon where pollutants are adsorbed onto the surface of an adsorbent via physical and/or chemical forces. It depends on many factors, such as temperature, solution pH, etc. In recent years, researchers have tried their best to develop various kinds of grapheme-based composites for adsorption of organic pollutants. Kim et al. (2015), Tang et al. (2013), Wang et al. (2013) reported the adsorption of isotherms and kinetics of dyes, pesticides on RGO materials. Diagboya and co-workers (2014) showed that graphene oxide–tripolyphosphate material (GPM) had higher potency for adsorption of cationic dyes than anionic dyes, and the adsorption process was through electrostatic and π–π interactions. Zhang et al. (2014), Zhou et al. (2014), Yang et al. (2013) reported that magnetic GO composite material could be used for the efficient removal of dyes, naphthalene and its derivatives from water. Chi and colleagues (2015) showed high adsorption capacities of HGAs for identical oils and organic solvents.

Table 1. Applications of graphene-based composites for adsorption of organic pollutants.

Adsorbents	Organic pollutant	pH	Temp.	Adsorption capacity	References
3D rGO	AC 1	7	353 K	277.01 mg/g	Kim et al. 2015
GPM	AM	10	288 K	1165.6 mg/g	Diagboya et al. 2014
GO/PPy	Aniline	6	338 K	7.57 mmol/g	Hu et al. 2015
GO-SA-M	AO	5	298 K	4.5 mmol/g	Sun et al. 2013
GO	AO	–	323 K	229.8 mg/g	Fiallos et al. 2015
GPM	AY	10	288 K	138.5 mg/g	Diagboya et al. 2014
Fe_3O_4@PDDA/GOx@DNA	BDE28	5–8	298 K	49.1 mg/g	Gan et al. 2014
GO/Fe_3O_4	Bisphenol A	6	298 K	123.2 mg/g	Zhang et al. 2014
GO/poly-dopamine	BF	–	–	1700 mg/g	Dong et al. 2014
GPM	BF	10	288 K	2722.7 mg/g	Diagboya et al. 2014
RGO/Fe_3O_4	Ciprofloxacin (CIP)	6.2	298 K	18.22 mg/g	Tang et al. 2013
G-KOH	CIP	~7	298 K	194.6 mg/g	Yu et al. 2014
GO	Chlorpyrifos (CP)	–	298 K	1200 mg/g	Maliyekkal et al. 2013
RGA	Crude oil	–	–	169 mg/g	Li et al. 2012
Fe_3O_4@rGO	Crystal violet	–	–	64.93 mg/g	Sun et al. 2014
RGO	4,4'-DCB	7	–	1552 mg/g	Wang et al. 2013
3D HGAs	Diesel oil	5–7	–	25 times	Chi et al. 2015
GO	Endosulfan (ES)	–	295 K	1100 mg/g	Maliyekkal et al. 2013
3D HGAs	Gasoline	5–7	–	21 times	Chi et al. 2015
(RGO@ZnO)10/ PET Membrane	Lubricant oil	–	–	up to 23 times its own weight	Wang et al. 2014
GO/Fe_3O_4	Methylene Blue (MB)	12	293 K	246 mg/g	Zhou et al. 2014
3D rGO	MB	7	353 K	302.11 mg/g	Kim et al. 2015
RL-GO	MB	11	318 K	581.40 mg/g	Wu et al. 2014
rGO-30	MB	6	298 K	746 mg/g	Russo et al. 2015
GO-SA	MB	–	303 K	833.3 mg/g	Ma et al. 2014
r-GO-PIL	MB	–	298 K	1910 mg/g	Zhao et al. 2015
GPM	MB	10	288 K	2761.5 mg/g	Diagboya et al. 2014
Fe_3O_4/β-CD/GO	Malachite green	7	318 K	990.10 mg/g	Wang et al. 2015
RGO	Methyl green	9	298 K	5.167 mmol/g	Sharma et al. 2014
GO	Malathion (ML)	–	298 K	800 mg/g	Maliyekkal et al. 2013
GPM	MO	10	288 K	142.9 mg/g	Diagboya et al. 2014
GO/poly-dopamine	MV	–	–	2100 mg/g	Dong et al. 2014
GPM	MV	10	288 K	2627.5 mg/g	Diagboya et al. 2014
GO/FeO·Fe_2O_3	1-Napthol	7	283.15 K	2.70 mmol/g	Yang et al. 2013
GO/FeO·Fe_2O_3	1-Napthylamine	7	283.15 K	2.85 mmol/g	Yang et al. 2013
GO/FeO·Fe_2O_3	Naphthalene	7	303.15 K	5.72 mmol/g	Yang et al. 2013
RGO/Fe_3O_4	Norfloxacin (NOR)	6.2	298 K	22.2 mg/g	Tang et al. 2013
RGO-SO_3H/Fe_3O_4	Neutral Red (NR)	6	298 K	216.8 mg/g	Wang et al. 2013
ss-GF	Oil	7	–	~196 times	Zhu et al. 2015
Grapheme-CNT aerogel	Oil	5.6	298 K	21–35 times	Kabiri et al. 2014
GO/PPy	Phenol	6	338 K	3.31 mmol/g	Hu et al. 2015
MCG	PPD	8	318 K	1102.6 mg/g	Wang et al. 2015
S2P1 hydrogel	P2R	6.2	298 K	1.05 mg/g	Li et al. 2015
3D HGAs	Pump oil	5–7	–	45 times	Chi et al. 2015
S2P1 hydrogel	RB	6.2	298 K	0.98 mg/g	Li et al. 2015
GO-magnetic chitosan	Reactive Black 5	3	298 K	391 mg/g	Travlou et al. 2013
Fe_3O_4/RGO(M2)	RhB	5.3	303 K	432.91 mg/g	Qin et al. 2014
RGO-SO_3H/Fe_3O_4	Safranine T (ST)	6	298 K	199.3 mg/g	Wang et al. 2013
3D HGAs	Soya oil	5–7	–	50 times	Chi et al. 2015
RGO-SO_3H/Fe_3O_4	Victoria Blue (VB)	6	298 K	200.6 mg/g	Wang et al. 2013

Table 1 gives a brief summary of recent reports on the adsorption of organic pollutants by graphene-based composites.

3 CONCLUSION

The removal of various organic pollutants from the wastewater is a challenge. Many graphene-based materials have been extensively explored for adsorption applications. Different materials are with different adsorption capacities. Compared with different adsorbents or different study conditions, we will have a large adsorption capacity, high selectivity, low-cost way. Efforts should be devoted to developing methods like this. Therefore, we need to correctly understand the adsorption mechanisms, assess the adsorption performance under multi-component pollutants, and recover the adsorbents for some economic purposes.

ACKNOWLEDGMENT

This research was supported by National Natural Science Foundation of China (No. 21275131), and Zhejiang Environmental Protection Bureau (No. 2013A025).

REFERENCES

Chi, C.X., K. Zhang, Y.B. Wang, S.H. Zhang, X.S. Liu, & X. Liu (2015). 3D hierarchical porous graphene aerogels for highly improved adsorption and recycled capacity. *Mater. Sci. Eng., B.* 194.

Diagboya, P.N., B.I. Olu-Owolabi, D. Zhou, & B.H. Han (2014). Graphene oxide–tripolyphosphate hybrid used as a potent sorbent for cationic dyes. *Carbon.* 79, 174–182.

Dong, Z.H., D. Wang, X. Liu, X.F. Pei, L.W. Chen, & J. Jin (2014). Bio-inspired surface-functionalization of graphene oxide for the adsorption of organic dyes and heavy metal ions with a superhigh capacity. *J. Mater. Chem., A.* 2, 5034–5040.

Fiallos, D.C., C.V. Gómez, G.T. Usca, D.C. Pérez, P. Tavolaro, & G. Martino (2015). Removal of acridine orange from water by graphene oxide. *International Conferences and Exhibition on Nanotechnologies and Organic Electronics (NANOTEXNOLOGY 2014).* 1646, 38–45.

Gan, N., J.B. Zhang, S.C. Lin, N.B. Long, T.H. Li, & Y.T. Cao (2014). A novel magnetic graphene oxide composite absorbent for removing trace residues of polybrominated diphenyl ethers in water. *Materials.* 7, 6028–6044.

Hu, R., S.Y. Dai, D.D. Shao, A. Alsaedi, B. Ahmad, & X.K. Wang (2015). Efficient removal of phenol and aniline from aqueous solutions using graphene oxide/polypyrrole composites. *J. Mol. Liq.* 203, 80–89.

Kabiri, S., D.N. Tran, T. Altalhi, & D. Losic (2014). Outstanding adsorption performance of graphene–carbon nanotube aerogels for continuous oil removal. *Carbon.* 80, 523–533.

Kim, H., S.O. Kang, S. Park, & H.S. Park (2015). Adsorption isotherms and kinetics of cationic and anionic dyes on three-dimensional reduced graphene oxide macrostructure. *J. Ind. Eng. Chem.* 21, 1191–1196.

Li, J., F. Wang, & C.Y. Liu (2012). Tri-isocyanate reinforced graphene aerogel and its use for crude oil adsorption. *J. Colloid Interface Sci.* 382, 13–16.

Li, H., J.C. Fan, Z.X. Shi, M. Lian, M. Tian, & J. Yin (2015). Preparation and characterization of sulfonated graphene-enhanced poly (vinyl alcohol) composite hydrogel and its application as dye absorbent. *Polymer.* 60, 96–106.

Lü, K., G.X. Zhao, & X.K. Wang (2012). A brief review of graphene-based material synthesis and its application in environmental pollution management. *Chin. Sci. Bull.* 57, 1223–1234.

Ma, T.T., P.R. Chang, P.W. Zheng, F. Zhao, & X.F. Ma (2014). Fabrication of ultra-light graphene-based gels and their adsorption of methylene blue. *Chem. Eng. J.* 240, 595–600.

Maliyekkal, S.M., T. Sreeprasad, D. Krishnan, S. Kouser, A.K. Mishra, & U.V. Waghmare (2013). Graphene: a reusable substrate for unprecedented adsorption of pesticides. *Small.* 9, 273–283.

Qin, Y.L., M.C. Long, B.H. Tan, & B.X. Zhou (2014). RhB Adsorption Performance of Magnetic Adsorbent Fe3O4/RGO Composite and Its Regeneration through A Fenton-like Reaction. *Nano-Micro Lett.* 6, 125–135.

Russo, P., L. D'Urso, A. Hu, N. Zhou, & G. Compagnini (2015). In liquid laser treated graphene oxide for dye removal. *Appl. Surf. Sci.* 348, 85–91.

Sharma, P., B.K. Saikia, & M.R. Das (2014). Removal of methyl green dye molecule from aqueous system using reduced graphene oxide as an efficient adsorbent: Kinetics, isotherm and thermodynamic parameters. *Colloids Surf., A.* 457, 125–133.

Shi, Y.P., S.X. Zhong, M.L. Wu, & J.R. Chen (2014). Comparing Different Kinds of Materials for Adsorption of Methylene Blue. *Appl. Mech. Mater.* 651–653, 1331–1334.

Sun, L., & F. Bunshi (2013). Effect of encapsulated graphene oxide on alginate-based bead adsorption to remove acridine orange from aqueous solutions. *Arxiv.* 10, 1–22.

Sun, J.Z., Z.H. Liao, R.W. Si, G.P. Kingori, F.X. Chang, & L. Gao (2014). Adsorption and removal of triphenylmethane dyes from water by magnetic reduced graphene oxide. *Water Sci. Technol.* 70, 1663–1669.

Tang, Y.L., H.G. Guo, L. Xiao, S.L. Yu, N.Y. Gao, & Y.L. Wang (2013). Synthesis of reduced graphene oxide/magnetite composites and investigation of their adsorption performance of fluoroquinolone antibiotics. *Colloids Surf., A.* 424, 74–80.

Travlou, N.A., G.Z. Kyzas, N.K. Lazaridis, & E.A. Deliyanni (2013). Functionalization of graphite oxide with magnetic chitosan for the preparation of a nanocomposite dye adsorbent. *Langmuir: the ACS journal of surfaces and colloids.* 29, 1657–1668.

Wang, D.X., L.L. Liu, X.Y. Jiang, J.G. Yu, & X.Q. Chen (2015). Adsorption and removal of malachite green from aqueous solution using magnetic β-cyclodextrin-graphene oxide nanocomposites as adsorbents. *Colloids Surf., A.* 466, 166–173.

Wang, D.X., L.L. Liu, X.Y. Jiang, J.G. Yu, X.H. Chen, & X.Q. Chen (2015). Adsorbent for p-phenylenediamine adsorption and removal based on graphene oxide functionalized with magnetic cyclodextrin. *Appl. Surf. Sci.* 329, 197–205.

Wang, J.F., T. Tsuzuki, B. Tang, L. Sun, X.J. Dai, & G.D. Rajmohan (2014). Recyclable textiles functionalized with reduced graphene oxide@ ZnO for removal of oil spills and dye pollutants. *Aust. J. Chem.* 67, 71–77.

Wang, Q., J. Li, Y. Song, & X. Wang (2013). Facile synthesis of high-quality plasma-reduced graphene oxide with ultrahigh 4,4'-dichlorobiphenyl adsorption capacity. *Chem. Asian J.* 8, 225–231.

Wang, S.B., H.Q. Sun, H.M. Ang, & M.O. Tadé (2013). Adsorptive remediation of environmental pollutants using novel graphene-based nanomaterials. *Chem. Eng. J.* 226, 336–347.

Wang, S., J. Wei, S.S. Lv, Z.Y. Guo, & F. Jiang (2013). Removal of Organic Dyes in Environmental Water onto Magnetic-Sulfonic Graphene Nanocomposite. *CLEAN–Soil, Air, Water.* 41, 992–1001.

Wu, Z.B., H. Zhong, X.Z. Yuan, H. Wang, L.L. Wang, & X.H. Chen (2014). Adsorptive removal of methylene blue by rhamnolipid-functionalized graphene oxide from wastewater. *Water Res.* 67, 330–344.

Yang, X., J.X. Li, T. Wen, X.M. Ren, Y.S. Huang, & X.K. Wang (2013). Adsorption of naphthalene and its derivatives on magnetic graphene composites and the mechanism investigation. *Colloids Surf., A.* 422, 118–125.

Yu, F., J. Ma, & D.S. Bi (2014). Enhanced adsorptive removal of selected pharmaceutical antibiotics from aqueous solution by activated graphene. *Environ. Sci. Pollut. Res.* 22, 1–10.

Zhang, Y.X., Y.X. Cheng, N.N. Chen, Y.Y. Zhou, B.Y. Li, & W. Gu (2014). Recyclable removal of bisphenol A from aqueous solution by reduced graphene oxide–magnetic nanoparticles: Adsorption and desorption. *J. Colloid Interface Sci.* 421, 85–92.

Zhao, W.F., Y.S. Tang, J. Xi, J. Kong (2015). Functionalized graphene sheets with poly (ionic liquid) s and high adsorption capacity of anionic dyes. *Appl. Surf. Sci.* 326, 276–284.

Zhou, C.J., W.J. Zhang, H.X. Wang, H.Y. Li, J. Zhou, & S.H. Wang (2014). Preparation of Fe$_3$O$_4$-Embedded Graphene Oxide for Removal of Methylene Blue. *Arabian J. Sci. Eng.* 39, 6679–6685.

Zhu, H.G., D.Y. Chen, N.J. Li, Q.F. Xu, H. Li, & J.H. He (2015). Graphene Foam with Switchable Oil Wettability for Oil and Organic Solvents Recovery. *Adv. Funct. Mater.* 25, 597–605.

Advances in Energy, Environment and Materials Science – Wang & Zhao (Eds)
© *2016 Taylor & Francis Group, London, ISBN 978-1-138-02931-6*

Corrosion inhibition of nitrite in existing reinforced concrete structures

Jing Liu, Jiali Yan, Yanhua Dai & Yushun Li
Ningbo University, Ningbo, P.R. China

ABSTRACT: This paper presents the diffusion model of nitrite ion in concrete and predicts a model for inhibition effect of reinforced concrete based on the diffusion equation of Fick's second law when the concrete surface is brushed with nitrite solution. This study confirms that nitrite ion can diffuse well in concrete to reach an effective mol ratio from the surface of the concrete to the surface of the reinforcement, and have an effective protection for the bars against corrosion. It will help increase the tendency of inhibiting duration and prolong the life of reinforced concrete. The nitrite solution amount is used to predict the corrosion inhibiting duration of concrete that contains chloride salt by the developed prediction model based on the Magge equation.

1 INTRODUCTION

Corrosion in reinforced concrete structures is one of the most serious durability problems. The resistance against corrosion can be improved by improving the compaction density etc. Although methods like dechlorination, cathodic protection and such other methods have theoretical meanings, when it comes to practical applications in the field there are certain restrictions. The existing method is mending the concrete cover using mortar containing corrosion inhibiting compositions at places where the crack is relatively large in the concrete surface and places where the reinforcement is exposed due to the delamination of the cover concrete. The disadvantage with this method is that the damage caused to the cover concrete is large to structures that have only small cracks or are sometimes not rust bloated. Further, due to this method only a part of the whole steel bars (affected portion) is mended and because the surface of the steel bars after being mended is in different environment, it is easy to become the anode of corrosion cell to accelerate the macro cell corrosion. This may further deteriorate the structure. The most effective method shall therefore be without damaging the cover concrete and putting the steel bars in the inhibiting environment to restrain from corrosion. Nitrite solution is one such material that has good inhibiting nature corrosion.

2 EXPERIMENTAL

2.1 Raw materials

The cement was Ordinary Portland cement with a specific gravity of 3.2. Crushed lime stone with 15 mm maximum size, specific gravity 2.65 and water absorption 1.0% was used as coarse aggregate. Fine aggregate is river sand with a specific gravity of 2.6, water absorption 1.0% and fineness modulus 2.7. Mix proportion of concrete are cement: coarse aggregate: sand: water = 310 kg: 1200 kg: 600 kg: 150 kg. Prismatic specimens of 200 mm × 200 mm × 100 mm were used as test specimens. Standard moulds were used for casting the cube blocks. The moulds were demoulded after 48 hours and cured in water for 28 days. 30% sodium nitrite solution is used in this investigation. The bottom side of the specimen is brushed with the solution after treating with the abrasive paper. This has ensured easy permeation of the nitrite solution. The four sides of the specimen are brushed with the epoxy resin dope to prevent the loss of nitrite solution applied on the bottom side. The top most surface of the specimens were cured with 1 mm polymer cement mortar and 2 mm thick acrylic rubber on the surface of smearing side. After the specimens were smeared with the solution, they were kept for curing for six months at 21°C and 65 percent relative humidity.

2.2 Determination of the concentration of nitrite ion

The procedure for the determination of the nitrite ion in concrete is based on JCI standards. The dimension of the sample was $\varphi 60$ mm by 150 mm which was sampled from concrete at 12 mm intervals from the concrete surface to the inner with drill machine. The determination of the concentration of the nitrite ion was conducted based on JCI-SC4.

3 THE MODEL OF DIFFUSION OF THE NITRITE IN CONCRETE

Assuming that concrete is a one-dimensional, half-infinite and homogeneous substance, the nitrite permeates into concrete by the diffusion effect. This process of diffusion can be described using the Fick's second law as

$$\frac{\partial C}{\partial t} = D \frac{\partial^2 C}{\partial x^2}$$

where, C is the concentration of nitrite ion at a distance x from the surface of concrete, t is time (months), x is the distance from the surface of concrete, D is the diffusion coefficient of nitrite ion in concrete.

In order to solve the equation (1), a boundary condition is proposed. Assuming the thickness of nitrite solution is H, then the concentration of nitrite ion $C(X, t)$ is the only function of time t. The boundary condition can be assumed as:

$$C(0, t) = C_0(t)$$

$$C(\infty, t) = 0$$

where, $C_0(t)$ is the concentration of the cover with time.

Assuming that the diffusive ion weight is in direct proportion with the nitrite ion concentration square, the concentration of the cover is given by

$$\frac{dC_0(t)}{dt} = -K \cdot C_0(t)^2$$

$$\frac{1}{C_0(t)} = K \cdot t + \frac{1}{C_0}$$

where, K is the coefficient of the outflow, C_0 is the initial concentration of the cover. Solving the Equations and we obtain

$$C = \frac{2}{\sqrt{\pi}} \frac{C_0}{C_0 \cdot K \cdot t + 1} \int_{\sqrt{4Dt}}^{\infty} \frac{x}{\sqrt{4Dt}} \exp(-\eta^{-2}) d\eta$$

4 NUMERICAL ANALYSIS OF THE DIFFUSION MODEL

4.1 Solution of K and C_0

The initial concentration C_0 of the brushing layer can be obtained from concrete specific gravity and amount of nitrite solution.

$$C_0 = C_{NO_2^-} \frac{M}{\rho_0 \cdot d} \times 100\%$$

where, M is the amounts of nitrite calcium solution, d is the thickness of the cover, $C_{Cl_0^-}$ is the concentration of nitrite solution (%).

Assuming that the thickness of the cover is 10 mm, then the initial concentration C_0 of the cover is 0.25%, 0.5%, 1% and 1.5%, respectively.

$$\frac{2}{\sqrt{\pi}} \frac{C_0}{C_0 \cdot K \cdot t + 1} \int_{\sqrt{4Dt}}^{\infty} \frac{x}{\sqrt{4Dt}} \exp(-\eta^{-2}) d\eta$$

Equation is a linear regression based on the measured concentration of the cover and initial concentration C_0, the equation of linear regression is:

$$Y = 0.6X + 1.2, 6K = 1.2 \text{ then } K = 0.2\% \text{ month.}$$

5 NUMERICAL ANALYSIS OF THE DIFFUSION MODEL

The diffusion coefficient can be obtained from the least-squares procedure. If K, C_0 and D in equation are known, $C_T(X_i, 6)$ at the sixth month from 0 to 50 mm can be obtained as

$$M = \sum_1^5 [C_T(X_i, 6) - C_M(X_i, 6)]^2$$

where, $C_M(X_i, 6)$ is the measured concentration of nitrite ion, $C_T(X_i, 6)$ is the theoretical concentration of nitrite ion in equation, M is the minimum sum of squares of the difference between $C_T(X_i, 6)$ and $C_M(X_i, 6)$.

The diffusion coefficients of 76 mm²/month, 70 mm²/month, 70 mm²/month and 65 mm²/month were obtained from the numerical analysis of equation, the average diffusion coefficient is $D = 70$ mm²/month. The concentration of nitrite ion in concrete at the sixth month can be obtained from D. The results indicate that the theoretical values are consistent with the measured experimental values.

6 PREDICTION MODEL

The true concentration of chloride, $C_{Cl_0^-}$ the true nitrite concentration $C_{NO_2^-}$ and the thickness of the protecting covering δ are three main effective factors, which determine if the surface of reinforcement is protected against corrosion or not according to the corrosion resisting mechanics of nitrite. The prediction model is given by

$$C_{Cl^-} = C_{Cl_0^-} + (C_{Cl_s^-} - C_{Cl_0^-}) \left[1 - erf \frac{\delta}{2\sqrt{D_0 t_0^m t^{1-m}}} \right]$$

where, $C_{Cl_0^-}$ is initial chloride concentration in concrete (%), $C_{Cl_s^-}$ is the superficial chloride concentration in concrete (%), erf is the error function, and D_0 is the diffusion coefficient of chloride in concrete when $t = t_0$, m is a constant, δ is the thickness of the protecting covering (mm). In model predicted, $C_{NO_2^-}$ is calculated by initial concentration and diffusion equation:

$$C_{NO_2^-} = C_{NO_2^-} +$$

$$C_{Cl_s^-} = C_{Cl_0^-};$$

where is the initial nitrite concentration in concrete, Due to

$$C_{NO_2^-} = \frac{2}{\sqrt{\pi}} \frac{C_0}{C_0 \cdot K \cdot t + 1} \int_{\frac{x}{\sqrt{4Dt}}}^{\infty} \exp(-\eta^{-2}) d\eta$$

It is known that nitrite is effective in concrete containing chloride NO_2^-/Cl^- when is 0.5. The concentration of nitrite, which permeates through the minimum protective cover is transferred to the maximum concentration of chloride according to the effective mol ratio, and the results are shown in equation below.

$$C_{Cl^-} = 1.54 C_{NO_2^-}$$

Then

$$C_{Cl^-} = \frac{3.08}{\sqrt{\pi}} \frac{C_0}{C_0 \cdot K \cdot t + 1} \int^{\infty} \frac{x}{\sqrt{4Dt}} \exp(-\eta^{-2}) d\eta$$

7 APPLICATION OF MODEL

The concentration of nitrite ion that permeates the minimum protective cover is transferred to the maximum concentration of chloride according to the effective molecular ratio. The corrosion resistance time increase when the amount of solution is more, because a large amount of solution accelerates the time of threshold value of mol ratio on the surface of rebar, and retards the time for losing effective threshold value of mol ratio. If the chloride content in concrete is low, there is a less influence of the amount of solution on the duration of inhibition effect, whereas a larger effect on the amounts of solution on duration of inhibiting effect.

8 CONCLUSIONS

The diffusion of Nitrite ion into concrete from the surface of concrete to the surface of reinforcement by brushing nitrite solution on the surface of concrete is an effective method of inhibition for corrosion in reinforced concrete elements. The diffusion concentration of nitrite in concrete can be obtained by using the diffusion model. The corrosion resistance time increase when the amount of solution is more, because a larger amount of brushing solution accelerates the time of threshold value of mol ratio on surface of rebar, and retards the time for losing effective threshold value of mol ratio. The theoretical model of duration of inhibiting effect in concrete is presented for predicting the inhibiting duration and the amount of nitrite solution is determined for inhibiting duration in concrete containing chloride salt. If the chloride content in concrete is low, the influence of the amount of solution on the duration of inhibiting effect is less, whereas much more influence on the amount of solution on duration of inhibiting effect can be observed.

REFERENCES

Abate C., B.E. Scheetz. Aqueous phase equilibria in the system CaO-Al2O3-CaCl2-H2O: The significance and stability of Friedel's salt. Journal of the American Ceramic Society, 1995, 78(4):939–944.

Álvarez-Bustamante R., G. Negrón-Silva, M. Abreu-Quijano. Electrochemical study of 2-merca- ptoimidazole as a novel corrosion inhibitor for steels. Electro chimica Acta. 2009, 54(23), 5393–5399.

Antonio A. Nepomuceno. Steel protection capacity of polymeric based cement mortars against chloride and carbonation attacks studied using electrochemical polarization resistance. Cement Concrete Comp, 2006, 28:716–721.

Bentz D.P., Mizell S., Satterfield S., et al. The visible cement data set. J Res Natl Inst Stand Technol, 2002, 107(2):137–148.

Burlion N., Bernard D., Chen D. X-ray microtomography: Application to microstructure analysis of a cementitious material during leaching process. Cem Concr Res, 2006, 36(2):346–357.

Castellote M., Andrade C., Turrillas X., et al. Accelerated carbonation of cement pastes in situ monitored by neutron diffraction. Cem Concr Res, 2008, 38(2): 1365–1373.

Chotard T.J., Martel M.P.B., Smith A. Application of X-ray computed tomography to characterise the early hydration of calcium aluminate cement. Cem Concr Compos, 2003, 25(1):145–152.

Erdogan S.T., NIE X., Stutzman P.E., et al. Micrometer-scale 3-D shape characterization of eight cements: Particle shape and cement chemistry, and the effect of particle shape on laser diffraction particle size measurement. Cem Concr Res, 2010, 40(5): 731–739.

Gallucci E., Scrivener K., Groso A., et al. 3D experimental investigation of the microstructure of the cement pastes using synchrotron X-ray micro tomography (µCT). Cem Concr Res, 2007, 37(3):360–368.

Ganesha Achary, Y. Arthoba Naik, S. Vijay Kumar, T.V. Venkatesha, B.S. Sherigara. An electroactive co-polymer as corrosion inhibitor for steel in sulphuric acid medium. Applied Surface Science. 2008, 254, 5569–5573.

Gonzalez J.A., E. Ramirez, A. Bautista. Protection of steel embedded in chloride containing concrete by means of inhibitors. Cem Concr Res. 2013, 28(4), 577–589.

Han Jiande, Pan Ganghua, Sun Wei. Investigation on carbonation induced meso-defects changes of cement mortar using 3D X-Ray computed tomography. J Chin Ceram Soc, 2011, 399(10):75–79.

Haque M.N., Kawamura M. Carbonation and chloride-induced corrosion of reinforcement in fly ash concretes. ACI Mater J, 1992, 89(6):602–605.

Kwon S.J., Song H.W. Analysis of carbonation behavior in concrete using neural network algorithm and carbonation modeling. Cem Concr Res, 2010, 40(1): 119–127.

Liu Junzhe, He Zhimin. Predicting the threshold values based on diffusion model of nitrite ions in reinforced concrete. J. Wuhan University of Technology-Mater. Sci. Ed.. 2010,25(2): 308–311.

Liu Zanqun, Deng Dehua, De Schutte G, et al. Chemical sulfate attack performance of partially exposed cement and cement fly ash paste. Constr Build Mater, 2012, 28(1): 230–237.

Marco Ormellese, Luciano Lazzari, Sara Goidanich, Gabriele Fumagalli, Andrea Brenna. A study of organic substances as inhibitors for chloride-induced corrosion in concrete. Corrosion Science. 2009, 51, 2959–2968.

Marques P.F., Costa A. Service life of RC structures: carbonation induced corrosion. Constr Build Mater, 2010, 24(3):258–265.

Martz H.E., Schnebeck D.J., Roberson G.P., et al. Computerized tomography analysis of reinforced concrete. ACI Mater, 1993, 90(3):259–264.

Monteiro P.J.M., Kurtis K.E. Time to failure for concrete exposed to severe sulfate attack. Cem Concr Res, 2003, 33(7): 987–993.

Morgan I.L., Ellinger H., Klinksiek R., et al. Examination of concrete by computerized tomography [J]. J Am Concr Inst, 1980, (1–2):23–27.

Ngala V.T., Page C.L. Effects of carbonation on pore structure and diffusion properties of hydrated cement paste. Cem Concr Res, 1997, 27(7):995–1007.

Pedro Faustino Marques, Carlos Chastre. Carbonation service life modelling of RC structures for concrete with Portland and blended cements. Cement Concrete Comp, 2013, 37(6):171–184.

Promentilla M.A.B., Sugiyama T., Hitomi T., et al. Quantification of tortuosity in hardened cement pastes using synchrotron-based X-ray computed microtomography. Cem Concr Res, 2009, 39(6):548–557.

Rougelot T., Burlion N., Bernard D., et al. About microcracking dueto leaching in cementitious composites: X-ray microtomography description and numerical approach. Cement and Concrete Research, 2010, 40(2):271–283.

Shah S.P., Choi S. Nondestructive techniques for studying fracture processes in concrete. Int J Fract, 1999, 98(3–4):351–359.

Shayan A. Effects of seawater on AAR expansion of concrete [J]. Cem Concr Res, 2010, 40(4):563–568.

Song H.W., Kwon S.J. Permeability characteristics of carbonated concrete considering capillary pore structure. Cem Concr Res, 2007, 37(6):909–915.

Song H.W., Lee Ch, Ann K.Y. Factors influencing chloride transport in concrete structures exposed to marine environments. Cement Concrete Comp, 2008, 30(2): 113–121.

Stock S.R., Naik N.K., Wilkinson A.P., et al. X-ray micro tomography (micro CT) of the progression of sulfate attack of cement paste. Cem Concr Res, 2002, 32(10):1673–1675.

Sugiyama T., Promentilla M.A.B., Hitomi T., et al. Application of synchrotron micro tomography for pore structure characterization of deteriorated cementitious materials due to leaching. Cem Concr Res, 2010, 40(8):1265–1270.

Suryavanshi A.K., R. Narayan Swamy. Stability of Friedel's salt in carbonated concrete structural elements. Cement Concrete Res, 1996, 26(5):729–741.

Valcarce M.B., M. Vázquez. Carbon steel passivity examined in solutions with a low degree of carbonation: The effect of chloride and nitrite ions. Materials Chemistry and Physics. 2014, 115(1), 313–321.

Verbeck G. Carbonation of hydrated Portland cement. PCA Bull. 1958, 87:17–36.

Yang T., Keller B., Magyari E. Dirtect observation of the carbonation process on the surface of calcium hydroxide crystals in hardened cement paste using an atomic force microscope. Journal of Materials Science, 2003, 38:1909–1916.

Zitrou E., J. Nikolaou, P.E. Tsakiridis. Atmospheric corrosion of steel reinforcing bars produced by various manufacturing process. Construction and Building Materials, 2007, 21 (6):1161–1169.

Advances in Energy, Environment and Materials Science – Wang & Zhao (Eds)
© 2016 Taylor & Francis Group, London, ISBN 978-1-138-02931-6

Binding mechanism of chloride ions in mortar

Jing Liu, Jiali Yan, Yanhua Dai & Yushun Li
Ningbo University, Ningbo, P.R. China

ABSTRACT: The adsorption and binding mechanisms of chloride ion in sea sand mortars were investigated through measuring the water-soluble chloride ion content in this paper. In addition, the microstructure of Friedel's salt were characterized by X-ray diffraction, scanning electron microscopy, and thermal analysis method. The results show that water soluble chloride ion concentration in the sea sand mortar is closely related to the extraction temperature. The higher the extraction temperature, the higher the water-soluble chloride concentration is. When extraction temperature is 65°C, the water soluble chloride ion concentration is about twice as mortars at 15°C. Physically bound chloride ion in fly ash mortar shows low dissolution in lower temperatures. The endothermic peak of the Friedel's salt in the sea sand mortar does not appear on the TG/DTA curve, while the existence of Friedel's salt was indicated by XRD and SEM picture. This is probably due to the small amount of instable Friedel's salt content in the sea sand mortar.

1 INTRODUCTION

Coastal areas are rich in sea sand. The sea sand washed by freshwater to remove impurity has already been widely used in coastal concrete industry in China due to its convenience in mining and transportation. Chloride-induced corrosion of rebars in concrete structures is a major concern to the concrete professionals. The residual chloride in the desalted sea sand, potentially threatens the durability of reinforced concrete structures.

Generally, there are three types of chloride ions in the sea sand concrete: free ions, physically bound and chemically bound chloride ions. Only the free chloride is responsible for the corrosion of rebars. Even though the pH of concrete is more than 12, the free chloride ions accumulated on the surface of rebars can cause or aggravate the corrosion.

Base on the survey of the utilization of the desalted sea sand in Ningbo city, this paper studied the binding mechanism of chloride ion and the effect of fly ash used as a partial replacement of cement on the concentration of free chloride ion in the regular sea sand and desalted sea sand mortars. The characteristic of Friedel's salt and microstructure of mortars contained chloride were clarified in order to provide the theory base for durable sea sand concrete structures.

2 EXPERIMENTAL PROCEDURE

Grade 42.5 ordinary Portland cement produced by Ningbo shunjiang Cement Plant was used in this study; Grade II Fly Ash (FA) used in this research was supplied by Ningbo Power Plant. River Sand (RS) was used as fine aggregate, and its fineness modulus is 2.27. Regular Sea Sand (SS) and Desalted Sea Sand (DSS) came from five plants and its chloride ion contents are shown in Table 1.

The mortar cube specimens of 100 mm × 100 mm × 100 mm were prepared and tested after being cured in standard curing room for 28 days.

3 TEST METHOD

The sand was quartered to 1500 g and dried in an oven with 105°C. Then the sand was cooled to room temperature. 500 g sand was put in a reagent bottle with rubber stopper then 500 ml distilled water was added. For the complete extraction of chloride in sand, the bottle was vibrated once in 24 h and thereafter three times every five minutes. After a certain time, the clear solution in the bottle was filtered and let the filtrate flow into a glass beaker. 50 ml filtrate was transferred into a triangular flask by pipette. Then the filtrate was titrated by 0.01 mol/L standard silver nitrate solution with 5% potassium chromate as indicator until the solution become red and the red color can be maintained for 5–10 s. In

Table 1. Chloride ion contents in sea sand.

Sample	1	2	3	4	5
DSS	0.006	0.002	0.005	0.01	0.01
SS	0.029	0.028	0.03	0.06	0.054

DSS—Desalted sea sand; SS—Sea sand.

Table 2. Mix proportions of the mortar.

Sample	Sand type	Cl⁻contents in sand (%)	Fly ash (%)	Mix proportion (C: S: W)
HCD0	RS	0.1	0	1:2.5:0.45
D0	DSS	0.01	0	1:2.5:0.45
W0	SS	0.054	0	1:2.5:0.45

the course of titration, the amount of standard silver nitrate solution consumed was recorded and the total chloride content in sea sand can be calculated according to the amount. Cl⁻ contents in sand was 0, 0.02, 0.03, 0.06, 0.1, 0.054 respectively.

The motar cubes were cured at the standard condition for 28 days. The mole method was adopted to test the free chloride ion content. The pH of the solution under the test must be near neutral according to the requirement of mole method, however, the solution of mortar is alkaline. The dilute sulfuric acid was used for the neutralization in this study. To investigate the effect of extraction temperature on the total or free chloride concentration of mortar, the tests of chloride ion content were carried out at 15°C and 65°C, respectively.

The proportions of three samples HCD0, W0, D0 for TG/DTA and XRD tests were shown in Table 2. The mortar used for the SEM test was mixed with regular sea sand. The scanning electron microscope was produced by Japan Hitachi Company. The polycrystalline X-ray diffraction and TG/DTA thermal analyzer were produced by German Brueck and America Perkin Amelmer Company. The inner products, microstructure and its content in mortars can be determined by use of these analyzers.

4 EXPERIMENTAL RESULTS AND DISCUSSION

Table 1 shows chloride ion contents of samples derived from the five plants are different. The chloride ion content in desalted sea sand is significantly lower than regular sea sand. It also shows that the the higher the original chloride ion content, the higher the residual content is after the sea sand is desalted.

Results shows the soluble and bonded chloride ion contents at 15°C and 65°C in mortars. It can be seen from Table 3 that there is a close correlation between free chloride concentration and extraction temperatures. At 15°C, free chloride concentration content varies from 22%–34% and at 65°C, it varies from 52%~65%. This indicates that the free chloride concentration of mortars at extraction temperature 65°C is about twice as mortars at 15°C. This may be due to that the extraction rate

of physically bound chloride ion increase with an increase in temperature. Therefore, taking the differential between total chloride ion content and free chloride ion content as the amount of chemically bound chloride is somewhat inaccurate.

There are physically and chemically bound chloride ions in mortars. Chemical bonding is generally the result of reaction between chlorides and C3 A to form Friedel's salt or the reaction with C4 AF to form a Friedel's salt analogue. Physical binding is due to the adsorption of chloride ion to the C-S-H surfaces.

Most of the chloride ion is physically bound to ion exchange sites of C-S-H gel and there exists a significant degree of reversibility. In fact, for a physical adsorption, an elevated temperature increases the thermal vibration of absorbates, resulting in more free chloride.

In addition, Figure 1 shows the free chloride extraction rate of the mortar with 20% fly ash is higher than that of the cement mortar without fly

Table 3. Existing status of chloride ions.

Sample	Soluble ratio of Cl⁻/%		Bonding ratio of Cl⁻/%		TCl⁻/% *
	15°C	65°C	15°C	65°C	
HCA0	22.1	54.7	77.9	45.3	0.0071
HCA20	28.3	57.8	71.7	42.2	0.0071
HCB0	22.1	57.8	77.9	42.2	0.0214
HCB20	24.1	52.6	75.9	47.4	0.0214
HCC0	26.4	60.1	73.6	39.9	0.0428
HCC20	31.2	63.5	68.8	36.5	0.0428
HCD0	22.9	58.8	77.1	41.2	0.0714
HCD20	27.8	57.3	72.2	42.7	0.0714
D0	29.5	63.0	70.5	37.0	0.0077
D20	33.4	64.7	66.6	35.3	0.0077
W0	30.3	56.7	69.7	43.3	0.0385
W20	32.0	58.8	68.0	41.2	0.0385

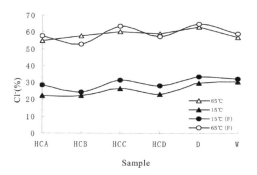

Figure 1. Relationship between the temperature and soluble chloride ion concentrations.

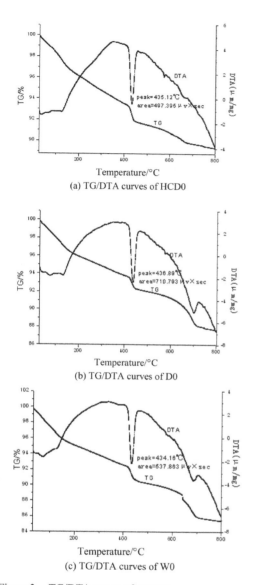

(a) TG/DTA curves of HCD0

(b) TG/DTA curves of D0

(c) TG/DTA curves of W0

Figure 2.   TG/DTA curves of mortar.

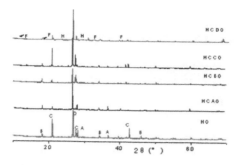

(a) XRD patterns of mortars with river sand

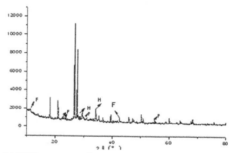

(b) XRD patterns of mortars with desalted sea sand

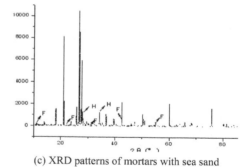

(c) XRD patterns of mortars with sea sand

Figure 3.   XRD patterns of mortars.

ash at extraction temperature15°C, however, the free chloride extraction rate of the mortars with or without fly ash is almost the same at extraction temperature 65°C. This may be due to less chemical bound and more physical bound chloride ions at lower extraction temperature.

As chloride enters the cementitious material, it may be converted to Friedel's salt due to chemical binding. Chloride ion can be introduced into concrete through two ways: (1) as an admixture (internal chloride); (2) penetration from external environment (external chloride). In the literature

review, the chloride was frequently dissolved in the mixing water and then entered the mixture. At the same time, the amount of chloride introduced varies 2%~10% by weight of the cementitious material. Many researchers investigated the Friedel's salt with the abovementioned condition. However, because the chloride contents in the regular sea sand and desalted sea sand are much lower than that of abovementioned researches, it has not been reported if the Friedel's salt exists in regular sea sand or desalted sea sand mortars and concretes and if it can be observed.

Generally, Friedel's salt yields an endothermal effect at about 360°C. Figure 2 shows the TG/DTA curves of mortars with river sand, regular sea sand and desalted sea sand, and the chloride content of these sands were 0.1%, 0.054% and 0.01%, respectively. Unfortunately, no clear endothermal peaks corresponding to the Friedel's salt appear for the three TG/DTA curves of mortars.

To further clarify weather Friedel's salt exists in regular sea sand and desalted sea sand mortar, the XRD tests were carried out for the grounded mortars samples at 28 days of age and the results are shown in Figure 3. From Figure 3, regardless of the mortars with regular sea sand or desalted sea sand, several intensity peaks of Friedel's salt appear at the corresponding position, however, the intensity of the Friedel's salt peak is much lower than the other compositions (for example Ca(OH)$_2$) of the mortar. This indicates that the chloride ions introduced by regular sea sand or desalted sea sand still form some Friedel's salt in the mortars.

Because the total chloride content in regular sea sand or desalted sea sand is relatively low, the intensity peak of Friedel's salt is very low. Moreover, the abovementioned TG/DTA curves of mortars do not show the endothermal peak of Friedel's salt. This is because the small amount of chloride ions introduced by regular sea sand or desalted sea sand formed a very small amount of unstable Friedel's salt. In this paper, the intensity peak of Friedel's salt can be clearly observed in the mortars with river sand containing more than 0.03% chloride content, and not in the mortars with river sand containing less than 0.03% chloride content.

5 CONCLUSIONS

Regardless of the mortars with river sand, regular sea sand or desalted sea sand, the free chloride content correlates closely to the extraction temperature. The free chloride concentration increases with the increase of extraction temperature, and the free chloride concentration increases by about two times when the extraction temperature varies from 15°C to 65°C. The free chloride concentration of the mortar with fly ash is higher than that of the mortar without fly ash at 15°C extraction temperature. This may be due to the chemically bound chloride ion content is low and the physically bound chloride ion content is relatively high. Small amount of unstable Friedel's salts exists in the mortars, which explains why the TG/DTA curves of regular sea sand and desalted sea sand mortars do not show endothermal peaks. Thus, it is unreasonable to use TG/DTA curves only to determine the existence of Friedel's salt, and the investigation of XRD and SEM should be carried out.

REFERENCES

Haque, M.N., Kawamura, M. (1992). Carbonation and chloride-induced corrosion of reinforcement in fly ash concretes. ACI Mater J, 89, 602–605.
Marques, P.F., Costa, A. (2010). Service life of RC structures: carbonation induced corrosion. Constr Build Mater, 24, 258–265.
Ngala V.T., Page C.L. (1997). Effects of carbonation on pore structure and diffusion properties of hydrated cement paste. Cem Concr Res, 27, 995–1007.
Pedro Faustino Marques, Carlos Chastre (2013). Carbonation service life modelling of RC structures for concrete with Portland and blended cements. Cement Concrete Comp, 37, 171–184.
Zitrou, E. J. Nikolaou, P.E. Tsakiridis (2007). Atmospheric corrosion of steel reinforcing bars produced by various manufacturing process. Construction and Building Materials, 21, 1161–1169.

Computer applications

Advances in Energy, Environment and Materials Science – Wang & Zhao (Eds)
© *2016 Taylor & Francis Group, London, ISBN 978-1-138-02931-6*

Multi-channel PM$_{2.5}$ sampler based on intelligent embedded software control system

Daozhu Hua
College of Environmental and Resource Sciences of Zhejiang University, Hangzhou, China
Focused Photonics (Hangzhou) Inc., Hangzhou, China

Wei Huang, Zhonghua Liu, Xu Cao & Huajun Ye
Focused Photonics (Hangzhou) Inc., Hangzhou, China

Weiping Liu
College of Environmental and Resource Sciences of Zhejiang University, Hangzhou, China

ABSTRACT: As an air pollutant, the particle with aerodynamic diameters less than 2.5 μm (PM$_{2.5}$) has been widely concerned in recent years due to its impacts on visibility, climate, and human health. It is significant to analyze source apportionment of PM$_{2.5}$ for controlling the particulate pollution. However, many particle matter samplers with one sample channel only achieve one sample within a specified time. In this paper, we have developed a multi-channel PM$_{2.5}$ sampler based on intelligent control system that includes sampling control module, instrument temperature control module, instrument maintenance module, and long-distance monitoring module. The results demonstrate that the instrument based on the control system can achieve the intelligent multi-channel sampling and provide one of potential tools to study the source apportionment of the particulate matter.

1 INTRODUCTION

Currently, atmospheric pollution has become a severe environmental issue in China for the rapid industrialization and motorization (Ye, H.L. et al., 2014; Yang, L.X. et al., 2012; Wang, J. et al., 2013; Zhou, J.M. et al., 2012). As one of the important air pollutants, PM$_{2.5}$ includes much nocuous components, such as sulfate, nitrate, organic carbon, inorganic carbon and heavy metals (Judith, C.C. et al., 2008; Admir, C.T. et al., 2012; Holler, R., 2002). Therefore, PM$_{2.5}$ has adverse effect on the human health, except visibility and climate (Brook, R. et al., 2010; Ramanathan, V. et al., 2001; Chakra, O. R. A. et al., 2007; Doğen, M., 2002; Schwartz, J. et al. 2004).

There have been more and more researches on the source apportionment of particulate matter since the study on relationship between the human morbidity and the aerosol particles published (He, X. et al., 2009; Wang, C.M., 2008). The laboratory analysis is the normal method for the source apportionment. It includes sampling the PM$_{2.5}$ on the membrane, taking the samples back to laboratory and analyzing the sample using laboratory instrument (Wang, G. et al., 2015). During the laboratory analyzing process,

each component checking need one PM$_{2.5}$ sample which cannot be reusable. For many chemical components in PM$_{2.5}$, source apportionment requires large numbers of samples. However, most PM$_{2.5}$ sampler with one sampling channel can only obtain one sample at one time. It is significant to develop a multi-channel sampler to meet the requirement of the source apportionment of the particulate matter.

In this paper, we design a multi-channel PM$_{2.5}$ sampler based on intelligent embedded software control system. The control system has four parts, including sampling control module, instrument temperature control module, instrument maintenance module, and long-distance monitoring module. Owing to the sampling control module, the sampler can be sampling four samples simultaneously with 5.00 L/min or 16.67 L/min. For supporting by its temperature control module, it can be employed in the temperature of −30°C~50°C. In addition, the sampler can be monitored by the internet terminal far from the sampling site using the long-distance monitoring module. The results of performance testing validate that the control system can effectively support the multi-channel PM$_{2.5}$ sampler intellectualized working.

Figure 1. The feature of the multi-channel PM$_{2.5}$ sampler.

2 INSTRUMENT

The feature of the instrument is shown in the Figure 1. There are four parts including canopy, sampling unit, host and fixtures. The function of canopy is keeping out the rains to protect the instrument. Sampling unit is constituted by PM$_{2.5}$ cutter and cassettes. The PM$_{2.5}$ cutter is used to separate the PM$_{2.5}$ from the Total Suspended Particulates (TSP) and the cassettes are used to hold the filter membrane that is employed to filter and enrich the PM$_{2.5}$ when the air flow passes the membrane. Host is the major part of the sampler and supports the sampler to achieve specific function. There are some flow devices, sensors of temperature and pressure, circuit board, and display screen in the host. The fixtures can maintain the sampling unit and the host at the certain height to avoid the interference of raise dust from platform and conveniently operate.

3 CONTROL SYSTEM

In order to ensure the multi-channel PM$_{2.5}$ sampler to work well, the intelligent embedded software control system has been developed. The control system is constituted by sampling control module, instrument temperature control module, maintenance module, and long-distance monitoring module.

3.1 Sampling control module

3.1.1 Sampling type
In order to meet the sampling requirements in different situation, sampling type has been developed. The sampling type provides different available options for programming the multi-channel PM$_{2.5}$ sampler.

The default option is suited for the sampling in the whole day. The sampling starts at the 0:00 and stops at 24:00. It supports sampler starting at the fixed time with unattended operation.

Considering the requirements of specific sampling time, the sampling control module has another sampling type that the user can set the start time and stop time by self.

3.1.2 Sampling flow control module
The accuracy of PM$_{2.5}$ cutter is closely related with the flow rate. Therefore, it must accurately control the flow rate. In the instrument, the closed loop feedback controlling method is employed to control the flow rate. The regulating valve has been used to adjust the flow rate, for the opening degree of the valve is linearly dependent on the supplying voltage. The flow sensor has been assembled into the flow path, and dynamically monitors the flow rate. When the flow rate deviates from the fixed value, the system will reversely adjust the supplying voltage of the valve. This method can effectively maintain the fluctuation of flow rate within ±2% of the fixed flow rate.

In addition, the concentration of the PM$_{2.5}$ is defined as (1).

$$C = \frac{\Delta m}{Q_{STD}} \qquad (1)$$

C is the concentration of the PM$_{2.5}$, μg/m^3;
Δm is the increased weight after sampling, μg;
Q_{STD} is the standard sampling volume, m^3.

Note that the sampling volume used to calculate the concentration of PM$_{2.5}$ is the standard sampling volume. The control system can also accumulate the standard sampling volume automatically. The volume can be computed by the formula, as shown in (2). The temperature and the pressure on the operating condition are measured by the temperature and pressure sensors in real time.

$$Q_{STD} = Q \times \frac{P}{P_{STD}} \times \frac{T_{STD}}{T} \qquad (2)$$

Q is the volume on the operating condition, m^3;
P is the pressure on the operating condition, KPa;
T is the temperature on the operating condition, K;
P_{STD} is the standard pressure, 101.325 KPa;
T_{STD} is the standard temperature, 273.15 K.

3.2 Instrument temperature control module

The multi-channel sampler is an instrument using outside. It requests that the instrument has to be suited for the environment with extreme temperature −30°C~50°C. In order to protect the inside devices, the temperature control module has been designed based on the temperature sensors, heater, and refrigeration equipment. When the inner temperature sensor indicates that the temperature is lower than the fixed value T_1, the control system will start the heater and not stop the heater until the temperature higher than T_1. Contrarily, when the inner temperature sensor indicates that the temperature is higher than the fixed value T_2 ($T_2 > T_1$), the refrigeration equipment will work till the inner temperature lower than T_2. Supporting by the temperature control module, the temperature of inside can maintain at the range of T_1~T_2. It can completely support the sampler applying in the extreme temperature environment.

3.3 Instrument maintenance module

The performance parameters of the instrument may be degeneration after long time running. Therefore, it is very necessary to support maintenance routine for a favorable instrument.

3.3.1 Leak check

Leak check is very important for the particulate matter sampler. Because it can mortally affect the weight of the $PM_{2.5}$ filtered and enriched on the membrane. The maintenance module can provide automatic leak check procedure. It can distinguish the tightness of the sampler though the change of flow pressure in a specific time period under the situation of approximate vacuum in the flow path.

3.3.2 Calibration of the sensors

The precision of the sensors, especially of the flowmeter, can determine the accuracy of the instrument. Maintenance module provides calibration function for all the sensors in the instrument.

3.3.3 Long-distance monitoring module

In order to meet the detection limit of laboratory analysis instrument, each time sampling often last several hours. Therefore, it is significant to monitor the operation status from long-distance site under the unattended situation.

According to the requirement, the long-distance monitoring module based on wireless network data transmission technique has been applied in the sampler, as shown in the Figure 2. Through General Packet Radio Service (GPRS) net, the sampler can upload the information of the instrument to a specified server by the inner

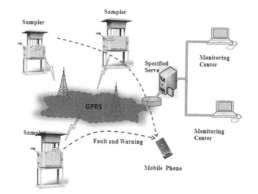

Figure 2. Schematic diagram of long-distance monitoring.

Table 1. The performance of the sampler.

Performance	Working parameters	Accuracy
Flow rate	5 L/min 16.67 L/min	±2% with the working flow rate
Atmospheric temperature	/	±2°C
Atmospheric pressure	/	±1 KPa
Atmospheric humidity	/	±5% RH
Operation temperature	−30°C~50°C	/

GPRS module. The other computer can visit the specified server to check the running status of the sampler. At the same time, user also can control the sampler reversely, such as parameters setting, start sampling or stop sampling. In addition, the information of fault and warning can be directly sent to the mobile phone which number has been set into the instrument previously.

4 RESULTS AND DISCUSSION

Owing to the hardware and the intelligent embedded software control system, the multi-channel $PM_{2.5}$ sampler reveals well performance and applicability.

The test results of the performance are shown in Table 1. The departure from the working flow rate is less than ±2% at the working flow rate is 5.00 L/min or 16.67 L/min. After calibration, the accuracy of the atmospheric temperature, pressure, and humidity is less than ±2°C, ±1 KPa and ±5% RH, respectively.

5 CONCLUSION

In this paper, the multi-channel $PM_{2.5}$ sampler based on intelligent embedded software control system has been demonstrated. The intelligent control system has major four parts, such as sampling control module, instrument temperature control module, maintenance module, and long-distance monitoring module. The test results validate that the multi-channel $PM_{2.5}$ sampler based on the intelligent control system can achieve excellent performance and applicability, which can provide one of the powerful tool to study the source apportionment of the particulate matter.

ACKNOWLEDGMENT

This work was supported with funding from two Major National Scientific Instrument Projects (No. 2012YQ060147 and No. 2013YQ060569).

REFERENCES

Admir C. Targino, Particia Krecl, Christer Johansson, Erik Swietlick, Andreas Massling, Cuilherme C. Coraiola and Heikki Lihavainen. 2012. Deterioration of air quality across Sweden duo to transboundary agricultural burning emissions. *Bormal Environment Reseach* 18:19–36.

Brook R., Rajagopalan S., Pope A., Brook J.R., Bhatnagar A., Diez-Roux A.V., Holguin F., Hong Y., Luepker R.V., Mittleman M.A., Peters A, Siscovick D., Smith S.C., Whitsel L., Kaufman J. 2010. Particulate matter air pollution an cardiovascular diseasse: an update to the scientific statement from the American Heart Association. *Circulation* 12:2331–2387.

Chakra O.R.A., Joyeux M., Nerriere E., Strub M.P., Zmirou-Navier D. 2007. Genotoxicity of organic extracts of urban airbrone particulate matter: an assessment within a personal exposure study. *Chemosphere* 66:1375–1381.

Doğen M. 2002. Environmental pulmonary health problems related to mineral dust: Examples from central Anatolia, Turkey. *Environmental Geology* 41:571–578.

He X., Li C.C., Lau A.K., et al. 2009. An intensive study of aerosol optical properties in Beijing urban area. *Atmospheric Chemistry and Physics* 9:8903–8915.

Holler R., Tohno S., Kasahara M., Hitzenberger R. 2002. Long-term characterization of carbonaceous aerosol in Uji, Japan. *Atmospheric Environment* 36:1267–1275.

Judith C. Chow, Parkash Doraiswamy, John. G. Watson, L.W., Antony Chen, Steven Sai Hang Ho and David A. Sodeman. 2008. Advances in integrated an dcontinuoys measurements for particle mass and chemical composition. *Journal of The Air & Waste Management Association* 58:141–163.

Nastos T., Athanasios G., Michael B., Eleftheria S.R., Kostas N.P. 2010. Outdoor particulate matter and childhood asthma admission in Athens, Greece: a time-series study. *Environmental Health* 9:1–9.

Ramanathan V., Crutzen P.J., Kiehl J.T., et al. 2001. Aerosol, climate, and the hydrologic cycle. *Science* 294:2119–2124.

Schwartz J. 2004. Air pollution and childeern's health. *Pediatrics*, 113:1037–1043.

Wang Gang, Cheng Shuiyuan, Li Jianbing, Lang Jianlei, Wen Wei, Yang Xiaowen, Tian Liang. 2015. Source apportionment and seasonal variation of $PM_{2.5}$ carbonaceous aerosol in the Beijing-Tianjin-Hebei Region of China. *Environmental Monitoring and Assessment* 187:143.

Wang Jun, Hu Zimei, Chen Y.Y., Chen Zhenlou, Xu Shiyuan. 2013. Contamination characteristics and possible sources of PM_{10} and $PM_{2.5}$ in different functional areas of Shanghai, China. *Atmospheric Environment* 68:221–229.

Wong C.M., Vichit-Vadakan N., Kan H.D., et al. 2008. Public health and air pollution in Asia (PAPA): A multi-city study of short-term effects of air pollution on mortality. *Environmental Health Perspectives* 116:1195–1202.

Yang Lingxiao, Zhou Xuehua, Wang Zhe, Zhou Yang, Cheng Shuhui, Xu Pengju, Gao Xiaomei, Nie Wei, Wang Xinfeng, Wang Wenxing. 2012. Airborne fine particulate pollution in Jinan, China: concentrations, chemical compositions and influence on visibility impairment. *Atmospheric Environment* 55:506–514.

Ye Huajun, Yang Kai, Jiang Xuejiao, Niu Lili, Hua Daozhu, Li Dan, Shi Shuihe. 2014. Hourly variations and potential sources of airborne trace elements in PM_{10} in four representative regions of southeastern China. *Aerosol and Air Quality Research* 14:1986–1997.

Zhou Jiamao, Zhang Renjian, Cao Junji, Chow J.C., Watson J. 2012. Carbonaceous and Ionic Components of Atmospheric Fine Particles in Beijing and Their Impact on Atmospheric Visibility. *Aerosol and Air Quality Research* 12:492–502.

Advances in Energy, Environment and Materials Science – Wang & Zhao (Eds)
© 2016 Taylor & Francis Group, London, ISBN 978-1-138-02931-6

Structure optimization design based on the numerical simulation of subway station deep foundation in open cut

Linli Tan, Ming Li & Mingli Wu
School of Civil Engineering, University of South China, Hengyang, Hunan, China

ABSTRACT: In the subway station deep excavation in open cut, there are many factors affecting its envelope stability for engineering geology and environmental variability. Now, taking Zhengzhou line 2 subway station foundation pit engineering as an example, taking measures monitors the horizontal displacement of retaining piles for deep foundation pits around the retaining structure stability retaining structure, at the same time uses different softwares for numerical simulation. First, in the same location, the datum from the different softwares and construction monitoring are compared. Then, as the project safe, feasible and economy under the premise, the structural optimization design is put forward. Namely, the date obtained software simulation and the corresponding are compared in different position of the inner support measured. Last, the analysis of the retaining structural deformation law and the impact on the structural stability of deep foundation pit in the different support position are necessary, so that the design has been optimized.

1 INTRODUCTION

In recent years, with the gradual increase in the degree of urban density and the widening of the transportation demand, developing underground subway engineering has become trend of ease traffic congestion in the cities in China. Subway Station Deep Foundation Pit construction method is widely used in subway construction. For the construction of the metro station open-cut method, the level of stability envelope is undoubtedly a key issue to the success of its construction, but also is the difficult point in design and construction of ditches (Xie et al. 2007). Deep foundation pit retaining structure is not only to satisfy the demands of structure strength (GB50010-2010.2011), and at the same time to satisfy the requirement of the components deformation, and deep foundation pit engineering of underground construction, but deep excavation for the underground construction in general, can be variable, and factors affecting are many and complex (GB50157-2013.2013). Taking Zhengzhou open-cut metro station deep foundation pit retaining structure as an example, with actual construction monitoring data of row pile retaining structure deformation is given priority to, at the same time for numerical simulation analysis. The datum of Beijing Lizheng simulation program, ANSYS program and the actual construction excavation monitoring were compared and analyzed for the low of deformation (Sun et al. 2004), thus optimizing the structure design.

2 ENGINEERING PROJECTS

2.1 Engineering situation

The station is Zhengzhou City Rail Transit Line 2 project a station, located at the junction of Zhenghua road and three national railway, north-south along Zhenghua road layout. Station starting mileage is ZAK11 + 304.0, the station end mileage is ZAK11 + 490.5, and effective platform centers mileage is ZAK11 + 418.0. The station is underground two island standard station, the station's full-length is 186.5 m, the station's width is 18.7 m, effective platform width is 10.0 m, main structure of the standard segment size is 18.7 × 12.96 m and bottom depth is about 17 m. Shield interval is 23.9 × 12.96 m, and embedded depth of floor is about 17 m, mileage station center turns the soil is about 3.0 m. Based on end hole MBZ3 XLL − 003 as an example, the drilling of the cross section of the retaining structure is calculated, the drill hole in each layer soil distribution and soil physical and mechanical parameters is seen in Table 1.

2.2 Inner support design of foundation pit retaining structure

The station's body is double span on the second floor underground closed box frame structure, using the open-cut method of construction along the construction method. Using bored piles combined with outside the foundation pit precipitation

Table 1. Physical and mechanical parameters of soils.

Soil layer number	Name of the soil	Natural density (kg/m³)	Modulus of compression/MPa	Coefficient of earth pressure at rest	Internal friction angel/(°)	Cohesionc/ KPa	Soil depth/m
1–0	Miscellaneous fill soil	18.0			15.0	15	2.6
2–3	Fine sand	20.3	11.40	0.34	24.6		13.9
4–3	Fine sand	21.2	11.97	0.32	28.5		7.2
4–5	Floury soil	20.5	10.00	0.43	21.0	13	4.5
4–3	Fine sand	21.2	11.97	0.32	28.5		3.0
4–5	Floury soil	20.5	10.00	0.43	21.0	13	4.8

Comment: According to the geological exploration reports.

scheme. Foundation pit retaining structure use 1 m distance of 1.2 m diameter bored piles and inside-pipe-steel support. Outward from the 300 mm diameter is 600 mm distance of 850 mm diameter spacing of three-axis mixing pile waterproof curtain. Set three supports and a transformation in the foundation pit support, four supports adopt Φ 609 steel pipes, specific arrangement as shown in the Figure 1.

2.3 Finite element modeling and simulation

The deep foundation pit retaining structure mainly includes the whole retaining piles, steel support, reinforced concrete structure, the surrounding soil and much more. Because the subway deep foundation pit construction process is relatively complex, when using finite element method analysis of soil mass is regarded as elastic or elastic-plastic material, and make the following assumptions: (1) Rock mass is regarded as continuous homogeneous and isotropic medium, using D-P yield criterion. (2) Only the effect of gravity stress of soil is considered (Huo et al. 2011). Interactions between the parts in the process of engineering construction, during the establishment of the station model, the horizontal and vertical direction of soil from each boundary is about three times the width and depth of foundation pit excavation. In ANSYS software, entity elements Solid45 is used to simulate soil and steel support use beam element beam188 simulated (Liu et al. 2010). Retaining pile adopts concrete C30. Modulus of elasticity is 25 Gpa. Poisson's ratio is 0.20, natural gravity is 25 KN/m² of concrete C30. Elastic modulus is 25 Gpa. Poisson's ratio is 0.20, natural gravity is 25 KN/m² of steel support. Before the excavation has been underway for foundation pit precipitation process, so the effect of groundwater on foundation pit retaining structure deformation is not considered. Deep foundation pit integral finite element model are shown in Figure 2.

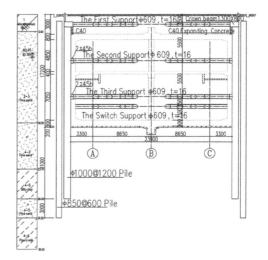

Figure 1. Cross section of the main station building envelope.

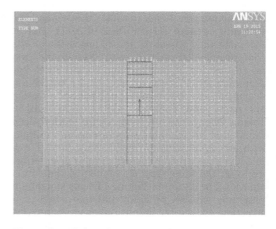

Figure 2. Finite element model of the foundation excavation.

3 STRUCTURE ANALYSIS

3.1 Compared with the results of numerical simulation of construction monitoring

This paper selects the station at the end of a section of the well cross-section as a model, using three steel supports and a conversion support, the first line of the horizontal spacing is 6 m, the remaining three steel supports are 3 m. Each vertical distance of supports is 1.5 m, 5.5 m, 5.5 m and 1.5 m. Statistics by monitoring the actual construction of the horizontal displacement of the pile, and the resulting simulation model data and the corresponding ANSYS, Beijing rationale for deep excavation F-SPW7.0PB1 version of the numerical simulation results were compared and the specific comparison results are below Figure 3.

According to the construction of foundation pit supporting technology regulation (JGJ120-2012.2012), this engineering foundation pit safety level is level 1, so that the deformation control protection level is level 1. The maximum outer surface subsidence is 25.8 mm or less (0.15% or less H), maximum horizontal displacement of retaining structures of 25.8 mm or less (0.15% or less H and 30 mm or less), H for foundation pit excavation depth. The deformations meet the deep foundation pit horizontal displacement deformation. By the statistics and computer simulation results analysis, the overall deformation of the pile tends to "bow" type, deformation roughly three overall trends are similar, at about 2/3 of maximum deformation of the pile. Therefore, in practical engineering, numerical simulation program can be used to simulate the actual project, calculate the approximate amount of deformation of pile construction is more, by continuously adjusting the parameters and then making the structure further optimized design (Li et al. 2012).

3.2 Results analysis

Different design scheme in actual construction engineering structure causes different degree of the deformation (Yang.2011). In this paper, on the basis of the original scheme, by changing the location of original design, Lizheng and ANSYS numerical simulation program is simulated, specific as follows as shown in Table 2 and Figure 4.

The results of comparative analysis:

1. According to the data in Table 2 and Figure 4 showed, changing location within the support have a major impact on the horizontal displacement of the pile among other factors under the same conditions. In practical engineering design, it is better to take its optimum design position of support after repeated calculation simulation.
2. With the support and excavation of pit, the maximum horizontal displacement of the position of the pile is about place on the partial excavation depth position and its deformation overall tend to "bow" type. So we can take corresponding measures to prevent the other near the location where more large deformation takes place.

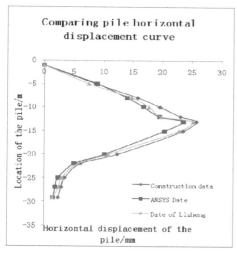

Figure 3. Correlation curve of horizontal deformation about the pile.

Table 2. Supporting position of the optimization design.

Construction conditions	Design 1	Design 2	Design 3	Design 4	Design 5	Design 6
Vertical spacing 1 (m)	1.5	1.5	1.5	1.5	1.5	1.5
Vertical spacing 2 (m)	5.3	5.5	5.7	5.9	6.1	6.3
Vertical spacing 3 (m)	5.7	5.5	5.3	5.1	4.9	4.7
Vertical spacing 4 (m)	1.5	1.5	1.5	1.5	1.5	1.5
Maximum displacement deformation of pile (mm)	26.15	25.2	24.98	23.07	22.01	22.55

Comment: The difference of each design is the size of the vertical distance between 2 and 3.

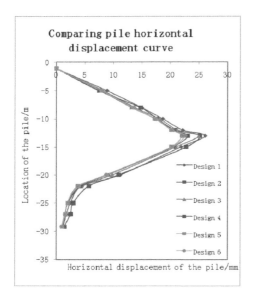

Figure 4. Correlation curve of horizontal deformation about the pile.

3. Because of the complex underground engineering, construction error, so the actual project we should try to make deep pit design optimization, making it the amount of deformation to meet regulatory requirements, enable the project to achieve security, stability effect.

4. Based on the above data showed, the change trend of retaining structure horizontal displacement of the pile and the actual construction monitoring data is similar by Lizheng and ANSYS numerical simulation software. So the calculation results can be appropriately used as a design basis for reference.

REFERENCES

Code for the design of metro (GB50157-2013).Beijing, China Planning Press. 2013.

Code for Design of Concrete Structures (GB50010-2010). Beijing, China Building Industry Press. 2011.

Construction of foundation pit supporting technology procedures (JGJ120-2012). Beijing, China Planning Press, 2012.

Huo Runke, Yan Mingyuan, Song Zhanping. The Monitoring and Numerical Analysis of Deep Foundation in Subway Station [J]. Journal of Railway Engineering Society, 2011(5), 81–85.

Li Shu, Zhang Dingli, Fang Qian, Lu Wei. Research on Characteristics of Ground Surface Deformation during Deep Excavation in Beijing Subway [J]. Chinese Journal of Rock Mechanics and Engineering, 2012, 31(1): 189–198.

Liu Jie, Yao Hailin, Ren Jianxi. Monitoring and numerical simulation of deformation of retaining structure in subway station foundation pit [J]. Rock and Soil Mechanics, 2010, 31(S2): 456–461.

Sun Kai, Xu Zhengang, Liu Tingjin, Fang Zongxin. Construction Monitoring and Numerical Simulation Foundation of a Analysis Pit [J]. Chinese Journal of Rock Mechanics and Engineering, 2004, 24(2): 293–298.

Xie Xiu-dong, Liu Guo-bin, Li Zhi-gao, Guo Zhij-ie. Analysis of Soil Layers Displacement Characteristics in Foundation Pit Adjacent to Subway Station [J]. Chinese Journal of Underground Space and Engineering, 2007, 3(4): 742–744, 757.

Yang Lei. Subway Deep Foundation Support Programs of Optimization Discussion [D]. Wuhan University of Technology, 2011.

Advances in Energy, Environment and Materials Science – Wang & Zhao (Eds)
© 2016 Taylor & Francis Group, London, ISBN 978-1-138-02931-6

Numerical calculation on Stress Intensity Factor in rock using Extended Finite Element Method

Bo Sun, Bo Zhou & Shifeng Xue

College of Pipeline and Civil Engineering, China University of Petroleum, Qingdao, China

ABSTRACT: The Extended Finite Element Method (XFEM) is a new numerical method for modeling discontinuities such as inclusions, holes, cracks etc. within a standard finite element framework. Compared with conventional finite element method, it has a unique advantage that the problem domain is modelled without explicitly meshing the discontinuities. Thus it can improve the computational efficiency and has broad use in the field of the fracture mechanics. The basic theory of XFEM is presented and the discretization scheme is derived. Numerical examples are carried out to quantitatively study the influencing factors-mesh size, crack length and domain size of interaction integral—of Stress Intensity Factor (SIF) of rock crack. Results show that the mesh size and the domain size of interaction integral will affect the computational accuracy. The constant value c associated with the domain size of interaction integral is suggested.

1 INTRODUCTION

As a kind of special geological body, rock has many multi-scaled inner defects, such as pores and cracks. Lots of engineering practice show that the occurrence, growth and link-ups of cracks inside the rock will directly affect the stability of structures. Thus, it's of great importance in practice to predict the crack initiation and crack path in order to keep them safe and reliable.

To analyze discontinuity problems, Belytschko (1999) proposed a new kind of numerical method-Extended Finite Element Method (XFEM). Based on the partition of unity method, XFEM adds enrichment functions to the standard FEM displacement approximation to accurately describe the discontinuous boundary. XFEM has inherited the advantages of FEM and has favorable transplantation. Besides, XFEM owns a unique merit: discontinuities are independent of the finite element mesh. Therefore, there is no need of remeshing when simulating discontinuity (crack) evolution. So it improves the computational efficiency.

Due to the special advantage in dealing with discontinuity problems, XFEM has been a focus in the area of computational mechanics and fracture mechanics and rapidly developed for more than a decade. Daux et al. (2000) constructed enrichment functions according to the discontinuous geometric feature, and analyzed arbitrary branched and intersecting cracks. Sukumar et al. (2000) adopted two-dimensional asymptotic crack tip functions to model three-dimensional crack. Stress Intensity

Factors (SIFs) are in good agreement with benchmark solutions. Li & Wang (2005) systematically introduced basic theory, implementation procedures and formulations and practical applications of XFEM. Fang & Jin (2007) presented a virtual node method to implement XFEM in commercial finite element software ABAQUS. Sukumar et al. (2001) used the level set method to represent the location of holes and material interfaces and coupled the level set method to XFEM. Ying et al. (2008) simulated the displacement field of a plate with multi-circular inclusions under uniaxial tension by XFEM and compared results with conventional finite element method to verify the extended finite element method. Song et al. (2006) described the discontinuity by superposed elements and phantom nodes and presented a new method for modelling of arbitrary dynamic crack and shear band propagation. Gong and Yu (2013) established the XFEM model of initiation and evolution of shear band by a simple growth algorithm.

This paper elaborates in detail the basic theory of XFEM and derives the discretization scheme of XFEM. Through numerical experiments, we quantitatively analyze the three influencing factors of SIFs of the rock crack: mesh size, crack length and domain size of interaction integral; we also compare the numerical results with theoretical solutions to verify the robustness of XFEM. Besides, the optimal domain size of interaction integral is determined by choosing proper value of constant c from results of numerical examples.

2 BASIC THEORY OF XFEM

2.1 Displacement approximation

The finite element approximation for a single crack in a two-dimensional body can be written as

$$u^h = \sum_{i \in I} u_i \phi_i + \sum_{j \in J} a_j \phi_j H(x) + \sum_{k \in K} \phi_k \left(\sum_{l=1}^{4} b_k^l F_l(x) \right) \quad (1)$$

where I is the set of all nodes; J is the set of nodes whose shape function support is cut by a crack; K is the set of nodes whose shape function support contains the crack front; ϕ_i is the shape function associated to node i; u_i are classical degrees of freedom (i.e. displacement) for node i; a_i account for the jump in the displacement field across the crack at node j; b_k^l are the additional degrees of freedom associated with the crack-tip enrichment functions F_l; $H(x)$ is the Heaviside function to represent the discontinuity across the crack; $F_l(x)$ are four asymptotic functions associated with crack-tip displacement field in a linear elastic solid (Fig. 1).

The near-tip functions $F_l(x)$ are given by

$$\{F_l(r, \theta)\} \equiv \left\{ \sqrt{r} \sin\left(\frac{\theta}{2}\right), \sqrt{r} \cos\left(\frac{\theta}{2}\right), \right.$$
$$\left. \sqrt{r} \sin\left(\frac{\theta}{2}\right)\sin(\theta), \sqrt{r} \cos\left(\frac{\theta}{2}\right)\sin(\theta) \right\} \quad (2)$$

where (r, θ) are local polar co-ordinates at the crack tip.

2.2 Governing equations

Consider the problem domain Ω and its boundary Γ as shown in Figure 2. The boundary Γ is consisted of the outer boundary Γ_u, Γ_t and inner boundary Γ_c, such that $\Gamma = \Gamma_u \cup \Gamma_t \cup \Gamma_c$. Γ_u is the

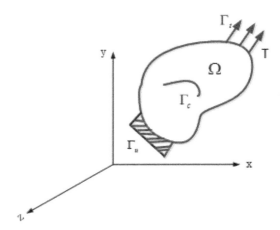

Figure 2. Cracked body subjected to loads.

displacement boundary while Γ_t is the traction boundary. The crack surface Γ_c is composed of Γ_c^+, and Γ_c^- is assumed to be traction-free.

The strong form of equilibrium and boundary conditions are

$$\nabla \cdot \sigma + b = 0 \quad \text{in } \Omega \quad (3a)$$

$$\sigma \cdot n = T \quad \text{on } \Gamma_t \quad (3b)$$

$$\sigma \cdot n = 0 \quad \text{on } \Gamma_c^+ \quad (3c)$$

$$\sigma \cdot n = 0 \quad \text{on } \Gamma_c^- \quad (3d)$$

$$u = U \quad \text{on } \Gamma_u \quad (3e)$$

where n is the unit outward normal; σ is the Cauchy stress, u is the displacement, b is the body force per unit volume. T and U represent the stress and displacement on the boundary respectively.

2.3 Discretization scheme

Under the assumption of small strains, the constitutive relation is given by Hooke's law

$$\sigma = C : \varepsilon \quad (4)$$

where C is the Hooke's tensor.

The space of admissible displacement fields is defined by

$$\Omega_d = \{u = U \text{ on } \Gamma_u, u \in C^0 \text{ and discontinuous on } \Gamma_c\}$$

The space of test function is defined similarly as

$$\Omega_{d0} = \{v = 0 \text{ on } \Gamma_u, v \in C^0 \text{ and discontinuous on } \Gamma_c\}$$

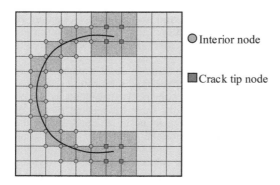

○ Interior node

■ Crack tip node

Figure 1. Computational mesh and enriched nodes for a cracked body.

The weak form of the equilibrium equations is given by the principle of virtual work for arbitrary $v \in \Omega_{d0}$

$$\int_{\Omega} \varepsilon(\mathbf{u}) : C : \varepsilon(v) d\Omega = \int_{\Omega} b \cdot v d\Omega + \int_{\Gamma_t} T \cdot v d\Gamma \quad (5)$$

On substituting Equation (1) into Equation (5), the following discrete system of linear equations is obtained

$$KU = f \quad (6)$$

where U is the vector of unknown degrees, K is the global stiffness matrix and f is the external force vector. Element stiffness K_{ij} and element external force f_i are defined in the following

$$K_{ij}^e = \begin{bmatrix} K_{ij}^{uu} & K_{ij}^{ua} & K_{ij}^{ub} \\ K_{ij}^{au} & K_{ij}^{aa} & K_{ij}^{ab} \\ K_{ij}^{bu} & K_{ij}^{ba} & K_{ij}^{bb} \end{bmatrix} \quad (7)$$

$$f_i^e = \left\{ f_i^u, f_i^a, f_i^{b1}, f_i^{b2}, f_i^{b3}, f_i^{b4} \right\}^T \quad (8)$$

$$U = \left\{ u, a, b_1, b_2, b_3, b_4 \right\}^T \quad (9)$$

In Equation (7) and (8), the components of stiffness matrix K and external force vector f are defined respectively as

$$K_{ij}^{rs} = \int_{\Omega} (B_i^r)^T D B_j^s d\Omega \quad (r, s = u, a, b) \quad (10a)$$

$$f_i^u = \int_{\Gamma_t} N_i T d\Gamma + \int_{\Gamma_t} N_i b d\Omega \quad (10b)$$

$$f_i^a = \int_{\Gamma_t} N_i H T d\Gamma + \int_{\Gamma_t} N_i H b d\Omega \quad (10c)$$

$$f_i^{bj} = \int_{\Gamma_t} N_i F_\alpha T d\Gamma + \int_{\Gamma_t} N_i F_\alpha b d\Omega \quad (\alpha = 1, 2, 3, 4) \quad (10d)$$

In above equations, D is the constitutive matrix for an isotropic linear elastic material, B_i^u, B_i^a, and B_i^b are the matrix of shape function derivatives which are given by

$$B_i^u = \begin{bmatrix} N_{i,x} & 0 \\ 0 & N_{i,y} \\ N_{i,y} & N_{i,x} \end{bmatrix}, \quad B_i^a = \begin{bmatrix} (N_i H)_{,x} & 0 \\ 0 & (N_i H)_{,y} \\ (N_i H)_{,y} & (N_i H)_{,x} \end{bmatrix}$$

$$B_i^b = \begin{bmatrix} B_i^{b1}, B_i^{b2}, B_i^{b3}, B_i^{b4} \end{bmatrix},$$

$$B_i^{b\alpha} = \begin{bmatrix} (N_i F_\alpha)_{,x} & 0 \\ 0 & (N_i F_\alpha)_{,y} \\ (N_i F_\alpha)_{,y} & (N_i F_\alpha)_{,x} \end{bmatrix} (\alpha = 1, 2, 3, 4)$$

3 INFLUENCING FACTORS OF ACCURACY OF SIF OF ROCK CRACK

3.1 Mesh size

Consider an edge crack in rock under uniaxial tension as shown in Figure 3. The geometry size of the model is given by $L/W = 6/2$ and $a/W = 1/2$. The model uses structured mesh as shown in Figure 4.

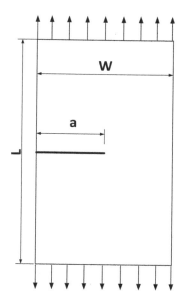

Figure 3. Edge crack under uniaxial tension.

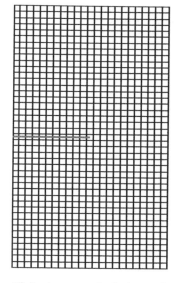

Figure 4. Finite element mesh of edge crack.

The theoretical solution of SIF can be found in the SIF handbook which is given by.

$$K_I^t = (1.12 - 0.231\lambda + 10.55\lambda^2$$
$$-21.72\lambda^3 + 30.39\lambda^4)\sigma_0\sqrt{\pi a} \quad (11)$$

where $\lambda = a/W$ and σ_0 is applied stress.

After calculations, normalized stress intensity factors are given by $K_I^n = K_I^{XFEM}/K_I^t$ and the results are listed in Table 1.

The results in the Table 1 show that when the mesh size decreases, the computational accuracy of SIFs increases. From this numerical example, good accuracy will be achieved when the mesh size is 1/15. It also demonstrates that XFEM has the same good solutions with coarser mesh compared to the conventional finite element method, thus improving computational efficiency.

3.2 Crack length

In this section, we use the same numerical example in previous section to consider effect of crack length on stress intensity factor. For simplicity, let the dimensionless crack length be a/W and take $a/W = 0.2, 0.3, 0.4, 0.5$ respectively.

It can be seen in the Table 2 that when the mesh size is fixed, the increase of crack length has little effect on the accuracy of SIFs, and the accuracy is good when proper mesh size is chosen; when the crack length is fixed, the accuracy of SIFs improves with the increase of the mesh density. So we can conclude that the change of crack length would

Table 1. Normalized stress intensity factors of different mesh sizes.

Mesh size	Node number	K_I^n
1/5	11×31	0.8816
1/10	21×61	0.9883
1/15	31×91	0.9953
2/35	36×72	0.9957
1/20	41×121	0.9961
1/30	61×181	0.9971
1/40	81×241	0.9973

Table 2. Normalized SIFs of different crack length and mesh.

Mesh size	Dimensionless crack length a/W			
	0.2	0.3	0.4	0.5
1/10	0.9772	0.9839	0.9919	0.9990
1/15	0.9964	0.9986	1.0017	0.9966

not affect the accuracy of SIFs while the accuracy would increase with high mesh density.

3.3 Domain size of interaction integral

The interaction integral method (Nikishkov, G.P. & Atluri, S.N. 1987) is utilized by XFEM to compute SIFs. The domain of interaction integral is centered at crack tip (Fig. 5) and the radius of the domain is r which is given by

$$r = c\sqrt{A} \quad (12)$$

where c is constant and A is the area of an element.

Now we use the same case in section 3.1 to analyze the effect of domain radius on the accuracy of SIFs. Define relative error of K_I as

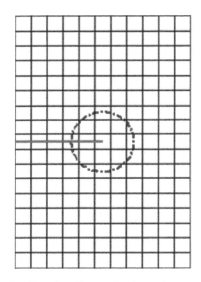

Figure 5. Domain of interaction integral.

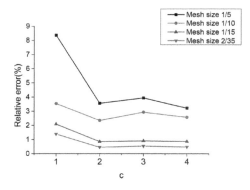

Figure 6. The relation between c and relative error.

$\left|\left(K_I^{XFEM} - K_I^t\right)\middle/K_I^t\right|$. We can see from the Figure 6 that the relative error of K_I computed by XFEM decreases with the increase of constant c. We can also draw a conclusion that when the value of c is larger than 2 the relative error is low and changes slowly. The computational requirements are met when the value of c is 2.

4 CONCLUSIONS

In above analysis, we can conclude that in the XFEM, the crack is treated as a separate geometric entity and the finite element meshes are independent of the crack. Only a unique set of mesh is introduced to compute the SIFs without remeshing in the vicinity of crack tips. Thus the computational efficiency is improved. The accuracy of SIFs will increase as the mesh size decreases and the crack length has little impact on the accuracy of SIFs. In addition, the domain size of interaction integral will affect the accuracy of SIFs and the proper constant c associated with the domain size is 2 by numerical studies.

XFEM has the exceptional advantage that the mesh is independent of the discontinuity. It avoids the trouble of remeshing during the crack propagation and reduces the complexity of computation. The only drawback of XFEM is the need for a variable number of degrees of freedom per node. However, XFEM still has promising applications in both static and dynamic discontinuity problems.

REFERENCES

Belytschko, T. & Black, T. 1999. Elastic crack growth in finite elements with minimal remeshing. *International Journal for Numerical Methods in Engineering* 45(5):601–620.

Belytschko, T. & Gracie, R. 2007. On XFEM applications to dislocations and interfaces. *International Journal of Plasticity* 23:1721–1738.

Daux, C., Moës, N., Dolbow J., Sukumar, N., Belytschko, T. 2000. Arbitrary branched and intersecting cracks with extended finite element method. *International Journal for Numerical Methods in Engineering* 48:1741–1760.

Fang, X.J. & Jin, F. 2007. Extended finite element method based on ABAQUS. *Engineering Mechanics* 24(7):6–10.

Gracie, R. & Belytschko, T. 2009. Concurrently coupled atomistic and XFEM models for dislocations and cracks. *International Journal for Numerical Methods in Engineering* 78:354–378.

Gong, Z.W. & Yu, T.T. 2013. Modeling of shear band evolution in soils by XFEM. *Chinese Journal of Underground Space and Engineering* S2:1817–1821+1826.

Li, L.X. & Wang, T.J. 2005. The extended finite element method and its applications. Advances in Mechanics 35(1): 5–20.

Melenk, J.M. & Babuska, I. 1996. The partition of unity finite element method: Basic theory and applications. *Computer Methods in Applied Mechanics and Engineering* 139:289–314.

Nikishkov, G.P. & Atluri, S.N. 1987. Calculation of fracture mechanics parameters for an arbitrary three-dimensional crack by the 'equivalent domain integral method'. *International Journal for Numerical Methods in Engineering* 24(9):1801–1821.

Song, J.H., Areias, P. & Belytschko, T. 2006. A method for dynamic crack and shear band propagation with phantom nodes. *International Journal for Numerical Methods in Engineering* 67:868–893.

Sukumar, N., Moës, N., Moran, B., Belytschko, T. 2000. Extended finite element method for three-dimensional crack modelling. *International Journal for Numerical Methods in Engineering* 48:1549–1570.

Sukumar, N., Chopp, D.L., Moës, N., Belystchko, T. 2001. Modelling holes and inclusions by level sets in extended finite-element method. *Computer Methods in Applied Mechanics and Engineering* 190: 6183–6200.

Ying, Z.Q., Du, C.C. & Cheng, L. 2008. Application of extended finite element method in heterogeneous materials with inclusions. *Journal of Hohai University (Natural Sciences)* 04:546–549.

Advances in Energy, Environment and Materials Science – Wang & Zhao (Eds)
© *2016 Taylor & Francis Group, London, ISBN 978-1-138-02931-6*

Yinger Learning Dynamic Fuzzy Neural Network algorithm for the three stage inverted pendulum

Ping Zhang
College of Electrical and Information Engineering, Lanzhou University of Technology, Lanzhou, China

Guodong Gao
University Hospital of Gansu Traditional Chinese Medicine, Lanzhou, China

Xin Zhang
State Grid Gansu Maintenance Company, Lanzhou, China

Wei Chen
College of Electrical and Information Engineering, Lanzhou University of Technology, Lanzhou, China

ABSTRACT: In order to avoid the over fitting and training and solve the knowledge extraction problem in fuzzy neural networks system. The Yinger Learning Dynamic Fuzzy Neural Network (YL-DFNN) algorithm is proposed. The Learning Set based on Yinger Learning is constituted from message. Then the framework of Yinger Leaning Dynamic Fuzzy Neural Network is designed and its stability is proved. Finally, Simulation results of the three stage inverted pendulum system indicates that the novel Lazy Learning Dynamic Fuzzy Neural Network is fast, compact, and capable in generalization.

1 INTRODUCTION

The fuzzy neural network control in the control domain has already become a hot topic at present. Applying the neural network in fuzzy systems may solve the fuzzy systems knowledge extraction problem; applying the fuzzy system in neural networks, the neural network is no Longer a black box, and humanity's knowledge is very easy to fuse in the neural network. It is apparent that a fuzzy neural network derives its computing power through, first, massively parallel distributed structure and, second, ability to learn and therefore generalize.

1.1 *Dynamic fuzzy neural network*

The most useful property of the FNN is the ability to arbitrarily approximate linear or nonlinear mappings through learning. To improve short-term load forecasting accuracy, a Modified Particle Swarm Optimizer (MPSO) and Fuzzy Neural Network (FNN) hybrid optimization algorithm is proposed in paper (Huang, 2015). In which the FNN is trained by MPSO to implement the optimization of FNN parameters. The short term load forecasting accuracy is improved in Guizhou power system, whose average percentage error is not more than 1.2%. A fault-tolerant control method based on fuzzy neural networks was presented for nonlinear

systems in paper (Zhang, 2013). The fault parameters were designed to detect the fault, adaptive updating method was introduced to estimate and track fault, and fuzzy neural networks were used to adjust the fault parameters and construct automated fault diagnosis. The simulation results in induction motor show that it is still able to work well with high dynamic performance and control precision under the condition of motor parameters' variation fault and load torque disturbance. In paper (Meng, 2014), a fuzzy diagonal regression neural networks recurrent forecast model is proposed based on analyzing influential factors of passenger traffic volume and the experimental results show that the simulation system has well application prospect than the promoted value.

However, all of them do not think about input messages. Input-output specimens with a lot of futile noise have been collected by sensors in the control system (David, 2013). All these useless messages make nonlinear systems more complicated. In order to solve this problem, Yinger Learning Dynamic Fuzzy Neural Network (YL-DFNN) algorithm is first proposed in this paper. It can extract similar messages form input-output specimens to structure identification. Form theoretical analysis and factual examination, we can see that the LL-DFNN is fast, compact, capable in generalization.

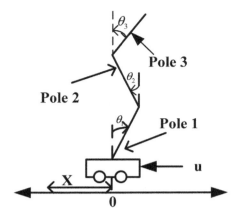

Figure 1. Model of the three stage inverted pendulum system.

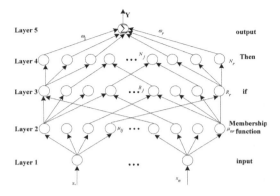

Figure 2. Network topology.

1.2 Three stage inverted pendulum

Set up three stage inverted pendulum system in MATLAB. Figure 1 is simplified model of it. Where u is the input, x is displacement of the car, the quality of the car is $0.7\ kg$, the quality of poles are $0.1\ kg$, the length of poles are $0.25\ m$, friction factor is $22.9\ N \cdot s/m$. Let $K = 60$ and $J \leq 0.01$.

2 YINGER LEARNING DYNAMIC FUZZY NEURAL NETWORK

2.1 Network topology

In the paper, the dynamic fuzzy neural network put in Figure 1 includes 5 layers:

Layer 1: Each node in layer 1 represents an input $X = [x_1 \cdots x_n]^T$

Layer 2: Each node in layer 2 represents a member ship function (MF), which is in the form of Gaussian

$$\mu_{ij} = e^{-(x_i - c_i)^2 / \sigma_j^2} \tag{1}$$

where $i = 1,\ldots,n$, $j = 1,\ldots,r$, μ_{ij} is the membership function of the input variable x_i, σ_j is the center of the Gaussian membership of, is the width of the Gaussian membership of x_i.

Layer 3: Each node in layer 3 represents a possible rule for fuzzy rules.

$$\phi_j = e^{-\sum\limits_{i=1}^{k}(x_i - c_i)^2 / \sigma_j^2} \tag{2}$$

where $i = 1,\ldots,n$, $j = 1,\ldots,r$.

Layer 4: Each node in layer 4 represents the then

$$\varphi_j = \phi_j \Big/ \sum_{k=1}^{r} \phi_k \tag{3}$$

where $j = 1,\ldots,r$.

Layer 5: Each node in layer 4 represents the output variable as a weighted summation of incoming signals.

$$y(x) = \frac{\sum\limits_{k=1}^{r}[(a_{k_0} + a_{k_1}x_1 + \ldots a_{k_n}x_n)]e^{-\sum\limits_{i=1}^{k}(x_i - c_i)^2 / \sigma_j^2}}{\sum\limits_{i=1}^{r} e^{-\sum\limits_{i=1}^{k}(x_i - c_i)^2 / \sigma_j^2}} \tag{4}$$

where $k = 1,\ldots,r$.

2.2 Yinger Learning Dynamic Fuzzy Neural Network algorithm

1. Criterion of neuron generation

$$\text{let } \|E_i\| = \|Y_i - \hat{Y}_i\| \tag{5}$$

$$\text{if } \|E_i\| = \|Y_i - \hat{Y}_i\| > E_e \tag{6}$$

add one rule.while $E_e = cD(X_1, X_d)$ and c is coefficient of error which is decided by system.

2. Allocation of RBF unit parameters

σ_j is the width of the Gaussian membership and it is important to generalization. Using lazy leaning results to allocation of these parameters.

$$\text{Let } C_1 = X_1, \ \sigma_1 = D(X_1, X_d) \tag{7}$$

when more messages were trained in the network, new parameters are defined as follows:

$$C_i = X_i, \ \sigma_1 = E_i \tag{8}$$

3. Weight adjustment

By local space, fuzzy rules can be defined as:

$$\varphi = \begin{bmatrix} \phi_{11} \cdots \phi_{1k} \\ \vdots \quad \vdots \\ \phi_{r1} \cdots \phi_{rk} \end{bmatrix} \tag{9}$$

The output: $Y = W\psi$ (10)
input: $X_j = [x_{1j}, ..., x_{nj}]$,

$$W = [a_{10} \cdots a_{r0} \cdots a_{1n} \cdots a_{rn}],$$

$$\Psi = \begin{bmatrix} \phi_{11} \cdots \phi_{1q} \\ \vdots \quad \vdots \\ \phi_{r1} \cdots \phi_{rq} \\ \vdots \quad \vdots \\ \phi_{11} x_{n1} \cdots \phi_{1q} x_{nq} \\ \vdots \quad \vdots \\ \phi_{r1} x_{n1} \cdots \phi_{rq} x_{nq} \end{bmatrix}, \ W \in R^r, \ \Psi \in R^{r \times q}$$

Let perfect output:

$$\hat{Y}_j = [\hat{y}_1, ..., \hat{y}_q] \in R^K \tag{11}$$

$$J = \min \sum_{i=1}^{k} \left\| \hat{Y}_i - Y_i \right\| \tag{12}$$

In order to reduce computing using recursive-least-squares get:

$$W_i = W_{i-1} + E_i \Psi_i^T (\hat{Y}_i - \Psi_i W_{i-1}) \tag{13}$$

$$E_i = E_{i-1} - \frac{E_{i-1} \Psi_i^T \Psi_i E_{i-1}}{1 + \Psi_i E_{i-1} \Psi_i^T} \tag{14}$$

where, E_i is covariance matrix of error, Ψ_i is column vector of Ψ, W_i is coefficient matrix after i time iterative learning algorithm. The local space is dynamic with input, so this method can avoid over-saturation.

4. Stability of system

1. equilibrium point

Define1. Let $Y_j(X) = \sum_{k=1}^{r} \varpi_k \phi_k, j = 1, \cdots q$

where $\phi_k, k = 1, \cdots q$ and:

1. $|\phi_k(X_c)| \le M_j, X_c \in \Omega, k = 1, \cdots q$
2. $|\phi_k(X_1) - \phi_k(X_2)| \le K_k |X_1 - X_2|, X_1, X_2 \in \Omega,$
   $k = 1, \cdots q$,

While $M_j \ge 0, K_k \ge 0$ the system at least has the equilibrium point.

Prove: if $X_s = [x_{1s}, \cdots, x_{ns}]^T$ is equilibrium point of system, then $Y = W\Psi$, let mapping $F(X) = W\Psi$, from Define 3 the $\phi(X)$ is a mapping of $\Omega^n \to \Omega^n$, so $F(X)$ also is a mapping of $\Omega^n \to \Omega^n$.

$$\|X\| = \sqrt{\sum_{i=1}^{n} x_i^2} \text{ then } \|F(X)\| = \|W\Psi\| \le \|W\| \cdot \|\Psi\|.$$

From Brouwer theory, $\exists X_w \in \Omega$, to $F(X_s) = X_s$, so X_s is a equilibrium point of system.

2. Stability of equilibrium point

Let $G = X - X_S = [g_1, \cdots, g_n]^T$, using Lyapunov function

Define 2 exist Lyapunov function V, meet demand: $V \ge 0$ and $V' \le 0$, then equilibrium point is steady.

Prove: form YL-DFNN algorithm get the Lyapunov function

$$V = \sum_{i=1}^{n} E(X_i, X_d) G_i^2,$$

$$E(\cdot, \cdot) \in (0,1) \text{ so } V \ge 0$$

$$\frac{dV(t)}{d(t)} = \frac{d \sum_{i=1}^{n} D(X_i, X_d) G_i^2}{d(t)}$$

$$\le \sum_{i=1}^{n} [G_i[-G_i + \sum_{i=1}^{n} D(X_i, X_d) G_i]]$$

$$\le \sum_{i=1}^{n} [[-1 + D(X_1, X_d)] G_i^2] \le 0 \tag{15}$$

5. Steps of YL-DFNN

Step 1. Form new message X_d to local space Ω_k with;

Step 2. Let local space Ω_k to networks;

Step 3. Criterion of neuron generation by (6);

Step 4. Allocation of RBF units parameters by (8);

Step 5. If error meets the system's demand at (12), to step 6, otherwise gets $i + 1$ to step 3.

Step 6. Keep down W and ψ, calculate output of $Y = W\psi$

3 SIMULATION

3.1 *Network parameters comparison*

Table 1 is the contrast of YL-DFNN and DFNN, and YL-DFNN has less training parameters than DFNN.

The new network only has 7 rules to the same system, but the old network has 15 rules. On

Table 1. Contrust of YL-DFNN and DFNN.

	YL-DFNN	DFNN
Rule	7	15
Parameter number	64	120
Root-mean-square error	0.0095	0.0175
Runtime/S	5.16	6.55

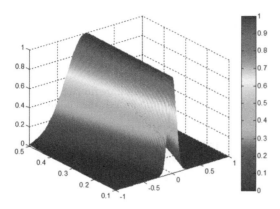

Figure 3. Curved surface of one membership function.

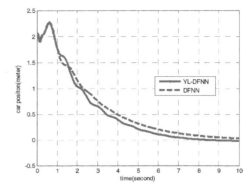

Figure 4. Position of car.

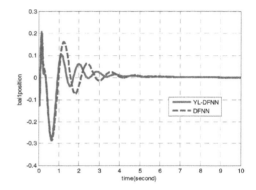

Figure 5. Angle of θ_1.

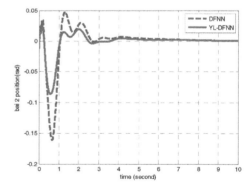

Figure 6. Angle of θ_2.

the other hand, the Root-mean-square error of new network is 0.0095 and the Root-mean-square error of old network is 0.0175, so we can see that the new system is better than traditional system.

Trajectory adjustment.

3.2 Membership function

The Figure 3 is the curved surface of one membership function.

From the Figure 3, we can see that the width of anyone membership function is gradually shrinking with the process of learning. So the controller based on YL-DFNN is gradually improving into optimal controller with optimal capacity and simple structure.

3.3 System trajectory

Figure 4 is the car position, and the car position is adjusted to the equilibrium point under the YL-DFNN in 8.667 s but the DFNN is 9.945 s. Figure 5 is the angle of pole 1, and its position is adjusted to the equilibrium point under YL-DFNN in 3.867 s but the DFNN is 4.945 s.

Figure 6 is the angle of pole 2, and its position is adjusted to the equilibrium point under YL-DFNN in 4.056 s but the DFNN is 5.835 s. Figure 7 is the

angle of pole 3, and its position is adjusted to the equilibrium point under YL-DFNN in 2.952 s but the DFNN is 6.295 s.

From them, we get the result that new algorithm is quicker and the system is steadier.

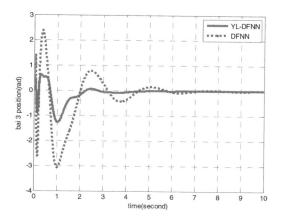

Figure 7. Angle of θ_1.

4 CONCLUSION

The YL-DFNN algorithm was found to be fast due to its dynamic local space and computational simplicity in determining the optimum weights. In practice, noise will invariably be present. Hence, it is necessary to dispose of rubbish message from input sets. YL-DFNN algorithm uses current data and history data, so it accords with the trait of dynamic channel. It adjusts the coefficient of the equalizer dynamically according to the received signals to enhance the equalization performance. The simulation results with three stage inverted pendulum system shows that the new algorithm has a faster convergence speed, lower error rate, smaller root-mean-square error than DFNN algorithm.

ACKNOWLEDGEMENT

This work was financed by the research fund of the National Natural Science Foundation of China (No. 61263008), (No. 2013 A-028), No. 1208RJYA070), (No. 1106B-10).

REFERENCES

Cai Xiushan. "Global asymptotic stability of a class of nonlinear systems with parametric uncertainty". *Journal of Systems Engineering and Electronics*. vol. 20, pp.168–173, January 2012.

David Hand. "Principle of Date Mining". *Citic Publishing House*. pp.36–45, August 2013.

Dietrich Wettschereck, "A Review and Empirical Evaluation of Feature Weighting Methods for a Class of Lazy Learning Algorithms". *Artificial Intelligence Review*, vol. 25, pp.273–314. January 2014.

Huang Xiaoping R; Nanning University. "The Three Level Inverted Pendulum Control Model Based on Genetic Neural Network". Bulletin of Science and Technology. vol. 5, pp. 157–166, 2015.

Ji Pengcheng. "Infinitesimal dividing modeling method for dual suppliers inventory model with random lead times". *Journal of Systems Engineering and Electronics*. vol. 20, pp. 527–536, March 2009.

Liu De-you. Generalized. "LMI-based approach to global asymptotic stability of cellular neural networks with delay". *Applied Mathematics and Mechanics*. vol. 20, pp.811–819, June 2008.

Ma Li. "A Bio-Engineering Algorithm Research Based on Dynamic Fuzzy Neural Networks". *Computer Engineering and Scinence*.vol. 32, pp.137–143, March 2014.

Meng Jian-jun. "Research on civil aviation logistics forecasting based on fuzzy neural networks and simulation analysis".*Computer Engineering and Design*. vol. 31, pp.1056–1061, March 2014.

Pan Tianhong, LI Shaoyuan, Wang Xin. "A multi-model modeling approach to nonlinear systems based on lazy Learning"*Proceeding of the 24th Chinese Control Conference*. pp.268–273. November 2005.

Sun bin. "The Study on the Application of DFNN in Stock Index Prediction and Financial Nonlinear System Identification". *China Management Information-ization*. vol. 21, pp. 89–93, November 2013.

Thomas H. Cormen. "Introduction to Algorithms". *Higher Education Press*. pp.197–205, August 2006.

Wu S.Q. "Dynamic Neural Fuzzy Networks: Principle, Algorithm and Application". *School of EEE NTU*. pp.658–66, September 1998.

Zhang DeFeng. "Dynamic Fuzzy Neural Network Method Research of the Glide Window and Pruning Technology". *Acta Scientiarum Naturalium Universitatis Sunyatseni*. vol. 49, pp.48–56, January 2010.

Zhang Longlong, He Liping. "Fluid—solid Coupling Analysis of Blade Strength and Deformation in Screw Extruder of Straw". *Journal of Agricultural Mechanization*. vol. 6 pp.105–124, January 2015.

Zhang Yong-li, Cheng Hui-feng, Li Hong-xing. "The swing-up and stabilization of the triple inverted pendulum". *Control Theory & Applications*. vol. 26, pp. 634–641, June 2013.

Advances in Energy, Environment and Materials Science – Wang & Zhao (Eds)
© 2016 Taylor & Francis Group, London, ISBN 978-1-138-02931-6

Simulation of ET_0 based on PSO and LS-SVM methods

Bin Ju
School of Hydrology and Water Resources, Hohai University, Nanjing, China

Hao Liu
Suqian Hydrologic and Water Resources Survey Branch Bureau of Jiangsu Province, Suqian, China

Dan Hu
The Bureau of Forestry and Water Resources of Yuhang District of Hangzhou Municipality, Hangzhou, China

ABSTRACT: Different meteorological factors combinations were used as the input data, the results calculated by the FAO Penman-Monteith equation were used as the calibration value, and the model based on Least Square-Support Vector Machine (LS-SVM) and Particle Swarm Optimization (PSO) was established. The monthly observation data from 1986 to 2013 in the Habahe Meteorological Station over Irtysh River Basin were used to train and test the model, and the results calculated by PSO-$LSSVM$ and other commonly ET_0 calculation formula were compared. It shows that the PSO-$LSSVM$ model can well reflect the non-linear relationships between ET_0 and the meteorological factors with high accuracy of simulation, but the simulation accuracy will become low with the decrease of meteorological factor number, and temperature is the most important factor. The accuracy of the PSO-$LSSVM$ is higher than that of Priestley-Taylor and Hargreaves-Samani formulas. It provides a new approach for ET_0 study in the area of lacking forecast data.

1 INTRODUCTION

Crop water requirement is the basic foundation to formulate area irrigation system scientifically, and to set the reasonable layout of the irrigation and drainage project, and the reference crop evapotranspiration (ET_0) is the key index to calculate the crop water requirement (Kang, 1998). At present, there are many formulas to calculate ET_0, such as FAO Penman-Monteith, Priestley-Taylor (Liu, 2003), Hargreaves-Samani (Peng, 2004), and so on. Among them, the FAO Penman-Monteith formula is widely used in the world which has been recommended by the United Nations Food and Agriculture Organization, the research shows that it has a high computational accuracy and regional applicability (Mao, 2000). But the FAO penman Monteith formula is quite complex, which has too many parameters. It not only needs the daily meteorological data, also needs the elevation, latitude, solar declination, and geographic parameters that need specialized meteorological sites to observe (Wang, 2004). It causes inconvenience to apply the FAO penman Monteith formula.

The reference crop evapotranspiration (ET_0) is a complex non-linear system that was strongly influenced by meteorological factors. In recent years, with the development of the machine learning

theory, there appears many models to analyze the complex non-linear relationship, such as artificial neural network model (Wang, 2000), the support of vector machine (Liao, 2006), and the Bayesian network model (Mu, 2000). Xu et al. explores the model of artificial neural network and its application in the ET_0 simulation (Xu, 2006); Hou et al. calculated the daily ET_0 by using the least squares support vector machine in the Hetao area (Hou, 2011). However, the learning sample size of artificial neural network model is large, the convergence speed is slow, and the local extremum and overfitting problems exist. Besides, the model parameters of least squares support vector machine are very sensitive and model prediction accuracy is affected by the parameters obviously. Therefore, it has important theoretical significance and application value to search the ET_0 forecasting model with the breadth and speed of search, which has the high prediction accuracy and can express the ET_0 features.

This paper selects the daily meteorological data of Habahe meteorological station over Irtysh River Basin in Xinjiang from 1986 to 2013 year, and constructs the prediction model of ET_0 (PSO-$LSSVM$) based on particle swarm optimization algorithm and least squares support vector machine. Using the FAO Penman-Monteith formula calculation results as expected output value to precede model simulation training. It explores the relative influence degree of different meteorological factors on the prediction accuracy of ET_0 model, and compares the simulation results with other commonly ET_0 calculation formulas, and analyzes the accuracy and applicability of *PSO-LSSVM* model. It provides a new idea and method for the future prediction of ET_0.

2 STUDY AREA

Irtysh river basin ($85°35'$~$90°30'$E, $46°52'$~$49°15'$N) is located in the northeast of the Xinjiang Altai area. Its length is 546 km in China, and the annual runoff is 11.9 billion m^3, only after the Ili River. Irtysh River is the second longest river in Xinjiang and the only international river to flow into the Arctic Ocean.

Irtysh river basin has a typical continental arid climatic characteristic. It has less precipitation and more evaporation. The climatic characteristics are significantly different in time and space (Ju, 2014). Xinjiang is one of the five Chinese major pastoral areas, the Altai region where Irtysh River basin located in is one of the major pastoral areas in Xinjiang, so it has important significance to study ET_0 simulation model for the development of water-saving irrigation and reasonable irrigation system (Liao 2009).

This paper selects the meteorological data from 1986 to 2013 years of Habahe meteorological station ($48.05°$ N, $86.4°$E) over the Irtysh river basin to study ET_0 simulation. The meteorological data come from the National Meteorological Bureau Information Center.

3 MODEL CONSTRUCTION

3.1 *Research method*

3.1.1 *Least square support vector machine*
Support Vector Machine (*SVM*) is a new statistical method, which is proposed by Vapnik to solve regression and pattern recognition problem (Vapnik, 1995). Support Vector Machine (*SVM*) is based on the VC dimension theory of statistics and structural risk minimization principle, it can better solve the practical problems of overfitting, nonlinear, local extreme point and high dimensional number (Vapnik, 1999).

Least squares support vector machine algorithm (*LS-SVM*) is proposed by Suykens (2002) due to the optimization index uses the square item and changes the inequality constraints of traditional Support Vector Machine (*SVM*) into equality

constraints, it greatly simplifies the complexity of calculation model and changes the quadratic programing problem into the problem of solving linear equations. Least squares support vector machine principle is as follows:

A sample of n-dimensional vector, a region of the l samples (x_1, y_1), (x_2, y_2), ..., $(x_l, y_l) \in R^n \times R$, map samples from raw space R^n to the feature space $\varphi(x_i)$ by using nonlinear mapping, and structure optimal decision function in the high dimensional feature space:

$$y(x) = \omega \cdot \varphi(x) + b \tag{1}$$

Define the following optimization problems:

$$\begin{cases} \min J(\omega, e) = \dfrac{1}{2}\|\omega\|^2 + \dfrac{1}{2}c\sum_{i=1}^{l} e_i^2 \\ s.t. \quad y_i = \varphi(x_i) \cdot \omega + b + e_i \end{cases} \tag{2}$$

where C is the penalty factor; e_i is the relaxation factor of the insensitive loss function.

Use Lagrange method for solving the above optimization problem.

$$L(\omega, b, e, \alpha) = j(\omega, e) - \sum_{i=1}^{l} \alpha_i(\varphi(x_i) \cdot \omega + b + e - y_i) \tag{3}$$

where α_i is a lagrange multiplier.

According to the optimal conditions, the kernel function is defined as $K(x_i, x_j) = \varphi(x_i)^T \cdot \varphi(x_j)$, and the programing problem is transformed into a linear equation group:

$$\begin{pmatrix} 1 & 1 & \cdots & 1 \\ 1 & K(x_1,x_1)+1/c & \cdots & K(x_1,x_l) \\ \vdots & \vdots & \ddots & \vdots \\ 1 & K(x_l,x_1)+1/c & \vdots & K(x_l,x_l)+1/c \end{pmatrix} \cdot \begin{pmatrix} b \\ \alpha_1 \\ \vdots \\ \alpha_l \end{pmatrix} = \begin{pmatrix} 0 \\ y_1 \\ \vdots \\ y_l \end{pmatrix} \tag{4}$$

Using the least square method to solve α, b, LS-SVM model is as follows:

$$f(x) = \sum_{i=1}^{l} \alpha_i K(x, x_i) + b \tag{5}$$

The kernel function is *RBF* function:

$$K(x_i, x_j) = \exp\left(-\frac{|x_i - x_j|^2}{2\sigma^2}\right) \tag{6}$$

where σ is kernel width parameter.

3.1.2 Particle swarm optimization

Level of the precision of the prediction model has a close relationship of the parameter value of the model, this paper uses the particle swarm optimization algorithm (*PSO*) to optimize the c, σ of *LS-SVM* model, *PSO-LSSVM* model algorithm flow chart as shown in Figure 1.

Particle Swarm Optimization (*PSO*) is a swarm intelligence algorithm, proposed by Eberhart and Kennedy (1995), which has a strong global optimization ability. Particle swarm optimization algorithm principle is as follows:

The particle swarm algorithm is a randomly generated particle when it is initialized, it is the random solution of optimization problem, where the space position of the i particle in the k iteration is represented as $X_i = (x_{i1}, x_{i2}, ..., x_{id})^T$, $V_i = (v_{i1}, v_{i2}, ..., v_{id})^T$, and ($i = 1, 2, ..., m$). Particle swarm algorithm update their velocity and position according to the formula (7) and (8) while searching for the optimal solution, measure the pros and cons of x_i^k when each particle is substituted into the optimization objective function. In every iteration, the particles update their own by two "extreme": one is the particle found in the optimal solution, known as individual extremum (p_i), another is the whole population found in the optimal solutions, known as the global extremum (p_g).

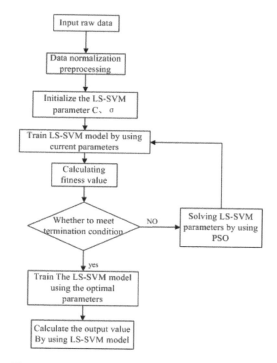

Figure 1. Algorithm flow chart of *PSO-LSSVM* model.

$$v_{id}(k+1) = \omega \cdot v_{id}(t) + c_1 \cdot rand(p_{id} - x_{id}(k))$$
$$+ c_2 \cdot rand(p_{gd} - x_{id}(k)) \qquad (7)$$

$$x_{id}(k+1) = x_{id}(k) + v_{id}(k) \qquad (8)$$

where c_1 and c_2 are the learning factors, $c_1 = c_2$, and the range of c_1 and c_2 are 0~4; *rand*( ) is a random number between (0, 1); ω is non-negative constant, which is called inertia weight.

3.2 Combination scheme

Select six meteorological factors what are closely related to reference crop evapotranspiration (ET_0), namely: monthly maximum temperature, monthly lowest temperature, monthly average temperature, average wind speed, sunshine time number and the relative humidity. There are 10 kinds of combination scheme through selecting 2 to 6 factors in six meteorological factors (Table 1), select parameters c and σ of least squares support vector machine model by using particle swarm optimization algorithm, construct the *PSO-LSSVM* model for ET_0 simulation, and analyze the precision of the ET_0 simulation of different meteorological factor combination.

In this paper, the calculation results of FAO penman Montieth formula that recommended by international society as the expected output value of *PSO-LSSVM* model, and compare it with the other two commonly used ET_0 calculation formula (Hargreaves-samani, Priestley-Taylor) in this study, the form of expression of the FAO Penman-Monteith, Hargreaves-samani, Priestley-Taylor equation are as followed:

$$ET_0(PM) = \frac{0.408\Delta(R_n - G) + \gamma \dfrac{900}{T+273} u_2(e_a - e_d)}{\Delta + \gamma(1 + 0.34u_2)}$$

$$(9)$$

where ET_0 (*PM*) is reference crop evapotranspiration that was calculated by FAO penman Montieth formula, *mm/d*; Δ is the slope of the tangent in T form of temperature ~ saturated vapor pressure curve, *kPa/°C*; γ is gamma humidity table constant, *kPa/°C*; R_n is the crop surface net radiation, *MJ/($m^2 \cdot d$)*; G is soil heat flux, *MJ/($m^2 \cdot d$)*; u_2 is the wind speed at 2 m height, *m/s*; the e_a is the saturation vapor pressure, *kPa*; e_d is the actual water vapor pressure, *kPa*.

$$ET_0(HS) = 0.0023(T_{mean} + 17.8)(T_{max} - T_{min})^{0.5} \frac{R_a}{\lambda}$$

$$(10)$$

where ET_0 (*HS*) is the reference crop evapotranspiration calculated by Hargreaves-samani formula,

Table 1. Combination mode of different meteorological factors.

Meteorological factors number	Number	Monthly mean temperature ($T/°C$)	Monthly maximum temperature ($T_{max}/°C$)	Average relative humidity ($T_{min}/°C$)	Average relative humidity ($RH/\%$)	Average wind speed ($u_h/(m/s)$)	Sunshine hours (n/h)
6	1	√	√	√	√	√	√
5	2	√	√	√		√	√
	3	√	√	√	√		√
	4	√	√	√	√	√	
4	5	√	√	√	√		
	6	√	√	√		√	
	7	√	√	√			√
3	8	√	√	√			
	9				√	√	√
2	10					√	√

mm/d; R_a is the top atmosphere radiation, $MJ/(m^2 \cdot d)$; T_{max} is the highest temperature, °C; T_{min} is the lowest temperature, °C; T_{mean} is average temperature, °C.

$$ET_0(PT) = \frac{\alpha}{\lambda} \frac{\Delta}{\Delta + \gamma}(R_n - G) \qquad (11)$$

where $ET_0(PT)$ is the reference crop evapotranspiration calculated by Priestley-Taylor formula; $\alpha = 1.26$; and the other parameters were consistent with the Penman-Monteith FAO method.

3.3 Evaluation indexs

The fitting precision of the model is measured by the mean relative error (RME), correlation coefficient (R^2), and the certainty factor (D_y). The specific formulas of the statistics are as followed:

$$R^2 = \frac{\left[\sum(x_i - \bar{x})(y_i - \bar{y})\right]^2}{\sum(x_i - \bar{x})^2 \sum(y_i - \bar{y})^2} \qquad (12)$$

$$RME = \frac{1}{n}\sum\left|(x_i - y_i)/y_i\right| \qquad (13)$$

$$D_y = 1 - \frac{\sum_{i=1}^{n}(x_i - y_i)^2}{\sum_{i=1}^{n}(x_i - \bar{x})^2} \qquad (14)$$

where x_i is the forecast value; the y_i is the actual value; $\bar{x}$, $\bar{y}$ are the mean of the forecast value and the actual value of the data; n is the forecast sample.

4 RESULTS AND ANALYSIS

4.1 Analysis of PSO-LSSVM model simulation results

The daily meteorological data of Habahe meteorological station form 1986 to 2013 years were studied, among them, 240 samples from 1986 to 2005 years are used in model training, 96 samples from 2006 to 2013 years are used in model prediction. Bring the σ and c that was optimized by particle swarm algorithm into LS-SVM prediction model. Prediction results of different meteorological factors are shown in Table 2.

From Table 2, we can see that when all six meteorological factors were used to simulate ET_0 (scheme 1), PSO-LSSVM model simulation calculation results have higher fitting degree with the FAO penman Montieth calculation results, and a good correlation, RME is only 13.52%, and the number of samples that the relative error less than twenty percent is 83.33%, R^2 up to 0.98. It can be seen that when the six meteorological factors are used to simulate the PSO-LSSVM model, the prediction accuracy is highest, and it can reflect the nonlinear relationship between the meteorological factors and ET_0.

When the five meteorological factors were used in the simulation of ET_0 (scheme 2, 3, 4), the PSO-LSSVM model is still able to predict the ET_0 value. Under the combination of 3 kinds of different meteorological factors, the R^2 and D_y of the sample simulation results are more than 0.9. In the case of lacking only the relative humidity of the sample, the average relative error is 14.86%, accounting for 85.42% of the number of samples with a relative error less than twenty percent, the simulation accuracy is a slight decline compare to scheme 1, so, the relative humidity has visible influence on ET_0 simulation calculation. When the lack of average

Table 2. Simulation results under different meteorological factors combination.

Scheme	C	σ^2	R^2	D_y	RME (%)	Error < 20% proportion (%)
1	318.76	11.93	0.981	0.980	13.52	83.33
2	763.12	8.31	0.970	0.967	14.86	85.42
3	511.90	6.07	0.969	0.955	18.33	77.90
4	571.42	16.99	0.981	0.966	17.22	73.96
5	805.31	6.73	0.962	0.960	17.90	79.00
6	511.90	6.07	0.972	0.960	16.77	84.38
7	668.80	20.69	0.974	0.969	17.80	69.79
8	668.80	21.69	0.959	0.956	17.08	68.75
9	834.36	1.57	0.782	0.779	46.39	53.12
10	574.14	5.22	0.862	0.8041	91.90	39.58

wind speed, ET_0 simulation accuracy have the biggest decline, the mean relative error is 18.33%, the samples number of relative error less than twenty percent is 77.9%. When there is lack of sunshine, the average relative error is 17.22%, the number of samples in the simulation results that relative error is less than twenty percent only accounted for 73.96%. We can see that the visible average wind speed and sunshine hours have the same influence degree on ET_0 simulation, more than the relative humidity.

When the four meteorological factors were used into ET_0 simulation calculation (scheme 5, 6, and 7), the simulation accuracy both decreased slightly. In the case of lacking two factors of average wind speed, sunshine duration and relative humidity, scheme 5, 6, 7 have same ET_0 simulation accuracy. Among of scheme 8 and 9, when lack of the average wind speed, sunshine days, relative humidity, the average relative error of *PSO-LSSVM* simulation results is 17.08%, small decline comparing to scheme 1, but the sample number of the simulation results that the relative error less than twenty percent is only 68.75%; when due to lack of temperature conditions, ET_0 simulation accuracy is worse, the correlation coefficient R^2 is only 0.782, average relative error up to 46.39%, and the sample number that relative error less than twenty percent accounted for only 53.12%. It shows that the temperature condition is a very important factor.

When only two meteorological factors were calculated (scheme 10), the average relative error of simulation results up to 91.9%, the simulation results that relative error is less than twenty percent accounted for only 39.58% is significantly lower than the results of other combinations.

4.2 *Comparison with other ET_0 calculation formulas*

When the study area lack of meteorological data so that the FAO penman Montieth formula can not be used, general use of the semi empirical method based on radiation and temperature data to estimate ET_0, such as Hargreaves-Samani method based on average temperature and temperature difference is recommended by FAO as an alternative replace method; and the Priestley-Taylor equation that based on radiation data. In this paper, comparative analysis of the *PSO-LSSVM* model and the two common formulas mentioned above.

Figure 2 shows the monthly ET_0 calculation results based on Priestley-Taylor equation, FAO penman Montieth formula and *PSO-LSSVM* model from 2006 to 2013 years; Figure 3 shows the monthly ET_0 calculation results based on Hargreaves-Samani formula, FAO penman Montieth formula and *PSO-LSSVM* model from 2006 to 2013 years; Table 3 shows the evaluation indexes of Priestley-Taylor equation, Hargreaves-Samani formula and *PSO-LSSVM* model.

You can see from Figure 2, the calculation results of the Priestley-Taylor formula, the *PSO-LSSVM* model and FAO Penman-Montieth have the same trend based on the radiation data, and the calculation results of *PSO-LSSVM* model is closer with FAO Penman-Montieth method, fitting precision is better than the Priestley-Taylor formula. The *PSO-LSSVM* model has good precision in the minimum value, and the maximum point of calculation results is too large, but the value of Priestley-Taylor formula is too small, and the minimum point is too small even more serious. Analysis reason, we find the Priestley-Taylor formula will not take wind speed into account, which leads to the calculating values often small when wind speed is greater. From the calculation results of Table 3, we can also see, *MRE* value of Priestley-Taylor equation calculation results is 31.63%, the sample number of the relative error less, than 20%, is 48.96%, far below than the PSO-LSSVM model simulation accuracy.

From Table 3 we can see that *PSO-LSSVM* model simulation accuracy is still better than

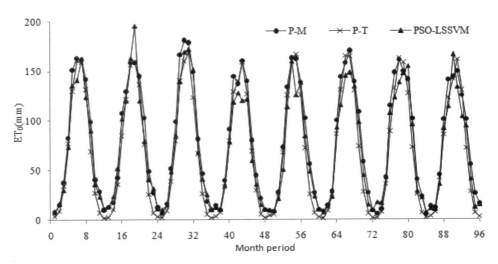

Figure 2. The calculation results of different methods based on the radiation conditions.

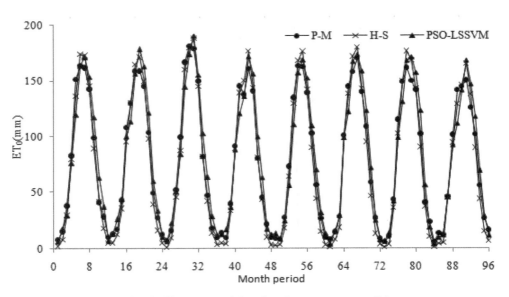

Figure 3. The calculation results of different methods based on the temperature condition.

Table 3. Comparison of results of different methods.

Calculation method	R^2	D_y	MRE (%)	Error <20% proportion (%)
PSO-LSSVM	0.963	0.946	17.80	69.79
P-T	0.973	0.936	31.63	48.96
PSO-LSSVM	0.959	0.9563	17.08	69.79
H-S	0.989	0.978	25.03	59.37

Hargreaves-Samani formula when there are only temperature conditions, but R^2 and D_y calculated from Hargreaves-Samani formula are better than PSO-LSSVM model. From Figure 3, we can see that Hargreaves-Samani formula is influenced by the temperature, the calculation results are often too small when the low temperature in winter and spring, and in June and July, Hargreaves-Samani formula and PSO-LSSVM calculation results

712

showed a bias trend, the gap between FAO penman Montieth method, Hargreaves samani formula and *PSO-LSSVM* model are lesser in the rest of month, fitting precision are very high.

5 CONCLUSION

In the paper, *PSO-LSSVM* model was constructed based on the meteorological data at Habahe weather station over Irtysh river basin from 1986 to 2013 years, and *PSO-LSSVM* model simulation results were studied when different meteorological factors combination scheme as model input conditions. Results show that ET_0 prediction model (*PSO-LSSVM*) based on particle swarm optimization algorithm and least squares support vector machines can well reflect the nonlinear relationship between different meteorological factors and the ET_0, and the accuracy of simulation model is reduced with the meteorological factors reduce. Among these meteorological factors, the temperature conditions have great influence on the simulation accuracy of ET_0, while the relative humidity has little effect, and the average wind speed and sunshine duration are in the middle.

When you calculate the ET_0 based on temperature conditions, Hargreaves-Samani formula calculation results will appear too small when the temperature is lower and need to modify, and at the time the *PSO-LSSVM* obtained simulation results are more accurate, and when the temperature is high fitting accuracy of both appear little difference. When calculated ET_0 based on the radiation data, Priestley-Taylor equation calculation results will appear too small, and little worse than calculation results by *PSO-LSSVM*. When the existing formula could not be used due to lack of the meteorological data, the *PSO-LSSVM* can be used to calculate the results of ET_0.

The ET_0 prediction model based on multi factors quantified index, reflects the relative degree of different meteorological factors on affecting ET_0, realize the unity of precision and practicality of model, provides a new point of view and approach for ET_0 prediction research in areas lacking of data.

REFERENCES

Hou Zhi-qiang, Yang Pei-ling, Su Yan-ping. et al. Simulation of ET0 based on LS-SVM method[J]. Shuili xuebao, 2011, 42(6):743–749.

Ju Bin, Hu Dan. Research on adaptability of estimated method of different reference crop evapotranspiration in Irtysh river basin[J]. Journal of Water Resources & Water Engineering, 2014, 25(5):106–111.

Kennedy, Eberhart. Particle Swarm Optimization[C]. In: Proceeding of IEEE International Conference on Neural Networks, Piscataway, NJ: IEEE CS, 1995:1942–1948.

Kang Shao-zhong. New Agricultural Sci-Technological Revolut ion and Development of Chinese Water-Saving Agriculture in 21st Century[J]. Agriculturai research in the arid areas, 1998,16(1):11–17.

Liu Xiao-ying, Lin Er-da, Liu Pei-jun. Comparative study on Priestley-Taylor and Penman methods in calculating reference crop evapotranspiration[J]. Transactions of the CSAE, 2003,19(1):32–36.

Liao Jie, Wang Wen-sheng, Li Yue-qing. et al. Support vector machine method and Its application to prediction of runoff[J]. Journal of Sichuan university (engineering science edition), 2006, 38(6):24–28.

Liao Xian-qin, Li Yi. Adaptability Research of Different Reference Crop Evapotranspiration estimated methods in Shanxi[J]. Journal of irrigation and drainage, 2009, 28(6):14–17.

Mao Fei, Zhang Guang-zhi, Xu Xiang-de. Several methods of calculating the reference evapotranspiration and comparison of the results[J]. Quarterly journal of applied meteorology, 2000, 6(Z1):128–136.

Mu Chun-di, Dai Jian-bin, Ye Jun. Bayesian network for data mining[J]. Journal of software, 2000, 11(5): 660–666.

Peng Shi-zhang, Xu Jun-zeng. Comparison of reference crop evapotranspiration computing methods[J]. Journal of irrigation and drainage, 2004, 23(6):5–9.

Suykens Jak, Gestel TV, Brabanter JD, et al. Least Squares Support Vector Machines [M]. Singapore: World Scientific Publishing Co, 2002.

Vapnik VN. The Nature of Statistical Learning Theory [M]. New York: Springer, 1995.

Vapnik VN. An overview of Statistical Learning Theory [J]. IEEE Trans Neural Network, 1999, 10(5): 988–999.

Wang Wen-sheng, Ding Jing, Liu Guo-dong. Application of artificial neural network model with nonlinear time series in hydrologic forecast[J]. Sichuan Water Power, 2000, 19:8–10.

Wang Yu-bao, Wang Zhi-nong, Shang Hu-jun. et al. Developing an instrument to measure crop evapotranspiration[J]. Journal of irrigation and drainage, 2004, 23(3): 61–64.

Xu Jun-zeng, Peng Shi-zhang, Zhang Rui-mei.et al. Neural network model for reference crop evapotranspiration prediction based on weather forecast[J]. Shuili xuebao, 2006, 37(3):376–379.

Advances in Energy, Environment and Materials Science – Wang & Zhao (Eds)
© 2016 Taylor & Francis Group, London, ISBN 978-1-138-02931-6

Evapotranspiration (ET) prediction based on least square support vector machine

Junping Liu, Wei Wang & Junjie Zhou
College of Civil Engineering and Architecture, Zhejiang University of Technology, Hangzhou, China

ABSTRACT: Monthly ET monitoring data of Haihe river basin in 2002~2007 have been obtained by remote sensing, which composites a time series. The Monthly one-dimension ET time series input space was mapped to a high dimension input space for obtaining the data which is adapted to SVM modeling. Using radial kernel function and grid search, the support vector machine of ET series is set up by learning 48 training samples. The model is used to predict 12 samples, the mean squared error of predicting samples is 12.01%, and the deterministic coefficient between predicting values and real values is 0.985. The results show that SVM possesses a stronger generalization ability and higher prediction accuracy and is a useful tool for evapotranspiration time series analysis.

1 INTRODUCTION

The World Bank aided China in Haihe river basin water resources and water environment comprehensive management project (refer to as the GEF Haihe river project) by Global Environment Facility, introduced the new concept of ET management firstly. It was used to alleviate the shortage of water resources, improve the utilization efficiency of water resource and realize the "real" water-saving. ET calls evapotranspiration as the general term of Evaporation and Transpiration, which includes vegetation interception evaporation, plant transpiration, soil evaporation and water surface evaporation and mainly affected by meteorological factors, land use, soil moisture, and vegetation status and agricultural technical measures, etc.[1–2].

The invalid and ineffective ET can be reduced through the monitoring and management of ET and the purpose of real water saving can be received. The prerequisite of achieving efficient management of water resources is to obtain the ET data. The ET monitoring provides a technical support for the realization of regional water resources management with ET management as the core.

2 ET MONITORING METHOD

The GEF Haihe River project introduces ET remote sensing monitoring model developed by Water Watch company in Holland which also known as the Surface Energy Balance Algorithm for Land (SEBAL). The model has a relatively stable physical found and has many successful applications in some European, Asian, and African countries in recent years. This model conducts parameter inversion of the surface albedo, land surface temperature, surface emissivity with the remote sensing images and meteorological data and obtains each energy subitem in the surface energy balance equation[3–5]. The ET value is calculated by

$$R_n - G = H + \lambda \cdot ET \qquad (1)$$

where R_n = net solar radiation, w/m²; G = soil heat flux, w/m²; H = Sensible heat flux, w/m²; $\lambda \cdot ET$ = Latent heat flux (λ = latent heat of evaporation of water), w/m², it expresses the part of the energy for evapotranspiration.

Remote sensing monitoring of ET also had the deficiency. For example, it is more suitable for use when sunny, if the satellite transit or cloudy, the parameters required for ET can't be estimated from the satellite images and other methods must been adopted to determine ET. Some parameters which used to the calculation of solar radiation, soil heat flux, and sensible heat flux in the inversion model are obtained through regression analysis or simplification of the actual physical process, it is applicable only under specific conditions; how to extrapolate the instantaneous ET transit to the day or even longer period of time still need to be improved through the experimental study[5].

The ET monitoring data of the Haihe River Basin from 2002 to 2007 are obtained through the remote sensing monitoring and SEBAL satellite model, 72 data forms out of a time series.

The ET time series arrange according to the order of time, these series often show a random due to the impact of various accidental factors and have a statistical dependencies exist between each other. Although the value of each moment is random and may not be completely and accurately predict the future with historical values, but the value of the correlation before and after the moment often shows some trend or cyclical changes, so the future of ET can been predicted by analyzing the historical data of ET time series and revealing the inherent law. The study establishes the ET time series prediction model which based on support vector machine, the model can provide a new idea for ET prediction and service for the management of water resources better with combining with the remote sensing monitoring of ET.

3 LEAST SQUARES SUPPORT VECTOR MACHINE

3.1 *The introduction of least squares support vector machine*

Supporting vector machine prediction, namely supporting vector regression. The basic idea of SVR is mapping the data into a high dimensional feature space by a nonlinear mapping, and conducting a linear regression in this space[6-8].

Least square support vector machine is a kind of algorithm developed by the standard support vector machine. The loss function is defined as the sum of square error, which is obtained by the inequality constrained in the standard support vector machine algorithm getting into the equality constraints. LSSVM replaces two quadratic programming problems in the standard support vector machine by solving a set of linear equations, which reduces the computation complexity and improves the speed of convergence. LSSVM only need to determine the shape parameters and the penalty coefficient of kernel function and don't need to choose the insensitive loss function values, so it is easier to use than the standard support vector machine[9-12].

For nonlinear regression problems, assuming that the training samples is $(x_i, y_i), ..., (x_n, y_n) \in R^n \times R, i = 1, 2, ..., l$, the nonlinear regression function is

$$y_i = w^T \varphi(x_i) + b + e_i \qquad (2)$$

S-SVM uses the insensitive loss function, LSSVM defines loss function as two items of error e_i.

According to the principle of structural risk minimization, the risk function for least squares support vector machine is

$$J_l = \frac{1}{2} \|W\|^2 + \gamma \sum_{i=1}^{l} e_i^2 \qquad (3)$$

The original optimization problem was defined as

$$Min \quad \frac{1}{2} \|w\|^2 + \frac{1}{2} \gamma \sum_{i=1}^{l} e_i^2 \qquad (4)$$

The constraint condition is

$$y_i = w^T \varphi(x_i) + b + e_i \quad i = 1, 2, ..., l \qquad (5)$$

where $\varphi(\cdot): R^n \to R^{nh}$ = nonlinear function mapping data from the original space to the high-dimensional Hilbert feature space; ω = weight vector; $e_i \geq 0$, = error variance; amounts to the slack variable in S-SVM; $\gamma > 0$ = parameter adjustment factor.

To solve the above optimization problem and get the constrained optimization problem into unconstrained optimization problem, Lagrange function was used

$$L = \frac{1}{2} \|w\|^2 + \frac{1}{2} \gamma \sum_{i=1}^{l} e_i^2 - \sum_{i=1}^{l} \alpha_i (w^T \varphi(x_i) + b + e_i - y_i) \qquad (6)$$

where α_i = Lagrange multiplier; the optimal α_i and b can be obtained by KKT (Karush-Kuhn-Tucker) conditions.

$$\left.\begin{array}{l}
\dfrac{\partial L}{\partial w} = 0 \to w = \sum_{i=1}^{l} \alpha_i \varphi(x_i) \\[3mm]
\dfrac{\partial L}{\partial b} = 0 \to \sum_{i=1}^{l} \alpha_i = 0 \\[3mm]
\dfrac{\partial L}{\partial e_i} = 0 \to \alpha_i = \gamma e_i \\[3mm]
\dfrac{\partial L}{\partial \alpha_i} = 0 \to w^T \varphi(x_i) + b + e_i - y_i = 0 \\[3mm]
i = 1, 2, ..., l
\end{array}\right\} \qquad (7)$$

By formula (4), the optimization problem can be transformed into solving linear equations as follows

716

$$
\begin{bmatrix}
0 & 1 & 1 & \cdots & 1 \\
1 & K(X_1,X_1)+1/\gamma & K(X_1,X_2) & \cdots & K(X_1,X_n) \\
1 & K(X_2,X_1) & K(X_2,X_2)+1/\gamma & \cdots & K(X_2,X_n) \\
\vdots & \vdots & \vdots & & \vdots \\
1 & K(X_n,X_1) & K(X_n,X_2) & \cdots & K(X_n,X_n)+1/\gamma
\end{bmatrix}
\begin{bmatrix}
b \\ \alpha_1 \\ \alpha_2 \\ \vdots \\ \alpha_l
\end{bmatrix}
=
\begin{bmatrix}
0 \\ y_1 \\ y_2 \\ \vdots \\ y_l
\end{bmatrix}
\tag{8}
$$

where $K(x_i,x_j)$ = kernel function.

α and b and the solution for nonlinear regression function can be obtained by solving the formula (7)

$$
f(x) = w^T \phi(x) + b = \sum_{i=1}^{l} \alpha_i K(x_i,x) + b \tag{9}
$$

3.2 The best parameters choose of LSSVM modeling

The Radial Basis Function (RBF) was thought to be suitable for the problem, which is studied in this paper by comparison. By determining the parameters γ and σ must be determined when the kernel function uses the RBF function. The two parameters can be determined by Cross-Validation (CV).

The CV methods have k-fold Cross-Validation (k-CV) and Leave-one-out Cross-Validation (Loo-CV), the k-CV method is used in this paper. The k-CV method is

1. Assume the number of cycles $t = 1$;
2. Randomly rearrange the original input and output matrix.
3. Divide Samples $(x_i,y_i), i = 1, 2, ..., n$ into disjoint and equal k portions, namely k-fold $S_1, S_2, ..., S_k$.

Training and testing of k times to the data, namely iteration of k times. The way of i time of iteration is that the S_i is selected as the test set and the collection of $S_1, ...S_{i-1}, S_{i+1}, ...S_k$ are as the training set. The corresponding regression model will be got after training set is used for modeling training, and then testing the test set S_i by the regression model. The error of actual S_i value and predicted S_i value is represented by square error

$$
\mathrm{cost}_i = mse(f(x_j) - y_j), \ j \in S_i \tag{10}
$$

4. The k errors $\mathrm{cost}_1, \mathrm{cost}_2, ..., \mathrm{cost}_k$ are got after k calculations, the average error of the k errors is

$$
error(t) = \frac{\sum_{i=1}^{k} \cos t(i)}{k} \tag{11}
$$

5. $error(1), error(2), ..., error(T)$ can be got by repeating the second step to the fourth step T times, the average error for T cycles is

$$
error = \frac{\sum_{t=1}^{T} error(t)}{T} \tag{12}
$$

Formula (12) can be as the estimation of modeling error, namely the error of the k-fold cross validation.

3.3 Network search method

The grid can be given in a given choice of interval for the parameters which need to choose the best, and then calculating the value of the objective function in the grid. The parameter of the objective function meeting the modeling function optimal properties is selected as the optimal parameter. The initial value of γ and σ must firstly be identified for the LSSVM in this paper. The parameters of 10×10 grid can be got if the initial value interval is divided into 10 portions in the corresponding range of these two parameters. The error value can be used cross validation algorithm to calculate in each network point, which can be as the target function of corresponding points. Each error which corresponds to 10×10 mesh point value can be drawn by contour line. The γ and σ can be obtained as the optimal parameters in the contour meeting of the conditions.

4 THE ET PREDICTION MODEL OF LEAST SQUARES SUPPORT VECTOR MACHINE

4.1 The establishment of ET prediction model

The GEF Haihe river project uses remote sensing technology to remote sensing monitoring for ET. The monthly remote sensing ET values from 2002 to 2007 are shown in Figure 1.

The first 48 samples after reconstructing phase space can be seen as the training samples to train the network in order to obtain the parameters of the model the ET with the remote sensing data of the Haihe river basin in 7 years (from 2002 to 2007). The ET data of 12 months in 2007 can be

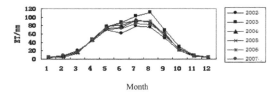

Figure 1. Changes of remote sensing ET (2002–2007).

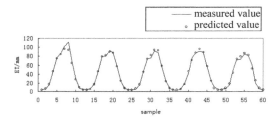

Figure 2. The fitting and predicting value of LSSVM model.

Table 1. The result of water demand prediction.

Month	Actual value/mm	SVM Predictive value/mm	Absolute error/mm	Relative error/%
1	5.0	3.83	1.17	23.45
2	8.1	7.93	0.17	2.07
3	15.8	18.72	2.92	18.45
4	47.0	42.80	4.20	8.93
5	72.4	73.65	1.25	1.72
6	73.0	79.61	6.61	9.06
7	85.0	86.96	1.96	2.30
8	81.6	82.78	1.18	1.44
9	61.0	52.53	8.47	13.89
10	21.2	26.04	4.84	22.83
11	6.3	8.25	1.95	30.95
12	3.7	4.03	0.33	9.01

seen as the test sample to examine the model's ability of prediction[13].

4.2 The prediction results and analysis

Taking $T = 10$, $k = 10$, the optimization parameters of the LSSVM model $\gamma = 736.4295$, $\sigma^2 = 16.2782$ after 10 cycles of the 10-fold cross validation. According to this set of parameters, the study on 48 samples can get the conclusion of the ET prediction model of least squares support vector machine that has a good effect on the fitting result of historical samples. The mean absolute error is 2.0083 mm, the average relative error is 6.8149%, and the mean square error is 3.4284 mm. The fitting results can be shown in Figure 2.

The ET value from January to December in 2007 was predicted by applying LSSVM model which has been trained with 12 prediction samples. The mean absolute error of 12 prediction samples is 2.92 mm, the average relative error is 12% and the mean square error is 3.84 mm, the specific results can be shown in Table 1 and Figure 2. The samples of relative error less than 20% account for 75%, the samples which relative error is more than 20% are January, October and November, which account for 25% of the total number of prediction samples. The deterministic coefficient is 0.985. The result shows the predictive ET value of support vector machine has a high coincide degree with the remote sensing ET value, the accuracy of Prediction model of support vector machine can meet the requirements and has a certain reference value. The ET prediction model of support vector machine can be improved with accumulation of the ET remote sensing data.

The result shows that the ET prediction model of least squares support vector machine not only has high fitting accuracy, the prediction precision is also very high, which has a very strong generalization ability.

5 CONCLUSION

The ET prediction model which was based on least squares support vector machine is established, the results show that the model has a high prediction precision, the deterministic coefficient can reach 0.985 and can provide a reference for remote sensing monitoring ET value. The ET prediction model of support vector machine can be further improved with the accumulation of the ET remote sensing data which provides a basic condition for the realization of water resources management "ET management" as the core.

ACKNOWLEDGMENTS

This work was financially supported by the Natural Science Foundation of Zhejiang Province (LY14E090007) and the Open Foundation of the Most Important Subjects of Zhejiang University of Technology (20150306).

REFERENCES

1. Qin, D.Y., Lv., J.Y., Liu, J.H., et al. 2008. The theory and calculation method of regional ET target. Chinese Science Bulletin 53(19):1168–1175.

2. Wang, H., You, J.J., 2008. The progress of studying on water resources rational allocation. Journal of hydraulic engineering 39(10):2384–2390.
3. Bastiaanssen, W.G.M., Menenti, M., Feddes, R.A., et al. 1998. The Surface Energy Balance Algorithm for Land (SEBAL): Part 1 formulation [J]. Journal of Hydrology(212–213):198–212.
4. Sun, M.Z., Liu, Z.X., Wu, B.F., et al. 2005. Application of satellite remote sensing monitoring of ET method in terms of water management. Advances in Water Science 16(3):468–474.
5. Wang, J.M., Gao, F., Liu, S.M. 2003. Remote sensing Inversion of basin scale ET, Remote sensing technology and Application 18(5):332–338.
6. Vapnik, V.N. 1995. The nature of statistical learning theory [M]. Springer: 70–256. New York.
7. Nello, C., John, S.T. 2000. An Introduction to Support Vector Machines and other Kernel-based Learning Methods [M]. London: Cambridge University Press.
8. Shevade, S.K., Keerthi, S.C., Bhattacharyy, et al. 2000. Improvements to SMO algorithm for SVM regression [J]. IEEE Trans on Natural Networks 11(5):1188–1193.
9. Liu, Z.X., Zhong, H.L., Zhang, D.Y. 2005. Multi scale of short-term load forecasting model of least squares support vector machine. Journal of Xi'an Jiao Tong University 39(6):620–623.
10. Guo, H., Liu, H.P., Wang, L. 2006. The selection method and application research of least squares support vector machine parameter. Journal of system simulation18(7):2033–2036.
11. Lin, J.Y., Cheng, C.T. 2006. Application of support vector machine in the middle and long-term runoff forecast. Journal of hydraulic engineering 37(6):681–686.
12. Li, X.D., Xi, S.Y., Pan, L. 2007. Study on prediction model of grain yield Chinese based on least squares support vector machine. Study on Soil and water conservation 14(6):329–331.
13. Fen, H.Z., Chen, Y.Y. 2004. A new method to process nonlinear classification and regression problems (II)—Application of support vector machine method in weather forecast. Journal of Applied Meteorology15(3):355–364.

Advances in Energy, Environment and Materials Science – Wang & Zhao (Eds)
© 2016 Taylor & Francis Group, London, ISBN 978-1-138-02931-6

A method of high maneuvering target dynamic tracking

Jin-pei Yang, Wei-tai Liang & Jun Wang
Science and Technology on Information System Engineering Key Laboratory, Qinhuai, Nanjing, China

ABSTRACT: This paper studies the problem of high-maneuvering target dynamic tracking based on particle filtering. The basic principle of particle filtering are firstly discussed, then the model of a high-maneuvering target dynamic tracking is set up, and finally the computer simulations are used to compare particle filtering with other tracking methods, the results show that the particle filtering have a good dynamic tracking performance of high-maneuvering target.

1 INTRODUCTION

The objection of Radar detection is usually to form a target trajectory, such as parameters estimation of the target trajectory (delay, Doppler, etc.). But the target track is a typical problem of uncertainty; its uncertainty is mainly as follows: the target (signal source) of the uncertainty is caused by the movement itself, the uncertainty of the measurement noise, the uncertainty of measuring data fuzzy and the uncertainty of deceptive, especially for high maneuvering move air targets. The Target trajectory is given by resolved target dynamic tracking model (state equation and measurement equation). As the tracking model have the uncertainty factors, which are non-linear and non-Gaussian noise, the common tracking methods (Zhou, 1991) achieve the desired results with difficulty. The paper using the particle filter to establish the high maneuvering target tracking model, and the performance of target tracking are computer simulated and analyzed, and some target tracking results are given.

2 PARTICLE FILTER

Particle Filter is based on a sample of Bayesian estimation theory approximation algorithm that Monte Carlo and Bayesian theory are combined together. The basic idea is to find a random sample in the state space to approximate the conditions posterior probability density $p(x_{0:n}/y_{1:n})$, which use the sample mean to substitute $E[g(x_{0:n})/y_{1:n}]$ so as to obtain the minimum variance estimation of a state. The key is to find a distributed random sample which obeys $p(x_{0:n}/y_{1:n})$, and these samples are often called particles. In a sense, the particle filter uses random adaptive grid method to approximate Bayesian.

2.1 Basic equation

Assume the particles $\{x_{0:n}^{(i)}\}_{i=1}^{N}$ can be obtained from the random sampling independently of conditional posterior probability density $p(x_{0:n}/y_{1:n})$, there are [2]:

$$p(x_{0:n}/y_{1:n}) = \frac{1}{N}\sum \delta_{x_{0:n}^{(i)}}(x_{0:n}) \tag{1}$$

where: $\delta_x(\cdot)$ is the Dirac function at the point x, and $g(x_{0:n})$ expectations can be expressed as:

$$E(g(x_{0:n})/y_{1:n}) = \int g(x_{0:n}) p(x_{0:n}/y_{1:n}) dx_{0:n} \tag{2}$$

Available approximation is expressed as:

$$\overline{g(x_{0:n})} = \frac{1}{N}\sum_{i=1}^{N} g(x_{0:n}^{(i)}) \tag{3}$$

According to the law of large numbers, because of the mutual independence between the particles $\{x_{0:n}^{(i)}\}_{i=1}^{N}$, when $N \to \infty$, $g(x_{0:n})$ with probability 1, convergence moves to $E(g(x_{0:n})/y_{1:n})$.

However, directly sampling $p(x_{0:n}/y_{1:n})$ is more difficult, now it is based on sampling theory to solve this important problem.

2.2 Bayesian important sampling (IS)

Assume the important density function is $q(x_{0:n}/y_{1:n})$, then:

$$E(g(x_{0:n})/y_{1:n}) = \int g(x_{0:n}) \frac{p(x_{0:n}/y_{1:n})}{q(x_{0:n}/y_{1:n})} q(x_{0:n}/y_{1:n}) dx_{0:n}$$

$$= \int g(x_{0:n}) \frac{p(y_{1:n}/x_{0:n}) p(x_{0:n})}{p(y_{1:n}) q(x_{0:n}/y_{1:n})} q(x_{0:n}/y_{1:n}) dx_{0:n} \tag{4}$$

Remember $w(x_{0:n}) = \dfrac{p(y_{1:n}/x_{0:n})p(x_{0:n})}{q(x_{0:n}/y_{1:n})}$, that is the important weights function. And:

$$p(y_{1:n}) = \int \frac{p(y_{1:n}/x_{0:n})p(x_{0:n})}{q(x_{0:n}/y_{1:n})} q(x_{0:n}/y_{1:n}) dx_{0:n}$$
$$= \int w(x_{0:n})q(x_{0:n}/y_{1:n}) dx_{0:n} \qquad (5)$$

So the equation (5) becomes:

$$E(g(x_{0:n})/y_{1:n})$$
$$= \frac{\int [g(x_{0:n})w(x_{0:n})]q(x_{0:n}/y_{1:n}) dx_{0:n}}{\int w(x_{0:n})q(x_{0:n}/y_{1:n}) dx_{0:n}} \qquad (6)$$

When the N individual distribution particles $\{x_{0:n}^{(i)}\}_{i=1}^{N}$ are obtained from sampling $q(x_{0:n}/y_{1:n})$, the expected value approach as:

$$\overline{g(x_{0:n})} = \frac{\dfrac{1}{N}\sum_{i=1}^{N} g(x_{0:n}^{(i)}) w(x_{0:n}^{(i)})}{\dfrac{1}{N}\sum_{i=1}^{N} w(x_{0:n}^{(i)})} = \sum_{i=1}^{N} g(x_{0:n}^{(i)}) \tilde{w}(x_{0:n}^{(i)})$$

$$\tilde{w}(x_{0:n}^{(i)}) = \frac{w(x_{0:n}^{(i)})}{\sum_{j=1}^{N} w(x_{0:n}^{(j)})} \qquad (7)$$

where: $\tilde{w}(x_{0:n}^{(i)})$ is a normalized weight.

From this, important sampling is used as easy sampling $q(x_{0:n}/y_{1:n})$ instead of difficult sampling $p(x_{0:n}/y_{1:n})$, which is a Monte Carlo integration method that is easy to implement. However, this simple method is not enough to solve the state posterior probability density recursive estimation. Because every time a new data y_{n+1} arrives, will need to recalculate the importance weights including a past time, and the entire state sequence, the estimated value at each time has to be updated, so as to spend a lot of computing time, therefore to provide an improved strategy—sequential importance sampling.

2.3 Sequential importance sampling (SIS)

SIS algorithm is a recursive important sample algorithm, and is the basis for particle filter. So far, a variety of particle filters are formed of different forms SIS transformation (Sanjeev, 2002).

Assume the important density function is $q(x_{0:n}/y_{1:n})$; in order to obtain a recursive form, the importance probability density function can be broken down as follows:

$$q(x_{0:n}/y_{1:n}) = q(x_n/x_{0:n-1}, y_{1:n})q(x_{0:n-1}/y_{1:n-1}) \qquad (8)$$

Because the state has the Markov property, the measurement is independent in the given state conditions, that is:

$$p(x_{0:n}) = p(x_0)\prod_{j=1}^{n} p(x_j/x_{j-1}),$$
$$p(y_{1:n}/x_{0:n}) = \prod_{j=1}^{n} p(y_j/x_j)$$

So important weights can be written as:

$$w_n = \frac{p(y_{1:n}/x_{0:n})p(x_{0:n})}{q(x_n/x_{0:n-1}, y_{1:n})q(x_{0:n-1}/y_{1:n})}$$
$$= w_{n-1}\frac{p(y_{1:n}/x_{0:n})p(x_{0:n})}{p(y_{1:n-1}/x_{0:n-1})p(x_{0:n-1})q(x_n/x_{0:n-1}, y_{1:n})}$$
$$= w_{n-1}\frac{p(y_n/x_n)p(x_n/x_{n-1})}{q(x_n/x_{0:n-1}, y_{1:n})} \qquad (9)$$

Equation (9) is form of an important weight recursive function, and it is called sequential importance sampling. Assuming the sample particles $\{x_{0:n-1}^{(i)}\}_{i=1}^{N}$ before $n-1$ moments and normalized weights $\tilde{w}_{n-1}^{(i)}$ at $n-1$ moments can be known, it is denoted as $\{x_{0:n-1}^{(i)}, \tilde{w}_{n-1}^{(i)}\}_{i=1}^{N}$. When new data arrives at n time, recursive algorithm can be summarized as follows:

1. The particles $\{x_{0:n}^{(i)}\}_{i=1}^{N}$ are obtained from the important function of sample $q(x_n/x_{0:n-1}^{(i)}, y_{1:n})$;
2. According to equation (9), the weights at n time are calculating:

$$w_n^{(i)} = \tilde{w}_{n-1}^{(i)}\frac{p(y_n/x_n^{(i)})p(x_n^{(i)}/x_{n-1}^{(i)})}{q(x_n^{(i)}/x_{n-1}^{(i)}, y_{1:n})} \qquad (10)$$

3. Normalized weights:

$$\tilde{w}_n^{(i)} = \frac{w_n^{(i)}}{\sum_{j=1}^{N} w_n^{(j)}} \qquad (11)$$

Particle and weights at n time are denoted $\{x_{0:n-1}^{(i)}, \tilde{w}_n^{(i)}\}_{i=1}^{N}$;
4. According to equation (7), the estimates are obtained:

$$\bar{x}_n = \sum_{i=1}^{N} x_n^{(i)} \tilde{w}_n^{(i)} \qquad (12)$$

2.4 Eliminate degradation of key technologies

However, there is a problem of weights degradation in SIS particle filter. It refers to the algorithm after several iterations, the variance of the weights will gradually increase with time, so that the weight of a small number of the particles will became larger, but most of the weight of the particles are negligible, which means that a large number of computing work is wasted on the update of solving $p(x_{0:n}/y_{1:n})$, with almost no effect of particles. Therefore, degradation of the particle filter has a serious negative impact on how to eliminate degradation as a vital part of the particle filter. Geweke proposed a measure of the degree of importance of sampling degradation—the relative efficiency (Sanjeev, 2002). It is defined as follows:

$$(RNE)^{-1} = \frac{\text{var}_q\left(E_N\left(g(x_{0:n})/y_{1:n}\right)\right)}{\text{var}_p\left(E_N\left(g(x_{0:n})/y_{1:n}\right)\right)} \approx \left(1 + \text{var}_\pi(\tilde{w})\right) \tag{13}$$

where, $\text{var}_p(E_N(g(x_{0:n})/y_{1:n}))$ is obtained by sampling posterior probability density $p(x_{0:n}/y_{1:n})$, $\text{var}_q(E_N(g(x_{0:n})/y_{1:n}))$ is obtained by sampling the importance sampling density $q(x_{0:n}/y_{1:n})$ of SIS filter.

In general, the relative efficiency can often be equivalent to another measurement scales, that effective sampling scales N_{eff} is as follows:

$$N_{eff} = \frac{N}{1 + \text{var}_q(\tilde{w}_n)} = \frac{N}{E_q\left((\tilde{w}_n)^2\right)} \leq N \tag{14}$$

In practical calculations, it can be used as the approximate calculation:

$$\hat{N}_{eff} = \frac{1}{\sum\limits_{i=1}^{N}\left(\tilde{w}_n^{(i)}\right)^2} \tag{15}$$

The smaller the effective sampling scale, the more serious is the degradation. From equation (15) it is apparent, as long as the particles can be increased to increase the sample, the degradation can also be improved. However, the increase will inevitably affect the calculation of particles in real time; usually it is not practical; it generally uses other strategies to eliminate degradation. Currently, there are two main eliminations for the degradation of key technologies, important density function properly chosen and important sampling.

2.5 Particle filter algorithm with important sampling

Although important sampling can reduce the algorithm degradation in some extent, but in fact, it has some disadvantages: increase the amount of calculation, limiting the parallel implementation of filter and sample impoverishment problem may occur because after important sampling with more particle power values are multiplied, particles' diversity is lost to a certain extent. So while selecting the important sampling is very important, under normal circumstances, only when degradation SIS filter reaches a certain limit, consider using important sampling, the most common method is to use a threshold to determine the use of important sampling.

Set threshold N_{th}, when the effective sampling scale $N_{eff} < N_{th}$, on the contrary, do not use important sampling. Thus, particle filter algorithm with important sampling process can be summarized in (Sanjeev, 2002) as follows:

1. Initialize the generation of particles $\{x_0^{(i)}\}_{i=1}^{N}$, all particles weights is $1/N$; then $n = 1, 2, ...,$ start the cycle;
2. Important sampling
   1. The particles $\{x_0^{(i)}\}_{i=1}^{N}$ can be obtained from sampling of the important function $q(x_n/x_{0:n-1}^{(i)}, y_{1:n})$;
   2. According to equation (10), the weight of the moment can be calculated:

$$w_n^{(i)} = \tilde{w}_{n-1}^{(i)} \frac{p\left(y_n/x_n^{(i)}\right)p\left(x_n^{(i)}/x_{n-1}^{(i)}\right)}{q\left(x_n^{(i)}/x_{0:n-1}^{(i)}, y_{1:n}\right)} \tag{16}$$

   3. Normalized weights:

$$\tilde{w}_n^{(i)} = w_n^{(i)} \bigg/ \sum_{j=1}^{N} w_n^{(j)} \tag{17}$$

3. Important sampling
   1. Calculate the effective sampling scale N_{eff};
   2. When the effective sampling scale $N_{eff} < N_{th}$, on the use of important sampling;
   3. Get a new set of particles $\{\tilde{x}_{0:n}^{(l)}\}_{i=1}^{N}$, their weights are $\tilde{w}_n^{(l)} = 1/N$.
   4. Get the estimated value is

$$\overline{x_n} = \frac{1}{N} \sum_{l=1}^{N} \tilde{x}_n^{(l)} \tag{18}$$

3 TARGET MOVING MODEL

Nonlinear filtering algorithm of motion model can provide better tracking performance only if it depends on the model and the actual model has consistent conditions. Combined with high maneuvering air target geometry, build its sports model, that is state equation and measurement equation of targets.

3.1 Targets state equation

The state equation model of air maneuvering target generally has a uniform acceleration that can be expressed as (Zhou, 1991; Li, 2014; Lou, 2012):

$$X(k) = F(k/k-1)X(k-1) + v(k-1) \quad (19)$$

where:
The state vector is taken as:

$$X = [x, y, z, \dot{x}, \dot{y}, \dot{z}, \ddot{x}, \ddot{y}, \ddot{z}] \quad (20)$$

State transition matrix is:

$$F = \begin{bmatrix} 1 & 0 & 0 & T & 0 & 0 & T^2/2 & 0 & 0 \\ 0 & 1 & 0 & 0 & T & 0 & 0 & T^2/2 & 0 \\ 0 & 0 & 1 & 0 & 0 & T & 0 & 0 & T^2/2 \\ 0 & 0 & 0 & 1 & 0 & 0 & T & 0 & 0 \\ 0 & 0 & 0 & 0 & 1 & 0 & 0 & T & 0 \\ 0 & 0 & 0 & 0 & 0 & 1 & 0 & 0 & T \\ 0 & 0 & 0 & 0 & 0 & 0 & 1 & 0 & 0 \\ 0 & 0 & 0 & 0 & 0 & 0 & 0 & 1 & 0 \\ 0 & 0 & 0 & 0 & 0 & 0 & 0 & 0 & 1 \end{bmatrix} \quad (21)$$

State process noise covariance matrix:

$$Q_{k-1} = \begin{bmatrix} T^4/4 & 0 & 0 & T^3/2 & 0 & 0 & T^2/2 & 0 & 0 \\ 0 & T^4/4 & 0 & 0 & T^3/2 & 0 & 0 & T^2/2 & 0 \\ 0 & 0 & T^4/4 & 0 & 0 & T^3/2 & 0 & 0 & T^2/2 \\ T^3/2 & 0 & 0 & T^2 & 0 & 0 & T & 0 & 0 \\ 0 & T^3/2 & 0 & 0 & T^2 & 0 & 0 & T & 0 \\ 0 & 0 & T^3/2 & 0 & 0 & T^2 & 0 & 0 & T \\ T^2/2 & 0 & 0 & T^3/2 & 0 & 0 & T^2 & 0 & 0 \\ 0 & T^2/2 & 0 & 0 & T^3/2 & 0 & 0 & T^2 & 0 \\ 0 & 0 & T^2/2 & 0 & 0 & T^3/2 & 0 & 0 & T^2 \end{bmatrix} \cdot \sigma_v^2 \quad (22)$$

The initial value of the covariance matrix:

$$P = [p_{ij}]_{i=1, 2, \ldots, 9; \, j=1, 2, \ldots, 9} \quad (23)$$

where: T is the radar scan cycle, σ_v^2 is the state process noise variance, the value of the initial covariance matrix see (Zhou, 1991).

3.2 Target measurement equation

Target measurement equation is:

$$Z(k) = H(k)X(k) + n_k \quad (24)$$

where: H is the observation matrix. If the measured is position, the H is linear:

$$H = \begin{bmatrix} 1 & 0 & 0 & 0 & 0 & 0 & 0 & 0 & 0 \\ 0 & 1 & 0 & 0 & 0 & 0 & 0 & 0 & 0 \\ 1 & 0 & 1 & 0 & 0 & 0 & 0 & 0 & 0 \\ 0 & 0 & 0 & 1 & 0 & 0 & 0 & 0 & 0 \\ 0 & 0 & 0 & 0 & 1 & 0 & 0 & 0 & 0 \\ 0 & 0 & 0 & 0 & 0 & 1 & 0 & 0 & 0 \\ 0 & 0 & 0 & 0 & 0 & 0 & 1 & 0 & 0 \\ 0 & 0 & 0 & 0 & 0 & 0 & 0 & 1 & 0 \\ 0 & 0 & 0 & 0 & 0 & 0 & 0 & 0 & 1 \end{bmatrix} \quad (25)$$

If the measurement is both position and radial velocity, and H is non-linear, the equation (24) becomes:

$$Z(k) = f(X_k) + n_k \quad (26)$$

$$H = \begin{bmatrix} 1 & 0 & 0 & 0 & 0 & 0 & 0 & 0 & 0 \\ 0 & 1 & 0 & 0 & 0 & 0 & 0 & 0 & 0 \\ 1 & 0 & 1 & 0 & 0 & 0 & 0 & 0 & 0 \\ \frac{\partial f}{\partial x} & \frac{\partial f}{\partial y} & \frac{\partial f}{\partial z} & \frac{\partial f}{\partial \dot{x}} & \frac{\partial f}{\partial \dot{y}} & \frac{\partial f}{\partial \dot{z}} & \frac{\partial f}{\partial \ddot{x}} & \frac{\partial f}{\partial \ddot{y}} & \frac{\partial f}{\partial \ddot{z}} \end{bmatrix} \quad (27)$$

If angle information (azimuth φ, pitch angle θ) of passive tracking are only measured, the equation (24) becomes:

$$Z(k) = H(k)X(k) + n_k$$

$$Z(k) = \begin{bmatrix} \theta(k) \\ \varphi(k) \end{bmatrix}$$

$$H(k) = \begin{bmatrix} tg^{-1}(y/x) \\ tg^{-1}\left(z/\sqrt{x^2+y^2}\right) \end{bmatrix} \quad (28)$$

Measurement error covariance matrix [1] is as follows:

$$R_n = \begin{bmatrix} \sigma_x{}^2 & \sigma_{xy} & \sigma_{xz} & \sigma_{x\dot{x}} & \sigma_{x\dot{y}} & \sigma_{x\dot{z}} & \sigma_{x\ddot{x}} & \sigma_{x\ddot{y}} & \sigma_{x\ddot{z}} \\ \sigma_{xy} & \sigma_y{}^2 & \sigma_{yz} & \sigma_{y\dot{x}} & \sigma_{y\dot{y}} & \sigma_{y\dot{z}} & \sigma_{y\ddot{x}} & \sigma_{y\ddot{y}} & \sigma_{y\ddot{z}} \\ \sigma_{xz} & \sigma_{yz} & \sigma_z{}^2 & \sigma_{\dot{x}z} & \sigma_{\dot{y}z} & \sigma_{\dot{z}z} & \sigma_{\ddot{x}z} & \sigma_{\ddot{y}z} & \sigma_{\ddot{z}z} \\ \sigma_{x\dot{x}} & \sigma_{y\dot{x}} & \sigma_{z\dot{x}} & \sigma_{\dot{x}}{}^2 & \sigma_{\dot{y}\dot{x}} & \sigma_{\dot{z}\dot{x}} & \sigma_{\ddot{x}\dot{x}} & \sigma_{\ddot{y}\dot{x}} & \sigma_{\ddot{z}\dot{x}} \\ \sigma_{x\dot{y}} & \sigma_{y\dot{y}} & \sigma_{z\dot{y}} & \sigma_{\dot{x}\dot{y}} & \sigma_{\dot{y}}{}^2 & \sigma_{\dot{z}\dot{y}} & \sigma_{\ddot{x}\dot{y}} & \sigma_{\ddot{y}\dot{y}} & \sigma_{\ddot{z}\dot{y}} \\ \sigma_{x\dot{z}} & \sigma_{y\dot{z}} & \sigma_{z\dot{z}} & \sigma_{\dot{x}\dot{z}} & \sigma_{\dot{y}\dot{z}} & \sigma_{\dot{z}}{}^2 & \sigma_{\ddot{x}\dot{z}} & \sigma_{\ddot{y}\dot{z}} & \sigma_{\ddot{z}\dot{z}} \\ \sigma_{x\ddot{x}} & \sigma_{y\ddot{x}} & \sigma_{z\ddot{x}} & \sigma_{\dot{x}\ddot{x}} & \sigma_{\dot{y}\ddot{x}} & \sigma_{\dot{z}\ddot{x}} & \sigma_{\ddot{x}}{}^2 & \sigma_{\ddot{y}\ddot{x}} & \sigma_{\ddot{z}\ddot{x}} \\ \sigma_{x\ddot{y}} & \sigma_{y\ddot{y}} & \sigma_{z\ddot{y}} & \sigma_{\dot{x}\ddot{y}} & \sigma_{\dot{y}\ddot{y}} & \sigma_{\dot{z}\ddot{y}} & \sigma_{\ddot{x}\ddot{y}} & \sigma_{\ddot{y}}{}^2 & \sigma_{\ddot{z}\ddot{y}} \\ \sigma_{x\ddot{z}} & \sigma_{y\ddot{z}} & \sigma_{z\ddot{z}} & \sigma_{\dot{x}\ddot{z}} & \sigma_{\dot{y}\ddot{z}} & \sigma_{\dot{z}\ddot{z}} & \sigma_{\ddot{x}\ddot{z}} & \sigma_{\ddot{y}\ddot{z}} & \sigma_{\ddot{z}}{}^2 \end{bmatrix} \quad (29)$$

where: $\sigma_{x^2}, \sigma_{y^2}, \sigma_{z^2}, \sigma_{\dot{x}^2}, \sigma_{\dot{y}^2}, \sigma_{\dot{z}^2}, \sigma_{\ddot{x}^2}, \sigma_{\ddot{y}^2}, \sigma_{\ddot{z}^2}$ are, respectively, the position, velocity, and acceleration variance, and others are the interaction variance for the ones between them.

4 COMPUTER SIMULATIONS

Simulation 1: set up a two-dimensional moving target, the initial position (20 m, 0), radar scan rate 1 s, $v_k \in N(0,10)$ and $n_k \in N(0,1)$ are independent Gaussian white noise. The trajectory model of target in the two-dimensional plane is as follows:

$$x(k) = [x]$$

$$x(k) = 0.5x(k-1) + [20x(k-1)]/[(1+x(k-1))^2]$$
$$+ 10\cos(0.08(k-)) + v_{k-1}$$

$$z(k) = x(k)^2/2 + n_k \tag{30}$$

The model is non-linear, using the particle SIS filter track for target dimensional tracking.

Using the equation (8) (10) (11) (12) (13) (18) to simulate particle SIS filter, tracking results are shown in Figure 1 to 3.

The figure shows that the nonlinear target tracking model, SIS particle filter is superior to the maximum peak (maximum likelihood ratio) filtering.

Simulation 2: set up a two-dimensional moving target, the initial position (20 m, 0), radar scan rate

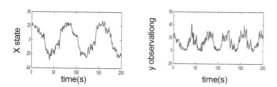

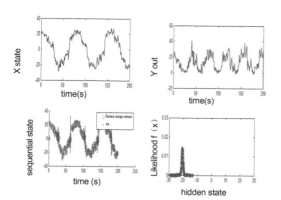

Figure 1. Target trajectory and noise pattern.

Figure 2. Sequential estimate.

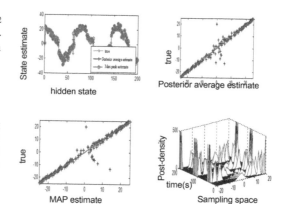

Figure 3. Particle filter (SIS) renderings.

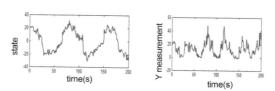

Figure 4. Target trajectory and noise pattern.

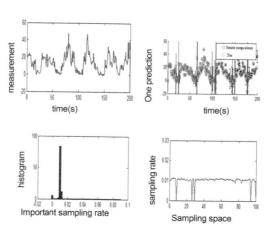

Figure 5. Mixed important sampling estimate.

1 s, $v_k \in N(0,10)$ and $n_k \in N(0,1)$ that are independent Gaussian white noise; the target in the two-dimensional plane of trajectory model is as follows:

$$x(k) = [x]$$
$$x(k) = 0.5x(k-1) + [20x(k-1)]/[(1+x(k-1))^2]$$
$$+ 10\cos(0.08(k-)) + v_{k-1}$$
$$z(k) = x(k)^2/2 + n_k \tag{31}$$

725

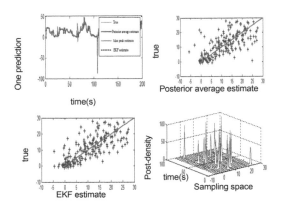

Figure 6.   Particle filter result.

The model is non-linear, using particle hybrid SIR to filter for target dimensional tracking.

Using the equation (16) (18) (28) to simulate mixed SIR filter, and tracking the results shown in Figures 4 to 6.

The figure shows that the target model for non-linear tracking, SIR particle filter is superior to the maximum peak (maximum likelihood ratio) filter and EKF filtering.

5   CONCLUSION

In summary, the tracking system state estimation of particle filter and posterior probability distribution estimate of the state and its true value is almost a complete match; the variance of state estimation error is quite small, so the particle filter for high maneuvering target tracking on the non-Gaussian background has a very good track performance.

REFERENCES

Li Meng, Xie, et al. "Mean Shift tracking algorithm based on aerial background weights [J], Command Information Systems and Technology," 2014, chapter 6, pp. 22–26.

Lou Xiao Xiang, et al. "multi-target tracking algorithm based on multi-assumptions," Command Information Systems and Technology," 2012, chapter 6, pp. 36–39.

Sanjeev M., et al. "A tutorial on particle filters for online nonlinear/non-gaussian Bayesian tracking. IEEE Trans. on Signal Processing, " Vol. 50, No. 2, 2002, pp. 174–187.

Xiaolong Deng, Jianying Xie. "Nonlinear target tracking based on particle filter. Proceedings of the Five World Congress on Intelligence Control and Automation," Hangzhou, P.R China, 2004, pp. 1618–1620.

Zhou Hong-ren, et al. "Maneuvering Target Dynamic Tracking," Beijing: National Defense Industry Publish, 1991.

Advances in Energy, Environment and Materials Science – Wang & Zhao (Eds)
© 2016 Taylor & Francis Group, London, ISBN 978-1-138-02931-6

A novel semantic data model for big data analysis

Yi Li, Kang Li, Xi-Tao Zhang & Shou-Biao Wang
Equipment Academy, Beijing, China

ABSTRACT: Based on the application requirements of big data analysis, this paper proposed a novel semantic data model for a big data analysis. First, the advantages of RDF model for data modeling are presented by analyzing the relevant research about data model. Second, the RDF model based on Key/Value is studied, which gives the related definitions and procedures of model. Finally, the effectiveness and feasibility of the prototype system based on HBase are verified through the experiment test on massive data.

1 INTRODUCTION

Big data analysis, which is a new research model, has produced a profound influence on the scientific researching and technological development. Peter Norvig, chief technology officer for Google's, has proposed a view on the new IT summit, that is "All models are wrong, and increasingly you can succeed without them" (Anderson, 2008); pointing out that large data analysis has no longer focused on the causal relationship building in the data model, but the new knowledge and new law discovered from a large numbers of analyses on the relationships among data, in contrast to the traditional research methods based on mathematical models.

The data analysis and processing techniques in the field of traditional database research have become more and more mature. Traditional data model is no longer suitable for solving large data storage and management. The RDF-based semantic data provides a good foundation for semantic data modeling in large data analysis, with its advantages in structured, semi-structured and unstructured data representation and the characteristics of "self-describing". Then, the semantic data model for large data analysis based on RDF is built (Chebotko, 2013).

2 RDF MODEL BASED ON KEY/VALUE

Data model (Data Model) is a describing collection of the data itself, data relationships, data semantics, and consistency constraints of conceptual tools, which provides a description of the method to design mode data in physical layer, logical layer and view layer (Abraham, 2012). Here are some common data models for solving the data integration problem, such as the relationship model (Relation Model) (Zhang, 2014), OEM model (Object Exchange Model)

(Papakonstantinou, 1995), XML model (eXtensible Markup Language Model) (Suciu, 1998), and RDF model (Resource Description Framework Model) (Beckett, 2014) and so on.

RDF model is proposed by W3C, which gets the ability of self-describing. The advantages of RDF model lies in the description of data, which is a triple about subject, predicate and object composition, and clear representation of the semantic relationships between data and data. Currently, the open source big data analytics platform, represented by Hadoop, has been widely used, as well as the developed SimpleDB, Big Table and other cloud-based data management system on this basis. The cores of these products and technologies are Map Reduce and Key/Value. The construction of Big Data analytics platform in this section is in line with the current mainstream Map Reduce processing framework; thus, the issues of semantic data model building for large data can be converted into a research on the RDF model based on Key/Value. Here is the Semantic data mode.

The so-called semantic data (Semantic Data), the narrow concept of body of data (Ontology Data), means the data itself and the rules described by using of body language, which can be queried and reasoned. Compared to the concept of the Ontology Data, semantic data is not limited to the description of the main body of data, but focuses more on establishing the relationship between the data and rules. By reference to the Ontology model defined by Gomez Perez and the expounding of Semantic Data in literature (Vagner, 2013), this paper formalizes the semantic data as follows: In this paper, "the body" as a manifestation of the semantic data model, Gomez Perez is defined with by reference to the body, as well as the literature (Vagner, 2013) semantic data set forth in the formal definition of the semantics of data is as follows: the semantic data can be formalized as follows:

Definition 1. Semantic data. It is defined as the quintuple of classes, relationships, attributes, instances and rules, and expressed as:.

$$S_d = \{C, R, P, I, L\} \qquad (1)$$

Which C represents the class, whose concept is corresponding to the concept of class in object-oriented model and it is a collection of objects. R describes the relationship between the concepts of correlation and constraints; it can be described as a subset of the Cartesian product dimension. P indicates the property, which is used to describe the type's features, generally including the DataTypeProperty and ObjectProperty. I represents Individual, showing the specific cases that belong to a class.

According to the definition 1, five concepts, concepts, relationships, attributes, individual and rules, are used to depict the conceptual model of the objective world. In coding implementations, semantic data will be described by using RDF, and RDF's definitions are given below.

Definition 2. RDF Tuple. Show as below:

$$\tau = (s, p, o) \in (U \cup B) \times U \times (U \cup B \cup Ls) \qquad (2)$$

U represents the collection of URI (Uniform Resource Identifier, Uniform Resource Identifier) B represents the set of blank nodes (BlankNodes) Ls shows a collection of string Literals, (s, p, o), respectively, represents the subject, predicate, and object of RDF.

The method of model construction is given as follows: (1) build ontology class, the default class is the most basic class, named Thing. All instances in model are totally the subclasses of Thing. (2) construct model properties ontology, it can describe the relationship between the instance or instances and values, by defining the attributes. The relationship between the various entities called ObjectProperty (object properties), which reflects the concept of the extrinsic properties. (3) construct the model instance. The instances are the specific things that belong to a class of the body. Semantic data model defines the URI system of all things, the system of each local ontology file can be named with a unique URI. Local ontology file size refinement to the instance level, the resources for each body contained in the document are in Key / Value model stored in NoSQL. In this example, the URI of resource (the file name is also the body) is set as Key, and Value is the file which the URI corresponds to.

3 EXPERIMENT

HBase is a distributed database based on Apache Hadoop. The character of open source has implemented all kinds of functions and features of BigTable (Chang, 2006) to support large data storage and management. According to the RDF-based ontology model previously discussed, two HTable tables are created, HBClass table and HBProperty table, which are used to store information about classes and properties. HBClass table is used to store information about the class, setting Row-Key as the class name; additionally, two subclasses of column families, SubClass and Property, and their attributes are set. Data of column family can be added and updated in real time via a timestamp, which meets the data's dynamically changing needs. The HBProperty table is used to store attribute information and the property name is defined as Row-Key. Two column families, SubProperty and ValProperty, are set to store sub-attribute information and information associated.

On the basis of HBClass table and HBProperty table instances of table of the system and HBIndividual is created to record specific information for each instance. Set the URI of each instance as Row-Key and four column families: HBClass, HBProperty, Deal, and Participant to, respectively, store the information of the instances, class information, property information, transaction information, and evaluation information. Six groups of test data sets with different sizes are randomly generated in "Database Experiment", which is the functional module based on UTPB's, and each RDF graph contains 400 tuples, the specific information is shown in Table 1.

Experiments using the sample query statement UTPB provided on Graph, Dependencies and Artifacts of the experiments, nine query statements, {Q1,Q2,...,Q9} are used to make a test, which is the sample query statement provided by UTPB, and the core code of the query statement is shown in Table 2.

The results are shown in Figure 1, where axis-x represents the test data set, the axis-y represents the test run time in ms. What can be derived from the analysis of experimental results is that the running time of statements Q1 and Q2 to query Graph class is significantly longer than other statements, that is because the Q1 statement is for

Table 1. RDF test data set.

Data set	RDF num	RDF turple	Data size
DS1	10,000	400,00,000	3.2 GB
DS2	20,000	800,00,000	6.4 GB
DS3	30,000	1200,00,000	9.6 GB
DS4	40,000	1600,00,000	12.8 GB
DS5	50,000	2000,00,000	16.0 GB
DS6	60,000	2400,00,000	19.2 GB

Table 2. Q1~Q9.

Q1. Discover all RDF graph node ID.
SELECT *
WHERE { ?graph rdf:type owl:Thing . }

Q3. Discover all labor export dependence of particular RDF graph.
SELECT ?causeArtifact ?effectArtifact
FROM NAMED WHERE {
   GRAPH utpb:opmGraph {
   ?wdf rdf:type opmo:WasDerivedFrom .
   ?wdf opmo:cause ?causeArtifact .
   ?wdf opmo:effect ?effectArtifact .}}

Q5. Discover all labor use (artifact use) dependency particular RDF graph.
SELECT ?process ?artifact
FROM NAMED WHERE {
   GRAPH utpb:opmGraph {
   ?used rdf:type opmo:Used .
   ?used opmo:cause ?artifact .
   ?used opmo:effect ?process .}}

Q7. Discover all controllable (controlled-by) dependency particular RDF graph.
SELECT ?process ?agent
FROM NAMED
<http://cs.panam.edu/utpb#opmGraph>
WHERE {
   GRAPH utpb:opmGraph {
   ?wcb rdf:type opmo:WasControlledBy .
   ?wcb opmo:cause ?agent .
   ?wcb opmo:effect ?process . }}

Q9. RDF query a specific figure for all initial input node or an output node Workflow
SELECT ?inputArtifact ?outputArtifact
FROM NAMED WHERE {
   GRAPH utpb:opmGraph {
   { ?inputArtifact rdf:type opmv:Artifact .
   OPTIONAL { ?wdf1 rdf:type
opmo:WasDerivedFrom .
   ?wdf1 opmo:effect ?inputArtifact . } .
   OPTIONAL { ?wgb rdf:type
opmo:WasGeneratedBy.

Q2. Query each node to identify an RDF graph.
SELECT * FROM NAMED
WHERE { GRAPH utpb:opmGraph { ?s ?p ?o .}}

Q4. Discover all processes triggered dependency particular RDF graph.
SELECT ?causeProcess ?effecProcess
FROM NAMED WHERE {
   GRAPH utpb:opmGraph {
   ?wtb rdf:type opmo:WasTriggeredBy .
   ?wtb opmo:cause ?causeProcess .
   ?wtb opmo:effect ?effecProcess .}}

Q6. Discover all artificially produced (artifact generation) dependency particular RDF graph.
SELECT ?process ?artifact
FROM NAMED WHERE {
   GRAPH utpb:opmGraph {
   ?wgb rdf:type opmo:WasGeneratedBy .
   ?wgb opmo:cause ?process .
   ?wgb opmo:effect ?artifact .}}

Q8. Query RDF graphs specific type of human-built node and values.
SELECT ?artifact ?value
FROM NAMED WHERE {
   GRAPH utpb:opmGraph {
   ?artifact rdf:type opmv:Artifact .
OPTIONAL { ?artifact opmo:annotation ?annotation .
?annotation opmo:property ?property .
?property opmo:value ?value . } .
PTIONAL { ?artifact opmo:avalue ?artifactValue .
opmo:content ?value . }.}}

?wgb opmo:effect ?inputArtifact . } .
OPTIONAL { ?used rdf:type opmo:Used .
?used opmo:cause ?inputArtifact . } .
OPTIONAL { ?wdf2 rdf:type
opmo:WasDerivedFrom .
?wdf2 opmo:cause ?inputArtifact . } .
FILTER (!bound(?wdf1) && !bound(?wgb) &&
(bound(?used) ||
bound(?wdf2))) } }

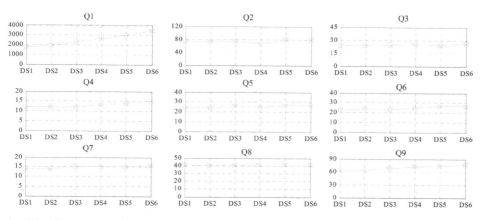

Figure 1. RDF Query test results.

all RDF graph traversal of each node, Q2 carries a query to each RDF graph. However, the query statements of Dependencies class, Q3~Q7, check centralized correlation dependence relationship of RDF data, a limited number of such relationship exists and therefore running time is shorter; Q8 and Q9, the query statements of Artifacts, are used to traverse non-empty resource, and the running time between the former; the query statements Q1~Q9 have little effect on the magnitude of data sets. As data volume grows, the running time increases only slightly, which meets the access requirements to a large scale of data. The results show that semantic data model for large data, presented in this paper, can accomplish to read and query semantic data processing in a short time, confirming the effectiveness and feasibility of the model, which provides a model-base for the realization of the equipment assessment system based on large data analysis.

4 CONCLUSION

In this paper, the research for large data semantic data model is analyzed, and a semantic data model based on it is proposed, so that it can adapt computing and storage MapReduce framework. This paper also gives the relevant definitions of semantic data and constructing method of semantic data model. Finally, the feasibility and effectiveness of the model is validated through system implementation and experimental.

REFERENCES

Abraham Silberschat, Henry F. Korth, S. Sudarshan. Data System Concepts (Sixth Edition) [M]. McGraw-Hill Companies, 2012.

Anderson Chris. The end of theory: the data deluge makes the scientific method obsolete [N]. Wired. 2008–6–23(7).

Beckett D, Berners-Lee, T. Turtle-Terse RDF Triple Language. W3C Team Submission[EB/OL]. http://www.w3.org/standards/techs/rdf#w3c_all/. 2014–6–30.

Chebotko Artem, Abraham John, Brazier Pearl. Storing, indexing and querying large provenance data sets as RDF graphs in apache HBase [C]. In: Proceedings of 2013 IEEE 9th World Congress on Services, 2013: 1–8.

Fay Chang, Jeffrey Dean, Sanjay Ghemawat. Bigtable: A Distributed Storage System for Structured Data [J]. Operating Systems Design and Implementation. 2006: 1–14.

Papakonstantinou Y, Garcia-Molina H, Widom J. Object Exchange Across Heterogeneous Information Sources [C]. In: Proceedings of the Eleventh International Conference on Data Engineering, 1995, 251–260.

Science. Special online collection: Dealing with data [EB/OL]. http://www.sciencemag.org/site/spec-ial/data/, 2014–6–30.

Suciu D. Semistructured Data and XML [C]. In: Proceedings of The 5th International Conference of Foundations of Data Organization (FODO'98), 1998: 1–12.

Utpb. University of Texas Provenance Benchmark [EB/OL]. http://faculty.utpa.edu/chebotkoa/utpb/. 2015–3–30.

Vagner Nascimento, Daniel Schwabe. Semantic Data Driven Interfaces for Web Applications [J]. Lecture Notes in Computer Science. 2013, 7977(1): 22–36.

Zhang Ke, Liu Tao, Li Zhong. A new directional relation model [J]. Science and Engineering Research Support Society, 2014, 7(2): 237–248.

Advances in Energy, Environment and Materials Science – Wang & Zhao (Eds)
© 2016 Taylor & Francis Group, London, ISBN 978-1-138-02931-6

Research on the power allocation algorithm in MIMO-OFDM systems

Xuchen Lin & Yunxiao Zu
School of Electronic Engineering, Beijing University of Posts and Telecommunications, Beijing, China

ABSTRACT: MIMO (Multiple-Input-Multiple-Output) and OFDM (Orthogonal Frequency Division Multiplexing) are the key techniques for the new-generation wireless communication systems. With the typical resource allocation optimization model, the capacity performance of the power average allocation and water-filling allocation algorithm for MIMO-OFDM systems were simulated and compared in this paper. The simulation results show that when the SNR (Signal-to-Noise Ratio) is low, the water-filling allocation algorithm earns more system capacity than the power average allocation, and when the SNR is high, the capacity performance of these two algorithms are almost the same, while the power average allocation lowers the system complexity greatly. Furthermore, in view of high computational complexity of the classical water-filling algorithm, we propose an improved water-filling algorithm with low complexity whose performance comes close to the classical water-filling algorithm.

1 INTRODUCTION

In the field of modern communications, wireless mobile communication technology has become a hot technology with a great development potential. Along with social progress and development of communication technology, higher requirements for the transmission rates and quality of service are expected. The goal of the fourth generation mobile communication is higher data rate, better quality of service, higher spectrum efficiency and higher security.

However, in order to achieve this goal, we need to overcome many challenges. Wireless mobile communication system is facing a very bad mobile wireless channel, in which signals broaden in time domain and cause frequency selective fading. The time-varying characteristics of channel will cause the spectrum spreading of signals when the mobile terminal is moving (Stuber et al., 2004, Jang et al., 2003). Hence, the new generation of mobile communication systems will use OFDM (Orthogonal Frequency Division Multiplexing) as the core technology to provide services. OFDM converts a frequency-selective channel into a parallel collection of frequency flat sub-channels. The sub-carriers have the minimum frequency separation required to maintain the orthogonality of their corresponding time domain waveforms, yet the signal spectrum corresponding to the different subcarriers overlap in frequency (Rey et al., 2005, Shen et al., 2005). Hence, the available bandwidth is efficiently used.

Meanwhile, multiple antennas can be used as the transmitter and receiver, an arrangement called a MIMO (Multiple-Input Multiple-Output) system.

A MIMO system takes advantage of the spatial diversity that is obtained by spatially separated antennas. And it may be implemented in a number of different ways to obtain either a diversity gain to combat signal fading or to obtain a capacity gain (Wang et al., 2006). Recent developments in MIMO techniques promise a significant boost in performance for OFDM systems. Research on wireless resource management in MIMO-OFDM systems is of great significance, which is directly related to the capacity performance of MIMO-OFDM systems. Wireless resources mainly consist of the spectrum resources and power resources, this paper discuss the power resources allocation.

The rest of this paper is arranged as following: in Section II, we study two classical power allocation algorithms in MIMO-OFDM systems, including the power average allocation algorithm and the water-filling algorithm, and then we propose a new low complexity water-filling algorithm. In Section III, we simulate the algorithms discussed and compare their system performances. In Section IV, we give the conclusions.

2 POWER ALLOCATION ALGORITHM IN MIMO-OFDM SYSTEMS

2.1 The power average allocation algorithm

In MIMO-OFDM systems, each orthogonal sub-carrier can be seen as a sub-channel. Hence, the system capacity is the sum of the channel capacity of all subcarriers. According to the Shannon formula defined in information theory, the equation of system capacity can be represented as

$$C = MB \log\left(1 + \frac{P_{AV}}{BN_0}\right) \qquad (1)$$

where C is the channel capacity, M is the number of sub-channels, B is the bandwidth of sub-channel, P_{AV} is the average signal power and N_0 is the unilateral noise spectral power density. The power average allocation algorithm allocates the system power to every subcarrier channel evenly. Equation (1) shows that in the power average algorithm, the system capacity is determined by SNR (Signal-to-Noise Ratio). When the channel condition is poor, which means low SNR, the system capacity is comparatively small. On the other hand, when the channel condition is good, which means high SNR, the system capacity is large.

The power average allocation algorithm is a static allocation algorithm with low complexity and easy to implement, but it does not consider the varying condition of channel state of sub-channel and the user demand, so it can't offer stable and high system capacity performance.

2.2 The water-filling allocation algorithm

OFDM channel has the characteristic of frequency selective fading, the water-filling theorem in information theory allocates the limited power to this kind of channel in order to achieve the goal of faster and more reliable data transmission and maximize the channel capacity. The basic idea of the water-filling theorem is that by dividing the available bandwidth into a number of narrow channels, the transmission characteristics of each sub-channel can come close to the ideal one (Qi et al., 2012, Jiang et al., 2010).

Assume $H(f)$ is the transmission function of a channel whose bandwidth is W, and the power spectral density of additive white Gauss noise in channel is $N(f)$. The signal's power should meet

$$\int_w P(f)df \le P_{av} \qquad (2)$$

where P_{av} represents the average transmit power of the transmitter, $P(f)$ represents the signal power density spectrum. Through a series of deduction we can get the channel capacity is

$$C = \int_w \log_2\left[1 + \frac{P(f)|H(f)|^2}{N(f)}\right]df \qquad (3)$$

And also we can get

$$P(f) = \begin{cases} K - \dfrac{N(f)}{|H(f)|^2} & f \in W \\ 0 & f \notin W \end{cases} \qquad (4)$$

From the equations above we can know that the corresponding channel power is comparatively large when the signal-to-noise ratio $|H(f)|^2/N(f)$ is high, and the corresponding channel power is comparatively small when the signal-to-noise ratio $|H(f)|^2/N(f)$ is low. The water-filling algorithm allocates more power to the sub-channels whose channel gain fading is low. However, whether to apply the water-filling algorithm is determined by the channel transmission function. When the channel transmission function represents constant average fading, the water-filling allocation algorithm earns less system capacity than the power average allocation; on the other hand, when the channel transfer function represents the selective fading, the water-filling algorithm can earn higher system capacity.

2.3 Low complexity water-filling algorithm

The classical water-filling algorithm which uses iterative method can achieve the theoretical upper bound of the system channel capacity, but it needs to make several times of iterative computations to achieve the optimization goal (Jindal et al., 2005). In other words, the computational complexity of the classical water-filling algorithm is quite large, and its real-time ability is poor. Hence, this paper proposes a low complexity water-filling algorithm whose capacity performance comes close to the classical water-filling algorithm.

Let H_n represents the channel gain of each sub-channel, P_n represents the power allocated to each sub-channel, and sort both of them in ascending order. Assume that $H_1 \le H_2 \le \ldots \le H_N$, so $P_1 \le P_2 \le \ldots \le P_N$.

We can get the value of P_1 through

$$P_1 = \frac{1}{N}\left(P - \frac{N}{H_1} + \sum_{n=1}^{N}\frac{1}{H_n}\right) \qquad (5)$$

If $P_1 \le 0$, we consider the power allocated to that sub-channel is zero. After knowing the value of P_1, we calculate the power allocated to the next sub-channel, and delete the sub-channel whose allocated power is zero previously in the later procedure of power allocation, that is

$$P_2 = \frac{1}{N}\left(P - \frac{N-1}{H_2} + \sum_{n=2}^{N}\frac{1}{H_n}\right) \qquad (6)$$

Continue this procedure above until a sub-channel whose allocated power is greater than 0 or equal to 0 is found. Assume $P_m \ge 0$, calculate the powers that allocated to the later sub-channels according to

$$P_n = P_m + \frac{1}{H_m} - \frac{1}{H_n} \qquad (7)$$

where $m, n = \{1, 2 ..., N\}$.

This algorithm only sorts the states of sub-channels, so the computational complexity is $O(N \log_2 N)$; then it rejects the sub-channels which do not need to allocate power, finally calculates the allocated power of later sub-channels one by one according to the order, the computational complexity is $O(N)$. Thus in contrast to the classical water-filling algorithm, this improved water-filling algorithm reduces the computational complexity greatly.

3 SIMULATION RESULTS

In this research, we build the simulation platform of MIMO-OFDM systems, and then simulate the algorithms discussed above, including the power average allocation, the classical water-filling algorithm and the improved water-filling algorithm with low complexity. In this simulation, the systems are single cell MIMO-OFDM downlinks, the bandwidth is 1 MHz, the number of the subcarriers is 64, and the channel is Rayleigh channel with frequency selective fading.

3.1 *Simulation 1: the capacity performance of power average allocation and water-filling allocation algorithm in MIMO-OFDM systems*

Run the power average allocation and water-filling allocation algorithm simultaneously on the platform. The simulation result is shown in Figure 1, the solid line represents the result of the water-filling power

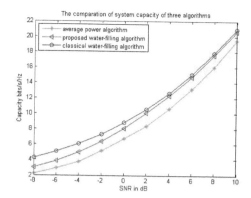

Figure 2. The capacity performance of the three algorithms discussed above.

allocation algorithm, and the dotted line represents the result of the average power allocation.

As is shown in Figure 1, we can draw a clear conclusion that the system capacity increases as the number of the transmitting and receiving antennas increase. Besides, we also can draw the following conclusions: 1) When the SNR is low, the water-filling allocation algorithm earn more system capacity than the power average allocation. 2) The SNR of the water-filling algorithm is approximately 3 dB less than that of the power average allocation when the system capacity is the same. 3) As the SNR increases to a certain degree, the channel capacity earned by the water-filling algorithm will converge to the channel capacity earned by the power average allocation. In other words, when the SNR is high, the capacity performance of this two algorithms are almost same, while the power average allocation lowers the system complexity greatly.

3.2 *Simulation 2: the comparison between the low complexity water-filling algorithm and the other two classical algorithms*

The simulation result is shown in Figure 2, it indicates that the proposed water-filling algorithm earns more capacity performance than the simple power average algorithm, but earns slightly less than the classical water-filling power allocation algorithm. However, in contrast to the classical water-filling algorithm, the proposed algorithm reduces the computational complexity greatly.

4 CONCLUSIONS

With the typical resource allocation optimization model, this paper presents two classical power allocation algorithms in MIMO-OFDM systems,

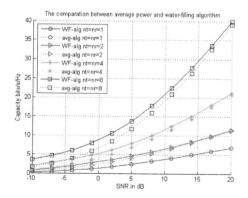

Figure 1. The capacity performance of the power average allocation and the water-filling allocation algorithm in MIMO-OFDM systems.

including the power average allocation algorithm and the water-filling algorithm. And the system capacity and the computational complexity of these algorithms have been discussed and compared through simulation. The power average allocation is a static algorithm with low computational complexity which don't have real-time adjusting ability, and it earns satisfactory system capacity only when the channel condition is very good. And the water-filling algorithm can earn high system capacity, but it needs multiple iterations to achieve the optimization goal of the system capacity maximization. Hence, in view of high computational complexity of the classical water filling algorithm, this paper proposes an improved water-filling algorithm with low complexity whose performance comes close to the classical water-filling algorithm. And the simulation results show that the proposed water-filling algorithm earns slightly less system capacity than the classical water-filling power allocation algorithm, but it reduces the computational complexity greatly.

ACKNOWLEDGMENT

This work is supported by the Beijing Key Laboratory of Work Safety Intelligent Monitoring (Beijing University of Posts and Telecommunications).

REFERENCES

Jang, J., Lee, K.B. & Lee, Y.-H. Transmit power and bit allocations for OFDM systems in a fading channel. Global Telecommunications Conference, 2003. GLOBECOM'03. IEEE, 2003. IEEE, 858–862.

Jiang, Y., Shen, M. & Zhou, Y. 2010. Two-dimensional water-filling power allocation algorithm for MIMO-OFDM systems. Science China Information Sciences, 53, 1242–1250.

Jindal, N., Rhee, W., Vishwanath, S., Jafar, S.A. & Goldsmith, A. 2005. Sum power iterative water-filling for multi-antenna Gaussian broadcast channels. Information Theory, IEEE Transactions on, 51, 1570–1580.

Qi, Q., Minturn, A. & Yang, Y. An efficient water-filling algorithm for power allocation in OFDM-based cognitive radio systems. Systems and Informatics (ICSAI), 2012 International Conference on, 2012. IEEE, 2069–2073.

Rey, F., Lamarca, M. & Vazquez, G. 2005. Robust power allocation algorithms for MIMO OFDM systems with imperfect CSI. Signal Processing, IEEE Transactions on, 53, 1070–1085.

Shen, Z., Andrews, J.G. & Evans, B.L. 2005. Adaptive resource allocation in multiuser OFDM systems with proportional rate constraints. Wireless Communications, IEEE Transactions on, 4, 2726–2737.

Stuber, G.L., Barry, J.R., Mclaughlin, S.W., Li, Y., Ingram, M.A. & Pratt, T.G. 2004. Broadband MIMO-OFDM wireless communications. Proceedings of the IEEE, 92, 271–294.

Wang, J., Love, D.J. & Zoltowski, M.D. User selection for MIMO broadcast channel with sequential water-filling. Allerton Conference on Communication, Control and Computing, 2006. 27–34.

Analysis and research of android security system

Ling Zheng & Yanjiao Liu
Control and Computer Engineering College, North China Electric Power University, Beijing, China

ABSTRACT: Android platform is an intelligent mobile phone operating-system based on Linux 2.6 kernel, which is launched by Google. Due to its openness and scalability, more and more mobile terminal venders and developers have been attracted to join in, so the Android systems gradually occupies most of the market both in smart phone and tablets. For this reason, its security has received more and more wide attention. When Google designed the Android system, it has considered the safety and has set up a relatively security system. However, because of its openness, the security risks are also more likely to be found and exploited, so now it is necessary to analyze and study its safety system, we need to analyze the shortcomings, and propose a solution to security issues of Android programs.

1 INTRODUCTION

With the widespread use of smart phones, mobile security problem is increasingly highlighted. Computers' virus security issues also appear on the phone, and the information is more personal privacy on phone. It is more dangerous after it is stolen. Thus the safety of the mobile phone has caused more and more concern. Android is an open source system based on Linux kernel, its openness and freedom plays a big role in the promotion of its popularity. But it also brings security risks, since it is easy to get Android system source code, its security risks can be easily found and exploited. In order to prevent Android security issues, we must analyze the Android security system, we need to analyze its internal security operation mechanism, how secure communications are between the contents of each module, and found insufficient safety system, that is to identify safety risks. This article briefly describes the Android system architecture, based on this it takes further analysis to Android's security mechanisms and security risks.

2 ANDROID SYSTEM ARCHITECTURE

Android system is a mobile operating system developed by Google in 2007. The entire platform framework consists of four parts from bottom to top in turn is Linux Kernel, Libraries and Runtime, System Framework, and Application. The architecture is shown in Figure 1. Here are the layers function and key technologies are in detail.

2.1 Android kernel

Android kernel based on Linux 2.6 kernel is a kind of porting and developing on Linux. The latest version is based on the 2.6.31, which is an enhanced version of the kernel, it made a cut and added some device drivers for mobile devices such as power management, network management, memory management, process management, driver model, etc. The main role of the Linux kernel is used as an abstraction layer between the hardware and software that can hide specific hardware details to provide a unified service for the upper layer. Android kernel is different from standard Linux kernel; the differences are reflected in several aspects like the

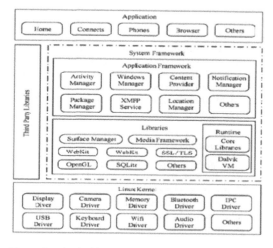

Figure 1. Android system architecture.

file system, inter-process communication mechanism and memory management.

2.2 Libraries and runtime

This part includes a number of core class libraries and an Android virtual machine—Dalvik—this is the middle part of the system, playing a connecting role between the program framework and the core. Android includes a set of C/C ++ libraries for various components of the Android system, these functions are provided to developers through the Android application framework. Libraries implement a number of important functions in the system. It primarily includes databases, graphics processing, Web browsers, and media libraries. Dalvik is a Java virtual machine designed for Google's own Android platform; each Android application is an instance of the Dalvike, where each application runs in its own process. When the virtual machine collapses, separated process can prevent all programs from being closed. The Executable file format in Dalvik is in .dex,.dex format, and is in a compression format designed specifically for Dalvik, which is suitable for low-memory and limited-speed processor system.

2.3 System framework

Android framework provides a variety of API to application developers. With an open development platform, Android developers can write extremely rich and innovative applications. Some important features have been achieved such as accessing to location information, the package manager, running background services, setting the alarm, adding notification to the status bar, they are all interfaces. Application only needs to call these interfaces to achieve the corresponding functions. Android framework architecture is designed to improve component reuse; it makes the use of the framework of the function inside easily, and allows us to develop our own components in order to replace the original components inside. Different mobile phone manufacturers can modify them to form their own unique function, which is an important manifestation of the openness of Android.

2.4 Application

Android applications are located in the top level of the whole system framework. It is a series of applications developed by Java language, including system programs like e-mail client, SMS program, calendar, maps, browser, contacts, and a variety of install applications. Android system application is a signature generated Apk file package to be installed, including code and non-coded files and AndroidManifest. xml file, the XML layout file contains the application's basic information, such as the package name, component descriptions, permission notice, and so on.

3 ANALYSIS OF ANDROID SECURITY SYSTEM

Android system inherited part of the security features of Linux by using the Linux kernel, and established a security mechanism by improvement and innovation. It has greatly improved the Android system's security, and established a security system, which makes android became a more secure mobile operating system. The security architecture is shown in Table 1.

1. Sandbox mechanism
   When an application (.apk) is installed, Android will give the application a unique ID. At runtime, the system will allocate a memory space for each program, in which the application runs to achieve the isolation between programs. However, in some cases, some applications need to share data or processes, different applications can be run in the same process, but they need to use the same private key to sign the application, and then assign them the same Linux user ID. When installing or running an application, Android will examine UserId and signature of the application. If the system UserId and signature of an application are the same, the Dalvik would know that they belong to the same publisher, and will give them the same UID, so these two programs can share a Linux process and the Dalvik virtual machine, they can visit each other. The sandbox mechanism of Android allows each application to run independently in its own process space. Such as the contacts application open the message editor, the editor can only edit and send text messages, regardless of the permissions contact.
2. File Access Control
   By using the Linux kernel, file access control in Android system is same as the Linux system; each file is closely associated with a user ID and group ID as well as three sets of Read/Write/Execute (RWX) permissions. When you create a

Table 1. Android security mechanism.

System architecture	Security mechanism
Linux kernel	Sandbox mechanism File Access Control
Libraries and runtime	Root Management
System framework	Authority Management Digital Signatures
Application	Security Encryption Digital rights management

file with a application ID, it cannot be accessed by other applications unless they have the same ID or the file is set to be readable and writeable.

There is another enhanced security design–mounting the system image as read-only. All important executable files and configuration files are stored in Firmware or system image, they are read-only mode, so even when the attacker got the file system, it cannot change the key file. But you can reload the system image to circumvent this restriction, the requirement is to get the system root privileges. Also it will bring a lot of potential threats after the phone root.

3. Root Management

In the Linux system, the user is divided into ordinary user and root user, only Root user can do some important operations to protect the security of the system. The Android system also has Root mechanism. Some important or related to system security operations, such as deleting system files and uninstall system applications, etc., must obtain the phone Root privileges, similarly to jailbreak in IOS system. By using Android Root privileges, it can prevent users and applications to modify the operating system kernel and other applications.

4. Authority Management

There are about one hundred kinds of behavior or service permissions controls in Android, including phone calls, send text messages, access the Internet, and so on. When installing an application, the required permissions must declare. Android give the appropriate permissions by checking signatures and interacting with the user. Application permissions can only be approved or rejected during the installation, and cannot apply to any authority while running. Once the application has been successfully installed, the application has the appropriate authority functions.

5. Digital Signatures

Signature mechanism plays a very important role in the security of Android applications, the signature mechanism can indicate the issuer or the developer of the Apk installer, determine the source of the application, determine whether this Apk is officially issued or a pirated software by comparing the Apk's signature. In the Android system, all programs must have a signed certificate, otherwise it is not allowed to install. Android system allows developers to use a self-signed certificate to sign the application, and give Apk file stamped digital certificates by digital certificates Java-related mechanisms.

6. Security Encryption

In the Android3.0 and later, Android provides a full file system encryption. All users' data are available in the kernel level, and be encrypted by using AES encryption algorithm. The secret key in

AES encryption algorithm is related to the user's password, it prevents stored data from unauthorized accessing. Encryption system can bear a lot of password guessing attacks, such as brute-force guessing; the password is a random number combination. In order to bear the dictionary-based password guessing attacks, Android provides a set of password complexity rules.

File encryption can protect user data in the system. It does not matter even if the phone is infected with malicious software or the user's information is stolen. It is difficult to break out of the encrypted data. But this is only 3.0, later system has the encryption mechanism.

7. Digital rights management

Android provides a scalable DRM—Digital Rights Management that allows applications to manage and protect their own content, DRM mechanism can use the media content on the phone to the user, such as e-books, mp3, etc. copy to a third party, limiting the frequency of use, limiting the time, so as to protect the rights of content providers and security, prevent the data from unauthorized use. DRM framework supports a number of rights management scheme, there are equipment based programs or based on the producer.

Android's DRM framework consists of two key parts, namely, DRM framework interface and DRM management. DMR frame interface provides a connection between standard applications and application frameworks and Dalvik; DMR management unit achieves a variety of digital rights management and decryption programs, and controls the access of interfaces framework to the underlying DMR digital copyright plug.

4 ANDROID SECURITY RISKS

Android security system is not perfect; there are some security risks. After the analysis of Android security system, we can find some of these security mechanisms are somewhat flawed, which may lead to security risks. After the analysis of security mechanisms and the prevalence of Android security issues, we can know that Android system security risks are mainly in the following aspects: Android permissions defect, Android system vulnerabilities, Root and Brush risk. Here are introductions, analyses, and proposed solutions.

4.1 *Android permissions defect*

4.1.1 *The performance*
Authority mechanism is very important in Android system. It can control the application access to system functions. After analyzing the implementation of a permission mechanism, we can find some flaws

in permission mechanism, for example the permission cannot be modified during installation. Users do not understand a wide range of privileges. The defects mainly show in the following aspects:

1. Cannot be dynamically modified. When installing the application, the user cannot choose the permission of the application. In some cases, in order to install the software, even if the users know that the application applies for an unrelated permission, they will also install. This leads to a result of applying unrelated authority in the application.
2. Cannot descript the permission. During the installation, description of the permission is not very clear, especially for the description of the degree of risk is not intuitive, so that the majority of users cannot understand.
3. Apply the permission without displaying the interface. It cannot show the permission interface by using adb installation to install the application or loading the application from Google Market. In this case, it can install the application in default mode.

Because of the drawbacks mentioned above, it can easily lead to some dangerous operations when applications obtain permission, and after the application is installed the user cannot modify permissions. So the user's privacy is in a serious threat, a lot of the software will collect your personal information, even though these software are some of the regular application.

4.1.2 The solution

For the above problems, we may first consider encrypt to the user's privacy information. The user's information is safe even if the malware have full access, only if the data is encrypted, they cannot break the encrypted-data. Present common data encryption algorithm is MD5 algorithm, RSA algorithm, DES algorithm and so on. RSA algorithm can be used to encrypt for contact information, text messages or other information. RSA algorithm is a relatively sophisticated algorithm; it is a mathematical function based on asymmetric algorithm, which uses a different secret key in encryption and decryption, and though the secret encryption key and algorithm are public, the user only needs to know the secret key to decrypt on it.

4.2 Android system vulnerabilities

4.2.1 The performance

Due to the imperfections, there are some loopholes in Android system; Android system has issued a total number of versions from the beginning until now, each version there may be some loopholes. In the Android application development process, you

can access components of other programs to break the control authority, so as to enhance their own procedures' privileges, which can be a privileged attack conspiracy, which is a loophole in system security. For example, the program 1 does not have an access to the Internet and SD card, but its networking components want to download the file and save it to the SD card, which is not allowed under normal circumstances, but in the same time the program 2 has a component 1 which has permissions for networking and operating SD card, the program 1 can access the component 1 in program 2 to get the permission of networking and operating SD card access procedures to break Android access control.

4.2.2 The solution

A number of shortcomings and deficiencies in system and hardware lead to the Android system loopholes. We can improve the system for the reason of its appearance or take response measures to handle the existing loopholes. However, in order to reduce the impact of system vulnerabilities fundamentally, we can enhance the underlying system and virtual machines.

The first is to strengthen the system kernel, we can control the access with the key Linux process to the underlying data, such that the above program1 component cannot easily access the program 2 components. Reference [10] proposed the implementation of SELinux in Android, if the Android run up SELinux and we can work out the right strategy, we can enhance the security level of the system. The second is to strengthen the system virtual machine, it can be carried out to strengthen security by virtualization, virtualization technology can achieve the separation of public and private platform consolidation, and because the virtual machine monitor has a higher authority, you can use it to complete the system of monitoring and control platform behavior, which can improve the security of the entire system.

4.3 Root and Brush risk

4.3.1 The performance

Due to the openness in Android system, everyone can easily get the source code, the mirror produced after compiling is ROM file, you can re-install the phone, it is equivalent to the computer to reinstall the system, which is Brush. Since the brush can change the system interface, add new features and become even more personality, there are more and more brush machine ROMs in the market, people will get root privileges and Brush the mobile phones easily. But there are some random Brush risks, especially if the brush ROM is not provided by the official. Driven by the interests, some of the ROM creators prefabricated the malicious software inside to push advertising and illegal SP services. When a user uses the configuration ROM games, chat software, or other built-in businesses, it will automatically rebates without user's confirmation. These charges are divided by the phones ROM makers, software companies and mobile phone operators. After the program gains root privileges, it can access system files, cut low-level bar, automatically install applications and do other dangerous operations.

4.3.2 *The solution*

For security risks due to the Root Brush machines will result to malware intrusions. It doesn't relate to the security mechanisms. It mainly relates to the openness of the Android system. So the good way to prevent security issues from arising is mainly intrusion detection and discovery Brush ROM if it contains malware.

5 SUMMARY

Research on safety aspects Android system has been widespread concern. Whether in the system level or application-level studies have got better results, but this does not mean that security researchers have been completed. With the development of Android system, more and more new capability is introduced, where the security research issues will also be endless.

Taking the system level under consideration, SELinux has been added to the Android 4.4 version, which means a more secure Android system, therefore, security research in Android system architecture will have a broader development. In addition, the design and specification in API interface, the user authentication mechanism low user participation, are all worth to have a great research.

Taking the application level under consideration, in order to protect the security of Android applications, malicious application detection technology has been great a progress, but it still has more research space with the development of the system; and there are less vulnerability mining technology breakthrough points in the Android system, and other advanced symbolic execution vulnerability mining technology is not sufficient; it needs researchers to take a broader concern.

REFERENCES

Android. A New Android Market for Phones [EB/OL]. http://www.android.com/.

Burns J. Developing Secure Mobile Applications for Android [R]. Technical Report, iSEC, 2008.

Wang Yang, Wang Qian. Development of the sandbox security technology [J]. Software Guide. 2009, 8(8):152.

Enck W., OngTang M., and McDaniel P. Understanding Android Security [J]. IEEE Security and Privacy, 2009, 7(01):50–57.

http://source.android.com/tech/security/.

Burns, J. Exploratory Android Surgery [R], Black Hat Technical Security Conference: USA, Jul. 2009.

Schlegel, R. Zhang, K., Zhou, X., Intwala, M? Kapadia, A., and Wang, X. Soundcomber: A Stealthy and Context-Aware Sound Trojan for Smartphones. In Proceedings of the Networkand Distributed System Security Symposium (2011).

Shabtai A. Fledel Y. Elovici Y Securing Android-powered mobile devices using SE Linux 2010.

W. Enck, M. Ongtang and P. McDaniel, Mitigating Android Software Misuse Before It Happens [R], Network and Security Research Center, Sep. 2008.

Zhang Yuqing, Wang Kai, Yang Huan. Android Security Review. Journal of Computer Research and Development. 2014, 51(7).

Feature fused multi-scale segmentation method for remote sensing imagery

T.Q. Chen, J.H. Liu, Y.H. Wang & F. Zhu
Laboratory of Remote Sensing and Intelligent Information System, Xi'an Institute of Optics and Precision Mechanics of CAS, Xi'an, China

J. Chen & M. Deng
Department of Geo-informatics, Central South University, Changsha, China

ABSTRACT: Image segmentation is a key step in Object-Based Image Analysis (OBIA) for High Spatial Resolution Remote Sensing Imagery (HSRI). Previous segmentation method used image spectrum and structure information only, so that it is difficult to obtain good segmentation results. A novel segmentation method based on multiple feature fusion and multi-scale region merging was proposed in this paper. Firstly, initial over-segmentation objects were obtained by marked watershed transform. Then, the spectrum, texture, shape, area, and edge intensity information were combined to build the merging cost of adjacent objects, and all objects were mapped to adjacent graph nodes with their merging cost. Finally, the nodes in Region Adjacency Graph (RGA) were merged to get multi-scale segmentation result. Experiment results indicate that the proposed method can produce nested multi-scale segmentation results with accurate boundaries, and achieve a better performance compared to Mean-shift and ENVI segmentation method.

1 INTRODUCTION

Image segmentation is an essential and important process of Object-Based Image Analysis (OBIA) for High Spatial Resolution Remote Sensing Imagery (HRSI). It aims to partition the image into perceptual meaningful regions, which influences the effectiveness of OBIA significantly, such as, object-oriented land classification, objects extraction, and so on. It is still a difficult task to develop an effective method to obtain the perfect segments for HRSI. The segmentation results are influenced by the complex and various features, such as spectrum, texture, shape, size, and spatial relationship of the objects in HRSI. The existing image segmentation algorithms are difficult to take full advantage of these features to obtain ideal segmentation results.

In order to solve this problem, a lot of researches on image segmentation based on feature fusion have been done. Wu et al (2008) proposed a cloud-model-based image segmentation method connected with geometrical features. Gaetano et al (2009) proposed a hierarchical texture-based segmentation method of multi-resolution remote sensing images to make full use of texture information. Wuest et al (2009) proposed a region based segmentation method through band ratios and fuzzy comparison for QuickBird multispectral

imagery. Zhang et al (2013) proposed a boundary-constrained multi-scale segmentation method for HRSI. Zhou et al (2013) proposed an adaptive multi-scale image segmentation method incorporating multiple features. Chen et al (2013) proposed a field theory based multi-level clustering method for high-spatial resolution remote sensing imagery segmentation. These methods can obtain good segmentation results in certain conditions; however, there are still some disadvantages in making use of features of images.

In order to improve segmentation precision, a feature fused multi-scale segmentation method is proposed in this work. The proposed method is illustrated in Section II, and followed with experimental results in Section III. Finally, conclusions and discussions are presented in Section IV.

2 THE PROPOSED METHOD

The three steps of the proposed feature fused multi-scale segmentation method are displayed in Figure 1: 1) Texture feature extraction and initial segmentation; 2) Region Adjacency Graph (RAG) construction based on initial segmentation using features of spectrum, texture, shape, area and edge intensity; 3) Hierarchy Correlated Merging Process (HCMP) based on RGA.

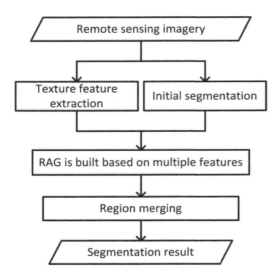

Figure 1. Flowchart of the proposed method.

2.1 Texture feature extraction and initial segmentation

HRSI contains a lot of texture information, and the textures of different ground objects are various. The segmentation method using texture information can get better segments than the method not used (Wu et al., 2008). In this paper, the texture features are extracted by Gabor filters, which have good discrimination properties for different textures and is widely used in remote sensing image analysis. The equation of 2D Gabor filters consists of a sinusoidal plane wave and orientation listed as below:

$$G_{\theta,\lambda,\varphi,\sigma,\gamma}(x,y) = \exp\left(-\frac{(x')^2 + (\gamma y')^2}{2\sigma^2}\right) \times \cos\left(2\pi\frac{x'}{\lambda} + \varphi\right) \tag{1}$$

where $x' = x\cdot\cos\theta + y\cdot\sin\theta$, $y' = -x\cdot\sin\theta + y\cdot\cos\theta$, x and y are coordinates of the pixel in the spatial domain; φ is phase angle of the Gabor filters; γ is the ratio of spatial aspect; λ is the wavelength of the cosine wave; σ is the standard deviation of the Gaussian factor; and σ/λ is the half-response spatial frequency bandwidth. In this paper, the parameters are set as: $\gamma = 1$, $\sigma/\lambda = 1.3$, and $\varphi = 0$, $\theta = \left\{0, \frac{\pi}{6}, \frac{\pi}{3}, \frac{\pi}{2}, \frac{2\pi}{3}, \frac{5\pi}{6}\right\}$. The value of λ is set in the same way as in (Chen et al., 2014). We use this Gabor filters to get a 36D (six orientations and six wavelengths) vector of texture features.

The precision of multi-scale merging segmentation method proposed in this paper is influenced by over-segmented initial segmentation, so the precise marker-based watershed transformation is used for initial segmentation. Before watershed transformation, the median filter is used to reduce noise. The value of a pixel is replaced by the median of the spectral values in the neighborhood of a 3 × 3 window. Then the horizontal I_h and vertical gradient I_v images are produced by Sobel operator, which can get precise edge and tolerate noise well. The final gradient image can be defined as:

$$I_g = \sqrt{I_v + I_h} \tag{2}$$

After we get gradient image, marker-based watershed algorithm (Chen et al., 2013) is applied to get initial segments, the steps are list as follows: 1) H-minima transform and minima imposition techniques are applied to calculate local minima and eliminate the meaningless local minima in the gradient image, in this study the H is set as $I_g/3$; 2) the watershed algorithm based on FIFO technique is used to get initial segments.

2.2 Build RGA based on multiple features

Merging initial segments is an important step to obtain accurate segmentation result, therefore, a good merging criterion plays a crucial role in the final segmentation result. In this paper, merging criterion is built based on the spectrum and texture homogeneity, the compactness of shape, area, and edge intensity.

The homogeneity of spectrum and texture is an objective measurement criterion of segmentation, therefore, the change of homogeneity after merge of two neighbors are to be integrated in the merging criterion. When the homogeneity change is small, there is a greater merging possibility of the two neighbors. The variance is a useful criterion, which can indicate the homogeneity of a region. Thus, the change of variance (CVar) is applied to evaluate the change of homogeneity of spectrum and texture after mergence of two neighbors (Zhang et al., 2013). CVar is defined as:

$$CVar = Var - \left(a_1 \cdot Var_1 + a_2 \cdot Var_2\right)/\left(a_1 + a_2\right) \tag{3}$$

where a_1 and a_2 are the sizes of two neighbors, which are evaluated by the number of pixels, Var, Var_1, and Var_2 representing the variance of the newly created region and two original neighbors.

The shape is also an important measurement criterion of segmentation. According to the compactness proposed by Haralick (Zhang et al., 2013), the boundaries of good segments should be simple. Similar to CVar, the change of compactness (CComp) is applied to evaluate the change of the shape after a visual mergence of the two neighbors. When the *CComp* is small, there is a greater merging possibility. *CComp* is defined as:

$$CComp = L/\sqrt{a} - \left(\sqrt{a_1} \cdot L_1 + \sqrt{a_2} \cdot L_2\right)/\left(a_1 + a_2\right) \quad (4)$$

where a is the size of the newly created region, L, L_1, and L_2 are the edge lengths of the new region and two original neighbor regions, respectively.

The *CVar* and *CComp* are used to evaluate region features, which are not imposed on local structure information. The segments in HRSI also have the similar spectrum, texture, and shapes, however, they are not parts of the same objects, such as the adjacent farmlands, the adjacent buildings, the adjacent road and building, and so on. The edge information of the segments is important for splitting different objects. Hence, the feature of edge intensity is introduced as a merging criterion. When the edge intensity between neighbors is small, there is a greater merging possibility. The gradient image is calculated by Sobel operator of equation (2), which can reflect the edge intensity, the edge intensity of neighbors is calculate using the mean value of the gradient, which is defined as:

$$EI = \sum_{i=1}^{l} Sobel_i/l \quad (5)$$

where l is the edge length of two neighbors, i is the identifier of the edge, and *Sobel* is the gradient of the image. When the *EI* is small, there is a greater merging possibility.

Region Adjacency Graph (RAG) is constituted of nodes and arcs. A node represents a segment and an arc represents adjacency between two nodes. The arc weight represents the merging possibility of two nodes, there are five features: spectrum, texture, shape, area, and edge intensity integrated as merging criterion of arc weight. Normalization needs to be done before incorporating different merging criterions to constitute Merging Criterion (*MC*) of arc weight. Finally, the *MC* of arc weight is defined as:

$$MC = \left(a_1 + a_2\right) \cdot \left(\omega_1 \cdot CVar\_t + \omega_2 \cdot CVar\_s + \omega_3 \cdot CComp\right) \cdot EI \quad (6)$$

where a_1 and a_2 are the size of two neighbors, $CVar\_t$ is the change of variance of texture, $CVar\_s$ is the change of variance of spectrum, *CComp* is the change of compactness, ω_1, ω_2 and ω_3 are weight of $CVar\_t$, $CVar\_s$ and *CComp*, $\omega_1 + \omega_2 + \omega_3 = 1$, *EI* is edge intensity. The small *MC* means that the corresponding adjacent regions are more easily to be merged.

2.3 Multi-scale region merging based on RAG

Threshold of arc weight is used to control region merging process, and is considered as the scale parameter of merging process. Only if the arc weight of the neighbors is smaller than the threshold, the two neighbors are allowed to be merged. If the threshold is set smaller, less merges of neighbors are permitted, resulting in segmentation with small region size or finer scale. If the threshold is set larger, more mergers are permitted, and segmentation with a coarser scale would be produced. Hence, multi-scale segmentation results are produced by setting different thresholds or scale parameters. However, in this case, the boundaries of different scale segments are not matching. In order to get boundary-matching multi-scale segmentations, a Hierarchy Correlated Merging Process (HCMP) is proposed in this paper, which is described as below:

Step 1. The threshold of the merging scale is set.
Step 2. MC is sorted according to ascending order.
Step 3. If the minimum MC is smaller than the threshold, the two adjacent segments are merged, else, the merging process is terminated.
Step 4. Refresh the RGA and MC, and go to Step 2.

The multi-scale segmentation results with matched boundaries are obtained using HCMP method, by setting different thresholds or multi-scale parameters.

3 EXPERIMENTAL RESULTS

In this article, an aerial image is used to evaluate the proposed method, which contains red, green, and blue visible spectral bands. The area is 256×256 pixels, which includes different types of land cover, as shown in Figure 2(a). Figure 2(b), 2(c), and 2(d), segment results of the MI, are equal to 0.05, 0.1 and 0.2, respectively; the weight of texture, spectral and shape are set as 0.1, 0.3, and 0.6. It can be seen that the result of Figure 2(b) contains over-segmentation phenomena, the result of Figure 2(c) has precise boundaries, and the result of Figure 2(d) has some under-segmentation phenomenon. Based on the experiment, MI = [0.05, 0.15] is a good parameter for getting good segmentation result using the proposed method.

In order to prove the efficiency of proposed method in this paper, the segmentation method of EDISON software based on Mean-shift and multi-scale merging method of commercial software ENVI are used for comparison. Figure 3 shows the segmentation results of three methods. Figure 3(a) shows the segmentation results of the proposed method while the MI is set as 0.15. Figure 3(b)

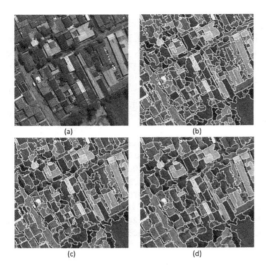

Figure 2. The segmentation results: (a) original image; (b) The segmentation result of *MI* equal 0.05; (c) The segmentation result of *MI* equal 0.1; and (d) The segmentation result of *MI* equal 0.2.

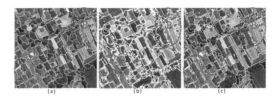

Figure 3. The comparison of segment results: (a) The segment result of the proposed method; (b) The segment result of EDISON; and (c) The segment result of ENVI.

ACKNOWLEDGMENTS

We would like to thank the anonymous reviewers for their detailed and constructive comments. The work was supported by the Open Fund of Key Laboratory of spectral imaging technology CAS [grant number LSIT201406], the Major State Basic Research Development Programme of China (973 Programme) [grant number 2012CB719906], the National Natural Science Foundation of China [grant number 41201428], the Open Fund of State Key Laboratory of Information Engineering in Surveying, Mapping and Remote Sensing [grant number13R01].

shows the segmentation results of the software EDISON, while the spatial bandwidth *hs* is set at 3, the frequency bandwidth *hr* is set at 3.25, and the area threshold is set at 20. Figure 3(c) shows the segmentation results of the software ENVI, while the segment parameter is set at 40, and merging parameter is set at 90. In the oval area A, it can be see that the boundaries of segments in Figure 3(a) are better preserved than Figure 3(b) and Figure 3(c). In the oval area B, Figure 3(a) and Figure 3(c) have less over-segmentation phenomenon than Figure 3(b), and the shape are better preserved. Compared with the segmentation method of EDISON and ENVI, the proposed method has better performance in HSRI segmentation.

4 CONCLUSION

In this study, a novel multi-scale image segmentation method based on multi-scale merging, which combines spectrum, texture, shape, area, and edge intensity is presented for High Spatial Remote Sensing Imagery (HSRI). It includes three steps: texture features extraction and initial segmentation, Region Adjacency Graph (RAG) built based on multiple features, and multi-scale merging based on RAG. The experimental results show that the proposed method is effective in obtaining precise boundaries of the objects and it can reduce the over-segmentation phenomena effectively. However, this method also cannot obtain the scale parameters automatically. The next study will focus on scale parameter selection.

REFERENCES

Chen T.Q., Chen J., Mei X.M., Shao Q.B., Zhang T., Deng M. 2013. Field theory based multi-level clustering for high-spatial resolution remote sensing imagery segmentation, *Geography and Geo-Infmation Science* 29(6): 10–13.

Chen J., Deng M., Mei X.M., Chen T.Q., Shao Q.B., Hong L. 2014. Optimal segmentation of a high-resolution remote-sensing image guided by area and boundary, *International Journal of Remote Sensing* 35(19): 6914–6939.

Gaetano R., Scarpa G, Poggi G. 2009. Hierarchical texture-based segmentation of multiresolution remote sensing images, *IEEE Transactions on Geoscience and Remote Sensing* 47(7): 2129–2141.

Wu Z.C., Qin M.Y., Zhang X. 2008. A cloud-model-based remote sensing image segmentation connected with geometrical features, *Geomatics and Information Science of Wuhan University* 33(9): 939–942.

Wuest B., Zhang Y. 2009. Region based segmentation of QuickBird multispectral imagery through band Ratios and fuzzy comparison, *ISPRS Journal of Photogrammetry and Remote Sensing* 64(1): 55–64.

Zhou Y.N., Luo J.C., Cheng X., Shen Z.F. 2013. Adaptive multi-scale remote sensing image segmentation incorporated multiple features, *Geomatics and Information. Science of Wuhan University* 38(1): 19–23.

Zhang X.L., Xiao P.F., Song X. Q, She J.F. 2013. Boundary-constrained multi-scale segmentation method for remote sensing images, *ISPRS Journal of Photogrammetry and Remote Sensing* 78: 15–25.

Advances in Energy, Environment and Materials Science – Wang & Zhao (Eds)
© *2016 Taylor & Francis Group, London, ISBN 978-1-138-02931-6*

Performance study on PWM rectifier of electric vehicle charger

L.X. Qu

The State Administration of Production Safety Supervision and Management Information Communication Center, Beijing, China

ABSTRACT: In order to make an electric vehicle charger with good performance and no electricity pollution, this paper gives a detailed analysis of the quasi-Proportional Resonant (PR) strategy used in PWM rectifier, focusing on its principle analysis and its parameters design. On the basis this, the control system is simulated using MATLAB, and compared with PI strategy. The result indicates that the dynamic and state performances of controller using quasi-PR are better than that of PI controller.

1 INTRODUCTION

As we all know, to make the electric vehicle charger (Chen, 2012) green, that is to say, the power factor should be 1 and the power harmonics less, the key part of which is rectifier, with the rapid development of power electronics technology, PWM rectifier is widely used in daily life, for example, AC variable speed, APF, transportation, household appliances, and so on. Due to its advantages of the adjustable power factor, any dc voltage could be got; therefore, PWM is applied in the charger in this paper. In order to make the charger obtain an excellent performance, the PWM rectifier should adopt a good control strategy. In the paper, the author took the current loop Quasi-PR controller method. This method can not only achieve steady astatic characteristic, but also has fast dynamic response.

If the three-phase voltage type PWM rectifier (VSR) wants to have good functions, it should have the two merits as followed. First, the dc voltage should be stable. Second, the ac current should be high sinusoidal, which is to say the harmonics less. Between the two aspects, the control of current is more important, which is the key of the paper. At present, the PI control is most common, but according to the control theory we know that the PI control can not achieve deadbeat control. So, this paper uses the Quasi-PR controller to control the ac current. The paper discussed the structure principle and the mathematical model of the VSR. Based on the math mode, it expounds the principles and the parameters design of the Quasi-PR controller.

2 THE TOPOLOGY AND OPERATION PRINCIPLE OF VSR

2.1 The main circuit topology

Figure 1 shows the topology of the VSR. U_a, U_b, and U_c represents the three phase voltage, respectively; i_a, i_b, and i_c are the currents; ac filter inductor is L and its internal resistance is R_s; the dc capacitor is C and RL is the load resistor.

2.2 The math model of the VSR

Define the function as the (1):

$$\beta_k = \begin{cases} 1 \\ 0 \end{cases} \quad k = a, b, c \tag{1}$$

β_k is the switch function, which represents the switch of power tube in Figure 1.

When $\beta_k = 1$, it shows that the upper arm of the phase bridge arm is on. Conversely when $\beta_k = 0$, the lower arm is on. According to KCL and KVL, the state equation of the VSR can be got where u_{dc} is the dc voltage and i_L is the load current.

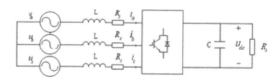

Figure 1. Main circuit topology of three-phase VSR.

$$\begin{cases} L\,di_k/dt + Ri_k = u_k - u_{dc}\left(\beta_k - \frac{1}{3}\sum_{j=a,\,b,\,c}\beta_j\right) \\ C\,du_{dc}/dt = \sum_{k=a,\,b,\,c} i_k\beta_k - i_L \end{cases} \quad (2)$$

From the (2) could get the math model under two phase rotating coordinate system:

$$\begin{cases} L\,di_q/dt = U_{sq} - Ri_q - \omega Li_d - u_{dc}\beta_q \\ L\,di_d/dt = U_{sd} - Ri_d + \omega Li_q - u_{dc}\beta_d \\ C\,du_{dc}/dt = \frac{3}{2}\left(i_q\beta_q + i_d\beta_d\right) - i_L \end{cases} \quad (3)$$

The transformation matrix between two phase stable and rotating coordinate system is as follows:

$$\begin{cases} C_{2r/2s} = \begin{bmatrix} \cos\theta & -\sin\theta \\ \sin\theta & \cos\theta \end{bmatrix} \\ C_{2s/2r} = \begin{bmatrix} \cos\theta & \sin\theta \\ -\sin\theta & \cos\theta \end{bmatrix} \end{cases} \quad (4)$$

According to (3) and (4), the voltage equation of the VSR under two phase stable coordinate system is as shown in (5):

$$\begin{cases} \begin{bmatrix} e_\alpha \\ e_\beta \end{bmatrix} = \begin{bmatrix} L\dfrac{di_\alpha}{dt} \\ L\dfrac{di_\beta}{dt} \end{bmatrix} + \begin{bmatrix} R_s & 0 \\ 0 & R_s \end{bmatrix}\begin{bmatrix} i_\alpha \\ i_\beta \end{bmatrix} + \begin{bmatrix} u_\alpha \\ u_\beta \end{bmatrix} \\ C\dfrac{du_{dc}}{dt} = \frac{3}{2}\left(i_\alpha\beta_\alpha + i_\beta\beta_\beta\right) - i_L \end{cases} \quad (5)$$

We can see from the (5), there did not exist any coupling relationship between the axis α-β. So there is no need to decouple this model, which greatly simplifies the design of control system (Fan, 2010).

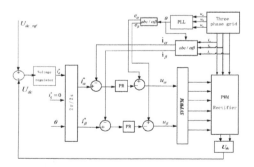

Figure 2. Overall control block diagram of three-phase PWM rectifier based on PR.

But the PI control strategy as most studiers used was uneasy to control the ac variables exactly. The Quasi-PR control, which will be mentioned next, can resolve the problem. Based on the Introduction above, the control diagram of PWM rectifier is shown in Figure 2.

3 THE QUASI-PR CONTROL

3.1 Principle analysis

From Figure 2, the PR control applied to the current control of VSR which avoided the coordinate transform and inverse transform could control the grid current under two-phase static coordinate system. Meanwhile, it omits the decoupling part, which is helpful to the design of the control system and can simplify the structure of the system.

Resonant controller is refined from the internal model principle. Its internal controller is made up of the corresponding mathematical model of given signal which is a control system with zero steady-state static error system. In the case of the given instruction which is with the form of $A\sin(\omega t+\varphi)$, the traditional PI controller can not trace it quickly with zero error. The control system which can replace it should have the regulator model of ω^2/s^2 or $k_{rs}/s^2 + \omega^2$ in the transfer function. Thus, it can make sure that the frequency is invariable when the amplitude of controlled object changed. Then, we can quickly and accurately obtain the instructions wanted. Next, deriving the PR model under the stationary and rotating coordinate system as is shown in Figure 3.

Taking the transform relationship of the two coordinate systems and the uncoupling control into consideration, the output expression of control amount is as shown in (6).

$$\begin{bmatrix} y_\alpha(t) \\ y_\beta(t) \end{bmatrix} = C_{2r/2s}\left(G_{dq}(t) * \left(C_{2r/2s}^{-1}\begin{bmatrix} x_\alpha(t) \\ x_\beta(t) \end{bmatrix}\right)\right) \quad (6)$$

* is convolution operation; $C_{2r/2s}^{-1}$ is the inverse operation of the $C_{2r/2s}$.

According to the Euler equations and Laplace transform, (7) can be formed.

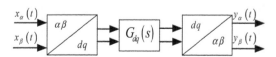

Figure 3. Control block diagram of two coordinate systems.

$$\begin{bmatrix} y_\alpha(s) \\ y_\beta(s) \end{bmatrix} = \frac{1}{2} \begin{bmatrix} G_{dq}(s+j\omega)+G_{dq}(s-j\omega) & jG_{dq}(s+j\omega)-jG_{dq}(s-j\omega) \\ -jG_{dq}(s+j\omega)+jG_{dq}(s-j\omega) & G_{dq}(s+j\omega)+G_{dq}(s-j\omega) \end{bmatrix} \begin{bmatrix} x_\alpha(s) \\ x_\beta(s) \end{bmatrix} \tag{7}$$

After the rotating coordinate transform, the regulator $k_p + k_i/s$ which contains the integral term could track the ac instruction accurately.

Such as (8) shows the math model under α-β static coordinate system.

$$\begin{bmatrix} y_\alpha(s) \\ y_\beta(s) \end{bmatrix} = \begin{bmatrix} k_p + \dfrac{k_i s}{s^2+\omega^2} & \dfrac{k_i \omega}{s^2+\omega^2} \\ -\dfrac{k_i \omega}{s^2+\omega^2} & k_p + \dfrac{k_i s}{s^2+\omega^2} \end{bmatrix} \begin{bmatrix} x_\alpha(t) \\ x_\beta(t) \end{bmatrix} \tag{8}$$

Known from (8), the PI controller could be replaced by the proportional resonant controller with $k_p + k_i s/(s^2 + \omega^2)$, which could track the ac signal accurately. It is consistency to the Internal Model Control what is mentioned above.

$$G_{\alpha\beta}(s) = \begin{bmatrix} k_p + \dfrac{k_r s}{s^2+\omega^2} & 0 \\ 0 & k_p + \dfrac{k_r s}{s^2+\omega^2} \end{bmatrix} \tag{9}$$

$$G_{\alpha\beta}^{-1}(s) = \begin{bmatrix} k_p + \dfrac{\omega}{s^2+\omega^2} & 0 \\ 0 & k_p + \dfrac{\omega}{s^2+\omega^2} \end{bmatrix} \tag{10}$$

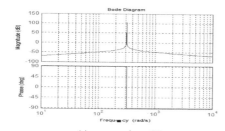

(a) correspondence (9)

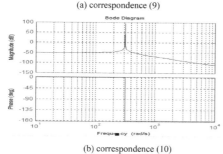

(b) correspondence (10)

Figure 4. Bode diagram of the two PR controllers.

(9) and (10) shows the general form of the PR controller. Then we can get the BODE figure of the two transfer functions.

From (4) we could find the two controllers that have the feature—zero steady-state error characteristic, but the phase margin of Figure 4(b) is 0, which will reduce the stability of system. The PR controller above has a good phase margin, so in the paper chooses the Figure 4(a) that is the (9) model.

The ideal PR controller, which has the feature of infinite gain will cause the unsteady to the system. On the other hand, it is difficult to realize in practice. Therefore, this paper will improve the ideal PR controller which is easy to realize. The method replaces the proportional integral with the Low pass filter $H_{dq}(s) = k_p + k_r s/(s^2 + \omega^2)$, which has the high gain property, if $\omega_c \ll \omega_0$ can get the PR controller expression.

$$H_{dq}(s) = k_p + \frac{k_r \omega_c s}{s^2 + 2\omega_c s + \omega_0^2} \tag{11}$$

3.2 Parameter design

The (11) shows that the controller is mainly related to the factors: k_p, k_r, and ω_c. To analyze the impacts the three factors cause, we could make the two of them invariant to analyze the only variate. The Figure 5(a), 5(b), and 5(c) show the influence of the three factors produced. (1) $\omega_c = 8$, $k_r = 3.5$, k_p is changed as is shown in Figure 5(a). (2) $\omega_c = 8$, $k_p = 100$, k_r is changed as is shown in Figure 5(b). (3) $k_p = 100$, $k_r = 3.5$, ω_c is changed as is shown in Figure 5(c).

From the BODE above, in the 5(a) the gain kept increasing with the k_p increasing. And the harmonics content decreased gradually when the k_p is big enough. If the value is appropriate, it can reach optimum performance and improve the anti-interference capacity. In the 5(b), when k_r increases, the gain in resonance point gets big, which would not affect the bandwidth. But the frequency range will widen leading to the useless signal getting big. The stability of the system will be badly affected, so the k_r should not be too big. In the 5(c), when ω_c changed, it would cause both the gain and phase margin to change. But it also causes the bandwidth to widen and the resonance point gets smaller. As a result, it is unfavorable to weaken the steady error and not good to add the dynamic response speed (Wanchak, 2006). According to all the factors, in this paper $k_p = 100$, $k_r = 3.5$, $\omega_c = 8$ rad/s, and $\omega_0 = 314$ rad/s.

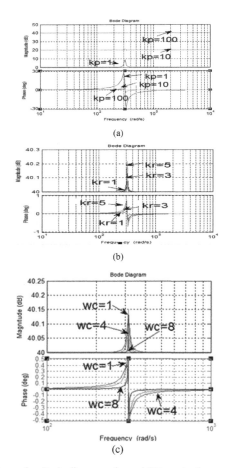

(a)

(b)

(c)

Figure 5.   Bode diagram of quasi-PR controller.

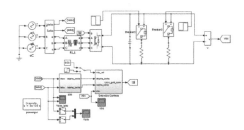

Figure 6.   Simulation of three-phase VSR based on PR.

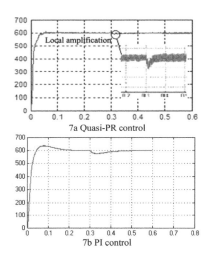

7a Quasi-PR control

7b PI control

Figure 7.   Output waveform of DC voltage at load dip instant.

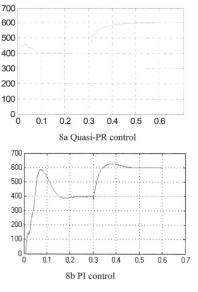

8a Quasi-PR control

8b PI control

Figure 8.   Output waveform of DC voltage at DC given dip instant.

4   SIMULATION AND ANALYSIS

After the theoretical analysis previously put the Quasi-PR and PI controller into the system, and make simulations in Matlab, that is shown in Figure 6. In order to obtain good grid current wave, we take the SVPWM (Wang, 2010) as modulation. The parameters are as followed (Zhu, 2009):

AC side: voltage 220 v/50 Hz, filter inductor L = 0.09 mH, Rs = 8.5e-3Ω, ω = 314 rad/s;

DC side: C = 0.012F, Udc = 600 V; RL = 50Ω; switching frequency: f = 5 kHz.

The simulation model of the Quasi-PR control is as shown in Figure 7.

At the time 0.3 second, the RL decreased from 50Ω to 40Ω. The Figure 7a and Figure 7b show the diversifications of dc voltage under two different control strategies. The diversifications of dc voltage are given in the Figure 8a and 8b when the voltage reference changes.

748

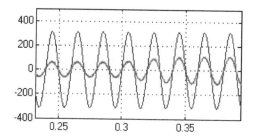

Figure 9. A-phase current and voltage waveforms at load dip instant.

Compared to the two sets of simulations, it is obvious that the Quasi-PR control is superior to the PI control when the load is changed. The voltage of Quasi-PR control could reach the given value quickly. But the PI control needs a long fluctuation time. In addition, the overshoot amount of the Quasi-PR control is smaller than the PI control. So it is known as the performance of the Quasi-PR control, which is obviously superior to the PI control.

The current and voltage waves of A phase are shown in Figure 9.

5 CONCLUSION

The theoretical PR control and the simulation are first introduced in this paper. And it demonstrates the feasibility of control strategy of the current loop of VSR that is applied in the electric vehicle charging machine. Without decoupling control, it could well control the current of VSR, and simplify the control system of VSR. Then the steady-state and dynamic characteristic is good, which has a great reference value to the analysis of the correlation problems of the power electronics technology in the future.

REFERENCES

Chen Li-ping. 2012. Zhang Jian-wen et al. Study on Control of Three-phase Rectifiers Based on PR [J]. Advanced Technology of Electrical Engineering and Energy, 31; 24–27.

Fan Yi-hui, Xu Hong-bing, et al. 2010. Three-phase PWM Rectifier Applied in the Electric Vehicle Charger [J]. The World of Inverters, 08: 90–95.

Geng Cui-hong, Cao Yi-long. 2014. Simulation of SVPWM Based on Matlab/Simulink [J]. Control and Instruments in Chemical Industry, 41(06): 707–710.

Wanchak Lenwari, Mark Summer, Pericle Zanchetta, et al. 2006. A high performance harmonic current control for shunt active filters based on resonant compensators [A]. IECON2006 [C], 2109–2114.

Wang Wan-wei, Yin Hua-jie, Guan Lin. 2010. Parameter Setting for Double Closed-loop Vector Control of Voltage Source PWM Rectifier [J]. Transactions of China Electro-technical Society, 25(2): 67–72.

Zhu Rong-wu, Dai Peng, et al. 2009. The Research of Three-phase Voltage Source PWM Rectifier Based on Two-phase Stationary Coordinate [J]. The World of Inverters, 4(4): 42–45.

Advances in Energy, Environment and Materials Science – Wang & Zhao (Eds)
© 2016 Taylor & Francis Group, London, ISBN 978-1-138-02931-6

Analysis and research on iOS security system

Ling Zheng & Dandan Li
Control and Computer Engineering College, North China Electric Power University, Beijing, China

ABSTRACT: With the popularity of mobile intelligent devices, hardware performance continues to improve. Mobile intelligent terminals are increasingly used to handle a variety of important data which is followed by a large number of new security issues, which is closely linked with mobile intelligent terminals. These security issues include privacy leak, the phone Trojan virus, malicious programs, and so on. Today's mainstream, smart, and terminally operating systems are, the most talked about, the iOS and Android. Apple's iOS operating system developed for iPhone, iPod Touch, iPad, and other and other products supports Wi-Fi, video telephone, GPS positioning and installation of third-party developers of applications, and other functions. This paper describes the overall architecture of iOs operating system, and focuses on the analysis of a number of important security mechanisms introduced in iOS operating systems, including code signing, sandboxing, memory address layout randomization strategy, and data encryption mechanisms.

1 INTRODUCTION

IOS operating systems are developed by Apple for its mobile devices. iOS security features include: sandboxing, ASLR, DEP, Code signing, Data Protect API, and so on. In addition, Apple launched its own official electronic market AppStore; the AppStore is the only channel for all iOS devices to download and install the software. All software must be reviewed before it goes on the shelves. The above-mentioned security mechanisms and AppleStore well protected security iOS system greatly limits the spread of malicious software in the device. With the advent of iOS system is "Jailbreak" and the third-party electronic market, piracy and cracking software, plug-jailbreak began to spread among iOS devices. After the iOS device jail breaks, the system loses the protection of many security mechanisms; applications running on the device can obtain root privileges; you can use the high-risk API that can be run with an unsigned code that has not been making the devices face enormous security threats and challenges.

This paper researching on the iOS operating system is mainly for security of the operating systems. Since the iOS is a non-open-source operating system, so existing research for iOS safety literature and work are extremely scarce. This paper combines the existing research results and data, and notes on the iOS security mechanisms including trusted boot, code signing, sandbox technology, and data protection. And by means of reverse analysis of the above security level from the realization of the principle of in-depth analysis, and focus on analysis and summary of the iOS security architecture of the key points and weaknesses.

2 iOS SECURITY ARCHITECTURE

2.1 iOS operating system architecture

1. Core OS is located in the bottom layer of the system architecture, which is the core of the operating system layer, which includes memory management, file systems, power management, and other operating system tasks. It can directly interact with hardware devices as app developers do not need to deal with this layer.
2. Core Services is a core service layer; it can be used to access the iOS operating system for some services.
3. Media is the media layer, through which we can use in the application of various media files, record audio and video, graphics rendering, and the production of basic animations.
4. Core Touch is a touch layer, which provides a variety of useful framework for our application development, and most of these frameworks are related to user interface, in essence, it is responsible for up user touch interaction on iOS devices.
5. iOS is a Unix-like operating systems, its core is XNU, mainly by Mach, BSD, lOKit, and other components.

2.2 iOS operating system hardware security features

IOS devices insert a AES 265 encryption engine operation between the running system memory and Flash memory in the middle of DMA path, which makes the SHA-1 operation complete in the

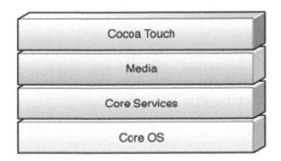

Figure 1.　iOS operating system architecture diagram.

middle of the hardware; it greatly improves the efficiency of the cryptographic operation. Each iOS device has a unique device ID (UID) and a group ID (GID), which are cured in the application processor chip when the device is produced; there is no software or firmware that can read them, but only see the result after using them to encrypt or decrypt the operation. UID is a data which can ensure that each device is unique, and any hardware providers, including Apple's own, will not be recorded and save the data. GID is a data which is related to the equipment hardware processor; the same processor has the same GID, the data can be used to correctly identify the hardware processor during software installation and restore time; this gives iOS devices in the installation and restore the operating system firmware that provides an additional layer of security.

UID ensures that the data which is encrypted by UID is not easily stolen, because even if the Flash memory chip moves from one device to another device, due to the different UID, file is inaccessible. Except UID and GID, other keys required by systems to use are produced by system random number generator based on Yarrow algorithms. And the system entropy needed by the algorithm can also be obtained from the interrupt timer, as well as some of the sensors inside the device.

Except secure produce key, the system also needs a safe destroyer, a key, since the flash memory using the average consumption technology, which means that a data erasure needs to erase multiple copies of it, in order to solve this problem, iOS devices introduced something called erasable storage security to erase data technology; erasable memory using the technology can be directly addressed and erase the data within a small area on the lower level.

2.3　iOS operating system software security features

iOS is derived by its desktop version of Mac OSX; it inherits many Mac OSX excellent security features.

In the process of designing the security system, as opposed to Mac OSX, Apple made a lot of improvements for the iOS mobile operating system characteristics, the most important ones are the following aspects.

For security reasons, iOS has less feature support respect to the desktop operating system. It includes: the current popular java and flash programs. Because they have a lot of security problems, reducing support for such a program will greatly reduce the way iOS attacks.

More streamlined components, as opposed to desktop system. iOS removes a lot of unnecessary existing system components such as iOS has not integrated shell, therefore, iOS can't use shell scripts or some shell commands, which makes an attacker to attack the system more difficult. An attacker who wants to achieve their desired operating in its own process has to prepare their own range of environmental tools needed to attack. iOS takes a good separation mode design between system service and application process, it has advanced means of communication. Various system services are based on the Unix Daemon implements, among these system services, system services communicate with the upper application process through the XPC, this design has two major benefits: make the system more robust and privilege separation. Separating design separate the error-prone modules from the entire system to run on the Daemon process, which allows codes to appear without any exception; even collapsing will not affect the normal operation of other parts of the system. Through this practice of separating program into different modules, it can be well separated from authority, so that each module only functions strictly responsible for its part in the whole system, and each module can simultaneously and strictly limit its permissions, which makes a system service appear even more secure and the issue does not completely undermine the security of the entire system. Also in the method called between processes through messaging time, iOS introduces entitlements to permissions audit of the caller, to ensure the legitimacy of the method call.

iOS also introduces a number of important security mechanisms including:

1. The code signing mechanism. This mechanism ensuring that all running on iOS binaries and runtime must be signed by Apple or Apple-authorized third party credible mechanism to run. In fact, the kernel signature to check the memory page which will be sent to execute, in fact, the kernel signature check will be sent to execute the memory page, if you find an error signature or no signature, the kernel will be rejected.
2. Data Execution Prevention. Using data execution prevention mechanism, the processor can

identify which areas of memory are executing code which is common data. The data execution protection is to ensure that the processor executes code only and not performs data.

3. Address space layout randomization. For many open, and data execution protection, operating systems, attackers often cleverly re-use existing code fragments in memory to achieve a certain purpose, but it requires the attacker to accurately infer which memory address needs to use the code accurately. After introducing the address space layout randomization mechanism, in memory, of all the dynamic library, the dynamic linker, stack and address are made randomizing process, which allows an attacker to locate objects in memory address become very difficult for better protecting the system from attack.

4. Sandbox technology. Sandbox technology to limit application process runs in a separate environment, this process has no ability to access resources beyond the sandbox, but also can't detect any sandbox environment. Sandbox technology has two main functions, one greatly limits the damage of malicious programs on the system, and second, the system also makes hacking more difficult.

Except the above listed several software-based security mechanism to the operating system, Apple also offers an important protection, that is, App-Store. AppStore is Apple's official market, all the iOS applications must be submitted to Apple and audited. The application which has Apple's private key signature can be shelved to the AppStore where users are allowed to download. Once at any time the program at any time after the discovery of the shelves have any violations, including security issues, Apple has the right to force the software off the shelf.

3 ENTERPRISE SECURITY MANAGEMENT

3.1 Safe boot chain

iOS system every step of the boot process are included by Apple's signature encryption module to ensure correctness and the integrity of the procedure, and only when it is verified after the trust chain, the steps to be carried out, these components include encrypted boot loader, kernel, kernel extensions, as well as baseband firmware.

Whenever an iOS device is turned on, its application processor immediately executes a code on the read-only memory of Boot ROM, the code can't be changed when the hardware chip would be implanted in the manufacture, so obviously it

is trustworthy. Boot ROM code includes Apple's public root certificate, the public key is used before the Low-Level Boot loader (LLB) load to verify that it has the correct signature Apple. When the LLB executed its mission, it will validate, and run the next stage boot loader, iBoot. Finally iBoot verifies and starts the iOS kernel.

This secure boot chain ensures the bottom of the software system will not be tampered, and also ensures that the started iOS only runs on the proven iOS devices. If this starts, the process of any one of the steps to verify a problem, the boot process will be terminated and force the system into recovery mode, even if the Boot Rom is unable to successfully start LLB, then the device goes directly into the factory DFU mode.

3.2 Code signing

Once the iOS kernel boots up, it will control the processes or procedures, which may be taken to get the kernel running, in order to ensure that the application is not illegal to tamper with, iOS requires that all code execution must go through certificates issued by Apple signature.

Code signing mechanism controlled by the Mandatory Access Control Framework, the system framework inherited from the Trusted BSD MAC Framework of the FreeBSD. MACF allows additional access control policy, and the new strategy is loaded at the startup time.

By analyzing bsd/kern/mach loader.c of XUN kernel source, we can see that when the executable file is sent to the kernel to perform, the kernel calls parse-machfile () to parse the mach-o file, and calls load_code_signature () in the body of parse_machfile () to load signature. During loading, ubc_cs_blob_add function checks whether the signature is legitimate, and ultimately through Mach's Remote Method Invocation service to ask AMFID if the signature is not a valid signature. So we know that when an application is loaded successfully, the summary information lettered of the signed certificate is also being loaded, the

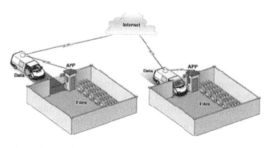

Figure 2. iOS sandbox mechanism.

summary information is stored in the cs_blob inside, and Correspondence with the actual execution memory address. Finally, while the program is running, the AMFI compares the digest information stored in the cs-blob and the digest value appearing in the virtual memory of the actual code. If the match results are inconsistent, then an error gets generated and terminates with the page memory address processes being related.

3.3 *Sandbox mechanism*

All third-party programs running under iOS system, must be in the sandbox environment; the program in the sandbox can't access any resources or files outside of the sandbox and can't make any changes to the system. When the application is installed, it will be assigned a random name directory by the system as the root directory of the application, when the application needs to access resources or files outside the directory it must go through the API of Apple.

Sandbox mechanism as well as AMFI has policy modules under the Mandatory Access Framework of TrustedBSD. Sandbox framework under its base adds some relatively large changes, including calling a user-space hooking and policy management engine above TmstedBSD system can be configured in the configuration file for each management processes that they have.

IOS's whole sandbox mechanism consists of the following components: a set of user space library function is used to configure and initialize the sandbox. A Mach service to handle kernel log and save preset configurations. A used kernel extension TrustedBSD API, to enforce a single access control policy. A kernel extension support providing a regular expression engine used to access control policies to match operations.

iOS applications to start the system sandbox by calling sandbox_init. Initialization function depends on libsandbox.dylib library, and the library can convert human-readable policy to kernel-readable binary format. The binary format of message is then passed to mac_syscall, so as to be handled by the TrustedBSD subsystem processes. Finally, TrustedBSD sends sandbox initialization request to Sandbox.kext process. This kernel process will set rules for the current process sandbox. After a rule is set up, the kernel will return a success flag. After sandbox initialization, TrustedBSD layer will monitor all function calls, and get a function call through the Sandbox.kext to check whether the resource is available. For different system calls, Sandbox. kext according AppleMatch.kext to records different process resource access rules. It uses the regular expression to match whether related

processes follow the expected resource access rules. Mach messages of the system store system information into log system.

3.4 *Address space layout randomization*

Address Space Layout Randomization security mechanism is introduced from iOS4.3, and its role is randomized each time the program load address is spaced in memory. It is able to configure the important data to the memory location that the malicious code is difficult to guess, and the other attackers difficult to attack.

Depending on whether open Position Depend Executables, Address Space Layout Randomization mechanism has two protected modes. If an application is not turned on PIE function at compilation time, it has only a limited function of ASLR protection. Specifically, its main program and the dynamic linker will be loaded at a fixed address in memory, the main thread's stack always starts at a fixed memory address. If one opens the PIE function when the program is compiled, it will turn on all the features of ASLR, and in this program, all memory regions are randomized. To better understand which memory area will receive ASLR effect, and how these memory addresses are changed, we can observe simple procedures to get the conclusion.

In order to better understand, which memory areas will be effected by ASLR, and how these memory addresses change, we can observe simple procedures for conclusion. To better understand which memory area will be, how they effect ASLR, and these memory addresses are changed, we can observe simple procedures for conclusion. Figure 3 shows a common procedure in the open PIE, and not open PIE, in the role of ASLR mechanism main load

PIE	No	Yes
Executable	Fixed	Randomized per execution
Data	Fixed	Randomized per execution
Heap	Randomized Per execution	Randomized per execution (more entropy)
Stack	Fixed	Randomized per execution
Libraries	Randomized Per device boot	Randomized per device boot
Linker	Fixed	Randomized per execution

Figure 3.   iASLR with PIE.

address, an address of a variable is allocated on the heap, a stack in the middle of the variable address, on a shared link library function address, as well as the dynamic linker (dyld) address memory to load before and after the change of system reboot.

3.5 Data encryption and data protection

Compared to the traditional desktop PC, mobile equipment is easily lost, stolen or may have been powered on and network status, therefore, Apple introduced the Data Protection (Data Protection) technology in iOS to protect user data stored on the device flash memory.

From the beginning of iOS4, Apple provides a data protection application interface to iOS application developers to bring more effective protection of data stored in a file and the keychain. Developers only need to declare which files or data keychain in the project is sensitive data and under what circumstances is the possible acquisition. In the program source code, developers using constants represent different protection classes, to mark the protected data and keychain items. Different protection classes are divided into different type of classes through protected file data or keychain items, as well as under what conditions was to enable protected area.

Data Protection Application Programming Interface is used to allow the application stated in the project file, or at what time keychain is encrypted, and by adding a new definition of protection class to mark an existing application program interface which makes these encrypted files or keychain item at any time can be decrypted again. To protect a file, the application needs to use this class NSFile ProtectKey's NSFileManager to set a value.

4 SUMMARY

This paper studied and analyzed the overall structure of the iOS operating system security system. On the basis of this, it focuses on the important introduced security mechanisms, as well as studies the realization of the principle. These researches can better understand the pros and cons of iOS operating system in terms of security. Research in this paper can provide important direction and targeted reference for other research work of iOS system.

REFERENCES

Cedric Halbronn, Jean Sigwald: iPhone security model & vulnerabilities [J]. HITBSecConf 2010.

Information on http://developer.apple.com/library/ios/navigation/index.htmlt2014.5.20].

Information on http://www.cydiasubstrate.com/inject/darwi3Q/[2014.5.20].

Jonathan Levin: Mac OS X and iOS Internals: To the Apple!s Core [M], Wrox, 2012.

Jonathan Zdziarski: Hacking and Securing iOS Applications [M]. O'Reilly, 2012.

Keen Team exploits Safari in mobile browser category [EB/OL]: 2013–12–13 [2014–05–20]. http://www.pwn2own.com/2013/1r/keen-team-exploits-safari-in-mobile-browser-category/.

Manuel Egele, Christopher Kruegel, Engin Kirda, Giovanni Vigna: PiOS: Detecting Privacy Leaks in iOS Applications. NDSS 2011.

Sean MorsseyrOS Forensic Analysis: for iPhone5 iPad, and iPod touch [M], Apress,2010_.

Shub-Nigxirrath.: Primer on Reversing Jailbroken iPhone Native Applications. 2008.

Design and implementation of the anti-cancer diet system based on .net

Qiang Li
Department of General Surgery, Zhujiang Hospital, Southern Medical University, Guangzhou, China

Zheng Xiang
College of Medical Information Engineering, Guangdong Pharmaceutical College, Guangzhou, China

ABSTRACT: China is currently in the high incidence of cancer, which is why prevention and treatment of cancer is a very urgent task. This design of anticancer medicated diet system will provide a comprehensive knowledge of medicated therapeutic aspects in treatment and rehabilitation of patients. The system includes information editing module, information query module, and message module. It can add, delete, and update the data and information of the diet through the information editing module. It can query the diet information including board materials, medicated formula, and other related information through information query module. Message module provides the user with a mutual communication platform. System implementation uses ASP.net, Transact-SQL, HTML, and B/S three-tier architecture, using a SQL Server 2008 as the database.

1 INTRODUCTION

Diet is a traditional product combined with Chinese medical knowledge and cooking experience (Liu, 2005). We can change the public's unreasonable eating habits by anticancer diet, and use science and a healthy diet to reduce the risk of cancer (Zhao, 2006). System of this study is to provide users with targeted anti-cancer medicated diet information to facilitate the users to find relevant medicated diet, guide the public towards a reasonable diet, and improve people's health. (Hao, 2005; Hong, 2007).

2 RELATED TECHNOLOGY

2.1 B/S modue

In this paper, the development of anti-cancer medicated therapeutic system's architecture is based on B/S Mod. Architecture B/S mode of the data processing is divided into three parts. The first layer is the client (user interface) that provides user-friendly access to the system. The second layer is the application server, responsible for business logic implementation. The third layer is the data server, responsible for storing data information access and optimization. B/S three-layer structure (see Fig. 1). (Liu, 2003).

2.2 ASP.Net technology

ASP.Net application framework is built on the common language runtime. It is used to build powerful web applications on the server side. ASP. Net framework consists of a rich toolbox. ASP. Net is language-neutral, so you can choose the language you are most familiar with, or in several languages to complete an application. The system selects the C # language as the primary development language, with a small portion of the JavaScript scripting language. (Dong, 2004).

2.3 Database technology

The system uses SQL Server2008 e to store data. SQL Server2008 is a Web database product with full support, providing core Extensible Markup Language (XML) support, providing check ability on the internet and outside the firewall, and providing a Web standards-based extensions database programming. Rich support for XML and internet standards allows the use of the built-in stored procedure to XML format and easily store and retrieve data. SQL Server provides a powerful development tools and various development characteristics; it greatly improves development efficiency and further expands the application space bringing new opportunities for commercial applications. (Yang, 2007).

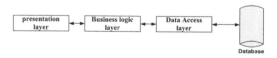

Figure 1. B/S three-layer structure.

3 SYSTEM ANALYSIS AND DESIGN

3.1 *The system functional requirements*

1. Member login functions. An illegal user is not a registered user. He cannot operate the message. According to the requirements of the customer registration module, a customer can become a member. At the same time the system administrator can delete user registration information, and can manage user data.
2. Medicinal herbs management functions. This section include medicinal herbs and herbal medicines management query. Medicinal herbs are for system administrators' management, divided into two parts of medicinal herbs and medicinal herbs management. These two portions include adding their new data, deleting their data, modifying their data, and querying their functions. For tourists, medicinal herbs membership query and search functions, mainly conditional query and multi-page displays a category of all commodities.
3. Message management functions. Message management is for system administrators, including website message, the new member message, delete, and query. Members are mainly oriented Web site message, personal message and reply to add, query function.
4. Administrator management functions. Administrators can query administrator information, add a new administrator, or delete the administrator.
5. Membership management functions. Administrators can freely view member information, and remove the member.

3.2 *Overall system function module*

Function modules include membership management module, medicated management module, medicine management module, query module, message management module, and administrator management module. Reception system module as Figure 2 shows. System background function as Figure 3 shows.

3.3 *System data table design*

According to the database requirements analysis and conceptual design of the above, we can design well-known database called Diet by SQL Server 2008 database. Diet database is composed of the multiple tables below, Administrator information table (Table 1), Medicinal material information table (Table 2), Diet information table (Table 3), Member information table (Table 4), Disease information table, Reply information table, Peer Review information table, Message table, and Information table. Each table is shown below. Each table corresponds to a table in the database.

4 SYSTEM PROCESS

System operation flowchart diagram describes the system to run the program process. Flowchart can help users familiarize with the operation of the system, System background operation processes as Figure 4 shows.

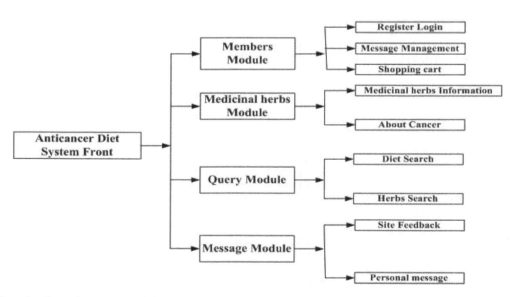

Figure 2. Reception system module.

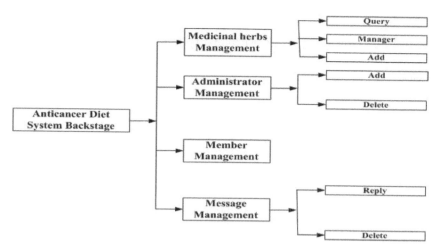

Figure 3. System background function module.

Table 1. Administrator information table.

Name	Type	Is empty	Primary key/foreign key	Explanation
AdminID	int	No	Primary key	Administrator ID
AdminName	varchar(50)			Administrator name
Password	varchar(50)			Password
RealName	varchar(50)			Real name
Email	varchar(50)			Mail
LoadDate	datetime			Load time

Table 2. Medicinal material information table.

Name	Type	Is empty	Primary key/foreign key	Explanation
MaterialID	int	No	Primary key	Medicinal material ID
Material	varchar(50)			Medicinal material name
Taste	varchar(Max)			Taste
Application	varchar(Max)			Application
Effect	varchar(Max)			Effect
Attention	varchar(Max)			Precautions
Note	varchar(Max)			Remark

Table 3. Diet information table.

Name	Type	Is empty	Primary key/foreign key	Explanation
MedID	int	No	Primary key	Diet ID
TypeID	int			Type ID
MedName	varchar(50)			Diet name
Usage	varchar(Max)			Usage amount
Method	varchar(Max)			Usage method
Prescription	varchar(Max)			Prescription
MedIntroduce	varchar(Max)			Remark

Table 4. Member information table.

Name	Type	Is empty	Primary key/foreign key	Explanation
MemberID	int	No	Primary key	Member ID
UserName	varchar(50)			Member name
Password	varchar(50)			Password
RealName	varchar(50)			Real name
Sex	bit			Sex
Address	varchar(200)			Address
PostCode	varchar(20)			Zip code
LoadDate	datetime			Load time
Phonecode	varchar(20)			Telephone
Email	varchar(50)			Mail

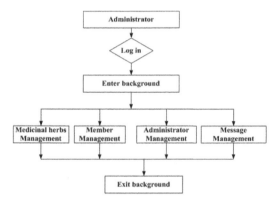

Figure 4. System background operation processes.

5 DISTRIBUTED DATABASE SYSTEM LAYOUT

This design will be a distributed system. Distributed database system is a subset of the database system, which includes not only distributed database and distributed database management systems, but also other parts. It is a storage medium, complex processing target, management systems and network environments. Distributed systems have two characteristics.

1. Distribution. Database data is not stored in the same space, more precisely, the data is not stored on the same server; it is stored in multiple independent database server.
2. Logical integrity. Distributed database systems are linked; it is a unified whole logic, that from the outside looks like a centralized database system. Take metadata and data separation designed in a manner distributed as a storage system.

6 COPYRIGHT FORMS AND REPRINT ORDERS

This paper analyzes the research and development process of a medicated diet system. We use Visual C # as the programming language, Microsoft SQL Server 2008 as the database in Microsoft Visual Studio 2008 platform, and system based on B/S. We can query the information and communication through the system.

ACKNOWLEDGMENT

Author Qiang Li, and Zheng Xiang contribute equally. I would like to declare on behalf of my co-authors that the work described was original research that has not been published previously, and not under consideration for publication elsewhere, in whole or in part. All the authors listed have approved the manuscript that is enclosed. This work is supported in part by the science and technology plan of Guangdong province (NO. 2012B060500063), China.

REFERENCES

Dong Fang. "ASP.NET Database Development". Beijing: Tsinghua University Press, 2004.
Hao jianxin. "New Chinese medicated diet recipe book". Beijing: Science and Technology Literature Publishing Press, 2005.
Hong shanggang. "Symptomatic Diet Health". Beijing: China Textile Press, 2007.
Liu zhenyan. ".NET-Based programming". Beijing: Electronic Industry Press, 2003.
Liu zhenyan. "Life is nutrition". Guangzhou: Guangdong People's Publishing Press, 2005.
Yang Tianqi. "ASP.NET Network programming". Machinery Industry Press, 2007.
Zhao lin. "Talk about a balanced diet and health advice". Beijing: People's Health Publishing Press, 2006.

Advances in Energy, Environment and Materials Science – Wang & Zhao (Eds)
© 2016 Taylor & Francis Group, London, ISBN 978-1-138-02931-6

Non-Intrusive Load Identification based on Genetic Algorithm and Support of Vector Machine

Yanli Ma, Bo Yin, Xiaopeng Ji & Yanping Cong
Ocean University of China, Qingdao, Shandong Province, China

ABSTRACT: Load identification is a major concept in the field of smart homes and smart grids. Non-Intrusive Load Monitoring (NILM) method is applied to solve this problem, which is performed by analyzing the total current and voltage signal of the main distribution board to estimate the energy consumption of individual appliances and turning on/off or other operations. In this paper, we have used the theory of NILM to identify household electric load. By analyzing the total current signal, extracting related features and using Genetic Algorithm (GA) and Support Vector Machine (SVM) to identify different electric load. This paper has used the BLUED dataset (Anderson et al. 2012) as the experimental data. In the end, the rationality and effectiveness of the proposed method was verified by MATLAB Simulation.

1 INTRODUCTION

In smart grids, electric load monitoring is significant. It will help to improve the load composition and guide rational consumption to reduce electricity costs. Therefore, the study of electric load monitoring and identification is necessary for building the electric monitoring system, which has practical significance and economic value of the electricity industry.

In recent years, automatic electricity load monitoring and decomposition method based on measuring sensor technology has obvious advantages than manual investigation, which is widely concerned. There are two implementations: 1). Intrusive residential Load Monitoring (ILM): Each appliance is equipped with sensors, to collect and send consumption information. This way has a high cost and is difficult to promote. 2). Non-Intrusive Load Monitoring and Decomposition (NILMD): By installing a set of current and voltage sensors at the power service entry in a residence and analyzing the measured electrical signals, the electricity energy consumption of individual appliances and working status can be identified. Using this way, the financial costs are less.

For non-intrusive monitoring of electricity, the load identification is a very important part. In this context, many researchers proposed many different theories and algorithms. Hart et al. (Hart 1992) proposed the method of active and reactive power variation sequence, using cluster analysis algorithm, for load identification; Lee et al. (Leeb et al. 1995) used the steady-state and transient information of load, with the method of multistage analysis to get the

power information; JG Roos et al. (Roos & Lane 1994) have developed the method of multi-layer neural network to analyze the features of current, power, and harmonic characteristics for the effective recognition of electrical equipment; Jiaming Li et al. (Jiang et al. 2012) extracted the current harmonics of load as parameter, and employed SVM for classification; H. Chang et al. (Chang 2010) analyzed the transient electrical energy when turned on, and used artificial neural network algorithm for electrical identification.

This paper combines genetic algorithm and supports vector machine (Wu et al. 2009) to identify resident electric load. The five relevant parameters of current are extracted as features, then they use the SVM optimized by GA to identify load. The result of MATLAB simulation showed the validity of the method.

2 GENETIC ALGORITHM AND SUPPORT VECTOR MACHINE

2.1 Genetic algorithm

Generally, Genetic Algorithm (GA) (Hong-bo 2004) consists of four parts: encoding mechanism, control parameters, fitness function, and genetic operators. Encoding mechanism is the foundation of GA. Using fitness function to describe the suitability of each individual. For the optimization problem, the fitness function is the objective function. The purpose of the introduction of the fitness function is that it can be evaluated according to its fitness to determine the merits of the individual than the degree. The most important part is a genetic operator, there are three kinds: selection, crossover, and variation.

Genetic algorithm, firstly, puts a problem solving as genotype (such as binary encoded string), which constitute a chromosomal group. According to the principle of the fittest, which chose to adapt to individual environment and eliminate the bad individual, the individual copy retained regeneration. By cross, genetic mutation operator group to a new generation of chromosomes. Based on some kind of convergence conditions, evolving from generation to generation, and finally converge to the most to adapt to the environment, so as to obtain the optimal solution to the problem.

2.2 *Support vector machine*

Support of Vector Machine (SVM) is based on a small sample of statistical learning theory proposed by Vapnik, which focuses on statistical learning rule in small sample conditions. The main idea of SVM is to create an optimal decision hyperplane, make the plane on both sides of the plane recently to maximize the distance between the two types of samples, thus provide good generalization ability for classification issues.

For classification problem is to find a computable classification function: $y = f(x)$, $x \in R^n$, $y \in \{-1,1\}$, for k given samples:

$$(x_1, y_1), (x_2, y_2), ..., (x_k, y_k), x_i \in R^n, y_i \in \{-1, 1\}.$$

To find the hyperplane (decision surface) of samples can be separated, that is

$$(\mathrm{w} \cdot x) + b = 0, w \in R^n, b \in R,$$

the corresponding identification function is $f(x) = sign((w \cdot x) + b)$

2.3 *Support vector machine based on genetic algorithm*

Support vector machine is an effective tool based on a small learning sample, which is considered high promotion performance as a classifier without prior knowledge, self-learning, self-organizing, adaptive, and strong nonlinear mapping ability. Its structure is simple and easy to operate, can simulate any non-linear input-output relationship. But the classification performance of SVM is related to kernel parameters σ2 and penalty factor C. According to this case, GA is chosen to optimize the parameters of SVM. With specific steps of this process of optimization (Wu et al. 2009) (see Fig. 1) as follows:

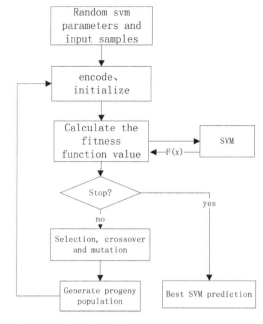

Figure 1. Flow chart of GA-SVM algorithm.

1. Randomly generated SVM binary encoding parameters, and then constructed the initial population;
2. Calculate its error function in order to determine their fitness value, generally return errors after running the reciprocal of the square and evaluation function as a chromosome error, the smaller fitness;
3. Select a large number of fitness values of the individual, directly to the next generation;
4. The use of crossover and mutation genetic operator groups are processed on the current generation and the next generation of population;
5. Repeat 2) to a group of support vector machine parameters initial determination evolving until the training objective condition is met.

3 IMPLEMENTATION OF THE PROPOSED METHOD OF LOAD IDENTIFICATION

NILM (Yu et al. 2013) is used to identify the operation status of different load, which includes four parts: data collection, data preprocessing, event monitoring and feature extraction, load analysis, and identification. The workflow is as shown in Figure 2.

To verify the feasibility of this method, we used a public dataset, BLUED, as the raw experimental data, and used the GA-SVM algorithms to classify loads.

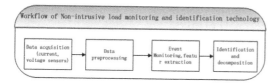

Figure 2. The workflow of NILM.

Table 1. Four appliances in the experiment.

Label	Name	PowerCoAvg (W)	Phrase
129	TV	190	B
140	Monitor	40	B
152	Hallway stairs light	110	B
156	Bathroom upstar light	65	A

*PowerCoAvg means the average power of each appliance.

3.1 Processing data

BLUED dataset were released by Carnegie Mellon University, collected the changes in voltage, current and power consumption of an American family during a week. The basic frequency is 60 Hz and the sampling frequency is 12 KHz. In this paper, four electrical appliances (see Table 1) were selected as the experimental subject from the BLUED dataset: TV (label 129), Monitor (label 140), and Hallway stairs light (label 152), Bathroom upstar light (label 156).

According to the file of event list from the BLUED dataset, the load energizing and de-energizing events were acquired. Extract one second mixed current data before and after the acquired event as the raw data. Then use different methods to get the transient current waveform for single load. The processing results are shown as below (see Fig. 3).

3.2 Feature extraction

In this paper, five features were extracted as the experiment feature, which were calculated from the transient current data of 10 cycles, $i(n)$, $n = 1, 2, ... N$ as above result. The five features (I_{peak}, I_{avg}, I_{rms}, FF, CF) (Tsai & Lin 2011) can be, respectively, computed by

$$I_{peak} = max(i(n)) - min(i(n)) \quad n = 1, 2, ..., N \quad (1)$$

$$I_{avg} = \frac{\sum_{n=1}^{N} i(n)}{N} \quad (2)$$

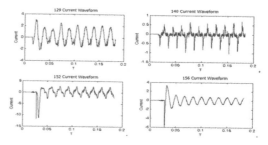

Figure 3. The current waveform of single load.

Table 2. Results of SVM and GA-SVM load identifiers.

	SVM		GA-SVM	
Label	Right number (right/total)	Recognition rate (%)	Right number (right/total)	Recognition Rate (%)
129	19/20	95	19/20	95
140	20/20	100	20/20	100
152	17/20	85	19/20	95
156	15/20	75	16/20	80
Total	71/80	88.75	74/80	92.5

$$I_{rms} = \sqrt{\frac{\sum_{n=1}^{N} i(n)^2}{N}} \quad (3)$$

$$FF = \frac{I_{rms}}{I_{avg}} \quad (4)$$

$$CF = \frac{I_{peak}}{I_{rms}} \quad (5)$$

where $i(n)$ is the samples of 10 cycles, which is acquired in section 3.1.

3.3 Load identification

To identify the operation status of different types of appliances, the SVM identifier with GA to optimize the parameters is used in this paper. The classification accuracy of SVM associated with the selected parameters, using GA to choose parameters, can effectively improve the classification performance of SVM.

3.4 Result analysis

For the selected four appliances, each appliance has 40 groups of features as the train and test data. Using the method of cross-valind, 20 groups of each load were selected randomly as training data set, the rest as the test data. The training and test data were brought into the SVM and GA-SVM prediction recognition, the highest recognition rate of SVM was 88.5%; and GA-SVM recognition rate was up to 92.5%. Recognition result of each appliance is in Table 2.

GA-SVM had a better recognition rate for small sample classification and linearly inseparable appliances. From the result, the use of GA-SVM classifier properly improved the recognition rate.

4 CONCLUSION

The Non-intrusive load identification technology is mainly composed by event detection, feature extraction and load identification. The problem of choosing effective features and classification algorithm can limit the evaluation process. In this paper, transient related features and GA-SVM as classifier for small samples are used. This method is capable of identifying the operation of different loads, and employs genetic algorithms to optimize the parameters for support vector machines that can improve identification precision. Finally, although the simulation result is high, the proposed method needs more research in large number of measured data of more appliances in the future.

ACKNOWLEDGMENT

This work was financially supported by Qingdao innovation and entrepreneurship leading talent project (13-cx-2), Qingdao strategic industry development project (13-4-1-15-HY), and Shandong province science and technology project (2013GHY11519).

REFERENCES

Anderson, K., Ocneanu, A. & Benitez, D., 2012. BLUED: A fully labeled public dataset for event-based non-intrusive load monitoring research. Proceedings of the 2nd ....

Chang, H., 2010. Load identification of non-intrusive load-monitoring system in smart home. WSEAS Transactions on Systems.

Hart, G., 1992. Nonintrusive appliance load monitoring. Proceedings of the IEEE.

Hong-bo, Z., 2004. A study on evolutionary support vector machine based on genetic algorithm. Journal of Shaoxing College of Arts and Sciences.

Jiang, L., Luo, S. & Li, J., 2012. An approach of household power appliance monitoring based on machine learning. 2012 Fifth International Conference on Intelligent Computation Technology and Autormation.

Leeb, S., Shaw, S. & Jr, J.K., 1995. Transient event detection in spectral envelope estimates for nonintrusive load monitoring. IEEE Transactions on Power Delivery.

Roos, J. & Lane, I., 1994. Using neural networks for non-intrusive monitoring of industrial electrical loads. ... in I & M., 1994 IEEE.

Tsai, M. & Lin, Y., 2011. Development of a non-intrusive monitoring technique for appliance'identification in electricity energy management. 2011 The International Conference on Advanced Power System Automation and Protection.

Wu, J., Yang, S. & Liu, C., 2009. "Parameter Selection for Support Vector Machines Based on Genetic Algorithms to Short-Term Power Load Forecasting,." Journal of Central South University (Science and Technology).

Yu, Y., Liu, B. & Luan, W., 2013. Nonintrusive Residential Load Monitoring and Decomposition Technology. Southern Power System Technology.

Advances in Energy, Environment and Materials Science – Wang & Zhao (Eds)
© *2016 Taylor & Francis Group, London, ISBN 978-1-138-02931-6*

A fast algorithm combined with SSIM and adjacent coding information for HEVC intra prediction

T. Yan, X.X. Xiao, X.C. Zhang & H. Liu
School of Physical Science and Technology, Central China Normal University, Wuhan, China

X.S. Zhang
Department of Electrical Engineering, College of Engineering, University of Arkansas, USA

ABSTRACT: A fast algorithm for intra mode decision in High Efficiency Video Coding (HEVC) is presented in this paper. This algorithm takes full use of both the advantages of SSIM of the adjacent PU block and the regular of probability of MPM, PLANAR mode and DC model, which are selected in the candidate modes to reduce the number of intra prediction modes and cut down the computational complexity. The experimental results show that complexity reduction from HM12.1 is over 35.6% and stable for various sequences. The corresponding bit rate is decreased by 1.34% and PSNR-Y is increased by 0.5058 dB.

1 INTRODUCTION

As the new generation standard of video coding is compared to H.264/AVC, the High Efficiency Video Coding (HEVC) is developed by a Joint Collaborative Team on Video Coding (JCT-VC) (Bross, 2013), including the ITU-T Video Coding Experts Group (VCEG) and ISO/IEC Moving Picture Experts Group (MPEG). It is intended to provide a better coding compression efficiency, to enable the use of the coded video representation in a flexible manner for a wide variety of network environments, and to enable the use of multicore parallel encoding and decoding devices. However, to improve compression performance better, it is made more complex so that HEVC cannot supply in the real time communication system. In the intra coding processing (Bross, 2011), computational complexity is especially intensive due to a unit structure of recursive quad-tree coding and the best mode was selected from 35 intra prediction modes by an enumeration method. Therefore, to reduce algorithm complexity, to quicken coding speed, and satisfy actual application without sacrificing its subjective and objective video quality already became an important topic of HEVC research. Kim J., et al (2011) proposes a method under its hierarchical structure including Coding Unit (CU), Prediction Unit (PU), and Transform Unit (TU) (Kim, 2011). Matsuo S., et al (2012) proposes a new intra angular prediction method exploiting the conventional 2-tap filter and a 4-tap DCT-based interpolation filter. Motra A.S., et al

(2012) uses direction information of the co-located neighboring block of previous frame along with neighboring blocks of current frame to speed up intra mode decision. To reduce coding complexity, all the above fast algorithms add extra coding information without considering its own rule. After analyzing its own rule, this paper proposes an intra prediction method based on image correlation and a high probability rule of the best prediction mode selected from PLANAR mode, DC mode and Most Probable Mode (MPM) finally. Experimental results show that this proposed method could reduce the computational complexity of prediction mode decision further, and meet practical applications better, with a lower cost.

2 INTRA PREDICTION IN HEVC

2.1 *Intra prediction mode in HEVC*

As similar as H.264/AVC is, intra prediction in HEVC is divided into luma prediction and chroma prediction. HEVC provides 35 prediction modes for luma prediction as well as 6 prediction modes for chroma prediction (Sugimoto, 2010). As time passes, the best mode is also selected by rate-distortion cost. However, the greatest advantage of HEVC is that up to 35 prediction directions are available, compared to only 9 prediction directions in H.264/AVC (Wiegand, 2012). Besides, HEVC has another non-directed prediction mode, called PLANNR mode (McCann, 2011), which has a good effect on an area of texture smoothing

Table 1. The number of candidate mode in the set of candidate mode.

PU size	N
64 × 64	3
32 × 32	3
16 × 16	3
8 × 8	8
4 × 4	8

and certain gradient trend. In addition, compared with H.264/AVC, prediction unit in HEVC could be matched with texture better, due to more shape, size, and types of PU block, so that it receives better prediction results.

2.2 Mode prediction proceeding

According to the HM12.1, intra mode prediction proceeding in HEVC is roughly divided into three stages.

The first stage: First, all of 34 or 17 prediction modes are for Rough Mode Decision by cost function based on Sum of Absolute Transformed Difference (SATD). Second, their Rate-Distortion Cost (RDCost) (Corrêa, 2012) is calculated. Finally, the cost is sorted in an ascending order. Therefore, candidate modes are the N kinds, in order of the least cost in modes; the value of N depending on the size of PU block.

For 4 × 4 and 8 × 8 blocks, N is equal to 8. For other blocks, N is equal to 3. The number of modes in the list of candidate modes is illustrated in Table 1.

The second stage: Whether MPM (Zhao, 2011) has been included in the candidate mode is to be validated. If not, it will be added in the end.

The third stage: The best prediction mode is the minimum cost mode, which is selected from the candidate modes calculated by the cost function of Rate-Distortion Optimization (RDO) (Bross, 2011).

3 OPTIMIZATION ALGORITHM

3.1 Structural Similarity (SSIM)

The structure information of image is suggested by correlation among pixels of an image. Eyes are more sensitive for the change of structure information than that of the value of a single pixel while observing an image (Yang, 2011). Wang Z., et al at (2014) proposed a method of quality evaluation for SSIM of image, in which content of image structure information remaining approximately image perception quality. On the basis of the feature, SSIM of current PU block x and current PU block

y are modeled to be of a combination of luma, contrast and structure information (Yang, 2011).

$$SSIM(x,y) = l(x,y) \cdot c(x,y) \cdot s \quad (1)$$

In the above expression, luma, contrast and structure information are related to a mean of x and y, a standard deviation of x and y and a covariance of x and y. Their expressions are as follows:

$$l(x,y) = \frac{2u_x u_y + c_1}{u_x^2 + u_y^2 + c_1} \quad (2)$$

$$c(x,y) = \frac{2\sigma_x \sigma_y + c_2}{\sigma_x^2 + \sigma_y^2 + c_2} \quad (3)$$

$$s(x,y) = \frac{\sigma_{xy} + c_3}{\sigma_x \sigma_y + c_3} \quad (4)$$

In the above expression, c_1, c_2, and c_3 are small constants for avoiding the denominator to be zero. $c_1 = (k_1 L)^2$, $c_2 = (k_2 L)^2$, $c_3 = c_2/2$, $k_1 k_2 \ll 1$, where L is the dynamic range of pixel value, generally $k_1 = 0.01$, $k_2 = 0.03$, and $L = 255$.

First, lines of paragraphs are indented 5 mm (0.2″) except for paragraphs after a heading or a blank line (First paragraph tag).

3.2 Condensed candidate

In the list of candidate modes, cost is sorted in an ascending order. The paper, respectively, counted the percentages of all candidate modes selected as the best mode. The percentages of all candidate modes of the file named RaceHorses.yuv in the kind of D video sequences are reflected in the Figure 1.

From the Figure 1, the cost is less which is the front of the list of candidate modes, the probability of the best prediction is selected from it is higher.

To reduce algorithm complex and ensure the quality of video prediction, the number of RDO decision models is cut. RDO decision models for PU block is bigger than 8 × 8 that are selected from the first several prediction modes in the list of candidate modes whose probabilities are over 80%. For other PU block whose information is more precise, those modes are selected from the first several prediction modes whose probabilities are over 65% and then both PLANAR modes and DC modes which are adopted frequently in the intra prediction of HEVC are added in the list of candidate modes.

The RDO decision model number for the various blocks is obtained by the statistic analysis as following Table 2.

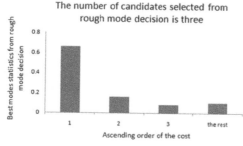

The number of candidates selected from rough mode decision is three

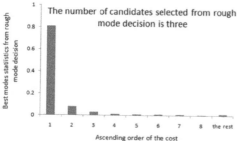

The number of candidates selected from rough mode decision is three

Figure 1. The percentage of each candidate mode.

Table 2. RDO decision model number after cut the mode.

The size of prediction unit	Number of intra prediction mode	Number of RDO modes
64×64	5	1
32×32	34	2
16×16	34	2
8×8	17	2~4
4×4	17	2~4

3.3 Algorithm description

W Algorithm process proposed in the paper is shown as following:

1. The four SSIM between the current PU block with its neighbor to the left block, the top of block, the top left block, and the top right block are calculated before rough mode decision. If certain SSIM of these is bigger than 0.000001, the prediction mode of current PU block is selected from the mode of the PU block.
2. If certain SSIM of these is lesser than or equal to 0.000001, SATD of MPM of the current PU block is counted. The prediction mode of current PU block is selected from the mode of the PU block if the SATD is the minimum.
3. If the SATD of MPM is not the minimum, the prediction mode of current PU block is selected from the mode of PU block whose the cost

Rate—Distortion (RD) is the minimum in the cut candidate modes.

4 EXPERIMENT RESULTS AND ANALYSE

The experiment platform used is Pentium (R) Dual-Core CPU with 2.6 GHZ and 2 GB RAM

Table 3. Experiment performance comparison of fast algorithm of mode selection.

Video sequences	Δ		
	Time (%)	BD-psnr (dB)	BD-rate (%)
PeopleOnStreet	−36.43	−0.091	1.61
Traffic	−34.18	−0.0966	1.26
BasketballDrive	−46.37	5.9064	−33.44
BQTerrace	−36.12	−0.0968	0.27
BasketballDrill	−31.46	−0.0811	1.35
BQMall	−32.77	−0.1048	0.89
BlowingBubbles	−33.91	0.1218	0.15
BasketballPass	−33.03	−0.0948	−1.09
Vidyo1	−36.23	−0.0866	3.34
Vidyo4	−35.45	−0.0745	2.53
The average	−35.60	0.5058	−2.29

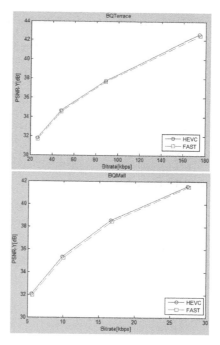

Figure 2. RD curve of algorithm in this paper and the RD algorithm.

under Windows 7, which was implemented in the model HM12.1 under VS2010 with encoder_intra_main.cfg of the configuration file and Quantization Parameters (QP) is 27.

In the proceeding of the test, cycle-time-setting of I-frames is 1 with every 100 frames sequences, of which all coding frames are intra coding. The results are shown in Table 3.

From the Table 3, the proposed method saves encoding time of 35.6% on average compared with HM12.1 while the increase of Peak Signal Noise Ratio (PSNR) is 0.5058% and the decrease of coding rate is 2.29%. The method could reduce nearly complex under different resolution or sequence. Therefore, the proposed algorithm has potential in the coding application of HEVC.

The RD curves of BQTerrace sequence, BQMall sequence are compared by the proposed method and algorithm of HEVC, respectively, under QP 22, 27, 32 and 37 due to more loss of the coder performance. From the Figure 2 it can be seen that the nearer the two curves are, the better the performance of the proposed algorithm is.

5 CONCLUSION

First, we propose a fast decision method used by SSIM among image pixel based on both intra prediction algorithm in HEVC and the existing methods. Second, intra prediction candidate modes are cut to decide prediction mode of PU fast under calculating the probabilities of MPM, PLANAR mode, and DC mode selected as the best prediction mode, respectively. Finally, the proposed method is tested compared with that of HM12.1. The proposed algorithm saves encoding time of 35.6% on an average while rate is reduced slightly and PSNR is a little increased. As a whole, the proposed fast mode decision algorithm for intra prediction in HEVC has obvious practical values and significance of research.

ACKNOWLEDGMENTS

This work was financially supported by self-determined research funds of CCNU from the colleges' basic research and operation of MOE (CCNU14A05044), the Central China Normal University Students' innovative undertaking training program (2014).

REFERENCES

Bross B., K. McCann, W. Han, et al. HEVC Draft and Test Model editing[C]// ITU-T SG 16 WP 3 and ISO/IEC JTC 1/SC 29/WG 11 12th Meeting. Jan: 2013. 3–58.

Bross B., W. Han, J. Ohm, et al. WD5: Working Draft 5 of High-Efficiency Video Coding[C]//ITU-T SG16 WP3 and ISO/IEC JTC1/SC29/WG11 7th Meeting, Geneva: 2011. 31–158.

Bross B., W. Han, J. Ohm, et al. WD5: Working Draft 5 of High-Efficiency Video Coding[C]// ITU-T SG16 WP3 and ISO/IEC JTC1/SC29/WG11 7th Meeting, Geneva: 2011. 401–438.

Corrêa G., Assuncao P., da Silva Cruz L.A., et al. Adaptive coding tree for complexity control of high efficiency video encoders[C]//Picture Coding Symposium (PCS), 2012. IEEE: 2012. 425–428.

Kim J., Yang J., Lee H., et al. Fast intra mode decision of HEVC based on hierarchical structure[C]//Information, Communications and Signal Processing (ICICS) 2011 8th International Conference on. IEEE: 2011. 1–4.

Matsuo S., Takamura S., Jozawa H. Improved intra angular prediction by DCT-based interpolation filter[C]//Signal Processing Conference (EUSIPCO), 2012 Proceedings of the 20th European. IEEE: 2012. 1568–1572.

McCann K., B. Bross, I. Kim, et al. HM5: High Efficiency Video Coding (HEVC) Test Model 5 Encoder Description[C]//ITU-T SG16 WP3 and ISO/IEC JTC1/SC29/WG11 7th Meeting, Geneva: 2011. 89–147.

Motra A.S., Gupta A., Shukla M., et al. Fast intra mode decision for HEVC video encoder[C]//Software, Telecommunications and Computer Networks (SoftCOM), 2012 20th International Conference on. IEEE: 2012. 1–5.

Sugimoto K., Itani Y., Isu Y., et al. Description of video coding technology proposal by Mitsubishi Electric[J]. JCTVC-A107, 2010: 29–59.

Wiegand T., J. Ohm, G.J. Sullivan, et al. Special Section on the Joint Call for Proposals on High Efficiency Video Coding (HEVC) Standardization[J]. IEEE Transactions on Circuits and Systems for Video Technology, 2012: 1661–1666.

Wang Z., Bovik A.C., Sheikh H.R., et al. Image quality assessment: from error visibility to structural similarity[J]. Image Processing, IEEE Transactions on, 2004, 13(4): 600–612.

Yang Chun-ling, Xiao Dong-1 in. Improvements for H. 264 Intra Mode Selection Based on SSE and SSIM[J]. Journal of Electronics & Information Technology, 2011, 33(2): 1–6.

Zhao L., Zhang L., Ma S., et al. Fast mode decision algorithm for intra prediction in HEVC[C]//Visual Communications and Image Processing (VCIP), 2011. IEEE: 2011. 1–4.

Advances in Energy, Environment and Materials Science – Wang & Zhao (Eds)
© 2016 Taylor & Francis Group, London, ISBN 978-1-138-02931-6

Persuasive tech in keeping chronic patients' willingness in health self-management

Yongyan Guo & Minggang Yang
East China University of Science and Technology, Shanghai, China

Jun Hu & Linkai Tao
Eindhoven University of Technology, The Netherlands

ABSTRACT: Chronic disease is a long and very harmful illness to patient. With the improvement of e-health technology, many users can monitor and manage their health condition at home. At the same time, recording health data is a time-consuming and boring work. Without the supervision of health care professionals, many patients feel very difficult to persist on it by themselves. This platform with persvasive tech is designed to test the effectiveness of peer pressure in the health self-management in the e-health.

1 INTRODUCTION

Chronic illnesses are rising faster in Asia than globally. Asia will be home to half of the world's elderly population and half of the global burden of chronic conditions by 2030. Asians increasing life expectancy and growing economic affluence will add tremendous pressure on the region's already-stretched health-care systems and create ever higher demand for quality care. So health self-management can improve patients' medical experience and reduce the societal burden of chronic patients.

Self-management is providing services that are person-centered (Marie McGill). (Norm, 2010) there are many ways in which technology could assist with the management process. In almost all, the monitoring physical data by sensor and user's self-entering everyday data to the system are the mainly two methods. Physical data can be stored, reviewed, and analyzed to detect by the professionals and be set back to the patients for their feedback. Because this is a extensive area for e-health, this paper is priority focus on the user-inputting health data by themselves. According to the user interview, one of the most difficult parts is user's persistent in the health journal of self-management. Normally, user doesn't know what these data means for him and he can't have the perception of his body. People are born with inertia. In view of the technology still can not solve the problem of automatic recording daily health behaviors, we should turn to the persuasive tech to improve chronic patient's willingness in persistent health self-management. Although chronic disease is a irreversible lifelong disease and it is an important cause of death, it is easy to be neglected for many people. Without records of patients health condition and relative behaviors for long term, the diagnosis of the chronic patients can't be detect timely and accurately. To pull on that thread a bit further, the number of chronic patients will became huge in the next several decades. If more patients can monitor and record their physical conditions at home, a lot of medical resources will be saved, that is helpful for the aging society.

2 GETTING STARTED

2.1 Model of persuasive tech

Model of persuasive tech is a tool to describe the user's behavior change, which is studied in the Persuasive Technology Lab of Stanford University. According to BJ Fogg's article, there are three important events in the model: motivation, ability and triggers. A trigger is something that tells people to perform a behavior now. In fact, for behaviors where people are already above the activation threshold—meaning they have sufficient motivation and ability—a trigger is all that's required (Fig. 1). Triggers are divided into three types: sparks, facilitators, and signals. A spark is a trigger that motivates behavior. A facilitator makes behavior easier. And a signal indicates or reminds.

In our research, we pay more attention to the behavior triggers. The reason is that for the patients who suffer the chronic diseases, they have the strong motivation to keep their health and maintain their lifelong. Because the young patients

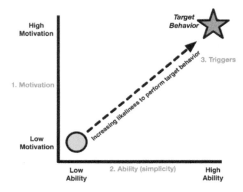

Figure 1.   The Fogg Behavior Model.

are growing fast in the recently ten years, they have the ability to operate the mobile phone application and other medical instruments at home. According to our interview, the most difficult thing for them to interrupt recording health conditions is the lack of behavior triggers. They may forget, lazy or feel too boring to do this task. Some patients think they can't see any feedback on the recording data and their health data are almost the same everyday, which makes the task more difficult to maintain. For this reason we introduce the social psychology to the FBM model as a behavior trigger and design an experiment to test the effectiveness.

2.2   Peer pressure theory

Peer effects can arise for a variety of reasons. In their theoretical contribution, Kandel and Lazear (1992) underline how the positive effects of peer pressure on effort can overcome free riding in environments with profit sharing. They distinguish between internal pressure (or guilt) and external pressure (or shame), with observability being the discriminant between the two. (Sotiris, 2013) Peer pressure is influence on a peer group, observers or individual exerts that encourages others to change their attitudes, values, or behaviors to conform to groups. Social groups affected include membership groups, in which individuals are "formally" members, or social cliques in which membership is not clearly defined. Peer pressure theory has been verified in psychology labs, but it is about the users in real physical environment. The effect of online peer pressure between the strangers has not been tested yet. Some mobile phone APPs are designed according to the peer pressure theory, but they are more about the competition social interactions among acquaintance. Very few is about the co-work social interaction between the strangers.

In this paper, the aim of the experiment is to improve the chronic patients' behavior of recording their healthy data everyday. Peer pressure will

be a good choice to be the triggers in the FBM. The hypothesis are: 1. online co-work can be a group dynamics that improves users' motivation & behavior. 2. obscure identity social relationship can form the trigger to motivation & behavior.

3   CASE STUDY: HEALTH SELF-MANAGEMENT WITH PEER PRESSURE

3.1   Experiment design

This experiment conception is divided into two parts. First, test the the effectiveness of peer presser theory to improve the user's motivation in e-health self-management. Second, test the matrix dot graphics as a trigger for long term effectiveness of this prototype. As motivation includes award and fear emotions, this prototype will designed to full-fill this psychology needs. System's Producing graphic is the trigger in this prototype (Figs. 2, 3).

3.2   Research methods

Test 1:
The prototype is a tangible social board with a dot matrix display. Two group of participant are invited to take part in the test in the lab. Group A is the control group who only do test separately. Group B is the experimental group without knowing who are the other peers. He is showed the social board and explained that every dot represents a person. At first, the participant is asked to do a trouble and time-consuming task (for example, a set of mathematical problems). One dot will be lighted when someone in the group is finished. The more peers are finished, the more dots will be lighted (Fig. 4). The time-consuming will be recorded and the two group data will be compared. If the Group B's time is shorter than the Group A, that means the peer pressure theory is valid in the experiment.

Test 2:
Group A and Group B are doing test 2 as the test 1.When all the test finished, the dots will become a dot matrix graphics. (This dot matrix graphic is used to be the feedback of stranger's

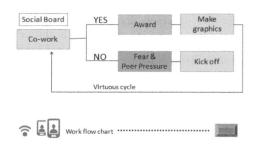

Figure 2.   Flow chart 1.

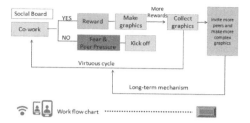

Figure 3.　Flow chart 2.

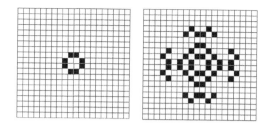

Figure 4.　Dot matrix board.

interaction online and it can be changed to different forms, such as a picture, a jigsaw puzzle, etc.) This experiment is asked to do several rounds. The number of rounds will be recorded and compared. If the Group B do more rounds than Group A, that means the co-work is valid to the endurable behavior of group members.

As the members in group are always separated in the tests, the peer pressure is come from the dot matrix on the social board. This experiment is designed perfectly to test the hypothesis goal.

3.3　*Implementation*

This experiment design is a platform for testing the effectiveness of social computing in changing user's behavior. As social forces is difficult to test in the lab, this platform is brought up to resolve this problem. The advantage are low cost, convenient and quick. With the use of group dynamic theory, the peers are abstracted to dots (or other icons), the feed back of group goal (it can be any difficult or troublesome work) is represented by the completed graphic (or other jigsaws) that is also a trigger to all users. If somebody of the group is failed to complete his daily goal, the group graphic will not completed display, then the other peers will urge him to do the project as soon as possible. When all the group numbers are insist on completing the group goal, the system will produce more complex graphics for them as awards. Users can collect these graphics and print them on different products or their skins as badges. This will be supposed to improve user's motivation of persistent health record (Figs. 5, 6).

Figure 5.　Complex dots matrix feedback.

Figure 6.　Implication of matrix graphic.

4　DISCUSSION

Chronic disease is a long and very harmful illness to patient. With the improvement of e-health technology, many users can monitor and manage their health condition at home. At the same time, recording health data is a time-consuming and boring work. Without the supervision of health care professionals, many patients feel very difficult to persist on it by themselves. This platform is designed to test the effectiveness of peer pressure in the e-health. In the next step, we will collect the data and testify the hypotheses.

5　CONCLUSION

Given the popularity of chronic disease, applications designed to promote healthy living are promising for helping users set and achieve their health-related goals, but have not yet proven themselves for long-term adoption and behavior change (Sajanee, 2010). Thus, other measures should be used. This platform can be used to compare the implication in the cross-culture and the effectiveness of different trigger patterns to users' behaviors. This prototype can be used as a trigger in the smart interactions for health self management.

ACKNOWLEDGEMENTS

Grateful acknowledgement is made to my supervisor M. Yang and J. Hu who gave me considerable help by means of suggestion, comments and criticism. I also owe my sincere gratitude to my friends and my fellow classmates who gave me their help and time in listening to me and helping me work out my problems during the difficult course of the thesis.

REFERENCES

Fogg BJ. A Behavior Model for Persuasive Design. Persuasive Technology Lab, Stanford University. www.bjfogg.com

http://appcrawlr.com/ios-apps/best-apps-peer-pressure

http://captology.stanford.edu/

http://en.wikipedia.org/wiki/Peer_pressure

Marie McGill. Self-Management—Health Professional Transcript

Norm Archer. Smart Interactions for Health Self Management. Invited Presentation at CASCON Conference, November 3, 2010.

Sajanee Halko, Julie A. Kientz. Personality & Persuasive Technology: An Exploratory Study on Health-Promoting Mobile Applications. Persuasive Technology, 5th International Conference, PERSUASIVE 2010, Copenhagen, Denmark, June 7–10, 2010.

Sotiris Georganas, Mirco Tonin, Michael Vlassopoulos. Peer Pressure and Productivity: The Role of Observing and Being Observed. IZA Discussion Paper No. 7523 July 2013.

Modeling and simulation of the air fuel ratio controlling the biogas engine based on the fuzzy PID algorithm

Zuhua Fang, Ying Sun, Kang Wang, Gengjuan Guo & Jiale Wang
Shanghai Normal University, Shanghai, China

ABSTRACT: According to the research of the air fuel ratio control characteristics of biogas engine, the paper proposes a control scheme of biogas engine, builds a mean value model of biogas engine and designs the fuzzy PID control of the algorithm. While the above has been done, a mean simulation model of the air fuel ratio controlling the biogas engine has been built, then the paper also simulates the control of the air fuel ratio of biogas engine under different load and speed conditions by using the Matlab/Simulink software. The simulation results show that while taking the fuzzy PID control algorithm, the control of the air fuel ratio of the biogas engine is more accurate, the response time is shorter, the fluctuations are smaller and the static and dynamic characteristics of the system are more stable.

1 INTRODUCTION

With the increasing emphasis on environmental protection, and the shortage of oil, coal, and other fossil energies looking for a new, clean, and highly effective energy has become the most critical issue in the world (Wang, 2015). Biogas, which is renewable, distributes large heat but has lower emissions, can be used as an engine fuel, which has a great prospect of development aroused a lot of attention. In order to obtain a higher power, fewer fuel consumption and lower emissions, we need to control the air fuel ratio precisely (Zhou, 2006; Wu, 2013; Huang, 2013; Li, 2014). While the engine is a typical, multi-input and output, nonlinear, time-varying, and complex system, and the components of biogas vary greatly, which is not as stable as liquid fuel, the control of air fuel ratio of biogas engine is more difficult. Therefore, the control system of engine should have the self-learning ability to adjust the control parameters adaptively.

2 AIR FUEL CONTROL STRATEGY OF BIOGAS ENGINE BASED ON FUZZY PID ALGORITHM

Combined with the PID control and fuzzy control algorithms, this paper has proposed the fuzzy PID control strategy to control the air fuel ratio of biogas power. The fuzzy PID controller of the biogas engine is shown in Figure 1.

2.1 Establishment of mean value model of biogas engine

Based on the mathematical model of four cylinder engine proposed by Crossley and Cook (Lang, 2006), combined with the common mean model, the paper has set up a mean value model of biogas engine, as is shown in Figure 2, which includes the air-intake sub model, the compression model, the detection model of air-intake time and crankshaft position, and the dynamics model.

The parameter selection of the mean value model of biogas engine:

The chosen biogas engine is the 465 type engine, and the initial values of the mean value: the outside temperature is 300 K, the outside pressure is 1 bar, the engine initial load is 0 Nm, the engine initial speed is 300 r/min, the rotational inertia is 0.14 Kg*m², bore/stroke is 65.5/74 mm, Desired air fuel value is 17.4.

2.2 Air fuel ratio controller

The inputs of the air fuel ratio controller are the desired value and actual value of air fuel ratio, then

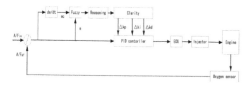

Figure 1. The fuzzy PID controller.

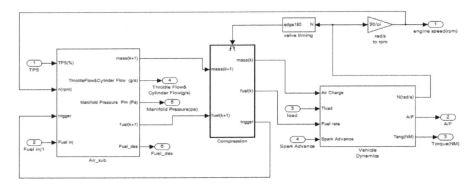

Figure 2. The mean value model of biogas engine.

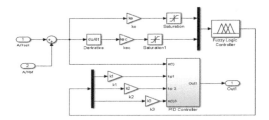

Figure 3. The air fuel ratio controller.

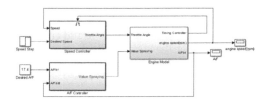

Figure 4. Control model of the air fuel ratio.

the output is the adjustment after calculated by the fuzzy PID control strategy, the air fuel ratio controller is as shown in Figure 3. The fuzzy controller and the conventional PID controller are packaged together in the air fuel controller, and the proportion coefficient of k1, k2, and k3 are valued as 0.08, 00043, and 0.0198, which are obtained by a lot of experiments, then the output is reacted fast, the fluctuation are small and the control effect is ideal.

2.3 Speed controller

The speed controller is used in the following situation: when the operating mode of biogas engine is varying, the speed is obstructed, we can adjust the speed by adjusting the opening throttle, which is the output of the controller, to ensure the speed is stable in the desired value.

2.4 Control model of air fuel ratio of biogas engine

The control model of the air fuel ratio of biogas power is shown in Figure 4, which is combined with the air fuel ratio controller and the speed controller.

3 SIMULATION AND ANALYSIS

The paper has taken the fuzzy PID algorithm to control the air fuel ratio, simulated the control effect in different conditions, and then compared it to the control effect while taking the normal PID algorithm. The analysis is as below:

3.1 Control of air fuel ratio when engine load changes

Set engine speed is 1000 r/min, the desired air fuel ratio is 17.4, and engine loads are 0, 20, 50 Nm. The control effects of the air fuel ratio are as shown in Figure 5.

1. The two control algorithms (fuzzy PID and normal PID) can also control the air fuel ratio of biogas power;
2. While taking the normal PID control algorithm, the system is sensitive to engine load, the response is slow and the fluctuation is large;
3. While taking the fuzzy PID control algorithm, along with the larger of engine load, the injected fuel is larger, and the fluctuation and response time also becomes larger; when the load is 0 Nm, the fluctuation of the air fuel ratio value is the smallest and the response time is shortest, about 2.2 s, when the load is 50 Nm, the fluctuation of the air fuel ratio value is the largest and the response time is longest, about 3.1 s.

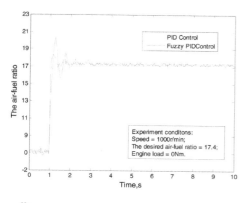

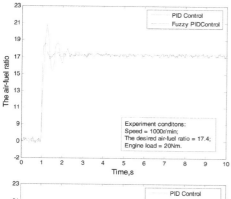

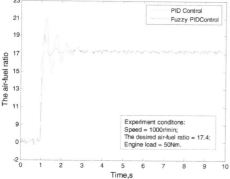

Figure 5. The control effect of the air fuel ratio when engine's loads are 0, 20, 50 Nm.

3.2 Control of air fuel ratio when engine speed changes

1. The engine becomes stable after acceleration
The biogas engine starts at the speed of 300 r/min, then accelerates to 700 r/min at the time of 10 s and keeps it stable, then we can consider that the biogas engine is in stable condition, as is shown in Figure 6. Compare the control effect by using the two kinds of control algorithms, which is shown in Figure 7.

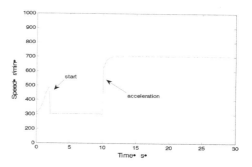

Figure 6. Speed curve when acceleration.

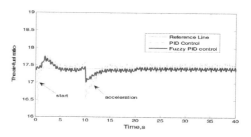

Figure 7. Control effects of the air fuel ratio when acceleration.

From Figure 7, we can find that the two kinds of control algorithms can control the air fuel ratio effectively; but while taking the normal PID control algorithm during acceleration, the response time is longer, the fluctuation of the air fuel ratio value is larger, and the minimum air fuel ratio value is 16.5, which means the control effect is not as ideal as the fuzzy PID control algorithm.

2. The engine becomes stable after deceleration
The biogas engine starts at the speed of 300 r/min, speeds up to 700 r/min at the time of 10 s and then decelerates to 500 r/min at the time of 20 s, as is shown in Figure 8. Compare the control effect by using the two kinds of control algorithms, which is shown in Figure 9.
We can see from Figure 9 that the two kinds of control algorithms can control the air fuel ratio effectively; but while taking the normal PID control algorithm, the varying of the engine speed would cause a slower response and the larger control error, the maximum and minimum air fuel ratio values are 18.3 and 16.5, which indicates the normal PID control algorithm is not as good as the fuzzy PID control algorithm.

3. The engine becomes stable after obstruction
The biogas engine starts at the speed of 300 r/min and accelerates to 700 r/min at the time of 10 s,

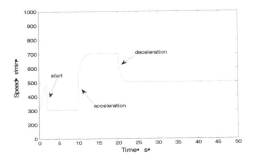

Figure 8. Speed curve when deceleration.

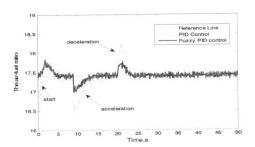

Figure 9. Control effects of air fuel ratio when deceleration.

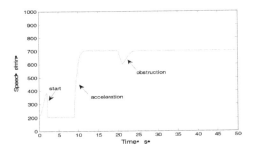

Figure 10. Speed curve when obstruction.

then adds a speed step signal at the time of 15 s to simulate the obstruction of the speed, make the speed jump to 600 r/min, the speed step signal stops after 2 s, and finally the speed changes back to 700 r/min, as is shown in Figure 10. Compare the control effect by using the two kinds of control algorithms, which is shown in Figure 11.

We can find that the two kinds of control algorithms can control the air fuel ratio effectively; but while taking the normal PID control algorithm, the response is slower, the fluctuation of the air fuel ratio value is larger, and its anti-interference performance is not as good as the fuzzy PID control algorithm.

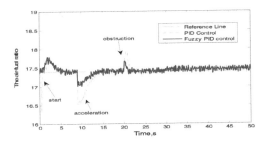

Figure 11. Control effects of air fuel ratio when obstruction.

4 CONCLUSION

Based on the mean value model of biogas engine and the fuzzy PID control strategy, this paper has built the simulation model of the air fuel model of biogas power that uses the Matlab/Simulink software to simulate the control effect of the air fuel ratio in different loads and speeds. The result shows that while taking the fuzzy PID control algorithm to control the air fuel ratio in different operating conditions, the response time is shorter, the fluctuation of the air fuel ratio value is smaller, then the static and dynamic performance of the system is more stable and the robustness is better.

ACKNOWLEDGMENT

This work was financially supported by the Research Program of Science and technology Committee of Shanghai (12160503000).

REFERENCES

Hua Lang. 2006. The research on the key technology of the air fuel ratio of gas engine [D]. Shandong University of Technology.

Kang Wang. 2015. The research on control technology of air fuel control of biogas power [D]. Shanghai Normal University.

Lvdong Wu. 2013. The simulation of gas supply system and the research on injection control of CNG engine [D]. Xihua University.

Naijun Zhou, Shengchong Bao, Hailing Pei, Hongde Chen. 2006. The simulation research of the air fuel ratio based on the model gasoline engine [J]. Journal of Chongqing Institute of Technology, 2006, 20(8):15–20.

Shuai Huang. 2013. The research on the technology of closed-loop control of gas injection of natural gas engine [D]. Haibin Engineering University.

Zimiao Li.2014. The research on the PID control technology of pumping station [D]. Changchun University of Technology.

Advances in Energy, Environment and Materials Science – Wang & Zhao (Eds)
© 2016 Taylor & Francis Group, London, ISBN 978-1-138-02931-6

Evaluation of software realization algorithms of industrial building operation life

Zolina Tatyana Vladimirovna & Sadchikov Pavel Nikolaevich
Astrakhan Institute of Civil Engineering, Astrakhan, Russia

ABSTRACT: There is a presented calculation software complex «DINCIB-new» that is capable to realize summary algorithm on data processing of technical surveys of an industrial building, including the one equipped with overhead cranes in order to find out a term of its problem-free operation. The complex is made as an instrument for probability calculations organization and conduction of object's accommodation of exciting influences combinations at known offset values in frame nodes.

1 INTRODUCTION

One of the most important issues that require a solution within the design is the erection and operation of a building or a structure, which is immune to a number of influences. It's possible to enlist a large number of factors influencing the change of proper operation of separate structural elements and the object in whole. These include:

- squally wind blast,
- snow deposits,
- ground subsidence during scouring,
- rapid temperature change,
- temporary impacts on inaccessible roof, etc.

In the case of industrial facilities, combination of environmental factors resulted due to changes in climatic conditions for the functioning of the technical system, which is complemented with technological components. Structures' accommodation of the influences combination leads to a reduction of bearing capacity of industrial building frame. The fact points out to the need of constant monitoring of the stress-strain state of the operated facility due to its significant changes as well as the accumulation of defects and damages in the node joints of the structures, which are one of the main accidents causes (Tamrazyan, 2009).

2 MODELS AND ALGORITHMS

A mathematical model of the research object with an operation live of an industrial building at the specific moment of its operation as an objective function allows evaluating the loss degree of the object's operation condition. For the ability to detect real operation conditions of a building, the model must take into account as many possible number of load factors through the action, which the frame is involved in the oscillatory process. Taking into account the simultaneous influences accommodation by structures and the random nature of their symptoms (Pichugin, 1995; Nikolayenko, 1988), there is a need to attract automated means of calculation and control.

Several computer-aided design systems providing tools for settling similar problems are known. However, the authors have developed calculation software complex «DINCIB-new» (Zolina, 2012), which has a number of advantages in comparison with analogues. Its specification includes calculation organization and conduction of industrial building structures equipped with overhead cranes at a whole complex of influences expanding the range of issues involved. The source code is written in high level language Pascal in IDE Delphi can be implemented on a PC compatible with the class IBM PC.

Application performance is determined with a set of interconnected tasks. Each of them is presented as a separate object. DINCIB-new includes the following modules formalizing the performance:

Unit 1—prescribes work with file structures and coordinates invocation variability to the main menu options;

Unit 2—organizes work with software dialog windows;

Unit 3—shows the results of lateral forces calculation based on already known formulas for the possibility of carrying out a comparative analysis;

Unit 4—organizes the input of initial data of the building;

Unit 5—fills in stiffness matrices and building inertial characteristics;

Vvod_kran—organizes the input of initial data for overhead cranes;

Vvod_3_mat_og—realizes the calculation of loads numerical coefficients;

Vvod_3_rama—calculates the coefficients of loaded frame stiffness matrix;

Obr_zad—formalizes solution algorithm of an inverse problem at offsets change in separated frame points;

Ost_resurs—determines the combination of disturbance influence and estimates remaining lifetime value;

Sob_Zn_Vec—realizes calculation algorithms of stochastic crane and seismic loads;

St_nagr—organizes the search for bending moments and stresses under designed scheme static load;

Info_zas—provides reference information about software package.

The package DINCIB-new is introduced to the PC Programs State Register of The Russian Federation.

UI control is provided with the selection of main menu options or project tree that includes all calculation stages.

Consolidated algorithm (Zolina, 2015) being a basis of program implementation is designed to find the values of bending moments and stresses in the separated structural elements of the frame. The main criterion for the technical system reliability at this approach is the condition of elements' impossibility of their limit states excess at the most negative combinations of design loads at a specific period of time.

3 METHOD'S APPLICATIONS

In assessing the ability of frame bearing capacity, maximum stresses and deformations in the bearing structures are found out with the crane components in overall load value. Operation of overhead cranes for lifting and moving loads of different masses along and across the shop sends the disturbing influences to the object's frame. Depending on the choice of a design diagram for fixing its geometrical and stiffness characteristics (Fig. 1), placement of spans and crane equipment parameters

(Fig. 2) in the program are calculated for individual components of crane load.

Among them are:

- Vertical pressure when lifting the load of a certain weight (Fig. 3)
- Performance of horizontal forces caused with crane truck braking (Fig. 4)
- Performance of lateral forces as a consequence of truck transverse hopping at crane movement on the railway tracks (Fig. 5).

Figure 2. Enter the initial data for crane equipment.

Figure 3. The image of calculation window with crane vertical pressure.

Figure 4. The image of the calculation window for crane truck braking.

Figure 1. Entering of initial data of industrial buildings.

Figure 5. The image of the calculation window for lateral forces influence.

4 METHOD'S IMPLEMENTATION

Generating random variable values of applied crane loads that were ordered with normal distribution law with known parameters of a mathematical expectation and average mean-square deviation at which each of the appropriate folded calculation parameters was run at 500 times each, which allows the achieving of highly precise results. An initial data of in-situ measurements is offset data in node points of design diagram of an industrial facility. Results comparison of several chronologically arranged surveys indicate an increase in an appropriate offset and as a consequence a reduction of stiffness of properties.

SMatrixGEVDReduce and SMatrixGEVD described in the module SpdGEVD are used to allow implementing algorithms aimed to find the eigenvalues and eigenvectors of linear transformations defined by the large dimension matrices. Solving systems of linear equations and finding corresponding inverse matrices (Fig. 6) are realized with connection to program code of procedures and functions of modules Ssolve and MatInv being members of numerical analysis cross-platform library ALGLIB.

Having offset experimental data in node frame points recorded in various time intervals using module "Inverse Problem" has made it possible:

- to trace dynamics of frame stiffness properties of industrial buildings;
- to create a voltage values' regression dependence of the stress on time factor;
- to predict resistance degree of the building to internal and external influences after a certain period of its operation.

Proposed software also allows automating values calculation: the conditional and complete seismic risk, system dynamic coefficient, the effective period of operation, and the remaining life of the technical system. Embedded algorithms implementation of option "Seismic Impacts" allows obtaining results for each specific object

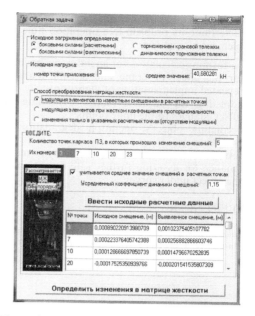

Figure 6. Window of module "Inverse Problem" of «DINCIB-new» program complex.

Figure 7. The window image of seismic influences calculation.

and conduct an analysis of its operation under the influence of seismic wave to a system (Fig. 7).

Within the research they have conducted probable deviations comparison of the designed values of similar algorithms with separated elements that were earlier realized in math packages MathCAD and Maple, and CAD-system SCAD. There is a proven high validity of results and conclusions of developed methodologies taken as the basis of the software package.

5 CONCLUSIONS

Presented software package «DINCIB-new» is considered as a tool for technical survey data processing of a specific industrial building in order to determine its problem-free operation time. Its implementation allows:

– performing calculations of an industrial building for the effect of crane and seismic loads at repeated implementation of algorithms;
– adjusting the stiffness matrix depending on offset changes frame node points;
– predicting object's operation degradation in perspective;
– setting terms, directions, and difficulty of maintenance for assuring further safe operation of the object under survey.

REFERENCES

Nikolayenko N.A., Nazarov Y.U.P. Dynamics and seismic durability of structures. M.: Stroyizdat, 1988. p. 312.

Pichugin S.F. Stochastic presentation of load acting on construction structures // Building.—1995.—№ 4.—pp. 12–18.

Tamrazyan A.G. Risk and reliability evaluation of structure and node elements is a necessary condition of building and facilities safety // Newsletter of Central Research Institute of Construction Structures. 2009. № 1. pp. 160–171.

Zolina T.V., Sadchikov P.N. Automated calculation system of industrial building on crane and seismic loads // Industrial and civil building. 2012. № 8. pp. 14–16.

Zolina T.V., Sadchikov P.N. Revisiting the Reliability Assessment of frame constructions of Industrial Building Applied Mechanics and Materials Vols. 752–753 (2015) pp. 1218–1223© (2015) Trans Tech Publications, Switzerland doi:10.4028/www.scientific. net/AMM.752–753. 1218.

Advances in Energy, Environment and Materials Science – Wang & Zhao (Eds)
© *2016 Taylor & Francis Group, London, ISBN 978-1-138-02931-6*

Anomaly detection based on the dynamic feature of network traffic

Yaxing Zhang, Shuyuan Jin, Yuanzhuo Wang & Yanxia Wang
Institute of Computing Technology, Chinese Academy of Sciences, Beijing, China

ABSTRACT: Anomaly detection is always one of the most important issues that network security researchers focus on. For the complexity of network structure and the diversity of the attack, detecting abnormal traffic in the network accurately becomes much harder. This paper proposes a method of anomaly detection based on the dynamic feature of network traffic, which is called DF-IPEAD. Firstly, this method constructs the dynamic feature of network traffic, and then builds a detection model of Elman neural network with the history data of dynamic feature. To optimize the Elman detection model, an improved Particle Swarm Optimization (PSO) algorithm is proposed in this paper. We implemented DF-IPEAD in this paper and performed experiments to compare it with other methods. Experimental results show that the proposed method can detect abnormal traffic accurately.

1 INTRODUCTION

With the rapid development of attack techniques, the Internet is becoming more and more insecure. Tremendous attacks on the Internet always cause abnormal network traffic which is malicious. Except malicious network traffic, there is unintentional network traffic for network failure or configuration error (Ramah, 2006). No matter what kind of abnormal network traffic that afflicts it, the availability and reliability of network will both be influenced to some extent. It can also cause the network to get disabled or show a denial of service. This calls for effective methods to detect abnormal network traffic. If abnormal traffic is detected timely in the network, there is a great opportunity to prevent potential attacks and avoid greater losses.

The existing anomaly detection methods are grossly divided into two groups, misuse detection and anomaly detection. Snort is a classic tool with the schema of misuse detection. It detects abnormal traffic through matching with pre-specified rules and receives high accuracy. However, it just works while detecting known attacks. If the pattern is not matched, there may be a false negative. The expert system also has the same limitation. Moreover, it is difficult to build a huge library to restore patterns. Even after library built, the problem of quick search in the library for pattern matching is also a challenged.

The basic idea of anomaly detection is to build the distribution of normal traffic, and then calculate the deviation between observed traffic and normal traffic. If the deviation exceeds the threshold, it shows that there is abnormal traffic in the current network. One class of anomaly detection method adopts statistical analysis. Daniela Brauckhoff (Roesch, 1999) applied Principal Component Analysis (PCA) in the detection of abnormal traffic and obtained good results. Due to high complexity, the approach has not been widely used. Another detection method with statistical analysis uses time series analysis models. It firstly selects an appropriate time series model to fit the network traffic, and then predicts the traffic of next moment with the model chosen. It detects abnormal traffic by comparing the predicted value with the normal traffic baseline (Brauckhoff, 2009; Bianco, 2001; Galeano, 2006). The ways described above are easy to utilize, but there is no universally accepted standard for the threshold or parameter setting.

Other kinds of anomaly detection methods are on the basis of machine learning including Support Vector Machine (SVM) (Yaacob, 2010), Artificial Neural Network (ANN) (George, 2012), Naïve Bayes (Shah, 2012), K-means (Sharma, 2012), decision tree (Yasami, 2010), K-neighbor (Guan, 2006) and so on. These methods have high sensitivity, strong adaptability and are able to detect unknown abnormal traffic. Appropriate parameters are set by training models. Sometimes, the false alarm rate of these methods is a little high.

The above methods are not so effective as to satisfy the demand of abnormal traffic detection. Anomaly detection approaches needs a stable traffic to train the detection model, while the current research shows that the network traffic is not stable with dynamic features, such as self-similarity, long-term correlation, recurrence, and so on (Willinger, 1998; Su, 2011). Yuan indicated that multi-scale and recurrence are the two important characteristics of network traffic and proposed a WRC model for anomaly detection (Grossglauser, 1999). On the one hand, network traffic has different statistical behaviors on different time scales. It shows stability and periodic change in the long term, and in the short time it shows randomness and volatility (Yuan, 2014). Different attacks may exist on the different time scales (Yuan, 2014). On the other hand, recurrence is a basic feature of dynamic system that indicates the evolution rule of traffic states (Grossglauser, 1999). Jin applied covariance in the detection of abnormal traffic by analyzing correlativity of traffic characteristics (Barford, 2002). Compared to the other statistical features of network traffic, recurrence is a relatively stable feature, as well as correlativity. Therefore, this paper presents an anomaly detection method DF-IPEAD based on the dynamic feature including correlativity and recurrence of network traffic. Simulation experiments are performed to verify the effectiveness of this method. And so, we compare our approach with other detection methods. Experimental results show that the approach proposed in this paper has a better performance.

The main contributions are shown as followed.

Dynamic infers to a process or system characterized by constant change, activity, or progress. The traffic of network shows dynamic features which are much more stable than other known statistical features. This paper proposes a dynamic feature, including covariance and recurrence of two aspects. In the anomaly detection process, the dynamic feature of normal traffic is stable and it almost has nothing to do with the scale of network. The details of the dynamic feature constructing were described in section II.

This paper presented an anomaly detection approach DF-IPEAD based on the dynamic feature of network traffic. The DF-IPEAD method adopts Elman neural network to build the detection model with the dynamic feature of network traffic, and then it utilizes an improved PSO algorithm to optimize the Elman network detection model to obtain better performance.

This paper is structured as follows. Section II introduces dynamic feature proposed in this paper. The detection method DF-IPEAD based on the dynamic feature of traffic is described in the Section III. Section IV presents the simulation experiments to show the effectiveness of the proposed method. Section V draws conclusions with a summary and future work.

2 DYNAMIC FEATURE

This paper presents the dynamic feature, including covariance and recurrence of the two aspects.

They both reflect the relations about the first-order characteristics of network traffic. The former represents the relation between the first-order characteristics and the latter interprets the inner relation of each first-order characteristic. The first-order characteristics of the network can be calculated from the raw packets with an interval of 2 seconds.

2.1 Covariance

In the statistical analysis, covariance represents the tendency in the linear relationship between two variables. If the covariance is greater than zero, it shows that the two variables have the same trend. That is to say, if one variable increases, the other increases as well. On the contrary, if the covariance is less than zero, it means that the two variables have converse trend. They are mutually independent when the covariance is zero. The covariance is calculated as shown as in equation 1.

$$\text{cov}(X, Y) = E((X - a)(Y - b)) \qquad (1)$$

In matrix X, each column can be recognized as a vector X_i, as shown in equation 2.

$$X_i = \begin{bmatrix} x_{1i} \\ x_{2i} \\ \vdots \\ x_{ni} \end{bmatrix} \qquad (2)$$

Covariance matrix is generated by calculating the covariance between any two vectors in the matrix. It extends the random variables to high dimensions. The covariance matrix of matrix X is cov X. cov $X_{i,j}$ can be obtained by equation 3.

$$\text{cov}\,X_{i,j} = \text{cov}(X_i, X_j) = E[(X_i - a_i)(X_j - a_j)^T] \quad (3)$$

where a_i represents the expectation of vector X_i. And cov X can be obtained from equation 4.

$$\text{cov}\,X = E[(X - E(X))(X - E(X))^T] \qquad (4)$$

Covariance matrix of traffic shows the correlations between the first-order characteristics of traffic. This paper adopted covariance matrix of traffic as a part of dynamic feature.

2.2 Recurrence

Generally, RQA (Recurrence Quantification Analysis) is a common method to analyze the recurrence of network traffic. It is proposed by Eckman (Yeung, 2007). This method describes the recurrence property in the dynamic system with

Recurrence Plot (RP). RP is a two-dimensional square chart. The x axis in the chart represents time, as well as the y axis. Black spots in the plot indicate that there are re-currency states of traffic and white spots have the opposite meaning. Figure 1 is a case of RP.

Suppose that there is a time series $[S_Y = [Y]_1, Y_2, ..., Y_N]$ with an interval of T, then the sub sequences are generated with the schema of sliding time window, as shown in equation 5.

$$\left[Sub_{Y_i} = [Y]_i, Y_{i+1}, ..., Y_{i+m-1} \right], i = 1, 2, ..., N \quad (5)$$

where Sub_{Y_i} represents the sub sequence of S_Y, and m is the size of sliding time window, which is also called embedding dimension. Then, we calculate the similarity of the two sub sequences of S_Y and compare the similarity with threshold pre-defined to get the spot of RP, which is shown in equation 6.

$$R_{i,j} = \Theta\left(\varepsilon - sim\left(Sub_{Y_i}, Sub_{Y_j} \right) \right), i, j = 1, 2, ..., N \quad (6)$$

where $R_{i,j}$ represents the element in recurrence plot, ε is the threshold, $sim(\Omega)$ is the similarity function, for example, which may be a distance function, and $\Theta(\Omega)$ is Heaviside function, as shown in equation 7.

$$\Theta(k) = \begin{cases} 0, k \le 0 \\ 1, k > 0 \end{cases} \quad (7)$$

If $R_{i,j}$ is 1, that is to say the similarity between Sub_{Y_i} and Sub_{Y_j} is larger than the threshold and there will be a black spot in the corresponding locations of RP. On the contrary, there will be a white spot.

To deal with network traffic, we can also use RP because it describes the recurrence of traffic states in space. Unlike covariance feature, recurrence property of traffic demonstrates the inner relation of each of the traffic characteristic. The RPs of normal traffic are usually quite different from those of abnormal traffic, as shown as in Figure 1 and Figure 2.

Recurrence Quantification Analysis (RQA) with the recurrence plot is firstly proposed by Zbilut (Eckmann, 1987) and Webber Jr (Zbilut, 1992), and extended by Marwan (Webber, 2005). There are quantification parameters given as followed.

Recurrence Ratio (RR) is the ratio of recurrent spots in RP, as shown in equation 2–8.

$$RR = \frac{1}{N^2} \sum_{i,j=1}^{N} R_{i,j}(i, j = 1, 2, ..., N) \quad (8)$$

where N denotes the number of spots and $R_{i,j}$ denotes as shown in equation 6.

Determinism Ratio (DETR) is the percentage of recurrent spots in the lines parallel to the diagonal of RP. It is the ratio of spots constituting a slash in RP, indicating the certainty of system.

$$DETR = \frac{\sum_{len=lT}^{N} len * P(len)}{\left(\sum_{i,j=1}^{N} R_{i,j} \right)(i, j = 1, 2, ..., N)} \quad (9)$$

where len is the length of slash composed of recurrent points, and its frequency is $P(len)$. lT is the threshold of slash length and N is the number of spots.

Recurrence Entropy (RENT) is the Shannon entropy of the distribution of slash in RP, indicating

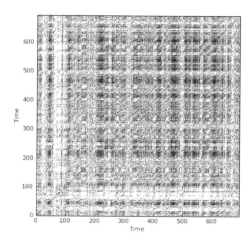

Figure 1.　Recurrence plot of normal traffic.

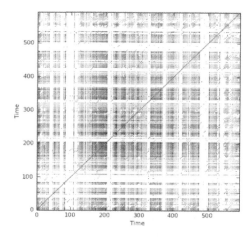

Figure 2.　Recurrence plot of abnormal traffic.

the complexity of system. Equation 10 shows the calculation of Recurrence Entropy.

$$RENT = -\sum_{len=lMin}^{N} p(len) * \log_2 p(len)(i, j = 1, 2, ..., N) \quad (10)$$

$$p(len) = \frac{P(len)}{\sum_{len=lMin}^{N} P(len)} \quad (11)$$

where $p(len)$ is the distribution of slash length in RP.

This paper also presents a method CR-E for extracting the dynamic feature of traffic. First, it uses Lift Wavelet Transform (LWT) to analyze network traffic on multi-scale. This step splits the network traffic on the time scales to discover various abnormal traffic. And then it calculates the covariance matrix and the recurrence parameters, respectively, to generate the dynamic feature.

The network traffic here is a matrix, composed of first-order characteristics of network traffic. The first-order characteristics of traffic are simple statistical characteristics of raw packets during a period, such as total number of packets in an interval. A sample of traffic $Tmatrix_t$ described above is shown in equation 12 at time t.

$$Tmatrix_t = \begin{bmatrix} x_1^{t-n+1} & x_2^{t-n+1} & \cdots & x_p^{t-n+1} \\ x_1^{t-n+2} & x_2^{t-n+2} & \cdots & x_p^{t-n+2} \\ \vdots & \vdots & \vdots & \vdots \\ x_1^{t-1} & x_2^{t-1} & \cdots & x_p^{t-1} \\ x_1^{t} & x_2^{t} & \cdots & x_p^{t} \end{bmatrix} \quad (12)$$

where n and p are the dimensions of the traffic matrix from time $t - (n - 1)$ to t. p is the number of first-order characteristics. After LWT procedure, $Tmatrix_t$ is divided into two matrixes, which are $Tmatrix_{tL}$ and $Tmatrix_{tH}$. Then the CR-E method calculates the covariance matrixes and recurrence parameters of $Tmatrix_{tL}$ and $Tmatrix_{tH}$, respectively, to build the dynamic feature.

3 THE ANOMALY DETECTION METHOD DF-IPEAD

Artificial Neural Network (ANN), also called Neural Network (NN), is a kind of artificial intelligence approach, which was developed in 1980s. In recent years, NN has a huge breakthrough both in theory and practice (Marwan, 2007). It has become an emerging field of interdisciplinary frontier including computer science, artificial

intelligence, brain science, information science and intelligent control, etc. In theory, NN can approximate any nonlinear functions with enough neurons and training time. Existing study about anomaly detection methods with neural network are mostly based on feed-forward network, for example, Backward Propagation (BP) network. However, a feedback network may be more suitable for modeling a dynamic system, such as Elman neural network. With memory of historical data and the feedback mechanism, Elman neural network has a good ability for nonlinear dynamic mapping. In consideration of the dynamic characteristic of traffic, this paper presents a detection method DF-IPEAD based on the Elman neural network with an improved Particle Swarm Optimization (PSO) algorithm. This model will be fed to the dynamic feature of network traffic extracted by CR-E as described in chapter 2.

3.1 Elman network for anomaly detection

Elman network was firstly proposed by Jeffrey L. Elman in 1990 (Simon, 1998). It is a kind of feedback neural network with strong computing power. Generally, an Elman network is composed of an input layer, hidden layer, undertake layer, and output layer. In the training process of Elman network, the previous output of the neurons in the hidden layer is passed to the neurons in the undertake layer. The feedback signal will be another part of input to the hidden layer. In this way, the Elman network becomes sensitive to the historical data and has the ability to deal with dynamic information. The Figure 3 displays the architecture of Elman network for anomaly detection proposed in this paper.

The Elman network has one more layer than the BP network, called the undertake layer, so the output of Elman network will consider the output of undertaker layer.

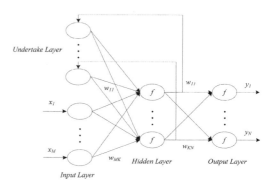

Figure 3. The architecture of Elman neural network for anomaly detection.

$$x_u(t) = x(t-1) + a * x_u(t-1) \tag{13}$$

$$h(t) = f(w^{IH} * x(t-1) + w^{UH} * x_u(t) + \theta_j^H),$$
$$j = 1, 2, ..., N_H \tag{14}$$

$$y(t) = W^{HO} * h(t) + \theta_k^o, k = 1, 2, ..., N_o \tag{15}$$

where $x_u(t)$ represents the output of undertake layer at time t, and $a(0 \le a \le 1)$ is the feedback factor. $h(t)$ denotes the output of hidden layer and $y(t)$ is the result. The weight matrices between the input layer, hidden layer, and output layer are W^{IH}, W^{UH} and W^{HO}, respectively. θ_j^H and θ_k^O are the thresholds of the hidden layer and output layer. N_H and N_O are the neuron numbers of the hidden layer and output layer. $f(\Omega)$ is the activation function. In this paper, we have adopted a backward propagation algorithm to train the Elman neural network.

3.2 DF-IPEAD based on improved PSO algorithm

As a BP network, Elman network has the limitation of easy to fall into local minimum. In order to solve this problem, this paper presents an improved PSO algorithm to optimize Elman network. And then, proposes the anomaly detection method DF-IPEAD based on the dynamic feature of network traffic.

PSO algorithm was presented by Eberhart (Elman, 1990) and Kermedy (Kennedy, 1995) in 1995. It indicates the regularity of birds cluster activity, utilizing the sharing information among the individuals in a population to search the optimal solution. The algorithm takes birds in population with speed and position. Each particle will share its n nearest position to food with other particles. After sharing, each particle will know the global nearest position. And then it will change its direction toward the global optimal solution. At last, all particles in population are gathered to the same place nearest to food roughly. As the basic idea is described above, PSO algorithm firstly initializes the position and speed of each particle randomly. And then it updates each particle's speed and position as shown in equation 16 and equation 17.

$$v_i^d = v_i^d + c_1 * rand_1^d * (pBest_i^d - x_i^d)$$
$$+ c_2 * rand_2^d * (gBest^d - x_i^d) \tag{16}$$

$$x_i^d = x_i^d + v_i^d \tag{17}$$

where $rand_1^d$ and $rand_2^d$ are random variables between 0 and 1. c_1 and c_2 are learning factors. d represents the dimension of search space. $pBest_i^d$ is the best historical position of a particle and

$gBest^d$ is the best position of the whole population. This is the basic PSO algorithm. Shi and Eberhart proposed a standard PSO algorithm based on the basic one (Ebethart, 1995), adding inertia weight in the speed and position updating as shown as in equation 18. This paper takes PSO algorithm as the standard PSO algorithm as follows.

$$v_i^d = w * v_i^d + c_1 * rand_1^d * (pBest_i^d - x_i^d)$$
$$+ c_2 * rand_2^d * (gBest^d - x_i^d) \tag{18}$$

Due to tracking single particle of population and using information sharing to approximate the optimal space, PSO algorithm is easy to tap into the local minimal without concerning with inter relationships among the individuals in population. To solve this problem, this paper presents an improved PSO algorithm by introducing the generic operators in Generation Algorithm (GA). In this way, particles' relationships will be considered and this is helpful to avoiding falling into local minimal or jumping out of local minimal when it happens. The improved PSO algorithm is shown as Figure 4.

The anomaly detection method DF-IPEAD proposed in this paper uses the improved PSO algorithm described above and its framework is indicated in Figure 5.

Algorithm 1 Improved PSO Algorithm

Input:
  S: the size of population
  C: the iteration
  w: inertia weight
  $rand_i^d, rand_2^d$: random value from 0-1
  c_1, c_2: learning factor
  F: the fitness function
  pcross: cross probability
  pmutation: mutation probability
Output: best solution
Approach:
1: initialize the speed and position of particles in population
2: **while** not meet terminal conditions {
  //satisfied solution or the iterations greater than C
3: **for** each particle i in population {
  //update speed and position of particle
4:   $v_i^d = w * v_i^d + c_1 * rand_i^d * (pBest_i^d - x_i^d) + c_2 * rand_2^d * (gBest^d - x_i^d)$
5:   $x_i^d = x_i^d + v_i^d$
  // calculating the fitness of each particle
6:   $Fitness_{x_i^d} = F(x_i^d)$
7:   Selection Operator
8:   Cross Operator (pcross)
9:   Mutation Operator (pmutation)
10:   Update the best historical position of particles
11:   Update the best position of population
12:   Update the inertia weight
13: }
14: best salutation = $gBest^d$
15: }

Figure 4. Improved PSO algorithms.

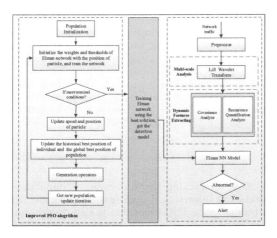

Figure 5.   The framework of DF-IPEAD.

4   SIMULATION

4.1   *Network environment and sample data*

To verify the validity of DF-IPEAD, a simulated network environment is set up in this paper, as shown in Figure 6. They have three subnets. In subnet 1, a detection machine and a traffic pretreatment server are the prime components. Machines in subnet 2 are used to simulate attacks. Some common machines and servers are in subnet 3. DF-IPEAD generates a record of traffic dynamic feature every three minutes and $24 * 60/3 = 480$ records will be obtained every day. Training and testing data used in this paper are the records of traffic dynamic feature collected from the above environment.

4.2   *Performance indicators*

In order to evaluate the performance of the proposed method, we use four indicators calculated by confusion matrix shown in Table 1.

1. Precision is the percentage of actual normal samples in the normal samples detected.

$$\text{Precision} = \frac{TP}{TP + FP} \tag{19}$$

2. Detection Accuracy Rate (DA Rate) is the percentage of actual abnormal samples in the abnormal samples detected.

$$\text{DA Rate} = \frac{TN}{TN + FN} \tag{20}$$

3. False Positive Rate (FP Rate) is the rate of actual abnormal samples detected as normal samples incorrectly.

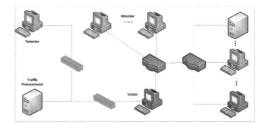

Figure 6.   Simulated network environment.

Table 1.   Confusion matrix.

	Normal detected	Abnormal detected
Normal actually	TP	FN
Abnormal actually	FP	TN

$$\text{FP Rate} = \frac{FN}{TP + FN} \tag{21}$$

4. False Negative Rate (FN Rate) is the rate of actual normal samples detected as abnormal samples incorrectly.

$$\text{FN Rate} = \frac{FP}{TN + FP} \tag{22}$$

4.3   *Results*

In the experiment, we selected the row of traffic matrix used in CR-E 70, and the embedding dimension 3. The fitness function used in DF-IPEAD detects accuracy rate. To compare with other methods, such as DF-PEAD (using Elman NN based on PSO algorithm), DF-GEAD (using Elman NN based on Generation algorithm), DF-EAD (only using Elman NN) and so on, some initial parameters are shown in Table 2.

Table 3 shows the detection results of DF-IPEAD. The average of precision and detection accuracy rate is around 95%, and that of false positive rate and false negative rate is less than 5%.

We also implemented a group experiments with the method DF-IPEAD, DF-PEAD and DF-GEAD. The comparison of different methods is shown as in Figure 7. From the picture, we can see the method DF-IPEAD gets the best performance than other methods.

We got the fitness curve of different methods in iterative process, as shown in Figure 4–3. We can see that the method DF-IPEAD receives the best fitness at last. The best fitness curve of DF-PEAD is increasing in the optimal process until

Table 2. Initial parameters.

	DF-IPEAD	DF-PEAD	DF-GEAD	DF-EAD
Iteration	20	20	20	
Population size	10	10	10	
Cross probability	0.3		0.3	
Mutation probability	0.1		0.1	
Inertia weight	0.1	0.1		
c1	1.49445	1.49445		
c2	1.49445	1.49445		
Neurons in hidden layer	9	9	9	9
Training epoch	20	20	20	20
Training goal	0.01	0.01	0.01	0.01
Training data records	480	480	480	480
Testing data records	480*7	480*7	480*7	480*7

Table 3. The detection results of DF-IPEAD.

	Precision	Detection accuracy rate	False positive rate	False negative rate
Day 1	97.06%	93.75%	6.25%	2.94%
Day 2	95.24%	96.30%	3.70%	4.76%
Day 3	90.90%	97.22%	2.70%	9.10%
Day 4	96.97%	93.33%	6.67%	3.03%
Day 5	100%	96.50%	0.00%	4.90%
Day 6	92.25%	93.24%	5.41%	1.23%
Day 7	97.06%	100%	2.94%	0.00%
Average	95.64%	95.76%	3.95%	3.71%

DF-PEAD obtains the convergence finally. From the curve of DF-GEAD, the best fitness rises at the beginning, and then it tends gently from the fourth iteration to the tenth iteration. However, there is a big steep after iteration 10 and then the method DF-GEAD converges to a stable state. This means that GA has much help to jump out of the local minimum. As a result, DF-IPEAD has a much better effectiveness due to introducing the generation operators based on PSO algorithm, combining the advantages of both. The curve of DF-IPEAD in Figure 8 describes the same meaning.

Experiments with DF-EAD and DF-BPAD (using BP network to build detection model) were also performed in this paper. The results are shown in Figure 9. This figure shows that the average precision and detection of accuracy rate of DF-EAD is a little higher than that of DF-BPAD. And the average of false positive rate and false negative rate is much lower than that of DF-BPAD, which indicates that Elman neural network is more suitable to deal with dynamic information.

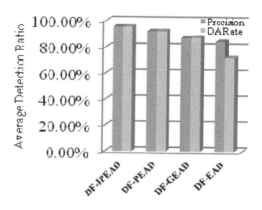

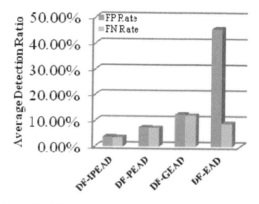

Figure 7. The comparison of different methods.

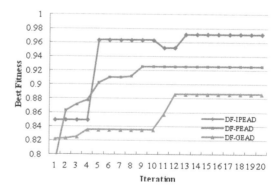

Figure 8. The fitness curve of DF-IPEAD, DF-PEAD and DF-GEAD.

We also make experiments only with covariance and recurrence, respectively. As shown in Table 4, DF-IPEAD_cov represents detection only with covariance, and DF-IPEAD_rqa represents detection only with recurrence. DF-IPEAD denotes

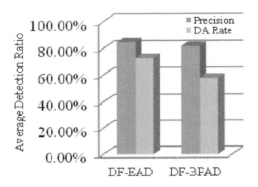

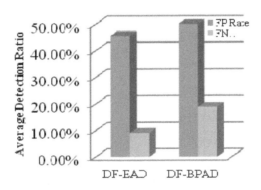

Figure 9.   The comparison of DF-EAD and DF-BPAD.

Table 4.   The detection results of DF-IPEAD.

	Precision	Detection accuracy rate	False positive rate	False negative rate
DF-IPEAD_cov	88.06%	89.75%	8.25%	9.9%
DF-IPEAD_rqa	93.24%	92.30%	7.83%	6.7%
DF-IPEAD	95.64%	95.76%	3.95%	3.7%

detection with both. From the table, we can see that DF-IPEAD has a better performance than the other two methods due to more fully dynamic feature that is taken into account.

5   CONCLUSIONS AND FUTURE WORKS

In order to detect the abnormal traffic in the network accurately, this paper proposes DF-IPEAD, a detection method based on Elman neural network with an improved PSO algorithm. The model is fed with the dynamic feature of traffic, thus CR-E is presented to extract dynamic feature in DF-IPEAD. After preprocessing network

traffic with lift wavelength transform, CR-E extracts the covariance and recurrence of traffic. Experimental results show that the proposed method has a good performance in the anomaly detection of network traffic. Since the training and testing data used in this paper is simulation data in a small network, we need do more experiments to verify the effectiveness of our methods in the future.

ACKNOWLEDGEMENTS

This work was supported by the National Natural Science Foundation of China under Grant No. 61402437. The authors would like to thank our reviewers for the feedbacks.

REFERENCES

Barford P., Kline J., Plonka D., et al. A signal analysis of network traffic anomalies[C]// Proceedings of the 2nd ACM SIGCOMM Workshop on Internet measurement. ACM, 2002: 71–82.

Bianco A.M., Garcia Ben M., Martinez E.J., et al. Outlier detection in regression models with arima errors using robust estimates[J]. Journal of Forecasting, 2001, 20(8): 565–579.

Brauckhoff D., Salamatian K., May M. Applying PCA for traffic anomaly detection: Problems and solutions[C]// INFOCOM 2009, IEEE. IEEE, 2009: 2866–2870.

Ebethart R.C., Kermedy J. A new optimizer using particle swarm theory In: Proc of the 6th Int'l Symposium on Micro Machine and Human Science[J]. Piscataway NJ: IEEE Service Center, 1995: 39–43.

Eckmann, J.P. Kamphorst, S.O. Ruelle, D. Recurrence plots of dynamical systems, Europhysics Letters, 4(9): 973–977, 1987.

Elman J.L. Finding structure in time[J]. Cognitive science, 1990, 14(2): 179–211.

Galeano P., Peña D., Tsay R.S. Outlier detection in multivariate time series by projection pursuit[J]. Journal of the American Statistical Association, 2006, 101(474): 654–669.

George, Annie. Anomaly detection based on Machine learning dimens ionality reduction using PCA and classification using SVM[J]. International Journal of Computer Applications, 2012, Vol. 47, No. 21: 5–8.

Grossglauser M., Bolot J.C. On the relevance of long-range dependence in network traffic[J]. IEEE/ACM Transactions on Networking (TON), 1999, 7(5): 629–640.

Guan X.Q. Research on the classifying algorithm based on decision tree[M]. Taiyuan: Shanxi University, 2006.

Kennedy J., Ebethart R. Particle Swarm Optimization[C]. In: Proceeding of IEEE International Conference on Neural Networks, Piseataway, NJ: IEEECS, 1995:1942–1948.

Marwan N., Carmen Romano M., Thiel M., et al. Recurrence plots for the analysis of complex systems[J]. Physics Reports, 2007, 438(5): 237–329.

Ming-Yang Su. Using clustering to improve KNN-based classifiers for online anomaly network traffic identification[J]. Journal of Network and Computer Applications, 2011, 34: 722–730.

Performance of Computer and Communication Networks; Mobile and Wireless Communications Systems. Springer Berlin Heidelberg, 2006: 136–147.

Ramah K.H., Ayari H., Kamoun F. Traffic anomaly detection and characterization in the tunisian national university network[M]//NETWORKING 2006. Networking Technologies, Services, and Protocols.

Roesch M. Snort: Lightweight Intrusion Detection for Networks[C]//LISA. 1999, 99(1): 229–238.

Shah Bhavin, H. Trivedi Bhushan. Artificial Neural Network based Intrusion Detection System: A survey[J]. International Journal of Computer Applications, 2012, Vol. 39, No. 6: 13–18.

Sharma Sanjay Kumar, Pendey Pankaj, Susheel Kumar, Sisodia Mahendra Singh. An improved network intrusion detection technique based on K-means clustering via Naïve bayes classification[C]. International conference on Advances in Engineering, Science and management, 2012: 417–422.

Shi, Y.H., R.C. Eberhart. A Modified Particle Swarm Optimizer. IEEE International Conference on Evolutionary Computation. Anchorage, 1998: 69–73.

Simon Haykin, Neural Networks, a Comprehensive Foundation, Second Edition, Prentice Hall, 1998: 161~175, 183~221, 400~438.

Webber Jr C.L., Zbilut J.P. Recurrence quantification analysis of nonlinear dynamical systems[J]. Tutorials in contemporary nonlinear methods for the behavioral sciences, 2005: 26–94.

Willinger W., Paxson V., Taqqu M.S. Self-similarity and heavy tails: Structural modeling of network traffic[J]. A practical guide to heavy tails: statistical techniques and applications, 1998, 23: 27–53.

Yaacob A.H., Tan I.K.T., Chien S.F., et al. ARIMA based network anomaly detection[C]//Communication Software and Networks, 2010. ICCSN'10. Second International Conference on. IEEE, 2010: 205–209.

Yasami Y., Mozaffari S.P. A novel unsupervised classification approach for network anomaly detection by k-Means clustering and ID3 decision tree learning methods[J]. The Journal of Supercomputing, 2010, 53(1): 231–245.

Yeung D.S., Jin S., Wang X. Covariance-matrix modeling and detecting various flooding attacks[J]. Systems, Man and Cybernetics, Part A: Systems and Humans, IEEE Transactions on, 2007, 37(2): 157–169.

Yuan J., Yuan R., Chen X. Network Anomaly Detection based on Multi-scale Dynamic Characteristics of Traffic[J]. International Journal of Computers Communications & Control, 2014, 9(1): 101–112.

Zbilut, J.P.; Webber, C.L.; Embedding and delays as derived from quantification of recurrence plots, Physics Letter A, 171: 199–203, 1992.

Advances in Energy, Environment and Materials Science – Wang & Zhao (Eds)
© *2016 Taylor & Francis Group, London, ISBN 978-1-138-02931-6*

Application of computer simulation in the elastic-plastic seismic response of bottom frame structure

H.Y. Deng
College of Space Technology and Civil Engineering, Harbin Engineering University, Harbin, China

B.T. Sun
Key Laboratory of Earthquake Engineering and Engineering Vibration, China Earthquake Administration, Harbin, China

ABSTRACT: The earthquake response of structure is much more complicated. The finite element simulation is an effective method through computer software. In order to study the seismic performance of masonry buildings with frame and seismic wall in the lower stories, continuum element model and constitutive relationship of non-reinforced masonry materials are used in this paper. Elasto-plastic time—history analysis and numerical simulation were carried out on the use of ABAQUS finite element software. On this basis, a series of different condition calculation models were designed to analyze and compare. The computer simulation results showed that: In site classification I to III, the seismic capacity of the masonry buildings with frame and seismic wall in the lower-two stories was higher than the masonry buildings with frame and seismic wall in first story, and the latter had better aseismic ability than the full masonry buildings. In site classification IV, the seismic capacity of the three was quite close.

1 INTRODUCTION

Due to the development process of the brittle and crack formation of the non-reinforced masonry materials, the nonlinear characteristics of the material are unique. So it is difficult to establish the constitutive relationship of masonry materials and simulate the nonlinear through the finite element. After continuous technological improvement, many scholars have obtained good calculation results in the failure of masonry structure, the development process of the cracks and the collapse of the structure. The continuous element model has obvious advantages in both computational efficiency and computational accuracy. In this paper, the continuous element model is used to simulate the finite element calculation of the bottom frame structures under earthquake action. Under the ABAQUS finite element software platform, using ABAQUS's own concrete damage plasticity model to define the nonlinearity of materials. In this model, the nonlinear response of the whole structure under earthquake action is simulated by defining the stress-strain relationship of the material under the compression and tensile behavior.

2 CONSTITUTIVE RELATION OF MATERIALS

2.1 *Constitutive relation of concrete*

The constitutive relation of concrete is described in the literature (China Architecture & Building Press, 2010). The uniaxial compression stress-strain (Fig. 1) and the uniaxial tension stress-strain

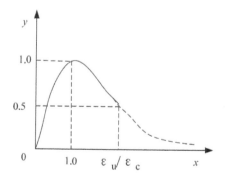

Figure 1.   Concrete uniaxial compression constitutive relation model.

relationship (Fig. 2) of the reinforced concrete members are presented in the paper.

The following formula is used to calculate the compression stress-strain curve of single axis.

Curve rising section (When $x \leq 1$)

$$y = \alpha_a x + (3 - 2\alpha_a)x^2 + (\alpha_a - 2)x^3 \quad (1)$$

Curve down section (When $x \geq 1$)

$$y = \frac{x}{\alpha_d (x-1)^2 + x} \quad (2)$$

$$x = \varepsilon / \varepsilon_c \quad (3)$$

$$y = \sigma / f_c^* \quad (4)$$

where α_a and α_d are parameters related to the rising and falling of the stress-strain curves of uniaxial compression, f_c^* is uniaxial compressive strength of concrete, ε_c is the peak compressive strain of concrete corresponding to f_c^*.

The following formula is used to calculate the tension stress-strain curve of single axis.

Curve rising section (When $x \leq 1$)

$$y = 1.2x - 0.2x^6 \quad (5)$$

Curve down section (When $x \geq 1$)

$$y = \frac{x}{\alpha_t (x-1)^{1.7} + x} \quad (6)$$

$$x = \varepsilon / \varepsilon_t \quad (7)$$

$$y = \sigma / f_t^* \quad (8)$$

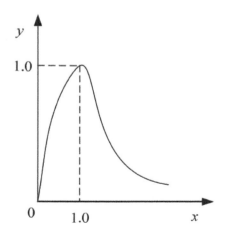

Figure 2. Concrete uniaxial tension constitutive relation model.

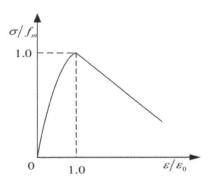

Figure 3. Masonry compression constitutive relation model.

where α_t is the parameter related to the falling of the stress-strain curves of uniaxial tension, f_t^* is uniaxial tensile strength of concrete, ε_t is the peak tensile strain of concrete corresponding to f_t^*.

2.2 Constitutive relation of masonry

The mechanical properties of masonry materials are complex, so a lot of different masonry constitutive relations are proposed by domestic and foreign scholars. This paper uses the masonry constitutive relationship which is put forward by C.X. Shi and G.Q (Liu, 2005). It is based on previous experimental results and combined with the masonry material characteristics in our country. The relationship of the stress and strain of masonry is descripted as the increasing section is a parabola and the falling section is linear (Fig. 3). The expressions are shown as formula (9) and (10). The tensile stress-strain curve of masonry is basically consistent with that of the compressive stress-strain curve, but the compressive strength of masonry is replaced by the tensile strength.

$$\frac{\sigma}{f_m} = 1.96\left(\frac{\varepsilon}{\varepsilon_0}\right) - 0.96\left(\frac{\varepsilon}{\varepsilon_0}\right)^2 \quad 0 \leq \frac{\varepsilon}{\varepsilon_0} \leq 1 \quad (9)$$

$$\frac{\sigma}{f_m} = 1.2 - 0.2\frac{\varepsilon}{\varepsilon_0} \quad 1 \leq \frac{\varepsilon}{\varepsilon_0} \leq 1.6 \quad (10)$$

where f_m is average compression strength of masonry, ε_0 is the peak compressive strain of masonry materials.

3 ESTABLISHMENT OF FINITE ELEMENT MODEL

A six-story residential bottom frame structure is taken as an example. The model is shown in Figure 4.

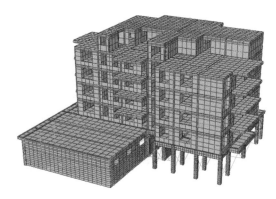

Figure 4. Finite element model.

The model is established through the finite element software ASBAQUS, and finite element simulation is carried out of the prototype structure. Frame column, beam and reinforced concrete shear wall of the first layer uses a solid unit (C3D20R), floor and common brick masonry wall uses shell unit (S4R), steel bar uses surface unit (SF3D4R), infilled wall of the first layer is simplified as equivalent diagonal strut model, using the truss unit (T3D2). The tie connection in constraint is used between the floors and walls, joint connections in the connector are used between the truss rod, which simulates the infilled wall and uses the truss unit (T3D2) and frame columns, the Embedded connection is used between steel bars of frame columns, beams and steel reinforced concrete shear walls and the main part to simulate joint coordinating role of the steel and concrete, slab reinforcement is added to shell unit (S4R), which is used in the floor by using rebar in the pre CAE processing.

4 SEISMIC RESPONSE CALCULATION OF BOTTOM FRAME STRUCTURE

4.1 Election of seismic wave

The site classification according to mean shear wave velocity and thickness of site soil layer is divided into four types in our country (China Architecture & Building Press, 2010). For the study of the damage of bottom frame structure in different types of ground conditions, 16 different types of ground seismic waves are selected and adjusted the input peak acceleration to 0.6 g recorded, then input to prototype structure for numerical calculation respectively. Each kind of site classification chooses four seismic waves and a total of 16 seismic waves are selected.

4.2 Analysis of structural dynamic response

For comparing the seismic capacity of the masonry buildings with frame and seismic wall in the lower two stories, the masonry buildings with frame and seismic wall in first story, and the full masonry buildings on earthquake performance in different site classification. Three contrast models were established for elastic-plastic time history analysis. Model 1: The prototype structure was changed into a full masonry building, the storey height of the first and the sixth layers is 3 m, and plane layout was same as other masonries layer. The masonry materials of 1 to 6 floors were made of MU10/M5. Model 2: The masonry buildings with frame and seismic wall in first story, compared with the prototype, the podium and surrounding ancillary components were removed and adjusted to the lateral stiffness ratio of 2. Only reinforced concrete seismic wall was added to the bottom without the addition of infill wall. The storey height of first layer was 3.8 m, and the sixth layer was 3 m, the plane was same as the fifth layer. Beam, plate, and column cast in place from foundation top surface to the first floor slab was C30, and the masonry materials of 2 to 6 floor were made of MU10/M5. Model 3: The masonry buildings with frame and seismic wall in the lower-two stories, compared with Model 2, the bottom two layers were changed into frame-shear wall structure. The total number of layers was also six, the second layer and the first layer had the same height 3.8 m. Height of the sixth layer was 3 m, which had the same plane with fifth floor. The material grade of beams, columns and anti-seismic wall in first two floors were made of C30, and the masonry materials of 3 to 6 floors were made of MU10/M5. After the adjustment, the plane of the frame-shear wall stories is shown in Figure 5.

Sixteen seismic waves which were applicable to different site classification in above were put into

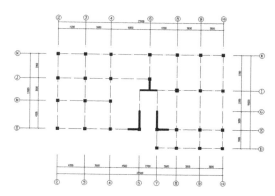

Figure 5. The plane of the frame-shear wall stories after adjustment.

the structure, respectively, and the peak acceleration was adjusted to 0.6 g. The average values of the floor transcendence strength rate in four site classifications are calculated. Judging floor damage state can use the concept of the floor transcendence strength rate E (Yin & Yang, 2004). The E equals the ratio of the actual earthquake shear and resistance. The structure is basically integral when $E \leq 1.0$, is slightly damaged when $1.0 < E \leq 1.35$, is medium damaged when $1.35 < E \leq 2.10$, is seriously damaged when $2.10 < E \leq 2.5$, and is destructively damaged when $E > 2.5$. To save space, only the floor transcendence strength rate of the first and transitional masonry layer in X orientation were given as shown in Table 1.

The structure of damaged can be summarized in that: In site classification I to III, the seismic capacity of the masonry buildings with frame and seismic wall in the lower two stories was higher than the masonry buildings with frame and seismic wall in first story, and the latter had better aseismic ability than the full masonry buildings. In site classification IV, the seismic capacity of the three was quite close. The reason for bottom frame structure has better seismic ability than full masonry building can be understood as: after the cracking of seismic wall in frame-shear wall stories, the layer stiffness of bottom layer is descended and a relatively flexible bottom layer is formed, which can absorb more earthquake energy and play a damping effect. So that the upper block layer response is reduced. Also from the periodic characteristic angle, bottom frame structure has a longer natural vibration period than the full masonry buildings. The natural vibration period increase with the building damage, so it is getting more and more far away from the predominant period of hard ground. At the meanwhile, the earthquake response of the structure gets decreased. So in the

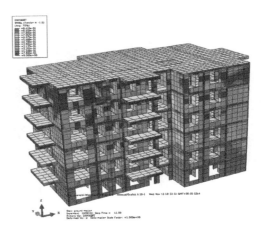

Figure 6. Tensile failure nephogram of Model 1.

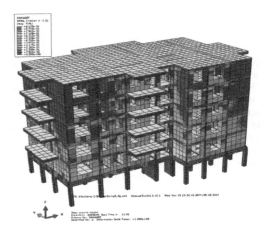

Figure 7. Tensile failure nephogram of Model 2.

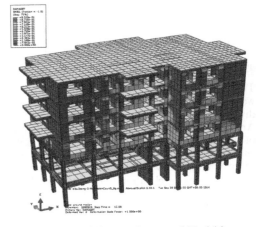

Figure 8. Tensile failure nephogram of Model 3.

Table 1. The results of the floor transcendence strength rate in X orientation.

Site classification	Layer	Model 1	Model 2	Model 3
I	First layer	2.11	1.10	0.89
	Transitional layer	1.92	1.84	1.75
II	First layer	2.30	0.74	0.63
	Transitional layer	2.19	1.88	1.67
III	First layer	2.25	1.21	0.86
	Transitional layer	2.21	1.90	1.78
IV	First layer	1.44	0.96	1.08
	Transitional layer	1.35	1.94	2.05

same earthquake wave input, its seismic capacity is better than the full masonry buildings.

The damage of masonry is mainly in tensile failure, in ABAQUS, the concrete damage plasticity model is used to calculate the tensile failure nephogram of structure, and the position and degree of structural damage can be judged from it. The tensile failure nephograms of these three models are shown in Figure 6–8.

5 CONCLUSIONS

This paper is based on the large general purpose finite element software ABAQUS, which contains rich elements and materials library, powerful nonlinear calculation function. An analysis of dynamic elastic plastic time history of the bottom frame structure under the earthquake is taken. It provides a useful reference for the calculation of elastic plastic deformation of the structure under rare earthquake. It can be seen that the large finite element software ABAQUS is suitable for the analysis of the complex structure of the dynamic elastic plastic analysis.

The building design and anti-seismic ability are closely related with the site classification where the building structure site. In site classification I to III, the seismic capacity of the masonry buildings with frame and seismic wall in the lower two stories was higher than the masonry buildings with frame and seismic wall in first story, and the latter had better aseismic ability than the full masonry buildings. In site classification IV, the seismic capacity of the three was quite close. From the look on the whole, the masonry buildings with frame and seismic wall in the lower two stories performs better than the other two types of structure.

REFERENCES

China Architecture & Building Press. GB 50010-2010. Code for design of concrete structures, Beijing, 2010.
China Architecture & Building Press. GB 50011-2010. Code for seismic design of buildings, Beijing, 2010.
Dorn, M. Computer prediction of the damage to and collapse of complex masonry structures from explosions, Structures and Materials, 2000, 8(4): 227–286.
Liu, G.Q. The research on the basic mechanical behavior of masonry structure, thesis of Master of Philosophy, Hunan University. 2005. Changsha: (in Chinese).
Yin Z.Q. & S.W. Yang, Analysis of earthquake loss and fortification criterion, Beijing: seismological press, 2004: (in Chinese).

Advances in Energy, Environment and Materials Science – Wang & Zhao (Eds)
© 2016 Taylor & Francis Group, London, ISBN 978-1-138-02931-6

The application of simulation method in one dimension pipeline flow based on elastic pipeline

Xiangdong Xue
Petrochina Pipeline R&D Center, Langfang, China

Pengfei Cao
China University of Petroleum (East China), Qingdao, China

Jian Shao
Petrochina Pipeline Jinan Oil Transportation Sub-Company, Jinan, China

Linjie Duan, Jingnan Zhang & Hao Lan
Petrochina Pipeline R&D Center, Langfang, China

ABSTRACT: This paper describes a complete method to simulate the one dimension elastic pipeline flow. The pipe is controlled horizontally and initial conditions and boundary conditions are set in order to close the calculation equation. Governing equations are chosen in the set condition. A new interactive method is used to calculate the mathematic model of elastic pipeline. Finally, the flow data in one dimension pipeline is obtained.

1 INTRODUCTION

China has a large amount of oil and gas transportation pipelines, in which the elastic pipe is widely used. Compared with the application of elastic pipes, the research to the elastic pipe doesn't receive a widespread attention and is still in the initial stage.

As the gas or oil flows through the elastic pipe, the structure and characteristic of the elastic pipe changes too, and hence, the flow characters inside the pipeline change distinctly. Therefore, for elastic pipe care must be taken to make sure that the flow inside the pipe.

Former researches mainly focus on the flow of rigid pipe and elasticity of wall; viscosity and non-constant flow were not mentioned in the research of pipe flow. However, in the practical oil & gas transmission pipelines all of the characters are included. Kuchar and Ostrach gave a detailed research on the coupling problem of wall deformation of elastic pipe and oil/gas flow. Flow distribution formula was given in the condition of constant flow of oil/gas. So, non-constant flow of elastic pipe is still a problem to solve.

This paper presents a method in dealing with the problem of variable diameter of elastic pipe. On the basis of flow conservation and control equation for continuous medium flow, combined with AGA state equation, differential form of conservation equation of one dimensional flow of compressible variable section pipe flow is brought out.

2 MATHEMATICAL MODEL

Assuming elastic pipe is an unlimited long cylindrical pipe with variable section, diameter is $D(x)$ and cross-sectional area is $A(x)$. We propose the pipe is a one dimensional pipe, where fluid flow direction is defined as X axis, and flow velocity is replaced by sectional average velocity, which is defined as u.

The elastic pipe flow model is built as follows: the velocity of inlet flow is uniform, which is defined as $u(0)$ and inlet pressure is defined as Ps. Due to the effect of viscosity, velocity and pressure reduce gradually. So the magnitude of elastic pipe expansion is also gradually reduced. At the outlet of the elastic pipe, pipe flow turns to a constant flow as in rigid pipe.

For this one dimension unsteady case, if the transaction surface in the pipeline is changeable, and we assume that the elastic pipe is straight, and semi-infinite pipe, the governing equation for the entire pipeline segment can be defined as follows:

The continuity equation

$$A\frac{\partial \rho}{\partial t} + \frac{\partial(\rho u A)}{\partial x} = 0$$

The momentum equation

$$A\frac{\partial(\rho u)}{\partial t} + \frac{\partial(\rho u^2 A)}{\partial x} = -g\rho A\theta - \frac{\partial(PA)}{\partial x} - \frac{\lambda}{D}\frac{u^2}{2}\rho A$$

$$A\frac{\partial(\rho u)}{\partial t} + \frac{\partial(\rho u \cdot u A)}{\partial x} = -A\frac{\partial P}{\partial x} - A\rho g\theta - A\lambda\frac{\rho\omega^2}{2D}$$

The energy equation

$$A\frac{\partial\left(\rho(U + u^2)\right)}{\partial t} + \frac{\partial\left(\rho Au\cdot\left(U + \frac{u^2}{2}\right)\right)}{\partial x}$$

$$= -\frac{\partial(PAu)}{\partial x} - \rho Aug\theta - \pi Dh_1(T - T_0)$$

And the frictional resistance coefficient is computed by F. ColeBrook-White formula:

$$1/\sqrt{\lambda} = 1.7385 - 2\lg\left(2e/D + 18.574\Big/\sqrt{\lambda * R_e}\right)$$

where R_e refers to Reynolds number, e/D denotes the ratio of roughness with respect to diameter.

$$Z = 1 + \frac{D_r B}{K^3} - D_r\sum_{n=13}^{18}C_n^*T^{-u_n} + \sum_{n=13}^{58}C_n^*T^{-u_n}\sqrt{2}$$
$$\cdot\left(b_n - c_n k_n D_r^{k_n}\right)D_r^{b_n}\exp\left(-c_n D_r^{k_n}\right)$$

3 NUMERICAL MODEL

Using (ρ, u, and T) as the basic variable, assume the pipeline as a horizontal pipeline, the transaction surface is not changeable with position, the pipeline is changed in spatial position can be obtained by processing the following contents.

Discretization governing equations:

Discretization of continuity equation of one dimension flow:

$$A\frac{\partial\rho}{\partial t} + \frac{\partial(\rho u A)}{\partial x} = 0$$

Take Infinitesimal pipe as the control body, the transaction surface acreage is specified as the average acreage of length and step of each point. The discretization form is as follows:

$$\frac{A_{i+1}^{k+1} + A_{i+1}^k + A_i^{k+1} + A_i^k}{4}\times\frac{\rho_{i+1}^{k+1} - \rho_{i+1}^k + \rho_i^{k+1} - \rho_i^k}{2\Delta t}$$

$$\rho_{i+1}^{k+1}u_{i+1}^{k+1}A_{i+1}^{k+1} - \rho_{i+1}^k u_{i+1}^k A_{i+1}^k$$

$$+\frac{+\rho_i^{k+1}u_i^{k+1}A_i^{k+1} - \rho_i^k u_i^k A_i^k}{2\Delta x} = 0$$

Discretization of momentum equation of one dimension flow:

$$A\frac{\partial(\rho u)}{\partial t} + \frac{\partial(\rho u^2 A)}{\partial x} = -g\rho A\theta - \frac{\partial(PA)}{\partial x} - \frac{\lambda}{D}\frac{u^2}{2}\rho A$$

Analogy of discretization of continuity equation, we can get the discretization of momentum equation is as follows:

$$\frac{A_{i+1}^{k+1} + A_{i+1}^k + A_i^{k+1} + A_i^k}{4}$$

$$\times\frac{\rho_{i+1}^{k+1}u_{i+1}^{k+1} - \rho_{i+1}^k u_{i+1}^k + \rho_i^{k+1}u_i^{k+1} - \rho_i^k u_i^k}{2\Delta t}$$

$$+\frac{\rho_{i+1}^{k+1}(u_{i+1}^{k+1})^2 A_{i+1}^{k+1} - \rho_{i+1}^k(u_{i+1}^k)^2 A_{i+1}^k}{2\Delta x}$$

$$+\frac{\rho_i^{k+1}(u_i^{k+1})^2 A_i^{k+1} - \rho_i^k(u_i^k)^2 A_i^k}{2\Delta x}$$

$$+ g\theta\left(\frac{\rho_{i+1}^{k+1} + \rho_{i+1}^k + \rho_i^{k+1} + \rho_i^k}{4}\right)\times\left(\frac{A_{i+1}^{k+1} + A_{i+1}^k + A_i^{k+1} + A_i^k}{4}\right)$$

$$+\frac{P_{i+1}^{k+1}A_{i+1}^{k+1} - P_{i+1}^k A_{i+1}^k + P_i^{k+1}A_i^{k+1} - P_i^k A_i^k}{2\Delta x}$$

$$+\frac{2\lambda}{D_{i+1}^{k+1} + D_{i+1}^k + D_i^{k+1} + D_i^k}\left(\frac{\rho_{i+1}^{k+1} + \rho_{i+1}^k + \rho_i^{k+1} + \rho_i^k}{4}\right)$$

$$\times\left(\frac{u_{i+1}^{k+1} + u_{i+1}^k + u_i^{k+1} + u_i^k}{4}\right)^2\left(\frac{A_{i+1}^{k+1} + A_{i+1}^k + A_i^{k+1} + A_i^k}{4}\right) = 0$$

Discretization of energy equation of one dimension flow:

$$A\frac{\partial(\rho(U + u^2/2))}{\partial t} + \frac{\partial(\rho Au(U + u^2/2))}{\partial x}$$

$$= -g\rho A\theta - \frac{\partial(PAu)}{\partial x} - \pi Dh_1(T - T_0)$$

Analogy of discretization of continuity equation, we can also get the discretization of energy equation is as follows:

$$\frac{A_{i+1}^{k+1} + A_{i+1}^{k} + A_{i}^{k+1} + A_{i}^{k}}{4}$$

$$\times \frac{\rho_{i+1}^{k+1} h_{i+1}^{k+1} + \frac{1}{2}\rho_{i+1}^{k+1}\left(u_{i+1}^{k+1}\right)^2 - \left(\rho_{i+1}^{k} h_{i+1}^{k} + \frac{1}{2}\rho_{i+1}^{k}\left(u_{i+1}^{k}\right)^2\right)}{\Delta t}$$

$$+ \frac{A_{i+1}^{k+1} + A_{i+1}^{k} + A_{i}^{k+1} + A_{i}^{k}}{4}$$

$$\times \frac{\rho_{i}^{k+1} h_{i}^{k+1} + \frac{1}{2}\rho_{i}^{k+1}\left(u_{i}^{k+1}\right)^2 - \left(\rho_{i}^{k} h_{i}^{k} + \frac{1}{2}\rho_{i}^{k}\left(u_{i}^{k}\right)^2\right)}{2\Delta t}$$

$$+ \frac{\rho_{i+1}^{k+1} h_{i+1}^{k+1} A_{i+1}^{k+1} u_{i+1}^{k+1} + \frac{1}{2}A_{i+1}^{k+1}\rho_{i+1}^{k+1}\left(u_{i+1}^{k+1}\right)^3}{2\Delta x}$$

$$- \frac{\rho_{i+1}^{k} h_{i+1}^{k} A_{i+1}^{k} u_{i+1}^{k} + \frac{1}{2}A_{i+1}^{k}\rho_{i+1}^{k}\left(u_{i+1}^{k}\right)^3}{2\Delta x}$$

$$+ \frac{\rho_{i}^{k+1} h_{i}^{k+1} A_{i}^{k+1} u_{i}^{k+1} + \frac{1}{2}A_{i}^{k+1}\rho_{i}^{k+1}\left(u_{i}^{k+1}\right)^3}{2\Delta x}$$

$$- \frac{\rho_{i}^{k} h_{i}^{k} A_{i}^{k} u_{i}^{k} + \frac{1}{2}A_{i}^{k}\rho_{i}^{k}\left(u_{i}^{k}\right)^3}{2\Delta x}$$

$$\frac{P_{i+1}^{k+1} A_{i+1}^{k+1} u_{i+1}^{k+1} - P_{i+1}^{k} A_{i+1}^{k} u_{i+1}^{k} + P_{i}^{k+1} A_{i}^{k+1} u_{i}^{k+1} - P_{i}^{k} A_{i}^{k} u_{i}^{k}}{2\Delta x}$$

$$+ g\theta\left(\frac{\rho_{i+1}^{k+1} + \rho_{i+1}^{k} + \rho_{i}^{k+1} + \rho_{i}^{k}}{4}\right)$$

$$\times \left(\frac{A_{i+1}^{k+1} + A_{i+1}^{k} + A_{i}^{k+1} + A_{i}^{k}}{4}\right)$$

$$+ \pi \frac{4\alpha_1}{D_{i+1}^{k+1} + D_{i+1}^{k} + D_{i}^{k+1} + D_{i}^{k}}$$

$$\left(T_{i+1}^{k+1} + T_{i+1}^{k} + T_{i}^{k+1} + T_{i}^{k} - 4T_w\right) = 0$$

4 ITERATIVE METHOD

The discrete control equations with boundary conditions and initial conditions can be closed, then form the nonlinear equations after closed, if matrix form can be expressed as $C(x)x = b$, $C(x)$ is the coefficient matrix, $X = (x_1, x_2, x_3, ..., x_n)^T$ are the solved variables, $b = (b_1, b_2, b_3, ..., b_N)^T$ is the right of the vector equation. Suppose that we use the basic NEWTON method to solve the nonlinear equations, first we introduce the function $F = (F_1, F_2, F_3, ..., F_n)^T$

$$F_i = r_i = C_i(x) \cdot x - b_i \quad i = 1, 2, 3, ..., N$$

Among them, namely F_i is the residual vector. With the introduction of the function F, we will solve the nonlinear equations into seeking the problems; the formula was established as follow:

$$F = (F_1, F_2, F_3, ..., F_n)^T = 0$$

For arbitrary point X_0 and its adjacent points/ neighborhood $X_0 + \delta_X$, By TAYLOR expansion we can achieve that:

$$F_i(X_0 + \delta_x) = F_i(X_0) + \sum_{j=1}^{N}\frac{\partial F_i}{\partial x_j}\delta_{x_j} + O(\delta x^2)$$
$$i = 1, 2, 3..., N$$

If use the matrix form:

$$F(X_0 + \delta_x) = F(X_0) + J\delta_x + O(\delta x^2)$$

The J is N*N Jacobian matrix and $J_{ij} = \frac{\partial F_i}{\partial x_j}$.

If we omit one of the higher order term, and demand $F(X_0 + \delta_x)$ as 0, we can obtain:

$$\delta_x = -J^{-1} \cdot F$$

This condition is called as the Newton Condition. Who solves method based on this condition can be called as Newton method, the basic Newton method (Newton-Raphson method) use the iterative method as follow:

$$X_{k+1} = X_k + \delta_x$$

Obviously Newton-Raphson method is not a method of global convergence, and it is not even a drop algorithm. Overcoming the defect of Newton Raphson method has a lot of measures, we introduce the one below. This method combined with Newton method of quick convergence and global convergence strategy.

We first introduce the objective function

$$f = 0.5F \cdot F$$

So through simple mathematical operation, we can know the Newton iteration along the gradient direction of the objective function ∇f, α Can always decline to make an objective function.

$$\nabla f \cdot \delta x = FJ \cdot (-J^{-1}F) = -F \cdot F < 0$$

So we can adopt the new iteration method: $X_{k+1} = X_k + \alpha \delta_x$.

α is the step length of the gradient direction; the initial value of α must be 1. If the value cannot satisfy the drop conditions, we can use the back-tracking algorithm (backtracking).

The method is very similar to the optimization algorithm of quasi-Newton algorithm; in fact the two are not consistent. First, convergence condition is different, second, it does not solve the HESSIAN matrix.

5 CALCULATION RESULT

In order to solve the above mathematical model, set the initial condition and boundary condition equations to close:

a. The initial conditions:
   Inlet pressure: P_{IN}
   Inlet temperature: T_{IN}
   Outlet pressure: P_{OUT}
   Mass flow rate: M
b. Pipeline import and export of boundary conditions:
   The pipe diameter at the end of pipe up hypothesis and cross-sectional area is kept constant, for the first kind of boundary condition, the expression is as follows:

$$D_0 = D_N = D_o$$

$$A_0 = A_N = \frac{\pi D_o^2}{4}$$

c. Pipe wall boundary conditions:
   For this model, the effective thickness of soil's calculation is as follow:

$$R_2 - R_1 = R_1 \left(\frac{2h}{D} + \left(\left(\frac{2h}{D} \right)^2 - 1 \right)^{0.5} - 1 \right)$$

Among them:
$R_2 - R_1$—The model of soil thickness;
R_1—The radius of the center from the pipe to soil layer (including coating and filling layer);
h—At the actual embedded depth of pipe center;
D—Pipe diameter.
Pipeline transient thermal model and the surrounding environment are as follows:

$$k(rT_r)/r = C_P \cdot \rho T_t$$

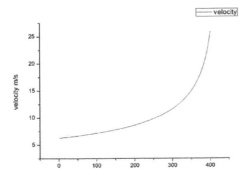

Figure 1. Velocity distribution of elastic pipeline.

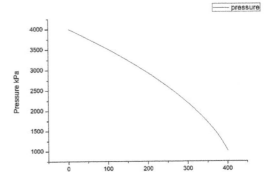

Figure 2. Pressure distribution of elastic pipe.

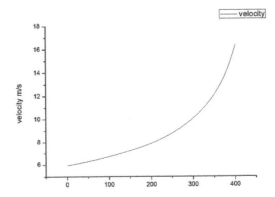

Figure 3. Velocity distribution of elastic pipeline.

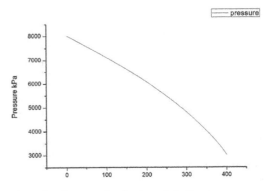

Figure 4. Pressure distribution of elastic pipe.

The heat flow per unit length can be calculated for the third kind of boundary conditions, expression is as follows:

$$\phi = \frac{2\pi\lambda(T_w - T_0)}{\ln((R_2 - R_1)/R_1)} = \pi h_1 D(T - T_w)$$

Calculated by programming the following result:
1. Set the initial condition: $P_{IN} = 4$ MPa, $P_{OUT} = 1$ MPa, $T_{IN} = 20°C$, $M = 59$ kg/s.
2. Set the initial condition: $P_{IN} = 8$ MPa, $P_{OUT} = 3$ MPa, $T_{IN} = 20°C$, $M = 115$ kg/s.

6 CONCLUSION

In this article, described in detail from the establishment of mathematical model and the whole process of mathematical modeling design flow process, using the model of elastic circular tube flow velocity field can make one dimension numerical simulation research for temperature field and stress field in the process of the change process. Through programming the results of simulation analysis, we can get the following conclusion:

1. As the elastic fluid flow in the pipe, pipe wall on each point of the elastic expansion of quantity increases with the reduction of horizontal direction gradually. In this paper, on the surface the, parameters related to the elastic pipe is not introduced, in fact it indirectly considers the control equations and boundary heat transfer conditions, that simplified the complicated problem. But simulation in the pipe of unsteady flow at the entrance of area is not in consideration; the change of elastic tube in the length of steady flow is ignored, this is a follow-up study that needs to be considered.

2. In terms of temperature field, we present the elastic tube when the steady flow of temperature distribution, at the same time, considering the influence of soil heat exchanger. But for one dimensional pipe flow, fluid flow process of the influence of different flow regime is not considered.

To further study the flow of the elastic tube, we must further consider different factors such as fluid flow, consider the higher precision of different formats and better algorithms, large-scale numerical calculations, but all this is the important research content of our the next phase, and from a one dimensional flow extending to the two dimensional flow are the focus of future research.

By the introduction of fluid state equation, Darcy friction factor empirical formula and wall heat transfer formula gives models closed and simplified.

REFERENCES

Bo Xu, Guoqun Chen, Zhao jianting, Deji Wang, A Hybrid BFGS-Based Method And Its Applications in Pipeline Steady Simulation, ICCET 2010–2010 International Conference on Computer Engineering and Technology, Proceedings, v 4, p V465–V468, 2010.

Miao Qing, Xu Bo, Oil/Gas Pipeline Simulation Using A Segregated Method, Proceedings of the Biennial International Pipeline Conference, IPC, 2010, 485–490.

Advances in Energy, Environment and Materials Science – Wang & Zhao (Eds)
© 2016 Taylor & Francis Group, London, ISBN 978-1-138-02931-6

On designing MAC—ROUTE cross layer to support hybrid antennas in wireless ad hoc networks

Yongjiang Zhang & Ying Li
School of Computer Science and Technology, Soochow University, Suzhou, China

ABSTRACT: In the mode of hybrid antennas wireless ad hoc networks, due to the directivity of directional antenna and the directional antenna and the layered structure of a wireless protocol stack and the lack of coordination between Medium Accesses Control (MAC) and routing protocols. These limitations result in long processing delays in a forwarding node. In order to make full use of the advantages of long distance transport of directional antenna and alleviate these issues, we propose a solution based on cross-layer MAC design, which improves the coordination between MAC and routing layers using an idea about control router which designed in MAC layer. Experimental results show that the proposed cross-layer design significantly improves the performance in terms of reduced round trip time, reducing processing time in the intermediate forwarding nodes and increased throughput compared to the legacy architecture.

1 INTRODUCTION

Currently there has been a lot of interest in terms of using directional antennas in wireless ad hoc networks. The broadcast nature of omni-directional antennas is one of the major causes of excessive multi-user interference. Traditional MAC protocols such as IEEE 802.11 cannot achieve high throughput in wireless ad hoc network because it blocks a large portion of the spectrum for each transmission. To address this problem, directional antennas or smart antennas can be used. They strongly reduce signal interferences in unnecessary directions and also significantly improve spatial reuse of the wireless channel by allowing several nodes to communicate simultaneously without interfering with each other, thereby significantly improving the capacity of ad hoc networks. To best utilize the benefit of directional antennas, a suitable Medium Access Control (MAC) protocol must be used. IEEE 802.11 MAC protocol assumes the use of omni-directional antenna at its physical layer; hence it does not exploit the maximum potential when used with directional antennas. In this paper, we propose a cross layer MAC design enabling virtual link for reducing processing time in routing for forwarding nodes. At the same time, the energy consumption for every data frame is also decreased. This solution has been successfully implemented and tested in a test bed based on the IEEE 802.11 MAC. The test bed consists of Soekris NET5501 devices. With this setup we experimentally evaluate and compare the performances of the proposed virtual link enabled MAC with legacy MAC/routing policy.

With the advances in wireless and mobile technologies and the ever increasing demand from the first responders and commercial users for ubiquitous connectivity, wireless ad hoc and mesh networks are gaining prominence and playing a major role in many applications. For example, during a natural disaster in a region, while it will be inconvenient or even impossible to create an infrastructure-based network, a multi-hop wireless ad-hoc network for first responder communications could be established quickly with the help of low–cost, low–power, multi–functional radio nodes. Unlike infrastructure based networking, multi-hop ad hoc architecture can create wide-area back-haul networks where traffic can flow among the peers directly using relay/forwarding via multiple hops resulting in higher capacity, ubiquitous connectivity and increased coverage.

Efficient routing in multi-hop wireless ad hoc networks is a fundamental issue to be addressed, especially in the mode of hybrid antennas wireless ad hoc networks. Within multi-hop ad hoc network, a data packet must traverse through multiple hops from source node to destination node. However, the current large number of MAC protocols and Router protocols in the mode of hybrid antennas can't work properly, or poor performance, the basic reason is that protocols tend to assume that the network exist only design the same type of antenna. To date, a great deal of literature has focused on routing issues for multi-hop ad-hoc network, primarily in three directions: (1) shortest route is discovery/maintenance; (2) developing link metrics to analyze a wide variety of performance objectives and (3) taking advantages of long distance transport of directional antenna in the mode of hybrid antennas.

At the very beginning, proactive routing protocols like Dynamic Destination-Sequenced Distance Vector (DSDV) and reactive routing protocols such as Ad hoc On-Demand Distance Vector Routing (AODV), were designed for solving routing issues. When these were implemented in real wireless networks several issues were encountered, such as: the destination never learns of a route to the source node, the manner in which RREP packets are forwarded and a rebooted node will lead to routing loops, etc. These issues have been discussed in (Royer & Perkins 2000, Chakeres & Belding-Royer 2004). In (Yang et al. 2009), a performance comparison between AODV and Dynamic Source Routing Protocol (DSR) is discussed. Here, "routing efficiency" is used for evaluating the performance of these routing protocols. In (Bansal et al. 2003), the energy overhead of DSDV, AODV and DSR are evaluated. This work shows that with different transmission ranges there are significant differences in energy overhead between the ad-hoc routing protocols. In (Niezen, et al. 2007), researchers compare energy consumption between different routing protocols including AODV, LEACH. AODV is observed to be not feasible for low energy-limited environment. A novel routing protocol is proposed to improve energy efficiency. The main idea of this protocol is to reduce message transmissions for routing information update. Cluster-based infrastructure creation are proposed for saving energy. There have also been few cross-layer attempts. The researchers investigate reducing handshaking/control frames before data transmission and reserving channel resource beyond more than one hop. In (Manjeshwar & Agrawal, 2001), a novel routing protocol is proposed to improve energy efficiency. The main idea of this protocol is to reduce message transmissions for routing information update. Cluster-based infrastructure creation are proposed for saving energy in (Sheltami & Mouftah, 2003). There have also been few cross-layer attempts. In (Raguin, 2004) the researchers investigate reducing handshaking/control frames before data transmission and reserving channel resource beyond more than one hop.

None of the above-mentioned works focuses on the long processing delays in a relay/forwarding node due to the layer structure in wireless communication, the lack of coordination between MAC and routing protocols and tend to assume that all nodes are configured in the system has the same type of antenna. In our experiments, for a three hops wireless ad-hoc setup, we observe that 16% of the Round Trip Time (RTT) is spent due to these processing delays; for a four hop ad-hoc network

this is around 17% of the total RTT. Moreover, it is also observed that at a routing node, an average of 30–40% of the processing time is used in transferring data packet between MAC layer and routing layer. Clearly, these delays result in battery power wastage, higher latency, reduced system efficiency and data rate. In this paper, we propose a cross layer MAC design enabling control router and taking advantages of long distance transport of directional antenna for reducing processing time in routing for forwarding nodes.

2 THE PROPOSED PROTOCOL

2.1 *The protocol*

In order to resolve the delays in a relay/forwarding node, we design a CLMRCR (Cross Layer MAC-Router with Control Route) protocol utilizes multi-hop RTS/CTS. Routing protocols in the conventional wireless layer structure are typically implemented in the network layer. Under this structure, the network layer has three major functions: (i) maintaining the routing table, including route discovery, creating corresponding routing entry in the routing table, modifying or deleting routing entries when a route is broken, etc.; (ii) computing the suitable next hop for the packet and (iii) re-encapsulating the packets according to the corresponding route entry. In the traditional routing structure, packets traversing over multiple hops, experience long unwanted processing delays and energy wastage at intermediate routers. Such delays may not be acceptable, especially in critical wireless applications such as first responders' networks, delay-sensitive real-time media applications etc. Moreover, if devices in a multi-hop wireless ad-hoc network are resource-limited, a traditional routing structure with such unwanted delays will make the system lifetime much shorter and inefficient to operate. In the mode of hybrid antennas, due to the directional antenna has a specific direction, how to make full use of the advantages of long distance transport of directional antenna is a problem. In order to break away from such inefficiencies, we introduce a new "control route" concept and propose a new cross layer MAC design which makes use of a virtual link to forward data packet much more efficiently. Under this proposal the routing layer implements the first function as mentioned above while the second and third functions are implemented in the data link layer.

2.2 *Control route*

Consider an ad-hoc network example as shown in Figure 1. Nodes A and F in this example are within

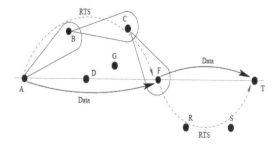

Figure 1. Control router from node A to F.

the communication range of each other, i.e., a physical link between nodes A and B exists. Then, these two nodes can communicate with each other over this physical link by using legacy DCSMAC. In contrast to the physical link between node A and B, the nodes A and F are not within the communication range of each other as shown in the Figure 1. However, note that physical links exist between A-B, B-C and C-F. If these two physical links could be combined to obtain a link then we obtain a logical link between A and F. It is then possible for nodes A and F to communicate with each other directly minimizing intermediate router processing thereby reducing the delays discussed before. In the mode of hybrid antennas, in order to make full use of the advantages of long distance transport of directional antenna, find some directional nodes as forwarding nodes in control route. For example, in Figure 1 nodes A and T are within the communication, assume node F is a directional node. Through node F forward data between nodes A and T, make full use of the directional antenna, reducing forward time. If there are many routers between nodes A and F, according to control route algorithm, choose a shortest route as nodes A and F to communication. By extending this idea, any two nodes in Figure 1 could communicate with each other. The proposed MAC protocol will establish and maintain links automatically and efficiently. With the proposed method, the routing layer will be responsible for the task of maintaining the routing table; however, the task for data frame re-encapsulation will be done by the control router enabled MAC. In the following subsection, we present the details of this system.

2.3 Analytical model of the protocol

The steps for creating a control route are then as follows. When the wireless MAC starts to run in a node, its IP address is noted. The node checks for the destination IP address on each frame. If the destination IP is equal to its own IP address, this is treated as a normal frame. Otherwise, the

node will look up the corresponding control router entry for this frame. If a suitable control router is located successfully, this frame will be re-encapsulated according to this control route entry and sent to the physical layer immediately for forwarding purpose. If no corresponding control route is found, the node will be scanning around to find some node to rebuild a suitable control route according to the new MAC header of this frame. When other data frames arrive at this node, the node will re-encapsulate the MAC header according to corresponding control route entry.

3 PERFORMANCE EVALUATION

To evaluate our protocols, we performed simulations using an extended version of the UCB/LBNL network simulator, ns-2. The ns-2 simulator is a discrete event network simulator that was developed as part of VINT project at the Lawrence Berkeley National Laboratory. The extensions implemented by the CMU Monarch project—which enable it to accurately simulate mobile nodes connected by wireless network interfaces and multi-hop wireless ad hoc networks—were used. We modified the ns-2 to implement the directional antennas which could transmit in a particular direction.

3.1 Simulation model

The simulation parameters are listed in Table 1. Directional antennas have been incorporated into this simulator. All nodes are randomly located in a 200*200 area. When compares different number of antennas beam, the proportion of dir-nodes and omni-nodes is n1:n2 = 1:1, and total number of nodes is 20. The communication hop-count from omni-directional node to directional node is 2. To get the saturation performance, we assume each node always has packets to send.

3.2 Simulation result

System throughput and delay not only relate to channel rate, interval time slot, frame size, number of nodes, packet byte and other related numerical value, but also the proportion of the directional

Table 1. Simulation parameters.

Parameter	Value	Parameter	Value
Channel rate	2 Mb/s	DIFS	50 us
Packet size	1024 byte	(D) RTS	32 bytes
Slot time	20 us	(D) CTS	14 bytes
SIFS	10 us	ACK	14 bytes

antenna with omni-directional antenna, directional antenna beam bandwidth have relations. Above all, Hop count has important influence on delay.

In this section, we compared the CLMRCR protocol with AODV and DSR protocol. And then we analyzed the delay and throughput in different situations.

The result in Figure 2 indicates that the performance of the protocols is improving as the number of the nodes increased. CLMRCR's throughput is as similar as AODV and DSR. But the delay of CLMRCR in Figure 3 is bigger than other protocols. As observed from the Figure 2 and Figure 3, with the proposed approach, we get a 15% delay improvement with 100 byte long packets. This clearly demonstrates that cross-layer MAC design enabled control router the efficiency of multi-hop routing by sufficiently reducing the processing time consumption at the intermediate relay nodes.

We focus on the delay performance comparison for data packets with different nodes and hop counts. Figure 4 show delay comparison CLMRCR,

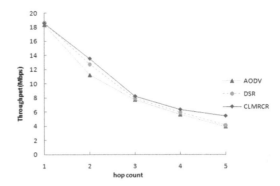

Figure 4.   Throughput with different hop count.

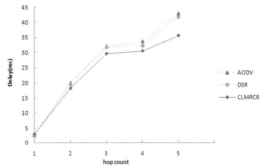

Figure 5.   Delay with different hop count.

DSR and AODV. As shown in the results, proposed method clearly reduces the delay over the legacy approach with 100 byte long packets. Moreover, it is also interesting to note that with the increasing hop counts the performance improvements are also increasing thus indicating that the proposed method to be useful for multi-hop wireless ad hoc networks. Figure 5 shows throughput difference with DSR, CLMRCR, and AODV. The throughput is measured from the client side which runs in destination node. As shown in the Figure 5, the CLMRCR's throughput improved by the proposed approach and better than DSR and AODV protocol.

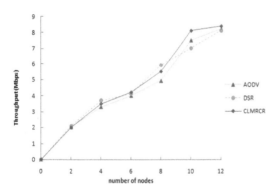

Figure 2.   Throughput with different node.

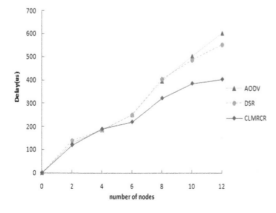

Figure 3.   Delay with different node.

4   CONCLUSION

In this work, we proposed a Cross Layer MAC-Router with Control Router protocol utilizes multi-hop RTS/CTS suitable for networks with directional antennas and omni-directional antennas to deal with delays in a relay/forwarding node. The protocol makes full use of the advantages of long distance transport of directional antenna as forward node and designs control router to improve the delay. The tests performed in the simulation clearly demonstrate that cross-layer MAC design employing the proposed virtual link concept reduces the processing time at the intermediate nodes by approximate 30% while the throughput increases by 6–9% when compared with the legacy routing algorithm.

ACKNOWLEDGMENT

This research is supported by the National Science Foundation of China (61103236).

REFERENCES

Bansal, S., Shorey, R. & Misra, A. 2003. Comparing the routing energy overheads of ad-hoc routing protocols. *In IEEE Wireless Communications and Networking (WCNC)*, 2:1155–1161.

Chakeres, I. & Belding-Royer, E. 2004. AODV Routing Protocol Implementation Design. *In Proceedings of the 24th International Conference on Distributed Computing Systems Workshops (ICDCSW'04)*, 7:698–703.

Manjeshwar, A. & Agrawal, D. 2001. Teen: a routing protocol for enhanced efficiency in wireless sensor networks. In 15th International Proc. of Parallel and Distributed Processing Symposium. 1:2009–2015.

Niezen, G., Hancke, G., Rudas, I., et al. 2007. Comparing wireless sensor network routing protocols. 1:1–7.

Raguin, D., Kubisch, M., Karl, H., et al. 2004. Queue-driven cutthrough medium access in wireless ad hoc networks. In *IEEE Wireless Communications and Networking Conference (WCNC)*, 3:1909–1914.

Royer, E. & Perkins, C. 2000. An implementation study of the AODV routing protocol. *In IEEE Wireless Communications and Networking Conference (WCNC)*, 3: 1003–1008.

Sheltami, T. & Mouftah, H. 2003. An efficient energy aware clusterhead formation infrastructure protocol for manets. In Eighth IEEE Intl. Symp. on Computers and Communication (ISCC), 1:203–208.

Yang, T., Barolli L., Ikeda M., et al. 2009. Performance evaluation of a wireless sensor network considering mobileevent. *In International Conference on Complex, Intelligent and Software Intensive Systems (CISIS)*, 1:1169–1174.

Advances in Energy, Environment and Materials Science – Wang & Zhao (Eds)
© 2016 Taylor & Francis Group, London, ISBN 978-1-138-02931-6

Detecting anomalies in outgoing long-wave radiation data by a window average martingale method

Jing Zeng & Liping Chen
School of Mathematics and Computer Science, Fujian Normal University, Fuzhou, China

ABSTRACT: A Window Average Martingale method is developed based on the martingale framework. The method can be used on unlabeled data in statistical signal processing. The analyses of the Outgoing Long-Wave Radiation data reveal that the method can been successfully used in detecting anomalies.

1 INTRODUCTION

Recently data mining has become a popular issue. It has been widely concerned and extensively researched in the field of machine learning and earthquake science research. It is important to find the changes in the high-dimensional multimedia data, such as Outgoing Long-wave Radiation data (OLR) and so on (Domingos, 2001). Wald (1947) advocated the sequential analysis, as a method of changes detection, for the first time. Page (1954) further developed the cumulative sum method.

In order to detect seismic precursors, we need the data which can reflect the reality. Researchers found that there are temperature anomalies, geothermal phenomena, electromagnetic perturbations before many earthquakes, and this information can be found in the satellite remote sensing data. There are a lot of advantages that ground-based observations don't possess, such as objectivity and high density information. So in order to identify the potential earthquake zones, it is important to detect seismic precursors within the satellite data.

Some studies have been concerned with ionospheric anomalies appearing a few days before the seismic shock, such as (Molchanov, 2006; Parrot, 2006; Zhu, 2008). Electromagnetic perturbations possibly connected with seismic activity were reported by (Asada, 2001; Hobara, 2005; Parrot, 1994) and so on. The phenomena of satellite thermal infrared radiation before major earthquakes were found by Bi (2009), Ouzounov (2006), Saraf (2005).

The key challenge for detecting seismic precursor research is to process massive volumes of satellite data properly and rapidly. It means we need to find the effective data processing methods to deal with the remote sensing data. Cervone et al. (Cervone, 2004) introduced an innovative data mining technique to identify precursory signals associated with earthquakes. Song et al. (Song, 2007) proposed a log-likelihood method relying on kernel density approximation to detect changes. Marzocchi et al. (Marzocchi, 2012) used the Bayesian estimate value method and integrated analysis to investigate earthquake data.

The data mining method–Martingale, can detect abnormal events embedded in the satellite data. Ho (2005) proposed a martingale method for change detecting of high-dimensional labeled data streams. At the same year, an adaptive support vector machine for time-varying data streams based on the martingale method was proposed (Ho, 2005). Ho and Wechsler (2010) used the martingale theory to detect changes in data streams, and their results have showed the effectiveness and feasibility of the martingale methodology.

The main contribution of this paper is the Window Average Martingale method (WAM). It combined the Martingale method (Ho, 2005) and the Window Average method (Pierre, 2008). The WAM method works as a data mining tool to detect the OLR data, and the WAM value can reflect the earthquake zone. Our results show the significant abnormality is found in the OLR data by using the WAM method. In this paper the WAM method are proposed first. Then the WAM values of OLR data are calculated. At last we analyze the results.

2 THE WINDOW AVERAGE MARTINGALE METHOD

In this section we propose the Window Average Martingale method base on the Martingale method (Ho, 2005) and the Window Average method (Pierre, 2008). The WAM method is described in detail below.

Let $\{z_i; 1 \leq i < \infty\}$ be a sequence of unlabeled or labeled random variables. A Martingale is a sequence of random variables $\{M_i: 0 \leq i < \infty\}$ such that M_n is a measurable function of $z_1, ..., z_n$ for all $n = 0, 1, ...$ ($M_0 = 1$). The conditional expectation of M_{n+1} is given by

$$E(M_{n+1} \mid M_1, ... M_n) = M_n. \tag{1}$$

A strangeness value s_i of z_i (for $i = 1, ..., n$) is defined the difference between a data point and the others,

$$s_i(z_1, ..., z_{n-1}, z_n) = \|z_i - C(\{z_1, \cdots, z_{n-1}, z_n\})\|, \tag{2}$$

where $C(\cdot)$ is a strange measure, it is the cluster center. We use the K-means cluster method (Kanungo, 2002) to compute the center $C(\cdot)$. $\|\cdot\|$ is the Euclidean distance. It turns out that if a data point is further away from the cluster center, the strangeness value will be high.

Next, the p-value of z_n is constructed to rank the strangeness,

$$p_i = \frac{\#\{i : s_i > s_n\} + \theta_n \cdot \#\{i : s_i = s_n\}}{n}, \tag{3}$$

where "#" is a function which counts the number of i satisfying the condition $s_i = s_n$, and s_i is the strangeness measure mentioned in formula (2). θ_n is randomly chosen from $[0,1]$. The power martingale is defined as

$$M_n^\varepsilon = \prod_{i=1}^n (\varepsilon p_i^{\varepsilon-1}), \tag{4}$$

where p_i ($i = 1, ..., n$) are the p-value in formula (3), $\varepsilon \in [0,1]$. The initial martingale is $M_0^\varepsilon = 1$. We also defined the stopping rule:

$$M_n \geq h, \tag{5}$$

where h is a chosen threshold. If the computed M_n exceeds some value such as h, it is regarded as an abnormal condition. That is if $M_n \geq h$, we stop the computation and start a new one by initializing the

algorithm. According to the definition of stopping rule, we pick out the point n satisfying $M_n \geq h$, and define n as a change point.

In order to find the abnormal change, Bertrand and Fleury (Bertrand, 2008) use the Finite Moving Average method to detect the small shift on the mean. We defined the WAM value as the following:

$$\text{WAM}_m = \frac{\sum_{n=m-L}^m M_n^\varepsilon}{L} = \frac{\sum_{n=m-L}^m \prod_{i=1}^n (\varepsilon p_i^{\varepsilon-1})}{L}, \tag{6}$$

where L is the size of window, M_n^ε is defined in formula (4). Formula (6) indicates that the WAM value reflects the abnormal change degree of the current data, and it computes the mean of abnormal change degree by the current data and $L-1$ data points before current data.

3 THE WAM VALUE AND DISCUSSION

Outgoing Long-wave Radiation data (OLR) are the data of the energy released by the Earth as infrared radiation at low energy. It is a critical component of the Earth's radiation and represents the total radiation going to space emitted by the atmosphere (Susskind, 2011). The OLR data we adopted are collected by the National Oceanic and Atmosphere Administration (NOAA) satellites. NOAA uses two polar-orbiting satellites, a "day satellite" and a "night satellite", to ensure every part of the Earth is observed at least twice every 24 hours. The OLR data have been recorded twice daily, making up time series over different time zone along with the spatial coverage of the entire Earth (90°N ~ 90°S, 0°E ~ 357.5°E) (N.C.A.R., 2008). The data are stored in 144×72 arrays in ASCII, and each value represents the OLR flux on 2.5 degree latitude × 2.5 degree longitude (units are watts/(m*m)).

The data we adopted cover the range of 23.75°N ~ 36.25°N and 97.5°E ~ 110°E. The areas are divided into 25 grids as in Table 1, and recorded as grid (1) ~ grid (25). We take $h = 1000$ in formula (5), $\varepsilon = 0.8$ in formula (4) and $L = 30$ in formula (6).

Table 1. The WAM value of grid (1)~(25) on 19 April 2013, the day before the Lushan earthquake, units are acd (abnormal change degree).

	97.5°E~100°E	100°E~102.5°E	102.5°E~105°E	105°E~107.5°E	107.5°E~110°E
33.75°N~36.25°N	(1) 26.44888	(2) 14.20673	(3) 1.899750	(4) 103.0629	(5) 32.74593
31.25°N~33.75°N	(6) 1.841195	(7) 0.736587	(8) 3.212655	(9) 178.6314	(10) 87.29493
28.75°N~31.25°N	(11) 13.75534	(12) 105.5608	(13) 307.1419	(14) 286.0259	(15) 53.38453
26.25°N~28.75°N	(16) 6.892492	(17) 12.27177	(18) 229.4337	(19) 70.40291	(20) 0.290468
23.75°N~26.25°N	(21) 106.1383	(22) 227.3611	(23) 4.890695	(24) 0.165424	(25) 0.075593

Lushan and Wenchuan earthquake's OLR data are selected for our data minting examples. The main reason why we choose these earthquakes is that they are geographically close, but also the WAM value of them exhibit different states.

3.1 The OLR data of Lushan earthquake

Lushan earthquake of magnitude 7.0 occurred on 20 April 2013, the depth of focal is 13 km (C.S.I.N., 2013). The epicenter is Lushan County (30.3°N, 103.0°E), which is in the southern section of the Longmenshan tectonic zone. The maximum seismic intensity is 9 degrees, and the earthquake affected an area of approximately 18,682 square kilometers. Lushan earthquake has given people a huge shock, not only for its magnitude, but also for it is so near in distance to Wenchuan, only 69 kilometers. And it is not long in time after the Wenchuan earthquake, about 5 years.

By the WAM method, the WAM values of the surrounding area of the Lushan epicenter on the day before Lushan earthquake are calculated (see Table 1). The epicenter of Lushan earthquake is in grid (13) (28.75°N ~31.25°N, 102.5°E ~ 105°E), whose WAM value reaches the global maximum 307.1419 (see Table 1) in the region. This is proved to be a harbinger of Lushan earthquake, which occurred at the next day, 20 April 2013.

Is there an abnormal change brewing already before the earthquake? To answer this question, the graphs of the original OLR data (see Fig. 1(A)) and the WAM value (Fig. 1(B)) of grid (13) are drown. The original OLR data are varying every day, and we connect these data dots with straight lines. So Figure 1(A) is a polyline, but a continuous curve. This explains spurious peaks in Figure 1(A). There are minimums on three days, 19 September, 30 October and 4 December 2013, and no

earthquake phenomena are reported around the three days. No obvious abnormality was observed in Figure 1(A), including on the day of earthquake, 20 April 2013. We can not draw conclusion from the original OLR data.

But how about the graph of WAM value (Fig. 1(B))? Figure 1(B) and Figure 1(A) are in the same time span (September 19, 2012 to April 26, 2013). The WAM value (Fig. 1(B)) is very small from 19 September 2012 to 18 March 2013. It is gradually bigger from then on, and keeps the stable high level from 15 April 2013. There is an abnormal change brewing already before Lushan earthquake (20 April 2013).

Next we will show that the abnormal change of WAM value have some connection with the earthquake zone. Now we focus on the area around the grid (13), that is grid (7), (8), (9), (12), (14), (17), (18) and grid (19). The WAM value of these grids are showed in Figure 2(A) and Figure 2(B) in the time span of a year, and the day of Lushan earthquake (20 April 2013) is marked by a vertical line. All the WAM value in the nine grids is very small from 19 September 2012 to 18 March 2013 (see Fig. 2(A) and Fig. 2(B)). From then on, these WAM values become bigger gradually, and show the abnormal change. We take particular note of the curves of grid (13), (14), (18), whose WAM values are greater than 200 on 19 April 2013 (see Table 1 and Fig. 2(B)). It shows a significant change in these grids.

3.2 The OLR data of Wenchuan earthquake

Wenchuan earthquake (magnitude 8.0) is the largest earthquake of China in the past sixty years.

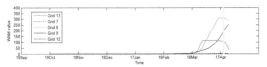

Figure 2(A). The Lushan WAM value of grid (13, 7, 8, 9, 12) from 19 September 2012 to 26 April 2013. We mark the day that Lushan earthquake occurred (20 April 2013) with the vertical yellow line.

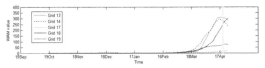

Figure 2(B). The Lushan WAM value of grid (13, 14, 17, 18, 19) from 19 September 2012 to 26 April 2013. We mark the day that Lushan earthquake occurred (20 April 2013) with the vertical yellow line.

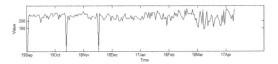

Figure 1(A). The Lushan original NOAA data of grid (13) from 19 September 2012 to 26 April 2013.

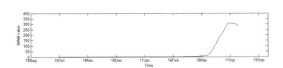

Figure 1(B). The Lushan WAM value of grid (13) from 19 September 2012 to 26 April 2013.

It occurred at the southeast edge of the Qinghai-Tibet Plateau, Wenchuan-Maowen large fault zone on 12 May 2008 (C.S.I.N., 2013). We take the same parameters as above, and the OLR data are also from the NOAA satellites. By the WAM method, we get the WAM value of the surrounding area of the earthquake (23.75°N~36.25°N, 97.5°E~110°E) on 11 May 2008, the day before Wenchuan earthquake (see Table 2).

The epicenter of the earthquake is in grid (13) (28.75°N~31.25°N, 102.5°E~105°E), with the highest WAM value 188.0978 (see Table 2). Original OLR data and the WAM value of grid (13) are showed in Figure 3(A) and Figure 3(B) respectively. The original OLR data dots are varying every day, and be connected with straight lines. So Figure 3(A) is a polyline, but a continuous curve. There is no earthquake phenomena reported around 16 June 2008, whose original OLR data reaches global minimum in Figure 3(A). So we cannot find any connection between the original OLR data and the earthquake.

Let us pay attention to the WAM value in Figure 3(B), whose time span is the same as Figure 3(A). The WAM value is very small from 28 September 2007 to 28 March 2008, then gradually bigger and shows the abnormal change (see Fig. 3(B)). The data reach its peak 189.6828 on 12 May 2008 (the Wenchuan earthquake). Next it gradually decreases, and reaches its normal value after 18 June 2008. It is no doubt that there is abnormal change brewing already before the earthquake, and the time that the earthquake occurs is exactly the time that the WAM value reaches its peak.

Next we want to show that the WAM value reflect the earthquake zone in some sense. The WAM data around the epicenter (grid (7), (8), (9), (12), (14), (17), (18) and grid (19)) are analyzed, in the time span of a year (28 September 2007 to 27 September 2008) (see Fig. 4(A) and Fig. 4(B)). The day of Wenchuan earthquake (12 May 2008) is marked with the vertical line in these figures. Among these curves in Figure 4, we take particular note of the curves of grid (13), (14), (18) and grid (19), whose WAM value are greater than 100 and are found significant abnormalities around earthquake day.

These show that the WAM value have some links with the earthquake zone. Compared with

Table 2. The Wenchuan WAM value of grid (1)–(25) on 11 May 2008, the day before the Wenchuan earthquake (units are acd).

	97.5°E~100°E	100°E~102.5°E	102.5°E~105°E	105°E~107.5°E	107.5°E~110°E
33.75°N~36.25°N	(1) 4.383995	(2) 30.88060	(3) 10.97135	(4) 129.7492	(5) 171.9776
31.25°N~33.75°N	(6) 0.005727	(7) 0.012335	(8) 64.96195	(9) 3.057392	(10) 21.73620
28.75°N~31.25°N	(11) 115.3304	(12) 0.894902	(13) 188.0978	(14) 108.8898	(15) 7.538662
26.25°N~28.75°N	(16) 0.139102	(17) 49.29567	(18) 157.7295	(19) 128.4422	(20) 0.287400
23.75°N~26.25°N	(21) 0.942082	(22) 9.976640	(23) 14.37489	(24) 38.71420	(25) 5.344377

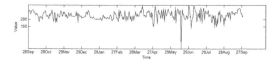

Figure 3(A). The Wenchuan original NOAA data of grid (13) from 28 September 2007 to 27 September 2008.

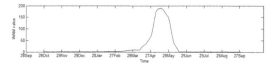

Figure 3(B). The Wenchuan WAM value of grid (13) from 28 September 2007 to 27 September 2008.

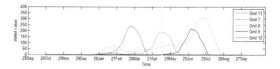

Figure 4(A). The Wenchuan WAM value of grid (13, 7, 8, 9, 12) from 28 September 2007 to 27 September 2008. We mark the day that the earthquake occurred (12 May 2008) with the vertical yellow line.

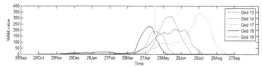

Figure 4(B). The Wenchuan WAM value of grid (13, 14, 17, 18, 19) from 28 September 2007 to 27 September 2008. We mark the day that the earthquake occurred (12 May 2008) with the vertical yellow line.

other experimental results, ours are relatively good. Xiong et al. (Xiong, 2011) used wavelet maxima to analyze the singularities of seismic precursors in DEMETER satellite data. But their prominent singularities do not consistent with the earthquake zone and the time of the earthquake.

ACKNOWLEDGMENT

The authors would like to acknowledge the NOAA center for making NOAA satellite data available for various research communities. And we thank the useful advices given by Kong Xiangzeng. The first author is a postdoctoral in Fujian Normal University, and is supported by the China National Postdoctoral Science Foundation (2014M551830). Corresponding author: Zeng Jing.

REFERENCES

Asada, T., Baba, H., Kawazoe, M. & Sugiura, M. 2001. An attempt to delineate very low frequency electromagnetic signals associated with earthquakes. *Earth Planets Space*, 53: 55–62.

Bertrand, P.R. & Fleury, G. 2008. Detecting small shift on the mean by finite moving average. *International Journal of Statistics and Management System*, 3(1): 56–73.

Bi, Y., Wu, S., Xiong, P. & Shen, X. 2009. A comparative analysis for detecting seismic anomalies in data sequences of Outgoing Long-wave Radiation. *Knowledge Science, Engineering and Management Lecture Notes in Computer Science*, 5914: 285–296.

Cervone, G., Kafatos, M., Napoletani, D. & Singh, R.P. 2004. Wavelet maxima curves of surface latent heat flux associated with two recent Greek earthquakes. *Natural Hazards and Earth System Sciences*: 359–374.

C.S.I.N. 2013. *China seismic information network*. http://www.csi.ac.cn.

Domingos, P. & Hulton, G. 2001. Catching up with the data: research issues in mining data streams. *In SIGMOD workshop on research issues in data mining and knowledge discovery*, Santa Barbara, CA.

Ho, S.S. 2005. A martingale framework for concept change detection in time-varying data streams. *In Luc De Raedt, Stefan Wrobel, Proceedings of the 22nd Annual International Conference on Machine Learning, AC M*: 321–327.

Ho, S.S. & Wechsler, H. 2005a. Adaptive support vector machine for time-varying data streams using martingale. *In Leslie Pack Kaelbling and Alessandro Saffiotti, IJCAI*: 1606–1607.

Ho, S.S. & Wechsler, H. 2005b. On the detection of concept change in time-varying data streams by testing exchangeability. *In Proc. 21st Conference on Uncertainty in Artificial Intelligence*: 267–274.

Ho, S.S. & Wechsler, H. 2010. A martingale framework for detecting changes in data streams by testing exchangeability. *IEEE Transactions on Pattern Analysis and Machine Intelligence*, 32(12): 2113–2127.

Hobara, Y., Lefeuvre, F., Parrot, M. & Molchanov, O.A. 2005. Low-latitude ionospheric turbulence observed by Aureol-3 satellite. *Annales Geophysicae*, 23: 1259–1270.

Kanungo, T., Mount, D.M., Netanyahu, N.S., Piatko, C.D., Silverman, R. & Wu, A.Y. 2002. An efficient k-Means clustering algorithm: Analysis and implementation. *IEEE Transactions on pattern analysis and machine intelligence*, 24(7): 881–892.

Marzocchi, W., Zecha, J. & Jordan, T.H. 2012. Bayesian forecast evaluation and ensemble earthquake forecasting. *Bulletin of the Seismological Society of America*, 6: 2574–2584.

Molchanov, O., Rozhnoi, A. & Solovieva, M. 2006. Global diagnostics of the ionospheric perturbations related to the seismic activity using the VLF radio signals collected on the DEMETER satellite. *Natural Hazards and Earth System Sciences*, 6: 745–753.

N.C.A.R. & N.O.A.A. 2008. ftp ftp.cpc.ncep.noaa.gov; cd precip/noaa* for OLR directories.

Ouzounov, D., Bryant, N., Logan, T. & Pulinets, S. 2006. Satellite thermal IR phenomena associated with some of the major earthquakes in 1999–2003. *Physics and Chemistry of the Earth*, 31: 154–163.

Page, E.S. 1954. *Continuous inspection schemes*, *Biometrika*, 41: 100–115.

Parrot, M. 1994. Statistical study of ELF/VLF emissions recorded by a low-altitude satellite during seismic events. *Journal of geophysical Research*, 99: 23339–23347.

Parrot, M., Berthelier, J.J., Lebreton, J.P. & Sauvaud, J.A. 2006. Examples of unusual ionospheric observations made by the DEMETER satellite over seismic regions. *Physics and Chemistry of the Earth*, 31: 486–495.

Pierre, R.B. & Fleury, G. 2008. Detecting small shift on the mean by finite moving average. *International Journal of Statistics and Management System*, 3(1–2): 56–73.

Saraf, A.K. & Choudhury, S. 2005. Satellite detects surface thermal anomalies associated with the Algerian earthquakes of May 2003. *International Journal of Remote Sensing*, 26(13): 2705–2713.

Song, X., Wu, M., Jermaine, C. & Ranka, S. 2007. Statistical change detection for multi-dimensional data. *In KDD'07, Proceedings of the 13th ACM SIGKDD international conference on Knowledge discovery and data mining*, San Jose, California, USA, 667–676.

Susskind, J., Molnar, G. & Iredell, L. 2011. Contributions to climate research using the AIRS science team Version-5 products. *NASA. Goddard Space Flight Center*: 1–13.

Wald, A. 1947. *Sequential analysis*, John Wiley and Sons, Inc., New York, 1947.

Xiong, P., Gu, X., Shen, X., Zhang, X., Kang, C. & Bi, Y. 2011. A wavelet-based method for detecting seismic anomalies in DEMETER satellite data. *Knowledge Science, Engineering and Management Lecture Notes in Computer Science*, 7091: 1–11.

Zhu, R., Yang, D., Jing, F., Yang, J. & Ouyang, X. 2008. Ionospheric perturbations before Pu'er earthquake observed on DEMETER, *Acta Seismologica Sinica*, 21(1): 77–81.

Advances in Energy, Environment and Materials Science – Wang & Zhao (Eds)
© 2016 Taylor & Francis Group, London, ISBN 978-1-138-02931-6

Human detection system based on Android

Qing Tian & Shan Lin
North China University of Technology, Beijing, China

Yun Wei
Beijing Urban Construction Design and Development Group Co. Ltd., Beijing, China

Weiwei Fei
Systems Engineering Research Institute, CSSC, Beijing, China

ABSTRACT: In this paper, combined with the existing algorithm of human detection and the depth camera, a human detection system under the Android system was set. Depth information and color information of the scene can be obtained by depth camera. According to depth information and threshold segmentation method, the area of head and shoulder is determined. Then the histogram of oriented gradient is extracted from the head and shoulder region. And the support of vector machine is used to classify and identify the target.

1 INTRODUCTION

As the core of intelligent transportation and intelligent vehicles, the research of pedestrian detection technology has been started in the 90's in the twentieth century, and a large number of research results appear every year. With the development of computer hardware and video capture devices, many representative algorithms of pedestrian detection have been proposed. (Xu & Cao 2008) In 2005's CVPR, the French researchers Navneet Dalal and Bill Triggs gave an excellent algorithm based on histograms of oriented gradients combined with support vector machine. This algorithm has become one of the classical algorithms of human detection. (Navneet & Bill 2005) (Dollar 2012).

In recent years, with the development of video capture devices, a new method based on depth map was used in the area of human detection. The method can obtain the depth information of the real scene through the camera, and can describe the real scene from the three-dimensional perspective. Depth information is not affected by illumination changes and shadows. At the same time, the distance between different targets can effectively overcome the phenomenon of occlusion in the image. (Tang & Lin 2015) (Rauter 2013) (Ikemura 2010).

Android is a mobile Operating System (OS) based on the Linux kernel and is currently developed by Google. With a user interface based on direct manipulation, Android is designed primarily for touchscreen mobile devices such as smartphones and tablet computers, with specialized user interfaces for televisions (Android TV), cars (Android Auto), and wrist watches (Android Wear).

In this paper, the development of pedestrian detection algorithm is successfully transplanted into the Android. After testing, the results are satisfied. Then we will introduce the system hardware in the second section, the third section is the algorithm in Android system and the fourth section of the system include test results.

2 HARDWARE SYSTEM

According to open and free services to these two major advantages, android obtains the support of manufacturers and a large number of developers. According to its share in the domestic mobile phone market, it is sufficient to prove the degree of Android system. Recently, more than 90% of the new products have the Android system. It can be seen that android systems in the field of mobile terminal are still rising.

Kinect (codenamed in development as Project Natal) is a line of motion sensing input device by Microsoft for Xbox 360 and Xbox One video game consoles and Windows PCs. Based around a webcam-style add-on peripheral, it enables users to control and interact with their console/computer without the need for a game controller, through a natural user interface using gestures and spoken commands. (Mircrosoft 2009) The device features an "RGB camera, depth sensor and multi-array

Figure 1.   The aspect of hardware system.

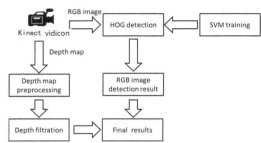

Figure 2.   The flow chart of pedestrian detection algorithm.

microphone running proprietary software", which provides full-body 3D motion capture, facial recognition and voice recognition capabilities. (Totilo 2010) The depth sensor consists of an infrared laser projector combined with a monochrome CMOS sensor, which captures video data in 3D under any ambient light conditions. (Totilo 2009) The sensing range of the depth sensor is adjustable.

Tiny4412 is a high performance quad core Cortex-A9 core board, which uses Samsung Exynos4412 as the main processor. Exynos4412 running frequency can be reached 1.5GHz. Exynos4412 internal integrated MP Mali-400 high performance graphics engine, support 3D graphics smooth running.

In this paper, we develop of pedestrian detection using the TINY4412 of development board with Android 4.4 and Kinect.

3   ALGORITHM OF SOFTWARE

3.1   *Algorithm design*

The pedestrian detection system using Kinect depth camera gets depth information and color information. Then use the open source computer vision library to filter depth in depth image and detect human in color image.

The detailed flow chart of the pedestrian detection algorithm is as Figure 2.

3.2   *The processing procedure of depth map*

First, explain the meaning of depth map. Kinect has three cameras, the camera in the middle is RGB color camera, the other two sides of the camera lens and sensing objects were composed of infrared transmitter and the infrared CMOS camera depth of 3D structured light sensor. Kinect calculates the distance between an object and the

camera by emitting an infrared and sensing object. A black and white depth map is formed by the gray value. The luminance of each pixel in the picture represents the distance from the camera.

The error information at the four edges of the depth image is removed at the first. As the depth map is the distance value of the object presented as the gray value, and the effective distance of Kinect depth camera is 1M~7M. In order to filter out the background, set a reasonable threshold to turn gray image into binary image. Afterward, remove noise by processing the prior result with morphological method (Opening). Opening removes small objects from the foreground (usually taken as the dark pixels) of an image, placing them in the background, while closing removes small holes in the foreground, changing small islands of background into foreground.

We can see the pedestrian connectivity area, after the edge extraction. And extract rectangular area according to the size of the contour. Finally in the background removing rectangular area, process the pedestrian detection of color image.

3.3   *The processing procedure of RGB image*

In this paper, a pedestrian detection algorithm based on histogram of oriented gradients and support vector machine was proposed by Dalal Navneet and Triggs Bill in 2005. The Histogram of Oriented Gradients (HOG) is a feature descriptor used in computer vision and image processing for the purpose of object detection. The technique counts occurrences of gradient orientation in localized portions of an image. This method is similar to that of edge orientation histograms, scale-invariant feature transforms descriptors, and shape contexts, but differs as it is computed on a dense grid of uniformly spaced cells and uses overlapping local contrast normalization for improved accuracy. Navneet Dalal and Bill Triggs, researchers for the French National Institute for Research in Computer Science and Control (INRIA), first

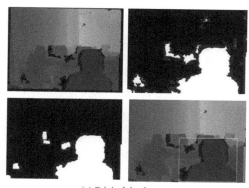

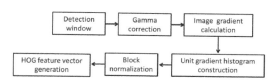

(a) Original depth map
(b) Remove edge and convert to binary image
(c)Morphology(opening)
(d)Extract edges and select rectangular area

Figure 3.   The processing procedure of depth map.

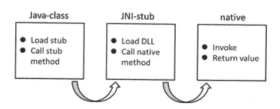

Figure 4.   The calculation process of HOG feature.

Figure 5.   The training process of SVM classifier.

Figure 6.   The diagram of Java Native Interface call process.

(a)Result of background filtering in depth map
(b)Final result of human detection in RGB image

Figure 7.   The result of human detection.

In machine learning, Support Vector Machines (SVM) are supervised learning models with associated learning algorithms that analyze data and recognize patterns, used for classification and regression analysis. (Cortes & Vapnik 1995) Given are a set of training examples, each marked for belonging to one of the two categories, an SVM training algorithm builds a model that assigns new examples into one category or the other, making it a non-probabilistic binary linear classifier.

The function of the SVM classifier is to calculate the optimal partition of these features according to certain rules, and to realize the classification and recognition of the target. In this paper we select 5,000 positive samples including head and shoulder feature and 10,000 negative samples for SVM training. Training process is as shown in Figure 5.

3.4   Algorithm optimization in Android project

Due to the limitation of Android development board, this paper uses JNI to call the native layer of pedestrian detection algorithm. In computing, the Java Native Interface (JNI) is a programming framework that enables Java code running in a Java Virtual Machine (JVM) to call and be called by native applications (programs specific to a hardware and operating system platform) and libraries written in other languages such as C, C++ and assembly. (Liang 2009) In other words, JNI can make the Android project in the Java layer and the native layer to play their respective roles and cooperate with each other.

4   CONCLUSION

Figure 7 (a) is the result of filtering depth map. The area in the white rectangle is the result of setting the threshold segmentation. The red rectangle in figure 7 (b) is the result of the histogram of the oriented gradients detection combined with the support vector machines classifier in color image. The green rectangle in color image is the final detection result. The final detection result is the intersection

described HOG descriptors at the 2005 Conference on Computer Vision and Pattern Recognition (CVPR). In this work they focused on pedestrian detection in static images, although since then they expanded their tests to include human detection in videos, as well as to a variety of common animals and vehicles in static imagery. HOG features are computed as shown in Figure 4.

After extracting the HOG features, they classified the features using support vector machine.

of depth filtration area and HOG detection results. It can be seen that human detection system runs normally. And it can reach the requirement of the pedestrian detection system.

In this paper, according to the actual needs of passenger flow detection system in real life, we combine the existing pedestrian detection algorithm with the Kinect depth camera. A relatively accurate pedestrian detection system is designed and implemented on the Android system. Pedestrian detection algorithm is optimized based on the characteristics of the Android system. We use the method of threshold segmentation based on depth image, extracting HOG feature of human head and shoulder area and combining the SVM classifier to identify the pedestrian detection. In the Android project, the JNI technology is used to ensure that the detection algorithm can deal with the image in real time. In the detection algorithm, through the collection of a large number of field images, positive and negative samples can be formed after collection and marking. Then the HOG features of these positive and negative samples are used to train the classifier. Effectively, it reduces the occurrence of the false detection of the classifier.

REFERENCES

Cortes, C, Vapnik, V. 1995. Support-vector networks, *Machine Learning* 20 (3): 273. doi:10.1007/BF00994018.

Dalal Navneet, Triggs Bil. 2005. Histograms of oriented gradients for human detection [C]. *Computer Vision and Pattern Recognition, 2005. CVPR 2005. IEEE Computer Society Conference on. IEEE*, 1: 886–893.

Dollar, P., C. Wojek, B. Schiele, et al. 2012. Pedestrian detection: an evaluation of the state of the art [J]. *IEEE Transactions on Pattern Analysis and Machine Intelligence*, 34(4): 743–761.

Ikemura S, Fujiyoshi H. 2010. Real-time human detection using relational depth similarity features [C]. *Computer Vision–ACCV 2010*, Lecture Notes in Computer Science. Los Alamitos, CA, United States: IEEE, 2011: 25–38.

Michael Rauter. Reliable. 2013. Human Detection and Tracking in Top-View Depth Images [C]. *Computer Vision and Pattern Recognition Workshops (CVPRW), 2013 IEEE Conference on*, 2013: 529–534.

Microsoft, 2009, Project Natal 101, Archived from the original on June 1, 2009. Retrieved June 2, 2009.

Nick C. Tang, Yen-Yu Lin, Ming-Fang Weng, Hong-Yuan Mark Liao.2015. Cross-Camera Knowledge Transfer for Multiview People Counting. *IEEE Transactions on image processing*, 24(1): 80–93.

Sheng Liang, Role of the JNI. 2009. *The Java Native Interface Programmer's Guide and Specification.* Retrieved 2008–02–27.

Totilo, Stephen. 2009. Microsoft: Project Natal Can Support Multiple Players, See Fingers. *Kotaku. Gawker Media.* Retrieved June 6, 2009.

Totilo, Stephen. 2010. Natal Recognizes 31 Body Parts, Uses Tenth of Xbox 360 Computing Resources. *Kotaku, Gawker Media.* Retrieved November 25, 2010.

Xu yanwu, Cao xianbin, Qiao hong. 2008. Survey on the Latest Development of Pedestrian Detection System and Its Key Technologies Expectation, *ACTA ELECTRONICA SINICA*, 36(5): 368–376.

Advances in Energy, Environment and Materials Science – Wang & Zhao (Eds)
© *2016 Taylor & Francis Group, London, ISBN 978-1-138-02931-6*

Research on attack model and security mechanism of RFID system

Tian Wang & Shisong Wu
Information Center, Guangdong Power Grid Corporation, Guangzhou, China

Qian Li, Hengfeng Luo & Ruiqi Liu
The 5th Electronics Research Institute of the Ministry of Information Industry of China, Guangdong, China

ABSTRACT: In recent years, RFID technology has developed rapidly, and at the same time, the security of the RFID system has emerged. This paper introduces three security domains, summarizing and analyzing the attack model of the RFID system. Based on security domain and attack model, this paper proposes a collaborative work model of security and privacy challenges in RFID system, which is an effective way of information security.

1 INTRODUCTION

There are three security domains in the RFID information systems: wireless data collection areas by tags and readers, internal systems including data centers and application management platforms, information systems and application platforms for data exchange and network inquiry such as e-commerce service platform, e-port interface, and e-government service platform interface. Security risks are more likely to occur in the first security domain among tag wireless transmissions and reader, which could cause privacy information leakage and tracking, etc. In addition, security risks may also occur in the third security domain, which could cause privacy from unauthorized access or misuse, if poorly managed. A great variety of security and privacy risks exist in security domains as follows (Alavi, 2015):

1. Security domain of wireless data collection areas by tags and readers. The attack technologies through the interface of tags and readers including Spoofing, Insert, Replay, and DOS (Denial of service), which leads to the security risks including tag of forged, tag of illegal access and tampered, eavesdropping through wireless interface, and tag information obtained and tag for track and monitoring.
2. Security domain of internal systems includes data centers and application management platform. In this domain the security risk is having the same enterprise network. Strengthening management at the same time, it must prevent the unlawful or unauthorized access and use of internal staff, and prevents unauthorized readers access to the internal network.
3. Security domain of information systems and application platforms for data exchange and network inquiry such as e-commerce service platform, e-port interface, and e-government service platform interface. Through the authentication and authorization mechanisms, as well as in accordance with the relevant privacy laws and regulations, ensure that data is not used for other than normal, commercial use and disclosure, and ensure that legitimate users can query and monitor the information for other purposes.

Through analyzing the principle of RFID system, the security risks of RFID systems mainly exist in three aspects: the tag itself, the middleware between the reader and the back-end database for processing, and the back-end data processing system. This paper will introduce the various attacks on front-end data collection areas, middleware, back-end system back-end communication.

2 ATTACK ON FRONT-END DATA COLLECTION AREAS

2.1 *Tag encoding attack*

Due to the small size and low power RFID chip designed to reduce operational costs, it has also become a major RFID security flaws, for it does not have enough computing power to encrypt. System security is based on a cryptographic algorithm, in front of the encryption algorithm that has not been cracked, potential security attacks or clones that have the correct label could not be confirmed; if the attacker cracked the encryption algorithm, it could capture the tag's query and response signals to get the needed results (Arbit, 2015).

2.2 Tag application attack

2.2.1 Man-in-the-middle attack

The man-in-the-middle attack is capable of winning the trust of both the sides and then transferring information back in communication system so that both sides regard the attacker as the other side (Kato, 2014).

Most RFID system reads information right in the reader's distance. Readers read tag information for any intermediate steps that the user intervention is not required for. So when near RFID tags, it can be transferred the RFID tag reader frequency to read tag information, and communicate with the tag event.

2.2.2 Chip cloning attack

RFID technology is a form of contactless cards. RFID card itself provides access to energy or gets energy from RF reader. RFID RF reader is also contactless, when the tag is close to the reader, the reader queries and checks the database information from the background of the tags, and then the tag's ID is passed to the reader. In the case of no certification, if any device can send the correct code to the reader, it will be allowed to access the system (Calderoni, 2014). So if the people clone the tag, it could get the same permissions and lead to a lot of security issues. Avoid eavesdropping and limit the application permissions, log check, card encryption, which is a method for ensuring system security.

3 ATTACKS ON MIDDLEWARE

There are three parts of security risks of RFID middleware mainly: data transmission, authentication, and authorization management; so middleware attacks may occur in any aspects among the reader to the back-end data processing system.

3.1.1 Data transmission

When the RFID data is transmitted to all network layers, it will likely be a security risk that illegal intruders intercepted RFID tags, such as eavesdropping, tampering, and DDOS. The transmission interference signal of an illegal user would block communication link, which is overloaded with the reader and can not receive the tag data.

3.1.2 Authentication

Sometimes illegal users (such as industry competitors or hackers) use the middleware to obtain confidential information and business secrets, which will cause a great deal of risk to legitimate users. And an attacker can use the analog tag to send the data to the reader, so that the reader processes false

data, rather than the real data (hidden). Therefore, it is necessary to verify the authentication of tags.

3.1.3 Authorization management

Users who are not authorized may try to protect the RFID middleware services, so it must provide security service for the user. According to the different needs of the user, it must be limited to the legitimate scope of user rights. For instance, the business needs of different users of different industries, they are also different in the middleware functions, so they have no right to use each other's business functions.

4 BACK-END SYSTEM ATTACKS

Back-end systems are vulnerable, especially in the process of data, such as the back-end system is full of errors of data, read false data, copy the malicious duplicate tag, so the system will stop (Niu, 2014).

4.1.1 Data attack

RFID middleware collects RFID event data, and then transfers them to the back-end system, but the transmission of the data may cause security risks to the system, such as data overflow and false data.

Data overflow, if placed in a large number of readers, it will have a very large number of data to the back-end system, and then it may lead to data overflow.

False data, an attacker may make a false tag; it is similar to the forgery or copy of the credit card at the same time.

4.1.2 Buffer overflow attack

In fact, the tag has a unique ID. RFID reader can read and write data, and the data is connected to the system for further processing. So the design of the error tag data and back-end system can lead to security problems (Liu, 2015).

Middleware system is designed to receive data marked with a certain scale, and if the buffer size is not big enough, enough to store data, attackers use labeled data over the writing of the buffer size, it will lead to buffer overflow and achieve the goal of the attacker.

4.1.3 SQL injection attack

SQL injection is through the SQL command to insert the Web table to submit or enter the domain name or page request query string, and ultimately deceive the server to execute malicious SQL command (Hung, 2014). When the application uses the input to construct dynamic SQL statements to access the database, SQL injection attacks occur.

If the code uses a stored procedure, the stored procedure can be injected into a string that is not filtered by the user's input. SQL injection may cause an attacker to use an application to log in to the database for execution of commands. In the application, the high privilege account is connected to the database and the problem will become very serious. In a number of tables, the user's input is directly used to construct (or impact) the dynamics of the SQL commands or store the procedure input parameters. These tables are particularly vulnerable to SQL injection attacks. In code, a lot of programs do not need to judge the legitimacy of the user input variable or the variable itself, so that the application has a security risk. In this way, an attacker can submit database query code, and according to the results obtained by the return of the program, some of them may sensitize the information or control the entire server. So SQL injection attack may arise.

SQL is injected from normal WWW ports, whose access seemingly is with no difference between the ordinary Web page views, so SQL injection is not aware of the current availability of the firewall. If the administrator did not check the IIS log, the system could be invaded and not found for a long time. SQL injection technology is very flexible, when the injection will encounter many unexpected situations. Therefore, according to the analysis of the specific circumstances, the attacker should skillfully construct the SQL statement to successfully obtain the required data.

In the RFID system, an attacker may implant the attack data in the back-end middleware of the SQL database. The restriction of RFID tag data storage is not limited to this attack, so a few of the SQL command can cause a lot of risk, such as "ShutOff" command can immediately close the SQL server. As long as the database runs in the root directory, there must be SQL injection vulnerability.

4.2 Back-end communication attack

RFID systems typically use JMS, HTTP and other communication protocols between middleware and back-end systems. And the man-in-the-middle attack, application layer attack and TCP rebroadcast attack may occur in these protocols.

4.3 Man-in-the-middle attack

The man-in-the-middle attack existing back-end systems, to information query, that is, in the implementation of the communication middleware and back-end system, because both sides are unable to recognize who is this, so an attacker could forge proof of identity at the same time they exchange data. As the wireless network system has less physical control, it is more prone to the man-in-the-middle attack.

4.4 Application layer attack

Application layer attacks can cause server operating system errors or application program errors, so the attacker can access and control system. An attacker may use the application layer system to

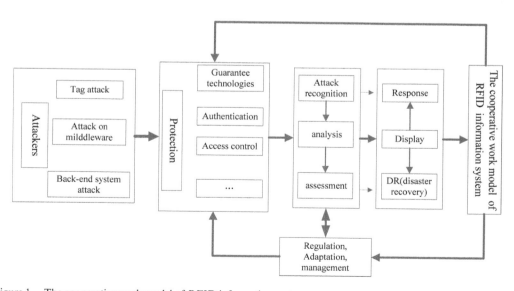

Figure 1. The cooperative work model of RFID information system.

easily obtain the control's application, system, or network (Niu, 2014).

4.5 *TCP rebroadcast attack*

TCP rebroadcast attacks is that the attackers use sniffer that can come from the data packets and data packets, then obtained from data software package and authentication information. After that, an attacker puts the packet back to the network, or rebroadcasts it (Benssalah, 2014).

5 COLLABORATIVE WORK MODEL

This paper proposed a cooperative work model of security and privacy protection for the security and privacy issues of RFID systems, which is shown in Figure 1. In this collaborative work model, which consists of the protection, detection, response recovery, and counter, these four process organizations constitute a more complete cooperative security model. In Figure 1, the protection part includes detection analysis and monitoring control; regulation of learning (knowledge accumulation), automatic adjustment of security policies (adaptation) and security management. Safety protection is to prevent and delay the attack that may be through a variety of security mechanisms and security measures. Response recovery is enabled to respond to attacks, where the attack is to be minimized the harm. The detection is used to identity and analyze the attack; the display is designed to monitor the protection, response, recovery, and regulatory. Scheduling and monitoring is the core to ensure that the key problem of the safe is a reliable operation of the network.

The design of the security and privacy of RFID information system actually is not a simple technical problem. It is related to organizational structure, security mechanism, disaster recovery, security protocol, security protection, and so on. Any aspect of a system will not necessarily lead to security problem. Therefore, the security of RFID public service platform is a system engineering, which needs to establish a flexible security system. So this is a dynamic and collaborative work process.

6 CONCLUSION

First, this paper classifies the information system based on RFID technology into three kinds of security domain, and then introduces the attack model, including attack on front-end data collection areas, attack on middleware and back end communication attack. Finally, based on these analyses, this paper proposes a cooperative work model of RFID system security mechanism, which is an effective way of information security.

REFERENCES

Alavi, Seyed Mohammad, et al. "Traceability Analysis of Recent RFID Authentication Protocols." Wireless Personal Communications (2015).

Alex Arbit, S et al. "Implementing public-key cryptography on passive RFID tags is practical." International Journal of Information Security 14.1(2015):85–99.

Ben Niu, et al. "Privacy and Authentication Protocol for Mobile RFID Systems." Wireless Personal Communications 77.3(2014):1713–1731.

Benssalah, Mustapha, M. Djeddou, and K. Drouiche. "Security enhancement of the authenticated RFID security mechanism based on chaotic maps." Security & Communication Networks 7.12(2014):2356–2372.

Calderoni L, Maio D. Cloning and tampering threats in e-Passports [J]. Expert Systems with Applications, 2014.

Hermans, J., R. Peeters, and B. Preneel. "Proper RFID Privacy: Model and Protocols." Mobile Computing IEEE Transactions on 13.12(2014):1–1.

Hung, Lun Ping, T.T. Lu, and C.L. Chen. "Application of Active RFID Technology to Build an Infection Control System in Seniors Villages." Lecture Notes in Electrical Engineering (2014).

Kato, Zoltan. "Knowledge And Acceptance Of RFID Traceability Solutions." SEA—Practical Application of Science (2014):595–602.

Liu, Zilong, et al. "Implementation of a New RFID Authentication Protocol for EPC Gen2 Standard." Sensors Journal IEEE 15(2015).

Niu, Ben, et al. "Privacy and Authentication Protocol for Mobile RFID Systems." Wireless Personal Communications 77.3(2014):1713–1731.

Advances in Energy, Environment and Materials Science – Wang & Zhao (Eds)
© 2016 Taylor & Francis Group, London, ISBN 978-1-138-02931-6

Wear feature extraction for diamond abrasives based on image processing techniques

Y.F. Lin & F. Wu
Xiamen Institute of Technology, Xiamen, Fujian, China

C.F. Fang
College of Mechanical Engineering and Automation, Huaqiao University, Xiame, Fujian, China

ABSTRACT: Diamond wear morphology has a heavy influence on tool life besides workpiece quality and machining efficiency, which statistical results have a great significance to while revealing machining mechanism. For the complexity of diamond abrasive wear during machining, few previous studies were focused on diamond wear features except diamond measurement and 3D reconstruction with image processing methods. Therefore, wear feature extraction of diamond abrasives was studied using multiple algorithms of Snake model and Mathematical morphology in the present paper. The results show that the global contour can be obtained by Snake model, but the local contour only be precisely obtained comprehensively using Snake model and Mathematical morphology; the big difference is mainly due to the complexity of diamond abrasive wear regions. Combining the global and local contours, wear features of diamond abrasives can be better described, and it can be further used to automatic image identification for diamond abrasive wear with the purpose of improving efficiency and evaluation accuracy compared with traditional manual sampling observation method.

1 INTRODUCTION

Image processing technology is an important high and new technology, which is widely used in machinery, medical, industrial testing, aviation remote sensing, artificial intelligence, and many other fields (Li et al., 1995; Atkociunas, 2005; Zhang et al., 2009). In grinding and sawing, tools used are often made with diamond abrasives. Diamond abrasives in tools will gradually wear during machining process, and it will reduce tool life and processing efficiency, it will also seriously reduce workpiece surface quality and even lead to loss tool machining performance. Some early studies indicate that tool wear is directly related to diamond wear morphology (Luo, 1996). Therefore, it is of obvious significance to reveal diamond abrasive wear, which will help to deeply understand the mechanism of diamond wear and guide machining process such as tool optimal design, processing parameters selection and etc.

In view of diamond abrasive wear in sawing, the essence of wear failure was expected to be revealed by observing the wear of diamond abrasives. Some early studies were carried on the classification of diamond wear form, such as broken, flat and fall-off forms of diamond wear, which is based on the purpose of better describing the machining process (Bailey, 1979). However, those studies were carried based on tracking a certain amount of diamond abrasives with microscope. With microscopic observation and qualitative evaluation, the wear characteristics of diamond abrasives were finally obtained based on the classification of diamond wear form. Although, the wear characteristics of diamond abrasives can be obtained by manual sampling observation method, it has the shortcomings of low efficiency of artificial observation, less simple size and great error of estimation, especially low comparative between different results.

Recently, many researches has been carried out on measurement, evaluation and 3D reconstruction of diamond abrasives in grinding and sawing based on image processing technology (Huang et al., 2010; Cui et al., 2013). However, few studies have been focused on diamond abrasive wear in machining due to the complexity of wear state. In the present paper, multiple methods of image processing are adopted in the attempt to analyze diamond abrasive wear, which will give guidance to automatic image identification of diamond abrasive wear.

2 ABRASIVE WEAR MORPHOLOGY

In sawing, diamond abrasives gradually wear due to the sawing force impact, sawing thermal impact, sawing swarf erosion and etc. The main wear morphology forms of diamond abrasives include whole form, broken form, and flat form and fall-off form (Luo et al., 1995, 1997). The greatest complexity of diamond abrasive wear occurs in sawing when it is broken into flat forms, which is shown in Figure 1.

As shown in Figure 1, diamond abrasives located in the tool has two features: one is the global contour which reflects the boundary between diamond abrasive and tool matrix; the other is the local contour, which reflects wear degree on diamond abrasive surface. In different diamond wear states, the local contour is much different shown in Figure 1. The big difference of global and local contours is illustrated in Figure 2.

As shown in Figure 2, the global contour of single diamond abrasive is the boundary between diamond abrasive and tool matrix, which is mainly dependent on the shape of diamond abrasive. But the local contour is the boundary between wear and non-wear areas on diamond abrasive surface, which is mainly dependent on the

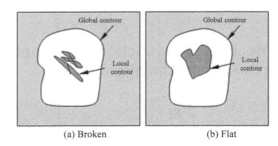

(a) Broken   (b) Flat

Figure 2. Illustrations of different local contours for diamond wear.

machining process. Therefore, the main challenge is the wear extraction of local contour. Taking the complexity of wear states, multiple methods of image processing technology must be adopted and studied in order to better extract wear contours in the present paper.

3 METHODS OF ABRASIVE WEAR FEATURE EXTRACTION

3.1 Global contour extraction

In order to analyze the diamond wear feature, the global contour of diamond abrasive must be first recognized and extracted. The typical method used in contour extraction is the Snake model (Kass et al., 1987), which uses a set of control points to describe the contour and the model that can be given as:

$$v(s) = [x(s), y(s)] \quad s \in [0, 1] \quad (1)$$

where $x(s)$ and $y(s)$ are the coordinate positions of each of the control points in the image. s is the boundary variable based on the Fourier transform.

A target energy function is usually defined in typical Snake model in order to prevent falling into the non-expected local minimum energy and accelerate convergence. The energy function for control points is defined as:

$$E_{snake}(v) = \frac{1}{2} \int \Omega \left[E_{int}(v) + E_{image}(v) + E_{con}(v) \right] ds \quad (2)$$

where E_{int} is the inner energy, E_{image} is the image energy, and E_{con} is the additional energy.

The inner energy E_{int} and image energy E_{image} can be, respectively, defined as:

$$E_{int} = \alpha(s)|v_s(s)|^2 + \beta(s)|v_{ss}(s)|^2 \quad (3)$$

(a) Broken

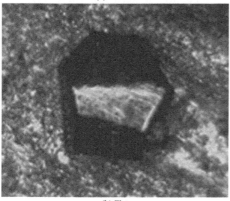

(b) Flat

Figure 1. Typical abrasive wear morphology.

$$E_{image} = \omega_{gray}E_{gray} + \omega_{edge}E_{edge} + \omega_{line}E_{line} \qquad (4)$$

where ω_{gray}, ω_{edge} and ω_{line} is the coefficient related to position in the image. E_{gray} is the image gray energy, E_{edge} is the image edge energy, and E_{line} is the image line energy.

3.2 *Local contour recognition*

With the Snake model, the global contour of diamond abrasive can be extracted, but it cannot be directly used in the local contour extraction of diamond abrasive within the global contour, which is due to the complexity of diamond wear states and the limitation of initial position in the Snake model. In order to overcome the above problem, mathematical morphology is adopted in the present paper.

The basic ideals of mathematical morphology is to analyze and recognize images by using a certain shape of structure elements to measure and extract the corresponding shape in images. Mathematical morphology has four basic operations, which is expansion, corrosion, opening, and closing. Different algorithms can be obtained through different combinations of basic operations. A new combination form is adopted in order to better extract the local contour, and it can be written as:

$$\beta(A) = A - (A \ominus B) + A \circ B \qquad (5)$$

where A is the target image, B is the structure element image, and β is the boundary set for A. $\circ$ is the opening operation.

Different diamond wear states contain different non-connected regions, and then connected components are adopted here in order to recognize and identify the local contour. Connected components can be described as:

$$X_k = (X_{k-1} \oplus B) \cap A \qquad k = 1, 2, 3, \dots \qquad (6)$$

where X is the array with the same size of A and k is the iteration number.

In consideration of many connected components in processed image when using mathematical morphology, the contour is very messy; this makes it unclear for us to better obtain contours. Therefore, watershed image segmentation method is adopted, and it can be written as:

$$g(x, y) = grad\,(f(x, y))$$
$$= \{[f(x, y) - f(x-1, y)]^2 [f(x, y) - f(x, y-1)]^2\}^{0.5} \qquad (7)$$

where $f(x, y)$ is the original image and $grad$ {.} is the gray operation.

4 RESULT AND DISCUSSION

Based on the above methods, diamond abrasives with broken and flat forms shown in Figure 1 were processed after pretreatment of gray transform, filtering and threshold processing, respectively. The results were processed with Snake model shown in Figure 3.

As shown in Figure 3, the global and local contours for flat state of diamond abrasive are clearer than that of broken state of diamond abrasive, which indicates that more wear features are there in broken abrasive. Compared to the processed results of broken and flat abrasives, it can be found that the global contour for both wear states of broken and flat can be better extracted than that of the local contour.

In order to well extract wear features on diamond abrasive surface, it is not suitable to extract wear feature just through using the Snake model. Therefore, mathematical morphology was used to process the wear images, and the results obtained, using mathematical morphology, are shown in Figure 4.

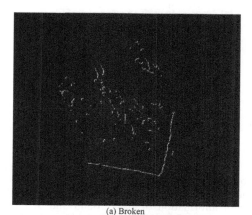

(a) Broken

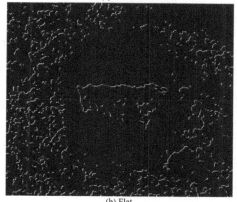

(b) Flat

Figure 3. Processed results with Snake model.

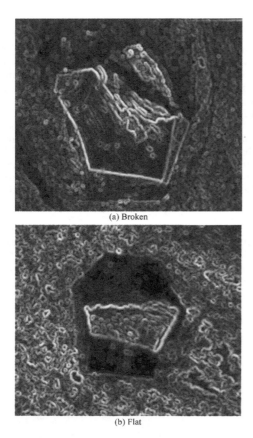

(a) Broken

(b) Flat

Figure 4. Processed results with Snake model.

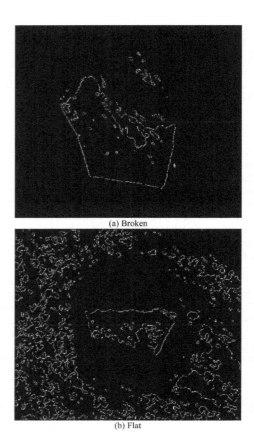

(a) Broken

(b) Flat

Figure 5. Processed results with Snake model and mathematical morphology.

As shown in Figure 4, the global and local contours for both broken and flat states of diamond abrasives can be easily obtained, but the detail information of background and diamond surface wear features in images is missed. Therefore, it is also not suitable to extract wear features just through using mathematical morphology.

Compared with Figure 3 and 4, it can be found that the detail wear features of diamond abrasives can be better obtained by the Snake model except the contour, but the contour can be better obtained by mathematical morphology, except the detail wear features. Therefore, the method of combining the Snake model with mathematical morphology was put forward, with the purpose of using their, respective, advantages in extracting contour and detail wear features.

The results, by using the combining method, are shown in Figure 5. From Figure 5, it can be seen that the global and local contours are both well obtained, which can reflect the shape of diamond abrasives and detail wear features on diamond abrasives surface.

However, the global and local contours are not continuous, which is the disadvantage of using the Snake model and mathematical morphology. It will be bad for further analysis. In order to overcome the shortcomings of the combining method, watershed image segmentation method was adopted and used to process the image. The results are shown in Figure 6.

As shown in Figure 6, the global and local contours for diamond abrasives are well obtained, the global contour, which reflects the boundary between diamond abrasive and tool matrix is clearly extracted. In addition, the local contour, which reflects the wear features caused during machining, is also clearly extracted. The results indicate that wear features can be well extracted based on the methods put forward in the present paper.

Based on above analysis, it can be seen that the detail wear features on diamond abrasive surface cannot be well extracted by adopting the Snake model or mathematical morphology due to the complex wear states on diamond abrasive surface. With the purpose of improving efficiency and evaluation

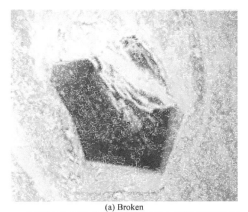

(a) Broken

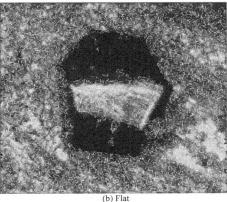

(b) Flat

Figure 6. Processed results with watershed image segmentation method based on Figure 5.

accuracy compared with traditional manual sampling observation method, the method of combining the Snake model with mathematical morphology put forward in the present paper can be further used to analyze the wear features of diamond abrasive, such as wear state recognition and wear degree of diamond abrasives, which is of significance for automatic image recognition and analysis.

5 CONCLUSION

The statistical results of diamond wear morphology are of obvious significance as they deeply reveal and understand the machining mechanism for sawing process. The global and local contours of diamond abrasives can be well extracted through by using the watershed image segmentation method based on the Snake model and mathematical morphology; the processed results can be further used to recognize the wear state and wear degree of diamond abrasives.

The original images of wear diamond abrasives processed just by the Snake model or mathematical morphology has the shortcomings of unclear overall contours or missing detail information of background and wear features on diamond abrasive surface in images. Through combining the Snake model with mathematical morphology, the global and local contour can be better obtained, but the contours are discontinuous in general. After further using the watershed image segmentation method, the contours can be well extracted, and detail wear features on diamond abrasive surfaces can be well presented.

Further work will be carried out on automatic wear forms recognition and wear degree analysis.

ACKNOWLEDGMENTS

The research was supported by the Project of Fujian Provincial Department of Education, China (Grant No. JA13465) and the National Natural Science Foundation of China (Grant No. 51305145).

REFERENCES

Atkociunas, E., Blake, R. & Juozapavicius, A. 2005. Image processing in road traffic analysis. *Nonlinear Analysis: Modeling and Contro* 10(4): 315–332.
Bailey, M.W. 1979. Sawing in the stone and civil engineering industries. *Industrial Diamond Review* 39(2): 56–60.
Cui, C.C., Xu, X.P., Huang, H., Hu, J., Ye, R.F., Zhou, L.J. & Huang, C.H. 2013. Extraction of the grains topography form grinding wheels. *Measurement* 46: 484–490.
Huang, S.G., Duan, N., Huang, H. & Xu, X.P. 2010. Reconstructing the topography of grinding wheel surface by image mosaic method. *Diamond & Abrasives Engineering* 6(30): 29–32.
Kass, M., Witkin, A. & Terzopoulos, D.S. 1987. Active Contour Models. *International Journal of Computer Vision* 321–331.
Li, H., Manjunath, B. & Mitra, S.K. 1995. A contour-based approach tomultisensor image registration. *IEEE Transactions on Image Processing* 4(3): 320–334.
Luo, S.Y. 1996. Characteristics of diamond sawblade wear in sawing. *International Jounal of Machine Tools & Manufacture* 36(6): 661–672.
Luo, S.Y. 1997. Investigation of the worn surfaces of diamond sawblades in sawing granite. *Journal of Materials Processing Technology* 70: 1–8.
Luo, S.Y. & Liao, Y.S. 1995. Study of the behaviour of diamond saw-blades in stone processing. *Journal of Materials Processing Technology* 51: 296–308.
Zhang, L.P., Li, G.Y. & Bao, Y. 2009. The improvement of tissue contour extraction method in medical image. Canada: Acta Press.

Detection the deviation from process target using VP loss chart

Chung-Ming Yang
Ling Tung University, Taiwan, R.O.C.

Su-Fen Yang
National Chengchi University, Taiwan, R.O.C.

ABSTRACT: A single chart, instead of $\overline{X}$-bar and R charts or $\overline{X}$-bar and S charts, to monitor simultaneously the process mean and variability if found would cut down the time and effort. Some researches have been done in finding such charts. Moreover, some researches have showed that the adaptive control charts could detect process shifts faster than the traditional Shewhart charts. A much easier average loss chart with Variable design Parameters (VP) is proposed here to detect the deviation from the process target and the increase in the process variability. Furthermore, a more efficient optimal VP average loss chart is proposed and performs better than the average loss chart with fixed design parameters and the Shewhart $\overline{X}$-bar and S^2 charts. A numerical example is given to illustrate the findings.

1 INTRODUCTION

In industry, quality of products and loss of productivities are important factors among competing companies. Loss function is widely used in industry to measure the loss due to poor quality (Spiring and Yeung (1998)). From Taguchi's philosophy (Gopalakrishnan, Jaraiedi, Iskander and Ahmad (2007)), the target value is a vital process measurement. Sullivan (1984) gave examples that stress the importance of monitoring the target value and variability. In practice, the in-control process mean may not be the process target. Hence, a single control chart is effective if it could detect the deviation from the process target and the shift in process variability with smallest loss. So far, no single Variable design Parameters (VP) control chart detecting both the deviation from the process target and the increases in process variability has ever been proposed. A single VP average loss chart is thus proposed to improve the performance of the single average loss chart with Fixed design Parameters (FP), and/or the Shewhart $\overline{X} - S^2$ charts. In section 2, a Taguchi loss function is introduced and the distribution of its average loss is derived. The FP and VP average loss charts are constructed in section 3. The performance measurement of the VP average loss chart is described in section 4. Finally, the application of the optimal VP average loss chart is illustrated by an example and the data analysis compares the performance among the VP, FP loss charts and the Shewhart $\overline{X} - S^2$ charts.

2 THE DISTRIBUTION OF TAGUCHI LOSS FUNCTION

Taguchi quadratic loss function is defined as $AL = k'E(X - T)^2$, where AL = loss, k' = coefficient of the loss, X = the quality variable of interest and T = the process target.

Let $X \sim N(\mu, \sigma^2)$ when the process is in-control. When μ and σ^2 are unknown, the average loss AL is estimated by $L = \sum_{i=1}^{n}(X_i - T)^2/n$, where n = the sample size. Statistic L is an unbiased estimator of the average loss; its statistical property is superior to the weighted loss function in Wu and Tian (2006), Zhang and Wu (2006), and Wu, Wang, and Wang (2009).

The S^2 and $\overline{X}$ are independent under normality assumption, the in-control distribution of the statistic L is derived as a non-central chi-square distribution with n degrees of freedom and non-centrality λ_0.

3 DESIGN OF THE LOSS CHART

A FP loss chart have fixed sample size (n_0), fixed sampling interval (t_0) and a fixed control factor (k_0) of upper control limit determined by a fixed false alarm rate (α_0). Based on the in-control distribution of the statistic L the Upper Control Limit (UCL) and the Lower Control Limit (LCL) of the FP loss chart are $UCL = k_0\sigma^2$, $LCL = 0$, where k_0 is the critical value of a non-central chi-square distribution with n degrees of freedom, non-centrality λ_0 and cumulative probability $1 - \alpha_0$.

To quickly detect the change in deviation from the process target and increase in variance simultaneously, a VP loss chart is designed. The design parameters—the sample size (n_q), the sampling interval (t_q), the warning factor (w_q) and the control factor (k_q) determined by the false alarm rate (α_q) are all variables. The limits of the VP loss chart are: $UCL = k_q\sigma^2$, $UWL = w_q\sigma^2$ and $LCL = 0$, where (w_q) is the warning factor of Upper Warning Limit (UWL), $0 < w_q < k_q$. The region between LCL and UWL is called the 'Central Region' (CR), that between UWL and UCL the 'Warning Region' (WR) and that above UCL the 'Action Region' (AR). Two variable combinations of the design parameters, (n_1, t_1, w_1, k_1) and (n_2, t_2, w_2, k_2), are adopted. The small sample size (n_1), long sampling interval (t_1), the corresponding warning factor (w_1) and the corresponding control factor (k_1) determined by the corresponding false alarm rate (α_1) are adopted when the plotted point fell into the CR; the large sample size (n_2), the short sampling interval (t_2), the corresponding warning factor (w_2) and the corresponding control factor (k_2) determined by the corresponding false alarm rate (α_2) are adopted when the plotted point fell into the WR, where $n_1 < n_2$, $t_2 < t_1$, $w_1 < k_1$ and $w_2 < k_2$. When the plotted point fell into the AR, the process was stopped to search, diagnose, and eliminate the special cause.

Two variable combinations of the design parameters, (n_1, t_1, w_1, k_1) and (n_2, t_2, w_2, k_2), are adopted. The small sample size (n_1), long sampling interval (t_1), the corresponding warning factor (w_1) and the corresponding control factor (k_1) determined by the corresponding false alarm rate (α_1) are adopted when the plotted point fell into the central region; the large sample size (n_2), the short sampling interval (t_2), the corresponding warning factor (w_2) and the corresponding control factor (k_2) determined by the corresponding false alarm rate (α_2) are adopted when the plotted point fell into the warning region, where $n_1 < n_2$, $t_2 < t_1$, $w_1 < k_1$ and $w_2 < k_2$. When the plotted point fell into the action region, the process was stopped to search, diagnose, and eliminate the special cause.

The proposed VP loss chart may detect the out-of-control process. When $k_1 = k_2 = k_0$ the VP loss chart is reduced to a VSSI loss chart; while $n_1 = n_2 = n_0$, $t_2 < t_1$, $w_q \neq 0$ and $k_1 = k_2 = k_0$ the VP loss chart is reduced to a VSI loss chart. And when $n_1 < n_2$, $t_1 = t_2 = t_0$ and $w_q \neq 0$ the VP loss chart is reduced to a VSS loss chart; while $n_1 = n_2 = n_0$, $t_1 = t_2 = t_0$, $w_1 = w_2 = 0$ and $k_1 = k_2 = k_0$ the VP loss chart is reduced to a FP loss chart.

4 PERFORMANCE MEASUREMENT OF THE VP LOSS CHART

The speed with which a control chart detects process shifts measures the chart's statistical efficiency. For a VP chart, the detection speed is measured by the average time from out-of-control process until loss chart signals, which is known as the Adjusted Average Time to Signal (AATS). The larger AATS indicates the better detection performance.

Assuming the occurrence time until the special cause is an exponential distribution with mean $1/\gamma_1$. That is, $T_{SC} \sim \exp(-\gamma_1 t)$, $t > 0$.

Hence, AATS = ATC $- 1/\gamma_1$.

The Average Time of the Cycle (ATC) is the average time from the start of the in-control process until a true signal is obtained from the proposed charts (see Duncan (1956)). The Markov chain approach is used to compute the ATC using the memory-less property of the exponential distribution. Thus, at the end of each sampling, one of the 6 states is assigned based on whether the process step is in- or out-of-control and the position of samples (see Table 1 for the 6 possible states of the process). The status of the process when the ($i + 1$)th sample is taken and the position of the ith samples on the loss chart define the transition states of the Markov chain. The loss chart produces a signal when a sample fell in the AR. If the current state is any one of the States 1, 2, 4 and 5, then there is no signal. If the current state is State 3, it indicates that a false signal came from the process and that the process instantly becomes either State 1 with probability p_{01} or State 2 with probability $1 - p_{01}$. If the current state is State 6, it indicates that a true signal occurs and that the process stops to be adjusted and brought back to an in-control state. If the current state is any one of the States 1–5, then it may transit to any other state, hence States 1–5 are transient states. The absorbing state (State 6) is reached when a true signal occurs.

When the quality engineers are unable to determine the specific combinations (n_1, t_1, w_1, k_1) and

Table 1. Possible states of the process.

State	Does SC occurs	In which region	Process is adjusted?	Transient or absorbing state
1	No	Central	No	Transient
2	No	Warning	No	Transient
3	No	Action	No	Transient
4	Yes	Central	No	Transient
5	Yes	Warning	No	Transient
6	Yes	Action	Yes	Absorbing

(n_2, t_2, w_2, k_2), the optimal VP loss chart with minimal AATS and the optimal values of (n_1, t_1, w_1, k_1) and (n_2, t_2, w_2, k_2) is recommended.

The procedure in determining the optimal VP L chart is described below:

Step 1: Specify the sample size, sampling interval and false alarm rate (n_0, t_0, α_0) of the FP L chart and the values of $\alpha_1, \delta_1, \delta_2, \delta_3, \gamma_1, T_r$ and T_f.

Step 2: Determine the available range of the variable false alarm rates, variable sample sizes and variable sampling intervals of the VP L chart.

Step 3: The optimal VP's $((n_1, t_1, w_1, k_1)$ and $(n_2, t_2, w_2, k_2))$ are determined using optimization technique, Fortran IMSL BCONF subroutine, to minimize AATS under the constraints in Step 2 and parameters as described in Step 1.

Step 4: Using the optimal parameters (n_1, t_1, w_1, k_1), the wide L chart is constructed; using the optimal parameters (n_2, t_2, w_2, k_2), the narrow L chart is constructed.

5   AN EXAMPLE

Consider 35 samples from duplicate measurements of Braverman (1981). Let the quality variable X follow a normal distribution. One machine could only fail in the process and shift the mean and variance of X distribution. Presently, the joint FP loss chart is used to monitor the deviation from the process target and the increase in process variance every hour ($t_0 = 1$). When the control chart indicates that the process is out-of-control, it requires adjustment. To construct the control chart, 35 samples of size $n_0 = 4$ are taken. The estimated in-control process mean and standard deviation are 9.73 and 48.42, respectively. The target value of X is 7.46, T = 7.46. Consequently, $\delta_3 = 0.326$. From historical data, the estimated failure frequency of the machine is 0.05 times per hour (or $\gamma_1 = 0.05$). The failure machine would shift the mean and variance to $N(\mu + \delta_1\sigma, \delta_2^2\sigma^2)$ with $\delta_1 = 0.5$ and $\delta_2 = 1.1$. The average correct adjustment time of the process is 0.5 h (or $T_r = 0.5$) when a true signal occurs. The time to search a false alarm on the process is $T_f = 0.1$ h. To construct the optimal VP loss chart we set $n_L = 2, n_U = 30, t_L = 0.1, t_U = 2, \alpha_U = 0.0059$ and $\alpha_L = 0.0027$. Using Fortran IMSL BCONF subroutine to minimize AATS under the constraints, the determined optimal VPs are $(n_1 = 2, t_1 = 1.14, w_1 = 4.14, k_1 = 12.58)$ and $(n_2 = 17, t_2 = 0.12, w_2 = 1.59, k_2 = 2.14)$, and the minimum AATS = 5.85 h. Hence, the optimum VP loss chart with wide control limits are constructed as $UCL_1 = 609.12, UWL_1 = 213.53, LCL_1 = 0$; the optimum VP loss charts with narrow control limits are constructed as $UCL_2 = 103.62$,

$UWL_2 = 76.99, LCL_2 = 0$. The AATS of the VP loss chart is 5.85.

The FP loss chart has control limits UCL = 200.46 and LCL = 0 with average run length = 185 (or false alarm rate = $\alpha_0 = 0.0054$) when $T_f = T_r = 0$, and AATS = 15.57. The AATS of the VP loss chart represents a saving time of 62.41% in out-of-control process detection.

The strategy of the process monitoring is

1. The combination of the variable design parameters, $(n_1 = 2, t_1 = 1.14, w_1 = 4.14, k_1 = 12.58)$ or $(n_2 = 17, t_2 = 0.12, w_2 = 1.59, k_2 = 2.14)$, is determined randomly when the process starts.
2. When the statistic L fell into CR, the combination $(n_1 = 2, t_1 = 1.14, w_1 = 4.14, k_1 = 12.58)$ is adopted next time.
3. When the statistic L fell into WR, the combination $(n_2 = 17, t_2 = 0.12, w_2 = 1.59, k_2 = 2.14)$ is adopted next time.
4. When the statistic L fell into AR, the process stops to search and discard the assignable cause(s) and bring the process back into control.

6   CONCLUSIONS

The proposed VP scheme to monitor the deviation from the process target and increase in variance using a single chart improves the performance of the FP scheme by increasing the speed with small to medium shifts in the mean and variance. Furthermore, the optimum VP loss chart always works better (in the cases examined) than the FP loss chart and the Shewhart $\bar{X} - S^2$ charts. Hence, the VP loss chart is recommended.

We recommend to use the optimum VP scheme in monitoring a process when quality engineers were unable to specify the VPs. This paper proposed a Shewhart-type single chart. However, the studies of the variables control scheme could be extended to study CUSUM/EWMA charts or the multiple steps of cascade processes.

REFERENCES

Braverman, J.D. 1981. *Fundamentals of Statistical Quality Control*, Reston Publishing Co., Inc., pp. 144.

Duncan, A.J. 1956. The economic design of charts used to maintain current control of a process. *Journal of American Statistical Association* 51: 228–242.

Gopalakrishnan, B., Jaraiedi, M., Iskander, W.H. and Ahmad, A. 2007. Tolerance synthesis based on Taguchi philosophy. *International Journal of Industrial and Systems Engineering* 2(3): 311–326.

Spring, F.A. and Yeung, A.S. 1998. A general class of loss functions with individual applications. *J. Qual. Technol.* 30:152–162.

Sullivan, L.P. July 1984. Reducing variability: A new approach to quality. *Quality Progress*, 15–21.

Wu, Z. and Y. Tian. 2006. Weighted-loss-function control charts. *International Journal of Advanced Manufacturing Technology* 31:107–115.

Wu, Z., Wang, P. and Wang, Q. 2009. A loss function-based adaptive control chart for monitoring the process mean and variance. *Int. J. Adv. Manuf. Technol.* 40: 948–959.

Zhang, S. and Wu, Z. 2006. Monitoring the process mean and variance using a weighted loss function CUSUM scheme with variable sampling intervals. *IIE Transactions* 38(4): 377–387.

Advances in Energy, Environment and Materials Science – Wang & Zhao (Eds)
© *2016 Taylor & Francis Group, London, ISBN 978-1-138-02931-6*

Optical rotation calculations on chiral compounds with HF and DFT methods

L.R. Nie, J. Yu, X.X. Dang, W. Zhang, S. Yao & H. Song
Department of Pharmaceutical and Biological Engineering, Sichuan University, Chengdu, P.R. China

ABSTRACT: The *ab initio* Hartree-Fock (HF) and the Density Functional Theory (DFT) methods were used to calculate the optical rotation of 24 chiral compounds. Their structures were optimized using DFT method. Then the frequency analysis was performed to obtain chemical thermodynamic data of molecules, which provided theoretical data for the latter calculations of the optical rotation. The calculated results indicated that the deviations of six compounds in eight heterocyclic compounds were less than 10° compared with experimental data. But, the calculation results were not satisfactory for benzene compounds and straight-chain compounds. Thus, it can be seen that HF and DFT methods were more suitable for optical rotation calculations of heterocyclic compounds and could give acceptable predictions.

1 INTRODUCTION

Optical rotation is one of important physical properties to chiral compounds. There has been increasing interest in the relationships between molecular absolute configuration and its optical rotation. Most scholars believe that there is a certain connection between molecular absolute configuration and its optical rotation (Yun 1990; Zhang 1990). Related researches were limited because of the limitations of research methods, and most calculations were concentrated on small and rigid molecules for accuracy. Ruud et al. (Ruud and Helgaker 2002) calculated the optical rotation of small molecules: H_2O_2 and H_2S_2. Until recent years, calculations of optical rotation have been improved significantly with the development of quantum chemistry and algorithms. More calculations were directed to large and flexible molecules or natural product molecules.

Nowadays, the calculation of optical rotation with quantum chemical can be achieved through a variety of methods. Such as Coupled Cluster (CC) theory, Complete Active Space Self-Consistent Filed (CASSCF) theory, Density Functional Theory (DFT) and Hartree-Fock (HF) theory. CC theory and CASSCF theories have not been popularly used in the calculation of optical rotation after they were first reported by Ruud and Polavarapu et al., respectively. DFT is widely applied for the calculation of optical rotation, and the first *ab initio* calculation of optical rotation was reported by Polavarapu and coworkers (Polavarapu 1997) in 1997 and subsequently more attention has been paid to its application. Compared with the DFT,

the accuracy of HF theory is slightly worse because it ignores the electronic effects. However, the calculation results of HF theory can be used to evaluate the accuracy of the other calculation methods.

Optical activity and stereochemistry of compounds have been the focus of attention in drug synthesis. Previous studies for optical rotation mainly focused on small and rigid compounds. Few literatures have reported and compared the optical rotations of a series of compounds. In this paper, our focus was committed to investigate the structural features of the compounds, which are suitable for the optical rotation calculations with HF or DFT method. DFT was used to optimize the structures of chiral compounds which were often chosen as raw materials for the pharmaceutical synthesis. The frequency analysis was calculated to obtain the energy and chemical thermodynamic data of molecules. In addition, the optical rotation of 24 molecules was calculated by HF and DFT methods.

2 EXPERIMENTAL SECTION

2.1 *Optimization of the molecular structure*

The 3D structures of these compounds were obtained from the Cambridge Crystallographic Data Centre (CCDC). All molecules were optimized by the DFT method at the B3-LYP/6-31G (d) level (Yang 2001; Hodgson et al. 2010; Tsang et al. 2012). Moreover, frequency calculations of these molecules were performed on the basis of the optimization. The same calculation method and basis set were used in the frequency analysis. More information about molecules, such as IR spectra, Raman spectra and thermodynamic data were obtained from frequency calculations. The errors between the calculated values and experimental values are inevitable, and the accuracy can be improved through error correction by using the correction factor (0.9613) of B3 LYP/6-31G (d). Finally, the relationship between these thermodynamic data and the experimental values of optical rotation of the compounds was also discussed.

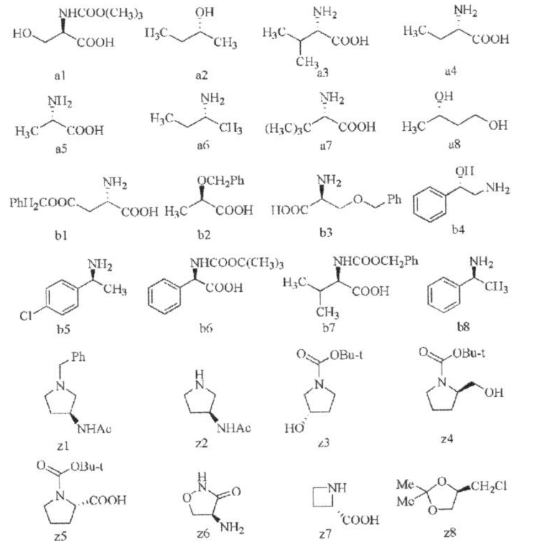

2.2 Calculation of optical rotation

The optical rotation was calculated by the HF and DFT methods. Two exchange-correlation functional, B3-LYP (hybrid three-parameter Becke-Lee-Yang-Parr functional) and B3PW91 (hybrid three-parameter Becke-Perdew-Wang91 functional) were applied in the DFT method. In addition, three basis sets were used in both the HF and DFT methods: 6-31G (d), 6-311+G (d) and 6-311++G (3df, 2pd). In order to characterize the accuracy of the two methods, it is necessary to choose sufficient compounds. In this study, 24 chiral compounds, including eight straight-chain compounds, eight benzene compounds and eight heterocyclic compounds were studied. Their structures are detailed in Figure 1 and the values of their experimental optical rotation are listed in Table 1. All calculations were performed using the Gaussian 03 program in this investigation.

Figure 1.   Structures of the three kinds of compounds.

Table 1.   Optical rotation of three kinds of compounds and determination conditions.

Compound	Configuration	t/°C	c/g·mL^{-1}	Solvent	$[\alpha]_D$/dm^{-1} g^{-1} cm^3
a1	R	20	2	Acetic acid	3.5
a2	R	20	10	Methanol	−13
a3	S	20	5	5 mol/LHCl	27.6
a4	S	20	4.8	6 mol/LHCl	20.5
a5	S	20	5	5 mol/LHCl	14
a6	R	20	–	–	7.5
a7	R	20	5	5 mol/LHCl	7
a8	S	20	–	–	30
b1	S	20	1	1 mol/LHCl	27.5
b2	R	20	1	IMS	90
b3	R	20	1	1 mol/LHCl	7
b4	S	20	2	Ethanol	42
b5	S	20	–	–	−31
b6	R	20	1	Ethanol	−144
b7	R	20	2	Acetic acid	3.8
b8	R	20	10	Ethanol	−30
z1	S	20	1.00	Chloroform	19
z2	S	20	1	Methanol	46.5
z3	S	27	1.02	Chloroform	22.75
z4	R	20	3.0	Chloroform	44
z5	S	24	1.40	Chloroform	72.7
z6	S	22	–	Water	115
z7	S	20	3.6	Water	118
z8	R	25	5.03	Benzene	35.9

3 RESULTS AND DISCUSSION

3.1 Optimization results

Figure 2 illustrated the 3D structure of the compound **z3** before (left) and after (right) optimization calculation. As can be seen, the imperceptible changes in the aspects of bond lengths and bond angles were observed. In order to investigate these changes more clearly, these data were listed in Table 2.

From the Table 2, the bond length changed slightly after optimization calculation and only the change of bond length of 28O–29H reached 0.12915 Å. Moreover, only the change of one in six bond angles around the chiral C atom was above 1°.

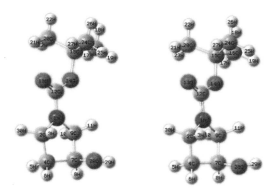

Figure 2. Change of the structure of the compound before (left) and after (right) optimization calculation.

The fact indicated that the compound structure obtained from CCDC was relatively close to the lowest energy conformation. So, the lowest energy conformation could be found through slight adjustment for the structure after optimization.

3.2 Frequency analysis

The investigation was focus on discussing the relationship between these thermodynamic data and the experimental values of optical rotation of the compounds. Thermodynamic data were obtained from frequency calculations. In order to understand more clearly the relationship between thermodynamic data and experimental values of optical rotation, the graphs were plotted with thermodynamic data versus the experimental values of optical rotation, and the fitting plots of heterocyclic compounds were obtained. The results were shown in Figure 3.

According to the results, for heterocyclic compounds, it can be seen that the zero-point energy, internal energy, heat capacity at constant volume and entropy presented an ascending trend, while electronic energy and heat energy had obvious trend of decline with the increase of experiment values of optical rotation. For benzene compounds and straight-chain compounds, there was no significant change for their thermodynamic data. In addition, the electronic energy and heat energy of straight-chain compounds almost were above the heterocyclic compounds, while the zero-point energy, internal energy, heat capacity at constant volume and entropy showed the

Table 2. Change of the bond length and bond angle of compounds before and after optimization calculation.

Bond length/ bond angle	Before optimization	After optimization	Change before and after optimization
12C–14O	1.33963 Å	1.36181 Å	0.02218
12C=13O	1.22676 Å	1.22333 Å	0.00343
1N–9C	1.47369 Å	1.46266 Å	0.01103
9C–7C	1.52997 Å	1.54017 Å	0.01020
7C–4C	1.52186 Å	1.53259 Å	0.01073
4C–2C	1.53814 Å	1.53538 Å	0.00276
2C–1N	1.46262 Å	1.46657 Å	0.00395
7C–28O	1.41871 Å	1.42606 Å	0.00735
28O–29H	0.84061 Å	0.96976 Å	0.12915
7C–8H	1.00018 Å	1.09974 Å	0.09956
∠9C–7C–4C	103.00015°	103.10149°	0.10134
∠8H–7C–28O	111.57871°	110.67489°	0.90382
∠4C–7C–8H	111.62454°	112.64749°	1.02295
∠9C–7C–8H	111.59637°	111.32054°	0.27583
∠9C–7C–28O	111.91006°	112.22277°	0.31271
∠4C–7C–28O	106.72426°	106.58996°	0.13430

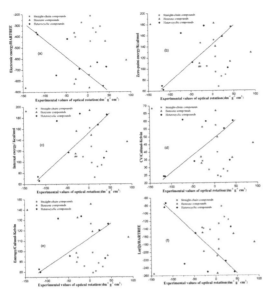

Figure 3. The relationship between experimental data of optical rotation and electronic energy (a), zero-point energy (b), internal energy (c), CV (d), entropy (e), heat energy (f), respectively.

reverse distribution. These experimental results were expected to lay the foundation for the following calculations of optical rotation.

3.3 Calculation values of OR with the HF and DFT methods

The calculation values of $[\alpha]_D$ were obtained utilizing HF and DFT methods and same basis sets. Absolute differences calculated and experimental $[\alpha]_D$ values were also given in Table 3.

First, the HF and DFT methods gave poor calculation results overall for straight-chain compounds. In the DFT method, only the deviation of compound **a6** was less than 10° compared with experimental value. The HF method gave relatively accurate results for molecules **a4** and **a8** with their derivations of calculated values were less than 10° compared with experimental value. Second, the best calculation results were belonged to compound **b5** among benzene compounds and the derivation was less than 10° with HF method. The calculated values of all other compounds both in the HF and DFT methods were poor and their derivations were more than 50°. Compared with straight-chain compounds, the calculation results

Table 3. Comparison of the $[\alpha]_D$ (dm^{-1} g^{-1} cm^3) obtained using HF and DFT methods.

Compd.	HF	$[\alpha]_D$	DFT	$[\alpha]_D$
a1	6-31G (d)	29.6	B3LYP/6-31G (d)	63.5
a2	6-311+G (d)	55.21	B3PW91/6-31 1++G (3df, 2pd)	59.62
a3	6-31G (d)	13.62	B3LYP/6-31G (d)	23.04
a4	6-31G (d)	4.08	B3PW91/6-31G (d)	28.09
a5	6-31G (d)	38.37	B3PW91/6-31G (d)	94.77
a6	6-31G (d)	33.08	B3PW91/6-31G (d)	6.87
a7	6-31G (d)	38.7	B3LYP/6-31G (d)	80.67
a8	6-31G (d)	7.61	B3LYP/6-31G (d)	18.21
b1	6-311++G (3df, 2pd)	112.3	B3PW91/6-31G (d)	100.6
b2	6-311++G (3df, 2pd)	151.1	B3PW91/6-311++G (3df, 2pd)	249.2
b3	6-311++G (3df, 2pd)	60.51	B3PW91/6-31 1++G (3df, 2pd)	95.6
b4	6-311++G (3df, 2pd)	80.91	B3PW91/6-311++G (3df, 2pd)	108.8
b5	6-311++G (3df, 2pd)	1.26	B3PW91/6-311++G (3df, 2pd)	15.03
b6	6-311++G (3df, 2pd)	183.2	B3PW91/6-31G (d)	177.8
b7	6-311++G (3df, 2pd)	63.53	B3LYP/6-31G (d)	65.12
b8	6-31G (d)	63.43	B3LYP/6-31G (d)	66.96
z1	6-311++G (3df, 2pd)	20.49	B3PW91/6-311++G (3df, 2pd)	17.38
z2	6-311+G (d)	4.05	B3PW91/6-311++G (3df, 2pd)	7.12
z3	6-311++G (3df, 2pd)	5.14	B3PW91/6-311++G (3df, 2pd)	0.92
z4	6-31G (d)	0.60	B3LYP/6-31G (d)	5.90
z5	6-31G (d)	1.87	B3LYP/6-31G (d)	5.60
z6	6-311++G (3df, 2pd)	4.68	B3PW91/6-311++G (3df, 2pd)	7.50
z7	6-31G (d)	128.3	B3LYP/6-31G (d)	79.05
z8	6-311++G (3df, 2pd)	4.74	B3LYP/6-311++G (3df, 2pd)	7.45

of optical rotation for benzene compounds were worse, which indicated that the effect of benzene ring on the optical rotation could be greater than the effect of molecular conformation. The introduction of the benzene ring, should have contributed to the change in the electron density distribution throughout the compound, or the change of whole molecule conformation, which caused a significant change in the specific rotation of the compound. Thereby, the determination of the specific rotation of benzene compounds is a challenging task. At last, compared with straight-chain and benzene compounds, the calculation results of the optical rotation of heterocyclic compounds were better. All derivations between the calculated values and experimental values were within 10° except compounds **z1** and **z7**. As can be seen from Figure 1, all compounds are five-membered nitrogen containing heterocyclic compounds except compound **z7**, and the chiral centers are on the ring, which have contribution to the better calculations. Although the molecular structure of compound **z7** contains nitrogen atom on the ring, the poor calculations may be caused by the four-membered ring of instability. It is worth mentioning that the structure of compound **z1** is in accordance with the above assumption, but the derivation between calculated value and experimental value is more than 20° because of the presence of benzene ring. In a word, the HF and DFT methods gave more accurate calculations for heterocyclic compounds than benzene compounds and straight-chain compounds.

4 CONCLUSIONS

First of all, the structure of 24 molecules was optimized using DFT method, and found that the change in the structure of compounds was quite subtle before and after optimization. Secondly, frequency calculations were performed on the basis of the optimization to obtain zero-point energy, electronic energy, internal energy, heat energy, entropy and enthalpy of compound molecules, which laid theoretical foundation for the calculations of

optical rotation. At last, the accuracy of the HF and DFT methods was evaluated for the calculation of the optical rotation of 24 compounds. The HF method with three basis sets was employed for the calculations. In parallel, B3LYP and B3PW91 functional with the same three basis sets have been performed in the DFT method. The calculation results indicated that there were many influencing factors for the calculations of the optical rotation utilizing the HF and DFT methods. Only for heterocyclic compounds, a good agreement between experimental and calculated values of the optical rotation has been observed. This can be attributed to the limit of molecular conformation because chiral center of these compounds was just on the ring. In conclusion, the HF and DFT methods can give acceptable prediction for the optical rotation of heterocyclic compounds, and further research is needed.

REFERENCES

Hodgson, J.L., L.B. Roskop, M.S. Gordon, C.Y. Lin, and M.L. Coote. 2010. Side Reactions of Nitroxide-Mediated Polymerization: N-O versus O-C Cleavage of Alkoxyamines. *Journal of Physical Chemistry A* 114: 10458–10466.

Polavarapu, P.L. 1997. Ab initio molecular optical rotations and absolute configurations. *Molecular Physics* 91: 551–554.

Ruud, K., T. Helgaker. 2002. Optical rotation studied by density-functional and coupled-cluster methods. *Chemical Physics Letters* 352: 533–539.

Tsang, Y., C.C.L. Wong, C.H.S. Wong, J.M.K. Cheng, N.L. Ma, and C.W. Tsanga. 2012. Proton and potassium affinities of aliphatic and N-methylated aliphatic α-amino acids: Effect of alkyl chain length on relative stabilities of K⁺ bound zwitterionic complexes. *International Journal of Mass Spectrometry* 316–318: 273–283.

Yang, D.C. 2001. *In Computational Chemistry: A Practical Guide for Applying Techniques to Real World Problems*. New York, USA: John Wiley & Sons Inc.

Yun, K.G. 1990. *Organic Chemistry*. Beijing, China: Higher Education Press.

Zhang, A. 1990. *Organic Chemistry Course*. Beijing, China: Higher Education Press.

Advances in Energy, Environment and Materials Science – Wang & Zhao (Eds)
© 2016 Taylor & Francis Group, London, ISBN 978-1-138-02931-6

Decision-making method of strategy intelligence based on the lifecycle model for technology equipages

Nan Ma, Li Zhang, Jian Zhao & Yuan Xue
Air Force Engineering University, Science College, Xi'an, Shaanxi, China

Huinan Guo
Xi'an Institute of Optics and Precision Mechanics Chinese Academy of Sciences, Xi'an, Shaanxi, China

ABSTRACT: Technology literatures contain useful information regarding strategy intelligence, which can be utilized through modern computer-aided technique and digital data mining method. In order to provide an effective decision support for military equipage strategy, we propose a novel technology lifecycle model, which can clearly describe the trend of development of a core technology. Meanwhile, a decision-making method is proposed that provides a new approach to analyze literature intelligence. To approve the proposed method, an actual case about FSS is selected, and simulation analysis shows that the proposed method can accurately analyze transformation relationship between technology and new equipages, which is a useful decision support mode to policymakers.

1 INTRODUCTION

War potential is the basic guarantee of the national security, and military power is one of the core indexes for evaluating this potential. Since World War II, the demand for military around the world brings a gradual increase of combination between the new technology theories and the weapons (Yuan et al. 2010). The national military power depends not only on the domestic industrial productivity and the scale of military, but also relies on the transforming ability of new technology equipage application, which stimulates the governments to pay more attention to develop their information search and analyze their capability for technology strategy intelligence.

As an important information carrier for development of science and technology, literature intelligence play a significant role in science and technology strategy and military strategic policy in the new period (Nerur et al. 2005). With the popularization and development of network technology, open source information in website and literature intelligence, which embodies the attributes of many advantageous aspects such as lower cost for search, lower legal and moral issues, and the like have become a research hotspot in scientific and technological intelligence field (Valverde et al. 2007). Scientific and technological literatures, which contain the latest and most comprehensive technical information, have some distinct characteristics: novel ideas, fresh methods, innovative

technology, and coherence. It is a kind of special composite resources of technology, law and economy. The World Intellectual Property Organization show that 90%~95% of the new technology, which is directly related to technological innovation in the world, is firstly expressed in literature; even though 70%~80% have only appeared in them (Cantrell 1997). Therefore, the literature intelligence can only reflect the trend of development and the procedure of evolution from a low level to a high stage for a new technology, which makes it possible to predict the development of technology by analyzing literature intelligence.

In this paper, we utilize some open literature database as the data source, by summarizing and analyzing the literature index information, to establish a lifecycle estimation model for a core technology. This model, based on multivariate information of literature index intelligence, represents the development of core technology in a series of related literatures that could provide a novel sight or a new method to intelligence analysts for strategy decision-making.

2 INTELLIGENCE ANALYSIS BASED ON CITATION DATABASE

The intelligence analysis based on the citation database grows up gradually after the digital citation network is wildly used. Science and technology literatures (including natural science and social science) are the important components of the intelligence source that are the carriers of the new technology, new methods, and new ideas in social and technology development. Actually the citation data of the kinds of literatures contains a deeply and indirect relation, which as important intelligence information provides a novel perspective and analysis method for analysts from different point of views.

The intelligence analysis methods based on citation database can be divided into two categories: on one hand, according to some limited number or limited correlative information elements, to explore and analyze the essential law in mass data (Zhang 2008), such as using the authors' country, author affiliation, year, language and other information of some target literatures to dig deeply useful information by means of the probability and mathematical statistics method. Meanwhile, combined with professional knowledge in some technical field, the complex internal relationship among data sources are looked for. On the other hand, using the statistical advantages of the citation database, to find the complex relationship between the citation and original literature, and explore the coupling and competition relationship between the citations in different point of views (Lv et al. 2012). When the chained interconnections between the citations are enlarged, the core technology development in literatures can be represented to a network or topology. Therefore, the intelligence subject—analysts could make strategy decision and even dig out more valuable and deeper information by making use of the comparability and related analysis to citation topology.

With the development of computer-aided calculation and mass data analysis, a series of computer analysis techniques, such as data mining, information extraction, and clustering technology, are widely used in technology prediction. The analysis methods of technology prediction in intelligence field mainly include the Delphi method (Kong et al. 2009), scenario analysis method (White 2000), literatures and key technology research (Zhao et al. 2002) and technology roadmap method (Bauer et al. 2005).

3 DECISION SUPPORT

The traditional intelligence analyses mostly focus on the process of technical innovation and application, which cannot directly provide help for intelligence analysts and decision makers to make a long-term strategy policy. For this limitation, we propose a novel analysis method based on technology lifecycle model, which could provide an effective decision support.

3.1 *Period analysis of technology lifecycle model*

As a technology is born, it culminates, and it decays always through a series of complex processes, and we could divide the whole development process into several characteristic stages. According to the technical maturity, the whole technology lifecycle can be defined as four stages: embryonic stage (E), growth stage (G), mature stage (M), and decline stage (D). And as we can see in Figure 1, we call it "E-G-M-D" mode.

However, in practice, it is difficult to accurately judge the development stage of some technology, and it usually involves in horizontal comparison by a number of related information so as to estimate the vitality and practical potential. At the same time, for many core technologies, the lifecycle is not completely compliant to this division of stages mentioned above.

For example, when a new technology in planning its period, it is gradually concerned by related fields. At this time, the technology is in the embryonic stage (E) in its lifecycle. After years of research and innovation, this core technology has transformed gradually from the theoretical experimental stage to the actual application stage, which indicates that the technology is in the growth stage (G). With the development of application on practical situations, its applications have expanded in industrial, military and other aspects which means that this technology has been in its mature stage (M). Nevertheless, for some continuous technology, the trend of development and applying foreground do not directly continue to decline (in its decline stage (D)), but go into a kind of new mode—the technology innovation. We could call it innovation stage (I), which could be involved in much deeper research and attract more investment. In fact, see in Figure 2, we proposed a new lifecycle mode—"E-G-M-D-I" mode.

A technology that in innovation stage mainly features basic characteristics as follow: First of all, a new generation of technical products is born from the original technology and is gradually concerned

Figure 1. The "E-G-M-D" mode.

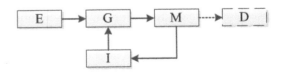

Figure 2. The "E-G-M-D-I" mode.

by researchers in some field. Besides that, the new technology in innovation stage is usually based on the original mature methods, which grows powerful and rapid, and over to a new generation of growth stage at a rapid pace. Last but not least, the new technology could be with higher application transform efficiency, which attracts extensive investment from the government and society.

Therefore, it is of great important significance theoretically and practically to extend the traditional lifecycle model from embryonic stage, growth stage, mature stage, and decline stage ("E-G-M-D" mode) to the closed loop structure—embryonic stage, growth stage, mature stage, decline stage, and innovation stage ("E-G-M-D-I" mode). Meanwhile, it is significant for analyzing and forecasting the inheritance and variability of core technology.

3.2 Decision-making method based on lifecycle model

From the different stages of technology lifecycle mentioned above we can see that the vitality and practical potential of a technology is changing along with time. As shown in Figure 3, the different structures of lifecycle models are with different parameter curves. Therefore, in order to estimate the development level for some core technology, we can introduce some quantitative parameters to describe the independent features of different stages in lifecycle.

Let, in a certain time interval $[t_1, t_2]$, there be N relevant literatures of a target technology collected in a citation database, and here let the n_t ($t \in [t_1, t_2]$), as the number of citation in each year, the citation index statistical distribution $S(t)$ for key words can be represented as:

$$\left\{ S \mid S(t) = \sum_{i=1}^{n_t} X_i(t); \sum_{t=t_1}^{t_2} n_t = N \right\} \tag{1}$$

where $X(t)$ denotes the literature sample which is indexed in the t year. For the discrete samples, $X(t) \equiv 1$.

In order to avoid the local disturbance in statistical distribution caused by some years, the statistical values in five years of continuous samples are used to replace the parameter $X(t)$, and the citation index statistical distribution $S(t)$ can be reshaped as:

$$S(t) = \sum_{i=1}^{n_t} \frac{X_{i-2}(t) + X_{i-1}(t) + X_i(t) + X_{i+1}(t) + X_{i+2}(t)}{5} \tag{2}$$

And now, we can estimate the technical state in some time point by using the distribution character in some interval. At the same time, let $t_x \in [t_1, t_2]$, and the expectation $\overline{E}(S)$ and variance $\overline{D}(S)$ can be represented as:

$$\begin{cases} \overline{E}(S) = \dfrac{1}{t_x - t_1} \cdot \sum_{t=t_1}^{t_x} S(t) \\ \overline{D}(S) = \dfrac{1}{t_x - t_1} \cdot \sum_{t=t_1}^{t_x} \left(S(t) - \overline{E}(S) \right)^2 \end{cases} \tag{3}$$

Let the variance threshold be α, and the judgment rule for technical state is shown in Table 1.

Where $Grad(t_x)$ is the gradient at the t_x year, which depicts the disparity between two of the nearest interval data. Let the neighborhood step be λ, and the gradient is expressed as follow:

$$Grad(t_x) = S(t_x + \lambda) - S(t_x - \lambda) \tag{4}$$

And the parameter β is a gradient threshold.

According to the judgment rules mentioned above, we propose a decision-making method for technical strategy intelligence. For different technical development stage, we usually focus on three aspects: occasion, strategy, and investment. Specifically, occasion means that we should step in some technology field at the right time, which may bring the maximized interests. Strategy here mainly depicts how we taking part in this field—research core technology from the start or purchase mature products. At last, investment involves in market input policy and industrial structure.

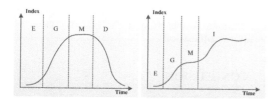

Figure 3. The lifecycle model curves: (a) is open loop structure; (b) is closed loop structure.

Table 1. Judgment rule for technical state.

Constrains		Judgment
$\overline{D}(S) \leq \alpha$	-----	E
$\overline{D}(S) > \alpha$	$Grad(t_x) < -\beta$	D
$\overline{D}(S) > \alpha$	$\|Grad(t_x)\| \leq \beta$	M
$\overline{D}(S) > \alpha$	$Grad(t_x) > \beta$	G I

Table 2. Decision-making method and strategy.

Stage	Strategy (Research □ vs. Purchase ■)	Investment level
E	□□□□□	1
G	□□□□■	2
M	□■■■■	5
I	□□□■■	4
D	-----	1

Table 3. The contrast schedules about F22 and F35.

Time	F22	F35
1981	Initiation	---
1993	---	Initiation
1997	First flight	---
2005	Fitted out	---
2006	---	First light
2008	Phase out	---
2012	---	Fitted out

The decision-making strategy is shown in Table 2. We can see that there are different input strategies in each stage. For the "Strategy", we use □ depicts "Research" and ■ for "Purchase", and these show clearly the ratio of strategy formulation. Meanwhile, the investment scale can be graded from 1 to 5, and the society financial input is increasing with the investment level.

4 SIMULATION AND DATA ANALYSIS

In order to prove the effectiveness and correctness of the proposed method, we choose an actual case to simulate decision results.

It is reported that FSS (Frequency Selection Surface) is core technology for stealth aircraft in the American air force, which has already been used in F22 and F35 jet fighters. The contrast schedules between F22 and F35 are shown in Table 3.

According to the statistical parameters mentioned above, we select Ei and SCI database as data source, and the statistical distribution of related literatures about FSS from 1981 to 2012 are shown in Figure 4. Each stage in the lifecycle is judged by rules shown in Table 1. Through comparing stage markers and strategy schedules of FSS, we can see that at the primary stage (E) of research for FSS, it is used in F22 by the US air force. And it costs 16 years to achieve their first flight and 24 years to fit out. While at the growth stage (G), it is adopted for F35, it costs 13 years to achieve its first flight and 19 years to fit out. It costs less by adopting such a new technology at the growth stage (G) than at the primary stage (E).

For the Chinese air force, the stealth technologies for fighters have been widely studied. According to the material processing level and related technological sophistication of stealth technology in China, there are many ways for the Chinese air force to choose. One of possible selection is adopting the FSS pattern. As shown in Figure 5, through analyzing literatures data by the proposed method in this paper we can see that the technology lifecycle of FSS is just in its growth stage (G). The statistical curve expresses that around the year

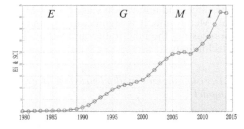

Figure 4. The statistical distribution about FSS of the US.

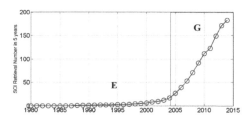

Figure 5. The statistical distribution about FSS of China.

of 2005 the research about FSS in China gradually became a hotspot. According to the proposed decision strategy, it is implied that it would shorten transformation process between research and application if the FSS technology used in stealth fighter by Chinese air force after the year of 2005. It means that an effective investment in the mature stage by Chinese military could shorten product development cycle as well as increase paces of weapons application.

5 CONCLUSION

By using computer-aid in network and data mining for digital literature database, a novel technology lifecycle model and decision-making strategy have been proposed in this paper, which provides a new approach to analyze literature intelligence. It is effective to forge industrial and military development strategy of some new technology. Meanwhile, by using this model it can offer an efficacious decision support about core technology in new military equipages to policymaker.

REFERENCES

Bauer K., Bakkalbasi N. (2005). An Examination of Citation Counts in a New Scholarly Communication Environment. DLib Magazine, 11(9), 1–8.

Cantrell R. (1997). Patent intelligence from legal and commercial perspective. World Patent Information, 19(4), 251–264.

Kong Li-sha, Liu Wen. (2009). Simply Discussing on the Technology of Data Mining and the Military Decision Support. Equipment Manufacturing Technology, 10, 117–118.

Lv Cao-fang, Hou Zhi-bin. (2012). Data Mining Realization Technology Based on Text Intelligence Data. Computer and Information Technology, 20(6), 32–34.

Nerur S., Sikora R., Mangalaraj G., et al. (2005). Assessing the relative influence of journals in a citation network. Communication of the ACM, 48(11), 71–74.

Valverde S., Sole R.V., Bedau M.A., et al. (2007). Topology and evolution of technology innovation networks. Physical Review, 76(5).

White, H.D. (2000). Toward ego-centered citation analysis. Medford, NJ: Information Today, Inc. 475–496.

Yuan Bin-cheng, Fang Shu, Liu Qing. (2010). Overview on Progress in Citation Analysis at Home and Abroad. Intelligence Science, 28(1), 147–153.

Zhang Yong-hua. (2008). The Application of Data Mining Technology in Military Information Acquisition. Sci-Tech Information Development & Economy, 18(8), 160–161.

Zhao D., Logan E. (2002). Citation analysis using scientific publication on the web as data source: A case study in the XML research area. Scientometrics, 54(3), 449–472.

Advances in Energy, Environment and Materials Science – Wang & Zhao (Eds)
© 2016 Taylor & Francis Group, London, ISBN 978-1-138-02931-6

Logical model of outdoor lighting soft interface TALQ protocol

Jin Liang & DongHui Jia
School of Information Technology in Education, South China Normal University, Guangzhou, China

ABSTRACT: Different outdoor lighting systems block the development of the intelligent control lighting systems. An identical interface protocol of outdoor lighting is an important way to solve the problem. TALQ protocol is an authoritative protocol recently established by many famous international companies including Philiphs, GE Lighting, Zumtobel etc. This paper introduces the newest technology of the protocol how to solve the key problem for different outdoor lighting control systems. First, some solution methods are discussed and the idea resource is described; Second, the specification protocol and information description are introduced, especially two important parts the central management system and TALQ bridge. According to the framework of the TALQ protocol, the logical model of one core technology is founded in the protocol, then some concepts including TALQ functions and the logical device address are introduced, and the data synchronization is introduced between the central management system and TALQ bridge. Lastly a XML file is exemplified to describe the logical model, and it is a good way to describe different outdoor lighting control systems.

1 INTRODUCTION

With the development demand of the smart city, it is important to save energy and decrease the carbon dioxide emissions. The electrical energy consumption of street and outdoor lighting constitutes an important part of total energy consumption (Zotos et al. 2012). There are two important ways to save energy and decrease the carbon dioxide emissions on the street and outdoor lighting: one way is to replace all the traditional lightings to the LED lightings, according to the statistics from TALQ consortium, if all the outdoor lightings in US are replaced to the LED lightings, then it could decrease the emissions of the CO_2 as much as 90 million metric tons and save much energy, and the number of those outdoor lightings needed to replace in some districts are exemplified as follows: over 50 millions outdoor lightings in US, over 80 millions outdoor lightings in Europe and over 17 millions outdoor lightings in china (Jin & Nim 2014, TALQ Consortium 2015), so it is an important way and will be realized in the near futur; the other way is to control the outdoor lighting systems intelligently, which will save energy and decrease the carbon dioxide emissions on outdoor lighting control systems through the adaptive and interoperable controls (Mohsen et al. 2014, Siddiqui et al. 2012, Schaeper et al. 2013, SangCheol et al. 2013), so the outdoor lighting intelligent control system becomes an important part of the smart city, especially to the LED outdoor lighting control system.

Then many outdoor lighting control systems are developed individually, such as london westminister roadway lighting system, delft university of technology intelligent roadway lighting system in the netherlands, san jose intelligent roadway lighting system in American and kingsun intelligent roadway lighting system in china etc (Jin & Nim 2014). Most outdoor lighting systems are composed of three parts: A centralized control center, a remote concentrator and street lamps control terminals (Zotos et al. 2012). However, it is difficult to combine these outdoor lighting control systems to work together without standards, which prohibits the development of LED outdoor lighting control systems and smart lighting. In order to solve the problem, many organizations and companies try to draw up some standards around the upper computer protocol and the lower computer protocol for the outdoor lighting systems, such as csa018 by china soled state lighting alliance, which is mainly according to the lower computer protocols. But the hardware will sometimes be limited and not flexible for the lower computer protocols in the outdoor lighting systems, which will be possibly uncertain to be accepted by many other organizations and companies in the world, so whether the standard is accepted by most organizations or companies and the market will be decided by its authority and flexibility. Apparently, the outdoor lighting interface standard is supported by more famous companies, and then the standard will be more authority.

To solve the problem of the outdoor lighting control interface standard, a great international consortium named TALQ is founded in 2012, which is initiated by famous lighting and electric companies in the world, including Philips, Zumtobel and Schneider etc. The TALQ Consortium aims to set a globally accepted standard for management software interface to control and monitor heterogeneous outdoor lighting networks (TALQ Consortium 2015). TALQ Consortium began to establish the standard at the June 2012, and try best to establish the standard. After more than two years hard working and development, more and more famous companies joined TALQ Consortium, such as Osram, GE Lighting etc, and the total membership has grown to 13 Regular and 15 Associate members, which mostly are famous important companies from all over the world. Now the standard named Approved TALQ Specification Version 1.0.2 has been finished in January 2015, and the corresponding certification standard named Approved TALQ Certification Test Specification Version 1.0 has been finished in January 2015 too. Some countries, such as England, Germany etc, have admitted the authority of TALQ standard acted as the outdoor lighting interface standard. So there is no doubt that TALQ standards are authoritative standards for the outdoor lighting control interface in the future, and it is necessary to introduce and discuss the TALQ standards.

This paper introduces an important part of TALQ specification, and describes the standard how to solve the problems and its merits; some cases are discussed to reflect the flexibility of the standard. Because of the length limitation of an article, only the logical model of the TALQ specification is mainly introduced and discussed here.

2 RANGE OF TALQ PROTOCOL

2.1 *Idea resource*

In order to solve the above problems, TALQ consortium has definitely defined the range of the TALQ specification. TALQ protocol is an upper computer protocol and the range of its definition is between the centralized control center and remote concentrator, which belongs to the application layer based on TCP/IP protocol. How the TALQ specification combines and manages different outdoor lighting systems, as exemplified like the Figure 1. Outdoor lighting network is abbreviated to OLN, and OLNs represents the outdoor lighting network system here. In the Figure 1, outdoor lighting networks 1 to outdoor lighting network n are different outdoor lighting networks.

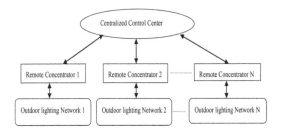

Figure 1. Combining different outdoor lighting systems.

In order to control and coordinate these different OLNs, the centralized control center has to understand those messages sent from OLNs and sent those understandable messages or commands to configure and manage different OLNs. So it is an important responsibility for the centralized control center to translate those different messages on different protocol and message format from different OLNs. But when all these responsibilities and tasks are put in the centralized control center, it would leads to too burden for the centralized control center, otherwise, it is almost impossible to unify the protocol between the remote concentrator and OLN, which would limit the flexibility of hardware so much and possibly be unacceptable for many producers. So TALQ consortium presents a flexible soft standard interface scheme to solve the problem after carefully discussion. The TALQ protocol is discussed as follows.

2.2 *TALQ specification protocol*

In TALQ specification, the centralized control center is denoted as the central management system, abbreviated to CMS, and a new part named TALQ Bridge is defined as the interface to translate different messages, then the solution of TALQ specification is shown in the Figure 2. From the Figure 2, the TALQ Bridge is the key part to bridge different OLNs in TALQ specification. The TALQ specification is defined in the application layer according to the TCP/IP protocol stack or the OSI network reference model. Message types, data format, parameters and behavior of the application end-points are defined both at the TALQ Bridge and the CMS, and the TALQ protocol applies the services presented by underlying data transport and network layer to establish communication between the TALQ end points.

2.3 *TALQ information description*

From the Figure 2, protocols of OLNs are always different, but the information or commands of different

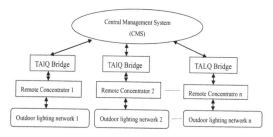

Figure 2. TALQ solution for different outdoor lighting systems.

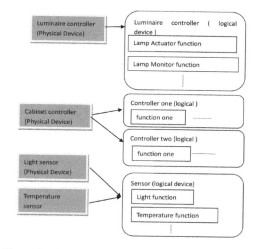

Figure 3. Map example between physical device and logical device.

OLNs have to be identified and transformed to the identical language and format by TALQ bridge, which are certainly understood by CMS. Then the XML language is selected as the identical language for TALQ specification and used to describe the information or commands. So the information among OLNs and CMS have to be formalized firstly, and models of different OLNs hardwares need to be defined or created by the XML language.

It is necessary to set up data models to describe different OLNs, so many physical devices are described through the logical model defined in the TALQ specification.

3 TALQ LOGICAL MODEL

3.1 Logical device model

In order to keep the flexibility of different OLNs physical characteristics, TALQ logical models need to be defined flexibly, and easy to be extended and suitable for different physical characteristics. One complex physical device can be described in several logical devices, or some physical devices are combined to be defined in one logical model. Physical devices' characteristics can be described in logical functions including attributes and events.

Some typical physical devices such as luminaire controller, cabinet controller etc could be described in logical devices, exemplified as follows.

From the Figure 3, physical devices can be described flexibly through functions of logical devices, and some functions are currently defined in the TALQ specification, which include basic, talq bridge, communication, lamp actuator, lamp monitor, electrical meter, photocell, light sensor, binary sensor, generic sensor and time. A logical device is defined by the set of functions and a TALQ Address.

3.2 TALQ function

In order to describe a functionality of a logical device, a group of attributes and events are defined in a TALQ function. The configuration, operation, and management of physical devices are realized by the logical devices, and the corresponding configuration, operation, and management of the logical devices are implemented by the manipulation of function attributes according to the corresponding TALQ Service (TALQ Consortium 2015). Then some function attribute types are defined in the TALQ function including configuration, operational, measurement and status. Configuration attributes are defined to describe device's capabilities and characteristics; Operational attributes are used in the TALQ services, which is in order to control the device's behavior as the reaction to a request from the CMS during the system operations; The performance data of specific device functions are provided by the measurement attributes; Status attributes are used to described the status of some specific event or the result of the preceding event.

3.3 Logical device address

In order to identify the logical device, an unique TALQ Address shall be assigned to each logical device, which is assigned by the TALQ bridge, and the format of logical address should be defined as follows:

<address-domain>:<LogicalDeviceEntity-
    Address>

The <address-domain> represents the type of TALQ address, which identifies the addressable resources in the context of a service. The function and attribute could be assigned "id" to identify their addresses, which is like the id attribute of element in HTML or XML page.

3.4 Data synchronization

It is necessary to define the data synchronization, which certify the consistency of the logical devices and other entities between the CMS and TALQ Bridge side. Most entities are ruled to be created and modified by only one application end point (CMS or TALQ Bridge). Logical devices are created by the TALQ bridge, but CMS can have their configuration attributes modified, and not permitted to delete them. Entity sequence number is defined to information synchronization between the CMS and TALQ bridge, and each side have two sequence numbers to record its own attribute and the other side attribute. The sequence numbers are defined in the "since" and "seq" xml attribute in some manipulation information such as refresh, update etc.

4 APPLICATION CASE

The logical device class of a luminaire controller are described by XML language as an example here. Suppose a luminaire controller contains a lamp actuator function and a lamp monitor function, then a XML file example could be used to describe the logical model of the luminaire controller as follows:

```
<?xml version = "1.0" encoding = "UTF-8"?>
<LogicalClass
xmlns = http://lightingcompany.org/schemas/
core/2015/6
xmlns:xsi = "http://www.w3.org/2001/
XMLSchema-instance">
  <LuminaireController class = "dev:Luminaire"
name = "xsi:LuminaireController">
<type> xsi:LogicalDevice</type>
<address > dev:LC001</address>
<functions>
  <lampactuator functionId = "xsi:la0001">
   <type > xsi:lampactuator</type>
   <attributes>
    <lampTypeId > dev:lamp001</lampTypeId>
    <defaultLightState > 40</defaultLightState>
    .....
   </attributes>
   <events>
    <lightStateChange>
     <lightState > 60</lightState>
    </lightStateChange>
    .....
   </events>
  </lampactuator>
   <lampmonitor functionId = "xsi:lm0001">
    ......
   </lampmonitor>
</functions>
  </LuminaireController>
</LogicalClass>
```

5 CONCLUSION

TALQ protocol is a very important outdoor lighting control interface protocol in the world, which is established recently by many famous international companies together and accepted by many advanced countries. It is necessary to introduce the protocol and promote the development of LED outdoor lighting control and intelligent lighting in the city. Because the length limitation of article, only the logical device model is discussed here and one case is exemplified. From the paper, TALQ protocol is a good way to combine different outdoor lighting systems together, and we can found the TALQ protocol how to solve the problem of combination of different outdoor lighting systems.

ACKNOWLEDGMENT

This research was financially supported by the 2014 Guangzhou Scientific Plan Project, which number is 2014J4100079, and the project name is LED outdoor lighting intelligent network control based on TALQ international standard and advanced intelligent method.

REFERENCES

Jin, L. & Nim C.K. 2014. LED Roadway Lighting Control Scheme Based on TALQ Protocol. Advanced Materials Research. (1044–1045):1541–1544.

Mohsen, M. & Tooraj, A.N., etc. 2014. A smart street lighting control system for optimization of energy consumption and lamp life. The 22nd Iranian Conference on Electrical Engineering: 1290–1294.

SangCheol, M., Gwan-Bon, K. & Gun-Woo, M. 2013. A New Control Method of Interleaved Single-Stage Flyback AC-DC Converter for Outdoor LED Lighting Systems. IEEE Tansactions on Power Electronics, Vol. 28:4051–4062.

Schaeper, A. & Palazuelos, C., etc. 2013. Intelligent lighting control using sensor networks. 2013 10th IEEE International Conference on Networking, Sensing and Control: 170–175.

Siddiqui, A.A. & Ahmad, A.W., etc. 2012. ZigBee based energy efficient outdoor lighting control system. 2012 14th International Conference on Advanced Communication Technology, PyeongChang: 916–919.

TALQ Consortium. 2015. Information on http://www.talq-consortium.org.

TALQ Consortium. 2015. Approved TALQ Specification Version 1.0.2, unpublished.

Zotos, N. & Pallis, E., etc. 2012. Case Study of a dimmable outdoor lighting system with intelligent management and remote control. 2012 International Conference on Telecommunications and Multimedial: 43–48.

Advances in Energy, Environment and Materials Science – Wang & Zhao (Eds)
© 2016 Taylor & Francis Group, London, ISBN 978-1-138-02931-6

The matlab simulation of the frequency hopping transmission station

Na Wang & Zhigang Song
92941 Troop, China & Post-Doctoral Scientific Research Workstation, 92493 Troop, China

ABSTRACT: The frequency hopping transmission station model overall construction technique and the principle of its main module design were studied. The Matlab/Simulink simulation platform was used to construct the frequency hopping transmission station model. Through the error rate computation and statistic module, the model is verified by a simulation example finally.

1 INTRODUCTION

The frequency hopping transmission station is applied widely for its high anti-jamming performance (Shen, 2011; Yan, 2011; Liu, 2009; Yan, 2011). Compared with the common communication system (Zhou, 2012), the frequency hopping modulation is added in transmission end, and the frequency hopping demodulation added in reception end. Firstly, the overall construction model technique based on matlab was studied. Then, the main modules were given and Matlab/Simulink simulation platform was used to construct the frequency hopping transmission station model. Through the error rate computation and statistic module, the model is verified by a simulation example finally.

2 SYSTEM OVERALL MODELING

The performance of the frequency hopping transmission station is decided by its technique system and parameters. To improve the simulation precision and obtain the more creditable simulation results to support decision, the simulation is given in physics level.

Simulink is a simulation technique based on time flow, which can make the simulation modeling and frame design in engineering together (Men, 2012; Liu, 2014). By this method, it can be convenient to design the system, debug the simulation and verify the model. The frequency hopping transmission station simulink model is constructed in Figure 1.

3 MAIN MODULE DESIGN

3.1 Source module

Commonly, the text information is transmitted by the frequency hopping transmission station. In order to improve the transmission validity, the text information should be encoded in binary bit flow to transmit. In the receiver, the bit flow should be decoded into the text information.

3.2 Encoder/Decoder module

To improve the transmission quality of the channel and correct the transmission mistake, the channel encoding should be used. Channel encoding technique is to construct code structure, to insert the least redundance code and to get the best anti-jamming performance.

The channel encoding technique with certain corrective ability, such as parity check code, the rank of supervision code, Hamming code, cyclic code and so on. Communication toolbox in matlab simulink can provide many encapsulated Encoder/Decoder module given in Figure 2 and Figure 3 to construct the simulation system rapidly and conveniently.

3.3 Modulation and demodulation module

The transmission performance of the communication system can be effected by the modulation

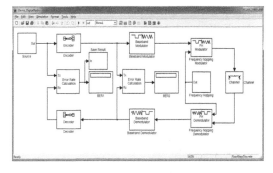

Figure 1. Frequency hopping transmission station Simulink model.

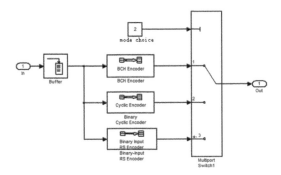

Figure 2. Channel encoder module.

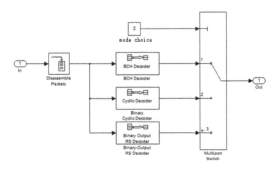

Figure 3. Channel decoder module.

methods greatly, sometimes, definitively. Frequency Shift Keying (FSK) modulation method is commonly used in frequency hopping system, and GFSK modulation system is commonly used in frequency hopping transmission station.

FSK modulation can be expressed as,

$$s(t) = \sqrt{\frac{2E_b}{T}} \cos\left(2\pi f_c t + 2\pi I_n \Delta f t\right) \qquad (1)$$

where $I_n = \pm 1, \pm 3, \ldots \pm (M-1)$.

In FSK modulation method, the information is transmitted by the amount of frequency shift $I_n \Delta f$, which may introduce larger frequency sidelobe spectrum in the continue signal transmitting interval.

In order to reduce the sidelobe spectrum, continue Phase FSK (CPFSK) is introduced as

$$s(t) = \sqrt{\frac{2E_b}{T}} \cos\left(2\pi f_c t + \phi(t, I) + \phi_0\right) \qquad (2)$$

where $\phi(t, I) = 4\pi T f_d \int_{-\infty}^{t} \left[\sum I_n g(\tau - nT)\right] d\tau$, f_c is carrier wave frequency, f_d is peak value frequency shift, ϕ_0 is initial phase of the carrier wave.

To reduce the signal sidelobe spectrum futherly, Gaussian filer can be used before CPFSK, i.e. the GFSK modulation method.

The transfer function of Gaussian filer is

$$H(f) = \exp\left(-a^2 f^2\right) \qquad (3)$$

where a is a variable.

Let B is 3dB bandwidth of the filer, i.e., $H(B) = 1/\sqrt{2}$ it can be

$$a = \frac{\sqrt{\ln 2}}{\sqrt{2} B} = \frac{0.5887}{B} \qquad (4)$$

GFSK modulation can be accomplished in two steps in Figure 4. Firstly, the input information bit flow is filtered by Gaussian filer. Then, The preprocessed wave is frequency modulated to get GFSK signal.

GFSK demodulation can be accomplished in two steps in Figure 5 in commonly.

3.4 Frequency hopping modulation and demodulation module

The output frequency of the radio oscillator can be discretely controlled by the frequency hopping system, which can make the carrier wave frequency and its center frequency changing with the frequency hopping. By this method, the anti-jamming and anti-interception capability of the communication system can be improved. The frequency hopping system is constructed by frequency synthesizer and random code generator. Random code is generated synchronously with the frequency of the carrier wave changing by the frequency synthesizer. The random code is used to choose the channel not to transmit the signal. The carrier wave of the frequency hopping system can be regarded as Multiple Frequency Shift Keying

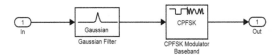

Figure 4. GFSK modulation module.

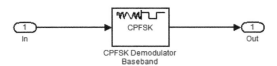

Figure 5. GFSK demodulation module.

(MFSK). In simulation, PN sequence generator module is regarded as random code generator, the text information is coded into binary bit flow, modulated and then frequency hopping modulated with the pseudo random sequence. In the frequency hopping demodulation module, the received signal conjugate multiply with the frequency hopping signal to accomplish the frequency hopping demodulation. The models are given in Figures 6–8.

3.5 Channel module

When analyzing the performance of the communication system, commonly ideal Additive White Gaussian Noise channel given in Figure 9 is used as the basis. The signal power and SNR (signal-to-noise) set can be realized in the channel mode.

3.6 Error rate computation and statistic module

The error rate is an index to evaluate the data transmitting accuracy, it can be expressed as

$$R_e = \frac{C_e}{C_o} \times 100\% \tag{5}$$

where R_e is the error rate, C_e is error codes in transmission, C_o is the overall codes in transmission.

The error rate computation mode is shown in Figure 10. There are two inputs and one output,

where one input (Tx) is to receive the information from the transmission and the other one is to receive the information from the receiver. Through the error rate calculation mode, the output is put in the BER statistic mode, where the first line is error rate, the second line is error code and the third line is the overall code.

4 SIMULATION EXAMPLE

The relation between SNR and error rate for different codes are compared. On the given simulation condition, the transmission performance of the simulation communication system with different codes are compared. The simulation results are given in Table 1. Parameter is set as follow, frequency hopping speed is 200/s, frequency hopping number is 64, frequency interval is 100 KHz, baseband information speed is 9600 bps, FSK is 2.

It can be concluded from the Table 1 that on the same SNR condition, between the above encode modes, the error rate of the RS code is the lowest and that of the original signal not encoded is the highest.

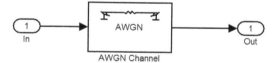

Figure 9. AWGN channel module.

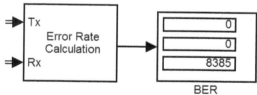

Figure 10. Error rate computation and statistic module.

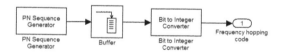

Figure 6. Frequency hopping code module.

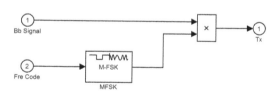

Figure 7. Frequency hopping modulation module.

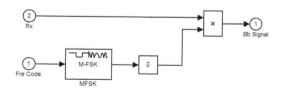

Figure 8. Frequency hopping demodulation module.

Table 1. The relation between SNR and error rate for different codes.

Setting	BCH code	RS code	CRC	Original data
0	0	0	0	0
−3	0	0	0	2.8e-5
−6	0	0	0	2.1e-3
−7	2.341e-5	1e-6	1.096e-5	0.00563
−8	0.00019	9e-5	0.0001	0.01209
−9	0.00102	0.00085	0.00113	0.0223
−10	0.00415	0.00465	0.00421	0.03754

5 CONCLUSION

Only the relationship between the SNR and the error rate of the common encode methods are given in this paper, the simulation method proposed can be used as a method to test the anti-jamming performance given communication system.

REFERENCES

Andong Zhou, Cheng Fan, Lugang Yang, in: Computer and Digital Engineering. 2012, 40(1): p63–66 In Chinese.

Kefei Liu, Dongkai Yang and Jiang Wu, in: Journal of System Simulation. 2009. 21(24):7969–7973 In Chinese.

Lanning MEN and Chiheng GE in: Electronic Science and Technology. 2012, 5(3): p35–37 In Chinese.

Nan Shen, Jun He and Feng Qi, in: Modern defence technology. 2011, 39(1): p88–91 In Chinese.

Siyuan Liu, Jianjun Chen and Xin Zhang, in: Journal of Nanjing University of Information Science & Technology. 2014, 3): 226–230 In Chinese.

Yunbin Yan, Houde Quan, and Peizhang Cui, in: Computer Measurement & Control. 2011,19(12):3082–3084,3088 In Chinese.

Yunbin Yan, Houde Quan, Peizhang Cut. Research on Follower jamming of FH Communication[c] 2011 International Conference on Communication and Electronics Information. 2011:244–247. In Chinese.

Author index

An, H.M. 459

Bai, A.L. 619, 629
Bai, S.Y. 101
Bao, X.K. 299
Bao, Z.B. 213, 227
Bao, Z.Y. 237

Cai, C.L. 331
Cao, C.Y. 161
Cao, J.Y. 525
Cao, P.F. 797
Cao, W.P. 641
Cao, X. 687
Chai, S.X. 237
Chang, J. 15
Chao, M. 9
Chen, C. 61
Chen, E.J. 155, 613
Chen, F.Y. 49
Chen, G.P. 555
Chen, J. 741
Chen, J.R. 223, 673
Chen, K.N. 291
Chen, L.P. 475
Chen, L.P. 809
Chen, P. 635
Chen, R.H. 561
Chen, R.Y. 151
Chen, S.-t. 543
Chen, S.Y. 573
Chen, T.Q. 741
Chen, W. 701
Chen, X. 421
Chen, X.D. 45
Chen, X.L. 155, 613
Chen, X.M. 493
Chen, Y. 73
Chen, Y. 321
Chen, Z.H. 119
Cheng, H.X. 451
Cheng, L. 451
Cheng, X.W. 481
Chu, X.B. 665
Cong, Y.P. 761
Cui, M.N. 665

Cui, Y.H. 45
Cui, Y.X. 195
Cui, Z.J. 581

Dai, X.D. 653
Dai, Y.H. 677, 681
Dang, X.X. 833
Deng, G.B. 155, 613
Deng, H.Y. 791
Deng, J. 349
Deng, M. 741
Deng, Q. 155
Ding, K.W. 487
Ding, X.H. 253, 469
Ding, Y.L. 101
Dong, G.Q. 247
Dong, K. 169
Dong, S. 61
Du, X.-l. 543
Duan, J. 493
Duan, L.J. 797

Fan, H. 169
Fan, M.Y. 91
Fang, C.F. 823
Fang, X.Z. 607
Fang, Y. 411, 669
Fang, Z.H. 357, 773
Fei, W.W. 815
Feng, K. 531
Feng, L.T. 619
Fu, D. 569
Fu, G.Q. 303
Fu, X.S. 505

Gao, C.C. 9
Gao, G.D. 701
Gao, G.L. 397
Gao, H.B. 549
Gao, H.-h. 97
Gao, Q. 77
Ge, Y.Y. 3
Gu, C. 303
Gu, D.S. 465
Gu, X.Y. 427
Guo, G.J. 357, 773

Guo, H.N. 839
Guo, M. 327
Guo, R.C. 145, 361
Guo, Y.Y. 769

He, G.J. 475
He, H. 635
He, H.L. 379
He, Y.J. 613
Hou, J. 531
Hou, X.J. 421
Hu, D. 707
Hu, J. 769
Hu, J.H. 181, 367
Hu, J.W. 91
Hu, S.G. 401
Hu, X.M. 653
Hu, X.Q. 303
Hu, X.X. 597, 625
Hu, Z.M. 195
Hua, D.Z. 687
Huang, C.H. 299
Huang, H. 37
Huang, H.Y. 57
Huang, J.H. 625
Huang, J.Q. 597
Huang, J.W. 181, 367
Huang, L.-l. 97
Huang, W. 687
Huo, W. 635

Ji, X.P. 761
Jia, D.H. 845
Jia, H.L. 401
Jia, Q. 265
Jia, W.H. 405
Jiang, C.X. 119
Jiang, J.Y. 375
Jiang, S. 45
Jiang, Y.L. 131
Jin, D.C. 213, 227
Jin, G.Y. 161
Jin, S.Y. 781
Jin, X.M. 45
Jin, Z. 379
Ju, B. 707

Kang, H. 593
Kang, K. 531
Khachatryan, G.E. 191
Kobayashi, N. 57
Kong, L.L. 451
Kuronuma, H. 57

Lai, X.P. 309
Lan, H. 797
Lan, J.P. 339
Lang, Q.C. 581
Lei, L.R. 573
Li, A. 25
Li, C. 597, 625
Li, C.C. 91
Li, D.D. 751
Li, D.Y. 275
Li, G. 131
Li, G.-h. 441
Li, G.M. 165
Li, H.H. 291
Li, H.Y. 653
Li, J. 441
Li, J. 451
Li, J. 641
Li, J. 657
Li, J.L. 91
Li, J.Q. 587
Li, K. 437
Li, K. 727
Li, L. 169
Li, L. 475
Li, L.C. 345
Li, L.Y. 91
Li, M. 691
Li, Q. 195
Li, Q. 757
Li, Q. 819
Li, Q.S. 619, 629
Li, S.H. 3
Li, S.M. 375
Li, S.N. 175
Li, T. 45
Li, T.Y. 531
Li, X. 497
Li, X.B. 303
Li, X.F. 101
Li, X.H. 145, 361
Li, X.P. 233
Li, Y. 213
Li, Y. 417
Li, Y. 497
Li, Y. 727
Li, Y. 803
Li, Y.B. 349
Li, Y.F. 401
Li, Y.G. 493

Li, Y.M. 573
Li, Y.S. 677, 681
Lian, E.X. 555
Liang, J. 845
Liang, L.Y. 135
Liang, W.J. 109,
  187
Liang, W.-t. 721
Liang, X.D. 647
Liang, Y. 653
Liang, Y.-p. 97
Liang, Y.P. 135, 141
Liao, J.B. 513
Lin, F.-T. 391
Lin, G. 9
Lin, S. 815
Lin, T. 41
Lin, W.-C. 391
Lin, X.C. 731
Lin, Y.F. 823
Liu, B. 41
Liu, B. 275
Liu, C.Q. 587
Liu, D.-h. 259
Liu, D.N. 9
Liu, H. 109
Liu, H. 405
Liu, H. 707
Liu, H. 765
Liu, H.C. 151
Liu, H.H. 299
Liu, H.L. 513
Liu, J. 677, 681
Liu, J.H. 741
Liu, J.P. 715
Liu, K. 653
Liu, L. 49
Liu, P. 445
Liu, R.Q. 819
Liu, S.W. 61
Liu, W.M. 49
Liu, W.P. 687
Liu, W.Q. 481
Liu, X.K. 169
Liu, X.X. 481
Liu, X.Y. 213
Liu, Y.F. 161
Liu, Y.J. 735
Liu, Y.-x. 353
Liu, Z.H. 687
Long, L.D. 505
Lu, B. 561
Lu, J.C.-C. 391
Lu, M.Y. 465
Lu, W.D. 481
Lu, Y.N. 233
Lu, Z.H. 309

Luo, H.F. 819
Lv, D.P. 481

Ma, E.W. 581
Ma, N. 839
Ma, X.Y. 385
Ma, Y.L. 555
Ma, Y.L. 761
Meng, H.L. 459
Meng, J. 125
Meng, X.J. 531
Meng, Y. 41
Meng, Y.T. 519
Mi, C.L. 569
Min, S.-S. 199
Mkrtchyan, N.I. 191
Mo, L.-Y. 97

Ni, D. 657
Nie, L.R. 833
Nikolaevich, S.P. 777
Niu, M.F. 109

Ouyang, S.J. 549

Pan, B. 417
Peng, F. 199
Peng, J.P. 49
Peng, J.X. 227
Peng, Q. 321

Qi, H. 253, 469
Qian, F.F. 481
Qiao, H.Y. 265
Qiao, X.N. 219
Qiao, Y. 241
Qin, J. 203
Qin, L.-t. 97
Qiu, B.Y. 525
Qiu, W. 37
Qiu, Y.Z. 73
Qu, L.X. 745

Ren, A.L. 593
Ren, H.Q. 331
Rong, X. 445
Ruan, Y.J. 327

Shan, P.F. 309
Shao, J. 797
Shen, H.X. 209
Shi, J.J. 459
Shi, L. 271
Shi, X.D. 91
Shi, X.D. 161
Shi, Y. 105
Si, W.H. 569

Song, D.S. 549
Song, G.F. 345
Song, H. 619, 629
Song, H. 833
Song, S.W. 151
Song, Z.G. 849
Su, W. 151
Sun, B. 695
Sun, B.T. 791
Sun, F. 3
Sun, F. 635
Sun, G. 411, 669
Sun, G.W. 151, 607
Sun, J.J. 3, 19
Sun, J.P. 647
Sun, K. 657
Sun, L. 101
Sun, L. 537
Sun, W.T. 45
Sun, Y. 357, 773
Sun, Y. 397

Tan, H. 597
Tan, L.L. 691
Tan, Y.-z. 353
Tang, G.W. 513
Tang, J. 321
Tang, L.M. 619
Tao, L.K. 769
Tao, Z.Y. 165
Teng, H.J. 213, 227
Teng, L. 77
Tian, Q. 815
Tong, C.Y. 601

Vladimirovna, Z.T. 777

Wang, B. 497
Wang, C. 19
Wang, D.X. 455
Wang, F.T. 601
Wang, G. 19
Wang, G.J. 537
Wang, H.J. 33
Wang, J. 105
Wang, J. 721
Wang, J.G. 331
Wang, J.L. 773
Wang, J.M. 9
Wang, K. 271
Wang, K. 357, 773
Wang, L. 25, 29, 33
Wang, M.-S. 317
Wang, M.X. 77
Wang, N. 849
Wang, P. 223
Wang, P.-y. 353

Wang, Q.M. 291
Wang, R. 37
Wang, S.B. 181, 367
Wang, S.-B. 727
Wang, S.G. 345
Wang, S.Z. 433
Wang, T. 819
Wang, W. 715
Wang, W.H. 497
Wang, X. 339
Wang, X.B. 67
Wang, Y. 77
Wang, Y. 161
Wang, Y. 597, 625
Wang, Y.G. 487
Wang, Y.H. 741
Wang, Y.J. 151
Wang, Y.Q. 339
Wang, Y.Q. 505
Wang, Y.Q. 629
Wang, Y.X. 781
Wang, Y.Z. 781
Wei, X. 459
Wei, Y. 815
Wu, F. 823
Wu, J.-p. 441
Wu, M.L. 691
Wu, N. 227
Wu, Q.S. 581
Wu, S.L. 349
Wu, S.S. 819
Wu, X. 105
Wu, X. 441
Wu, X.H. 233
Wu, X.Y. 331
Wu, X.Y. 581
Wu, Z. 657
Wu, Z.B. 641
Wu, Z.F. 339
Wu, Z.J. 141

Xi, C.Y. 665
Xi, M.Y. 665
Xia, F.J. 345
Xiang, Z. 757
Xiao, H.Z. 475
Xiao, J. 73
Xiao, X.X. 765
Xie, C.J. 3
Xing, W. 653
Xing, Z.C. 77
Xu, B.S. 61
Xu, D.D. 445
Xu, J.X. 181
Xu, W.Q. 105
Xu, X. 37
Xu, X.L. 283

Xu, Y. 265
Xu, Y.F. 555
Xu, Y.P. 169
Xu, Y.Y. 241
Xue, C.G. 601
Xue, G. 519
Xue, H.F. 613
Xue, S.F. 455, 695
Xue, X.D. 797
Xue, Y. 839

Yan, D.M. 175
Yan, J.L. 677, 681
Yan, Q. 61
Yan, T. 765
Yan, Y. 109, 187
Yan, Y.X. 421
Yang, C. 619, 629
Yang, C.-M. 829
Yang, F. 213
Yang, G.-D. 317
Yang, H. 91
Yang, J.-p. 721
Yang, L. 227
Yang, M. 339
Yang, M.G. 769
Yang, M.J. 175, 247
Yang, S.-F. 829
Yang, S.Z. 85
Yang, Y.J. 219
Yang, Y.Z. 525
Yang, Z.R. 537
Yang, Z.Y. 175, 247
Yao, S. 833
Ye, H.J. 687
Yin, B. 761
Yin, C.B. 405
Yin, Y.H. 321
You, J.J. 101
Yu, J. 833
Yu, S. 271
Yu, Y.D. 175, 247
Yu, Y.M. 619, 629
Yu, Z.L. 175, 247
Yuan, B.Q. 113
Yun, Y. 53
Yue, X.J. 309

Zeng, H.H. 135, 141
Zeng, H.-h. 97
Zeng, H.L. 85
Zeng, J. 809
Zha, M. 119
Zhai, J.Q. 291
Zhai, L. 259
Zhang, D.Y. 537
Zhang, G.F. 3

Zhang, G.Q. 451
Zhang, G.S. 155
Zhang, H.B. 219
Zhang, H.J. 237
Zhang, H.S. 561
Zhang, J. 45
Zhang, J. 169
Zhang, J.-K. 317
Zhang, J.N. 797
Zhang, L. 49
Zhang, L. 839
Zhang, M. 67
Zhang, M.L. 241
Zhang, N. 657
Zhang, P. 701
Zhang, P.F. 379
Zhang, Q. 19
Zhang, Q.W. 379
Zhang, S.Y. 175
Zhang, T. 19
Zhang, T. 199
Zhang, W. 379
Zhang, W. 833

Zhang, W.-x. 543
Zhang, W.Y. 665
Zhang, X. 701
Zhang, X.C. 765
Zhang, X.-Q. 317
Zhang, X.Q. 613
Zhang, X.S. 765
Zhang, X.-T. 727
Zhang, Y.G. 113
Zhang, Y.H. 15
Zhang, Y.-h. 543
Zhang, Y.J. 803
Zhang, Y.X. 781
Zhang, Y.Z. 33
Zhang, Y.Z. 309
Zhao, D. 29
Zhao, D.D. 253, 469
Zhao, H.X. 555
Zhao, J. 401
Zhao, J. 839
Zhao, Q.L. 587
Zhao, Q.Y. 561
Zhao, R.S. 165

Zhao, S. 165
Zhao, W.X. 593
Zhao, X. 327
Zhao, X.G. 25,
  29, 33
Zhao, Y. 291
Zhao, Y.B. 253,
  469
Zheng, L. 735, 751
Zheng, X.L. 125
Zhou, B. 455, 695
Zhou, J.J. 715
Zhou, K.J. 587
Zhou, M.Y. 673
Zhou, Q.Y. 161
Zhou, S.Y. 41
Zhou, T.Q. 427
Zhou, X. 105
Zhu, D.S. 405
Zhu, F. 741
Zhu, H.L. 505
Zhu, Z.J. 607
Zu, Y.X. 731